T0140166

Studies in Fuzziness and Soft Computing

Volume 306

Series Editor

Janusz Kacprzyk, Polish Academy of Sciences, Warsaw, Poland
e-mail: kacprzyk@ibspan.waw.pl

For further volumes:
http://www.springer.com/series/2941

About this Series

The series "Studies in Fuzziness and Soft Computing" contains publications on various topics in the area of soft computing, which include fuzzy sets, rough sets, neural networks, evolutionary computation, probabilistic and evidential reasoning, multi-valued logic, and related fields. The publications within "Studies in Fuzziness and Soft Computing" are primarily monographs and edited volumes. They cover significant recent developments in the field, both of a foundational and applicable character. An important feature of the series is its short publication time and world-wide distribution. This permits a rapid and broad dissemination of research results.

Zongmin Ma · Fu Zhang
Li Yan · Jingwei Cheng

Fuzzy Knowledge Management for the Semantic Web

Zongmin Ma
Fu Zhang
Li Yan
Jingwei Cheng
Northeastern University
Shenyang
People's Republic of China

ISSN 1434-9922 ISSN 1860-0808 (electronic)
ISBN 978-3-662-50767-4 ISBN 978-3-642-39283-2 (eBook)
DOI 10.1007/978-3-642-39283-2
Springer Heidelberg New York Dordrecht London

Softcover reprint of the hardcover 1st edition 2014

Printed on acid-free paper

Springer is part of Springer Science+Business Media (www.springer.com)

Preface

The Web is a huge information resource depository all around the world and the huge amount of information on the Web is getting larger and larger every day. Nowadays, the Web is the most important means for people to acquire and publish information. Under this context, it is becoming very crucial for computer programs to deal with information on the Web automatically and intelligently. But most of today's Web content is suitable for human consumption. The Semantic Web has emerged as an extension of the current Web, in which Web resources are given computer-understandable semantics, enabling computers and people to better work in cooperation. The central idea of the Semantic Web is to make the Web more understandable to computer programs so that people can make more use of this gigantic asset.

The development of the Semantic Web proceeds in several steps, each step building a layer on top of another. The first layer is Uniform Resource Identifier (URI) and Unicode. The second layer is Extensible Markup Language (XML) and the third layer is the data representation format for Semantic Web, i.e., Resource Description Framework (RDF). The core layer is the ontology layer, in which RDF Schema (RDFS) and Web Ontology Language (OWL) represent the ontologies. Description Logics (DLs) are essentially the theoretical counterpart of OWL. Moreover, rule languages (e.g., RIF) are being standardized for the Semantic Web as well. In addition, to query RDF data as well as RDFS and OWL ontologies, Simple Protocol, and RDF Query Language (SPARQL) are proposed. On top of the ontology layer is the logic layer, which is the logical foundation of the Semantic Web. The rest of the Semantic Web layers are the Proof and Trust layers. On top of these layers, application with user interface can be built. From the layers of the Semantic Web, one of the important and declared goals of the Semantic Web is to reason about knowledge and data which are pervaded on the Web. So knowledge representation and reasoning have become very important research topics of the Semantic Web.

As it may be known, in the real world, human knowledge and natural language have a great deal of imprecision and vagueness. Fuzzy logic has been applied in a large number and a wide variety of applications, with real-world impact across a wide array of domains with human-like behavior and reasoning. Fuzzy logic has been a crucial means of implementing machine intelligence. So, in order to bridge the gap between human-understandable soft logic and machine-readable hard logic, fuzzy logic cannot be ignored. None of the usual logical requirements can be guaranteed: there is no centrally defined format for data, no guarantee of truth for assertions made, and no guarantee for consistency. It can be believed that fuzzy knowledge management can play an important and positive role in the development of the Semantic Web. It should be noted, however, that fuzzy logic may not be assumed to be the (only) basis for the Semantic Web, but its related concepts and techniques will certainly reinforce the systems classically developed within the W3C. Currently, the research of fuzzy logic in the area of the Semantic Web for fuzzy knowledge representation and reasoning are attracting increased attention.

This book goes to great depth concerning the fast-growing topic of technologies and approaches of fuzzy logic in the Semantic Web. The topics of this book include fuzzy description logics and fuzzy ontologies, queries of fuzzy description logics and fuzzy ontology knowledge bases, extraction of fuzzy description logics and ontologies from fuzzy data models, storage of fuzzy ontology knowledge bases in fuzzy databases, fuzzy Semantic Web ontology mapping, and fuzzy rules and their interchange in the Semantic Web. Concerning the fuzzy description logics, three kinds of fuzzy description logics are introduced, which are the tractable fuzzy description logics, the expressive fuzzy description logics, and the fuzzy description logics with fuzzy data types. The syntax, semantics, knowledge base, and reasoning for the fuzzy description logics are investigated, and a fuzzy description logic reasoner called FRESG supporting fuzzy data information with customized fuzzy data types *G* is introduced. Concerning the queries of fuzzy description logics and fuzzy ontology knowledge bases, conjunctive query answering for the tractable fuzzy description logics, expressive fuzzy description logics and fuzzy description logics with data type support is investigated, and a fuzzy SPARQL query language is proposed. Concerning the extraction of fuzzy description logics and ontologies from fuzzy data models, extracting fuzzy description logics and ontologies from the fuzzy conceptual data models (including fuzzy ER and UML data models) and the fuzzy database models (including fuzzy relational and fuzzy object-oriented databases) is discussed, respectively. Concerning the storage of fuzzy ontology knowledge bases in fuzzy databases, storing fuzzy knowledge base structure information, and instance information in the fuzzy relational databases are presented. Concerning the fuzzy Semantic Web ontology mapping, based on fuzzy ontology concept comparison and conceptual graph, mappings among multiple fuzzy ontologies are investigated. Concerning the fuzzy rules and their interchange in the Semantic Web, fuzzy rule languages are introduced and a fuzzy RIF framework f-RIA (fuzzy Rule Interchange Architecture) is proposed based on the fuzzy rule interchange format RIF-FRD (RIF Fuzzy Rule Dialect).

This book aims to provide a single record of current research in the fuzzy knowledge representation and reasoning for the Semantic Web. The objective of the book is to provide the state-of-the art information to researchers, practitioners, and graduate students of the Web intelligence and at the same time serve the knowledge and data engineering professional faced with nontraditional applications that make the application of conventional approaches difficult or impossible. Researchers, graduate students, and information technology professionals interested in the Semantic Web and fuzzy knowledge representation and reasoning will find this book a starting point and a reference for their study, research, and development.

We would like to acknowledge all of the researchers in the area of the Semantic Web and fuzzy knowledge representation and reasoning. Based on both their publications and the many discussions with some of them, their influence on this book is profound. The materials in this book are the outgrowth of research conducted by the authors in recent years. The initial research work were supported by the *National Natural Science Foundation of China* (61073139, 61202260, 61370075, 60873010), and in part by the *Program for New Century Excellent Talents in University* (NCET-05-0288). We are grateful for the financial support from the *National Natural Science Foundation of China* and the *Ministry of Education of China* through research grant funds. Additionally, the assistances and facilities of Northeastern University, China, are deemed important and highly appreciated. Special thanks go to Janusz Kacprzyk, the series editor of *Studies in Fuzziness and Soft Computing*, and Thomas Ditzinger, the senior editor of Applied Sciences and Engineering of Springer-Verlag, for their advice and help to propose, prepare and publish this book. This book will not be completed without the support from them. Finally, we wish to express deep thanks to Dr. Hailong Wang, Dr. Xing Wang, and Dr. Lingyu Zhang for their excellent research work and their great contributions to the book.

May 2013

Zongmin Ma
Fu Zhang
Li Yan
Jingwei Cheng

Contents

Chapter 1
Knowledge Representation and Reasoning in the Semantic Web

1.1 Introduction

The Semantic Web (Berners-Lee et al. 2001) is an extension of the current Web, in which Web resources are given computer-understandable semantics, better enabling computers and people to work in cooperation. The development of the Semantic Web proceeds in steps, each step building a layer on top of another. The layers of the Semantic Web are shown in Fig. 1.1 (Berners-Lee et al. 2006). As shown in Fig. 1.1, the first layer, URI and Unicode, follows the important features of the existing WWW. Unicode is a standard of encoding international character sets and it allows that all human languages can be used (written and read) on the web using one standardized form. Uniform Resource Identifier (URI) is a string of a standardized form that allows to uniquely identify resources (e.g., documents). The second layer, Extensible Markup Language (XML) layer with XML namespace and XML schema definitions makes sure that there is a common syntax used in the Semantic Web. The third layer is the data representation format for Semantic Web, i.e., Resource Description Framework (RDF). RDF is a framework for representing information about resources in a graph form. It was primarily intended for representing metadata about WWW resources. All data in the Semantic Web use RDF as the primary representation language. RDF itself serves as a description of a graph formed by triples. Anyone can define vocabulary of terms used for more detailed description. To allow standardized description of taxonomies and other ontological constructs, a RDF Schema (RDFS) is created together with its formal semantics within RDF. RDFS can be used to describe taxonomies of classes and properties and use them to create lightweight ontologies. More detailed ontologies can be created with Web Ontology Language OWL, which is the core layer of the Semantic Web. The OWL offers more constructs over RDFS. It is syntactically embedded into RDF, so like RDFS, it provides additional standardized vocabulary. Moreover, to provide rules beyond the constructs available from these languages, rule languages (e.g., RIF) are being standardized for the Semantic Web as well. In addition, for querying RDF data as well as RDFS and OWL ontologies, a Simple Protocol and RDF Query Language

Z. Ma et al., *Fuzzy Knowledge Management for the Semantic Web*,
Studies in Fuzziness and Soft Computing 306, DOI: 10.1007/978-3-642-39283-2_1,

(SPARQL) is available. SPARQL is SQL-like language, but uses RDF triples and resources for both matching part of the query and for returning results of the query. Since both RDFS and OWL are built on RDF, SPARQL can be used for querying ontologies and knowledge bases directly as well. Note that SPARQL is not only query language, but also a protocol for accessing RDF data. On top of the ontology layer is the logic layer, which is the logical foundation of the Semantic Web. In particular, Description Logics (DLs) (Baader et al. 2003) are essentially the theoretical counterpart of OWL. Since OWL is based on Description Logic, it is not surprising that a formal semantics is defined for OWL. RDFS and OWL have semantics defined and this semantics can be used for reasoning within ontologies and knowledge bases described using these languages. Last, the rest layers of the Semantic Web are the Proof and Trust layers. On top of these layers, application with user interface can be built. The detailed introduction about the architecture of Semantic Web can be found in Obitko (2007); Antoniou and van Harmelen (2008).

From the architecture of the Semantic Web, one of the important and declared goals of the Semantic Web is to reason about knowledge and data which are pervaded on the Web. Therefore, how to represent these knowledge and data and how to reason with them have become very important research topics of the Semantic Web. From this point of view, some methods which can describe the semantic meaning of the web should be developed. Fortunately, "ontology" (Staab and Studer 2009) can do this. As shown in Fig. 1.1, the core of the Semantic Web is "ontology" which is a formal and explicit specification of a shared conceptualization and can enable semantic interoperability. With ontologies, it is possible to formally represent the knowledge of an application domain in terms of a set of vocabulary, the fundamental concepts in the domain and relations between them.

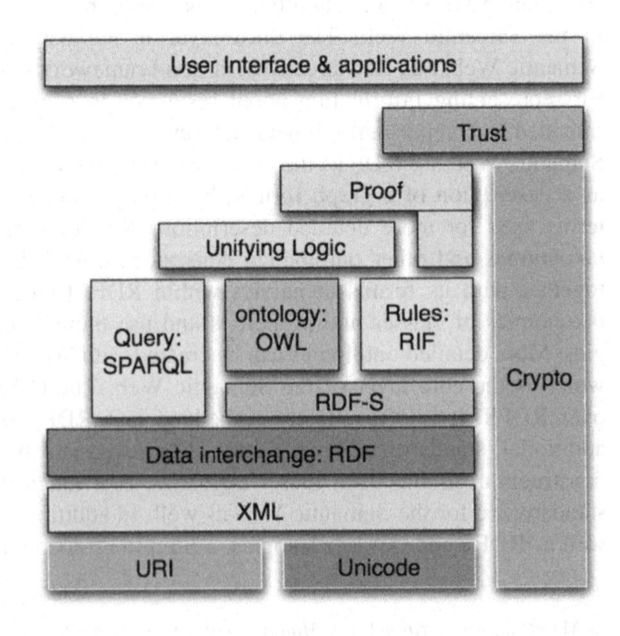

Fig. 1.1 The architecture of the semantic web

Over the past few years, several ontology definition languages for the Semantic Web have emerged, including RDF(S), OIL, DAML, DAML+OIL, and OWL (Berners-Lee et al. 2001). Among them, Web Ontology Language (OWL) is the newly released standard recommended by W3C. As the Semantic Web expects, OWL has the reasoning nature because Description Logics (DLs) (Baader et al. 2003) are essentially the theoretical counterpart of OWL and play a crucial role in this context. DLs, which provide a logical reconstruction of object-centric and frame-based knowledge representation languages, are the subset of first-order logic that can provide sound and decidable reasoning support (Baader et al. 2003). Today, there are lots of researches on OWL and DLs, which are used widely in the context of the Semantic Web and other applications such as medical, image analysis, databases and more (Staab and Studer 2009; Baader et al. 2003).

Although OWL adds considerable expressive power to the Semantic Web it does have expressive limitations, particularly with respect to what can be said about properties. In particular, there is no composition constructor, so it is impossible to capture relationships between a composite property and another (possibly composite) property. The standard example here is the obvious relationship between the composition of the "parent" and "brother" properties and the "uncle" property (Horrocks et al. 2005). Fortunately, rules can represent and reason with them, and they can enhance the expressive power of the Semantic Web ontologies (Horrocks 2008). As an important tool for representation and reasoning of knowledge, rules possess many features including high expressiveness, terseness, easy understanding and requiring small memory space, which let rules play important roles in many areas including Artificial Intelligence, and have been applied to markup the constructions of E-Business and the Semantic Web. With the development of rules in the Semantic Web, rule languages came into being successively.

Based on the observations mentioned above, knowledge representation and reasoning play a key role in the context of the Semantic Web. The Semantic Web ontologies and rules can together represent and reason with the knowledge in lots of applications and tasks both of the Semantic Web as well as of applications using ontologies and rules. The DLs are the main logical foundations of the ontologies and rules, and they are the key of reasoning with knowledge in the Semantic Web. In this chapter, we brief introduce DLs, ontologies, and rules in the Semantic Web, which can provide a general understanding of them.

1.2 Description Logics

In the last decade a substantial amount of work has been carried out in the context of Description Logics (DLs). DLs are a logical reconstruction of the so-called frame-based knowledge representation languages, with the aim of providing a simple well-established Tarski-style declarative semantics to capture the meaning of the most popular features of structured representation of knowledge. Nowadays,

DLs have gained even more popularity due to their applications in the context of the Semantic Web (Baader et al. 2003).

The recent researches on DLs can be divided into three categories:

- Introducing the *theoretical foundations* of DLs, addressing some of the most recent developments in theoretical research in the area;
- Focusing on the *implementation* of knowledge representation systems based on DLs, describing the basic functionality of a DL system, surveying the most influential systems based on DLs, and addressing specialized implementation techniques;
- Addressing the use of DLs and of DL-based systems in the design of several *applications* of practical interest.

In the following statements, we mainly focus on the first category, especially the theoretical formalism of a DL, with respect to the balance between its expressive power and its computational complexity. Indeed, the tradeoff between the expressiveness of a DL language and its complexity of reasoning shows that a careful selection of language constructs is needed and that the reasoning services provided by the DL system are deeply influenced by the set of constructs. On this basis, we can characterize three different approaches to the implementation of reasoning services. The first can be referred to as *limited* + *complete* systems, which are designed by restricting the set of constructs in such a way that subsumption would be computed efficiently, possibly in polynomial time. The second approach can be denoted as *expressive* + *incomplete* systems, since the idea is to provide both an expressive language and efficient reasoning. The drawback is, however, that reasoning algorithms turn out to be incomplete in these systems. After some of the sources of incompleteness are discovered, often by identifying the constructs—or, more precisely, combinations of constructs—that would require an exponential algorithm to preserve the completeness of reasoning, the systems with complete reasoning algorithms are designed. Systems of this sort are therefore characterized as *expressive* + *complete*, they provide a test bed for the implementation of reasoning techniques developed in theoretical investigations, and they play a key role in stimulating comparison and benchmarking with other systems.

In the following, we survey the languages of DLs according to their expressive power with the beginning of *AL* (i.e., *attributive language*) (Baader et al. 2003). The language *AL* has been introduced as a minimal language that is of practical interest, and the different DL languages are distinguished by the constructors they provide. In the sequel we shall discuss various languages from the family of *AL*-languages. The other expressive languages from the extensions of this family are also introduced.

1.2.1 The Basic Description Logic AL

Elementary descriptions are atomic concepts and atomic roles (also called concept names and role names), and complex descriptions can be built from them inductively with concept constructors and role constructors. In abstract notation, we use the letters A and B for atomic concepts, the letter R for atomic roles, and the letters C and D for concept descriptions.

Concept descriptions in AL are formed according to the following syntax rule:

$$
\begin{aligned}
C, D \rightarrow \quad & \top | && \text{(universal concept)} \\
& \bot | && \text{(bottom concept)} \\
& A | && \text{(atomic concept)} \\
& C \sqcap D | && \text{(intersection)} \\
& \forall R.C | && \text{(value restriction)} \\
& \exists R.\bot && \text{(limited existential quantification)}
\end{aligned}
$$

In AL, negation can only be applied to atomic concepts, and only the top concept is allowed in the scope of an existential quantification over a role. In order to define a formal semantics of AL-concepts, we consider an interpretations I that consist of a non-empty set Δ^I (the domain of the interpretation) and an interpretation function $\bullet^I$, which assigns to every atomic concept A a set $A^I \subseteq \Delta^I$ and to every atomic role R a binary relation $R^I \subseteq \Delta^I \times \Delta^I$. The interpretation function is extended to concept descriptions by the following inductive definitions:

$$
\begin{aligned}
T^I &= \Delta^I \\
\bot^I &= \varnothing \\
(\neg A)^I &= \Delta^I \backslash A^I \\
(C \sqcap D)^I &= C^I \cap D^I \\
(\forall R.C)^I &= \{a \in \Delta^I | \forall b.\,(a,b) \in R^I \rightarrow b \in C^I\} \\
(\exists R.\bot)^I &= \{a \in \Delta^I | \exists b.\,(a,b) \in R^I\}
\end{aligned}
$$

The above constructors can be used to build DL knowledge bases for representing the knowledge of an application domain (the "world") by first defining the relevant concepts of the domain (i.e., its terminology), and then using these concepts to specify properties of objects and individuals occurring in the domain (i.e., the world description) (Baader et al. 2003). Generally speaking, a DL knowledge base (*KB*) comprises two components, the TBox and the ABox. The TBox introduces the terminology, i.e., the vocabulary of an application domain, while the ABox contains assertions about named individuals in terms of this vocabulary. In brief, a TBox T is a finite set of terminology axioms of the form $C \subseteq D$ or $C = D$. An interpretation I satisfies $C \subseteq D$ if $C^I \subseteq D^I$, and similarly for $C = D$. An interpretation I is a model of TBox T iff I satisfies all axioms in T.

An ABox *A* is a finite set of assertions of the form *C*(*d*) or $R(d_1, d_2)$. By the first kind, called concept assertions, one states that *d* belongs to (the interpretation) *C*, by the second kind, called role assertions, one states that d_2 is a filler of the role *R* for d_1. An interpretation *I* is a model of ABox *A* iff *I* satisfies all assertions in *A*. Moreover, some DL *KBs* also include the RBox, which is a finite set of transitive role axioms (*Trans*(*R*)) and role inclusion axioms ($R \sqsubseteq S$). Please refer to (Baader et al. 2003) in detail. Furthermore, an interpretation *I* satisfies a *DL KB* if it satisfies all axioms and assertions in *KB*.

1.2.2 The Family of AL-languages

We obtain more expressive languages if we add constructors to *AL*. The following introduces several constructors which are often added to *AL*.

The *union* of concepts (indicated by the letter ***U***) is written as $C \sqcup D$, and interpreted as

$$C \sqcup D^I = C^I \cup D^I$$

Full existential quantification (indicated by the letter ***E***) is written as $\exists R.C$, and interpreted as

$$(\exists R.C)^I = \{a \in \Delta^I | \exists b.\,(a,b) \in R^I \wedge b \in C^I\}$$

Note that $\exists R.C$ differs from $\exists R.\bot$ in that arbitrary concepts are allowed to occur in the scope of the existential quantifier.

Number restrictions (indicated by the letter ***N***) are written as $\geq nR$ (at-least restriction) and as $\leq nR$ (at-most restriction), where *n* ranges over the nonnegative integers. They are interpreted as

$$(\geq nR)^I = \{a \in \Delta^I | \#\{b | (a,b) \in R^I\} \geq n\}$$
$$(\leq nR)^I = \{a \in \Delta^I | \#\{b | (a,b) \in R^I\} \leq n\}$$

where “#{}” denotes the cardinality of a set.

The *negation* of arbitrary concepts (indicated by the letter ***C***, for “complement”) is written as $\neg C$, and interpreted as

$$(\neg C)^I = \Delta^I \setminus C^I$$

Extending *AL* by any subset of the above constructors yields a particular *AL*-language. We name each *AL*-language by a string of the form:

$$AL[U][E][N][C]$$

where a letter in the name stands for the presence of the corresponding constructor. For instance, *ALEN* is the extension of *AL* by full existential quantification and number restrictions. Moreover, since the union (*U*) and full existential quantification (*E*) can be expressed using negation (*C*), the letters *UE* in language names will be replaced by the letter *C*. For instance, we shall write *ALC* instead of *ALUE* and *ALCN* instead of *ALUEN*.

1.2.3 The More Expressive Description Logics

There are several possibilities for extending *AL* in order to obtain a more expressive DL. The three most prominent are adding additional concept constructors, adding role constructors, and formulating restrictions on role interpretations. For these extensions, we introduce a naming scheme. Basically, each extension is assigned a letter or symbol:

- For *concept constructors*, the letters are written after the starting *AL*;
- For *role constructors*, we write the letters/symbols as superscripts;
- For *restrictions* on the interpretation of roles as subscripts.
- As an example, a DL called $ALCQI_{R+}$, extends *AL* with the concept constructors negation (*C*) and qualified number restrictions (*Q*), the inverse role (*I*), and the restriction that some roles are transitive($_{R+}$).

Below, we start with the third case, since we need to refer to restrictions on roles when defining certain concept constructors.

Restrictions on role interpretations. These restrictions enforce the interpretations of roles to satisfy certain properties, such as functionality and transitivity. We consider these two prominent examples in more detail. Others would be symmetry or connections between different roles.

- *Functional roles*. Here one considers a subset N_F of the set of role names N_R, whose elements are called *features*. An interpretation must map features *f* to functional binary relations $f^I \subseteq \Delta^I \times \Delta^I$. *AL* extended with features is denoted by AL_f.
- *Transitive roles*. Here one considers a subset N_{R+} of N_R. Role names $R \in N_{R+}$ are called *transitive roles*. An interpretation must map transitive roles $R \in N_{R+}$ to transitive binary relations $R^I \subseteq \Delta^I \times \Delta^I$. *AL* extended with transitive roles is denoted by AL_{R+}.

All the DLs mentioned until now contain the concept constructors such as intersection ($C \sqcap D$) and value restriction ($\forall R.C$) as a common core. DLs that allow for intersection of concepts and existential quantification (but not value restriction) are collected in the *EL-family*. The only constructors available in *EL* are intersection of concepts ($C \sqcap D$) and existential quantification ($\exists R.\bot$). Extensions of *EL* are again obtained by adding appropriate letters/symbols. Moreover, in order to avoid very long names for expressive DLs, the abbreviation *S* was introduced for ALC_{R+}, i.e., DL that extends *ALC* by transitive roles.

Prominent members of the *S*-family are *SIN* (which extends ALC_{R+} with number restrictions and inverse roles), *SHIF* (which extends ALC_{R+} with role hierarchies, inverse roles, and number restrictions of the form $\leq 1R$), and *SHIQ* (which extends ALC_{R+} with role hierarchies, inverse roles, and qualified number restrictions). Actually, DLs *SIN*, *SHIF*, and *SHIQ* are somewhat less expressive than indicated by their name since the use of roles in number restrictions is restricted: roles that have a transitive sub-role must not occur in number restrictions.

1.2.4 Description Logics with Data Type Representation

A drawback that all DLs introduced until now share is that all the knowledge must be represented on the abstract logical level. In many applications, one would like to be able to refer to concrete domains and predefined predicates on these domains when defining concepts, such as one wants to define the concept "Minor" using the axiom Minor = Person $\sqcap$ $\exists$age. $\leq$ 18. To solve the problem, Baader and Hanschke (1991a) integrated concrete domains into concept languages, and the definition of *concrete domain* was given and a tableau-based algorithm for deciding consistency of *ALC*(*D*)-ABoxes for admissible *D* was introduced. The algorithm has an additional rule that treats existential predicate restrictions according to their semantics. The main new feature is that, in addition to the usual "abstract" clashes, there may be concrete ones, i.e., one must test whether the given combination of concrete predicate assertions is non-contradictory. This is the reason why we must require that the satisfiability problem for *D* is decidable. As described in Baader and Hanschke (1991b), the algorithm is not in *PSpace*. Using techniques similar to the ones employed for *ALC* it can be shown, however, that the algorithm can be modified such that it needs only polynomial space (Lutz 1999), provided that the satisfiability procedure for *D* is in *PSpace*. In the presence of acyclic TBoxes, reasoning in *ALC(D)* may become *NExpTime-hard* even for rather simple concrete domains with a polynomial satisfiability problem (Lutz 2001).

The more expressive DLs *SHN(D)*$^-$ (Haarslev and Mölleret 2001) and *SHOQ*(*D*) (Horrocks and Sattler 2001), which can represent data information, are proposed. Although *SHN(D)*$^-$ and *SHOQ*(*D*) are rather expressive, they also have a very serious limitation on data types, i.e., they do not support customized data types. It has been pointed out that many potential users will not adopt them unless the limitation is overcome. Pan and Horrocks release a series of papers about data types to solve the problem (Pan 2007; Pan and Horrocks 2006). In the two papers, they summarize the limitations of OWL datatyping and propose the data type approach. For example, *SHIQ*(*G*) and *SHOQ*(*G*) DLs presented in Pan (2007) and Pan and Horrocks (2006) can support user-defined data type and user-defined data type predicates.

1.3 Semantic Web Ontologies

Ontology is a W3C standard knowledge representation model for the Semantic Web (Berners-Lee et al. 2001). The vocabulary "ontology" often appears in various applications. While having its roots in *philosophy*, the term ontology is today popular also in computer science. Ontology helps people and machines to communicate concisely by supporting information exchange based on semantics rather than just syntax. In general terms, ontologies are a formal, explicit specification of a shared conceptualization (Studer et al. 1998):

- *Conceptualization* refers to an abstract model of some part of the world which identifies the relevant concepts and relations.
- *Explicit* means that the type of concepts, the relations between the concepts, and the constraints on their usage, are explicitly defined.
- *Formal* refers to the fact that ontology should be machine readable.
- *Shared* means that the ontology should reflect the understanding of a community and should not be restricted to the comprehension of some individuals.

In particular, ontology allows the semantics of a domain to be expressed in a language understood by computers, enabling automatic processing of the meaning of shared information. Therefore, ontologies are a key element in the context of the Semantic Web, an effort to make information on the Internet more accessible to agents and other software (Tommila et al. 2010). In general, ontologies are based on knowledge representation formalisms such as the Description Logics (Baader et al. 2003), which is the logical counterpart of some popular ontology representation languages, e.g. Web Ontology Language (OWL). Currently, *ontology is used as a standard (W3C recommendation) knowledge representation model for the Semantic Web*. Some ontology definitions have already been presented sparsely in various papers, which will not be listed here. The detailed introduction about ontologies can be found in Chandrasekaran et al. (1999); Ding et al. (2005); Nieto et al. (2003); Staab and Studer (2009). From these definitions, we can identify some essential aspects of ontologies:

- Ontologies are used to describe a domain. The terms (including concepts, properties, objects, and relations) are clearly defined.
- There is a mechanism to organize the terms commonly as a hierarchical structure.
- There is an agreement between users of an ontology in such a way the meaning of the terms is used consistently.

An ontology can be defined by the *ontology representation languages* such as RDF(S), SHOE, OIL, DAML, DAML+OIL, and OWL (Horrocks et al. 2003). Among them, the most widely used is OWL, which is a W3C standard for expressing ontologies in the Semantic Web. OWL has three increasingly expressive sublanguages OWL Lite, OWL DL, and OWL Full. OWL Full has the highest expressiveness while it has been proved to be undecidable (Motik 2005). OWL

Lite and OWL DL are *almost* equivalent to the *DLs* SHIF(D) and SHOIN(D). It is "*almost*" since OWL Lite and OWL DL can provide annotation properties while DLs do not. Moreover, OWL has two interchangeable syntactic forms. One is the *exchange syntax*, i.e., the RDF/XML syntax. Another form is the frame-like style *abstract syntax*. For the detailed introduction about OWL, including its syntax and semantics, please refer to (http://www.w3.org/2004/OWL/).

In the following we summarize a formal definition of ontologies in order to provide a general understanding of the components of ontologies. An ontology O can be considered as a couple $O = <O_S, O_I>$ which consists of the ontology structure O_S and the ontology instances O_I. O_S is a set of *OWL identifiers* and *class/property axioms* used to describe concepts, properties and their relationships in a domain, and O_I is a set of *individual axioms* used to describe objects, concepts and their relationships in the domain. Definition 1.1 gives a formal definition of ontologies.

Definition 1.1 (*ontologies*) An ontology may be defined as a couple $O = (O_S, O_I) = (ID_O, Axiom_O)$, where:

(1) $ID_O = CID_O \cup IID_O \cup DRID_O \cup OPID_O \cup DPID_O$ is an OWL identifier set (see Table 1.1 in detail):

- a subset CID_O of class identifiers, which models the concepts of a domain. A concept is often considered as a class in an ontology.
- a subset IID_O of individual identifiers.
- a subset $DRID_O$ of data range identifiers; each data range identifier is a predefined XML Schema datatype, which models the data types of properties of an object in a domain, such as integer, string, and etc.
- a subset $OPID_O$ of object property identifiers, which models the relationships between concepts in a domain. Each object property has its characteristics (e.g., Symmetric, Functional, Transitive, etc.) and its restrictions (e.g., someValuesFrom, allValuesFrom, minCardinality, etc.) as will be shown in Table 1.1.
- a subset $DPID_O$ of datatype property identifiers, which models the properties of concepts in a domain. Each datatype property has its characteristics (e.g., Functional) and its restrictions (e.g., someValuesFrom, allValuesFrom, minCardinality, etc.) as will be shown in Table 1.1.

(2) $Axiom_O$ is an OWL axiom set (see Table 1.1 in detail):

- a subset of class/property axioms, used to represent the ontology structure, which models the structure information of a domain.
- a subset of individual axioms, used to represent the ontology instance, which models the object instance information of a domain.

Table 1.1 gives *the OWL abstract syntax*, *the respective Description Logic syntax*, and *the semantics*. The semantics for OWL is based on the interpretation of SHOIN(D) (Horrocks et al. 2003). In detail, the semantics is provided by an interpretation $I = (\Delta^I, \Delta_D, \bullet^I, \bullet^D)$, where Δ^I is the abstract domain and Δ_D is the

Table 1.1 OWL abstract syntax, the corresponding DL syntax, and the semantics

OWL abstract syntax	DL syntax	Semantics
class description		
Class(*A*)	A	$A^I \subseteq \Delta^I$
owl:Thing	T	$T^I = \Delta^I$
owl:Nothing	$\bot$	$\bot^I = \varnothing$
intersectionOf ($C_1 \ldots C_n$)	$C_1 \sqcap \ldots \sqcap C_n$	$(C_1 \sqcap \ldots \sqcap C_n)^I = C_1^I \cap \ldots \cap C_n^I$
unionOf ($C_1 \ldots C_n$)	$C_1 \sqcup \ldots \sqcup C_n$	$(C_1 \sqcup \ldots \sqcup C_n)^I = C_1^I \cup \ldots \cup C_n^I$
complementOf (*C*)	$\neg C$	$(\neg C)^I = \Delta^I \backslash C^I$
oneOf ($w_1 \ldots w_n$)	$\{w_1,\ldots,w_n\}$	$(w_1,\ldots,w_n)^I = \{w_1^I,\ldots,w_n^I\}$
restriction (*P* some ValuesFrom(*E*))	$\exists P.E$	$(\exists P.E)^I = \{a \mid \exists w. < a, w > \in P^I \wedge w \in E^I\}$
restriction (*P* allValuesFrom(*E*))	$\forall P.E$	$(\forall P.E)^I = \{a \mid \forall w. < a, w > \in P^I \rightarrow w \in E^I\}$
restriction (*P* hasValue(*w*))	$\exists P.\{w\}$	$(\exists P.w)^I = \{a \mid < a, w^I > \in P^I\}$
restriction (*P* minCardinality(n))	$\geq nP$	$(\geq nP)^I = \{a \mid \#\{w \mid < a, w > \in P^I\} \geq n\}$
restriction (*P* maxCardinality(n))	$\leq nP$	$(\leq nP)^I = \{a \mid \#\{w \mid < a, w > \in P^I\} \leq n\}$
restriction (*P* Cardinality(n))	$=nP$	$(\geq nP \sqcap \leq nP)^I$
ObjectProperty(*R*)	R	$R^I \subseteq \Delta^I \times \Delta^I$ $P \in \{R, T\}$ $W \in \{a, v\}$
DatatypeProperty(*T*)	T	$T^I \subseteq \Delta^I \times \Delta_D$
AbstractIndividual(*a*)	a	$a^I \in \Delta^I$
DatatypeIndividual(*v*)	v	$v^I \in \Delta_D$
class axioms		
Class(*A* partial $C_1 \ldots C_n$)	$A \sqsubseteq C_1 \sqcap \ldots \sqcap C_n$	$A^I \sqsubseteq C_1^I \cap \ldots \cap C_n^I$
Class(*A* complete $C_1 \ldots C_n$)	$A \equiv C_1 \sqcap \ldots \sqcap C_n$	$A^I = C_1^I \cap \ldots \cap C_n^I$
SubClassOf (C_1 C_2)	$C_1 \sqsubseteq C_2$	$C_1^I \subseteq C_2^I$
EquivalentClasses (C_1 C_2)	$C_1 \equiv \ldots \equiv C_n$	$C_1^I = \ldots = C_n^I$
DisjointClasses($C_1 \ldots C_n$)	$C_1 \sqcap C_2 \sqsubseteq \bot$	$C_1^I \cap C_2^I = \varnothing, i \neq j$
EnumeratedClasses(*A* $a_1 \ldots a_2$)	$A^I \equiv \{a_1,\ldots,a_n\}$	$A^I = \{a_1^I,\ldots,a_n^I\}$
property axioms		
Datatype Property (*T*		
domain(C_1)...domain(C_m)	$\leq 1T \sqsubseteq Ci$	$T^I \subseteq C_i^I \times \Delta_D$
range(d_1)...range(d_k)	$T \sqsubseteq \forall T.di$	$T^I \subseteq \Delta^I \times d_i^D$
[Functional])	$T \sqsubseteq \leq 1T$	$\forall a \in \Delta^I.\#\{v \mid < a,v > \in T^I\} \leq 1$
ObjectProperty (*R*		
domain(C_1)...domain(C_m)	$\geq 1R \sqsubseteq Ci$	$R^I \subseteq C_i^I \times \Delta^I$
range(C_1)...range(C_k)	$T \sqsubseteq \forall R.Ci$	$R^I \subseteq \Delta^I \times C_i^I$
[Functional]	$T \sqsubseteq \leq 1 R$	$\forall a \in \Delta^I.\#\{a' \mid < a,a' > \in R^I\} \leq 1$
[InverseOf($E_1,\ldots E_2$)	$R = R_0^-$	$R^I = (R_0^-)^I$
[Symmetric]	$R = R^-$	$R^I = (R^-)$
[InverseFunction]	$T \sqsubseteq \leq 1 R^-$	$\forall a \in \Delta^I.\#\{a' \mid < a,a' > \in (R_0^-)^I\} \leq 1$
[Transitive])	Trans(*R*)	$\{< a,a_1 >, < a_1,a_2 >\} \subseteq R^I \rightarrow < a,a_2 > \in R^I$
SubPropertyOf(E_1,E_2)	$E_1 \sqsubseteq E_2$	$E_1^I \subseteq E_2^I$
EquivalentProperties($E_1,\ldots,E_2$)	$E_1 \equiv \ldots \equiv E_n$	$E_1^I \equiv \ldots \equiv E_n^I$
individual axioms		
Individual (*a* type(C_i)...	$a: C_i$	$a^I \in C_i^I$ $1 \leq i \leq n$
value(R_i,a_i)...	$(a,a_i):R_i$	$< a^I,a_i^I > \in R_i^I$ $1 \leq i \leq n$
value(T_i,v_i)...)	$(a,v_i):T_i$	$< a^I,v^I > \in T_i^I$ $1 \leq i \leq n$
SameIndividual($a_1 \ldots a_n$)	$a_1 = \ldots = a_n$	$a_1^I = \ldots = a_n^I$
DifferentIndividuals($a_1 \ldots a_n$)	$a_i \neq a_j$	$a_i^I \neq a_j^I$

datatype domain (disjoint from Δ^I), and $\bullet^I$ and $\bullet^D$ are two interpretation functions, which map:

- An abstract individual a to an element $a^I \in \Delta^I$;
- A concrete individual v to an element $v^D \in \Delta_D$;
- A concept name A to a subsut $A^I \subseteq \Delta^I$;
- An abstract role name R to a relation $R^I \subseteq \Delta^I \times \Delta^I$;
- A concrete datatype D to a subset $D^D \subseteq \Delta_D$;
- A concrete role name T to a relation $T^I : \Delta^I \times \Delta_D$.

1.4 Rules in the Semantic Web

As an important representation of knowledge, rules play an important role in the Semantic Web. They possess many features including high expressiveness, terseness and requiring small memory space, and they can enhance the expressive power of the Semantic Web ontologies, which let rules become indispensable and widely used in the Semantic Web. With the development of rules in the Semantic Web, rule languages came into being successively. Horrocks et al. (2005) presented the rule language Semantic Web Rule Language (SWRL), which are used to construct different rule systems applied in Web and Semantic Web. It is unavoidable that these systems communicate with each other (i.e., interchange rules). However, because of the heterogeneity of different rule languages in the respect of syntaxes and semantics, creating a generally accepted interchange format is by no means a trivial task (Boley et al. 2007). Rule Markup Language (RuleML) (Boley et al. 2001) proposed by the RuleML Initiative tries to encompass different kinds of rules. REWERSE Rule Markup Language (R2ML) (Wagner et al. 2005) which integrates Object Constraint Language (OCL), SWRL and RuleML has strong markup ability. Especially, Rule Interchange Format (RIF), (http://www.w3.org/2005/rules/wg/charter.html#w3c-xml), a recommendation of the WWW Consortium (W3C), aims at implementing rule interchange between different rule languages and thus between different rule systems in the Semantic Web.

Here, instead of introducing all kinds of the rule languages, we only recall the rule language Semantic Web Rule Language (SWRL) (Horrocks et al. 2005) and the rule makeup language REWERSE Rule Markup Language (R2ML) (Wagner et al. 2005), in order to provide a general understanding of the rules in the Semantic Web.

1.4.1 Semantic Web Rule Language-SWRL

The form of the SWRL rule (Horrocks et al. 2005) is: *antecedent* → *consequent*. For example, the following SWRL rule denotes that the brother of parent of a person is his uncle: *HasParent(?x,?y)* ∧ *HasBrother(?y,?z)* → *HasUncle(?x,?z)*, where *HasParent*, *HasBrother*, *HasUncle* are individual-valued property, and ?*x*, ?*y* and ?*z* are individual variables.

A SWRL rule axiom consists of an antecedent (body) and a consequent (head), each of which consists of a (possibly empty) set of atoms. Just as for class and property axioms, rule axioms can also have annotations. These annotations can be used for several purposes, including giving a label to the rule by using the rdfs:label annotation property.

rule :: = '*Implies*(' {*annotation*} *antecedent consequent* ')'
antecedent :: = '*Antecedent*(' {*atom*} ')'
consequent :: = '*Consequent*(' {*atom*} ')'

Atoms in rules can be of the form *C*(*x*), *P*(*x*,*y*), *Q*(*x*,*z*), *sameAs*(*x*,*y*) or *different From*(*x*,*y*), where *C* is an OWL DL description, *P* is an OWL DL individual-valued Property, *Q* is an OWL DL data-valued Property, *x*, *y* are either variables or OWL individuals, and *z* is either a variable or an OWL data value. In the context of OWL Lite, descriptions in atoms of the form *C*(*x*) may be restricted to class names.

atom ::= *Description* '(' *i-object* ')'
| *DataRange* '(' *d-object* ')'
| *individualvaluedProperty* '(' *i-object*, *i-object* ')'
| *sameAs* '(' *i-object*, *i-object* ')'
| *differentFrom* '(' *i-object*, *i-object* ')'
| *builtIn* '(' *builtinID*, *d-object* ')'
builtinID :: = *URI reference*
i-object ::= *i-variable* | *individualID*
d-object ::= *d-variable* | *dataLiteral*
i-variable ::= '*I-variable* (' *URI reference* ')'
d-variable ::= '*D-variable* (' *URI reference* ')'

The model-theoretic semantics for SWRL is a straightforward extension of the semantics for OWL DL given in (Patel-Schneider et al. 2004). The basic idea is that we define bindings—extensions of OWL interpretations that also map variables to elements of the domain in the usual manner. A rule is satisfied by an interpretation iff every binding that satisfies the antecedent also satisfies the consequent. The semantic conditions relating to axioms and ontologies are unchanged, so an interpretation satisfies an ontology iff it satisfies every axiom (including rules) and fact in the ontology.

An abstract OWL interpretation is a tuple of the form $I = < R, EC, ER, L, S, LV >$, where *R* is a set of resources, $LV \subseteq R$ is a set of literal values, *EC* is a mapping from classes and datatypes to subsets of *R* and *LV* respectively, *ER* is a mapping from properties to binary relations on *R*, *L* is a mapping from typed literals to elements of *LV*, and *S* is a mapping from individual names to elements of *EC*(owl : Thing). Given an abstract OWL interpretation *I*, a binding *B*(*I*) is an abstract OWL interpretation that extends *I* such that *S* maps *i*-variables to elements of *EC*(owl : Thing) and *L* maps *d*-variables to elements of *LV* respectively. A binding *B*(*I*) satisfies an antecedent *A* iff *A* is empty or *B*(*I*) satisfies every atom in *A*. A binding *B*(*I*) satisfies a consequent *C* iff *C* is not empty and *B*(*I*) satisfies every atom in *C*. A rule is satisfied by an interpretation *I* iff for every binding *B* such that *B*(*I*) satisfies the antecedent, *B*(*I*) also satisfies the consequent.

The detailed introduction about the rule language Semantic Web Rule Language (SWRL) can be found in Horrocks et al. (2005).

1.4.2 REWERSE Rule Markup Language-R2ML

R2ML (Wagner et al. 2005) is an important rule markup language, which is the combination of SWRL, RuleML and OCL, and has strong markup capability. R2ML adopts logic and non-logic form to express rules in different systems and tools, and it support strict differentiation of concepts. R2ML does not acquire that users transform their rule representation to a different form. The content of R2ML is expressed by data type language, user-defined vocabulary, term, atom, action and rules.

The construct of R2ML rules is atom formula. There are nine kinds of R2ML atoms. They are object description atom, nonequivalent atom, object class atom, association atom, property atom, reference property atom, equivalent atom, data class atom and datatype predicate atom. R2ML has two kinds of formulas, i.e., AndOrNafNegFormula and Logical Formula, which represent free quantifier logical formula including weak negation and strong negation, and first-order logic formula, respectively. To express production rules and reaction rules, R2ML defines action which can be InvokeActionExpression, AssignActionExpression, CreatActionExpression or DeleteActionExpression. R2ML includes three kinds of rules. They are integrity rules, derivation rules and production rules, where, integrity rules include alethic rules and deontic rules, derivation rules include condition and conclusion, and production rules include condition, post condition and action. To enable computers automatically dealing with R2ML rules, REWERSE defines R2ML XML Schema which includes all the elements defined in R2ML, which lays a solid foundation for rule interchange based on XML.

The following codes are an XML snippet which means that a discount for a customer buying a product is 7.5 percent if the customer is premium and the product is luxury.

```
<r2ml:DerivationRule>
    <r2ml:condition>
  <r2ml:ObjectClassificationAtom r2ml: classID = “premium-Customer”>
     <r2ml:ObjectVariable r2ml:name=“customer” r2ml:classID = “Customer”>
  <r2ml:ObjectClassificationAtom/>
  <r2ml:ObjectClassificationAtom r2ml: classID = “Luxury-Product”>
     <r2ml:ObjectVariable r2ml:name=“product” r2ml:classID = “Product”>
  <r2ml:ObjectClassificationAtom/>
  <r2ml:AssociationAtom r2ml: associationPredicateID = “buy”>
      <r2ml:objectArgument>
        <r2ml:ObjectVariable r2ml:name=“customer”/>
        <r2ml:ObjectVariable r2ml:name=“product”/>
      </r2ml:objectArgument>
  </r2ml:AssociationAtom>
 </r2ml:condition>
 <r2ml:conclusion>
  <r2ml:AttributeAtom r2ml: attributeID = “discount”>
      <r2ml:subject>
        <r2ml:ObjectVariable r2ml:name=“customer”/>
      </r2ml:subject>
      <r2ml:value>
        <r2ml:TypeLiteral r2ml:datatype=“xs:decimal” r2ml:lexicalValue
         =“7.5”/>
      </r2ml:value>
  </r2ml:AttributeAtom>
 </r2ml:conclusion>
</r2ml:DerivationRule>
```

1.5 Summary

Knowledge representation and reasoning plays an essential role in creating machine-processing content in the context of the Semantic Web. Ontology, which can capture the knowledge in a domain in a formal and machine-processable way, is a W3C standard knowledge representation model for the Semantic Web. Also, as an important representation of knowledge, rules, which possess many features including high expressiveness, terseness and requiring small memory space, can enhance the expressive power of the Semantic Web ontologies. In particular, ontology has the reasoning nature because of the existence of Description Logics, which are a family of logic-based knowledge representation formalisms designed to represent and reason about the knowledge of an application domain in a structured and well-understood way. Therefore, all of these features let ontologies, Description Logics, and rules become indispensable and widely used in the context of the Semantic Web. In this chapter, we briefly introduce some basic notions of Description Logics, ontologies, and rules.

However, the traditional ontologies, Description Logics, and rules feature limitations, mainly with what can be said about fuzzy information that is commonly found in many application domains. In order to provide the necessary means to handle such information and knowledge there are today a huge number of proposals for fuzzy extensions to ontologies, Description Logics, and rules. In particular, Zadeh's *fuzzy set theory* (Zadeh 1965) has been identified as a successful technique for modelling the fuzzy information in many application areas, especially in the Semantic Web. In the next chapter, we will briefly introduce the fuzzy set theory.

References

Antoniou G, van Harmelen F (2008) A semantic web primer, 2nd edn. MIT Press, Cambridge

Baader F, Hanschke P (1991a) A scheme for integrating concrete domains into concept languages. In: Proceedings of the 12th international joint conference on artificial intelligence (IJCAI'91), pp 452–457

Baader F, Hanschke P (1991b) A scheme for integrating concrete domains into concept languages. Research report RR-91-10, DFKI, Kaiserslautern

Baader F, Calvanese D, McGuinness D, Nardi D, Patel-Schneider PF (2003) The description logic handbook: theory, implementation and applications. Cambridge University Press, Cambridge

Berners-Lee T, Hendler J, Lassila O (2001) The semantic web. Sci Am 284(5):34–43

Berners-Lee T, Hall W, Hendler J, O'Hara K, Shadbolt N, Weitzner DJ (2006) A framework for web science. Found Trends Web Sci 1(1):1–130

Boley H, Tabet S, Wagner G (2001) Design rationale for RuleML: a markup language for semantic web rules. In: Proceedings of the first Semantic web working symposium (SWWS'01), pp 381–401

Boley H, Kifer M, Patranjan PL, Polleres A (2007) Rule interchange on the web. In: Proceedings of reasoning web, LNCS 4636, pp 269–309

Chandrasekaran B, Josephson J, Benjamins V (1999) What are ontologies, and why do we need them? IEEE Intell Syst 14(1):20–26

Ding L, Kolari P, Ding Z, Avancha S, Finin T, Joshi A (2005) Using ontologies in the semantic web: a survey. In: TR-CS-05-07, UMBC

Haarslev V, Möller R (2001) The description logic ALCNHR+ extended with concrete domains: a practically motivated approach. In: Proceedings of the international joint conference on automated reasoning (IJCAR 2001). Lecture Notes in Artificial Intelligence, vol 2083, pp 29–44

Horrocks I, Sattler U (2001) Ontology reasoning in the SHOQ(D) description logic. In: Proceedings of the 17th international joint conference on artificial intelligence (IJCAI 2001), pp 199–204

Horrocks I, Patel-Schneider PF, van Harmelen F (2003) From SHIQ and RDF to OWL: the making of a web ontology language. J. Web Semant 1(1):7–26

Horrocks I, Patel-Schneider PF, Bechhofer S, Tsarkov D (2005) OWL rules: a proposal and prototype implementation. J Web Semant 3(1):23–40

Horrocks I (2008) Ontologies and the semantic web. The Communication of ACM 51:58–67

Lutz C (1999) Reasoning with concrete domains. In: Proceedings of the 16th international joint conference on artificial intelligence (IJCAI'99), Stockholm, Sweden, pp 90–95

Lutz C (2001) NEXPTIME-complete description logics with concrete domains. In: Proceedings of the international joint conference on automated reasoning (IJCAR 2001). Lecture notes in artificial intelligence, vol 2083. Springer, pp 45–60

Motik B (2005) On the properties of metamodeling in OWL. In: Proceedings of the fourth international semantic web conference (ISWC 2005). LNCS, vol 3729. Springer, pp 548–562

Nieto MAM (2003) An overview of ontologies. Technical report, Center for Research in Information and Automation Technologies and Interactive and Cooperative Technologies Lab, Universidad De Las Americas Puebla, http://www.starlab.vub.ac.be/teaching/ontologies_overview.pdf

Obitko M (2007) Semantic web architecture. http://www.obitko.com/tutorials/ ntologies-semantic-web/semantic-web-architecture.html

Pan JZ, Horrocks I (2006) OWL-Eu: adding customized data types into OWL. J Web Semant 4(1):29–49

Patel-Schneider PF, Hayes P, Horrocks I (2004) OWL web ontology language semantics and abstract syntax, W3C Recommendation. http://www.w3.org/TR/owl-semantics/. Accessed 10 Feb 2004

Pan JZ (2007) A flexible ontology reasoning architecture for the Semantic Web. IEEE Trans Knowl Data Eng 19(2):246–260

Staab S, Studer R (2009) Handbook on ontologies, 2nd edn. Springer, Dodrecht

Studer R, Benjamins R, Fensel D (1998) Knowledge engineering: principles and methods. Data Knowl Eng 25(1–2):161–198

Tommila T, Hirvonen J, Pakonen A (2010) Fuzzy ontologies for retrieval of industrial knowledge-a case study. VTT Working Papers, pp 1459–7683

Wagner G, Viegas Damásio C, Giurcaj A (2005) Towards a general web rule language. Int J Web Eng Technol 2:181–206

Zadeh LA (1965) Fuzzy sets. Inf Control 8(3):338–353

Chapter 2
Fuzzy Sets and Possibility Theory

2.1 Introduction

In many real-world applications, information is often imprecise and uncertain. Many sources can contribute to the imprecision and uncertainty of data or information. We face, for example, increasingly large volumes of data generalized by non-traditional means (e.g., sensors and RFID), which is almost always imprecise and uncertain. It has been pointed out that in the future, we need to learn how to manage data that is imprecise or uncertain, and that contains an explicit representation of the uncertainty (Dalvi and Suciu 2007).

Over the years, imprecise and uncertain information has been widely investigated in lots of application domains, including database, information system, and so on (Klir and Yuan 1995; Ma and Yan 2008). Also, this is a well-known problem especially for semantics-based applications of the Semantic Web, such as knowledge management, e-commerce, and web portals (Calegari and Ciucci 2006). As the examples mentioned by Stoilos et al. (2010), a task like a "holiday organization" could involve a request like: "Find me a *good* hotel in a place that is *relatively hot* and with *many* attractions", or a "doctor appointment" could look like: "Make me an appointment with a doctor *close to* my home not *too early* and of *good* references". Moreover, several intermediate processes, like information extraction or retrieval, matching user preferences with data and more, might involve imperfect information due to their automatic nature (Stoilos et al. 2010). Last but not least, several modern applications that have adopted Semantic Web technologies in order to enhance their performance and "connect" with the Semantic Web require the management of such knowledge (Stoilos et al. 2010). For example there are schemes for using Semantic Web technologies in multimedia applications for multimedia analysis, in Semantic Portals, ontology alignment and semantic interoperability, Semantic Web Services matching, word-computing based systems and many more, all of which require the management of some form of fuzzy information. For example, in image analysis one has to map low-level numerical values that are extracted by analysis algorithms for the color, shape and texture of a region into more high-level symbolic features like concepts.

Z. Ma et al., *Fuzzy Knowledge Management for the Semantic Web*,
Studies in Fuzziness and Soft Computing 306, DOI: 10.1007/978-3-642-39283-2_2,

For example, values of the RGB color model would need to be mapped into concepts like *Blue*, *Green*, etc. or values of special shape and texture (signal) transforms need to be mapped to concepts like *Rectangular_Shaped*, *Coarse_Textured*, *Smooth_Textured* and more, all of which are obviously fuzzy concepts and need proper handling (Stoilos et al. 2010).

The conceptual formalism supported by typical Semantic Web techniques (such as ontologies, Description Logics, and rules mentioned in Chap. 1) may not be sufficient to represent such information and knowledge as mentioned above. In particular, the problem to deal with fuzzy information has been addressed in several decades ago by Zadeh (1965), who gave bird in the meanwhile to the so-called fuzzy set and fuzzy logic theory and a huge number of real life applications are based on it. Fuzzy set theory (Zadeh 1965), which is interchangeably referred as fuzzy logic, is a generalization of the set theory and provides a means for the representation of imprecision and vagueness. The *fuzzy set theory* has been identified as a successful technique for modeling the fuzzy information in many application areas such as text mining, multimedia information system, database (Klir and Yuan 1995; Bosc et al. 2005; Ma and Yan 2008). Also, in the context of the Semantic Web as well as applications using ontologies, Description Logics, and rules, the fuzzy set theory has been considered as the main theory basis for extensions to them to handle fuzzy information (Sanchez 2006; Ma et al. 2012). In this chapter, we briefly introduce some notions about imperfect information and fuzzy sets and possibility theory, which are of interest or relevance to the discussions of successive chapters.

2.2 Imperfect Information

In order to satisfy the need for modeling fuzzy information, the following several kinds of imperfect information have been extensively introduced into real-world applications (Bosc et al. 2005; Klir and Yuan 1995; Ma and Yan 2008; Smets 1996):

- Inconsistency is a kind of semantic conflict, meaning the same aspect of the real world is irreconcilably represented more than once in a database or in several different databases. For example, the *age* of *George* is stored as 34 and 37 simultaneously.
- Imprecision is relevant to the content of an attribute value and means that a choice must be made from a given range (interval or set) of values without knowing which one to choose. For example, the *age* of *Michael* is a set {18, 19, 20, 21}.
- Vagueness is like imprecision but which is generally represented with linguistic terms. For example, the *age* of *John* is a linguistic value "*old*".
- Uncertainty is related to the degree of truth of its attribute value, meaning that we can apportion some, but not all, of our belief to a given value or group of values. For example, the possibility that the *age* of *Chris* is 35 right now should

be 0.98. The random uncertainty, described using probability theory, is not considered here.

- Ambiguity means that some elements of the model lack complete semantics, leading to several possible interpretations.

Generally, several different kinds of imperfection can co-exist with respect to the same piece of information. For example, the *age* of *John* is a set {18, 19, 20} and their possibilities are 0.70, 0.95, and 0.98, respectively. Imprecision, vagueness, and uncertainty are three types of imperfect information (Ma and Yan 2008; Smets 1996). Imprecision are essential properties of the information itself, whereas uncertainty is a property of the relation between the information and our knowledge about the world.

Moreover, imprecise values generally denote a set of values in the form of [ai1, ai2,..., aim] or [ai1, ai2] for the discrete and continuous universe of discourse, respectively, meaning that exactly one of the values is the true value for the single-valued attribute, or at least one of the values is the true value for the multivalued attribute. So, imprecise information here has two interpretations: disjunctive information and conjunctive information.

One kind of imprecise information that has been studied extensively is the well-known null values (Codd 1986, 1987; Motor 1990; Parsons 1996; Zaniolo 1984), which were originally called incomplete information. The possible interpretations of null values include: (a) "*existing but unknown*", (b) "*nonexisting*" or "*inapplicable*", and (c) "*no information*". A null value on a multivalued object, however, means an "open null value" (Gottlob and Zicari 1988), i.e., the value may not exist, has exactly one unknown value, or has several unknown values. Null values with the semantics of "existent but unknown" can be considered as the special type of partial values that the true value can be any one value in the corresponding domain, i.e., an applicable null value corresponds to the whole domain.

The notion of a partial value is illustrated as follows (Grant 1979). A partial value on a universe of discourse *U* corresponds to a finite set of possible values in which exactly one of the values in the set is the true value, denoted by $\{a_1, a_2, \ldots, a_m\}$ for discrete *U* or $[a_1, a_n]$ for continua *U*, in which $\{a_1, a_2, \ldots, a_m\} \subseteq U$ or $[a_1, a_n] \subseteq U$. Let η be a partial value, then $sub(\eta)$ and $sup(\eta)$ are used to represent the minimum and maximum in the set.

Note that crisp data can also be viewed as special cases of partial values. A crisp data on discrete universe of discourse can be represented in the form of {p}, and a crisp data on continua universe of discourse can be represented in the form of [p, p]. Moreover, a partial value without containing any element is called an *empty partial value*, denoted by $\perp$. In fact, the symbol $\perp$ means an inapplicable missing data (Codd 1986, 1987). Null values, partial values, and crisp values are thus represented with a uniform format.

2.3 Representation of Fuzzy Sets and Possibility Distributions

In 1965, L. A. Zadeh published his innovating paper "Fuzzy Set" in the journal of *Information and Control* (Zadeh 1965). Since then fuzzy set has been infiltrating into almost all branches of pure and applied mathematics that are set-theory-based. This has resulted in a vast number of real applications crossing over a broad realm of domains and disciplines. Over the years, many of the existing approaches dealing with imprecision and uncertainty are based on the theory of fuzzy sets (Zadeh 1965) and possibility theory (Zadeh 1978).

Fuzzy data is originally described as fuzzy set (Zadeh 1965). A fuzzy set, say {0.6/18, 0.7/19, 0.8/20, 0.9/21} for the age of *Michael*, is more informative because it contains information imprecision (the age may be 18, 19, 20, or 21 and we do not know which one is true) and uncertainty (the degrees of truth of all possible age values are respectively 0.6, 0.7, 0.8, and 0.9) simultaneously.

Let U be a universe of discourse. A fuzzy value on U is characterized by a fuzzy set F in U. A membership function

$$\mu_F : U \rightarrow [0, 1]$$

is defined for the fuzzy set F, where $\mu_F(u)$, for each $u \in U$, denotes the degree of membership of u in the fuzzy set F. For example, $\mu_F(u) = 0.75$ means that u is "likely" to be an element of F by a degree of 0.75. For ease of representation, a fuzzy set F over universe U is organized into a set of ordered pairs:

$$F = \{\mu_F(u_1)/u_1, \mu_F(u_2)/u_2, \ldots, \mu_F(u_n)/u_n\}$$

When the membership function $\mu_F(u)$ above is explained to be a measure of the possibility that a variable X has the value u in this approach, where X takes values in U, a fuzzy value is described by a possibility distribution π_X (Zadeh 1978).

$$\pi_X = \{\pi_X(u_1)/u_1, \pi_X(u_2)/u_2, \ldots, \pi_X(u_n)/u_n\}$$

Here, $\pi_X(u_i)$, $u_i \in U$, denotes the possibility that u_i is true.

Moreover, the extension principle introduced by Zadeh (1975) has been regarded as one of the most basic ideas of fuzzy set theory. By providing a general method, the extension principle has been extensively employed to extend non-fuzzy mathematical concepts. The idea is to induce a fuzzy set from a number of given fuzzy sets through a mapping.

Zadeh's extension principle can also be referred to maximum-minimum principle sometimes. Let X_1, X_2, …, X_n and Y be ordinary sets, f be a mapping from $X_1 \times X_2 \times \ldots \times X_n$ to Y such that $y = f(x_1, x_2, \ldots, x_n)$, P ($X_i$) and P (Y) be the power sets of X_i and Y ($0 \leq i \leq n$), respectively. Here, P (X_i) = $\{C|C \subseteq X_i\}$ and P (Y) = $\{D|D \subseteq Y\}$. Then f induces a mapping from P (X_1) × P (X_2) × … × P (X_n) to P (Y) with

$$f(C_1, C_2, \ldots, C_n) = \{f(X_1, X_2, \ldots, X_n) | X_i \in C_i, 0 \leq i \leq n\}$$

Here $C_i \subseteq X_i$ and $0 \leq i \leq n$. Now, let F (X_i) be the class of all fuzzy sets on X_i, i.e., F $(X_i) = \{\}$, $0 \leq i \leq n$ and F (Y) be the class of all fuzzy sets on Y, i.e., F (Y) = { }, then *f* induces a mapping from F $(X_1) \times$ F $(X_2) \times \ldots \times$ F (X_n) to F (Y) such that for all $A_i \in$ F (X_i), *f* $(A_1, A_2, \ldots, A_n)$ is a fuzzy set on Y with

$$f(A_1, A_2, \ldots, A_n)(y) = \begin{cases} \sup\limits_{f(x1,x2,\ldots,xn)=y} (\min(\mu_{A1}(x1), \mu_{A2}(x2), \ldots, \mu_{An}(xn)), f^{-1}(y)) \neq \Phi \\ xi \in Xi\,(i = 1, 2, \ldots, n) \\ 0, f^{-1}(y) = \Phi \end{cases}$$

2.4 Operations on Fuzzy Sets

In order to manipulate fuzzy sets (as well as possibility distributions), several operations, including *set operations*, *arithmetic operations*, *relational operations*, and *logical operations*, should be defined. The usual *set operations* (such as union, intersection and complementation) have been extended to deal with fuzzy sets (Zadeh 1965). Let *A* and *B* be fuzzy sets on the same universe of discourse *U* with the membership functions μ_A and μ_B, respectively. Then we have

Union. The union of fuzzy sets *A* and *B*, denoted *A* U *B*, is a fuzzy set on *U* with the membership function $\mu_{A \cup B}$: $U \rightarrow [0, 1]$, where

$$\forall u \in U, \mu_{A \cup \bar{B}}(u) = \max(\mu_A(u), \mu_B(u)).$$

Intersection. The intersection of fuzzy sets *A* and *B*, denoted $A \cap B$, is a fuzzy set on *U* with the membership function $\mu_{A \cap B}$: $U \rightarrow [0, 1]$, where

$$\forall u \in U, \mu_{A \cap B}(u) = \min(\mu_A(u), \mu_B(u)).$$

Complementation. The complementation of fuzzy set $\bar{A}$, denoted by $\bar{A}$, is a fuzzy set on *U* with the membership function $\mu_{\bar{A}}$: $U \rightarrow [0, 1]$, where

$$\forall u \in U, \mu_A(u) = 1 - \mu_A(u).$$

Based on these definitions, the *difference* of the fuzzy sets B and A can be defined as:

$$B - A = B \cap \overline{A}.$$

Also, most of the properties that hold for classical set operations, such as DeMorgan's Laws, have been shown to hold for fuzzy sets. The only law of ordinary set theory that is no longer true is the law of the excluded middle, i.e.,

$$A \cap \overline{A} \neq \emptyset \, \text{and} \, A \cup \overline{A} \neq U.$$

Let A, B, and C be fuzzy sets in a universe of discourse U, then the operations on fuzzy sets satisfy the following conditions:

- Commutativity laws: $\mathrm{A} \cup \mathrm{B} = \mathrm{B} \cup \mathrm{A}, \mathrm{A} \cap \mathrm{B} = \mathrm{B} \cap \mathrm{A}$
- Associativity laws: $(\mathrm{A} \cup \mathrm{B}) \cup \mathrm{C} = \mathrm{A} \cup (\mathrm{B} \cup \mathrm{C}), (\mathrm{A} \cap \mathrm{B}) \cap \mathrm{C} = \mathrm{A} \cap (\mathrm{B} \cap \mathrm{C})$
- Distribution laws: $(\mathrm{A} \cup \mathrm{B}) \cap \mathrm{C} = (\mathrm{A} \cap \mathrm{C}) \cup (\mathrm{B} \cap \mathrm{C}), (\mathrm{A} \cap \mathrm{B}) \cup \mathrm{C} = (\mathrm{A} \cup \mathrm{C}) \cap (\mathrm{B} \cup \mathrm{C})$
- Absorption laws: $(\mathrm{A} \cup \mathrm{B})\ \mathrm{A} = \mathrm{A}, (\mathrm{A} \cap \mathrm{B}) \cup \mathrm{A} = \mathrm{A}$
- Idempotency laws: $\mathrm{A} \cup \mathrm{A} = \mathrm{A}, \mathrm{A} \cap \mathrm{A} = \mathrm{A}$
- De Morgan laws: $\overline{\mathrm{A} \cup \mathrm{B}} = \overline{\mathrm{A}} \cap \overline{\mathrm{B}}, \overline{\mathrm{A} \cap \mathrm{B}} = \overline{\mathrm{A}} \cup \overline{\mathrm{B}}$

Given two fuzzy sets A and B in U, B is a fuzzy subset of A, denoted by $B \subseteq A$, if

$$\mu_B(u) \leq \mu_A(u)$$

for all $u \in U$.

Two fuzzy sets A and B are said to be equal if $A \subseteq B$ and $B \subseteq A$.

In order to define Cartesian product of fuzzy sets, let $U = U_1 \times U_2 \times \ldots \times U_n$ be the Cartesian product of n universes and $A_1, A_2, \ldots, A_n$ be fuzzy sets in $U_1, U_2, \ldots, U_n$, respectively. The Cartesian product $A_1 \times A_2 \times \ldots \times A_n$ is defined to be a fuzzy subset of $U_1 \times U_2 \times \ldots \times U_n$, where

$$\mu_{A1 \times \ldots \times An}(u_1 \ldots u_n) = \min(\mu_{A1}(u_1), \ldots, \mu_{An}(u_n))$$

and $u_i \in U_i, i = 1, \ldots, n$.

Moreover, Let U be a universe of discourse and F a fuzzy set in U with the membership function μ_F: $U \rightarrow [0, 1]$. We have then the following notions related to fuzzy sets.

Support. The set of the elements that have non-zero degrees of membership in F is called the support of F, denoted by

$$supp(F) = \{u | \, u \in U \, \text{and} \, \mu_F(u) > 0\}$$

Kernel. The set of the elements that completely belong to F is called the kernel of F, denoted by

$$ker(F) = \{u | \, u \in U \, \text{and} \, \mu_F(u) = 1\}.$$

α-Cut. The set of the elements which degrees of membership in F are greater than (greater than or equal to) α, where $0 \leq \alpha < 1$ ($0 < \alpha \leq 1$), is called the strong (weak) α-cut of F, respectively denoted by

$$F_{\alpha+} = \{u |\, u \in U \,\text{and}\, \mu_F(u) > \alpha\}$$

and

$$F_{\alpha} = \{u |\, u \in U \,\text{and}\, \mu_F(u) \geq \alpha\}.$$

The relationships among the support, kernel, and α-cut of a fuzzy set can be illustrated in Fig. 2.1.

Consider the example of a preliminary product design. The value of the performance parameter *capacity* is "about 2.5e + 03", which is represented by the following fuzzy set

$$F = \{1.0/2.5\text{e} + 03,\ 0.96/5.0\text{e} + 03,\ 0.88/7.5\text{e} + 03,\ 0.75/1.0\text{e} + 04,\ 0.57/1.25\text{e} + 04,\ 0.32/1.5\text{e} + 04,\ 0.08/1.75\text{e} + 04\}.$$

Then, we have

$$\text{supp}(F) = \{2.5\text{e} + 03, 5.0\text{e} + 03, 7.5\text{e} + 03, 1.0\text{e} + 04, 1.25\text{e} + 04, 1.5\text{e} + 04, 1.75\text{e} + 04\},$$
$$\ker(F) = \{2.5\text{e} + 03\},$$
$$F_{0.88+} = \{2.5\text{e} + 03, 5.0\text{e} + 03\},\ \text{and}\ F_{0.88} = \{2.5\text{e} + 03,\ 5.0\text{e} + 03,\ 7.5\text{e} + 03\}.$$

Moreover, utilizing Zadeh's extension principle, which can also be referred to maximum-minimum principle sometimes, *arithmetic operations* can be defined. Let A and B be fuzzy sets on the same universe of discourse U with the membership functions μ_A and μ_B, respectively, and "θ" be an infix operator. $A\ \theta\ B$ is a fuzzy set on U with the membership function $\mu_{A\ \theta\ B}$: $U \rightarrow [0, 1]$, where

$$\forall z \in U, \mu_{A\theta B}(z) = \max_{z=x\theta y}(\min(\mu_A(x), \mu_B(y))).$$

In addition, *fuzzy relational operations* are various kinds of comparison operations on fuzzy sets, namely, *equal* (=), *not equal* ($\neq$), *greater than* (>), *greater*

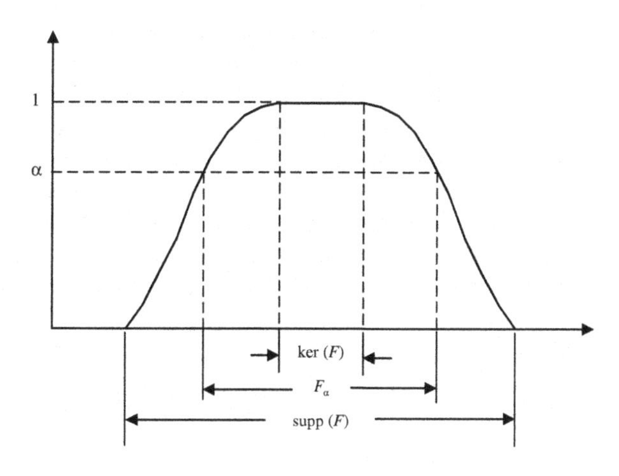

Fig. 2.1 Support, kernel, and α-cut of fuzzy sets

than or equal (≥), *less than* (<), and *less than or equal* (≤). The definitions of these fuzzy relational operations are essentially related to the closeness measures between fuzzy sets and the given thresholds.

Let A and B be fuzzy sets on the same universe of discourse U with the membership functions μ_A and μ_B, respectively, and β be a given threshold value. Then we have

(a) $A \approx_\beta B$ if $SE(A, B) \geq \beta$,
(b) $A \not\approx_\beta B$ if $SE(A, B) < \beta$,
(c) $A \succ_\beta B$ if $SE(A, B) < \beta$ and max (supp (A)) > max (supp (B)),
(d) $A \prec_\beta B$ if $SE(A, B) < \beta$ and max (supp (A)) < max (supp (B)),
(e) $A \succeq_\beta B$ if $A \succ_\beta B$ or $A \approx_\beta B$, and
(f) $A \preceq_\beta B$ if $A \prec_\beta B$ or $A \approx_\beta B$.

Now let us have a close look at the fuzzy sets being operators. Three kinds of fuzzy sets can be identified: simple (atomic) fuzzy set, modified (composite) fuzzy set, and compound fuzzy set.

Simple (atomic) fuzzy set. A simple fuzzy set such as *young* or *tall* is defined by a fuzzy number with membership function.

Modified (composite) fuzzy set. A modified fuzzy set such as *very young* or *more or less tall* is described by a fuzzy number with membership function. Note that its membership function is not defined but computed through the membership function of the corresponding simple fuzzy set. In order to compute the membership function of modified fuzzy set, some semantic rules should be used. Let F is a simple fuzzy set represented by a fuzzy number in the universe of discourse U and its membership function is μ_F: $U \rightarrow [0,1]$, then we have the following rules.

Concentration rule: $\mu_{\text{very } F}(\text{u}) = (\mu_F(\text{u}))^2$
More generally, $\mu_{\text{very very } \ldots \text{ very } F}(\text{u}) = (\mu_F(\text{u}))^{2 \times \text{(times of very)}}$
Dilation rule: $\mu_{\text{more or less } F}(\text{u}) = (\mu_F(\text{u}))^{1/2}$.

Compound fuzzy set. A compound fuzzy set such as *young* ∩ *very tall* is represented by simple fuzzy sets or modified fuzzy sets connected by union (∪), intersection (∩) or complementation connectors.

Generally speaking, the results of fuzzy relational operations are fuzzy Boolean. They can be combined with logical operators such as *not* (¬), *and* (∧), and *or* (∨) to form complicated logical expression. Such expression can be used to represent logical conditions for information retrieval and so on. In the above definitions of the fuzzy relational operations, classical two-valued logic (2VL), namely, true (T) and false (F), is used because of the use of threshold values.

In the relational operations of fuzzy sets, there may be some fuzzy relations such as (*not*) *close to*/*around*, (*not*) *at lease*, and (*not*) *at most* etc. with crisp values in addition to the traditional operators such as >, <, =, ≠, ≥, and ≤. Now consider fuzzy relations as operators and crisp values as operands. For $A\,\tilde{\theta}Y$, where A is an attribute, $\tilde{\theta}$ is a fuzzy relation, and Y is a crisp value, $\tilde{\theta}Y$ is a fuzzy number.

First, let's focus on fuzzy relation "*close to (around)*". According to (Chen and Jong 1997), the membership function of the fuzzy number "*close to Y (around Y)*" on the universe of discourse can be defined by

$$\mu_{close\ to\ Y}(\mathrm{u}) = \frac{1}{1 + \left(\frac{\mathrm{u-Y}}{\beta}\right)^2}$$

The membership function of the fuzzy number "*close to Y*" is shown in Fig. 2.2.

It should be noted that the fuzzy number above is a simple fuzzy term. Based on it, we have the following modified fuzzy terms: "*very close to Y*", "*very very … very close to Y*", and "*more or less close to Y*", which membership functions can be defined as

$$\mu_{\text{very close to Y}}(\mathrm{u}) = (\mu_{\text{close to Y}}(\mathrm{u}))^2,$$

$$\mu_{\text{very very...very close to Y}}(\mathrm{u}) = (\mu_{\text{close to Y}}(\mathrm{u}))^{2\times(\text{times of very})}, \text{ and}$$

$$\mu_{\text{more or less close to Y}}(\mathrm{u}) = (\mu_{\text{close to Y}}(\mathrm{u}))^{1/2}.$$

Based on fuzzy number "*close to Y*", a compound fuzzy term "*not close to Y*" can be defined. Its membership function is as follows.

$$\mu_{\text{not close to Y}}(\mathrm{u}) = (1 - \mu_{\text{close to Y}}(\mathrm{u}))$$

Second, the membership function of the fuzzy number "*at least Y*" on the universe of discourse (Bosc and Pivert 1995) can be defined by

$$\mu_{at\ least\ Y}(\mathrm{u}) = \begin{cases} 0, u \le \omega \\ \dfrac{u-\omega}{Y-\omega}, \omega < u < Y. \\ 1, u \ge Y \end{cases}$$

The membership function of the fuzzy number "*at least Y*" is shown in Fig. 2.3.

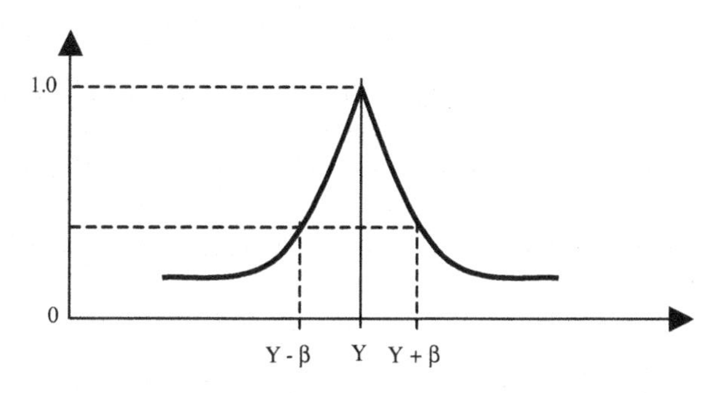

Fig. 2.2 Membership function of the fuzzy number "*close to Y*"

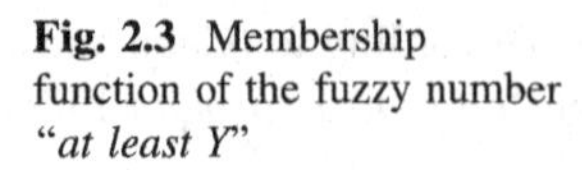
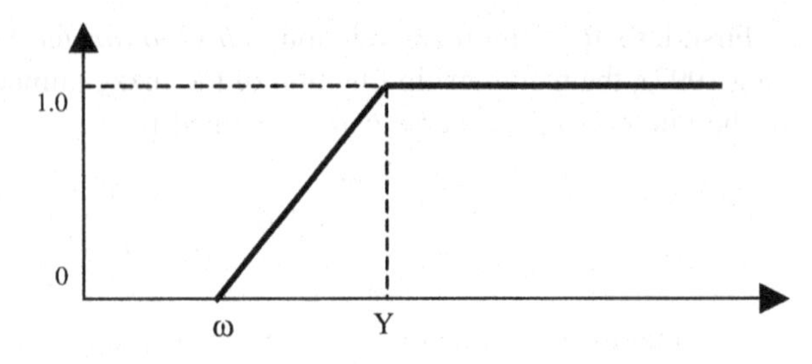

Fig. 2.3 Membership function of the fuzzy number "*at least Y*"

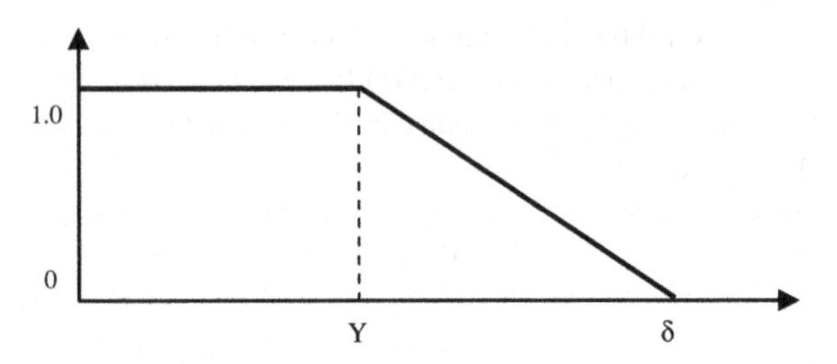

Fig. 2.4 Membership function of the fuzzy number "*at most Y*"

Based on fuzzy number "*at least Y*", a compound fuzzy term "*not at least Y*" can be defined. Its membership function is as follows.

$$\mu_{not\ at\ least\ Y}(\mathrm{u}) = (1 - \mu_{\mathrm{at\ least\ Y}}(\mathrm{u}))$$

Finally let's focus on fuzzy relation "*at most*". The membership function of the fuzzy number "*at most Y*" on the universe of discourse can be defined by

$$\mu_{at\ most\ Y}(\mathrm{u}) = \begin{cases} 1, u \leq Y \\ \dfrac{\delta - u}{\delta - Y}, Y < u < \delta \\ 0, u \geq \delta \end{cases}.$$

The membership function of the fuzzy number "*at most Y*" is shown in Fig. 2.4.

Based on fuzzy number "*at most Y*", a compound fuzzy term "*not at most Y*" can be defined. Its membership function is as follows.

$$\mu_{\mathrm{not\ at\ most\ Y}}(\mathrm{u}) = (1 - \mu_{\mathrm{atmostY}}(\mathrm{u}))$$

The fuzzy relations "close to", "not close to", "at least", "at most", "not at least", and "not at most" can be viewed as "fuzzy equal to", "fuzzy not equal to", "fuzzy greater than and equal to", "fuzzy less than and equal to", "fuzzy greater than", and "fuzzy less than", respectively. Using these fuzzy relations and crisp values, the fuzzy query condition with fuzzy operators, which has the form $\mathrm{A}\tilde{\theta}\mathrm{Y}$, is formed.

Fuzzy logical operations dependent on the representation of fuzzy Boolean values as well as fuzzy logic. Three logical operations *fuzzy not* ($\tilde{\neg}$), *fuzzy and* ($\tilde{\wedge}$), and *fuzzy or* ($\tilde{\vee}$), which operands are fuzzy Boolean value(s) represented by fuzzy sets, are defined in the following.

Fuzzy and. The result of *fuzzy and* is a fuzzy Boolean value. *Fuzzy and* can be defined with "intersection" kinds of operations such as "min" operation. Let A: μfi$_A$ (u) and B: μ_B (u) be two fuzzy Boolean values represented by fuzzy sets on the same universe of discourse U. Then

$$A\tilde{\wedge}B : \min(\mu_A(u), \mu_B(u)), u \in U.$$

Fuzzy or. The result of *fuzzy or* is a fuzzy Boolean value. *Fuzzy or* can be defined with "union" kinds of operations such as "max" operation. Let A: μ_A (u) and B: μ_B (u) be two fuzzy Boolean values represented by fuzzy sets on the same universe of discourse U. One has

$$A\tilde{\vee}B : \max(\mu_A(u), \mu_B(u)), u \in U.$$

Fuzzy not. The result of *fuzzy not* is a fuzzy Boolean value. *Fuzzy not* can be defined with "complementation" kinds of operations such as "subtraction" operation. Let A: μ_A (u) be a fuzzy Boolean values represented by fuzzy sets on the universe of discourse U. One has

$$\tilde{\neg}A : (1 - \mu_A(u)), u \in U.$$

A fuzzy set A w.r.t U is called *convex*, iff for all $u_1, u_2 \in U$, $\mu_A(\lambda u_1 + (1-\lambda)u_2) \geq min(\mu_A(u_1), \mu_A(u_2))$, where $\lambda \in [0,1]$. A fuzzy set A w.r.t U is called *normal*, if $\exists u \in U$, s.t. $\mu_A(u) = 1$. A *fuzzy number* is a convex and normal fuzzy set. The set of elements whose membership degrees in A are greater than (greater than or equal to) α, where $0 \leq \alpha < 1\,(0 < \alpha \leq 1)$, is called the *strong (weak)* α*-cut* of A, denoted by $A_{\alpha+} = \{u \in U | \mu_A(u) > \alpha\}$ and $A_\alpha = \{u \in U | \mu_A(u) \geq \alpha\}$. The α-cut of a fuzzy number corresponds to an interval of U. Let A and B be two fuzzy numbers of U, and $A_\alpha = [x_1, y_1]$ and $B_\alpha = [x_2, y_2]$ the α-cuts of A and B, respectively. Then we have $(A \cup B)_\alpha = A_\alpha \underline{\cup} B_\alpha$, $(A \cap B)_\alpha = A_\alpha \underline{\cap} B_\alpha$, where $\underline{\cup}$ and $\underline{\cap}$ denote the union and the intersection operators between two intervals, respectively. They are defined as follows,

$$A_\alpha \underline{\cup} B_\alpha = \begin{cases} [x_1, y_1] \underline{\cup} [x_2, y_2], & if\, A_\alpha \cap B_\alpha = \emptyset \\ [\min(x_1, x_2), \max(y_1, y_2)], & if\, A_\alpha \cap B_\alpha \neq \emptyset \end{cases}$$

$$A_\alpha \underline{\cap} B_\alpha = \begin{cases} \emptyset, & if\, A_\alpha \cap B_\alpha = \emptyset \\ [\max(x_1, x_2), \min(y_1, y_2)], & if\, A_\alpha \cap B_\alpha \neq \emptyset \end{cases}$$

2.5 Summary

In real-world applications, information is often imprecise or uncertain. Many sources can contribute to the imprecision and uncertainty of data or information. It is particular true in the knowledge representation and reasoning in the Semantic Web as well as applications using Semantic Web techniques such as ontologies, Description Logics, and rules. In particular, many of the existing approaches dealing with imprecision and uncertainty are based on the theory of fuzzy sets and possibility theory. Zadeh's fuzzy set theory has been identified as a successful technique for modelling the fuzzy information in many application areas. For example, as we have known, fuzzy set theory has been extensively applied to extend various database models and resulted in numerous contributions as will be introduced in the next Chap. 3.

References

Bosc P, Pivert O (1995) SQLf: a relational database language for fuzzy querying. IEEE Trans Fuzzy Syst 3(1):1–17

Bosc P, Kraft DH, Petry FE (2005) Fuzzy sets in database and information systems: status and opportunities. Fuzzy Sets Syst 156(3):418–426

Calegari S, Ciucci, D. (2006) Integrating fuzzy logic in ontologies. In: Proceedings of the 8th international conference on enterprise information systems, pp. 66–73

Chen SM, Jong WT (1997) Fuzzy query translation for relational database systems. IEEE Trans Syst Man Cybern Part B: Cybern 27(4):714–721

Codd EF (1986) Missing information (applicable and inapplicable) in relational databases. SIGMOD Rec 15:53–78

Codd EF (1987) More commentary on missing information in relational databases (applicable and inapplicable information). SIGMOD Rec 16(1):42–50

Dalvi N, Suciu D (2007) Management of probabilistic data: foundations and challenges. In: Proceedings of the ACM SIGACT-SIGMOD-SIGART symposium on principles of database systems, pp. 1–12

Gottlob G, Zicari R (1988) Closed world databases opened through null values. In: Proceedings of the 1988 international conference on very large data bases, pp. 50–61

Grant J (1979) Partial values in a tabular database model. Inf Process Lett 9(2):97–99

Klir GJ, Yuan B (1995) Fuzzy sets and fuzzy logic: theory and applications. Prentice-Hall, Upper Saddle River

Ma ZM, Yan L (2008) A literature overview of fuzzy database models. J Inf Sci Eng 24(1):189–202

Ma ZM, Zhang F, Wang HL, Yan L (2012) An overview of fuzzy description logics for the Semantic Web. Knowl Eng Rev 28(1):1–34

Motor A (1990) Accommodation imprecision in database systems: issues and solutions. ACM SIGMOD Rec 19(4):69–74

Parsons S (1996) Current approaches to handling imperfect information in data and knowledge bases. IEEE Trans Knowl Data Eng 8:353–372

Stoilos G, Stamou G, Pan JZ (2010) Fuzzy extensions of OWL: logical properties and reduction to fuzzy description logics. Int J Approx Reason 51:656–679

Sanchez E (2006) Fuzzy logic and the Semantic Web. Elsevier, Amsterdam

Smets P (1996) Imperfect information: imprecision and uncertainty. In: Motro A, Smets P (eds) Uncertainty management in information systems: from needs to solution. Kluwer Academic Publishers, Boston, pp. 225–254

Zaniolo C (1984) Database relations with null values. JCSS 21(1):142–162

Zadeh LA (1975) The concept of a linguistic variable and its application to approximate reasoning. Inf Sci 8:119–249 and 301–357; 9:43–80

Zadeh LA (1965) Fuzzy sets. Inf Control 8(3):338–353

Zadeh LA (1978) Fuzzy sets as a basis for a theory of possibility. Fuzzy Sets Syst 1(1):3–28

Chapter 3
Fuzzy Data Models and Formal Descriptions

3.1 Introduction

Information is often imprecise and uncertain in real-world applications. It is also true in the context of the Semantic Web. Therefore, in order to realize the fuzzy knowledge management for the Semantic Web, some knowledge management techniques such as query, extraction and storage should be developed. In this case, it is not surprising that fuzzy databases may be useful for providing some technique supports for managing fuzzy knowledge in the Semantic Web. As we have known, a significant body of research in the area of fuzzy databases has been developed over the past thirty years and tremendous gain is hereby accomplished in this area. Therefore, on the basis of the widespread studies and the relatively mature techniques of fuzzy databases, the existing numerous contributions of fuzzy databases may facilitate the fuzzy knowledge management for the Semantic Web. For example, fuzzy databases might be the more promising alternative for storing fuzzy knowledge in the Semantic Web because of the efficient storage mechanisms of fuzzy databases; or much valuable information and implicit knowledge in the fuzzy databases may be considered as the main data resources for supporting fuzzy ontology development; or the query techniques in the area of fuzzy databases may be applied for querying fuzzy knowledge bases in the Semantic Web.

On this basis, in this chapter, we introduce fuzzy data models. As have been reviewed in Ma and Yan (2008, 2010), since the early 1980s, Zadeh's fuzzy logic (Zadeh 1965) has been used to extend various data models. The purpose of introducing fuzzy logic in databases is to enhance the classical models such that uncertain and imprecise information can be represented and manipulated. This has resulted in numerous contributions, mainly with respect to the popular fuzzy conceptual data models (fuzzy ER/EER model, fuzzy UML model, and etc.) and fuzzy logical database models (fuzzy relational database model, fuzzy object-oriented database model, and etc.). All of these fuzzy data models will be introduced in the following sub-chapters in detail.

Z. Ma et al., *Fuzzy Knowledge Management for the Semantic Web*,
Studies in Fuzziness and Soft Computing 306, DOI: 10.1007/978-3-642-39283-2_3,

3.2 Fuzzy Conceptual Data Models

Conceptual data modeling involves conceptual (semantic) data models. The conceptual data models, e.g., ER/EER and UML, provide the designers with powerful mechanisms in generating the most complete specification from the real world. They represent both complex structures of entities and complex relationships among entities as well as their attributes. Hence, the conceptual data models play an important role in conceptual data modeling and database conceptual design.

A major goal for database research has been the incorporation of additional semantics into the data model. Classical data models often suffer from their incapability to represent and manipulate imprecise and uncertain information that may occur in many non-traditional applications in the real world (e.g., decision-making and expert systems). In order to deal with complex objects and imprecise and uncertain information in conceptual data modeling, one needs fuzzy extension to conceptual data models, which allow imprecise and uncertain information to be represented and manipulated at a conceptual level. Various fuzzy conceptual data models have been proposed in the literature such as fuzzy ER model, fuzzy EER model, fuzzy IFO model, and fuzzy UML model. Please refer to Ma and Yan (2010) in detail. In this sub-chapter, we will focus on the common fuzzy ER model and fuzzy UML model.

3.2.1 *Fuzzy ER Models*

Zvieli and Chen (1986) first applied the fuzzy set theory to some of the basic ER concepts, in which fuzzy entity sets, fuzzy relationship sets and fuzzy attribute sets were introduced in addition to fuzziness in entity and relationship occurrences and in attribute values. The fuzzy extension to the ER data model is carried out in three levels:

- At the first level, entity types, relationship types and attributes may all be fuzzy sets. That means we only have the partial knowledge that they possibly belong to the ER model.
- At the second level, for each entity type and relationship type, the sets of their instances can be fuzzy sets. That is, they reflect possible partial belonging of the corresponding values to their types. Formally, let *e* and *r* be the fuzzy instance sets of an entity type, say *E*i, and a relationship type, say *R*i, respectively, and μ_e and μ_r be their membership functions. Then
  - for an entity instance of fuzzy instance set *e* to the (fuzzy) entity type, say *e*i, we have $\mu_e(ei)/ei$, where $\mu_e(ei)$ is the degree of *e*i belonging to *e* and $0 \le \mu_e(ei) \le 1$, and

– for a relationship instance of fuzzy instance set r to the (fuzzy) relationship type, say ri, we have $\mu_r(ri)/ri$, where $\mu_r(ri)$ is the degree of ri belonging to r and $0 \le \mu_r(ri) \le 1$.

- At the third level, for each attribute, any of its values can be a fuzzy set. Suppose we have attribute "*Age*" and its values may be "*young*" or "*old*".

The detailed introduction about fuzzy ER models and the fuzzy extensions to the ER algebra can be found in Zvieli and Chen (1986). Other efforts to extend the ER model with fuzzy features can be found in Ma et al. (2001), Bouaziz et al. (2007), Chen and Kerre (1998), Ruspini (1986), Vandenberghe (1991) and Vert et al. (1986).

In the following we provide *a formal definition of fuzzy ER models* in Ma et al. (2010); Zhang et al. (2013). The formal definition includes most of general and important features of fuzzy ER models such as fuzzy entities, fuzzy relationships, fuzzy attributes, and some constraints. Note that, the fuzziness at the first level is not included in the formal definition. Moreover, *we provide an example of a full fuzzy ER model including its graphical model and formal representation* (Ma et al. 2010; Zhang et al. 2013). Further, we give the semantics interpretation method of fuzzy ER models. All of these will be used in the later chapters when we introduce how to manage fuzzy knowledge for the Semantic Web.

Here we first define a function as follows: for two finite sets X and Y, we call a function from a subset of X to Y an X-labeled tuple over Y. The labeled tuple T that maps $x_i \in X$ to $y_i \in Y$ ($i \in \{1,\ldots,k\}$) is denoted $[x_1: y_1, \ldots, x_k: y_k]$, and we also write $T[x_i]$ to denote y_i.

Definition 3.1 (*fuzzy ER models*) A fuzzy ER model can be defined as a tuple $FS = (L_{FS}, \le_{FS}, att_{FS}, rel_{FS}, card_{FS})$, where L_{FS} is a set of symbols; $\le_{FS}$ is a binary relation denoting the inheritance relationship (i.e., ISA) between two fuzzy entities or two fuzzy relationships; att_{FS} is a function that associates each fuzzy entity to its corresponding attributes; rel_{FS} is a function that associates each fuzzy relationship to its corresponding participation entities; $card_{FS}$ is a function that specifies cardinality constraints that instances of a fuzzy entity may participate in a fuzzy relationship:

- $L_{FS} = E_{FS} \cup A_{FS} \cup U_{FS} \cup R_{FS} \cup D_{FS}$ is a finite *alphabet*, where E_{FS} is a set of *fuzzy entity* symbols; A_{FS} is a set of *fuzzy attribute* symbols; U_{FS} is a set of *fuzzy role* symbols; R_{FS} is a set of *fuzzy relationship* symbols; and D_{FS} is a set of *fuzzy domain* symbols. Each fuzzy domain symbol $FD \in D_{FS}$ has an associated pre-defined basic domain D^{FB}, and we assume the various basic domains to be pairwise disjoint.
- $\le_{FS} \subseteq ER_{FS} \times ER_{FS}$ is a relation over ER_{FS}. Here $ER_{FS} = E_{FS} \cup R_{FS}$. $\le_{FS}$ denotes the *inheritance relationship* (i.e., *ISA*) between two fuzzy entities or two fuzzy relationships.
- att_{FS}: $E_{FS}/R_{FS} \rightarrow T(A_{FS}, D_{FS})$ is a function that maps each *fuzzy entity/relationship* symbol in E_{FS}/R_{FS} to an A_{FS}-labeled tuple over D_{FS}. The function att_{FS} associates each fuzzy entity/relationship to its corresponding attributes. For

simplicity we assume that attributes are single-valued and mandatory, and we could easily handle multi-valued attributes with associated cardinalities.

- rel_{FS}: $R_{FS} \rightarrow T(U_{FS}, E_{FS})$ is a function that maps each *fuzzy relationship* symbol in R_{FS} to an U_{FS}-labeled tuple over E_{FS}. Without loss of generality we assume that:
  - Each fuzzy role is specific to exactly one fuzzy relationship;
  - For each fuzzy role $FU \in U_{FS}$, there is a fuzzy relationship FR and a fuzzy entity FE such that $rel_{FS}(FR) = [\ldots, FU{:}FE, \ldots]$. The function rel_{FS} combines a fuzzy relationship with fuzzy entities and fuzzy roles, and it decides the arity of the fuzzy relationship.
- $card_{FS}$ is a function from $E_{FS} \times R_{FS} \times U_{FS}$ to $N_0 \times (N_0 \cup \{\infty\})$ that satisfies the condition: for a fuzzy relationship $FR \in R_{FS}$ such that $rel_{FS}(FR) = [\ldots, FU{:}FE, \ldots]$, defining $card_{FS}(FE, FR, FU) = (cmin_{FS}(FE, FR, FU), cmax_{FS}(FE, FR, FU))$, if not stated otherwise, $cmin_{FS}(FE, FR, FU)$ is assumed to be 0 and $cmax_{FS}(FE, FR, FU)$ is assumed to be ∞. The function $card_{FS}$ specifies *cardinality constraints*, i.e., constraints on the minimum and maximum number of times an instance of a fuzzy entity may participate in a fuzzy relationship via some fuzzy role.

Figure 3.1 shows a *fuzzy ER model* FS_1 modeling a part of the reality at a company. The corresponding *formal representation* of the fuzzy ER model FS_1 is given in Fig. 3.2. In the fuzzy ER model:

- In a fuzzy ER model, an instance may belong to an entity with a degree of [0, 1]. An attribute $u \in [0, 1]$ is introduced into an entity to represent the membership degree of an instance to the entity.
- The attributes of entities may be fuzzy. A fuzzy keyword *FUZZY* appears in front of attributes indicating the attributes are fuzzy attributes. For example, the attribute "*Age*" of the fuzzy entity *Young-Employee* is a fuzzy attribute.

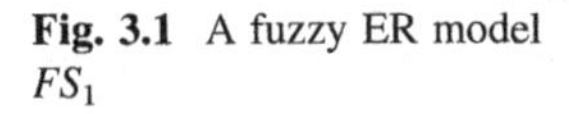
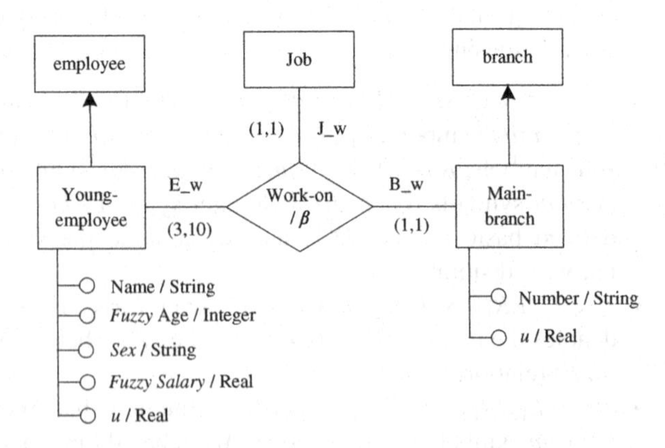

Fig. 3.1 A fuzzy ER model FS_1

According to Definition 3.1, the formal syntax of the fuzzy ER model FS_1 in Fig. 3.1 is as follows:

$FS_1 = (L_{FS}, \leq_{FS}, att_{FS}, rel_{FS}, card_{FS})$, where:

$L_{FS} = E_{FS} \cup A_{FS} \cup U_{FS} \cup R_{FS} \cup D_{FS}$ is a set of symbols partitioned into:
E_{FS} = {employee, Young-employee, Job, branch, Main-branch}
A_{FS} = {Name, Fuzzy Age, Sex, Fuzzy Salary, u, Number}
U_{FS} = {E_w, J_w, B_w} R_{FS} = {Work-on}
D_{FS} = {String, Integer, Real}

and the set of functions/relations over the symbols in L_{FS} consists of:
Young-employee$\leq_{FS}$ employee; Main-branch$\leq_{FS}$ branch;
att_{FS} (Young-employee) = [Name: String, FUZZY Age: Integer, Sex: String, Fuzzy Salary: Real, u: Real];
att_{FS} (Main-branch) = [Number: String, u: Real];
att_{FS} (Work-on) = [β: Real];
rel_{FS} (Work-on) = [E_w: Young-employee, J_w: Job, B_w: Main-branch];
$card_{FS}$ (Young-employee, Work-on, E_w) = (3, 10);
$card_{FS}$ (Job, Work-on, J_w) = (1, 1);
$card_{FS}$ (Main-branch, Work-on, B_w) = (1, 1).

Fig. 3.2 The formal representation of the fuzzy ER model FS_1 in Fig. 3.1

- The relationships among entities may be fuzzy. The symbol $\beta \in [0, 1]$ is also introduced into a fuzzy relationship to denote the membership degree of the fuzzy relationship occurring in several entities.

Being similar to the semantics of ER models (Baader et al. 2003; Calvanese et al. 1999), the semantics of fuzzy ER models can be given by fuzzy database states. That is, by specifying which fuzzy database state is consistent with the information structure of a fuzzy ER model.

Definition 3.2 (*semantics of fuzzy ER models*) Formally, a fuzzy database state FB corresponding to a fuzzy ER model $FS = (L_{FS}, \leq_{FS}, att_{FS}, rel_{FS}, card_{FS})$ is constituted by a nonempty finite set Δ^{FB}, assumed to be disjoint from all basic domains, and a function $\cdot^{FB}$ that maps:

- Each domain symbol $FD \in D_{FS}$ to the corresponding basic domain D^{FB}, that is $FD^{FB} \in D^{FB}$.
- Each fuzzy entity $FE \in E_{FS}$ to a subset FE^{FB} of Δ^{FB}, that is $FE^{FB} \subseteq \Delta^{FB}$.
- Each fuzzy attribute $FA \in A_{FS}$ to a set $FA^{FB} \subseteq \Delta^{FB} \times \cup FD^{FB}$, where $FD \in D_{FS}$.
- Each fuzzy relationship $FR \in R_{FS}$ to a set FR^{FB} of U_{FS}-labeled tuples over Δ^{FB}.

Furthermore, a fuzzy database state is considered acceptable if it satisfies all integrity constraints of the fuzzy ER model. This is captured by the following definition.

Definition 3.3 A fuzzy database state FB is legal for a fuzzy ER model $FS = (L_{FS}, \leq_{FS}, att_{FS}, rel_{FS}, card_{FS})$ if the following conditions are satisfied:

- For each pair of fuzzy entities $FE_1, FE_2 \in E_{FS}$ such that $FE_1 \leq_{FS} FE_2$, $FE_1^{FB} \subseteq FE_2^{FB}$ holds.

- For each pair of fuzzy relations FR_1, $FR_2 \in R_{FS}$ such that $FR_1 \leq_{FS} FR_2$, $FR_1^{FB} \subseteq FR_2^{FB}$ holds.
- For each fuzzy entity $FE \in E_{FS}$, if $att_{FS}(FE) = [FA_1{:}\ FD_1,\ldots,FA_k{:}\ FD_k]$, there is exactly one element $Fa_i = < Fe, Fd_i > \in FA_i^{FB}$, $Fd_i \in D_i^{FB}$ for each instance $Fe \in FE^{FB}$ and for each $i \in \{1,\ldots,k\}$, where the first component is Fe and the second component is an element of D_i^{FB}.
- For each fuzzy relationship $FR \in R_{FS}$ and $rel_{FS}(FR) = [FU_1{:}\ FE_1,\ldots,FU_n{:}\ FE_n]$, all instances of FR are of the form $[FU_1{:}\ Fe_1,\ldots,FU_n{:}\ Fe_n]$, where $Fe_i \in FE_i^{FB}$, and $i \in \{1,\ldots,n\}$.
- For each fuzzy relationship $FR \in R_{FS}$ such that $rel_{FS}(FR) = [\ldots,FU{:}\ FE,\ldots]$, $cmin_{FS}(FE,FR,FU) \leq \#\{Fr \in FR^{FB} \mid Fr[FU] = Fe\} \leq cmax_{FS}(FE,FR,FU)$ holds for each instance $Fe \in FE^{FB}$, where $\#\{\}$denotes the base of set $\{\}$.

It is worth emphasizing that the definition of fuzzy database state reflects the usual assumption that fuzzy database states are finite structures. In fact, the basic domains are not required to be finite, but for each legal fuzzy database state for a fuzzy ER model, only the finite set of values of attributes is actually of interest.

3.2.2 Fuzzy UML Models

The fuzzy UML model is a fuzzy extension of unified modeling language (UML), which is a set of object-oriented (OO) modeling notations that has been standardized by the object management group (OMG). UML provides a collection of models to capture the many aspects of a software system (Booch et al. 1998). While the UML reflects some of the best OO modeling experiences available, it suffers from some lacks of necessary semantics. One of the lacks can be generalized as the need to handle imprecise and uncertain information although such information exist in knowledge engineering and databases and have extensively being studied. Consequently, fuzzy UML models were widely investigated for modeling imprecise and uncertain information (Ma and Yan 2007; Ma et al. 2011a; Haroonabadi and Teshnehlab 2007, 2009; Sicilia et al. 2002).

In the following, we first introduce some basic notions of fuzzy UML models, including fuzzy class, fuzzy generalization, fuzzy aggregation, fuzzy dependency, and fuzzy association (Ma et al. 2011b; Zhang and Ma 2013). Then, we provide a formal definition of fuzzy UML models, give the semantic interpretation of fuzzy UML models, and show an example of a fuzzy UML model (Ma et al. 2011b; Zhang and Ma 2013). All of these are the basic of extracting fuzzy Description Logics and fuzzy ontologies from fuzzy UML models as will be introduced in the later chapters.

3.2.2.1 Fuzzy Class

An object, which is an entity of the real world, is fuzzy because of a lack of information. Formally, objects that have at least one attribute whose value is a fuzzy set are fuzzy objects. The objects with the same structure and behavior are grouped into a class, and classes are organized into hierarchies. Theoretically, a class can be considered from two viewpoints:

- An *extensional* class, where the class is defined by a list of its object instances;
- An *intensional* class, where the class is defined by a set of attributes and their admissible values.

In a fuzzy UML model, a class is fuzzy because of the following several reasons (Ma and Yan 2007):

- A class is extensionally defined, where some objects with similar properties are fuzzy ones. The class defined by these objects may be fuzzy, and these objects belong to the class with degree of [0, 1].
- When a class is intensionally defined, the domains of some attributes may be fuzzy, and thus a fuzzy class is formed.
- The subclass produced by a fuzzy class by means of specialization, and the superclass produced by some classes (in which there is at least one class who is fuzzy) by means of generalization, are fuzzy.

Figure 3.3 shows a fuzzy UML class *Old-Employee*, where a fuzzy class is denoted by using a *dashed-outline rectangle* to differentiate a classical class. Here:

- A fuzzy keyword *FUZZY* is appeared in front of an attribute indicating the attribute may take fuzzy values. For example, *FUZZY* Age or *FUZZY* Email.
- For an attribute a of type T in a class C, an optional multiplicity $[i...j]$ for a specifies that a associates to each instance of C at least i and most j instances of T. When the multiplicity is missing, [1...1] is assumed, i.e., the attribute is single-valued. For example, the attribute "*FUZZY* Email [1...∞]" in Fig. 3.3 means that each object instance of the class *Old-Employee* has at least one email, and possibly more.

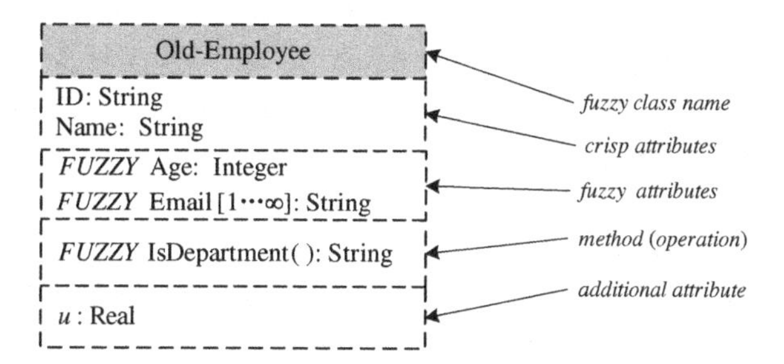

Fig. 3.3 Representation of a *fuzzy class Old-Employee* in a fuzzy UML model

- The method *IsDepartment():String* denotes the dynamic aspect of fuzzy UML models. It returns a possibility distribution value. The type of the parameter is *null*.
- An additional attribute $u \in [0, 1]$ in the class is defined for representing the object instance membership degree to the class.

3.2.2.2 Fuzzy Generalization

The concept of *generalization* is one of the basic building blocks of the fuzzy UML model. A *generalization* is a taxonomic relationship between a more general classifier named *superclass* and a more specific classifier named *subclass*. The subclass is produced from the superclass by means of inheriting all attributes and methods of the superclass, overriding some attributes and methods of the superclass, and defining some attributes and methods.

A class produced from a fuzzy class must be fuzzy. If the former is still called subclass and the latter superclass, the subclass/superclass relationship is fuzzy. If a fuzzy class is a subclass of another fuzzy class, for any object, say o, let the membership degree that it belongs to the subclass, say B, be $u_B(o)$ and the membership degree that it belongs to the superclass, say A, be $u_A(o)$. Then $u_B(o) \leq u_A(o)$. This characteristic can determine if two classes have a subclass/superclass relationship.

Several generalizations can be grouped together to form a *class hierarchy* as shown in Fig. 3.4. Figure 3.4 shows a fuzzy generalization relationship, where a *dashed triangular arrowhead* is used to represent a fuzzy generalization relationship. The *disjointness* and *completeness* constraints, which are optional, can be enforced on a class hierarchy. The *disjointness* means that all the specific classes are mutually disjoint, and *completeness* means that the union of the more specific classes completely covers the more general class. That is, the union of object instances of several subclasses completely covers the object instances of the superclass, and the membership degree that any object belongs to the subclass must be less than or equal to the membership degree that it belongs to the superclass.

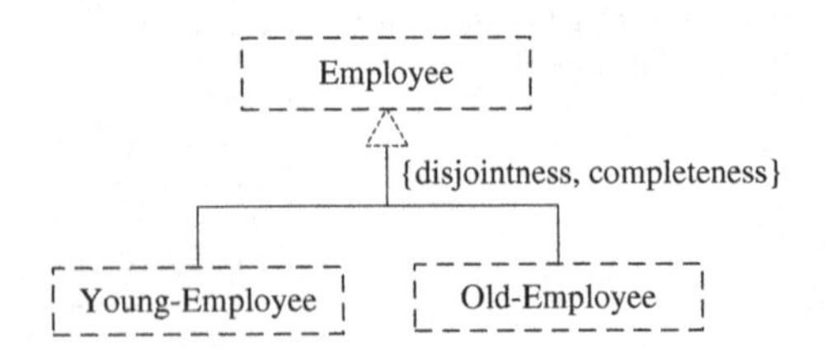

Fig. 3.4 Representation of a *fuzzy generalization* in a fuzzy UML model

3.2.2.3 Fuzzy Aggregation

An *aggregation* captures a *whole-part* relationship between a class named aggregate and a group of classes named constituent parts. The constituent parts can exist independently. Aggregate class *New-Computer*, for instance, is aggregated by constituent parts *Monitor*, *New CPU box*, and *Keyboard*. Each object of an aggregate can be projected into a set of objects of constituent parts. Formally, Let C be an aggregate of constituent parts $C_1, C_2, \ldots, C_n$. For $o \in C$, the projection of o to C_i is denoted by $o\downarrow_{Ci}$. Then we have

$(o\downarrow_{C1}) \in C_1$, $(o\downarrow_{C2}) \in C_2$, …, and $(o\downarrow_{Cn}) \in C_n$.

A class aggregated from fuzzy constituent parts must be fuzzy. For any object, say o, let the membership degree that it belongs to the aggregate, say C, be $u_C(o)$. Also, let the projection of o to the constituent parts, say $C_1, C_2, \ldots, C_n$, be $o\downarrow_{C1}$, $o\downarrow_{C2}, \ldots, o\downarrow_{Cn}$. Let the membership degrees that these projections belong to C_1, $C_2, \ldots, C_n$ be $u_{C1}(o\downarrow_{C1})$, $u_{C2}(o\downarrow_{C2})$, …, $u_{Cn}(o\downarrow_{Cn})$, respectively. Then it follows $u_C(o) \le u_{C1}(o\downarrow_{C1})$, $u_C(o) \le u_{C2}(o\downarrow_{C2})$,…, and $u_C(o) \le u_{Cn}(o\downarrow_{Cn})$.

This characteristic can be used to determine if several classes have a fuzzy aggregation relationship.

Figure 3.5 shows a fuzzy aggregation relationship, where a *dashed open diamond* is used to denote a fuzzy aggregation relationship. Here:

- The class *New CPU box* is a fuzzy class, and thus the class *New-Computer* aggregated by *Monitor*, *New CPU box*, and *Keyboard* is also fuzzy.
- The *multiplicity* $[m_i, n_i]$ specifies that each instance of the aggregate class consists of at least m_i and at most n_i instances of the i-th constituent class. For example, a New-Computer may contain at least one Monitor, and possibly more.

Fuzzy Dependency

A *dependency*, which is a relationship between a source class and a target class, denotes that the target class exists dependently on the source class. In addition, the dependency between the source class and the target class is only related to the classes themselves and does not require a set of instances for its meaning.

A fuzzy dependency relationship is a dependency relationship with a degree of possibility η as shown in Fig. 3.6, which can be indicated explicitly by the designers or be implied implicitly by the source class based on the fact that the target class is decided by the source class.

Figure 3.6 shows a fuzzy dependency relationship, which is denoted by a *dashed line with an arrowhead*. It is clear that *Employee Dependent* is dependent on fuzzy class *Employee* with membership degree $\eta \in [0, 1]$.

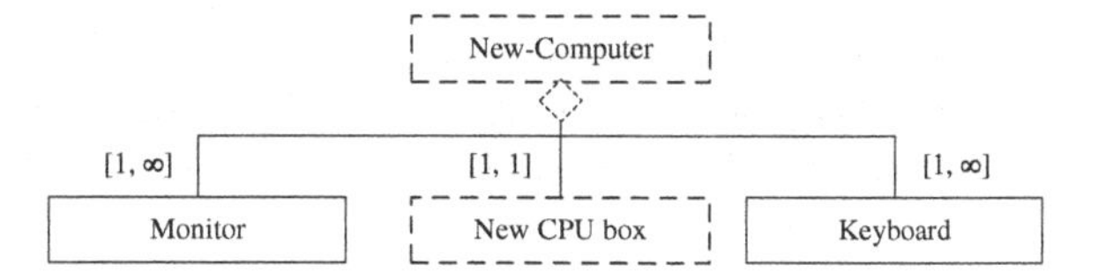

Fig. 3.5 Representation of a *fuzzy aggregation* in a fuzzy UML model

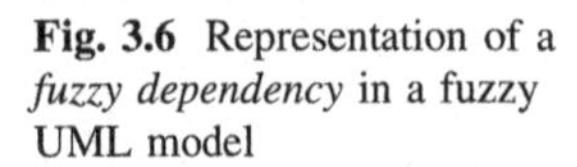

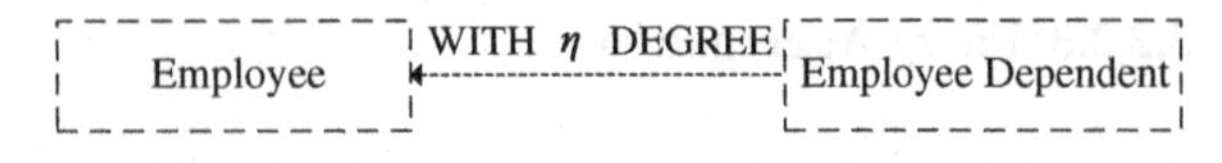

Fig. 3.6 Representation of a *fuzzy dependency* in a fuzzy UML model

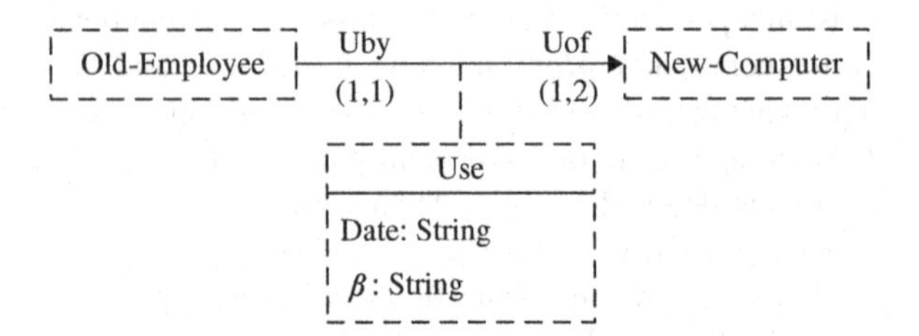

Fig. 3.7 Representation of a *fuzzy association* (*class*) in a fuzzy UML model

Fuzzy Association

An *association* is a relation between the instances of two or more classes. Names of associations are unique in a fuzzy UML model. An association has a related *association class* that describes properties of the association. Three kinds of fuzziness can be identified in an association relationship:

(i) The association is fuzzy itself, it means that the association relationship fuzzily exists in n classes, namely, this association relationship occurs with a degree of possibility.
(ii) The association is not fuzzy itself, i.e., it is known that the association relationship must occur in n classes. But it is not known for certain if n class instances (i.e., n objects) respectively belonging to the n classes have the given association relationship.
(iii) The association is fuzzy caused by such fact that (*i*) and (*ii*) occur in the association relationship simultaneously.

A fuzzy association relationship is an association relationship with a degree of possibility. We introduce a symbol β, as shown in Fig. 3.7, into a fuzzy UML model to denote the degree of possibility of a fuzzy association, and the calculating methods of β with respect to the three kinds of fuzziness above have been introduced in Ma et al. (2004).

Figure 3.7 shows a fuzzy association class *Use* between two classes *Old-Employee* and *New-Computer*. A *single line with an arrowhead* is used to denote a fuzzy association, and the association class is connected with the association by a *dashed-outline*. Here:

- *Date* is an attribute of the association class *Use*, which describes the start date that an *Old-Employee* uses a *New-Computer*.
- The additional symbol β denotes the membership degree of the fuzzy association occurring in several classes as mentioned above.
- The participation of a class in a fuzzy association is called a *role* which has a unique name. For example, *Uby* and *Uof* in Fig. 3.7.

- The cardinality constraint (m, n) on an association S specifies that each instance of the class C_i can participate at least m times and at most n times to S. For example, in Fig. 3.7, (1, 1) and (1, 2) denote that each *Old-Employee* can use at least 1 and at most 2 *New-Computers* and each *New-Computer* can be used by exactly one *Old-Employee*.

Formalization of Fuzzy UML Models

The main notions of a fuzzy UML model are introduced above. Based on these notions, a formal definition and semantic interpretation method of fuzzy UML models will be provided in the following.

Firstly, for two finite sets X and Y we call a function from a subset of X to Y an *X-labeled tuple over Y*. The labeled tuple T that maps $x_i \in X$ to $y_i \in Y$, for $i \in \{1, \ldots, k\}$, is denoted $[x_1 : y_1, \ldots, x_k : y_k]$. We also write $T[x_i]$ to denote y_i.

Definition 3.4 (*fuzzy UML models*) A fuzzy UML model is a tuple $\mathcal{F}_{UML} = (\mathcal{Q}\mathcal{F}, \leq_{\mathcal{F}}, att_{\mathcal{F}}, agg_{\mathcal{F}}, dep_{\mathcal{F}}, ass_{\mathcal{F}}, card_{\mathcal{F}}, mult_{\mathcal{F}}, mult'_{\mathcal{F}})$, where:

- $\mathcal{Q}\mathcal{F} = FO_{\mathcal{F}} \cup FA_{\mathcal{F}} \cup FM_{\mathcal{F}} \cup FT_{\mathcal{F}} \cup FC_{\mathcal{F}} \cup FH_{\mathcal{F}} \cup FG_{\mathcal{F}} \cup FD_{\mathcal{F}} \cup FS_{\mathcal{F}} \cup FR_{\mathcal{F}}$ is a finite *alphabet* partitioned into a set $FO_{\mathcal{F}}$ of *fuzzy object* symbols, a set $FA_{\mathcal{F}}$ of *fuzzy attribute* symbols (i.e., static attributes), a set $FM_{\mathcal{F}}$ of *fuzzy method* symbols (i.e., dynamic attributes), a set $FT_{\mathcal{F}}$ of *datatype* symbols, a set $FC_{\mathcal{F}}$ of *fuzzy class* symbols, a set $FH_{\mathcal{F}}$ of *fuzzy hierarchy* symbols, a set $FG_{\mathcal{F}}$ of *fuzzy aggregation* symbols, a set $FD_{\mathcal{F}}$ of *fuzzy dependency* symbols, a set $FS_{\mathcal{F}}$ of *fuzzy association* symbols, and a set $FR_{\mathcal{F}}$ of *role* symbols. Here it should be noted that:
  - A method in $FM_{\mathcal{F}}$ is the form of $f(P_1, \ldots, P_m) : R$, where f is the name of the method, $P_1, \ldots, P_m$ are types of m parameters, and R is the type of the result.
  - Each attribute $FA \in FA_{\mathcal{F}}$ is associated with a domain $FT \in FT_{\mathcal{F}}$, and each fuzzy domain symbol FT has an associated predefined basic domain and the various basic domains are usually assumed to be pairwise disjoint. A fuzzy keyword *FUZZY* is appeared on the front of attributes indicating these attributes are fuzzy attributes.
- $\leq_{\mathcal{F}}(FC) = FC_1 \times FC_2 \times \ldots \times FC_n$ is a relation that models the *hierarchy* (several generalizations can be grouped together to form a class hierarchy) between a superclass FC and several subclasses $FC_1, \ldots, FC_n$. Moreover, there may be an optional constraint (*disjointness*, *completeness*) in the hierarchy relation.
- $att_{\mathcal{F}} : FC \rightarrow T(FA, FT)$ is a function that maps each *fuzzy class* symbol in $FC_{\mathcal{F}}$ to a *$FA_{\mathcal{F}}$-labeled tuple over $FT_{\mathcal{F}}$*, i.e., $att_{\mathcal{F}}(FC) \rightarrow [FA_1 : FT_1, \ldots, FA_n : FT_n]$.
- $agg_{\mathcal{F}}(FG) = FC \times (FC_1 \cup FC_2 \cup \ldots \cup FC_m)$ is a relation that models the *aggregation* between an aggregate class FC and a group of constituent classes FC_i, $i = 1, 2, \ldots, m$.
- $dep_{\mathcal{F}} \subseteq FC_1 \times FC_2$ is a binary relation over $FC_{\mathcal{F}}$, which models the *dependency* between a source class FC_1 and a target class FC_2.

- $ass_{\mathcal{F}} : FS \rightarrow T(FR, FC)$ is a function that maps each *fuzzy association* symbol in $FS_{\mathcal{F}}$ to a $FR_{\mathcal{F}}$*-labeled tuple over* $FC_{\mathcal{F}}$, i.e., $ass_{\mathcal{F}}(FS) = [FR_1: FC_1, \ldots, FR_k: FC_k]$. The function $ass_{\mathcal{F}}$ actually associates a set of *roles* to each fuzzy association, determining implicitly also the arity of the fuzzy association. We assume without loss of generality that:
  - Each role is specific to exactly one fuzzy association, i.e., for two fuzzy associations $FS, FS' \in FS_{\mathcal{F}}$ with $FS \neq FS'$, if $ass_{\mathcal{F}}(FS) = [FR_1 : FC_1, \ldots, FR_k : FC_k]$ and $ass_{\mathcal{F}}(FS') = [FR_1' : FC_1', \ldots, FR_k' : FC_k']$, where roles $FR_i, FR_i' \in FR_{\mathcal{F}}$, fuzzy classes $FC_i, FC_i' \in FC_{\mathcal{F}}$, and $i \in \{1, \ldots, k\}$, then $\{FR_1, \ldots, FR_k\} \cap \{FR_1', \ldots, FR_k'\} = \varnothing$.
  - For each role $FR \in FR_{\mathcal{F}}$ there is a fuzzy association $FS \in FS_{\mathcal{F}}$ and a fuzzy class $FC \in FC_{\mathcal{F}}$ such that $ass_{\mathcal{F}}(FS) = [\ldots, FR : FC, \ldots]$.
- $card_{\mathcal{F}}$ is a function from $FC_{\mathcal{F}} \times FS_{\mathcal{F}} \times FR_{\mathcal{F}}$ to $\mathbb{N}_0 \times (\mathbb{N}_0 \cup \{\infty\})$ that satisfies the following condition (where $\mathbb{N}_0$ denotes non-negative integers): for a fuzzy association $FS \in FS_{\mathcal{F}}$ such that $ass_{\mathcal{F}}(FS) = [FR_1 : FC_1, \ldots, FR_k : FC_k]$, defining $card_{\mathcal{F}}(FC_i, FS, FR_i) = (card_{min}(FC_i, FS, FR_i), card_{max}(FC_i, FS, FR_i))$. The function $card_{\mathcal{F}}$ is used to specify the *cardinality constraints* on the minimum and maximum number of times an object instance of a fuzzy class may participate in a fuzzy association via some role. If not stated otherwise, $card_{min}(FC_i, FS, FR_i)$ is assumed to be 0 and $card_{max}(FC_i, FS, FR_i)$ is assumed to be ∞.
- $mult_{\mathcal{F}}$ is a function from $FT_{\mathcal{F}} \times FC_{\mathcal{F}} \times FA_{\mathcal{F}}$ to $\mathbb{N}_1 \times (\mathbb{N}_1 \cup \{\infty\})$ (where $\mathbb{N}_1$ denotes positive integers). The function $mult_{\mathcal{F}}$ is used to specify *multiplicities*, i.e., constraints on the minimum and maximum number of values that an attribute of an object may have. In detail, for an attribute $FA_i \in FA_{\mathcal{F}}$ of type $FT_i \in FT_{\mathcal{F}}$ in a class $FC \in FC_{\mathcal{F}}$ such that $att_{\mathcal{F}}(FC) \rightarrow [FA_1 : FT_1, \ldots, FA_n : FT_n]$, defining $mult_{\mathcal{F}}(FT_i, FC, FA_i) = (mult_{min}(FT_i, FC, FA_i), mult_{max}(FT_i, FC, FA_i))$. If not stated otherwise, $mult_{min}(FT_i, FC, FA_i)$ is assumed to be 1 and $mult_{max(FT_i}, FC, FA_i)$ is also assumed to be 1, i.e., the attribute is mandatory and single-valued.
- $mult'_{\mathcal{F}}$ is a function from $FC_{\mathcal{F}} \times FC_{\mathcal{F}}$ to $\mathbb{N}_0 \times (\mathbb{N}_0 \cup \{\infty\})$, which is used to specify *multiplicities*, i.e., constraints on the minimum and maximum number of times that an object instance of a constituent class may participate in a fuzzy aggregation. In detail, for a fuzzy aggregation $agg_{\mathcal{F}}(FG) = FC \times (FC_1 \cup FC_2 \cup \ldots \cup FC_m)$, where $FC \in FC_{\mathcal{F}}$ is an aggregate class, $FC_i \in FC_{\mathcal{F}}$ is a constituent class, and $i \in \{1, \ldots, m\}$, defining $mult'_{\mathcal{F}}(FC_i, FC) = (mult'_{min}(FC_i, FC), mult'_{max}(FC_i, FC))$. If not stated otherwise, $mult'_{min}(FC_i, FC)$ is assumed to be 0 and $mult'_{max}(FC_i, FC)$ is assumed to be ∞.

In the following we give the semantic interpretation of fuzzy UML models. Being similar to the fuzzy ER models in Sect. 3.2.1, the *semantics* of a fuzzy UML model can be given by *fuzzy object states* (see Definition 3.5), i.e., by specifying which fuzzy object state is consistent with the information structure of the fuzzy UML model.

Definition 3.5 A fuzzy object state $\mathscr{F}\mathscr{B}$ with respect to a fuzzy UML model $\mathscr{F}_{UML}$ is constituted by a nonempty finite set $\Delta^{\mathscr{F}\mathscr{B}}$, and a function $\bullet^{\mathscr{F}\mathscr{B}}$ that maps:

- Every domain symbol $FT \in FT_{\mathscr{F}}$ to a set $FT^{\mathscr{F}\mathscr{B}} \in FT_{\mathcal{F}}^{\mathcal{FB}}$, where $FT_{\mathcal{F}}^{\mathcal{FB}}$ is a set of domains. For non-fuzzy attributes, each domain is associated with the basic types such as Integer and String; for fuzzy attributes, each fuzzy domain is a fuzzy set or a possibility distribution.
- Every fuzzy class $FC \in FC_{\mathscr{F}}$ to a membership degree function $FC^{\mathscr{F}\mathscr{B}}$: $\Delta^{\mathscr{F}\mathscr{B}} \rightarrow [0, 1]$. That is, each fuzzy class FC is mapped into a possibility distribution $\{o_1/u_1, \ldots, o_n/u_n\}$, where o_i is an object identifier denoting a real world object belonging to FC, and each o_i is associated with a membership degree u_i denoting the object o_i membership degree to the fuzzy class FC.
- Every fuzzy attribute $FA \in FA_{\mathscr{F}}/FM_{\mathscr{F}}$ to a set $FA^{FB} \subseteq \Delta^{FB} \times \cup_{FT \in FT_{\mathcal{F}}} FT^{\mathcal{FB}}$.
- Every fuzzy association $FS \in FS_{\mathscr{F}}$ to a set $FS^{\mathscr{F}\mathscr{B}}$ of $FR_{\mathscr{F}}$-labeled tuples over $\Delta^{\mathscr{F}\mathscr{B}}$, i.e., $FS^{\mathscr{F}\mathscr{B}} \subseteq T(FR, \Delta^{\mathscr{F}\mathscr{B}})$.

The elements of $FC^{\mathscr{F}\mathscr{B}}$, $FA^{\mathscr{F}\mathscr{B}}$ and $FS^{\mathscr{F}\mathscr{B}}$ are called *instances* of FC, FA, and FS respectively.

A fuzzy object state is considered *acceptable* if it satisfies all integrity constraints of the fuzzy UML model. This is captured by the definition of legal fuzzy object state (see Definition 3.6).

Definition 3.6 A fuzzy object state $\mathscr{F}\mathscr{B}$ is said to be legal for a fuzzy UML model $\mathscr{F}_{UML}$, if it satisfies the following conditions:

- For each pair of fuzzy classes FC_1, $FC_2 \in FC_{\mathscr{F}}$ such that $\leq_{\mathscr{F}}(FC_2) = FC_1$, it holds that $FC_1^{\mathscr{F}\mathscr{B}} \subseteq FC_2^{\mathscr{F}\mathscr{B}}$.
- For each fuzzy hierarchy $\leq_{\mathscr{F}}(FC) = FC_1 \times FC_2 \times \ldots \times FC_n$, it hold that $FC_i^{\mathscr{F}\mathscr{B}} \subseteq FC^{\mathscr{F}\mathscr{B}}$; if there are disjointness and completeness constraints, then it follows $FC^{\mathscr{F}\mathscr{B}} = FC_i^{\mathscr{F}\mathscr{B}} \cup \ldots \cup FC_n^{\mathscr{F}\mathscr{B}}$ and $FC_i^{\mathscr{F}\mathscr{B}} \cap FC_j^{\mathscr{F}\mathscr{B}} = \varnothing$, $i \neq j$, $i, j \in \{1, \ldots, n\}$.
- For each fuzzy class $FC \in FC_{\mathscr{F}}$, if $att_{\mathscr{F}}(FC) = [FA_1 : FT_1, \ldots, FA_n : FT_n]$, then for each instance $c \in FC^{\mathscr{F}\mathscr{B}}$ and for each $i \in \{1, \ldots, n\}$ the following holds: (*i*) there is at least one element $a_i \in FA_i^{\mathscr{F}\mathscr{B}}$ whose first component is c, and the second component of a_i is an element $t_i \in FT_i^{\mathscr{F}\mathscr{B}}$; (*ii*) $mult_{\min}(t_i, c, a_i) \leq \#\{c \in FC^{\mathscr{F}\mathscr{B}} \mid c[a_i] = t_i\} \leq mult_{\max}(t_i, c, a_i)$, where #{ }denotes the base of set { }.
- For each fuzzy aggregation $FG \in FG_{\mathscr{F}}$ such that $agg_{\mathscr{F}}(FG) = FC \times (FC_1 \cup FC_2 \cup \ldots \cup FC_m)$, for each instance $c \in FC^{\mathscr{F}\mathscr{B}}$ and $c_i \in FC_i^{\mathscr{F}\mathscr{B}}$, where $i \in \{1, \ldots, m\}$, it holds that: $mult'_{\min}(c_i, c) \leq \#\{c \in FC^{\mathscr{F}\mathscr{B}} \mid c[FG] = c_i\} \leq mult'_{\max}(c_i, c)$, where #{ }denotes the base of set { }.
- For each pair of fuzzy classes FC_1, $FC_2 \in FC_{\mathscr{F}}$ such that $dep_{\mathscr{F}} \subseteq FC_1 \times FC_2$, it holds that: if $FC_1^{\mathscr{F}\mathscr{B}} = \varnothing$, then $FC_2^{\mathscr{F}\mathscr{B}} = \varnothing$.
- For each fuzzy association $FS \in FS_{\mathscr{F}}$ such that $ass_{\mathscr{F}}(FS) = [FR_1 : FC_1, \ldots, FR_k : FC_k]$, all instances of FS are of the form $[r_1 : c_1, \ldots, r_k : c_k]$, where $r_i \in FR_i^{\mathscr{F}\mathscr{B}}$, $c_i \in FC_i^{\mathscr{F}\mathscr{B}}$, $i \in \{1, \ldots, k\}$.

- For each fuzzy association $FS \in FS_{\mathscr{F}}$ such that $ass_{\mathscr{F}}(FS) = [\ldots, FR : FC, \ldots]$, for each instance $c \in FC^{\mathscr{F}\mathscr{B}}$, it holds that: $card_{\min}(FC, FS, FR) \leq \#\{s \in FS^{\mathscr{F}\mathscr{B}} \mid s[FR] = c\} \leq card_{\max}(FC, FS, FR)$, where $\#\{\}$denotes the base of set $\{\}$.

In the following we provide an example of a full fuzzy UML model, including the structure and object instance information, which will be used to construct fuzzy ontology in the later sections.

Figure 3.8 shows a graphic fuzzy UML model $\mathscr{F}_{\mathrm{UML1}}$ modeling the situation at a company. Furthermore, Fig. 3.9 gives formal representation of the fuzzy UML model $\mathscr{F}_{\mathrm{UML1}}$ in Fig. 3.8 according to the formal definition of fuzzy UML models (i.e., Definition 3.4).

The detailed instruction about the fuzzy UML model $\mathscr{F}_{\mathrm{UML1}}$ in Figs. 3.8 and 3.9 is as follows:

- A *fuzzy class* is denoted by using a *dashed-outline rectangle* to differentiate a classical class, e.g., *Old-Employee* as shown in Fig. 3.8. A fuzzy class may contain four parts, i.e., crisp attributes, fuzzy attributes, methods, or an additional attribute u as shown in Fig. 3.8, where:
  - A fuzzy keyword *FUZZY* is appeared in front of an attribute indicating the attribute may take fuzzy values. For example, *FUZZY* Age or *FUZZY* Email. Moreover, the *multiplicity* $[1\ldots\infty]$ of the attribute "*FUZZY* Email $[1\ldots\infty]$" means that each object instance of the fuzzy class *Old-Employee* has at least one email, and possibly more.
  - An additional attribute $u \in [0, 1]$ in a fuzzy class is defined for representing an object membership degree to the fuzzy class.
  - The method *IsDepartment():String* returns a possibility distribution value $\{Department/u_i\}$, which denotes that an *Old-Employee* works in the *Department* with degree $u_i \in [0, 1]$. The type of the parameter is *null*.
- A *fuzzy generalization* is denoted by using a *dashed triangular arrowhead*, e.g., the class *Employee* is a generalization of classes *Young-Employee* and *Old-Employee*, and two classes *Young-Employee* and *Old-Employee* are *disjoint* and *the union of them* completely covers the class *Employee*.

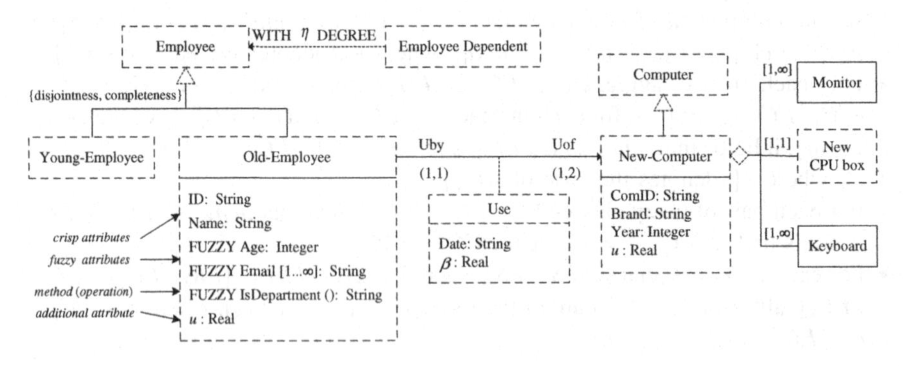

Fig. 3.8 A graphic fuzzy UML model $\mathscr{F}_{\mathrm{UML1}}$

The formal syntax of the fuzzy UML model $\mathcal{F}_{UML1}$ in Fig. 3.8 is as follows:

$\mathcal{F}_{UML1} = (\mathcal{L}_{\mathcal{F}}, \preccurlyeq_{\mathcal{F}}, att_{\mathcal{F}}, agg_{\mathcal{F}}, dep_{\mathcal{F}}, ass_{\mathcal{F}}, card_{\mathcal{F}}, mult_{\mathcal{F}}, mult'_{\mathcal{F}})$, where

$\mathcal{L}_{\mathcal{F}}$ is a set of symbols partitioned into:

$FC_{\mathcal{F}}$ = {Employee, Young-Employee, Old-Employee, Computer, New-Computer, ..., Keyboard}

$FA_{\mathcal{F}}$= {ID, Name, FUZZY Age, FUZZY Email, Date, *u*, *β*, ComID, Brand, Year}

$FT_{\mathcal{F}}$ = {String, Integer, Real}

$FM_{\mathcal{F}}$ = {FUZZY IsDepartment()}

...

$FR_{\mathcal{F}}$ = {Uby, Uof}

The set of functions/relations over the above symbols consists of:

$att_{\mathcal{F}}$(Old-Employee) = [ID: String, Name: String, FUZZY Age: Integer, FUZZY Email: String, FUZZY IsDepartment(): String, *u*: Real]

$att_{\mathcal{F}}$(New-Computer) = [ComID: String, Brand: String, Year: Integer, *u* : Real]

$att_{\mathcal{F}}$(Use) = [Date: String, *β*: Real]

$\preccurlyeq_{\mathcal{F}}$(Employee) = Young-Employee × Old-Employee (*disjointness, completeness*)

$\preccurlyeq_{\mathcal{F}}$(Computer) = New-Computer

$agg_{\mathcal{F}}(FG)$ = New-Computer × (Monitor ∪ New CPU box ∪ Keyboard)

$dep_{\mathcal{F}} \subseteq$ Employee × Employee-Department

$ass_{\mathcal{F}}$(Use) = [Uby : Old-Employee, Uof : New-Computer]

$mult_{\mathcal{F}}$(String, Old-Employee, FUZZY Email) = (1,∞)

$mult'_{\mathcal{F}}$(Monitor, New-Computer) = (1,∞)

$mult'_{\mathcal{F}}$(New CPU box, New-Computer) = (1, 1)$mult'_{\mathcal{F}}$(Keyboard, New-Computer) = (1,∞)

$card_{\mathcal{F}}$(Old-Employee, Use, Uby) = (1, 1)

$card_{\mathcal{F}}$(New-Computer, Use, Uof) = (1, 2).

Fig. 3.9 The formal syntax of the fuzzy UML model $\mathscr{F}_{UML1}$ in Fig. 3.8

- A *fuzzy dependency* is denoted by using a *dashed line with an arrowhead*, e.g., there is a fuzzy dependency relationship between the target class *Employee Dependent* and the source class *Employee*.
- A *fuzzy association* is denoted by using a *single line with an arrowhead* and the association class is connected with the association by a *dashed-outline*, e.g., *Use* is a fuzzy association class between two classes *Old-Employee* and *New-Computer*. Here:
  - *Date* is an attribute of the association class *Use*, which describes the start date that an *Old-Employee* uses a *New-Computer*.
  - The additional attribute *β* denotes the degree of possibility that an association relationship occurs in *n* classes.
  - The participation of a class in a fuzzy association is called a *role* which has a unique name. For example, *Uby* and *Uof*. The cardinality constraints (1, 1) and (1, 2) denote that each *Old-Employee* can use at least 1 and at most 2 *New-Computers* and each *New-Computer* can be used by exactly one *Old-Employee*.
  - A *fuzzy aggregation* is denoted by using a *dashed open diamond*, e.g., the class *New-Computer* is aggregated by *Monitor*, *New CPU box*, and *Keyboard*.

The *multiplicity* [m_i, n_i] specifies that each instance of the aggregate class consists of at least m_i and at most n_i instances of the *i*-th constituent class. For example, a *New-Computer* may contain at least one *Monitor*, and possibly more.

3.3 Fuzzy Database Models

In order to manage fuzzy data in the databases, a significant body of research in the area of fuzzy database has been developed over the past thirty years. Various fuzzy database models (e.g., fuzzy relational and object-oriented databases) were proposed, and some major issues related to the models have been investigated.

Regarding modeling fuzzy information in relational databases, generally speaking, several basic approaches can be classified: (*i*) one of fuzzy relational databases is based on possibility distribution (Chaudhry et al. 1999; Prade and Testemale 1984; Umano and Fukami 1994); (*ii*) the other one is based on the use of similarity relation (Buckles and Petry, 1982), proximity relation (De et al. 2001; Shenoi and Melton 1999), resemblance relation (Rundensteiner and Bic 1992), or fuzzy relation (Raju and Majumdar 1988); (*iii*) another possible extension is to combine possibility distribution and similarity (proximity or resemblance) relation (Chen et al. 1992; Ma et al. 2000; Ma and Mili 2002). In this book, the fuzzy relational database is based on the possibility distribution.

Regarding modeling fuzzy information in object-oriented databases, Zicari and Milano (1990) first introduced incomplete information, namely, null values, where incomplete schema and incomplete objects can be distinguished. From then on, the incorporation of imprecise and uncertain information in object-oriented databases has increasingly received attention. A fuzzy object-oriented database model was defined in Bordogna and Pasi (2001) based on the extension of a graphs-based object model. Based on similarity relationship, uncertainty management issues in the object-oriented database model were discussed in George et al. (1996). Based on possibility theory, vagueness and uncertainty were represented in class hierarchies in Dubois et al. (1991). In more detail, also based on possibility distribution theory, Ma et al. (2004) introduced fuzzy object-oriented database models, some major notions such as objects, classes, objects-classes relationships and subclass/superclass relationships were extended under fuzzy information environment. Moreover, other fuzzy extensions of object-oriented databases were developed. In Marín et al. (2000, 2001), fuzzy types were added into fuzzy object-oriented databases to manage vague structures. The fuzzy relationships and fuzzy behavior in fuzzy object-oriented database models were discussed in Cross (2001) and Gyseghem and Caluwe (1995). Several intelligent fuzzy object-oriented database architectures were proposed in Koyuncu and Yazici (2003), Ndouse (1997) and Ozgur et al. (2009). The other efforts on how to model fuzziness and uncertainty in object-oriented database models were done in Lee et al. (1999),

Majumdar et al. (2002) and Umano et al. (1998). The fuzzy and probabilistic object bases (Cao and Rossiter 2003 and Nam et al. 2007), fuzzy deductive object-oriented databases (Yazici and Koyuncu 1997), and fuzzy object-relational databases (Cubero et al. 2004) were also developed. In addition, an object-oriented database modeling technique was proposed based on the level-2 fuzzy sets in de Tré and de Caluwe (2003), where the authors also discussed how the object Data Management Group (ODMG) data model can be generalized to handle fuzzy data in a more advantageous way. Also, the other efforts have been paid on the establishment of consistent framework for a fuzzy object-oriented database model based on the standard for the ODMG object data model (Cross et al. 1997). More recently, how to manage fuzziness on conventional object-oriented platforms was introduced in Berzal et al. (2007). Yan and Ma (2012) proposed the approach for the comparison of entity with fuzzy data types in fuzzy object-oriented databases. Yan et al. (2012) investigated the algebraic operations in fuzzy object-oriented databases, and discussed fuzzy querying strategies and gave the form of SQL-like fuzzy querying for the fuzzy object-oriented databases.

In the following, we introduce the fuzzy relational database models and fuzzy object-oriented database models in detail.

3.3.1 Fuzzy Relational Database Models

Basically, a *fuzzy relational database* (*FRDB*) is based on the notions of *fuzzy relational schema*, *fuzzy relational instance*, *tuple*, *key*, and *constraints*, which are introduced briefly as follows (Ma et al. 2011c):

- A *fuzzy relational database* consists of a set of *fuzzy relational schemas* and a set of *fuzzy relational instances* (i.e., simply *fuzzy relations*).
- The set of *fuzzy relational schemas* specifies the structure of the data held in a database. A fuzzy relational schema consists of a fixed set of attributes with associated domains. The information of a domain is implied in forms of schemas, attributes, keys, and referential integrity constraints.
- The set of *fuzzy relations*, which is considered to be an instance of the set of fuzzy relation schemas, reflects the real state of a database. Formally, a fuzzy relation is a two-dimensional array of rows and columns, where each column represents an attribute and each row represents a tuple.
- Each *tuple* in a table denotes an *individual* in the real world identified uniquely by primary key, and a foreign key is used to ensure the data integrity of a table. A column (or columns) in a table that makes a row in the table distinguishable from other rows in the same table is called the *primary key*. A column (or columns) in a table that draws its values from a primary key column in another table is called the *foreign key*. As is generally assumed in the literature, we assume that the primary key attribute is always crisp and all fuzzy relations are in the third normal form.

- An *integrity constraint* in a schema is a predicate over relations expressing a constraint; by far the most used integrity constraint is the referential integrity constraint. A *referential integrity constraint* involves two sets of attributes S_1 and S_2 in two relations R_1 and R_2, such that one of the sets (say S_1) is a key for one of the relations (called primary key). The other set is called a foreign key if $R_2[S_2]$ is a subset of $R_1[S_1]$. Referential integrity constraints are the glue that holds the relations in a database together.

In summary, in a fuzzy relational database, the structure of the data is represented by a set of *fuzzy relational schemas*, and data are stored in *fuzzy relations* (i.e., *tables*). Each table contains *rows* (i.e., *tuples*) and *columns* (i.e., *attributes*). Each tuple is identified uniquely by the *primary key*. The relationships among relations are represented by the referential integrity constraints, i.e., *foreign keys*. Moreover, here, two types of fuzziness are considered in fuzzy relational databases, one is the fuzziness of attribute values (i.e., attributes may be fuzzy), which may be represented by possibility distributions; another is the fuzziness of a tuple being a member of the corresponding relation, which is represented by a membership degree associated with the tuple. For a more comprehensive review about fuzzy relational databases, please refer to Ma and Yan (2008).

In the following, we give a simple formal definition of FRDBs (Zhang et al. 2012a), followed by an example to well illustrate the definition.

Definition 3.7 (*fuzzy relational databases*) A fuzzy relational database $FRDB = < FS, FR >$ consists of a set of fuzzy relational schemas *FS* and a set of fuzzy relations *FR*, where:

- Each *fuzzy relational schema FS* can be represented formally as $FR(A_1/D_1, A_2/D_2, \ldots, A_n/D_n, \mu_{FR}/D_{FR})$, which denotes that a fuzzy relation *FR* has attributes $A_1, A_2, \ldots, A_n$ and μ_{FR} with associated datatypes $D_1, D_2, \ldots, D_n$ and D_{FR}. Here, μ_{FR} is an additional attribute for representing the membership degree of a tuple to the fuzzy relation.
- Each *fuzzy relation FR* on a fuzzy relational schema $FR(A_1/D_1, A_2/D_2, \ldots, A_n/D_n, \mu_{FR}/D_{FR})$ is a subset of the Cartesian product of $Dom(A_1) \times Dom(A_2) \times \ldots \times Dom(A_n) \times Dom(\mu_{FR})$, where $Dom(A_i)$ may be a fuzzy subset or even a set of fuzzy subset and $Dom(\mu_{FR}) \in [0, 1]$. Here, $Dom(A_i)$ denotes the domain of attribute A_i, and each element of the domain satisfies the constraint of the datatype D_i. Formally, each *tuple* in *FR* has the form $t = < \pi_{A1}, \pi_{A2}, \ldots, \pi_{Ai}, \ldots, \pi_{An}, \mu_{FR} >$, where the value of an attribute A_i may be represented by a possibility distribution π_{Ai}, and $\mu_{FR} \in [0, 1]$.

To provide the intuition on the fuzzy relational database we show an example. The following gives a fuzzy relational database modeling parts of the reality at a company, including fuzzy relational schemas in Table 3.1 and fuzzy relations in Table 3.2. The detailed introduction is as follows:

- The attribute underlined stands for primary key *PK*. The foreign key (*FK*) is followed by the parenthesized relation called referenced relation. A relation can have several candidate keys from which one primary key, denoted *PK*, is chosen.
- An '*f*' next to an attribute means that the attribute is fuzzy.
- In Table 3.1, there are the inheritance relationships *Chief-Leader* "is-a" *Leader* and *Young-Employee* "is-a" *Employee*. There is a *1-many relationship* between *Department* and *Young-Employee*. The relation *Supervise* is a *relationship relation*, and there is *many–many relationship* between *Chief-Leader* and *Young-Employee*.
- Note that, a relation is different from a relationship. A relation is essentially a table, and a relationship is a way to correlate, join, or associate the two tables.

3.3.2 Fuzzy Object-oriented Database Models

Generally speaking, a fuzzy object-oriented database (FOODB) is based on the basic notions of fuzzy object, fuzzy class, and fuzzy inheritance (Ma et al. 2004).

3.3.2.1 Fuzzy Objects

Objects model real-world entities or abstract concepts. Objects have properties that may be attributes of the object itself or relationships also known as associations between the object and one or more other objects. An object is fuzzy because of a lack of information. For example, an object representing a part in preliminary design for certain will also be made of *stainless steel*, *moulded steel*, or *alloy steel* (each of them may be connected with a possibility, say, 0.7, 0.5 and 0.9,

Table 3.1 The fuzzy relational schemas of a fuzzy relational database

Relation name	Attribute and datatype	Foreign key and referenced relation
Leader	leaID (String), lNumber (String), μ_{FR}(Real)	No
Employee	empID (String), eNumber (String), μ_{FR}(Real)	No
Chief-Leader	leaID (String), clName (String), f_clAge (Integer), μ_{FR}(Real)	leaID (Leader(leaID))
Young-Employee	empID (String), yeName (String), f_yeAge (*Integer*), f_yeSalary (Integer), dep_ID (String), μ_{FR}(Real)	empID (Employee(empID)) dept_ID (Department(depID))
Supervise	supID (String), lea_ID (String), emp_ID (String), μ_{FR}(Real)	lea_ID (Chief-Leader(leaID)) emp_ID(Young-Employee(empID))
Department	depID (String), dName (String), μ_{FR}(Real)	No

Table 3.2 The fuzzy relations of a fuzzy relational database

Leader

leaID	*lNumber*	μ_{FR}
L001	1	0.7
L002	2	0.9
L003	3	0.8

Employee

empID	*eNumber*	μ_{FR}
E001	1	0.8
E002	2	0.9

Chief-Leader

leaID	*clName*	*f_clAge*	μ_{FR}
L001	Chris	{35/0.8, 39/0.9}	0.65
L003	Billy	37	0.7

Young-Employee

empID	*yeName*	*f_yeAge*	*f_yeSalary*	*dep_ID*	μ_{FR}
E001	John	{24/0.7,25/0.9}	{2000/0.3, 3000/0.4}	D001	0.75
E002	Mary	23	{4000/0.5, 4500/0.7, 5000/1.0}	D003	0.85

Department

depID	*dName*	μ_{FR}
D001	HR	0.8
D002	Finance	0.9
D003	Sales	0.7

Supervise

supID	*lea_ID*	*emp_ID*	μ_{FR}
S001	L001	E001	0.78
S002	L001	E002	0.8
S002	L003	E002	0.9

respectively). Formally, objects that have at least one attribute whose value is a fuzzy set are fuzzy objects.

3.3.2.2 Fuzzy Classes

The fuzzy classes in fuzzy object-oriented databases are similar to the notion of the fuzzy classes in fuzzy UML models as introduced in Sect. 3.2.2.

The objects having the same properties are gathered into classes that are organized into hierarchies. Theoretically, a class can be considered from two different viewpoints (Dubois et al. 1991): (*a*) an extensional class, where the class is defined by the list of its object instances, and (*b*) an intensional class, where the class is defined by a set of attributes and their admissible values. In addition, a subclass defined from its superclass by means of inheritance mechanism in the object-oriented database (OODB) can be seen as the special case of (*b*) above.

Therefore, a class is fuzzy because of the following several reasons. First, some objects are fuzzy ones, which have similar properties. A class defined by these objects may be fuzzy. These objects belong to the class with membership degree of [0, 1]. Second, when a class is intensionally defined, the domain of an attribute may be fuzzy and a fuzzy class is formed. For example, a class *Old equipment* is a fuzzy one because the domain of its attribute *Using period* is a set of fuzzy values such as *long*, *very long*, and *about* 20 *years*. Third, the subclass produced by a fuzzy class by means of specialization and the superclass produced by some classes (in which there is at least one class who is fuzzy) by means of generalization are also fuzzy.

The main difference between fuzzy classes and crisp classes is that the boundaries of fuzzy classes are imprecise. The imprecision in the class boundaries is caused by the imprecision of the values in the attribute domain. In the FOODB, classes are fuzzy because their attribute domains are fuzzy. The issue that an object fuzzily belongs to a class occurs since a class or an object is fuzzy. Similarly, a class is a subclass of another class with membership degree of [0, 1] because of the class fuzziness. In the OODB, the above-mentioned relationships are certain. Therefore, the evaluations of fuzzy object-class relationships and fuzzy inheritance hierarchies are the cores of information modeling in the FOODB.

In the FOODB, the following four situations can be distinguished for object-class relationships (Ma et al. 2004).

(a) *Crisp class* and *crisp object*: This situation is the same as the OODB, where the object belongs or not to the class certainly. For example, the objects *Car* and *Computer* are for a class *Vehicle*, respectively.
(b) *Crisp class* and *fuzzy object*: Although the class is precisely defined and has the precise boundary, an object is fuzzy since its attribute value(s) may be fuzzy. In this situation, the object may be related to the class with the special degree in [0, 1]. For example, the object whose *position* attribute may be *graduate*, *research assistant*, or *research assistant professor* is for the class *Faculty*.
(c) *Fuzzy class* and *crisp object*: Being the same as the case in (*b*), the object may belong to the class with the membership degree in [0, 1]. For example, a Ph.D. student is for *Young student* class.
(d) *Fuzzy class* and *fuzzy object*: In this situation, the object also belongs to the class with the membership degree in [0, 1].

The object-class relationships in (*b*)–(*d*) above are called fuzzy object-class relationships. In fact, the situation in (*a*) can be seen as the special case of fuzzy object-class relationships, where the membership degree of the object to the class is one. It is clear that estimating the membership of an object to the class is crucial for fuzzy object-class relationship when classes are instantiated. In the OODB, determining if an object belongs to a class depends on if its attribute values are respectively included in the corresponding attribute domains of the class. Similarly, in order to calculate the membership degree of an object to the class in a fuzzy object-class relationship, it is necessary to evaluate the degrees that the attribute domains of the class include the attribute values of the object. The details about how to calculate the membership degree of an object to the class in a fuzzy object-class relationship can be found in Ma et al. (2004).

3.3.2.3 Fuzzy Inheritance Hierarchies

In the object-oriented database (OODB), a new class, called subclass, is produced from another class, called superclass by means of inheriting some attributes and methods of the superclass, overriding some attributes and methods of the superclass, and defining some new attributes and methods. Since a subclass is the specialization of the superclass, any one object belonging to the subclass must belong to the superclass. This characteristic can be used to determine if two classes have subclass/superclass relationship.

In the fuzzy object-oriented database (FOODB), however, classes may be fuzzy. A class produced from a fuzzy class must be fuzzy. If the former is still called subclass and the later superclass, the subclass/superclass relationship is fuzzy. In other words, a class is a subclass of another class with membership degree of [0, 1] at this moment. Correspondingly, the method used in the OODB for determination of subclass/superclass relationship is modified as:

(a) for any (fuzzy) object, if the member degree that it belongs to the subclass is less than or equal to the member degree that it belongs to the superclass, and
(b) the member degree that it belongs to the subclass is greater than or equal to the given threshold.

The subclass is then a subclass of the superclass with the membership degree, which is the minimum in the membership degrees to which these objects belong to the subclass.

Let $C1$ and $C2$ be (fuzzy) classes and β be a given threshold. We say $C2$ is a subclass of $C1$ if $(\forall o)$ $(\beta \leq \mu_{C2}(o) \leq \mu_{C1}(o))$.

It can be seen that by utilizing the inclusion degree of objects to the class, one can assess fuzzy subclass/superclass relationships in the FOODB (Ma et al. 2004). It is clear that such assessment is indirect. If there is no any object available, this method is not used. In fact, the idea used in evaluating the membership degree of an object to a class can be used to determine the relationships between fuzzy subclass and superclass. One can calculate the inclusion degree of a (fuzzy) subclass with respect to the (fuzzy) superclass according to the inclusion degree of the attribute domains of the subclass with respect to the attribute domains of the superclass as well as the weight of attributes. The method for evaluating the inclusion degree of fuzzy attribute domains is introduced in Ma et al. (2004) in detail. Moreover, more concepts and operations about fuzzy object-oriented databases (such as the comparison of entity with fuzzy data types and the algebraic operations in fuzzy object-oriented databases) can be found in Yan and Ma (2012) and Yan et al. (2012).

3.3.2.4 Formalization of Fuzzy Object-oriented Database Models

In the following, a formal definition and semantic interpretation method of fuzzy object-oriented database (FOODB) model (Zhang et al. 2012b) are provided.

Definition 3.8 (*fuzzy object-oriented databases*) A FOODB model is a set of class declarations, which consists of a tuple $FS = (FC_{FS}, FA_{FS}, FD_{FS})$, where:

- FC_{FS} is a set of fuzzy class names, denoted by the letter FC;
- FA_{FS} is a set of fuzzy attribute names, denoted by the letter FA;
- FD_{FS} is a set of fuzzy class declarations. For each fuzzy class FC ∈ FC_{FS}, FD_{FS} contains exactly one such declaration:

$$\text{Class FC } is - a \text{ FC}_1, \ldots, \text{FC}_n \; type - is \, \text{FT},$$

FT denotes a type expressive built as the following syntax:

$$\begin{aligned} &\text{FT} \rightarrow \text{FC}| \\ &\quad \text{Union} FT_{1,}\ldots, FT_k(disjointness, covering)End| \\ &\quad Set - ofFT\,[\,(m_1, n_1),\ (m_2, n_2)\,]| \\ &\quad \text{Record} FA_1 : FT_{1,}\ldots,\ FA_k : FT_k End| \\ &\quad f(P_1,\ \ldots, P_m)\ : R \end{aligned}$$

where:

(a) The *is-a* part, which is *optional*, denotes the inheritance relationship between fuzzy classes;
(b) The *type-is* part specifies the structure of the class *FC* through a type expressive *FT*;
(c) The *Union…End* part denotes a generalization relationship between a general class and several specific classes; The (*disjointness*, *covering*) part is *optional*, *disjointness* means that all the specific classes are disjoint, and *covering* means that the union of specific classes completely covers the general class;
(d) $[(m_1, n_1), (m_2, n_2)]$ denotes the cardinality constraints of a set;
(e) $f(P_1, \ldots, P_m) : R$ represents the *method*, where *f* is the name of the method, $P_1,\ldots,P_m$ are types of *m* parameters, and *R* is the type of the result.

Figure 3.10 shows a simple FOODB model FS_1 modeling parts of the reality at a company.

Fig. 3.10 A FOODB model FS_1 modeling parts of the reality at a company

```
Class Employee type-is
   Union Young-Employee, Old-Employee
   End

Class Young-Employee is-a Employee type-is
   Record
      Name: String
      FUZZY Age: Integer
      FUZZY Salary: Integer
      FUZZY IsDepartment( ): String
   End

Class Chief-Leader is-a Leader type-is
   Record
      Number: String
      FUZZY Manage: Set-of Young-Employee [(2, 6), (1, 1)]
   End
```

The detailed instructions about Fig. 3.10 are as follows:

- A fuzzy keyword *FUZZY* is appeared in front of the attributes indicating attributes may take fuzzy values.
- The method *IsDepartment():String* returns a possibility distribution value {*Department/u*}, which denotes the possibility $u \in [0, 1]$ that a *Young-Employee object* works in the *Department*.
- [(2, 6), (1, 1)] denotes that each *Chief-Leader* manages at least 2 and at most 6 *Young-Employees*, and each *Young-Employee* is managed by exactly one *Chief-Leader*.

Figure 3.11 gives a *database instance* (i.e., *object diagram*) w.r.t. the FOODB model FS_1 in Fig. 3.10 (only parts of data):

- $\beta = 0.7$ denotes that the possibility which the object O_1 manages the object O_2 is 0.7.
- The attributes such as *Age* and *Salary* are fuzzy, e.g., the age of *Chris* is represented by a possibility distribution: {20/0.9, 25/0.8}.
- $u = 0.85$ denotes that the possibility which an object O_1 belongs to the class *Chief-Leader* is 0.85.

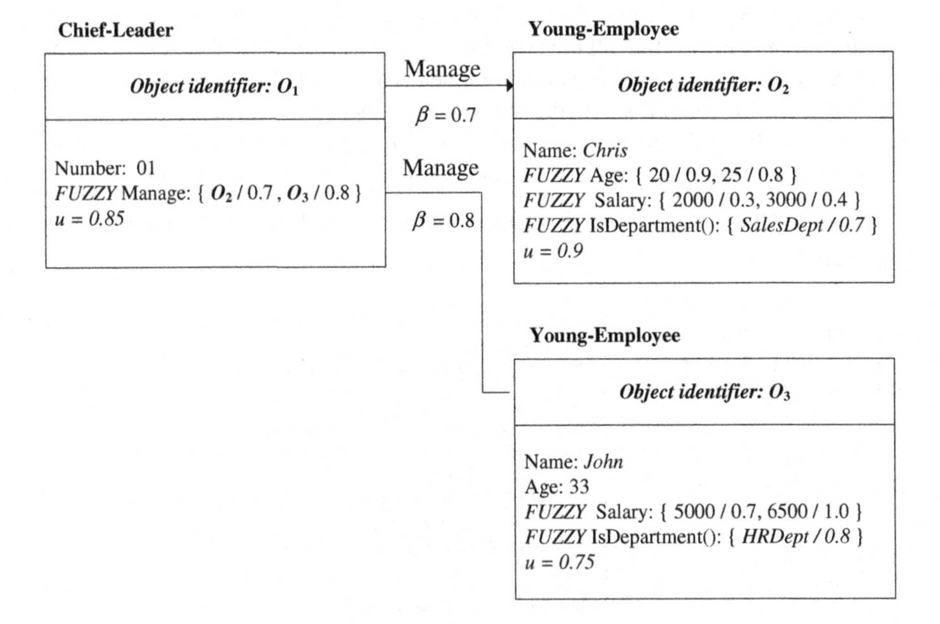

Fig. 3.11 A fragment of the instance w.r.t. the FOODB model FS_1 in Fig. 3.10

The semantics of a FOODB model can be given by the fuzzy database state (e.g., the database instance in Fig. 3.11), i.e., by specifying which fuzzy database state/instance is consistent with the information structure of the FOODB model (see Definition 3.9).

Definition 3.9 The definition of fuzzy database state *FJ* w.r.t. the FOODB model *FS* is given by a fuzzy interpretation $FI_{FD} = (FV_{FJ}, \pi^{FJ}, \rho^{FJ}, \bullet^{FJ})$, which consists of:

(1) A set $FV_{FJ} = FD_{FS} \cup FO^{FJ} \cup FV_{FR} \cup FV_{FS}$ of fuzzy values is inductively defined as follows:

- $FD_{FS} = \cup_{i=1}^{n} FD_i$, where FD_i is a set of domains, $FD_i \cap FD_j = \emptyset$, $i \neq j$. For fuzzy attributes, each fuzzy domain is associated with possibility distributions.
- $FO^{FJ} = \{o_1/u_1, \ldots, o_n/u_n\}$, where o_i is a symbol denoting the real world object called object identifier, and each o_i is associated with a degree of possibility u_i.
- FV_{FR} is the set of record values. A record value is denoted by $[FA_1{:}FV_1, \ldots, FA_k{:}FV_k]$, where FA_i is the attributes of the class, $FV_i \in FV_{FJ}$, $i \in \{1, \ldots, k\}$.
- FV_{FS} is the set of set-values. A set-value is denoted by $\{FV_1, \ldots, FV_k\}$, where $FV_i \in FV_{FJ}$, $i \in \{1, \ldots, k\}$.

(2) A mapping π^{FJ} assigning to each class $FC \in FC_{FS}$ a subset of FO^{FJ}.
(3) A mapping ρ^{FJ} assigning to each object in FO^{FJ} a value of FV_{FJ}.
(4) An function $\bullet^{FJ}$ that maps each type expression FT to a set FT^{FJ} in such a way that:

- If FT is a class FC, then $FT^{FJ} = FC^{FJ} = \pi^{FJ}(FC)$;
- If FT is a union type Union $FT_1, \ldots, FT_k$ End, then $FT^{FJ} = FT_1^{FJ} \cup \ldots \cup FT_k^{FJ}$;
- If FT is a record type Record $FA_1{:}FT_1, \ldots, FA_k{:}FT_k$ End (resp. set type Set-of FT), then FT^{FJ} is a set of record values FV_{FR} (resp. set values FV_{FS}). Note that the method is included in the record type.

A fuzzy database state is considered *acceptable* if it satisfies all the constraints of the FOODB model. This is captured by the following definition.

Definition 3.10 A fuzzy database state FJ is said to be legal w.r.t. a FOODB model FS, if for each class declaration: Class FC is-a $FC_1, \ldots, FC_n$ type-is FT, it satisfies:

- $FC^{FJ}(o) \leq FC_i^{FJ}(o)$ for each object $o \in \pi^{FJ}(FC)$, $i \in \{1, \ldots, n\}$;
- $\rho^{FJ}(FC^{FJ}) \subseteq FT^{FJ}$.

3.4 Summary

In real-world applications, information is often imprecise or uncertain. For modelling fuzzy information in the area of databases, Zadeh's fuzzy logic (Zadeh 1965) is introduced into databases to enhance the classical databases such that uncertain and imprecise information can be represented and manipulated. This resulted in numerous contributions, mainly with respect to the popular fuzzy conceptual data models and fuzzy logical database models.

In this chapter, we mainly introduce several popular fuzzy database models, including two fuzzy conceptual data models (fuzzy ER models and fuzzy UML models) and two fuzzy logical database models (fuzzy relational database models and fuzzy object-oriented database models). As mentioned at the beginning of this chapter, based on the widespread studies and the relatively mature techniques of fuzzy databases, it is not surprising that fuzzy databases may be useful for providing some technique supports for managing fuzzy knowledge in the Semantic Web. In particular, over the years, fuzzy information has been found in many applications in the context of the Semantic Web and resulted in numerous contributions as will be introduced in the next chapter.

References

Baader F, Calvanese D, McGuinness D, Nardi D, Patel-Schneider PF (2003) The description logic handbook: theory, implementation, and applications. Cambridge University Press, Cambridge

Berzal F, Marín N, Pons O, Vila MA (2007) Managing fuzziness on conventional object-oriented platforms. Int J Intell Syst 22(7):781–803

Booch G, Rumbaugh J, Jacobson I (1998) The unified modeling language userguide. Addison-Welsley Longman, MA

Bouaziz R, Chakhar S, Mousseau V, Ram S, Telmoudi A (2007) Database design and querying within the fuzzy semantic model. Inf Sci 177(21):4598–4620

Bordogna G, Pasi G (2001) Graph-based interaction in a fuzzy object oriented database. Int J Intell Syst 16:821–841

Buckles B, Petry F (1982) A fuzzy representation for relational databases. Fuzzy Sets Syst 7:213–226

Cao TH, Rossiter JM (2003) A deductive probabilistic and fuzzy object-oriented database language. Fuzzy Sets Syst 140:129–150

Calvanese D, Lenzerini M, Nardi D (1999) Unifying class-based representation formalisms. J Artif Intell Res 11(2):199–240

Chaudhry N, Moyne J, Rundensteiner EA (1999) An extended database design methodology for uncertain data management. Inf Sci 121(1–2):83–112

Chen GQ, Kerre EE (1998) Extending ER/EER concepts towards fuzzy conceptual data modeling, In: Proceedings of the 1998 IEEE international conference on fuzzy systems, pp. 1320–1325

Chen GQ, Vandenbulcke J, Kerre EE (1992) A general treatment of data redundancy in a fuzzy relational data model. J Am Soc Inform Sci 43:304–311

Cross V (2001) Fuzzy extensions for relationships in a generalized object model. Int J Intell Syst 16:843–861

Cross V, Caluwe R, Vangyseghem N (1997) A perspective from the fuzzy object data management group (FODMG). In: Proceedings of fuzzy systems, pp. 721−728

Cubero JC, Marín N, Medina JM, Pons O, Vila MA (2004) Fuzzy object management in an object-relational framework. In: Proceedings of the 10th international conference on information processing and management of uncertainty in knowledge-based systems, IPMU'2004, pp. 1767−1774

de Tré G, de Caluwe R (2003) Level-2 fuzzy sets and their usefulness in object-oriented database modeling. Fuzzy Sets Syst 140:29–49

De SK, Biswas R, Roy AR (2001) On extended fuzzy relational database model with proximity relations. Fuzzy Sets Syst 117:195–201

Dubois D, Prade H, Rossazza JP (1991) Vagueness, typicality, and uncertainty in class hierarchies. Int J Intell Syst 6:167–183

George R, Srikanth R, Petry FE, Buckles BP (1996) Uncertainty management issues in the object-oriented data model. IEEE Trans. Fuzzy Syst 4(2):179–192

Gyseghem NV, Caluwe RD (1995) Fuzzy behavior and relationships in a fuzzy OODB-model. In: Proceedings of the tenth annual ACM symposium on applied computing, Nashville, pp. 503−507

Haroonabadi A, Teshnehlab M (2007) Applying Fuzzy-UML for uncertain systems modeling. In: Proceedings of the first joint congress on fuzzy and intelligent systems, Ferdowsi University of Mashhad, Iran

Haroonabadi A, Teshnehlab M (2009) Behavior modeling in uncertain information systems by Fuzzy-UML. Int J Soft Comput 4(1):32–38

Koyuncu M, Yazici A (2003) IFOOD: an intelligent fuzzy object-oriented database architecture. IEEE Trans Knowl Data Eng 15(5):1137–1154

Lee J, Xue NL, Hsu KH, Yang SJH (1999) Modeling imprecise requirements with fuzzy objects. Inf Sci 118:101–119

Majumdar AK, Bhattacharya I, Saha AK (2002) An object-oriented fuzzy data model for similarity detection in image databases. IEEE Trans Knowl Data Eng 14:1186–1189

Marín N, Pons O, Vila MA (2001) A strategy for adding fuzzy types to an object oriented database system. Int J Intell Syst 16:863–880

Marín N, Vila MA, Pons O (2000) Fuzzy types: a new concept of type for managing vague structures. Int J Intell Syst 15:1061–1085

Ma ZM, Zhang WJ, Ma WY (2000) Semantic measure of fuzzy data in extended possibility-based fuzzy relational databases. J Intell Syst 15:705–716

Ma ZM, Zhang WJ, Ma WY, Chen GQ (2001) Conceptual design of fuzzy object-oriented databases using extended entity-relationship model. Int J Intell. Syst. 16(6):697–711

Ma ZM, Mili F (2002) Handling fuzzy information in extended possibility-based fuzzy relational databases. Int J Intell Syst 17(10):925–942

Ma ZM, Zhang WJ, Ma WY (2004) Extending object-oriented databases for fuzzy information modeling. Inf Syst 29(5):421–435

Ma ZM, Yan L (2007) Fuzzy XML data modeling with the UML and relational data models. Data Knowl Eng 63(3):970–994

Ma ZM, Yan L (2008) A literature overview of fuzzy database models. J Inform Sci Eng 24(1):189–202

Ma ZM, Yan L (2010) A Literature overview of fuzzy conceptual data modeling. J Inform Sci Eng 26:427–441

Ma ZM, Zhang F, Yan L, Lv Y (2010) Formal semantics-preserving translation from fuzzy ER model to fuzzy OWL DL ontology. Web Intell Agent Syst Int J 8(4):397–412

Ma ZM, Zhang F, Yan L (2011a) Fuzzy information modeling in UML class diagram and relational database models. Appl Soft Comput 11(6):4236–4245

Ma ZM, Zhang F, Yan L (2011b) Representing and reasoning on fuzzy UML models: a description logic approach. Expert Syst Appl 38(3):2536–2549

Ma ZM, Zhang F, Yan L, Cheng JW (2011c) Extracting knowledge from fuzzy relational database with Description Logic. Integr Comput Aided Eng 18:1–19

Nam M, Ngoc NTB, Nguyen H, Cao TH (2007) FPDB40: a fuzzy and probabilistic object base management system. In: Proceedings of the FUZZ-IEEE 2007, pp. 1–6

Ndouse TD (1997) Intelligent systems modeling with reusable fuzzy objects. Int J Intell Syst 12:137–152

Ozgur NB, Koyuncu M, Yazici A (2009) An intelligent fuzzy object-oriented database framework for video database applications. Fuzzy Sets Syst 160:2253–2274

Prade H, Testemale C (1984) Generalizing database relational algebra for the treatment of incomplete or uncertain information and vague queries. Inf Sci 34:115–143

Raju K, Majumdar A (1988) Fuzzy functional dependencies and lossless join decomposition of fuzzy relational database systems. ACM TODS 13(2):129–166

Ruspini E (1986) Imprecision and uncertainty in the Entity-Relationship model. In: Prade H, Negoita CV (eds) Fuzzy logic in knowledge engineering, Verlag TüV Rheinland GmbH, Köln, pp. 18–28

Rundensteiner E, Bic L (1992) Evaluating aggregates in possibilistic relational databases. Data Knowl Eng 7:239–267

Shenoi S, Melton A (1999) Proximity relations in the fuzzy relational database model. Fuzzy Sets Syst. (Supplement) 100:51–62

Sicilia MA, Garcia E, Gutierrez JA (2002) Integrating fuzziness in object oriented modeling language: towards a fuzzy-UML. In: Proceedings of international conference on fuzzy sets theory and its applications, pp. 66–67

Umano M, Fukami S (1994) Fuzzy relational algebra for possibility-distribution-fuzzy-relational model of fuzzy data. J Intell Inf Syst 3:7–27

Umano M, Imada T, Hatono I, Tamura H (1998) Fuzzy object-oriented databases and implementation of its SQL-type data manipulation language. In: Proceedings of the 7th IEEE international conference on fuzzy systems, pp. 1344–1349

Vandenberghe RM (1991) An extended entity-relationship model for fuzzy databases based on fuzzy truth values. In: Proceedings of the 4th international fuzzy systems association world congress, pp. 280–283

Vert G, Morris A, Stock M, Jankowski P (1986) Extending entity-relationship modeling notation to manage fuzzy datasets. In: Proceedings of the 8th international conference on information processing and management of uncertainty in knowledge-based systems, pp. 320–327

Yazici A, Koyuncu M (1997) Fuzzy object-oriented database modeling coupled with fuzzy logic. Fuzzy Sets Syst 89:1–26

Yan L, Ma ZM (2012) Comparison of entity with fuzzy data types in fuzzy object-oriented databases. Integr Comput Aided Eng 19(2):199–212

Yan L, Ma ZM, Zhang F (2012) Algebraic operations in fuzzy object-oriented databases, Inf Syst Front. in press. doi:10.1007/s10796-012-9359-8

Zadeh LA (1965) Fuzzy sets. Inf Control 8(3):338–353

Zvieli A, Chen PP (1986) Entity-relationship modeling and fuzzy databases. In: Proceedings of IEEE international conference on data engineering, pp. 320–327

Zicari R, Milano P (1990) Incomplete information in object-oriented databases. ACM SIGMOD Rec 19(3):5–16

Zhang F, Yan L, Ma ZM (2012a) Reasoning of fuzzy relational databases with fuzzy ontologies. Int J Intell Syst 27(6):613–634

Zhang F, Ma ZM, Yan L, Wang Y (2012b) A Description Logic approach for representing and reasoning on fuzzy object-oriented database models. Fuzzy Sets Syst 186(1):1–25

Zhang F, Ma ZM, Yan L (2013) Representation and reasoning of fuzzy ER model with description logic. J Intell Fuzzy Syst, in press. doi:10.3233/IFS-120754

Zhang F, Ma ZM (2013) Construction of fuzzy ontologies from fuzzy UML models. Int J Comput Intell Syst 6(3):442–472

Chapter 4
Fuzzy Description Logics and Fuzzy Ontologies

4.1 Introduction

In the context of the Semantic Web, ontology (Berners-Lee et al. 2001) is a W3C standard knowledge representation model for the Semantic Web, and Description Logics (DLs) (Baader et al. 2003) are essentially the theoretical counterpart of the ontology representation languages (e.g., OWL), both of them are the key of representing and reasoning with knowledge in the Semantic Web as has been introduced in Chap. 1. Currently, there are lots of researches on ontologies and DLs, which are used widely in the context of the Semantic Web and other applications such as medical, image analysis, databases and more (Staab and Studer 2009; Baader et al. 2003).

However, information imprecision and uncertainty exist in many real world applications, and such information would be retrieved, processed, shared, reused, and aligned in the maximum automatic way possible. Nowadays, in ontology-based and many applications information is often vague and imprecise. This is a well-known problem especially for semantics-based applications of the Semantic Web, such as knowledge management, e-commerce, and web portals (Calegari and Ciucci 2006). For example, a task like a "doctor appointment" could look like: "Make me an appointment with a doctor *close* to my home not *too early* and of *good* references" (Stoilos et al. 2010). The classical ontologies and DLs are quite expressive logical formalisms, but they feature limitations and may not be sufficient to represent such information and knowledge. Therefore, a requirement that naturally arises in many practical applications of knowledge-based systems, in particular the Semantic Web because ontologies and DLs are the representation and logical foundation of the Semantic Web. As a popular family of formally well-founded knowledge representation formalisms, fuzzy ontologies and fuzzy Description Logics (fuzzy DLs), which extend ontologies and DLs with fuzzy logic, are very well suited to cover for representing and reasoning with imprecision and uncertainty. Up to now, the literature on fuzzy ontologies and fuzzy DLs has

Z. Ma et al., *Fuzzy Knowledge Management for the Semantic Web*,
Studies in Fuzziness and Soft Computing 306, DOI: 10.1007/978-3-642-39283-2_4,

been flourishing. In this chapter, we focus on the knowledge representation models (i.e., fuzzy ontologies and fuzzy DLs) for representing and reasoning with imprecision and uncertainty in the Semantic Web, some main fuzzy DLs and the notions of fuzzy ontologies, and the related issues of them will be introduced.

4.2 Fuzzy Description Logics

As a popular family of formally well-founded and decidable knowledge representation languages, fuzzy Description Logics (fuzzy DLs), which extend Description Logics (DLs) with fuzzy logic, have been proposed as languages able to represent and reason about imprecise and vague knowledge. One of the most important research lines in (fuzzy) DLs is the balance between expressivity and computational complexity. Using expressive (fuzzy) DLs to model complex knowledge definitely brings about higher complexity, while reducing reasoning complexity will lead to the limitation of expressive power. In some "toy" system (or trial system), knowledge bases are relatively small, so the reasoning complexity is not of the main concern, whereas in the overwhelming majority of the practical application-oriented domain knowledge bases, as they involve a large number of individual assertions and relationship assertions, which make the ABoxes in large scale, the requirements for a low complexity are imperative. Therefore, according to the different requirements, many works proposed various fuzzy DLs from weaker to stronger in expressive power, including the *tractable* fuzzy DLs, the *expressive* fuzzy DLs, and fuzzy DLs with *fuzzy data types*. Also, in order to implement the automatic reasoning of fuzzy DLs, kinds of reasoners based on different fuzzy DLs have been put forward. A comprehensive literature overview of fuzzy DLs can be found at (Ma et al. 2013). In this sub-chapter, several types of fuzzy DLs and fuzzy DL reasoners will be introduced.

4.2.1 Tractable Fuzzy Description Logics

The tractable fuzzy DLs are fuzzy extensions of tractable DLs, which are rich enough to capture significant ontology languages but keeping low complexity of reasoning. The first kind of tractable fuzzy DLs is *fuzzy DL-Lite* family, i.e., fuzzy extensions of *DL-Lite* family languages (Calvanese et al. 2005). A fuzzy extension of *DL-Lite* called *fuzzy DL-Lite* was proposed by Straccia (2006). Furthermore, Pan et al. (2007) proposed two new expressive query languages accompanied with query answering algorithms over *fuzzy DL-Lite*, and also developed a prototype implementation for querying *fuzzy DL-Lite* ontologies. In order to model *n*-ary relations among more than two objects in some real world situations, a *fuzzy DLR-Lite* with fuzzy concrete domains was introduced in Straccia and Visco (2007). Cheng et al. (2008) proposed a fuzzy DL called $f\text{-}DLR\text{-}Lite_{F,\cap}$, which allows for the presence of *n*-ary relations and the occurrence of concept conjunction on the left land of inclusion axioms.

Table 4.1 The tractable fuzzy DLs

	Fuzzy DLs	*References*	*Mainly discussed issues*
Fuzzy DL-Lite family	fuzzy DL-Lite	Straccia 2006	Syntax, semantics, reasoning services, and query
	fuzzy DL-Lite	Pan et al. 2007	Syntax, semantics, reasoning services, and query
	fuzzy DLR-Lite(D)	Straccia and Visco (2007)	Syntax, semantics, reasoning services, and query
	f-DLR-Lite$_{F,\cap}$	Cheng et al. 2008	Syntax, semantics, reasoning services, and query
Fuzzy EL-family	*fuzzy EL*	VojtÄsõ 2006; Gursky¨ et al. 2008	Syntax, semantics, reasoning, and query
	fuzzy EL$^+$, *fuzzy EL*$^{++}$	Stoilos et al. 2008a; Mailis et al. 2008	Syntax, semantics, reason, and classification algorithms

Moreover, as we have known that DLs that allow for intersection of concepts and existential quantification (but not value restriction) are collected in the *EL*-family of DLs, which can also be considered as the tractable DLs. Some fuzzy extensions to the *EL*-family languages are presented, e.g., *fuzzy EL* (VojtÄsõ 2006; Gursky¨ et al. 2008), *fuzzy EL*$^+$ (Stoilos et al. 2008a), and *fuzzy EL*$^{++}$ (Mailis et al. 2008). Some existing tractable fuzzy DLs can be summarized in Table 4.1.

Here, we do not intend to introduce all the tractable fuzzy DLs mentioned above. In the following, we introduce the fuzzy DL *f-DLR-Lite*$_{F,\cap}$ proposed in Cheng et al. (2008), which may provide a general understanding of some notions of the tractable fuzzy DLs.

f-DLR-Lite$_{F,\cap}$ concepts are defined as follows:

$$B ::= A|\exists i : R$$
$$C ::= B|\neg B|C1 \sqcap C2$$

where A denotes an atomic concept, R denotes an n-ary relation, $\exists i : R$ denotes the projection of n-ary relation R on its ith element, $1 \le i \le n$. B denotes a basic concept that can be either an atomic concept or a concept of the form $\exists i : R$. Note that negation is only allowed to be used on basic concepts other than general concepts, and disjunction is not in consideration.

An *f-DLR-Lite*$_{F,\cap}$ knowledge base (KB) K can be partitioned into a terminological part called TBox and an assertional part called ABox, denoted by $K = < T, A >$.

An *f-DLR-Lite*$_{F,\cap}$ TBox T is a finite set of concept axioms and functionality axioms which are of the form:

$$Cl \sqsubseteq C(\text{inclusion axioms})$$
$$\text{funct}(i{:}R)\ (\text{functionality axiom})$$

where $Cl ::= B|C1 \sqcap C2$, denotes the concepts allowed to present in the left hand of inclusion axioms. A functionality axiom shows the functionality of ith component of n-ary relation. An *f-DLR-Lite*$_{F,\sqcap}$ ABox A is of the form $B(o) \geq n$, $\exists i : R(o) \geq n$.

As in other DLs, the semantics of *f-DLR-Lite*$_{F,\sqcap}$ is given in terms of interpretation. An interpretation of *f-DLR-Lite*$_{F,\sqcap}$ is $FI = (\Delta^{FI}, \bullet^{FI})$, where Δ^{FI} is a nonempty object set named interpretation domain and $\rfloor^{FI}$ is an interpretation function, which maps different individuals into different elements in Δ^{FI}, concept A into membership function A^{FI}: $\Delta \to [0, 1]$, role R into membership function R^{FI}: $\Delta^{FI} \times \cdots \times \Delta^{FI} \to [0, 1]$, while $(\exists i : R)^{FI}(o) = \sup(R^{FI}(o^{[i:o]}))$, where o denotes an n-tuple of individuals whose ith element is o. According to the conjunctive and negative operations in fuzzy set theory, the semantics of other concepts are given as $\neg B^{FI}(o) = 1 - B^{FI}(o)$ and $(C_1 \sqcap C_2)^{FI}(o) = \min(C_1^{FI}(o), C_2^{FI}(o))$.

Given an interpretation FI and inclusion axiom$B \sqsubseteq C$, FI is a model of $B \sqsubseteq C$, if $B^{FI}(o) \leq C^{FI}(o)$ for any $o \in \Delta^{FI}$, written as $FI| \approx B \sqsubseteq C$. Note that if concept disjunctions are interpreted as *max*(), then concept disjunctions in the form of $C_1 \sqcup C_2$ are allowed in the right hand of inclusion axioms. For functionality axiom funct(i: R), given an interpretation FI and o_1, $o_2 R$, if $o_1[i] = o_2[i]$ entails $o_1[j] = o_2[j]$ (for all $1 \leq j \leq n$), then we have $FI| \approx$ funct(i:R). Here $o[i]$ denotes ith element of o. Similarly, for ABox assertions, $FI| \approx B(o) \geq n(resp.\ FI| \approx R(o, orsquor;) \geq n)$, iff $B^{FI}(o^{FI}) \geq n(resp. R^{FI}(o^{FI}, o'^{FI}) \leq n)$.

If interpretation FI is a model of all the axioms and assertions in a KB K, we call it a model of K. A KB K is satisfiable iff it has at least one model. A KB K entails (logically implies) a fuzzy assertion α, iff all the models of K are also models of α, written as $K| \approx \alpha$.

4.2.2 Expressive Fuzzy Description Logics

The tractable fuzzy DLs mentioned in Sect. 4.2.1 can effectively express fuzzy knowledge, but they provide limited representation and reasoning ability. For example, they can not represent and reason about more complex fuzzy knowledge (e.g., the number restrictions, the transitive roles, the nominals, and etc.). Therefore, some scholars have carried out researches on more expressive fuzzy DLs. The initial idea combining fuzzy logic and DLs was presented by Yen (1991), where a construct called *membership manipulators* was introduced, and a structural subsumption algorithm was also provided in order to perform reasoning. A later approach was presented by Tresp and Molitor (1998), and the authors also developed a tableaux calculus for the proposed fuzzy DL called ALC_{FM} (*ALC* extended with fuzzy set theory and the membership manipulator constructor). In particular, the fuzzy extension of the basic DL *ALC*, i.e., *f-ALC*, was presented in Straccia (1998). From then on, the fuzzy extensions of the *f-ALC* by adding different constructors (e.g., number restrictions, inverse roles, and etc.) were proposed. All of them form the family of *f-ALC* languages, such as ALC_{FH} (HÎlldobler et al. 2002), *f-ALCN* (Li et al. 2005c; Lu et al. 2005), $ALCQ_F^+$ (SÄnchez and Tettamanzi 2004), *f-ALCIQ* (Stoilos et al. 2008b), and etc. Further, for representing and reasoning about more complex fuzzy knowledge, many works further extended the family of *f-ALC* languages with transitive roles or other constructors. Many expressive fuzzy DLs were thus presented, such as *f–SI* (Stoilos et al. 2005a), *f-SHIN* (Stoilos et al. 2005b), *f-SHOIN* (Stoilos et al. 2005c), *f-SHOIQ* (Stoilos et al. 2006), and *f-SROIQ* (Stoilos and Stamou 2007), and so on. Here, to avoid very long names for more expressive fuzzy DLs, the abbreviation *S* was introduced for ALC_{R+}, i.e., DL that extends *ALC* by transitive roles. In addition, the other fuzzy DLs with fuzzy cut sets, fuzzy rough sets, and fuzzy truth values were developed. A comprehensive review about many expressive fuzzy DLs can be found at (Ma et al. 2013). Table 4.2 gives some existing fuzzy DLs. Notice that, however, it does not means that Table 4.2 covers all publications in the research area and gives complete descriptions.

In the following, we briefly recall the basic fuzzy DL *f-ALC* (Straccia 1998), in order to well understand the notions of fuzzy DLs.

Let N_I, N_C, and N_R be three disjoint sets: N_I is a set of individual names, N_C is a set of fuzzy concept names, and N_R is a set of fuzzy role names. *f-ALC*-concepts are defined as follows (where $A \in N_C$, $R \in N_R$):

$$C, D \rightarrow \bot|\top|A|\neg C|C \sqcap D|C \sqcup D|\exists R.C|\forall R.C$$

The semantics of *f-ALC* is defined by a fuzzy interpretation $FI = < \Delta^{FI}, \rfloor^{FI} >$ where Δ^{FI} is a nonempty set and $\rfloor^{FI}$ is a function which maps every $d \in N_I$ to an element $d^{FI} \in \Delta^{FI}$, maps every $A \in N_C$ into a function A^{FI}: $\Delta^{FI} \rightarrow [0, 1]$, and maps every $R \in N_R$ into a function R^{FI}: $\Delta^{FI} \times \Delta^{FI} \rightarrow [0, 1]$. Further, for *f-ALC*-concepts *C* and *D*, $R \in N_R$, and $d, d' \in \Delta^{FI}$, we have:

$$\begin{aligned}
\top^{FI}(d) &= 1 \bot^{FI}(d) = 0 \\
(\neg C)^{FI}(d) &= 1 - C^{FI}(d) \\
(C \sqcap D)^{FI}(d) &= \min\{C^{FI}(d), D^{FI}(d)\} \\
(C \sqcup D)^{FI}(d) &= \max\{C^{FI}(d), D^{FI}(d)\} \\
(\forall R.C)^{FI}(d) &= \inf_{d' \in \Delta^{FI}} \{\max\{1 - R^{FI}(d, d'), C^{FI}(d')\}\} \\
(\exists R.C)^{FI}(d) &= \sup_{d' \in \Delta^{FI}} \{\min\{R^{FI}(d, d'), C^{FI}(d')\}\}
\end{aligned}$$

An *f-ALC* knowledge base *K* is composed of a TBox *T* and an ABox *A*:

- A TBox *T* is a finite set of terminology axioms of the form $C \sqsubseteq D$ or $C = D$. An interpretation *FI* satisfies $C \sqsubseteq D$ iff for any $d \in \Delta^{FI}$, $C^{FI}(d) \leq D^{FI}(d)$, and similarly for $C = D$. *FI* is a model of TBox *T* iff *FI* satisfies all axioms in *T*.
- An ABox *A* is a finite set of assertions of the form $< \alpha \bowtie n >$, Here $\bowtie \in \{>, \geq, <, \leq\}$, $n \in [0, 1]$, α is either of the form d: C or (d_1, d_2): R. Especially, in order to give a uniform format of the ABox, we define: when $n = 1$, the form $< \alpha \geq 1 >$ is equivalent to $< \alpha = 1 >$. Concretely speaking, $< d$: $C \geq 1 >$ means that *d* is determinately an individual of *C*; $< (d_1, d_2)$: $R \geq 1 >$ means that (d_1, d_2)

Table 4.2 Some expressive fuzzy DLs

	Representation of fuzzy terminologies and concepts							
	f-TSL	f-ALC	ALC_{FH}	$ALCQ_F^+$	f-ALCIQ	f-SHIN	f-SHOIN	*f-SROIQ*
Syntax and semantics	Yen, 1991	Straccia, 1998	HÎlldobler et al., 2002	SÄnchez and Tettamanzi, 2004	Stoilos et al., 2008b	Stoilos et al., 2005b	Stoilos et al., 2005c	Stoilos and Stamou, 2007; Bobillo et al., 2007
Tableau reasoning algorithm	Yen, 1991	Straccia, 2001	HÎlldobler et al., 2002	SÄnchez and Tettamanzi, 2006	Stoilos et al., 2008b	Stoilos et al., 2007	Stoilos et al., 2005c	Bobillo et al., 2007
Decid-ability	Yen, 1991	Straccia, 2001	HÎlldobler et al., 2002	SÄnchez and Tettamanzi, 2006	Stoilos et al., 2008b	Stoilos et al., 2007	Stoilos et al., 2005c	Bobillo et al., 2007

determinately has the relationship R. An interpretation FI satisfies $< d\colon C \bowtie n >$ iff $C^{FI}(d^{FI}) \bowtie n$ and satisfies $< (d_1, d_2)\colon R \bowtie n >$ iff $R^{FI}(d_1^{FI}, d_2^{FI}) \bowtie n$. FI is a model of ABox A iff FI satisfies all assertions in A.
- A fuzzy interpretation FI satisfies a knowledge base K if it satisfies all axioms and assertions in K.

Now, we introduce the reasoning problems of *f-ALC*, including satisfiability, subsumption, consistency, entailment, and so on. An *f-ALC* knowledge base K is *satisfiable* iff there exists a fuzzy interpretation FI which satisfies all axioms and assertions in K. An *f-ALC*-concept C is *satisfiable* w.r.t. a Tbox T iff there exists some model FI of T for which there is some $d \in \Delta^{FI}$ such that $C^{FI}(d) = n$, and $n \in [0, 1]$. Let C and D be two *f-ALC*-concepts, we say that C is *subsumed* by D w.r.t. T if for every model FI of T it holds that, $\forall d \in \Delta^{FI}$, $C^{FI}(d) \leq D^{FI}(d)$. An *f-ALC* Abox A is *consistent* w.r.t. T if there is a model FI of T which is also a model of A. Furthermore, given an axiom or assertion ψ, an *f-ALC* knowledge base K *entails* ψ, written $K \mid = \psi$, iff all models of K also satisfy ψ. Moreover, given an *f-ALC* knowledge base K and a fuzzy assertion α, the *greatest lower bound* of α w.r.t. K is $glb(K, \alpha) = sup\{n : K \mid = \alpha \geq n\}$, where $sup\ \varnothing = 0$.

The problem of concept satisfiability can be reduced to the problem of ABox consistency, and the problems of subsumption and entailment can be reduced to the problem of knowledge base satisfiability (Straccia 2001). Furthermore, in the presence of only *simple TBoxes* (A Tbox T is called *simple* if it neither includes cyclic nor general concept inclusions, i.e., axioms are the form $A \sqsubseteq C$ or $A \equiv C$, where A is a concept name that is never defined by itself either directly or indirectly, and A appears at most once at the left hand side.), all the problems above can be reduced to ABox consistency w.r.t. an empty TBox. The reasoning algorithm and its decidability for *f-ALC* have been introduced in detail in (Straccia 2001).

Note that, although the fuzzy DLs mentioned in Table 4.2 are quite expressive logical formalism they feature limitations, and they can represent only the binary relations and cannot represent the n-ary relations among more than two objects. A fuzzy extension of *DLR* called *FDLR*, which can directly represent the n-ary relations, was proposed in (Zhang et al. 2008c; Ma et al. 2011). In the following, the syntax, semantics, knowledge base, and reasoning for FDLR are introduced.

The FDLR, which is a fuzzy extension of DLR (Calvanese et al. 1997), has the same syntax with the DLR. The basic elements of FDLR are atomic relations and atomic concepts, denoted by P and A, respectively. The arbitrary relations (denoted by R) of arity between 2 and n_{max} and the arbitrary concepts (denoted by C) are formed according to the following *syntax*:

$$R \rightarrow \top_n |P| (\$i/n : C) |\neg R| R_1 \sqcap R_2$$

$$C \rightarrow \top_1 |A| \neg C | C \sqcap D | \exists [\$i] R | \leq k [\$i] R$$

where $i \in \{1, n_{\max}\}$ denotes the i-th component of a relation; $n \in \{2, n_{\max}\}$ denotes the arity of a relation, $i \leq n$; and k denotes a nonnegative integer. $\top_1$ denotes the interpretation domain, $\top_n$ covers all relations of arity n. The concepts and relations must be well-typed, which means that only the relations of the same arity n can be combined to form an expression such as $R_1 \sqcap R_2$ (which inherits the arity n).

The *semantics* of FDLR are provided by a fuzzy interpretation, which is a pair $FI = (\Delta^{FI}, \cdot^{FI})$ consisting of the interpretation domain Δ^{FI} and an interpretation function $\cdot^{FI}$ that mapping:

- an individual d to an element in Δ^{FI}.
- a concept C to a membership degree function C^{FI}: $\Delta^{FI} \to [0, 1]$.
- a role R to a membership degree function R^{FI}: $(\Delta^{FI})^n \to [0, 1]$.
- the syntax expressions of FDLR mentioned above to the following formulas:

$$\top_1^{FI} = \Delta^{FI} \qquad A^{FI} \subseteq \Delta^{FI} \qquad \top_1^{FI}(d_i) = 1$$

$$(\neg C^{FI})(d_i) = 1 - C^{FI}(d_i)$$
$$(C_1 \sqcap C_2)^{FI}(d_i) = \min\{C_1^{FI}(d_i), C_2^{FI}(d_i)\}$$
$$(\exists[\$i]R)^{FI}(d_i) = \sup_{d_i \in \Delta^{FI}} R^{FI}(d_1 \ldots d_i \ldots d_n)$$
$$(\leq K[\$i]R)^{FI}(d_i) = \inf_{D_1 \ldots D_{K+1} \in T_n^{FI}} \vee_{j=1}^{K+1} \neg R^{FI}(D_j)$$

$$\top_n^{FI} \subseteq (\Delta^{FI})^n \qquad P^{FI} \subseteq \top_n^{FI} \qquad \top_n^{FI}(d_1 \ldots d_n) = \begin{cases} 1 & (d_1 \ldots d_n) \in \top_n^{FI} \\ 0 & (d_1 \ldots d_n) \notin \top_n^{FI} \end{cases}$$

$$(\neg R)^{FI}(d_1 \ldots d_n) = 1 - R^{FI}(d_1 \ldots d_n)$$
$$(R_1 \sqcap R_2)^{FI}(d_1 \ldots d_n) = \min\{R_1^{FI}(d_1 \ldots d_n), R_2^{FI}(d_1 \ldots d_n)\}$$
$$(\$i/n : C)^{FI}(d_1 \ldots d_i \ldots d_n) = \min\{C^{FI}(d_i), \top_n^{FI}(d_1 \ldots d_i \ldots d_n)\}$$

where $(d_1 \ldots d_i \ldots d_n)$ denotes an individual $d_i \in (d_1 \ldots d_i \ldots d_n)$, i.e., $(d_1 \ldots d_i \ldots d_n)$ contains the appointed individual d_i; D_j is a tuple $(d_1 \ldots d_i \ldots d_n)$ that contains the appointed individual d_i (j does not depend on i). Moreover, the semantics of the following abbreviations can be obtained according to the formulas mentioned above:

$$\begin{array}{rcc} C_1 \sqcup C_2 & \text{for} & \neg(\neg C_1 \sqcap \neg C_2) \\ \forall[\$i]R & \text{for} & \neg\exists[\$i]\neg R \\ \geq (k+1)[\$i]R & \text{for} & \neg(\leq k[\$i]R) \\ = k[\$i]R & \text{for} & (\leq (k+1)[\$i]R) \sqcap (\geq k[\$i]R) \end{array}$$

A FDLR *knowledge base* (FDLR *KB*) comprises two components: the *TBox T* and the *ABox A*. The TBox introduces terminology, while the ABox contains assertions about the named individuals in terms of this vocabulary. A *TBox T* is a finite set of axioms such as inclusions $C \sqsubseteq D$ ($R \sqsubseteq S$) or equalities $C = D$ ($R = S$), where C and D are concepts, and R and S are roles. To simplify the exposition, we only deal with the axioms involving concepts. The semantics are given by a fuzzy interpretation FI as follows: $C \sqsubseteq D$, iff $\forall d_i \in \Delta^{FI}$, $C^{FI}(d_i^{FI}) \leq D^{FI}(d_i^{FI})$; $C \equiv D$, iff $C \sqsubseteq D$ and $D \sqsubseteq C$, that is to say, $\forall d_i \in \Delta^{FI}$, $C^{FI}(d_i^{FI}) = D^{FI}(d_i^{FI})$. An *ABox A* is a finite set of fuzzy assertions of the form $< \alpha \bowtie n >$. More precisely, $< C(d_i) \bowtie n >$ or $< R(d_1 \ldots d_n) \bowtie n >$, where $\bowtie$ stands for $\geq$, $>$, $\leq$, and $<$, and $n \in [0, 1]$. The semantics are given by a fuzzy interpretation *FI* such that a fuzzy interpretation *FI* satisfies the assertion formula $< C(d_i) \bowtie n >$, iff $C^{FI}(d_i^{FI}) \bowtie n$; A fuzzy interpretation *FI* satisfies the assertion formula $R(d_1 \ldots d_n) \bowtie n >$, iff $R^{FI}(d_1^{FI} \ldots d_n^{FI}) \bowtie n$. A fuzzy interpretation *FI* satisfies a FDLR *KB* $\sum$ iff $\forall \eta \in \sum$ such that *FI* satisfies η. If a fuzzy interpretation *FI* satisfies a FDLR *KB* $\sum$, then we say that it is a model of $\sum$.

The fundamental tasks are considered when reasoning over a FDLR knowledge base including the concept (relation) satisfiability, concept (relation) subsumption, and logical implication. A FDLR-concept *C* is *satisfiable* (*unsatisfiable*) w.r.t. a TBox *T* iff there exists (does not exist) some model *FI* of *T* for which there is some $d \in \Delta^{FI}$, such that $C^{FI}(d^{FI}) = n$, $n \in [0, 1]$. Let *C* and *D* be two FDLR-concepts. We say that *C* is *subsumed* by *D* ($C \sqsubseteq D$) w.r.t. a TBox *T* if for every model *FI* of *T* it holds that, $\forall\, d \in \Delta^{FI}$, $C^{FI}(d^{FI}) \leq D^{FI}(d^{FI})$. For simplicity the relation satisfiability and relation subsumption are omitted, which are similar to the concept satisfiability and concept subsumption, respectively. Moreover, given a fuzzy axiom or a fuzzy assertion η, a FDLR *KB* $\sum$ *logically implies* η, written $\sum \models \eta$, iff each model of $\sum$ satisfies η. The satisfiability and subsumption problems can be reformulated by means of logical implication, e.g., satisfiability of concept *C* in a FDLR *KB* $\sum$ can be reformulated as $\Sigma \not\models C \sqsubseteq \bot$, i.e., deciding whether $\sum$ admits a model *FI* for which there is some $d \in \Delta^{FI}$, such that $C^{FI}(d^{FI}) = n$, where $n \in [0, 1]$. Interestingly, logical implication can, in turn, be reformulated in terms of satisfiability. Observe that FDLR allows for Boolean constructs on relations as well as relation inclusion axioms $R \sqsubseteq S$ and relation equality axioms $R = S$. In fact, FDLR can be viewed as a proper generalization of fuzzy ALCIQ (Stoilos et al. 2008b). The traditional DL fuzzy ALCIQ constructs can be expressed in FDLR as follows:

$$
\begin{array}{ccc}
\exists \mathrm{P.C} & \text{as} & \exists[\$1](\mathrm{P} \sqcap \ominus(\$2/2:\ \mathrm{C})) \\
\exists \mathrm{P^-.C} & \text{as} & \exists[\$2](\mathrm{P} \sqcap (\$1/2:\ \mathrm{C})) \\
\forall \mathrm{P.C} & \text{as} & \neg\exists[\$1](\mathrm{P} \sqcap (\$2/2:\ \neg\mathrm{C})) \\
\forall \mathrm{P^-.C} & \text{as} & \neg\exists[\$2](\mathrm{P} \sqcap (\$1/2:\ \neg\mathrm{C})) \\
\leq \mathrm{k\ P.C} & \text{as} & \leq \mathrm{k}\ [\$1](\mathrm{P} \sqcap (\$2/2:\ \mathrm{C})) \\
\leq \mathrm{k\ P^-.C} & \text{as} & \leq \mathrm{k}\ [\$2](\mathrm{P} \sqcap (\$1/2:\ \mathrm{C}))
\end{array}
$$

On this basis, reasoning on FDLR can be reduced to reasoning on fuzzy ALCIQ as presented in (Ma et al. 2011). The detailed reasoning algorithm for the fuzzy ALCIQ, please refer to (Stoilos et al. 2008b).

4.2.3 Fuzzy Description Logics with Fuzzy Data Types

It can be found that both the tractable and expressive fuzzy DLs mentioned above cannot represent and reason about data information (i.e., concrete knowledge). The

Semantic Web is expected to process data information in an intelligent and automatic way. For this purpose, some fuzzy DLs with *fuzzy concrete domains* (*D*) and *fuzzy data type group* (*G*) have been proposed. Straccia (2005a) first proposed a fuzzy extension of *ALC*(*D*) (the basic DL *ALC* extended with concrete domains), where the representation of concept membership functions and fuzzy modifiers are allowed, together with an inference procedure based on a mixture of a tableaux and bounded mixed integer programming. A reasoning algorithm for fuzzy *ALCF*(*D*) with general concept inclusions and explicit membership functions was proposed in Bobillo and Straccia (2009a). Moreover, the more expressive DL which can support fuzzy concrete domains was shown in Straccia (2005b), where the language is a fuzzy extension of *SHOIN*(*D*), which is the corresponding DL of the ontology description language OWL DL. A fuzzy extension of OWL 2 language, i.e., fuzzy *SROIQ*(*D*) was proposed in Bobillo (2008); Bobillo and Straccia (2009b, 2010, 2011). In addition, in order to support customized fuzzy data types and customized fuzzy data type predicates *G*, a new kind of fuzzy DL called *F-ALC*(*G*) was proposed in (Wang and Ma 2008), which can not only support the representation and reasoning of fuzzy concept knowledge, but also support fuzzy data information with customized fuzzy data types and customized fuzzy data type predicates. Similarly, a fuzzy DL called *F-SHOIQ*(*G*) and *f-SROIQ*(*G*) were developed in Wang et al. (2008); Yan et al. (2012). Table 4.3 provides some existing fuzzy DLs with fuzzy data types.

In the following, we will first recall the fuzzy extension of *SHOIN*(*D*) (Straccia 2005b), which is the main logical foundation of the fuzzy ontology representation language (e.g., fuzzy OWL language). Then, we will introduce the fuzzy DL called *F-ALC*(*G*) (Wang and Ma 2008), which is a fuzzy DL supporting customized fuzzy data types and customized fuzzy data type predicates *G*.

As usual we have an alphabet of distinct concepts names $\boldsymbol{C}$, abstract role names $\boldsymbol{R}_{\mathrm{A}}$, concrete role names $\boldsymbol{R}_{\mathrm{D}}$, abstract individuals $\boldsymbol{I}_{\mathrm{A}}$, concrete individuals $\boldsymbol{I}_{\mathrm{D}}$, and concrete datatypes $\boldsymbol{D}$. The set of *fuzzy SHOIN*(*D*)-roles is defined by $\boldsymbol{R}_{\mathrm{A}} \cup \{R^{-} | R \in \boldsymbol{R}_{\mathrm{A}}\} \cup \boldsymbol{R}_{\mathrm{D}}$. Let $A \in \boldsymbol{C}$, $R \in \boldsymbol{R}_{\mathrm{A}}$, $S \in \boldsymbol{R}_{\mathrm{A}}$ is a simple role, $T \in \boldsymbol{R}_{\mathrm{D}}$, $d \in \boldsymbol{D}$, $o \in \boldsymbol{I}_{\mathrm{A}}$, $c \in \boldsymbol{I}_{\mathrm{D}}$ and $p \in \mathbb{N}$, then *fuzzy SHOIN*(*D*)-concepts are defined inductively by the production rule:

$$\begin{aligned} C, D &\rightarrow \top | \bot | A | \neg C | C \sqcap D | C \sqcup D | \forall R.C | \exists R.C | \geq pS | \leq pS | \forall T.u | \exists T.u | \\ &\quad \geq pT | \leq pT | \{o\} \\ u &\rightarrow d | \{c\} \end{aligned}$$

The semantics of *fuzzy SHOIN*(*D*) are provided by a fuzzy interpretation $FI = (\Delta^{\mathrm{FI}}, \Delta_D, \cdot^{\mathrm{FI}}, \cdot^{\mathrm{D}})$, where Δ^{FI} is the abstract domain and Δ_D is the datatype domain (disjoint from Δ^{FI}), and $\cdot^{\mathrm{FI}}$ and $\cdot^{\mathrm{D}}$ are two fuzzy interpretation functions, which map:

- An abstract individual o to an element $o^{\mathrm{FI}} \in \Delta^{\mathrm{FI}}$,
- A concrete individual c to an element $c^{\mathrm{D}} \in \Delta_D$,
- A concept name A to a membership degree function $A^{\mathrm{FI}} : \Delta^{\mathrm{FI}} \rightarrow [0, 1]$,

Table 4.3 Some existing fuzzy description logics with fuzzy data types

	Representation of fuzzy data information						
	f-ALC(D)	f-ALCF(D)	f-SHOIN(D)	f-SROIQ(D)	F-ALC(G)	F-SHOIQ(G)	*f-SROIQ(G)*
Syntax and semantics	Straccia (2005a)	Bobillo and Straccia (2009a)	Straccia (2005b)	Bobillo, 2008; Bobillo and Straccia, 2009b	Wang and Ma, 2008	Wang et al., 2008	Yan et al., 2012
Reasoning algorithm	Straccia (2005a)	Bobillo and Straccia (2009a)	Lukasiewicz and Straccia, 2008	Bobillo, 2008	Wang and Ma, 2008	Wang et al., 2008	Yan et al., 2012
Decid-ability	Straccia (2005a)	Bobillo and Straccia (2009a)	Lukasiewicz and Straccia, 2008	Bobillo, 2008	Wang and Ma, 2008	Wang et al., 2008	Yan et al., 2012

- An abstract role name R to a membership degree function $R^{\mathrm{FI}}: \Delta^{\mathrm{FI}} \times \Delta^{\mathrm{FI}} \rightarrow [0, 1]$,
- A datatype d to a membership degree function $d^{\mathrm{D}}: \Delta_D \rightarrow [0, 1]$,
- A concrete role name T to a membership degree function $T^{\mathrm{FI}}: \Delta^{\mathrm{FI}} \times \Delta_D \rightarrow [0, 1]$,
- The complete semantics of *fuzzy SHOIN*(*D*)-concepts are depicted as follows:

$$\top^{\mathrm{FI}}(d) = 1$$

$$\bot^{\mathrm{FI}}(d) = 0$$

$$c^{\mathrm{FI}} = c^{\mathrm{D}}$$

$$d^{\mathrm{FI}}(c) = d^{\mathrm{D}}(c)$$

$$(\neg C)^{\mathrm{FI}}(o) = 1 - C^{\mathrm{FI}}(o)$$

$$\{o\}^{\mathrm{FI}}(o') = 1 \quad \text{if} \quad o^{\mathrm{FI}} = o', \{o\}^{\mathrm{FI}}(o') = 0\,\text{otherwise}$$

$$\{c\}^{\mathrm{FI}}(c') = 1 \quad \text{if} \quad c^{\mathrm{FI}} = c', \{c\}^{\mathrm{FI}}(c') = 0 \ \text{otherwise}$$

$$(C \sqcap D)^{\mathrm{FI}}(o) = \min\{C^{\mathrm{FI}}(o), D^{\mathrm{FI}}(o)\}$$

$$(C \sqcup D)^{\mathrm{FI}}(o) = \max\{C^{\mathrm{FI}}(o), D^{\mathrm{FI}}(o)\}$$

$$(\forall R.C)^{\mathrm{FI}}(o) = \inf_{o' \in \Delta^{\mathrm{FI}}} \{\max\{1 - R^{\mathrm{FI}}(o, o'), C^{\mathrm{FI}}(o')\}\}$$

$$(\exists R.C)^{\mathrm{FI}}(o) = \sup_{o' \in \Delta^{\mathrm{FI}}} \{\min\{R^{\mathrm{FI}}(o, o'), C^{\mathrm{FI}}(o')\}\}$$

$$(\geq pS)^{\mathrm{FI}}(o) = \sup_{o_1, \ldots, o_p \in \Delta^{\mathrm{FI}}} \wedge_{i=1}^{p} S^{\mathrm{FI}}(o, o_i)$$

$$(\leq pS)^{\mathrm{FI}}(o) = \inf_{o_1, \ldots, o_{p+1} \in \Delta^{\mathrm{FI}}} \vee_{i=1}^{p+1} (1 - S^{\mathrm{FI}}(o, o_i))$$

$$(\geq pT)^{\mathrm{FI}}(o) = \sup_{c_1, \ldots, c_p \in \Delta_D} \wedge_{i=1}^{p} T^{\mathrm{FI}}(o, c_i)$$

$$(\leq pT)^{\mathrm{FI}}(o) = \inf_{c_1, \ldots, c_{p+1} \in \Delta_D} \vee_{i=1}^{p+1} (1 - T^{\mathrm{FI}}(o, c_i))$$

$$(\forall T.d)^{\mathrm{FI}}(o) = \inf_{c \in \Delta_D} \{\max\{1 - T^{\mathrm{FI}}(o, c), d^{\mathrm{FI}}(c)\}\}$$

$$(\exists T.d)^{FI}(o) = \sup_{c \in \Delta_D} \{\min\{T^{\mathrm{FI}}(o, c), d^{\mathrm{FI}}(c)\}\}$$

A *fuzzy SHOIN(D)* knowledge base *KB* is a triple $< T, R, A >$, which contains a fuzzy TBox *T*, a fuzzy RBox *R*, and a fuzzy ABox *A*, where: A TBox *T* is a set of fuzzy concept axioms, including fuzzy concept inclusion axioms $C \sqsubseteq D$ and fuzzy concept equivalence axioms $C \equiv D$. A RBox *R* is a set of fuzzy role axioms, including transitive role axioms *Trans*(*R*) and fuzzy role inclusion axioms $R \sqsubseteq S$ or $T \sqsubseteq U$ ($U \in R_D$). An ABox *A* is a set of fuzzy assertions, including concept assertions $< C(o) \bowtie k >$, role assertions $< R(o_1, o_2) \bowtie k >$ and $< T(o, c) \bowtie k >$, and individual (in)equality assertions $o_1 \approx o_2$ and $o_1 \not\approx o_2$, where $\bowtie \in \{\geq, >, \leq, <\}$, $k \in [0, 1]$. The main reasoning problems for *fuzzy SHOIN(D)* has been provided in (Lukasiewicz and Straccia 2008).

Also, in order to support customized fuzzy data types and customized fuzzy data type predicates *G*, Wang and Ma (2008) proposed a fuzzy DL called *F-ALC* (*G*). Before introducing the fuzzy DL *F-ALC*(*G*), we first introduce some basic notions of data types.

- A *data type* is characterized by a lexical space, *L*(*d*), which is a nonempty set of Unicode strings; a value space, *V*(*d*), which is a nonempty set, and a total mapping *L*2 *V*(*d*) from the lexical space to value space.
- A *base data type d* is a special kind of data type which is built-in in XML Schema, such as *xsd*: *integer*, *xsd*: *string*, and etc.
- A *data type predicate p* is used to constrain and limit the corresponding data types. It is characterized by an arity *a*(*p*). The semantic of a *fuzzy data type predicate p* can be given by (Δ_D, J^D), where Δ_D is the fuzzy data type interpretation domain and J^D is the fuzzy data type interpretation function. Here J^D maps each *n*-ary fuzzy predicate *p* to a function $p^D: \Delta_D^n \rightarrow [0, 1]$ which is an *n*-ary fuzzy relation over Δ_D. It means that the relationship of concrete variables $v_1, \ldots, v_n$ satisfies predicate *p* in a degree belonging to [0, 1]. We also use *a*(*p*) to represent the arity of fuzzy predicate *p*.
- A fuzzy data type group G is a triple (ϕ_G, D_G, *dom*), where φ_G is a set of predicates which have been defined by corresponding known predicate URIrefs, D_G is a set of base data types, and *dom*(*p*) is the domain of a fuzzy data type predicate,

$$dom(p) = \begin{cases} p & if \quad p \in D_G \\ (d_1, \ldots, d_n), \quad where\ d_1, \ldots, d_n \in D_G & if \quad p \in \phi_G \backslash D_G\ and\ a(p) = n \end{cases}$$

Given a fuzzy data type group $G = (\phi_G, D_G, dom)$ and a base data type $d \in D_G$, the *sub-group* of *d* in *G*, abbreviated as *sub-group*(*d*, *G*), is defined as:

$$sub\text{-}group(d, G) = \{p \in \phi_G | dom(p) = \{d, \ldots, d\}(a(p) times)\}$$

Let *G* be a fuzzy data type group. Valid *fuzzy G-data type expression E* is defined by the following abstract syntax:

$$E ::= \top_D|\bot_D|p|\bar{p}|\{l_1,\ldots,l_s\}|(E_1 \wedge \ldots \wedge E_s)|(E_1 \vee \ldots \vee E_s)$$
$$|[u_{1,\ldots,}u_s]$$

Here $\top_D$ represents the fuzzy top data type predicate while $\bot_D$ the fuzzy bottom data type predicate; p is a predicate URIref; $\bar{p}$ is the relativized negated form of p; $l_1,\ldots, l_s$ are typed literals; u_i is a unary predicate in the form of $\top_D$, $\bot_D$, p, or$\bar{p}$. The semantic of fuzzy data type expressions can be defined as follows:

$$\top_D^D(v_1,\ldots,v_n) = 1;$$

$$\bot_D^D(v_1,\ldots,v_n) = 0$$

$$p^D(v_1,\ldots,v_n) \to [0,1];$$

$$\bar{p}^D(v_1,\ldots,v_n) = 1 - p^D(v_1,\ldots,v_n)$$

$$\{l_1,\ldots,l_s\}^D(v) = \vee_{i=1}^s(l_i^D = v)$$

$$(E_1 \wedge \ldots \wedge E_s)^D(v,\ldots,v_n) = (E_1)^D(v_1,\ldots,v_n) \wedge \ldots \wedge (E_s)^D(v,\ldots,v_n)$$

$$(E_1 \vee \ldots \vee E_s)^D(v,\ldots,v_n) = (E_1)^D(v_1,\ldots,v_n) \vee \ldots \vee (E_s)^D(v,\ldots,v_n)$$

$$([u_1,\ldots,u_s])^D(v_1,\ldots,v_s) = u_1^D(v_1) \wedge \ldots \wedge u_s^D(v_s)$$

In the fuzzy data type group, the supported data type predicates have been defined in ϕ_G, while the unsupported ones can be defined based on the supported ones used fuzzy data type expressions.

A *fuzzy data type group G* is *conforming* iff: (*i*) for any $p \in_G \backslash D_G$ with $a(p) = n$ $(n \geq 2)$, $dom(p) = (w,\ldots,w)$ (n times) for some $w \in D_G$; (*ii*) for any $p \in \phi_G \backslash D_G$, there exists $p' \in_G \backslash D_G$ such that $p'^D = \bar{p}^D$; and (*iii*) the satisfiability problems for finite fuzzy predicate conjunctions of each sub-group of G is decidable.

Based on these notions mentioned above, the *syntax* of the fuzzy DL *F-ALC*(*G*) can be defined as follows: *F-ALC*(*G*) consists of an alphabet of distinct concept names (C), role names ($R = R_A \cup R_D$) and individual names (I). *F-ALC*(*G*)-roles are simply role names in R, where $R \in R_A$ is called an abstract role and $T \in R_D$ is called a data type role. Let A be an atomic concept in C, $R \in R_A$, $T_1, \ldots, T_n \in R_D$, E is fuzzy G-data type expression. Then the valid *F-ALC*(*G*)-concepts are defined as follows:

$$C ::= \top|\bot|A|\neg C|C \sqcap D|C \sqcup D|\exists R.C|\forall R.C|\exists T_1,\ldots,T_n.E|\forall T_1,\ldots,T_n.E|$$
$$\geq mT_1,\ldots,T_n.E| \leq mT_1,\ldots,T_n.E$$

The *semantics* of *F-ALC*(*G*) is given by an interpretation of the form $FI = (\Delta^{FI}, \rfloor^{FI}, \Delta_D, \rfloor^D)$. Here $(\Delta^{FI}, \rfloor^{FI})$ is an interpretation of the object domain, $(\Delta_D, \rfloor^D)$ is an interpretation of the fuzzy data type group G, and Δ^{FI} and Δ_D are disjoint each other. $\rfloor^{FI}$ is an individual interpretation function and the semantics of abstract

F-ALC(*G*)-concepts like ⊤, ⊥, *A*, ¬*C*, *C* ⊓ *D*, *C* ⊔ *D*, ∃*R*.*C*, and ∀*R*.*C* can be referred to the *fuzzy SHOIN*(*D*) as introduced above. J^D is an interpretation of the fuzzy data type group *G*, which respectively assigns each concrete individual to an element in Δ_D, each simple data type role $T \in R_D$ to a function T^{FI}: $\Delta^{FI} \times \Delta_D \rightarrow [0, 1]$, and each *n*-ary fuzzy predicate *p* to a function p^D: $\Delta_D^n \rightarrow [0,1]$ which is an *n*-ary fuzzy relation over Δ_D. We have the following semantics for concrete *F-ALC* (*G*)-concepts with fuzzy data type expressions-related constructors:

$$(\exists T_1, \ldots, T_n.E)^{FI}(x) = \sup_{v_1,\ldots,v_n \in \Delta_D} \left(\left(\wedge_{i=1}^n T_i^{FI}(x, v_i) \right) \wedge E^{FI}(v_1, \ldots, v_n) \right)$$

$$(\forall T_1, \ldots, T_n.E)^{FI}(x) = \inf_{v_1,\ldots,v_n \in \Delta_D} \left(\left(\wedge_{i=1}^n T_i^{FI}(x, v_i) \right) \rightarrow E^{FI}(v_1, \ldots, v_n) \right)$$

$$(\geq mT_1, \ldots, T_n.E)^{FI}(x) = \sup_{v_{11},\ldots,v_{1n},\ldots,\ldots,v_{m1},\ldots,v_{mn} \in \Delta_D} \wedge_{i=1}^m \left(\left(\wedge_{j=1}^n T_{ij}^{FI}(x, v_{ij}) \right) \wedge E^{FI}(v_{i1}, \ldots, v_{in}) \right)$$

$$(\leq mT_1, \ldots, T_n.E)^{FI}(x) = 1 - (\geq (m+1)T_1, \ldots, T_n.E)^{FI}(x)$$

An *F-ALC*(*G*) knowledge base *k* is in the form of $< T, A >$, where *T* is a fuzzy *TBox* and *A* a fuzzy *ABox*. A fuzzy *TBox* is a finite set of fuzzy concept axioms of the form $C \sqsubseteq D$, where *C* and *D* are *F-ALC*(*G*)-concepts. A fuzzy *ABox* is a finite set of fuzzy assertions of the form $< \alpha \bowtie n >$. Here $\bowtie \in \{>, \geq, <, \leq\}$, $n \in [0, 1]$, and α is the form either *x*: *C*, (*x*, *v*): *T* (*v* is concrete variable) or (*x*, *y*): *R* ($x, y \in$ I).

An *F-ALC*(*G*)-concept *C* is satisfiable iff there exists some fuzzy interpretation *FI* for which there is some $x \in \Delta^{FI}$ such that $C^{FI}(x) = n$, and $n \in (0,1]$. A fuzzy interpretation *FI* satisfies a fuzzy *TBox T* iff $\forall x \square \Delta^{FI}$, $C^{FI}(x) \leq D^{FI}(x)$ for each $C \sqsubseteq D$. *FI* is a model of *TBox T* iff *FI* satisfies all axioms in *T*. An interpretation *FI* satisfies $< x: C \bowtie n >$ iff $C^{FI}(x) \bowtie n$, satisfies $< (x, v): T \bowtie n >$ iff $T^{FI}(x, v) \bowtie n$, satisfies $< (x, y): R \bowtie n >$ iff $R^{FI}(x, y) \bowtie n$. Then *FI* is a model of *ABox A* iff *FI* satisfies all assertions in *A*. Finally, a fuzzy knowledge base *k* is satisfiable iff there exists an interpretation *FI* which satisfies the *TBox T* and *ABox A*.

In the following, a tableaux algorithm for *F-ALC*(*G*) was provided. For ease of presentation, assuming that all concepts to be in *negation normal form* (*NNF*) (Hollunder et al. 1990), i.e., negations occur in front of concept names only. In addition, we use the symbols $\rhd$ as a placeholder for the inequalities $>$, $\geq$ and $\lhd$ for $<$, $\leq$, use the symbol $\bowtie$ as a placeholder for all types of inequations. Also we use the symbols $\bowtie^{-1}$, $\rhd^{-1}$, $\lhd^{-1}$ to denote their reflections, respectively. If ψ is an assertion in *F-ALC*(*G*), then ψ^C is the conjugation of ψ. The tableaux algorithm for *F-ALC*(*G*) tries to prove the satisfiability of a concept expression *D* by constructing a model of *D*. The model is represented by a so-called completion forest. Its nodes correspond to either individuals (labeled nodes) or variables (nonlabeled nodes), each labeled node being labeled with a set of triples of the form $< C, X, k >$, which respectively denote the concept, the type of inequality (*X* the set $\{\geq, >, <, \leq\}$), and the membership degree that the individual of the node has been asserted to belong to *C*. We call such triples *membership triples*. While testing the satisfiability of an *F-ALC*(*G*)-concept *D*, the sets of concepts *C* appearing in membership triples are restricted to subsets of *cl*(*D*), which denotes the set of all sub-concepts of *D*. We use $cl_G(D)$ to denote the set of all the fuzzy *G*-data type expressions and their negations occurring in these sub-concepts.

Let *D* be a *F-ALC*(*G*)-concept in *NNF*, R_A^D the set of abstract roles occurring in *D*, R_D^D the set of concrete roles occurring in *D*, *G* a fuzzy data type group, $E \in cl_G(D)$ a (possibly negated) fuzzy data type expression, and *X* the set $\{\geq, >, <, \leq\}$. *A tableau* $T = (S_A, S_D, L, DC, _A, _D)$ for *D* is defined as follows:

- S_A is a set of individuals; S_D is a set of variables,
- L: $S_A \rightarrow 2^{cl(D)} \times X \times [0,1]$ maps each individual in S_A to membership triples,
- *DC*: is a set of fuzzy data type constraints of the form $< E(v_1,\ldots,v_n), X, k >$ or inequations of the form $(< v_{i1},\ldots,v_{in} >, < v_{j1},\ldots,v_{jn} >, \neq)$, where $E \in cl_G(D)$, $v_1,\ldots,v_n, v_{i1},\ldots,v_{in}, v_{j1},\ldots,v_{jn} \in S_D$, $k \in [0, 1]$,
- $\varepsilon_A : R_A^D \rightarrow 2^{S_A \times S_A} \times X \times [0,1]$ maps abstract role in R_A^D to membership triples,
- $\varepsilon_D : R_D^D \rightarrow 2^{S_A \times S_D} \times X \times [0,1]$ maps data type role in R_D^D to membership triples.

A tableaux algorithm for *F-ALC*(*G*) works on a completion forest *F* for *D*. There are two kinds of nodes in the completion forest: abstract nodes (the normal labeled nodes) and data type nodes (nonlabeled leaves of *F*). Each abstract node *x* is labeled by a set of triples *L*(*x*), which contains membership triples. We define $L(x) = \{<C, X, k>\}$, where $C \in cl(D) \cup \{\uparrow(R, \{o\}) \mid R \in R_A^D$ and $\{o\} \in I^D\}$. Each edge $< x, y >$ is labeled with a set of $L(< x, y >)$ defined as, $L(< x, y >) = \{< R, X, k >\}$ where *R* is either abstract roles occurring in R_A^D, or data type roles in R_D^D. In the first case, *y* is a node and called an abstract successor of *x*; in the second case, *y* is a data type node, and called a data type successor of *x*. For each $\{o\} \in I^D$, there is a distinguished node $z_{\{o\}}$ in *F* such that $\{o\} \in L(z)$.

We use a set *DC*(*x*) to store the fuzzy data type expressions that must hold with regard to data type successors of a node *x* in *F*. Each element of *DC*(*x*) is either of the form $< E(v_1,\ldots,v_n), X, k >$, or of the form $(< v_{i1},\ldots,v_{in} >, < v_{j1},\ldots,v_{jn} >, \neq)$, where $E \in cl_d(D)$, $v_1,\ldots,v_n, v_{i1},\ldots,v_{in}, v_{j1},\ldots,v_{jn} \in S_D$, and $k \in [0, 1]$. Here $< v_1,\ldots, v_n >$, $< v_{i1},\ldots,v_{in} >$ and $< v_{j1},\ldots,v_{jn} >$ are tuples of fuzzy data type successors of *x*. The tableaux algorithm calls a fuzzy data type reasoner as a subprocedure for the satisfiability of *DC*(*x*). We say that *DC*(*x*) is satisfiable if the fuzzy data type query

$$\wedge_{<E(v_{r1},\ldots,v_{rn}),X,k>} \in DC(x) < E(v_{r1},\ldots,v_{rn}), X, k > \wedge$$
$$\wedge_{(<v_{i1},\ldots,v_{in}>,<v_{j1},\ldots,v_{jn}>,\neq) \in DC(x)} \neq (v_{i1},\ldots,v_{in}; v_{j1},\ldots,v_{jn})$$

is satisfiable. *DC*(*x*) plays as the interface of the fuzzy *DL* concept reasoner and the fuzzy data type reasoner.

A node *x* of a completion forest *F* is said to contain a *clash* if (at least) it contains either an *F-ALC-clash* (Straccia 2001) or a G_f*-clash*. A node *x* of a completion forest *F* contains a G_f*-clash* if it contains one of the following:

(i) for some fuzzy data type roles $T_1,\ldots,T_n$, $< \leq mT_1,\ldots,T_n.E, \rhd, k > \in L(x)$ and there are $m+1 < T_1,\ldots,T_n >$ -successor $< v_{i1},\ldots,v_{in} > (1 \leq i \leq m+1)$ of x with $< E<v_{i1},\ldots,v_{in}>, \rhd, k > \in DC(x) (1 \leq i \leq m+1)$, and $(< v_{i1},\ldots,v_{in} >, < v_{j1},\ldots,v_{jn} >, \neq) \notin DC(x)$ $(1 \leq i < j \leq m+1)$;
(ii) for some abstract node x, $DC(x)$ is unsatisfiable.

In the following, we give a tableaux algorithm that can construct a fuzzy tableau for an *F-ALC(G)*-concept D, which is an abstraction of a model of the concept.

Input: *F-ALC(G)*-concept D;

Output: Boolean;

Algorithm

Step 1 Initialization for the complete forest F: If $\{o_1\}, \ldots, \{o_l\} = I^D$, the algorithm initializes the completion forest F to contain $l+1$ root nodes $x_0, z_{\{o1\}},\ldots, z_{\{ol\}}$ with $L(x_0) = \{< D, X, k >\}$ and $L(z_{\{oi\}}) = < \{o_i\}, \geq, 1 >$. The $DC(x)$ of any abstract node x is initialized to the empty set

Step 2 F is then expanded by repeatedly applying the completion rules, including *F-ALC*-complete rules (Straccia 2001) and G_f-complete rules shown in the following, until no rules can be applied any more

Step 3 If the completion rules can be applied in such a way that they yield a complete, clash-free completion forest, then D is satisfiable, and the algorithm returns TRUE; otherwise, D is unsatisfiable and the algorithm returns FALSE

Step 4 Terminate

(1) $\exists_{p\rhd}$**-rule**: if 1. $< \exists T_1,\ldots,T_n.\ E, \rhd, k > \in L(x)$, and 2. x has no $< T_1,\ldots,T_n >$ -successor $< v_1,\ldots,v_n >$ s.t.: $L(< x, v_i >) = \{< T_i, \rhd, k >\}$ $(1 \leq i \leq n)$, and $< E(v_1,\ldots,v_n), \rhd, k > \in DC(x)$,
then create a $< T_1,\ldots,T_n >$ - successor $< v_1,\ldots,v_n >$ of x, such that: $L(< x, v_i >) = \{< T_i, \rhd, k >\}$ $(1 \leq i \leq n)$, and $DC(x) = DC(x) \cup \{< E(v_1,\ldots,v_n), \rhd, k >\}$.
(2) $\forall_{p\lhd}$**-rule**: if 1. $< \forall T_1,\ldots,T_n.\ E, \lhd, k > \in L(x)$, and 2. x has no $< T_1,\ldots,T_n >$ - successor $< v_1,\ldots,v_n >$ s.t.: $L(< x, v_i >) = \{< T_i, \lhd^{-1}, 1\text{-}k >\}$ $(1 \leq i \leq n)$, and $< E(v_1,\ldots,v_n), \lhd, k > \in DC(x)$,
then create a $< T_1,\ldots,T_n >$ - successor $< v_1,\ldots,v_n >$ of x, such that: $L(< x, v_i >) = \{< T_i, \lhd^{-1}, 1\text{-}k >\}$ $(1 \leq i \leq n)$, and $DC(x) = DC(x) \cup \{< E(v_1,\ldots,v_n), \lhd, k >\}$.
(3) $\exists_{p\lhd}$**-rule**: if 1. $< \exists T_1,\ldots,T_n.\ E, \lhd, k > \in L(x)$, and 2. x has a $< T_1,\ldots,T_n >$ - successor $< v_1,\ldots,v_n >$ s.t.: $L(< x, v_i >) = \{< *, \rhd, r >\} (1 \leq i \leq n)$, and $< E(v_1,\ldots,v_n), \lhd, k > \notin DC(x)$, where $< *, \rhd, r >$ is conjugated with $< *, \lhd, k >$, * stands for any binary relationship of the form $< x, v_i >$, then $DC(x) = DC(x) \cup \{< E(v_1,\ldots,v_n), \lhd, k >\}$.
(4) $\forall_{p\rhd}$**-rule**: if 1. $< \forall T_1,\ldots,T_n.\ E, \rhd, k > \in L(x)$, and 2. x has a $< T_1,\ldots,T_n >$ - successor $< v_1,\ldots,v_n >$ s.t.: $L(< x, v_i >) = \{< *, \rhd, r >\} (1 \leq i \leq n)$, and $< E(v_1,\ldots,v_n), \rhd, k > \notin DC(x)$, where $< *, \rhd, r >$ is conjugated with $< *, \rhd^{-1}, 1\text{-}k >$, * stands for any binary relationship of the form $< x, v_i >$, then $DC(x) = DC(x) \cup \{< E(v_1,\ldots,v_n), \rhd, k >\}$.

(5) $\geq_{p\rhd}$**-rule**: if 1. $< \geq mT_1,\ldots,T_n.E, \rhd, k > \in L(x)$, and 2. there are no $m < T_1,\ldots,T_n >$ -successors $< v_{i1},\ldots,v_{in} >$ $(1 \leq i \leq m)$ connected to x s.t.: $L(< x, v_{ij} >) = \{< T_j, \rhd, k >\}$ $(1 \leq i \leq m, 1 \leq j \leq n)$, and $< E(v_{i1},\ldots,v_{in}), \rhd, k > \in DC(x)$ $(1 \leq i \leq m)$ and $(< v_{i1},\ldots,v_{in} >, < v_{j1},\ldots,v_{jn} >, \neq) \in DC(x)$ $(1 \leq i < j \leq m)$,
then create m new $< T_1,\ldots,T_n >$ -successors $< v_{i1},\ldots,v_{in} >$ $(1 \leq i \leq m)$ connected to x s.t.: 1. $L(< x, v_{ij} >) = \{< T_j, \rhd, k >\}$ $(\leq i \leq m, 1 \leq j \leq n)$, and 2. $DC(x) = DC(x) \cup < E(v_{i1},\ldots,v_{in}), \rhd, k >$ $(1 \leq i \leq m)$, and 3. $DC(x) = DC(x) \cup \{(< v_{i1},\ldots,v_{in} >, <v_{j1},\ldots,v_{jn} >, \neq)\}$ $(1 \leq i < j \leq m)$.

(6) $\leq_{p\lhd}$**-rule**: if $< \leq mT_1,\ldots,T_n.E, \lhd, k > \in L(x)$,
then apply $\geq_{p\rhd}$-rule for the triple $< \geq (m + 1)T_1,\ldots,T_n.E, \lhd^{-1}, 1\text{-}k >$.

(7) $\leq_{p\rhd}$**-rule**: if 1. $< \leq mT_1,\ldots,T_n.E, \rhd, k > \in L(x)$, and 2. x has $m + 1 < T_1,\ldots, T_n >$ -successors $< v_{i1},\ldots,v_{in} >$ $(1 \leq i \leq m + 1)$ s.t.:

   (a) $L(< x, v_{ij} >) = \{< *, \rhd, r >\}$ $(\leq i \leq m + 1, 1 \leq j \leq n)$, where $< *, \rhd, r >$ is conjugated with $< *, \rhd^{-1}, 1\text{-}k >$, * stands for any binary relationship of the form $< x, v_{ij} >$, and
   (b) $<E(v_{i1},\ldots,v_{in}), \rhd, k > \in DC(x)$ $(1 \leq i \leq m + 1)$, and
   (c) In the above $m + 1 < T_1,\ldots,T_n >$ -successors, there exists two different $< T_1,\ldots,T_n >$ -successors $< v_{s1},\ldots,v_{sn} >$ and $< v_{t1},\ldots,v_{tn} >$ $(s \neq t)$ s.t. $\{(< v_{s1},\ldots,v_{sn} >, < v_{t1},\ldots,v_{tn} >, \neq)\} \notin DC(x)$, then for each pair v_{sj}, v_{tj}, where v_{sj} and v_{tj} are not the same data type T_j-successor of x, do: 1. $L(< x, v_{sj} >) = L(< x, v_{sj} >) \cup L(< x, v_{tj} >)$, and 2. replace v_{tj} with v_{sj} in $DC(x)$, remove v_{tj}, and 3. $L(< x, v_{tj} > = \varnothing$.

(8) $\geq_{p\lhd}$**-rule**: if $< \geq mT_1,\ldots,T_n.E,\lhd,k > \in L(x)$,
then apply $\leq_{p\rhd}$-rule for the triple $< \leq (m\text{-}1)T_1,\ldots,T_n.E, \lhd^{-1}, 1\text{-}k >$.

In the following, we discuss the decidability of the *F-ALC(G)* tableau algorithm.

Theorem 4.1 *If G* $= (\phi_G, D_G, dom)$ *is a conforming fuzzy data type group, then the satisfiability problem for finite fuzzy predicate conjunctions of G is decidable.*

Proof Let the fuzzy predicate conjunction be $\zeta = \zeta_{w1} \wedge \ldots \wedge \zeta_{wk} \wedge \zeta_U$, where $D_G = \{w_1,\ldots,w_k\}$, ζ_{wi} is the fuzzy predicate conjunction for *sub-group*(ζ_{wi}, *G*), and ζ_U is the sub-conjunction of ζ and only unsupported fuzzy predicates appear. According to the notions mentioned above, $\zeta_S = \zeta_{w1} \wedge \ldots \wedge \zeta_{wk}$ is decidable. ζ_U is unsatisfiable iff there exist $< p(v_1,\ldots,v_n), \rhd, k >$ and $< p(v_1,\ldots,v_n), \rhd', k' >$ $(k \geq k')$ for some $p \notin_G$ appearing in ζ_U. The satisfiability of ζ_U is clearly decidable. □

Theorem 4.2 *Let G be a conforming fuzzy data type group. The fuzzy G-data type expression conjunctions of formula are decidable.*

Proof We can reduce the satisfiability problem for fuzzy *G*-data type expressions to the satisfiability problem for fuzzy predicate conjunctions over *G*; (*i*) $[u_1,\ldots,u_s]$ constructor simply introduces fuzzy predicate conjunctions. Similarly, its negation introduces disjunctions. According to the notions mentioned above, we can eliminate relativised negations in $[u_1, \ldots, u_s]$; (*ii*) The *and* and *or* constructors simply introduce disjunctions of fuzzy predicate conjunctions of *G*; (*iii*) We can transform the tuple-level constraints to a disjunction of conjunctions of variable-level constraints in form $\neq (v_i, v_j)$. The conjunctions is clearly decidable. □

According to Theorem 4.1, the satisfiability problem of fuzzy predicate conjunctions of *G* is decidable. So a fuzzy *G*-data type expression conjunction is satisfiable if one of its disjunctions is satisfiable.

Theorem 4.3 *For each F-ALC(G)-concept D in NNF, its satisfiability is decidable.*

Proof The reasoning in fuzzy concept and fuzzy data type group can be divided separately. The tableau algorithm of *F-ALC(G)* is decidable. Let $m = |sub(A)|$, $k = |\mathrm{R}_{\mathrm{A}}^{A}|$, $w = |\mathrm{I}|$, $p_{\max} = max\{p| \geq pT_1,\ldots,T_n.E\}$, l be the number of different membership degrees appearing in *A*. The new nodes are generated by the $\exists_{\rhd}$-, $\forall_{\lhd}$-, $\exists_{p\rhd}$-, $\forall_{p\lhd}$-, $\geq_{p\rhd}$-, and $\leq_{p\lhd}$-rules as successors of an abstract node *x*. For *x*, each of these concepts can trigger the generation of successors once at most, even if the node(s) generated are later removed by either the $\leq_{p\rhd}$- or $\geq_{p\lhd}$-rules. Since *sub* (*A*) contains a total of m $\exists_{\rhd}$-, $\forall_{\lhd}$-, $\exists_{p\rhd}$-, $\forall_{p\lhd}$-, $\geq_{p\rhd}$-, $\leq_{p\lhd}$-concepts at most, the out-degree of the forest is bounded by $2lmp_{max}$. The nodes are labeled with the sets of triples of the form $< C, \bowtie, k >$, where $C \in sub(A) \cup \{\uparrow(R, \{o\}) \mid R \in \mathrm{R}_{\mathrm{A}}^{A}$ and $\{o\} \in \mathrm{I}\}$. Since the concrete nodes have no labels and are always leaves, there are at most $2^{4l(m+kw)}$ different abstract node labels. Hence paths are of length at most $2^{4l(m+kw)}$. The reasoning of fuzzy data type group is actually to check the satisfiability of a set of *DC*(*x*), which is decidable according to Theorem 4.2. So the procedure of checking the satisfiability of *F-ALC*(*G*)-concept *D* in *NNF* is decidable. □

4.2.4 Fuzzy Description Logic Reasoners

Description logics (DLs) serve as the theoretical counterpart of the Semantic Web and provide reasoning supports for it, and the DL reasoners are the basic supporting bodies for the Semantic Web coming into use. Similarly, in order to implement the automatic reasoning of fuzzy DLs, kinds of reasoners based on different fuzzy DLs have been put forward.

Table 4.4 lists the existing fuzzy reasoners that can support the processing of fuzzy and uncertain information. In brief, for the implementation of fuzzy reasoners, there are two strategies: one is to translate the fuzzy DL into the classic DL and then call the classic DL reasoner for reasoning, e.g., DeLorean in Table 4.4. Another strategy is to directly implement fuzzy reasoners based on the tableau algorithm of the corresponding fuzzy DL, e.g., the reasoners in Table 4.4 except for DeLorean and SoftFacts. The latter strategy has the advantage that the optimized technology can be adopted according to the specific fuzzy DL, and their efficiency in the implementation can be enhanced.

Table 4.4 The existing fuzzy DL reasoners and their comparisons

Fuzzy reasoners	Fuzzy DLs	Features				
		Reasoning technology		*Data type*		*Unique features*
		Tableau algorithm	Reducing to crisp DLs	*D*	*G*	
FiRE (Stoilos et al. 2006b)	*f-SHIN*	√				Supporting graphical interface GUI
FuzzyDL (Bobillo and Straccia 2008)	*f-SHIF(D)*	√		√		Supporting that the degree of a fuzzy assertion is not only a constant, but also a variable
DeLorean (Bobillo et al. 2012)	*f-SROIQ(D)*		√	√		Reducing to crisp *DLs* to solve, and supporting fuzzy concrete domains *D*
GURDL (Haarslev et al. 2007)	*f-ALC*	√				Proposing some interesting techniques of optimization
GERDS (Habiballa 2007)	*f-ALC*	√				Adding role negation, top role, and bottom role to *f-ALC*
FRESG (Wang et al. 2009)	*f-ALC(G)*	√			√	Supporting fuzzy data information with customized fuzzy data types *G*
YADLR (Stasinos and Georgios 2007)	*SLG algorithm*	√				Allowing to deal with unknown degrees of truth in the fuzzy assertions of the knowledge base
SoftFacts (Straccia 2009)	*SoftFacts*					An ontology mediated top-k information retrieval system over relational databases

In the following, we introduce the fuzzy DL reasoner FRESG (Wang et al. 2009), and the design and implementation of a fuzzy DL reasoner is shown from the introduction.

FRESG (Fuzzy Reasoning Engine Supporting Fuzzy Data Type Group) is a prototype reasoner, which is based on the *ABox* consistency checking algorithm of the fuzzy DL *F-ALC*(*G*) as has been introduced above. The current version of FRESG is named FRESG 1.0. FRESG 1.0 can support the consistency checking of a knowledge base (KB) with empty *TBox*, and the future versions will provide supports for the consistency checking of a KB with simple *TBox* and general *TBox* restrictions. In fact, the consistency checking of a KB with empty *TBox* (equivalent to *ABox*) is the most basic reasoning problem in DL because the other inference services can be reduced to this case. FRESG 1.0 provides the following "standard" set of inference services, including:

- *ABox consistency checking*, which is used to check whether an *ABox* is "consistent" or "inconsistent";
- *Concept satisfiability*, which checks if it is possible for a concept to have any instances;

- *Entailment*. It is said that an interpretation *FI* satisfies (is a model of) a KB *k* iff *FI* satisfies each element in *k*. A KB *k* entails an assertion α (denoted by $k \mid = \alpha$) iff every model of *k* also satisfies α. The problem of determining if $k \mid = \alpha$ holds is called *entailment* problem;
- *Retrieval*. According to the given query conditions, FRESG 1.0 returns all the instances satisfying the conditions;
- *Realization*. Given an *ABox A* and a concept *C*, to find all the individuals *a* such that $\langle a\text{: } C, \bowtie, k\rangle \in A$.
- The other reasoning services, like *subsumption* and *classification*, can be reduced to consistency checking of the corresponding *ABox*.

FRESG 1.0 is a reasoner based on fuzzy DL *F-ALC(G)*, so its grammatical forms are in correspondence with the *F-ALC(G)* abstract syntax. Users can define their *ABox* according to the syntax of FRESG 1.0 and then use FRESG 1.0 reasoner to reason or query over the *ABox*. The detailed FRESG 1.0 syntax can be referred to FRESG 1.0 user manual. Example 4.1 shows an *ABox* A_1 described in valid FRESG 1.0 syntax.

Example 4.1 (instance *x* (some *R D*) > = 0.7); (instance *x* (all *R C*) > = 0.4);
(instance *y C* > = 0.2); (instance *y D* > = 0.3);
(related *x y R* > = 0.5); (instance *x* (some *R* (and *C D*)) < 0.5);
(instance *y* (dtsome T_1, (dtor p_1 p_2)) > = 0.7);
(instance *y* (dtatleast 3, T_2 T_3, p_3) > = 0.6);
(instance *y* (dtall T_2, p_1) > = 0.9);|

Its corresponding DL KB described in *F-ALC(G)* abstract syntax is as follows: $A_1' = \{< x\text{: } \exists R.D, \geq, 0.7, < x\text{:}\forall R.C, \geq, 0.4, < y\text{:}C, \geq, 0.2, < y\text{:}D, \geq, 0.3, < (x,y)\text{:}R, \geq, 0.5, < x\text{: } \exists R.(C \sqcap D), <, 0.5, < y\text{: } \exists T_1.p \vee p_2, \geq, 0.7, < y\text{: } \geq 3T_2,T_3.p_3, \geq, 0.6, < y\text{:}\forall T_2.p_1, \geq, 0.9\}$. Here *x*, *y* are individuals; *C*, *D* are concepts; *R* is an abstract role; T_1, T_2, T_3 are fuzzy data type roles; p_1, p_2, p_3 are fuzzy data type predicates based on base data type *xsd*: *integer*. Then, FRESG 1.0 can provide reasoning services over the compiled A_1'.

The core of FRESG 1.0 is a Tableaux reasoner and a fuzzy data type reasoner. The former is implemented based on the *F-ALC(G) ABox* consistency checking algorithm, and the latter is used to decide the satisfiability of fuzzy data type predicates conjunctions. The main design goal of FRESG 1.0 is that it has a small core reasoning engine, which is able to reason with customized fuzzy data type predicates and is suitable for extensions. Based on this goal, FRESG 1.0 has the overall architecture shown in Fig. 4.1. In the design, it requires that the fuzzy data type reasoner is independent of the Tableaux reasoner. On the other hand, if users want to add the construction operators of fuzzy data type expressions, they only need to modify the fuzzy data type manager, without having to change the Tableaux reasoner; If users want to add some basic data types supported by the reasoner, they only need to add the corresponding fuzzy data type checkers on the base of the current fuzzy data type reasoner. Such a design philosophy makes FRESG 1.0 have highly modular structure and be easy extended. As a result, a

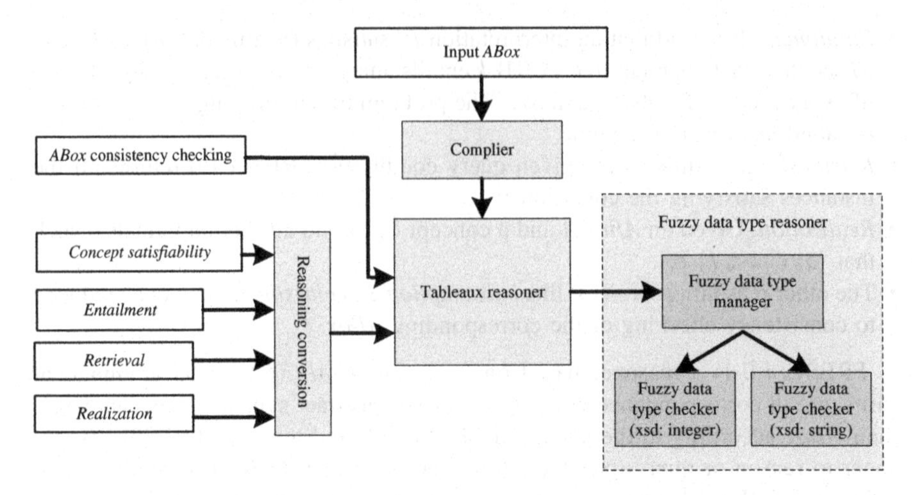

Fig. 4.1 The architecture of FRESG 1.0 reasoner

solid foundation is laid for implementing more expressive DL reasoner and the further research. If the expression capacity of a certain part (e.g., the Tableaux reasoner, fuzzy data type manager or fuzzy data type checker) is amended and strengthened independently, a new fuzzy DL reasoner is formed, which is suitable for applications in different backgrounds.

The Tableaux reasoner checks the consistency of an *ABox*. The FRESG 1.0 compiler translates the *ABox* described in FRESG 1.0 syntax into the file described in abstract *F-ALC(G)* syntax which can be identified by the Tableaux reasoner. The Tableaux reasoner conducts the decision or query over the compiled file according to the "*ABox* consistency checking" command, or the query requirements converted by the unit of "Reasoning conversion". If there are some sub-queries about the customized fuzzy data type predicates in the query, the Tableaux reasoner invokes the fuzzy data type reasoner to decide the satisfiability of the fuzzy data type predicate conjunctions, and finally returns the corresponding results.

The Tableaux reasoner has only one function: checking the consistency of an *ABox*. According to the model-theoretic semantics, an *ABox* is consistent if there is an interpretation that satisfies all the assertions in it. Such an interpretation is called a *model* of the *ABox*. The tableaux reasoner searches for such a model according to the *ABox* consistency checking algorithm.

In the process of implementation, the Tableaux reasoner firstly starts by constructing an initial forest F_A for *ABox*. There are two kinds of nodes in the constructed forest F_A: abstract nodes (the normal labeled nodes) and data type nodes (unlabeled leaves of F_A). Each abstract node x is labeled by a set of triples and each triple is in the form of $\langle C, \bowtie, k\rangle$, which indicates the membership that the individual x belongs to $C \bowtie k$. The edges in the forest F_A represent the relationship between nodes. An edge between two abstract nodes represents a fuzzy abstract role, and an edge between an abstract node and a data type node represents a fuzzy data type role. The initial forest F_A is shown in "Initial Status" window in the form of tree.

According to the *ABox* consistency checking algorithm, the Tableaux reasoner repeatedly applies the tableaux expansion rules until a clash is detected in the label of a node, or until a clash-free forest is found, to which no more rules are applicable. FRESG 1.0 applies *F-ALC*(*G*) Tableaux expansion rules in the following order: (*i*) For an abstract node, *F-ALC*-rules are used first and then *G*-rules are used; (*ii*) *F-ALC*-rules are used according to the order of $\neg_{\bowtie}$-, $\sqcup_{\rhd}$-, $\sqcap_{\lhd}$-, $\sqcap_{\rhd}$-, $\sqcup_{\lhd}$-, $\exists_{\rhd}$-, $\forall_{\lhd}$-, $\forall_{\rhd}$-, and $\exists_{\lhd}$-rules; (*iii*) *G*-rules are used according to the order of $\exists_{p\rhd}$-, $\forall_{p\lhd}$-, $\geq_{p\rhd}$-, $\leq_{p\lhd}$-, $\forall_{p\rhd}$-, $\exists_{p\lhd}$-, $\leq_{p\rhd}$-, and $\geq_{p\lhd}$-rules;

As a result of the use of uncertainty rules $\sqcup_{\rhd}$- and $\sqcap_{\lhd}$-, a number of complete forest F_{A1}, F_{A2},…, F_{An}($n \geq 1$) may be yielded. Finally, the Tableaux reasoner checks if these forests contain clashes or not. If one of these forests does not contain any clashes, there is a model about *ABox* and the input *ABox* is "consistent". If all these forests contain clashes, there is no model for *ABox* and the input *ABox* is "inconsistent". When the Tableaux reasoner checks the satisfiability of the forests, if there are some queries related to fuzzy data type, the Tableaux reasoner invokes the fuzzy data type reasoner to decide the satisfiability of the conjunction queries with fuzzy data type expression. The Tableaux reasoner returns the final results according to the results returned by fuzzy data type reasoner.

The fuzzy data type reasoner is used to decide the satisfiability of the conjunction query with fuzzy data type. In other words, the tableaux reasoner calls the fuzzy data type reasoner for the satisfiability of the fuzzy data type expression conjunction. The fuzzy data type reasoner includes one fuzzy data type manager and two fuzzy data type checkers. Among them, one checker is based on the basic data type *xsd*: *integer* and another one is based on the basic data type *xsd*: *string*.

The fuzzy data type manager works between the tableaux reasoner and the two fuzzy data type checkers. The design of the fuzzy data type manager is based on the algorithms in Wang and Ma (2008). The Tableaux reasoner reduces the fuzzy data information in *ABox* to the form of fuzzy data type expressions conjunctions (Wang and Ma 2008) using Tableaux expansion rules, and then call the fuzzy data type reasoner to check the satisfiability of the conjunction. The fuzzy data type manager decomposes the conjunction into disjunctions of fuzzy data type predicate conjunctions and then divides each of the predicates conjunction into two sub-conjunctions. One of the sub-conjunctions is based on *xsd*: *integer* and another one is based on the basic data type *xsd*: *string*. Then, the fuzzy data type manager sends these two sub-conjunctions to appropriate fuzzy data type checkers to decide their satisfiabilities. If the two sub-conjunctions of a fuzzy predicate conjunction are satisfiable, it is satisfiable. If any one of the fuzzy predicate conjunction is satisfiable, the input fuzzy data type expressions conjunctions is satisfiable. The fuzzy data type checkers decide the satisfiability problem of fuzzy data type predicate conjunctions, where the fuzzy data type predicates are defined over a base data type in a fuzzy data type group. If some new base data types are needed to add into FRESG 1.0, only corresponding fuzzy data type checkers are need to add. The design of the fuzzy data type checker is based on the algorithms in Wang and Ma (2008).

FRESG 1.0 reasoning engine is designed to transform other reasoning tasks into *ABox* consistency checking tasks. The current version of FRESG provides not only

ABox consistency checking services, but also the reasoning services such as concepts satisfaction checking, entailment reasoning function and querying. The FRESG 1.0 provides these three kinds of reasoning services through transforming them into *ABox* consistency checking. Firstly, it is easy to transform concepts satisfaction checking into *ABox* consistency checking. If users want to check the satisfaction of the *F-ALC*(*G*)-concept *C*, FRESG 1.0 transforms it into checking consistency of the fuzzy *ABox* $\{(x: C) > 0\}$, and calls the Tableaux reasoner to decide its consistency (x is an abstract individual); Secondly, the entailment reasoning function can also be transformed into *ABox* consistency checking task; Thirdly, according to the query definition, FRESG 1.0 returns all the individuals satisfying the query condition ($C \bowtie k$), so it is only required to do entailment reasoning for each individual x in the current *ABox* A. If $A \mid \approx < x: C, \bowtie, k$, x satisfies the querying condition and the result is displayed in the "Result Window". Otherwise, x does not satisfy the querying conditions. FRESG 1.0 needs to decide every individual in *ABox* when querying, so the querying is inefficient. How to make the querying tasks more efficient is one of our future works.

Example 4.2 In order to check the consistency of *ABox* A_1 in Example 4.1, we can run it in FRESG 1.0, and the result is partly shown in Fig. 4.2. According to the Tableaux expansion rules, the "Reasoning Details" window shows how the F_A is extended into four different kinds of completion-forests by the Tableaux reasoner. Then, the Tableaux reasoner calls the fuzzy data type reasoner, which checks the

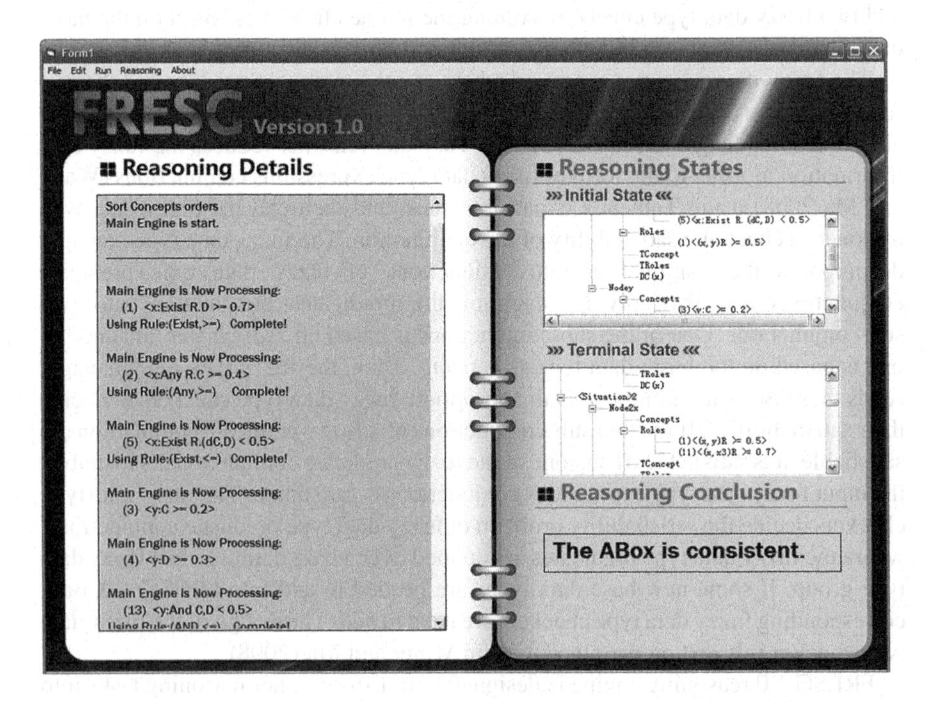

Fig. 4.2 The running results of FRESG 1.0 reasoner

satisfiability of the conjunction query with customized fuzzy data type predicates. As a result, the first and second situations are satisfiable, but the third and fourth situations are unsatisfiable. So the "Final Result" window gives "The *ABox* A_1 is consistent" as a conclusion.

Example 4.3 Now we check the satisfiability of the concept (and (some *R* (all *S* (and *C D*))) (some *S* (dtsome T_1 T_2 T_3, (dtand E_1 E_2)))). Firstly, FRESG∧ 1.0 compiler transforms the concepts that follow the FRESG 1.0 syntax into the concepts that follow the *F-ALC(G)* syntax, i.e., $(\exists R.(\forall S.(C \sqcap D))) \sqcap (\exists S.(\exists T_1,T_2, T_3.(E_1 \wedge E_2)))$. Secondly, using the unit of "reasoning conversion", FRESG 1.0 transforms the task that checks the satisfiability of concepts into the task that checks the consistency of the *ABox* A_2: $\{x : (\exists R.(\forall S.(C \sqcap D))) \sqcap (\exists S.(\exists T_1,T_2, T_3.(E_1 \wedge E_2))) > 0\}$.

We can get useful information from the results: the initial forest that corresponds to A_2 is shown in the "Initial States" window. The "reasoning detail" window shows how F_A is expanded to a complete forest by the Tableaux reasoner. Then the Tableaux reasoner calls the fuzzy data type reasoner to check the satisfiability of the conjunction query with customized fuzzy data type predicates. Finally, because no clash is contained in the complete forest, A_2 is "consistent", and the "Final Result" window also gives the corresponding conclusion.

We tested performance of the FRES0047 1.0 reasoner by checking the consistencies of different *ABoxes* with different sizes (see Fig. 4.3). The size of *ABox* is measured in instance number (both concept and role instances). In the graph, the reasoning time is in scale to the *ABox* size.

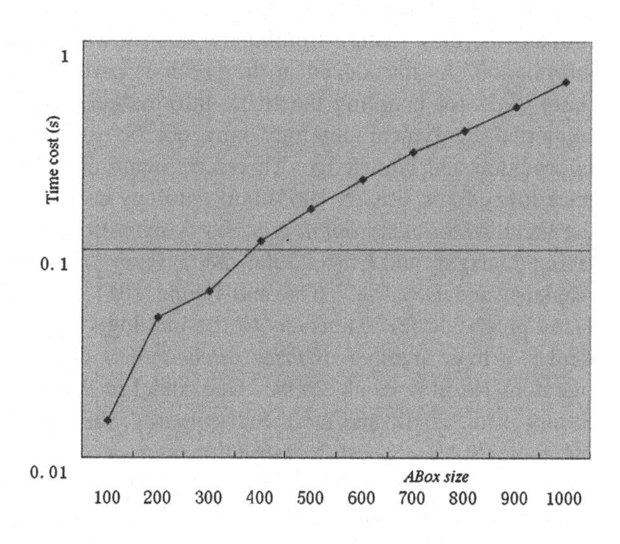

Fig. 4.3 Performance of the FRESG 1.0 reasoner

	*ABox*1	*ABox*2	*ABox*3	*ABox*4	*ABox*5	*ABox*6	*ABox*7	*ABox*8	*ABox*9	*ABox*10
ABox size	100	200	300	400	500	600	700	800	900	1000
Time cost(s)	0.015	0.047	0.063	0.110	0.157	0.218	0.296	0.375	0.484	0.640

4.2.5 Fuzzy Ontologies

Ontology is a W3C standard knowledge representation model for the Semantic Web (Berners-Lee et al. 2001). In general terms, ontology, which is an explicit formal specification of a shared domain conceptualization, can be used to describe the objects, properties, concepts, and their relationships existing in the domain. A number of ontology definition languages, such as RDF(S), SHOE, OIL, DAML, DAML + OIL, and OWL, have been developed over the past years (Horrocks et al. 2003). The most widely used among them is the W3C recommended standard language *OWL* (*Web Ontology Language*). In particular, OWL has the reasoning nature because of the existence of Description Logics (DLs). The logical underpinnings of OWL are mainly very expressive DLs, like SHOIN(D) and SHIF(D). More recently, OWL 2, an extension to and revision of OWL, is being developed within the W3C OWL Working Group (Cuenca Grau et al. 2008). Today, there are lots of researches on OWL and DLs, which are used widely in the Semantic Web and other applications such as medical, image analysis, databases and more (Staab and Studer 2009; Baader et al. 2003).

Although ontology and its representation languages are quite expressive logical formalisms, they feature limitations, mainly with what can be said about fuzzy information. As introduced in the previous parts, DLs have been extended with fuzzy logic for handling the fuzzy information. Also, the conceptual formalism supported by typical ontology may not be sufficient to represent such fuzzy information and knowledge. Therefore, many proposals have attempted to introduce fuzzy logic (Zadeh 1965) into ontology definition and reasoning, up to now the literature on *fuzzy ontologies* (fuzzy extensions of ontologies) has been flourishing (Calegari and Ciucci 2006, 2007; Parry 2004; Sanchez and Yamanoi 2006; Abulaish and Dey 2007; Cai and Leung 2011; and etc.). Also, there are today many proposals for extensions to the ontology representation languages. In particular, a huge number of *fuzzy extensions of OWL* have been presented in the literature (Stoilos et al. 2005c; Gao and Liu 2005; Calegari and Ciucci 2007; Stoilos et al. 2010; and etc.). Subsequently, the other important issues on fuzzy ontologies such as query, construction, storage, and mapping increasingly receive attention (Pan et al. 2008; Lv et al. 2009; Zhang et al. 2011; De Maio et al. 2009; Quan et al. 2006; Zhang et al. 2008a, 2008b, 2009, 2010; Zhang and Ma 2012). Also, for representing and managing such fuzzy ontologies, fuzzy OWL language and several tools (such as KAON and Fuzzy ProtÅgÅ) (Calegari and Ciucci 2008; Ghorbel et al. 2009) were developed. A comprehensive review about fuzzy ontologies can be found at (Sanchez 2006; Bobillo 2008). In this chapter, to provide a general understanding of fuzzy ontologies, we introduce fuzzy ontology language (i.e., fuzzy OWL) and some notions of fuzzy ontologies.

4.2.5.1 Fuzzy OWL Language

In order to represent fuzzy information of fuzzy ontologies, a fuzzy ontology representation language called *fuzzy OWL* (a fuzzy extension of the W3C recommended language OWL) (Stoilos et al. 2005c; Gao and Liu 2005; Calegari and Ciucci 2007; Stoilos et al. 2010; and etc.) was developed. In the following, we will recall fuzzy OWL language.

Being similar to OWL, fuzzy OWL has three increasingly expressive sublanguages fuzzy OWL Lite, fuzzy OWL DL, and fuzzy OWL Full. Fuzzy OWL Lite and fuzzy OWL DL are basically Description Logic (DLs); there are almost equivalent to the *f-SHIF(D)* and *f-SHOIN(D)* DLs. Furthermore, fuzzy OWL has two interchangeable syntactic forms: the RDF/XML syntax and the frame-like style abstract syntax. For example, the axiom “SubClassOf (*AdminStaff Staff*)” written in fuzzy OWL abstract syntax may be equivalently represented as the axiom “<fowl:Class rdf:ID = "*AdminStaff*" > < fowl:SubClassOf rdf:resource = *#Staff*/ > < fowl:ineqType fowl:degree = 1.0/> </fowl:Class>” written in RDF/XML syntax in (Calegari and Ciucci 2007). Moreover, OWL DL is the language chosen by the major ontology editors because it supports those users who want the maximum expressiveness without losing computational completeness and decidability of reasoning systems, and OWL Full has been proved to be undecidable. Therefore, when we mention fuzzy OWL in the following, we usually mean fuzzy OWL DL.

Table 4.5 gives *the fuzzy OWL abstract syntax*, *the respective fuzzy Description Logic syntax*, and *the semantics*. The semantics for fuzzy OWL is based on the interpretation of *f-SHOIN(D)* (Straccia 2005) as introduced in Sect. 4.2.3. In brief, the semantics is provided by a fuzzy interpretation $FI = (\Delta^{FI}, \Delta_D, J^{FI}, J^{D})$, where Δ^{FI} is the abstract domain and Δ_D is the datatype domain (disjoint from Δ^{FI}), and J^{FI} and J^{D} are two fuzzy interpretation functions, which map: An abstract individual d to an element $d^{FI} \in \Delta^{FI}$; A concrete individual v to an element $v^{D} \in \Delta_D$; A concept name A to a membership degree function A^{FI}: $\Delta^{FI} \rightarrow [0, 1]$; An abstract role name R to a membership degree function R^{FI}: $\Delta^{FI} \times \Delta^{FI} \rightarrow [0, 1]$; A concrete datatype D to a membership degree function D^{D}: $\Delta_D \rightarrow [0, 1]$; A concrete role name T to a membership degree function T^{FI} : $\Delta^{FI} \times \Delta_D \rightarrow [0, 1]$. In Table 4.5, $a \in \{d, v\}$; $\#S$ denotes the cardinality of a set S, and $\bowtie \in \{\geq, >, \leq, <\}$.

Fuzzy OWL has the reasoning nature because of the existence of fuzzy DLs, since the logical underpinnings of fuzzy OWL are mainly very expressive fuzzy DLs as mentioned above. Currently, some researchers proposed several kinds of languages based on different fuzzy DLs, such as fuzzy DL-Lite (Pan et al. 2007), fuzzy SHIN (Stoilos et al. 2005b; Cheng et al. 2009), fuzzy OWL (Stoilos et al. 2005c; Gao and Liu 2005; Calegari and Ciucci 2007; Stoilos et al. 2010), and etc. All of these languages can be used to express fuzzy knowledge of fuzzy ontologies with

Table 4.5 Fuzzy OWL abstract syntax, fuzzy DL syntax, and semantics

Fuzzy OWL abstract syntax	Fuzzy DL syntax	Semantics
Fuzzy class descriptions	*Fuzzy concepts*	
Class (A)	A	$A^{FI}: \Delta^{FI} \rightarrow [0,1]$
owl: Thing	$\top$	$\top^{FI}(d) = 1$
owl: Nothing	$\bot$	$\bot^{FI}(d) = 0$
intersectionOf ($C_1 \ldots C_n$)	$C_1 \sqcap \ldots \sqcap C_n$	$(C_1 \sqcap \ldots \sqcap C_n)^{FI}(d) = \min\{C_1^{FI}(d), \ldots, C_n^{FI}(d)\}$
unionOf ($C_1 \ldots C_n$)	$C_1 \sqcup \ldots \sqcup C_n$	$(C_1 \sqcup \ldots \sqcup C_n)^{FI}(d) = \max\{C_1^{FI}(d), \ldots, C_n^{FI}(d)\}$
complementOf(C)	$-C$	$(C)^{FI}(d) = 1 - C^{FI}(d)$
oneOf ($o_1 \ldots o_n$)	$\{o_1\} \sqcup \ldots \sqcup \{o_n\}$	$(\{o_1\} \sqcup \ldots \sqcup \{o_n\})^{FI}(d) = 1 \text{ if } d \in \{o_1^{FI}, \ldots, o_n^{FI}\}, 0 \text{ otherwise}$
restriction (P someValuesFrom(E))	$\exists P.E$	$(\exists P.E)^{FI}(d) = \sup_{a \in \Delta^{FI}}\{\min\{P^{FI}(d,a), E^{FI}(a)\}\}$
restriction (P allValuesFrom(E))	$\forall P.E$	$(\forall P.E)^{FI}(d) = \inf_{a \in \Delta^{FI}}\{\max\{1 - P^{FI}(d,a), E^{FI}(a)\}\}$
restriction (P hasValues(a))	$\exists P.\{a\}$	$(\exists P.\{a\})^{FI}(d) = P^{FI}(d,a)$
restriction (P minCardinality(n))	$\geq nP$	$(\geq nP)^{FI}(d) = \sup_{a_1,\ldots,a_n \in \Delta^{FI}} \wedge_{i=1}^{n} P^{FI}(d, a_i)$
restriction (P maxCardinality(n))	$\leq nP$	$(\geq nP)^{FI}(d) = \inf_{a_1,\ldots,a_{n+1} \in \Delta^{FI}} \vee_{i=1}^{n+1} (1 - P^{FI}(d, a_i))$
restriction (P cardinality(n))	$= nP$	$(\geq nP \sqcap \leq nP)^{FI}(d)$
Fuzzy class axioms	*Fuzzy axioms*	
Class (A partial $C_1 \ldots C_n$)	$A \sqsubseteq C_1 \sqcap \ldots \sqcap C_n$	$A^{FI}(d) \leq \min\{C_1^{FI}(d), \ldots, C_n^{FI}(d)\}$
Class (A complete $C_1 \ldots C_n$)	$A \equiv C_1 \sqcap \ldots \sqcap C_n$	$A^{FI}(d) = \min\{C_1^{FI}(d), \ldots, C_n^{FI}(d)\}$
SubClassOf(C_1 C_2)	$C_1 \sqsubseteq C_2$	$C_1^{FI}(d) \leq C_2^{FI}(d)$
EquivalentClasses ($C_1 \ldots C_n$)	$C_1 \equiv \ldots \equiv C_n$	$C_1^{FI}(d) = \ldots = C_n^{FI}(d)$
DisjointClasses ($C_1 \ldots C_n$)	$C_1 \sqcap C_2 \sqsubseteq \bot$	$\min\left\{C_i^{FI}(d), C_j^{FI}(d)\right\} = 0 \quad 1 \leq i < j \leq n$
EnumeratedClass ($A\, o_1 \ldots o_n$)	$A = \{o_1 \ldots o_n\}$	$A^{FI}(d) = 1 \text{ if } d \in \{o_1^{FI}, \ldots, o_n^{FI}\}, A^{FI}(d) = 0 \text{ otherwise}$

(continued)

Table 4.5 (continued)

Fuzzy OWL abstract syntax *Fuzzy class descriptions*	Fuzzy DL syntax *Fuzzy concepts*	Semantics
Fuzzy property axioms	*Fuzzy axioms*	
DatatypeProperty (T		
domain (C_1)…domain (C_m)	$\geq 1T \sqsubseteq C_i$	$T^{FI}(d,v) \leq C_i^{FI}(d) \quad i = 1,\ldots,m$
range (D_1)…range (D_k)	$\top \sqsubseteq \forall T.D_i$	$T^{FI}(d,v) \leq D_i^{FI}(v) \quad i = 1,\ldots,k$
[Functional])	$\top \sqsubseteq \leq 1T$	$\forall d \in \Delta^{FI} \# \{v \in \Delta_D : T^{FI}(d,v) \geq 0\} \leq 1$
ObjectProperty (R		
domain (C_1)…domain(C_m)	$\geq 1R \sqsubseteq C_i$	$R^{FI}(d_1,d_2) \leq C_j^{FI}(d_1) \quad i = 1,\ldots,m$
range (C_1)…range(C_k)	$\top \sqsubseteq \forall R.C_i$	$R^{FI}(d_1,d_2) \leq C_j^{FI}(d_2) \quad i = 1,\ldots,k$
[Functional])	$\top \sqsubseteq \leq 1R$	$\forall d_1 \in \Delta^{FI} \# \{d_2 \in \Delta^{FI} : R^{FI}(d_1,d_2) \geq 0\} \leq 1$
[InverseOf(R_0)]	$R = (R_0)^-$	$R^{FI}(d_1,d_2) \leq R_0^{FI}(d_2,d_1)$
[Symmetric]	$R = R^-$	$R^{FI}(d_1,d_2) = (R^-)^{FI}(d_1,d_2)$
[InverseFunctional]	$\top \sqsubseteq \leq 1 R^-$	$\forall d_1 \in \Delta^{FI} \# \left\{d_2 \in \Delta^{FI} : (R^-)^{FI}(d_1,d_2) \geq 0\right\} \leq 1$
[Transitive]	Trans(R)	$\sup_{d \in \Delta^{FI}} \{\min\{R^{FI}(d_1,d), R^{FI}(d,d_2)\}\} \leq R^{FI}(d_1,d_2)$
SubPropertyOf(E_1,E_2)	$E_1 \sqsubseteq E_2$	$E_1^{FI}(d,a) \leq E_2^{FI}(d,a)$
EquivalentProperties ($E_1,\ldots,E_n$)	$E_1 \equiv \ldots \equiv E_n$	$E_1^{FI}(d,a) = \ldots = E_n^{FI}(d,a)$
Fuzzy individual axioms	*Fuzzy assertions*	
Individual (o type (C_i)[$\ltimes m_i$]…	$o : C_i \ltimes m_i$	$C_i^{FI}(o) \ltimes m_i \quad m_i \in [0,1] \quad 1 \leq i \leq n$
value (R_i, o_j)[$\ltimes k_i$]…	$(o, o_i) : R_i \ltimes k_i$	$R_i^{FI}(o,o_i) \ltimes k_i \quad k_i \in [0,1] \quad 1 \leq i \leq n$
Value (T_i, v_j)[$\ltimes l_i$]…)	$(o, v_i) : T_i \ltimes l_i$	$T_i^{FI}(o,v_i) \ltimes l_i \quad l_i \in [0,1] \quad 1 \leq i \leq n$
SameIndividual ($o_1 \ldots o_n$)	$o_1 = \ldots = o_n$	$o_1^{FI} = \ldots = o_n^{FI}$
DifferentIndividuals ($o_1 \ldots o_n$)	$o_i \neq o_j$	$o_i^{FI} \neq o_j^{FI}$

Comments: where $P \in \{R, T\}$ is an abstract role R or a concrete role T; $E \in \{C, D\}$ is a concept C or a concrete datatype D; d and o are abstract individuals, v is a concrete individual, $a \in \{d, v\}$, and $\ltimes \in \{\geq, >, \leq, <\}$.

different expressive power. Most of the approaches provide reasoning support by means of fuzzy DLs. Here, we will not attempt to introduce the reasoning details of these languages based on different fuzzy DLs, but we rather going to start with the fuzzy OWL which is enough for our goals to illustrate the reasoning methods of fuzzy ontologies, since that OWL is the W3C standard ontology language and fuzzy OWL ontologies are common in many application domains.

In (Horrocks and Patel-Schneider 2004) a translation from OWL entailment to DL satisfiability was provided. Based on the idea in Horrocks and Patel-Schneider (2004), the authors in Stoilos et al. (2010) presented a translation method which reduces inference problems of fuzzy OWL into inference problems of fuzzy DLs, thus one can make use of fuzzy DL reasoners to support reasoning for fuzzy OWL. In brief, two steps may be needed for reasoning on fuzzy OWL:

(i) Translating fuzzy OWL into fuzzy DL syntax, thus fuzzy OWL entailment can be further reduced to fuzzy DL entailment.
(ii) Reducing fuzzy DL entailment to fuzzy DL (un)satisfiability, and the latter can be reasoned through the reasoning techniques or the fuzzy DL reasoners as mentioned in the previous parts.

For the first step (*i*), the translation from a fuzzy OWL to the corresponding fuzzy DL syntax can be defined by a function over the mappings between fuzzy OWL syntax and the respective fuzzy DL syntax as shown in Table 4.5. For instance, the fuzzy OWL axioms *SubClassOf*(C_1, C_2) and *EquivalentClasses* (C_1, …, C_n) are mapped to fuzzy DL axioms $C_1 \sqsubseteq C_2$ and $C_1 \equiv \cdots \equiv C_n$ respectively. Besides the straightforward translation above, it shousld be noted that the translation of individual axioms from fuzzy OWL to fuzzy DL syntax is complex as mentioned in (Stoilos et al. 2010), because the fuzzy OWL syntax supports anonymous individuals, which fuzzy DL syntax does not. For example, the individual axiom *Individual*(type(*C*) value(*R Individual*(type(*D*) $\geq$ 0.9)) > 0.8) will be translated into the fuzzy *DL* assertions $C(a) = 1$, $R(a, b) > 0.8$ and $D(b) \geq 0.9$, where *a* and *b* are new individuals. Based on the translation from a fuzzy OWL to the corresponding fuzzy DL syntax, thus fuzzy OWL entailment can be reduced to fuzzy DL entailment. That is, a fuzzy OWL axiom is satisfied if and only if the translated fuzzy *DL* axiom or assertion is satisfied. For the second step (*ii*), it is a common issue in the area of fuzzy DLs, and many proposals have been presented for reducing some reasoning problems of fuzzy DLs (e.g., concept subsumption and entailment problems) to the knowledge base (un)satisfiability. Ones can refer to (Lukasiewicz and Straccia 2008; Stoilos et al. 2007, 2010; Straccia 2001) for more details about the two steps above.

Overall, the translation approaches from fuzzy OWL to fuzzy DL together with the fuzzy DL reasoning techniques and fuzzy DL reasoners as introduced in the previous parts, may provide reasoning support over fuzzy OWL, which will be helpful to handle and manage fuzzy information in many applications and tasks both of the Semantic Web as well as of applications using fuzzy OWL.

4.2.5.2 Fuzzy Ontologies

Over the years, the literature on *fuzzy ontologies* (fuzzy extensions of ontologies) has been flourishing, and kinds of fuzzy ontology definitions were proposed (a list of fuzzy ontology definitions are summarized in Bobillo (2008)). By summarizing the characteristics of fuzzy ontologies, in the following we try to answer the question "what is a fuzzy ontology" for providing a general understanding of fuzzy ontologies. Of course, it should be noted that as we do not expect our fuzzy ontology definition to become a universal and standard definition, because we understand that a universal fuzzy ontology definition is difficult since the different application requirements.

In a more general sense, the short answer for this question "what is a fuzzy ontology" is: a fuzzy ontology is a shared model of some domain which is often conceived as a hierarchical data structure containing all concepts, properties, individual, and their relationships in the domain, where these concepts, properties and so on may be defined imprecisely.

In the following we provide a formal definition of fuzzy ontologies which summarizes the main notions of fuzzy ontologies, so that ones may have a general understanding of fuzzy ontologies.

Definition 4.1 (*fuzzy ontologies*) A fuzzy ontology can be formally defined as a tuple $\mathcal{FO} = (FID_0, FAxiom_0) = \{\mathcal{I}, \mathcal{A}, \mathcal{C}, \mathcal{R}, \mathcal{X}\}$, where FID_0 is a set of identifiers including $\mathcal{I}, \mathcal{A}, \mathcal{C}$, and $\mathcal{R}$, and $FAxiom_0$ is a set of axioms including $\mathcal{X}$:

- $\mathcal{I}$ is a set of *individuals*. Each individual is an instance of a fuzzy concept $\mathcal{C}$ with a membership degree of [0, 1].
- $\mathcal{A}$ is a set of concept properties, and a property may have crisp as well as fuzzy values. A fuzzy value such as "young" or "cheap" is defined through fuzzy sets which allow for fuzzy attributes to be modeled to sets using a gradual assessment of a membership. Currently, there are some membership functions for fuzzy sets membership specification (Oliboni and Pozzani 2008). Formally, a property a $\in \mathcal{A}$ can be defined as an instance of a ternary relation of the form a (c, v, f), where $c \in \mathcal{C}$ denotes a fuzzy ontology concept as will be mentioned below, v is a property value associated with c, and f denotes the restriction facets on v such as the types of the property (e.g., integer or string) or the cardinality constraints of the property (i.e., the upper and lower limits on the number of values for the property). More precisely, in a fuzzy ontology, a property can be classified into two kinds of properties:
  - $\mathcal{A}_R$ is a set of *fuzzy object properties*. Each fuzzy object property link individuals to individuals, and each fuzzy object property has its characteristics (e.g., Symmetric, Functional, Transitive, etc.) and its restrictions (e.g., someValuesFrom, allValuesFrom, minCardinality, etc.) as shown in Table 4.5;
  - $\mathcal{A}_T$ is a set of *fuzzy datatype properties*. Each fuzzy datatype property links individuals to data values, and the domains of data values may be the fuzzy datatypes (Oliboni and Pozzani 2008). Each fuzzy datatype property also has its characteristics and restrictions as shown in Table 4.5.

- $\mathcal{C}$ is a set of *fuzzy concepts* (also called fuzzy classes). A fuzzy concept $c \in \mathcal{C}$ may be defined from several different viewpoints:
  - A concept may be defined by a list of its object instances, i.e., individuals. In this case, the concept can be considered as an enumeration class, where some individuals with similar properties are fuzzy ones. The concept or class defined by these individuals may be fuzzy, and these individuals belong to the concept or class with membership degree of [0, 1], i.e., the fuzzy concept or class $c \in \mathcal{C}$ is a fuzzy set on the domain of individuals $c{:}\mathcal{I} \rightarrow [0, 1]$.
  - A concept may be defined by a set of attributes and their admissible values. The domains of some attributes may be fuzzy, and thus a fuzzy concept is formed.
  - A concept defined using concept forming expressions (e.g., concept conjunction or inheritance) may be a fuzzy one, e.g., a subclass produced from a fuzzy superclass may be fuzzy.

- $\mathcal{R}$ is a set of *fuzzy relations* denoting a type of interaction between concepts of the fuzzy ontology's domain. A fuzzy relation $r \in \mathcal{R}$ can be formally defined as $r \subset C \times C \times [0, 1]$, and the membership degree in [0, 1] denotes the strength of such relation. In a fuzzy ontology, a fuzzy relation may be classified into two types:
  - *Taxonomic relations*: such relations are usually used to represent the hierarchical structure of concepts, such as *equivalence* relation, *generalization* relation, and *part_of* relation. Here, *equivalence* relation denotes that two concepts are equivalent; *generalization* relation equally as is_a relation is the most adopted taxonomic relation and denotes the subclass/superclass relation, and if a fuzzy class is a subclass of another fuzzy class, for any individual, say o, let the membership degree that it belongs to the subclass, say C_i, be $u_{Ci}(o)$ and the membership degree that it belongs to the superclass, say C_j, be $u_{Cj}(o)$. Then $u_{Ci}(o) \leq u_{Cj}(o)$. This characteristic can be used to determine if two classes/concepts have a subclass/superclass relationship; *part_of* relation denotes that one concept is a part of another concept, and in a fuzzy ontology a fuzzy concept which is defined as aggregation of other fuzzy concepts is expressed using this relation.
  - *Description relations*: such relations denote the non-taxonomic relations between two concepts and are commonly used to define relations between instances. For example, we may define a description relation named *Take* between the concept *Student* and the concept *Course*. We can equally associate a degree in [0, 1] to the instantiated relations, e.g., the Student *John* is related to the Course *physics* with the relation *Take* and this relation has a degree of membership equal to 0.85. The main difference between a membership degree associated to fuzzy taxonomic relations and a membership degree associated to description ones is that the first quantifies relations between concepts while the second quantifies relations between instances.

- $\mathcal{X}$ is a set of fuzzy axioms defined over $\mathcal{I} \sqcup \mathcal{A} \sqcup \mathcal{C} \sqcup \mathcal{R}$, which includes fuzzy class/property/individual axioms as shown in Table 4.5. The fuzzy axioms are used to represent the relationships among individuals, concepts, properties and relations in the domain. The fuzzy class and property axioms are used to represent the fuzzy ontology structure information, and the fuzzy individual axioms are used to represent the fuzzy ontology instance information.

A fuzzy ontology, which may be formulated in fuzzy OWL language, is called fuzzy OWL ontology. That is, the elements in the tuple $\mathcal{FO} = \{\mathcal{I}, \mathcal{A}, \mathcal{C}, \mathcal{R}, \mathcal{X}\}$ are expressed by the fuzzy OWL syntax in Table 4.5.

4.3 Summary

Based on the introduction in this chapter, it is shown that lots of fuzzy Description Logics (DLs) and fuzzy ontologies have been investigated in order to handle fuzzy information in real-world applications. In particular, fuzzy DLs and fuzzy ontologies are the acknowledged key of representing and reasoning with knowledge in the Semantic Web. In this chapter, we focus our attention on the recent research achievements on fuzzy extension approaches of DLs and ontologies based on fuzzy set theory, and we provide a brief review of them. We introduce several fuzzy DLs in detail, including the tractable fuzzy DL *f-DLR-Lite*$_{\mathrm{F},\cap}$ proposed in Cheng et al. (2008); the expressive fuzzy DLs *f-ALC* (Straccia 1998) and *FDLR* (Zhang et al. 2008c; Ma et al. 2011); and the fuzzy DLs with fuzzy data types *f-SHOIN(D)* (Straccia 2005b) and *F-ALC*(*G*) (Wang and Ma 2008). A fuzzy DL reasoner called FRESG supporting fuzzy data information with customized fuzzy data types *G* (Wang et al. 2009) is introduced. Also, a definition of fuzzy ontologies is given in order to provide a general understanding of fuzzy ontologies. All of these are of interest or relevance to the discussions of successive chapters.

After reviewing most of the proposals of fuzzy extensions, it has been widely approved that fuzzy DLs and fuzzy ontologies could play a key role in the Semantic Web by serving as a mathematical framework for fuzzy knowledge representation and reasoning in applications. However, the researches on fuzzy DLs/ontologies are still in a developing stage and still the full potential of fuzzy DLs/ontologies has not been exhaustively explored. Some very important issues on fuzzy DLs/ontologies may be important in order for fuzzy DL/ontology technologies to be more widely adoptable in the Semantic Web and other application domains, such as extraction, mapping, query, and storage, which have increasingly received attention as will be discussed in detail in the later chapters.

References

Abulaish M, Dey L (2007) A fuzzy ontology generation framework for handling uncertainties and non-uniformity in domain knowledge description. In: Proceedings of the international conference on computing: theory and applications, Kolkata, India, pp 287–293

Baader F, Calvanese D, McGuinness D, Nardi D, Patel-Schneider PF (2003) The description logic handbook: theory, implementation and applications. Cambridge University Press, Cambridge

Berners-Lee T, Hendler J, Lassila O (2001) The semantic web. The Scientific American 284 (5):34–43

Bobillo F (2008) Managing vagueness in ontologies. PhD thesis, University of Granada, Spain

Bobillo F, Straccia U (2008) fuzzyDL: an expressive fuzzy description logic reasoner. In: Proceedings of the 2008 IEEE international conference on fuzzy systems, Hong Kong, China, pp 923–930

Bobillo F, Straccia U (2009a) Fuzzy description logics with general t-norms and datatypes. Fuzzy Sets Syst 160(23):3382–3402

Bobillo F, Straccia U (2009b) An OWL ontology for fuzzy OWL 2. In: Proceedings of ISMIS 2009, LNAI 5722, pp 151–160

Bobillo F, Straccia U (2010) Representing fuzzy ontologies in OWL 2. In: Proceedings of the 19th IEEE international conference on fuzzy systems (FUZZ-IEEE 2010), pp 2695–2700

Bobillo F, Straccia U (2011) Fuzzy ontology representation using OWL 2. J Approx Reason 52 (7):1073–1094

Bobillo F, Delgado M, Romero J (2007) Optimizing the crisp representation of the fuzzy description logic SROIQ. In: Proceedings of the 3rd ISWC workshop on uncertainty reasoning for the semantic web

Bobillo F, Delgado M, Romero J (2012) DeLorean: a reasoner for fuzzy OWL 2. Expert Syst Appl 39(1):258–272

Cai Y, Leung H (2011) Formalizing object membership in fuzzy ontology with property importance and property priority. In: Proceedings of 2011 IEEE international conference on fuzzy systems, pp 1719–1726

Calegari S, Ciucci D (2006) Integrating fuzzy logic in ontologies. In: Proceedings of the 8th international conference on enterprise information systems, pp 66–73

Calegari S, Ciucci D (2007) Fuzzy ontology, fuzzy description logics and fuzzy-owl. In: Proceedings of WILF 2007, LNCS 4578, pp 118–126

Calegari S, Ciucci D (2008) Towards a fuzzy ontology definition and a fuzzy-extension of an ontology. editor, ICEIS 2006 (Selected Papers). Lecture Notes in Business Information Processing, vol 3, pp 147–158

Calvanese D, De Giacoma G, Lenzerini M (1997) Conjunctive query containment in description logics with n-ary relations. In: Proceedings of the 1997 international workshop on description logic, pp 1–9

Calvanese D, De Giacoma G, Lembo D, Lenzerini M, Rosati R (2005) DL-Lite: tractable description logics for ontologies. In: Proceedings of AAAI 2005, Pittsburgh, pp 602–607

Cheng JW, Ma ZM, Yan L, Wang HL (2008) Querying over fuzzy description logic. Wuhan Univ J Nat Sci 13(4):429–434

Cheng JW, Ma ZM, Zhang F, Wang X (2009) Deciding query entailment for fuzzy SHIN ontologies. In: Proceedings of the 4th annual Asian semantic web conference (ASWC 2009), pp 120–134

Cuenca Grau B, Horrocks I, Motik B, Parsia B, Patel-Schneider P, Sattler U (2008) OWL 2: the next step for OWL. Web Semant Sci Serv Agents World Wide Web 6(4):309–322

De Maio C, Fenza G, Loia V, Senatore S (2009) Towards an automatic fuzzy ontology generation. In: Proceedings of the 2009 IEEE international conference on fuzzy systems, Jeju Island, Korea, pp 1044–1049

Gao M, Liu C (2005) Extending OWL by fuzzy description logic. In: Proceedings of the 17th IEEE international conference on tools with artificial intelligence, pp 562–567

Ghorbel H, Bahri A, Bouaziz R (2009) Fuzzy ProtÅgÅ for fuzzy ontology models. In: Proceedings of 11th international protÅgÅ conference IPC'2009, Academic Medical Center, University of Amsterdam, Amsterdam, Netherlands

Gursky¨ P, HorvÄth T, JirÄsek J, Novotny¨ R, PribolovÄ J, VanekovÄ V, VojtÄsõ P (2008) Knowledge processing for web search—an integrated model and experiments. Journal of Scalable comput pract experience 9(1):51–59

Habiballa H (2007) Resolution strategies for fuzzy description logic. In: Proceedings of the 5th conference of the European society for fuzzy logic and technology, pp 27–36

Haarslev V, Pai HI, Shiri N (2007) Optimizing tableau reasoning in ALC extended with uncertainty. In: Proceedings of the 2007 international workshop on description logics (DL-2007), pp 307–314

HÎlldobler S, Khang TD, StÎrr HP (2002) A fuzzy description logic with hedges as concept modifiers. In: Proceedings of Tech/VJFuzzy'2002, pp 5–34

Hollunder B, NuttW, Schmidt-Schaus M (1990) Subsumption algorithms for concept description languages. In: Proceedings of European conference on artificial intelligence, pp 348–353

Horrocks I, Patel-Schneider PF (2004) Reducing OWL entailment to description logic satisfiability. J Web Semant 1(4):345–357

Horrocks I, Patel-Schneider PF, van Harmelen F (2003) From SHIQ and RDF to OWL: the making of a web ontology language. J Web Semant 1(1):7–26

Li Y, Xu B, Lu J, Kang′ D, Wang P (2005) Extended fuzzy description logic ALCN. In: Proceedings of KES (4), pp 896–902

Lu JJ, Xu BW, Li YH, Kang DZ, Wang P (2005) Extended fuzzy ALCN and its tableau algorithm. In: Proceedings of the second international conference on fuzzy systems and knowledge discovery, pp 232–242

Lukasiewicz T, Straccia U (2008) Managing uncertainty and vagueness in description logics for the semantic web. Web Semant Sci Serv Agents World Wide Web 6(4):291–308

Lv YH, Ma ZM, Zhang XH (2009) Fuzzy ontology storage in fuzzy relational database. In: Proceedings of the international conference on fuzzy systems and knowledge discovery (FSKD 2009), pp 242–246

Ma ZM, Zhang F, Yan L (2011) Representing and reasoning on fuzzy uml models: a description logic approach. Expert Syst Appl 38(3):2536–2549

Ma ZM, Zhang F, Wang HL, Yan L (2013) An overview of fuzzy description logics for the semantic web. Knowl Eng Rev 28:1–34

Mailis T, Stoilos G, Simou N, Stamou G (2008) Tractable reasoning based on the fuzzy EL++algorithm. In: Proceedings of the fourth international workshop on uncertainty reasoning for the semantic web (URSW 2008)

Oliboni B, Pozzani G (2008) An XML Schema for managing fuzzy documents. Technical report, Department of Computer Science, University of Verona, Italy

Pan JZ, Stamou G, Stoilos G, Thomas E (2007) Expressive querying over fuzzy DL-Lite ontologies. In: Proceedings of 2007 international workshop on description logics (DL-2007)

Pan JZ, Stamou G, Stoilos G, Taylor S, Thomas E (2008) Scalable querying service over fuzzy ontologies. In: Proceedings of international world wide web conference (WWW 08), Beijing, pp 575–584

Parry D (2004) A fuzzy ontology for medical document retrieval. In: Proceedings of the Australian workshop on data mining and web intelligence (DMWI2004), pp 121–126

Quan TT, Hui SC, Fong ACM, Cao TH (2006) Automatic fuzzy ontology generation for Semantic Web. IEEE Trans Knowl Data Eng 18(6):842–856

Sanchez E (2006) Fuzzy logic and the semantic web. Elsevier, Amsterdam

SÄnchez D, Tettamanzi G (2004) Generalizing quantification in fuzzy description logic. In: Proceedings of the 8th fuzzy days in Dortmund

SÄnchez D, Tettamanzi G (2006) Reasoning and quantification in fuzzy description logics. Lect Notes Artif Intell 3846:81–88

Sanchez E, Yamanoi T (2006) Fuzzy ontologies for the semantic web. In: Proceedings of FQAS 2006, pp 691–699

Staab S, Studer R (2009) Handbook on ontologies, 2nd edn. Springer, Berlin

Stasinos K, Georgios A (2007) Fuzzy-DL reasoning over unknown fuzzy degrees. In: Proceedings of the 3rd international workshop on semantic web and web semantics, pp 1312–1318

Stoilos G, Stamou G (2007) Extending fuzzy description logics for the semantic web. In: Proceedings of the 3rd international workshop on OWL: experiences and directions (OWLED 2007)

Stoilos G, Stamou G, Tzouvaras V (2005a) A fuzzy description logic for multimedia knowledge representation. In: Proceedings of the 2005 International workshop on multimedia and the semantic web

Stoilos G, Stamou G, Tzouvaras V (2005b) The fuzzy description logic f-SHIN. In: Proceedings of the international workshop on uncertainty reasoning for the semantic web, pp 67–76

Stoilos G, Stamou G, Tzouvaras V (2005c) Fuzzy OWL: uncertainty and the semantic web. In: Proceedings of the 2005 international workshop on OWL: experience and directions

Stoilos G, Stamou G, Pan JZ (2006) Handling imprecise knowledge with fuzzy description logic. In: Proceedings of the 2006 international workshop on description logics

Stoilos G, Simous N, Stamou G, Kollias S (2006b) Uncertainty and the semantic web. IEEE Intell Syst 21(5):84–87

Stoilos G, Stamou G, Pan JZ, Tzouvaras V, Horrocks I (2007) Reasoning with very expressive fuzzy description logics. J Artif Intell Res 30(8):273–320

Stoilos G, Stamou G, Pan JZ (2008a) Classifying fuzzy subsumption in Fuzzy-EL+. In: Proceedings of the 21st international workshop on description logics (DL 2008)

Stoilos G, Stamou G, Kollias S (2008b) Reasoning with qualified cardinality restrictions in fuzzy description logics. In: Proceedings of the 17th ieee international conference on fuzzy systems (FUZZ-IEEE 2008), IEEE Computer Society, Hong Kong, China, pp 637–644

Stoilos G, Stamou G, Pan JZ (2010) Fuzzy extensions of OWL: logical properties and reduction to fuzzy description logics. Int J Approx Reasoning 51:656–679

Straccia U (1998) A fuzzy description logic. In: Proceedings of the 15th national conference on artificial intelligence (AAAI-98), pp 594–599

Straccia U (2001) Reasoning within fuzzy description logics. J Artif Intell Res 14(1):137–166

Straccia U (2005a) Description logics with fuzzy concrete domains. In: Proceedings of UAI-2005, pp 559–567

Straccia U (2005b) Towards a fuzzy description logic for the semantic web. In: Proceedings of the 2nd European semantic web conference (ESWC 2005), pp 167–181

Straccia U (2006) Answering vague queries in Fuzzy DL-LITE. In: Proceedings of the 11th International conference on information processing and management of uncertainty in knowledge-based systems (IPMU-06), pp 2238–2245

Straccia U (2009) SoftFacts: a top-k retrieval engine for a tractable description logic accessing relational databases. Technical Report, http://www.straccia.info/software/SoftFacts/SoftFacts.html

Straccia U, Visco G (2007) DLMedia: an ontology mediated multimedia information retrieval system. In: Proceedings of the international workshop on description logics (DL 2007)

Tresp C, Molitor R (1998) A description logic for vague knowledge. In: Proceedings of the 13th European conference on artificial intelligence (ECAI-98)

VojtÄsõ P (2006) A fuzzy EL description logic with crisp roles and fuzzy aggregation for web consulting. In: Proceedings of information processing and management under uncertainty (IPMU), pp 1834–1841

Wang HL, Ma ZM (2008) A decidable fuzzy description logic F-ALC(G). In: Proceedings of the 19th international conference on database and expert systems applications (DEXA 2008), pp 116–123

Wang HL, Ma ZM, Yan L, Cheng JW (2008) A fuzzy description logic with fuzzy data type group. In: Proceedings of the 2008 IEEE international conference on fuzzy systems, pp 1534–1541

Wang HL, Ma ZM, Yin JF (2009) FRESG: a kind of fuzzy description logic reasoner. In: Proceedings of DEXA 2009, pp 443–450

Yan L, Zhang F, Ma ZM (2012) f-SROIQ(G): an expressive fuzzy description logic supporting fuzzy data type group. In: Proceedings of SAC 2012, pp 320–325

Yen J (1991) Generalising term subsumption languages to fuzzy logic. In: Proceedings of the 12th international joint conference on artificial intelligence (IJCAI-91), pp 472–477

Zadeh LA (1965) Fuzzy sets. Inf Control 8(3):338–353

Zhang LY, Ma ZM (2012) ICFC: a method for computing semantic similarity among fuzzy concepts in a fuzzy ontology. In: Proceedings of FUZZ-IEEE 2012, pp 1–8

Zhang F, Ma ZM, Lv YH, Wang X (2008a) Formal semantics-preserving translation from fuzzy er model to fuzzy OWL DL ontology. In: Proceedings of 2008 IEEE/WIC/ACM international conference on web intelligence (WI 2008), pp 503–509

Zhang F, Ma ZM, Wang HL, Meng XF (2008b) A formal semantics-preserving translation from fuzzy relational database schema to fuzzy OWL DL ontology. In: Proceedings of the 3rd Asian semantic web conference on the semantic web (ASWC 2008), pp 46–60

Zhang F, Ma ZM, Yan L (2008c) Representation and reasoning of fuzzy ER model with description logic. In: Proceedings of the 17th IEEE international conference on fuzzy systems (FUZZ-IEEE 2008), pp 1358–1365

Zhang F, Ma ZM, Cheng JW, Meng XF (2009) Fuzzy semantic web ontology learning from fuzzy UML model. In: Proceedings of the 18th ACM conference on information and knowledge management (CIKM 2009), pp 1007–1016

Zhang F, Ma ZM, Fan GF, Wang X (2010) Automatic fuzzy semantic web ontology learning from fuzzy object-oriented database model. In: Proceedings of the international conference on database and expert systems applications (DEXA 2010), pp 16–30

Zhang F, Ma ZM, Yan L, Cheng JW (2011) Storing fuzzy ontology in fuzzy relational database. In: Proceedings of the 22nd international conference on database and expert systems applications (DEXA 2011), pp 447–455

Chapter 5
Fuzzy Description Logic and Ontology Extraction from Fuzzy Data Models

5.1 Introduction

Knowledge Extraction (KE) means a process of nontrivial extraction of implicit, previously unknown and potentially useful information from data in many different fields (Chen et al. 1996; Fayyad et al. 1996; Jayatilaka and Wimalarathne 2011). In particular, over the years, there has been an enormous growth in the amount of data stored in database systems, and millions of databases have been used in various fields such as business management and scientific and engineering data management (Chang et al. 2004; Hilderman and Hamilton 1999). However, the problem of data rich but knowledge poor may exist in many database systems, and much useful information is hidden within databases and, as such, cannot be immediately utilized (Martin 2003). This situation has generated an urgent need for approaches and tools to assist humans in extracting useful knowledge from databases, which has received much attention in recent years. When extracting knowledge from databases, different techniques may be accepted to represent the extracted knowledge in different forms (Chen et al. 1996). Among several ways to approach knowledge representation, two Semantic Web techniques, i.e., Description Logics and ontologies, are gaining a privileged place in recent years as have been mentioned in Chap. 1:

(i) On one hand, *Description Logics* (*DLs*, for short), which is a family of knowledge representation languages, can be used to represent the knowledge of an application domain in a structured and formally well-understood way (Baader et al. 2003). Based on several advantages of DLs (e.g., high expressive power and effective reasoning service), lots of work has investigated how to merge databases (DBs) and DLs, in order that the reasoning power in DBs, e.g., instance classification and query optimization, may be increased and the data in DL knowledge bases may be efficiently managed (Borgida and Brachman 1993; Bresciani 1995; Roger et al. 2002). In particular, in recent years, a significant interest developed regarding the problem of *describing DBs with DLs*, which is a fundamental problem in many fields because it turns databases to logical systems with enriched semantics and enhanced reasoning mechanism

Z. Ma et al., *Fuzzy Knowledge Management for the Semantic Web*,
Studies in Fuzziness and Soft Computing 306, DOI: 10.1007/978-3-642-39283-2_5,

(Hao et al. 2007). As we have known, Database Management Systems are suited to manage data efficiently, but the formalism for organizing them in a structured way is quite absent, as well as the capability to infer new information from that already existing; Description Logic Management Systems, which can provide a more structured representation of the universe of discourse in which data are placed, are suited to the semantic organization and the reasoning of data (Bresciani 1995). Therefore, extracting knowledge from databases by means of DLs will facilitate the accessing and intelligent processing of data in databases. After extracting knowledge from databases with DLs, i.e., the databases are described by DLs, firstly, the reasoning services of databases (such as the satisfiability of a database schema and the subsumption of two schemas) may be reasoned automatically through the reasoning mechanism of DLs, and secondly, DLs can be used as a tractable query language for databases, and the query processing in database systems could be intelligently done in a thoroughly logical way (Bresciani 1995; Schild 1995). For the former, it is necessary to investigate the relationships between database models (e.g., ER model, relational model, object-oriented data model, UML model, and XML model) and DLs (Calvanese et al. 1999a, b; Calì et al. 2001).

(ii) On the other hand, as the knowledge representation model in the Semantic Web, ontologies, are being accepted to represent the extracted knowledge from different domains, i.e., the well-known issue of constructing ontologies. Construction of ontologies is a very important issue in the context of the Semantic Web. The Semantic Web aims at creating ontology-based and machine-processable Web content, and thus the success and proliferation of the Semantic Web largely depends on *constructing ontologies* (Berners-Lee et al. 2001). To this end, kinds of approaches and tools have been developed to construct ontologies from various data resources such as text, dictionary, XML, databases, and etc. (Corcho et al. 2003; Maedche and Steffen 2001).

It should be noted that, in real-world application, information is often imprecise and uncertain. The representation of fuzzy information has been introduced into data models and DLs/ontologies as have been introduced in Chaps. 3 and 4. This resulted in numerous contributions, mainly with respect to the popular fuzzy data models such as fuzzy conceptual data models (fuzzy ER models and fuzzy UML models) and fuzzy logical database models (fuzzy relational database models and fuzzy object-oriented database models) (see Chap. 3 in detail), and with respect to the fuzzy DLs and fuzzy ontologies (see Chap. 4 in detail). However, such information and knowledge in fuzzy data models cannot participate in the Semantic Web directly. The classical knowledge extraction approaches are not sufficient for handling such imprecise and uncertain information. Therefore, there is an urgent need for extracting knowledge from these fuzzy data models with the Semantic Web techniques, i.e., fuzzy DLs and fuzzy ontologies. By extracting knowledge from fuzzy data models with fuzzy DLs, interesting and useful information and knowledge can be applied to information management, query processing, database reasoning, data integration, and many other applications. Also, by extracting knowledge from fuzzy data models with fuzzy ontologies, i.e., constructing fuzzy ontologies, may facilitate the fuzzy ontology development and the information sharing in the context of the Semantic Web, and also act as a gap-bridge of the realization of semantic interoperations between the existing database applications and the Semantic Web. In this chapter, we mainly investigate fuzzy DL/ontology knowledge extraction from several typical fuzzy data models, including fuzzy conceptual data models (fuzzy ER models and fuzzy UML models) and fuzzy logical database models (fuzzy relational database models and fuzzy object-oriented database models) mentioned in Chap. 3.

5.2 Fuzzy Description Logic Extraction from Fuzzy Data Models

In recent years, several works have been done to extract fuzzy Description Logics knowledge bases from fuzzy data models. In Zhang et al. (2008a, b, and c) and its extended version Zhang et al. (2013a, b), how to represent and reason on fuzzy ER models with a fuzzy DL FDLR was investigated, where FDLR knowledge bases are extracted from fuzzy ER models, and the reasoning problems of fuzzy ER models are reasoned by means of the reasoning mechanism of the fuzzy DL FDLR. Also, the fuzzy DL FDLR was used to represent and reason on fuzzy UML models in Ma et al. (2011a). Moreover, how to extract knowledge from fuzzy relational databases and fuzzy object-oriented databases were investigated in Ma et al. (2011b) and Zhang et al. (2012a, b), and the extracted knowledge was used to reason about the fuzzy relational databases and fuzzy object-oriented databases. In this subchapter, fuzzy DL extraction from these fuzzy data models will be introduced in detail. Also, how to apply the extracted fuzzy DL knowledge for reasoning about the fuzzy data models will be discussed.

5.2.1 Fuzzy Description Logic Extraction from Fuzzy Conceptual Data Models

The common conceptual data models (e.g., entity-relationship (ER) model and UML model) have played a crucial role in database design and information systems analysis. In spite of its wide applications, the classical data models often suffer from their incapability of representing and manipulating imprecise and uncertain information that may occur in many real-world applications. Since the early 1980s, Zadeh's fuzzy logic (Zadeh 1965) has been used to extend various data models. The purpose of introducing fuzzy logic in data models is to enhance the classical models such that uncertain and imprecise information can be represented and manipulated. This resulted in numerous contributions, mainly with respect to the popular fuzzy ER model (Chen and Kerre 1998; Ruspini 1986;

Zvieli and Chen 1986; etc.) and fuzzy UML model (Haroonabadi and Teshnehlab 2009; Ma and Yan 2007; Ma et al. 2011c; Motameni et al. 2008). A comprehensive literature overview of fuzzy data models can be found at Ma and Yan (2008). In particular, the fuzzy ER model and fuzzy UML model are the typical fuzzy conceptual data models as have been introduced in Sect. 3.2. In the following, we will introduce how to extract fuzzy DL knowledge bases from fuzzy conceptual data models including fuzzy ER model (Zhang et al. 2008c, 2013a) and fuzzy UML model (Ma et al. 2011a).

5.2.1.1 Fuzzy Description Logic Extraction from Fuzzy ER Model

In order to realize fuzzy DL extraction from fuzzy ER model, first it is necessary to choose an appropriate fuzzy DL which can well represent the extracted knowledge and has effective reasoning service. Based on the discussions in (Zhang et al. 2013a), it is shown that the DL DLR, which is powerful enough to account for the essential features of ER models, is suitable to represent the ER models. Considering all the nice features of DLR, it is natural to try to extend it through fuzzy logic for satisfying the need of fuzzy ER modeling. To this end, a fuzzy description logic called FDLR (i.e., the fuzzy extension of DLR) was proposed in (Zhang et al. 2008c) and has been introduced in Sect. 4.2.2. In the following, we introduce how to extract fuzzy DL FDLR knowledge bases from fuzzy ER models in part A and reason about the fuzzy ER models with the extracted knowledge in part B (Zhang et al. 2008c, 2013a).

(A) *Extracting FDLR Knowledge Bases from Fuzzy ER Models*

Based on the formal representation of fuzzy ER models introduced in Sect. 3.2.1 and the fuzzy DL FDLR introduced in Sect. 4.2.2, a formal approach and an example for extracting FDLR knowledge bases from fuzzy ER models will be introduced in the following.

For simplicity, for each fuzzy relationship *FR* of arity *n* in the fuzzy ER model *FS* as introduced in Sect. 3.2.1, we denote with μ_{FR} a mapping from the set of fuzzy ER-roles associated with *FR* to the integers 1, …, *n*. For example, $\mu_{FR}(FU_1) = 1, \ldots, \mu_{FR}(FU_n) = n$.

Definition 5.1 Given a fuzzy ER model $FS = (L_{FS}, \leq_{FS}, att_{FS}, rel_{FS}, card_{FS})$ mentioned in Sect. 3.2.1, the corresponding FDLR knowledge base $\varphi\ (FS) = (FA, FP, FT)$ can be extracted as the following rules:

(1) The set *FA* of atomic fuzzy concepts of $\varphi\ (FS)$ contains the following elements:
   - Each fuzzy entity $FE \in E_{FS}$ is mapped into an atomic fuzzy concept $\varphi\ (FE)$;
   - Each fuzzy domain symbol $FD \in D_{FS}$ is mapped into an atomic fuzzy concept $\varphi\ (FD)$
(2) The set *FP* of atomic fuzzy relations of $\varphi\ (FS)$, which is obtained from the relationships and attributes in *FS*, contains the following elements:

- Each fuzzy relationship $FR \in R_{FS}$, denoting a relation of arity n, is mapping into an atomic fuzzy relation $\varphi(FR)$;
- Each fuzzy attribute symbol $FA \in A_{FS}$ is mapped into a fuzzy binary relation $\varphi(FA)$ in $\varphi(FS)$.

(3) The set FT of fuzzy inclusion axioms of $\varphi(FS)$ contains the following elements:
For each pair of fuzzy entities $FE_1, FE_2 \in E_{FS}$ such that $FE_1 \leq_{FS} FE_2$, the fuzzy axiom:

$$\varphi(FE_1) \subseteq \varphi(FE_2);$$

- For each pair of fuzzy relations $FR_1, FR_2 \in R_{FS}$ such that $FR_1 \leq_{FS} FR_2$, the fuzzy axiom:

$$\varphi(FR_1) \subseteq \varphi(FR_2);$$

- For each fuzzy attribute FA with fuzzy domain FD of a fuzzy entity FE, the fuzzy axiom:

$$\varphi(FE) \subseteq (\forall[\$1]\,(\varphi(FA) \cap (\$2/2 : \varphi(FD))))\cap = 1[\$1]\varphi(FA);$$

- For each fuzzy relationship FR of arity n with roles $FU_1, \ldots, FU_n$ in which each FU_i is associated with fuzzy entity FE_i, the fuzzy axiom:

$$\varphi(FR) \subseteq (\$\mu_{FR}(FU_1)\,/n : FE_1) \cap \ldots \cap (\$\mu_{FR}(FU_n)\,/n : FE_n);$$

- For each fuzzy ER-role FU of fuzzy relationship FR associated with fuzzy entity FE, with cardinality constraints $m = cmin_{FS}(FE, FR, FU)$ and $n = cmax_{FS}(FE, FR, FU)$:
- If $m \neq 0$, the fuzzy inclusion axiom:

$$(\varphi(A_i))^{\alpha_{FS}(FDBI)}$$

- If $n \neq \infty$, the fuzzy inclusion axiom:

$$\varphi(FE) \subseteq \leq n[\$\mu_{FR}(FU)]\varphi(FR);$$

- For each pair of symbols $X, Y \in E_{FS} \cup R_{FS} \cup D_{FS}$, such that $X \neq Y$ and $X \in R_{FS} \cup D_{FS}$, a fuzzy inclusion axiom: $\varphi(X) \subseteq \neg\varphi(Y)$.

On the basis of the semantics of fuzzy ER models and FDLR introduced in Sects. 3.2.1 and 4.2.2, the correctness of the extracted approach mentioned above can be proved by establishing the correspondences between the fuzzy database states of fuzzy ER models and the models of FDLR knowledge bases (i.e., Theorem 5.1).

Theorem 5.1 *For every fuzzy ER model* $FS = (L_{FS}, \leq_{FS}, att_{FS}, rel_{FS}, card_{FS})$ *mentioned above, let FB be a fuzzy database state with respect to FS as shown in* Sect. 3.2.1 *and* $\varphi(FS)$ *be the FDLR knowledge base obtained from FS. Then there are two mappings* α_{FS}*, from the fuzzy database states w.r.t. FS to the fuzzy* *interpretations of $\varphi(FS)$, and β_{FS}, from the fuzzy interpretations of φ (FS) to the fuzzy database states corresponding to FS, such that:*

- *For each legal fuzzy database state FB for FS, $\alpha_{FS}(FB)$ is a model of φ (FS), and for each symbol $X \in E_{FS} \cup A_{FS} \cup R_{FS} \cup D_{FS}$, $X^{FB} = (\varphi(X))^{\alpha_{FS}(FB)}$ holds.*
- *For each fuzzy model FI of φ (FS), $\beta_{FS}(FI)$ is a legal fuzzy database state for FS, and for each symbol $X \in E_{FS} \cup A_{FS} \cup R_{FS} \cup D_{FS}$, $(\varphi(X))^{FI} = X^{\beta_{FS}(FI)}$ holds.*

Proof In the following, we prove the first part of Theorem 5.1. Note that for each legal database state *FB*, we only consider the finite set of fuzzy values Δ^{FB}, and a fuzzy interpretation $\alpha_{FS}(FB)$ of φ (*FS*) can be defined as follows:

- α_{FS} is a mapping from the fuzzy database state corresponding to *FS* to the fuzzy interpretation of its translation φ (*FS*). So the mapping of their corresponding elements also exists. That is, the domain elements $\Delta^{\alpha_{FS}(FB)}$ of fuzzy interpretation of φ (*FS*) is constituted by the fuzzy values of fuzzy database state *FB*, i.e., $\Delta^{\alpha_{FS}(FB)} = \Delta^{FB} \cup \bigcup_{FR \in R_{FS}} FR^{FB}$. Here, we know that each fuzzy relation *FR* of *FS* is assigned with a U_{FS}-labeled tuples over Δ^{FB}, thus we use $\cup_{FR \in R_{FS}} FR^{FB}$ to explicitly represent in FDLR knowledge base the type structure of fuzzy relations.
- The set of atomic fuzzy concepts *FA* and the set of atomic fuzzy relations *FP* are interpreted as follows: $(\varphi(X))^{\alpha_{FS}(FB)} = \{X^{FB} \mid X \in D_{FS} \cup E_{FS} \cup A_{FS} \cup R_{FS}\}$.

Based on the definition of $\alpha_{FS}(FB)$ mentioned above, now we prove the first part of Theorem 5.1 by considering each case of fuzzy inclusion axiom set *FT* of φ (*FS*) in the extracted approach. □

Case 1 For $FE_1 \leq_{FS} FE_2$ and $FR_1 \leq_{FS} FR_2$ in *FS*, the extracted fuzzy inclusion axioms: $\varphi(FE_1) \subseteq \varphi(FE_2)$ and $\varphi(FR_1) \subseteq \varphi(FR_2)$.

Proof Firstly, according to the semantics of fuzzy ER models, there are $FE_1^{FB} \subseteq FE_2^{FB}$ and $FR_1^{FB} \subseteq FR_2^{FB}$. Then, by definition of $\alpha_{FS}(FB)$ above, we have $(\varphi(FE_1))^{\alpha_{FS}(FB)} \subseteq (\varphi(FE_2))^{\alpha_{FS}(FB)}$ and $(\varphi(FR_1))^{\alpha_{FS}(FB)} \subseteq (\varphi(FR_2))^{\alpha_{FS}(FB)}$. That is, $\alpha_{FS}(FB)$ satisfies the inclusion axioms $\varphi(FE_1) \subseteq \varphi(FE_2)$ and $\varphi(FR_1) \subseteq \varphi(FR_2)$. □

Case 2 For each fuzzy attribute *FA* with fuzzy domain *FD* of a fuzzy entity *FE*, i.e., $att_{FS}(FE) = [\ldots, FA{:}FD, \ldots]$, the extracted fuzzy inclusion axiom: $\varphi(FE) \subseteq (\forall [\$1] (\varphi(FA) \sqcap (\$2/2 : \varphi(FD)))) \sqcap (= 1[\$1] \varphi(FA))$.

Proof Firstly, for a fuzzy instance $Fe \in (\varphi(FE))^{\alpha_{FS}(FB)}$, by definition of $\alpha_{FS}(FB)$ above, we have $Fe \in FE^{FB}$. Then, according to the semantics of fuzzy ER models, there is exactly one element $Fa = \langle Fe, Fd \rangle \in FA^{FB} = (\varphi(FA))^{\alpha_{FS}(FB)}$, $Fd \in D^{FB} = (\varphi(FD))^{\alpha_{FS}(FB)}$, where the first component of *Fa* is *Fe*, and the second component of *Fa* is an element of D^{FB}, i.e., $\forall [\$1] (\varphi(FA) \sqcap (\$2/2{:}\varphi(FD)))$ holds. Additionally, all fuzzy relations φ (*FA*) corresponding to fuzzy attribute names *FA* are functional, i.e., $= 1[\$1]\ \varphi\ (FA)$ holds. Therefore, $\alpha_{FS}(FB)$ satisfies the fuzzy axiom $\varphi(FE) \subseteq (\forall[\$1]\ (\varphi\ (FA) \cap (\$2/2 : \varphi\ (FD)))) \cap = 1[\$1]\ \varphi\ (FA)$. □

Case 3 For each fuzzy relationship *FR* of arity *n* with fuzzy ER-roles FU_1, …, FU_n in which each FU_i is associated with fuzzy entity FE_i, i.e., $rel_{FS}(FR) = [FU_1{:}FE_1,\ldots, FU_n{:}FE_n]$, the extracted fuzzy inclusion axiom: $\varphi\ (FR) \subseteq (\$\ \mu_{FR}(FU_1)/n : FE_1) \cap \ldots \cap (\$\ \mu_{FR}(FU_n)/n : FE_n)$.

Proof Firstly, for a fuzzy relation instance $Fr \in (\varphi(FR))^{\alpha_{FS}(FB)}$, by definition of $\alpha_{FS}(FB)$ above, we have $Fr \in FR^{FB}$. Then, according to the semantics of fuzzy ER models, we know that all instances of *FR* are of the form $[FU_1{:}Fe_1,\ldots, FU_n{:}Fe_n]$, where $Fe_i \in FE_i^{FB}$, $i \in \{1,\ldots,n\}$, i.e., $Fr = [FU_1{:}Fe_1, \ldots, FU_n{:}Fe_n]$, where $Fe_i \in FE_i^{FB}$, $i \in \{1, \ldots, n\}$. Furthermore, by definition of $\alpha_{FS}(FB)$ above once again, we have $Fe_i \in (\varphi(FE_i))^{\alpha_{FS}(FB)} = FE_i^{FB}$ for $i \in \{1, \ldots, n\}$. That is, $\alpha_{FS}(FB)$ satisfies the fuzzy inclusion axiom $\varphi\ (FR) \subseteq (\$\mu_{FR}(FU_1)/n : FE_1) \cap \ldots \cap (\$\mu_{FR}(FU_n)/n : FE_n)$. □

Note that, the two parts of Theorem 5.1 are a mutually inverse process, and the proof of the second part of Theorem 5.1, which can be treated analogously according to the first part mentioned above, is omitted here. According to Theorem 5.1, it is shown that the transformation from a fuzzy ER model to a FDLR knowledge base is semantics-preserving.

In the following, we illustrate the extraction procedure with an example.

Example 5.1 Given the fuzzy ER model FS_1 in Fig. 3.1 in Chap. 3, the extracted FDLR knowledge base from FS_1 by using the rules in Definition 5.1 is shown in Fig. 5.1.

Following the proposed extraction approach in Definition 5.1, we implemented a prototype extraction tool called *FER2FDLR*, which can automatically transform a fuzzy ER model into a FDLR knowledge base.

The core of *FER2FDLR* is that it can first read in an XML-coded fuzzy ER model file (a conceptual data model file produced from CASE tool PowerDesigner), and then transform automatically it into a FDLR knowledge base. In the following, we briefly introduce the design and implementation of *FER2FDLR*. The implementation of *FER2FDLR* is based on Java 2 JDK 1.5 platform, and the Graphical User Interface (GUI) is exploited by using the java.awt and javax.swing packages. Figure 5.2 shows the overall architecture of *FER2FDLR*. *FER2FDLR* includes four main modules, i.e., *input module*, *parse module*, *transformation module*, and *output module*: The input module first read in an XML-coded fuzzy ER model file produced from PowerDesigner; The parsing module uses the regular expression to parse the input file and store the parsed information as Java ArrayList classes. The features of the fuzzy ER model in the XML-coded file (such as fuzzy ER entities, attributes, roles, and so on) can be extracted and represented as the formalization of the fuzzy ER model; The transformation module transforms the parsed results into the corresponding FDLR knowledge base according to

Firstly, according to Definition 5.1, the fuzzy ER-roles in Fig. 3.1 in Chapter 3 can be mapped into the integers: $\mu_{FR}(E\_w) = 1$, $\mu_{FR}(J\_w) = 2$, $\mu_{FR}(B\_w) = 3$.
The FDLR knowledge base $\varphi(FS_1) = (FA, FP, FT)$ transformed from FS_1 is as follows (for the sake of simplicity, the symbol $\varphi()$ is omitted in the following descriptions):
FA = {*employee, Young-employee, Job, branch, Main-branch, String, Integer, Real*}
FP = {*Work-on, Number, Name, Fuzzy_Age, Sex, Fuzzy_Salary, u*}
FT = {
Young-employee $\subseteq$ *employee* ;
Main-branch $\subseteq$ *branch* ;
Young-employee $\subseteq$ ≥ 3 [\$1] *Work-on* $\cap \leq 10$ [\$1] *Work-on* $\cap$ =1[\$1] *Name* $\cap$ ($\forall$[\$1] (*Name* $\cap$ (\$2/2:*String*))) $\cap$ =1[\$1] *Fuzzy_Age* $\cap$ ($\forall$[\$1] (*Fuzzy_Age* $\cap$ (\$2/2:*Integer*))) $\cap$ =1[\$1] *Sex* $\cap$ ($\forall$[\$1] (*Sex* $\cap$ (\$2/2:*String*))) $\cap$ =1[\$1] *Fuzzy_Salary* $\cap$ ($\forall$[\$1] (*Fuzzy_Salary* $\cap$ (\$2/2:*Real*))) $\cap$ =1[\$1] *u* $\cap$ ($\forall$[\$1] (*u* $\cap$ (\$2/2:*Real*))) ;
Job $\subseteq$ =1[\$2] *Work-on* ;
Main-branch $\subseteq$ =1[\$3] *Work-on* $\cap$ =1[\$1] *Number* $\cap$ ($\forall$[\$1] (*Number* $\cap$ (\$2/2:*String*))) $\cap$ =1[\$1] *u* $\cap$ ($\forall$[\$1] (*u* $\cap$ (\$2/2:*Real*))) ;
Work-on $\subseteq$ (\$1/3:*Young-employee*) $\cap$ (\$2/3:*Job*) $\cap$ (\$3/3:*Main-branch*) $\cap$ =1[\$1]

Fig. 5.1 The FDLR knowledge base $\varphi(FS_1)$ extracted from the fuzzy ER model FS_1 in Fig. 3.1 in Chap. 3

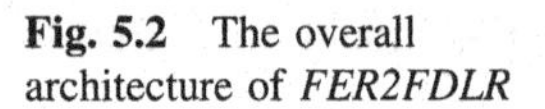
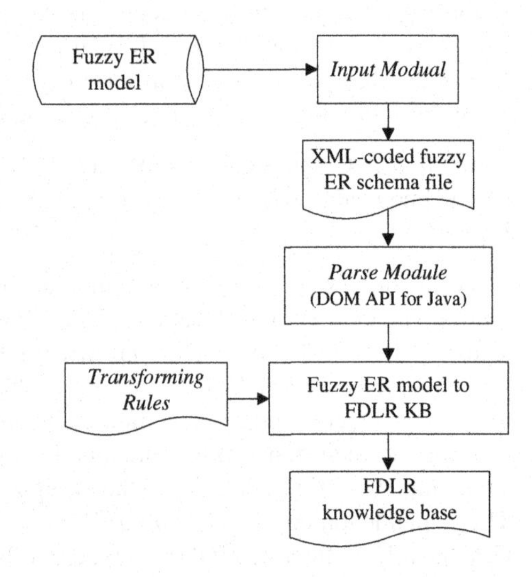

Fig. 5.2 The overall architecture of *FER2FDLR*

Definition 5.1; The output module finally produces the resulting FDLR knowledge base which is saved as a text file and displayed on the tool screen.

We carried out transformation experiments of some fuzzy ER models using the implemented tool *FER2FDLR*, with a PC (CPU P4/3.0 GHz, RAM 3.0 GB and Windows XP system). The sizes of the fuzzy ER models range from 0 to 800. Case studies show that our approach is feasible and *FER2FDLR* tool is efficient. Figure 5.3 shows the actual execution time routines in the *FER2FDLR* tool running ten fuzzy ER models, where the *preprocessing* denotes the operations of parsing and storing fuzzy ER models, i.e., parsing the fuzzy ER models and

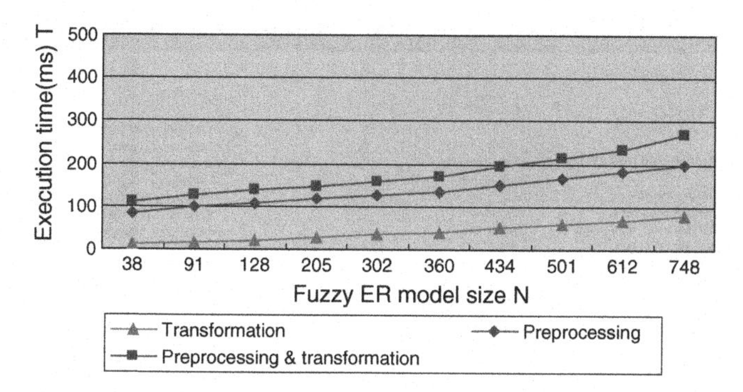

Fig. 5.3 The execution time of *FER2FDLR* routine on several fuzzy ER models

preparing the element data in computer memory for the usage in the transformation procedure.

Here we provide an example of *FER2FDLR*. Figure 5.4 is the screen snapshot of *FER2FDLR* tool running one of case studies, which displays a fuzzy ER model representation procedure with the fuzzy DL FDLR, i.e., the transformation from the fuzzy ER model FS_1 to the FDLR knowledge base as shown in Fig. 5.1. In Fig. 5.4, the XML-coded fuzzy ER model file, the parsed results, and the transformed FDLR knowledge base are displayed in the left, middle and right areas, respectively. The complete transformed FDLR knowledge base from the fuzzy ER model can be found in Fig. 5.1.

Until now, on the basis of the proposed approach and the implemented tool, fuzzy ER models can be represented by FDLR knowledge bases. Therefore, it is possible to take advantage of the associated reasoning techniques of FDLR to reason on the fuzzy ER models as will be discussed in the following.

(B) *Reasoning about the Fuzzy ER Models with the Extracted FDLR Knowledge Bases*

As we have known, using conventional techniques in reasoning on fuzzy ER models would mean to manually check the reasoning tasks of fuzzy ER models (e.g., whether a fuzzy entity is a subclass of another fuzzy entity or whether there is redundancy in a fuzzy ER model), which is a complex and time-consuming task. Using knowledge representation techniques such as the fuzzy Description Logic, reasoning on fuzzy ER models may be done more effectively or automatically by means of the reasoning ability of the fuzzy Description Logic. In the following we will introduce how to reason on fuzzy ER models with FDLR. Firstly, we introduce some familiar reasoning problems of fuzzy ER models and give their formal definitions. Then, we discuss how to reason on fuzzy ER models with FDLR, and propose an approach that can reduce reasoning on fuzzy ER models to reasoning on FDLR knowledge bases, so that the reasoning tasks of fuzzy ER models may be detected by checking the reasoning problems of FDLR.

Generally speaking, being similar to the classical data models such as ER models, UML models, and object-oriented data models (Baader et al. 2003;

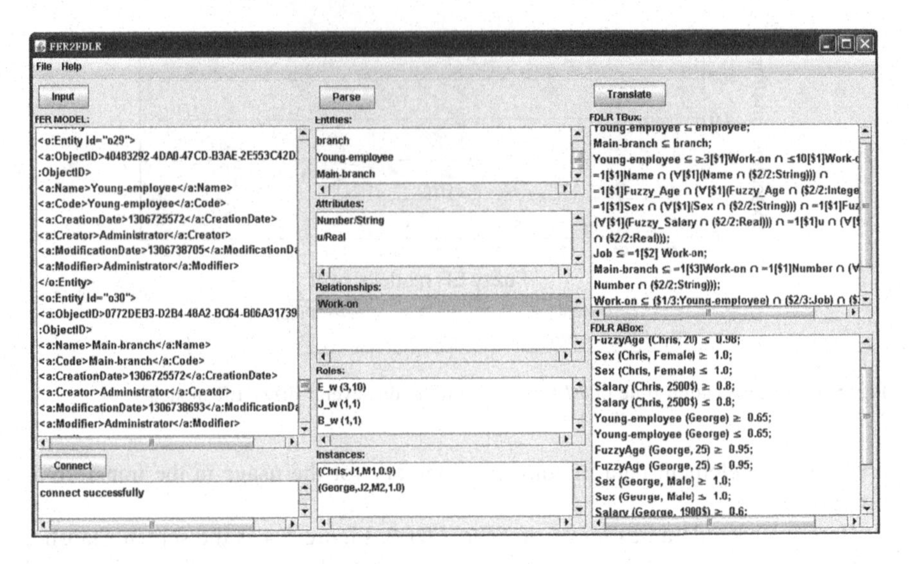

Fig. 5.4 Screen snapshot of *FER2FDLR*

Calvanese et al. 1999a, b; Calì et al. 2001), the familiar reasoning problems of fuzzy ER models include *consistency*, *satisfiability*, *subsumption*, and *redundancy*:

- The *consistency* task is to check whether a fuzzy ER model is consistent, i.e., whether the elements altogether in the fuzzy ER model are contradictory;
- The *satisfiability* task is to determine whether a fuzzy entity or relationship stands for a non-empty set;
- The *subsumption* task is to determine whether a fuzzy entity or relationship is the subclass of another fuzzy entity or relationship;
- The *redundancy* task is to check whether there is *redundancy* in a fuzzy ER model, e.g., if there are two equivalent fuzzy entities in a fuzzy ER model, then we know that the fuzzy ER model is *redundant*.

In the following we give formal definitions of these reasoning problems.

Definition 5.2 (*consistency*) A fuzzy ER model is *consistent*, if there is at least a set of data instances defined on the fuzzy ER model.

If a fuzzy ER model is not consistent, the entities altogether are contradictory. That is to say, the fuzzy ER model does not allow that any set of data instances can be populated without violating any of the requirements imposed by the fuzzy ER model. This may be due to a design error or over-constraining. Observe that the interaction of various types of constraints may make it difficult to detect inconsistencies of fuzzy ER models.

Definition 5.3 (*satisfiability*) A fuzzy entity *FE* of a fuzzy ER model *FS* is *satisfiable*, if there is at least one fuzzy database state *FB* of *FS* such that $FE^{FB} \neq \emptyset$.

Definition 5.4 (*satisfiability*) A fuzzy relationship *FR* of a fuzzy ER model *FS* is *satisfiable*, if there is at least one fuzzy database state *FB* of *FS* such that $FR^{FB} \neq \varnothing$.

A fuzzy entity *FE* or a fuzzy relationship *FR* of a fuzzy ER model is satisfiable, i.e., the fuzzy ER model admits a set of data instances in which *FE* or *FR* has a non-empty set of instances. An unsatisfiable fuzzy entity/relationship weakens the understandability of a fuzzy ER model, since the fuzzy entity/relationship stands for an empty set, and thus, at the very least, it is inappropriately named. In this case, to increase the quality of the fuzzy ER modeling, designers may remove the unsatisfiability by correcting errors or relaxing some constraints.

Definition 5.5 (*subsumption*) In a fuzzy ER model *FS*, a fuzzy entity FE_1 is subsumed by another fuzzy entity FE_2 if for each fuzzy database state *FB* of *FS* it holds $FE_1^{FB} \subseteq FE_2^{FB}$.

Definition 5.6 (*subsumption*) In a fuzzy ER model *FS*, a fuzzy relationship FR_1 is subsumed by another fuzzy relationship FR_2 if for each fuzzy database state *FB* of *FS* it holds $FR_1^{FB} \subseteq FR_2^{FB}$.

In a fuzzy ER model, a fuzzy entity FE_1 (or a fuzzy relationship FR_1) is subsumed by another fuzzy entity FE_2 (or another fuzzy relationship FR_2), i.e., for any instance, the membership degree of it to FE_1 (or FR_1) must be less than or equal to the membership degree of it to FE_2 (or FR_2). In a fuzzy ER model, subsumption is the basis for a classification of all the fuzzy entities and fuzzy relationships. If all instances of a more specific entity/relationship are not supposed to be instances of a more general entity/relationship in a fuzzy ER model, then something may be wrong with the fuzzy ER model, since it is forcing an undesired conclusion.

Definition 5.7 (*redundancy*) Afuzzy ER model is redundant, if there is any of the following cases: (i) there is some fuzzy entity/relationship standing for an empty set of instances, i.e., it is inappropriately named; (ii) there are two fuzzy entities FE_1 and FE_2 (or two fuzzy relationships FR_1 and FR_2) have same instances with same membership degrees, that is, FE_1 subsumes FE_2 and FE_2 subsumes FE_1 (similarly for FR_1 and FR_2); (iii) there are two fuzzy entities FE_1 and FE_2 (or two fuzzy relationships FR_1 and FR_2) which do not have subsumption relationship but they have same instances with different membership degrees.

Removing redundancy can reduce the complexity of a fuzzy ER model. In fact, when two fuzzy entities (or two fuzzy relationships) have same instances with same membership degrees, the two fuzzy entities (or two fuzzy relationships) are strictly equivalent. In this case, one of the fuzzy entities (or fuzzy relationships) can be removed and replaced by another. For the case (iii) in Definition 5.7, we take two fuzzy entities for example. Given two fuzzy entities FE_1 and FE_2, and assume that there is an instance *Fe* belonging to FE_1 and FE_2 with different membership degrees $u_{FE1}(Fe) \in [0, 1]$ and $u_{FE2}(Fe) \in [0, 1]$. In this case, which one in FE_1 and FE_2 is the corresponding entity of the instance *Fe* depends on the case: if $u_{FE1}(Fe) > u_{FE2}(Fe)$, then FE_1 is considered as the entity of the instance *Fe* and we say FE_1 fuzzily includes FE_2, else FE_2 is considered as the entity of the instance *Fe* and we say FE_2 fuzzily includes FE_1. Correspondingly, the fuzzy entity which is included by another fuzzy entity is removed, and the fuzzy entity which includes the removed fuzzy entity is retained.

In the fuzzy ER modeling, the designers usually need to check the above reasoning problems by hand. It not only is a complex and time consuming task, but also has some disadvantages such as a low reasoning efficiency and reliability. In the following, we introduce how to reason on the problems of fuzzy ER models by means of the reasoning mechanism of FDLR.

The following theorems allow us to reduce reasoning on a fuzzy ER model to reasoning on the transformed FDLR knowledge base, so that the reasoning tasks of the fuzzy ER model mentioned in Definitions 5.2-5.7 may be checked by detecting the reasoning problems of FDLR.

Theorem 5.2 *Let FS be a fuzzy ER model and φ (FS) be the FDLR knowledge base derived by Definition 5.1. FS is consistent iff: φ (FS) is satisfiable, i.e., φ (FS) admits at least a model.*

Theorem 5.2 is the straightforward consequences of the transformation procedures presented in Definition 5.1 and Definition 5.2, and thus the proof of Theorem 5.2 is omitted here.

Theorem 5.3 *Let FS be a fuzzy ER model, φ (FS) be the FDLR knowledge base derived by Definition 5.1, and FE be a fuzzy entity of FS. FE is satisfiable iff:* $\varphi\ (FS) \not\models \varphi\ (FE) \sqsubseteq \bot$.

Proof "⇒": If *FE* is satisfiable, then there is a legal fuzzy database state *FB* with $FE^{\mathrm{FB}} \neq \varnothing$. By part 1 of Theorem 5.1, $\alpha_{\mathrm{FS}}(FB)$ is a model of $\varphi\ (FS)$ with $FE^{FB} = (\varphi(FE))^{\alpha_{FS}(FB)}$, so $(\varphi(FE))^{\alpha_{FS}(FB)} \neq \varnothing$. That is, $\varphi\ (FS) \not\models \varphi\ (FE) \sqsubseteq \bot$;

"⇐": If $\varphi\ (FS) \not\models \varphi\ (FE) \sqsubseteq \bot$, then $\varphi\ (FE)$ is consistent in $\varphi\ (FS)$, i.e., there is a fuzzy model *FI* of $\varphi\ (FS)$ with $(\varphi(FE))^{\mathrm{FI}} \neq \varnothing$. By part 2 of Theorem 5.1, $\beta_{\mathrm{FS}}(FI)$ is a legal fuzzy database state for *FS* with $(\varphi(D_i))^{\alpha_{FS}(FDBI)}$, so$FE^{\beta_{FS}(FI)} \neq \varnothing$. That is, *FE* is satisfiable. □

Theorem 5.4 *Let FS be a fuzzy ER model, φ (FS) be the FDLR knowledge base derived by Definition 5.1, and FR be a fuzzy relationship of FS. Then FR is satisfiable iff:* $\varphi\ (FS) \not\models \varphi\ (FR) \sqsubseteq \bot$.

Proof "⇒": If *FR* is satisfiable, then there is a legal fuzzy database state *FB* with $FR^{\mathrm{FB}} \neq \varnothing$. By part 1 of Theorem 5.1, $\alpha_{\mathrm{FS}}(FB)$ is a model of $\varphi\ (FS)$ with $FR^{FB} = (\varphi(FR))^{\alpha_{FS}(FB)}$, and so $(\varphi(FR))^{\alpha_{FS}(FB)} \neq \varnothing$. That is, $\varphi\ (FS) \not\models \varphi\ (FR) \sqsubseteq \bot$.

"⇐": If $\varphi\ (FS) \not\models \varphi\ (FR) \sqsubseteq \bot$, then $\varphi\ (FR)$ is consistent in $\varphi\ (FS)$, i.e., there is a fuzzy model *FI* of $\varphi\ (FS)$ with $(\varphi(FR\varphi)^{\mathrm{FI}} \neq \varnothing$. By part 2 of Theorem 5.1,

$\beta_{FS}(FI)$ is a legal fuzzy database state for FS with $(\varphi(FR))^{FI} = FR^{\beta_{FS}(FI)}$, so $FR^{\beta_{FS}(FI)} \neq \varnothing$. That is, FR is satisfiable. □

Theorem 5.5 *Let FS be a fuzzy ER model, φ (FS) be the FDLR knowledge base derived by Definition 5.1, and* FE_1 *and* FE_2 *be two fuzzy entities of FS, then* $FE_1 \preceq_{FS} FE_2$ *iff:* φ $(FS) \vDash \varphi$ $(FE_1) \sqsubseteq \varphi$ (FE_2).

Proof "⇒": If φ $(FS) \nvDash \varphi$ $(FE_1) \sqsubseteq \varphi$ (FE_2), then φ $(FE_1) \sqcap \neg\, \varphi$ (FE_2) is consistent in φ (FS), i.e., there is a fuzzy model FI of φ (FS) with individual $d_i \in (\varphi\ (FE_1))^{FI}$ and $d_i \notin (\varphi\ (FE_2))^{FI}$, for some $d_i \in \Delta^{FI}$. By part 2 of Theorem 5.1, $\beta_{FS}(FI)$ is a legal fuzzy database state for FS with $(\varphi(FE_1))^{FI} = FE_1^{\beta_{FS}(FI)}$ and $(\varphi(FE_2))^{FI} = FE_2^{\beta_{FS}(FI)}$, so $d_i \in FE_1^{\beta_{FS}(FI)}$ and $d_i \notin (\varphi(A_i))^{\alpha_{FS}(FDBI)}$, i.e., FE_1 is not subsumed by FE_2, and thus there is a contradiction. That is, φ $(FS) \vDash \varphi$ $(FE_1) \sqsubseteq \varphi$ (FE_2).

"⇐": If FE_1 is not subsumed by FE_2, then there is a fuzzy instance $Fe \in FE_1^{FB}$ and $Fe \notin FE_2^{FB}$, where FB is a legal fuzzy database state of FS. By part 1 of Theorem 5.1, $\alpha_{FS}(FB)$ is a model of φ (FS) with $(\varphi(FR))^{\alpha_{FS}(FDBI)}$ and $FE_2^{FB} = (\varphi(FE_2))^{\alpha_{FS}(FB)}$, so $Fe \in \Delta^{\alpha_{FS}(FDBI)}$ and $Fe \notin (\varphi(FE_2))^{\alpha_{FS}(FB)}$, i.e., φ $(FS) \nvDash \varphi$ $(FE_1) \sqsubseteq \varphi$ (FE_2), and thus there is contradiction. That is, $FE_1 \preceq_{FS} FE_2$. □

Theorem 5.6 *Let FS be a fuzzy ER model, φ (FS) be the FDLR knowledge base derived by Definition 5.1, and* FR_1 *and* FR_2 *be two fuzzy relationships of FS, then* $FR_1 \preceq_{FS} FR_2$ *iff:* φ $(FS) \vDash \varphi$ $(FR_1) \sqsubseteq \varphi$ (FR_2).

Proof "⇒": If φ $(FS) \nvDash \varphi$ $(FR_1) \sqsubseteq \varphi$ (FR_2), then φ $(FR_1) \sqcap \neg\, \varphi$ (FR_2) is consistent in φ (FS), i.e., there is a fuzzy model FI of φ (FS) with a fuzzy relation instance $Fr \in (\varphi\ (FR_1))^{FI}$ and $Fr \notin (\varphi\ (FR_2))^{FI}$, for some $Fr \in \Delta^{FI}$. By part 2 of Theorem 5.1, $\beta_{FS}(FI)$ is a legal fuzzy database state for FS with $(\varphi(A_i))^{\alpha_{FS}(FDBI)}$ and $(\varphi(FR_2))^{FI} = FR_2^{\beta_{FS}(FI)}$, so $Fr \in FR_1^{\beta_{FS}(FI)}$ and $Fr \notin (\varphi(FR))^{\alpha_{FS}(FDBI)}$, i.e., FR_1 is not subsumed by FR_2, and thus there is a contradiction. That is, φ $(FS) \vDash \varphi$ $(FR_1) \sqsubseteq \varphi$ (FR_2).

"⇐": If FR_1 is not subsumed by FR_2, then there is a fuzzy relation instance $Fr \in FR_1^{FB}$ and $Fr \notin FR_2^{FB}$, where FB is a legal fuzzy database state. By part 1 of Theorem 5.1, $\alpha_{FS}(FB)$ is a model of $\varphi(FS)$ with ${}^{\alpha_{FS}(FDBI)}$ and $FR_2^{FB} = (\varphi(FR_2))^{\alpha_{FS}(FB)}$, so $Fr \in (\varphi(FR_1))^{\alpha_{FS}(FB)}$ and $Fr \notin (\varphi(FR_2))^{\alpha_{FS}(FB)}$, i.e., φ $(FS) \nvDash \varphi$ $(FR_1) \sqsubseteq \varphi$ (FR_2), and thus there is a contradiction. That is, $FR_1 \preceq_{FS} FR_2$. □

Theorem 5.7 *Let FS be a fuzzy ER model, φ (FS) be the FDLR knowledge base derived by Definition 5.1, FE,* FE_1 *and* FE_2 *be three fuzzy entities of FS, and FR,* FR_1 *and* FR_2 *be three fuzzy relationships of FS. Then FS is redundant if and only if at least one of the following conditions is satisfied:*

- φ *(FS)* $\models$ φ *(FE)* $\sqsubseteq \bot$*, or* φ *(FS)* $\models$ φ *(FR)* $\sqsubseteq \bot$*;*
- φ *(FS)* $\models$ φ *(FE_1)* $\sqsubseteq$ φ *(FE_2) and* φ *(FS)* $\models$ φ *(FE_2)* $\sqsubseteq$ φ *(FE_1), or* φ *(FS)* $\models$ φ *(FR_1)* $\sqsubseteq$ φ *(FR_2) and* φ *(FS)* $\models$ φ*(FR_2)* $\sqsubseteq$ φ *(FR_1);*
- φ *(FS)* $\not\models$ φ *(FE_1)* $\sqsubseteq$ φ *(FE_2),* φ *(FS)* $\not\models$ φ *(FE_2)* $\sqsubseteq$ φ *(FE_1), and* φ *(FE_1) and* φ *(FE_2) contain same instances, or* φ*(FS)* $\not\models$ φ *(FR_1)* $\sqsubseteq$ φ *(FR_2),* φ *(FS)* $\not\models$ φ *(FR_2)* $\sqsubseteq$ φ *(FR_1), and* φ *(FR_1) and* φ *(FR_2) contain same instances.*

Proof "$\Rightarrow$": Let *FS* be redundant, according to Definition 5.7, there is some fuzzy entity/relationship standing for an empty set of instances, by Theorems 5.3 and 5.4, then φ (*FS*) $\models$ φ(*FE*) $\sqsubseteq \bot$ and φ (*FS*) $\models$ φ (*FR*) $\sqsubseteq \bot$; or there are fuzzy entities FE_1 and FE_2 with $FE_1 \leq_{FS} FE_2$ and $FE_2 \leq_{FS} FE_1$, by Theorem 5.5, then φ (*FS*) $\models$ φ (FE_1) $\sqsubseteq$ φ (FE_2) and φ (*FS*) $\models$ φ (FE_2) $\sqsubseteq$ φ (FE_1); or there are fuzzy relationships FR_1 and FR_2 with $FR_1 \leq_{FS} FR_2$ and $FR_2 \leq_{FS} FR_1$, by Theorem 5.6, then φ(*FS*) $\models$ φ (FR_1) $\sqsubseteq$ φ (FR_2) and φ (*FS*) $\models$ φ (FR_2) $\sqsubseteq$ φ (FR_1); For the third case, the constraint that two entities FE_1 and FE_2 do not have subsumption relationship in Definition 5.7 can be reasoned by checking whether φ (*FS*) $\not\models$ φ (FE_1) $\sqsubseteq$ φ (FE_2) and φ (*FS*) $\not\models$ φ (FE_2) $\sqsubseteq$ φ (FE_1) (similarly for fuzzy relation ships FR_1 and FR_2). Furthermore, whether two entities contain same instances can also be checked by the retrieval reasoning service of FDLR. Given a FDLR knowledge base, a fuzzy concept *C* and $n' \in [0, 1]$, the retrieval reasoning service of FDLR can return all the individuals *d* satisfying the condition $C(d) \geq n'$. Therefore, given two FDLR concepts φ (FE_1) and φ (FE_2), and the $n' \in [0, 1]$ which may be a positive infinite small value or be specified by the designers in the fuzzy ER modeling activities, two sets of individuals respectively belonging to φ (FE_1) and φ (FE_2) can be created by the retrieval reasoning service. Then, whether the two sets of individuals are same can be easily detected by checking clashes among individual (in)equality assertions. The case "$\Leftarrow$" can also be proved following the similar procedures above. □

On the basis of the theorems, reasoning on fuzzy ER models can be reduced to reasoning on FDLR knowledge bases, so that the reasoning tasks of fuzzy ER models can be detected by checking the reasoning problems of FDLR. Moreover, by exploiting the correspondences between fuzzy ER models and FDLR, the richness of DL constructs makes it possible to add to the fuzzy ER models several features and modeling primitives that are currently missing. Notable examples are the possibility to refine properties of entities along ISA hierarchies, to specify arbitrary Boolean combinations of entities, and so on.

5.2.1.2 Fuzzy Description Logic Extraction from Fuzzy UML Model

In order to extract fuzzy Description Logic (DL) knowledge bases from fuzzy UML models, the fuzzy DL FDLR introduced in Sect. 4.2.2 was used to represent and reason on fuzzy UML models in Ma et al. (2011a). In the following, we introduce how to extract fuzzy DL FDLR knowledge bases from fuzzy UML models (Ma et al. 2011a).

Based on the formalization of fuzzy UML models introduced in Sect. 3.2.2 and the fuzzy DL FDLR introduced in Sect. 4.2.2, let $\mathcal{F}_{UML} = (\mathcal{Q}_{\mathcal{F}}, \preccurlyeq_{\mathcal{F}}, att_{\mathcal{F}}, agg_{\mathcal{F}}, dep_{\mathcal{F}}, ass_{\mathcal{F}}, card_{\mathcal{F}}, mult_{\mathcal{F}}, mult'_{\mathcal{F}})$ be a fuzzy UML model, the FDLR knowledge base $KB = \varphi(\mathcal{F}_{UML}) = (FA, FP, FT)$ is defined by function φ shown in Table 5.1 (here, for each association FS of arity n in the fuzzy UML model, we denote with μ_S a mapping from the set of fuzzy UML-roles associated with FS to the integers 1, …, n, for example, $\mu_S(FR_1) = 1$, …, $\mu_S(FR_n) = n$.):

The correctness of translation function φ in Table 5.1 can be proved following the similar procedures in the previous Theorem 5.1. Note that, in the extraction process, the dynamic attributes of a fuzzy class in a fuzzy UML model as mentioned in Sect. 3.2.2 are not considered. In the following, we illustrate the extraction procedure in Table 5.1 with an example.

Example 5.2 Given the fuzzy UML model $\mathcal{F}_{\mathrm{UML1}}$ in Fig. 3.8 in Sect. 3.2.2, the extracted FDLR knowledge base from $\mathcal{F}_{\mathrm{UML1}}$ by using the rules in Table 5.1 is shown in Fig. 5.5. Here, the fuzzy UML-roles in $\mathcal{F}_{\mathrm{UML1}}$ can be mapped to the integers as follows: $\mu_S(Uby) = 1$, $\mu_S(Uof) = 2$. Here $n = 2$ denotes that the arity of association *Use* is 2.

The formalization in FDLR of fuzzy UML models provides the ability to reason on fuzzy UML models. This represents a significant improvement and it is the first step towards developing the modeling tools that offer an automated reasoning support to the designers in the modeling activity. The reasoning of fuzzy UML models with the extracted FDLR knowledge bases is similar to the reasoning of fuzzy ER models as has been introduced in the previous parts, and thus will not be included here. As for how to implement the reasoning systems, two methods can be considered: (i) Encoding a fuzzy UML model in a FDLR knowledge base allows the designs for exploiting the reasoning services offered by DL reasoners. In particular, the existence of the decidable reasoning procedures for FDLR makes it more possible to exploit a FDLR reasoner. (ii) Establishing the correspondences between fuzzy UML models and FDLR, then extending the existing fuzzy DL reasoners for checking the relevant reasoning problems of fuzzy UML models.

5.2.2 Fuzzy Description Logic Extraction from Fuzzy Database Models

Over the years, there has been an enormous growth in the amount of data stored in database systems. Also, in order for databases to handle fuzzy information, the representation of fuzzy information has been introduced into databases and fuzzy databases (including fuzzy relational databases and fuzzy object-oriented

Table 5.1 Extracting rules of a FDLR knowledge base from a fuzzy UML model

Fuzzy UML model $\mathcal{F}_{UML}$	FDLR $KB = \varphi\ (\mathcal{F}_{UML}) = (FA, FP, FT)$
Alphabet $\mathcal{D}_{\mathcal{F}}$	*The set FA of atomic concepts of KB*
Each class $FC \in FC_{\mathcal{F}}$	An atomic concept $\varphi(FC)$
Each datatype $FT \in FT_{\mathcal{F}}$	An atomic concept $\varphi(FT)$
Each association class $FS_c \in FC_{\mathcal{F}}$	An atomic concept $\varphi(FS_c)$
Alphabet $\mathcal{D}_{\mathcal{F}}$	*The set FP of atomic relations of KB*
Each association $FS \in FS_{\mathcal{F}}$	An atomic *n*-ary relation $\varphi(FS)$
Each role symbol $FR \in FR_{\mathcal{F}}$	An atomic binary relation $\varphi(FR)$
Each attribute $FA \in FA_{\mathcal{F}}$	An atomic binary relation $\varphi(FA)$
Constructs	*The set FT of inclusion axioms of KB*
Each class $FC \in FC_{\mathcal{F}}$ such that $att_{\mathcal{F}}(FC) = [\ldots, FA_i\ [m_i, n_i]: FT_i, \ldots]$, where $FA_i \in FA_{\mathcal{F}}$, $FT_i \in FT_{\mathcal{F}}$, $[m_i, n_i]$ is *multiplicity*, $m_i = mult_{min}(FT_i, FC, FA_i)$, $n_i = mult_{max}(FT_i, FC, FA_i)$, $i \in \{1\ldots k\}$	$\varphi(FC) \sqsubseteq \ldots \sqcap (\forall[\$1]\ (\varphi(FA_i) \sqcap (\$2/2 : \varphi(FT_i)))) \sqcap \geq m_i\ [\$1]\ \varphi(FA_i) \sqcap \leq n_i\ [\$1]\ \varphi(FA_i) \sqcap \ldots$
Each association $FS \in FS_{\mathcal{F}}$ such that $ass_{\mathcal{F}}(FS) = [\ldots, FR_i\ (p_i, q_i) : FC_n, \ldots]$, where $FR_i \in FR_{\mathcal{F}}$, $FC_i \in FC_{\mathcal{F}}$, (p_i, q_i) is *cardinality constraints*, $p_i = card_{min}(FC_i, FS, FR_i)$, $q_i = card_{max}(FC_i, FS, FR_i)$, and $i \in \{1\ldots n\}$	$\varphi(FS) \sqsubseteq \ldots \sqcap (\$\mu_S(FR_i)/n : FC_i) \sqcap \ldots$ The association *FS* with *cardinality constraints* (p_i, q_i), if $p_i \neq 0$ and $q_i \neq \infty$, then creating axiom: $\varphi(FC_i) \sqsubseteq\ \geq p_i\ [\$\mu_S(FR_i)]\ \varphi(FS) \sqcap \leq q_i\ [\$\mu_S(FR_i)]\ \varphi(FS)$
Each association class $FS_c \in FC_{\mathcal{F}}$ such that: (i) $att_{\mathcal{F}}(FS_c) = [\ldots, FA_i\ [m_i, n_i]: FT_i, \ldots]$, where $FA_i \in FA_{\mathcal{F}}$, $FT_i \in FT_{\mathcal{F}}$, $[m_i, n_i]$ is *multiplicity*, $m_i = mult_{min}(FT_i, FC, FA_i)$, $n_i = mult_{max}(FT_i, FC, FA_i)$, and $i \in \{1\ldots k\}$. (ii) $ass_{\mathcal{F}}(FS_c) = [FR_1\ (p_1, q_1): FC_1, \ldots, FR_n\ (p_n, q_n): FC_n]$, $FR_j \in FR_{\mathcal{F}}$, $FC_j \in FC_{\mathcal{F}}$, (p_j, q_j) is *cardinality constraints*, $p_j = card_{min}(FC_j, FS, FR_j)$, $q_j = card_{max}(FC_j, FS, FR_j)$, $j \in \{1\ldots n\}$	$\varphi(FS_c) \sqsubseteq \ldots \sqcap (\forall[\$1]\ (\varphi(FA_i) \sqcap (\$2/2 : \varphi(FT_i)))) \sqcap \geq m_i\ [\$1]\ \varphi(FA_i) \sqcap \leq n_i\ [\$1]\ \varphi(FA_i) \sqcap \ldots$ $\varphi(FS_c) \sqsubseteq (\forall[\$1]\ (\varphi(FR_1) \sqcap (\$2/2 : \varphi(FC_1)))) \sqcap = 1\ [\$1]\ \varphi(FR_1) \sqcap \ldots \sqcap (\forall[\$1]\ (\varphi(FR_n) \sqcap (\$2/2 : \varphi(FC_n)))) \sqcap = 1\ [\$1]\ \varphi(FR_n)$ The association class FS_c with *cardinality constraints* (p_j, q_j), if $p_j \neq 0$ and $q_j \neq \infty$, then creating axiom: $\varphi(FC_j) \sqsubseteq\ \geq p_j\ [2]\ (\varphi(FR_j) \sqcap (\$1/2 : \varphi(FS_c))) \sqcap \leq q_j\ [2]\ (\varphi(FR_j) \sqcap (\$1/2 : \varphi(FS_c)))$
For classes $FC_1, FC_2 \in FC_{\mathcal{F}}$ such that $\preccurlyeq_{\mathcal{F}}(FC_1) = FC_2$	$\varphi(FC_2) \sqsubseteq \varphi(FC_1)$
Each a class $FC \in FC_{\mathcal{F}}$ generalizing *n* classes $FC_1, \ldots, FC_n \in FC_{\mathcal{F}}$ with the disjoint and complete constraints, i.e., $\preccurlyeq_{\mathcal{F}}(FC) = FC_1 \times FC_2 \times \ldots \times FC_n$	$\varphi(FC_i) \sqsubseteq \varphi(FC)$ $\Delta^{\alpha_{FS}(FJ)}$ Δ_{fid}
Each aggregation $FG \in FG_{\mathcal{F}}$ such that $agg_{\mathcal{F}}(FG) = FC \times (FC_1 \cup \ldots \cup FC_m)$, where $FC \in FC_{\mathcal{F}}$ is an aggregate class, $FC_i \in FC_{\mathcal{F}}$ is a constituent class, and $i \in \{1, \ldots, m\}$. The *multiplicity* of the *i*th constituent class FC_i is $[m_i', n_i']$, $m_i' = mult'_{min}(FC_i, FC)$, and $n_i' = mult'_{max}(FC_i, FC)$	Add *m* symbols <is_part_of_g_1, …, is_part_of_g_m>, which are mapped into *m* atomic binary relations φ(is_part_of_g_m), …, φ(is_part_of_g_m) Create axioms, $i \in \{1\ldots m\}$: φ (is_part_of_g_m) $\sqsubseteq (\$1/2 : \varphi\ (FC_i)) \sqcap (\$2/2 : \varphi\ (FC))$ $\varphi(FC_i) \sqsubseteq\ \geq m_i'\ [1]\ \varphi$(is_part_of_$g_i$) $\sqcap \leq n_i'\ [1]\ \varphi$(is_part_of_$g_i$)
Each pair of classes $FC_1, FC_2 \in FC_{\mathcal{F}}$ such that $dep_{\mathcal{F}} \subseteq FC_1 \times FC_2$	The dependency between FC_1 and FC_2 is only related to the classes themselves, and the fuzzy DL FDLR does not support this type of property, which can be indicated explicitly by the designers

The extracted FDLR knowledge base $\varphi(\Phi_{UML1}) = (FA, FP, FT)$ from the fuzzy UML model Φ_{UML1} in Fig. 3.8 in Sect. 3.2.2 is as follows:
FA = {New-Computer, Computer, Monitor, New CPU box, Keyboard, Employee, Young-Employee, Old-Employee, Use, …}
FP = {ComID, Brand, ID, Name, FUZZY Age, FUZZY Email, Uby, Uof, is_part_of_ g_1, is_part_of_ g_2, is_part_of_g_3, …}
FT = {
New-Computer ⊆ (∀ [$1] (ComID ∩ ($2/2:String))) ∩ = 1 [$1] ComID ∩ (∀ [$1] (Brand ∩ ($2/2:String))) ∩ = 1 [$1] Brand ∩ …
Old-Employee ⊆ (∀ [$1] (ID ∩ ($2/2:String))) ∩ = 1 [$1] ID ∩ (∀ [$1] (Name ∩ ($2/2:String))) ∩ ≥ 1 [$1] Name ∩ …
Use ⊆ (∀ [$1] (Date ∩ ($2/2:String))) ∩ = 1 [$1] Date ∩ (∀ [$1] (Uby ∩ ($2/2: Old-Employee))) ∩ = 1 [$1] Uby ∩ (∀ [$1] (Uof ∩ ($2/2: New-Computer))) ∩ = 1 [$1] Uof
Old-Employee ⊆ = 1 [$2] (Uby ∩ ($1/2: Use))
New-Computer ⊆ ≥1 [$2] (Uof ∩ ($1/2: Use)) ∩ ≤2 [$2] (Uof ∩ ($1/2:Use))
is_part_of_ g_1 ⊆ ($1/2 : Monitor) ∩ ($2/2 : New-Computer)
is_part_of_ g_2 ⊆ ($1/2 : New CPU box) ∩ ($2/2 : New-Computer)
is_part_of_ g_3 ⊆ ($1/2 : Keyboard) ∩ ($2/2 : New-Computer)
Monitor ⊆ = 1 [$1] is_part_of_$g_1$
New CPU box ⊆ = 1 [$1] is_part_of_$g_2$
Keyboard ⊆ ≥ 1 [$1] is_part_of_$g_3$
Young-Employee ⊆ Employee
Old-Employee ⊆ Employee
Old-Employee ⊆ ¬ Young-Employee
Employee ⊆ Young-Employee ∪ Old-Employee
…
}

Fig. 5.5 The FDLR knowledge base $\varphi(\mathcal{F}_{UML1})$ extracted from the fuzzy UML model $\mathcal{F}_{UML1}$ in Fig. 3.8 in Sect.3.2.2

databases) have been extensively investigated as have introduced in Chap. 3. Therefore, there is an urgent need for techniques and tools that can extract useful knowledge from fuzzy databases. To this end, how to extract knowledge from fuzzy relational databases and fuzzy object-oriented databases were investigated in Ma et al. (2011b) and Zhang et al. (2012a, b), and also the extracted knowledge was used to reason about the fuzzy relational databases and fuzzy object-oriented databases. In this chapter, fuzzy Description Logic (DL) extraction from fuzzy relational databases and fuzzy object-oriented databases will be introduced.

5.2.2.1 Fuzzy Description Logic Extraction from Fuzzy Relational Databases

In Ma et al. (2011b), in order to represent the extracted knowledge, a fuzzy DL called *f-ALCNI* is introduced after considering the characteristics of fuzzy relational databases (FRDB). On this basis, an approach which can extract the *f-ALCNI* knowledge base from the FRDB, i.e., transform the FRDB (including schema and data information) into the *f-ALCNI* knowledge base (i.e., TBox and ABox), was proposed. Furthermore, a prototype extraction tool called *FRDB2DL* was designed and implemented. In addition, to further demonstrate how the DLs are useful for improving some database applications, based on the extracted knowledge, how to reason about FRDB (e.g., consistency, satisfiability, subsumption, equivalence, and redundancy) by means of the reasoning mechanism of *f-ALCNI* was investigated. In the following, we introduce how to extract *f-ALCNI* knowledge bases from fuzzy relational databases (Ma et al. 2011b).

(A) *Extracting f-ALCNI Knowledge Bases from Fuzzy Relational Databases*

As introduced in Sect. 3.3.1, a fuzzy relational database *FRDB* constists of a set of fuzzy relational schemas *FS* (i.e., a set of $FR(A_1/D_1, A_2/D_2, \ldots, A_n/D_n, \mu_{FR}/D_{FR})$) and a set of fuzzy relations *FR*. Therefore, to extract *f-ALCNI* knowledge base from a fuzzy relational database, two steps need to be done, i.e., extracting *f-ALCNI* TBox from the schema information of the fuzzy relational database (i.e., *fuzzy relational schemas*), and extracting *f-ALCNI* ABox from the data information of the fuzzy relational database (i.e., *fuzzy relation instances*).

A fuzzy relational schema specifies a set of integrity constraints, which restrict the data that can be stored in the database. Given a fuzzy relational schema $FR(A_1/D_1, A_2/D_2, \ldots, A_n/D_n, \mu_{FR}/D_{FR})$, in order to establish the precise conceptual correspondences between fuzzy relational databases and fuzzy DLs, based on the appearance of primary key (PK) and foreign keys (FK) of *FR*, the relation *FR* is classified as one of the following:

(a) The relation *FR* has not foreign keys.
(b) The relation *FR* has exactly one foreign key that references the primary key of relation *FR′*, i.e., $|FR[\text{FK}]| = 1$, then:
    - If the foreign key *FR*[FK] is one of the candidate keys of *FR*, then the relationship between *FR* and *FR′* may be a *1–1*;
    - If the foreign key *FR*[FK] is not one of the candidate keys of *FR*, then the relationship between *FR* and *FR′* is a *many–1*.
(c) The relation *FR* has more than one foreign keys that respectively reference the primary key PK_i of relation FR_i, $i \in \{1,\ldots,n\}$, and $n \geq 2$, i.e., $|FR[\text{FK}]| > 1$, then *FR* is a *many–many* relationship relation. Moreover, the relation *FR* maybe has its own attributes.
(d) The relation *FR* has a primary key PK that is also a foreign key FK, and FK references the primary key of relation *FR'*, then an *inheritance* relationship exists, i.e., *FR* "*is-a*" *FR'*.

Based on the classifications of relations defined above, the following Definition 5.8 gives the formal approach for transforming a set of fuzzy relational schemas into an *f-ALCNI* TBox. Starting with the construction of the atomic concepts and roles, the approach induces a set of axioms from the fuzzy relational schemas, which forms the body of *f-ALCNI* TBox.

Definition 5.8 Given a set of fuzzy relational schemas $FR(A_1/D_1, A_2/D_2, \ldots, A_n/D_n, \mu_{FR}/D_{FR})$ of a fuzzy relational database as mentioned in Definition 3.7 in Sect. 3.3.1. The *f-ALCNI* TBox φ $(FR) = (FA, FP, FT)$ is obtained by defining a transformation function φ as follows:

(1) The set *FA* of atomic fuzzy concepts of φ (FR) contains the following elements:

- Each fuzzy relation *FR* is mapped into an atomic fuzzy concept φ (FR);
- Each fuzzy domain symbol D_i $(D_{FR} \in D_i)$ is mapped into an atomic fuzzy concept $\varphi(D_i)$.

(2) The set *FP* of atomic fuzzy roles of φ (FR) contains the following elements:

- Each fuzzy attribute A_i $(\mu_{FR} \in A_i)$ is mapped into a fuzzy atomic role φ (A_i);
- Each foreign key attribute A_i is mapped into a pair of inverse fuzzy roles φ (A_i) and $(\varphi(A_i))^-$.

(3) The set *FT* of fuzzy inclusion axioms of φ (FR) contains the following elements:

- *Case 1 FR* has not foreign keys. Creating the following fuzzy axiom: $\varphi\ (FR) \sqsubseteq \forall\varphi\ (A_1).\ \varphi\ (D_1) \sqcap \ldots \sqcap \forall\varphi\ (A_n).\ \varphi\ (D_n) \sqcap \forall\varphi\ (\mu_{FR}).\ \varphi\ (D_{FR}) \sqcap (= 1\ \varphi(A_1)) \sqcap \ldots \sqcap (= 1\ \varphi\ (A_n)) \sqcap (= 1\ \varphi\ (\mu_{FR}))$;
- *Case 2 FR* has exactly one foreign key attribute *FK* that references the primary key of *FR*′. Creating the following fuzzy axioms:

Case 2.1 FR and *FR*′ is a *1-1* relationship. Creating the fuzzy axioms:
$\varphi\ (FR) \sqsubseteq \forall\varphi\ (A_1).\ \varphi\ (D_1) \sqcap \ldots \sqcap \forall\varphi\ (A_n).\ \varphi\ (D_n) \sqcap \forall\varphi\ (\mu_{FR}).\ \varphi\ (D_{FR}) \sqcap (= 1\ \varphi(A_1)) \sqcap \ldots \sqcap (= 1\ \varphi\ (A_n)) \sqcap (= 1\ \varphi\ (\mu_{FR}))$;
$\varphi\ (FR) \sqsubseteq \forall\varphi\ (FK).\ \varphi\ (FR') \sqcap (= 1\ \varphi\ (FK))$;
$\varphi\ (FR') \sqsubseteq \forall(\varphi\ (FK))^-.\ \varphi\ (FR) \sqcap (= 1\ (\varphi\ (FK))^-)$.
Case 2.2 FR and *FR*′ is a *many-1* relationship. Creating the fuzzy axioms:
$\varphi\ (FR) \sqsubseteq \forall\varphi\ (A_1).\ \varphi\ (D_1) \sqcap \ldots \sqcap \forall\varphi\ (A_n).\ \varphi\ (D_n) \sqcap \forall\varphi\ (\mu_{FR}).\ \varphi\ (D_{FR}) \sqcap (= 1\ \varphi(A_1)) \sqcap \ldots \sqcap (= 1\ \varphi\ (A_n)) \sqcap (= 1\ \varphi\ (\mu_{FR}))$;
$\varphi\ (FR) \sqsubseteq \forall\varphi\ (FK).\ \varphi\ (FR') \sqcap (= 1\ \varphi\ (FK))$;
$\varphi\ (FR') \sqsubseteq \forall(\varphi\ (FK))^-.\varphi\ (FR) \sqcap (\geq 1(\varphi\ (FK))^-)$.

- *Case 3 FR* has more than one foreign key FK_i that respectively reference the primary key of relation FR_i, $i \in \{1, \ldots, n\}$, and $n \geq 2$. Creating the following fuzzy axioms:

- $\varphi\ (FR) \sqsubseteq \forall\varphi\ (A_1).\ \varphi\ (D_1) \sqcap \ldots \sqcap \forall\varphi\ (A_n).\ \varphi\ (D_n) \sqcap \forall\varphi\ (\mu_{FR}).\ \varphi\ (D_{FR}) \sqcap (= 1\ \varphi(A_1)) \sqcap \ldots \sqcap (= 1\ \varphi\ (A_n)) \sqcap (= 1\ \varphi\ (\mu_{FR}))$;
- $\varphi\ (FR) \sqsubseteq \forall\varphi\ (FK_1).\ \varphi\ (FR_1) \sqcap \ldots \sqcap \forall\varphi\ (FK_n).\ \varphi\ (FR_n) \sqcap (= 1\ \varphi\ (FK_1)) \sqcap \ldots \sqcap (= 1\ \varphi\ (FK_n))$;
- $\varphi\ (FR_i) \sqsubseteq \forall(\varphi\ (FK_i))^-.\ \varphi\ (FR) \sqcap (\geq 1\ (\varphi\ (FK_i))^-)$.

- *Case 4 FR inherits FR*′, i.e., *FR "is-a" FR*′. Creating the following fuzzy axiom:

- $\varphi\ (FR) \sqsubseteq \varphi\ (FR')$.

The following introduces how to extract the *data information* of a fuzzy relational database by establishing the mappings from *the fuzzy relations w.r.t. fuzzy relational schemas* to *the f-ALCNI ABox*, i.e., from the database assertions to the *f-ALCNI* ABox assertions. Firstly, let us briefly sketch the assertional formalisms used in the two cases.

In the *f-ALCNI* ABox, one describes a specific state of affairs of an application domain in terms of concepts and roles. Being similar to the fuzzy DL *f-ALC* (Straccia 1998) as recalled in Sect. 4.2, the assertions used in an *f-ALCNI* ABox contain concept assertions $< C(d) \bowtie n >$, and role assertions $< R(d_1, d_2) \bowtie n >$, where $\bowtie \in \{\geq, >, \leq, <\}$, $n \in [0, 1]$.

A fuzzy relational database describes the real world by means of a set of fuzzy relations (i.e., a set of tuples, including individuals, values, and their mutual relationships), which can be considered as a finite set of assertions. The assertional formalisms of the fuzzy relations with respect to a set of fuzzy relational schemas specify that: (i) An individual *I* (i.e., a tuple that is identified uniquely by primary key) is an instance of a fuzzy relation *FR* with membership degree of *n* by means of fuzzy assertion *I*: *FR*: *n*, where $n \in [0, 1]$; (ii) The value associated with the individual (tuple) *I* by means of fuzzy assertion $I: [A_1{:}V_1{:}n_1, \ldots, A_k{:}V_k{:}n_k]$, where A_i denotes the attribute of *FR*, $V_i \in \mathrm{Dom}(A_i)$ is the attribute value, $n_i \in [0, 1]$ denotes the membership degree, and $i \in \{1 \ldots k\}$. Here, since the value of an attribute A_i may be a possibility distribution, for simplicity, $V_i : n_i$ only denotes one element of the possibility distribution; (iii) Since the fuzzy inheritance relationships in a fuzzy relational database can be assessed by utilizing the inclusion degree of individuals to the fuzzy relation, the assertional formalism of fuzzy inheritance relationships can be re-expressed by the above assertional formalism of individuals to the fuzzy relation.

Based on the discussion above, given a fuzzy relational database $FRDB = < FS, FR >$, the mappings from the fuzzy relations *FR* w.r.t. fuzzy relational schemas *FS* to the *f-ALCNI* ABox can be established as follows:

- Each individual *I* is mapped into an *f-ALCNI* individual *I*;
- Each fuzzy assertion *I*: *FR*: *n* corresponds to the ABox assertion $< FR(I) \bowtie' n >$, where $\bowtie' \in \{\geq, \leq\}$;
- Each fuzzy assertion $I : [A_1{:}V_1{:}n_1,\ldots,A_k{:}V_k{:}n_k]$ corresponds to the ABox assertions $< A_i(I, V_i) \bowtie' n_i >$, where A_i is viewed as a fuzzy binary role with membership degree of n_i, $i \in \{1,\ldots,k\}$, $\bowtie' \in \{\geq, \leq\}$.

Below we discuss the effectiveness of the translation from fuzzy relational database models to *f-ALCNI*, which can be sanctioned by establishing the mappings between the fuzzy relational database instance w.r.t. fuzzy relational schemas (i.e., a set of fuzzy relations) and the model of extracted *f-ALCNI* knowledge base (see Theorem 5.8).

Theorem 5.8 *Given a fuzzy relational database* $FRDB = < FS, FR >$, *where FS is the set of fuzzy relational schemas (i.e., a set of* $FR(A_1/D_1, A_2/D_2, \ldots, A_n/D_n, \mu_{FR}/D_{FR})$*) and FR is the set of fuzzy relations, and* φ *(FRDB) is the extracted f-ALCNI knowledge base. Then:*

- *There exists the mapping α_{FS} from the fuzzy relational database instance FDBI (i.e., the set of fuzzy relations FR) w.r.t. FS to the model of φ (FRDB) such that, for each fuzzy relational database instance FDBI, there is α_{FS}(FDBI) which is a model of φ (FRDB).*
- *There exists the mapping β_{FS} from the model of φ (FRDB) to the fuzzy relational database instance such that, for each model FI of φ (FRDB), there is β_{FS}(FI) which is a fuzzy relational database instance on FS.*

Proof In the following, we prove the first part of Theorem 5.8. Note that a fuzzy relational database instance *FDBI* is an interpretation (function) that maps each fuzzy relation *FR* into a set FR^{FDBI} of total functions from the set of attributes of *FR* to Δ^{FDBI}. Formally, $FDBI = (\Delta^{FDBI}, \bullet^{FDBI})$, where Δ^{FDBI} is a non-empty finite set and $\bullet^{FDBI}$ is a function that maps each domain D_i into the corresponding basic domain D_i^{FDBI}, each relation *FR* into a set FR^{FDBI} of A_i-labeled tuples over Δ^{FDBI}, and each attribute A_i into a set $A_i^{FDBI} \sqsubseteq \Delta^{FDBI} \times D_i^{FDBI}$. Therefore, given a fuzzy relational database instance *FDBI* over *FS*, we define the fuzzy interpretation $\alpha_{FS}(FDBI)$ of φ (*FRDB*) as follows:

- The domain $\Delta^{\alpha_{FS}(FDBI)}$ of $\alpha_{FS}(FDBI)$ of φ (*FRDB*), which can be constituted by values of *FDBI*, is the union of Δ^{FDBI}, the set of tuples, and the set of domain values that appear in *FDBI* as a value of some attribute.
- The atomic fuzzy concepts of set *FA* and atomic fuzzy relations of set *FP* in Definition 5.8 are interpreted as follows:

For each symbol $X \in D_i \sqcup A_i \sqcup R_i$, $(\varphi(X))^{\alpha_{FS}(FDBI)} = X^{FDBI}$;

For each FK attribute A_i (i.e., FK_i), $\Delta^{\alpha_{FS}(FDBI)} = \{(FR, FR_i) \in \Delta^{\alpha_{FS}(FDBI)} \times (\varphi(FR))^{\alpha_{FS}(FDBI)} | \; FR \in FR^{FDBI}, FR_i \in FR_i^{FDBI}\}$;

For each non-FK attribute A_i, ${}^{\alpha_{FS}(FDBI)} = \{(FR, D_i) \in \Delta^{\alpha_{FS}(FDBI)} \times \Delta^{\alpha_{FS}(FDBI)} | \; FR \in FR^{FDBI}, D_i \in D_i^{FDBI}\}$;

To prove the first part of Theorem 5.8 it is sufficient to show that $\alpha_{FS}(FDBI)$ satisfies every fuzzy axiom in Definition 5.8. Based on definition of $\alpha_{FS}(FDBI)$ above, the following proves the first part of Theorem 5.8.

(i) If $FR(A_1/D_1, A_2/D_2, \ldots, A_n/D_n, \mu_{FR}/D_{FR})$ has not foreign key attributes. Firstly, for an instance $FR \in (\varphi(FR))^{\alpha_{FS}(FDBI)}$, based on definition of $\alpha_{FS}(FDBI)$ above, we have $FR \in FR^{FDBI}$. That is, there is a non-empty A_i-labeled tuple in relation *FR*, by definition of $\alpha_{FS}(FDBI)$ above, there is one element $a_i \in A_i^{FDBI} = \Delta^{\alpha_{FS}(FDBI)}$ whose first component is *FR*, and second component of a_i is an element $D_i \in D_i^{FDBI} = (\varphi(D_i))^{\alpha_{FS}(FDBI)}$, i.e., $(FR, D_i) \in (\varphi(A_i))^{\alpha_{FS}(FDBI)}$. Moreover, based on the characteristics of fuzzy relational databases presented in Sect. 3.3.1, all fuzzy roles φ (A_i) corresponding to fuzzy attributes A_i are functional. Therefore, for an instance $FR \in (\varphi(FR))^{\alpha_{FS}(FDBI)}$, we can show that for each A_i there is exactly one element $D_i \in \Delta^{\alpha_{FS}(FDBI)}$ such that $(FR, D_i) \in (\varphi(A_i))^{\alpha_{FS}(FDBI)}$. That is, $\alpha_{FS}(FDBI)$ satisfies the fuzzy axioms φ (*FR*) $\sqsubseteq \forall\varphi$ (A_1). φ (D_1) $\sqcap \ldots \sqcap \forall\varphi$

(A_n). φ (D_n) ⊓ ∀φ (μ_{FR}). φ (D_{FR}) ⊓ (= 1 φ (A_1)) ⊓ … ⊓ (= 1 φ (A_n)) ⊓ (= 1 φ (μ_{FR})) in Definition 5.8.

(ii) If $FR(A_1/D_1, A_2/D_2, \ldots, A_n/D_n, \mu_{FR}/D_{FR})$ has foreign key attributes FK_i. Firstly, for an instance $FR \in (\varphi(FR))^{\alpha_{FS}(FDBI)}$, by definition of $\alpha_{FS}(FDBI)$ above, we have $FR \in FR^{FDBI}$. Then, from Sect. 3.3.1, we know that all tuples in a relation are of the form $[A_1{:}D_1,\ldots, A_n{:}D_n, \mu_{FR}{:}D_{FR}]$, thus FR is the following FK_i-labeled tuples over Δ^{FDBI} of the form $[FK_1{:}\ FR_1,\ldots, FK_n{:}\ FR_n]$, where the non-FK attributes are omitted, $FR_i \in FR_i^{FDBI}$, and $i \in \{1,\ldots,n\}$. Furthermore, by definition of $\alpha_{FS}(FDBI)$ above, we have $FR_i \in \varphi\ (FR_i)^{\alpha_{FS}(FDBI)} \in (\varphi(FK_i))^{\alpha_{FS}(FDBI)}$. That is, for an instance $FR \in (\varphi(FR))^{\alpha_{FS}(FDBI)}$, it can be shown that for each FK_i there is exactly one element $FR_i \in \varphi(FR_i)^{\alpha_{FS}(FDBI)} \in \Delta^{\alpha_{FS}(FDBI)}$ such that $(FR, FR_i) \in (\varphi(FK_i))^{\alpha_{FS}(FDBI)}$. In addition, according to definition of $\alpha_{FS}(FDBI)$ above it follows $(\varphi(FK_i))^{\alpha_{FS}(FDBI)} \sqsubseteq FR^{FDBI} \times FR_i^{FDBI}$, and thus we have $(\varphi(FK_i))^{\alpha_{FS}(FDBI)} \sqsubseteq (\varphi(FR))^{\alpha_{FS}(FDBI)} \times \varphi(FR_i)^{\alpha_{FS}(FDBI)}$ and $((\varphi(FK_i))^{-})^{\alpha_{FS}(FDBI)} \sqsubseteq \varphi\ (FR_i)(\varphi(FK_i))^{\alpha_{FS}(FDBI)} \times (\varphi(FR))^{\alpha_{FS}(FDBI)}$, i.e., the first component of each element of $(\varphi(FK_i))^{\alpha_{FS}(FDBI)}$ is always an element of $FR^{FDBI} = (\varphi(FR))^{\alpha_{FS}(FDBI)}$. Moreover, the cardinalities of relations (e.g., *1–1*, *1–many*, etc.) denote that there are one or at least one FK_i-labeled tuples over Δ^{FDBI} in FR^{FDBI} whose FK_i component is equal to FR_i. That is, $\alpha_{FS}(FDBI)$ satisfies other fuzzy axioms except the axiom in case 4.

(iii) If FR_1 *inherits* FR_2. Firstly, if FR_1 inherits FR_2, then we know that for any individual (identified uniquely by primary key), the membership degree of it to FR_1 must be less than or equal to the membership degree of it to FR_2, i.e., $FR_1^{FDBI} \sqsubseteq FR_2^{FDBI}$. Then, by definition of $\alpha_{FS}(FDBI)$ above, we have $\varphi\ (FR_1)\ (\varphi(FK_i))^{\alpha_{FS}(FDBI)} \sqsubseteq \varphi\ (FR_2)\ (\varphi(FR))^{\alpha_{FS}(FDBI)}$. That is, $\alpha_{FS}(FDBI)$ satisfies fuzzy axiom $\varphi(FR_1) \sqsubseteq \varphi(FR_2)$. □

Theorem 5.8 shows that the extraction procedure in Definition 5.8 is an equivalence preserving transformation. Moreover, the proof of the second part of Theorem 5.8 are omitted, which can be treated analogously according to the proof of the first part, they are mutually inverse process.

In the following, we further illustrate the extraction procedure in Definition 5.8 with Example 5.3.

Example 5.3 Given the fuzzy relational database in Sect. 3.3.1 (including the set of fuzzy relational schemas in Table 3.1 and the fuzzy relations in Table 3.2), the extracted *f-ALCNI* knowledge base from the fuzzy relational database is shown in Fig. 5.6.

In the following, we implemented a prototype extraction tool called *FRDB2DL*, which can extract automatically the *f-ALCNI* knowledge bases (i.e., TBox and ABox) from the fuzzy relational databases (fuzzy relational schemas and fuzzy relation instances). The following briefly introduces the design and implementation

The *f-ALCNI* knowledge base $\varphi(FRDB) = (FA, FP, FT)$ extracted from the fuzzy relational database in Sect. 3.3.1 is as follows:

FA = {Leader, Chief-Leader, Supervise, Young-employee, Employee, Department, String, Integer, Real}

FP = {leaID, lNumber, μ_{FR}, clName, f_clAge, empID, eNumber, yeName, f_yeAge, f_yeSalary,dep_ID, dep_ID$^-$, supID, lea_ID, lea_ID$^-$, emp_ID, emp_ID$^-$, depID, dName}

FT = {

Leader $\sqsubseteq$ $\forall$leaID. String $\sqcap$ (=1 leaID) $\sqcap$ $\forall$lNumber. String $\sqcap$ (=1 lNumber) $\sqcap$ $\forall\mu_{FR}$. Real $\sqcap$ (=1μ_{FR}) ;

Chief-Leader $\sqsubseteq$ Leader $\sqcap$ $\forall$leaID. String $\sqcap$ (=1 leaID) $\sqcap$ $\forall$clName. String $\sqcap$ (=1 clName) $\sqcap$ $\forall$f_clAge. Integer $\sqcap$ (=1 f_clAge) $\sqcap$ $\forall$lea_ID$^-$. Supervise $\sqcap$ ($\geq$1 lea_ID$^-$) $\sqcap$ $\forall\mu_{FR}$. Real $\sqcap$ (=1μ_{FR}) ;

Employee $\sqsubseteq$ $\forall$empID. String $\sqcap$ (=1 empID) $\sqcap$ $\forall$eNumber. String $\sqcap$ (=1 eNumber) $\sqcap$ $\forall\mu_{FR}$. Real $\sqcap$ (=1μ_{FR}) ;

Young-Employee $\sqsubseteq$ Employee $\sqcap$ $\forall$empID. String $\sqcap$ (=1 empID) $\sqcap$ $\forall$yeName. String $\sqcap$ (=1 yeName) $\sqcap$ $\forall$f_yeAge. Integer $\sqcap$ (=1 f_yeAge) $\sqcap$ $\forall$f_yeSalary. Integer $\sqcap$ (=1 f_yeSalary) $\sqcap$ $\forall$dep_ID. Department $\sqcap$ (=1 dep_ID) $\sqcap$ $\forall$emp_ID$^-$. Supervise $\sqcap$ ($\geq$1 emp_ID$^-$) $\sqcap$ $\forall\mu_{FR}$. Real $\sqcap$ (=1μ_{FR}) ;

Supervise $\sqsubseteq$ $\forall$supID. String $\sqcap$ (=1 supID) $\sqcap$ $\forall\mu_{FR}$. Real $\sqcap$ (=1μ_{FR}) $\sqcap$ $\forall$lea_ID. Chief-Leader $\sqcap$ (=1 lea_ID) $\sqcap$ $\forall$emp_ID. Young-Employee $\sqcap$ (=1 emp_ID) ;

Department $\sqsubseteq$ $\forall$depID. String $\sqcap$ (=1 depID) $\sqcap$ $\forall$dName. String $\sqcap$ (=1 dName) $\sqcap$ $\forall$dep_ID$^-$. Young-Employee $\sqcap$ (>1 dep_ID$^-$) $\sqcap$ $\forall\mu_{FR}$. Real $\sqcap$ (=1μ_{FR}) ;

Leader (L001) ≥ 0.7 | Young-Employee (E001) ≥ 0.75
Leader (L001) ≤ 0.7 | Young-employee (E001) ≤ 0.75
Leader (L002) ≥ 0.9 | yeName (E001, John) ≥ 1
Leader (L002) ≤ 0.9 | yeName (E001, John) ≤ 1
Leader (L003) ≥ 0.8 | f_yeSalary (E001, 2000) ≥ 0.3
Leader (L003) ≤ 0.8 | f_ yeSalary (E001, 2000) ≤ 0.3
Chief-Leader (L001) ≥ 0.65 | f_yeAge (E002, 23) ≥ 1
Chief-Leader (L001) ≤ 0.65 | f_yeAge (E002, 23) ≤ 1

⋮ }

Fig. 5.6 The *f-ALCNI* knowledge base $\varphi(FRDB)$ extracted from the fuzzy relational database *FRDB* in Sect. 3.3.1

of *FRDB2DL*. The implementation of *FRDB2DL* is based on Java 2 JDK 1.5 platform. *FRDB2DL* includes three main modules: *connecting and parsing module*, *translation module*, and *output module*. The connecting and parsing module uses the Java JDBC/ODBC API to parse the fuzzy relational database stored in MySQL 5.0 and store the parsed information as Java ArrayList classes; translation module transforms the fuzzy relational database into the *f-ALCNI* knowledge base by using the previous extraction approach; output module produces the resulting *f-ALCNI* knowledge base which is saved as a text file and displayed on tool screen.

As for how to store the fuzzy relational databases in MySQL, the following introduces an approach presented in (Chaudhry et al. 1999). As we all know, the current database systems cannot provide the possibility distribution (or equivalently

Table 5.2 The original relation Young-Employee

empID	f_yeAge
E001	{24/0.7, 25/0.9}
E002	23

Table 5.3 The new relation Young-Employee

empID	f_yeAge
E001	1
E002	2

Table 5.4 The new relation f-yeAge-YE

f_yeAge	Value	μ
1	24	0.7
1	25	0.9
2	23	1.0

the fuzzy set) as a datatype. Therefore, in order that the knowledge extraction tool can automatically extract knowledge from fuzzy relational databases, we use the approach presented to store fuzzy data of fuzzy relational databases in MySQL 5.0.

Given a fuzzy relation *FR* on a fuzzy relational schema $FR(\ldots, A_i/D_i, \ldots, \mu_{FR}/D_{FR})$. For each fuzzy attribute A_i (a symbol '*f*' next to an attribute means that the attribute is fuzzy), create a new relation FR' with attributes $< K_i$, *value*, $\mu >$ and modify $FR(\ldots, A_i/D_i, \ldots \mu_{FR}/D_{FR})$ to $FR(\ldots, K_i/D_i, \ldots, \mu_{FR}/D_{FR})$ by replacing A_i with K_i, where K_i is an atomic identifier in *FR'* for A_i, and *value* and μ respectively model the elements and associated grades of the fuzzy attribute A_i. The approach will be explained directly with the following example.

Consider the fuzzy relation *Young-Employee* and part of representative data of which is repeated listed in Table 5.2. The attribute *f-yeAge* is fuzzy and its value is a possibility distribution. By the above approach, the fuzzy relation *Young-Employee* is replaced by the new relations *Young-Employee* in Table 5.3 and *f-yeAge-YE* in Table 5.4.

Based on the implemented tool *FRDB2DL*, we have carried out some case studies. Here we give an example which can show that the proposed approach is feasible and the implemented tool is efficient. Figure 5.7 shows the screen snapshot of *FRDB2DL*, which displays the *f-ALCNI* knowledge base extracted from the fuzzy relational database in Example 5.3.

After the fuzzy relational databases are described by *f-ALCNI*, the reasoning problems of fuzzy relational databases may be reasoned by means of *f-ALCNI*, which is similar to the reasoning of fuzzy ER models as has been introduced in Sect. 5.2.1, and thus will not be introduced here. Please refer to Ma et al. (2011b).

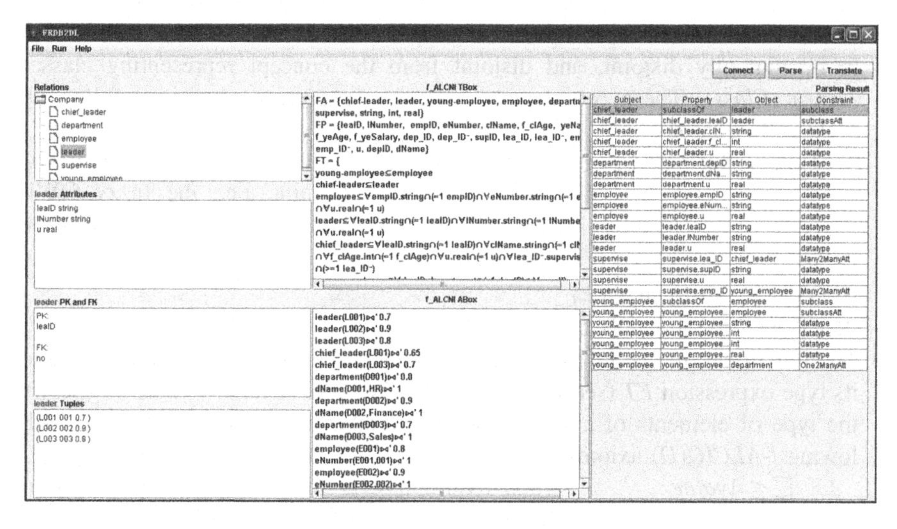

Fig. 5.7 Screen snapshot of *FRDB2DL*

5.2.2.2 Fuzzy Description Logic Extraction from Fuzzy Object-oriented Databases

Being similar to the extraction of fuzzy Description Logic (DL) from fuzzy relational databases, Zhang et al. (2012a, b) proposed an approach for extracting fuzzy DL *f-ALCIQ(D)* knowledge bases from fuzzy object-oriented databases (see the following part *A*), and also discussed how to reason about fuzzy object-oriented databases with the extracted knowledge (see part *B*).

(A) *Extracting f-ALCIQ(D) Knowledge Bases from Fuzzy Object-oriented Databases*

Based on the characteristics of fuzzy object-oriented databases, a fuzzy DL called *f-ALCIQ(D)* is chosen to represent the extracted knowledge in (Zhang et al. 2012b). To extract *f-ALCIQ(D)* knowledge base from a fuzzy object-oriented database, two steps need to be done, i.e., extracting *f-ALCIQ(D)* TBox from the schema information of the fuzzy object-oriented database, and extracting *f-ALCIQ(D)* ABox from the data information of the fuzzy object-oriented database. Here, it should be noted that the fuzzy DL *f-ALCIQ(D)* is similar to the fuzzy DLs *f-ALC* and *f-SHOIN(D)* as have been introduced in Chap. 4, and thus the syntax, semantics, knowledge base, and reasoning of *f-ALCIQ(D)* are not included here.

Given a fuzzy object-oriented database (FOODB) model $FS = (FC_{FS}, FA_{FS}, FD_{FS})$ as given in Definition 3.8 in Sect. 3.3.2, in order to explicitly represent in *f-ALCIQ(D)* the type structure of classes in FOODB models, based on (Baader et al. 2003; Calvanese et al. 1999b; Stoilos et al. 2010), we introduce in the *f-ALCIQ(D)* knowledge base concepts and roles with a specific meaning:

- The concepts *AbstractClass*, *RecType*, and *SetType* are used to denote fuzzy classes, record types, and set types, respectively.

- Furthermore, the concepts representing types *RecType* and *SetType* are assumed to be mutually disjoint, and disjoint from the concept representing classes *AbstractClass*. These constraints can be expressed by the following *f-ALCIQ(D)* axioms:

  RecType $\sqcap$ SetType $\sqsubseteq$ $\perp$ //this axiom asserts that two concepts *RecType* and *SetType* are disjoint, i.e., the fuzzy OWL axiom of the form *DisjointClasses*(RecType SetType).

  SetType $\sqcap$ AbstractClass $\sqsubseteq$ $\perp$
  RecType $\sqcap$ AbstractClass $\sqsubseteq$ $\perp$
- The functional role *value* models the association between a fuzzy class *FC* and its type expression *FT* (see Definition 3.8). The role *member* is used to specify the type of elements of a set. These constraints can be expressed by the following *f-ALCIQ(D)* axioms:

$\sqsubseteq$ $\perp\top \sqsubseteq \leq 1 value$ //this axiom asserts the intuitive definition *Func* (*value*), i.e., defines that *value* is a *functional* role.

$\exists value.\top \sqsubseteq$ AbstractClass //this axiom asserts the intuitive definition *domain*(*value*) = AbstractClass with a fuzzy *DL f-ALCIQ(D)* axiom form, i.e., defines that the *domain* of role *value* is AbstractClass.

AbstractClass $\sqsubseteq = 1 value$ //this axiom asserts that a fuzzy class has and only has one type expression.

$\exists member.\top \sqsubseteq$ SetType //this axiom asserts the intuitive definition *domain*(*member*) = SetType with an *f-ALCIQ(D)* axiom, i.e., defines that the *domain* of role *member* is SetType.

All of these *f-ALCIQ(D)* axioms above will be part of the *f-ALCIQ(D)* knowledge base that we are going to define in Definition 5.9.

Definition 5.9 (*Schema extraction*) Given a FOODB model $FS = (FC_{FS}, FA_{FS}, FD_{FS})$ mentioned in Sect. 3.3.2, the *f-ALCIQ(D)* TBox $\varphi(FS) = (FA, FP, FT)$ can be derived by function φ as shown in Table 5.5.

In the following we further introduce how to translate a *fuzzy database instance* (i.e., a set of object instances) w.r.t. the FOODB model into an *f-ALCIQ(D) ABox*. Firstly, let us briefly introduce the representation forms of instances in *f-ALCIQ(D)* and FOODB models.

Being similar to the fuzzy DL *f-SHOIN(D)* introduced in Sect. 4.2, in *f-ALCIQ(D)*, one describes a specific state of affairs of an application domain in terms of assertions. The assertions used in an *f-ALCIQ(D)* ABox contain concept assertions $< C(a) \bowtie k >$, role assertions $< R(a, b) \bowtie k >$ and $< T(a, v) \bowtie k >$, and individual (in)equality assertions $a \approx b$ and $a \not\approx b$, where $a, b \in \Delta^{FI}$, $v \in \Delta_D$, $\bowtie \in \{\geq, >, \leq, <\}$, $k \in [0, 1]$.

A fuzzy object-oriented database, which describes the real world by means of objects, values, and their mutual relationships, can be considered as a finite set of assertions (Artale et al. 1996; Beneventano et al. 2003; Roger et al. 2002). The assertion formalisms of a fuzzy database instance (i.e., a set of object instances) with respect to a FOODB model include:

- The assertion of the form *FO*: *FC*: *u*, which denotes that a fuzzy object *FO* is an instance of a fuzzy class *FC* with membership degree of $u \in [0, 1]$.
- The assertion of the form $FO : [FA_1{:}FV_1{:}n_1,\ldots,FA_k{:}FV_k{:}n_k]$, which denotes the fuzzy structured value associated with *FO*, where $FV_i \in \mathrm{FV}_{FJ}$ is the value of the attribute, FA_i denotes the attribute of *FO*, $n_i \in [0, 1]$ denotes the membership degree, and $i \in \{1\ldots k\}$. Here, since the value of an attribute FA_i may be a possibility distribution, for simplicity, $FV_i : n_i$ only denotes one element of the possibility distribution.
- Since the fuzzy subclass/superclass relationship in a FOODB model can be assessed by utilizing the inclusion degree of objects to the class (George et al.1996; Koyuncu and Yazici 2003; Ma et al. 2004), the assertion formalism of fuzzy subclass/superclass relationship can be re-expressed by the above assertion formalism of objects to the class.

On this basis, Definition 5.10 gives a translation approach from fuzzy database instances to *f-ALCIQ(D)* ABoxes.

Definition 5.10 (*Instance extraction*) Given a fuzzy database instance (i.e., a set of object instances) w.r.t. a FOODB model *FS*, the corresponding *f-ALCIQ(D)* ABox (w.r.t. the *f-ALCIQ(D)* TBox $\varphi(FS)$ in Definition 5.9) can be derived as the following rules:

- Each fuzzy object symbol *FO* is mapped into an *f-ALCIQ(D)* abstract individual $\varphi(FO)$;
- Each fuzzy class symbol $FC \in FC_{FS}$ and each fuzzy attribute symbol $FA \in FA_F$ are mapped into a fuzzy concept $\varphi(FC)$ and a fuzzy role $\varphi(FA)$, respectively. Here, to improve readability, $\varphi(FC)$ stands for all the fuzzy concepts, and $\varphi(FA)$ stands for all the fuzzy abstract and concrete roles (see Table 5.5 in detail);
- Each fuzzy assertion *FO*: *FC*: *u* is mapped into an *f-ALCIQ(D)* assertion $< \varphi(FC)\ (\varphi(FO)) \bowtie' u >$, where $\bowtie' \in \{\geq, \leq\}$;
- Each fuzzy assertion FO : $[\mathrm{FA}_1{:}\mathrm{FV}_1{:}n_1,\ldots,\mathrm{FA}_k{:}\mathrm{FV}_k{:}n_k]$ is mapped into the *f-ALCIQ(D)* assertions $< \varphi(FA_i)(\varphi(FO), FV_i) \bowtie' n_i >$, where $FV_i \in \mathrm{FV}_{FJ}$ as mentioned in Sect. 3.3.2, which is mapped into an abstract individual or a concrete individual, $i \in \{1,\ldots,k\}$, $\bowtie' \in \{\geq, \leq\}$.

Based on (Baader et al. 2003; Calvanese et al. 1999b; Artale et al. 1996), below we prove the correctness of the translation φ in Definitions 5.9 and 5.10, which can be sanctioned by establishing mappings between *fuzzy database states* w.r.t. the

Table 5.5 Extracting rules of *f-ALCIQ(D)* TBox from FOODB model

FOODB model *FS*	*f-ALCIQ(D)* TBox $\varphi(FS) = (FA, FP, FT)$
Fuzzy class FC_{FS}, *fuzzy attribute* FA_{FS}	*Atomic fuzzy concept set FA*
Each fuzzy class *FC*	An atomic fuzzy concept $\varphi(FC)$
Each result type *R* in method *f*():*R* (the parameter is *null*)	A fuzzy domain predicate $\varphi(R)$
Each method $f(P_1,\ldots,P_m) : R$	$m + 1$ atomic fuzzy domain predicates $\varphi(P_1), \ldots, \varphi(P_m), \varphi(R)$, denoting the types of *m* parameters and a result, respectively.
Each domain *FB* of fuzzy attributes *FA*	A fuzzy domain predicate $\varphi(FB)$
	Additional atomic fuzzy concepts *AbstractClass*, *RecType*, and *SetType*
Fuzzy class FC_{FS}, *fuzzy attribute* FA_{FS}	*Atomic fuzzy role set FP*
Each fuzzy attribute *FA* ($u \in FA$)	An atomic fuzzy role $\varphi(FA)$, which may be a abstract role or concrete role
Each method $f() : R$	An atomic fuzzy role $\varphi(R_{f()})$, denoting that an object invokes the method by $\varphi(R_{f()})$
Each method $f(P_1,\ldots,P_m) : R$	$1 + m{+}1$ atomic fuzzy roles $\varphi(R_{f(P)}), \varphi(R_1), \ldots, \varphi(R_m), \varphi(R_R)$, where role $\varphi(R_{f(P)})$ denotes that an object invokes the method by $\varphi(R_{f(P)})$, and the rest of roles denote the types of *m* parameters and a result
	Atomic fuzzy roles *value* and *member*
Fuzzy class declaration FD_{FS}	*Fuzzy axioms set FT*
Each class declaration in FD_{FS}:	RecType $\sqcap$ SetType $\sqsubseteq \bot$
Class FC *is-a* $FC_1,\ldots,FC_n$ *type-is* FT	SetType $\sqcap$ AbstractClass $\sqsubseteq \bot$
FT $\rightarrow$ $FC \mid FB$	RecType $\sqcap$ AbstractClass $\sqsubseteq \bot$
Union $FT_1,\ldots, FT_k$ (*disjointness, covering*) *End* $\mid$	$\top \sqsubseteq\ \leq 1 value\ \exists value.\top \sqsubseteq$ AbstractClass
Set-of FT $[(m_1, n_1), (m_2, n_2)] \mid$	AbstractClass $\sqsubseteq\ = 1 value$
Record $FA_1{:}FT_1,\ldots, FA_k$:	$\exists member.\top \sqsubseteq$ SetType
FT_k, *u*:Real *End* $\mid$	$\varphi(FC) \sqsubseteq$ AbstractClass $\sqcap\ \varphi(FC_1) \sqcap \ldots \sqcap \varphi(FC_n) \sqcap \forall value.\ \varphi(FT)$
$f(P_1, \ldots, P_m) : R$	where a function φ maps type expression FT to a fuzzy concept expression $\varphi(FT)$ as follows:

(continued)

Table 5.5 (continued)

FOODB model *FS*	*f-ALCIQ(D)* TBox $\varphi(FS) = (FA, FP, FT)$
Each type expression FT	*A fuzzy concept expression* $\varphi(FT)$
Each fuzzy class FC	An atomic fuzzy concept $\varphi(\mathrm{FC})$
Each expression *Union* $\mathrm{FT}_1,\ldots,\mathrm{FT}_k$ (*disjointness, covering*) *End*	A concept expression $(\varphi(\mathrm{FT}_1) \sqcup \ldots \sqcup \varphi(\mathrm{FT}_k))$; If there is the constraint (*disjointness, covering*), then adding the fuzzy axioms: $\varphi(\mathrm{FT}_i) \sqcap \varphi(\mathrm{FT}_j) \sqsubseteq \perp,\ i \neq j,\ i, j \in \{1,\ldots,k\}$ $\varphi(\mathrm{FT}_1) \sqcup \ldots \sqcup \varphi(\mathrm{FT}_k) \sqsubseteq \varphi(\mathrm{FT})$
Each expression FA *Set-of* FT $[(m_1, n_1), (m_2, n_2)]$	A fuzzy concept expression: (SetType $\sqcap \forall$member. $\varphi(\mathrm{FT}) \sqcap \geq m_1$ member $\sqcap \leq n_1$ member); Moreover, adding the fuzzy axiom: $\varphi(\mathrm{FT}) \sqsubseteq \geq m_2\ (\varphi(\mathrm{FA}))^-.\ \varphi(\mathrm{FC}) \sqcap \leq n_2\ (\varphi(\mathrm{FA}))^-.\ \varphi(\mathrm{FC})$
Each record type expression may include the following three cases: *Record* $\mathrm{FA}_1{:}\mathrm{FT}_1,\ldots,\mathrm{FA}_k{:}\mathrm{FT}_k$, u:Real $f() : R$ $f(P_1, \ldots, P_m) : R$ *End*	A fuzzy concept expression $\varphi(\mathrm{FT})$ is formed as a conjunction of the following three concept expressions, i.e., $\varphi(\mathrm{FT})$ = (i) $\sqcap$ (ii) $\sqcap$ (iii) *The concept expression* (i): RecType $\sqcap \forall\varphi(\mathrm{FA}_1).\ \varphi(\mathrm{FT}_1) \sqcap = 1\ \varphi(\mathrm{FA}_1) \sqcap \ldots \sqcap \forall\varphi(u).\ \varphi(\mathrm{Real}) \sqcap = 1\ \varphi(u) \sqcap \forall\varphi(\mathrm{FA}_k).\ \varphi(\mathrm{FT}_k) \sqcap = 1\ \varphi(\mathrm{FA}_k)$ *The concept expression* (ii): $\forall\varphi(R_{f()}).\varphi(R) \sqcap \leq 1\ \varphi(R_{f()}).\top$ *The concept expression* (iii) is also a conjunction of the following three parts, i.e., (iii) = $part_1 \sqcap part_2 \sqcap part_3$ $part_1$: $\exists\varphi(R_{f(P)}).\top \sqcap \leq 1\ \varphi(R_{f(P)}).\top \sqcap \exists\varphi(R_1).\top \sqcap \leq 1\ \varphi(R_1).\ \varphi(P_1) \sqcap$ $\ldots \sqcap \exists\varphi(R_m).\top \sqcap \leq 1\ \varphi(R_m).\ \varphi(P_m)$ $part_2$: $\forall\varphi(R_1).\ \varphi(P_1) \sqcap \ldots \sqcap \forall\varphi(R_m).\ \varphi(P_m)$ $part_3$: $\forall(\varphi(R_{f(P)}))^-.\ (\neg\varphi(\mathrm{FC}) \sqcup \forall\varphi(R_R).\ \varphi(R))$ In the concept expression (iii), the $part_1$ states that each instance of $\varphi(\mathrm{FC})$, representing a tuple, correctly is connected to exactly one object for each of the roles $\varphi(R_{f(P)})$, $\varphi(R_1)$, ..., $\varphi(R_m)$; The other two parts impose the correct typing of the parameters and of the return value

FOODB model *FS* and *models* of the translated *f-ALCIQ(D)* TBox $\varphi(FS)$ (see Theorem 5.9).

Theorem 5.9 *For every FOODB model FS* = *(*FC_{FS}*,* FA_{FS}*,* FD_{FS}*), let FJ be a fuzzy database state w.r.t. FS as introduced in Sect. 3.3.2*, and φ(FS) be the f-ALCIQ(D) TBox derived from FS *by Definition 5.9. There are mappings:* α_{FS} *is a mapping from FJ to a fuzzy interpretation of* $\varphi(FS)$*, and* α_{FV} *is a mapping from values of FJ to domain elements of the fuzzy interpretation of* $\varphi(FS)$*; conversely,* β_{FS} *is a mapping from a fuzzy interpretation of* $\varphi(FS)$ *to FJ, and* β_{FV} *is a mapping from domain elements of fuzzy interpretation of* $\varphi(FS)$ *to values of FJ, such that:*

- *For each legal fuzzy database state FJ for FS, there is* $\alpha_{FS}(FJ)$ *which is a model of* $\varphi(FS)$*, and for each type expression FT of FS and each* $V \in FV_{FJ}$*,* $V \in FT^{FJ}$ *if and only if* $\alpha_{FV}(V) \in {}^{\alpha_{FS}(FDBI)}$;
- *For each model FI of* $\varphi(FS)$*, there is* $\beta_{FS}(FI)$ *which is a legal fuzzy database state for FS, and for each concept expression* $\varphi(FT)$ *and each* $d \in \Delta^{FI}$*,* $d \in (\varphi(FT))^{FI}$ *if and only if* $\beta_{FV}(d) \in FT^{\beta_{FS}(FI)}$.

Proof The following gives the proof of the first part of Theorem 5.9. Here, since Definition 5.9 aims at translating a FOODB model to an *f-ALCIQ(D)* TBox at terminological level, the following proof takes no account of the membership degrees which occur at three levels of fuzziness in a FOODB model. Firstly, let us define a fuzzy interpretation $\alpha_{FS}(FJ)$ of $\varphi(FS)$ as follows:

- α_{FV} is a mapping from values FV_{FJ} of the fuzzy database state *FJ* w.r.t. *FS* to domain elements $((\varphi(FK_i))^-)^{\alpha_{FS}(FDBI)}$ of the fuzzy interpretation of $\varphi(FS)$, i.e., $\Delta^{\alpha_{FS}(FJ)}$ is constituted by the set of elements $\alpha_{FV}(V)$ for each $V \in FV_{FJ}$. Since each fuzzy object of *FS* is assigned a structured value, in order to explicitly represent in *f-ALCIQ(D)* the type structure of fuzzy classes, we denote with Δ_{fid}, Δ_{frec}, and Δ_{fset} the elements of $\Delta^{\alpha_{FS}(FJ)}$ corresponding to fuzzy object identifiers, fuzzy record values, and fuzzy set values, respectively.
- The atomic fuzzy concepts of set *FA* in Definition 5.9 are interpreted: $AbstractClass^{\alpha_{FS}(FJ)} = \Delta_{fid}$; $RecType^{\alpha_{FS}(FJ)} = \Delta_{frec}$; $SetType^{\alpha_{FS}(FJ)} = \Delta_{fset}$; $(\phi(FC))^{\alpha_{FS}(FJ)} = \{\alpha_{FV}(v) | v \in \pi^{FJ}(FC),\ FC \in FC_{FS}$
- The atomic fuzzy relations of set *FP* in Definition 5.9 are interpreted: $member^{\alpha_{FS}(FJ)} == \{(d_1, d_2) \mid d_1 \in \Delta_{fset}$ and $\alpha_{FV}^{-1}(d_1) = \{\ldots, \alpha_{FV}^{-1}(d_2), \ldots\}\}$; $value^{\alpha_{FS}(FJ)} = \{(d_1, d_2) \mid (\alpha_{FV}^{-1}(d_1) =, \alpha_{FV}^{-1}(d_2)) \in \rho^{FJ}\}$;$(\phi(FA))^{\alpha_{FS}(FJ)} = \{(d_1, d_2) \mid d_1 \in \Delta_{frec}$ and $\alpha_{FV}^{-1}(d_1) = [\ldots, FA{:}\alpha_{FV}^{-1}(d_2), \ldots]\}$, where $FA \in FA_{FS}$, $\alpha_{FV}^{-1}(d_1)$ is the inverse function (mapping) of $\alpha_{FV}(d_1)$.

Now, based on the definition of $\alpha_{FS}(FJ)$ above, we prove the first part of Theorem 5.9 by considering each case of type expressions *FT* in Definition 5.9.

Case 1 FT → FC. By Definition 5.9, FC is mapped into φ(FC).

Proof "⇒" : If $v \in FT^{FJ} = FC^{FJ}$, by Definition 3.9 in Sect. 3.3.2, we have $v \in \pi^{FJ}(FC)$. Then, according to the definition of $\alpha_{FS}(FJ)$ above it follows $\alpha_{FV}(v) \in (\varphi(FC))^{\alpha_{FS}(FJ)}$, and vice versa.

Case 2 FT → Set-of FT′, for simplicity, the cardinality constraint of a set is omitted. By Definition 5.9, FT is mapped into SetType ⊓ ∀*member*.φ(FT′). We first give an induction assumption that $v \in (FT')^{FJ}$ iff $\alpha_{FV}(v) \in (\varphi(FT'))^{\alpha_{FS}(FJ)}$, and it shows that $v \in FT^{FJ}$ iff $\alpha_{FV}(v) \in (\varphi(FT))^{\alpha_{FS}(FJ)}$.

Proof "⇒" : If $v \in FT^{FJ}$, according to Definition 3.9 in Sect. 3.3.2, we have $v = \{FV_1',\ldots, FV_k'\}$ and $FV_i' \in (FT')^{FJ}$ for $i \in \{1,\ldots, k\}$. Then, by the induction assumption above it follows $\alpha_{FV}(FV_i') \in (\varphi(FT'))^{\alpha_{FS}(FJ)}$ for $i \in \{1,\ldots, k\}$. Furthermore, by the definition of $\alpha_{FS}(FJ)$ above, we have $\alpha_{FV}(v) \in SetType^{\alpha_{FS}(FJ)}$ and $(\alpha_{FV}(v), \alpha_{FV}(FV_i')) \in member^{\alpha_{FS}(FJ)}$, $i \in \{1,\ldots, k\}$. Therefore, $\alpha_{FV}(v) \in (\varphi(FT))^{\alpha_{FS}(FJ)}$.

Proof "⇐" : If $d = \alpha_{FV}(v) \in (\varphi(FT))^{\alpha_{FS}(FJ)}$, according to φ(FT) = SetType ⊓ ∀*member*. φ(FT'), then there is exactly one $d_i \in \Delta^{\alpha_{FS}(FJ)}$ with $(d, d_i) \in member^{\alpha_{FS}(FJ)}$ and $d_i \in (\varphi(FT'))^{\alpha_{FS}(FJ)}$, $i \in \{1,\ldots, k\}$. Then, by the definition of $\alpha_{FS}(FJ)$ above, we have $v = \{FV_1',\ldots, FV_k'\}$ and $FV_i' = \alpha_{FV}^{-1}(d_i)$ for $i \in \{1,\ldots, k\}$. Furthermore, by the induction assumption above it follows $FV_i' \in (FT')^{FJ}$ for $i \in \{1,\ldots, k\}$. Therefore, $v \in FT^{FJ} = (\text{Set-of } FT')^{FJ}$.

Case 3 FT → Record FA_1:FT_1,..., FA_k:FT_k End, note that $u \in FA_i$ and the method is included in the record type. By Definition 5.9, FT is mapped into Rec-
Type ⊓ ∀$\varphi(FA_1).\varphi(FT_1)$ ⊓ = 1$\varphi(FA_1)$ ⊓ ... ⊓ ∀$\varphi(FA_k).\varphi(FT_k)$ ⊓ = 1$\varphi(FA_k)$.
We first give an induction assumption that $v \in FT_i^{FJ}$ iff $\alpha_{FV}(v) \in (\varphi(FT_i))^{\alpha_{FS}(FJ)}$, for $i \in \{1,\ldots, k\}$, and it shows that $v \in FT^{FJ}$ iff $\alpha_{FV}(v) \in (\varphi(FT))^{\alpha_{FS}(FJ)}$.

Proof "⇒" : If $v \in FT^{FJ}$, by Definition 3.9 in Sect. 3.3.2 it follows $v = [FA_1{:}FV_1,\ldots,FA_k{:}FV_k]$ and $FV_i \in FT_i^{FJ}$ for $i \in \{1,\ldots, k\}$. Furthermore, according to the induction assumption above, we have $\alpha_{FV}(FV_i) \in \Delta^{\alpha_{FS}(FJ)}$ for $i \in \{1,\ldots, k\}$. Then, by the definition of $\alpha_{FS}(FJ)$ above we have $\alpha_{FV}(v) \in RecType^{\alpha_{FS}(FJ)}$, $(\alpha_{FV}(v), \alpha_{FV}(FV_i)) \in (\varphi(FA_i))^{\alpha_{FS}(FJ)}$ for $i \in \{1,\ldots, k\}$, and all fuzzy relations φ(FA) corresponding to fuzzy attribute names FA are functional. Therefore, $\alpha_{FV}(v) \in (\varphi(FT))^{\alpha_{FS}(FJ)}$.

Proof "⇐" : If $d = \alpha_{FV}(v) \in (\varphi(FT))^{\alpha_{FS}(FJ)}$, according to φ(FT) = RecType ⊓ ∀$\varphi(FA_1)$. $\varphi(FT_1)$ ⊓ = 1$\varphi(FA_1)$ ⊓ ... ⊓ ∀$\varphi(FA_k)$. $\varphi(FT_k)$ ⊓ = 1$\varphi(FA_k)$, then there is exactly one $d_i \in \Delta^{\alpha_{FS}(FJ)}$ with $(d, d_i) \in (\varphi(FA_i))^{\alpha_{FS}(FJ)}$ and $d_i \in (\varphi(FT_i))^{\alpha_{FS}(FJ)}$, $i \in \{1,\ldots, k\}$. Furthermore, by the definition of $\alpha_{FS}(FJ)$ above, we have $v = [FA_1{:}FV_1,\ldots, FA_k{:}FV_k]$ with $FV_i = \alpha_{FV}^{-1}(d_i)$ for $i \in \{1,\ldots, k\}$. Then, by the induction assumption above, we have $FV_i \in FT_i^{FJ}$ for $i \in \{1,\ldots, k\}$. Therefore, $v \in FT^{FJ} = (\text{Record } FA_1{:}FT_1,\ldots, FA_k{:}FT_k \text{ End})^{FJ}$.

Case 4 The case for FT → Union $FT_1,\ldots, FT_k$ (disjointness, covering) End can be treated analogously. □

The two parts of Theorem 5.9 are a mutually inverse process. The proof of the second part of Theorem 5.9, which can be treated analogously according to the first part above, is omitted here. In the following, we further illustrate the translation procedure in Definitions 5.9 and 5.10 with Example 5.4.

Example 5.4 Given the fuzzy object-oriented database mentioned in Sect. 3.3.2 (including the schema information in Fig. 3.10 and the object instance information in Fig. 3.11), by applying Definitions 5.9 and 5.10, we can extract the corresponding *f-ALCIQ(D)* knowledge base as shown in Fig. 5.8.

Following the proposed approaches, we developed a prototype extraction tool called *FOOD2DL*, which can automatically extract FOODB models and object instances stored in databases and translate them into *f-ALCIQ(D)* knowledge bases. The fuzzy object-oriented databases (including the schema information and the object instances) are stored in the object-oriented database *db4o* (http://www.db4o.com). *db4o*, which is an open source object database developed in Java, is extensively used. *db4o* enables Java and .NET developers to store and retrieve any application object with only one line of code, eliminating the need to predefine or maintain a separate, rigid data model. In the following, we briefly introduce the implementation of *FOOD2DL*.

The implementation of *FOOD2DL* is based on Java 2 JDK 1.5 platform. *FOOD2DL* includes three main modules: *connecting and parsing module*, *translation module*, and *output module*. The connecting and parsing module uses the regular expression to parse the FOODB model file and stores the parsed results as Java ArrayList classes; translation module translates the FOODB model and data instances into the *f-ALCIQ(D)* knowledge bases based on the proposed approaches; output module produces the resulting *f-ALCIQ(D)* knowledge base which is saved as a text file and displayed on the tool screen.

Here we give an example to show that the proposed approach is feasible and the implemented tool is efficient. Figure 5.9 shows the screen snapshot of *FOOD2DL*, which displays the translations from the FOODB model to the *f-ALCIQ(D)* knowledge base mentioned in Example 5.4, where the FOODB model information, the *f-ALCIQ(D)* knowledge base, and parsed results are displayed in the left, middle and right areas, respectively.

So far, on the basis of the proposed approach and the implemented tool, FOODB models can be represented by *f-ALCIQ(D)* knowledge bases. Therefore, it is possible for us to take advantage of the associated reasoning techniques of *f-ALCIQ(D)* to reason on the FOODB models. In the following we will further introduce how to reason on FOODB models with *f-ALCIQ(D)*.

(B) *Reasoning about the Fuzzy Object-oriented Databases with the Extracted f-ALCIQ(D) Knowledge Bases*

The *f-ALCIQ(D)* knowledge base *KB* = (*FA*, *FP*, *FT*) extracted from the FOODB model mentioned in Sect. 3.3.2 is as follows:

FA = {AbstractClass, RecType, SetType, Chief-Leader, Leader, Old-Employee, Young-Employee, Employee, String, Integer, Real}

FP = {*value*, *member*, *u*, Name, Age, Salary, IsDepartment, Number, Manage}

FT = { SetType $\sqcap$ AbstractClass $\sqsubseteq \bot$

RecType $\sqcap$ AbstractClass $\sqsubseteq \bot$

RecType $\sqcap$ SetType $\sqsubseteq \bot$

$\top \sqsubseteq\ \leq 1 value$ $\qquad \exists value.\top \sqsubseteq$ AbstractClass

AbstractClass $\sqsubseteq\ =1 value$ $\qquad \exists member.\top \sqsubseteq$ SetType

Employee $\sqsubseteq$ AbstractClass $\sqcap\ \forall$ *value*.(Young-Employee $\sqcup$ Old-Employee)

Young-Employee $\sqcap$ Old-Employee $\sqsubseteq \bot$

Young-Employee $\sqcup$ Old-Employee $\sqsubseteq$ Employee

Young-Employee $\sqsubseteq$ AbstractClass $\sqcap$ Old-Employee $\sqcap\ \forall value$.(RecType $\sqcap\ \forall$Name. String $\sqcap$ =1 Name $\sqcap\ \forall u$. Real $\sqcap$ =1 u $\sqcap\ \forall$Age. Integer $\sqcap$ =1 Age $\sqcap\ \forall$Salary. Integer $\sqcap$ =1 Salary $\sqcap$ =1 Manage^{-}. Chief-Leader $\sqcap\ \forall$IsDepartment. String $\sqcap\ \leq 1$ IsDepartment.$\top$)

Chief-Leader $\sqsubseteq$ AbstractClass $\sqcap$ Leader $\sqcap\ \forall value$.(RecType $\sqcap\ \forall$ Number. String $\sqcap$ =1 Number $\sqcap\ \forall u$. Real $\sqcap$ =1 u $\sqcap\ \forall$Manage.(SetType $\sqcap\ \forall member$. Young-Employee $\sqcap \geq 2$ *member* $\sqcap \leq 6$ *member*))

Young-Employee (O_1) ≥ 0.9 $\qquad$ Salary (O_2, 6000) ≥ 0.8

Young-Employee (O_1) ≤ 0.9 $\qquad$ Salary (O_2, 6000) ≤ 0.8

Name (O_1, Chris) $\qquad$ Chief-Leader (O_3) ≥ 0.85

Age (O_1, 20) ≥ 0.5 $\qquad$ Chief-Leader (O_3) ≤ 0.85

Age (O_1, 20) ≤ 0.5 $\qquad$ Number (O_3, 03)

IsDepartment (O_1, SalesDept) ≥ 0.7 $\qquad$ Old-Employee (O_1') ≥ 0.6

IsDepartment (O_1, SalesDept) ≤ 0.7 $\qquad$ Old-Employee (O_1') ≤ 0.6

Young-Employee (O_2) ≤ 0.95 $\qquad$ Employee (O_1) ≥ 0.93

Name (O_2, John) $\qquad$ Employee (O_1) ≤ 0.93

...}

Fig. 5.8 The *f-ALCIQ(D)* knowledge base extracted from the FOODB model in Sect. 3.3.2

In the following we introduce how to reason on fuzzy object-oriented databases (FOODB) models based on the extracted *f-ALCIQ(D)* knowledge bases, including: (*i*) we first provide an example to well motivate the later reasoning work; (*ii*) we then introduce several familiar reasoning tasks considered in FOODB models and give their formal definitions; (*iii*) we investigate how to reason on FOODB models by *f-ALCIQ(D)*, i.e., how to turn the reasoning problems of FOODB models to the reasoning of *f-ALCIQ(D)* knowledge bases. Further, reasoning on *f-ALCIQ(D)* knowledge bases can be automatically done by means of the existing fuzzy reasoner DeLorean (Bobillo et al. 2012).

The following brief example will be helpful to understand the reasoning tasks of FOODB models and can show that it is possible and meaningful to represent and reason on FOODB models with fuzzy DLs.

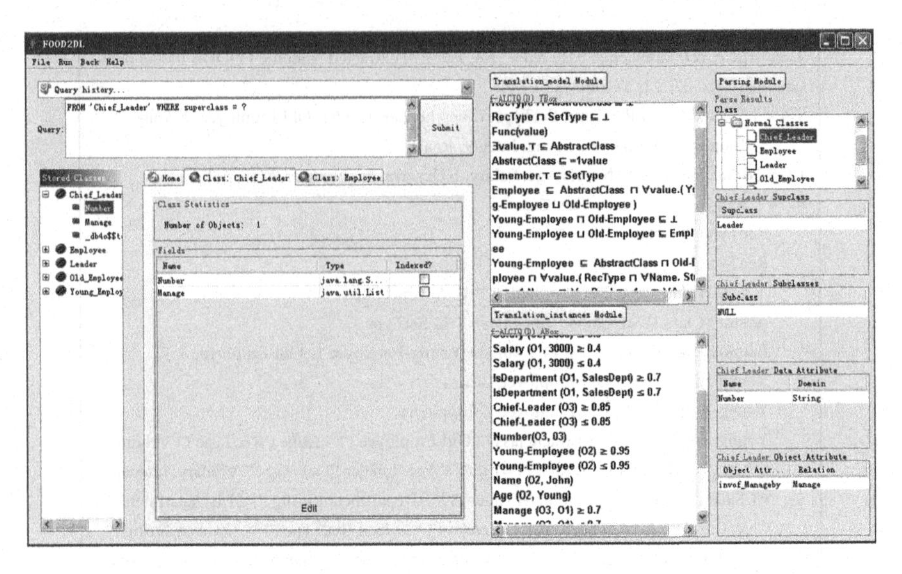

Fig. 5.9 Screen snapshot of *FOOD2DL*

Example 5.5 We take the FOODB model in Fig. 5.10 (including the schema and object instance information) for example to introduce the reasoning tasks of FOODB models. From Fig. 5.10, there are unsatisfiability and redundancy in the FOODB model since the following reasons:

(a) Since *Young-Employee* is a subclass of *Old-Employee*, i.e., for any object, the membership degree that it belongs to the subclass *Young-Employee* is *less than or equal to* the membership degree that it belongs to the superclass *Old-Employee*, we have $\{o_1/\geq 0.9, o_2/\geq 0.95, o_1'/0.6, o_2'/0.8\} \in$ *Old-Employee*;
(b) Since *Young-Employee* and *Old-Employee* are disjoint, i.e., there is no instance which belongs to the two classes, we have $\{o_1/0, o_2/0\} \in$ *Old-Employee*;
(c) From (*a*) to (*b*), it is shown that the two instance sets of *Old-Employee* are conflictive. Therefore, we know that the instance set of *Young-Employee* is an *empty set*, because the *empty set* is the only set that can be *at the same time disjoint from* and *contained in* the class *Old-Employee*. That is, *Young-Employee* is unsatisfiable since it is an empty class;
(d) The class *Young-Employee* is an empty class, and *Employee* is the union of *Young-Employee* and *Old-Employee*. Therefore, *Employee* is equivalent to *Old-Employee*, i.e., there is redundancy in the FOODB model;
(e) The above cases may result in that other undesirable problems occur in a more complete FOODB model.

All the reasoning problems mentioned in Example 5.5 may occur in the FOODB modeling activities, and the burden of checking these problems is left to the designers. Therefore, it would be highly desirable to improve the ability of

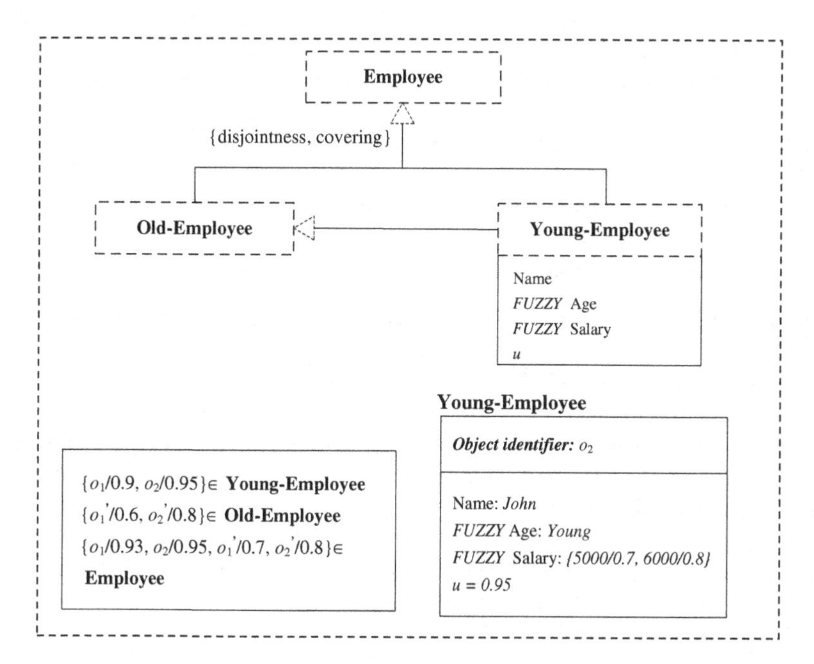

Fig. 5.10 A brief fuzzy object-oriented database model

reasoning on database models. Based on the previous translation work, in the following we make an attempt to resolve the problem by means of the fuzzy DL *f-ALCIQ(D)*. The familiar reasoning problems considered in FOODB models include *consistency*, *satisfiability*, *subsumption*, and *redundancy*. The following gives their formal definitions.

Definition 5.11 (*consistency of fuzzy object-oriented databases*) A fuzzy object-oriented database contains its schema information (i.e., a FOODB model) and its instance information (i.e., a set of object instances). A fuzzy object-oriented database is *consistent*, if the set of object instances satisfies all the constraints of the FOODB model.

In particular, when no instance exists in the fuzzy object-oriented database, the consistency problem of the database above is reduced to the satisfiability problem of the FOODB model in Definition 5.12.

Definition 5.12 (*satisfiability of FOODB models*) A FOODB model is *satisfiable*, if it admits at least one fuzzy database instance (i.e., a set of object instances as introduced in Sect. 3.3.2).

If a FOODB model is not satisfiable, the classes altogether are contradictory, i.e., it does not allow that any class can be populated without violating any of the requirements imposed by the FOODB model. This may be due to a design error or over-constraining.

Definition 5.13 (*satisfiability of fuzzy classes*) A fuzzy class FC in a FOODB model FS is *satisfiable*, if there is at least one fuzzy database state/instance FJ of FS such that $FC^{FJ} \neq \varnothing$ (see Sect. 3.3.2), i.e., FS admits a fuzzy database instance in which FC has a non-empty set of objects.

An unsatisfiable fuzzy class weakens the understandability of a FOODB model, since the fuzzy class stands for an empty class, and thus, at the very least, it is inappropriately named. In this case, designers should modify or delete the fuzzy class to increase understandability.

Definition 5.14 (*subsumption of fuzzy classes*) Let FS be a FOODB model, and FC_1, FC_2 be two fuzzy classes in FS. For each fuzzy database state/instance FJ of FS, if FC_1^{FJ} is a subset of FC_2^{FJ}, then FC_1 is a subclass of FC_2.

In short, a fuzzy class FC_1 is a subclass of another fuzzy class FC_2 if, for any object, the membership degree of it to FC_1 is *less than or equal to* the membership degree of it to FC_2. Class subsumption is the basis for a classification of all the classes in a FOODB model.

As mentioned in Example 5.5, a FOODB model is redundant if there is an empty class or two equivalent fuzzy classes. Therefore, before giving the definition of redundancy of FOODB models, the concept of equivalence of fuzzy classes should be introduced first.

As we have known, in the classical object-oriented database, two classes are equivalent if they denote the same set of object instances. In a FOODB model, however, an object may belong to a class with membership degree of [0, 1], and thus two fuzzy classes may have same objects with same/different membership degrees. In this case, the definition of equivalent classes should be extended under fuzzy environment as will be shown in Definition 5.15.

Definition 5.15 (*equivalence of fuzzy classes*) A fuzzy class FC_1 is equivalent to another fuzzy class FC_2, if at least one of the following conditions is satisfied: (*i*) two classes have same objects with same membership degrees; (*ii*) two classes do not have subclass/superclass relationship but they have same objects with different membership degrees.

Note that, when two fuzzy classes have same objects with same membership degrees, the two fuzzy classes are strictly equivalent. In this case, one of the fuzzy classes can be removed and replaced by another. Now let us focus on how to handle the case that two classes have same objects with different membership degrees. Let FC_1 and FC_2 be two fuzzy classes, and assume that there is an object o belonging to the two fuzzy classes with different membership degrees $u_{FC1}(o) \in [0, 1]$ and $u_{FC2}(o) \in [0, 1]$. At this moment, which one in FC_1 and FC_2 is the class of object o depends on the following case: if $u_{FC1}(o) > u_{FC2}(o)$, then FC_1 is considered as the class of object o and we say FC_1 fuzzily include FC_2, else FC_2 is considered as the class of object o and we say FC_2 fuzzily include FC_1. Correspondingly, the class which is included by another class is removed, and the class which includes the removed class is retained.

Based on Definition 5.15, the following gives the formal definition of redundancy of FOODB models.

Definition 5.16 (*redundancy of FOOD models*) A FOODB model is *redundant*, if there is some class standing for an empty class, i.e., it is inappropriately named; or there are two equivalent fuzzy classes.

Removing redundancy can reduce the complexity of a FOODB model. The reasoning problems mentioned above may occur in the FOODB modeling activities, the following section will investigate how to reason on these problems of FOODB models by means of the fuzzy *DL f-ALCIQ(D)*.

The following theorems allow us to reduce reasoning on FOODB models to reasoning on *f-ALCIQ(D)* knowledge bases, so that the reasoning problems of FOODB models may be checked through the reasoning mechanism of*f-ALCIQ(D)*.

Theorem 5.10 (consistency of fuzzy object-oriented databases) *Given a fuzzy object-oriented database (including a FOODB model and a fuzzy database instance), and $\Sigma = <TBox, ABox>$ is the extracted f-ALCIQ(D) KB according to the previous proposed approach. The fuzzy object-oriented database is consistent iff Σ is satisfiable, i.e., the ABox is consistent w.r.t. the TBox.*

Theorem 5.11 (satisfiability of FOODB models) *Given a FOODB model FS, and $\varphi(FS)$ is the extracted f-ALCIQ(D) TBox. FS is satisfiable iff $\varphi(FS)$ is satisfiable, i.e., there is a fuzzy interpretation FI which is a model of $\varphi(FS)$.*

Note that, when no instance exists in a fuzzy object-oriented database (i.e., the translated *f-ALCIQ(D)* KB contains only the TBox), the consistency problem of the fuzzy object-oriented database (Theorem 5.10) is reduced to the satisfiability problem of the FOODB model (Theorem 5.11). Moreover, Theorems 5.10-5.11 are the straightforward consequences of Definitions 5.11-5.12, and thus their proofs are omitted here.

Theorem 5.12 (satisfiability of fuzzy classes) *Given a FOODB model FS, $\varphi(FS)$ is the translated f-ALCIQ(D) TBox, and FC is a fuzzy class in FS. FC is satisfiable iff: $\varphi(FS) \nvDash \varphi(FC) \sqsubseteq \bot$.*

Proof "$\Rightarrow$": If *FC* is consistent, then there is a legal fuzzy database state *FJ* with $FC^{FJ} \neq \emptyset$, i.e., $\exists v.\ v \in FV_{FJ}$ and $v \in FC^{FJ}$. By part 1 of Theorem 5.9, $\alpha_{FS}(FJ)$ is a model of $\varphi(FS)$ with $\alpha_{FV}(v) \in (\varphi(FC))^{\alpha_{FS}(FJ)}$, i.e., $(\varphi(FC))^{\alpha_{FS}(FJ)} \neq \emptyset$. That is, $\varphi(FS) \nvDash \varphi(FC) \sqsubseteq \bot$.

"$\Leftarrow$": If $\varphi(FS) \nvDash \varphi(FC) \sqsubseteq \bot$, i.e., $\varphi(FC)$ is consistent in $\varphi(FS)$, then there is a fuzzy interpretation *FI* of $\varphi(FS)$ with $(\varphi(FC))^{FI} \neq \emptyset$, i.e., $\exists d.\ d \in \Delta^{FI}$ and $d \in (\varphi(FC))^{FI}$. By part 2 of Theorem 5.9, $\beta_{FS}(FI)$ is a legal fuzzy database state for *FS* with $\beta_{FV}(d) \in FC^{\beta_{FS}(FI)}$, i.e., $FC^{\beta_{FS}(FI)} \neq \emptyset$. That is, *FC* is consistent. □

Theorem 5.13 (*subsumption of fuzzy classes*) *Given a FOODB model FS, $\varphi(FS)$ is the translated f-ALCIQ(D) TBox, and FC_1, FC_2 are two fuzzy classes in FS. FC_1 is a subclass of FC_2 iff: $\varphi(FS) \vDash \varphi(FC_1) \sqsubseteq \varphi(FC_2)$.*

Proof "⇒": Let $\varphi(FS) \nvDash \varphi(FC_1) \sqsubseteq \varphi(FC_2)$, i.e., $\varphi(FC_1) \sqcap \neg\varphi(FC_2)$ is consistent in $\varphi(FS)$, then there is a fuzzy interpretation FI of $\varphi(FS)$ with $(\varphi(FC_1) \sqcap \neg\varphi(FC_2))^{FI} \neq \emptyset$, i.e., $\exists d \in \Delta^{FI}$ such that $d \in (\varphi(FC_1))^{FI}$ and $d \notin (\varphi(FC_2))^{FI}$. By part 2 of Theorem 5.9, $\beta_{FS}(FI)$ is a legal fuzzy database state for FS with $\beta_{FV}(d) \in FC_1^{\beta_{FS}(FI)}$ and $\beta_{FV}(d) \notin FC_2^{\beta_{FS}(FI)}$, i.e., FC_1 is not the subclass of FC_2, and thus there is a contradiction. That is, $\varphi(FS) \vDash \varphi(FC_1) \sqsubseteq \varphi(FC_2)$.

"⇐": If FC_1 is not the subclass of FC_2, i.e., there is a legal fuzzy database state FJ for FS and $v \in FV_{FJ}$ with $V \in FC_1^{FJ}$ and $V \notin FC_2^{FJ}$. By part 1 of Theorem 5.9, $\alpha_{FS}(FJ)$ is a model of $\varphi(FS)$ with $\alpha_{FV}(v) \in (\varphi(FC_1))^{\alpha_{FS}(FJ)}$ and $\alpha_{FV}(v) \notin (\varphi(FC_2))^{\alpha_{FS}(FJ)}$, i.e., $\varphi(FS) \nvDash \varphi(FC_1) \sqsubseteq \varphi(FC_2)$, and thus there is a contradiction. That is, FC_1 is the subclass of FC_2. □

Before giving Theorem 5.14, we first introduce a notion of having the same individuals for two different *f-ALCIQ(D)*-concepts. Given an *f-ALCIQ(D)* knowledge base, a fuzzy concept $\varphi(FC)$ and $n' \in [0, 1]$, the *retrieval* reasoning service of *f-ALCIQ(D)* can return all the individuals d, which satisfy the condition $\varphi(FC)(d) \geq n'$. Therefore, given two *f-ALCIQ(D)*-concepts $\varphi(FC_1)$ and $\varphi(FC_2)$, two sets of individuals respectively belonging to $\varphi(FC_1)$ and $\varphi(FC_2)$, written as *Retrieval*($\varphi(FC_1)$, n) and *Retrieval*($\varphi(FC_2)$, n), can be created by the retrieval reasoning service above. Here, the degree of truth $n \in [0, 1]$ may be a positive infinite small value or be specified by the designers in the practical FOODB modeling activities. Furthermore, we say that two concepts $\varphi(FC_1)$ and $\varphi(FC_2)$ have the same individuals if the following conditions hold:

(a) For every x in *Retrieval*($\varphi(FC_1)$, n), one of the following holds: (i) x in *Retrieval*($\varphi(FC_2)$, n), or (ii) exists y in *Retrieval*($\varphi(FC_2)$, n) such that $x \approx y$;
(b) For every x in *Retrieval*($\varphi(FC_2)$, n), one of the following holds: (i) x in *Retrieval*($\varphi(FC_1)$, n), or (ii) exists y in *Retrieval*($\varphi(FC_1)$, n) such that $x \approx y$.

Based on the discussion above, the following gives a theorem of checking the redundancy of a FOODB model.

Theorem 5.14 (*redundancy*) *Given a FOODB model FS*, $\varphi(FS)$ is the translated *f-ALCIQ(D)* TBox, and FC, FC_1, FC_2 *are fuzzy classes in FS. FS is redundant if and only if at least one of the following conditions is satisfied: (i)* $\varphi(FS) \vDash \varphi(FC) \sqsubseteq \bot$; (ii) $\varphi(FS) \vDash \varphi(FC_1) \equiv \varphi(FC_2)$; (iii) $\varphi(FS) \nvDash \varphi(FC_1) \sqsubseteq \varphi(FC_2)$, $\varphi(FS) \nvDash \varphi(FC_2) \sqsubseteq \varphi(FC_1)$, and $\varphi(FC_1)$ and $\varphi(FC_2)$ contain same individuals.

Note that, since two concepts are called equivalent in fuzzy DLs when they have same individuals with same membership degrees, the equivalence problem of fuzzy classes in the case (1) of Definition 5.16 can be reasoned by checking whether $\varphi(FS) \vDash \varphi(FC_1) \equiv \varphi(FC_2)$, i.e., $\varphi(FS) \vDash \varphi(FC_1) \sqsubseteq \varphi(FC_2)$ and $\varphi(FS) \vDash \varphi(FC_2) \sqsubseteq \varphi(FC_1)$. However, for the case (2) of Definition 5.16, i.e., when two classes do not have subclass/superclass relationship but they have same objects with different membership degrees, it cannot be handled directly by checking whether $\varphi(FS) \models \varphi(FC_1) \equiv \varphi(FC_2)$. In this case, the constraint that two classes do not have subclass/superclass relationship can be reasoned by checking whether $\varphi(FS) \not\models \varphi(FC_1) \sqsubseteq \varphi(FC_2)$ and $\varphi(FS) \not\models \varphi(FC_2) \sqsubseteq \varphi(FC_1)$. Furthermore, whether two classes contain same objects can also be checked by detecting whether two concepts $\varphi(FC_1)$ and $\varphi(FC_2)$ have the same individuals as mentioned above. Moreover, Theorem 5.14 is the straightforward consequences of Definitions 5.16, and the proof is hereby omitted here.

5.3 Fuzzy Ontology Extraction from Fuzzy Data Models

Ontology extraction (i.e., construction of ontologies) is a very important issue in the context of the Semantic Web. The Semantic Web aims at creating ontology-based and machine-processable Web content, and thus the success and proliferation of the Semantic Web largely depends on *constructing ontologies* (Berners-Lee et al. 2001). To this end, kinds of approaches and tools have been developed to construct ontologies from various data resources such as Text, Dictionary, XML documents, database models, etc. (Corcho et al. 2003; Maedche and Steffen 2001).

However, classical ontology construction approaches are not sufficient for handling imprecise and uncertain information that is commonly found in many application domains. Therefore, great efforts on construction of fuzzy ontologies have been made in recent years. Constructing fuzzy ontologies may facilitate the fuzzy ontology development and the information sharing in the context of the Semantic Web, and also act as a gap-bridge of the realization of semantic interoperations between the existing applications and the Semantic Web. For this purpose, several approaches have been developed for constructing fuzzy ontologies based on the Formal Concept Analysis (FCA) theory (Chen et al. 2009; De Maio et al. 2009; Quan et al. 2006). Some approaches aim at constructing fuzzy ontologies from the data sources, such as fuzzy narrower terms (Widyantoro and Yen 2001), fuzzy relations (Gu et al. 2007), and etc.

Recently, constructing fuzzy ontologies from *fuzzy database models* has received much attention. In particular, as mentioned in Chap. 3, since the early 1980s, Zadeh's fuzzy logic (Zadeh 1965) has been used to extend various data models. This resulted in numerous contributions, mainly with respect to the popular fuzzy data models such as fuzzy ER models, fuzzy UML models, fuzzy relational database models, and fuzzy object-oriented database models. However, such information and knowledge in fuzzy data models cannot participate in the Semantic Web directly. To this end, Ma et al. (2010) and Zhang et al. (2008a) presented a fuzzy ontology construction approach from *fuzzy ER model*. A formal approach for constructing fuzzy ontology from *fuzzy UML model* was developed in (Zhang et al. 2009; Zhang and Ma 2013), where the constructed fuzzy ontologies are formulated in the fuzzy OWL language. Moreover, Zhang et al. (2008b) translated the fuzzy relational schema into the fuzzy ER model and then constructed the fuzzy ontology from the fuzzy ER model at schema level. Similarly, how to construct fuzzy ontologies from *fuzzy relational databases* was also investigated in (Ma et al. 2008), where the authors first implicitly translated the fuzzy relational schema into the fuzzy ER model by means of reverse engineering, and then translated the fuzzy ER model and database instances into the fuzzy ontology structure and fuzzy RDF data model, respectively. Also, an approach for reasoning on fuzzy relational databases with fuzzy ontologies was proposed in Zhang et al. (2012a), where the precondition is that the fuzzy relational databases are transformated into fuzzy ontologies. In Zhang et al. (2010), a formal approach and a tool for constructing fuzzy ontologies from fuzzy object-oriented database models were developed. In adiition, recently Zhang et al. (Zhang et al. 2013a) investigated how to construct fuzzy ontologies from fuzzy XML models. In this chapter, we will introduce fuzzy Semantic Web ontology extraction (i.e., learning) from fuzzy data models, i.e., constructing fuzzy ontologies from fuzzy data models, including from fuzzy ER and UML conceptual data models, fuzzy relational and object-oriented logical database models introduced in Chap. 3.

5.3.1 Fuzzy Ontology Extraction from Fuzzy Conceptual Data Models

As mentioned in Chap. 3, since the early 1980s, Zadeh's fuzzy logic (Zadeh 1965) has been used to extend various conceptual data models (such as the common ER and UML conceptual data models). This resulted in numerous contributions, mainly with respect to the popular fuzzy ER model and fuzzy UML model (see Chap. 3 in detail). In the following, we will focus on fuzzy ontology extraction from fuzzy conceptual data models including fuzzy ER model (Ma et al. 2010; Zhang et al. 2008a) and fuzzy UML model (Zhang et al. 2009; Zhang and Ma 2013).

5.3.1.1 Fuzzy Ontology Extraction from Fuzzy ER Model

In the following, we focus on fuzzy ontology extraction from fuzzy ER model (Ma et al. 2010; Zhang et al. 2008a), i.e., transforming fuzzy ER model into fuzzy ontology structure by a semantics-preserving translation algorithm.

Based on the formal representation of fuzzy ER models introduced in Sect. 3.2.1 and the fuzzy ontology introduced in Sect. 4.3, Definition 5.17 gives the approach for translating fuzzy ER model into fuzzy ontology, following a set of mapping rules.

Definition 5.17 Given a fuzzy ER schema $FS = (L_{FS}, \leq_{FS}, att_{FS}, rel_{FS}, card_{FS})$ mentioned in Sect. 3.2.1, a fuzzy ontology $FO = \varphi(FS) = (FID_0, FAxiom_0)$ is defined by a translation function φ as follows:

(1) The fuzzy ontology identifier set FID_0 of $\varphi(FS)$ contains the elements:

- Each fuzzy entity FE $\in E_{FS}$ is mapped into a fuzzy class identifier φ(FE) $\in$ $FCID_0$.
- Each fuzzy relationship FR $\in R_{FS}$ is mapped into a fuzzy class identifier φ(FR) $\in FCID_0$.
- Each fuzzy attribute symbol FA $\in A_{FS}$ is mapped into a fuzzy datatype property identifier φ(FA) $\in FDPID_0$.
- Each fuzzy domain symbol FD $\in D_{FS}$ is mapped into a fuzzy data range identifier φ(FD) $\in FDRID_0$, where each φ(FD) is predefined XML Schema fuzzy datatype.
- Each fuzzy ER-role FU of fuzzy relationship FR associated with fuzzy entity FE is mapped into a pair of inverse fuzzy object property identifiers φ(FU) $\in$ $FOPID_0$ and FV = invof_φ(FU) $\in FOPID_0$.

(2) The fuzzy ontology axiom set $FAxiom_0$ of $\varphi(FS)$ as follows:

- Each pair of fuzzy entities FE_1, $FE_2 \in E_{FS}$ such that $FE_1 \leq_{FS} FE_2$, a fuzzy class axiom:
  Class ($\varphi(FE_1)$ partial $\varphi(FE_2)$).
- Each fuzzy entity FE $\in E_{FS}$ with att_{FS}(FE) = [FA_1:FD_1,…,FA_k:FD_k], the following fuzzy axioms:
  A fuzzy class axiom:
  Class (φ(FE) partial restriction ($\varphi(FA_1)$ allValuesFrom ($\varphi(FD_1)$) cardinality (x)) … restriction ($\varphi(FA_k)$ allValuesFrom ($\varphi(FD_k)$) cardinality (x)));
  FOR i = 1,2,…,k DO, A fuzzy property axiom:
  DatatypeProperty ($\varphi(FA_i)$ domain (φ(FE)) range ($\varphi(FD_i)$)).
- Each fuzzy relationship FR $\in R_{FS}$ with rel_{FS}(FR) = [FU_1: FE_1,…,FU_n:FE_n], the following fuzzy axioms:
  Class (φ(FR) partial restriction ($\varphi(FU_1)$ allValuesFrom ($\varphi(FE_1)$) cardinality (1)) … restriction ($\varphi(FU_n)$ allValuesFrom ($\varphi(FE_n)$) cardinality (1)));
  FOR i = 1, 2, …, n DO, the fuzzy axioms:
  ObjectProperty (FV_i domain ($\varphi(FE_i)$) range (φ(FR)) inverseOf $\varphi(FU_i)$);
  Class ($\varphi(FE_i)$ partial restriction (FV_i allValuesFrom (φ(FR)))).
- Each fuzzy ER-role FU of fuzzy relationship FR associated with fuzzy entity FE, rel_{FS}(FR) = […,FU:FE,…], the following fuzzy axioms:
  A fuzzy property axiom:
  ObjectProperty (φ(FU) domain (φ(FR)) range (φ(FE)));
  If m = $cmin_{FS}$(FE, FR, FU) $\neq$ 0, a fuzzy class axiom:
  Class (φ(FE) partial restriction (FV minCardinality (m)));
  If n = $cmax_{FS}$(FE, FR, FU) $\neq \infty$, a fuzzy class axiom:
  Class (φ(FE) partial restriction (FV maxCardinality (n))).
- Each pair of symbols X, Y $\in E_{FS} \cup R_{FS} \cup D_{FS}$, such that X $\neq$ Y and X $\in R_{FS} \cup D_{FS}$, a fuzzy class axiom: DisjointClasses (φ(X) φ(Y)).

The correctness of translation function φ can be proved by defining two mappings that from the fuzzy database state FB w.r.t. the fuzzy ER schema FS mentioned in Sect. 3.2.1 to the model of fuzzy ontology φ(FS) and from the model of φ(FS) to the fuzzy database state FB w.r.t. the fuzzy ER schema FS.

Theorem 5.15 *For every fuzzy ER schema FS, $\varphi(FS)$ be the fuzzy OWL ontology structure obtained according to Definition 5.17, there exist two mappings α_{FS}, from the fuzzy database state FB with respect to FS to the model of $\varphi(FS)$, and β_{FS}, from the model of $\varphi(FS)$ to FB for FS, such that*:

- For each legal fuzzy database state FB for FS, there is α_{FS}(FB) which is a model of φ(FS).
- For each model FI of φ(FS), there is β_{FS}(FI) which is a legal fuzzy database state for FS.

The proof of Theorem 5.15 can be proved analogously according to the proof of Theorem 5.1 in Sect. 5.2.1.1, and thus is omitted here.

In the following, we provide an extraction example to well illustrate the extraction procedure in Definition 5.17.

Example 5.6 According to the Definition 5.17, the following will translate the fuzzy ER schema FS_1 mentioned in Fig. 3.1 of Sect. 3.2.1 into the fuzzy OWL ontology structure $\varphi(FS_1)$ in Fig. 5.11.

5.3.1.2 Fuzzy Ontology Extraction from Fuzzy UML Model

Being similar to the fuzzy ontology extraction from fuzzy ER models, the following Definition 5.18 introduces how to extract fuzzy ontologies from fuzzy UML models (Zhang et al. 2009; Zhang and Ma 2013).

Definition 5.18 Given a fuzzy UML model $\mathcal{F}$UML = ($\mathcal{Q}\mathcal{F}$, $\preccurlyeq\mathcal{F}$, att$\mathcal{F}$, agg$\mathcal{F}$, dep$\mathcal{F}$, ass$\mathcal{F}$, card$\mathcal{F}$, mult$\mathcal{F}$, mult$'\mathcal{F}$) as mentioned in Definition 3.4 in Sect. 3.3.2. The fuzzy OWL ontology $\mathcal{FO} = \varphi(\mathcal{F}$UML) can be extracted by transformation function φ as shown in Table 5.6.

The following Theorem 5.16 proves the correctness of the transformation in Definition 5.18. The proof of Theorem 5.16 can be found in (Zhang and Ma 2013).

Theorem 5.16 *For each fuzzy UML model $\mathcal{F}_{UML}$, $\mathcal{FFB}$ is a legal fuzzy object state corresponding to $\mathcal{F}_{UML}$,* and *$\varphi(\mathcal{F}_{UML})$ is the fuzzy OWL ontology extracted from the $\mathcal{F}$ by Definition 5.18. Then there exist two mappingsUML $\alpha_{\mathcal{F}}$, from the fuzzy object state $\mathcal{FFB}$ to the model of $\varphi(\mathcal{F}_{UML})$,* and *$\beta_{\mathcal{F}}$, from the model of $\varphi(\mathcal{F}_{UML})$ to the fuzzy object state $\mathcal{FFB}$, such that*:

- For each legal fuzzy object state $\mathcal{FB}$ for $\mathcal{F}_{UML}$, there is $\alpha_{\mathcal{F}}(\mathcal{FI})$ which is a model of $\varphi(\mathcal{F}_{UML})$.

According to Definition 1, the fuzzy OWL ontology structure φ(FS_1) can be extracted from the fuzzy ER schema FS_1 in Fig. 3.1 in Sect. 3.2.1 as fo llows:
φ(FS_1) = (FID_0, $FAxiom_0$), where
FID_0 = {$FCID_0$ ∪ $FDRID_0$ ∪ $FOPID_0$ ∪ $FDPID_0$}
$FCID_0$ = {employee, Young-employee, Job, branch, Main-branch, Work-on}
$FDRID_0$ = {xsd: string, xsd: integer, xsd: real}
$FDPID_0$ = {Name, FUZZY Age, Sex, FUZZY Salary, Number, *u*}
$FOPID_0$ = {E_w, J_w, B_w, invof_E_w, invof_ J_w, invof_B_w}
$FAxiom_0$= {
Class (Young-employee partial employee); Class (Main-branch partial branch);
Class (Young-employee partial restriction (Name allValuesFrom (xsd: string) cardinality(1)) restriction (FUZZY Age allValuesFrom (xsd: integer) cardinality(1)) ...);
DatatypeProperty (Name domain (Young-employee) range (xsd: string) [Functional]);
DatatypeProperty (FUZZY Age domain (Young-employee) range (xsd: integer) [Functional]);
Class (Work-on partial restriction (E_w allValuesFrom (Young-employee) cardinality(1)) restriction (J_w allValuesFrom (Job) cardinality(1)) restriction (B_w allValuesFrom (Main-branch) cardinality(1)));
ObjectProperty (invof_E_w domain (Young-employee) range (Work-on) inverseOf E_w);
ObjectProperty (invof_J_w domain (Job) range (Work-on) inverseOf J_w);
ObjectProperty (invof_B_w domain (Main-branch) range (Work-on) inverseOf B_w);
Class (Young-employee partial restriction (invof_E_w allValuesFrom (Work-on)));
Class (Job partial restriction (invof_J_w allValuesFrom (Work-on)));
Class (Main-branch partial restriction (invof_B_w allValuesFrom (Work-on)));
Class (Young-employee partial restriction (invof_E_w minCardinality (3)));
Class (Young-employee partial restriction (invof_E_w maxCardinality (10)));
...
}

Fig. 5.11 The fuzzy OWL ontology structure $\varphi(FS_1)$ extracted from the fuzzy ER schema FS_1 in Fig. 3.1 in Sect. 3.2.1

- For each model $\mathcal{FI}$ of $\varphi(\mathcal{F}_{UML})$, there is $\beta_{\mathcal{F}}(\mathcal{FI}))$ which is a legal fuzzy object state for $\mathcal{F}_{UML}$.

Example 5.7 Given the fuzzy UML model $\mathcal{F}_{\mathrm{ML1}}$ in Fig. 3.8 in Sect. 3.2.2, by Definition 5.18, we can extract the corresponding fuzzy OWL ontology $\mathcal{F} = \varphi(\mathcal{F}_{\mathrm{UML1}})$ in Fig. 5.12.

5.3.2 Fuzzy Ontology Extraction from Fuzzy Database Models

Based on the proposed approaches in Sect. 5.3.1, the fuzzy conceptual data models (including fuzzy ER models and fuzzy UML models) can be transformed into fuzzy ontologies. However, the approaches cannot transform the data instances in fuzzy databases into fuzzy ontologies. In this chapter we will focus on fuzzy

Table 5.6 Transforming rules from a fuzzy UML model to fuzzy OWL ontology

Fuzzy UML model $\mathcal{F}_{UML}$	Fuzzy OWL ontology structure $\mathcal{F}_S = \varphi(\mathcal{F}_{UML}) = (FID_0, FAxiom_0)$
Alphabet set $\mathcal{Q}_\mathcal{F}$	*Fuzzy OWL class description set* FID_0
Each fuzzy class $FC \in FC_\mathcal{F}$	A fuzzy class identifier $\varphi(FC) \in FCID_0$
Each fuzzy attribute symbol $FA \in FA_\mathcal{F}$	A fuzzy datatype property identifier $\varphi(FA) \in FDPID_0$
Each method without parameters $f()$: $R \in FM_\mathcal{F}$	A fuzzy datatype property identifier $\varphi(f) \in FDPID_0$ A fuzzy data range identifier $\varphi(R) \in FDRID_0$
Each fuzzy datatype symbol $FT \in FT_\mathcal{F}$	A fuzzy data range identifier $(FT) \in FDRID_0$
Each method with parameters $f(P_1,\ldots,P_m)$: $R \in FM_\mathcal{F}$	A fuzzy class identifier $\varphi(FC_{f(P1, \ldots, Pm)}) \in FCID_0$ m fuzzy data range identifiers $\varphi(P_1)\ldots\varphi(P_m) \in FDRID_0$ A fuzzy data range identifier $\varphi(R) \in FDRID_0$ A fuzzy object property identifier $\varphi(r_1) \in FOPID_0$ $(m + 1)$ fuzzy datatype property identifiers $\varphi(r_2)\ldots\varphi(r_{m+2}) \in FDPID_0$ *Note that*: $\varphi(r_1)$ represents the object of invocation, the next m $\varphi(r_2)\ldots\varphi(r_{m+1})$ represent the parameters, and the last one $\varphi(r_{m+2})$ represents the return result
Each fuzzy association (class) symbol $FS \in FS_\mathcal{F}$	A fuzzy class identifier $\varphi(FS) \in FCID_0$
Each role symbol $FR \in FR_\mathcal{F}$	A pair of inverse fuzzy object property identifiers: $\varphi(FR) \in FOPID_0$ and $v = \text{invof\_}\varphi(FR) \in FOPID_0$
Each fuzzy aggregation $FG \in FG_\mathcal{F}$ such that $agg_\mathcal{F}(FG) = FC \times (FC_1 \cup \ldots \cup FC_m)$	m pairs of fuzzy object property identifiers: $\varphi(\text{is\_part\_of\_}g_1)$, $\varphi(\text{is\_whole\_of\_}g_1)$, …, $\varphi(\text{is\_part\_of\_}g_m)$, $\varphi(\text{is\_whole\_of\_}g_m)$
Each extensionally defined class $FC \in FC_\mathcal{F}$ such that $FC = \{o_1, \ldots, o_n\}$, where $o_i \in FO_\mathcal{F}$	n fuzzy individual identifiers $\varphi(o_1), \ldots, \varphi(o_n) \in FIID_0$
Constraints	*Fuzzy OWL axiom set* $FAxiom_0$
Each enumeration class $FC = \{o_1,\ldots,o_n\}$, where $FC \in FC_\mathcal{F}$, $o_i \in FO_\mathcal{F}$	Creating a fuzzy class axiom: EnumeratedClass ($\varphi(FC)$ $\varphi(o_1)\ldots\varphi(o_n)$) .
Each fuzzy generalization $\preccurlyeq_\mathcal{F}(FC_2) = FC_1$, where $FC_1, FC_2 \in FC_\mathcal{F}$	Creating a fuzzy class axiom: Class ($\varphi(FC_1)$ partial $\varphi(FC_2)$) or SubClassOf ($\varphi(FC_1)$ $\varphi(FC_2)$).
Each fuzzy class hierarchy $\preccurlyeq_\mathcal{F}(FC) = FC_1 \times FC_2 \times \ldots \times FC_n$, i.e., a fuzzy class FC generalizing n fuzzy classes $FC_1, \ldots, FC_n$ with the optional *disjointness* and *completeness* constraints	Creating fuzzy class axioms: Class ($\varphi(FC_i)$ partial $\varphi(FC)$) or SubClass of ($\varphi(FC_i)$ $\varphi(FC)$); If there are the *disjointness* and *completeness* constraints, then adding fuzzy class axioms: EquivalentClasses ($\varphi(FC)$ unionOf ($\varphi(FC_1)$, …, $\varphi(FC_n)$) DisjointClasses ($\varphi(FC_i)$, $\varphi(FC_j)$) $i, j \in 1, 2, \ldots, n$, $i \neq j$.

(continued)

Table 5.6 (continued)

Fuzzy UML model $\mathcal{F}_{UML}$	Fuzzy OWL ontology structure $\mathcal{F}_S = \varphi(\mathcal{F}_{UML}) = (FID_0, FAxiom_0)$
Each fuzzy dependency such that $dep_{\mathcal{F}} \subseteq FC_1 \times FC_2$	The dependency between FC_1 and FC_2 is only related to the classes themselves, and the current fuzzy OWL ontologies are not enough to support this type of property, which can be indicated by the designers
Each pair of symbols $x, y \in FC_{\mathcal{F}} \cup FS_{\mathcal{F}}$ with $x \neq y$ and $x \in FS_{\mathcal{F}}$	Creating a fuzzy class axiom: DisjointClasses ($\varphi(x)$, $\varphi(y)$)
Each fuzzy class $FC \in FC_{\mathcal{F}}$ such that $att_{\mathcal{F}}(FC) \rightarrow [\ldots, f() : R, \ldots]$	Creating a fuzzy class axiom: Class ($\varphi(FC)$ partial restriction ($\varphi(f)$ all valuesFrom ($\varphi(R)$) maxCardinality(1))).
Each fuzzy aggregation $FG \in FG_{\mathcal{F}}$ such that $agg_{\mathcal{F}}(FG) = FC \times (FC_1\ [w_1 \ldots z_1] \cup \ldots \cup FC_m\ [w_m \ldots z_m])$ Here, FC, $FC_i \in FC_{\mathcal{F}}$, $(w_i \ldots z_i)$ is the multiplicity $mult'_{\mathcal{F}}(FC_i, FC) = (mult'_{min}(FC_i, FC), mult'_{max}(FC_i, FC))$, and $i \in \{1 \ldots m\}$	Creating a fuzzy class axiom: Class (owl:Thing partial restriction (inverseOf φ(is_part_of_g_1) allValuesFrom ($\varphi(FC)$) Cardinality (1)) restriction (φ(is_whole_of_g_1) allValuesFrom ($\varphi(FC_1)$) minCardinality (w_1) maxCardinality (z_1)) … restriction (inverseOf φ(is_part_of_g_m) allValuesFrom ($\varphi(FC)$) Cardinality (1)) restriction (φ(is_whole_of_g_m) allValuesFrom ($\varphi(FC_m)$) minCardinality (w_m) maxCardinality (z_m))); Creating fuzzy property axioms: ObjectProperty (φ(is_part_of_g_i) domain ($\varphi(FC_i)$) range ($\varphi(FC)$)); ObjectProperty (φ(is_whole_of_g_i) domain ($\varphi(FC)$) range ($\varphi(FC_i)$)).
Each fuzzy class $FC \in FC_{\mathcal{F}}$ such that $att_{\mathcal{F}}(FC) \rightarrow [\ldots, f(P_1, \ldots, P_m) : R, \ldots]$. Here, $f(P_1, \ldots, P_m) : R \in FM_{\mathcal{F}}$ is a method with m parameters $P_1, \ldots, P_m$.	Creating fuzzy class axioms: Class ($\varphi(FC_{f(P1, \ldots, Pm)})$ partial restriction ($\varphi(r_1)$ someValuesFrom (owl:Thing) Cardinality(1)) … restriction ($\varphi(r_{m+1})$ someValuesFrom (owl:Thing) Cardinality(1))); Class ($\varphi(FC_{f(P1, \ldots, Pm)})$ partial restriction ($\varphi(r_2)$ allValuesFrom ($\varphi(P_1)$)) … restriction ($\varphi(r_{m+1})$ allValuesFrom ($\varphi(P_m)$))); Class ($\varphi(FC)$ partial restriction (inverseOf($\varphi(r_1)$) allValuesFrom (unionOf (complementOf ($\varphi(C_{f(P1, \ldots, Pm)})$) restriction ($\varphi(r_{m+2})$ allValuesFrom ($\varphi(R)$)))))). *Note that*: The first axiom states that each instance of $\varphi(FC_{f(P1, \ldots, Pm)})$, representing a tuple, correctly is connected to exactly one object for each of $r_1, \ldots r_{m+1}$; The other two axioms impose the correct typing of parameters and of the return value

(continued)

Table 5.6 (continued)

Fuzzy UML model $\mathcal{F}_{UML}$	Fuzzy OWL ontology structure $\mathcal{F}_S = \varphi(\mathcal{F}_{UML}) = (FID_0, FAxiom_0)$
Each fuzzy association $FS \in FS_{\mathcal{F}}$ such that $ass_{\mathcal{F}}(FS) = [FR_1 : FC_1, \ldots, FR_k : FC_k]$. Here, $FR_i \in FR_{\mathcal{F}}$, $FC_i \in FC_{\mathcal{F}}$, each role FR_i is associated with a *cardinality constraint* $card_{\mathcal{F}}(FC_i, FS, FR_i) = (card_{min}(FC_i, FS, FR_i), card_{max}(FC_i, FS, FR_i))$, and $i \in \{1 \ldots k\}$	Creating fuzzy class axioms: Class ($\varphi(FS)$ partial restriction ($\varphi(FR_1)$ allValuesFrom ($\varphi(FC_1)$) cardinality (1)) … restriction ($\varphi(FR_k)$ allValuesFrom ($\varphi(FC_k)$) cardinality (1))); Class ($\varphi(FC_i)$ partial restriction (v_i allValuesFrom ($\varphi(FS)$))), where v_i = invof_$\varphi(FR_i) \in FOPID_0$ denotes an inverse property of the fuzzy object property $\varphi(FR_i)$ as mentioned above; Do case $card_{\mathcal{F}}(FC_i, FS, FR_i)$ of: (i) $m = card_{min}(FC_i, FS, FR_i) \neq 0$, creating axiom: Class ($\varphi(FC_i)$ partial restriction (v_i minCardinality (m))); (ii) $n = card_{max}(FC_i, FS, FR_i) \neq \infty$, creating axiom: Class ($\varphi(FC_i)$ partial restriction (v_i maxCardinality (n))); (iii) $q = card_{\mathcal{F}}(FC_i, FS, FR_i)$, creating axiom: Class ($\varphi(FC_i)$ partial restriction (v_i Cardinality (q))); Creating fuzzy property axioms: ObjectProperty (v_i domain ($\varphi(FC_i)$) range ($\varphi(FS)$) inverseOf $\varphi(FR_i)$); ObjectProperty ($\varphi(FR_i)$ domain ($\varphi(FS)$) range ($\varphi(FC_i)$))

ontology extraction from fuzzy database models, including extracting fuzzy ontologies from fuzzy relational databases (Zhang et al. 2008b, 2012a) and fuzzy object-oriented databases (Zhang et al. 2010).

5.3.2.1 Fuzzy Ontology Extraction from Fuzzy Relational Databases

Fuzzy ontology extraction from fuzzy relational database (i.e., transforming fuzzy relational database into fuzzy ontology) is similar to the fuzzy Description Logic extraction from fuzzy relational database as introduced in Sect. 5.2.2. In order to transform fuzzy relational database into fuzzy ontology, we first recall a classification method of fuzzy relational schemas as given in Sect. 5.2.2. The fuzzy relational schemas specify a set of integrity constraints, which restricts the data that may be stored in the fuzzy relational database. Given a fuzzy relational schema $FS = FR(A_1/D_1, A_2/D_2, \ldots, A_n/D_n, \mu_{FR}/D_{FR})$ mentioned in Sect. 3.3.1, the

ΦO = φ(Φ_{UML1}) = <FID_0, $FAxiom_0$>, for brevity, FID_0, part of axioms, and the symbol φ( ) are omitted:
$FAxiom_0$ = { Class (Young-Employee partial Employee);
Class (Old-Employee partial Employee); Class (New-Computer partial Computer);
EquivalentClasses(φ(Employee), unionOf (φ(Young-Employee), φ(Old-Employee)));
DisjointClasses (Young-Employee, Old-Employee);
Class (Old-Employee partial restriction (ID allValuesFrom (xsd:String) Cardinality (1)) restriction (Name allValuesFrom (xsd:String) Cardinality (1)) … restriction (*u* allValuesFrom (xsd:Real) Cardinality (1)));
Class (se partial restriction (Date allValuesFrom (xsd:String) Cardinality (1)) restriction (*β* allValuesFrom (xsd:Real) Cardinality (1)) restriction (Uby allValuesFrom (Old-Employee) Cardinality (1)) restriction (Uof allValuesFrom (New-Computer) Cardinality (1))) ;
Class (owl:Thing partial restriction (inverseOf φ(is_part_of_g_1) allValuesFrom (New-Computer) Cardinality (1)) restriction (φ(is_whole_of_g_1) allValuesFrom (Monitor) minCardinality (1)) … restriction (inverseOf φ(is_part_of_g_m) allValuesFrom (New-Computer) Cardinality (1)) restriction (φ(is_whole_of_g_m) allValuesFrom (Keyboard) minCardinality (1))) ;
Class (Old-Employee partial restriction (invof_Uby allValuesFrom (Use)));
Class (New-Computer partial restriction (invof_Uof allValuesFrom (Use)));
Class (Old-Employee partial restriction (invof_Uby Cardinality (1)));
Class (New-Computer partial restriction (invof_Uof minCardinality (1) maxCardinality (2)));
ObjectProperty (is_part_of_g_1 domain (Monitor) range (New-Computer));
ObjectProperty (is_whole_of_g_1 domain (New-Computer) range (Monitor));
ObjectProperty (Uby domain (Use) range (Old-Employee));
ObjectProperty (Uof domain(Use) range (New-Computer));
ObjectProperty (invof_Uby domain (Old-Employee) range (Use) inverseOf Uby);
… }

Fig. 5.12 The fuzzy OWL ontology $\varphi(\mathcal{F}_{\text{UML1}})$ extracted from the fuzzy UML model $\mathcal{F}_{\text{UML1}}$ in Fig. 3.8 in Sect. 3.2.2

fuzzy relation *FR* defined on the fuzzy relational schema *FS*, is classified as one of the following:

(a) The fuzzy relation *FR* has not foreign keys.
(b) The fuzzy relation *FR* has one foreign key *FR*[FK] that references the primary key of a fuzzy relation *FR'*, i.e., |*FR*[FK]| = 1, then:

- If the foreign key *FR*[FK] is one of the candidate keys of *FR*, then the relationship between *FR* and *FR'* may be a *1–1* relationship;
- If the foreign key *FR*[FK] is not one of the candidate keys of *FR*, then the relationship between *FR* and *FR'* is a *many–1* relationship.

(c) The fuzzy relation *FR* has a primary key PK that is also a foreign key FK, and FK references the primary key of a fuzzy relation *FR*′, then *FR*′ subsumes *FR*, i.e., *FR* "is-a" *FR*′.
(d) The fuzzy relation *FR* has more than one foreign keys that respectively reference the primary keys PK_i of fuzzy relations FR_i, $i \in \{1,\ldots,n\}$, and $n \geq 2$, i.e., |*FR*[FK]| > 1, then *FR* is a *many–many* relationship relation. Moreover, the fuzzy relation *FR* may have its own attributes.

Moreover, as we have known, the fuzzy ontology structure differs from the fuzzy relational database schema in that: In the fuzzy ontology structure, everything is a concept; in the concept, there are no attributes and relationships; everything is a property; and the concepts and properties can be organized into axioms or inheritance hierarchies.

Based on the discussions above, we give a formal approach for transforming *the fuzzy relational schemas* of a fuzzy relational database into *fuzzy OWL ontology structure*. Starting with the construction of fuzzy OWL descriptions, the approach further induces a set of fuzzy axioms (i.e., fuzzy class and property axioms) from the fuzzy relational schemas.

Definition 5.19 Given a set of fuzzy relational schemas *FS* of a fuzzy relational database (where each fuzzy relational schema is the form of $FR(A_1/D_1, A_2/D_2, \ldots, A_n/D_n, \mu_{FR}/D_{FR})$). The fuzzy ontology structure $FO_S = \varphi(FS) = (FID_0, FAxiom_0)$ is obtained by transformation function φ:

(1) The fuzzy OWL class description set FID_0 of $\varphi(FS)$ contains following elements:

- Each fuzzy relation symbol *FR* is mapped into a fuzzy class $\varphi(FR) \in FCID_0$;
- Each fuzzy domain symbol D_i ($D_{FR} \in D_i$) is mapped into a fuzzy data range $\varphi(D_i) \in FDRID_0$;
- Each foreign key attribute A_i is mapped into a pair of inverse fuzzy object properties $\varphi(A_i) \in FOPID_0$ and invof_$\varphi(A_i) \in FOPID_0$;
- Each fuzzy attribute A_i ($\mu_{FR} \in A_i$) is mapped into a fuzzy datatype property $\varphi(A_i) \in FDPID_0$.

(2) The fuzzy class/property axiom set $FAxiom_0$ of $\varphi(FS)$ contains following elements:

- *Case 1 FR* has not foreign keys.
  Creating a fuzzy class axiom:

  Class ($\varphi(FR)$ partial restriction ($\varphi(A_1)$ allValuesFrom ($\varphi(D_1)$) cardinality (1)) … restriction ($\varphi(A_n)$ allValuesFrom ($\varphi(D_n)$) cardinality (1)) restriction ($\varphi(\mu_{FR})$ allValuesFrom ($\varphi(D_{FR})$) cardinality (1)));
  FOR $i = 1, 2, \ldots, n$ DO, Creating fuzzy datatype property axioms:

  DatatypeProperty ($\varphi(A_i)$ domain ($\varphi(FR)$) range($\varphi(D_i)$) [Functional]);
  DatatypeProperty ($\varphi(\mu_{FR})$ domain ($\varphi(FR)$) range($\varphi(D_{FR})$) [Functional]).
- *Case 2 FR* has exactly one foreign key attribute *FK* that references the primary key of a fuzzy relation *FR'* (where *FR'* is contained in the set of fuzzy relational schemas *FS*). Then:

  Case 2.1 FR and *FR′* is a *1-1* relationship, then:
  Creating fuzzy class axioms:

Class ($\varphi(FR)$ partial restriction ($\varphi(A_1)$ allValuesFrom ($\varphi(D_1)$) cardinality (1)) … restriction ($\varphi(A_n)$ allValuesFrom ($\varphi(D_n)$) cardinality (1)) restriction ($\varphi(\mu_{FR})$ allValuesFrom ($\varphi(D_{FR})$) cardinality (1)));
Class ($\varphi(FR)$ partial restriction ($\varphi(FK)$ allValuesFrom ($\varphi(FR')$) cardinality (1)));
Class ($\varphi(FR')$ partial restriction (invof_$\varphi(FK)$ allValuesFrom ($\varphi(FR)$) cardinality (1)));

Creating fuzzy datatype property axioms:

DatatypeProperty ($\varphi(A_i)$ domain ($\varphi(FR)$) range($\varphi(D_i)$) [Functional]);
DatatypeProperty($\varphi(\mu_{FR})$ domain ($\varphi(FR)$) range($\varphi(D_{FR})$) [Functional]);

Creating fuzzy object property axioms:

ObjectProperty ($\varphi(FK)$ domain ($\varphi(FR)$) range ($\varphi(FR')$) [Functional]);
ObjectProperty (invof_$\varphi(FK)$ domain ($\varphi(FR')$) range ($\varphi(FR)$) [Functional] inverseOf $\varphi(FK)$).

Case 2.2 FR and *FR'* is a *many-1* relationship, then:
Creating fuzzy class axioms:

Class ($\varphi(FR)$ partial restriction ($\varphi(A_1)$ allValuesFrom ($\varphi(D_1)$) cardinality (1)) … restriction ($\varphi(A_n)$ allValuesFrom ($\varphi(D_n)$) cardinality (1)) restriction ($\varphi(\mu_{FR})$ allValuesFrom ($\varphi(D_{FR})$) cardinality (1)));
Class ($\varphi(FR)$ partial restriction ($\varphi(FK)$ allValuesFrom ($\varphi(FR')$) cardinality (1)));
Class ($\varphi(FR')$ partial restriction (invof_$\varphi(FK)$ allValuesFrom ($\varphi(FR)$) mincardinality (1)));

Creating fuzzy object property axioms:

ObjectProperty ($\varphi(FK)$ domain ($\varphi(FR)$) range ($\varphi(FR')$) [Functional]);
ObjectProperty (invof_$\varphi(FK)$ domain ($\varphi(FR')$) range ($\varphi(FR)$) inverseOf $\varphi(FK)$);

Creating fuzzy datatype property axioms:

DatatypeProperty ($\varphi(A_i)$ domain ($\varphi(FR)$) range($\varphi(D_i)$) [Functional]);
DatatypeProperty ($\varphi(\mu_{FR})$ domain ($\varphi(FR)$) range($\varphi(D_{FR})$) [Functional]).

- *Case 3 FR* has a primary key *PK* that is also a foreign key *FK*, and *FK* references the primary key of a fuzzy relation *FR'*, i.e., *FR'* subsumes *FR*. Then creating a fuzzy class axiom:
  SubClassOf ($\varphi(FR)$ $\varphi(FR')$).
- *Case 4 FR* has more than one foreign key FK_i that respectively reference the primary keys of fuzzy relations FR_1, …, FR_n, $i \in \{1,\ldots,n\}$, and $n \geq 2$. Then:Creating fuzzy class axioms:

Class (φ (FR) partial restriction ($\varphi(A_1)$ allValuesFrom ($\varphi(D_1)$) cardinality (1)) … restriction ($\varphi(A_n)$ allValuesFrom ($\varphi(D_n)$) cardinality (1)) restriction ($\varphi(\mu_{FR})$ allValuesFrom ($\varphi(D_{FR})$) cardinality (1)));
Class ($\varphi(FR)$ partial restriction ($\varphi(FK_1)$ allValuesFrom ($\varphi(FR_1)$) cardinality (1)) … restriction ($\varphi(FK_n)$ allValuesFrom ($\varphi(FR_n)$) cardinality (1)));
Class ($\varphi(FR_i)$ partial restriction (invof_$\varphi(FK_i)$ allValuesFrom ($\varphi(FR)$) mincardinality (1))), $i = 1, 2, …, n$;
Creating fuzzy datatype property axioms:

DatatypeProperty ($\varphi(A_i)$ domain ($\varphi(FR)$) range($\varphi(D_i)$) [Functional]);
DatatypeProperty($\varphi(\mu_{FR})$ domain ($\varphi(FR)$) range($\varphi(D_{FR})$) [Functional]);
Creating fuzzy object property axioms:

ObjectProperty ($\varphi(FK_i)$ domain ($\varphi(FR)$) range ($\varphi(FR_i)$) [Functional]), $i = 1, 2, …, n$;
ObjectProperty (invof_$\varphi(FK_i)$ domain ($\varphi(FR_i)$) range ($\varphi(FR)$) inverseOf $\varphi(FK_i)$), $i = 1, 2, …, n$.

Based on the structure transformation mentioned above, here we further give an approach that can transform the fuzzy relational database instance into the fuzzy OWL ontology instance at instance level, i.e., from *fuzzy relations* (w.r.t. the fuzzy relational schemas) into the *fuzzy OWL ontology instance* (w.r.t. the fuzzy OWL ontology structure). In a fuzzy OWL ontology, the instances are represented by the fuzzy individual axioms: Individual (o type(C_1) [$\bowtie m_1$] … value(R_1, o_1) [$\bowtie k_1$] … value(T_1, v_1) [$\bowtie l_1$]…); DifferentIndividuals ($o_1…o_n$); SameIndividual ($o_1…o_n$) (see Table 4.5 in Sect. 4.3). A fuzzy relational database describes the real world by means of a set of fuzzy relations or tables, which can be considered as a finite set of assertions as have mentioned in Sect. 5.2.2.

On this basis, given a fuzzy relational database $FRDB = < FS, FR >$ as mentioned in Sect. 3.3.1, the set of fuzzy relational schemas FS can be transformed into the fuzzy OWL ontology structure $FO_S = \varphi(FS)$ according to Definition 5.19, and the following Definition 5.20 can further transform the fuzzy relational instance (i.e., a set of fuzzy relations FR) into the fuzzy OWL ontology instance $FO_I = \varphi(FR)$. The FO_S and FO_I form the body of a target fuzzy OWL ontology.

Definition 5.20 Given the *fuzzy relations* FR of a fuzzy relational database FRDB $= < FS, FR >$ as mentioned in Sect. 3.3.1, the corresponding *fuzzy OWL ontology instance* $FO_I = \varphi(FR) = (FIID_O, FAxiom_O)$ can be derived as the following rules:

- Each tuple t in FR (i.e., an individual identified uniquely by the primary key) is mapped to a fuzzy OWL individual $\varphi(t) \in FIID_O$;
- Each fuzzy relation symbol FR is mapped into a fuzzy class description $\varphi(FR) \in FCID_O$ (see Definition 5.19 in detail);
- Each non-foreign key attribute A_i ($\mu_{FR} \in A_i$) is mapped into a fuzzy datatype property $\varphi(A_i) \in FDPID_O$; Each foreign key attribute A_i is mapped into a pair of

inverse fuzzy object properties $\varphi(A_i) \in FOPID_O$ and invof_$\varphi(A_i) \in FOPID_O$ (see Definition 5.19 in detail);

- Each assertion t: FR: n is mapped into a fuzzy individual axiom *Individual* ($\varphi(t)$ *type*($\varphi(FR)$) [$\bowtie' n$]), where $\bowtie' \in \{\geq, \leq\}$;
- Each assertion $t : [\ldots, A_i : V_i : n_i, \ldots]$, if A_i is a non-foreign key attribute, the assertion is mapped into a fuzzy individual axiom *Individual* ($\varphi(t)$ *value*($\varphi(A_i)$, V_i) [$\bowtie$ss$' n_i$] ...); if $A_i = FK_i$ is a foreign key attribute referencing a fuzzy relation FR_i, i.e., V_i denotes a tuple t_i in FR_i and there is a corresponding fuzzy OWL individual $\varphi(t_i) \in FIID_O$, the assertion is mapped into fuzzy individual axioms *Individual* ($\varphi(t)$ *value*($\varphi(FK_i)$, $\varphi(t_i)$) [$\bowtie' n_i$]...) and *Individual* ($\varphi(t_i)$ *value* (invof_$\varphi(FK_i)$, $\varphi(t)$) [$\bowtie' n_i$]...), where $\bowtie' \in \{\geq, \leq\}$, and for the sake of simplicity $V_i : n_i$ in the above assertion only denotes one element of the possibility distribution.

The correctness of the transformations in Definitions 5.19 and 5.20 can be proved following the similar procedure mentioned in Theorem 5.8 in Sect. 5.2.2.

Example 5.8 Given the fuzzy relational database in Sect. 3.3.1 (including the set of fuzzy relational schemas in Table 3.1 and the fuzzy relations in Table 3.2), a fuzzy ontology can be extracted from the fuzzy relational database is shown in Fig. 5.13.

5.3.2.2 Fuzzy Ontology Extraction from Fuzzy Object-oriented Databases

Being similar to the extraction of fuzzy ontologies from fuzzy relational databases, Zhang et al. (2010) proposed a brief approach for extracting fuzzy ontologies from fuzzy object-oriented databases, i.e., transforming fuzzy object-oriented databases into fuzzy ontologies, which including the transformations at the structure and instance levels. The approach only considers a part of constraints of fuzzy object-oriented databases mentioned in Sect. 3.3.2. The following Definition 5.21 gives the formal approach for translating a fuzzy object-oriented database into a fuzzy ontology at structure level (Zhang et al. 2010). Starting with the construction of the *fuzzy OWL atomic identifiers* FID_O, the approach induces a set of *fuzzy class/ property axioms* $FAxiom_O$ from the fuzzy object-oriented database.

Definition 5.21 Given a fuzzy object-oriented database model $FS = (FC_{FS}, FA_{FS}, FD_{FS})$ as mentioned in Sect. 3.3.2. The fuzzy OWL ontology structure $FO_S = \varphi(FS) = (FID_O, FAxiom_O)$ can be derived by function φ as shown in Table 5.7.

On this basis, the mappings from the *database instances* w.r.t. a fuzzy object-oriented database model to the *fuzzy OWL ontology instances* can be established by Definition 5.22.

The following is the fuzzy OWL ontology $FO = (FO_S, FO_I) = (FID_0, FAxiom_0)$ derived from the fuzzy relational database in Sect. 3.3.1. Here, the part FID_0 and some axioms are omitted.

$FAxiom_0$ = {
Class (Department partial restriction (depID allValuesFrom (xsd:string) cardinality(1)) restriction (dName allValuesFrom (xsd:string) cardinality(1)) restriction (μ_{FR} allValuesFrom (xsd:real) cardinality(1)));
Class (Young-Employee partial restriction (empID allValuesFrom (xsd:string) cardinality(1)) restriction (yeName allValuesFrom (xsd:string) cardinality(1)) ...);
DatatypeProperty (depID domain(Department) range(xsd:string) [Functional]);
DatatypeProperty (dName domain(Department) range(xsd:string) [Functional]);
Class (Young-Employee partial restriction (dep_ID allValuesFrom (Department) cardinality(1))) ;
Class (Department partial restriction (invof_dep_ID allValuesFrom (Young-Employee) mincardinality(1))) ;
Class (Supervise partial restriction (supID allValuesFrom (xsd:string) cardinality(1)) restriction (lea_ID allValuesFrom (Chief-Leader) cardinality(1)) restriction (emp_ID allValuesFrom (Young-Employee) cardinality(1)) restriction (μ_{FR} allValuesFrom (xsd:real) cardinality(1))) ;
ObjectProperty (dep_ID domain (Young-Employee) range (Department) [Functional]) ;
ObjectProperty (invof_dep_ID domain (Department) range (Young-Employee) inverseOf dep_ID) ;
DifferentIndividuals (L001, L002, L003, E001, E002, ..., S001, S002, S003) ;
Individual (L001 type(Leader) [⋈′ 0.7]) ;
Individual (L001 value(leaID, L001) value(lNumber, 001) value(μ_{FR}, 0.7)) ;
Individual (E001 type(Employee) [⋈′ 0.8]) ;
Individual (L001 value(leaID, L001) value(clName, Chris) value(f_clAge, 35) [⋈′ 0.8] ... value(μ_{FR}, 0.65)) ;
Individual (L003 value(leaID, L003) value(clName, Billy) value(f_clAge, 37) value(μ_{FR}, 0.7)) ;
Individual (E001 value(empID, E001) value(yeName, John) value(f_yeAge, 24) [⋈′ 0.7] ... value(f_yeSalary, 3000) [⋈′ 0.4] value(dep_ID, D001) value(μ_{FR}, 0.75));
Individual (L001 value(invof_lea_ID, S001));
Individual (S002 value(supID, S002) value(lea_ID, L001) value(emp_ID, E002) value(μ_{FR}, 0.8)) ;
...}.

Fig. 5.13 The fuzzy OWL ontology extracted from the fuzzy relational database in Sect. 3.3.1

Definition 5.22 Given the database instances w.r.t. a fuzzy object-oriented database model (i.e., a finite set of assertions), the corresponding fuzzy OWL ontology instances (i.e., a set of fuzzy individual axioms) can be derived as the mapping rules shown in Table 5.8.

The correctness of the transformations in Definitions 5.21 and 5.22 can be proved following the similar procedure in Theorem 5.9 in Sect. 5.2.2.

Example 5.9 Given the fuzzy object-oriented database mentioned in Sect. 3.3.2 (including the schema information in Fig. 3.10 and the object instance information in Fig. 3.11), by applying Definitions 5.21 and 5.22, we can extract a fuzzy ontology as shown in Fig. 5.14.

Table 5.7 Transforming from a fuzzy object-oriented database model to a fuzzy OWL ontology at structure level

FOOD model $FS = (FC_{FS}, FA_{FS}, FD_{FS})$	Fuzzy ontology structure $FO_S = \varphi(FS) = (FID_0, FAxiom_0)$
Fuzzy class FC_{FS}*, fuzzy attribute* FA_{FS}	*Identifier set* FID_0
Each fuzzy class symbol FC	A fuzzy class identifier $\varphi(FC) \in FCID_0$
Each type expression *Record* FA : *Set-of* FT *End*	
Each *object attribute* symbol FA	A fuzzy class identifier $\varphi(FA) \in FCID_0$ Add two additional roles FU_1 and FU_2;
Each *object attribute* symbol FA	FU_1 and FU_2 are mapped into two fuzzy object property identifiers $\varphi(FU_1)$ and $\varphi(FU_2)$: $\varphi(FU_1) \in FOPID_0$ $\varphi(FU_2) \in FOPID_0$; In addition, we use symbols FV_1 and FV_2 denote the inverse properties of $\varphi(FU_1)$ and $\varphi(FU_2)$, respectively: $FV_1 = invof\_\varphi(FU_1) \in FOPID_0$ $FV_2 = invof\_\varphi(FU_2) \in FOPID_0$
Each type expression *Record* FA_1:FD_1, …, FA_k:FD_k *End*	
Each *datatype attribute* symbol FA_i	A fuzzy datatype property identifier $\varphi(FA_i) \in FDPID_0$
Each domain symbol FD_i	A fuzzy XML Schema datatype identifier $\varphi(FD_i)$ $FDRID_0$
Fuzzy class declarations FD_{FS}	*Fuzzy class/property axiom set* $FAxiom_0$
Each fuzzy class declaration: Class FC *is-a* FC_1,…, FC_n	Create a fuzzy class axiom: Class (φ(FC) partial $\varphi(FC_1)$,…,$\varphi(FC_n)$)
Each fuzzy class declaration: Class FC *type-is* *Union* FT_1,…, FT_k *End*	Create fuzzy class axioms: Class (φ(FC) partial unionOf($\varphi(FT_1)$,…, $\varphi(FT_k)$)); SubClassOf ($\varphi(FT_i)$ φ(FC)) If there are the *disjointness* and *covering* constraints, then adding fuzzy class axioms: DisjointClasses ($\varphi(FC_i)$, $\varphi(FC_j)$) $i, j \in 1, 2, \ldots, k, i \neq j$; EquivalentClasses ($\varphi(FC)$ unionOf ($\varphi(FC_1)$, …, $\varphi(FC_k)$)) (Comments: here the above axiom "Class (φ(FC) partial unionOf($\varphi(FT_1)$,…,$\varphi(FT_k)$))" is replace with the axiom "EquivalentClasses ($\varphi(FC)$ unionOf ($\varphi(FC_1)$, …, $\varphi(FC_n)$))"

(continued)

Table 5.7 (continued)

FOOD model $FS = (FC_{FS}, FA_{FS}, FD_{FS})$	Fuzzy ontology structure $FO_S = \varphi(FS) = (FID_O, FAxiom_O)$
Each fuzzy class declaration: Class FC_1 *type-is* *Record* FA: *Set-of* FC_2 *End* //here the cardinality constraints of a set $[(m_1, n_1), (m_2, n_2)]$ as mentioned in Sect. 3.3.2 is not considered in Zhang et al. (2010)	Create a fuzzy class axiom: Class (φ(FA) partial restriction (φ(FU_1) allValuesFrom (φ(FC_1)) cardinality(1)) restriction (φ(FU_2) allValuesFrom (φ(FC_2)) cardinality(1))); FOR i = 1, 2 DO: The fuzzy property axioms: ObjectProperty (φ(FU_i) domain (φ(FA)) range (φ(FC_i))); ObjectProperty (FV_i domain (φ(FC_i)) range (φ(FA)) inverseOf φ(FU_i)); The fuzzy class axioms: Class (φ(FC_i) partial restriction (FV_i allValuesFrom (φ(FA))))
Each fuzzy class declaration: Class FC *type-is* *Record* FA_1:FD_1,…, FA_k:FD_k *End* //here the method of an object mentioned in Sect. 3.3.2 is not considered in Zhang et al. (2010)	Create fuzzy class/property axioms: Class (φ(FC) partial restriction (φ(FA_1) allValuesFrom (φ(FD_1)) cardinality(1)) … restriction (φ(FA_k) allValuesFrom (φ(FD_k)) cardinality(1))); DatatypeProperty (φ(FA_i) domain (φ(FC)) range (φ(FD_i)) [Functional]) where i = 1, 2,…,k .

Table 5.8 Mapping rules from fuzzy object-oriented database instances to fuzzy OWL ontology instances

Database instances	Fuzzy OWL ontology instances FO_I
Each fuzzy object symbol *FO*	A fuzzy individual identifier $\varphi(FO) \in FIID_0$
Each fuzzy class symbol *FC*	A fuzzy class identifier $\varphi(FC) \in FCID_0$
Each fuzzy *datatype attribute* FA_i	A fuzzy datatype property identifier $\varphi(FA_i) \in FDPID_0$, denoted by U_i (Definition 5.21)
Each fuzzy *object attribute FA*	A fuzzy class identifier $\varphi(FA) \in FCID_0$; In addition, adding the additional fuzzy object property identifiers, denoted by R_i (see Definition 5.21)
The fuzzy assertion *FO*: *FC*: *n* (see Sect. 5.2.2.2)	The fuzzy individual axiom: Individual ($\varphi(FO)$ type($\varphi(FC)$) [$\bowtie'$ *n*]) where $\bowtie'$ denotes $\geq$ and $\leq$
The fuzzy assertion: *FO*:[FA_1:FV_1:n_1,…,FA_k:FV_k:n_k] (see Sect. 5.2.2.2)	The fuzzy individual axiom: Individual ($\varphi(FO)$ value(R_i, o_i) [$\bowtie'$ n_i]… value(U_i, v_i) [$\bowtie'$ n_i]…) where o_i, $v_i \in \varphi(FV_i)$, i $\in$ {1…k}

The following is the fuzzy OWL ontology $FO = (FO_S, FO_I) = (FID_0, FAxiom_0)$ extracted from the fuzzy object-oriented database in Sect. 3.3.2. Here, the part FID_0 and some axioms are omitted.

$FAxiom_0$ = {SubClassOf (Young-Employee Employee);
SubClassOf (Old-Employee Employee);
Class (Employee partial unionOf (Young-Employee, Old-Employee));
Class (Young-Employee partial restriction (Name allValuesFrom (xsd:String) cardinality (1)) restriction (FUZZY Age allValuesFrom (xsd: Integer) cardinality (1))…);
Class (Chief-Leader partial restriction (Number allValuesFrom (xsd:String) cardinality (1)));
Class (Manage partial restriction (Mby allValuesFrom (Young-Employee) cardinality(1)) restriction (Mof allValuesFrom (Chief-Leader) cardinality(1)));
Class (Young-Employee partial restriction (invof_Mby allValuesFrom (Manage)));
Class (Chief-Leader partial restriction (invof_Mof allValuesFrom (Manage)));
ObjectProperty (invof_Mby domain (Young-Employee) range (Manage) inverseOf Mby);
ObjectProperty (invof_Mof domain (Chief-Leader) range (Manage) inverseOf Mof);
DatatypeProperty (Name domain (Young-Employee) range (xsd:String) [Functional]);
DifferentIndividuals (o_1, o_2, o_3, o'); Individual (o' type(Manage));
Individual (o_1 type(Chief-Leader) [$\bowtie'$ 0.85]);
Individual (o_2 type(Young-Employee) [$\bowtie'$ 0.9]);
Individual (o_3 type(Young-Employee) [$\bowtie'$ 0.75]);
Individual (o' value(Mby, o_1) value(Mof, o_2) [$\bowtie'$ 0.7] value(Mof, o_3) [$\bowtie'$ 0.8]);
Individual (o_1 value(Number, *01*) value(invof_Mby, o'));
Individual (o_2 value(Name, *Chris*) value(Age, 20) [$\bowtie'$ 0.9]… value(invof_Mof, o') [$\bowtie'$ 0.7]); … }

Fig. 5.14 The fuzzy ontology extracted from the fuzzy object-oriented database in Sect. 3.3.2

5.4 Summary

With the requirement of fuzzy Description Logic (DL) and ontology extraction, it is necessary and meaningful to extract fuzzy DL and ontology knowledge from various data resources. In particular, with the widespread studies and the relatively mature techniques of fuzzy databases as introduced in Chap. 3, much valuable information and implicit knowledge in the fuzzy data models may be considered as the main data resources for supporting fuzzy DL and ontology extraction. Extracting fuzzy DL and ontology knowledge from the fuzzy data models may facilitate the development of the Semantic Web and the realization of semantic interoperations between the existing applications of fuzzy data models and the Semantic Web.

In this chapter, we introduce how to extract fuzzy DLs and ontologies from several fuzzy data models, including fuzzy ER and UML conceptual data models, fuzzy relational and object-oriented logical database models. The existing numerous contributions of fuzzy data models provide the rich data resource support for the fuzzy knowledge management in the Semantic Web. Besides of the extraction techniques of fuzzy knowledge, for realizing the fuzzy knowledge management in the Semantic We, the other issues about mapping, query and storage of fuzzy DLs/ontologies also have been attracting significant attention as will be introduced in the later chapters.

References

Artale A, Cesarini F, Soda G (1996) Describing database objects in a concept language environment. IEEE Trans Knowl Data Eng 8(2):345–351

Baader F, Calvanese D, McGuinness D, Nardi D, Patel-Schneider PF (2003) The description logic handbook: theory, implementation, and applications. Cambridge University Press, Cambridge

Beneventano D, Bergamaschi S, Sartori C (2003) Description logics for semantic query optimization in object-oriented database systems. ACM Trans Database Syst 28:1–50

Berners-Lee T, Hendler J, Lassila O (2001) The semantic web. Sci Am 284(5):34–43

Bobillo F, Delgado M, Romero J (2012) DeLorean: a reasoner for Fuzzy OWL 2. Expert Syst Appl 39(1):258–272

Borgida A, Brachman RJ (1993) Loading data into description reasoners. SIGMOD Record 22(2):217–226

Bresciani P (1995) Querying database from description logics. In: Proceedings of the 2nd workshop KRDB'95, pp 1–4

Calì A, Calvanese D, De Giacomo G, Lenzerini M (2001) Reasoning on UML class diagrams in description logics. In: Proceedings of IJCAR workshop on precise modelling and deduction for object-oriented software development

Calvanese D, Lenzerini M, Nardi D (1999a) Unifying class-based representation formalisms. J Artif Intell Res 11(2):199–240

Calvanese D, De Giacomo G, Lenzerini M (1999b) Representing and reasoning on XML documents: a description logic approach. J Logic Comput 9(3):295–318

Chang KC, He B, Li C, Zhang Z (2004) Structured databases on the web: observation and implications. SIGMOD Record 33(3):61–70

Chaudhry N, Moyne J, Rundensteiner EA (1999) An extended database design methodology for uncertain data management. Inf Sci 121(1–2):83–112

Chen GQ, Kerre EE (1998) Extending ER/EER concepts towards fuzzy conceptual data modeling. In: Proceedings of the 1998 IEEE international conference on fuzzy systems, pp 1320–1325

Chen MS, Han J, Yu PS (1996) Data mining: an overview from a database perspective. IEEE Trans Knowl Data Eng 8(6):866–883

Chen W, Yang Q, Zhu L, Wen B (2009) Research on automatic fuzzy ontology generation from fuzzy context. In: Proceedings of the 2nd international conference on intelligent computation technology and automation, pp 764–767

Corcho O, Fernández-López M, Gómez-Pérez A (2003) Methodologies, tools and languages for building ontologies. Where is their meeting point? Data Knowl Eng 46:41–64

De Maio C, Fenza G, Loia V, Senatore S (2009) Towards an automatic fuzzy ontology generation. In: Proceedings of the 2009 IEEE international conference on fuzzy systems, Jeju Island, Korea, pp 1044–1049

Fayyad UM, Piatetsky-Shapiro G, Smyth P, Uthurusamy R (1996) Advances in knowledge discovery and data mining. AAAI/MIT Press, Menlo Park

George R, Srikanth R, Petry FE, Buckles BP (1996) Uncertainty management issues in the object-oriented data model. IEEE Trans Fuzzy Syst 4(2):179–192

Gu H, Lv H, Gao J, Shi J (2007) Towards a general fuzzy ontology and its construction. In: Proceedings of ISKE a part of series: advances in intelligent system research

Hao G, Ma S, Sui Y, Lv J (2007) An unified dynamic description logic model for databases: relational data, relational operations and queries. In: Proceedings of the 26th international conference on conceptual modeling, pp 121–126

Haroonabadi A, Teshnehlab M (2009) Behavior modeling in uncertain information systems by fuzzy-UML. Int J Soft Comput 4(1):32–38

Hilderman RJ, Hamilton HJ (1999) Knowledge discovery and interestingness measures: a survey. Technical report, Computer Science, University of Regina

Jayatilaka ADS, Wimalarathne GDSP (2011) Knowledge extraction for Semantic Web using web mining. In: Proceedings of the international conference on advances in ICT for emerging regions—ICTer2011, pp 089–094

Koyuncu M, Yazici A (2003) IFOOD: an intelligent fuzzy object-oriented database architecture. IEEE Trans Knowl Data Eng 15(5):1137–1154

Ma ZM, Yan L (2007) Fuzzy XML data modeling with the UML and relational data models. Data Knowl Eng 63(3):970–994

Ma ZM, Yan L (2008) A literature overview of fuzzy database models. J Inf Sci Eng 24(1):189–202

Ma ZM, Zhang WJ, Ma WY (2004) Extending object-oriented databases for fuzzy information modeling. Inf Syst 29(5):421–435

Ma ZM, Lv YH, Yan L (2008) A fuzzy ontology generation framework from fuzzy relational databases. Int J Semant Web Inf Syst 4(3):1–15

Ma ZM, Zhang F, Yan L, Lv YH (2010) Formal semantics-preserving translation from fuzzy ER model to fuzzy OWL DL ontology. Web Intell Agent Syst Int J 8(4):397–412

Ma ZM, Zhang F, Yan L (2011a) Representing and reasoning on fuzzy UML models: a Description Logic approach. Expert Syst Appl 38(3):2536–2549

Ma ZM, Zhang F, Yan L, Cheng JW (2011b) Extracting knowledge from fuzzy relational database with Description Logic. Integr Comput Aided Eng 18:1–19

Ma ZM, Zhang F, Yan L (2011c) Fuzzy information modeling in UML class diagram and relational database models. Appl Soft Comput 11(6):4236–4245

Maedche A, Steffen S (2001) Ontology learning for the Semantic Web. IEEE Intell Syst 16(2):72–79

Martin TP (2003) ASK-acquisition of semantic knowledge. In: Proceedings of the 2003 joint international conference on artificial neural networks and neural information processing, pp 917–924

Motameni H, Movaghar A, Daneshfar I, Nemat ZH, Bakhshi J (2008) Mapping to convert activity diagram in fuzzy UML to fuzzy Petri Net. World Appl Sci J 3(3):514–521

Quan TT, Hui SC, Fong ACM, Cao TH (2006) Automatic fuzzy ontology generation for Semantic Web. IEEE Trans Knowl Data Eng 18(6):842–856

Roger M, Simonet A, Simonet M (2002) Bringing together description logics and database in an object oriented model. In: Proceedings of the database and expert systems applications, pp 504–513

Ruspini E (1986) Imprecision and uncertainty in the entity-relationship model. In: Prade H, Negoita CV (eds) Fuzzy logic in knowledge engineering. Verlag TüV Rheinland GmbH, Köln, pp 18–28

Schild K (1995) The use of Description Logics as Database Query Languages. In: Proceedings of the 2nd workshop KRDB'95

Stoilos G, Stamou G, Pan JZ (2010) Fuzzy extensions of OWL: logical properties and reduction to fuzzy Description Logics. Int J Approximate Reasoning 51:656–679

Straccia U (1998) A fuzzy description logic. In: Proceedings of the 15th national conference on artificial intelligence (AAAI-98), pp 594–599

Widyantoro DH, Yen J (2001) Using fuzzy ontology for query refinement in a personalized abstract search engine. In: Proceedings of joint 9th IFSA world congress and 20th NAFIPS international conference, Vancouver, Canada, pp 610–615

Zadeh LA (1965) Fuzzy sets. Inf Control 8(3):338–353

Zhang F, Ma ZM (2013) Construction of fuzzy ontologies from fuzzy UML models. Int J Comput Intell Syst 6(3):442–472

Zhang F, Ma ZM, Wang HL, Meng XF (2008a) Formal semantics-preserving translation from fuzzy ER model to fuzzy OWL DL ontology. In: Proceedings of the 2008 IEEE/WIC/ACM international conference on web intelligence (WI 2008), pp 503–509

Zhang F, Ma ZM, Wang HL, Meng XF (2008b) A formal semantics-preserving translation from fuzzy relational database schema to fuzzy OWL DL ontology. In: Proceedings of the 3rd Asian Semantic Web conference (ASWC 2008), pp 46–60

Zhang F, Ma ZM, Yan L (2008c) Representation and reasoning of fuzzy ER model with description logic. In: Proceedings of the 17th IEEE international conference on fuzzy systems, pp 1358–1365

Zhang F, Ma ZM, Cheng JW, Meng XF (2009) Fuzzy Semantic Web ontology learning from fuzzy UML model. In: Proceedings of the 18th ACM conference on information and knowledge management (CIKM 2009), pp 1007–1016

Zhang F, Ma ZM, Fan GF, Wang X (2010) Automatic fuzzy Semantic Web ontology learning from fuzzy object-oriented database model. In: Proceedings of the international conference on database and expert systems applications (DEXA 2010), pp 16–30

Zhang F, Yan L, Ma ZM (2012a) Reasoning of fuzzy relational databases with fuzzy ontologies. Int J Intell Syst 27(6):613–634

Zhang F, Ma ZM, Yan L, Wang Y (2012b) A Description Logic approach for representing and reasoning on fuzzy object-oriented database models. Fuzzy Sets Syst 186(1):1–25

Zhang F, Ma ZM, Yan L (2013a) Representation and reasoning of fuzzy ER model with description logic. J Intell Fuzzy Syst. doi:10.3233/IFS-120754

Zhang F, Ma ZM, Yan L (2013b) Construction of fuzzy ontologies from fuzzy XML models. Knowl Based Syst 42:20–39

Zvieli A, Chen PP (1986) Entity-relationship modeling and fuzzy databases. In: Proceedings of the 1986 IEEE international conference on data engineering, pp 320–327

Chapter 6
Fuzzy Semantic Web Ontology Mapping

6.1 Introduction

The need for sharing and reusing independently developed ontologies has become even more important and attractive. Ontology reuse is now one of the important research issues in the ontology field. Ontology mapping, integration, merging, alignment and versioning are some of its subprocesses. Ontology mapping is the effective method to solve the problems of knowledge sharing and reusing across the heterogeneous ontologies in the Semantic Web (Doan et al. 2002). Note that, one common issue to these subprocesses is the problem of defining similarity relations among ontology components (Zhao et al. 2007).

The current ontology mapping technologies are not sufficient for fuzzy ontologies. Therefore, with the growing number of heterogeneous fuzzy ontologies in the Semantic Web, the *fuzzy ontology mapping* that can handle fuzzy data becomes a research hotpot. In particular, the treatment of the semantic similarity between fuzzy concepts plays a major role in ontology mapping. On this basis, Bahri et al. (2007) proposed an approach for dealing with similarity relations in fuzzy ontologies by reducing the problem of determining similarity relations among concepts to an (un)satisfiability problem. Wang et al. (2009) proposed an approach to measure the similarity of fuzzy concepts for mapping fuzzy ontologies. This method applies fuzzy set to represent fuzzy concept, and computes fuzzy concept similarity according to the compatibility of fuzzy sets. Xu et al. (2005) proposed a framework of mapping fuzzy concepts between fuzzy ontologies. It applied the approximate concept mapping approach, extended atom fuzzy concept sets and defined the least upper bounds to reduce the searching space. It resolved the mapping problem of fuzzy concepts into finding the simplified least upper bounds for atom fuzzy concepts, and gave an algorithm for searching the simplified least upper bounds. Bakillah and Mostafavi (2011) proposed a solution to the problem of fuzzy geospatial ontology and fuzzy semantic mapping. They first provided a definition of fuzzy geospatial ontologies. Then, they proposed a new fuzzy semantic mapping approach which integrated fuzzy logic operators and predicates to reason with fuzzy concepts. Finally, they demonstrated a possible application of

Z. Ma et al., *Fuzzy Knowledge Management for the Semantic Web*,

Studies in Fuzziness and Soft Computing 306, DOI: 10.1007/978-3-642-39283-2_6,

the fuzzy semantic mapping, which is the propagation of fuzzy queries to the relevant sources of a network. Xu et al. (2010) presented a novel method based on the Fuzzy Formal Concept Analysis (FCA) theory in order to handle vague information of ontology mapping. The ontology was first defined to model vague data, and then a fuzzy formal context was constructed with the combination of two ontologies. At last, a concept similarity measure model was proposed for improving ontology mapping. Li et al. (2012) proposed a modularized multi-strategy fuzzy ontology mapping method. By using the approach of iterative correction, the mapping results are merged and repaired so that the mapping results are more reliable and the efficiency and accuracy of fuzzy ontology mapping can be improved. More recently, Zhang and Ma (2012) proposed a novel semantic similarity method, ICFC (Information Content of Fuzzy Concept), for fuzzy concepts in a fuzzy ontology. The experiment is done to prove the benefits of ICFC on a fuzzy ontology. Further, to realize the mappings among multiple fuzzy ontologies, Ma and Zhang (2013) proposed a new method for multiple fuzzy ontology mapping called FOM-CG (Fuzzy Ontology Mapping based on Conceptual Graph). To reduce unnecessary comparisons for multiple fuzzy ontologies in a domain, FOM-CG firstly creates or finds out a Reference Ontology that contains the most common and shared information. The other fuzzy ontologies in the domain are Source Ontologies. Then, these fuzzy ontologies are transformed into conceptual graph sets (i.e. R-set and S-sets). Next, some algorithms are presented to create mappings among conceptual graph sets. Finally, the obtained mappings are transformed into the mappings among fuzzy ontologies. Experimental results with some fuzzy ontologies from the real world indicate that FOM-CG performs encouragingly well. In this chapter, we introduce the problem of fuzzy ontology mapping. First, the method for computing semantic similarity among fuzzy concepts in a fuzzy ontology in (Zhang and Ma 2012) is introduced. Then, we further introduce the conceptual graph based approach for mappings among multiple fuzzy ontologies proposed in (Ma and Zhang 2013).

6.2 Fuzzy Ontology Concept Comparison

In (Zhang and Ma 2012), a novel semantic similarity method, ICFC (Information Content of Fuzzy Concept), for fuzzy concepts in a fuzzy ontology was proposed. In this method, a semantic similarity between two fuzzy concepts is computed by their information content values. But, it is difficult to obtain information content values for fuzzy concepts, because a fuzzy concept is considered as a fuzzy set, and all the instances belong to it with membership degrees. In this case, ICFC achieves two tasks: (1) determines whether or not an instance belongs to a fuzzy concept; (2) provides a method for computing the membership degree of instance to fuzzy concept, which exploits the degrees that the attribute domains of fuzzy concept include the attribute values of instance. Experimental results with a fuzzy ontology from the real world indicate that ICFC performs encouragingly well.

In the following, we introduce the fuzzy ontology concept comparison method in (Zhang and Ma 2012). Before that, we first briefly recall some preliminaries on the definition of fuzzy ontology and some computing methods of semantic similarity between concepts.

As mentioned in Sect. 4.3, although there are many fuzzy ontology definitions in the literature, a universal fuzzy ontology definition is difficult since the different application requirements. In (Zhang and Ma 2012), a fuzzy ontology is considered as a quintuple $FO = \{\boldsymbol{FC}, \boldsymbol{P}, \boldsymbol{FR}, \boldsymbol{I}, \boldsymbol{A}\}$, where:

- $\boldsymbol{FC}$ is the set of fuzzy concepts that are made up of properties. Each fuzzy concept $C \in \boldsymbol{FC}$ is a fuzzy set on the domain of instances C: $\boldsymbol{I} \rightarrow [0, 1]$.
- $\boldsymbol{P}$ is the set of fuzzy concept properties that are considered as the attribute domains of fuzzy concept. Each property $P \in \boldsymbol{P}$ is a basic unit for constructing fuzzy concept.
- $\boldsymbol{FR}$ is the set of fuzzy relations. Each $R \in \boldsymbol{FR}$ is a binary fuzzy relation with a membership degree [0, 1] between two fuzzy concepts.
- $\boldsymbol{I}$ is the set of instances that belong to fuzzy concepts with membership degrees [0, 1]. Each $I \in \boldsymbol{I}$ is constituted by attribute values with respect to the attribute domains of fuzzy concept.
- $\boldsymbol{A}$ is the set of axioms expressed in a proper logical language, for example asserting class subsumption, equivalence, more generally to (fuzzily) constrain the possible values of concepts or instances.

  Moreover, some methods have been presented for calculating the semantic similarity between two concepts in a classic ontology. They can be classified into three categories: edge based, node based and hybrid methods:
- The edge based method evaluates the semantic similarity between two concepts by the distance, which is the number of edges found on the path between them in a graph representing the ontology. Obviously, the shorter the path from a concept to the other, the more semantic similar they are. There are some edge-based methods that are similar. But, the most typical one is shown in the following formula, where C_1 and C_2 are concepts in ontology, D is the maximum depth of the ontology, and $len(C_1, C_2)$ is the length of the shortest path between concepts C_1 and C_2.

$$Sim_{EDGE}(C_1, C_2) = (2 * D) - len(C_1, C_2) \tag{6.1}$$

- The node based method applies information theory to deal with the semantic similarity calculation. It regards that: the more information two concepts share in common, the more similar they are, and the information shared by two concepts is indicated by the information content of the concepts that subsume them in the ontology. For instance, Resnik (1995) introduced a node-based method shown in the following formula, where $S(C_1, C_2)$ is the set of concepts that subsume both C_1 and C_2, $IC(C)$ is the information content of concept C defined as the negative log likelihood of the probability of encountering an instance of C, i.e. $IC(C) = -\log P(C)$.

$$Sim_{NODE}(C_1, C_2) = \underset{C \in S(C_1, C_2)}{Max}[IC(C)] \tag{6.2}$$

- The Hybrid method combines both methods defined above. Jiang and Conrath (1997) provided a hybrid method that computes semantic distance for two concepts with the edge-based method. Then, the semantic distance is represented as link strength that is the difference of information content between them. Specifically, the semantic distance is defined as follows:

$$Dist(C_1, C_2) = IC(C_1) + IC(C_2) - 2 * IC(LCA(C_1, C_2)) \tag{6.3}$$

In the following, a semantic similarity measure for fuzzy concepts based on the Information Content of Fuzzy Concept (i.e., ICFC) in (Zhang and Ma 2012) is introduced.

As shown in Fig. 6.1, ICFC computes the semantic similarity between two fuzzy concepts FC_1 and FC_2 in four basic steps. Firstly, the lowest common ancestor, FC = LCA(FC_1, FC_2), of fuzzy concepts FC_1 and FC_2 is found, based on the structure of fuzzy ontology. Next, we create three sets $\{I_{FC}\}$, $\{I_{FC1}\}$ and $\{I_{FC2}\}$, which are respectively made up of the instances for the fuzzy concepts FC, FC_1 and FC_2. Note that, the three sets are fuzzy sets whose elements belong to them with memberships. Therefore, we provide a measure for the membership of an instance belonging to a fuzzy concept by using the relations between attribute values of the instance and attribute domains of the fuzzy concept. Next, according to these instances with memberships, we respectively compute the probabilities of occurrence of fuzzy concepts FC, FC_1 and FC_2, and transform the probabilities (P(FC), P(FC_1) and P(FC_2)) into information content (IC(FC), IC(FC_1) and IC(FC_2)) by the function of the negative log likelihood. Finally, based on the information content, ICFC computes the semantic similarity of FC_1 and FC_2. Figure 6.1 shows the process of ICFC, and we will further illustrate each step in details.

Step 1: Finding Out the Lowest Common Ancestor of the Two Specified Fuzzy Concepts

As we known, a fuzzy ontology is a hierarchical semantic structure. More specifically, fuzzy concepts in the same fuzzy ontology are organized as a tree structure by hyponymy. A fuzzy concept inherits all the features of its super concept, and possesses at least one feature that distinguishes it from its superconcept. In general, the similarity between two fuzzy concepts is related to their common features that are included into the lowest common ancestor (LCA) of them. For this reason, it is necessary for us to find out the LCA of two fuzzy concepts in the same fuzzy ontology when calculating similarity.

To represent a semantic inclusion relationship between two fuzzy concepts in a fuzzy ontology, the binary relationship "is-kind-of" is used to connect them. For example, pine is a kind of tree, so "pine" is connected with "tree" by the relationship "is-kind-of". Due to the transitivity of "is-kind-of", we can find out all the ancestors (e.g. super concepts) for any fuzzy concept following the path that is

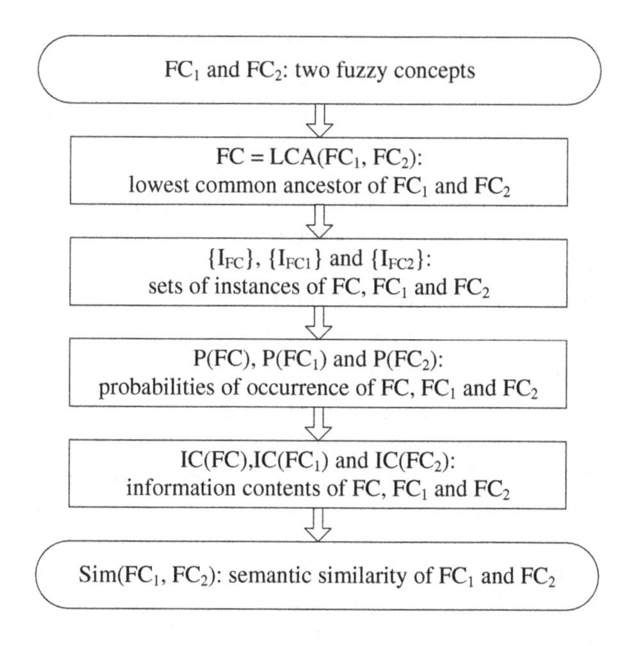

Fig. 6.1 The procedure of computing similarity for fuzzy concepts in ICFC

from it to the root concept in a fuzzy ontology. For two fuzzy concepts, they own at lest one common ancestor (i.e., root of the conceptual hierarchy tree). The most specific one from the common ancestors is the LCA of the two fuzzy concepts.

Step 2: Computing a Membership of an Instance to a Fuzzy Concept

Theoretically, a concept can be considered from two different viewpoints: an extensional concept, where the concept is defined by the list of its instances, and an intensional concept, where the concept is defined by a set of attributes and their admissible values. Since fuzzy concept is the fuzzy extension of concept, the viewpoints are also appropriate for fuzzy concept. To creating an instance set for a fuzzy concept, two steps should be done: (1) finding all the instances for a fuzzy concept based on its sub-concepts; (2) exploiting the relationship between attribute value of instance and attribute domain of fuzzy concept to compute the membership of instance to fuzzy concept.

After creating an instance set for a fuzzy concept, we need further to find out all the instances for a fuzzy concept. As all the fuzzy concepts in a fuzzy ontology are organized by hyponymy, some semantic information of a fuzzy concept is contained in its sup-concept. Obviously, all the attributes of the fuzzy concept are inherited by its sub-concepts, and all the instances contained in its sub-concepts belong to the fuzzy concept. In order to reduce data redundancy and obtain the most obvious information, a fuzzy concept is restored with an internal instance set in which the instances are used to define the fuzzy concept. But, the instances of its sub-concepts are not contained in it. Hence, to find all the instances belonging to a fuzzy concept, we should put its instances and the instances of its sub-concepts together and form an instance set.

Example 6.1 As shown in Fig. 6.2, a fuzzy ontology is composed by fuzzy concepts A to G, whose internal instance sets are respectively named as $\{I_A\}$ to $\{I_G\}$. Note that, an internal instance set for a fuzzy concept may be not equal to the instance set for the fuzzy concept, and the relation between them is that the former is contained in the latter. For example, fuzzy concept D in Fig. 6.2 is one of the leaf-nodes which do not have sub-concepts, and the instance set of it is the internal instance set $\{I_D\}$. But, for the fuzzy concept B which is a non-leaf node in a fuzzy ontology, the instance set of it is made up of its internal instance set and the instance sets of its sup-concepts (i.e. fuzzy concepts D and E). Therefore, the instance set of B is the union of $\{I_B\}$, $\{I_D\}$ and $\{I_E\}$.

When we have find out all the instances for a fuzzy concept, we can compute a membership of an instance to a fuzzy concept. In a fuzzy ontology, the instances having the same properties are gathered into fuzzy concepts with memberships. We regard that a fuzzy concept is composed of attributes. Therefore, in order to calculate the membership of an instance to a fuzzy concept, it is necessary to evaluate the degrees that the attribute domains of the fuzzy concept include the attribute values of the instance.

Let FC be a fuzzy concept with attributes $\{A_1, A_2, \ldots, A_n\}$, I be an instance on attribute set $\{A_1, A_2, \ldots, A_n\}$, and let $I(A_i)$ denote the attribute value of I on A_i $(1 \leq i \leq n)$. In FC, each attribute A_i is connected with a domain denoted $dom_{FC(Ai)}$. Therefore, the membership of the instance I to the fuzzy concept FC should be calculated using the memberships with which the attribute values $I(A_i)$ belong to the attribute domain $dom_{FC}(A_i)$. In (Ma et al. 2004), the membership of the attribute value $(I(A_i))$ to the attribute domain $(dom_{FC}(A_i))$ is defined by the inclusion degree $ID(dom_{FC}(A_i), I(A_i))$:

$$ID(dom_{FC}(A_i), I(A_i)) = \frac{SS(dom_{FC}(A_i)) \cap SS(I(A_i))}{SS(dom_{FC}(A_i))} \tag{6.4}$$

In this formula, SS() is a function that returns the semantic space (Ma et al. 1999) of a fuzzy data, and $ID(dom_{FC}(A_i), I(A_i))$ can also be considered as a probability of occurrence of $I(A_i)$ on $dom_{FC}(A_i)$, denoted by $P(I(A_i) \in dom_{FC}(A_i))$.

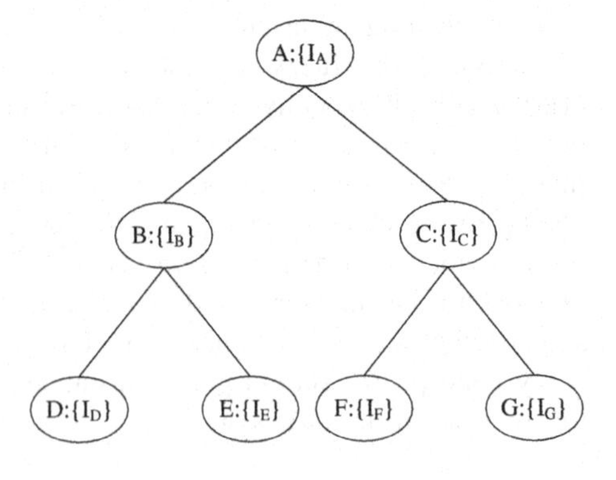

Fig. 6.2 Fuzzy concepts with their internal instance set

As we know, a fuzzy concept is an intersection of attributes, and the condition of I belonging to FC is that all the attribute values of I must be respectively on the attribute domains of FC. Therefore, we should use joint probability distribution function to define the formula for the membership of the instance I to the fuzzy concept FC as follows.

$$u_{FC}(I) = P(I(A_1) \in dom_{FC}(A_1), I(A_2) \in dom_{FC}(A_2), \ldots, I(A_n) \in dom_{FC}(A_n)) \tag{6.5}$$

Further, all the attributes A_i ($1 \leq i \leq n$) of a fuzzy concept FC are mutual independent, because the value of any attribute is not related to other attributes. We apply probability formula of mutual independent random variable to rewrite the above formula.

$$u_{FC}(I) = \prod_{1 \leq i \leq n} P(I(A_i) \in dom_{FC}(A_i)) \tag{6.6}$$

Example 6.2 Consider a fuzzy concept FC with attributes A_1 and A_2, and its instance I. Assume $dom_{FC}(A_1) = \{1/x\}$, and $dom_{FC}(A_2) = \{0.5/y\}$. Let $I(A_1) = \{0.8/x\}$, $I(A_2) = \{0.3/y\}$. According to the measure supported in (Ma et al. 2004), we can obtain that $P(I(A_1) \in dom_{FC}(A_1)) = 0.8$, $P(I(A_2) \in dom_{FC}(A_2)) = 0.6$. So the membership of I to C is 0.48 (0.8*0.6) computed by the formula (6.6).

However, there is a problem about the formula (6.6). When a fuzzy concept owns many attributes (e.g., 10), for an instance of the fuzzy concept, although the membership of each attribute value of the instance belonging to the attribute domain is very high (e.g., 0.9), the membership of the instance to the fuzzy concept may be very low (e.g., $0.9^{10} = 0.39$). The reason for this problem is that weights of attributes of the fuzzy concept are not taken into account. Generally speaking, a weight of an attribute stands for the importance of the attribute, and the sum of all the weights in a fuzzy concept should be 1 (if the sum of all the weights is not 1, we use the sum to divide every weight for standardization).

Based on the reasonable use of the weights of attributes, we provide the complete formula for the membership $u_{FC}(I)$ as follows, where $P(A_i)$ is the abbreviation for $P(I(A_i) \in dom_{FC}(A_i))$ and the weights standardized by the sum of all the weights are considered as the exponentials of the corresponding $P(A_i)$. In the Example 6.2, assume the weights of A_1 and A_2 are equal, that is to say $w_{A1} = w_{A2} = 0.5$. Then, the membership $u_{FC}(I)$ is $0.8^{1/2} * 0.6^{1/2} = 0.69$ according to the formula (6.7).

$$\mu_{FC}(I) = P(A_1)^{\frac{wA1}{\sum wA1+wA2+\ldots+wAn}} * P(A_2)^{\frac{wA2}{\sum wA1+wA2+\ldots+wAn}} * \ldots * P(A_n)^{\frac{wAn}{\sum wA1+wA2+\ldots+wAn}} \tag{6.7}$$

Step 3: Associating Fuzzy Concepts with Probabilities

Inspired by (Ross 1976), the semantic similarity between two fuzzy concepts (e.g. FC_1 and FC_2) in the same fuzzy ontology is calculated by the probability of their Least Common Ancestor (LCA:FC). Therefore, in order to obtain the semantic similarity of any two fuzzy concepts, it is necessary to compute the probability for every fuzzy concept in a fuzzy ontology. We use P(FC) to represent the probability of fuzzy concept FC, and it means the probability of encountering an instance of the fuzzy concept FC.

Note that, in a classic ontology, the membership of an instance belonging to a concept is 1, and the membership of the instance not belonging to the concept is 0. So, the probability of a concept is measured as the ratio between the number of the instances to the concept and the number of all the instances. That is to say P(FC) = Num(I(FC))/Num(I). However, in a fuzzy ontology, the membership of an instance to a fuzzy concept is a fraction between 0 and 1. To compute the probability of the fuzzy concept, we should use the memberships of its instances, but not the number of its instances. Therefore, the probability of a fuzzy concept is calculated by formula (6.8), where the numerator is the sum of the memberships of instances to fuzzy concept, and the denominator is the number of all the instances.

$$P(FC) = \frac{\sum_{I \in FC} u_{FC}(I)}{Num(I)} \tag{6.8}$$

The relation of two connected fuzzy concepts in a fuzzy ontology is hyponymy. The semantics of a fuzzy concept is included in the semantics of its sup-concept. Therefore, all the instances of the fuzzy concept are the instances of its sup-concept. It is easy to see that the instances of a fuzzy concept can be divided into two parts: one part is composed by its own instances; the other part is made up of the instances of its sub-concepts. The memberships of the instances in the first part are given by domain experts during the generation of fuzzy ontology. But, to decrease data restored in fuzzy ontology, the memberships of the instances in the second part are not provided. To this end, when an instance is not directly contained by a fuzzy concept, the formula (6.7) is needed to compute the membership of the instance to the fuzzy concept.

Step 4: Computing Semantic Similarity Between Two Fuzzy Concepts Based on Information Content of Fuzzy Concept

Intuitively, one key to the semantic similarity of two fuzzy concepts is the extent to which they share information in common, indicated in a fuzzy ontology by a common highly specific fuzzy concept that is called Least Common Ancestor (LCA). More precisely, the LCA of two fuzzy concepts is the first fuzzy concept upward that subsumes both fuzzy concepts in a fuzzy ontology. For classic ontology, Resnik (1995) provided an approach to measure the shared information of two concepts by evaluating the information content (IC) value of their LCA. The information content of a concept is defined as the negative log likelihood of the probability of encountering an instance of the concept (the probability of the concept). Since the calculation of probability of fuzzy concept is provided, information content can also be applied to compute the semantic similarity between fuzzy concepts. For example, given FC_1 and FC_2, and their LCA FC, the semantic similarity of FC_1 and FC_2 is computed by $Sim(FC_1, FC_2) = IC(FC) = -\log P(FC)$.

However, the method proposed in (Resnik 1995) considered the common information two fuzzy concepts shared as the only criterion for semantic similarity, and ignored the importance of the difference of the two fuzzy concepts. In order to accurately compute semantic similarity for two fuzzy concepts (i.e. FC1 and FC2), not only should shared information be computed, but the information contained by themselves must also be taken into account. Borrowing the idea from (Lin 1998), a formula is introduced to compute the semantic similarity of FC_1 and FC_2 as follows, where FC is the LCA of FC_1 and FC_2.

$$Sim(FC_1, FC_2) = \frac{2 * IC(FC)}{IC(FC_1) + IC(FC_2)} = \frac{2 * \log P(FC)}{\log P(FC_1) + \log P(FC_2)} \quad (6.9)$$

From the formula (6.9), it is easy to see that the similarity of two fuzzy concepts (FC_1 and FC_2) is in proportion to the information content value of their LCA (FC). As we know, for a fuzzy concept, the more the information content it contains, the less the probability of it. Therefore, the similarity of FC_1 and FC_2 is inversely proportional to the probability of FC. But, the sum of the information content values $IC(FC_1)$ and $IC(FC_2)$ also affect the similarity of FC_1 and FC_2.

Example 6.3 Take the fuzzy ontology shown in Fig. 6.2 as an example, the similarity of fuzzy concepts B and C is higher than the similarity of fuzzy concepts D and G, although the pairs of (B, C) and (D, G) possess the same LCA A. The reason is that D (G) is a sub-concept of B(C) and the information content value IC(D) (IC(G)) is higher than the information content value IC(B) (IC(C)). Therefore, the formula (6.9) does better in calculating the semantic similarity for fuzzy concepts.

To state the advantage of ICFC in evaluating the semantic similarity for fuzzy concepts, we provide an instance coming from the real world. As shown in Fig. 6.3, there is a fragment of a fuzzy ontology in some university. Each fuzzy concept possesses an internal instance set represented by a table adjacent to it. For the convenience of description of ICFC, we use capital letters (A to H) to label fuzzy concepts in the fuzzy ontology.

Following the process of ICFC shown in Fig. 6.1, we should firstly create instance sets for fuzzy concepts. Example 6.1 introduced how to create an instance set for a fuzzy concept. But, the instance set is also a fuzzy set, in which each instance has a membership. Therefore, we give the following example to emphasize how to calculate the membership of instance to fuzzy concept by the attribute values of instance to the attribute domains of fuzzy concept.

Example 6.4 We want to obtain the membership of an instance of D, noted by I_{D1}, to fuzzy concept B, based on the attribute values of I_{D1} and the attribute domains of B. The concrete data for I_{D1} and B are shown in Table 6.1. Note that, in fuzzy

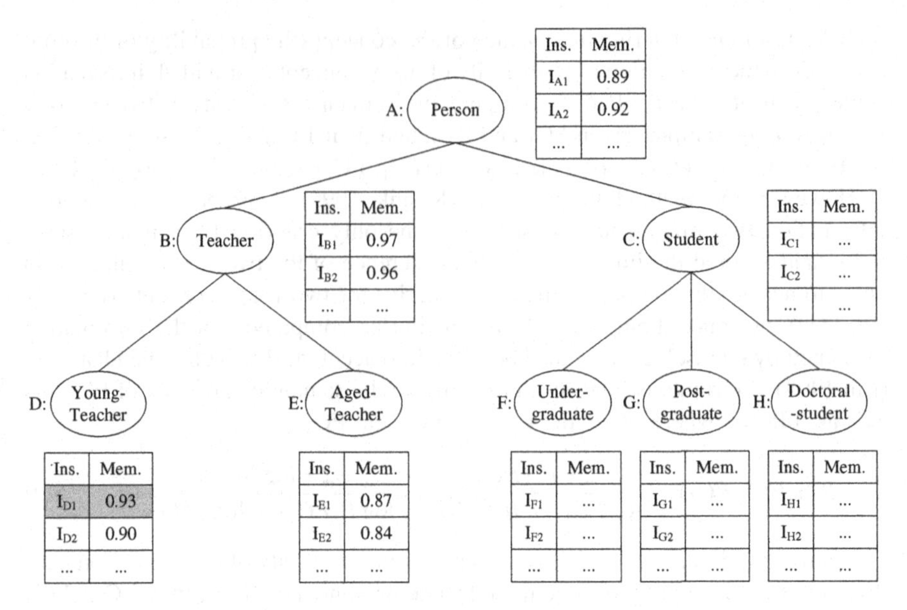

Fig. 6.3 A fragment of a fuzzy ontology in some university

Table 6.1 Concrete data for I_{D1} and B

B. Name	ID	Name	Age	Education	Salary
B. Domain	T0000 ~ T9999	String (0 ~ 30)	{0.97/27 0.97/28…0.87/60}	{0.90/bachel or 0.95/master 0.99/doctor}	[3,000, 10,000]
Weight	0.5	0	0.2	0.2	0.1
I_{D1}. Value	T0243	Tom	28	master	[2,700, 4,000]
Membership of I_{D1} to B	1.0	1.0	0.97	0.95	0.90

concept B, the attribute domains of "Age" and "Education" are fuzzy sets, and the others are crisp sets.

Now, we will obtain the memberships, according to the attribute values of I_{D1} (i.e. < T0243, Tom, 28, master, [2,700, 4,000] >). Because the attribute values "T0243" and "Tom" are crisp data, which belong to crisp sets {ID} and {Name}, the memberships of them are "1". As we known, attribute values "28" and "master" are also crisp data, but the corresponding attribute domains are fuzzy set. Therefore, the memberships of them are 0.97 and 0.95, respectively. Inspired by (Ma et al. 1999), the membership of set [2,700, 4,000] to set [3,000, 10,000] is 0.90. Based on these memberships above, the membership of instance I_{D1} to fuzzy concept B is calculated by formula (6.7). That is: $u_B(I_{D1}) = 1^{0.5} * 1^0 * 0.97^{0.2} * 0.95^{0.2} * 0.9^{0.1} = 0.974$.

After creating an instance set for a fuzzy concept, the probability of the fuzzy concept is computed by the formula (6.8), where the numerator is the sum of

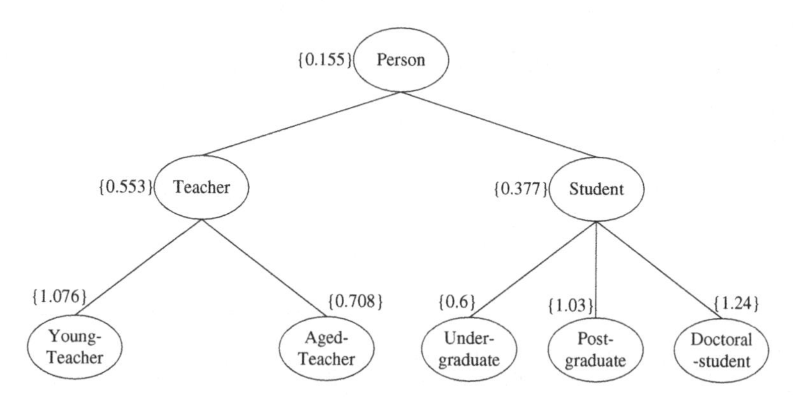

Fig. 6.4 Fuzzy concepts labeling with information contents

memberships of instances in the instance set. Next, we get an information content value for each fuzzy concept by IC(FC) = -log [P(FC)]. Figure 6.4 shows the fuzzy concepts labeling with their information content values. The number in the bracket of a fuzzy concept indicates the corresponding information content.

Now, we can easily calculate the semantic similarity between any two fuzzy concepts in a fuzzy ontology, using the information content values. For example, to compute the semantic similarity between fuzzy concepts "Young-Teacher" and "Aged-Teacher", the information contents (i.e. IC(Teacher) = 0.553, IC(Young-Teacher) = 1.076 and IC(Aged-Teacher) = 0.708) are substituted into formula (6.9), and the result is Sim(Young-Teacher, Aged-Teacher) = 0.57.

6.3 Fuzzy Semantic Web Ontology Mapping

Fuzzy ontology mapping is an important tool to solve the problem of interoperation among heterogeneous ontologies containing fuzzy information. At present, some researches have been done to expand existing mapping methods to deal with fuzzy ontology. However, these methods can not perform well when creating mappings among multiple fuzzy ontologies in a specific domain. To this end, Ma and Zhang (2013) proposed a method for fuzzy ontology mapping called FOM-CG (Fuzzy Ontology Mapping based on Conceptual Graph). This method is more suitable for multiple fuzzy ontologies composed of reference ontology and source ontologies in an application domain.

Before introducing the FOM-CG method, we first recall the notion of conceptual graph. A conceptual graph (CG) (Sowa 1984) is a visual knowledge representation formalism based on the semantic networks of artificial intelligence and the existential graphs of Charles Sanders Peirce. A CG can be considered as a bipartite graph where two kinds of nodes are used to stand for concepts and conceptual relations, respectively. It is a kind of knowledge model to represent

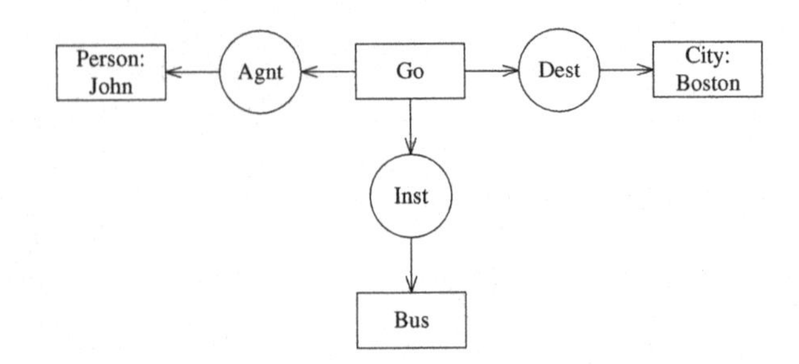

Fig. 6.5 A CG for the sentence "John is going to Boston by bus"

complex logic relations among concepts. For example, Fig. 6.5 shows a CG for the sentence "John is going to Boston by bus", in which the rectangles stand for concepts, and the circles stand for relations. An arc pointing toward a circle marks the first argument of the relation, and an arc pointing away from a circle marks the last argument. The arrowhead can be omitted, if the central word of a CG is identified.

Formally, a CG is a 3-tuple $\boldsymbol{CG} = \{\boldsymbol{S}, \boldsymbol{G}, \boldsymbol{\lambda}\}$, where:

- $\boldsymbol{S} = (T_C, T_R, I, *)$ is a support, where T_C is a finite, partially ordered set (poset) of concept types, TR is a finite set of relation types, I is a countable set of individual markers used to refer to specific concepts, and * is the generic marker that refers to an unspecified concept of a specified type;
- $\boldsymbol{G} = (N_C, N_R, EG)$ is an ordered bipartite graph, where N_C is a set of concept nodes labeled by the concept types $\in T_C$ or the individual markers $\in$ I, N_R a set of conceptual relation nodes labeled by the relation types $\in T_R$, and EG a set of edges;
- $\boldsymbol{\lambda}$ is a mapping function, which relates every node in the G to an element from the support S.

In the following, we introduce the method for Fuzzy Ontology Mapping based on Conceptual Graph, which is called FOM-CG. This method firstly translates multiple fuzzy ontologies (i.e., reference ontology and source ontologies) into sets of conceptual graphs (i.e. R-set and S-sets). In R-set and S-sets, a CG represents a fuzzy concept (FC) and all the other entities that are relative to the FC with fuzzy relationships. In this way, the problem of creating mappings among multiple fuzzy ontologies is translated into the problem of how to create mappings among the CG sets. Next, two mapping sub-processes for FOM-CG are provided. In the first sub-process, mappings between R-set and S-sets are created, and we name these mappings as R-S mappings. In the second sub-process, R-S mappings are used to generate the mappings among S-sets, which are named by FOM-CG as S-S mappings. Finally, FOM-CG obtains mappings among multiple fuzzy ontologies according to the mappings among CGs. The main steps of FOM-CG are shown in Fig. 6.6.

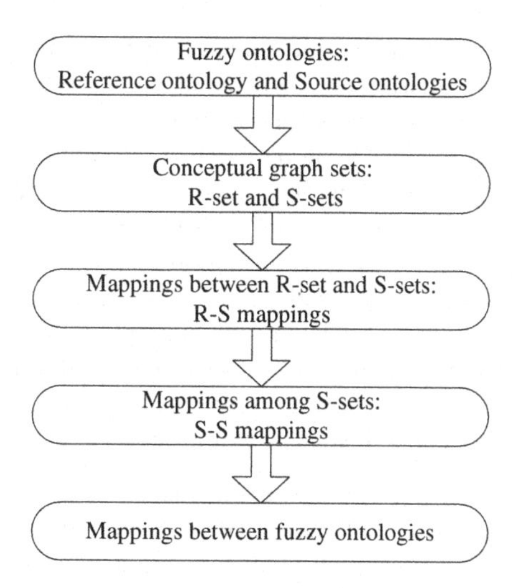

Fig. 6.6 The main steps of FOM-CG

6.3.1 Generating Conceptual Graph Sets: R-Set and S-Sets

The conceptual graph model has many advantages in application domains such as query and reasoning, especially in similarity calculation. In order to bring in the conceptual graph model to deal with the process of fuzzy ontology mapping, FOM-CG represents fuzzy ontologies by conceptual graph sets where each CG is represented by a FC and all the other entities connecting to the FC with relationships. However, the definitions for fuzzy ontology and the support of conceptual graph indicate that the two models have some similarities, although they represent knowledge by different ways. For this reason, we provide the transformation method from a fuzzy ontology to a support. Then, using the support and the structure of fuzzy ontology FOM-CG generates CGs. Therefore, the process of representation can be divided into two steps that are respectively used to generate the support and CGs based on structure of fuzzy ontology.

In the first step, a transformation function (φ) is usually used, and the specific transformation rules are as follows

(a) TC = φ(FC) v φ(P), i.e., fuzzy concepts (FC) and properties (P) in fuzzy ontology are all translated into concept types of the support, and organized by hyponymy to form a hierarchy (TC).
(b) TR = φ(FR) v φ(A), i.e., fuzzy relationships (FR) and axioms (A) are all translated into relation types of the support. Note that, there are five kinds of relation types: "Is-parent-of", "Is-child-of", "Is-relative-to", "Has-instance", "Has-property".
(c) I = φ(I), i.e., instances of fuzzy concepts are translated into instances of concept types.

The above-mentioned transformation method is complete. To explain it, let us summarize the correspondence between the elements of fuzzy ontology and the elements of conceptual graph. It is easy to find that the five elements (i.e., FC, FR, P, I, and A) can be mapped into the corresponding elements in the support using the transformation function φ. More specifically, the mapping relationships are composed by FC v P → TC, FR v A → TR, and I → I. To sum up, the translation does not cause information loss.

In the second step, FOM-CG generates a conceptual graph set for a fuzzy ontology. Each conceptual graph in this set is composed of an entry, some entities, and relation types connecting the entry and elements. Now, let us give an example to explain how to represent a fuzzy concept and its adjacent entities (i.e., fuzzy concepts, instances and properties) as a conceptual graph.

Example 6.5 Some information of a fuzzy concept *A* in a fuzzy ontology is shown in Fig. 6.7. Fuzzy concepts are represented by rectangles, the properties (P_1 to P_m) of *A* are represented by squares, and the instances (I_1 to I_n) belonging to *A* are represented by diamonds. Meanwhile, *A* is related to C_1 and C_2 with hyponym (Is-parent-of), related to *P* with hyponym (Is-child-of), and related to *D* with predicate-relationship (Is-relative-to). All the relations are labelled by the membership α_i, where the subscript "i" represents the corresponding element.

It is easy to see that *A* is a central entity (i.e., entry) that is connected by all the other entities. Hence, FOM-CG takes *A* as an entry of CG, the objects relating to the entry as entities, and the connections between *A* and objects as relationships. Besides, instances are used to enrich the information of fuzzy concept, and they are connected by the relation type "Has-instance". The representation result is shown in Fig. 6.8, where the rectangles are called concept nodes, and the ovals are called conceptual relation nodes. It is worthwhile to note that: entities in a CG are classified by the five kinds of relation types. This will bring a lot of convenience for FOM-CG to compare CGs and compute similarities for CGs in the next step.

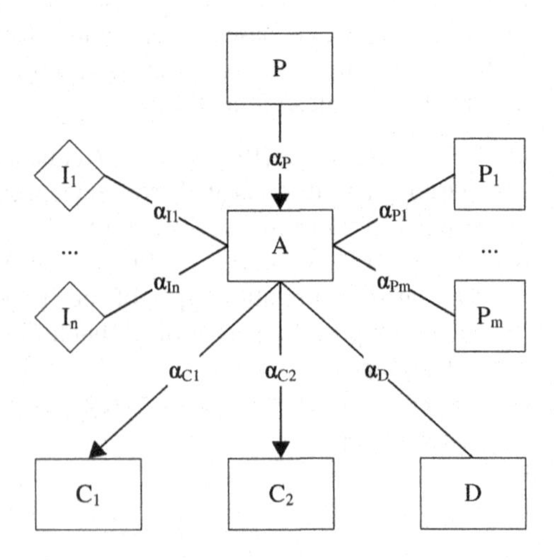

Fig. 6.7 A fuzzy concept A and its elements

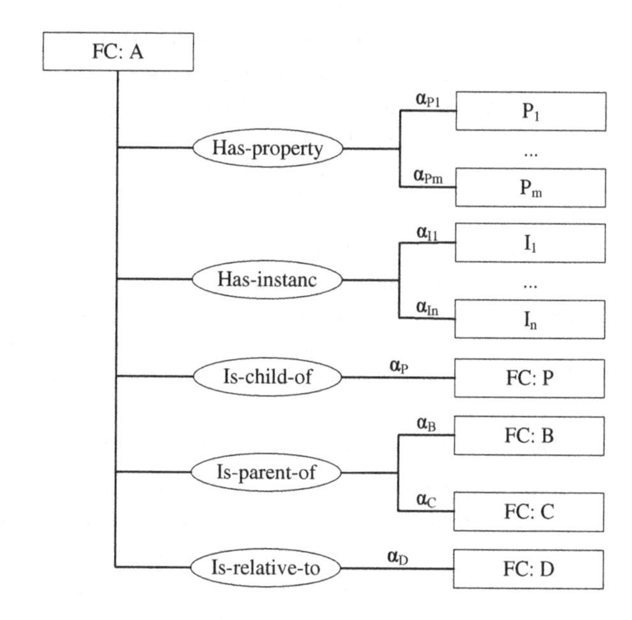

Fig. 6.8 The conceptual graph CGA translated from the fuzzy concept *A*

Following the above method of generating a CG based on a fuzzy concept, a fuzzy ontology can be translated into a CG set. According to reference ontology and source ontologies as input, FOM-CG generates conceptual graph sets R-set and S-sets, respectively.

6.3.2 Creating Mappings Between R-Set and S-Sets

There was a very strong feeling (average rating 4.4 on a 0–5 scale) that clients did not well understand the capabilities of the technologies. Similarly it was felt (average rating 4.2) that clients did not understand their own needs as they related to the technology. Perhaps surprisingly, anecdotal evidence indicated that respondents felt that clients had a low understanding of their own organisations and existing processes (most of the time undocumented) that need to be changed to allow for the effective integration of the new system. There was a majority consensus (i.e. 83 % responding as *Strongly Agree* or *Agree*) that there needed to be a process at the beginning of the projects focussed on educating their clients.

As is known to all, many source ontologies are derived from the reference ontology by modifying, adding, or deleting some information. Obviously, the mappings between R-set and S-sets can be considered as clues that are useful for creating mappings among S-sets. In fact, an R-S mapping is a CG-pair that is made up of two CGs, where a CG is from R-set and the other one from S-sets. To the end, the following provides a method to compare CGs and compute similarity for them.

In practice, the comparison of two CGs is difficult as many elements contained in the two CGs should be considered. To solve the problem, we divide each CG into fragments called Relation-Entity pairs (RE-pairs), and compare these RE-pairs to calculate the similarity for CGs. More specifically, a RE-pair is composed of a relation type and an entity, and the relation type is connected with the entity in CG.

Example 6.6 Figure 6.9 shows two sketch maps of CG and CG', where CG contains x entities, and CG' contains y entities. After decomposing them we obtain many RE-pairs, and each pair is represented as a pair (Relation-type, Entity). To compare effectively, two RE-pair sets (i.e. $\{RE\}_X$ and $\{RE'\}_Y$) are created shown in Table 6.2, where X and Y are respectively the numbers of entities belonging to the two RE-pair sets. Thus, the problem of mapping between fuzzy concepts is translated to the problem of how to calculate the similarity for RE-pair sets.

Before explaining how to calculate the similarity of two sets of RE-pairs, we firstly provide the function "sim_p-p" to calculate the similarity of two RE-pairs. Take the data shown in Table 6.2 as an example, we can use the following formula to calculate the similarity between $RE_i = (R_i{:}a_{Ei}\ E_i)$ belonging to $\{RE\}_X$ and $RE_j' = (R_j'{:}a_{Ej}', E_j')$ belonging to $\{RE'\}_Y$.

$$sim\_p - p(RE_i, RE_j') = \begin{cases} (\alpha_{Ei}/\alpha_{Ej}') * sim(R_i, R_j') * sim(E_i, E_j'), & \alpha_{Ei} < \alpha_{Ej}' \\ (\alpha_{Ej}'/\alpha_{Ei}) * sim(R_i, R_j') * sim(E_i, E_j'), & \alpha_{Ej}' \le \alpha_{Ei} \end{cases} \quad (6.10)$$

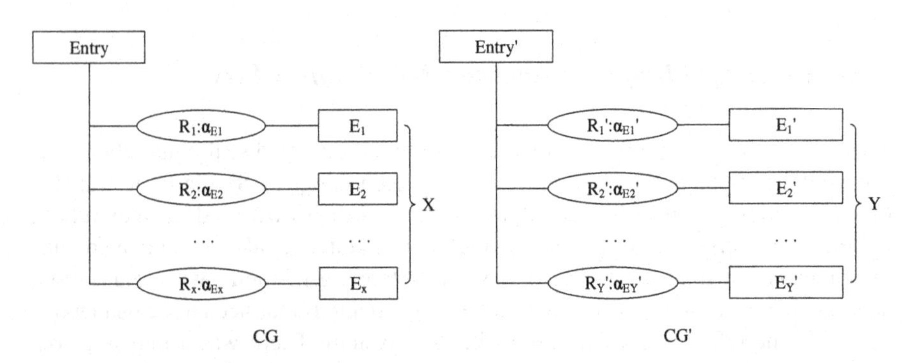

Fig. 6.9 Two CGs composed of RE-pairs

Table 6.2 RE-pair sets

{RE}	{RE'}
$(R_1{:}\ \alpha_{E1}, E_1)$	$(R_1'{:}\ \alpha_{E1}', E_1')$
$(R_2{:}\ \alpha_{E2}, E_2)$	$(R_2'{:}\ \alpha_{E2}', E_2')$
...	...
$(R_X{:}\ \alpha_{EX}, E_X)$	$(R_X'{:}\ \alpha_{EX}', E_X')$

Here, sim(R_i' R_j') is the semantic similarities of relation types, and sim(E_i' E_j') is the semantic similarities of entities. The calculation methods for sim(R_i' R_j') and sim(E_i' E_j') are respectively shown in formulas (6.11) and (6.12). In the formula (6.11), the value of sim(R_i' R_j') is 1 only if R_i is identical with R_j', otherwise, the value is 0. Previous transformation from fuzzy ontology to conceptual graph has provided the five kinds of relation types shown in Fig. 6.8. Therefore, it is easy for us to compute the semantic similarities of relation types. However, the entities are varied, which contain more semantic information. Thus, formula (6.12) appears more complex, and the notion of information content (IC) (Resnik 1995) is introduced into it to calculate to the semantic similarities of entities. In formula (6.12), E represents the least common ancestor of E_i and E_j' in semantic lexicon such as WorldNet (Miller et al. 1990), and IC(E) = -log P(E), where P(E) is the probability of E.

$$sim(R_i, R_j') = \begin{cases} 1\ , & \mathrm{R_i = R_j'} \\ 0\ , & \mathrm{R_i \neq R_j'} \end{cases} \tag{6.11}$$

$$sim(E_i, E_j') = \frac{2 * IC(E)}{IC(E_i) + IC(E_j')} \tag{6.12}$$

Next, we provide the definition of the "sim_p-s" algorithm for calculating the similarity of a RE-pair w.r.t. a set of RE-pairs. This algorithm compares the RE-pair (R_i:a_{Ei}, E_i) with all the RE-pairs in the set of {RE'}, and the objective of this algorithm is to find out the maximal value as the similarity of a RE-pair w.r.t. a set of RE-pairs.

```
Algorithm sim_p-s(RE-pair (Ri:aEi, Ei), RE-pair-set {RE'}Y
Similarity = 0
Foreach j = 1 to Y
 If Similarity < sim_p-p((Ri:aEi, Ei), (Rj':aEj', Ej'))
 Similarity = sim_p-p((Ri:aEi, Ei), (Rj':aEj', Ej'))
 EndIf
EndForeach
Return Similarity
```

Based on the above efforts, we propose the "sim_s-s" algorithm that is able to calculate the similarity of two sets of RE-pairs. This algorithm is implemented by calling the algorithm "sim_p-s". In detail, a similarity for each RE-pairs in $\{RE\}_X$ w.r.t. $\{RE'\}_Y$ are calculated by "sim_p-s", and the similarity between $\{RE\}_X$ and $\{RE'\}_Y$ are calculated by the formula $(\sum_{i=1}^{X} similarity_i)/X$.

```
Algorithm sim_s-s(RE-pair-set {RE}, RE-pair-set {RE'})
Sum = 0
Foreach i = 1 to X
 Sum = Sum + sim_p-s((R_i:a_Ei, E_i), {RE'})
EndForeach
Similarity = Sum/X
Return Similarity
```

Now, the CGs in the R-set can be compared with the CGs in S-sets. The similarities between these candidate pairs, where a CG is from R-set and the other one is from S-sets, is computed by the algorithm "sim_s-s". Finally, according to the threshold of similarity given by experts, the mappings between R-set and S-sets are created. Moreover, the threshold is alterable to ensure that the accuracy and recall of FOM-CG could meet the requirement proposed by experts.

6.3.3 *Creating Mappings Among Conceptual Graphs in S-Sets*

In the following we introduce the similarity calculation for CGs in S-sets. Concretely, the method can be divided into two parts. In the first part, all the CGs in S-sets are classified into groups according to the R-S mappings. The purpose of classifying is to put together the CGs that are similar to the same CG in R-set. The benefit of doing so is to avoid the comparisons for the CGs that have little common characteristics, and to move up the efficiency of comparison. During the grouping, FOM-CG labels all the RE-pairs of CGs in a group with the similar RE-pairs of CGs in R-set. In the second part, an algorithm "sim_c-c" is provided to compute the similarity of any two CGs belonging to the same group. The algorithm is implemented based on the comparisons of the labels of RE-pairs.

Firstly, we need to classify CGs in S-sets into groups by R-S mappings. For a CG (e.g. CG_A) in R-set, some CGs from different S-sets are mapped to it. To be convenience for explanation, we put these CGs mapping to CG_A into a group named as $Group_{CGA}$. According to the thinking of the "sim_s-s" algorithm, these CGs in the group are similar to CG_A in R-set. It is reasonable to suppose that these CGs may be similar to each other, as they possess some common characteristics hidden in CG_A. Conversely, given a CG (e.g. CG_B) in S-sets that is not mapped to CG_A, it is almost impossible that the CG_B is similar to any CG in the $Group_{CGA}$. To this end, in order to move up the efficiency of comparison for the CGs in S-sets, FOM-CG classifies the CGs in S-sets into groups according to the mappings from R-set to S-sets. More specifically, all the CGs in a group are mapped to the same CG of R-set, and the names of CGs in R-set are used to labeled these groups, respectively.

However, it is also a troublesome work to compare CGs in a group, because the group may contains many CGs, and any two CGs should be found out and compared. To solve this problem, FOM-CG labels all the RE-pairs of CGs in a group with pairs each of which contains an index and a similarity. The index can be used to find out the similar RE-pair of CGs in R-set. Thus, comparing RE-pairs can be realized by the comparison of their labels. More specifically, a label is composed of two parts, and represented as the pair (Index, Similarity). The following is an example to explain how to generate a label for a RE-pair.

Example 6.7 Suppose that CG_A that is a CG in R-set is similar to CG_B' that is a CG in S-sets, and a RE-pair (e.g. RE_i) belonging to CG_A is similar to a RE-pair (e.g. RE_j') belonging to CG_B'. The similarity of the two RE-pairs is represented by the capital letter "S" calculated by the algorithm "sim_p-p" mentioned above. When the CG_B' is put into the group $Group_{CGA}$, FOM-CG labels all the RE-pairs of CG_B' with the similar RE-pairs of CG_A and the similarity, such as RE_j': (A_i, S). Conversely, based on "Index" (i.e. A_i) of label (A_i, S) of RE_j', we know that the ith RE-pair in CG_A is similar to RE_j', and the similarity of two RE-pairs is "S".

Then, we can compare the labeled CGs in groups. After classifying, all the RE-pairs of CGs in a group are labeled by the same set of RE-pairs. It is easy to see that the labels of RE-pairs are not only simpler than RE-pairs but also easy to be compared. Therefore, the comparison of RE-pairs in groups can be replaced by the comparison of their labels. Next, FOM-CG illustrates how to compute the similarity of two CGs in a group by the comparison of labels of RE-pairs.

Suppose that two CGs (CG and CG') are respectively composed of two RE-pair sets $\{RE\}_X$ and $\{RE'\}_Y$, where X and Y are the cardinalities of sets. The following algorithm "sim_c-c" can be used to compute the similarity of CG and CG'. In the algorithm "sim_c-c", some functions should be stated. The function Group(CG) returns the name of group that contains CG. The function Index(RE) returns a value that is the serial number of the similar RE-pair of CG in R-set. The function S(RE) returns the similarity of RE-pairs.

```
Algorithm sim_c-c(CG, CG')
Sum = 0
If Group(CG) == Group(CG')//the CGs are from the same group
Foreach i = 1 to X
 Foreach j = 1 to Y
 If Index(RE_i) == Index(RE_j')
 If S(RE_i) < S(RE_j')
 Sum = Sum + S(RE_i)/S(RE_j')
 Else
 Sum = Sum + S(RE_j')/S(RE_i)
 EndIf
 EndForeach
EndForeach
Similarity = Sum/Max(X,Y)
EndIf
Return Similarity
```

The algorithm "sim_c-c" has two characteristics: (1) only the CGs in the same group are allowed to be compared; (2) the similarity between two CGs is calculated by comparing the labels of their RE-pairs. Finally, according to the similarities calculated by Algorithm sim_c-c, FOM-CG creates mappings among S-sets (S-S mappings).

In the following, we apply the method, i.e., FOM-CG, to create mappings among fuzzy ontologies that are on the domain of supermarket. Our goals are to introduce the mapping process of FOM-CG in details, to evaluate the matching accuracy of FOM-CG, and to verity that FOM-CG can be sufficient for creating mappings among multiple fuzzy ontologies in an application domain.

Figure 6.10 shows three fragments of fuzzy ontologies that are composed of a reference ontology and two source ontologies. Obviously, all the fuzzy concepts are connected by fuzzy relations (Is-child-of and Is-parent-of) with memberships, and properties and instances belong to fuzzy concepts by fuzzy relations (Has-property and Has-instance) with memberships. To be convenience for explanation, FOM-CG uses letters in round brackets to label the corresponding fuzzy concepts, instances and properties.

To creating mappings, FOM-CG firstly takes these fuzzy ontologies as inputs, and transforms them into conceptual graph sets: one R-set and two S-sets as shown in Fig. 6.11. Here, a conceptual graph is represented by a set of RE-pairs, and all the conceptual graphs belonging to the same CG set are stored in a table.

Next, the algorithm sim_s-s is applied to calculate similarities for CGs in which a CG belongs to R-set and the other one belongs to S-sets. Take Conceptual Graph B and B1 as an example, FOM-CG compares all the RE-pairs of B with the RE-pair set of B1 in turn. In every comparison, the maximum similarity is added to the variable "Sum". Finally, the similarity between B and B1 is calculated by Sum/X, where X is the number of RE-pairs of B. According to these similarities, FOM-CG creates R-S mappings shown in Fig. 6.12, where mappings are labeled with the similarities.

Next, FOM-CG puts CGs in S-sets into groups based on R-S mappings, and labels all RE-pairs of CGs in Sets. For example, CGs B^1 and B^2 in S-sets are mapped into B in R-set. Thus, they are put into $Group_B$, and their RE-pairs are labeled shown in Fig. 6.13. Here, a label of RE-pair is composed of an index and a similarity. According to the index, FOM-CG can easily find out the similar RE-pair belonging to the CG in R-set. Meanwhile, the similarity can be used to create S-S mappings in the next step.

Finally, FOM-CG applies the algorithm sim_c-c to compare the labeled CGs in the same group. In other words, any two CGs from different groups are not considered to be compared. Due to the more simple form of labels, the algorithm sim_c-c compares labels, not RE-pairs, to calculate similarities for the labeled CGs. Likewise S-S mappings are created by FOM-CG based on these similarities. Figure 6.14 shows the mappings among CG sets containing one R-set and Two S-sets, where solid lines represent R-S mappings and broken lines represent S-S mappings. On the basis of these mappings, FOM-CG can further generate the mappings among the fuzzy ontologies.

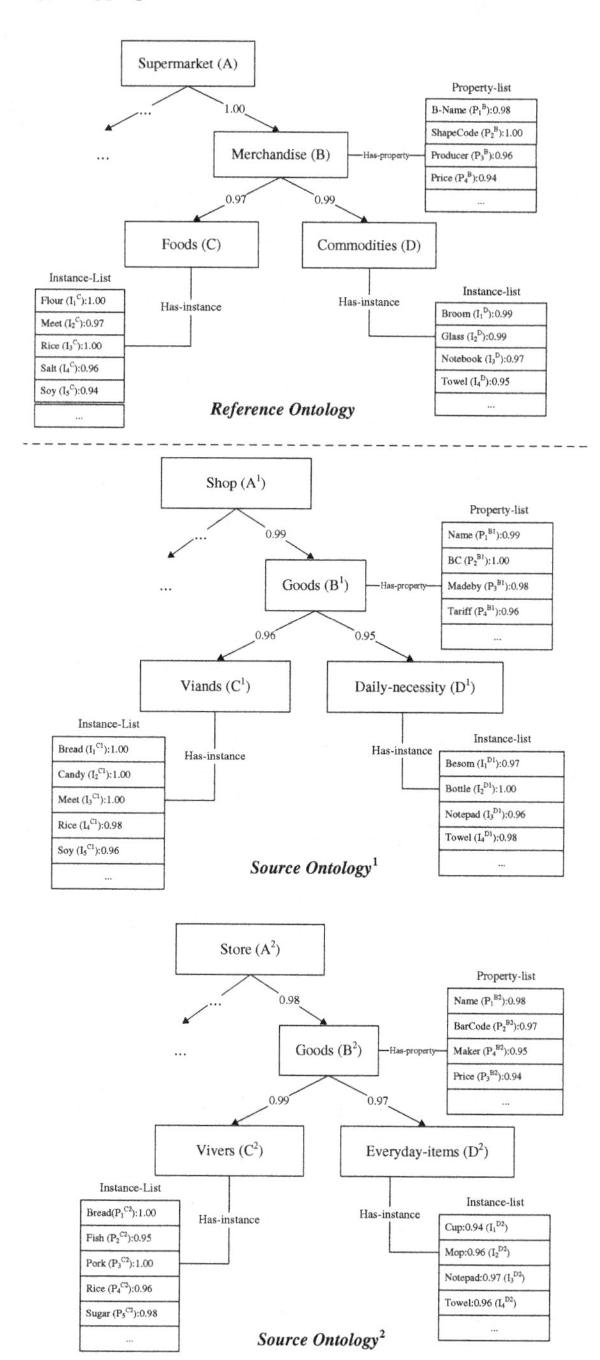

Fig. 6.10 Multiple fuzzy ontologies containing reference and source ontologies

R-set	
Is-parent-of: 1.00, B	CG: A
...	
Is-child-of: 1.00, A	CG: B
Is-parent-of: 0.97, C	
Is-parent-of: 0.99, D	
Has-property: 0.98, P_1^B	
Has-property: 1.00, P_2^B	
...	
Is-child-of: 0.97, B	CG: C
Has-instance: 1.00, I_1^C	
Has-instance: 0.97, I_2^C	
...	
Is-child-of: 0.99, B	CG: D
Has-instance: 0.99, I_1^D	
Has-instance: 0.99, I_2^D	
...	

$S\text{-set}^1$	
Is-parent-of: 0.99, B^1	CG: A^1
...	
Is-child-of: 0.99, A^1	CG: B^1
Is-parent-of: 0.96, C^1	
Is-parent-of: 0.95, D^1	
Has-property: 0.99, P_1^{B1}	
Has-property: 1.00, P_2^{B1}	
...	
Is-child-of: 0.96, B^1	CG: C^1
Has-instance: 1.00, I_1^{C1}	
Has-instance: 1.00, I_2^{C1}	
...	
Is-child-of: 0.95, B^1	CG: D^1
Has-instance: 0.97, I_1^{D1}	
Has-instance: 1.00, I_2^{D1}	
...	

$S\text{-set}^2$	
Is-parent-of: 0.98, B^2	CG: A^2
...	
Is-child-of: 0.98, A^2	CG: B^2
Is-parent-of: 0.99, C^2	
Is-parent-of: 0.97, D^2	
Has-property: 0.98, P_1^{B2}	
Has-property: 0.97, P_2^{B2}	
...	
Is-child-of: 0.99, B^2	CG: C^2
Has-instance: 1.00, I_1^{C2}	
Has-instance: 0.95, I_2^{C2}	
...	
Is-child-ofa0.97, B^2	CG: D^2
Has-instance: 0.94, I_1^{D2}	
Has-instance: 0.96, I_2^{D2}	
...	

Fig. 6.11 Conceptual graph sets translated from fuzzy ontologies

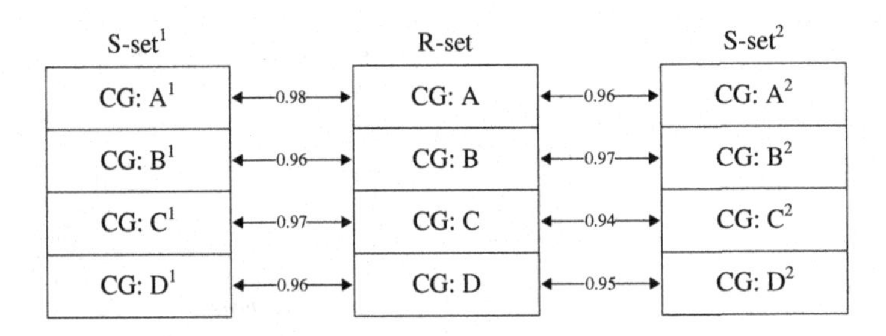

Fig. 6.12 R-S mappings between R-set and S-sets

The labeled CG B^1

RE-pairs in CG B^1	Labels
Is-child-of: 0.99, A^1	(B_1, 0.97)
Is-parent-of: 0.96, C^1	(B_2, 0.94)
Is-parent-of: 0.95, D^1	(B_3, 0.96)
Has-property: 0.99, P_1^{B1}	(B_4, 0.94)
Has-property: 1.00, P_2^{B1}	(B_5, 0.96)
...	...

The labeled CG B^2

RE-pairs in CG B^2	Labels
Is-child-of: 0.98, A^2	(B_1, 0.95)
Is-parent-of: 0.99, C^2	(B_2, 0.96)
Is-parent-of: 0.97, D^2	(B_3, 0.98)
Has-property: 0.98, P_1^{B2}	(B_4, 0.97)
Has-property: 0.97, P_2^{B2}	(B_5, 0.99)
...	...

Fig. 6.13 The labeled CGs B1 and B2 of S-sets in GroupB

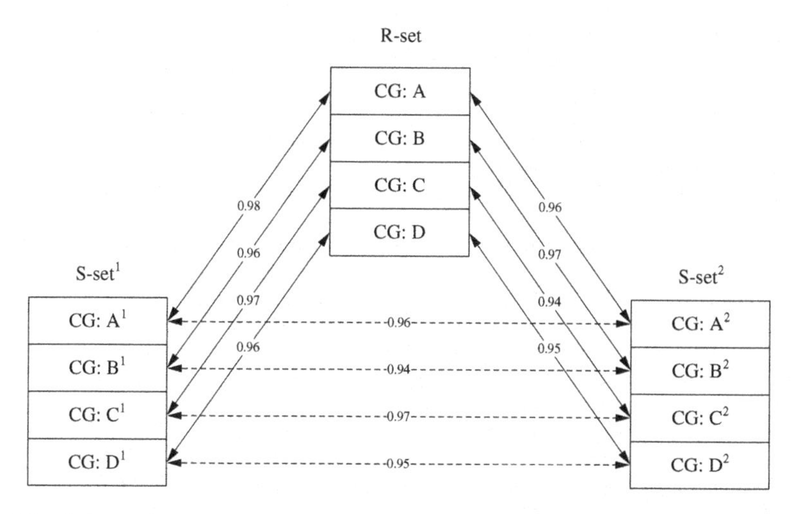

Fig. 6.14 S-S mappings among sets of CGs

6.4 Summary

With the growing number of heterogeneous fuzzy ontologies in the Semantic Web, fuzzy ontology mapping becomes an important problem to solve the interoperation among heterogeneous ontologies containing fuzzy information. In particular, defining similarity relations among fuzzy ontology components is the core of fuzzy ontology mapping. In this chapter, we introduce a method for computing semantic similarity among fuzzy concepts in a fuzzy ontology. In this method, a semantic similarity between two fuzzy concepts is computed by their information content values. Further, we introduce a conceptual graph based approach for mappings among multiple fuzzy ontologies. This method is suitable for multiple fuzzy ontologies composed of reference ontology and source ontologies in an application domain. Both of the methods of computing semantic similarity and creating mappings among multiple fuzzy ontologies introduced in this chapter may contribute to the fuzzy knowledge management for the Semantic Web.

References

Bahri A, Bouaziz R, Gargouri F (2007) Dealing with similarity relations in fuzzy ontologies. In: Proceedings of fuzzy systems conference, pp 1–6

Bakillah M, Mostafavi MA (2011) A fuzzy logic semantic mapping approach for fuzzy geospatial ontologies. In: Proceedings of SEMAPRO 2011, Lisbon, Portugal, pp 21–28

Doan A, Madhavan J, Domingos P, Halevy A (2002) Learning to map between ontologies on the semantic web. In: Proceedings of the eleventh international world wide web conference, pp 662–673

Jiang J, Conrath DW (1997) Semantic similarity based on corpus and lexical taxonomy. In: Proceedings of the 10th international conference on computational linguistics
Li GY, Zhao Y, Rao ZM (2012) A modularized multi-strategy fuzzy ontology mapping method. Inform Technol J 11:1730–1736
Lin D (1998) An information-theoretic definition of similarity. In: Proceedings of the 15th international conference on machine learning
Ma ZM, Zhang LY (2013) A conceptual graph based approach for mappings among multiple fuzzy ontologies. J Web Eng 26: 427–441 (in press)
Ma ZM, Zhang WJ, Ma WY (2004) Extending object-oriented databases for fuzzy information modeling. Inf Syst 29(5):421–435
Ma ZM, Zhang WJ, Ma WY (1999) Assessment of data redundancy in fuzzy relational databases based on semantic inclusion degree. Inf Process Lett 72(1–2):25–29
Miller GA, Beckwith R, Fellbaum C, Gross D, Miller KJ (1990) Introduction to wordNet: an online lexical database. Int J Lexicography 3(4):235–244
Ross S (1976) A first course in probability. University of Southern California Upper Saddle River, New Jersey 07458
Resnik P (1995) Using information content to evaluate semantic similarity. In Proceedings of the international joint conference on artificial intelligence (IJCAI95), pp 448–453
Sowa JF (1984) Conceptual structures—information processing in mind and machine. Addison-Welsey, Boston
Wang Y, Zhang RB, Lai JB (2009) Measuring concept similarity between fuzzy ontologies. Fuzzy Inf Eng 2:163–171
Xu BW, Kang DZ, Lu JJ, Li YH, Jiang JX (2005) Mapping fuzzy concepts between fuzzy ontologies. In: Proceedings of 9th international conference on knowledge-based intelligent information and engineering systems (KES 2005), pp 199–205
Xu X, Wu Y, Chen J (2010) Fuzzy FCA based ontology mapping. In: Proceedings of the first international conference networking and distributed computing, pp 181–185
Zhao Y, Wang X, Halang WA (2007) Ontology mapping techniques in information integration. In: Cruz-Cunha M, Cortes B, Putnik G (eds) Adaptive technologies and business integration: social, managerial and organizational dimensions, IGI Global, Hershey (Pennsylvania, USA), pp 306–326
Zhang LY, Ma ZM (2012) ICFC: a method for computing semantic similarity among fuzzy concepts in a fuzzy ontology. In: Proceedings of FUZZ-IEEE 2012, pp 1–8

Chapter 7
Querying Fuzzy Description Logics and Ontology Knowledge Bases

Abstract In order to represent the widespread imprecision and uncertainty in Semantic Web applications, there have been substantial amounts of work carried out in the context of fuzzy extensions of (Description Logics) DLs and ontologies, and corresponding reasoning algorithms and reasoners are thus developed. Also, some querying techniques are developed to provide users with querying services. However, some existing simple queries, such as instance retrieval, are relatively weak in expressive power and are unable to effectively express users' query intents. Therefore, with emergence of these large-scale fuzzy DL knowledge bases (KBs) and fuzzy ontologies, it is of particular importance to provide users with expressive querying services. In particular, conjunctive queries (CQs) originated from research in relational databases, and, more recently, have also been identified as a desirable form of querying DL knowledge bases. Conjunctive queries provide an expressive query language with capabilities that go beyond standard instance retrieval. In this chapter, we will focus on the querying techniques of fuzzy DL KBs and fuzzy ontologies. We aim at introducing the conjunctive querying techniques from tractable to more expressive fuzzy DL KBs. Also, querying over fuzzy ontologies and some flexible extensions to the W3C standard query language SPARQL will be included.

7.1 Introduction

As introduced in Chap. 1, Description logics (DLs, for short) (Baader et al. 2003) and ontologies (Berners-Lee et al. 2001) are the key techniques in the Semantic Web, which can support knowledge representation and reasoning. However, the reasoning services that aim at accessing and querying the data underlying ontologies, such as *retrieval*, *realization* and *instantiation*, are only in weak form and do not support complex queries (mainly *conjunctive queries*, CQs). CQs originated from research in relational databases, and, more recently, have also been identified as a desirable form of querying ontologies and DL knowledge bases (KBs). The

Z. Ma et al., *Fuzzy Knowledge Management for the Semantic Web*, Studies in Fuzziness and Soft Computing 306, DOI: 10.1007/978-3-642-39283-2_7,

first conjunctive query algorithm (Calvanese et al. 1998) over DLs was actually specified for the purpose of deciding conjunctive query containment for DLR_{reg}. Recently, query entailment and answering have also been extensively studied both for tractable DLs (Calvanese et al. 2007; Rosati 2007) and for expressive DLs (Levy and Rousset 1998; Glimm et al. 2007).

In order to capture and reason about vague and imprecise domain knowledge, there has been a substantial amount of work carried out in the context of fuzzy DLs and fuzzy ontologies as introduced in Chap. 4. With the emergence of a good number of large-scale knowledge bases, it is of particular importance to provide users with querying services. When querying over fuzzy DL KBs and fuzzy ontologies, as in the crisp case, the same difficulties emerge in the existing fuzzy DL reasoners such as fuzzyDL and FiRE (see Chap. 4 in detail), which are not capable of dealing with CQs either. Therefore, for providing users with expressive querying service over fuzzy DL and ontology knowledge bases, many efforts have been made. In this chapter, we focus on the query of fuzzy Semantic Web DL and ontology knowledge bases.

7.2 Querying Fuzzy Description Logic Knowledge Bases

With the development of fuzzy Description Logics (DLs) in the Semantic Web as has introduced in Chap. 4, some work were done to provide users with querying services over kinds of fuzzy DL knowledge bases. Some works of querying over tractable KBs were developed (Straccia 2006; Cheng et al. 2008a). Moreover, in order to support complex queries (mainly *conjunctive queries*, CQs), in (Mailis et al. 2007), a fuzzy extension of CARIN system is provided, along with a decision procedure for answering union of conjunctive queries. Also, several proposals for answering expressive and fuzzy conjunctive queries (CQs) were proposed (Cheng et al. 2008b; Cheng et al. 2009a). In addition, an approach for query answering in fuzzy DLs with data type support was presented in (Cheng et al. 2010a). In this chapter, querying from tractable to more expressive DL KBs will be introduced.

7.2.1 Conjunctive Query Answering of Tractable Fuzzy Description Logics

In order to capture significant ontology languages but to keep low complexity of reasoning, many tractable fuzzy DLs are presented as introduced in Sect. 4.2.1. In this case, some work focusing on tractable (or *lightweight*) *fuzzy DL languages* has been done. Straccia (2006) extended DL-Lite with fuzzy set theory and showed how to compute efficiently the top-*k* answers of a complex query (i.e., *conjunctive query*) over a huge set of instances. Further, in order to acquire more expressive but still tractable fuzzy DL, Cheng et al. (2008a) proposed a conjunctive query algorithm over *f-DLR-Lite*$_{F,\sqcap}$ knowledge base (it allows for the presence of n-ary relations and the occurrence of concept conjunction in the left land of inclusion axioms), and showed that the query answering algorithm over the extended fuzzy DL is still FOL reducible and shows polynomial data complexity.

In the following, we will introduce the conjunctive query answering of tractable fuzzy DLs presented in (Cheng et al. 2008a). The fuzzy DL *f-DLR-Lite*$_{F,\sqcap}$ in (Cheng et al. 2008a) has been introduced in Sect. 4.2.1 in detail, and thus we brief recall the notions of *f-DLR-Lite*$_{F,\sqcap}$. *f-DLR-Lite*$_{F,\sqcap}$ concepts will be redefined as $B ::= A|\exists i : R$ and $C ::= B|\neg B|C_1 \sqcap C_2$, where A denotes an atomic concept, R denotes an n-ary relation, $\exists i : R$ denotes the projection of n-ary relation R on its ith element, $1 \leq i \leq n$; B denotes a basic concept that can be either an atomic concept or a concept of the form $\exists i : R$. An *f-DLR-Lite*$_{F,\sqcap}$ knowledge base (KB) K can be partitioned into a terminological part called TBox and an assertional part called ABox, denoted by $K = <T, A>$. A TBox T is a finite set of concept axioms and functionality axioms which are of the form: $Cl \sqsubseteq C_1$ (inclusion axioms) and funct(i:R) (functionality axiom), where $Cl ::= B|C_1 \sqcap C_2$, denotes the concepts allowed to present in the left hand of inclusion axioms. A functionality axiom shows the functionality of ith component of n-ary relation. An *f-DLR-Lite*$_{F,\sqcap}$ ABox A is of the form $B(o) \geq n$, $\exists i : R\ (o) \geq n$. An interpretation of *f-DLR-Lite*$_{F,\sqcap}$ is $I = (\Delta^I, \bullet^I)$, where Δ^I is a nonempty object set named interpretation domain and $\bullet^I$ is an interpretation function.

In general, the present DL knowledge base oriented reasoners can support different kinds of queries over an ABox for obtaining assertional knowledge. These queries include *retrieval* (given an ABox A and a concept C, to find all individuals such that $A| = C(o)$, i.e. retrieve all the instances of a given concept), *realization* (determine the most specific concept an individual is an instance of) and *instantiation* (given an individual o and a concept C, to find if KB $K| = C(o)$). However, there is no support for queries that ask for n-tuples of related individuals or for the use of variables to formulate a query, just as conjunctive queries do.

Conjunctive queries stemmed from the domain of relational databases, and have attracted more attentions in Semantic Web recently. Through queries, especially conjunctive queries, we may extract extensional information from a KB. The *conjunctive queries* (CQs) over a KB are expressions of the form $q(\vec{x}) \leftarrow \exists \vec{y}.conj(\vec{x}, \vec{y})$, where vector $\vec{x}$ is constituted by distinguished variables, which can be bound with individuals in a given KB to answer the CQs. Vector $\vec{y}$ is constituted by non-distinguished variables, which are also existential quantified variables. $conj(\vec{x}, \vec{y})$ is a conjunction of atoms of the form $B(z)$ and $R(z_1,\ldots,z_n)$, where B and R are basic concept and role in a KB K respectively, z, z_1 and z_2 are constants in K or variables in $\vec{x}$ or $\vec{y}$.

In order to express our query intentions in a more efficient way, we extend the conjunctive language as follows: firstly, our new query language allows user to specify a threshold between 0 and 1 for every atom of a CQ. The specification is a generalization of original CQ as well as an extension of expressivity of query language. Secondly, various users may have various preferences to query conjunctive atoms and expect to assign each of them a weight which can reflect importance degree of the specific atom in whole query. Thus, the atom of improved fuzzy CQ language will be defined as $\langle conj_i(\vec{x},\vec{y}) \geq n, wi\rangle$, where $conj_i(\vec{x},\vec{y})$ is the i-th conjunctive atom of the form $B(z)$, $R(z_1,\ldots,z_n)$, n is the threshold between 0 and 1, w_i is the user-defined weight for i-th conjunctive atom. Furthermore, we extend an interpretation I of a given KB. For individual tuples $\vec{c}_x, \vec{c}_y \in \Delta^I \times \cdots \times \Delta^I$, $conj(\vec{x},\vec{y})$ is interpreted as the minimum of every conjunctive atoms, while $\exists\vec{y}\cdot conj(\vec{x},\vec{y})$ is interpreted as $sup_{\vec{c}_y \in \Delta^I \times \cdots \times \Delta^I}$ $\left\{(conj(\vec{c}_x,\vec{c}_y))^I\right\}$, thus $q(\vec{c}_x)$ is interpreted as $sup_{\vec{c}_y \in \Delta^I \times \cdots \times \Delta^I}\left\{(conj(\vec{c}_x,\vec{c}_y))^I\right\}$.

Just as the algorithm for crisp case, our fuzzy query answering algorithm for a given KB can be divided into four steps: firstly, we may normalize the given KB K, and store the normalized ABox into a relational tables; secondly, check the normalized KB for consistency; Thirdly, combine user queries with TBox inclusion axioms so as to reformulate queries; finally, send the reformulated queries to DBMS where normalized ABox stored to obtain answers. The realization details are given below.

7.2.1.1 KB Normalization

The object of this step is to transform present KB into appropriate form, and store the ABox in tables of relational database.

Firstly, ABox is expanded by adding assertions $(\exists i : R)(\vec{o}[i]) \geq n$ for every n-ary fuzzy relation assertion $R(\vec{o}) \geq n$. Then, for conjunctive concepts in TBox, following rules are recursively applied: if there exist $Cl \sqsubseteq C_1 \sqcap C_2$, then replace it with $Cl \sqsubseteq C_1$ and $Cl \sqsubseteq C_2$. TBox after expanding only exist inclusion axioms of the form: (i) $Cl \sqsubseteq B$, where B is a basic concept (that is to say, it is an atomic concept or an existential concept of the form $\exists i : R$). We call these inclusion axioms *positive inclusions* (*PIs*); (ii) $Cl \sqsubseteq \neg B$, which we call *negative inclusions* (*NIs*). Then, we expand TBox by calculating all the *NIs* entailed by TBox. The TBox is still closed with regard to following rules: if $B_1 \sqcap \ldots \sqcap B_n \sqsubseteq B$ occurs in TBox, and if $B'_1 \sqcap \ldots \sqcap B'_m \sqsubseteq \neg B$ occurs in TBox or there exists $i \in \{1,\ldots,m\}$, s.t. $B'_1 \sqcap \ldots \sqcap B'_{i-1} \sqcap B \sqcap B'_{i+1} \sqcap \ldots \sqcap B'_m \sqsubseteq \neg B'_i$ occurs in TBox, then add $B_1 \sqcap \ldots \sqcap B_n \sqcap \ldots \sqcap B'_{m-1} \sqsubseteq \neg B'_m$ to TBox.

After aforementioned closure operations, for every basic concepts sequence $B_1,\ldots,B_n$, $T| \approx B_1 \sqcap \ldots \sqcap B_{n-1} \sqsubseteq \neg B_n$ iff there exists $i \in \{1,\ldots,n\}$, s.t. $B_1 \sqcap \ldots \sqcap B_{i-1} \sqcap B_n \sqcap B_{i+1} \sqcap \ldots \sqcap B_{n-1} \sqsubseteq \neg B_i$ occurs in normalized KB. The normalized KB is equivalent to the original KB in this sense that they have the same models.

In order to take advantage of well-established and efficient data managing and indexing technique, we store normalized ABox into a database. The detailed steps are as follows: (i) we assume that every concept and relation in ABox is also in TBox. For every concept B in ABox, we define a 2-ary relational table tab_B, s.t.

$\langle o, n\rangle \in tab_B$ iff $B(o) \geq n$ occurs in ABox. We use *indname* and *memdegree* as the field names of individual name and its least member degree separately. (ii) for every n-ary relation R in ABox, we define a relational table tab_R of arity $n + 1$, s.t. $\langle \vec{o}, n\rangle \in tab_R$ iff $R(\vec{o}) \geq n$ occurs in ABox. Similarly, we use field names $indname_1$, ..., $indname_n$ to denote the n elements of vector $\vec{o}$, and *memdegree* to denote least member degree of $\vec{o}$ belongs to R. We denotes hereby established database *DB*(*A*).

7.2.1.2 KB Satisfiability Checking

It is no sense to query over a KB that is not satisfiable. For this reason, we will make a satisfiability checking of normalized KB aiming at following conditions: (i) there is *NI* $B_1 \sqcap \ldots \sqcap B_{m-1} \sqsubseteq \neg B_m$ in TBox, and there also exists individual o and a sequence of rational numbers $n_1,\ldots,n_m$, s.t. $B_i(o) \geq n_i (1 \leq i \leq m)$ occur in ABox, where $min(n_i) + n_m > 1 (1 \leq i \leq m - 1)$. (ii) there is functionality axiom funct(i:R) in TBox, there also $R^I(\vec{o}_1{}^{[i:o]}) \geq m$ and $R^I(\vec{o}_2{}^{[i:o]}) \geq n$ occur in ABox with $n,m > 0$, and $(\vec{o}_1 \backslash i) = (\vec{o}_2 \backslash i)$, where $(\vec{o} \backslash i)$ denotes a vector obtained by removing i-th element of vector $\vec{o}$. Condition (i) holds for checking if there are some contradictions between ABox and some *NIs* in TBox, condition (ii) holds for checking if there are some contradictions between ABox and some functionality axioms in TBox. Note that the least membership degrees of $R^I(\vec{o}_1{}^{[i:o]})$ and $R^I(\vec{o}_2{}^{[i:o]})$ need not to be equal, since once there exist two ABox assertions with regard to the same n-ary relation R, then the conflict between the two assertions generates. If any above condition is satisfiable, then the algorithm returns false, or true otherwise.

The practical process for KB satisfiability checking can be done by sending queries to the database where normalized ABox is stored.

For example, if we want to check whether condition (i) holds, we may send following query written in relational algebra (combining with relational calculus) to *DB*(*A*) for each *NI* in the TBox:

$$\exists o.\left(\begin{array}{r}\bigwedge_{1 \leq i \leq m} Count(\sigma_{indname} = \text{`}o\text{'}(B_i)) \neq 0 \wedge \min_{1 \leq i \leq m-1}\left(\prod_{B_{i.memdegree}}(\sigma_{indname} = \text{`}o\text{'}(B_i))\right) \\ +\prod_{B_{m.memdegree}}(\sigma_{indname} = \text{`}o\text{'}(B_m)) > 1\end{array}\right).$$

As for condition (ii), we may send following query written in relational algebra to *DB*(*A*) for each *functionality axiom* funct(i:R) in the TBox: $\exists o.Count(\sigma_{indname} = \text{`}o\text{'}(R)) \geq 2$ or $\exists o.Count(\sigma_{indname} = \text{`}o\text{'}(\exists i : R)) \geq 2$. The latter one is also a rational one if we additionally store fuzzy concept $\exists i : R$ in *DB*(*A*).

7.2.1.3 Query Reformulation

For a given KB *K* and a query *q*, we will reformulate the original query *q* into new query *qr* by binding *PIs* in TBox to *q*, and then send query *qr* to the database where normalized ABox locates.

Subsequently, we will analyze the query reformulation algorithm *QueryRef* from a technical point of view. If any argument in an atom of query *q* corresponds to a distinguished variable or a shared variable (i.e., a variable appears at least twice in query body), or a constant (an individual name), then we call it *bound*. On the contrary, if the argument corresponds to a non-distinguished non-shared variable, we call it *unbound*.

For example, given a query $q(x) \leftarrow B_1(x) \wedge R_1(y_1, y_2) \wedge B_2(y_2)$, where x is a distinguished variable, y_2 is a shared variable, thus they are all *bound*, while y_1 is *unbound* (for the sake of simplicity, we omit the thresholds and user weights in query atoms). In the following, we will use the symbol _ to represent non-distinguished non-shared variable, then the example above can be written as $q(x) \leftarrow \exists y_1, y_2.B_1(x) \wedge R_1(\_, y_2) \wedge B_2(y_2)$.

According to principle below, we apply *PI I* to query atom *g* to reformulate it and assign threshold and user weight to new atom. We denote hereby obtained atom *gr(g,I)*.

If *PI* $I = B_1 \sqcap \ldots \sqcap B_m \sqsubseteq A$ (resp. $I = B_1 \sqcap \ldots \sqcap B_m \sqsubseteq \exists j : R$), $g = \langle A(x) \geq n, w\rangle$ (resp. $g = \langle R(\_, \ldots, \_, x, \_, \ldots, \_) \geq n, w\rangle$, where x occurs in j-th position of relation R, and variables in other positions are *unbound*), we say *PI I* is applicable to atom g, and $gr(g, I) = \langle min(C_1(x), \ldots, C_m(x)) \geq n, w\rangle$, where for every $i \in \{1, \ldots, m\}$, if $B_i = A_i$, then $C_i(x) = A_i(x)$; if $B_i = \exists k : R_i$, then $C_i(x) = \exists z_1, \ldots, z_l.R_i(z_1, \ldots, z_{k-1}, x, z_{k+1}, \ldots, z_l)$, where l is the arity of R_i.

Here is the full algorithm of *QueryRef*:

Algorithm: *QueryRef(q, T)*
Input: CQ *q*, normalized f-DLR-Lite$_\sqcap$TBox T
Output: FOL query *qr*

```
1  P:={q};
2  repeat
3  P':=P;
4        for each q∈P' do
5        (a) for each g in q do
6        for each PI I in T do
7              if I is applicable to g
8              then P := P∪{q[g/gr(g, I)] };
9        (b) for each g1, g2 in q do
10             if g1 and g2 unify
11             then P := P ∪ {τ(reduce(q, g1, g2))};
12 until P'= P;
13 let qr be a query in P
14 for each q ∈ P do
15       body_qr = body_qr ∨ body_q;
16 return qr
17 end
```

At line 8 of the algorithm, we use $q[g/g']$ to represent the query obtained by replacing atom g with g'. In step (a), by applying *PI* axiom, we reformulate every atom g of each query q in P', and insert the new query into P. In step (b), given a pair of atom $g_1 = \langle R(x_1,\ldots,x_n) \geq n_1, w_1\rangle$ and $g_2 = \langle R(y_1,\ldots,y_n) \geq n_2, w_2\rangle$, if for, $i \in \{1,\ldots,n\}$, either $x_i = y_i$, or $x_i = \_$ or $y_i = \_$, then we call g_1, g_2 unify (Cali et al. 2003), and their unification is $\langle R(z_1,\ldots,z_n) \geq min(n_1,n_2), w_1 + w_2\rangle$. If x_i or y_i is distinguished variable, then $z_i = x_i$ or $z_i = y_i$. If $x_i = y_i$ or $y_i = \_$, then $z_i = y_i$, otherwise $z_i = x_i$. If g_1, g_2 unify, then algorithm *reduce*(q, g_1, g_2) returns new query by applying most general unification of g_1 and g_2 to q. Due to the unification, the *unbound* variables in query q may be *bound* in the new query q', thus the *PIs* not applicable to q may become applicable to q' in the next execution of step (*a*). Function $\tau(q)$ replaces the unbound variables in q with _. From line 13 to line 16, the algorithm transforms conjunctive query set into FOL query.

For example, given a normalized TBox $T = \{B_1 \sqsubseteq \exists 2 : R_2,\quad \exists 2 : R_1 \sqsubseteq B_1, B_3 \sqcap B_4 \sqsubseteq \exists 2 : R_2\}$ and query $q(x) \leftarrow \langle R_1(x,y,\_) \geq n_1, w_1\rangle \wedge \langle R_2(\_,y,\_) \geq n_2, w_2\rangle$, where atom $g_1 = \langle R_1(x,y,\_) \geq n_1, w_1\rangle, g_2 = \langle R_2(\_,y,\_) \geq n_2, w_2\rangle$.

At the first execution of line 8, the algorithm gets atom $gr_1(g_2, I_1) = \langle B_1(y) \geq n_2, w_2\rangle$ by applying *PI* $I_1 = B_1 \sqsubseteq \exists 2 : R_2$ to g_2, then obtains a new query $q_1 = g_1 \wedge gr_1$ and inserts it into *P*. At the second execution of line 8, the algorithm gets atom $gr_2(g_1, I_1) = \langle R_1(\_,y,\_) \geq n_2, w_2\rangle$ by applying *PI* $I_2 = \exists 2 : R_1 \sqsubseteq B_1$ to gr_1, then obtains a new query $q_2 = g_1 \wedge gr_2$ and inserts it into *P*. Since atom gr_2 and g_1 unify, the algorithm enters line 11 and gets a new query $q_3 = \tau(reduce(q_2, g_1, gr_2)) = \langle R_1(x,\_,\_) \geq min(n_1,n_2), w_1 + w_2\rangle$. At the third execution of line 8, the algorithm gets atom $gr_3(g_2, I_3) = \langle min(B_3(y), B_4(y)) \geq n_2, w_2\rangle$ by applying *PI* $I_3 = B_3 \sqcap B_4 \sqsubseteq \exists 2 : R_2$ to g_2, then obtains a new query $q_3 = g_1 \wedge gr_3$ and inserts it into *P*. The algorithm ends up the loop when it finds no *PI* applicable to any atom.

7.2.1.4 Top-k Answer

To take advantage of existing top-*k* query answering algorithms, we temporarily do not take user weights into consideration. Firstly, we convert obtained FOL queries into SQL queries and find top-*k* answers in database where ABox locates in virtue of e.g. RankSQL system (Li et al. 2005). For every (abbreviated) query q in query set *qr*, we compute the answer set and take the union, written as *ans*. We denote every tuple in set *ans* as $\vec{t_j} = \left\langle \vec{d_j}, \vec{c_j} \right\rangle$, where $\vec{d_j} = \left\langle d_{conj1}, \ldots, d_{conji}, \ldots \right\rangle$ is membership degree tuple of *ans*, whose every element corresponds to one of the query atom, $\vec{c_j}$ is the answer of querying. Then we compute the vector product of $\vec{d_j}$ and the transpose of user weight vector $\vec{w}$, then rank the scores to get top-*k* answer, written as *answer*.

7.2.1.5 Complete Algorithm

Finally, we present the complete algorithm *fQA* of query answering over *f-DLR-Lite*$_{F,\cap}$.

Algorithm: *fQA(q,K)*
Input: CQ *q*, *f-DLR-Lite*$_{F,\cap}$ KB *K*
Output: answer set *answer*
1 *A'=ExpandA(A)* //ABox Expanding
2 *T'=ExpandT(T)* //TBox axiom transformation&entailment computing
3 *DB(A)=Store(A)* //ABox storing into DB
4 **if** *Consistent(A', T')=true* **then** //Checking consistency for TBox and ABox
5 **return** *(SQLRef(QueryRef(q,T')),DB(A))* //Transforming reformulated query into SQL, then finding answers in ABox DB

The correctness of the query answering algorithm is similar to that of the crisp case in (Calvanese et al. 2005). Given a normalized KB *K* = <*T*, *A*>, we use *chase*(*K*) to denote the (possibly infinite) ABox obtained starting from *A* and closing it with respect to the following *chase* rule: (i) if $B_i(o) \geq n \in chase(K)$ for $i \in \{1,\ldots,m\}$ and $B_1 \sqcap \ldots \sqcap B_m \sqsubseteq A$, then add $A(o) \geq n$ to *chase*(*K*). (ii) if $\exists k :$ $R(o) \geq n \in chase(K)$ and there exists no vector $\vec{o}$ with o as its k-th element, s.t. $R(\vec{o}) > 0$, then add the assertion $R(\vec{o}^{[k:o]}) \geq n$ to *chase*(*K*). The correctness of query processing technique is based on a crucial property of *chase*(*K*): if *K* is satisfiable, then *chase*(*K*) is a representative of all models of *K*. This property implies that query answering can be in principle done by evaluating the query over *chase*(*K*) seen as a database. However, since *chase*(*K*) is in general infinite, we obviously avoid the construction of the *chase*. Rather, as we said before, we are able to compile the TBox into the query, thus simulating the evaluation of the query over the (in general infinite) *chase* by evaluating a finite reformulation of the query over the initial ABox.

From the concrete steps of the algorithm, we can intuitively make out that *f-DLR-Lite*$_{F,\cap}$ is still FOL reducible, i.e., the inputted CQs can be transformed into equivalent FOL queries. Meanwhile, algorithm *ExpandA*, *ExpandT*, *Consistent* and *QueryRef* are polynomial data complexity with regard to original ABox size, original TBox size, normalized KB size and normalized TBox size respectively. Hence, we can conclude that *fQA* is polynomial data complexity with regard to KB size, and this is the minimal data complexity to the best of our knowledge.

7.2.2 Conjunctive Query Entailment for Expressive Fuzzy Description Logics

The tractable fuzzy DLs and their relevant query techniques introduced above cannot realize the conjunctive query for expressive fuzzy DLs. Currently, there are lots of expressive fuzzy DL KBs as mentioned in Sect. 4.2.2. Correspondingly, querying in expressive fuzzy DLs has been carried out to provide users with expressive querying services. Mailis et al. (2007) described fuzzy *CARIN*, a knowledge representation language combining fuzzy Description Logics with Horn rules, and also proposed a fuzzy version of existential entailment algorithm for answering conjunctive queries in fuzzy *ALCNR*. However, it allows only *positive role atoms* in a query, while the *negative atoms* are not touched on. On this basis, Cheng et al. (2008b) presented the algorithm for answering expressive and fuzzy conjunctive queries, allowing the occurrence of both lower bound and the upper bound of threshold for a query atom (i.e., not only the $>$ and $\geq$ atoms, but the $<$ and $\leq$ atoms are allowed in a query), over the relative expressive fuzzy DL, *f-ALC*. The algorithm can easily be adapted to existing (and future) DL implementations. Similarly, Cheng et al. (2009a) proposed an algorithm for answering expressive fuzzy conjunctive queries, which allows the occurrence of both lower bound and the upper bound of thresholds in a query atom, over the relative expressive DL, namely *fuzzy ALCN*. In the following, we will introduce the algorithm for answering expressive and fuzzy conjunctive queries over the fuzzy DL *f-ALC* in (Cheng et al. 2008b).

Before introducing the algorithm for answering fuzzy conjunctive queries over *f-ALC*, we briefly reviews the necessary background knowledge of *f-ALC* as has been introduced in Sect. 4.2.2 in detail. Let N_C, N_R, N_I be countable infinite and pairwise disjoint sets of concept, role and individual names respectively. *f-ALC* concepts (denoted by C and D) are formed out of concept names according to the following abstract syntax:

$$C, D \rightarrow \top | \bot | A | C \sqcap D | C \sqcup D | \neg C | \forall R.C | \exists R.C$$

where $A \in N_C, R \in N_R$. Here symbols $\top$, $\bot$, $\sqcap$, $\sqcup$, and $\neg$ are used to represent top concept, bottom concept, concept conjunction, concept disjunction, and concept negation, respectively. Symbols $\forall$ and $\exists$ are used to represent universal quantification and existential quantification, respectively.

An *f-ALC* knowledge base $\mathcal{K}$ can be partitioned into a terminological part called TBox and an assertional part called ABox, denoted by $\mathcal{K} = \langle \mathcal{T}, \mathcal{A} \rangle$. An *f-ALC* TBox $\mathcal{T}$ is a finite set of concept inclusion axioms of the form $C \sqsubseteq D$. An *f-ALC* ABox $\mathcal{A}$ is of the form $B(o) \bowtie n, R(o, o') \bowtie n$, where $o, o' \in N_I$ and $\bowtie$ stands for one of the signs of inequality of $\geq$, $>$, $\leq$ and $<$. For example, the concept assertion $B(o) \geq n$ means that "o is B with the membership degree equal to or more than n", while role assertion $R(o,o') < n$ means that "o is related to o' via relation R with the membership degree less than n".

The semantics of *f-ALC* are provided by an interpretation, which is a pair $\mathcal{I} = (\Delta^{\mathcal{I}, \cdot \mathcal{I}})$. Here $\Delta^{\mathcal{I}}$ is a non-empty set of objects, called the domain of interpretation, and $\cdot^{\mathcal{I}}$ is an interpretation function which maps different individual names into different elements in $\Delta^{\mathcal{I}}$. Given an interpretation $\mathcal{I}$ and an inclusion axiom $A \sqsubseteq C, \mathcal{I}$ is a model of $A \sqsubseteq C$, if $A^{\mathcal{I}}(o) \leq C^{\mathcal{I}}(o)$ for any $o \in \Delta^{\mathcal{I}}$, written as $\mathcal{I}| \approx A \sqsubseteq C$. Similarly, for ABox assertions, $\mathcal{I} |\approx B(o) \bowtie n$(resp. $\mathcal{I} |\approx R(o, o') \bowtie n$), iff $B^{\mathcal{I}}(o^{\mathcal{I}}) \bowtie n$(resp.$R^{I}(o^{I}, o'^{I}) \bowtie n)$. If an interpretation $\mathcal{I}$ is a model of all the axioms and assertions in a KB $\mathcal{K}$, we call it a model of $\mathcal{K}$. A KB is *satisfiable* iff it has at least one model. A KB $\mathcal{K}$ *entails* (logically implies) a fuzzy assertion φ, iff all the models of $\mathcal{K}$ are also models of φ, written as $\mathcal{K} \models_{\approx} \varphi$.

7.2.2.1 Fuzzy Querying Language

We now formally define the syntax and semantics of the fuzzy querying language for *f-ALC*.

Definition 7.1 (*syntax*) Let N_C, N_R, N_I, and N_V be countable infinite and pairwise disjoint sets of concept, role, individual, and variable names respectively, and let FL be a fuzzy Description Logic (e.g. *f-ALC*). A *term t* is either an individual name from N_I or a variable name from N_V. Let C be an FL-concept, R an FL-role, and t, t' terms. An *fuzzy query atom* is an expression $C(t) \bowtie n$ or $R(t,t') \bowtie m$, where n denotes the threshold the lower bound (which corresponds to $>$ or $\geq$) or upper bound (which corresponds to $<$ or $\leq$) of membership of being the term t a member of the fuzzy set C, m denotes the degree of membership of being the term pair (t, t') member of the fuzzy role R. We refer to these two different types of atoms as *concept atoms* and *role atoms* respectively. A *fuzzy conjunctive query q* is considered conjunction of fuzzy query atoms, or a non-empty set of query atoms. The conjunction form of a fuzzy conjunctive query q is $q = \diamondsuit \leftarrow \wedge_{i=1}^{n} \langle at_1 \bowtie n_i \rangle$, where $\langle at_i \bowtie n_i \rangle$ denotes i-th query atom of q, and at_i stands for the corresponding crisp query atom of i-th fuzzy query atom. The set form of query q is $q = \{\langle at_1 \bowtie n_i \rangle, \ldots, \langle at_n \bowtie n_k \rangle\}$. Then for every fuzzy query atom, we can say $\langle at_i \bowtie n_i \rangle \in q$.

Note that, in a fuzzy query $q = \diamondsuit \leftarrow \wedge_{i=1}^{n} \langle at_i \bowtie n_i \rangle$, the fuzzy query atoms of the form $\langle at_i > 1 \rangle$ and $\langle at_i < 0 \rangle$ are not allowed. Furthermore, we assume that a fuzzy query q does not contain conjugated atoms, i.e., there exists in q no atom pair of $\langle \alpha \geq n \rangle$ and $\langle \alpha \leq m \rangle$ with n > m, which apparently makes q to be false.

We use **Var**(q) to denote the set of variables occurring in q, **Ind**(q) to denote the set of individual names occurring in q, and **Term**(q) for the set of terms in q, where **Term**(q) = **Var**(q) ∪ **Ind**(q). If all terms in q are individual names, i.e. **Term**(q) ⊆ **Ind**(q), we say that q is *ground*.

A Boolean query consisting of fuzzy query atoms is called a *fuzzy Boolean query*, which is of the form $\langle\rangle \leftarrow \langle C(x) \bowtie n_1 \rangle \wedge \langle R(x,y) \bowtie n_2 \wedge \langle D(y) \bowtie n_3 \rangle$, where the membership degree of i-th atom is $\bowtie n_i$.

The semantics of a fuzzy query is given in the same way as for the related fuzzy DL by means of interpretations consisting of an interpretation domain and a fuzzy interpretation function.

Definition 7.2 *(semantics)* Let $\mathcal{I} = (\Delta^{\mathcal{I}}, \cdot^{\mathcal{I}})$ be a fuzzy interpretation of a fuzzy Description Logic FL, q be a fuzzy conjunctive query and t, t' terms in q, and π : **Var**(q) ∪ **Ind**(q) $\rightarrow \Delta^{\mathcal{I}}$ a total function (also called an *assignment*) such that $\pi(a) = a^{\mathcal{I}}$ for each $a \in$ **Ind**(q). We say

- $\mathcal{I} \mid\approx^{\pi} \langle C(t) \bowtie n \rangle$if $C^{\mathcal{I}}(\pi(t)) \bowtie n$,
- $\mathcal{I} \mid\approx^{\pi} \langle R(t,t') \bowtie n \rangle$if $R^{\mathcal{I}}(\pi(t), \pi(t)') \bowtie n$.

If there is an assignment π, such that $\mathcal{I} \mid\approx^{\pi}$ *at* for all atom *at* $\in q$, then we say $\mathcal{I}$ entails q, written as $\mathcal{I} \mid\approx q$. If $\mathcal{I} \mid\approx q$ for each model $\mathcal{I}$ of a knowledge base $\mathcal{K}$, then we say $\mathcal{K}$ entails q, written as $\mathcal{K} \mid\approx q$.

Fuzzy Boolean queries can be classified into two categories: those only allowing individual names in knowledge base occurring in query atoms, i.e., ground query or variable-free query, and those with non-distinguished variables.

7.2.2.2 Ground Boolean Queries

Since the atoms occurring in a ground query are ABox assertions at all, the fuzzy conjunctive query entailment problem is actually a generalized fuzzy KB entailment problem. For example, given a fuzzy Boolean query including only one fuzzy query atom, e.g., $q = \langle\rangle \leftarrow \langle C(a) \geq n \rangle$, then the conjunctive query entailment, in fact, is to check whether or not $\mathcal{K} \mid\approx \langle C(a) \geq n \rangle$. Furthermore, given a fuzzy Boolean query with more than one query atom, denoted as $q = \wedge_{i=1}^{n} \langle at_i \bowtie n_i \rangle$, checking whether $\mathcal{K} \mid\approx q$, can be dealt with by checking whether $\mathcal{K} \mid\approx \langle at_i \bowtie n_i \rangle$ for every $\langle at_i \bowtie n_i \rangle \in q$. The fuzzy KB entailment problem, in turn, can be reduced to the undecidability problem:

- $\mathcal{K} \mid\approx \langle at_i \geq n_i \rangle$iff $\mathcal{K} \cup \{\langle at_i < n_i \rangle\}$ is not satisfiable,
- $\mathcal{K} \mid\approx \langle at_i \leq n_i \rangle$iff $\mathcal{K} \cup \{\langle at_i > n_i \rangle\}$ is not satisfiable,
- $\mathcal{K} \mid\approx \langle at_i > n_i \rangle$iff $\mathcal{K} \cup \{\langle at_i \leq n_i \rangle\}$ is not satisfiable,
- $\mathcal{K} \mid\approx \langle at_i < n_i \rangle$iff $\mathcal{K} \cup \{\langle at_i \geq n_i \rangle\}$ is not satisfiable.

7.2.2.3 Boolean Queries with Variables

A Boolean query with variables is of the form $q_1 = \langle\rangle \leftarrow \langle C(a) \geq n_1 \rangle \wedge \langle R(a,y_1) \geq n_2 \rangle \wedge \langle S(y_1,y_2) \geq n_3 \rangle \wedge \langle D(y_2) \geq n_4 \rangle$, we use $v(q) = (v_1,\ldots,v_m)$ to denote variable tuple occurring in a query, e.g. $v(q_1) = (y_1, y_2)$. Since the conjunctive query answering problem can be reduced to conjunctive query entailment by replacing all the distinguished variables with individual names from the knowledge base. If a query has only distinguished variables, then the obtained query is a ground query. However, for a query with non-distinguished variables, the obtained Boolean query still has variables. Substituting these variables with individual names occurring in the knowledge base is at odds with the principle of OWA (open world assumption) in DLs. Consider a knowledge base (with only an assertion) $\mathcal{K}_1 = \{(\exists R.(\exists S.D) \sqcap C)(a) \geq n\}$, where $n \geq \max(\{n_i\})$ with $1 \leq i \leq 4$. Apparently, $\mathcal{K}_1 \mid\approx q_1$ while the variables may need to be interpreted as elements of domain that are not the interpretations of any individual names, i.e., there is no tuple of domain elements which can be used to replace the variable tuple so as to make the query q_1 true. For this reason, we may resort to other methods.

To restrict our attentions within the method itself, we assume a KB is a *purely assertional* one (i.e., a KB without TBox) in that a KB can be transformed into a purely assertional one without loss of its reasoning characteristic (Straccia 2001). Since ABoxes and queries are all sets of assertions when ignoring the difference between individual names and variable names, it immediately occurs to us that we can check the *model containment* between ABox and query, i.e., if the model of a query is subsumed by the model of an ABox.

Furthermore, *f-ALC* has a so-called forest model property, i.e., a model of it has the form of a set of trees. The query also can be represented as a directed, labelled graph called a *query graph*. We assume a query q is connected, i.e., the query graph of q is a directed connected graph. This is w.l.o.g. since a non-connected query graph may be treated by treating each connected subgraph of it separately, and a knowledge base $\mathcal{K}$ entails q iff $\mathcal{K}$ entails every subquery of q depicted by a connected subgroup. There is a direct correspondence between the completion forest and models of the *f-ALC* knowledge base, also for the ease of illustration, the model containment problem can be deduced to syntactically checking the completion forest for a match of the query graph.

Initially, we start building the completion forest with a set of nodes, each of which corresponds to each individual name in the ABox. A node is labelled with a set of triples, each of which corresponds to each fuzzy concept assertion with regard to the individual name that the node represents and is defined, formally, as $tr = (cName, ineqSign, memDegree)$, where *cName*, *ineqSign* and *memDegree* denote the concept name, sign of equality and the membership degree in the concept assertion, respectively. For instance, given a concept assertion $\langle C(b) \leq 0.3 \rangle$ in ABox, the label of node b may contain a triple $(C, \leq, 0.3)$. Note that, for the sake of simplicity, we use the same name of an individual b to denote its corresponding node. Each role assertion is represented by an edge between two nodes in the initial completion graph. The edge between two nodes is labelled with a set of triples, each of which corresponds to a fuzzy role assertion to which the two nodes are related, and is defined similar as a concept triple. For instance, given a fuzzy role assertion $\langle R(a,b) \geq 0.6 \rangle$, the label of edge between node a and b contains a triple $(R, \geq, 0.6)$. An initial completion forest is expanded according to a set of expansion rules (Straccia 2001) that reflect the constructors allowed in *f-ALC*. The expansion stops when an obvious contradiction, called a *clash*, occurs, or when no more rules are applicable. In the latter case, the completion forest is called complete. Termination is guaranteed by a cycle-checking technique called *blocking*. The expansion and blocking rules are such that we can build a model for the knowledge base from each complete and clash-free completion forest. The model built from a complete and clash-free completion graph is called a *canonical* model.

Then, we build the query graph for a query q. The nodes in the query graph correspond to the terms in q, either individual names or variables, and are labeled with a triple corresponds to a related fuzzy concept atom. The edges correspond to the role atoms in q and are labeled accordingly.

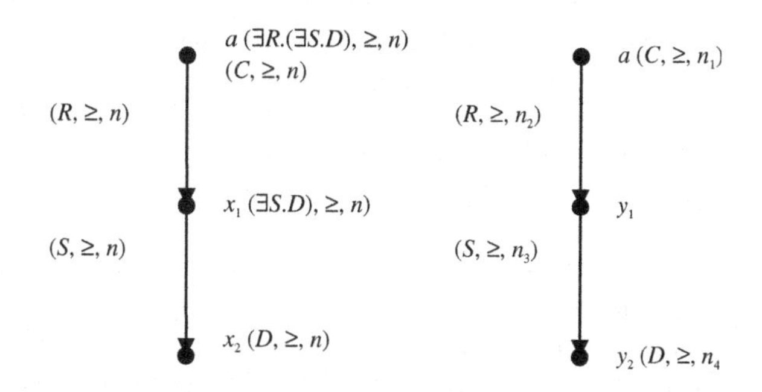

Fig. 7.1 The completion forest of K_1 and the query graph of q_1

Table 7.1 The conditions under which tr is more general than tr′

tr	*tr′*			
	(*C*, >,*n*)	(*C*, ≥,*n*)	(*C*, <,*n*)	(*C*, ≤,*n*)
(*C*, >,*m*)	$m \leq n$	$m < n$	NULL	NULL
(*C*, ≥,*m*)	$m \leq n$	$m \leq n$	NULL	NULL
(*C*, <,*m*)	NULL	NULL	$m \geq n$	$m > n$
(*C*, ≤,*m*)	NULL	NULL	$m \geq n$	$m \geq n$

After the completion forest and the query graph are built, we use a method parallel to the so-called *existential entailment* algorithm (Levy and Rousset 1998). The following example may be helpful in illustrating our method. Recall the query $q_1 = \Diamond \leftarrow \langle C(a) \geq n_1 \rangle \wedge \langle R(a, y_1) \geq n_2 \rangle \wedge \langle S(y_1, y_2) \geq n_3 \rangle \wedge \langle D(y_2) \geq n_4 \rangle$ and the KB $\mathcal{K}_1 = \{(\exists R.(\exists S.D) \sqcap C)(a) \geq n\}$ where $n \geq \max(\{n_i\})$ with $1 \leq i \leq 4$. We first build completion forest for $\mathcal{K}_1$. Initially, there is only one node *a* labelled with triple (∃*R*. (∃*S*.*D*) ⊓ *C*, ≥, *n*), then expansion rules are applied to label of nodes, and lead to the completion forest and the query graph of q_1 are depicted in Fig. 7.1.

Now, we can compare the completion forest of $\mathcal{K}_1$ and query graph of q_1 for a match. We start from the root node in the query graph of q_1 and, at the same time, locate in the completion forest of $\mathcal{K}_1$ for a node whose label (a set of triples) *is more general than* the label (also a set of triples) of the node in the query graph. We say a label l_1 is subsumed by a label l_2 if for every triple *tr* in l_1, there exists a triple *tr′* in l_2, s.t. (i) *tr′* share the same concept or role name with *tr*, (ii) and the range identified by the sign of equality and the membership degrees in *tr* is *more general than* that of *tr′*. In Table 7.1, we list the conditions under which *tr* is more general than *tr′*.

Then, we traverse the query graph along the outgoing edge to the next node, while in the completion forest we undeterministically select a branch to proceed (if any). The algorithm terminates when the leaf node in query graph is encountered. In according to what they represent, we distinguish the nodes, in the completion forest and query graph, into *constant nodes* which represent individual names and *variable nodes* which represent variable names. During the traversal, if (i) every constant node occurs in the query graph is mapped to a constant node in the completion forest that represents the same individual, and (ii) the label of every node and edge in query graph is more general than that of in completion graph, we say that there is a match for a query in the knowledge base. Given an $f{-}\mathcal{ALC}$ knowledge base $\mathcal{K}$ and a query q, the algorithm answers "$\mathcal{K}$ entails q" if a match for q exists in each complete and clash-free completion graph and it answers "$\mathcal{K}$ does not entail q" otherwise.

However, this method cannot extend directly to the case where $<$ and $\leq$ atoms are allowed in a query. For example, given a KB $\mathcal{K}_2 = \{\forall R.D(a) \geq 0.6, D(b) < 0.3\}$ and a query $q_2 = \lozenge \leftarrow \langle R(a,y) \leq 0.5\rangle \wedge \langle D(y) \leq 0.4\rangle$, we can instinctively recognize that $\mathcal{K}_2$ entails q_2. However, existing algorithm cannot build a completion forest for $\mathcal{K}_2$, in which there exists a match of the query graph of q_2.

We thus introduce some new rules for mending this. For the consistency with other rules in (Straccia 2001), we adopt the same form here.

$$(\forall_{\geq R})\langle w_1 : \forall R.C \geq n\rangle, \langle w_2 : C < m\rangle \rightarrow \langle (w_1, w_2) : R \leq 1 - n\rangle \text{ where } m \leq n.$$

$$(\forall_{> R})\langle w_1 : \forall R.C > n\rangle, \langle w_2 : C < m\rangle \rightarrow \langle (w_1, w_2) : R < 1 - n\rangle \text{ where } m < n.$$

The rule $\forall_{\geq R}$ says, if $\langle \forall R.C(w_1) \geq n\rangle$ and $\langle C(w_2) < n\rangle$ are in $\mathcal{K}$, then add $\langle R(w_1,w_2) \leq 1\text{-}n \rangle$ into $\mathcal{K}$. The rule $\forall_{>R}$ is defined similarly. After applying these rules, the otherwise isolated two nodes are related by a fuzzy role R.

We, in the following, prove the correctness of newly introduced rules. For each fuzzy interpretation $\mathcal{I}$ that is a model of $\mathcal{K}$, we show that $\mathcal{I} | \approx \mathcal{K}'$, where $\mathcal{K}'$ stands for the knowledge base obtained by applying $\forall_{\geq R}$ rule to $\mathcal{K}$, i.e. $\mathcal{K}' = \mathcal{K} \cup \{\langle (w_1,w_2){:}R \leq 1\text{-}n \rangle\}$. Given an assertion $\langle w_1{:}\forall R.C \geq n \rangle$ in $\mathcal{K}$, we have $\mathcal{I} | \approx \langle w_1{:}\forall R.C \geq n \rangle$ according to $\mathcal{I} | \approx \mathcal{K}$, i.e., $\inf_{w' \in \Delta^{\mathcal{I}}}. \{\max\{1 - R^{\mathcal{I}}(w_1, w'), C^{\mathcal{I}}(w')\}\} \geq n$. It follows that for every $w' \in \Delta^{\mathcal{I}}$, $\max\{1 - R^{\mathcal{I}}(w_1, w'), C^{\mathcal{I}}(w')\} \geq n$ holds. As w_2 is an element in $\Delta^{\mathcal{I}}$, we have $\max\{1 - R^{\mathcal{I}}(w_1, w_2), C^{\mathcal{I}}(w_2)\} \geq n$. Since there is an additional assertion of $\langle w_2{:}C < m \rangle$ with $m \leq n$, the $1 - R^{\mathcal{I}}(w_1, w_2) \geq n$ (or its equivalence $R^{\mathcal{I}}(w_1, w_2) \leq n$) holds.

In our example, before applying $\forall_{\geq R}$ rule, the completion forest of $\mathcal{K}_2$ consists of two isolated nodes a and b labelled with ($\forall R.D$,≥,0.6) and (D,<,0.3) respectively. The completion forest of $\mathcal{K}_2$ is shown in Fig. 7.2. The $\forall_{\geq R}$ rule then additionally adds an role assertion $\langle R(a,b) \leq 1{-}0.6 \rangle$ (simplified as $\langle R(a,b) \leq 0.4\rangle$) depicted in Fig. 7.2 as a dashed line, into $\mathcal{K}_2$ and therefore connects the isolated a and b, resulting in the completion forest of $\mathcal{K}_2'$, which is also shown in Fig. 7.2.

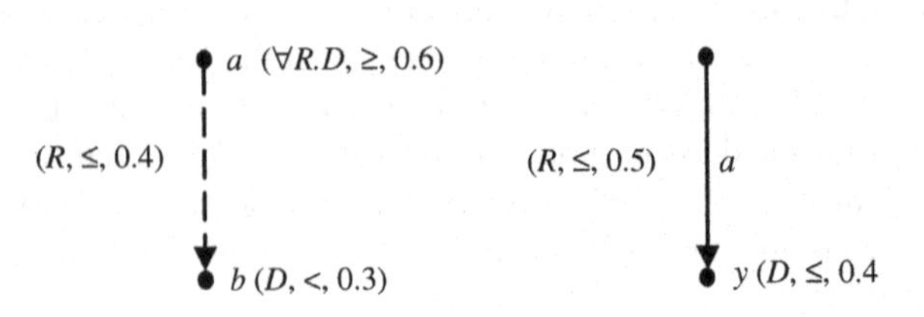

Fig. 7.2 The completion forest of K_2 and K_2' and the query graph of q_2

We can easily find a match of the query graph of q_2, which is shown in Fig. 7.2, in the completion forest of $\mathcal{K}_2'$.

Finally, we briefly discuss the complexity of query entailment algorithm.

Proposition 7.1 Let $\mathcal{K}$ be an f-ALC knowledge base and let q be a fuzzy Boolean query. The fuzzy query entailment is a PSpace-complete problem.

Proof (*sketch*) As mentioned previously, query entailment of a fuzzy ground Boolean query is a generalized fuzzy knowledge base entailment problem. The complexity result comes from the PSpace-complete of the knowledge base entailments (Straccia 2001). As for fuzzy Boolean queries with variables, since the newly introduced rules of $\forall_{\geq R}$ and $\forall_{>R}$ do not add any node or label of node into the completion forest, thus do not bring about additional computational cost. The procedure for checking the completion forest for a match of the query graph is polynomial w.r.t the size of knowledge base. So, fuzzy query entailment problem is a PSpace-complete problem totally. □

Note that our result is base on the assumption that a knowledge base is a purely assertional one, whereas the complexity of fuzzy query entailment w.r.t a knowledge base with TBox is dominated by the problem of expanding the KB into a purely assertional one, which is in ExpTime.

7.2.3 Conjunctive Query Answering in Fuzzy Description Logics with Data Type Support

Besides the query of tractable and expressive fuzzy DLs mentioned above, how to query data information in many fuzzy DL KBs is expected to be realized. As introduced in Sect. 4.2.3, data type support is one of the most useful features that Web Ontology Languages or DLs are expected to provide. There are many fuzzy DLs with data type support, such as *f-ALC(D)* (Straccia 2005), *f-ALC(G)* (Wang and Ma 2008), and etc. (see Sect. 4.2.3 in detail). Therefore, for realizing conjunctive query answering in fuzzy DLs with data type support, an approach for query answering in fuzzy DLs with data type support was presented in (Cheng et al. 2010a), where the authors dealt with fuzzy conjunctive query entailment for an expressive fuzzy DL *f-ALC(G)* (Wang and Ma 2008), presented a novel tableau-based algorithm for checking query entailment over *f-ALC(G)* KBs, and generalized the mapping conditions from a fuzzy query into a completion forest, reducing the times required for checking mapping in different completion forests. In the following we introduce the conjunctive query answering in fuzzy DLs with data type support (Cheng et al. 2010a).

In order to support customized fuzzy data types and customized fuzzy data type predicates *G*, Wang and Ma (2008) proposed a fuzzy DL called *f-ALC(G)* as have been introduced in Sect. 4.2.3 in detail, and thus the details about the fuzzy DL *f-ALC(G)* will not included here again. In the following, we will mainly focus on the conjunctive query entailment algorithm for *f-ALC(G)*.

First, for a fuzzy concept *D*, we denote by *sub*(*D*) the set that contains *D* and it is closed under sub-concepts of *D*, and define *sub* ($\mathcal{K}$) as the set of all the sub-concepts of the concepts occurring in the *f-ALC(G)* knowledge base $\mathcal{K}$. We abuse the notion *sub* ($\mathcal{D}$) to denote the set of all the data type predicates occurring in a knowledge base. Given a fuzzy KB $\mathcal{K}$, we can w.l.o.g assume that (i) all concepts are in their negative normal forms (NNFs), i.e. negation occurs only in front of concept names. (ii) all fuzzy concept assertions are in their positive inequality normal forms (PINFs). A negative concept assertion can be transformed into its equivalent PINF by applying fuzzy complement operation on it. For example, $C(o) < n$ is converted to $\neg C(o) > 1 - n$. (iii) all fuzzy assertions are in their normalized forms (NFs). By introducing a positive, infinite small value ε, a fuzzy assertion of the form can be normalized forms (NFs). By introducing a positive, infinite small value ε, a fuzzy assertion of the form $C(o) > n$can be normalized to $C(o) \geq n + \varepsilon$.

7.2.3.1 Fuzzy Querying Language

A conjunctive query over a classical DL knowledge base is of the form $q(\vec{x}) \leftarrow \exists\vec{y}.conj(\vec{x},\vec{y})$ (Horrocks and Tessaris 2002), where vector $\vec{x}$ is constituted by *distinguished variables* (also known as *answer variables*), which can be bound with individual names in a given knowledge base to answer the conjunctive queries, $\vec{y}$ is constituted by *non-distinguished* variables, which are treated as existential quantified, i.e., we just require the existence of a suitable element in the model, but this element does not have to correspond to an individual explicitly named in the ABox. $conj(\vec{x},\vec{y})$ is a conjunction of atoms of the form $A(z)$ or $R(z_1, z_2)$, where *A* and *R* are basic concept and role in $\mathcal{K}$ respectively. The concept and role atoms are syntactically equal to concept and role assertions except that z, z_1 and z_2 may be variables $\vec{x}$ in or $\vec{x}$, besides constants in $\mathcal{K}$. A CQ is called a Boolean conjunctive query (BCQ) if it has no distinguished variable. When querying fuzzy DL knowledge bases, however, the aforementio-ned query language bears a major limitation: the language does not allow users to specify for each query atom a threshold. The outcome for this is that every query holds to some degree. In addition, the language in (Horrocks and Tessaris 2002) does not provide for data type group querying. To this end, we provide the formal definition of the syntax and semantics of the fuzzy querying language used in this paper, extending Mallis's work (Mailis et al. 2007) to allow for querying fuzzy data type group.

Let N_v be a countable infinite set of variables and is disjoint from N_I, N_C and N_R. A *term t* is either an individual name from N_I, or a variable name from N_v. A *fuzzy query atom* is an expression of the form $\langle C(t) \geq n\rangle$, $\langle R(t,t') \geq n\rangle$, or $\langle T(t,t') \geq n\rangle$ *C* with a concept, *R* a simple abstract role *T*, a data type role, and *t*, *t'* terms. As with fuzzy assertions, we refer to these three different types of atoms as *fuzzy concept atoms*, *fuzzy abstract role atoms*, and *fuzzy data type role atoms*, respectively. The fuzzy abstract role atoms and the fuzzy data type role atoms are collectively referred to as *fuzzy role atoms*.

Definition 7.3 (*Fuzzy Boolean Conjunctive Queries*) A fuzzy boolean conjunctive query q is a non-empty set of fuzzy query atoms of the form $q = \{\langle at_1 \geq n_1 \rangle, \ldots, \langle at_k \geq n_k \rangle\}$. Then for every fuzzy query atom, we can say $\langle at_i \geq n_i \rangle \in q$.

We use *Vars*(q) to denote the set of variables occurring in q, *AInds*(q) and *CInds*(q) to denote the sets of abstract and data type individual names occurring in q, *Inds*(q) to denotes the union of *AInds*(q) and *CInds*(q), and *Terms*(q) for the set of terms in q, i.e. *Terms*(q) = *Vars*(q) ∪ *Inds*(q).

The semantics of a fuzzy query is given in the same way as for the related fuzzy DL by means of fuzzy interpretation consisting of an interpretation domain and a fuzzy interpretation function.

Definition 7.4 (*Models of Fuzzy Queries*) Let $I = (\Delta^I, \bullet^I)$ be a fuzzy interpretation of an *f-ALC(G) KB*, q a fuzzy boolean conjunctive query, and t, t', terms in q. We say $\mathcal{I}$ is a model of q, if there exists a mapping $\pi : Terms(q) \to \Delta^{\mathcal{I}} \cup \Delta_D$ such that $\pi(a) = a^{\mathcal{I}}$ for each $a \in Ind(q)$, $C^{\mathcal{I}}(\pi(t)) \geq n$ for each fuzzy concept atom $C(t) \geq n \in q$, $R^{\mathcal{I}}(\pi(t), \pi(t')) \geq n$ (resp. $T^{\mathcal{I}}(\pi(t), \pi(t')) \geq n$) for each fuzzy role atom $R(t, t') \geq n$ (resp. $T(t, t') \geq n$) $\in q$.

If $\mathcal{I} \models^{\pi} at$ for every atom $at \in q$, we write $\mathcal{I} \models^{\pi} q$. If there is a π, such that $\mathcal{I} \models^{\pi} q$, we say $\mathcal{I}$satisfies q, written as $\mathcal{I} \models q$. We call such a π a *match* of q in $\mathcal{I}$. If $\mathcal{I} \models q$ for each model of $\mathcal{I}$ a KB $\mathcal{K}$, then we say $\mathcal{K}$ entails q, written as $\mathcal{I} \models q$. The *fuzzy query entailment problem* is defined as follows: given a knowledge base $\mathcal{K}$ and a query q, decide whether $\mathcal{K} \models q$.

7.2.3.2 Fuzzy Query Entailment Algorithm

The intuition behind our algorithm for deciding fuzzy query entailment is that, we first (i) construct all the models for a given KB, and then (ii) check each of them for a match of the query. The first task can be fulfilled with the tableau reasoning algorithm for KB satifiability checking. However, a knowledge base may have infinitely many possibly infinite models, whereas tableau algorithms construct only a subset of finite models of the knowledge base. As is defined above, the query entailment holds only if the query is true in all models of the knowledge base, we thus have to show that inspecting only a subset of the models, namely the *canonical* ones, suffices to decide query entailment.

Our algorithm works on a data structure called *completion forest*. A completion forest is a finite relational structure capturing a set of models of a KB $\mathcal{K}$. Roughly speaking, models of are represented by an initial completion forest $\mathcal{F}_{\mathcal{K}}$. Then, by applying *expansion rules* repeatedly, new completion forests are generated. Since every model of $\mathcal{K}$ is preserved in some completion forest that results from the expansion, $\mathcal{K} \models q$ can be decided by considering a set $\mathbb{F}_{\mathcal{K}}$ of sufficiently expanded forests. From each such an $\mathcal{F}$, a single *canonical model* is constructed. Semantically, the finite set of these canonical models is sufficient for answering all queries of q bounded size. Furthermore, we prove that entailment in the canonical model obtained from can be checked effectively via a syntactic mapping of the terms into the nodes in $\mathcal{F}$.

Completion Forests

Definition 7.5 (*Completion Forest*) A completion tree T for an *f-ALC(G)* KB is a tree all of whose nodes are generated by expansion rules, except for the root node which might correspond to an abstract individual name in N_I. A completion forest $\mathcal{F}$ for an *f-ALC(G)* KB consists of a set of completion trees whose root nodes correspond to abstract individual names occurring in the ABox, an equivalent relation $\approx$ and an inequivalent relation $\not\approx$ among nodes, and a data structure $DC(x)$ to store the fuzzy data type expressions that must hold with regard to data type successors of a node x.

The nodes in a completion forest $\mathcal{F}$, denoted Nodes($\mathcal{F}$), can be divided into abstract nodes (denoted ANodes($\mathcal{F}$)) and data type nodes (or data type nodes, denoted CNodes($\mathcal{F}$)). Each abstract node o in a completion forest is labeled with a set $\mathcal{L}(o) = \{\langle C, \geq, n\rangle\}$, where $C \in sub(\mathcal{K}), n \in (0,1]$. The data type nodes v can only serve as leaf nodes, each of which is labeled with a set $\mathcal{L}(v) = \{\langle d, \geq, n\rangle\}$, where $d \in sub(D), n \in (0,1]$. Similarly, the edges in a completion forests can be divided into abstract edges and data type edges. Each abstract edge $\langle o, o'\rangle$ is labeled with a set $\mathcal{L}(\langle o, o'\rangle) = \{\langle R, \geq, n\rangle\}$, where $R \in N_R^A$. Each data type edge $\langle o, v\rangle$ is labeled with a set $\mathcal{L}(\langle o, v\rangle) = \{\langle T, \geq, n\rangle\}$,, where $T \in N_R^D$.

If $\langle o, o'\rangle$ is an edge in a completion forest with $\langle R, \geq, n\rangle \in \mathcal{L}(\langle o, o'\rangle)$, then o' is called an $R_{\geq,n}$-successor of$o$ and$o$ is called an $R_{\geq,n}$-predecessor of o'. Ignoring the inequality and membership degree, we can also call o' an R-successor of o and o an R-predecessor of o'. Ancestor and descendant are the transitive closure of predecessor and successor, respectively. The union of the successor and predecessor relation is the neighbor relation. The distance between two nodes o, o' in a completion forest is the shortest path between them.

Starting with an *f-ALC(G)* KB $\mathcal{K} = \langle \mathcal{T}, \mathcal{A}\rangle$, the completion forest $\mathcal{F}_{\mathcal{K}}$ is initialized such that it contains (i) a root node o, with $\mathcal{L}(o) = \{\langle C, \geq, n\rangle \mid C(o) \geq n \in \mathcal{A}\}$, for each abstract individual name o occurring in $\mathcal{A}$, (ii) an abstract edge $\langle o, o'\rangle$ with $\mathcal{L}(\langle o, o'\rangle) = \{\langle R, \geq, n\rangle \mid \langle R(o, o') \geq n\rangle \in \mathcal{A}\}$ for each pair $\langle o, o'\rangle$ of individual names for which the set $\{R \mid R(o, o') \geq n \in \mathcal{A}\}$ is non-empty, and (iii) a data type edge $\langle o, v\rangle$ with $\mathcal{L}(\langle o, v\rangle) = \{\langle T, \geq, n\rangle \mid \langle T(o, v) \geq n\rangle \in \mathcal{A}\}$. We initialize the relation $\not\approx$ as $\{\langle o, o'\rangle \mid o \not\approx o' \in \mathcal{A}\}$, the relation $\approx$ to be empty, the $DC(x)$ for each node x to be empty.

Now we introduce the blocking condition, called k-blocking (Levy and Rousset 1998), for fuzzy query entailment depending on a depth parameter $k > 0$.

Definition 7.6 (*k-tree equivalence*) The k-tree of a node v in T, denoted as T_v^k, is the subtree of T rooted at v with all the descendants of v within distance k. We use $Nodes\left(T_v^k\right)$ to denote the set of nodes in T_v^k. Two nodes v and w in T are said to be k-tree equivalent in T, if T_v^k and T_w^k are isomorphic, i.e., there exists a bijection $\psi : Nodes(T_v^k) \rightarrow Nodes(T_w^k)$ such that (i) $\psi(v) = w$, (ii) for every node $o \in Nodes\left(T_v^k\right)$, $\mathcal{L}(o) = \mathcal{L}(\psi(o))$, (iii) for every edge connecting two nodes o and o' in T_v^k, $\mathcal{L}(\langle o, o' \rangle) = \mathcal{L}(\langle \psi(o), \psi(o') \rangle)$.

Definition 7.7 (*k-witness*) A node w is a k-witness of a node v, if v and w are k-tree equivalent in T, w is an ancestor of v in T and v is not in T_w^k. Furthermore T_w^k, tree-blocks T_v^k and each node o in T_w^k tree-blocks node $\psi^{-1}(o)$ in T_v^k.

Definition 7.8 (*k-blocking* A node o is k-blocked in a completion forest $\mathcal{F}$ iff it is not a root node and it is either directly or indirectly k-blocked. Node o is directly k-blocked iff none of its ancestors is k-blocked, and o is a leaf of a tree-blocked k-tree. Node o is indirectly k-blocked iff one of its ancestors is k-blocked or it is a successor of a node o' and $\mathcal{L}(\langle o', o \rangle) = \varnothing$.

An initial completion forest is expanded according to a set of expansion rules that reflect the constructors in *f-ALC(G)*. The expansion rules, which syntactically decompose the concepts in node labels, either infer new constraints for a given node, or extend the tree according to these constraints (see Table 7.2). Termination is guaranteed by k-blocking. We denote by $\mathbb{F}_{\mathcal{K}}$ the set of all completion forests obtained this way.

For a node o, $\mathcal{L}(o)$ is said to contain a *clash*, if it contains either an *f-ALC*-clash (Straccia 2001) or a G-clash: (i) for some fuzzy data type roles $T_1, \ldots, T_n$, $\langle \leq mT_1, \ldots, T_n.E, \geq, n \rangle$ and there are $m + 1$ $\langle T_1, \ldots, T_n \rangle$-successor $\langle v_{i1}, \ldots, v_{in} \rangle$ of x with $\langle E(v_{i1}, \ldots, v_{in}), \geq, n \rangle \in DC(x)(1 \leq i \leq m + 1)$, and $(\langle v_{i1}, \ldots, v_{in} \rangle, \langle v_{j1}, \ldots, v_{jn} \rangle, \neq) \notin DC(x)(1 \leq i < j \leq m + 1)$; (ii) for some abstract node x, $DC(x)$ is unsatisfiable.

Definition 7.9 (*k-complete and clash-free completion forest*) A completion forest is called k-complete and clash-free if under k-blocking no rule can be applied to it, and none of its nodes and edges contains a clash. We denote by $\text{ccf}_k(\mathbb{F}_{\mathcal{K}})$ the set of k-complete and class-free completion forests in $\mathbb{F}_{\mathcal{K}}$.

Models of Completion Forests

We now show that every model of a KB $\mathcal{K}$ is preserved in some complete and clash-free completion tree $\mathcal{F}$. If we view all the nodes (either root nodes or generated nodes) in a completion forest $\mathcal{F}$ as individual names, we can define models of $\mathcal{F}$ in terms of models of over an extended vocabulary.

Table 7.2 Expansion rules

Rule	Description
$\sqcap_{\geq}$	if 1. $\langle C \sqcap D, \geq, n\rangle \in \mathcal{L}(x)$, x, is not indirectly k-blocked, and 2. $\{\langle C, \geq, n\rangle, \langle D, \geq, n\rangle\} \not\subseteq \mathcal{L}(x)$ then $\mathcal{L}(x) \rightarrow \mathcal{L}(x) \cup \{\langle C, \geq, n\rangle, \langle D, \geq, n\rangle\}$
$\sqcup_{\geq}$	if 1. $\langle C \sqcup D, \geq, n\rangle \in \mathcal{L}(x)$, x is not indirectly k-blocked, and 2. $\{\langle C, \geq, n\rangle, \langle D, \geq, n\rangle\} \cap \mathcal{L}(x) = \varnothing$ then $\mathcal{L}(x) \rightarrow \mathcal{L}(x) \cup \{C'\}$, where $C' \in \{\langle C, \geq, n\rangle, \langle D, \geq, n\rangle\}$
$\exists_{\geq}$	if 1. $\langle \exists R.C, \geq, n\rangle \in \mathcal{L}(x)$, x is not k-blocked. 2. x has no $R_{\geq,n}$-neighbory s.t. $\langle c, \geq, n\rangle \in \mathcal{L}(y)$, then create a new node y with $\mathcal{L}(x, y) = \{\langle R, \geq, n\rangle\}$ and $\mathcal{L}(y) = \{\langle C, \geq, n\rangle\}$
$\forall_{\geq}$	if 1. $\langle \forall R.C, \geq, n\rangle \in \mathcal{L}(x)$, x is not indirectly k-blocked. 2. x has an $R_{\geq,n'}$-neighbor y with $\langle C, \leq, n\rangle \notin \mathcal{L}(y)$, where $n' = 1 - n + \varepsilon$, then $\mathcal{L}(y) \rightarrow \mathcal{L}(y) \cup \{\langle C, \geq, n\rangle\}$
$\sqsubseteq$	if 1. $C \sqsubseteq D \in \mathcal{T}$ and 2. $\{\langle \neg C, \geq, 1 - n + \varepsilon\rangle, \langle D, \geq, n\rangle\} \cap \mathcal{L}(x) = \varnothing$ for $n \in N^{\mathcal{A}} \cup N^{q}$, then $\mathcal{L}(x) \rightarrow \mathcal{L}(x) \cup$ $\{C'\}$ for some $C' \in \{\langle \neg C, \geq, 1 - n + \varepsilon\rangle, \langle D, \geq, n\rangle\}$
$\forall_{p\geq}$	if 1. $\langle \forall T_1, \ldots, T_n \cdot E, \geq, n\rangle \in \mathcal{L}(x)$, x, is not blocked, and 2. x has a $\langle T_1, \ldots, T_n\rangle_{\geq,n'}$-successor $\langle v_1, \ldots, v_n\rangle$ with $\langle E(v_1, \ldots, v_n), \geq, n\rangle \notin DC(x)$, where $n' = 1 - n + \varepsilon$, then $DC(x) \rightarrow DC(x) \cup \{\langle E(v_1, \ldots, v_n), \geq, n\rangle\}$
$\exists_{p\geq}$	if 1. $\langle \exists T_1, \ldots, T_n \cdot E, \geq, n\rangle \in \mathcal{L}(x)$, x, is not blocked, and 2. x has no $\langle T_1, \ldots, T_n\rangle_{\geq,n}$-successor $\langle v_1, \ldots, v_n\rangle$ with $\langle E(v_1, \ldots, v_n), \geq, n\rangle \in DC(x)$,, where $n' = 1 - n + \varepsilon$, then 1. create a $\langle T_1, \ldots, T_n\rangle_{\geq,n}$-successor $\langle v_1, \ldots, v_n\rangle$ with $\mathcal{L}(x, v_i) = \{\langle T_i, \geq, T_n\rangle\}$ for all $1 \leq i \leq n$, and 2. $DC(x) \rightarrow DC(x) \cup \{\langle E(v_1, \ldots, v_n), \geq, n\rangle\}$
$\exists_{p\geq}$	if 1. $\langle \exists T_1, \ldots, T_n \cdot E, \geq, n\rangle \in \mathcal{L}(x)$, x, is not blocked, and 2. x has no $\langle T_1, \ldots, T_n\rangle_{\geq,n}$-successor $\langle v_1, \ldots, v_n\rangle$ with $\langle E(v_1, \ldots, v_n), \geq, n\rangle \in DC(x)$, where$n' = 1 - n + \varepsilon$, then 1. create a $\langle T_1, \ldots, T_n\rangle_{\geq,n}$-successor $\langle v_1, \ldots, v_n\rangle$ with $\mathcal{L}(x, v_i) = \{\langle T_i, \geq, T_n\rangle\}$ for all $1 \leq i \leq n$, and 2. $DC(x) \rightarrow DC(x) \cup \{\langle E(v_1, \ldots, v_n), \geq, n\rangle\}$
$\geq_{p\geq}$	if 1. $\langle \geq m T_1, \ldots, T_n \cdot E, \geq, n\rangle \in \mathcal{L}(x)$, x is not blocked, and 2. there are no m $\langle T_1, \ldots, T_n\rangle_{\geq,n}$-successor $\langle v_{11}, \ldots, v_{1n}\rangle, \ldots, \langle v_{m1}, \ldots, v_{mn}\rangle$ of x with $\langle E v_{i1}, \ldots, v_m, \geq n\rangle \in DC(x)$ and $(\langle v_{i1}, \ldots, v_{in}\rangle, \ldots, \langle v_{j1}, \ldots, v_{jn}\rangle, \neq) \in DC(x)$ for$1 \leq i < j \leq m$, then 1. create m $\langle T_1, \ldots, T_n\rangle_{\geq,n}$-successor $\langle v_{11}, \ldots, v_{1n}\rangle, \ldots, \langle v_{m1}, \ldots, v_{mn}\rangle$ of x with $\mathcal{L}(x, v_{ij}) = \{\langle T_j, \geq, n\rangle\}$ for all $1 \leq i \leq m$, $1 \leq j \leq n$, and 2. $DC(x) \rightarrow DC(x) \cup \{\langle E(v_{i1}, \ldots, v_{in}), \geq, n\rangle\}$ for $1 \leq i \leq m$, and 3. $DC(x) \rightarrow DC(x) \cup \{(\langle v_{i1}, \ldots, v_{in}\rangle,$ $\langle v_{j1}, \ldots, v_{jn}\rangle, \neq)\}$ for all $1 \leq i < j \leq m$

Definition 7.10 (*Models of completion forests*) An interpretation $\mathcal{I}$ is a model of a completion forest $\mathcal{F}$ for $\mathcal{K}$, denoted $\mathcal{I} \models \mathcal{F}$, if $\mathcal{I} \models \mathcal{K}$ and for all abstract nodes o, o' and data type nodes v in $\mathcal{F}$ it holds that (i) $C^{\mathcal{I}}(o^{\mathcal{I}}) \geq n$ if $\langle C, \geq, n\rangle \in \mathcal{L}(o)$, (ii) $d^{\mathcal{I}}(v^{\mathcal{I}}) \geq n$ if $\langle d, \geq, n\rangle \in \mathcal{L}(v)$, (iii) $R^{\mathcal{I}}(o^{\mathcal{I}}, o'^{\mathcal{I}}) \geq n$ if there exists an abstract

edge in $\langle o, o' \rangle$ in $\mathcal{F}$ and $\langle R, \geq, n \rangle \in \mathcal{L}(\langle o, o' \rangle)$, (iv) $T^{\mathcal{I}}(o^{\mathcal{I}}, v^{\mathcal{I}}) \geq n$ if there exists a data type edge $\langle o, v \rangle$ in $\mathcal{F}$ and $\langle T, \geq, n \rangle \in \mathcal{L}(\langle o, v \rangle)$, (v) $o^{\mathcal{I}} \neq o'^{\mathcal{I}}$ if $o \not\approx o' \in \mathcal{F}$.

We first show that the models of the initial completion forest $\mathcal{F}_{\mathcal{K}}$ and of $\mathcal{K}$ coincide.

Lemma 7.1 *$\mathcal{I} \models \mathcal{F}_{\mathcal{K}}$ iff $\mathcal{I} \models \mathcal{K}$.*

Proof The only if direction follows from Definition 7.10. For the if direction, we need to show that, for all nodes v, w. in $\mathcal{F}_{\mathcal{K}}$, Property (i)–(v) in Definition 7.10 hold. By Definition 7.5, each node in $\mathcal{F}_{\mathcal{K}}$ corresponds to an individual name in $\mathcal{K}$. For each abstract individual o in N_I, the label of node o in $\mathcal{F}_{\mathcal{K}}$ is $\mathcal{L}(o) = \{\langle C, \geq, n \rangle | C(o) \geq n \in \mathcal{A}\}$. Since $\mathcal{I} \models \mathcal{K}$, then we have $C^{\mathcal{I}}(o^{\mathcal{I}}) \geq n$ and Property (i) thus holds. Property (ii)–(v) can be proved in a similar way with (i).□

We then show that, each time an expansion rule is applied, all models are preserved in some resulting forest.

Lemma 7.2 *Let $\mathcal{F}$ be a completion forest in $\mathbb{F}_{\mathcal{K}}$, r a rule, F a set of completion forests obtained from $\mathcal{F}$ by applying r, then for each model $\mathcal{I}$ of $\mathcal{F}$, there exist an $\mathcal{F}'$ in F and an extension $\mathcal{I}'$ of $\mathcal{I}$, such that $\mathcal{I}' \models \mathcal{F}'$.*

Proof (sketch) $\exists_{\geq}$-rule. Since $\langle \exists R.C, \geq, n \rangle \in \mathcal{L}(x)$ and $\mathcal{I} \models \mathcal{F}$, then there exists some $o \in \Delta^{\mathcal{I}}$, such that $R^{\mathcal{I}}(x^{\mathcal{I}}, o) \geq n$ and $C^{\mathcal{I}}(o) \geq n$ hold. In the completion forest $\mathcal{F}'$ obtained from $\mathcal{F}$ by applying $\exists_{\geq}$-rule, a new node y is generated $\langle R, \geq, n \rangle \in \mathcal{L}(\langle x, y \rangle)$ and $\langle C, \geq, n \rangle \in \mathcal{L}(y)$. By setting $y^{\mathcal{I}'} = o$, we obtain an extension $\mathcal{I}'$ of $\mathcal{I}$, and thus $\mathcal{I}' \models \mathcal{F}'$. The case of $\geq_{\geq}$-rule is analogous to the $\exists_{\geq}$-rule. The proofs for other rules are in similar way. □

Since the set of k-complete and clash-free completion forests for $\mathcal{K}$ semantically captures $\mathcal{K}$ (modulo new individuals), we can transfer query entailment $\mathcal{K} \models q$ to logical consequence of q from completion forests as follows. For any completion forest $\mathcal{F}$ and any CQ q, let $\mathcal{F} \models q$ denote that $\mathcal{I} \models q$ for every model $\mathcal{I}$ of $\mathcal{F}$.

Theorem 7.1 Let $k \geq 0$ be arbitrary. Then $\mathcal{K} \models q$ iff $\mathcal{F} \models q$ for each $\mathcal{F} \in \mathrm{ccf}_{\mathcal{K}}(\mathbb{F}_{\mathcal{K}})$.

Proof By Lemma 7.1 and Lemma 7.2, for each model $\mathbb{I}$ of $\mathcal{K}$, there exists some $\mathcal{F} \in \mathbb{F}_{\mathcal{K}}$ and an extension $\mathcal{I}'$ of $\mathcal{I}$, such that $\mathcal{I}' \models \mathcal{F}$. Assume $\mathcal{F} \in \mathrm{ccf}_{\mathcal{K}}(\mathbb{F}_{\mathcal{K}})$, then there still are rules applicable to $\mathcal{F}$. We thus obtain an expansion $\mathcal{F}'$ from $\mathcal{F}$ and an extension $\mathcal{I}''$ of $\mathcal{I}'$, such that $\mathcal{I}'' \models \mathcal{F}'$, and so forth until no rule is applicable. Now we either obtain a complete and clash free completion forest, or encounter a clash. The former is in conflict with the assumption, and the latter is in conflict with the fact that $\mathcal{F}$ have models. □

Checking Query Entailment Within Completion Forests

We now show that, if k is large enough, we can decide $\mathcal{F} \models q$ for each $\mathcal{F} \in \mathrm{ccf}_{\mathcal{K}}(\mathbb{F}_{\mathcal{K}})$ by syntactically mapping the query q into $\mathcal{F}$.

Definition 7.11 (*Query mapping*) A fuzzy query q can be mapped into $\mathcal{F}$, denoted $q \rightarrowtail \mathcal{F}$, if there is a mapping $\mu : Terms(q) \rightarrow Nodes(\mathcal{F})$, such that (i) $\mu(a) = a$, if $a \in \text{Inds}(q)$, (ii) for each fuzzy concept atom $C(x) \geq n$ (resp. $d(x) \geq n$) in q, $\langle C(x), \geq, m\rangle$ (resp. $\langle d(x), \geq, m\rangle$) $\in \mathcal{L}(\mu(x))$ with $m \geq n$, (iii) for each fuzzy role atom $R(x, y) \geq n$ (resp. $T(x, y) \geq n$) in q, $\mu(y)$ is a $R_{\geq m}$-neighbor (resp. a $T_{\geq m}$-neighbor) of $\mu(x)$ with $m \geq n$.

In fact, for completion forests $\mathcal{F}$ of a KB $\mathcal{K}$, syntactic mapping $q \rightarrowtail \mathcal{F}$ implies semantic consequence $\mathcal{F} \models q$.

Lemma 7.3 *If $q \rightarrowtail \mathcal{F}$, then $\mathcal{F} \models q$.*

Proof If $q \rightarrowtail \mathcal{F}$, then there is a mapping $\mu : Terms(q) \rightarrow Nodes(\mathcal{F})$ satisfying Definition 7.11. For any model $\mathcal{I} = \left(\Delta^{\mathcal{I}}, \cdot^{\mathcal{I}}\right)$ of $\mathcal{F}$, it satisfies Definition 7.10. We construct a mapping $\pi : \text{Terms}(q) \rightarrow \Delta^{\mathcal{I}}$ such that, for each term $x \in \text{Terms}(q)$, $\pi(x) = (\mu(x))^{\mathcal{I}}$. It satisfies $C^{\mathcal{I}}(\pi(x)) = C^{\mathcal{I}}((\mu(x))^{\mathcal{I}}) \geq m \geq n$, for each fuzzy concept atom $C(x) \geq n \in q$. The proof for fuzzy role atoms can be shown in a similar way. Hence $\mathcal{I} \models$q, which by Theorem 7.1 implies $\mathcal{F}$. □

Lemma 7.3 shows the soundness of our algorithm. We prove, in the later part *D*, the converse (the completeness) also holds. We show that provided the completion forest $\mathcal{F}$ has been sufficiently expanded, a mapping from q to $\mathcal{F}$ can be constructed from a single canonical model.

Fuzzy Tableaux and Canonical Models

The construction of the canonical model $\mathcal{I}_{\mathcal{F}}$ for $\mathcal{F}$ is divided into two steps. First, $\mathcal{F}$ is unravelled into a fuzzy tableau. Then, the canonical model is induced from the tableau. The interpretation domain $\Delta^{\mathcal{I}_{\mathcal{F}}}$ of $\mathcal{I}_{\mathcal{F}}$ consists of a set of (maybe infinite) paths. The reason lies in that a KB $\mathcal{F}$ with GCIs may have infinite models, whereas the canonical model $\mathcal{I}_{\mathcal{F}}$ is constructed from $\mathcal{F}$, which is a finite representation decided by the termination of the algorithm. This requires some nodes in $\mathcal{F}$ represent several elements in $\Delta^{\mathcal{I}_{\mathcal{F}}}$. The paths is chosen to distinguish different elements represented in $\mathcal{F}$ by the same node. The definition of fuzzy tableau is based on the one in (Stoilos et al. 2006, 2007).

Definition 7.12 (*Fuzzy tableaux*) Let $\mathcal{K} = (\mathcal{T}, \mathcal{A})$ be a *f-ALC(G)* KB, $\mathbf{R}_{\mathcal{K}}$, $\mathbf{R_D}$ the sets of abstract and data type roles occurring in $\mathcal{K}$, $\mathbf{I}_A$ the set of individual names occurring in $\mathcal{A}$, and $\text{cl}_{\mathcal{G}}(\mathcal{K})$ the set of all the fuzzy $\mathcal{G}$-data type expressions and their negations occurring in $\mathcal{K}$. $T = \langle \mathbf{S}_A, \mathbf{S_D}, \mathcal{H}, \varepsilon_a, \varepsilon_d, DC\rangle$ is the fuzzy tableau of $\mathcal{K}$ if (i) $\mathbf{S}_A$ is a set of individuals; $\mathbf{S_D}$ is a set of variables, (ii) $\mathcal{H} : \mathbf{S}_A \times sub(\mathcal{K}) \rightarrow [0, 1]$ maps each element in $\mathbf{S}_A$ and each concept in $sub(\mathcal{K})$ to the membership degree the element belongs to the concept, (iii) $\varepsilon_a : \mathbf{S}_A \times \mathbf{S}_A \times \mathbf{R}_{\mathcal{K}} \rightarrow [0, 1]$ maps each pair of elements in $\mathbf{S}_A$ and each role in $\mathbf{R}_{\mathcal{K}}$ to the membership degree the pair belongs to the role, (iv) $\varepsilon_c : \mathbf{S}_A \times \mathbf{S_D} \times \mathbf{R_D} \rightarrow [0, 1]$ maps each pair of elements and data type values and each data type role in $\mathbf{R_D}$ to the membership degree the pair belongs to the role, (v) DC is a set of fuzzy data type constraints of the form $\langle E(v_1, \ldots, v_n), \geq, n \rangle$ or inequations of the form $(\langle v_{i1}, \ldots, v_{in} \rangle, \langle v_{j1}, \ldots, v_{jn} \rangle, \neq)$, where $E \in cl_{\mathcal{G}}(\mathcal{K})$, $v_1, \ldots, v_n, v_{i1}, \ldots, v_{in}, v_{j1}, \ldots, v_{jn} \in \mathbf{S_D}, n \in [0, 1]$. Additionally, for each $s, t \in \mathbf{S}_A$, $C, D \in sub(\mathcal{K})$, $R \in \mathbf{R}_{\mathcal{K}}$, and $n \in [0, 1]$, T satisfies:

(1) for each $s \in \mathbf{S}_A$, $\mathcal{H}(s, \bot) = 0$, $\mathcal{H}(s, \mathrm{T}) = 1$;
(2) if $\mathcal{H}(s, C \sqcap D) \geq n$, then $\mathcal{H}(s, C) \geq n$ and $\mathcal{H}(s, D) \geq n$;
(3) if $\mathcal{H}(s, C \sqcup D) \geq n$, then $\mathcal{H}(s, C) \geq n$ or $\mathcal{H}(s, D) \geq n$;
(4) if $\mathcal{H}(s, \forall R.C) \geq n$, then for all $t \in \mathbf{S}_A$, $\mathcal{E}(\langle s, t \rangle, R) \leq 1 - n$ or $\mathcal{H}(t, C) \geq n$;
(5) if $\mathcal{H}(s, \exists R.C) \geq n$, then there exists $t \in \mathbf{S}_A$, such that $\mathcal{E}_a(\langle s, t \rangle, R) \geq n$ and $\mathcal{H}(t, C) \geq n$;
(6) if $C \sqsubseteq D \in \mathcal{T}$, then for all $s \in \mathbf{S}_A$, $n \in N^{\mathcal{A}} \cup N^q$, $\mathcal{H}(s, \neg C) \geq 1 - n + \varepsilon$ or $\mathcal{H}(s, D) \geq n$;
(7) if $\mathcal{H}(s, \exists T_1, \ldots, T_n.E) \geq n$, then there exists $v_1, \ldots, v_n \in \mathbf{S_D}$, such that $\varepsilon_d(\langle s, v_i \rangle, T_i) \geq n$ $(1 \leq i \leq n)$ and $\langle E(v_1, \ldots, v_n), \geq, n \rangle \in DC(s)$;
(8) if $\mathcal{H}(s, \forall T_1, \ldots, T_n.E) \geq n$, and there exists $v_1, \ldots, v_n \in \mathbf{S_D}$, such that $\varepsilon_d(\langle s, v_i \rangle, T_i) \leq 1 - n$ $(1 \leq i \leq n)$, then $\langle E(v_1, \ldots, v_n), \geq, n \rangle \in DC(s)$;
(9) if $\mathcal{H}(s, \geq mT_1, \ldots, T_n.E) \geq n$, then there exists $v_{i1}, \ldots, v_{in} \in \mathbf{S_D}$ $(1 \leq i \leq m)$, such that $\mathcal{E}_d(\langle s, v_i \rangle, T_i) \geq n$ $(1 \leq i \leq n)$ and $\langle E(v_1, \ldots, v_n), \geq, n \rangle \in DC(s)$;
(10) if $\mathcal{H}(s, \leq mT_1, \ldots, T_n.E) \geq n$, then there exists no more than $m + 1$ pair-wise different no more than m + 1 pair-wise different $\langle v_1, \ldots, v_n \rangle \in \mathbf{S_D}$ such that $\varepsilon_d(\langle s, v_i \rangle, T_i) \leq 1 - n$ $(1 \leq i \leq n)$ and $\langle E(v_1, \ldots, v_n), \geq, n \rangle \in DC(s)$;

The process of inducing a fuzzy tableau from a completion forest $\mathcal{F}$ is as follows.

Each element in $\mathbf{S}_A$ corresponds to a path in $\mathcal{F}$. We can view a blocked node as a loop so as to define infinite paths. To be more precise, a path $p = [v_0/v_{0'}, \ldots, v_n/v_{n'}]$ is a sequence of node pairs in $\mathcal{F}$. We define $Tail(p) = v_n$, and $Tail'(p) = v_{n'}$. We denote by $[p|v_{n+1}/v_{n+1}{}']$ the path $[v_0/v_{0'}, \ldots, v_n/v_{n'}, v_{n+1}/v_{n+1}{}']$ and use $[p|v_{n+1}/v_{n+1'}, v_{n+2}/v_{n+2'}]$ as the abbreviation of $[[p|v_{n+1}/v_{n+1'}], v_{n+2}/v_{n+2'}]$. The set Paths($\mathcal{F}$) of paths in $\mathcal{F}$ is inductively defined as follows:

- if v is a root node in $\mathcal{F}$, then $[v/v] \in$ Paths($\mathcal{F}$);
- if $p \in$ Paths($\mathcal{F}$), and $w \in$ Nodes($\mathcal{F}$),
  - if w is the R-successor of $Tail(p)$ and is not k-blocked, then $[p|w/w] \in$ Paths($\mathcal{F}$);
  - if there exists $w' \in$ Nodes ($\mathcal{F}$) and is the R-successor of $Tail(p)$, and is directly k-blocked by w, then $[p|w/w'] \in$ Paths($\mathcal{F}$).

Definition 7.13 (*Induced fuzzy tableau*) The fuzzy tableau $T_{\mathcal{F}} = \langle \mathbf{S}_A, \mathbf{S_D}, \mathcal{H}, \varepsilon_a, \varepsilon_c, DC \rangle$ induced by $\mathcal{F}$ is as follows.

- $\mathbf{S}_A = \text{Paths}(\mathcal{F})$,
- $\mathbf{S_D} = \Delta_\mathbf{D}$,
- $\mathcal{H}(p, C) \geq \sup\{n_i | \langle C, \geq, n_i \rangle \in \mathcal{L}(Tail(p))\}$,
- $\varepsilon_a(\langle p, [p|w/w'] \rangle, R) \geq \sup\{n_i | \langle R, \geq, n_i \rangle \in \mathcal{L}(\langle Tail(p), w' \rangle)\}$,
- $\varepsilon_c(\langle p, [p|v_c/v_c] \rangle, T) \geq n$ with $\langle T, \geq, n \rangle \in \mathcal{L}(\langle Tail(p), v_c \rangle)\}$, where v_c is a data type node.

From the fuzzy tableau of a fuzzy KB $\mathcal{K}$, we can obtain the canonical model of $\mathcal{K}$.

Definition 7.14 (*Canonical model*) Let $T_\mathcal{F} = \langle \mathbf{S}_A, \mathbf{S_D}, \mathcal{H}, \varepsilon_a, \varepsilon_c, DC \rangle$ be a fuzzy tableau of $\mathcal{K}$, the canonical model of T, $\mathcal{I}_T = (\Delta^{\mathcal{I}_T}, {}^{\mathcal{I}_T})$, is defined as follows.

- $\Delta^{\mathcal{I}_T} = \mathbf{S}_A$;
- for each $s \in \mathbf{S}_A$ and each concept name A, $A^{\mathcal{I}_T}(s) = \mathcal{H}(s, A)$;
- for each $\langle s, t \rangle \in \mathbf{S}_A \times \mathbf{S}_A$, $R^{\mathcal{I}_T}(s, t) = \varepsilon_a(\langle s, t \rangle, R)$,
- for each $\langle s, v \rangle \in \mathbf{S}_A \times \Delta_\mathbf{D}$, $T^{\mathcal{I}_T}(s, v) = \varepsilon_c(s, v)$.

Lemma 7.4 *Let T be the fuzzy tableau of an f-ALC(G)* KB $\mathcal{K} = (\mathcal{T}, \mathcal{A})$, *then the canonical model* $\mathcal{I}_\mathrm{T}$ *of T is a model of* $\mathcal{K}$.

Proof Property 6 in Definition 7.12 ensures that $\mathcal{I}_\mathrm{T}$ is a model of $\mathcal{T}$. For a detailed proof, see Proposition 3 in (Stoilos et al. 2006). Property 1–5 and 7–10 in Definition 7.12 ensures that $\mathcal{I}_\mathrm{T}$ is a model of $\mathcal{A}$ (Wang and Ma 2008). □

From a complete and clash-free completion forest $\mathcal{F}$, we can obtain a fuzzy tableau $T_\mathcal{F}$, through which a canonical model $\mathcal{I}_\mathcal{F}$ is constructed.

Lemma 7.5 *Let* $\mathcal{F} \in ccf_k(\mathbb{F}_\mathcal{K})$, *then* $\mathcal{I}_\mathcal{F} | = \mathcal{K}$, *where* $k \geq 1$.

Proof It follows from Lemma 5.9 and 6.10 in (Stoilos et al. 2007) and Proposition 5 in (Stoilos et al. 2006) that the induced tableau in Definition 7.13 satisfies Property 1–10 in Definition 7.12. By Lemma 7.5, the canonical model $\mathcal{I}_\mathcal{F}$ constructed from $T_\mathcal{F}$ is a model of $\mathcal{K}$. □

Now we illustrate how to construct a mapping of q to $\mathcal{F}$ from a mapping of q to $\mathcal{I}_\mathcal{F}$.

Definition 7.15 (*Mapping graph*) Let $\mathcal{F} \in \text{ccf}_\mathcal{K}(\mathbb{F}_\mathcal{K})$ with $k \geq 0$, and a fuzzy query q, such that $\mathcal{I}_\mathcal{F} | = q$. π is a mapping in Definition 7.4, then the mapping graph $G_\pi = \langle V, E \rangle$ is defined as: $V(G_\pi) = \{\pi(x) \in^{\mathcal{I}_\mathcal{F}} \cup^\mathbf{D} | x \in Terms(q)\}$, $E(G_\pi) = \{\langle \pi(x), \pi(y) \rangle \in \Delta^{\mathcal{I}_\mathcal{F}} \times \Delta^{\mathcal{I}_\mathcal{F}} | R(x, y) \geq n \in q\} \cup \{\langle \pi(x), \pi(y) \rangle \in \Delta^{\mathcal{I}_\mathcal{F}} \times \Delta^\mathbf{D} | T(x, y) \geq n \in q\}$. $V(G_\pi)$ is divided into $Vr(G_\pi)$ and $Vn(G_\pi)$, i.e., $V(G_\pi) = Vr(G_\pi) \cup Vn(G_\pi)$ and $Vr(G_\pi) \cap Vn(G_\pi) = \varnothing$, where $Vr(G_\pi) = \{ [v/v] | v$ is a root node in $\mathcal{F}\}$.

Definition 7.16 (*Maximal q-distance*) For any $x, y \in Terms(q)$, if $\pi(x), \pi(y) \in Vn(G\pi)$, then we use $d^\pi(x, y)$ to denote the length of the shortest path between $\pi(x)$ and $\pi(y)$ in $G\pi$. If $\pi(x)$ and $\pi(y)$ are in two different connected components respectively, then $d^\pi(x, y) = -1$. We define maximal q-distance $d_q^\pi = max_{x, y \in Terms(q)}\{d^\pi(x, y)\}$.

We use n_{q} to denote the number of fuzzy role atoms in a fuzzy query q. We only consider fuzzy role atoms $R(x,y) \geq n$ with R a simple role, so $d^{\pi}(x,y) = 1$ and $d_{\mathrm{q}}^{\pi} \leq n_{\mathrm{q}}$. We show provided the k in the k-blocking condition is greater than or equal to n_{q}, it suffices to find a mapping from q to $\mathcal{F}$.

Lemma 7.6 *Let $\mathcal{F} \in \mathrm{ccf}_{\mathcal{K}}(\mathbb{F}_{\mathcal{K}})$ with $k \geq n_q$, and $\mathcal{I}_{\mathbb{F}}$ is the canonical model of $\mathcal{F}$. If $\mathcal{I}_{\mathcal{F}} |{=} q$, then $q \rightarrowtail \mathcal{F}$.*

Theorem 7.2 *Let $k \geq n_q$. Then $\mathcal{K} |{=} q$ iff for each $\mathcal{F} \in \mathrm{ccf}_{\mathcal{K}}(\mathbb{F}_{\mathcal{K}})$, it holds that $q \rightarrowtail \mathcal{F}$.*

Proof The if direction is trivial. By Lemma 7.5, for each $\mathcal{F} \in \mathrm{ccf}_{\mathcal{K}}(\mathbb{F}_{\mathcal{K}})$, $\mathcal{F} |{=} q$. Then, by Theorem 7.1, $\mathcal{K} |{=} q$. For the converse side, by Theorem 7.1, $\mathcal{F} |{=} q$ for each $\mathcal{F} \in \mathrm{ccf}_{\mathcal{K}}(\mathbb{F}_{\mathcal{K}})$. By Lemma 7.6, $q \rightarrowtail \mathcal{F}$. □

We can, from the only if direction of Theorem 7.2, establish our key result, which reduce query entailment $\mathcal{K} |{=} q$ to finding a mapping of q into every $\mathcal{F}$ in $\mathrm{ccf}_{\mathcal{K}}(\mathbb{F}_{\mathcal{K}})$.

Based on the above introduction, the issue of conjunctive querying over fuzzy DL KBs from tractable to more expressive fuzzy DL KBs is introduced. The relevant techniques can provide users with querying services to a good number of large-scale fuzzy DL KBs. In the following subchapter, we introduce how to query fuzzy ontologies.

7.3 Querying Fuzzy Semantic Web Ontologies

Being similar to the query of fuzzy DL knowledge bases introduced above, with the emergence a great number of large-scale fuzzy ontologies in the Semantic Web and other application domains, the efficient *query of fuzzy ontologies* gain more attention in recent years. To the best of our knowledge, the *query of fuzzy ontologies* mainly focuses on two aspects: on one hand, being similar to the querying over fuzzy DL knowledge bases as introduced in Sect. 7.2, for providing users with expressive querying service over fuzzy ontologies, many efforts have been made (Pan et al. 2007, 2008; Cheng et al. 2009b; Bahri et al. 2009; Bandini et al. 2006; Widyantoro and Yen 2001a, b). On the other hand, as we have known, SPARQL is the W3C standard query language for RDF, which is a standard model for data interchange on the Web. However, most of the existing query systems on RDF can not process queries with fuzzy information directly and thus can not adequately meet users' querying needs. Consequently, it becomes an important subject of Semantic Web to do research about fuzzy query on RDF. To this end, several proposals for fuzzy extension to SPARQL were proposed (Cheng et al. 2010b; De Maio et al. 2012; Wang et al. 2012). In this subchapter, we will focus on the query of fuzzy Semantic Web ontologies, and introduce some query techniques of fuzzy ontologies and the fuzzy extensions of SPARQL.

7.3.1 Query Answering of Fuzzy Ontologies

To query fuzzy ontologies, the work in (Pan et al. 2007, 2008) proposed two expressive query languages accompanied with query answering algorithms over fuzzy *DL-Lite* ontology, and presented a prototype implementation for querying fuzzy *DL-Lite* ontologies. Cheng et al. (2009b) developed a tableau-based algorithm for deciding query entailment over *f-SHIN* ontology, where the query also allowed the occurrence of both lower bound and the upper bound of threshold in a query atom, and the authors proved that the algorithm for query entailment is co3NExp-Time in the size of the knowledge base and the query. Similarly, Cheng et al. (2011) proposed an approach for deciding query entailment in fuzzy OWL Lite ontologies, i.e., fuzzy DL *SHIF(D)* ontologies. Bahri et al. (2009) addressed the problem of implementation and query answering of fuzzy ontologies on databases. They proposed a language to define fuzzy ontology schema and to query fuzzy ontology databases, and also developed an inferential engine to infer instances and subsumption relations between fuzzy concepts defined using concept forming expressions. Bandini et al. (2006) presented a solution to handle vague information in query processing into fuzzy ontology-based applications. They introduced the quality concept in the fuzzy ontology to better define the degree of truth of the fuzzy ontology entities. The constraint tree has been defined as a hierarchical indication of what qualities should be considered significant to evaluate an instance of a concept. Also, a strategy to parse a sentence in a set of constraints was proposed, and this will allow users to submit queries to the system using natural language requests. Widyantoro and Yen (2001a, b) used fuzzy ontology for query refinement, which is very promising to be useful to help users find the information they need. Additionally, the authors implemented and incorporated the fuzzy ontology of narrower and broader terms for query refinement in personalized abstract search services search engine.

In the following, the techniques of querying answering over fuzzy ontologies (Cheng et al. 2011) will be introduced. It should be noted that querying over fuzzy ontologies are basically based on the query techniques of fuzzy DLs as introduced in Sect. 7.2. Cheng et al. (2011) presented a novel tableau-based algorithm for checking query entailment over fuzzy OWL Lite ontologies. They generalize the mapping conditions from a fuzzy query into a completion forest, reducing the times required for checking mapping in different completion forests. Also, they close the open problem of the complexity for answering fuzzy conjunctive queries in expressive fuzzy DLs by establishing two complexity bounds: for data complexity, they prove a coNP upper bound, as long as only simple roles occur in the query. Regarding combined complexity, they prove a co3NExpTime upper bound in the size of the knowledge base and the query. Before introducing the techniques of querying answering over fuzzy OWL Lite ontologies, we first recall some basic notions, including the syntax and semantics of fuzzy OWL Lite ontologies, fuzzy conjunctive queries, and fuzzy query language.

7.3.1.1 Preliminaries

Fuzzy OWL Lite ontologies are expressed by fuzzy OWL Lite language. Note that, the *fuzzy DL SHIF(D)* is the logic counterpart of fuzzy OWL Lite language. Therefore, here we briefly recall the syntax and semantics of *f-SHIF(D)*, which is the sublanguage of fuzzy DL *f-SHOIN(D)* as has been introduced in Sect. 4.2.

Let $\mathbf{A}$, $\mathbf{R}$, $\mathbf{R_c}$, $\mathbf{I}$, and $\mathbf{I_c}$ be countable infinite and pairwise disjoint sets of concept names (denoted by A), abstract role names (denoted by R), concrete role names (denoted by T), abstract individual names (denoted by O), and concrete individual names (denoted by). We assume that the set of abstract role names $\mathbf{R}$ can be divided into two disjoint subsets, $\mathbf{R}_t$ and $\mathbf{R}_n$, which stands for the subset of *transitive role names* and *non-transitive role names*, respectively. *f-SHIF(D)* abstract roles (or abstract roles for short) are defined as $R \to R_N|R^-$, where $R_N \in \mathbf{R}$, R^- is called the *inverse role* of R. A role inclusion axiom is of the form $R \sqsubseteq S$, with R, S abstract roles. A role hierarchy (also called a RBox) $\mathcal{R}$ is a finite set of role inclusion axioms.

For the sake of brevity and clarity, we use following notations: (i) we use an abbreviation $Inv(R)$ to denote inverse role of R, (ii) for a RBox $\mathcal{R}$, we define $\sqsubseteq^*_{\mathcal{R}}$ as the reflexive transitive closure of $\sqsubseteq$ over $\mathcal{R} \cup \{Inv(R) \sqsubseteq Inv(S) | R \sqsubseteq S \in \mathcal{R}\}$, (iii) for a RBox $\mathcal{R}$ and a role S, we define the set $Trans_{\mathcal{R}}$ of transitive roles as $\{S|$ there is a role R with $R \equiv^*_{\mathcal{R}} S$ and $R \in \mathbf{R}_t$ or $Inv(R) \in \mathbf{R}_t\}$, (iv) a role S is called simple w.r.t. a RBox $\mathcal{R}$ if, for each role R such that $R \sqsubseteq^*_{\mathcal{R}} S$, $R \notin$ Trans $_{\mathcal{R}}$. The subscript $\mathcal{R}$ of $\sqsubseteq^*_{\mathcal{R}}$ and $Trans_{\mathcal{R}}$ is dropped if clear from the context.

f-SHIF(D) complex concepts (or simply concepts) are defined by concept names according to the following abstract syntax:

$$C \to \top|\bot|A|C_1 \sqcap C_2|C_1 \sqcup C_2|\neg C|\forall R.C|\exists R.C| \leq 1S| \geq 2S|.D|\exists T.D$$
$$D \to d|\neg d.$$

For decidability reasons, functional restrictions of the form $\leq 1S$ and their negation $\geq 2S$ are restricted to be simple abstract roles.

A fuzzy TBox is a finite set of fuzzy concept axioms. Fuzzy concept axiom of the form $A \equiv C$ are called *fuzzy concept definitions*, fuzzy concept axiom of the form $A \sqsubseteq C$ are called *fuzzy concept specializations*, and fuzzy concept axiom of the form $C \sqsubseteq D$ are called *general concept inclusion* (GCIs) axioms .

A fuzzy ABox consists of fuzzy assertions of the form $C(o) \bowtie n$ (*fuzzy concept assertions*), $R(o, o') \rhd n$ (*fuzzy abstract role assertions*), $T(o, v) \rhd n$ (*fuzzy data type role assertions*), or $.o \neq o'$. (*inequality assertions*), where $o, o' \in \mathbf{I}$, $v \in \mathbf{I}_c$, $\bowtie$ stands for any type of inequality, i.e., $\bowtie \in \{\geq, >, \leq, <\}$. We use $\rhd$ to denote $\geq$ or $>$, and $\lhd$ to denote $\leq$ or $<$. We call ABox assertions defined by $\rhd$ *positive assertions*, while those defined by $\lhd$ *negative assertions*. Note that, we consider only positive fuzzy role assertions, since negative role assertions would imply the existence of role negation, which would lead to undecidability (Mailis et al. 2007). An *f-SHIF(D)* knowledge base $\mathcal{K}$ is a triple $\langle \mathcal{T}, \mathcal{R}, \mathcal{A} \rangle$ with $\mathcal{T}$ a TBox, $\mathcal{R}$ a RBox and $\mathcal{A}$ an ABox.

For a fuzzy concept D, we denote by $sub(D)$ the set that contains D and it is closed under sub-concepts of D, and define $sub(\mathcal{K})$ as the set of all the sub-concepts of the concepts occurring in $\mathcal{K}$. We abuse the notion $sub(\mathbf{D})$ to denote the set of all the data type predicates occurring in a knowledge base.

The semantics of *f-SHIF(D)* are provided by a fuzzy interpretation $\mathcal{I}$ as being similar to the semantics of the fuzzy DL *f-SHOIN(D)* introduced in Sect. 4.2. A fuzzy interpretation $\mathcal{I}$ satisfies a fuzzy concept specification $A \sqsubseteq C$, if $A^{\mathcal{I}}(o) \leq C^{\mathcal{I}}(o)$ for any $o \in \Delta^{\mathcal{I}}$, written as $\mathcal{I} \models A \sqsubseteq C$. Similarly, $\mathcal{I} \models A \equiv C$ if $A^{\mathcal{I}}(o) = C^{\mathcal{I}}(o)$ for any $o \in \Delta^{\mathcal{I}}$, and $\mathcal{I} \models C \sqsubseteq D$, if $C^{\mathcal{I}}(o) \leq D^{\mathcal{I}}(o)$ for any $o \in \Delta^{\mathcal{I}}$. For ABox assertions, $\mathcal{I} \models C(o) \bowtie n$ (resp. $\mathcal{I} \models R(o, o') \rhd n$), iff $C^{\mathcal{I}}(o^{\mathcal{I}}) \bowtie n$ (resp $R^{\mathcal{I}}(o^{\mathcal{I}}, o'^{\mathcal{I}}) \rhd n$), and $\mathcal{I} \models o \not\approx o'$ iff $o^{\mathcal{I}} \neq o'^{\mathcal{I}}$. If an interpretation $\mathcal{I}$ satisfies all the axioms and assertions in a KB $\mathcal{K}$, we call it a model of $\mathcal{K}$. A KB is *satisfiable* iff it has at least one model. A KB $\mathcal{K}$ *entails* (logically implies) a fuzzy assertion φ, iff all the models of $\mathcal{K}$ are also models of φ, written as $\mathcal{K} \models \varphi$.

Given a KB $\mathcal{K}$, we can w.l.o.g assume that (i) all concepts are in their negative normal forms (NNFs), i.e. negation occurs only in front of concept names. Through de Morgan law, the duality between existential restrictions ($\exists R.C$) and universal restrictions ($\forall R.C$), and the duality between functional restrictions ($\leq 1S$) and their negations ($\geq 2S$), each concept can be transformed into its equivalent NNF by pushing negation inwards. (ii) all fuzzy concept assertions are in their positive inequality normal forms (PINFs). A negative concept assertion can be transformed into its equivalent PINF by applying fuzzy operation on it. For example, $C(o) < n$ is converted to $\neg C(o) > 1 - n$. (iii) all fuzzy assertions are in their normalized forms (NFs). By introducing a positive, infinite small value ε, a fuzzy assertion of the form $C(o) > n$ can be normalized to $C(o) \geq n + \varepsilon$. The model equivalence of a KB $\mathcal{K}$ and its normalized form was shown to justify the assumption (Stoilos et al. 2006). (iv) there are only fuzzy GCIs in the TBox. A fuzzy concept specification $A \sqsubseteq C$ can be replaced by a fuzzy concept definition $A \equiv A' \sqcap C$ (Stoilos et al. 2007), where A' is a new concept name, which stands for the qualities that distinguish the elements of A from the other elements of C. A fuzzy concept definition axiom $A \equiv C$ can be eliminated by replacing every occurrence of A with C. The elimination is also known as *knowledge base expansion*. Note that the size of the expansion can be exponential in the size of the TBox. But if we follow the principle of "Expansion is done on demand" (Baader et al. 2003), the expansion will have no impact on the algorithm complexity of deciding fuzzy query entailment.

Here, as a running example, we use the *f-SHIF(D)* KB $\mathcal{K} = \langle \mathcal{T}, \mathcal{R}, \mathcal{A} \rangle$ with $\mathcal{T} = \{C \sqsubseteq \exists R.C, \top \sqsubseteq \exists Td\}$, $\mathcal{R} = \varnothing$, and $\mathcal{A} = \{C(o) \geq 0.8\}$.

7.3.1.2 Fuzzy Query Language

We now provide the formal definition of the syntax and semantics of the fuzzy querying language, extending Mallis's work (Mailis et al. 2007) to allow for querying concrete domains.

Let **V** be a countable infinite set of variables and is disjoint from **A**, **R**, $\mathbf{R}_c$, **I**, and $\mathbf{I}_c$. A *term t* is either an individual name from **I** or $\mathbf{I}_c$, or a variable name from **V**. A *fuzzy query atom* is an expression of the form $\langle C(t) \geq n \rangle$, $\langle R(t, t') \geq n \rangle$, or $\langle T(t, t') \geq n \rangle$ with C a concept, R a simple abstract role, T a data type role, and t, t' terms. As with fuzzy assertions, we refer to these three different types of atoms as *fuzzy concept atoms*, *fuzzy abstract role atoms*, and *fuzzy data type role atoms*, respectively. The fuzzy abstract role atoms and the fuzzy data type role atoms are collectively referred to as *fuzzy role atoms*.

Definition 7.17 (*Fuzzy Boolean Conjunctive Queries*) A fuzzy boolean conjunctive query q is a non-emptorm $q = \{\langle \mathrm{at}_1 \geq n_1 \rangle, \ldots, \langle \mathrm{at}_k \geq n_k \rangle\}$. Then for every fuzzy query atom, we can say $\langle \mathrm{at}_i \geq n_i \rangle \in q$.

We use $Vars(q)$ to denote the set of variables occurring in q, $AInds(q)$ and $CInds(q)$ to denote the sets of abstract and concrete individual names occurring in q, $Inds(q)$ to denotes the union of $AInds(q)$ and $CInds(q)$, and $Terms(q)$ for the set of terms in q, i.e. $Terms(q) = Vars(q) \cup Inds(q)$.

The semantics of a fuzzy query is given in the same way as for the related fuzzy DL by means of fuzzy interpretation consisting of an interpretation domain and a fuzzy interpretation function.

Definition 7.18 (*Models of Fuzzy Queries*) Let $\mathcal{I} = (\Delta^{\mathcal{I}}, \cdot^{\mathcal{I}})$ be a fuzzy interpretation of an *f-SHIF(D)* KB, q a fuzzy boolean conjunctive query, and t, t' terms in q. We say $\mathcal{I}$ is a model of q, if there exists a mapping $\pi : Terms(q) \to \Delta^{\mathcal{I}} \cup \Delta^{\mathbf{D}}$ such that $\pi(a) = a^{\mathcal{I}}$ for each $a \in Ind(q)$, $C^{\mathcal{I}}(\pi(t)) \geq n$ for each fuzzy concept atom $C(t) \geq n \in q$, $R^{\mathcal{I}}(\pi(t), \pi(t')) \geq n$ (resp. $T^{\mathcal{I}}(\pi(t), \pi(t')) \geq n$) for each fuzzy role atom $R(t, t') \geq n$ (resp. $T(t, t') \geq n$) $\in q$.

If $\mathcal{I} \models^{\pi}$ *at* for every atom *at* $\in q$, we write $\mathcal{I} \models^{\pi} q$. If there is a π, such that $\mathcal{I} \models^{\pi} q$, we say $\mathcal{I}$ satisfies q, written as $\mathcal{I} \models q$. We call such a π a *match* of q in $\mathcal{I}$. If $\mathcal{I} \models q$ for each model $\mathcal{I}$ of a KB $\mathcal{K}$, then we say $\mathcal{K}$ entails q, written as $\mathcal{K} \models q$. The *query entailment problem* is defined as follows: given a knowledge base $\mathcal{K}$ and a query q, decide whether $\mathcal{K} \models q$.

Considering the following fuzzy boolean *CQ*: $q = \{R(x, y) \geq 0.6, R(y, z) \geq 0.8, T(y, y_c) \geq 1, C(y) \geq 0.6\}$

We observe that $\mathcal{K} \models q$. Given the GCI $C \sqsubseteq \exists R.C$, we have that, for each model $\mathcal{I}$ of $\mathcal{K}$, $\exists R.C^{\mathcal{I}}(o^{\mathcal{I}}) \geq C^{\mathcal{I}}(o^{\mathcal{I}}) \geq 0.8 > 0.6$ holds. By the definition of fuzzy interpretation, there exists some element b in $\Delta^{\mathcal{I}}$, such that $R(o^{\mathcal{I}}, b) \geq 0.8 > 0.6$ and $C(b) \geq 0.8 > 0.6$ holds. Similarly, there is some element c in $\Delta^{\mathcal{I}}$, such that $R^{\mathcal{I}}(b, c) \geq 0.8$ and $C^{\mathcal{I}}(c) \geq 0.8$ holds. Since $\mathrm{T} \sqsubseteq \exists T.d$, there is some element v in $\Delta^{\mathbf{D}}$, such that $T^{\mathcal{I}}(b,v) \geq 1 \geq 0.8$ and $d^{\mathcal{I}}(v) \geq 1$ holds. By constructing a mapping π with $\pi(x) = o^{\mathcal{I}}$, $\pi(y) = b$, $\pi(z) = c$, and $\pi(y_c) = v$, we have $\mathcal{I} \models q$.

7.3.1.3 Query Entailment Algorithm

As with the algorithms for basic inference services and simple query answering, our algorithm for deciding fuzzy query entailment is also based on tableau algorithms. However, the query entailment problem can not be reduced to the knowledge base satisfiability problem, since the negation of a fuzzy conjunctive query is not expressible with existing constructors provided by an *f-SHIF(D)* knowledge base. For this reason, tableau algorithms for reasoning over knowledge bases is not sufficient. A knowledge base $\mathcal{K}$ may have infinitely many possibly infinite models, whereas tableau algorithms construct only a subset of finite models of the knowledge base. As is defined in Sect. 7.2, the query entailment holds only if the query is true in all models of the knowledge base, we thus have to show that inspecting only a subset of the models, namely the *canonical* ones, suffices to decide query entailment.

Our algorithm works on a data structure called *completion forest*. A completion forest is a finite relational structure capturing a set of models of a KB $\mathcal{K}$. Roughly speaking, models of $\mathcal{K}$ are represented by an initial completion forest $\mathcal{F}_{\mathcal{K}}$. Then, by applying *expansion rules* repeatedly, new completion forests are generated. Since every model of $\mathcal{K}$ is preserved in some completion forest that results from the expansion, $\mathcal{K} \models q$ can be decided by considering a set $\mathbb{F}_{\mathcal{K}}$ of sufficiently expanded forests. From each such an $\mathcal{F}$, a single *canonical model* is constructed. Semantically, the finite set of these canonical models is sufficient for answering all queries q of bounded size. Furthermore, we prove that entailment in the canonical model obtained from $\mathcal{F}$ can be checked effectively via a syntactic mapping of the terms in q to the nodes in $\mathcal{F}$.

Definition 7.19 (*Completion Forest*) A completion tree T for an *f-SHIF(D)* KB is a tree all of whose nodes are generated by expansion rules, except for the root node which might correspond to an abstract individual name in **I**. A completion forest $\mathcal{F}$ for an *f-SHIF(D)* KB consists of a set of completion trees whose root nodes correspond to abstract individual names occurring in the ABox, an equivalent relation $\approx$ and an inequivalent relation $\neq$ among nodes.

The nodes in a completion forest $\mathcal{F}$, denoted Nodes($\mathcal{F}$), can be divided into abstract nodes (denoted ANodes($\mathcal{F}$)) and concrete nodes (or data type nodes, denoted CNodes($\mathcal{F}$)). Each abstract node o in a completion forest is labeled with a set $\mathcal{L}(o) = \{\langle C, \geq, n\rangle\}$, where $C \in sub(\mathcal{K})$, $n \in (0,1]$. The concrete nodes v can only serve as leaf nodes, each of which is labeled with a set $\mathcal{L}(v) = \{\langle d, \geq, n\rangle\}$, where $d \in sub(\mathbf{D})$, $n \in (0,1]$. Similarly, the edges in a completion forests can be divided into abstract edges and concrete edges. Each abstract edge $\langle o, o'\rangle$ is labeled with a set $\mathcal{L}(\langle o, o'\rangle) = \{\langle R, \geq, n\rangle\}$, where $R \in \mathbf{R}$. Each concrete edge $\langle o, v\rangle$ is labeled with a set $\mathcal{L}(\langle o, v\rangle) = \{\langle T, \geq, n\rangle\}$, where $T \in \mathbf{R}_c$.

If $\langle o, o' \rangle$ is an edge in a completion forest with $\langle R, \geq, n \rangle \in \mathcal{L}(\langle o, o' \rangle)$, then o' is called an $R_{\geq,n}$ *-successor* of o and o is called an $R_{\geq,n}$ *-predecessor* of o'. Ignoring the inequality and membership degree, we can also call o' an R-successor of o and o an R-predecessor of o'. *Ancestor* and *descendant* are the transitive closure of predecessor and successor, respectively. The union of the successor and predecessor relation is the *neighbor* relation. The *distance* between two nodes o, o' in a completion forest is the shortest path between them.

Starting with an *f-SHIF(D)* KB $\mathcal{K} = \langle \mathcal{T}, \mathcal{R}, \mathcal{A} \rangle$, the completion forest $\mathcal{F}_{\mathcal{K}}$ is initialized such that it contains (i) a root node o, with $\mathcal{L}(o) = \{\langle C, \geq, n \rangle | C(o) \geq n \in \mathcal{A}\}$, for each abstract individual name o occurring in $\mathcal{A}$, (ii) a leaf node v, with $\mathcal{L}(v) = \{\langle d, \geq, n \rangle | d(v) \geq n \in \mathcal{A}\}$, for each concrete individual name v occurring in $\mathcal{A}$, (iii) an abstract edge $\langle o, o' \rangle$ with $\mathcal{L}(\langle o, o' \rangle) = \{\langle R, \geq, n \rangle | \langle R(o, o') \geq n \rangle \in \mathcal{A}\}$, for each pair $\langle o, o' \rangle$ of individual names for which the set $\{R | R(o, o') \geq n \in \mathcal{A}\}$ is non-empty, and (iv) a concrete edge $\langle o, v \rangle$ with $\mathcal{L}(\langle o, v \rangle) = \{\langle T, \geq, n \rangle | \langle T(o, v) \geq n \rangle \in \mathcal{A}\}$. We initialize the relation $\neq$ as $\{\langle o, o' \rangle | o \neq o' \in \mathcal{A}\}$, and the relation $\approx$ to be empty.

In our running example mentioned above, $\mathcal{F}_{\mathcal{K}}$ contains only one node o labelled with $\mathcal{L}(o) = \{\langle C, \geq, 0.8 \rangle\}$.

Now we can formally define a new blocking condition, called *k-blocking*, for fuzzy query entailment depending on a depth parameter $k > 0$.

Definition 7.20 (*k-tree equivalence*) The k-tree of a node v in T, denoted as T_v^k, is the subtree of T rooted at v with all the descendants of v within distance k. We use $Nodes(T_v^k)$ to denote the set of nodes in T_v^k. Two nodes v and w in T are said to be k-tree equivalent in T, if T_v^k and T_w^k are isomorphic, i.e., there exists a bijection ψ : $Nodes(T_v^k) \rightarrow Nodes(T_w^k)$ such that (i) $\psi(v) = w$, (ii) for every node $o \in Nodes(T_v^k)$, $\mathcal{L}(o) = \mathcal{L}(\psi(o))$, (iii) for every edge connecting two nodes o and o' in T_v^k, $\mathcal{L}(\langle o, o' \rangle) = \mathcal{L}(\langle \psi(o), \psi(o') \rangle)$.

Definition 7.21 (*k-witness*) A node w is a k-witness of a node v, if v and w are k-tree equivalent in T, w is an ancestor of v in T and v is not in T_w^k. Furthermore, T_w^k tree-blocks T_v^k and each node o in T_w^k tree-blocks node $\psi^{-1}(o)$ in T_v^k.

Definition 7.22 (*k-blocking*) A node o is k-blocked in a completion forest $\mathcal{F}$ iff it is not a root node and it is either directly or indirectly k-blocked. Node o is directly k-blocked iff none of its ancestors is k-blocked, and o is a leaf of a tree-blocked k-tree. Node o is indirectly k-blocked iff one of its ancestors is k-blocked or it is a successor of a node o' and $\mathcal{L}(\langle o', o \rangle) = \varnothing$.

An initial completion forest is expanded according to a set of *expansion rules* that reflect the constructors in *f-SHIF(D)*. The expansion rules, which syntactically decompose the concepts in node labels, either infer new constraints for a given node, or extend the tree according to these constraints (see Table 7.3). Termination is guaranteed by k-blocking. We denote by $\mathbb{F}_{\mathcal{K}}$ the set of all completion forests obtained this way.

Table 7.3 Expansion rules

Rule	Description
$\sqcap_{\geq}$	if 1. $\langle C \sqcap D, \geq, n\rangle \in \mathcal{L}(x)$, x is not indirectly k-blocked, and 2. $\{\langle C, \geq, n\rangle, \langle D, \geq, n\rangle\} \not\subseteq \mathcal{L}(x)$ then $\mathcal{L}(x) \to \mathcal{L}(x) \cup \{\langle C, \geq, n\rangle, \langle D, \geq, n\rangle\}$
$\sqcup_{\geq}$	if 1. $\langle C \sqcup D, \geq, n\rangle \in \mathcal{L}(x)$, x is not indirectly k-blocked, and 2. $\{\langle C, \geq, n\rangle, \langle D, \geq, n\rangle\} \cap \mathcal{L}(x) = \varnothing$ then $\mathcal{L}(x) \to \mathcal{L}(x) \cup \{C'\}$, where $C' \in \{\langle C, \geq, n\rangle, \langle D, \geq, n\rangle\}$
$\exists_{\geq}$	if 1. $\langle \exists R.C, \geq, n\rangle$ (or $\langle \exists T.d, \geq, n\rangle$) $\in \mathcal{L}(x)$, x is not k-blocked. 2. x has no $R_{\geq,n}$-neighbor (resp. no $T_{\geq,n}$-neighbor) y s.t. $\langle C, \geq, n\rangle$ (resp. $\langle d, \geq, n\rangle$) $\in \mathcal{L}(y)$, then create a new node y with $\mathcal{L}(x, y) = \{\langle R, \geq, n\rangle\}$ and $\mathcal{L}(y) = \{\langle C, \geq, n\rangle\}$ (resp. with $\mathcal{L}(x, y) = \{\langle T, \geq, n\rangle\}$ and $\mathcal{L}(y) = \{d, \geq, n\}$)
$\forall_{\geq}$	if 1. $\langle \forall R.C, \geq, n\rangle$ (or $\langle \forall T.d, \geq, n\rangle$) $\in \mathcal{L}(x)$, x is not indirectly k-blocked. 2. x has an $R_{\geq,n'}$-neighbor y (resp. a $T_{\geq,n'}$-neighbor) with $\langle C, \geq, n\rangle$ (resp. $\langle d, \geq, n\rangle$) $\notin \mathcal{L}(y)$, where $n' = 1 - n + \varepsilon$, then $\mathcal{L}(y) \to \mathcal{L}(y) \cup \{\langle C, \geq, n\rangle\}$ (resp. $\mathcal{L}(y) \to \mathcal{L}(y) \cup \{\langle d, \geq, n\rangle\}$)
$\forall_{+}$	if 1. $\langle \forall R.C, \geq, n\rangle \in \mathcal{L}(x)$ with $Trans(R)$, x is not indirectly k-blocked, and 2. x has an $R_{\geq,n'}$-neighbor y with $\langle \forall R.C, \geq, n\rangle \notin \mathcal{L}(y)$, where $n' = 1 - n + \varepsilon$, then $\mathcal{L}(y) \to \mathcal{L}(yy \cup \{\langle \forall R.C, \geq, n\rangle\}$
$\forall_{+'}$	if 1. $\langle \forall S.C, \geq, n\rangle \in \mathcal{L}(x)$, x is not indirectly k-blocked, and 2. there is some R, with $Trans(R)$ and $R \sqsubseteq^* S$, 3. x has an $R_{\geq,n'}$-neighbor y with $\langle \forall R.C, \geq, n\rangle \notin \mathcal{L}(y)$, where $n' = 1 - n + \varepsilon$, then $\mathcal{L}(y) \to \mathcal{L}(yy \cup \{\langle \forall R.C, \geq, n\rangle\}$
$\geq_{\geq}$	if 1. $\langle \geq 2S, \geq, n\rangle \in \mathcal{L}(x)$, x is not k-blocked, 2. $\#\{x_i \in N_I \| \langle R, \geq, n\rangle \in \mathcal{L}(x, x_i)\} < 2$, then introduce new nodes, s.t. $\#\{x_i \in N_I \mid \langle R, \geq, n\rangle \in \mathcal{L}(x, x_i)\} \geq 2$
$\leq_{\geq}$	if 1. $\langle \leq 1S, \geq, n\rangle \in \mathcal{L}(x)$, x is not indirectly k-blocked, 2. $\#\{x_i \in N_I \mid \langle R, \geq, 1 - n + \varepsilon\rangle \in \mathcal{L}(x, x_i)\} > 1$ and 3. there exist x_l and x_k, with no $x_l \not\approx x_k$, 4. x_l is neither a root node nor an ancestor of x_k. then (i) $\mathcal{L}(x_k) \to \mathcal{L}(x_k) \cup \mathcal{L}(x_l)$ (ii) $\mathcal{L}(x, x_k) \to \mathcal{L}(x, x_k) \cup \mathcal{L}(x, x_l)$ (iii) $\mathcal{L}(x, x_l) \to \varnothing$, $\mathcal{L}(x_l) \to \varnothing$ (iv) set $x_i \not\approx x_k$ for all x_i with $x_i \not\approx x_l$.
$\leq_{r\geq}$	if 1. $\langle \leq 1S, \geq, n\rangle \in \mathcal{L}(x)$, 2. $\#\{x_i \in N_I \mid \langle R, \geq, 1 - n + \varepsilon\rangle \in \mathcal{L}(x, x_i)\} > 1$ and 3. there exist x_l and x_k, both root nodes, with no $x_l \not\approx x_k$, then 1. $\mathcal{L}(x_k) \to \mathcal{L}(x_k) \cup \mathcal{L}(x_l)$ 2. For all edges $\langle x_l, x'\rangle$, i. if the edge $\langle x_k, x'\rangle$, does not exist, create it with $\mathcal{L}(\langle x_k, x'\rangle) = \varnothing$, ii. $\mathcal{L}(\langle x_k, x'\rangle) \to \mathcal{L}(\langle x_k, x'\rangle) \cup \mathcal{L}(\langle x_l, x'\rangle)$. 3. For all edges $\langle x', x_l\rangle$, i. if the edge $\langle x', x_k\rangle$ does not exist, create it with $\mathcal{L}(\langle x', x_k\rangle) = \varnothing$, ii. $\mathcal{L}(\langle x', x_k\rangle) \to \mathcal{L}(\langle x', x_k\rangle) \cup \mathcal{L}(\langle x', x_l\rangle)$ 4. Set $\mathcal{L}(x_l) = \varnothing$ and remove all edges to/from x_l. 5. Set $x'' \not\approx x_k$ for all x'' with $x'' \not\approx x_l$ and set $x_l \approx x_k$
$\sqsubseteq$	if 1. $C \sqsubseteq D \in \mathcal{T}$ and 2. $\{\langle \neg C, \geq, 1 - n + \varepsilon\rangle, \langle D, \geq, n\rangle\} \cap \mathcal{L}(x) = \varnothing$ for $n \in N^{\mathcal{A}} \cup N^{q}$, then $\mathcal{L}(x) \to \mathcal{L}(x) \cup \{C'\}$ for some $C' \in \{\langle \neg C, \geq, 1 - n + \varepsilon\rangle, \langle D, \geq, n\rangle\}$.

For a node o, $\mathcal{L}(o)$ is said to contain a *clash*, if it contains one of the followings: (*i*) a pair of triples $\langle C, \geq, n\rangle$ and $\langle \neg C, \geq, m\rangle$ with $n + m > 1$, (*ii*) one of the triples: $\langle \bot, \geq, n\rangle$ with $n > 0$, $\langle C, \geq, n\rangle$ with $n > 1$, (*iii*) some triple $\langle \leq 1S, \geq, n\rangle$, and o has two $R_{\geq,n'}$-neighbors o_1, o_2 with $n + n' > 1$ and $o_1 \not\approx o_2$.

Definition 7.23 (*k-complete and clash-free completion forest*) A completion forest is called k-complete and clash-free if under k-blocking no rule can be applied to it, and none of its nodes and edges contains a clash. We denote by $\mathrm{ccf}_k(\mathbb{F}_{\mathcal{K}})$ the set of k-complete and class-free completion forests in $\mathbb{F}_{\mathcal{K}}$.

Figure 7.3 shows a 2-complete and clash-free completion forest $\mathcal{F}$ for $\mathcal{K}$, i.e., $\mathcal{F} \in \mathrm{ccf}_2(\mathbb{F}_{\mathcal{K}})$, where $\mathcal{L}_o = \{\langle C, \geq, 0.8\rangle, \langle \exists R.C, \geq, 0.8\rangle\}$, $\mathcal{L}_d = \{\langle d, \geq, 1\rangle\}$, $\mathcal{L}_R = \{\langle R, \geq, 0.8\rangle\}$, $\mathcal{L}_T = \{\langle T, \geq, 1\rangle\}$. In $\mathcal{F}$, o_1 and o_4 are 2-tree equivalent, and o_1 is a 2-witness of o_4. T^2_{o1} tree-blocks T^2_{o4}, and o_1 tree-blocks o_5. The node o_6 in T^2_{o4} is directly blocked by o_3 in T^2_{o1}-tree, indicated by the dashed line.

We now show that every model of a KB $\mathcal{K}$ is preserved in some complete and clash-free completion tree $\mathcal{F}$. If we view all the nodes (either root nodes or generated nodes) in a completion forest $\mathcal{F}$ as individual names, we can define models of $\mathcal{F}$ in terms of models of $\mathcal{K}$ over an extended vocabulary .

Definition 7.24 (*Models of completion forests*) An interpretation $\mathcal{I}$ is a model of a completion forest $\mathcal{F}$ for $\mathcal{K}$, denoted $\mathcal{I} \models \mathcal{F}$, if $\mathcal{I} \models \mathcal{K}$ and for all abstract nodes o, o' and concrete nodes v in $\mathcal{F}$ it holds that (i) $C^{\mathcal{I}}(o^{\mathcal{I}}) \geq n$ if $\langle C, \geq, n\rangle \in \mathcal{L}(\langle o\rangle)$, (ii) $d^{\mathcal{I}}(v^{\mathcal{I}}) \geq n$ if $\langle d, \geq, n\rangle \in \mathcal{L}(v)$, (iii) $R^{\mathcal{I}}(o^{\mathcal{I}}, o'^{\mathcal{I}}) \geq n$ if there exists an abstract edge $\langle o, o'\rangle$ in $\mathcal{F}$ and $\langle R, \geq, n\rangle \in \mathcal{L}(\langle o, o'\rangle)$, (iv) $T^{\mathcal{I}}(o^{\mathcal{I}}, v^{\mathcal{I}}) \geq n$ if there exists a concrete edge $\langle o, v\rangle$ in $\mathcal{F}$ and $\langle T, \geq, n\rangle \in \mathcal{L}(\langle o, v\rangle)$, (v) $o^{\mathcal{I}} \neq o'^{\mathcal{I}}$ if $o \neq o' \in \mathcal{F}$.

We first show that the models of the initial completion forest $\mathcal{F}_{\mathcal{K}}$ and of $\mathcal{K}$ coincide.

Lemma 7.7 $\mathcal{I} \models \mathcal{F}_{\mathcal{K}}$ *iff* $\mathcal{I} \models \mathcal{K}$.

Proof The only if direction follows from Definition 7.24. For the if direction, we need to show that, for all nodes v, w in $\mathcal{F}_{\mathcal{K}}$, Property (i)–(v) in Definition 7.24 hold. By Definition 7.19, each node in $\mathcal{F}_{\mathcal{K}}$ corresponds to an individual name in $\mathcal{K}$. For each abstract individual o in $\mathbf{I}$, the label of node o in $\mathcal{F}_{\mathcal{K}}$ is $\mathcal{L}(o) = \{\langle C, \geq, n\rangle | C(o) \geq n \in \mathcal{A}\}$. Since $\mathcal{I} \models \mathcal{K}$, then we have $C^{\mathcal{I}}(o^{\mathcal{I}}) \geq n$ and Property (i) thus holds. Property (ii)–(v) can be proved in a similar way with (i). □

We then show that, each time an expansion rule is applied, all models are preserved in some resulting forest.

Lemma 7.8 *Let* $\mathcal{F}$ *be a completion forest in* $\mathbb{F}_{\mathcal{K}}$, r a rule in Table 7.3, $\mathbf{F}$ *a set of completion forests obtained from* $\mathcal{F}$ *by applying* r*, then for each model* $\mathcal{I}$ *of* $\mathcal{F}$*, there exist an* $\mathcal{F}$ *in* $\mathbf{F}$ *and an extension* $\mathcal{I}$ *of* $\mathcal{I}$*, such that* $\mathcal{I}' \models \mathcal{F}'$.

Proof $\exists_{\geq}$-rule. Since $\langle \exists R.C, \geq, n\rangle \in \mathcal{L}(x)$ and $\mathcal{I} \models \mathcal{F}$, then there exists some $o \in \Delta^{\mathcal{I}}$, such that $R^{\mathcal{I}}(x^{\mathcal{I}}, o) \geq n$ and $C^{\mathcal{I}}(o) \geq n$ hold. In the completion forest $\mathcal{F}'$ obtained from $\mathcal{F}$ by applying $\exists_{\geq}$-rule, a new node y is generated such that $\langle R, \geq, n\rangle \in \mathcal{L}(\langle x, y\rangle)$ and $\langle C, \geq, n\rangle \in \mathcal{L}(y)$. By setting $y^{\mathcal{I}'} = o$, we obtain an extension $\mathcal{I}'$ of $\mathcal{I}$, and thus $\mathcal{I}' \models \mathcal{F}'$. The case of $\geq_{\geq}$-rule is analogous to the $\exists_{\geq}$-rule. The proofs for other rules are in similar way. □

Since the set of k-complete and clash-free completion forests for $\mathcal{K}$ semantically captures $\mathcal{K}$ (modulo new individuals), we can transfer query entailment $\mathcal{K} \models q$ to logical consequence of q from completion forests as follows. For any completion forest $\mathcal{F}$ and any CQ q, let $\mathcal{F} \models q$ denote that $\mathcal{I} \models q$ for every model $\mathcal{I}$ of $\mathcal{F}$.

Theorem 7.3 *Let $k \geq 0$ be arbitrary. Then $\mathcal{K} \models q$ iff $\mathcal{F} \models q$ for each $\mathcal{F} \in ccf_k(\mathbb{F}_{\mathcal{K}})$.*

Proof By Lemma 7.7 and Lemma 7.8, for each model $\mathcal{I}$ of $\mathcal{K}$, there exists some $\mathcal{F} \in \mathbb{F}_{\mathcal{K}}$ and an extension $\mathcal{I}'$ of $\mathcal{I}$, such that $\mathcal{I}' \models \mathcal{F}$. Assume $\mathcal{F} \notin \mathrm{ccf}_k(\mathbb{F}_{\mathcal{K}})$, then there still have rules applicable to $\mathcal{F}$. We thus obtain an expansion $\mathcal{F}'$ from $\mathcal{F}$ and an extension $\mathcal{I}''$ of $\mathcal{I}'$, such that $\mathcal{I}'' \models \mathcal{F}'$, and so forth until no rule is applicable. Now we either obtain a complete and clash free completion forest, or encounter a clash. The former is conflict with the assumption, and the latter is conflict with the fact that $\mathcal{F}$ have models. □

In the following, we check query entailment within completion forests. We now show that, if k is large enough, we can decide $\mathcal{F} \models q$ for each $\mathcal{F} \in ccf_k(\mathbb{F}_{\mathcal{K}})$ by syntactically mapping the query q into $\mathcal{F}$.

Definition 7.25 (*Query mapping*) A fuzzy query q can be mapped into $\mathcal{F}$, denoted $q \rightarrowtail \mathcal{F}$, if there is a mapping μ: *Terms*$(q) \to$ *Nodes*$(\mathcal{F})$, such that (i) $\mu(a) = a$, if $a \in$ Inds (q), (ii) for each fuzzy concept atom $C(x) \geq n$ (resp. $d(x) \geq n$) in q, $\langle C(x), \geq, m\rangle$ (resp. $\langle d(x), \geq, m\rangle$) $\in \mathcal{L}(\mu(x))$ with $m \geq n$, (iii) for each fuzzy role atom $R(x, y) \geq n$ (resp. $T(x, y) \geq n$) in q, $\mu(y)$ is a $R_{\geq m}$-neighbor (resp. a $T_{\geq m}$-neighbor) of $\mu(x)$ with $m \geq n$.

For example, by setting $\mu(x) = o_1$, $\mu(y) = o_2$, $\mu(z) = o_3$, and $\mu(y_c) = v_1$, we can construct a mapping μ of q into $\mathcal{F}_1$.

In fact, for completion forests $\mathcal{F}$ of a KB $\mathcal{K}$, syntactic mapping $q \rightarrowtail \mathcal{F}$ implies semantic consequence $\mathcal{F} \models q$.

Lemma 7.9 *If $q \rightarrowtail \mathcal{F}$, then $\mathcal{F} \models q$.*

Proof If $q \rightarrowtail \mathcal{F}$, then there is a mapping μ : Terms $(q) \to$ Nodes $(\mathcal{F})$ satisfying Definition 7.25. For any model $\mathcal{I} = (\Delta^{\mathcal{I}}, \cdot^{\mathcal{I}})$ of $\mathcal{F}$, it satisfies Definition 7.24. We construct a mapping π: Terms $(q) \to \Delta^{\mathcal{I}}$ such that, for each term $x \in$ Terms (q), $\pi(x) = (\mu(x))^{\mathcal{I}}$. It satisfies $C^{\mathcal{I}}(\pi(x)) = C^{\mathcal{I}}((\mu(x))^{\mathcal{I}}) \geq m \geq n$, for each fuzzy concept atom $C(x) \geq n \in q$. The proof for fuzzy role atoms can be shown in a similar way. Hence $\mathcal{I} \models q$, which by Theorem 7.3 implies $\mathcal{F} \models q$. □

Lemma 7.9 shows the soundness of our algorithm. We prove, in the following, the converse (the completeness) also holds. We show that provided the completion forest $\mathcal{F}$ has been sufficiently expanded, a mapping from q to $\mathcal{F}$ can be constructed from a single canonical model.

In the following we construct the canonical model. The construction of the canonical model $\mathcal{I}_{\mathcal{F}}$ for $\mathcal{F}$ is divided into two steps. First, $\mathcal{F}$ is unravelled into a fuzzy tableau. Then, the canonical model is induced from the tableau. The interpretation domain $\Delta^{\mathcal{I}_{\mathcal{F}}}$ of $\mathcal{I}_{\mathcal{F}}$ consists of a set of (maybe infinite) paths. The reason lies in that a KB $\mathcal{K}$ may have infinite models, whereas the canonical model $\mathcal{I}_{\mathcal{F}}$ is constructed from $\mathcal{F}$, which is a finite representation decided by the termination of the algorithm. This requires some nodes in $\mathcal{F}$ represent several elements in $\Delta^{\mathcal{I}_{\mathcal{F}}}$. The paths is chosen to distinguish different elements represented in $\mathcal{F}$ by the same node. The definition of fuzzy tableau is based on the one in (Stoilos et al. 2006, 2007).

Definition 7.26 (*Fuzzy tableaux*) Let $\mathcal{K} = \langle \mathcal{T}, \mathcal{R}, \mathcal{A} \rangle$ be an *f-SHIF(D)* KB, $\mathbf{R}_{\mathcal{K}}$, $\mathbf{R_D}$ the sets of abstract and concrete roles occurring in $\mathcal{K}$, $\mathbf{I}_{\mathcal{A}}$ the set of individual names occurring in $\mathcal{A}$, $T = \langle \mathbf{S}, \mathcal{H}, \mathcal{E}_a, \mathcal{E}_c, \mathcal{V} \rangle$ is the fuzzy tableau of $\mathcal{K}$ if (i) $\mathbf{S}$ is a nonempty set, (ii) $\mathcal{H} : \mathbf{S} \times sub(\mathcal{K}) \rightarrow [0, 1]$ maps each element in $\mathbf{S}$ and each concept in $sub(\mathcal{K})$ to the membership degree the element belongs to the concept, (iii) $\mathcal{E}_a : \mathbf{S} \times \mathbf{S} \times \mathbf{R}_{\mathcal{K}} \rightarrow [0, 1]$ maps each pair of elements in $\mathbf{S}$ and each role in $\mathbf{R}_{\mathcal{K}}$ to the membership degree the pair belongs to the role, (iv) $\mathcal{E}_c : \mathbf{S} \times \Delta^{\mathbf{D}} \times \mathbf{R_D} \rightarrow [0, 1]$ maps each pair of elements and concrete values and each concrete role in $\mathbf{R_D}$ to the membership degree the pair belongs to the role, (v) $\mathcal{V} : \mathbf{I}_{\mathcal{A}} \rightarrow \mathbf{S}$ maps each individual in $\mathbf{I}_{\mathcal{A}}$ to a element in $\mathbf{S}$. Additionally, for each $s, t \in \mathbf{S}$, $C, D \in sub(\mathcal{K})$, $R \in \mathbf{R}_{\mathcal{K}}$, and $n \in [0, 1]$, T satisfies:

1 for each $s \in \mathbf{S}$, $\mathcal{H}(s, \bot) = 0$, $\mathcal{H}(s, \top) = 1$;
2 if $\mathcal{H}(s, C \sqcap D) \geq n$, then $\mathcal{H}(s, C) \geq n$ and $\mathcal{H}(s, D) \geq n$;
3 if $\mathcal{H}(s, C \sqcup D) \geq n$, then $\mathcal{H}(s, C) \geq n$ or $\mathcal{H}(s, D) \geq n$;
4 if $\mathcal{H}(s, \forall R.C) \geq n$, then for all $t \in \mathbf{S}$, $\mathcal{E}(\langle s, t \rangle, R) \leq 1 - n$ or $\mathcal{H}(t, C) \geq n$;
5 if $\mathcal{H}(s, \exists R.C) \geq n$, then there exists $t \in \mathbf{S}$, such that $\mathcal{E}_a(\langle s, t \rangle, R) \geq n$ and $\mathcal{H}(t, C) \geq n$;
6 if $\mathcal{H}(s, \forall R.C) \geq n$, and Trans($R$), then $\mathcal{E}_a(\langle s, t \rangle, R) \leq 1 - n$ or $\mathcal{H}(t, \forall R.C) \geq n$;
7 if $\mathcal{H}(s, \forall S.C) \geq n$, Trans($R$), and $R \sqsubseteq^* S$, then $\mathcal{E}_a(\langle s, t \rangle, R) \leq 1 - n$ or $\mathcal{H}(t, \forall R.C) \geq n$;
8 if $\mathcal{H}(s, \geq 2S) \geq n$, then $\#\{t \in \mathbf{S} | \mathcal{E}_a(\langle s, t \rangle, R) \geq n$;
9 if $\mathcal{H}(s, \leq 1S) \geq n$, then $\#\{t \in \mathbf{S} | \mathcal{E}_a(\langle s, t \rangle, R) \geq 1 - n + \varepsilon$;
10 if $\mathcal{E}_a(\langle s, t \rangle, R) \geq n$, and $R \sqsubseteq^* S$, then $\mathcal{E}_a(\langle s, t \rangle, S) \geq n$;
11 $\mathcal{E}_a(\langle s, t \rangle, R) \geq n$ iff $\mathcal{E}_a(\langle t, s \rangle, Inv(R)) \geq n$;
12 if $C \sqsubseteq D \in \mathcal{T}$, then for all $s \in \mathbf{S}$, $n \in N^{\mathcal{A}} \cup N^q$, $\mathcal{H}(s, \neg C) \geq 1 - n + \varepsilon$ or $\mathcal{H}(s, D) \geq n$;
13 if $C(o) \geq n \in \mathcal{A}$, then $\mathcal{H}(\mathcal{V}(o), C) \geq n$;
14 if $R(o, o') \geq n \in \mathcal{A}$, then $\mathcal{E}_a(\langle \mathcal{V}(o), \mathcal{V}(o') \rangle, R) \geq n$;
15 if $o \not\approx o' \in \mathcal{A}$, then $\mathcal{V}(o) \neq \mathcal{V}(o')$.
16 if $\mathcal{H}(s, \forall T.d) \geq n$, then for all $t \in \Delta^{\mathbf{D}}$, $\mathcal{E}_c(\langle s, t \rangle, T) \leq 1 - n$ or $d^{\mathbf{D}}(t) \geq n$;
17 if $\mathcal{H}(s, \exists T.d) \geq n$, then there exists $t \in \Delta^{\mathbf{D}}$, such that $\mathcal{E}_c(\langle s, t \rangle, T) \geq n$ and $d^{\mathbf{D}}(t) \geq n$;

The process of inducing a fuzzy tableau from a completion forest $\mathcal{F}$ is as follows.

Each element in $\mathbf{S}$ corresponds to a path in $\mathcal{F}$. We can view a blocked node as a loop so as to define infinite paths. To be more precise, a path $p =$

$[v_0/v_{0'}, \ldots, v_n/v_{n'}]$ is a sequence of node pairs in $\mathcal{F}$. We define $Tail(p) = v_n$, and $Tail'(p) = v_{n'}$. We denote by $[p|v_{n+1}/v_{n+1}']$ the path $[v_0/v_{0'}, \ldots, v_n/v_{n'}, v_{n+1}/v_{n+1}']$ and use $[p|v_{n+1}/v_{n+1}', v_{n+2}/v_{n+2}']$ as the abbreviation of $[[p|v_{n+1}/v_{n+1}'], v_{n+2}/v_{n+2}']$. The set Paths($\mathcal{F}$) of paths in $\mathcal{F}$ is inductively defined as follows:

- if v is a root node in $\mathcal{F}$, then $[v/v] \in$ Paths ($\mathcal{F}$);
- if $p \in$ Paths ($\mathcal{F}$), and $w \in$ Nodes ($\mathcal{F}$),
  - if w is the R-successor of $Tail(p)$ and is not k-blocked, then $[p|w/w] \in$ Paths ($\mathcal{F}$);
  - if there exists $w' \in$ Nodes ($\mathcal{F}$) and is the R-successor of $Tail(p)$, and is directly k-blocked by w, then $[p|w/w'] \in$ Paths ($\mathcal{F}$).

Definition 7.27 (*Induced fuzzy tableau*) The fuzzy tableau $T_{\mathcal{F}} = \langle \mathbf{S}, \mathcal{H}, \mathcal{E}_a, \mathcal{E}_c, \mathcal{V} \rangle$ induced by $\mathcal{F}$ is as follows.

- $\mathbf{S} = \text{Paths}(\mathcal{F})$,
- $\mathcal{H}(p, C) \geq \sup\{n_i | \langle C, \geq, n_i \rangle \in \mathcal{L}(Tail(p))\}$,
- $\mathcal{E}_a(\langle p, [p|w/w'] \rangle, R) \geq \sup\{n_i | \langle R, \geq, n_i \rangle \in \mathcal{L}(\langle Tail(p), w' \rangle)\}$,
  $\mathcal{E}_a(\langle [p|w/w'], p \rangle, R) \geq \sup\{n_i | \langle Inv(R), \geq, n_i \rangle \in \mathcal{L}(\langle Tail(p), w' \rangle)\}$,
  $\mathcal{E}_a(\langle [v/v, w/w] \rangle, R) \geq \sup\{n_i | \langle R^*, \geq, n_i \rangle \in \mathcal{L}(\langle v, w \rangle)\}$, where v, w are root nodes, v is the R^*-neighbour of w, and R^* denotes R or $Inv(R)$,
- $\mathcal{E}_c(\langle p, [p|v_c/v_c] \rangle, T) \geq n$ with $\langle T, \geq, n \rangle \in \mathcal{L}(\langle Tail(p), v_c \rangle)\}$, where v_c is a concrete node,
- $\mathcal{V}(a_i) = \begin{cases} [a_i/a_i], & \text{if } a_i \text{ is a root node, and } \mathcal{L}(a_i) \neq \varnothing, \\ [a_j/a_j], & \text{if } a_i \text{ is a root node, and } \mathcal{L}(a_i) = \varnothing, \\ & \text{with } a_i \approx a_j \text{ and } \mathcal{L}(a_j) \neq \varnothing. \end{cases}$

From the fuzzy tableau of a fuzzy KB $\mathcal{K}$, we can obtain the canonical model of $\mathcal{K}$.

Definition 7.28 (*Canonical model*) Let $T = \langle \mathbf{S}, \mathcal{H}, \mathcal{E}_a, \mathcal{E}_c, \mathcal{V} \rangle$ be a fuzzy tableau of $\mathcal{K}$, the canonical model of T, $\mathcal{I}_T = (\Delta^{\mathcal{I}_T}, \cdot^{\mathcal{I}_T})$, is defined as follows.

- $\Delta^{\mathcal{I}_T} = \mathbf{S}$;
- for each individual name o in $\mathbf{I}_A$, $o^{\mathcal{I}_T} = \mathcal{V}(o)$;
- for each $s \in \mathbf{S}$ and each concept name A, $A^{\mathcal{I}_T}(s) = \mathcal{H}(s, A)$;
- for each $\langle s, t \rangle \in \mathbf{S} \times \mathbf{S}$,

$$R^{\mathcal{I}_T}(s,t) = \begin{cases} R_{\mathcal{E}}^{+}(s,t), & \text{if } Trans(R), \\ \max\limits_{S \sqsubseteq^* R, S \neq R} (R_{\mathcal{E}}(s,t), S^{\mathcal{I}_T}(s,t)), & \text{otherwise.} \end{cases}$$

- where $R_{\mathcal{E}}(s,t) = \mathcal{E}_a(\langle s, t \rangle, R)$, and $R_{\mathcal{E}}^{+}(s,t)$ is the sup-min transitive closure of $R_{\mathcal{E}}(s,t)$.
- for each $\langle s, t \rangle \in \mathbf{S} \times \Delta^{\mathbf{D}}$, $T^{\mathcal{I}_T}(s,t) = \mathcal{E}_c(s,t)$.

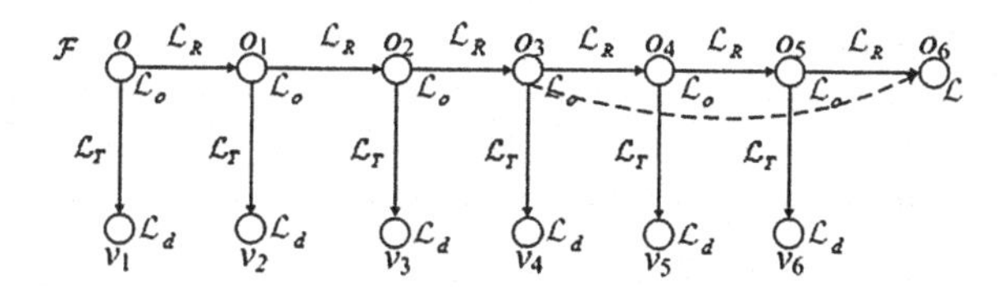

Fig. 7.3 A 2-complete and clash-free completion forest $\mathcal{F}$ of $\mathcal{K}$

Lemma 7.10 *Let T be the fuzzy tableau of an f-SHIF(D) KB* $\mathcal{K} = \langle \mathcal{T}, \mathcal{R}, \mathcal{A} \rangle$, *then the canonical model* $\mathcal{I}_T$ *of T is a model of* $\mathcal{K}$.

Proof Property 12 in Definition 7.26 ensures that $\mathcal{I}_T$ is a model of $\mathcal{T}$. For a detailed proof, see Proposition 3 in (Stoilos et al. 2006). Property 1–11 and 13–17 in Definition 7.26 ensures that $\mathcal{I}_T$ is a model of $\mathcal{A}$ and $\mathcal{R}$. For a detailed proof, see Lemma 5.2 and 6.5 in (Stoilos et al. 2007). □

For example, by unraveling $\mathcal{F}$ in Fig. 7.3, we obtain a model $\mathcal{I}_\mathcal{F}$ which has as domain the infinite set of paths from o to each o_i. Note that a path actually comprises a sequence of pairs of nodes, in order to witness the loops introduced by the blocked variables. When a node is not blocked, like o_1, the pair o_1/o_1 is added to the path. Since T^2_{o1} tree-blocks T^2_{o4}, each time a path reaches o_6, which is a leaf node of a blocked tree, we add o_3/o_6 to the path and 'loop' back to the successors of o_3. This set of paths constitute the domain $\Delta^{\mathcal{I}_\mathcal{F}}$. For each concept name A, we have $A^{\mathcal{I}_\mathcal{F}}(p_i) \geq n$, if $\langle A, \geq, n \rangle$ occurs in the label of the last node in p_i. For each role R, $R^{\mathcal{I}_\mathcal{F}}(p_i, p_j) \geq n$ if the last node in p_j is an $R_{\geq, n}$-successor of p_i. If role $R \in Trans$, the extension of R is expanded according to the sup-min transitive semantics. Therefore, $C^{\mathcal{I}_\mathcal{F}}(p_i) \geq 0.8$ for $i \geq 0$, and $R^{\mathcal{I}_\mathcal{F}}(p_i, p_j) \geq 0.8$ for $0 \leq i < j$.

From a complete and clash-free completion forest $\mathcal{F}$, we can obtain a fuzzy tableau $T_\mathcal{F}$, through which a canonical model $\mathcal{I}_\mathcal{F}$ is constructed.

Lemma 7.11 *Let* $\mathcal{F} \in ccf_k(\mathbb{F}_\mathcal{K})$, *then* $\mathcal{I}_\mathcal{F} \models \mathcal{K}$, *where* $k \geq 1$.

Proof It follows from Lemma 5.9 and 6.10 in (Stoilos et al. 2007) and Proposition 5 in (Stoilos et al. 2006) that the induced tableau in Definition 7.28 satisfies Property 1–15 in Definition 7.27. By Lemma 7.10, the canonical model R constructed from $\mathcal{R}$ is a model of $\sqsubseteq^*_\mathcal{R}$. □

Now we illustrate how to construct a mapping of q to $\mathcal{F}$ from a mapping of q to $\mathcal{I}_\mathcal{F}$.

Definition 7.29 (*Mapping graph*) Let $\mathcal{F} \in ccf_k(\mathbb{F}_\mathcal{K})$ with $k \geq 0$, and a fuzzy query q, such that $\mathcal{I}_\mathcal{F} \models q$. π is a mapping in Definition 7.18, then the mapping graph $G_\pi = \langle V, E \rangle$ is defined as: $V(G_\pi) = \{\pi(x) \in \Delta^{\mathcal{I}_\mathcal{F}} \cup \Delta^\mathbf{D} | x \in Terms(q)\}$, $E(G_\pi) = \{\langle \pi(x), \pi(y) \rangle \in \Delta^{\mathcal{I}_\mathcal{F}} \times \Delta^{\mathcal{I}_\mathcal{F}} | R(x, y) \geq n \in q\} \cup \{\langle \pi(x), \pi(y) \rangle \in \Delta^{\mathcal{I}_\mathcal{F}} \times \Delta^\mathbf{D} | T(x, y) \geq n \in q\}$. $V(G_\pi)$ is divided into $Vr(G_\pi)$ and $Vn(G_\pi)$, i.e., $V(G_\pi) = Vr(G_\pi) \cup Vn(G_\pi)$ and $Vr(G_\pi) \cap Vn(G_\pi) = \varnothing$, where $Vr(G_\pi) = \{[v/v] | v \text{is a root node in } \mathcal{F}\}$.

Definition 7.30 (*Maximal* **q-distance**) For any $x, y \in Terms(q)$, if $\pi(x)$, $\pi(y) \in V_n(G_\pi)$, then we use $d^\pi(x, y)$ to denote the length of the shortest path between $\pi(x)$ and $\pi(y)$ in G^π. If $\pi(x)$ and $\pi(y)$ are in two different connected components respectively, then $d^\pi(x, y) = -1$. We define maximal q-distance $d_q^\pi = \max_{x,y \in Terms(q)}\{d^\pi(x, y)\}$.

For example, Consider a mapping π such that $\pi(x) = p_6, \pi(y) = p_7$, $\pi(z) = p_8$, and $\pi(y_c) = v_5$. The mapping graph G_π contains the nodes p_6, p_7 and p_8, where $Vr(G_\pi) = \varnothing$, $Vn(G_\pi) = \{p_6, p_7, p_8, v_5\}$, and $E(G_\pi) = \{\langle p_6, p_7\rangle, \langle p_7, p_8\rangle, \langle p_7, v_5\rangle\}$. Moreover, $d^\pi(p_6, p_7) = 1$, $d^\pi(p_7, p_8) = 1$, $d^\pi(p_6, p_8) = 2$, $d^\pi(p_6, v_5) = 2$, and $d^\pi(v_5, p_8) = 2$, thus $d_q^\pi = 2$.

We use n_q to denote the number of fuzzy role atoms in a fuzzy query q. We only consider fuzzy role atoms $R(x, y) \geq n$ with R a simple role, so $d^\pi(x, y) = 1$ and $d_q^\pi \leq n_q$. We show provided the k in the k-blocking condition is greater than or equal to n_q, it suffices to find a mapping from q to $\mathcal{F}$.

Lemma 7.12 *Let* $\mathcal{F} \in ccf_k(\mathbb{F}_\mathcal{K})$ *with* $k \geq n_Q$, *and* $\mathcal{I}_\mathcal{F}$ *is the canonical model of* $\mathcal{F}$. *If* $\mathcal{I}_\mathcal{F} \models q$, *then* $q \rightarrowtail \mathcal{F}$.

Proof Since $\mathcal{I}_\mathcal{F} \models q$, then there exists a mapping π: $Terms(q) \rightarrow \Delta^{\mathcal{I}_\mathcal{F}} \cup \Delta^D$ such that $\pi(a) = a^{\mathcal{I}_\mathcal{F}}$ for each $a \in Ind(q)$, $C^{\mathcal{I}_\mathcal{F}}(\pi(t)) \geq n$ for each fuzzy concept atom $C(t) \geq n \in q$, $R^{\mathcal{I}_\mathcal{F}}(\pi(t), \pi(t')) \geq n$ (resp. $T^{\mathcal{I}_\mathcal{F}}(\pi(t), \pi(t')) \geq n$) for each fuzzy role atom $R(t, t') \geq n$ (resp. $T(t, t') \geq n$) $\in q$. □

We construct the mapping μ: $Terms(q) \rightarrow Nodes(\mathcal{F})$ from the mapping π. First, we consider G, a subgraph of G_π, which is obtained by eliminating the vertices of the form $[a/a]$ with a a individual name and the arcs that enter or leave these vertices. G consists of a set of connected components-written as $G_1, \ldots, G_m$. We define $Blocked(G_i)$ as the set of all the vertices p such that $Tails(p) \neq Tails'(p)$. Then, for every ancestor p' of p in G_i, $Tails(p) = Tails\prime(p)$. We use $AfterBlocked(G_i)$ to denote the set of the descendants of the vertices in $Blocked(G_i)$.

Recalling the definition of $Nodes(\mathcal{F})$, since $\mathcal{F}$ is k-blocked, if there are two node pairs v/v' and w/w' in a path (also a vertex) p with $v \neq v'$ and $w \neq w'$, then the distance between these two node pairs must greater than k. A path p always begins with v/v, if it contains a node pair w/w' with $w \neq w'$, then the distance between v/v and w/w' must greater than k.

We prove two properties of G_i as follows.

(1) If $\pi(x) \in AfterBlocked(G_i)$, then $Tail(\pi(x)) = Tail'(\pi(x))$.
If $\pi(x) \in AfterBlocked(G_i)$, then there exists some $y \in Vars(q)$, such that $\pi(y) \in Blocked(G_i)$, i.e., $Tail(\pi(y)) \neq Tail'(\pi(y))$, and $\pi(x)$ is a descendant of $\pi(y)$. $\pi(x)$ is of the form $[p|v_0/v_0', \ldots, v_m/v_m']$, where $Tail(p) \neq Tail'(p)$. Assume that $v_m \neq v_m'$, then the length of the path $\pi(x)$ is larger than k, which contradicts with the fact that $d^\pi(x, y) \leq d_q^\pi \leq n_q \leq k$.

(2) If $\pi(x) \in V(G_i)$ for some G_i with $afterblocked(G_i) \neq \varnothing$ and $\pi(x) \notin afterblocked(G_i)$, then $Tail'(\pi(x))$ is tree-blocked by $\psi(Tail'(\pi(x)))$.
If $afterblocked(G_i) \neq \varnothing$, then there exists some $y \in Vars(q)$, such that $\pi(y) \in Nodes(G_i)$ has some proper sub-path p such that $Tail(p) = Tail'(p)$. Since $\pi(x)$ and $\pi(y)$ are in the same G_i, then either $\pi(x)$ is an ancestor of $\pi(y)$ or there is some $z \in Terms(q)$ such that $\pi(z)$ is a common ancestor of $\pi(x)$ and $\pi(y)$ in $Nodes(G_i)$. In the first case, if $Tail'(\pi(x))$ was not tree-blocked, we would have that $d^{\pi}(x, y) > n \geq n_q$, which is a contradiction. In the second case, if $Tail'(\pi(x))$ was not tree-blocked, then $Tail'(\pi(z))$ would not be tree-blocked either, and thus we also derive a contradiction since $d^{\pi}(z, y) > n \geq n_q$.

We thus can construct the mapping μ: $Terms(q) \to Nodes(\mathcal{F})$ as follows.

- For each $a \in Inds(q)$, $\mu(a) = Tail(\pi(a)) = a$;
- For each $x \in Vars(q)$ with $\pi(x) \in afterblocked(G_i)$, $\mu(x) = Tail(\pi(x))$;
  - If $afterblocked(G_i) = \varnothing$, then $\mu(x) = Tail(\pi(x))$,
  - If $afterblocked(G_i) \neq \varnothing$, then

$$\mu(x) = \begin{cases} Tail\prime(\pi(x)) & if\, \pi(x) \in afterblocked(G_i), \\ \psi(Tail\prime(\pi(x))) & otherwise. \end{cases}$$

We now prove that μ satisfies Property 1–3 in Definition 7.25.

Property 1: follows from the construction of μ;

Property 2: For each fuzzy concept atom $C(x) \geq n \in q$, since $\mathcal{I}_{\mathcal{F}} \models q$, $\mathcal{H}(\pi(x), C) \geq n$ holds. It follows that $\langle C, \geq, n\rangle \in \mathcal{L}(Tail'(\pi(x)))$ or $\langle C, \geq, n\rangle \in \mathcal{L}(\psi(Tail'(\pi(x))))$, then we have $\langle C, \geq, n\rangle \in \mathcal{L}(\mu(x))$.

Property 3: For each fuzzy role atom $R(x, y) \geq n \in q$, $\mathcal{E}(\langle \pi(x), \pi(y)\rangle, R) \geq n$ holds. Then, either (1) $Tail'(\pi(y))$ is a $R_{\geq, n}$-successor of $Tail(\pi(x))$, or (2) $Tail'(\pi(x))$ is a $Inv(R)_{\geq, n}$-successor of $Tail(\pi(y))$.

Case (1): For each connected component G_i, if $AfterBlocked(G_i) = \varnothing$, then for each term x such that $\pi(x) \in G_i$, $Tail(\pi(x) = Tail'(\pi(x)$. If $Tail'(\pi(y))$ is a $R_{\geq, n}$-successor of $Tail(\pi(x))$, then $\mu(y) = Tail'(\pi(y))$ is a $R_{\geq, n}$-successor of $\mu(x) = Tail(\pi(x))$.

If $AfterBlocked(G_i) \neq \varnothing$, we make case study as follows.

(a) If $\pi(x), \pi(y) \in AfterBlocked(G_i)$, then $\mu(x) = Tail'(\pi(x))$ and $\mu(y) = Tail'(\pi(y))$. If $Tail'(\pi(y))$ is a $R_{\geq, n}$-successor of $Tail(\pi(x))$, then $\mu(y) = Tail'(\pi(y))$ is a $R_{\geq, n}$-successor of $\mu(x) = Tail'(\pi(x)) = Tail(\pi(x))$.
(b) If $\pi(x), \pi(y) \notin AfterBlocked(G_i)$, then $\mathrm{Tail}'(\pi(x))$. Otherwise, there will be $\pi(y) \in AfterBlocked(G_i)$. By Property (2), $\mathrm{Tail}'(\pi(x))$ is tree-blocked by $\psi(\mathrm{Tail}'(\pi(x)))$, $\mathrm{Tail}'(\pi(y))$ is tree-blocked by $\psi(\mathrm{Tail}'(\pi(y)))$. If $\mathrm{Tail}'(\pi(y))$ is a $R_{\geq, n}$-successor of $Tail(\pi(x))$, then $\mu(y) = \psi(\mathrm{Tail}'(\pi(y)))$ is a $R_{\geq, n}$-successor of $\mu(x) = \psi(\mathrm{Tail}'(\pi(x)))$.
(c) If $\pi(x) \notin AfterBlocked(G_i)$ and $\pi(y) \in AfterBlocked(G_i)$, then $\mathrm{Tail}(\pi(x)) \neq$ - $\mathrm{Tail}'(\pi(x))$ and $\mathrm{Tail}(\pi(x)) = \psi(\mathrm{Tail}'(\pi(x)))$. If $\mathrm{Tail}'(\pi(y))$ is a $R_{\geq, n}$-successor of $\mathrm{Tail}(\pi(x))$, then $\mu(y) = \mathrm{Tail}`(\pi(y))$ is a $R_{\geq, n}$-successor of $\mu(x) = \psi(\mathrm{Tail}'(\pi(x)))$.

The proof for case (2) is in a similar way with case (1).

Since μ has the Property 1–3, $q \rightarrowtail \mathcal{F}$ holds.

For example, since $Vr(G_\pi) = \varnothing$ and G_π is connected, the only connected component of G_π is itself. We have $Blocked(G_\pi) = \{p_6\}$, and $AfterBlocked(G_\pi) = \{p_7, p_8, v_5\}$. We obtain the mapping μ_1 from π by setting $\mu(x) = \psi(\text{Tail}'(p_6)) = o_3$, $\mu(y) = \psi(\text{Tail}'(p_7)) = o_4$, $\mu(z) = \psi(\text{Tail}'(p_8)) = o_5$, and $\mu_1(y_c) = v_5$. By Definition 7.25, $q_1 \rightarrowtail \mathcal{F}_1$.

Theorem 7.4 *Let $k \geq n_q$. Then $\mathcal{K} \vDash q$ iff for each $\mathcal{F} \in ccf_k(\mathbb{F}_\mathcal{K})$, it holds that $q \rightarrowtail \mathcal{F}$.*

Proof The if direction is easy. By Lemma 7.9, for each $\mathcal{F} \in \text{ccf}_k(\mathbb{F}_\mathcal{K})$, $\mathcal{F} \vDash q$. Then, by Theorem 7.3, $\mathcal{K} \vDash q$. For the converse side, by Theorem 7.3, $\mathcal{F} \vDash q$ for each $\mathcal{F} \in \text{ccf}_k(\mathbb{F}_\mathcal{K})$. By Lemma 7.12, $q \rightarrowtail \mathcal{F}$. □

We can, from the only if direction of Theorem 7.4, establish our key result, which reduce query entailment $\mathcal{K} \vDash q$ to finding a mapping of q into every $\mathcal{F}$ in $\text{ccf}_k(\mathbb{F}_\mathcal{K})$.

In the following, we discuss the termination and complexity.

For the standard reasoning tasks, e.g., knowledge base consistency, the combined complexity is measured in the size of the input knowledge base. For query entailment, the size of the query is additionally taken into account. The size of a knowledge base $\mathcal{K}$ or a query q is simply the number of symbols needed to write it over the alphabet of constructors, concept, role, individual, and variable names that occur in $\mathcal{K}$ or q, where numbers are encoded in binary. As for data complexity, we consider the ABox as the only input for the algorithm, i.e., the size of the TBox, the role hierarchy, and the query is fixed. When the size of the TBox, role hierarchy, and query is small compared to the size of the ABox, the data complexity of a reasoning problem is a more useful performance estimate since it tells us how the algorithm behaves when the number of assertions in the ABox increases.

Let $\mathcal{K} = \langle \mathcal{T}, \mathcal{R}, \mathcal{A} \rangle$ an *f-SHIF(D)* KB and q a fuzzy conjunctive query, we denote by $||\mathcal{K}, q||$ the string length of encoding $\mathcal{K}$ and q, $|\mathcal{A}|$ the sum of the numbers of fuzzy assertions and inequality assertions, $|\mathcal{R}|$ the number of role inclusion axioms, $|\mathcal{T}|$ the number of fuzzy GCIs, $\mathbf{c}$ the $sub(\mathcal{K}) \cup sub(C_q)$, where C_q is the set of concepts occurring in q, $\mathbf{r}$ the cardinality of $\mathbf{R}_\mathcal{K}$, $\mathbf{d} = |N^\mathcal{A}| + |N^q|$. Note that, when the size of q, TBox $\mathcal{T}$, and RBox $\mathcal{R}$ is fixed, $\mathbf{c}$, $\mathbf{d}$, and $\mathbf{r}$ is linear in the size of $||\mathcal{K}, q||$, and is constant in $\mathcal{A}$.

Lemma 7.13 *In a completion forest of $\mathcal{K}$, the maximal number of non-isomorphic k-trees is $T_k = 2^{p(\mathbf{c},\mathbf{d},\mathbf{r})^{k+1}}$, where $p(\mathbf{c}, \mathbf{d}, \mathbf{r})$ is some polynomial w.r.t. $\mathbf{c}$, $\mathbf{d}$, and $\mathbf{r}$.*

Proof Since $\mathcal{L}(x) \subseteq (sub(\mathcal{K}) \cup sub(C_q)) \times \{\geq\} \times (N^\mathcal{A} \cup N^q)$, there are at most $2^\mathbf{c}\mathbf{d}$ node labels in a k-complete clash free completion forest $\mathcal{F}$ of $\mathcal{K}$. Each successor of such a node v may be a root node of a $(k-1)$-tree. If a node label of a k-tree contains some tripes of the form $\langle \exists R.C, \geq, n \rangle$ or $\langle \geq 2S, \geq, n \rangle$, the $\exists_\geq$-rule or $\geq_\geq$-rule is triggered and new nodes are added. The generating rules can be applied to each node at most $\mathbf{c}$ times. Each time it is applied, it generates at most two R-successors (if the label of the node contains $\langle \geq 2S, \geq, n\rangle$) for each role R. This gives a bound of $2c$ R-successors for each role, and a bound of $2cr$ successors for each node. Since $\mathcal{L}(\langle x, y\rangle) \subseteq \mathbf{R}_{\mathcal{K}} \times \{\geq\} \times (N^{\mathcal{A}} \cup N^{q})$, there are at most $2^{\mathbf{r}}\mathbf{d}$ different edge labels, and thus can link a node to one of its successor in $2^{\mathbf{r}}\mathbf{d}$ ways. Hence, there can be at most $2^{\mathbf{r}}\mathbf{d}$ combinations from a single node in a completion forest to its successors. Thus, the upper bound of the number of non-isomorphic k-trees is $T_k = 2^{\mathbf{c}}\mathbf{d}(2^{\mathbf{r}}\mathbf{d}T_{k-1})^{2\mathbf{cr}}$. Let $y = 2\mathbf{cr}$, $x = \mathbf{c} + \mathbf{r}y$, then

$$\begin{aligned} T_k &= 2^x\mathbf{d}^{1+y}(T_{k-1})^y = 2^x\mathbf{d}^{1+y}(2^x\mathbf{d}^{1+y}(T_{k-2})^y)^y = \ldots \\ &= (2^x\mathbf{d}^{1+y})^{1+y+\ldots+y^{k-1}}(T_0)^{y^k} \leq (2^x\mathbf{d}^{1+y}T_0)^{y^k}. \end{aligned}$$

Since $T_0 = 2^{\mathbf{c}}\mathbf{d}$, we have for $y \geq 2$ (It also holds for $y = 1$) that

$$\begin{aligned} T_k &\leq (2^{\mathbf{c}+2\mathbf{cr}\,\mathbf{2}}\,\mathbf{d}^{1+2\mathbf{cr}}\,2^{\mathbf{c}})^{(2\mathbf{cr})^k} \leq (2^{2\mathbf{c}+2\mathbf{cr}\,\mathbf{2}+(\mathbf{1}+2\mathbf{cr})\mathbf{d}})^{(2\mathbf{cr})^k} \\ &\leq (2^{2\mathbf{c}\,\mathbf{2}+4\mathbf{c}\,2\mathbf{r}\,\mathbf{3}+4\mathbf{c}\,2\mathbf{r}\,2\mathbf{d}})^{k+1} = 2^{p(\mathbf{c},\mathbf{d},\mathbf{r})^{k+1}} \end{aligned}$$

where $p(\mathbf{c}, \mathbf{d}, \mathbf{r}) = 2\mathbf{c}\,\mathbf{2} + \mathbf{4c}\,\mathbf{2r}\,\mathbf{3} + \mathbf{4c}\,\mathbf{2r}\,\mathbf{2d}$.

Lemma 7.14 *The upper bound of the number of nodes in $\mathcal{F}$ is $O\,(\mathbf{I}_{\mathcal{K}}((\mathbf{cr})^{d+1}))$, where $d = (T_k + 1)k$.*

Proof By Lemma 7.13, there are at most T_k non-isomorphic k-trees. If there exists a path from v' to v with its length greater than $(T_k + 1)k$, then v would appear after a sequence of $T_k + 1$ non-overlapping k-trees, and one of them would have been blocked so that v would not be generated. The number of nodes in a k-tree is bounded by $(\mathbf{cr})^{d+1}$. The number of nodes in a $\mathcal{F}$ is bounded by $|\mathbf{I}_{\mathcal{K}}|(2\mathbf{cr})^{d+1}$. □

Corollary 7.1. *If k is linear in the size of $||\mathcal{K}, q||$, then the number of nodes in $\mathcal{F}$ is at most triple exponential in the size of $||\mathcal{K}, q||$; if the size of q, TBox $\mathcal{T}$, and RBox $\mathcal{R}$ is fixed, and k is a constant, then the number of nodes in $\mathcal{F}$ is polynomial in the size of $\mathcal{A}$.*

Theorem 7.5 (Termination) *The expansion of $\mathbb{F}_{\mathcal{K}}$ into a complete and clash free completion forest $\mathcal{F} \in_{\mathrm{ccf}_k} (\mathbb{F}_{\mathcal{K}})$ terminates in triple exponential time w.r.t. $||\mathcal{K}, q||$, and in polynomial time w.r.t. $|\mathcal{A}|$ if the size of q, TBox $\mathcal{T}$, and RBox $\mathcal{R}$ is fixed, and k is a constant.*

Proof By Lemma 7.14, the number of nodes in $\mathcal{F}$ is at most $|\mathbf{I}_{\mathcal{K}}|(2\mathbf{cr})^{d+1}$, written as M. We now make a case study of the numbers of the application of rules for expanding $\mathcal{F}_{\mathcal{K}}$ into $\mathcal{F}$. For each node, the $\sqcap$-rule and the $\sqcup$-rule may be used at most $\mathrm{O}(\mathbf{c})$ times, and the $\exists_{\geq}$, $\forall_{\geq}$, $\forall_{+}$, $\forall_{+'}$ and $\geq_{\geq}$-rules may be used at most $O(cr)$ times. The number of the application of these rules is at most $O(M\mathbf{cr})$ times. The number of the application of $\leq_{r\geq}$-rule is at most $O(|\mathbf{I}_{\mathcal{K}}|)$ times. For each node, The number of the application of $\sqsubseteq$-rule is at most $O(\mathbf{d}|\mathcal{T}|)$ times, and is at most $O(M\mathbf{d}|\mathcal{T}|)$ times for M nodes. To sum up, the total number for the application of rules is at most $O(Mcr + M\mathbf{d}|\mathcal{T}|)$ times. □

Theorem 7.6 *Let $\mathcal{K}$ be an f-SHIF(D) KB and q a fuzzy conjunctive query in which all the roles are simple, deciding whether $\mathcal{K} \models q$ is in co3NexpTime w.r.t. combined complexity, and is in coNP w.r.t. data complexity.*

Proof If $\mathcal{K} \not\models q$, there must exists a $\mathcal{F} \in_{ccf} (\mathbb{F}_{\mathcal{K}})$, such that $q \rightarrowtail \mathcal{F}$ does not hold. Due to the existence of the nondeterministic rules, and by Theorem 7.5, the construction of $\mathcal{F}$ can be done in nondeterministic tripe exponential time in the size of $||\mathcal{K}, q||$. By Corollary 7.1, the number of the nodes in $\mathcal{F}$ is at most triple exponential in the size of $||\mathcal{K}, q||$. For a fuzzy CQ q with k variables, checking $\mathcal{F}$ for a mapping takes M^k times, which is tripe exponential in the size of $||\mathcal{K}, q||$. Hence, deciding $\mathcal{K} \not\models q$ is in *3NexpTime*, and deciding $\mathcal{K} \models q$ is in *co3NexpTime*. Similarly, the data complexity deciding $\mathcal{K} \models q$ is in *coNP*. □

7.3.2 Fuzzy SPARQL Query Language

Besides the techniques of querying answering over fuzzy ontologies as introduced in Sect. 7.3.1, the well-known query language SPARQL also has been attracting significant attention in the context of the Semantic Web. The Semantic Web needs data models to support unified access to Web information source, Web service and intelligent applications. RDF, as a data model representing network resources object and the relations between them, defines the underlying model that can be exchanged between applications without loss of information semantic. The main RDF query languages adopted in Semantic Web consist of RDQL, RQL, SeRQL, N3QL, Triple, and SPARQL (Prud'hommeaux and Seaborne 2008). SPARQL is the W3C standard query language for RDF. Until now, however, standard SPARQL processes information only in a crisp way. In real-world applications, there exists uncertainty and vagueness in user querying intention. Therefore, SPARQL does not allow users to form queries with preferences or vagueness, which could be desirable for the following reasons (Kraft and Petry 1997): (i) to express soft query conditions; (ii) to control the size of the answers; (iii) to produce a discriminated answer.

For example. An advertisement company requires 30 models which are *close to* 175 cm and *not very young and not very old.*

Apparently, SPARQL can not efficiently express and answer such a request. Most of the existing query systems on RDF can not process queries with fuzzy information directly and thus can not adequately meet users' querying needs. Consequently, it becomes an important subject of Semantic Web to do research about fuzzy query on RDF. To address this problem, Cheng et al. (2010a, b) proposed an flexible extension of SPARQL by introducing fuzzy set theory and the

α-cut operation of fuzzy numbers into SPARQL. They showed how to efficiently compute the top-k answers of fuzzy queries with regard to membership degrees and user-defined weights. Also, a flexible query service was implemented and evaluated on the LUBM platform. Results of a preliminary user study demonstrated that the method for construction and translation of flexible queries and results ranking can capture the users's intension effectively. Moreover, De Maio et al. (2012) introduced an approach to context analysis and recognition that relies on f-SPARQL tool, that is a flexible extension of SPARQL. A JAVA implementation of f-SPARQL and the integrated support for fuzzy clustering and classification were discussed. This tool was exploited in the architecture that foresees some task oriented agents in order to achieve context analysis and recognition in order to identify critical situations. Finally, a simple application scenario and preliminary experimental results have been described. In addition, Wang et al. (2012) put forward a new fuzzy retrieval mechanism supporting user preference–fp-SPARQL, where Zadeh's type-II fuzzy set theory, the concepts of α-cut set, and linguistic variable were adopted to put forward the RDF fuzzy retrieval mechanism supporting user preference, which extended SPARQL language for further realizing fuzzy and preferred expression.

In the following we will introduce the fuzzy SPARQL query language (Cheng et al. 2010b). As in the case with databases (Kraft and Petry 1997; Ma and Yan 2007), the extension mainly takes place in the FILTER constraint part. For each such a flexible query, we declare with #FQ# before a SELECT query indicating that this is the case. In the FILTER part, the original expression of the form FILTER (?X op Y) is extended by allowing fuzzy terms and fuzzy operators.

7.3.2.1 Fuzzy Terms Extension

The basic form of FILTER constraint allowing fuzzy terms is FILTER (?X θ FT) [WITH α], where FT denotes a fuzzy term, e.g. "*tall*" or "*young*", which corresponds to a fuzzy number, and θ is one of the operators including >, <, =, >=, <=, !=, *between*, and *not between*. Note that if the operator θ is *between* or *not between*, then the syntax of FILTER constraint is a little different, i.e., FILTER (?X between/not between FT1 and FT2) [WITH α], where FT1 and FT2 are fuzzy numbers. The optional parameter [WITH α] indicates the condition that must be satisfied as the minimum membership degree threshold in [0,1]. Users choose an appropriate value of α to express his/her requirement. If not specified, then use 1 as default. There are three kinds of fuzzy terms: *simple (atomic) fuzzy term, modified (composite) fuzzy term,* and *compound fuzzy term.*

Simple fuzzy terms. A simple fuzzy term such as "*young*" or "*tall*" is defined in terms of a fuzzy number with a membership function.

Modified fuzzy terms. A modified fuzzy term, e.g. "*very young*" or "*more or less tall*", is described by a simple fuzzy term and a fuzzy modifier. Let A be a simple fuzzy term represented by a fuzzy number in the universe of discourse U and its membership function is, then we have the following rules.

- Concentration rule: $\mu_{veryA}(u) = (\mu_A(u))^2$. More generally,$\mu_{very...very}A(u) = (\mu_A(u))^{2\times(\textit{times of very})}$
- Dilation rule:$\mu_{more\ or\ less}A(u) = (\mu_A(u))^{1/2}$

Compound fuzzy terms. A compound fuzzy term such as "*young* or *very young*" is represented by simple fuzzy terms or modified fuzzy terms connected by *or* (union), *and* (intersection), or *not* (complement) connectors.

7.3.2.2 Fuzzy Operators

Now consider fuzzy relations as operators within FILTER constraint, formalized as FILTER (?X $\tilde{\theta}$ Y) [with α], where $\tilde{\theta}$ denotes the fuzzy operators, Y a string, a integer or any other data types allowed in RDF. The fuzzy operator $\tilde{\theta}$ and the crisp value Y constitute a fuzzy number $\tilde{\theta}$ Y. We will mainly discuss three types of fuzzy relations, which are "*close to (around)*", "*at least*", and "*at most*".

Close to. According to (Chen and Jong 1997), the membership function of the fuzzy number "*close to Y (around Y)*" on the universe of discourse *U* is defined as

$$\mu_{close\ to\ Y}(u) = \frac{1}{1 + \left(\frac{u-Y}{b}\right)^2}. \tag{7.1}$$

It is worth noting that the fuzzy number above is a simple fuzzy term. Based on it, we also have the following modified fuzzy terms: "*very close to Y*", "*very very ... very close to Y*", and "*more or less close to Y*".

At least. The membership function of the fuzzy number "*at least Y*" on the universe of discourse is defined by

$$\mu_{atleast\ Y}(u) = \begin{cases} 0, & if\ u \leq \omega \\ \frac{u-\omega}{Y-\omega}, & if\ \omega < u < Y \\ 1, & if\ u \geq Y \end{cases} \tag{7.2}$$

At most. The membership function of the fuzzy number "*at most Y*" on the universe of discourse is defined by

$$\mu_{atmost\ Y}(u) = \begin{cases} 1, & if\ u \leq Y \\ \frac{\delta-u}{\delta-Y}, & if\ u < Y < \delta \\ 0, & if\ u \geq \delta \end{cases} \tag{7.3}$$

Note that, the parameters of *b*, *ω*, and *δ* in the membership functions of "*close to Y*","*at least Y*", and "*at most Y*" respectively, are chosen according to the concrete value of and vary for different use cases.

7.3.2.3 User-Defined Weights

In the context of SPARQL, a user may have different preferences for different triple patterns, e.g. ?X ex:hasAge ?Age and ?X ex:hasHeight ?Height. In order to provide users a query language with adequate expressibility, we additionally allow users to specify, for each triple pattern, the weight of importance. The query criteria user specified can reflect the users's preferences on the triple. For each such a flexible query, we declare with #top-k FQ# with k before a SELECT query to indicate the query type.

For example. Given the query in the example mentioned above, the advertisement company believe models with similar heights (close to 175 cm) should be of the more important factor in their advertisement, so he/she specify a weight of 0.8 for the triple ?X ex:hasHeight ?Height, and therefore 0.2 for ?X ex:hasAge ?Age. The f-SPARQL query is formed as follows.

```
#top-k FQ# with 30
SELECT ?X ?Age ?Height WHERE {
?X rdf:type Model
?X ex:hasAge ?Age with 0.2.
FILTER (?Age = not very young && ?Age = not very old) with 0.9.
?X ex:hasHeight ?Height with 0.8.
FILTER (?Height close to 175 cm) with 0.8.}
```

7.3.2.4 Syntax of f-SPARQL

Table 7.4 presents the syntax of f-SPARQL. f-SPARQL extends two elements of SPARQL, i.e. the "Query" and the "Constrain". Each SELECT query is extended with the element *QueryType*, which can be further divide into #FQ# (flexible queries) and #top-k FQ# with k (top-k flexible queries). The "Constraint" element is extended with an additional "FlexibleExpression" element, as illustrated in the above subsections.

7.3.2.5 Query Translation

In order to still make use of existing SPARQL implementations for coping with fuzzy queries, a set of translation rules is needed.

Translation of Fuzzy Terms. We make a case study for different types of fuzzy terms.

Case 1 FT is a simple fuzzy term or a modified fuzzy term

Recall the query form FILTER (?X θ FT [WITH α]) mentioned above. Let α be a given threshold, FT_α the α-cut of FT. It is clear that FT_α is an interval, written as $FT_a = [a, b]$. Fuzzy queries are then translated into crisp ones according to the rules given in Table 7.5.

Table 7.4 f-SPARQL modifications to the SPARQL standard grammar

<table>
<tr><td>Query</td><td>::=</td><td>Prologue (QueryType SelectQuery | ConstructQuery | DescribeQuery | AskQuery)</td></tr>
<tr><td>QueryType</td><td>::=</td><td>‘#FQ#’ | #top-k FQ# with k</td></tr>
<tr><td>Constraint</td><td>::=</td><td>BrackettedExpression | BuiltInCall | FunctionCall | FlexibleExpression</td></tr>
<tr><td>FlexibleExpression</td><td>::=</td><td>FuzzyTermExpression | FuzzyOperatorExpression</td></tr>
<tr><td>FuzzyTermExpression</td><td>::=</td><td>(Var [‘=’,‘!=’,‘>’‘>=’,‘<’,‘<=’] FuzzyTerm)? [with threshold]</td></tr>
<tr><td>FuzzyOperatorExpression</td><td>::=</td><td>Var FuzzyOperator NumericLiteral</td></tr>
<tr><td>FuzzyOperator</td><td>::=</td><td>(Modifier)*FuzzyOperator</td></tr>
<tr><td>FuzzyTerm</td><td>::=</td><td>FuzzyTerm and FuzzyTerm | FuzzyTerm or FuzzyTerm | not FuzzyTerm | ModifiedFuzzyTerm</td></tr>
<tr><td>ModifiedFuzzyTerm</td><td>::=</td><td>(Modifier)* ModifiedFuzzyTerm | (Modifier)* SimpleFuzzyTerm</td></tr>
</table>

Table 7.5 Translation rules for fuzzy term FT with$FT_\alpha = [a, b]$

<table>
<tr><th>Flexible queries</th><th>Queries translated</th></tr>
<tr><td>FILTER (?X = FT WITH α)</td><td>FILTER (?X > = a && ?X <= b)</td></tr>
<tr><td>FILTER (?X > FT WITH α)</td><td>FILTER (?X > b)</td></tr>
<tr><td>FILTER (?X > = FT WITH α)</td><td>FILTER (?X > = b)</td></tr>
<tr><td>FILTER (?X < FT WITH α)</td><td>FILTER (?X < a)</td></tr>
<tr><td>FILTER (?X <= FT WITH α)</td><td>FILTER (?X <= a)</td></tr>
<tr><td>FILTER (?X ! = FT WITH α)</td><td>FILTER (?X < a || ?X > b)</td></tr>
</table>

Case 2 FT is a compound fuzzy terms

Firstly, we consider compound fuzzy terms FT of the form “FT1 and FT2”, where FT1 and FT2 are simple or modified fuzzy terms. As is shown in the equation at the end of Sect. 2.4, the α-cut of FT is of the form $[a, b] \cup [c, d]$. Queries of this kind can be translated into crisp SPARQL according to the rules shown in Table 7.6. Then, consider compound fuzzy terms FT of the form “FT1 or FT2”, where FT1 and FT2 are simple or modified fuzzy terms. Since the α-cut of FT is of the form when $A \cap B \neq \varnothing$ (see the equation at the end of Sect. 2.4), queries of this kind can be translated using rules in Table 7.5. For FILTER constraints of the form ?X θ not FT, they can firstly be translated into the form of ?X not θ FT. The not θ is defined as, e.g. not > = <=, and not = = ! = . Compound queries in more complex forms can be translated by iteratively using the aforementioned translation rules.

Translation of Fuzzy Operators. Given FILTER constraint with fuzzy relations as fuzzy operators, i.e., FILTER (?X $\tilde{\theta}$ Y), the *fuzzy* operator along with the crisp value Y is actually a fuzzy term, so the expression can be translated into an equivalent one of the form FILTER (?X = FT), where FT denotes $\tilde{\theta}$ Y.

Table 7.6 Translation rules for fuzzy term FT with $FT_{\alpha} = [a, b] \cup [c, d]$

<table>
<tr><th>Flexible queries</th><th>Queries translated</th></tr>
<tr><td>FILTER (?X = FT WITH α)</td><td>FILTER (?X > = a && ?X <= b || ?X > = c && ?X <= d)</td></tr>
<tr><td>FILTER (?X > FT WITH α)</td><td>FILTER (?X > max(b, d))</td></tr>
<tr><td>FILTER (?X > = FT WITH α)</td><td>FILTER (?X > = max(b, d))</td></tr>
<tr><td>FILTER (?X < FT WITH α)</td><td>FILTER (?X < min(a, c))</td></tr>
<tr><td>FILTER (?X <= FT WITH α)</td><td>FILTER (?X <= min(a, c))</td></tr>
<tr><td>FILTER (?X ! = FT WITH α)</td><td>FILTER (?X < a || ?X > b || ?X < c || ?X > d)</td></tr>
</table>

Table 7.7 Experimental queries

F_1: Find the *not famous* and **busy** teachers
F_2: Find the *famous* and **busy** teachers
F_3: Find the *not famous* and **busy** students
F_4: Find the *famous* and **busy** students

7.3.2.6 Ranking

A naive solution to the top-k retrieval problem is illustrated with the query in the example mentioned at the beginning of Sect. 7.3 as follows: (i) A flexible query is firstly translated into a crisp SPARQL query, then the latter is sent to a SPARQL implementation, e.g. ARQ. (ii) ARQ returns the results for the translated crisp query. (iii) The results are ranked ordered according to a scoring function.

Given an f-SPARQL query Q, the FILTER constraint conditions is $F = (F_1, F_2, \ldots, F_n)$, where F_i denotes the *i*-th constraint condition. We use $A = (A_1, A_2, \ldots, A_n)$ to denote one of the answers for F, then the score of A can be calculated with the following scoring function.

$$Score(A) = \sum_{i=1}^{n} memDegree(A_i) \times weight(F_i).$$

where $memDegree(A_i)$ denotes the membership degree of the answer for F_i, and $weight(F_i)$ denotes the user-defined weight for F_i.

7.3.2.7 Experiment

We generate, from the Lehigh University Benchmark LBUM (Guo et al. 2004), the data set which contains 6,000 k triples. The data is stored in and managed by Mysql 5.0.11. The translation and ranking algorithm are implemented using JAVA, while using Jena ARQ and SDB (which provides for large scale storage and query of RDF datasets using conventional SQL databases) for crisp queries processing. The applications run on a windows XP professional system with P4 3G CPU and with 1G RAM.

Table 7.8 Response time of queries

	F_1	F_2	F_3	F_4
User1	176	202	195	217
User2	199	231	189	272
User3	353	243	215	271
User4	256	277	316	317
User5	376	272	295	257

Table 7.9 Evaluation results

	F_1	F_2	F_3	F_4
User1	166	173	185	147
User2	176	155	195	177
User3	176	200	193	177
User4	156	182	168	198
User5	190	187	186	200

We developed four queries which are shown in Table 7.7, and request the top-100 results. The constraint with *italic* font denotes the first constraint, while the constraint with **bold** font denotes the second constraint.

We use the membership function of "*busy*" and "*famous*" in (Pan et al. 2008).

$$Busy(n)\frac{2}{1+exp(-0.4n)} \qquad Famous(n)\frac{2}{1+exp(-0.1n)}-1$$

where n in *Busy*(n) represents the number of courses taken by a student or taught by a teacher, and n in *Famous*(n) represents the number of papers published by a teacher or a student.

We employ five users to specify for every constraint the threshold and the weight. The response time of queries are listed in the Table 7.8. We invite the five testers to evaluate how closely the results satisfied their initial intentions. We define scores of 2, 1, and 0 for "very satisfying", "satisfying", and "dissatisfying", respectively. Table 7.9 shows the evaluation scores for top-100 results. The preliminary user study demonstrate that our query relaxation and results ranking methods can capture the users's intension effectively.

7.4 Summary

To capture and reason about vague and imprecise domain knowledge, there has been a substantial amount of work carried out in fuzzy Description Logics (DLs) and fuzzy ontologies as introduced in Chap. 4. With the emergence of the large-scale fuzzy DL knowledge bases (KBs) and fuzzy ontologies, it is of particular importance to provide users with expressive querying service.

In this chapter, the issues of querying over fuzzy DL KBs and fuzzy ontologies are introduced. Regarding the fuzzy DL KBs, conjunctive querying over fuzzy DL KBs from tractable to more expressive fuzzy DL KBs is introduced. Regarding the fuzzy ontologies, some efforts have been made for querying over lightweight and expressive fuzzy DL ontologies. Also, some flexible extensions to SPARQL were presented in order to provide users with an easy-to-grasp yet powerful way for querying the RDF data sets. The relevant techniques for querying fuzzy DL KBs and fuzzy ontologies can provide users with querying services to a good number of large-scale fuzzy KBs in the context of the Semantic Web.

References

Baader F, Calvanese D, McGuinness DL, Nardi D, Patel-Schneider PF (2003) The description logic handbook: theory, implementation, and applications. Cambridge University Press, New York

Bandini S, Calegari S, Radaelli P (2006) Towards fuzzy ontology handling vagueness of natural languages. In: Proceedings of the first international conference on rough sets and knowledge technology (RSKT 2006), LNAI 4062, pp 693–700

Bahri A, Bouaziz R, Gargouri F (2009) Fuzzy ontology implementation and query answering on databases. In: Proceedings of the 28th North American fuzzy information processing society annual conference, pp 1–6

Berners-Lee T, Hendler J, Lassila O (2001) The Semantic Web. Sci Am 284(5):34–43

Chen SM, Jong WT (1997) Fuzzy query translation for relational database systems. IEEE Trans Syst Man Cybern B 27(4):714–721

Calvanese D, De Giacomo G, Lenzerini M (1998) On the decidability of query containment under constraints. In: Proceedings of PODS 1998, pp 149–158

Calvanese D, De Giacomo G, Lembo D, Lenzerini M, Rosati R (2005) DL-lite: tractable description logics for ontologies. In: Proceedings of the 20th national conference on artificial intelligence (AAAI 2005), 602–607

Calvanese D, De Giacomo G, Lembo D, Lenzerini M, Rosati R (2007) Tractable reasoning and efficient query answering in description logics: the dl-lite family. J Autom Reason 39(3):385–429

Cali A, Lembo D, Rosati R (2003) Query rewriting and answering under constraints in data integration systems. In: Proceedings of the 18th international joint conference on artificial intelligence, pp 16–21

Cheng JW, Ma ZM, Yan L, Wang HL (2008a) Querying over fuzzy description logic. Wuhan Univ J Nat Sci 13(4):429–434

Cheng JW, Ma ZM, Zhang F, Wang X (2008b) Conjunctive query answering over an f-alc knowledge base. In: Proceedings of web intelligence/iat workshops, pp 279–282

Cheng JW, Ma ZM, Zhang F, Wang X (2009a) Deciding query entailment in fuzzy description logic knowledge bases. In: Proceedings of the 20th international conference on database and expert systems applications, pp 830–837

Cheng JW, Ma ZM, Zhang F, Wang X (2009b) Deciding query entailment for fuzzy SHIN ontologies. In: Proceedings of the 4th annual Asian semantic web conference (ASWC 2009), pp 120–134

Cheng JW, Ma ZM, Wang Y (2010a) Query answering in fuzzy description logics with data type support. In: Proceedings of web intelligence, pp 240–247
Cheng JW, Ma ZM, Yan L (2010b) f-SPARQL: a flexible extension of SPARQL. In: Proceedings of DEXA 2010, Part I. LNCS, vol 6261. pp 487–494
Cheng JW, Ma ZM, Yan L (2011) Deciding query entailment in fuzzy OWL lite ontologies. In: Yan L, Ma ZM (eds) Advanced database query systems: techniques, applications and technologies. IGI Global, Hershey
De Maio C, Fenza G, Furno D, Loia V (2012) f-SPARQL extension and application to support context recognition. In: Proceedings of 2012 IEEE international conference on fuzzy systems (FUZZ-IEEE), pp 1–8
Guo Y, Pan Z, Heflin J (2004) An evaluation of knowledge base systems for large owl datasets. In: Proceedings of international semantic web conference, pp 274–288
Glimm B, Horrocks I, Lutz C, Sattler U (2007) Conjunctive query answering for thedescription logic shiq. In: Proceedings of international joint conferences on artificial intelligence, pp 399–404
Horrocks I, Tessaris S (2002) Querying the semantic web: a formal approach. In: Proceedings of international semantic web conference, volume 2342 of Lecture notes in computer science, pp 177–191
Kraft DH, Petry FE (1997) Fuzzy information systems: managing uncertainty in databases and information retrieval systems. Fuzzy Sets Syst 90(2):183–191
Levy AY, Rousset MC (1998) Combining horn rules and description logics in CARIN. Artif Intell 104(1–2):165–209
Li C, Chang K, Ilyas IF, Song S (2005) RankSQL: query algebra and optimization for relational top-k queries. In: Proceedings of SIGMOD conference, pp 131–142
Li Y, Xu B, Lu J, Kang D (2006) Discrete tableau algorithms for shi. In: Proceedings of description logics 2006
Lukasiewicz T, Straccia U (2007) Top-k retrieval in description logic programs under vagueness for the semantic web. In: Proceedings of SUM 2007, LNCS (LNAI), vol 4772, pp 16–30
Mailis TP, Stoilos G, Stamou GB (2007) Expressive reasoning with horn rules andfuzzy description logics. In: Proceedings of RR 2007, LNCS, vol 4524, pp 43–57
Ma ZM, Yan L (2007) Generalization of strategies for fuzzy query translation in classical relational databases. Inf Softw Technol 49(2):172–180
Pan JZ, Thomas E, Sleeman D (2006) ONTOSEARCH2: searching and querying web ontologies. In: Proceedings of WWW/internet 2006, pp 211–218
Pan JZ, Stamou G, Stoilos G, Thomas E (2007) Expressive querying over fuzzy DL-lite ontologies. In: Proceedings of description logics 2007
Pan JZ, Stamou G, Stoilos G, Taylor S, Thomas E (2008) Scalable querying service over fuzzy ontologies. In: Proceedings of the international world wide web conference (WWW 08), pp 575–584
Prud'hommeaux E, Seaborne A (2008). Sparql query language for rdf. In: W3C Recommendation
Rosati R (2007) On conjunctive query answering in EL. In: Proceedings of description logics 2007
Stoilos G, Straccia U, Stamou GB, Pan JZ (2006) General concept inclusions in fuzzy description logics. In: Proceedings of the 17th international conference on artificial intelligence (ECAI 2006), pp 457–461
Stoilos G, Stamou G, Pan J, Tzouvaras V, Horrocks I (2007) Reasoning with very expressive fuzzy description logics. J Artif Intell Res 30(8):273–320
Straccia U (2001) Reasoning within fuzzy description logics. J Artif Intell Res 14:137–166
Straccia U (2005) Description logics with fuzzy concrete domains. In: Proceedings of UAI-2005, pp 559–567
Straccia U (2006) Answering vague queries in fuzzy dl-lite. In: Proceedings of the 11th international conference on information processing and management of uncertainty in knowledge-based systems (IPMU 2006), pp 2238–2245

Widyantoro DH, Yen J (2001a) A fuzzy ontology-based abstract search engine and its user studies. In: Proceedings of the 10th IEEE international conference on fuzzy systems, pp 1291–1294

Widyantoro DH, Yen J (2001b) Using fuzzy ontology for query refinement in a personalized abstract search engine. In: Proceedings of joint nineth IFSA world congress and 20th NAFIPS international conference.Vancouver, Canada, pp 610–615

Wang HL, Ma ZM (2008) A decidable fuzzy description logic F-ALC(G). In: Proceedings of the 19th international conference on database and expert systems applications, pp 116–123

Wang HR, Ma ZM, Cheng JW (2012) fp-Sparql: an RDF fuzzy retrieval mechanism supporting user preference. In: Proceedings of the 9th international conference on fuzzy systems and knowledge discovery, pp 443–447

Chapter 8
Fuzzy Ontology Knowledge Bases Storage in Fuzzy Databases

8.1 Introduction

The Semantic Web aims at extending the current Web so that the content of web pages is described with rich semantics and will not exclusively be meaningful for humans, but also be machine processable (Berners-Lee et al. 2001). As has mentioned in Chap. 1, Description Logics (DLs) and ontologies, as meaning providers, which enable a shared, explicit and formal description of the domain knowledge and have efficient reasoning ability, have been recognized to play an important role in the Semantic Web and many application domains. Currently, the Semantic Web is increasingly gaining popularity and importance with an increasing number of people using knowledge bases (Description Logic or ontology knowledge bases) to represent their information. Lots of knowledge bases have been created and real knowledge bases tend to become very large to huge (millions of items). Therefore, one problem is considered that has arisen from practical needs: namely, how to store these knowledge bases.

In general, there are several possible approaches to store the knowledge bases (Astrova et al. 2007; Gali et al. 2004). One is to use file systems for storing the knowledge bases, and the main problem with this technique is that file systems do not provide scalability or any query facility. Another is to store the knowledge bases in well-known relational, object or object-relational databases. In particular, among these kinds of databases, relational databases, which can provide maturity, reliability and availability, may be the widest model used in the world for storing the information of domains. On this basis, many proposals have been developed to store the knowledge bases in relational databases. An algorithm was developed to map OWL ontologies to relational schemas in (Gali et al. 2004). In (Astrova et al. 2007), the authors studied how to store ontologies in SQL relational databases. In (Vysniauskas and Nemuraite 2006), the authors analyzed the process how ontology of a particular domain named product configuration system may be stored in a relational database. In (Al-Jadir et al. 2010; Das et al. 2004), ontologies were stored in relational databases for reasoning and querying. A survey of current

Z. Ma et al., *Fuzzy Knowledge Management for the Semantic Web*,
Studies in Fuzziness and Soft Computing 306, DOI: 10.1007/978-3-642-39283-2_8,

implementations and future directions about ontology and database mapping was made in (Konstantinou et al. 2008).

The classical knowledge bases and the relevant storage techniques may not be sufficient to represent and store fuzzy information that is commonly found in many application domains. In particular, as have introduced in Chap. 4 in detail, there are lots of fuzzy Description Logic/ontology knowledge bases which have been created. Therefore, the efficient storage of fuzzy knowledge bases is of paramount importance. Being similar to the most common proposals that use relational databases to store classical knowledge bases, the fuzzy relational database may be also a good candidate for storing fuzzy knowledge bases because of the widespread investigation of the fuzzy relational databases as mentioned in Chap. 3. If fuzzy knowledge bases can be stored in the fuzzy relational database, the fuzzy relational database may solve the scalability issue raised by real fuzzy knowledge bases and also may facilitate for retrieving and manipulating the information of fuzzy knowledge bases within the existing fuzzy relational database techniques. To this end, currently, there are several proposals for storing fuzzy ontology knowledge bases in fuzzy databases. Barranco et al. (2007) presented a method to store a part of constructors of fuzzy ontology knowledge base in fuzzy object relation database. The work used fuzzy object relation databases to store fuzzy ontologies and presented a brief description of the framework. The authors also introduced several fuzzy datatypes in ontologies. Lv et al. (2009) developed a brief method to store fuzzy ontology knowledge base in fuzzy object relation database. The work only made a brief discussion about fuzzy ontology storage and provided three brief algorithms. The authors did not store many important constructors of fuzzy OWL in databases (including fuzzy class axioms, fuzzy property axioms, fuzzy individual axioms, and property restrictions such as someValuesFrom, minCardinality, Symmetric, Functional, etc.), and also did not consider many fuzzy datatypes in fuzzy OWL. More recently, Zhang et al. (2011) proposed an approach for storing fuzzy ontologies in fuzzy relational databases. The elements in fuzzy ontologies such as fuzzy classes, fuzzy properties, characters and restrictions of properties, data instances, and fuzzy axioms were considered, and an example was further provided throughout the paper to well explain the approach. In this chapter, we briefly discuss how to store fuzzy knowledge base in fuzzy relational database in (Zhang et al. 2011), give a storage mechanism and provide a storage example.

8.2 Fuzzy Ontology Knowledge Bases Storage in Fuzzy Databases

In (Zhang et al. 2011), we propose an approach for storing fuzzy ontologies in fuzzy relational databases. The elements of fuzzy ontologies are introduced first, where most of constructors of fuzzy ontologies are considered. On this basis, we propose an approach for storing all of these elements of fuzzy ontologies in fuzzy relational databases, and an example is provided throughout the paper to well explain the approach. Note that, the fuzzy ontology is almost equivalent to the fuzzy Description Logic *f-SHOIN*(*D*)) knowledge base as mentioned in Chap. 4, and thus the storage approach can store not only fuzzy ontologies but also fuzzy Description Logic knowledge bases.

Figure 8.1 shows a fuzzy knowledge base modeling parts of the reality at a university, which includes the structure information and the instance information. Therefore, to store a fuzzy knowledge base into fuzzy relational database, two steps need to be done: storing the structure information of the fuzzy knowledge base into fuzzy relational schema; and storing the instance information of the fuzzy knowledge base into fuzzy relation w.r.t. the fuzzy relational schema.

From Fig. 8.1, it is shown that the fuzzy knowledge base contains the following elements:

- six classes:

  {Department, Staff, AdminStaff, AcademicStaff, Student, Course};
- four object properties:

  {study_in, choosecourse, work_in, teach};
- four datatype properties:

  {staffname, title, email, age}, and the domains of these properties are as follows: {xsd:string, xsd:string, fuzzy:possibilitydistribution, fuzzy:label};

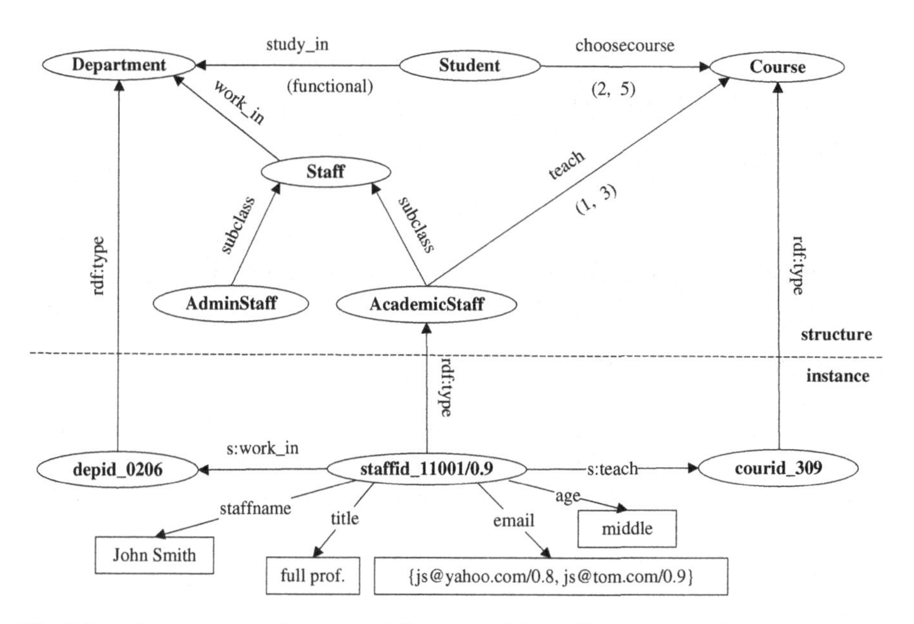

Fig. 8.1 A fuzzy ontology *Onto_1* modeling parts of the reality at a university

- three individuals:
  {staffid_110001, depid_0206, courid_309};
- some axioms:
  {subClassOf (AdminStaff, Staff);
  subClassOf (AcademicStaff, Staff);
  Class (Staff partial restriction (work_in allValuesFrom (Department)));
  ObjectProperty (work_in domain (Staff) range (Department));
  Class (Student partial restriction (study_in allValuesFrom (Department) cardinality (1)) restriction (choosecourse allValuesFrom (Course) minCardinality (2) maxCardinality (5)));
  ObjectProperty (study_in domain (Student) range (Department) [Functional]);
  ObjectProperty (choosecourse domain (Student) range (Course));
  Class (AcademicStaff partial restriction (teach allValuesFrom (Course) minCardinality (1) maxCardinality (3)));
  ObjectProperty (teach domain (AcademicStaff) range (Course));
  Individual (depid_0206 type(Department));
  Individual (courid_309 type(Course));
  Individual (staffid_110001 type(AcademicStaff) value(work_in, depid_0206) value(staffname, *John Smith*) value(title, *full prof.*) value(email, *js@yahoo.com*) [0.8] value(email, *js@tom.com*) [0.9] value(age, *middle*) value(teach, courid_309));}.

The following procedures will store the above fuzzy knowledge base information in the target fuzzy relational database.

8.2.1 Storing Fuzzy Knowledge Base Structure Information in Fuzzy Databases

In the following, we introduce how to store the structure information of the fuzzy knowledge base in Fig. 8.1 into the fuzzy relational database. For the sake of simplicity, some constraints (e.g., primary key, foreign key, check, and unique constraints) are not shown in tables of the target fuzzy relational database.

8.2.1.1 Storing the Resources

Table 8.1 shows Resource_Table in the target fuzzy relational database, which stores all the resources in Fig. 8.1, where: *OntologyName* is the fuzzy ontology name "Onto_1"; *ID* uniquely identifies a resource; *namespace* and *localname* describe the URIref of a resource; and *type* describes the type of a resource. The attribute "*ID*" is the primary key of Table 8.1.

Table 8.1 Resource_Table

OntologyName	ID	Namespace	Localname	Type
Onto_1	c_1	http://www...	Department	Class
Onto_1	c_2	http://www...	Staff	Class
Onto_1	c_3	http://www...	AdminStaff	Class
Onto_1	c_4	http://www...	AcademicStaff	Class
Onto_1	c_5	http://www...	Student	Class
Onto_1	c_6	http://www...	Course	Class
Onto_1	p_1	http://www...	study_in	Property
Onto_1	p_2	http://www...	work_in	Property
Onto_1	p_3	http://www...	teach	Property
Onto_1	p_4	http://www...	choosecourse	Property
Onto_1	p_5	http://www...	staffname	Property
Onto_1	p_6	http://www...	title	Property
Onto_1	p_7	http://www...	email	Property
Onto_1	p_8	http://www...	age	Property
Onto_1	i_1	http://www...	staffid_11001	Individual
Onto_1	i_2	http://www...	depid_0206	Individual
Onto_1	i_3	http://www...	courid_309	Individual

Table 8.2 Class_Table

$Class_1ID$	$Class_2ID$	Relationship	u
c_3	c_2	SubClassOf	1.0
c_4	c_2	SubClassOf	1.0

8.2.1.2 Storing the Relationships Between Classes

Table 8.2 shows Class_Table, which stores the subclass/superclass fuzzy inheritance hierarchies (or unionOf, intersectionOf, complementOf, and so on as mentioned in Chap. 4) relationships between classes. Here, u denotes the membership degree of a class belonging to the subclass of another class. The attributes "($Class_1ID$, $Class_2ID$)" is the primary key of Table 8.2.

For example, subClassOf (AdminStaff, Staff) is stored as the first tuple in Table 8.2.

8.2.1.3 Storing the Domains and Ranges of Properties

Table 8.3 shows Property_Field_Table, storing domains and ranges of properties. The attribute "*ProID*" is the primary key of Table 8.3.

For example, for the object property study_in (*p_1*), its domain is Student (*c_5*) and range is Department (*c_1*); for the datatype property staffname (*p_5*), its domain is AcademicStaff (*c_4*) and range is xsd:string.

Table 8.3 Property_Field_Table

ProID	Domain	Range
p_1	c_5	c_1
p_2	c_2	c_1
p_3	c_4	c_6
p_4	c_5	c_6
p_5	c_4	xsd:string
p_6	c_4	xsd:string
p_7	c_4	fuzzy: possdis
p_8	c_4	fuzzy: lable

Table 8.4 Property_Character_Table

ProID	Type	Character
p_1	object	Functional

Table 8.5 Property_Restriction_Table

ClassID	ProID	Type	Value
c_4	p_3	mincardinality	1
c_4	p_3	maxcardinality	3
c_5	p_4	mincardinality	2
c_5	p_4	maxcardinality	5

8.2.1.4 Storing the Characters and Restrictions of Properties

Tables 8.4 and 8.5 show Property_Character_Table and Property_Restriction_Table, which store the characters and restrictions of object and datatype properties. The attribute "*ProID*" is the primary key of Table 8.4, and the attributes "(*ClassID, ProID, type*)" is the primary key of Table 8.5.

For example, the functional object property study_in (*p_1*) can be stored in a tuple in Table 8.4; the cardinality constraints (2, 5) denotes that each student (*c_5*) can choose at least 2 and at most 5 courses, which can be stored in Table 8.5.

Since the fuzzy ontology in Fig. 8.1 does not contain all of the constructors in a fuzzy DL/ontology knowledge base (e.g., some class operations and property characters as introduced in Chap. 4), their corresponding tables in the target fuzzy relational database are not shown here, and the other constructors in a fuzzy DL/ontology knowledge base can be stored following the similar procedures above.

8.2.2 *Storing Fuzzy Knowledge Base Instance Information in Fuzzy Databases*

The following several procedures will store the fuzzy knowledge base instance information. From Fig. 8.1, there are three individuals in the fuzzy ontology as mentioned in Resource_Table (Table 8.1): staffid_110001 (*i_1*), depid_0206 (*i_2*), courid_309 (*i_3*).

8.2.2.1 Storing the Relationships of Individuals/Classes

Table 8.6 shows Individual_Class_Relation_Table in the target fuzzy relational database, which stores the relationships of individuals/classes in Fig. 8.1. Here, *u* denotes the membership degree of an individual to the class. The attribute "(*IndID, ClassID*)" is the primary key of Table 8.6.

For example, staffid_110001/0.9 in Fig. 8.1 denotes that the individual *i_1* is an instance of class AcademicStaff (*c_4*) with degree 0.9, which is stored in Table 8.6.

8.2.2.2 Storing the Values of Crisp Properties

Table 8.7 shows Individual_Crisp_Property_Table, which stores the values of crisp properties of individuals. The attribute "(*IndID, ProID*)" is the primary key of Table 8.7.

For example, the values of properties work_in (*p_2*), teach (*p_3*), staffname (*p_5*), and title (*p_6*) of the individual staffid_110001 (*i_1*), which are crisp, are stored in Table 8.7.

Table 8.6 Individual_Class_Relation_Table

IndID	ClassID	u
i_1	c_1	0.9
i_2	c_4	1.0
i_3	c_6	1.0

Table 8.7 Individual_Crisp_Property_Table

IndID	ProID	Value
i_1	p_2	i_2
i_1	p_3	i_3
i_1	p_4	John Smith
i_1	p_6	full prof

8.2.2.3 Storing the Values of Fuzzy Properties

From Fig. 8.1, there are fuzzy datatypes, such as the label (*middle*) and the possibility distribution {*js@ya…*}, where the linguistic label *middle* can be represented by a function *T*[30, 40, 50, 60] defined on a numerical domain (Oliboni and Pozzani 2008).

The following several sub-procedures introduce how to store these fuzzy property values.

Storing the fuzzy datatypes

Table 8.8 shows Fuzzy_Types_Table, storing the names of the fuzzy datatypes in a fuzzy knowledge base. The attribute "*fuzzytypeID*" is the primary key of Table 8.8.

For example, the fuzzy datatypes the label (*f_1*) and possibility distribution (*f_2*) are stored in Table 8.8.

Storing individuals with fuzzy properties and these fuzzy properties

Table 8.9 shows Individual_Fuzzy_Property_Value_Table, storing all those individuals with fuzzy properties and these fuzzy properties, where *FuzzyPropertyValueID* identifies the value of the fuzzy property *ProID* and it references the primary key attribute in Fuzzy_Property_Value_Table in the later Table 8.10. The attribute "(*IndID, ProID*)" is the primary key of Table 8.9.

For example, the property age (*p_8*) of the individual *i_1* is fuzzy and its value is identified by *fpv_1* which will be stored in the following Tables 8.10 and 8.11.

Storing the fuzzy values:

Table 8.10 shows Fuzzy_Property_Value_Table, storing *fuzzy typed literals* of the fuzzy property values, where *FuzzyPropertyValueID* uniquely identifies a fuzzy value, and *fuzzytypeID* denotes the datatype of a *fuzzytypeLiteral.*

Table 8.8 Fuzzy_Types_Table

FuzzytypeID	FuzzytypeName
f_1	Label
f_2	Possibility distribution

Table 8.9 Individual_Fuzzy_Property_Value_Table

IndID	ProID	FuzzyPropertyValueID
i_1	p_7	fpv_2
i_1	p_8	fpv_1

Table 8.10 Fuzzy_Property_Value_Table

FuzzyPropertyValueID	FuzzytypeLiteral	FuzzytypeID
fpv_1	middle	f_1
fpv_2	js@yahoo	f_2

Table 8.11 Fuzzy_Value_Table

Lable_Table				
FuzzyPropertyValueID	Alpha	Beta	Gamma	Delta
fpv_1	30	40	50	60
PossibilityDistribution_Table				
FuzzyPropertyValueID	Value_1	Poss_1	Value_2	Poss_2
fpv_2	js@yahoo.com	0.8	js@tom.com	0.9

Note that, since the current fuzzy database systems cannot support the fuzzy datatype storage, the field *fuzzytypeLiteral* stores the corresponding fuzzy typed literals of the fuzzy values in the form of *string* first, and then these fuzzy values will further be stored in the fuzzy database based on the approach in Table 8.11.

Table 8.11 shows Lable_Table and PossibilityDistribution_Table, storing all of the fuzzy property values in Fig. 8.1.

For example, the fuzzy typed literal *middle* is identified as *fpv_1*, and its datatype is "*Label*" identified by *f_1* and its value is stored in Lable_Table.

Until now, a fuzzy DL/ontology knowledge base, including classes, properties, individuals, axioms, datatypes, and characters and restrictions, can be stored in a fuzzy relational database. Moreover, our storage approach can be seen as a transformation. The evaluation of transforming or storing fuzzy DL/ontology knowledge base is an important issue, but it is also a difficult task. Currently there are still no any standard frameworks/metrics for DL/ontology evaluation, and many existing works used the standard information retrieval metrics to evaluate their approaches (Miller et al. 1993; Udrea et al. 2007; van Rijsbergen 1986). As mentioned in (Miller et al. 1993), when a transformation can preserve information capacities and it is considered as a correct transformation. Therefore, the correctness of the storage approach can be similarly proved based on the notion of information capacity (Miller et al. 1993). Moreover, note that, since it is well known that fuzzy DL/ontology fuzzy knowledge bases are more expressive than fuzzy relational databases in terms of semantic representation, being similar to (Astrova et al. 2007; Gali et al. 2004; Vysniauskas and Nemuraite 2006; Lv et al. 2009; Barranco et al. 2007; Konstantinou et al. 2008), in some cases the storage is not really lossless in the sense that there may be some loss of semantics, which may result in that reasoning on fuzzy DLs/ontologies becomes difficult after some axioms of fuzzy DLs/ontologies are stored in databases. Moreover, when a fuzzy knowledge base that imports another fuzzy knowledge base is stored in a fuzzy relational database, the naming strategy may be needed since SQL does not support namespaces. However, the information capacity preserving and correct storage of fuzzy knowledge bases in fuzzy relational databases proposed above makes it possible to solve the scalability issue raised by real fuzzy knowledge bases and also may facilitate for manipulating fuzzy knowledge bases within existing database techniques.

8.3 Summary

In the context of the Semantic Web, fuzzy extensions to OWL (the W3C standard ontology language) and Description Logics (DLs, the logical foundation of OWL) have been extensively investigated as introduced in Chap. 4, and many real knowledge bases based on fuzzy DLs and fuzzy OWL tend to become very large to huge. Therefore, how to store fuzzy knowledge bases has become an important issue. Based on the widespread investigation of fuzzy relational databases, in this chapter, we briefly introduce how to store fuzzy knowledge bases in fuzzy relational databases. Until now, there are a few papers discussing fuzzy DL or ontology knowledge base storage, which is still an open problem. Much work about fuzzy DL and ontology knowledge base storage may be needed for supporting the fuzzy knowledge management in the Semantic Web.

References

Astrova I, Korda N, Kalja A (2007) Storing OWL ontologies in SQL relational databases. Proceedings of world academy of science engineering and technology, pp 167–172

Al-Jadir L, Parent C, Spaccapietra S (2010) Reasoning with large ontologies stored in relational databases: the OntoMinD approach. Data Knowl Eng 69(11):1158–1180

Barranco CD, Campaña JR, Medina JM, Pons O (2007) On storing ontologies including fuzzy datatypes in relational databases. Proceedings of IEEE international conference on fuzzy systems, pp 1–6

Berners-Lee T, Hendler J, Lassila O (2001) The semantic web. Sci Am 284(5):34–43

Das S, Inseok Chong E, Eadon G, Srinivasan J (2004) Supporting ontology-based semantic matching in RDBMS. Proceedings of the 30th VLDB conference, pp 1054–1065

Gali A, Chen CX, Claypool KT, Uceda-Sosa R (2004) From ontology to relational databases. Proceedings of ER workshops 2004, LNCS, vol 3289, pp 278–289

Konstantinou N, Spanos DM, Nikolas M (2008) Ontology and database mapping: a survey of current implementations and future directions. J Web Eng 7(1):1–24

Lv YH, Ma ZM, Zhang X (2009) Fuzzy ontology storage in fuzzy relational database. Proceedings of international conference on fuzzy systems and knowledge discovery, pp 242–246

Miller RJ, Ioannidis YE, Ramakrishnan R (1993) The use of information capacity in schema integration and translation. Proceedings of the 19th VLDB conference, pp 120–133

Oliboni B, Pozzani G (2008) Representing fuzzy information by using XML schema. Proceedings of international conference on database and expert systems applications, pp 683–687

Udrea O, Getoor L, Miller RJ (2007) Leveraging data and structure in ontology integration. Proceedings of the 27th ACM SIGMOD international conference on management of data, pp 449–460

van Rijsbergen CJ (1986) A new theoretical framework for information retrieval. Proceedings of the 9th annual international ACM SIGIR conference on research and development in information retrieval, pp 194–200

Vysniauskas E, Nemuraite L (2006) Transforming ontology representation from OWL to relational database. Inf Technol Control 35(3):333–343

Zhang F, Ma ZM, Yan L, Cheng JW (2011) Storing fuzzy ontology in fuzzy relational database. Proceedings of the 22nd international conference on database and expert systems applications, pp 447–455

Chapter 9
Fuzzy Rules and Interchange in the Semantic Web

9.1 Introduction

In recent years, the Semantic Web (Berners-Lee et al. 2001) has gained considerable attention. One of the declared goals of the Semantic Web is to represent and reason about knowledge pervaded on the Web. As an important representation of knowledge, rules possess many features including highly expressive power and requiring small memory space, which let rules become indispensable and widely used in many areas of the Semantic Web.

With the development of rules in the Semantic Web, some important rule languages and rule markup languages, such as Semantic Web Rule Language (SWRL) (Horrocks et al. 2004) combining Web Ontology Language (OWL) and Horn rules, Rule Markup Language (RuleML) (Boley et al. 2005), are proposed to represent, reason and markup knowledge in the Semantic Web. Also, these rule languages may be used in different systems and tools. And it is unavoidable that these systems communicate with each other, i.e., interchange rules. And thus problems in sharing rules occur. Fortunately, rule markup languages can play the role of an intermediary language for rule interchange. There are some important evolving proposals for rule markup languages. The RuleML Initiative (http://ruleml.org/) has defined RuleML to try to encompass different kinds of rules. The Rule Interchange Format Working Group (RIF WG) has proposed Rule Interchange Format (RIF, http://www.w3.org/2005/rules/wg/charter.html) to interchange rules in the Semantic Web. What's more, Reasoning on the Web with Rules and Semantics (REWERSE) has proposed REWERSE Rule Markup Language (R2ML) (Wagner et al. 2006) which integrates Object Constraint Language (OCL), SWRL and RuleML, and it has strong markup capability. More recently, Ma and Wang (2012) propose and construct Rule Interchange Architecture based on XML Syntaxes (RIA_{XML})—a rule interchange structure centered on RIF, which enables bidirectional rule interchange between RIF and SWRL, RuleML, R2ML and Frame Logic, and further implement a prototype system of RIA_{XML}.

Z. Ma et al., *Fuzzy Knowledge Management for the Semantic Web*,

Studies in Fuzziness and Soft Computing 306, DOI: 10.1007/978-3-642-39283-2_9,

However, there is much imprecise and uncertain knowledge which cannot be represented and reasoned by the above languages. Fortunately, fuzzy logic (Zadeh 1965) is employed to handle imprecise and uncertain knowledge. Based on fuzzy sets, fuzzy SWRL (f-SWRL), a fuzzy extension of SWRL, is proposed to represent fuzzy knowledge in (Pan et al. 2006), where the general form, syntax and semantics of fuzzy rules are given. RuleML 0.9 (Hirtle et al. 2006) introduces <degree> to represent fuzzy knowledge, which makes the start of fuzzy RuleML (http://image.ntua.gr/FuzzyRuleML/). RIF has become a candidate recommendation on rule interchange of WWW Consortium (W3C), and the RIF Charter also proposes that RIF should support fuzzy rule interchange, but the W3C does not give RIF dialects which are capable of interchanging fuzzy rules. Zhao and Boley (2008) proposes RIF Uncertainty Rule Dialect (RIF-URD), an uncertain extension of RIF, to support a direct representation of uncertain knowledge in the Semantic Web.

Combining if–then rules, fuzzy sets and Ontology Web Language Description Logic (OWL DL), Wang et al. (2009a) proposes fuzzy rule language f-SW-if–then-RL, and based on unless rules, f-SW-if–then-unless-RL is also proposed. These two languages are the combination between AI rules and the Semantic Web, which enable representing fuzzy and nonmonotonic Semantic Web knowledge. Vague-SWRL combining vague sets and SWRL is also proposed in (Wang et al. 2008), which uses membership intervals to more precisely represent fuzzy knowledge. Moreover, Wang et al. (2009b) propose a fuzzy rule language, f-NSWRL, which combines SWRL and fuzzy sets, and can represent and reason with nonmonotonic and fuzzy knowledge. Wang et al. (2011) propose a fuzzy rule markup language named fuzzy R2ML (f-R2ML). In addition, Wang et al. (2012) propose RIF Fuzzy Rule Dialect (RIF-FRD), a rule interchange format based on fuzzy sets, and define its XML syntax and metamodel. Based on the metamodel, the authors propose a fuzzy RIF framework—fuzzy Rule Interchange Architecture (f-RIA), which supports fuzzy rule interchange in the Semantic Web. In this chapter, we introduce the fuzzy rules and interchange in the Semantic Web.

9.2 Fuzzy Rule Languages

As mentioned in Sect. 9.1, there have been some fuzzy rule languages, such as a fuzzy extension of SWRL (f-SWRL) (Pan et al. 2006), fuzzy Nonmonotonic Semantic Web Rule Language (f-NSWRL) (Wang et al. 2009b), f-SW-if–then-RL and f-SW-if–then-unless-RL (Wang et al. 2009a), and fuzzy R2ML (f-R2ML) (Wang et al. 2011). In the following, we introduce several fuzzy rule languages.

9.2.1 *f-SWRL*

To let rules capture imprecision and uncertainty, class and property in SWRL should be fuzzified. f-SWRL is proposed in (Pan et al. 2006), where each fuzzy individual axiom gets a membership degree with which one can assert that an individual (resp. a pair of individuals) is an instance of a given class (resp. property). Each atom has a weight which expresses the importance of the atom. For instance, the following fuzzy rule axiom, inspired from the field of housing purchase, asserts that if a house has a fairly low price, good quality, and sound environment (for example, there is no much noise around the house) then one may buy the house. And the condition that the house has a low price is more important than the others, and the condition that the house has good quality is a bit more important than the condition that the environment is harmonious.

$$\begin{gathered}\text{PriceLow }(?a) * 0.9 \wedge \text{QualityGood }(?a) * 0.8 \wedge \\ \text{EnvironmentSound}(?a) * 0.7 \rightarrow \text{Purchasing}(?a) * 0.95\end{gathered} \tag{9.1}$$

where PriceLow, QualityGood, EnvironmentSound and Purchasing are fuzzy class URIrefs, $?a$ is an individual-valued variable, and 0.9, 0.8, and 0.7 are weights of the corresponding fuzzy atoms. 0.95 represents the degree with which the consequent holds. The details about f-SWRL can be found in (Pan et al. 2006).

9.2.2 *f-NSWRL*

Fuzzy knowledge f-SWRL represents is monotonic. f-SWRL does not discuss the representation and reasoning of fuzzy knowledge including negation and negation as failure (NAF), i.e., it is incapable of expressing nonmonotonic fuzzy knowledge. We know that when expressing expert knowledge, exceptions are often encountered, and we usually employ "unless" to denote exceptions in natural language. To this end, Wang et al. (2009b) propose a fuzzy rule language, f-NSWRL, which combines SWRL and fuzzy sets, and can represent and reason with nonmonotonic and fuzzy knowledge. In the following, we first discuss the two kinds of negation.

9.2.2.1 Two Kinds of Negation

There are at least two kinds of negation, negation and negation as failure (NAF), in the real world. The former is also called strong negation, while the later is called weak negation. Negation is based on certain facts, gives a complete negation and expresses explicit falsity. NAF expresses the situation of non-truth which is like this: based on restricted conditions (incompleteness of knowledge), there is no information or there is no way to show that the positive form of a fuzzy atom holds. NAF is all-importance in nonmonotonic reasoning and its theoretical foundation is Closed World Assumption (CWA), and thus NAF is also called closed world negation.

Here, we offer an example to illustrate the difference between negation and NAF. Imagine a situation that Jack wants to buy a cake for his wife from an online shop. By coincidence, the shop will give a special discount if it is the birthday of the customer or the customer's spouse. An easy way to represent this rule in natural language is as follows:

r_1: If birthday, then special discount.
r_2: If not birthday, then regular price.

This solution works well in case that the shop knows Jack's or his wife's birthday. But imagine that Jack refuses to provide his and his wife's birthday information because of privacy concerns. The rule r_2 can not be applied to represent this case, because the shop does not know the birthday information rather than it knows that it is not their birthday that day. To capture this situation, we give the following rule:

r_3: If birthday is not known, then regular price.

The difference between rules r_2 and r_3 is exactly the difference between negation and NAF. The former rule represents the negation of class birthday, and it is an example of strong negation, the reason is that through the information Jack has registered the online shop definitely knows it is neither Jack's nor his wife's birthday that day (maybe it is Jack and his wife's wedding anniversary, and Jack wants to buy a cake for his wife to celebrate their wedding anniversary). The latter rule represents the NAF of the class birthday, and it is an example of weak negation. According to registered information, there is no way for the online shop to know whether it is the Jack's or his wife's birthday or not, because when the agent asks birthday information, Jack offers "unknown". Maybe it is the birthday of Jack or his wife that day, but because of lack of this piece of knowledge, Jack will not get a special discount. In a word, negation expresses the presence of explicit negative information, while NAF captures the absence of positive information. The rules r_1, r_2 and r_3 can be represented by the following rules:

$$\text{Birthday}(?x) \rightarrow \text{SpecialDiscount}(?x) \tag{9.2}$$

$$\neg\ \text{Birthday}(?x) \rightarrow \text{RegularPrice}(?x) \tag{9.3}$$

$$\text{not Birthday}(?x) \rightarrow \text{RegularPrice}(?x) \tag{9.4}$$

where Birthday, SpecialDiscount and RegularPrice are class URIrefs. Symbol '¬' represents negation of class Birthday, while 'not' denotes NAF of class Birthday.

9.2.2.2 Nonmonotonic Fuzzy Rules of f-NSWRL

f-NSWRL employs negation and NAF to extend fuzzy classes and properties. The extended rules can be of the following form:

$$\begin{aligned} A(x) * w_1 \wedge \neg B(x) * w_2 \wedge \text{not } C(x) * w_3 \rightarrow D(x)\ (\text{or} \neg D(x) \\ \text{or not } D(x)) * w \end{aligned} \tag{9.5}$$

where A(x), B(x), C(x) and D(x) are fuzzy classes, '¬' is strong negation, 'not' is weak negation, and w_1, w_2, w_3 and w are weights of the corresponding atoms.

Note that we call fuzzy rules crisp fuzzy rules, and call fuzzy rules capable of representing nonmonotonic knowledge fuzzy rules. Now, the extended rules can be used to represent nonmonotonic fuzzy knowledge in the Semantic Web. For example, default fuzzy rules we often encounter in expert systems are of the form "if x is A then y is B unless z is C", which can not be expressed by crisp fuzzy rules, the reason is that there is no element to represent unless part, but can be represented by fuzzy rules. According to concrete situations, we can transform this kind of fuzzy rules into the form "if x is A and z is not Z then y is B" or "if x is A and z is C then y is not B" or "y is B when x is A if nothing is known on z", which can be represented by the following forms, respectively:

$$A(x) * w_1 \wedge \neg C(z) * w_2 \rightarrow B(y) * w \tag{9.6}$$

$$A(x) * w_3 \wedge C(z) * w_4 \rightarrow \neg B(y) * w \tag{9.7}$$

$$A(x) * w_5 \wedge \text{not } C(z) * w_6 \rightarrow B(y) * w \tag{9.8}$$

We can see that the first two rules contain negation and the third rule is typically weak negation.

Now, we give an example of fuzzy rules with negation and negation as failure. Inspired from the field of medical science, generally speaking, the following rule holds: if a person has a fever, has a headache, and has a sore throat, then he or she possibly has a cold. Lucy feels bad and she guesses she may have a cold. And so she comes to see a doctor. The doctor gives her a preliminary examination. The situation is as follows: she does not have a high fever which is obtained through taking temperature and using the doctor's "fever function" (membership function of the fuzzy set 'fever'), has a sore throat, and has a bad headache. In the following, the doctor advises Lucy to do a blood routine examination to see whether there is a bronchial and pulmonary inflammation or not. But she refuses, because she always feels dizzy when seeing blood and she is afraid of pain caused by the blood examination. So there is no way for the doctor to get the information of Lucy's bronchi and lungs. The doctor has to make a decision that she gets viral influenza according to her external symptoms. The decision rule is as follows:

$$\begin{aligned} \neg\ \text{FeverHigh(Lucy)} * 0.7 \wedge \text{ThroatSore(Lucy)} * 0.8 \wedge \\ \text{HeadacheBad(Lucy)} * 0.7 \wedge \text{not Inflammation(Lucy)} * 0.5 \\ \rightarrow \text{ViralInfluenza(Lucy)} * 0.6 \end{aligned} \tag{9.9}$$

where FeverHigh, ThroatSore, HeadacheBad, Inflammation and ViralInfluenza are fuzzy class URIrefs, '¬' represents negation, and 'not' denotes NAF.

9.2.2.3 Nonmonotonic Extension for RuleML

The RuleML effort aims to establish a general rule language framework that supports different kinds of rules and various semantics. The Rule Markup Initiative has published a substantial amount of publications of RuleML (such as RuleML 0.8, RuleML tutorial and RuleML 0.9) which improve the expressive power of RuleML step by step and witness the development of RuleML. In order to represent imprecise and uncertain knowledge and information, fuzzy RuleML Technical Group extends RuleML with fuzzy sets theory and fuzzy logic (http://image.ntua.gr/FuzzyRuleML/). The element <degree> introduced in (Hirtle et al. 2006) makes the start of fuzzy RuleML. With the development of RuleML, it can be a suitable form for rules interchange. In order to make RuleML play the role of an intermediary language when f-NSWRL interchanges rules with other rule languages, we extend RuleML to represent nonmonotonic fuzzy rules of f-NSWRL.

The latest RuleML derivation rule is marked up as <Implies > <body> … </body> <head> … </head> </Implies>. And there are <and> … </and> and <or> …</or> to represent conjunctions and disjunctions of fuzzy atoms, respectively. The weights of atoms can be denoted by <degree> … </degree>. Strong negation and weak negation are denoted by <neg> … </neg> and <naf> … </naf>, respectively. RuleML codes of the above example (formula 9) are as follows:

```
<Implies>
 <body>
  <and>
    <atom>
      <degree><data>0.7</data></degree>
      <neg>
      <op><rel>FeverHigh</rel></op>
      <Ind>Lucy</Ind>
      </neg>
    </atom>
    <atom>
      <degree><data>0.8</data></degree>
      <op><rel> ThroatSore</rel></op>
      <Ind>Lucy</Ind>
```

```
      </atom>
      <atom>
        <degree><data>0.7</data></degree>
        <op><rel>HeadacheBad</rel></op>
        <Ind>Lucy</Ind>
      </atom>
      <atom>
        <degree><data>0.5</data></degree>
        <naf>
        <op><rel>Inflammation</rel></op>
        <Ind>Lucy</Ind>
        </naf>
      </atom>
    </and>
  </body>
  <head>
      <atom>
        <degree><data>0.6</data></degree>
        <op><rel>Inflammation</rel></op>
        <Ind>Lucy</Ind>
      </atom>
  </head>
</Implies>
```

9.2.2.4 Priorities of Fuzzy Rules in f-NSWRL

There are many imprecise and uncertain knowledge and information in the real world. What's more, facing different information sources or through different paths of propagation, available knowledge and information people captured may be incomplete. Therefore, on account of the incompleteness, imprecision and uncertainty of knowledge and information, facing the same facts, rules in fuzzy rule systems may offer different consequents and even conflicting consequents. We call these fuzzy rules competing fuzzy rules. Because of complexity and multiformity of knowledge and information, it is unavoidable that competing fuzzy rules exist simultaneously.

In nonmonotonic fuzzy rules bases, it worth noting that rules with contrary heads are competing rules, and there are many rules that do not have contrary heads are also competing rules. In our following example, we will illustrate this fact.

In face of these competing fuzzy rules, it would be wise to introduce the term priority into nonmonotonic fuzzy rules bases. Priorities are meant to resolve conflicts among competing fuzzy rules.

When new facts invoke rules to produce new conclusions, we need to guarantee the invoked rules in proper order. An efficient way is to compute priorities of these fuzzy rules, when adding them into fuzzy rule bases. And so we need a preprocessor to look for whether a rule conflicts with other rules, and if conflicting, the priorities of these fuzzy rules should be offered. The deciding principles of priorities of fuzzy rules are as follows:

1. search whether there is any conflicting rule or not;
2. if these is no competing rules, then add this rule into the fuzzy rule base and the value of the priority is assigned with a value of 0 (the value of 0 represents the highest priority. Only if no competing rules in fuzzy rule bases, 0 can be assigned to the rule, otherwise, the value of priority should be assigned from the value of 1. The advantage to this is that the preprocessor can immediately stop a searching process of one kind of rules, when it is searching competing rules and finds a rule with priority 0. The reason is that the value of 0 show that there is only one competing rule in the rule base. This method enhances execution efficiency of the processor.); else
3. compare with the competing rules (maybe the number of competing rules is more than two, for the sake of simplicity, we assume that there are only two), and perform matching function. If the atoms of antecedents of the two conflicting rules completely match, then through the comparison of weights (i.e., computing the multiplication of all the weights, comparing the two consequences, and assigning 1 to the priority of the rule whose multiplication is bigger and assigning 2 to the other), we can get the priorities of the two rules; else
4. in this case, all atoms of antecedent in one fuzzy rule present in antecedent of the other rule. We use r_{long} to represent the rule which has more atoms in its antecedent, and the other fuzzy rule is denoted by r_{short}. Because of the existence of weights in fuzzy rules, we should not take it for granted that the rule which have more atoms in its antecedent will get a higher priority. The preprocessor will give a function to compute these weights, through which the priorities can be obtained.

(1) if the weight of atomic consequent in r_{long} is equal to or greater than the weight of atomic consequent in r_{short}, then we give r_{long} a higher priority. i.e., because r_{long} has more atoms in its antecedent and the degree with which the consequent holds is greater, we give r_{long} a higher priority; else,
(2) if the weight of atomic consequent in r_{long} is less than the weight of atomic consequent in r_{short}, then we have to consider all the weights in the two fuzzy rules. The focal point is how to deal with the weights of the atoms which are not in r_{short} but in r_{long} (we call these atoms default atoms). Here we employ averaging method to treat weights of default atoms. We compute the mean value of the weights in r_{short} and consider it as the weights of default atoms. Then we compute the multiplications of weights of r_{long} and r_{short}, respectively. According to the two products, we can obtain the priorities of r_{long} and r_{short}.

In the following, an example will be given to illustrate the generating process of the priorities in a fuzzy rules base. Inspired also from the field of medical science, we assert that a person has acute appendicitis, if he or she has a high fever, has severe stomach pain (metastatic lower right abdominal pain) and may keep vomiting. The fuzzy rule already in the fuzzy rule base is as follows:

$$\begin{aligned} &\text{FeverHigh}(?a) * 0.8 \wedge \text{StomachPainSevere}(?a) * 0.95 \rightarrow \\ &\qquad \text{AcuteAppendicitis}(?\text{a}) * 0.9 \end{aligned} \tag{9.10}$$

where FeverHigh, StomachPainSevere and AcuteAppendicitis are fuzzy class URIrefs. *?a* is a fuzzy variable. 0.8 and 0.95 are the weights of FeverHigh(*?a*) and StomachPainSevere(*?a*). 0.9 represents the weight that is given to the rule axiom in determining the degree with which the consequent holds.

Now, we have fuzzy rule (9.11) which will be added into the fuzzy rule base. Through scanning the rule base by the preprocessor, a competing rule (9.10) is found. According to weights of the two fuzzy rules, we should determine the priorities of the two rules.

$$\begin{aligned} &\text{FeverHigh}(?a) * 0.8 \wedge \text{StomachPainSevere}(?a) * 0.9 \wedge \\ &\text{Vomiting}(?\text{a}) * 0.6 \rightarrow \text{AcuteAppendicitis}(?\text{a}) * 0.8 \end{aligned} \tag{9.11}$$

The decision procedure is as follows: according to deciding principles above, because of $0.9 > 0.8$, $0.8 \times 0.90 \times 0.6 \times 0.8 = 0.3456$, $0.8 \times 0.95 \times ((0.8 + 0.95)/2) \times 0.9 = 0.5985$ and $0.3456 < 0.5985$, we get the conclusion that the priority of (9.10) is higher than the priority of (9.11).

In what follows, we introduce priorities into RuleML to encode competing fuzzy rules. We note that there may be more than two competing rules in a fuzzy rule base, so already existing keywords (such as superior and inferior) in RuleML are no longer applicable. Here, tags <Priories> … </Priorities> and <xor> … </xor> are introduced into RuleML to support priority. The former tag represents the priorities of fuzzy rules, and the later which means exclusive disjunction shows the relationship of competing rules. The two competing rules above are encoded as follows (pl is priority level; for simplicity, we only offer the frame, the rest can be obtained analogously, according to the RuleML of the example mentioned above):

```
<Priorities>
  <xor>
   <Implies pl="0">
    <degree><data>0.9</data></degree>
      ........
   </Implies>
   <Implies pl="1">
    <degree><data>0.8</data></degree>
      ........
   </Implies>
  </xor>
</Priorities>
```

9.2.3 f-SW-if–then-RL and f-SW-if–then-unless-RL

Combining if–then rules, fuzzy sets and Ontology Web Language Description Logic (OWL DL), Wang et al. (2009a) proposes fuzzy rule language f-SW-if–then-RL, and based on unless rules, f-SW-if–then-unless-RL is also proposed.

9.2.3.1 f-SW-if–then-RL

$$\text{If } x_1 \text{ is in } \mathrm{A}_1 \text{ with weight } \omega_1,\ x_2 \text{ is in } \mathrm{A}_2 \text{ with weight } \omega_2, \ldots, x_n \text{ is in } \mathrm{A}_n \text{ with weight } \omega_n,\ \text{then } y \text{ is in B with weight } \omega \tag{9.12}$$

The above is the common form of f-SW-if–then-RL rules. Here A_i (i = 1,2…n) and B are fuzzy classes in an ontology, ω_i (i = 1,2…,n) and ω are weights (truth values in [0,1]) of corresponding fuzzy classes which are employed to represent the importance of the atoms.

A f-SW-if–then-RL rule axiom consists of an if-part and a then-part, each of which consists of a set of fuzzy atoms. Rule axioms can have annotations, and a rule axiom may be assigned a URIreference which could be used to uniquely identify it.

f-SW-if–then-RL rule ::='if–then ('[URIreference] {annotation} if-part then-part')'

if-part ::='If-part ('{fuzzy atom}')'

then-part ::='Then-part ('fuzzy atom')'

Figures 9.1 and 9.2 show the syntax of f-SW-if–then-RL rule and its if-part in the form of UML models, respectively. Then-part is a single fuzzy atom.

Fuzzy atoms in f-SW-if–then-RL rules can be of the form C(x), D(z), P(x,y), Q(x,z), sameAs(x,y), different- From(x,y), or builtIn(b,z_1,…z_n), where C is a fuzzy OWL DL description, D is a fuzzy OWL DL data range, P is a fuzzy OWL DL individual-valued property, Q is a fuzzy OWL DL data-valued property, x, y are either variables or fuzzy OWL DL individuals, and z,z_1,…,z_n are either variables or fuzzy OWL data literals.

```
Fuzzy atom::= fuzzyDescription'('i-object')'
            | fuzzyDataRange'('d-object')'
            | fuzzyIndividual-valuedPropertyID'('i-object, i-object')'
            | fuzzyData-valuedPropertyID'('i-object, d-object')'
            | sameAs'('i-object, i-object')'
            | differentFrom'('i-object, i-object')'
            | builtIn'('builtinID d-object')'
```

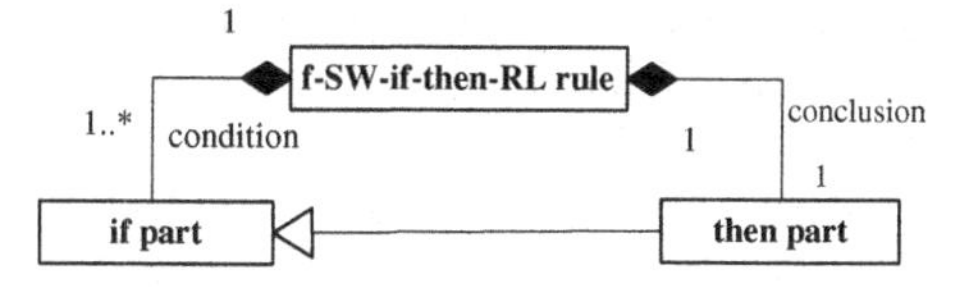

Fig. 9.1 Model of a f-SW-if–then-RL rule

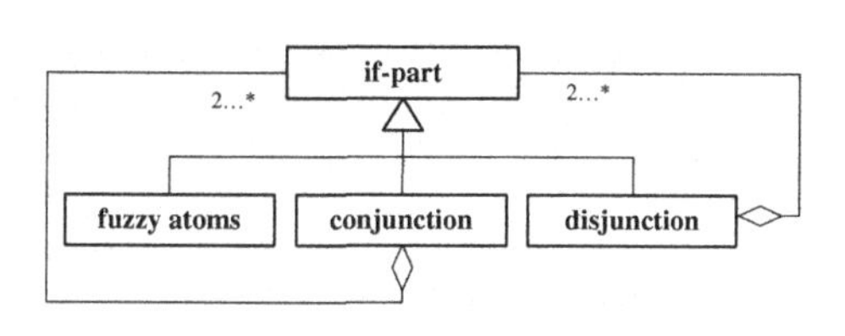

Fig. 9.2 Abstract syntax of if-part in a f-SW-if–then-RL rule

Fuzzy atoms refer to individuals, individual variables, data literals, or data variables, where the first two kinds are denoted by i-object, while the last two kinds by d-object. Variables are treated as universally quantified, with their scope limited to a given fuzzy rule. We do not require that fuzzy variables in if-part are present in then-part, which enhances the expressiveness of our fuzzy rules.

A fuzzy interpretation I ($I = (\Delta^I, \Delta_D, {}^I)$, where Δ^I is a non-empty set of abstract individuals, Δ_D is a non-empty set of concrete individuals and they are disjoint, and ^{I} is a fuzzy interpretation function) satisfies f-SW-if–then-RL* rule in Formula 12, if

$$\begin{aligned}\left(A_1^I(u_1)*\omega_1\right)\wedge\left(A_2^I(u_2)*\omega_2\right)\wedge\ldots\wedge\left(A_n^I(u_n)*\omega_n\right)\wedge\left(B^I(v)*\omega\right)\leq\\ \pi_{y|x1,\ldots,xn}(v,u_1,u_2,\ldots u_n)\leq n\left(\left(A_1^I(u_1)*\omega_1\right)\right)\vee n\left(\left(A_2^I(u_2)*\omega_2\right)\right)\\ \vee\ldots\vee n\left(\left(A_n^I(u_n)*\omega_n\right)\right)\vee\left(B^I(v)*\omega\right)\end{aligned}\tag{9.13}$$

Here $A_i^I(u_i)$ ($i = 1,\ldots,n$) and $B^I(v)$ are the interpretations of fuzzy classes, $u_1,\ldots u_n$ and v are the values of $x_1,\ldots x_n$ and y, respectively. $\wedge$ denotes conjunction operation, and $\vee$ denotes disjunction operation. $n((A_i^I(u_i)*\omega_i))$ denotes the negation of $A_i^I(u_i)*\omega_i$. $\Gamma\pi_{y|x1,\ldots,xn}(v,u_1,u_2,\ldots u_n)$ represents the possibility distribution of y for B, given that all the atoms in if-part hold.

The above model can be divided into two models:

$$\begin{aligned}\pi_{y|x1,\ldots,xn}(v,u_1,u_2,\ldots u_n)\geq\left(A_1^I(u_1)*\omega_1\right)\wedge\left(A_2^I(u_2)*\omega_2\right)\wedge\ldots\wedge\\ \left(A_n^I(u_n)*\omega_n\right)\wedge\left(B^I(v)*\omega\right)\end{aligned}\tag{9.14}$$

$$\begin{aligned}\pi_{y|x1,\ldots,xn}(v,u_1,u_2,\ldots u_n)\leq n\left(\left(A_1^I(u_1)*\omega_1\right)\right)\vee n\left(\left(A_2^I(u_2)*\omega_2\right)\right)\\ \vee\ldots\vee n\left(\left(A_n^I(u_n)*\omega_n\right)\right)\vee\left(B^I(v)*\omega\right)\end{aligned}\tag{9.15}$$

Inequality (9.14) is called conjunction-based model, while (9.15) is called implication-based model.

Based on implication-based model, there are two kinds of fuzzy rules, i.e., fuzzy certainty rule and fuzzy gradual rule. Fuzzy certainty rule corresponds to a rule of the type "the more x_i is in A_i with weight ω_i, the more certain y is in B with weight ω", whose interpretation is as follows:

$$\pi_{y|x1,\ldots,xn}(v, u_1, u_2, \ldots u_n) \le \max\left(1-\left(A_1^I(u_1) * \omega_1\right)\right), \left(A_2^I(u_2) * \omega_2\right), \ldots, \left(1-\left(A_n^I(u_n) * \omega_n\right)\right), \left(B^I(v) * \omega\right) \quad (9.16)$$

Fuzzy gradual rule corresponds to statement of the form "the more x_i is in A_i with weight ω_i, the more y is in B with weight ω". Its semantic interpretation is as follows:

$$\min\left(\left(A_1^I(u_1) * \omega_1\right), \left(A_2^I(u_2) * \omega_2\right), \ldots, \left(A_n^I(u_n) * \omega_n\right), \pi_{y|x1,\ldots,xn}(v, u_1, u_2, \ldots u_n) \le \left(B^I(v) * \omega\right) \quad (9.17)$$

Based on conjunction-based model, again, there are two cases, i.e., fuzzy possibility rule and fuzzy anti-gradual rule. The former corresponds to a rule of the type "the more x_i is in A_i with weight ω_i, the more possible y is in B with weight ω", whose interpretation is as follows:

$$\min\left(\left(A_1^I(u_1) * \omega_1\right), \left(A_2^I(u_2) * \omega_2\right), \ldots, \left(A_n^I(u_n) * \omega_n\right), \left(B^I(v) * \omega\right)\right) \le \pi_{y|x1,\ldots,xn}(v, u_1, u_2, \ldots u_n) \quad (9.18)$$

The latter is of the type "the more x_i is in A_i with weight ω_i and the less y is related to x_i, the less y is in B with weight ω". Its semantics is represented in the following:

$$\min\left(\left(A_1^I(u_1) * \omega_1\right), \left(A_2^I(u_2) * \omega_2\right), \ldots, \left(A_n^I(u_n) * \omega_n\right), 1-\pi_{y|x1,\ldots,xn}(v, u_1, u_2, \ldots u_n)\right) \le \left(1-\left(B^I(v) * \omega\right)\right) \quad (9.19)$$

9.2.3.2 f-SW-if–then-unless-RL

Because of incompleteness of knowledge, many conclusions may be reached according to some assumptions. When new knowledge appears, these conclusions may be defeated by new knowledge (such as some exceptions). These exceptions cannot be expressed by if–then rules. Therefore, it is necessary that if–then rules should be extended to express nonmonotonic knowledge. Based on the fact above, unless rules are introduced to express these exceptions, and thus f-SW-if–then-RL is extended to f-SW-if–then-unless-RL, whose rule is of the following form:

if x_1 is in A_1 with weight ω_1, x_2 is in A_2 with weight $\omega_2, \ldots, x_n$ is in A_n with weight ω_n, then y is in B with weight ω, unless z is in C with weight ω' (9.20)

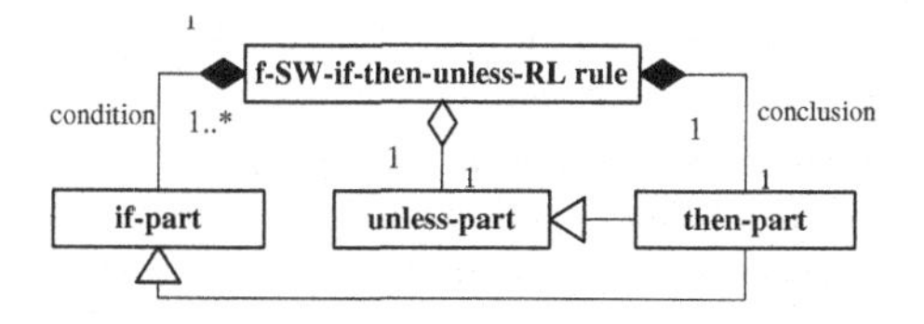

Fig. 9.3 Model of a f-SW-if–then-unless-RL rule

If-part is a conjunction or disjunction of atoms, then-part is an atomic consequent, and unless-part is an atomic condition. Its abstract syntax (cf. Fig. 9.3) is the same as that of f-SW-if–then-RL rule, except the following:

f-SW-if–then-unless-RL Rule ::='If–then-unless ('[URI-reference] {annotation} if-part then-part unless-part')'

if-part ::='If-part ('{fuzzy atom}')'

then-part ::='Then-part ('fuzzy atom')'

unless-part ::='unless-part ('fuzzy atom')'

Before giving the semantics of f-SW-if–then-unless-RL rules, we first discuss negation and NAF. Negation is also called strong negation, while NAF is essentially synonymous with weak negation. The former is based on certain facts, gives complete negation and expresses explicit falsity, while the latter expresses the non-truth fact: based on restricted conditions (i.e., incompleteness of knowledge), there is no information or there is no way to show that the positive form of a fuzzy atom holds.

Imagine a situation that a person Jack wants to buy a cake for his wife from an online shop. By coincidence, the shop will give a special discount if it is the birthday of the customer or the customer's spouse. An easy way to represent this rule in natural language is as follows:

r_1: if birthday, then special discount.
r_2: if not birthday, then regular price.

This solution works properly in case that the shop knows Jack's or his wife's birthday. But if Jack refuses to provide his and his wife's birthday information because of privacy concerns, rule r_2 can not be applied in this case, because the shop does not know the birthday information rather than it knows that it is not their birthday that day. To capture this situation we give a rule like this:

r_3: if birthday is not known, then regular price.

The difference between rules r_2 and r_3 is exactly the difference between negation and NAF. The former rule represents the negation of class birthday, and it is an example of strong negation, the reason is that through the information Jack has registered the online shop definitely knows it is neither Jack's nor his wife's birthday that day (maybe it is Jack and his wife's wedding anniversary. Jack wants to buy a cake for his wife to celebrate their wedding anniversary). The latter rule represents NAF of the class birthday. It is an example of weak negation. According to registered information, there is no way for the online shop to know whether it is the Jack's or his wife's birthday or not that day, because when the agent asks birthday information,

Jack offers "unknown". Maybe it is the birthday of Jack or his wife that day, but because of lack of this piece of knowledge, Jack cannot be given a special discount. In a word, a strong negation expresses the presence of explicit negative information, while a weak negation captures the absence of positive information. The rules r_1, r_2 and r_3 can be represented by the following rules, respectively:

$$\text{if } x \text{ is in } \mathit{Birthday} \text{ then } y \text{ is in } \mathit{SpecialDiscount}. \tag{9.21}$$

$$\text{if } x \text{ is in } \neg \mathit{Birthday} \text{ then } y \text{ is in } \mathit{RegularPrice}. \tag{9.22}$$

$$\text{if } x \text{ is in } \mathit{not\ Birthday} \text{ then } y \text{ is in } \mathit{RegularPrice}. \tag{9.23}$$

where *Birthday*, *SpecialDiscount* and *RegularPrice* are class URIrefs. Symbol '¬' represents strong negation of class *Birthday*, while '*not*' denotes NAF of class *Birthday*.

The method of interpreting our if–then-unless rules is that we use negation and NAF to substitute unless-part. Two kinds of negation equivalently substituting unless-part also does good to the rule representation in the form of rule markup language, because there is no unless elements in rule markup languages. The general rule forms containing two kinds of negation are as follows

$$\text{if } x_i \text{ is in } A_i \text{ with weight } \omega_i \text{ and } z \text{ is in } \neg C \text{ with weight } \omega' \text{ then } y \text{ is in B with weight } \omega \tag{9.24}$$

$$\text{if } x_i \text{ is in } A_i \text{ with weight } \omega_i \text{ and } z \text{ is in} \neg C \text{ with weight } \omega' \text{then } y \text{ is in } \neg B \text{ with weight } \omega \tag{9.25}$$

$$\text{if } x_i \text{ is in } A_i \text{ with weight } \omega_i \text{ and } z \text{ is in C with weight } \omega' \text{ then } y \text{ is unknown with weight } \omega \text{ (if } x_i \text{ is in } A_i \text{ with weight } \omega_i \text{ and } z \text{ is in C weight } \omega' \text{ then } y \text{ is in } \mathit{not} \text{ Bwith weight } \omega) \tag{9.26}$$

$$\text{if } x_i \text{ is in } A_i \text{ with weight } \omega_i \text{ and } z \text{ is unknown with weight } \omega' \text{ then } y \text{ is in B withweight } \omega \text{ (if } x_i \text{ is in } A_i \text{ with weight } \omega_i \text{ and } z \text{ is in } \mathit{not} \text{ C with weight } \omega' \text{ then } y \text{ is in Bwith weight } \omega) \tag{9.27}$$

It is clear that the first two models are based on negation, while the last two based on NAF. In a word, fuzzy unless rules can be represented by general rules with the two kinds of negation. The abstract syntax of if-part in f-SW-if–then-unless-RL rules is shown in Fig. 9.4.

The representation of f-SW-if–then-unless-RL rules (9.20) is the one and the only, but there are many interpretation models of the rules. The reason is that, in light of the different semantics of f-SW-if–then-RL rules which have been discussed above, it is clear there are many types of f-SW-if–then-unless-RL rules.

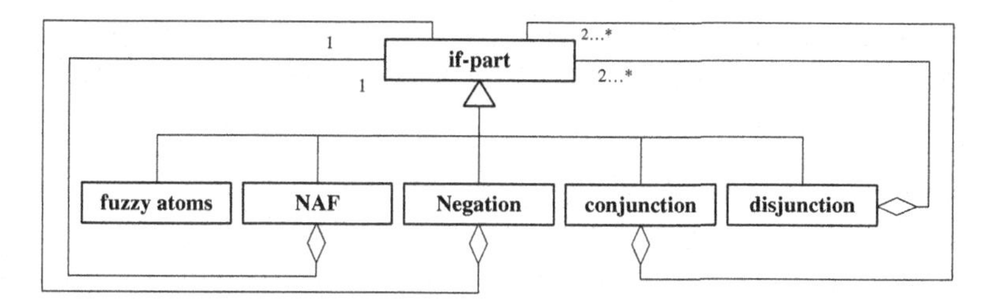

Fig. 9.4 Abstract syntax of if-part in a f-SW-if–then-unless-RL rule

According to representation of (9.24)–(9.27), for each type of if–then rules, we can imagine the associated if–then-unless rules.

Now we focus on the semantics of f-SW-if–then-unless-RL, which is provided by a *fuzzy interpretation I*. Here, we only discuss fuzzy rule axioms. Interpretations of other concepts are the same as the semantics of f-SW-if–then-RL. A fuzzy interpretation I satisfies (9.20), if

$$\begin{aligned}\pi_{y|x1,\ldots,xn,z}(\mathrm{v},\mathrm{u}_1,\mathrm{u}_2,\ldots\mathrm{u}_\mathrm{n},\mathrm{w}) \le \max\big(1-\big(\mathrm{A}_1^\mathrm{I}(\mathrm{u}_1)*\omega_1\big)\big),\ldots,\\ \big(1-\big(\mathrm{A}_\mathrm{n}^\mathrm{I}(\mathrm{u}_\mathrm{n})*\omega_\mathrm{n}\big)\big),\big(1-\big(1-\mathrm{C}^\mathrm{I}(\mathrm{w})\big)*\omega'\big),\big(\mathrm{B}^\mathrm{I}(\mathrm{v})*\omega\big)\end{aligned} \tag{9.28}$$

$$\begin{aligned}\min\big(\big(\mathrm{A}_1^\mathrm{I}(\mathrm{u}_1)*\omega_1\big),\ldots,\big(\mathrm{A}_\mathrm{n}^\mathrm{I}(\mathrm{u}_\mathrm{n})*\omega_\mathrm{n}\big),\big(1-\mathrm{C}^\mathrm{I}(\mathrm{w})\big)*\omega',\\ \pi_{y|x1,\ldots,xn,z}(\mathrm{v},\mathrm{u}_1,\mathrm{u}_2,\ldots\mathrm{u}_\mathrm{n},\mathrm{w})\big)\le\big(\mathrm{B}^\mathrm{I}(\mathrm{v})*\omega\big)\end{aligned} \tag{9.29}$$

$$\begin{aligned}\min\big(\big(\mathrm{A}_1^\mathrm{I}(\mathrm{u}_1)*\omega_1\big)\big),\ldots,\big(\mathrm{A}_\mathrm{n}^\mathrm{I}(\mathrm{u}_\mathrm{n})*\omega_\mathrm{n}\big),\big(1-\mathrm{C}^\mathrm{I}(\mathrm{w})\big)*\omega',\mathrm{B}^\mathrm{I}(\mathrm{v})*\omega\\ \le\pi_{y|x1,\ldots,xn,z}(\mathrm{v},\mathrm{u}_1,\mathrm{u}_2,\ldots\mathrm{u}_\mathrm{n},\mathrm{w})\end{aligned} \tag{9.30}$$

$$\begin{aligned}\min\big(\big(\mathrm{A}_1^\mathrm{I}(\mathrm{u}_1)*\omega_1\big),\ldots,\big(\mathrm{A}_\mathrm{n}^\mathrm{I}(\mathrm{u}_\mathrm{n})*\omega_\mathrm{n}\big),\big(1-\mathrm{C}^\mathrm{I}(\mathrm{w})\big)*\omega',\\ 1-\pi_{y|x1,\ldots,xn,z}(\mathrm{v},\mathrm{u}_1,\mathrm{u}_2,\ldots\mathrm{u}_\mathrm{n},\mathrm{w})\le\big(1-\mathrm{B}^\mathrm{I}(\mathrm{v})*\omega\big)\end{aligned} \tag{9.31}$$

Note that the above interpretation models are the models of (9.24), and they are fuzzy certainty rule, fuzzy gradual rule, fuzzy possibility rule, and fuzzy anti-gradual rule. The interpretation models of (9.25)–(9.27) can be obtained respectively, according to the interpretation models of (9.24).

Inspired from the field of medical science, generally speaking, the following rule holds: if a person has a fever, has a headache, and has a sore throat, then he or she possibly has a cold. Lucy feels bad and she guesses that she may have a cold. And so she comes to see a doctor. The doctor gives her a preliminary examination, and the situation is as follows: she does not have a high fever which is obtained through taking temperature and using the doctor's "fever function" (i.e., the membership function of the fuzzy set 'fever'), has a sore throat, and has a bad headache. In the following, the doctor advises her to do a blood routine examination to see whether there is a bronchial and pulmonary inflammation or not. But Lucy refuses to do it, because she always feels dizzy when seeing blood and she is afraid of pain caused by the blood examination. So there is no way for the doctor to get the information of

Lucy's bronchi and lungs. The doctor has to make a decision that she has got viral influenza according to Lucy's external symptoms. The decision rule is as follows:

$$\begin{gathered}\text{if Lucy is in} \neg \mathit{FeverHigh} \text{ with weight } 0.7, \text{ in } \mathit{ThroatSore} \\ \text{with weight } 0.8, \text{ in } \mathit{HeadacheBad}\ \mathit{with} \text{ weight } 0.7, \text{ and in not } \mathit{Inflammation} \\ \text{with weight } 0.5, \text{ then Lucy is in } \mathit{Viral} - \mathit{Influenza} \text{ with weight } 0.6\end{gathered} \tag{9.32}$$

where *FeverHigh*, *ThroatSore*, *HeadacheBad*, *Inflammation* and *ViralInfluenza* are all fuzzy class URIrefs.

Rule markup languages can play the role of a lingua franca for exchanging rules between different systems. As an important markup language, R2ML is more powerful than SWRL and RuleML, but it does not possess elements representing fuzzy information. To make R2ML play the role of an intermediary language when f-SW-if–then-unless-RL interchanges rules with other rule languages, we add fuzzy element—weight to R2ML. <Degree> is employed to represent weight, and DataValue is used to constrain its value. The DTD of weight is as follows:

<! ENTITY % Degree "Degree">

<! ENTITY % DataValue "DataValue">

The following Fig. 9.5 shows the R2ML codes of Formula 32.

Fig. 9.5 The R2ML codes of formula 32

```
<DerivationRule>
  <condition>
   <And>
    <Neg>
     <OClassAtom classRef="FeverHigh">
     <OConst name="Lucy"/>
     <Degree><DataValue>0.7</DataValue></Degree>
     </OClassAtom>
    </Neg>
     <OClassAtom classRef="ThroatSore">
     <OConst name="Lucy"/>
     <Degree><DataValue>0.8</DataValue></Degree>
     </OClassAtom>
     <OClassAtom classRef="HeadacheBad">
     <OConst name="Lucy"/>
     <Degree><DataValue>0.7</DataValue></Degree>
     </OClassAtom>
    <Naf>
     <OClassAtom classRef="Inflamation">
     <OConst name="Lucy"/>
     <Degree><DataValue>0.5</DataValue></Degree>
     </OClassAtom>
    </Naf>
   </And>
  </condition>
  <conclusion>
     <OclassAtom classRef="ViralInfluenza">
     <OConst name="Lucy"/>
     <Degree><DataValue>0.6</DataValue></Degree>
     </OClassAtom>
  </conclusion>
</DerivationRule>
```

9.2.4 *f-R2ML*

In order for the Semantic Web rule markup languages REWERSE Rule Markup Language (R2ML) to represent, markup and reason with fuzzy knowledge, a fuzzy rule markup language named fuzzy R2ML (f-R2ML) is proposed in Wang et al. (2011). The abstract syntax and concrete syntax of f-R2ML are presented, and the semantics and XML Schema of f-R2ML are also defined.

9.2.4.1 Syntax of f-R2ML

The syntaxes of f-R2ML include two parts, i.e., XML syntax and metamodel syntax. In the following, we will introduce these two parts.

XML Syntax off-R2ML. In order to represent f-R2ML rules in XML format, we define the XML schema of f-R2ML (which is presented in Appendix A). In the schema, rule set element is the root element, which is composed of integrity rule set, derivation rule set, production rule set, and reaction rule set. Because of the fuzziness of atoms, weights (truth values between 0 and 1) are used with atoms, and they represent importance of these atoms. Neg and naf are negation and negation as failure (Clark 1978) of fuzzy atoms.

Inspired from the field of medical science, generally speaking, the following rule holds: if a person has a fever, has a headache, and has a sore throat, then he or she possibly has a cold. Lucy feels bad and she guesses that she may have a cold. And so she comes to see a doctor. The doctor gives her a preliminary examination, and the situation is as follows: she does not have a high fever which is obtained through taking temperature and using the doctor's "fever function" (i.e., the membership function of the fuzzy set 'fever'), has a sore throat, and has a bad headache. In the following, the doctor advises her to do a blood routine examination to see whether there is a bronchial and pulmonary inflammation or not. But Lucy refuses to do it, because she always feels dizzy when seeing blood and she is afraid of pain caused by the blood examination. So there is no way for the doctor to get the information of Lucy's bronchi and lungs. The doctor has to make a decision that she has got viral influenza according to Lucy's external symptoms. The decision rule is as follows (here *FeverHigh*, *ThroatSore*, *HeadacheBad*, *Inflammation* and *ViralInfluenza* are all fuzzy class URIrefs).

$$\begin{aligned}&\text{if Lucy is in } \neg\, \mathit{FeverHigh} \text{ with weight } 0.7,\ \text{in } \mathit{ThroatSore}\\&\text{with weight } 0.8,\ \text{in } \mathit{HeadacheBad}\ \mathit{with} \text{ weight } 0.7,\ \text{ and in not } \mathit{Inflammation}\\&\text{with weight } 0.5,\ \text{then Lucy is in } \mathit{Viral}-\mathit{Influenza} \text{ with weight } 0.6\end{aligned} \tag{9.33}$$

```
<r2ml:ruleset>
 <r2ml:DerivationRuleSet>
  <r2ml:DerivationRule>
   <r2ml:condition>
    <r2ml:and>
     <r2ml:neg>
       <r2ml:OClassAtom classRef="FeverHigh">
        <r2ml:OConst name="Lucy"/>
         <r2ml:Degree><r2ml:DataValue>0.7
         </r2ml:DataValue></r2ml:Degree>
       </r2ml:OClassAtom>
     </r2ml:neg>
       <r2ml:OClassAtom classRef="ThroatSore">
        <r2ml:OConst name="Lucy"/>
         <r2ml:Degree><r2ml:DataValue>0.8
         </r2ml:DataValue></r2ml:Degree>
       </r2ml:OClassAtom>
       <r2ml:OClassAtom classRef="HeadacheBad">
        <r2ml:OConst name="Lucy"/>
         <r2ml:Degree><r2ml:DataValue>0.7
         </r2ml:DataValue></r2ml:Degree>
       </r2ml:OClassAtom>
     <r2ml:naf>
       <r2ml:OClassAtom classRef="Inflammation">
        <r2ml:OConst name="Lucy"/>
         <r2ml:Degree><r2ml:DataValue>0.5
         </r2ml:DataValue></r2ml:Degree>
       </r2ml:OClassAtom>
     </r2ml:naf>
    </r2ml:and>
   </r2ml:condition>
    <r2ml:conclusion>
       <r2ml:OclassAtom classRef="ViralInfluenza">
        <r2ml:OConst name="Lucy"/>
         <r2ml:Degree><r2ml:DataValue>0.6
         </r2ml:DataValue></r2ml:Degree>
       </r2ml:OClassAtom>
    </r2ml:conclusion>
  </r2ml:DerivationRule>
 </r2ml:DerivationRuleSet>
</r2ml:ruleset>
```

Metamodel of f-R2ML. Metamodels can simplify conceptual modelling, decrease syntactic and semantic errors, and increase readability (Brockmans et al. 2006). Therefore, it's necessary for f-R2ML to define its own rule metamodel. In the following, based on MOF (Meta Object Facility), we construct an f-R2ML rule metamodel.

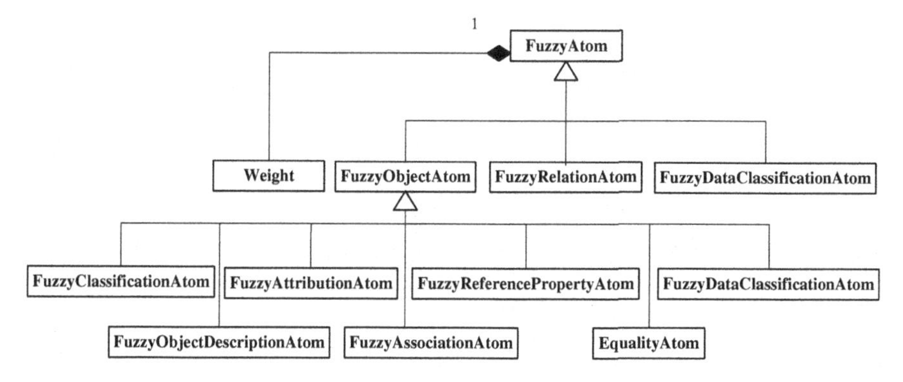

Fig. 9.6 The fuzzy atoms of f-R2ML

Fuzzy Atom. The basic constituent of a f-R2ML rule is the *FuzzyAtom* as shown in Fig. 9.6.

FuzzyAtom consists of weight, FuzzyObjectAtom, FuzzyRelationAtom and FuzzyData-ClassificationAtom, where weight represents the importance of fuzzy elements (they are of the values between 0 and 1). And FuzzyObjectAtom is the generalization of FuzzyClassificationAtom, FuzzyObjectDescriptionAtom, FuzzyAttributionAtom, FuzzyAssociationA tom, FuzzyReferencePropertyAtom, EqualityAtom and FuzzyData ClassificationAtom.

Fuzzy Formulas. f-R2ML defines two kinds of fuzzy formulas, i.e., *QF::AndOrNafNegFormula* (as shown in Fig. 9.7) and *LogicalFormula* (as shown in Fig. 9.8). FuzzyAtoms, QF::Negation, QF::Conjunction and QF::Disjunction inherit from *QF::AndOrNafNegFormula.*

Rule Metamodel. As shown in Fig. 9.9, we obtain the metamodel of f-R2ML derivation rules (the metamodels of the other three kinds of f-R2ML rules can be obtained analogously), through combining R2ML metamodels and fuzzy sets.

9.2.4.2 Semantics of f-R2ML

There are four kinds of rules in f-R2ML, i.e., integrity rules, derivation rules, production rules and reaction rules, where derivation rules are the most frequently used. Therefore, we only give the semantics of derivation rules, and the semantic of the other three kinds of rules can be obtained analogously.

A formal semantics for derivation rules is based on a Tarski-style theory (Wagner et al. 2005). Tarski-style model theory is not limited to classical first-order models, as employed in the semantics of OWL. The stable model semantics for normal and extended logic programs can be viewed as a Tarski-style model-theoretic semantics for nonmonotonic derivation rules. Fuzzy sets are added into the semantics. Therefore, the interpretation is a fuzzy interpretation.

A fuzzy Tarski-style model theory is a triple <L, I, |=>, such that

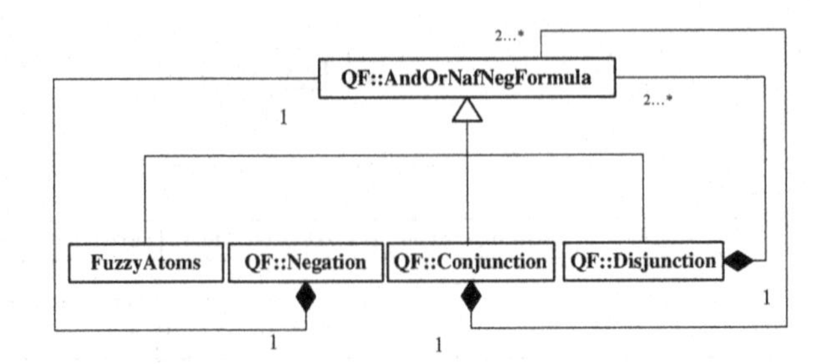

Fig. 9.7 AndOrNafNegFormula of f-R2ML

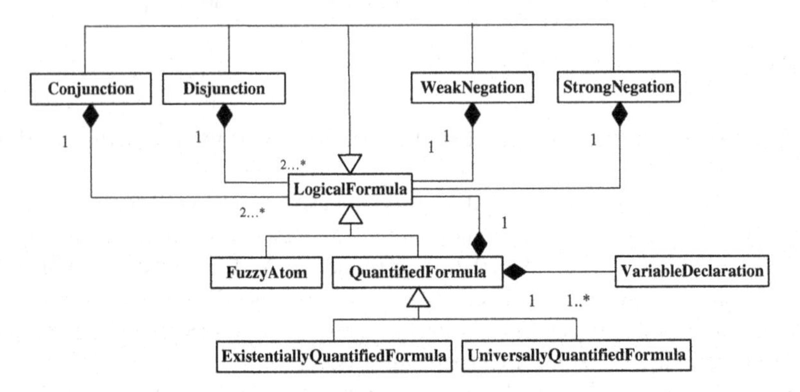

Fig. 9.8 Logical formula of f-R2ML

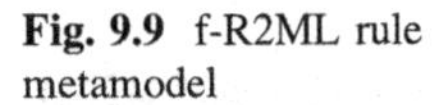
Fig. 9.9 f-R2ML rule metamodel

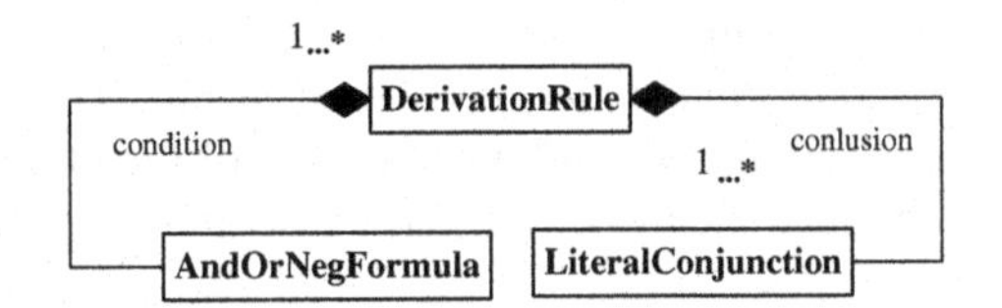

- L is a set of fuzzy formula, called language
- I is a set of fuzzy interpretation
- |= is a relation between fuzzy interpretation and fuzzy formula, called fuzzy model relation.

For each fuzzy Tarski-style model theory <L, I, |=>, we can define

- a notion of a derivation rule $F \rightarrow G$, where $F \in L$ is called "condition" and $G \in L$ is called "conclusion"
- $DR_L = \{F \rightarrow G : F, G \in L\}$, the set of derivation rules of L
- $(F)^I \leq (G)^I$, where F and G are both fuzzy formula
- A standard model operator

$Mod(X) = \{I \in \mathbf{I}: I \mid = Q, \text{ for all } Q \in X\}$, where $X \subseteq L \cup DR_L$, as a set of formula and/or derivation rules, is called a knowledge base.

9.3 Fuzzy Rule Interchange

Rule Interchange Format (RIF) has become W3C's candidate recommendation on rule interchange. To let RIF represent and interchange fuzzy knowledge in the Semantic Web, Wang et al. (2012) propose RIF Fuzzy Rule Dialect (RIF-FRD), a rule interchange format based on fuzzy sets, and define its XML syntax and metamodel. Based on the metamodel, the authors propose a fuzzy RIF framework—fuzzy Rule Interchange Architecture (f-RIA), which supports fuzzy rule interchange in the Semantic Web.

9.3.1 Rif-frd

9.3.1.1 RIF Dialects and RIF-BLD

RIF is an extensible framework for rule-based languages, called RIF dialects, which includes specifications of the syntax, semantics and XML serialization (Kifer 2008). To become a W3C's recommendation, the RIF Working Group has published five candidate recommendations for RIF, i.e., RIF-UCR, RIF Core, RIF-BLD, RIF-FLD, and RIF RDF and OWL Compatibility (http://www.w3.org/standards/techs/rif#w3c_all). RIF Basic Logic Dialect (RIF-BLD) is the first dialect of RIF dialects. It corresponds to the languages of definite Horn rules with equality and standard first-order semantics, and it has a number of extensions to support features such as objects and frames as in F-logic, Internationalized Resource Identifiers (IRI), and XML Schema datatypes. It should be noted that, the RIF-FRD metamodel is obtained according to RIF-BLD.

9.3.1.2 RIF-FRD Abstract Syntax

RIF-BLD is usually defined in terms of a presentation syntax which is defined in "mathematical English" (a special form of English for communicating mathematical definition) and XML syntax, and most care is spent on the formal semantics. However, RIF-BLD does not have a metamodel to describe its own features. What's more, RIF does not have elements expressing fuzzy knowledge. In this subsection, we construct the metamodel of RIF-FRD.

Fuzzy Terms

RIF-FRD defines the term *fuzzy term* (as shown in Fig. 9.10), which consists of constant, variable, fuzzy positional term, fuzzy terms with named arguments, equality, fuzzy membership, and external terms, where constant, variable are called *simple term*, and simple term, fuzzy positional term, fuzzy term with named arguments and external term are called *base term*.

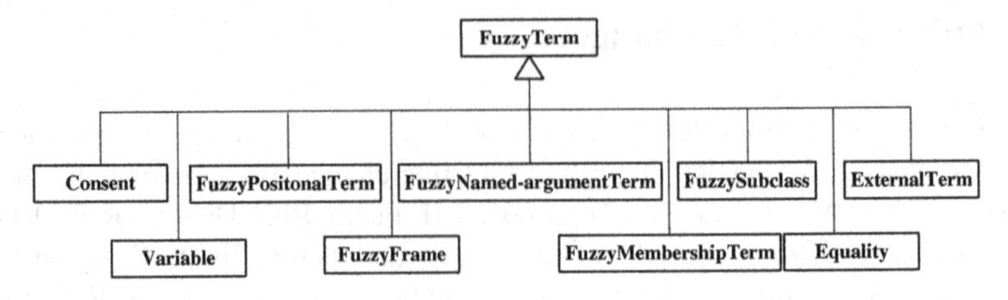

Fig. 9.10 The fuzzy terms of RIF-FRD

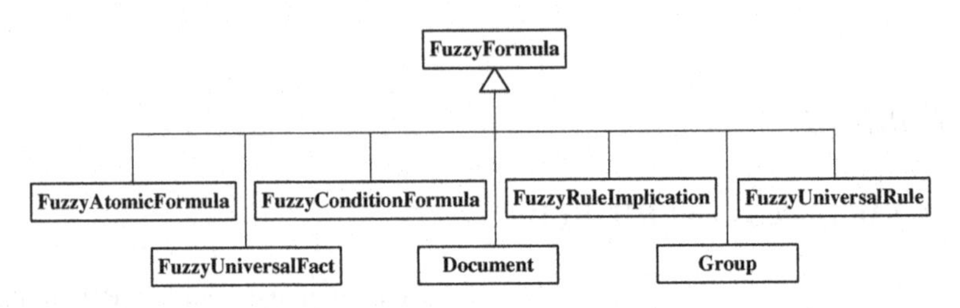

Fig. 9.11 The fuzzy formulas of RIF-FRD

Simple Term. If t∈Const or t∈Var, then t is a simple term, where Const and Var are countably infinite sets of constant symbols and variable symbols, respectively.

Fuzzy Positional Term. If t∈Const and $t_1,\ldots,t_n$ ($n \geq 0$) are base terms, then $t(t_1, \ldots, t_n)$ is a fuzzy positional term, i.e., it has a membership degree (a truth value between 0 and 1) with which $t_1, \ldots, t_n$ belong to t.

Fuzzy Term with Named Arguments. A fuzzy term with named arguments is of the form $t(s_1\text{-}> v_1 \ldots s_n\text{-}> v_n)$, where $n \geq 0$, t∈Const and $v_1, \ldots, v_n$ are base terms and $s_1, \ldots, s_n$ are pairwise distinct symbols from the set ArgNames (a countably infinite set of argument names). In fact, this fuzzy term has n degrees (truth values between 0 and 1) with which one can assert that the pair of t and v_i is an instance of s_i ($i = 1, \ldots, n$).

Equality Term. If t and s are base terms, then t = s is an equality term.

Fuzzy Membership Term. If t and s are base terms, then t # s is a fuzzy membership term, i.e., t is a member of s with a degree.

Fuzzy Subclass Term. If t and s are base terms, then t##s is a fuzzy subclass term, which also has a degree.

Fuzzy Frame Term. If t, $p_1, \ldots, p_n, v_1, \ldots, v_n$($n \geq 0$) are base terms, t $[p_1\text{-}> v_1 \ldots p_n\text{-}> v_n]$ is a fuzzy frame term. It has n degrees (truth values between 0 and 1) with which one can assert that the pair of t and v_i is an instance of P_i ($i = 1, \ldots, n$).

External Term. If t is a positional, named-argument, or a frame term then External(t) is an externally defined term.

Fuzzy Formulas

RIF-FRD defines many *fuzzy formulas* (cf. Fig. 9.11) including fuzzy atomic formula, fuzzy condition formula, fuzzy rule implication, fuzzy universal fact, group and document for representing fuzzy knowledge and information.

Fuzzy Atomic Formula. Fuzzy terms (i.e., fuzzy positional term, fuzzy named-argument term, equality, fuzzy membership, fuzzy subclass, and fuzzy frame) and External(Φ) (Φ is also a fuzzy atomic formula) are fuzzy atomic formulas.

Fuzzy Condition Formula. Fuzzy condition formulas are intended to be used inside the premises of fuzzy rules. A fuzzy condition formula is either a fuzzy atomic formula or a fuzzy formula that has one of the following forms:

Conjunction: And(Φ_1, .., Φ_n) (Φ_1, …, Φ_n are fuzzy condition formulas). As a special case, And () is always true.

Disjunction: Or(Φ_1, …, Φ_n) (Φ_1, …, Φ_n are fuzzy condition formulas). As a special case, Or () is always false.

Existentials: Φ is fuzzy condition formula, $?v_1$, …, $?v_n$ are variables, then Exists $?v_1,\ldots,?v_n$ (Φ) is a fuzzy existential formula.

Fuzzy Rule Implication. Φ:-Ψ is a fuzzy formula, called fuzzy rule implication, where Φ is a fuzzy atomic formula or a conjunction of fuzzy atomic formulas, and Ψ is a fuzzy condition formula.

Fuzzy Universal Rule. If Φ is a fuzzy rule implication, and $?v_1$, …, $?v_n$ are variables, then Forall$?v_1$, …, $?v_n$ (Φ) is called fuzzy universal rule.

Fuzzy Universal Fact. Fuzzy universal facts are fuzzy universal rules without premises.

Group. Groups are of the form Group (Φ_1, …, Φ_n), where Φ_1, …, Φ_n are fuzzy universal rules, fuzzy universal facts, fuzzy implication and fuzzy atomic formulas. Group is considered the sets of fuzzy rules and fuzzy facts.

Document. Documents are of the form Document (directive$_1$, …, directive$_n$, Γ), where directives are base directive, prefix directive and import directive, and Γ is a group.

9.3.1.3 Fuzzy Classes and Fuzzy Restrictions

Not only simple named classes but also classes with class constructors (cf. Fig. 9.12) are allowed to define in RIF-FRD. Class combination in RIF-FRD includes conjunction and disjunction. Negation is not allowed in RIF and RIF-FRD. The reason is that RIF is described to be simple dialect with limited expressiveness that lies within the intersection of first-order and logic-programming systems. Figure 9.12 also shows that fuzzy classes can be related with each other by employing fuzzy axioms. The axioms may include class subsumption (Subclass) and class equivalence (Equality) which are naturally modelled as associations. RIF-FRD provides membership (Membership) to relate fuzzy classes and their objects. With terms , Equality and Membership, we can easily represent objects and fuzzy class hierarchies.

Fuzzy restrictions use class constructors to restrict the range of a property for the context of the fuzzy class. In Fig. 9.13, restrictions are stated on datatype and fuzzy object property, as indicated by toClass and toDatatype. The datatype system

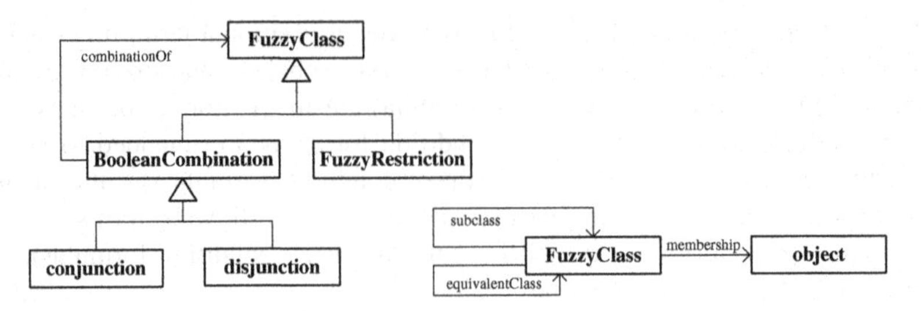

Fig. 9.12 The fuzzy class constructors and axioms of RIF-FRD

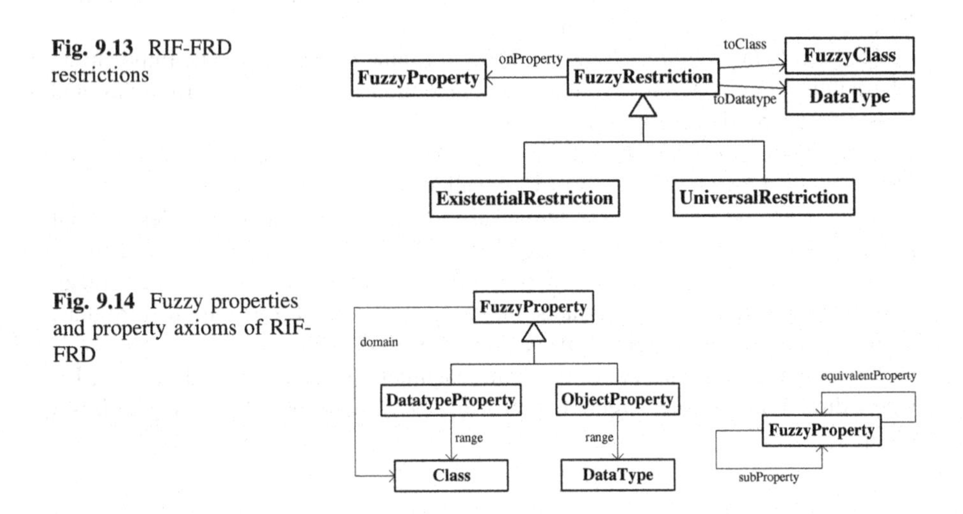

Fig. 9.13 RIF-FRD restrictions

Fig. 9.14 Fuzzy properties and property axioms of RIF-FRD

of RIF-FRD is provided by RIF-DTB document, which provides a predefined set of named datatypes, such as stings xsd:string. toClass and toDatatype are used to limit the value to a fuzzy class or a certain datatype. Exists and Forall are used to limit the variables to be used in the fuzzy rules. They require that all the variables used in fuzzy rules should be chosen from the variables declared with Exists and Forall.

Properties and Axioms

Generally speaking, in common rule languages, properties are used to represent two-ary relations, while properties in RIF-FRD can be employed to express n-ary (n = 2,3,...) relations, which lets RIF-FRD be more expressive to represent fuzzy knowledge and information in the real world. There are two kinds of fuzzy properties. They are datatype property and object property (cf. Fig. 9.14) whose domains are both classes. The range of the former is class, while the latter's range is datatype. The datatype is defined in the RIF-DTB document. When needing datatype, RIF-FRD can introduce it by using Import directive. Figure 9.14 also discusses fuzzy property axioms of RIF-FRD. As with fuzzy classes, fuzzy properties in RIF-FRD do not have explicit definitions. But we can obtain them from fuzzy terms and fuzzy formulas of RIF-FRD. The property axioms include property subsumption and property equivalence. These two axioms are implemented through the terms Subclass and Equality (in general, the two terms are used to represent class axioms), which may be a little counterintuitive. In fact, it is reasonable. In the sense of Meta Object Facility (MOF) metamodel, class and property are all modelled by metaclass. We can say that they are isogenous. The attributes that are fit for classes can be applied into properties analogically. Therefore, Subclass and Equality can be used to represent property axioms.

Rule Metamodel

A fuzzy rule implication (as shown in Fig. 9.15) is a kind of fuzzy formula, and it consists of two parts: antecedent and consequent, which are also called body and head of the rule, respectively. The relation between rule and its two parts can be represented by aggregation link, and so can antecedent, consequent and their atoms. The reason is that the same antecedent or consequent can appear in several fuzzy rules, and the same fuzzy atom is allowed in different antecedent or consequent. Both the antecedent and the consequent consist of a set of fuzzy atoms

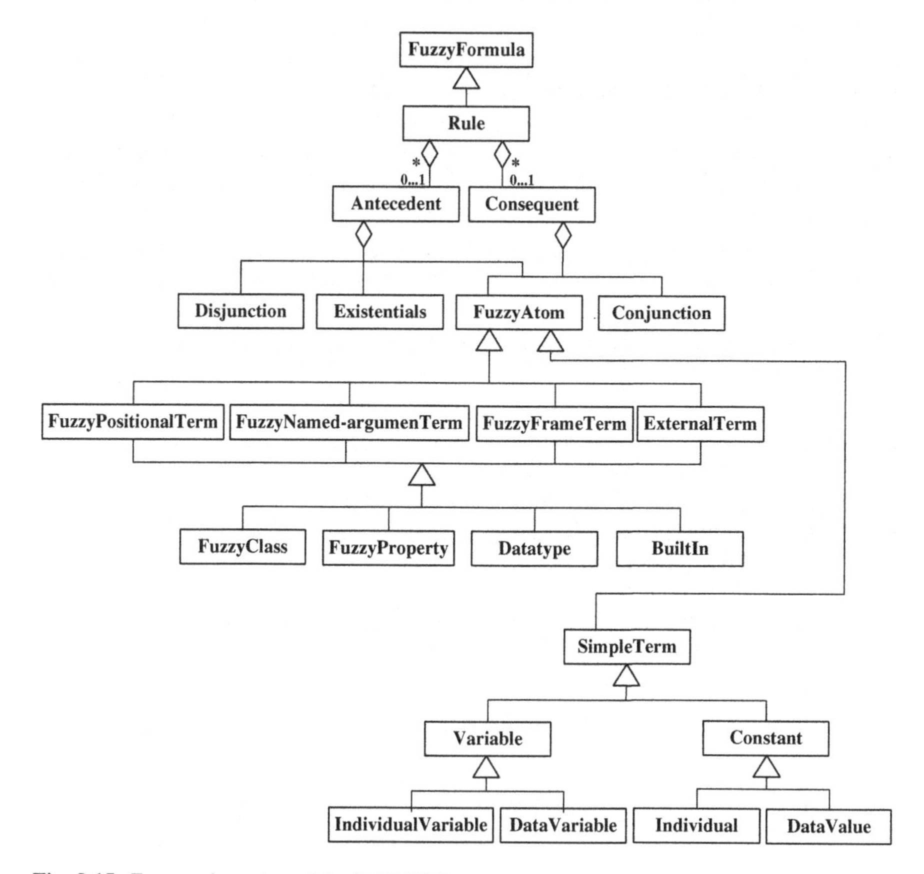

Fig. 9.15 Fuzzy rule metamodel of RIF-FRD

which can possibly be empty. A fuzzy rule means that if all fuzzy atoms of the antecedent hold, then the consequent holds. An empty antecedent is treated as trivially true, whereas an empty consequent is treated as trivially false. An antecedent is a set of conjunction, disjunction and existentials of fuzzy atoms, while a consequent is a fuzzy atomic formula or a conjunction of fuzzy atoms. Here, fuzzy atoms mean fuzzy atomic formulas and simple terms. Fuzzy atomic formulas used in rules consist of fuzzy positional formulas, fuzzy named-argument formulas, fuzzy frames and external terms. These atomic formulas are essentially composed of fuzzy class, fuzzy property, datatype and builtIn, where property can be n-ary, and datatype and builtIn can be introduced by Import directive from RIF-DTB document. Simple terms are variables and constants. And variables consist of individual variables and data variables, while constants are composed of individuals and data values. Again, data variables and data values are from RIF-DTB document.

RIF Concrete Syntax

To represent RIF-FRD rules in XML format, we define the XML schema of RIF-FRD. In the schema, document element is the root element, which is composed of group, annotation, formula, if-part, then-part, weight, and so on. Because of the fuzziness of formulas, weights (truth values between 0 and 1) are used with formulas, and they represent importance of formulas.

In the following, we give an example to illustrate elements in RIF-FRD XML schema. Inspired from the field of housing purchase, imagine that Jack wants to buy a house, and he asserts that if a house has a fairly low price, good quality, sound environment (for example, there is no much noise around the house), then he may buy the house, and the condition that the house has a low price is more important than the other conditions, and the condition that the house has good quality is a bit more important than the condition that the environment is harmonious. The presentation syntax of Jack's assertion is shown as follows:

```
Document (
Prefix (ppl  http://www.houseexample.com/people#)
Prefix (cpt  http://www.houseexample.com/concepts#)
Forall ?X (cpt: buy (ppl:Jack ?X)[0.95] :-
            And (cpt:priceLow(?X)[0.9]
                  cpt:qualityGood(?X)[0.8]
                  cpt:environmentSound(?X)[0.7])
          )
)
```

Compared with RIF, RIF-FRD has fuzzy terms (such as priceLow, such terms will be given a membership degree by some membership function), and every term has a weight (a truth value between 0 and 1) which is used to represent importance of the term in the whole rule. The following codes are corresponding XML form of the above fuzzy rule.

```
<!DOCTYPE Document [
<!ENTITY ppl "http:www.houseexample.com/people#">
<!ENTITY cpt"http:www.houseexample.com/concepts#">
]>
<Document>
xmlns=http://www.w3.org/2007/rif#
xmlns:xs=http://www.w3.org/2001/XMLSchema#>
<payload>
 <Group>
  <Forall>
   <declare><Var>X</Var></declare>
   <formula>
   <Implies>
    <if>
     <And>
      <formula>
       <Atom>
        <op><Const type="&rif;iri">&cpt;priceLow</Const></op>
        <args ordered="yes"><Var>X</Var></args>
       </Atom>
       <Const type="&xs;float">0.9</Const>
      </formula>
      <formula>
       <Atom>
          <op><Const type="&rif;iri">&cpt;qualityGood</Const> </op>
          <args ordered="yes"><Var>X</Var></args>
```

```
        </Atom>
        <Const type="&xs;float">0.8</Const>
       </formula>
       <formula>
       <Atom>
         <op><Const type="&rif;iri">&cpt;environmentSound
         </Const></op>
         <args ordered="yes"><Var>X</Var></args>
       </Atom>
       <Const type="&xs;float">0.7</Const>
      </formula>
     <And>
    </if>
    <then>
       <Atom>
         <op><Const type="&rif;iri">&cpt;buy</Const></op>
         <args ordered="yes">
         <Const type="&xs;string">Jack</Const>
         <Var>X</Var>
         </args>
       </Atom>
       <Const type="&xs;float">0.95</Const>
    </then>
   </Implies>
   </formula>
  </Forall>
 </Group>
</payload>
</Document>
```

9.3.2 f-RIA Architecture

Based on the metamodel of RIF-FRD in Sect. 9.3.1, we construct our architecture f-RIA (fuzzy Rule Interchange Architecture) in the following. As shown in Fig. 9.16, f-RIA is centered on RIF-FRD, and it supports rule interchange between RIF-FRD and f-SWRL, RuleML and f-R2ML (fuzzy R2ML). Certainly, pairwise transformations among the four languages can also be implemented. What's more, f-RIA suffices to interchange rules between languages which are centered on f-SWRL, RuleML and f-R2ML. Among others, all the transformations are

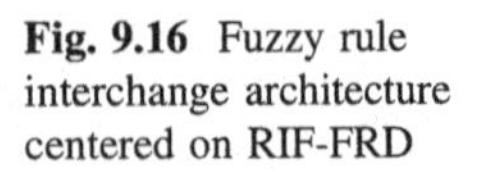
Fig. 9.16 Fuzzy rule interchange architecture centered on RIF-FRD

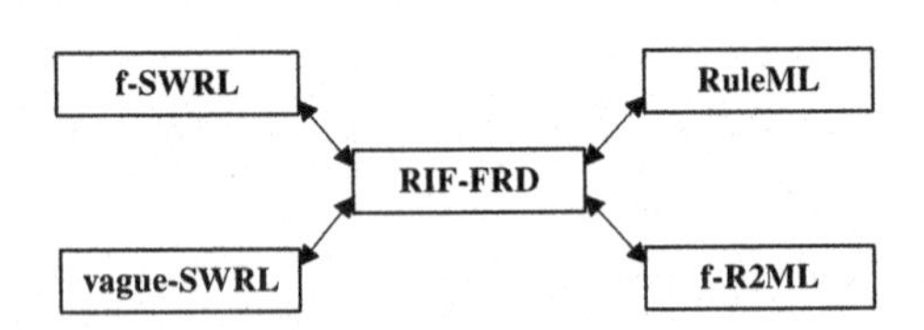

bidirectional and they are all based on the metamodels of the languages to be transformed, rather than based on XML-style concrete syntaxes. With the enhancement of expressiveness of RIF-FRD, f-RIA will be capable of transforming more rule languages. f-RIA is composed of two parts. The first part describes three pairs of transformations based on abstract syntaxes, i.e., transformations between RIF-FRD and f-SWRL, between RIF-FRD and RuleML, and between RIF-FRD and f-R2ML, while the second part depicts the mappings between the abstract syntaxes and the concrete syntaxes of the four fuzzy languages.

9.4 Summary

As an important representation of knowledge, rules possess many features including highly expressive power and requiring small memory space, which let rules become indispensable and widely used in many areas of the Semantic Web. However, there is much imprecise and uncertain knowledge in the context of the Semantic Web. Therefore, fuzzy extensions to rule languages and rule interchange architectures are needed to support the direct representation of uncertain knowledge in the Semantic Web.

In this chapter, some fuzzy rule languages such as f-SWRL, f-NSWRL, f-SW-if–then-RL, f-SW-if–then-unless-RL, and f-R2ML were introduced. Also, it is unavoidable that these languages communicate with each other, and thus fuzzy rule interchange becomes an important part of the Semantic Web. In particular, Rule Interchange Format (RIF) has become W3C's candidate recommendation on rule interchange. To let RIF represent and interchange fuzzy knowledge in the Semantic Web, a fuzzy RIF framework—fuzzy Rule Interchange Architecture (f-RIA) was described in this chapter. All of these fuzzy rule languages and the fuzzy rule interchange architecture will help to represent, reason and markup fuzzy knowledge in the Semantic Web.

References

Berners-Lee T, Hendler J, Lassila O (2001) The semantic web. Sci Am 284(5):34–43

Boley H, Grosof B, Tabet S (2005) RuleML tutorial. http://ruleml.org/papers/tutorial-ruleml-20050513.html

Brockmans S, Volz R, Eberhart A, Löffler P (2006) A metamodel and UML profile for rule-extended OWL DL ontologies. Proceedings of ESWC 2006, LNCS vol 4011, pp 303–316

Clark KL (1978) Negation as failure, logic and data bases. Plenum Press, New York, pp 293–322

Hirtle D, Boley H, Grosof B, Kifer M, Sintek M, Tabet S, Wagner G (2006) Schema specification of RuleML 0.9. http://www.ruleml.org/0.9/

Horrocks I, Patel-Schneider PF, Boley H, Tabet S, Grosof B, Dean M (2004) SWRL: a semantic web rule language | combining OWL and RuleML. http://www.w3.org/Submission/SWRL/

Kifer M (2008) Rule interchange format: the framework. Proceedings of RR2008, LNCS, vol 5341, pp 1–11

Ma ZM, Wang X (2012) Rule Interchange in the semantic web. J Inf Sci Eng 28(2):393–406
Pan JZ, Stoilos G, Stamou G, Tzouvaras V, Horrocks I (2006) f-SWRL: a fuzzy extension of SWRL. J Data Semant Emergent Semant 4090:28–46
Wagner G, Viegas Damásio C, Giurca A (2005) Towards a general web rule language. Int J Web Eng Technol 2(2–3):181–206
Wagner G, Giurca A, Lukichev S (2006) A usable interchange format for rich syntax rules integrating OCL, RuleML and SWRL. Reasoning on the Web Workshop at WWW (RoW'06)
Wang X, Ma ZM, Yan L, Meng XF (2008) Vague-SWRL: a fuzzy extension of SWRL. Proceedings of RR 2008, LNCS, vol 5341, pp 232–233
Wang X, Ma ZM, Yan L, Cheng JW (2009a) If-then and if-then-unless rules in the semantic Web. Proceedings of WI 2009, pp 357–360
Wang X, Ma ZM, Xu CM, Cheng JW (2009b) Nonmonotonic fuzzy rules in the semantic web. Proceedings of FSKD 2009, pp 275–279
Wang X, Sun J, Meng XF, Chen J (2011) f-R2ML: a fuzzy rule markup language. Proceedings of FSKD 2011, pp 1275–1279
Wang X, Yan L, Ma ZM, Zhao R (2012) RIF-FRD: a RIF dialect based on fuzzy sets. Proceedings of FSKD 2010, pp 1912–1916
Zadeh LA (1965) Fuzzy sets. Inf Control 8(3):338–353
Zhao JD, Boley H (2008) Uncertainty treatment in the rule interchange format: from encoding to extension. Proceedings of the fourth international workshop on uncertainty reasoning for the semantic web

Index

Z. Ma et al., *Fuzzy Knowledge Management for the Semantic Web*,
Studies in Fuzziness and Soft Computing 306, DOI: 10.1007/978-3-642-39283-2,

Printed in the United States
By Bookmasters

T0140057

Lecture Notes in Artificial Intelligence 13456

Subseries of Lecture Notes in Computer Science

More information about this subseries at https://link.springer.com/bookseries/1244

Honghai Liu · Zhouping Yin · Lianqing Liu ·
Li Jiang · Guoying Gu · Xinyu Wu ·
Weihong Ren (Eds.)

Intelligent Robotics and Applications

15th International Conference, ICIRA 2022
Harbin, China, August 1–3, 2022
Proceedings, Part II

Editors
Honghai Liu
Harbin Institute of Technology
Shenzhen, China

Zhouping Yin
Huazhong University of Science
and Technology
Wuhan, China

Lianqing Liu
Shenyang Institute of Automation
Shenyang, Liaoning, China

Li Jiang
Harbin Institute of Technology
Harbin, China

Guoying Gu
Shanghai Jiao Tong University
Shanghai, China

Xinyu Wu
Shenzhen Institute of Advanced Technology
Shenzhen, China

Weihong Ren
Harbin Institute of Technology
Shenzhen, China

ISSN 0302-9743 ISSN 1611-3349 (electronic)
Lecture Notes in Artificial Intelligence
ISBN 978-3-031-13821-8 ISBN 978-3-031-13822-5 (eBook)
https://doi.org/10.1007/978-3-031-13822-5

LNCS Sublibrary: SL7 – Artificial Intelligence

This Springer imprint is published by the registered company Springer Nature Switzerland AG
The registered company address is: Gewerbestrasse 11, 6330 Cham, Switzerland

Preface

With the theme "Smart Robotics for Society", the 15th International Conference on Intelligent Robotics and Applications (ICIRA 2022) was held in Harbin, China, August 1–3, 2022, and designed to encourage advancement in the field of robotics, automation, mechatronics, and applications. It aims to promote top-level research and globalize the quality research in general, making discussions, presentations more internationally competitive and focusing on the latest outstanding achievements, future trends, and demands.

ICIRA 2022 was organized by Harbin Institute of Technology, co-organized by Huazhong University of Science and Technology, Shanghai Jiao Tong University, and Shenyang Institute of Automation, Chinese Academy of Sciences, undertaken by State Key Laboratory of Robotics and Systems, State Key Laboratory of Digital Manufacturing Equipment and Technology, State Key Laboratory of Mechanical Systems and Vibration, and State Key Laboratory of Robotics. Also, ICIRA 2022 was technically co-sponsored by Springer. On this occasion, ICIRA 2022 was a successful event this year in spite of the COVID-19 pandemic. It attracted more than 440 submissions, and the Program Committee undertook a rigorous review process for selecting the most deserving research for publication. The advisory Committee gave advice for the conference program. Also, they help to organize special sections for ICIRA 2022. Finally, a total of 284 papers were selected for publication in 4 volumes of Springer's Lecture Note in Artificial Intelligence. For the review process, single-blind peer review was used. Each review took around 2–3 weeks, and each submission received at least 2 reviews and 1 meta-review.

In ICIRA 2022, 3 distinguished plenary speakers and 9 keynote speakers had delivered their outstanding research works in various fields of robotics. Participants gave a total of 171 oral presentations and 113 poster presentations, enjoying this excellent opportunity to share their latest research findings. Here, we would like to express our sincere appreciation to all the authors, participants, and distinguished plenary and keynote speakers. Special thanks are also extended to all members of the Organizing Committee, all reviewers for peer-review, all staffs of the conference affairs group, and all volunteers for their diligent work.

August 2022

Honghai Liu
Zhouping Yin
Lianqing Liu
Li Jiang
Guoying Gu
Xinyu Wu
Weihong Ren

Organization

Honorary Chair

Youlun Xiong	Huazhong University of Science and Technology, China

General Chairs

Honghai Liu	Harbin Institute of Technology, China
Zhouping Yin	Huazhong University of Science and Technology, China
Lianqing Liu	Shenyang Institute of Automation, Chinese Academy of Sciences, China

Program Chairs

Li Jiang	Harbin Institute of Technology, China
Guoying Gu	Shanghai Jiao Tong University, China
Xinyu Wu	Shenzhen Institute of Advanced Technology, Chinese Academy of Sciences, China

Publication Chair

Weihong Ren	Harbin Institute of Technology, China

Award Committee Chair

Limin Zhu	Shanghai Jiao Tong University, China

Regional Chairs

Zhiyong Chen	The University of Newcastle, Australia
Naoyuki Kubota	Tokyo Metropolitan University, Japan
Zhaojie Ju	The University of Portsmouth, UK
Eric Perreault	Northwestern University, USA
Peter Xu	The University of Auckland, New Zealand
Simon Yang	University of Guelph, Canada
Houxiang Zhang	Norwegian University of Science and Technology, Norway

Advisory Committee

Contents – Part II

Mechanism Design, Control and Sensing of Soft Robots

Intelligent Perception and Control of Rehabilitation Systems

Micro or Nano Robotics and Its Application

Biosignal Acquisition and Analysis

Rehabilitation and Assistive Robotics

Design and Control of a Bimanual Rehabilitation System for Trunk Impairment Patients

Lufeng Chen[1], Jing Qiu[2(✉)], Lin Zhou[3], Hongwei Wang[3], Fangxian Jiang[1], and Hong Cheng[1]

[1] Department of Automation Engineering, University of Electronic Science and Technology of China, Chengdu, China
[2] Department of Mechanical and Electric Engineering, University of Electronic Science and Technology of China, Chengdu, China
qiujing@uestc.edu.cn
[3] Beijing Aerospace Measurement and Control Technology Co., Ltd, Beijing, China

Abstract. Stroke is one of the leading causes of severe long-term disability worldwide. Past research has proved the effectiveness and importance of trunk rehabilitation in enhancing the locomotive capacities and balance of patients with hemiplegia. The current robot-assisted rehabilitation mainly focuses on limbs but less on trunk training. In this paper, we have developed a bimanual rehabilitation system for trunk impairment patients. The main contribution of the work is twofold. First, the synchronized training trajectory generation method of the two articulated robots is proposed by introducing a simplified yet effective trunk kinematic model and taking the past demonstrations as experience. Second, an improved admittance controller is embedded to modulate the training trajectories online in catering to subjects' different locomotion capacities. The physical experiments are conducted to validate the feasibility of our system. It is expected that, upon the availability of clinical trials, the effectiveness of our system in the locomotion restoring training of trunk impairment patients will be further confirmed.

Keywords: Stroke · Trunk rehabilitation · Robot system · Trajectory planning · Admittance control.

1 Introduction

Stroke is one of the most common global healthcare problems and the most common causes of adult disability [14]. Post-stroke patients suffer a lot daily due to the lack of locomotion capabilities and associated functions, especially

This work was supported in part by the National Natural Science Foundation of China (No. 52005082) and Natural Science Foundation of Sichuan (NSFSC)(No. 2022NSFSC1956).

H. Liu et al. (Eds.): ICIRA 2022, LNAI 13456, pp. 3–12, 2022.
https://doi.org/10.1007/978-3-031-13822-5_1

for those with impaired upper extremity [6]. The trunk plays a curial role in stabilizing the spine and upper extremity. Evidence in the literature suggests that patients with hemiplegia unavoidably witness a loss in trunk muscle strength, significantly weakening the sitting balance and mobilities [15].

Extensive work has been reported focusing on hemiplegic upper and lower limbs [7–9], while little attention has been received on the trunk rehabilitation. Recent literature asserts that trunk rehabilitation enhances the locomotion and balance of patients with hemiplegia [2,8]. Conventionally, the trunk impairment subjects receive simple passive exercises given by the therapists. Opposite to the passive one, the active training paradigm can be described as a challenging strategy that highly relies on the therapist to conduct the one-to-one training empirically, which is time inefficient and labor-intensive. More importantly, such a training pattern designates to provoke motor plasticity and relearning function through a guiding process and therefore improve motor recovery [10]. Conceivably, the effectiveness of the function recovery varies, thanks to therapists' experience substantially. Therefore, it calls for a standardized rehabilitation paradigm to improve therapeutic efficacy.

Rehabilitation robot is a newly emerged technology for rehabilitation intervention. The three primary characteristics of the robot-assisted rehabilitation are highly appreciated in the stroke recovery area, i.e., repetition, high intensity, and task specificity [5]. Research in developing such robots has been proliferating, and various robots have emerged in the last two decades. At the current stage, the recent research suggests that the robot-assist rehabilitation shows no significant difference in scores over other intensive comparison therapy for the upper limb [7]. Regarding the mechanical structures, rehabilitation robots usually fall into two categories: exoskeleton-type and end-effector type. The exoskeleton-type aligns the robot axes with the anatomical axes of the wearer, which controls the patients' joints locomotion in a passive pattern that has been successfully applied to upper extremity [4,7], hand [1], trunk [13], and lower extremity [17]. The end-effector type robot utilizes its end-effectors to guide the movement of the patient's one or multiple joints. Due to the complexity of the human musculoskeletal system, the design of the exoskeleton-type robot is a complicated task to resemble the human joints. The human-robot misalignment issue is still an open problem [16]. On the contrary, the end-effector type robot only connects the subject at some distal points without matching the human joints, which poses difficulties in controlling the isolated movement of a single joint [11]. Owing to the much smaller sample size, the battle for the more effective one has not been settled yet.

A systematic review of robot-assisted rehabilitation for stroke patients is reported in Ref. [3]. Most of the reviewed devices are expected to improve the locomotive function of some specific parts, such as the hand, to some extent. Maciejasz et al., [9] targeted the review of the rehabilitation systems exclusively for the upper limb. Regarding robot-assisted trunk rehabilitation, to the best of our knowledge, too little work has been devoted to developing a trunk rehabilitation system. The most relevant research we can find is a sitting balance robot [13] dedicated to post-stroke patients with the unbalanced sitting syndrome. They led pioneering work in developing a rehabilitation device that can perform a single-DOF reclining motion for trunk stability training.

To mimic the bimanual training paradigm of the therapist and to cater to the distinct body size of patients, we develop an end-effector-type trunk rehabilitation system integrating two collaborative robot arms for versatile and customized training for trunk impairment patients. The paper delivers the following main contributions:

- A bimanual rehabilitation system which combines two articulated 6-DoF robot is developed.
- The customized rehabilitation trajectory specific to the distinct subject is generated by introducing a simplified trunk kinematic model.
- An improved admittance control strategy is proposed to control the guidance force and to intend to invoke the self-movement of the subject during the trunk rehabilitation.

2 System Overview

The primary function of the trunk rehabilitation system is to replace therapists from repetitive and customized training trajectories for trunk impairment patients. The fundamental trunk training exercise (TTE) is the forward and backward bending, which is the result of the flexion and extension of the trunk [12]. Furthermore, the interaction with the human body is a delicate task wherein a physical therapist continually assesses the physical response of the impaired limb and adjusts the level of assistance to ensure the safety and training effectiveness of the subject. The therapist-mimic trunk training characteristics should have three functions: (1) Capable of executing various repetitive trunk training exercises; (2) Customized training trajectories; (3) Adjustable level of assistance.

Anatomically, the spinal column mainly takes charge of the trunk movements, consisting of a collection of vertebrae connected by ligaments, small muscles, vertebral joints, and intervertebral discs [12]. The mobile part of the trunk rooted at the sacrum can be organized into two main regions: 12 thoracic and 5 lumbar vertebrae, denoted as T1-T12 and L1-L5, respectively. Ideally, 18 revolute joints with up to 17 shape parameters are required to describe the trunk kinematics. Moreover, since each vertebra has a unique size and range of movement, it is tough to design an exoskeleton-type robot that resembles all these 18 distinct joints. Accordingly, the end-effector type becomes a better choice to fulfill the compound motions of the trunk by controlling the trajectories of some distal points.

The distal point that determines the trunk motion is described as the motion contact point (MCP) between the robot end and the human body. Ideally, 6-DoF is sufficient to push the MCP to move and rotate in the task space. To further perform the repetitive movement, such as forward and backward bending, we

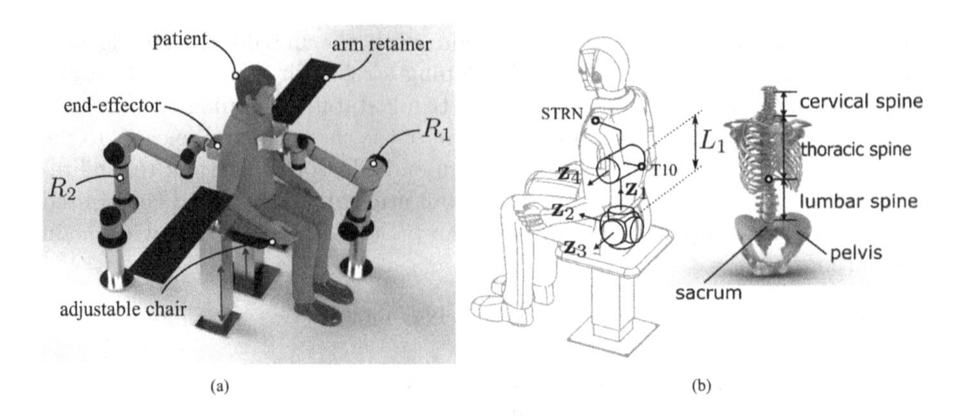

Fig. 1. System overview.

develop a bimanual trunk rehabilitation system consisting of two 6-DoF collaborative robots to replace the physical therapists to conduct training. The trunk training exercises can be conducted via driving the MCPs in a synchronized fashion performed by the two robots. Referring to Fig. 1(a), the rehabilitation system consists of two collaborative robots (denoted as R_1 and R_2), a height-adjustable chair, two arm retainers. The system is initialized by adjusting the chair and the arm retainers to match the sitting posture of the patient when he/she sits steadily on the chair with the two arms supported by the arm retainers. The two robots receive the move commands from a computing terminal which generates the online rehabilitation trajectories with feed information of the robot motion states and the contact forces between the robot and the patient.

To satisfy the three underlying functions described above, in this paper, we first introduce a simplified trunk kinematic model and integrate imitation learning to generate one tailored training trajectory. Then, an improved admittance controller that integrates the two robots is designed to modulate the training trajectories of the two robots in an online fashion.

3 Trunk Rehabilitation Trajectory Generation

Repetitive or periodic movements or motions are regularly seen in scenarios for the robot tasks, such as the repetitive rehabilitation for hemiplegia patients. If each past successful rehabilitation trajectory can be treated as one demonstration, the trajectories of each demonstration may vary a little due to the rehabilitation subject or environmental differences. In order to best utilize the past demonstrations as the experience and reflect the pattern behind these demonstrations, we differentiate the customized trunk training into a two-stage robot trajectory generation scheme. In the first stage, a simplified trunk model is introduced to generate a reference path for the patient with a distinct body size, followed by a customized training trajectory generated by the imitation learning of previous demonstrations in the second stage.

3.1 Reference Path Generation

The key to conducting customized training is to obtain the trajectory of the MCP of the given subject. Due to the distinct body size, trunk kinematics is highly desired to retrieve the specific trajectory. As previously asserted, the ideal motion of the distal point on the trunk is the compound motion of up to 18 revolute joints. Direct trunk modeling becomes an impossible task. To best reflect the anatomical characteristics of the trunk while reducing the computational complexity, we introduce a simplified 4-DOF serial chain model. Specifically, a spherical joint is mounted at the sacrum to match its high flexibility in all directions, while a revolute joint is placed at the vertebra connecting the lumbar spine and the thoracic spine, as shown in Fig. 1(b).

Before the representation of the trunk trajectory, we first define a global coordinate frame which served as the base frame for the trunk model. A subject-specific body coordinate frame (BCS) $O_b x_b y_b z_b$ centered at the intersection of the lumbar spine and the pelvis transverse plane is defined. The plane $y_b O_b z_b$ passes through the sagittal plane with y_b perpendicular to the coronal plane and z_b perpendicular to the transverse plane. The Denavit-Hartenberg (DH) convention is applied to attach a local frame on each rotational axis. The trunk kinematics can be described by the multiplication of the corresponding homogeneous matrix for each joint,

$$T_s(q,a) = \begin{bmatrix} R_s(q,a) & p_s(q,a) \\ \mathbf{0}^T & 1 \end{bmatrix} \tag{1}$$

where: $R_s(q,a)$ is the rotational matrix, $p_s(q,a)$ is the position vector, $q = [q_1, q_2, q_3, q_4]^T$ is the joint variable vector, $a = [a_1, a_2]^T$ are the body size specific variables representing the lengths of the lumbar spine and the effective thoracic spine, respectively. $R_s(q,a)$ and $p_s(q,a)$ are related to the orientation and position of the proximal chain (thoracic spine) of the trunk. Therefore, for a given subject, the position of any point on the thoracic spine can be easily described in BCS as,

$$p_e = T_s(q) p_l \tag{2}$$

where p_e and p_l are the positions of a point on the thoracic spine in the local coordinate frame originate at the last revolute joint and BCS, respectively. Note that $T_s(q)$ is the function of the generalized coordinates in the joint space obtained by the DH representation of the trunk model. When the joint variable vector is described by a sequence, i.e., $q_0 = q(t)$, $t \in [0,1]$, the resultant position sequence $p_e(q_0)$ becomes one parametric reference path of a point-of-interest (POI) on the trunk in the BCS.

Ideally, our developed trunk model can describe any trunk movement with an appropriate joint variable vector sequence, such as forward-leaning and reclining, side bending, and rotation. It is worth noting that the 4-DoF trunk model assumes that the thoracic and lumbar spine are relatively rigid, and the pseudo rotational joint achieves the relative motion of the two parts in between. The assumption holds when the rotation angle of the two parts is in a moderate range.

3.2 Customized Training Trajectory Generation

Due to the similarity of each successful repetitive training cycle, we intend to extract the features hidden among previous demonstrations while planning the subsequent training trajectory. The spatial trajectory of any point on the trunk computed from the simplified trunk kinematics is a function of two personalized parameters and the time-dependent generalized joint vector per 2. The similarity of past demonstrations encapsulates the probability distribution of the position of a set of task-related robot end trajectories, which is treated as the training data. The Gaussian mixture model (GMM) is utilized for encoding the training data, while the Gaussian mixture regression (GMR) is applied to imitate the demonstrations.

Given a set of demonstrations as $\{\{p_{m,h}, \kappa_{m,h}\}_{m=1}^{M}\}_{h=1}^{H}$, where $p_{m,h} \in \mathbb{R}^3$ and $\kappa_{m,h} \in \mathbb{R}^3$ are the input and output data, respectively, M and H represent the number of sample points and trajectories, respectively. We apply GMM to estimate the joint probabilistic distribution $\mathcal{P}(p, \kappa)$, which can be described as,

$$\mathcal{P}(p, \kappa) \sim \sum_{i=1}^{K} \pi_i \mathcal{N}(\mu_i, \Sigma_i) \tag{3}$$

where: π_i, μ_i and Σ_i are the prior probability, mean and covariance of the i-th Gaussian components, respectively, K is the number of Gaussians.

Once the given trajectories are encoded in GMM, a probabilistic reference trajectory $\{\hat{\kappa_i}\}_{i=1}^{N}$ can be retrieved by GMR which maximized the posterior probability $\mathcal{P}(\kappa|p)$ for each query points. The resultant trajectory follows the probability distribution as

$$\hat{\kappa_i}|p \sim \mathcal{N}(\hat{\mu_i}, \hat{\Sigma_i}) \tag{4}$$

where $\hat{\mu_i}$ and $\hat{\Sigma_i}$ are the mean and covariance of point $\hat{\kappa_i}$.

4 Robot arms integration and controls

4.1 Trunk Rehabilitation Trajectory Planning for the Bi-manunal Robots

In fulfilling the compound motions of the trunk while catering to our system configuration, two contact points are required on the patient's body. During the rehabilitation, the two articulated robots should always be in contact with the body and drive the whole trunk following the trajectories. Therefore, the corresponding trajectories of two MCPs become the robot trajectories. In this paper, we will take the forward-leaning and reclining motion of the trunk movement as an example to illustrate the trajectory generation. The rest training tasks will follow the same procedure. Referring to Figure 1(b), Two marker points, i.e., STRN and T10, which are often applied in the motion capture system, are chosen as the positions of the two MCPs. The trajectories of the two MCPs for the forward-leaning (or reclining) are two spatial curves generated by transforming a series of generalized coordinates in the joint space, i.e., $\{\Theta_i\}$ into the task space. The trajectory of the spatial curve starting from the initial position p_0 can be represented by Eq. 2.

To reveal the relationship between human movements and robot motions, we first define six Cartesian coordinate frames in the system - three fixed frames (world coordinate system (WCS) and two robot coordinate systems (RCS)) and three float frames (local coordinate system (LCS), tip coordinate system (TCS) and holder coordinate system (HCS)). The transformation matrix from WCS to HCS can be computed by the kinematic chain through the robot. Suppose a 6D pose is defined in the task space, i.e., $p = \{x, y, z, \rho, \varphi, \theta\}$, where $\{x, y, z\}$ is the position and $\{\rho, \varphi, \theta\}$ is the orientation represented in Euler angle, we say p is reachable by the robot arm if one or multiple generalized coordinates can be found in the joint space, which is also known as the inverse kinematic transformation (IKT). Since the reference point is defined on the human body, another kinematic chain starting from WCS, going through BCS, and ending at LCS is also built to describe the relationship between WCS and LCS. Once the path of the two MCPs is given in BCS, the path for the two robots in the correspondent RCS can be easily calculated.

4.2 Controls

An improved admittance controller is implemented to track the desired force trajectories of the sensor pair mounted between the robot wrists and the end-effectors. Thanks to the cushion-like design of the end-effector, the subject is always confined by the two end-effectors. Therefore, for each robot, the control can be modeled as a one-dimensional impedance model with a spring and a damper. Once the desired trajectories of the two MCPs are given, the actual trajectories are modulated by the external force pair on the contact points between the force sensors and the subject. The outputs of the system are the trajectories of the two robots, as follows,

$$M_d(\ddot{x}_d^i - \ddot{x}_0^i) + D_d(\dot{x}_d^i - \dot{x}_0^i) + K_d(x_d^i - x_0^i) = k_i F_i + k_{\hat{i}} F_{\hat{i}} \tag{5}$$

where: M_d, D_d and K_d are the inertial, damping and stiffness properties of the system, x_0 and x_d are the desire and actual trajectories, respectively, $i \in Q = \{1, 2\}$ represents the robot index and $\hat{i} = Q - i$ indicates the rest index, $\{F_i, F_{\hat{i}}\}$ denotes the external force pair.

5 Simulation and experiment

To test the feasibility of our system and algorithms, we have developed a physical prototype, as shown in Fig. 3(a). A suite of algorithms related to path planning and motion planning is implemented in C++. The data visualization is implemented in Matlab.

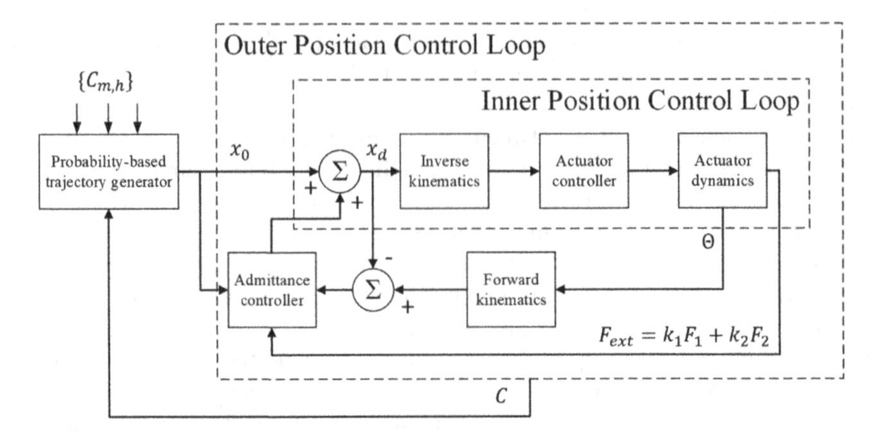

Fig. 2. Admittance control strategy for trunk rehabilitation training of the bi-manual robot.

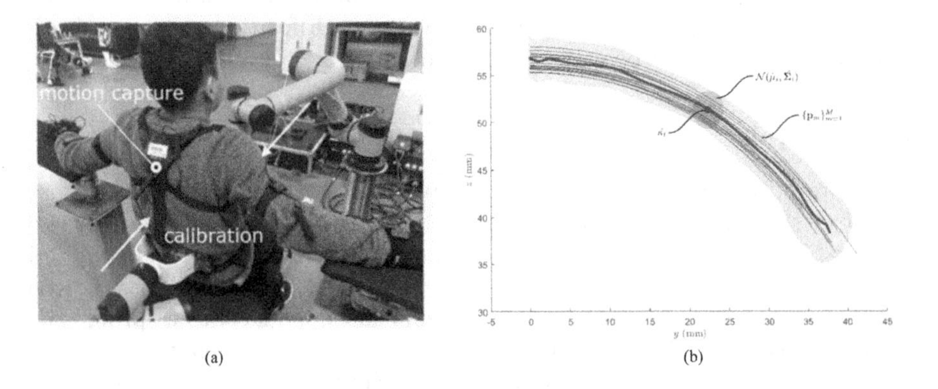

Fig. 3. (a) Physical experiment; (b) Motion capture from AxisNeuro.

5.1 Validation of the System

In the real system, apart from the maximally allowed velocity constraint, another safety precaution is also applied. The force sensors are mounted at the end-effector to avoid extreme interaction force between the patient and the robot arm. The threshold in our current experiment is fixed to 30 N. At the beginning of the test, the reference training path generated from Section 3.1 is sent to the controllers. The training trajectory is updated after each training cycle per Eq. 4.

5.2 Validation of the customized training trajectory generation and control strategy

Figure 3(b) shows an example of estimating the probability distribution given 10 trajectories for trunk forward-leaning and bending. After the trajectories

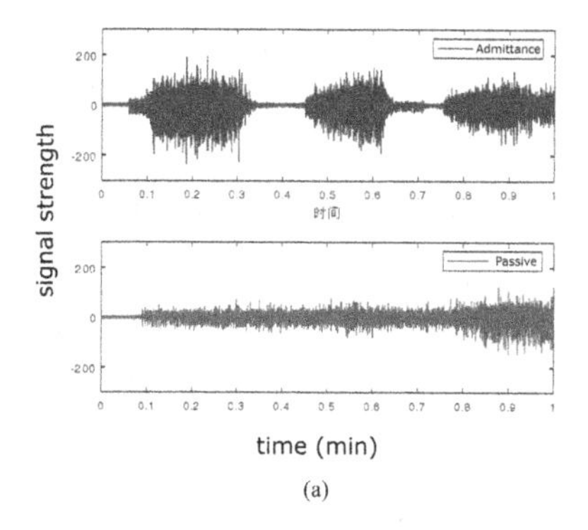

(a)

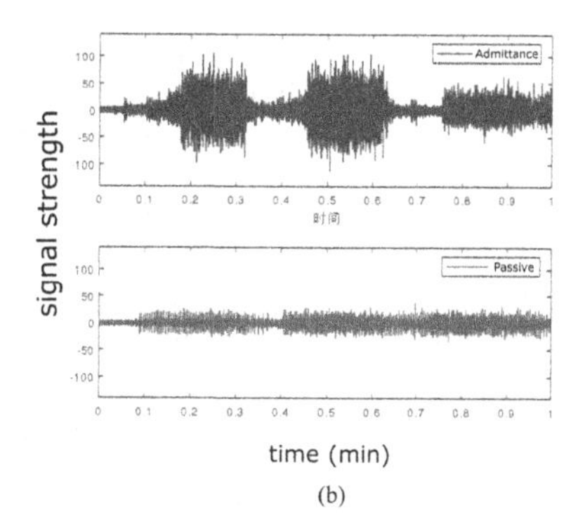

(b)

Fig. 4. Comparison of sEMG signals of the vastus lateralis of two legs between admittance mode and passive mode.

are encoded by GMM, the probabilistic reference trajectory $\{\hat{\kappa}_i\}, i = 1, 2, ..., N$ together with the corresponding probability distribution are computed by GMR.

To validate the effectiveness of the assistance level adjustment during the rehabilitation, we use the Noraxon-DTS sEMG acquisition system to collect the sEMG signals of the leg muscles during the training process. The rectus femoris, vastus lateralis and vastus medialis muscle of the thigh are selected to retrieve the sEMG data. We also implement a regular passive position controller for comparison. Figure 4 illustrates the comparison of sEMG signals between the two modes. It can be clearly seen that the sEMG signal has a periodic pattern during the repetitive training under the admittance control strategy, indicating a periodic activation of the contraction and relaxation of the muscles.

6 Conclusion and Future Work

In this study, we developed a novel training device for trunk rehabilitation of post-stroke patients. Explicitly, the customized training trajectory generation and cooperative control strategy related to the foundation of the system are resolved. To the best of our knowledge, it is the first trunk rehabilitation system integrating two collaborative robot arms to conduct various training actions.

Nevertheless, more work is needed, and the main caveat of the current work is the lack of clinical trials. It is expected that, upon the availability of the clinical trial data, the effectiveness of our system in the locomotion restoring training of trunk impairment patients will be further confirmed.

References

1. Agarwal, P., Fox, J., Yun, Y., O'Malley, M.K., Deshpande, A.D.: An index finger exoskeleton with series elastic actuation for rehabilitation: Design, control and performance characterization. Int. J. Robot. Res. **34**(14), 1747–1772 (2015)

2. Cabanas-Valdés, R., Cuchi, G.U., Bagur-Calafat, C.: Trunk training exercises approaches for improving trunk performance and functional sitting balance in patients with stroke: a systematic review. NeuroRehabilitation **33**(4), 575–592 (2013)
3. Chang, W.H., Kim, Y.H.: Robot-assisted therapy in stroke rehabilitation. J. Stroke **15**(3), 174–181 (2013)
4. Hunt, J., Lee, H.: Development of a low inertia parallel actuated shoulder exoskeleton robot for the characterization of neuromuscular property during static posture and dynamic movement. In: International Conference on Robotics and Automation, pp. 556–562 (2019)
5. Langhorne, P., Bernhardt, J., Kwakkel, G.: Stroke rehabilitation. The Lancet **377**(9778), 1693–1702 (2011)
6. Li, C., Rusak, Z., Horvath, I., Ji, L., Hou, Y.: Current status of robotic stroke rehabilitation and opportunities for a cyber-physically assisted upper limb stroke rehabilitation. In: Proceedings of the 10th international symposium on tools and methods of competitive engineering (TMCE 2014), pp. 358–366 (2014)
7. Lo, A.C., et al.: Robot-assisted therapy for long-term upper-limb impairment after stroke. N. Engl. J. Med. **362**(19), 1772–1783 (2010)
8. Lo, K., Stephenson, M., Lockwood, C.: Effectiveness of robotic assisted rehabilitation for mobility and functional ability in adult stroke patients: a systematic review. JBI Database System Rev. Implement. Rep. **15**(12), 3049–3091 (2017)
9. Maciejasz, P., Eschweiler, J., Gerlach-Hahn, K., Jansen-Troy, A., Leonhardt, S.: A survey on robotic devices for upper limb rehabilitation. J. Neuroeng. Rehabil. **11**, 3 (2014)
10. Marchal-Crespo, L., Reinkensmeyer, D.J.: Review of control strategies for robotic movement training after neurologic injury. J. Neuroeng. Rehabil. **6**(1), 1–15 (2009)
11. Molteni, F., Gasperini, G., Cannaviello, G., Guanziroli, E.: Exoskeleton and end-effector robots for upper and lower limbs rehabilitation: narrative review. PM and R **10**(9), 174–188 (2018)
12. Monheit, G., Badler, N.I.: A kinematic model of the human spine and torso. IEEE Comput. Graphics Appl. **11**(2), 29–38 (1991)
13. Song, Z., Ji, L., Wang, R., Wu, Q., Sun, X., Fan, S.: A sitting balance training robot for trunk rehabilitation of acute and subacute stroke patients. In: Proceedings of the 6th International Conference on Bioinformatics and Biomedical Science, Singapore, pp. 147–151. (2017)
14. Tyson, S.F., Hanley, M., Chillala, J., Selley, A., Tallis, R.C.: Balance disability after stroke. Phys. Ther. **86**(1), 30–38 (2006)
15. Wolfe, C.D.: The impact of stroke. Br. Med. Bull. **56**(2), 275–286 (2000)
16. Zanotto, D., Akiyama, Y., Stegall, P., Agrawal, S.K.: Knee joint misalignment in exoskeletons for the lower extremities: effects on user's gait. IEEE Trans. Rob. **31**(4), 978–987 (2015)
17. Zoss, A.B., Kazerooni, H., Chu, A.: Biomechanical design of the Berkeley lower extremity exoskeleton. IEEE/ASME Trans. Mechatron. **11**(2), 128–138 (2006)

Study on Adaptive Adjustment of Variable Joint Stiffness for a Semi-active Hip Prosthesis

Meng Fan[1,2,3], Yixi Chen[1,2,3], Bingze He[1,2,3], Qiaoling Meng[1,2,3], and Hongliu Yu[1,2,3](✉)

[1] Institute of Rehabilitation Engineering and Technology, University of Shanghai for Science and Technology, No. 516, Jungong Road, Yangpu District, Shanghai 200093, China
yhl_usst@outlook.com

[2] Shanghai Engineering Research Center of Assistive Devices, Shanghai 200093, China

[3] Key Laboratory of Neural-Functional Information and Rehabilitation Engineering of the Ministry of Civil Affairs, Shanghai 200093, China

Abstract. The moment and impedance of the hip antagonist muscle groups can change dynamically with movement, but existing passive hip prostheses are unable to achieve this function and limit the walking ability of hip amputees. This paper presents a novel adaptive adjustment stiffness system for a semi-active hip prosthesis that can simulate the function of hip antagonist muscles according to real-time detection of gait phase information and improve the walking ability of hip amputees. A method for predicting hip prosthesis kinematic information in advance is proposed to improve real-time performance of stiffness adjustment. The kinematic information at the amputee's healthy side is acquired from the posture sensor, and the kinematic information of the prosthetic side is predicted using a nonlinear autoregressive neural network model. To realize adaptive adjustment of variable stiffness joint, we have developed a stiffness controller based on PID algorithm. The target values for stiffness controller are obtained by analyzing the walking of the healthy person, and the actual stiffness of the hip joint on the prosthetic is determined using angle sensors and torque sensors from a customized embedded system. The hip joint stiffness tracking experiments show that the stiffness controller can effectively improve the regularity of the stiffness curve of the prosthetic joint, and the joint torque is significantly increased. The gait symmetry tests show that gait symmetry indices SI and R_{II} were improved by 86.3% and 85.1%, respectively. The average difference in bilateral gait length of walking amputees was reduced from 6.6 cm to 0.7 cm. The adjustable stiffness system proposed in this study can adaptively adjust the stiffness of the prosthetic joint to the amputee's gait, effectively improving the amputee's walking ability.

Keywords: Variable stiffness control · Hip prosthesis · NARX neural network · Gait phase division

H. Liu et al. (Eds.): ICIRA 2022, LNAI 13456, pp. 13–23, 2022.
https://doi.org/10.1007/978-3-031-13822-5_2

1 Introduction

Wearing lower limb prosthesis is a way to restore the walking function of lower limb amputees. Lower limb prostheses can be divided into hip prostheses, knee prostheses, and ankle prostheses depending on the amputation site [1]. Current research on lower limb prostheses mainly focuses on knee and ankle prostheses [1, 2]. Because the number of hip amputees is relatively small, there is less research on hip prostheses [3]. Wearing a hip prosthesis is an important way to help hip amputees regain normal walking ability, so it is important to develop a hip prosthesis.

Lower limb prostheses can be broadly divided into passive prostheses, semi-active prostheses, and powered prostheses based on the different types of joint adjustment [2]. Early research on prosthetic limbs mainly refers to passive prosthetic limbs. The joint damping and stiffness of passive prostheses cannot be adjusted adaptively to the amputee's motion and rely only on the prosthesis' own mechanical adjustment. With the advancement of electronic technology, lower limb prostheses have evolved towards semi-active and powered prostheses. The semi-active prosthesis contains microcontrollers, sensors and actuators that can adjust joint stiffness according to human posture, motion status and environment information [4]. It has the ability to adapt to different motion scenarios. The power prosthesis includes a sensor system and high-power motor system that can adjust the motor's rotation angle, speed, and output torque to match the amputee's gait speed and road conditions [5].

There are very limited studies related to hip prostheses available for search. In 2010, Ottobock developed a hydraulic hip joint, Helix3D [6], which uses a ball and socket joint and a four-link mechanism to allow three degrees of rotation in space and offers a stable mode with high joint damping and a flexible mode with low joint damping. In 2014, Orozco et al. [7] designed a multiaxial pneumatic damping hip prosthesis that provides damping through an air spring to support the prosthetic joint during walking. However, the air spring is poorly controllable and cannot achieve the damping required for walking in the hip joint. In 2019, Ueyama et al. [8] designed a fully active hip-knee prosthesis that was powered by electric motors but had only motor control and did not include a sensor system to sense environmental information. With the exception of the fully active hip-knee prosthesis by Ueyama et al., most studies on hip prostheses refer to passive prostheses.

Passive hip prostheses cannot dynamically adjust the joint stiffness of the prosthesis based on real-time gait information from the amputee [9]. This makes it difficult for most hip amputee to maintain stability through controlled swing and stance when using a passive hip prosthesis. To solve the above problems of passive hip prosthesis, our team has developed a semi-active bionic hip prosthesis with variable stiffness, where the hip prosthesis control system enables a dynamical adjustment of the joint stiffness according to the amputee's real-time gait information.

2 Hip Prosthesis and Variable Stiffness Joint Structure

Our team designed a semi-active hip prosthesis based on the Remote Center of Mechanism (RCM) [10], the structure of which is shown in Fig. 1. The whole lower limb prosthesis consists of three parts, the semi-active hip prosthesis, the knee prosthesis (Ottobock 3R60 dual hydraulic elastic flexion knee) and the prosthetic foot (Ottobock 1C10 lightweight carbon fiber prosthetic foot). The semi-active hip prosthesis consists of an RCM, an angle sensor, a torque sensor, a standing support plate, and a parallel actuator. The rotation range of the prosthetic joint is from $-30°$ to $95°$, including the swing phase from $-30°$ to $75°$ and the sit phase from $90°$. Angle sensors and torque sensors are used to measure information about the prosthetic joints in real time and provide feedback parameters for prosthetic control. Standing support plate supports human weight. Parallel actuators are used to dynamically adjust the stiffness of the hip joint to improve stability of the prosthesis during standing and flexibility during walking. The parallel actuator consists of an extension motor, a flexion motor, an extension spring, a flexion spring, and two ball screws as shown in Fig. 1. The ball screw converts the motor torque into a thrust on the spring. In this study, the variable stiffness of the semi-active hip prosthesis is achieved by controlling the motor to change the parallel spring extension.

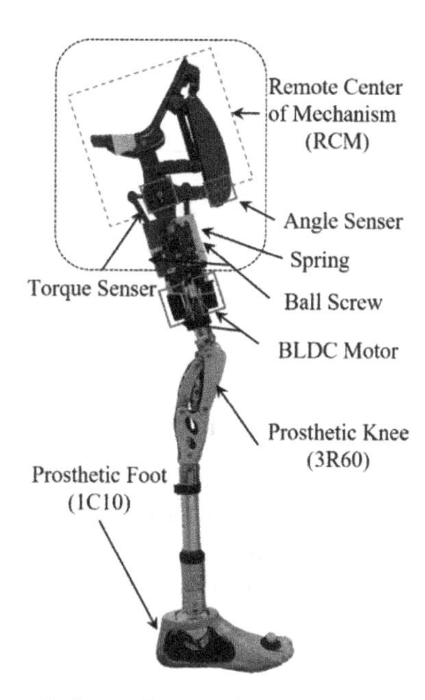

Fig. 1. The lower limb prostheses and variable stiffness joint structures.

3 Control Method

3.1 Overall Design

The overall design of the control system for the semi-active hip joint prosthesis is shown in Fig. 2. Including the target stiffness prediction system and the prosthesis control system. The posture sensor, which was placed on the healthy side of the amputee, records the hip joint angle x on the healthy side in real time. The above data was input to the trained NARX neural network model to predict the hip joint angle y on the prosthetic side. The gait phase was subdivided according to the predicted hip joint angle y. A model was created for the association of gait phase and stiffness. Combining the results of gait phase classification to derive the target stiffness of the hip prosthesis. A proportional-integral-derivative (PID) control algorithm was used to control the stiffness of the hip joint prosthesis.

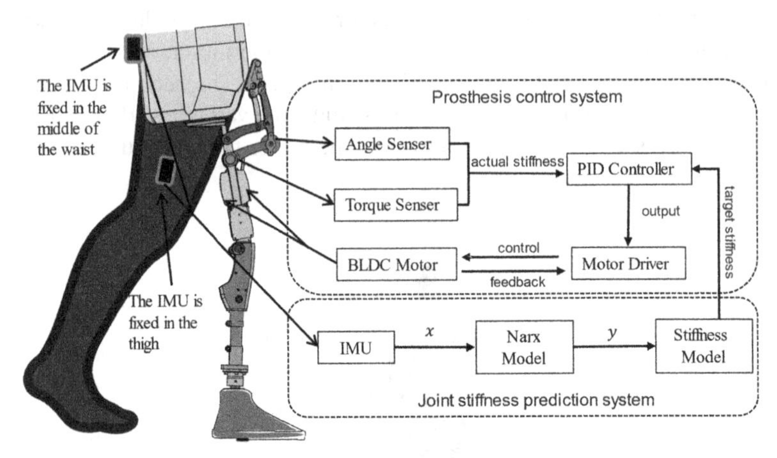

Fig. 2. Overall block diagram for the proposed control system.

3.2 Kinematics Mapping Model of Healthy Side to Prosthetic Side

The movement of the joints of the lower limbs in normal human walking shows periodic variation and a strong symmetry of bilateral movement. Therefore, the kinematic information of the hip joint on the healthy side can be obtained by the inertial measurement unit (IMU) and the kinematic information on the prosthetic side can be predicted by the neural network.

This study uses a 9-axis inertial measurement unit, model LPMS-ME1, developed by Alubi Inc. Ten volunteers were recruited to collect data on the hip angle of volunteers walking steadily at different speeds on a treadmill. Set the speed as 1.0 km/h, 1.5 km/h, 2.0 km/h, 2.5 km/h. The sampling frequency was set at 100 Hz. The hip joint angle belongs to the time series, so a neural network model suitable for time series prediction must be selected.

Nonlinear Autoregressive with Exogenous Inputs Model (NARX) is a class of models that describe nonlinear discrete systems [11]. At each sample time t, the model has an external input $x(t)$ that produces an output $y(t)$. The structure of NARX neural network contains a delay and feedback mechanism that records the last external inputs D_u and the last outputs D_y via a delay timer. Therefore, the NARX model has some short-term memory and is suitable for predicting time series. The defining equations of the NARX model are as follows.

$$y(t) = f\left(y(t-1), \ldots, y\left(t - D_y\right), x(t), x(t-1), \ldots, x(t - D_u)\right) \tag{1}$$

The kinematic mapping model based on NARX neural network architecture in this study for the healthy side to prosthetic side is shown in Fig. 3. The model consists of three parts: input layer, hidden layer, and output layer. The model consists of input layer, hidden layer and output layer. The input parameter is the hip joint angle of the healthy side. The hidden layer contains 10 neurons. The output is the angle of the hip joint on the amputation side. We set the input delay and the output delay as two time units.

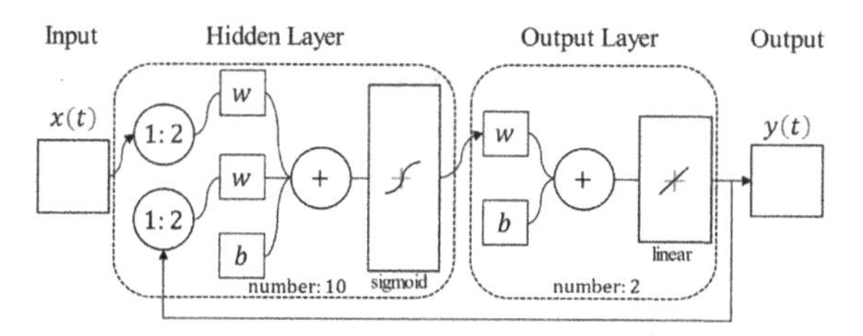

Fig. 3. Architecture of NARX neural network.

The sample data was 20,000 data points of bilateral lower limb hip angles during walking in ten healthy subjects, (70% as training, 15% as validation, 15% as test), and we used Bayesian regularization as the training mode. After 360 iterations, the correlation coefficient R of the model was 0.9937, and there was a significant correlation between input and output. Figure 4 shows the actual bilateral and neural network predicted curves of hip joint angle change. The correlation coefficient R was 0.977 when comparing the actual and predicted right hip curves, and the actual right hip angle was significantly correlated with the predicted hip angle.

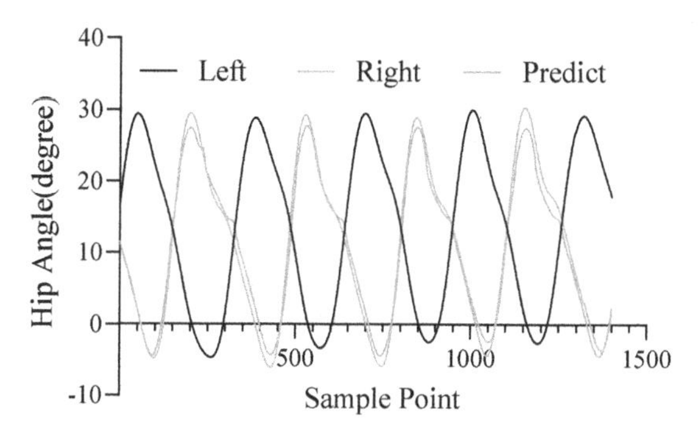

Fig. 4. A comparison curve of actual bilateral hip joint angle and neural network prediction.

3.3 Modeling of Gait Phase to Stiffness

The normal gait of the human body has a certain periodicity, and Fig. 5 shows a gait cycle motion phase of normal human walking. The gait phase can be divided into standing phase and swing phase, with the standing phase accounting for about 60% and the swing phase accounting for about 40%. According to the characteristics of hip flexors and extensors, the gait phase is divided into six phases, namely initial double stance (IDS), single stance (SS), double stance end (DSE), initial swing (ISW), mid swing (MSW) and swing end (SWE).

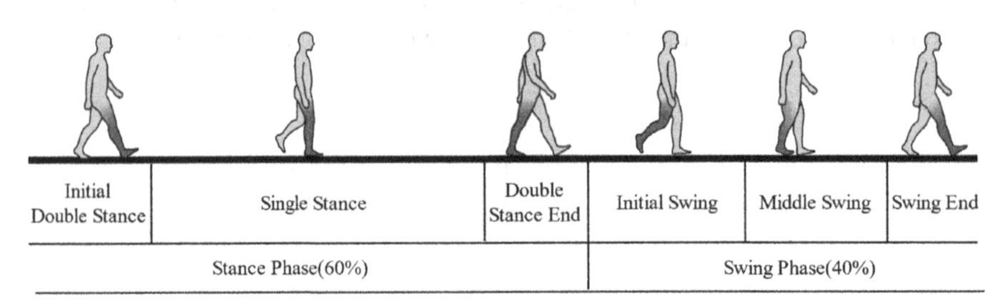

Fig. 5. Six phase division in a gait cycle of normal human walking.

Figure 6 (a) shows the hip joint angle and torque change curves during a gait cycle of normal walking in a healthy person. The stiffness is equal to the torque divided by the angle. When the slope of the hip angle curve is zero, the stiffness tends to infinity. However, the output of the adjustable stiffness system has a maximum saturation value, so the target stiffness is set to the maximum saturation value at the point. The stiffness change curve is shown in Fig. 6 (b).

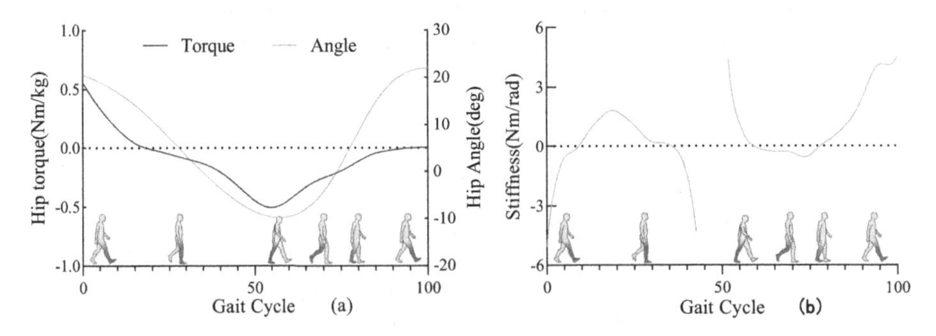

Fig. 6. (a) Hip joint torque and angle change curves during walking in healthy subjects; (b) Hip stiffness change curve during walking in healthy subjects.

3.4 Embedded System Design

The design of the embedded system is shown in Fig. 7. The hardware circuit can be divided into power supply module, data acquisition module, and communication module.

The main control chip was chosen as the STM32F407ZET6 with ARM Cortex-M4 core, 32-bit microcontroller, built-in ADC peripherals, and support for serial communication and CAN communication to meet the requirements of the semi-active hip joint prosthesis control system. The serial communication protocol is used to transmit the sensor data. A signal conditioning circuit is built to amplify the differential signal of the torque sensor, and the data is acquired peripherally in combination with the microcontroller ADC. The angle sensor uses resistance-angle correspondence, so that the voltage value of the common terminal can be detected, and the resistance value can be determined according to Ohm's law to obtain the actual angle value. The vector control of the motor is carried out by the CAN communication in conjunction with the motor driver. Wireless transmission of experimental data and remote control of the system are realized by Bluetooth. The software uses the real-time operating system RT-Thread to ensure real-time data acquisition and real-time control of the engine and minimize system latency.

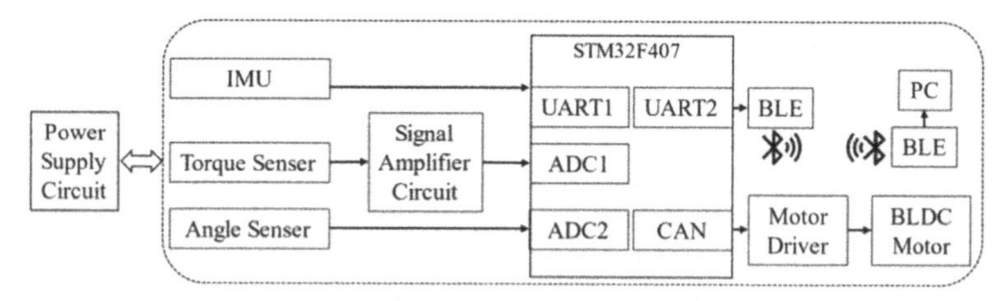

Fig. 7. Block diagram for embedded system design.

4 Experiments and Discussions

4.1 Experiments Protocol

In this study, a hip amputee was recruited to participate in the wearing experiment. The amputee was male, 33 years old, weighed 65 kg, and had been an amputee for 11 years. The amputation of the left hip was caused by an accident, and he rarely wore the prosthesis in daily life. The subjects were informed about the whole experiment before the experiment and signed a written informed consent. The experimental protocol was approved by the medical ethics committee of Shanghai Health Medical College.

To verify the effects of the variable stiffness control system on prosthesis stiffness when the amputee walks with the prosthesis. The change curves of hip joint stiffness were compared during walking at different speeds with the control system turned on and off. The prosthesis is actually a passive prosthesis when the control system is off and a semi-active prosthesis when the control system is on. According to the controller's conditions, prosthesis divided into two types: passive mode and active mode.

In the experiment, the Zebris gait analysis system (Zebris, Germany) was used to obtain data on the subject's gait speed and stride length during walking. To test the effect of the variable stiffness control system on the symmetry of human gait during walking, step length information was collected from amputees walking at different speeds with the variable stiffness control system turned on and off. With the control system off, the prosthesis was fixed in the middle position with the springs on both sides. Three preferable speeds were selected for the amputee walking with the prosthesis, namely 1.4 km/h, 1.8 km/h and 2.1 km/h. The results of this experiment were evaluated by the gait symmetry evaluation index [12].

$$\begin{cases} SI = \frac{2(X_R - X_L)}{X_R + X_L} \times 100\% \\ R_I = \frac{X_R}{X_L} \\ R_{II} = \frac{X_R - X_L}{max(X_R, X_L)} \end{cases} \tag{2}$$

where X_R and X_L represent the characteristic gait parameters of the left and right lower limbs, respectively, and the positive and negative signs indicate the dominance of the lower limbs. In this study, the step length parameter was used. When $SI = 0, R_I = 1, R_{II} = 0$, gait symmetry is good.

4.2 Results

Figure 8 shows the change curve of the stiffness of the hip prosthesis when the amputee walked at different speeds in active and passive modes. The curves are plotted from hip prosthesis angle to torque data for ten gait cycles. It can be seen from the figure that the regularity of the joint stiffness curve in the passive mode gradually decreases with increasing speed. The regularity of the stiffness curve in the active mode was not prominently correlated with speed, and it is preferable to walk at 1.8 km/h. The regularity of the active mode stiffness change curves was better than that of the passive mode in all three velocity cases, and the average joint torques were improved by 10 Nm, 12.5 Nm, and 9.3 Nm, respectively.

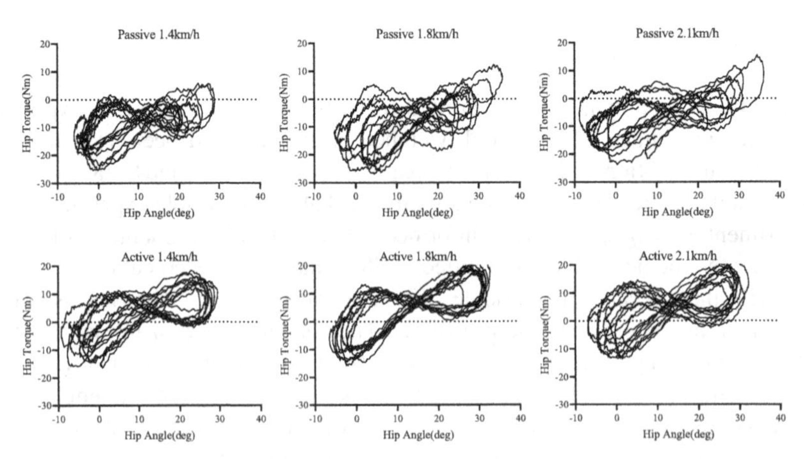

Fig. 8. Hip stiffness curve of the prosthesis when the amputee walks at different speeds in active and passive mode.

The calculated gait symmetry indices are shown in Table 1. From the table, it can be seen that in the active mode, SI, R_I, R_{II} are better than the passive mode in three different speed cases. Best results at 1.4 km/h, where SI improves from −26.0% to −2.8%, R_I improves from 0.77 to 0.97, and R_{II} improves from −0.23 to −0.03. Metrics of gait symmetry SI and R_{II} improved by in average 86.3% and 85.1%, respectively.

Table 1. Gait symmetry index at different speeds and modes.

Gait symmetry index	Passive mode			Active mode		
	SI	R_I	R_{II}	SI	R_I	R_{II}
1.4 km/h	−26.0%	0.77	−0.23	−2.8%	0.97	−0.03
1.8 km/h	−17.7%	0.83	−0.16	0	1	0
2.1 km/h	−8.9%	0.91	−0.08	4.3%	1.04	0.04

Figure 9 shows the step length data in the two modes under three different speeds. The step lengths of the left and right sides of the amputee were obviously asymmetric after the control system was turned off, and the average differences between the left and right sides were 9 cm, 7 cm, and 4 cm, respectively. With variable stiffness control turned on, the average step lengths difference between the two sides is 1 cm, 0 cm, and 1 cm, respectively.

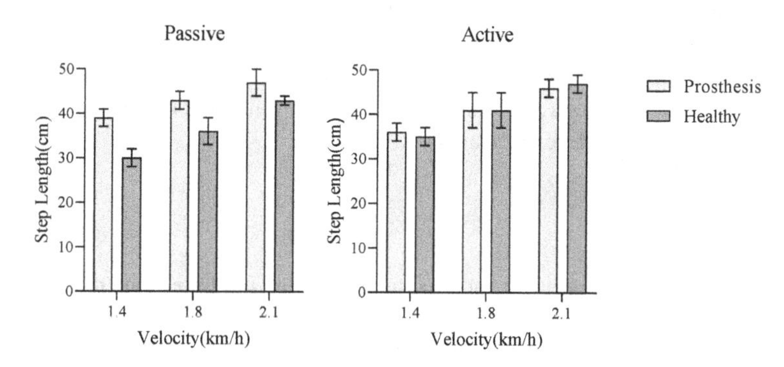

Fig. 9. Step length of amputees walking at different speeds and modes

4.3 Discussion

When the prosthesis is in passive mode, the stiffness adjustment of the prosthetic joint only depends on the spring, and the spring itself cannot adjust to achieve the stiffness required by the prosthesis during walking. When the prosthesis is in active mode, the variable stiffness control system can dynamically adapt the joint stiffness to the amputee's real-time gait information, which greatly improves the regularity of the joint stiffness.

The amputee will tend to rely on the healthy side for security and stability when walking with a passive prosthesis, resulting in a decrease in stride length on the healthy side and an increase in stride length on the prosthetic side. When the control system is turned on, it automatically adjusts the stiffness of the hip joint according to the amputee's current gait phase, which improves joint stability in the stance phase and joint flexibility in the swing phase. The amputee wearing the prosthesis can stand stably in the stance phase and swing the prosthesis more easily in the swing phase, which improves his stride length on the left and right sides.

5 Conclusion

In this study, a semi-active hip prosthesis variable stiffness control system is proposed. The NARX-based kinematic mapping model from the healthy side to the prosthetic side and the gait phase subdivision-based stiffness model are investigated to solve for the ideal stiffness. An embedded hardware and software control system was also developed to implement the prosthetic hip joint stiffness control. The effectiveness of this control system was verified by hip joint stiffness tracking experiments and gait symmetry tests. The hip joint stiffness tracking experiments showed that the control method proposed in this study can dynamically adjust the hip joint stiffness of the prosthesis according to the kinematic information of the healthy side of the lower limb when the amputee walks, which can effectively improve the regularity of the stiffness change curve of the prosthesis. Comparing the step length data and the symmetry indexes of the left and right sides in active and passive modes, the gait symmetry test experiment indicates that the control system developed in this study can effectively improve the gait symmetry of the left and right sides during walking. This study only explores the control strategy when walking on level ground. In our further work, we will investigate the control method in more complex situations such as walking up and down stairs and ramps, and realize the adaptive control across different terrains.

References

1. Asif, M., et al.: Advancements, trends and future prospects of lower limb prosthesis. IEEE Access **9**, 85956–85977 (2021)
2. Lara-Barrios, C.M., Blanco-Ortega, A., Guzman-Valdivia, C.H., Bustamante Valles, K.D.: Literature review and current trends on transfemoral powered prosthetics. Adv. Robot. **32**, 51–62 (2018)
3. Gholizadeh, H., et al.: Hip disarticulation and hemipelvectomy prostheses: A review of the literature. Prosthet. Orthot. Int. **45**, 434–439 (2021)
4. Lui, Z.W., Awad, M.I., Abouhossein, A., Dehghani-Sanij, A.A., Messenger, N.: Virtual prototyping of a semi-active transfemoral prosthetic leg. Proc. Inst. Mech. Eng. Part H-J. Eng. Med. **229**, 350–361 (2015)
5. Windrich, M., Grimmer, M., Christ, O., Rinderknecht, S., Beckerle, P.: Active lower limb prosthetics: a systematic review of design issues and solutions. Biomed. Eng. Online **15**, 140 (2016)
6. Gailledrat, E., et al.: Does the new Helix 3D hip joint improve walking of hip disarticulated amputees? Ann. Phys. Rehabil. Med. **56**, 411–418 (2013)

7. Orozco, G.A.V., Piña-Aragón, M., Altamirano, A.A. Diaz, D.R.: Polycentric mechanisms used to produce natural movements in a hip prosthesis. In: Braidot, A., Hadad, A. (eds.) VI Latin American Congress on Biomedical Engineering CLAIB 2014, Paraná, Argentina 29, 30 & 31 October 2014, pp. 289–292. Springer International Publishing (2015)
8. Ueyama, Y., Kubo, T., Shibata, M.: Robotic hip-disarticulation prosthesis: evaluation of prosthetic gaits in a non-amputee individual. Adv. Robot. **34**, 37–44 (2020)
9. Ebrahimzadeh, M.H., et al.: Long-term clinical outcomes of war-related hip disarticulation and transpelvic amputation. JBJS **95**, e114 (2013)
10. Li, X., et al.: Design and optimization of a hip disarticulation prosthesis using the remote center of motion mechanism. Technol. Health Care **29**, 269–281 (2021)
11. Lin, T., Horne, B.G., Tino, P., Giles, C.L.: Learning long-term dependencies in NARX recurrent neural networks. IEEE Trans. Neural Networks **7**, 1329–1338 (1996)
12. Karaharju-Huisnan, T., Taylor, S., Begg, R., Cai, J., Best, R.: Gait symmetry quantification during treadmill walking. In: The Seventh Australian and New Zealand Intelligent Information Systems Conference 2001, pp. 203–206. IEEE (2001)

A Hip Active Lower Limb Support Exoskeleton for Load Bearing Sit-To-Stand Transfer

Jinlong Zhou, Qiuyan Zeng, Biwei Tang, Jing Luo, Kui Xiang, and Muye Pang(✉)

School of Automation, Wuhan University of Technology, Wuhan, Hubei, China
pangmuye@whut.edu.cn

Abstract. Sit-to-stand (STS) transfer is a basic and important motion function in daily living. Most currently-existing studies focus on movement assistance for patients who lost mobility or have impaired their muscle strength. In this article, a hip-active lower limb support exoskeleton is designed to assist load bearing STS transfer for healthy persons. In order to provide effective assistance, a self-designed quasi-direct drive actuator is adopted to compose the actuation module and a load bearing stand up assistance strategy is designed based on virtual modwhutel control and gravity compensation. Control parameters are optimized in a musculoskeletal simulation environment with kinematic and kinetic data obtained from the wearer. The experimental results show that muscle activation levels of gluteus maximus and semimembranous are reduced with the help from the proposed exoskeleton during load bearing STS transfer.

Keywords: Sit-to-stand (STS) transfer · Stand up assistance · Exoskeleton

1 Introduction

The lower limb exoskeleton has been extensively studied and practically implemented on many occupational fields [1, 2], such as post-stroke rehabilitation training, industrial work supporting and military usefulness. Owing to its potential for releasing labour burden and protecting human health, exoskeleton is a compelling technology given the situation of increasing aging of population. Many of the present studies focus on locomotion assistance, such as walking [3], running [4] or upright standing push-recovery [5]. However, if only locomotion is supported, these devices are insufficient to perform agile and safe movements in human daily lives without monitor from technicians. Sit-to-stand (STS) transfer is one of the most common movements of daily-life activity [6]. It is essential for exoskeleton robots to assist STS function effectively in order to accelerate the process of exoskeleton implementation from laboratory environment to real-world applications.

The goal of STS transfer is commonly defined as moving the center of mass (CoM) of the body upward, from a sitting posture to upright standing, without loss of balance [6]. STS requires sophisticated coordination between trunk and lower limb muscles to remain stability of the entire body. For executing a thorough analysis on STS transfer, the movement is generally divided into different phases, ranging from two to five. Basically,

H. Liu et al. (Eds.): ICIRA 2022, LNAI 13456, pp. 24–35, 2022.
https://doi.org/10.1007/978-3-031-13822-5_3

two- (seat-off phase and ascending phase) or three-phase (add a stabilization phase after ascending phase) definition, is commonly adopted in many related studies. The seat-off phase is defined as the process from beginning to stand up to the moment where subject just leaves the chair. The characteristic of this phase is that most of the muscle work is transferred to a large enough momentum to push CoM leaning forward. The ascending phase describes the movement from leaving the chair to standing upright, in which lower limb muscles coordinate to move CoM upward to the desired position. The stabilization phase is the final step to ensure upright standing balance.

Most of the proposed exoskeletons developed for STS transfer aim to help individuals who are affected by muscle weakness or impair their legs because of stroke attack to perform successive and stable STS movement. In order to offer a proper STS transfer support for spinal cord injury (SCI) patients, Tsukahara et al. [7] proposed an intention estimation method based on preliminary motion which is before the desired STS transfer to construct a controller on Hybrid Assistive Limb. The support vector machine has also been implemented to recognize sit-to-stand and stand-to-sit motions in real-time [8]. Mefoued et al. [9] adopted a second order sliding mode controller to track a predefined movement trajectory to drive a knee joint orthosis to provide STS assistance. The proposed method can eliminate the chattering phenomenon of the classical sliding mode controller and ensure high performances. Song et al. [10] designed a wheelchair-exoskeleton hybrid robot which is able to transform from a wheelchair into an exoskeleton and be actuated by the same actuators. Misalignment between exoskeleton and wearer joints affects joint kinematics as well as wear comfort, even may cause injury. Junius et al. [11] discussed the metabolic effects induced by a kinematically compatible hip exoskeleton during STS transfer. They discovered that their proposed misalignment-compensating strategy is capable of reducing muscle activity. Indeed, the control method based on tracking STS kinematics trajectory is able to provide considerable assistant support to mobility impaired patients. However, it would not be the most proper option for 'assist-as-needed' mode since the induced undesired large driving force may become a disturbance for self-movable patients or intact persons when actual kinematics trajectory differs from the predefined trajectory. It is observed that every movement trajectory of a person is different from a predefined one, no matter how this predefined one is obtained. In order to provide a proper STS assistance for post-stroke persons, Shepherd et al. [12] proposed a series elastic actuator (SEA) based knee joint exoskeleton to provide torque assistance for STS transfer. The provided torque is prescribed as a function of knee angle which is obtained from the normalized torque-angle curve of able-bodied STS movement. As the proposed algorithm is impedance-control-like and the exoskeleton is composed by SEA, the physical interaction between subject and exoskeleton is complaint and safe. To modify the rigid control effect of predefined-trajectory-tracking method, Kamali et al. [13] implemented the dynamic movement primitives theory to predict a trajectory based on the initial joint angles of a wearer. The experimental results indicated that the proposed method is capable to perform a flexible control under different sitting postures. Apart from reference trajectory, impedance parameters have been modulated to provide appropriate power and balance assistance during STS transfer. Huo et al. [14] proposed an impedance modulation controller which includes balance reinforcement control and an impedance reduction ratio function.

Lower limb exoskeletons are not only designed for mobility impaired patients, but also worn by healthy persons to enhance their muscle force and protect from injury in industrial work. It requires capabilities of movement flexibility and load bearing for these exoskeletons which serve in daily living. However, based on our best knowledge, only few studies focus on load bearing STS transfer. Thus we design a hip active lower limb support exoskeleton (Hals-Exo) (as shown in the left column of Fig. 1) in this article to provide assistance for this common scene in daily life. The consideration to actuate hip joint is twofold. Firstly, hip joint is beneficial for both locomotion and STS transfer. As Hals-Exo is designed to be used by healthy persons in everyday tasks, it is essential to provide assistant force on hip joint. Secondly, it is not efficient and economical enough to provide full actuations for hip, knee and ankle joints due to the fact that the cooperation control for locomotion between exoskeleton and subject remains an issue. An inappropriate locomotion control may hinder healthy person voluntary locomotion rhythm, or even hurt the wearer.

2 The Design of Exoskeleton

The front view and side view of Hals-Exo is illustrated in the left column of Fig. 1. The main components of the exoskeleton structure are shown in the right column of Fig. 1. Hals-Exo mainly consists of a backboard and two hip-active support legs. The backboard, which is connected to the support legs via steel hollow rob, is designed to be sufficient to bear a 50 kg weighted load. There are six degrees of freedoms (DoFs) in each leg: three at hip joint, one at knee joint, and two at ankle joint. The three-DoF design at hip joint allows subjects to move flexibly while wearing the exoskeleton without being limited to the sagittal plane. In addition, the thigh and calf parts are designed to be able to slide, which provide fitness to individual lower limb differences. Therefore, the exoskeleton can be used to adapt to subjects with height ranged from 155 cm to 185 cm. The ergonomic design makes the wearer comfortable and allows subjects to move with minimal restrictions. Detailed information is shown in Table 1.

Table 1. Range of motion of Hals-Exo

DoF	Property	Range
Hip flexion/extension	Active	−30–150°
Hip adduction/abduction	Passive	0–90°
Hip internal/external rotation	Passive	−10–10°
Knee flexion/extension	Passive	−110–0°
Ankle dorsi/plantar flexion	Passive	−45–45°
Ankle eversion/inversion	Passive	−90–90°
Translation of two sliders	Passive	0–15 cm

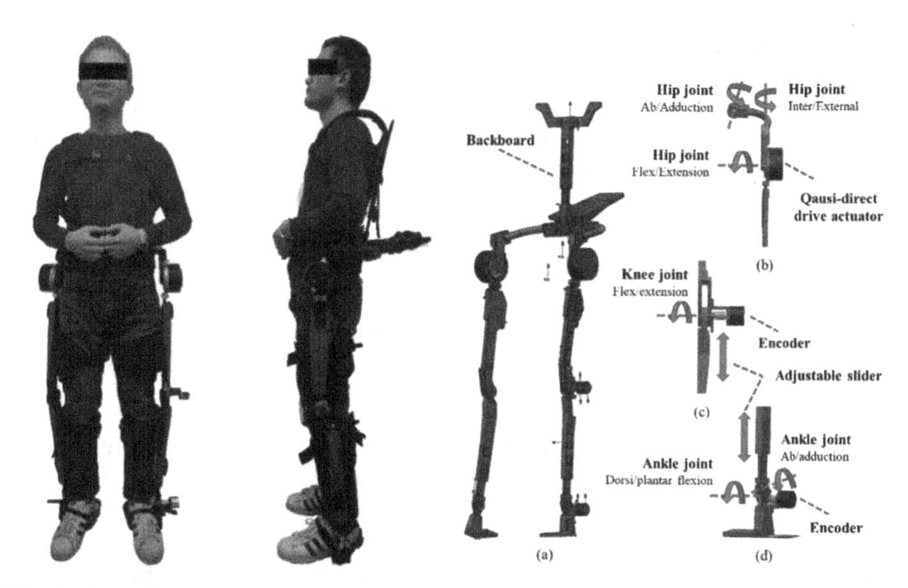

Fig. 1. Front view and side view of Hals-Exo (left column) and the main components of Hals-Exo mechanical system (right column). (a) The overview of assembled Hals-Exo, including two rigied legs and a backboard that supports the load. (b) Front view of the hip joint. The actuator provides torque to assist human STS movement. (c) Front view of the knee joint. The encoder records the bending angle of the knee joint. (d) Side view of the ankle joint.

The active rotational DoF in hip joint is driven by a self-designed quasi-direct drive actuator (reduced gear ratio 1:6, mass 0.5 kg). The actuator, which refers to an open source design [15], is capable to provide peak torque of 24 Nm and peak rotation velocity of 300 prm with a well-tuned torque control property. For the safety of subjects, a mechanical limitation is added to the hip joint, which works in conjunction with a programmed constrain in control to limit the motion range and maximum torque of actuator. Two encoders (OVW2–10-2MD, 1000P/R, Nemicon, China) are mounted on knee joint and ankle dorsi/plantar flexion joint respectively, to obtain the joint angles which are used to estimate the STS phase and to calculate the required assistant torque.

Compared with conventional single joint active assistant exoskeletons, Hals-Exo has full links from waist to ankle joint to transmit the load weight from backboard to the ground. This design is able to protect the wear from work-related musculoskeletal disorders caused by bearing load weight. Moreover, knee joint and ankle joint are passive which allows maximum flexibility in locomotion with the tradeoff of no active assistance for these joints.

3 Load Bearing STS Assistance Strategy Design

We adopt two-phase definition of STS movement to construct Hals-Exo control framework. A joint impedance control is applied in the first phase, i.e. the seat-off phase, to provide a rotation torque to push HAT (head-arm-torso) leaning forward. A stand up assistance algorithm based on virtual model control (VMC) [16] and gravity compensation is designed for the second phase, i.e. the ascending phase, to provide assistant torque from seat-off to upright standing. Phase transformation is estimated via hip joint angle with a pre-defined threshold.

The joint impedance controller for first phase follows a conventional form as follow:

$$\tau_h = K_h(\theta_d^h - \theta_r^h) + B_h\dot{\theta}_r^h \tag{1}$$

where ${\theta^h}_d$ is the desired hip joint angle and ${\theta^h}_r$ is the real hip joint angle read from the encoder. K_h and B_h are impedance control parameter of stiffness and damping, respectively.

The main idea of the stand-up assistance algorithm for the second phase is to provide an attractive force/torque (τ_a) to the CoM of HAT, driving it to move from the initial seat position to upright standing position. In order to decouple STS movement assistance and load weight bearing, we construct τ_a by two parts: the pulling force (F_p) which tries to pull HAT CoM to follow the preferred trajectory of the subject, and the compensation force (F_G) which offsets the effect of gravity.

3.1 VMC Based Pulling Force Calculation

In order to design an easy to adjust pulling force calculation algorithm and realize compliant human-robot physical interaction, we adopt VMC as the basic control framework. VMC, which is designed to try to realize locomotion control of bipedal robot, is well known in its characteristics of flexibility in planning compliant driving force under the situation of physical interaction. The driving point is attached at HAT CoM, as shown in Fig. 2. Model parameters are listed in Table 2.

Compared with the traditional VMC algorithm, we add a rotation driven term (the third line in Eq. 2) to represent body rotation during STS transfer:

$$F_p = \begin{bmatrix} f_x \\ f_y \\ \tau_r \end{bmatrix} = \begin{bmatrix} K_x(x_r - x) + B_x\dot{x} \\ K_y(y_r - y) + B_y\dot{y} \\ K_r(\theta_r - \theta) + B_r\dot{\theta} \end{bmatrix} \tag{2}$$

where subscript x, y and r denote forward direction, upward direction and body rotation respectively. f is the driving force in translation and τ is driving torque in rotation. Coefficients of K and B are complaint or impedance control parameters, representing physical interaction effect between subject and Hals-Exo. x_d, y_d and θ_d are referent trajectories. θ, which equals to $\theta_a + \theta_k + \theta_h$ (angle of ankle, knee and hip joint, respectively), is a synthetic scale variable indicating body gesture of the subject. As demonstrated by our previous study, the posture of the model will not be upright if the rotation driven term is lacked.

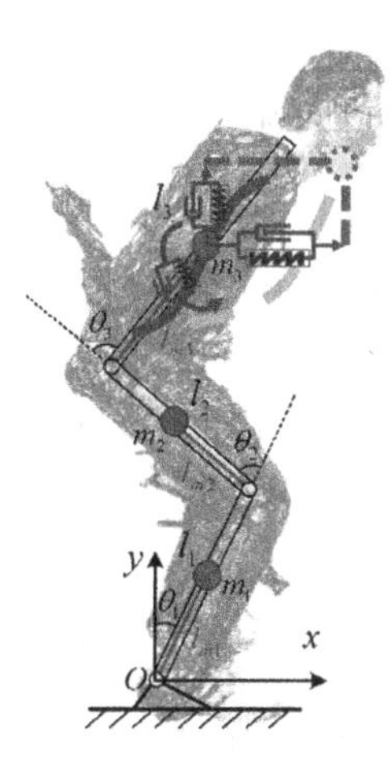

Fig. 2. Illustration of the proposed VMC-SUA. Means of symbols shown in figure are listed in Table 2. Red point is the CoM of HAT. Blue curve denotes the real trajectory of HAT CoM. Yellow dotted curve is the referent trajectory which is responsible for providing a proper pulling force.

Table 2. Model Parameters

Notation	Description
$\theta_a, \theta_k, \theta_h$	Angles of ankle, knee and hip joint
l_s, l_t, l_{hat}	Lengths of shank, thigh and HAT
l_{ms}, l_{mt}, l_{mhat}	Lengths from rotation center to CoM for ankle, knee and hip, respectively
m_s, m_t, m_{hat}	Masses of shank, thigh and HAT

After determination of the virtual force, the next step is to transfer the end-point virtual force into model joint space torque. The generalized position of HAT CoM can be represented as:

$$\hat{P}_{HAT} = \begin{bmatrix} x \\ y \\ \theta \end{bmatrix} = \begin{bmatrix} l_s \sin\theta_a + l_t \sin(\theta_a + \theta_k) + l_{mhat} \sin(\theta_a + \theta_k + \theta_h) \\ l_s \cos\theta_a + l_t \cos(\theta_a + \theta_k) + l_{mhat} \cos(\theta_a + \theta_k + \theta_h) \\ \theta_a + \theta_k + \theta_h \end{bmatrix} \quad (3)$$

The Jacobian matrix can be derived by partial differential $\hat{P}_{HAT}$ with θ_a, θ_k and θ_h:

$$J = \begin{bmatrix} y & l_t \cos(\theta_a + \theta_k) + l_{mhat} \cos(\theta_a + \theta_k + \theta_h) & l_{mhat} \cos(\theta_a + \theta_k + \theta_h) \\ -x & -l_t \sin(\theta_a + \theta_k) - l_{mhat} \sin(\theta_a + \theta_k + \theta_h) & -l_{mhat} \sin(\theta_a + \theta_k + \theta_h) \\ 1 & 1 & 1 \end{bmatrix} \quad (4)$$

Thus the pulling force expressed in joint space is given as $T = [\tau_a\ \tau_k\ \tau_h]^T = J^T F_p$. In real-time application to Hals-Exo, only τ_h is used as the command to hip joint actuators, as the other joints are passive and carried by the wearer. In applications, a scale factor is used to adjust the applied torque for the purposes of safety. The gravity effect is removed by adding a self-weight compensation torque, which is calculated from dynamic equation of static posture, on T.

3.2 Control Parameters Optimization

In order to tune the controller properly to individual subject, we use kinematic and kinetic data obtained from the corresponding subject STS movement to optimize the control parameters. The control parameters needed to be optimized are K_x, K_y, K_r, B_x, B_y, B_r, α and β.

We setup an optimization framework combining Opensim environment and MATLAB optimization toolbox. OpenSim [17] is an open source musculoskeletal simulation software which provides not only a good interface to motion capture system and force plate data, but also amount of powerful models to construct custom simulation environment. In this article, the human musculoskeletal model 'gait10dof18' which has 12 rigid bodies and 10 DoFs, is used as the basic model. The model is modified to be driven directly by joint torques. In order to perform the dynamic simulation under gravity condition, contact points (with stiffness coefficient of 108 and friction coefficient of 0.9) are added on spins, hips, thigh and feet (as shown in Fig. 3) to exert contact force between chair and ground. The initial state (joint angles, angular velocities and accelerations) is imposed from recording data obtained by motion capture system. The parameters of the model, for example the length and mass of each body, are properly scaled to the subject by OpenSim scale tools before simulation.

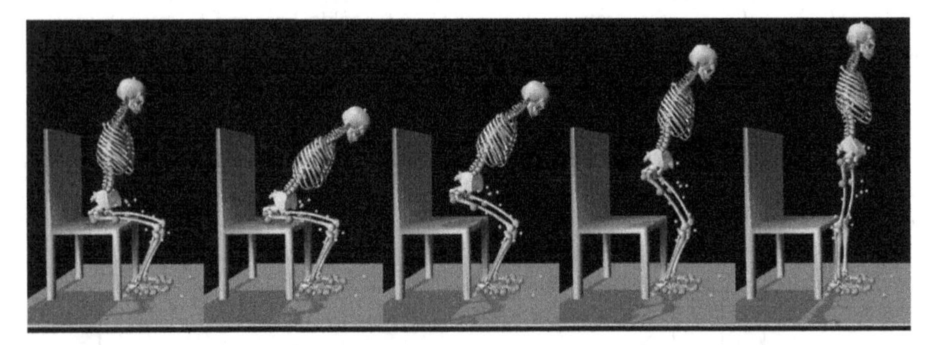

Fig. 3. The simulation environment built in OpenSim. Control parameter optimization is performed in this environment. Green spheres are contact points. Pink points denote marker positions recorded by motion capture system.

The fminsearch function in MATLAB optimization toolbox is used to seek for an optimized control parameter set. The cost function is the square of summation of HAT CoM trajectory distance between simulation model and subject experiment during STS transfer. Two key path points, which are the point connecting phase one and phase two of STS movement and the point referred to final upright standing state, are added to be the constraint condition, with the purpose of acquiring a suitable STS trajectory.

4 Experiment Setup and Data Process

Because these experiments serve as a proof-of-concept rather than a clinical validation, it is assumed to be sufficient that one subject (male, the age of 26 years, the height of 172 cm, and the body mass of 61.5 kg) participated in the experiment. The study was approved by the Wuhan University of Technology Institutional Review Board and informed written consent was obtained for the subject. Subject firstly sat straight on a chair (0.45 m height) with arms cross on chest, knee joint flexed about 100° and hip joint flexed 90°, kept torso as straight as possible. Then he was asked to perform STS movement with preferred speed. The experiment was considered to be ended when subject stood upright steadily. The kinematics data were recorded by a passive optical motion capture system (sampling frequency 50 Hz, Nokov, Beijing Metrology Technology Co., Ltd., China), with Helen Hayes Hospital marker setup method. The force and torque data of the feet in three directions during STS movement were obtained from a 6-dimensional force plate (sampling frequency 1000 Hz, BMS400600-1K, AMTI Company, USA). A 16-channel surface Electromyography device (sampling frequency 1000 Hz, ELONXI, China) was used to record muscle activity of gluteus maximus (GM), iliopsoas (IL), rectus femoris (RF), semimembranous (SM) and biceps femoris (BF) to verify the effectiveness of the proposed method.

The experiment consisted of two parts. In the first part, subject was asked to perform STS movement without Hals-Exo. The purpose of this part is to obtain original kinematics and kinetics data with no assistance from the exoskeleton. These data were used to optimize the control parameters and compare with exoskeleton assistance task. The second part was performed in the next day, in which subject wore Hals-Exo to perform STS movement. In each part, there were three tasks to be performed: without load, with a 4 kg load and with an 8 kg load. Subject was asked to perform STS transfer five times in each task and there was a 5 min break among each STS movement.

Paired-sample T-test (significant coefficient 5%) was applied to examine differences in muscle activation levels between load bearing STS with and without Hals-Exo.

5 Experimental Results

The optimized parameters for the subject are listed in Table 3 and Fig. 4 demonstrates the kinematic performance of Hals-Exo assisted load bearing STS movement. The red line and gray space denote mean kinematic value and stander deviation respectively. Reference values depicted by dotted blue lines are recorded from STS transfer without load and Hals-Exo. RMSEs between exoskeleton-on and exoskeleton-off are 0.09 ± 0.03 m, 0.03 ± 0.01 m, 0.12 ± 0.02 rad, 0.23 ± 0.08 rad, 0.38 ± 0.10 rad for HAT CoM x-direction trajectory, y-direction trajectory, ankle angle, knee angle and hip angle, respectively. The low RMSEs among kinematic data shows that the STS movements with and without Hals-Exo are similar, indicating that the interference from the proposed device to wearer voluntary movement is small.

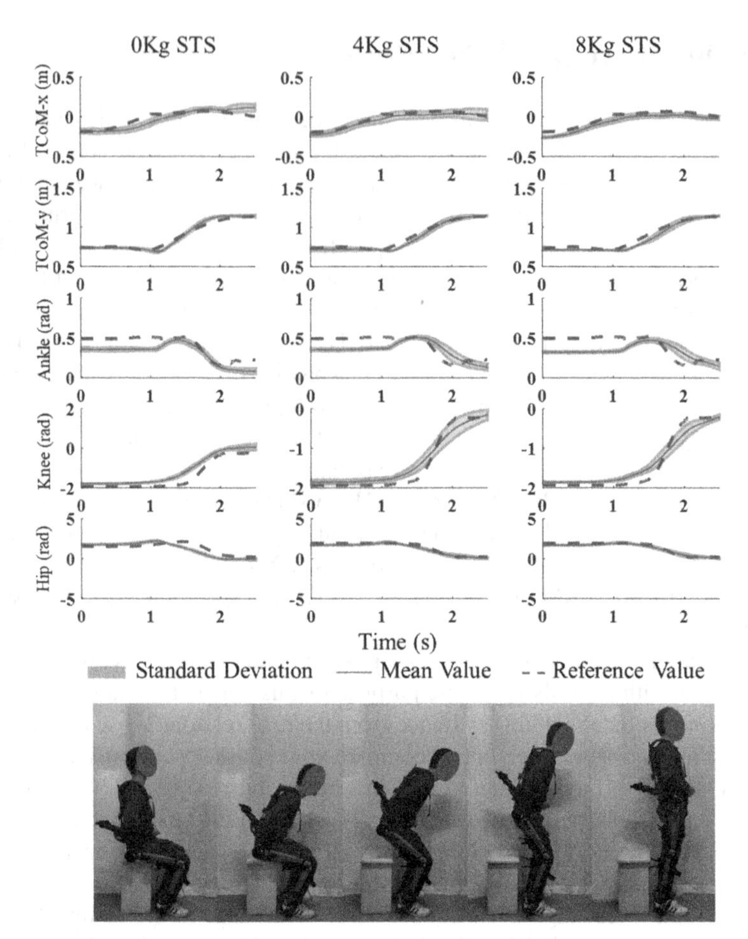

Fig. 4. Hals-Exo assisted load bearing STS experimental results. The three columns denote the kinematic data with load weight of 0 kg, 4 kg and 8 kg, respectively.

The experimental results of muscle activations during load bearing STS transfer with and without Hals-Exo are depicted in Fig. 5. Compared with no Hals-Exo assistant tasks, muscle activation levels of SM decrease significantly ($p < 0.05$) for all three cases (no load, 4 kg load and 8 kg load) when assistance is provided. Muscle activation levels of GM decrease during ascending phase (0.5 s to 1.5 s) with the help from Hals-Exo. As GM and SM are muscles response for hip extension which is the major motion of hip joint during STS, these experimental results indicate that Hals-Exo is able to reduce hip extensors effort during load bearing STS movement. For IL and RF, there is no significant differences ($p > 0.05$) in muscle activation levels with and without exoskeleton assistance.

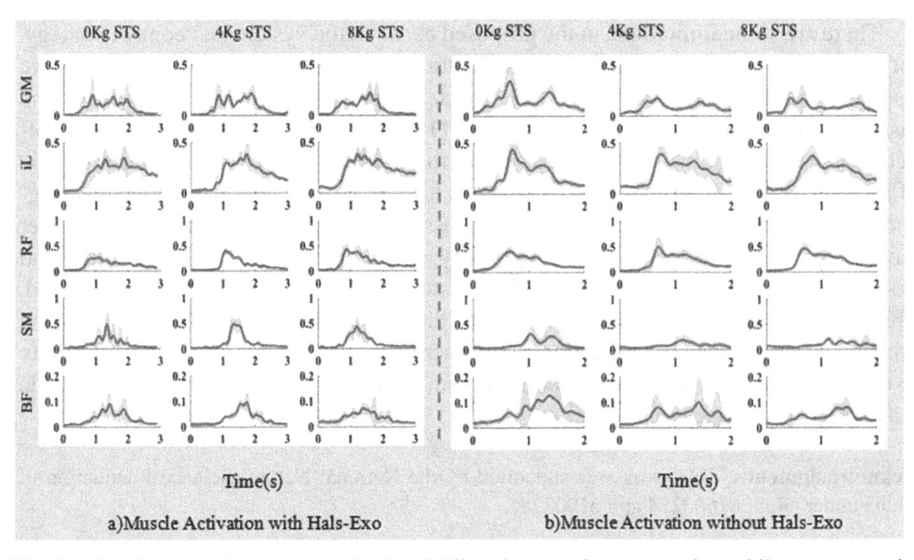

Fig. 5. Muscle activation response. Red solid line denotes the mean value while gray space is the standard deviation. a) Muscle activation performance when exoskeleton is worn, b) Muscle activation performance when subject executes STS movement without load and Hals-Exo.

Table 3. Control parameters

Parameter	Value	Unit
K_x	46.31	N/mm
B_x	1.54	N/mm/s
K_y	23.02	N/mm
B_y	1.52	N/(mm/s)
K_r	885.21	Nm/rad
B_r	15.17	Nm/(rad/s)

6 Discussion and Conclusions

The design and control of a lower limb exoskeleton aimed for load bearing STS transfer is described in this papaer Compared with other STS assistive exoskeleton, the proposed device has active hip joints and passive knee and ankle joint which provide it the abilities of transmission of the load weight to the ground, reduction of lumbar spinal compressive force and flexibility for lower limb movement. Quasi-direct drive actuators are used in the actuation modules to provide strong and fast response assistant torque. Simulation results indicate that our proposed STS assistive strategy is able to duplicate human sit to stand movement. sEMG data shows that Hals-Exo, controlled by the proposed strategy, is capable to reduce subject muscle activation and assist load bearing STS transfer.

There are some limitations in the proposed exoskeleton system and control strategy. First, the peak torque provided by the actuator is lower than human hip joint. As we implemented the quasi-direct drive type which has good performance on accurate and fast response torque control, rather than SEA [18] or high-gear-ration type, the tradeoff between the peak torque and total weight has to be accepted. Second, the consideration of adding passive spring-damping mechanism to Hals-Exo is lacked. Benefits of spring-damping unit for exoskeleton robot aimed to assist locomotion, lifting and STS transfer have been discussed. However, the hybrid-type exoskeleton consisted by active and passive unit for load bearing STS transfer assistance has not been thoroughly discussed. Third, we ignored the influence of the load on the position of HAT CoM as shapes of the load are similar during the experiment. In order to acquire a robust controller for daily living applications, a CoM estimation method has to be designed to obtain a reliable load weight compensation algorithm.

Acknowledgments. This work was supported by the National Natural Science Foundation of China under Grant 61603284 and 61903286.

References

1. Cao, W., et al.: A lower limb exoskeleton with rigid and soft structure for loaded walking assistance. IEEE Robotics and Automation Letters **7**, 454–461 (2022)
2. Ma, L., Leng, Y., Jiang, W., Qian, Y., Fu, C.: Design an underactuated soft exoskeleton to sequentially provide knee extension and ankle plantarflexion assistance. IEEE Robotics and Automation Letters **7**, 271–278 (2022)
3. Collins, S.H., Wiggin, M.B., Sawicki, G.S.: Reducing the energy cost of human walking using an unpowered exoskeleton. Nature **522**, 212–215 (2015)
4. Witte, K.A., Fiers, P., Sheets-Singer, A.L., Collins, S.H.: Improving the energy economy of human running with powered and unpowered ankle exoskeleton assistance. Science Robotics **5**, eaay9108 (2020)
5. Emmens, A.R., van Asseldonk, E.H.F., van der Kooij, H.: Effects of a powered ankle-foot orthosis on perturbed standing balance. J. Neuroeng. Rehabil. **15**, 50 (2018)
6. Galli, M., Cimolin, V., Crivellini, M., Campanini, I.: Quantitative analysis of sit to stand movement: experimental set-up definition and application to healthy and hemiplegic adults. Gait Posture **28**, 80–85 (2008)
7. Tsukahara, A., Kawanishi, R., Hasegawa, Y., Sankai, Y.: Sit-to-Stand and stand-to-sit transfer support for complete paraplegic patients with robot suit HAL. Adv. Robot. **24**, 1615–1638 (2010)
8. Liu, X., Zhou, Z., Mai, J., Wang, Q.: Real-time mode recognition based assistive torque control of bionic knee exoskeleton for sit-to-stand and stand-to-sit transitions. Robot. Auton. Syst. **119**, 209–220 (2019)
9. Mefoued, S., Mohammed, S., Amirat, Y., Fried, G.: Sit-to-Stand movement assistance using an actuated knee joint orthosis. In: 2012 4th IEEE RAS & EMBS International Conference on Biomedical Robotics and Biomechatronics (BioRob), pp. 1753–1758 (2012)
10. Song, Z., Tian, C., Dai, J.S.: Mechanism design and analysis of a proposed wheelchair-exoskeleton hybrid robot for assisting human movement. Mech. Sci. **10**, 11–24 (2019)
11. Junius, K., Lefeber, N., Swinnen, E., Vanderborght, B., Lefeber, D.: Metabolic effects induced by a kinematically compatible hip exoskeleton during STS. IEEE Trans. Biomed. Eng. **65**, 1399–1409 (2018)

12. Shepherd, M.K., Rouse, E.J.: Design and validation of a torque-controllable knee exoskeleton for sit-to-stand assistance. IEEE/ASME Trans. Mechatron. **22**, 1695–1704 (2017)
13. Kamali, K., Akbari, A.A., Akbarzadeh, A.: Trajectory generation and control of a knee exoskeleton based on dynamic movement primitives for sit-to-stand assistance. Adv. Robot. **30**, 846–860 (2016)
14. Huo, W., et al.: Impedance modulation control of a lower-limb exoskeleton to assist sit-to-stand movements. IEEE Transactions on Robotics, 1–20 (2021)
15. Wensing, P.M., Wang, A., Seok, S., Otten, D., Lang, J., Kim, S.: Proprioceptive actuator design in the MIT cheetah: impact mitigation and high-bandwidth physical interaction for dynamic legged robots. IEEE Trans. Rob. **33**, 509–522 (2017)
16. Pratt, J., Chew, C.-M., Torres, A., Dilworth, P., Pratt, G.: Virtual model control: an intuitive approach for bipedal locomotion. The Int. J. Roboti. Res. **20**, 129–143 (2001)
17. Delp, S.L., et al.: OpenSim: open-source software to create and analyze dynamic simulations of movement. IEEE Trans. Biomed. Eng. **54**, 1940–1950 (2007)
18. Sun, J., Guo, Z., Zhang, Y., Xiao, X., Tan, J.: A novel design of serial variable stiffness actuator based on an archimedean spiral relocation mechanism. IEEE/ASME Trans. Mechatron. **23**, 2121–2131 (2018)

Generative Adversarial Network Based Human Movement Distribution Learning for Cable-Driven Rehabilitation Robot

Zonggui Li, Chenglin Xie, and Rong Song(✉)

Sun Yat-sen University, Shenzhen 518107, China
songrong@mail.sysu.edu.cn

Abstract. Movement distribution analysis can reveal the body's changes from training with rehabilitation robotic assistance, and the distribution result has been used to develop robot control scheme. However, movement distribution modeling and further validation of the control scheme remain a problem. In this study, we propose a generative adversarial network (GAN) to learn the distribution of human movement, which will be used to design the control scheme for a cable-driven robot later. We preliminary collect a movement dataset of ten healthy subjects following a circular training trajectory, and develop a GAN model based on WGAN-GP to learn the distribution of the dataset. The distribution of the generated data is close to that of the real dataset (Kullback-Leibler divergence = 0.172). Ergodicity is also used to measure the movement trajectories generated by our GAN model and that of the real dataset, and there is no significant difference ($p = 0.342$). The results show that the developed GAN model can capture the features of human movement distribution effectively. Future work will focus on conducting further experiments based on the proposed control scheme, integrating human movement distribution into the control of real cable-driven robot, recruiting subjects for robot training experiments and evaluation.

Keywords: Rehabilitation robot · Human movement distribution · Generative adversarial network

1 Introduction

Stroke is the most common acute cerebrovascular disease that causes movement disabilities. Due to population growth and aging, the increasing number of stroke survivors creates a greater demand for rehabilitation services [1]. Rehabilitation robot can provide task-specific, repetitive and interactive movement training to help stroke patients accelerate neuroplasticity and regain partially lost movement functions. As a result, robot-assisted therapy technology has been developed and its effectiveness has been

This work was supported in part by the National Key Research and Development Program of China under Grant 2018YFC2001600, in part by the Guangdong Science and Technology Plan Project under Grant 2020B1212060077, and in part by the Shenzhen Science and Technology Plan Project Grant GJHZ20200731095211034.

H. Liu et al. (Eds.): ICIRA 2022, LNAI 13456, pp. 36–44, 2022.
https://doi.org/10.1007/978-3-031-13822-5_4

proven by clinical results [2]. To further understand the impact of robotics on human body, various methods have been used to describe the quality of movement in stroke patients undergoing robotic-assisted rehabilitation [3]. Nonetheless, characterizing the changes of a movement accurately is still challenging because of the inherent stochasticity in neuromotor commands and the resulting task executions [4], which in simple terms is that human movement is highly variable. Recent studies [5–7] have attempted to explain the body's changes from training with robotic assistance through movement distribution analysis and achieved advanced performance. In addition, ergodicity, which relates the movement trajectory to the target movement distribution, can also capture the changes of robotic assistance [4].

In this paper, we focus on movement distribution modeling and further development of rehabilitation robot control scheme based on movement distribution. In robotic training, an important question is how to apply force or movement so as to best facilitate the goals of producing new movement patterns with human body [8]. There have been studies applying movement distribution to promote the favorable changes of body with robot training by designing the force field [8–10]. In these studies, the weighted sum of multivariate Gaussian-normal components is used to fit the distribution of the movement data. Nevertheless, as is mentioned in [7], to accurately model movement distribution, increased multivariate normal components are needed. The fitting method is inconvenient in practice because of the fact that more consideration is required in the selection of parameters.

Generative adversarial network (GAN) is a framework that can capture the data distribution, with the advantage of representing very sharp, even degenerate distributions [11]. Figure 1 gives an overview of GAN. GAN consists of two parts: generator and discriminator, where the generator takes the random vectors as input and outputs generated samples, and the discriminator distinguishes the generated samples of the generator from the samples of the real dataset. Through adversarial training, the distribution of generated samples will tend to be consistent with the distribution of real samples. Compared to the method fitting with multivariate Gaussian normal component, the way to capture the data distribution in GAN is implicit, and the parameters of the generator and discriminator can be automatically learned by adversarial training, which is more convenient. The GAN-based distribution capture methods have been applied to human movement modeling [12–15], while the application in rehabilitation robot control needs to be further explored.

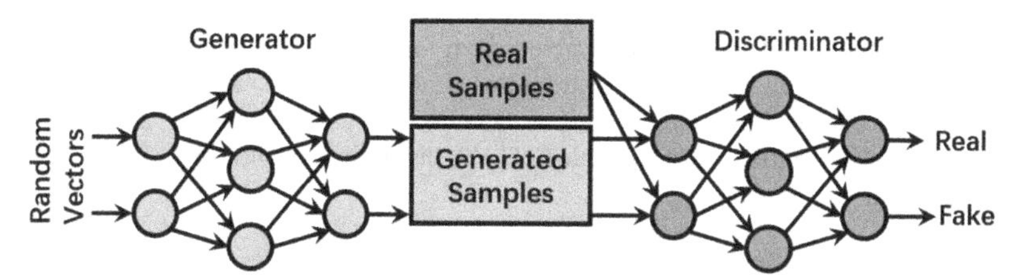

Fig. 1. Overview of GAN. It consists of two parts: generator and discriminator. The discriminator forces the generator to generated samples that tend to the real samples.

In this paper, we develop a GAN model based on Wasserstein GAN with a gradient penalty (WGAN-GP) [16] to capture the human movement distribution, and then the trained GAN model is used to design a control scheme for a cable-driven rehabilitation robot (CDRR). The results of our GAN show that it can capture the characteristics of human movement distribution effectively, which is meaningful for the control process later.

2 Method

2.1 Kinematic and Dynamic Analyses

As is shown in Fig. 2, we develop a GAN-based control scheme for a 3-DOF CDRR with 4 cables driving the end-effector. The end-effector is assumed as a point-mass platform in Cartesian coordinates. It tracks a reference trajectory, which in this study is a circular trajectory in the three-dimensional workspace.

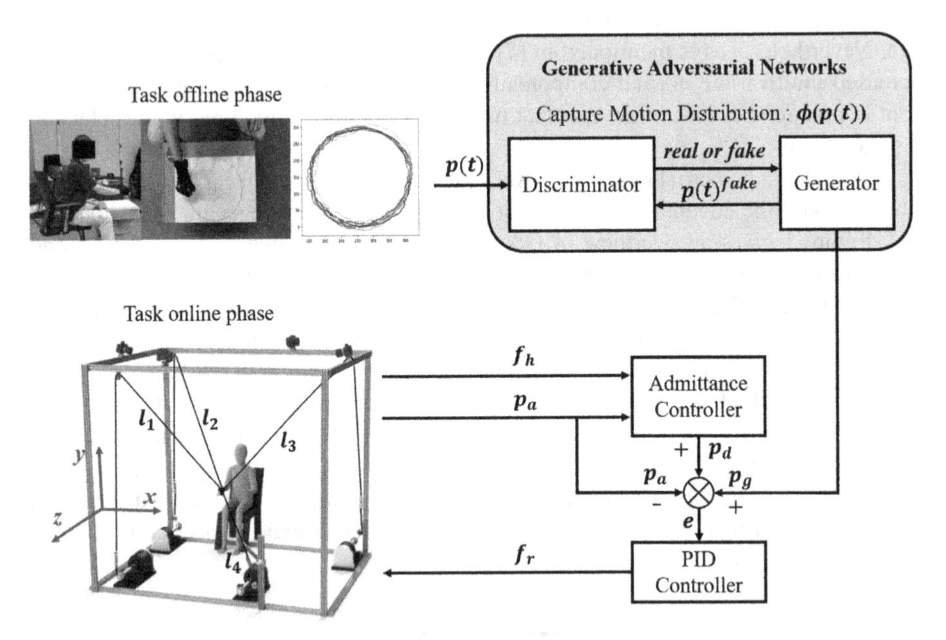

Fig. 2. GAN-based control scheme for a cable-driven rehabilitation robot (CDRR). In the task offline phase, GAN is used to learn the distribution of natural human movement $\phi(p(t))$. In the task online phase, the robot assistive force f_r is calculated by a PID controller, according to the error of the predicted position p_g of the trained GAN, the desired position p_d of the admittance controller, and real-time position p_a.

Referring to [17], the dynamic equation of the CDRR is as follows:

$$M(p)\ddot{p} + C(p, \dot{p})\dot{p} + G(p) = u + \tilde{G}(p) \tag{1}$$

where $\boldsymbol{M}(\boldsymbol{p})$ is the inertia matrix defined as (2), $\boldsymbol{C}(\boldsymbol{p}, \dot{\boldsymbol{p}})$ is the matrix of Coriolis and centripetal terms defined as (3), $\boldsymbol{G}(\boldsymbol{p})$ represents the gravity terms defined as (4), and $\tilde{\boldsymbol{G}}(\boldsymbol{p}) = \boldsymbol{G}(\boldsymbol{p})$ represents the gravity compensator of the end-effector.

$$\boldsymbol{M}(\boldsymbol{p}) = \begin{bmatrix} \boldsymbol{m} & 0 & 0 \\ 0 & \boldsymbol{m} & 0 \\ 0 & 0 & \boldsymbol{m} \end{bmatrix} \tag{2}$$

$$\boldsymbol{C}(\boldsymbol{p}, \dot{\boldsymbol{p}})\dot{\boldsymbol{p}} = \begin{bmatrix} 0 \\ 0 \\ 0 \end{bmatrix} \tag{3}$$

$$\boldsymbol{G}(\boldsymbol{p}) = \boldsymbol{M}(\boldsymbol{p}) \begin{bmatrix} 0 \\ \boldsymbol{g} \\ 0 \end{bmatrix} \tag{4}$$

where $\boldsymbol{m}$ is the mass of the end-effector, and $\boldsymbol{g}$ is the gravity constant. $\boldsymbol{u} \in \boldsymbol{R}^3$ can be expressed as:

$$\begin{cases} u = \boldsymbol{f}_h + \boldsymbol{f}_c \\ \boldsymbol{f}_c = \boldsymbol{J}^T \tau \end{cases} \tag{5}$$

where $\boldsymbol{f}_h \in \boldsymbol{R}^3$ is the active force applied by human, $\boldsymbol{f}_c \in \boldsymbol{R}^3$ is the actual force applied by the four cables, $\boldsymbol{\tau} = [\tau_1\ \tau_2\ \tau_3\ \tau_4]^T$ and τ_i is the tension of the ith cable. The tension τ_i can be measured by the tensor force sensor. $\boldsymbol{J}$ is the Jacobin matrix:

$$\boldsymbol{J} = [\boldsymbol{L}_1\ \boldsymbol{L}_2\ \boldsymbol{L}_3\ \boldsymbol{L}_4]^T \tag{6}$$

where $\boldsymbol{L}_i$ is the unit directional vector of the ith cable.

2.2 Generative Adversarial Network (GAN)

In this study, we develop a GAN model based on WGAN-GP for human movement distribution learning since WGAN-GP can realize stable learning while avoiding the problem of mode collapse. In the training, the generator $\boldsymbol{G}$ with parameter $\boldsymbol{\theta}_G$ takes the random vector z as input and is trained to generate fake samples $\boldsymbol{p}(\boldsymbol{t})^{\boldsymbol{fake}}$ to fool the discriminator $\boldsymbol{D}$. The discriminator $\boldsymbol{D}$ with parameter $\boldsymbol{\theta}_D$ is trained to distinguish between the real samples $\boldsymbol{p}(\boldsymbol{t})$ with data distribution $\boldsymbol{\phi}(\boldsymbol{p}(\boldsymbol{t}))$ and fake samples $\boldsymbol{p}(\boldsymbol{t})^{\boldsymbol{fake}}$ with data distribution $\boldsymbol{\phi}\left(\boldsymbol{p}(\boldsymbol{t})^{\boldsymbol{fake}}\right)$. $\boldsymbol{p}(\boldsymbol{t})$ is the movement trajectory which is a time series in the three-dimensional workspace. The training of GAN results in a minimax game, in which $\boldsymbol{D}$ forces $\boldsymbol{G}$ to produce fake samples asymptotically close to real samples, and finally $\boldsymbol{\phi}\left(\boldsymbol{p}(\boldsymbol{t})^{\boldsymbol{fake}}\right)$ tend to be consistent with $\boldsymbol{\phi}(\boldsymbol{p}(\boldsymbol{t}))$. The loss function of WGAN-GP $\boldsymbol{W}\left(\boldsymbol{\phi}\left(\boldsymbol{p}(\boldsymbol{t})^{\boldsymbol{fake}}\right), \boldsymbol{\phi}(\boldsymbol{p}(\boldsymbol{t}))\right)$ is a combination of the Wasserstein distance $\tilde{\boldsymbol{W}}\left(\boldsymbol{\phi}\left(\boldsymbol{p}(\boldsymbol{t})^{\boldsymbol{fake}}\right), \boldsymbol{\phi}(\boldsymbol{p}(\boldsymbol{t}))\right)$ and the gradient penalty term $\boldsymbol{W}_{\boldsymbol{gp}}$:

$$\boldsymbol{W}\left(\boldsymbol{\phi}\left(\boldsymbol{p}(\boldsymbol{t})^{\boldsymbol{fake}}\right), \boldsymbol{\phi}(\boldsymbol{p}(\boldsymbol{t}))\right) = \tilde{\boldsymbol{W}}\left(\boldsymbol{\phi}\left(\boldsymbol{p}(\boldsymbol{t})^{\boldsymbol{fake}}\right), \boldsymbol{\phi}(\boldsymbol{p}(\boldsymbol{t}))\right) + \boldsymbol{W}_{\boldsymbol{gp}} \tag{7}$$

The generator is trained to maximize the loss L_G with fixed parameter $\overline{\theta}_D$:

$$L_G\left(\theta_G, \overline{\theta}_D\right) = \mathbb{E}_{p(t)^{fake} \sim \phi\left(p(t)^{fake}\right)}\left[D_{\overline{\theta}_D}\left(p(t)^{fake}\right)\right] = \mathbb{E}_{p(t)^{fake} \sim \phi\left(p(t)^{fake}\right)}\left[D_{\overline{\theta}_D}\left(G_{\theta_G}(z)\right)\right] \tag{8}$$

The discriminator is trained to minimize the loss L_D with fixed parameter $\overline{\theta}_G$:

$$L_D\left(\theta_D, \overline{\theta}_G\right) = \tilde{W}\left(\phi\left(p(t)^{fake}\right), \phi(p(t))\right) = \mathbb{E}_{p(t) \sim \phi(p(t))}\left[D_{\theta_D}(p(t))\right] - \mathbb{E}_{p(t)^{fake} \sim \phi\left(p(t)^{fake}\right)}\left[D_{\theta_D}\left(p(t)^{fake}\right)\right] \tag{9}$$

The gradient penalty term W_{gp} is defined as follows:

$$W_{gp} = \lambda \mathbb{E}_{\hat{p}(t) \sim \phi(\hat{p}(t))}\left[\left(\nabla_{\hat{p}(t)} D(\hat{p}(t))_2 - 1\right)^2\right] \tag{10}$$

where $\mathbb{E}[\cdot]$ denotes the expectation, λ is a hyperparameter that balances the weight of the Wasserstein distance and the gradient penalty term, $\nabla[\cdot]$ is the gradient operator. $\hat{p}(t)$ is a subdivided point of a real sample $p(t)$ and a fake sample $p(t)^{fake}$:

$$\hat{p}(t) = \varepsilon p(t) + (1 - \varepsilon) p(t)^{fake} \tag{11}$$

where ε is a parameter from uniform distribution.

2.3 Admittance Controller

At each time step, the actual position p_a of the end-effector can be calculated using the movement capture system. In the training, the subject exerts the active force f_h on the end-effector to complete the circular trajectory tracking. Referring to [18], the admittance controller calculates a desired position p_d derived from p_a and f_h.

$$-M_h \ddot{p}_a - C_h \dot{p}_a + K_h(p_d - p_a) = f_h \tag{12}$$

where $M_h \in R^3$, $C_h \in R^3$, $K_h \in R^3$ is the mass, damper, and stiffness matrices of the human limb. In rehabilitation training, the damper and stiffness components are dominant. Therefore, Eq. (12) can be simplified as:

$$p_d = K_h^{-1}(f_h + C_h \dot{p}_a + K_h p_a) \tag{13}$$

2.4 PID Controller

As human movement distribution $\phi(p(t))$ is captured by GAN, the trained GAN is used to generated samples $p(t)^{fake}$. Since $p(t)^{fake}$ is a time series, the generator produces a predicted position p_g of the end-effector at each time step. In this paper, the position error $e(t)$ at time t is defined as follows:

$$e(t) = p_d + p_g - p_a \tag{14}$$

A PID controller is used to computes the robot force $\boldsymbol{f}_r \in \boldsymbol{R}^3$ on the end-effector:

$$\boldsymbol{f}_r = \boldsymbol{K}_p \boldsymbol{e}(t) + \boldsymbol{K}_i \int_0^t \boldsymbol{e}(\tau) d\tau + \boldsymbol{K}_d \dot{\boldsymbol{e}}(t) \tag{15}$$

where $\boldsymbol{K}_p, \boldsymbol{K}_i, \boldsymbol{K}_d$ is the control parameter of the proportional term, the integral term and the derivative term, respectively.

In summary, we use GAN to learn the distribution of human movement in the task offline phase. In the task online phase, the learned feature is used to design the control scheme for a CDRR. The trained generator is used to predict the movement position of an end-effector, on which the human-robot interface is located. A desired movement position of the end-effector is calculated by an admittance controller, according to the human-robot interaction force and the real-time position information. Finally, a position controller is used to calculate the robot force based on the error of the predicted position, desired position and real-time position of the end effector.

3 Experiment and Results of Human Movement Distribution Learning Based on GAN

We preliminary collect a movement dataset of ten healthy subjects. Each subject uses his/her usual hand to dangling move 10 laps, following a circular training trajectory with a radius of 0.15 m directly below. The movement trajectory and duration will be recorded by movement capture system. It is worth noting that the subjects move without interacting the robot during the dataset collection. As is mentioned in [19], some features of unconstrained human movement are a strong preference in the guided movement of physical human-robot interaction. Therefore, we use GAN to extract the natural human movement features, and then the extracted features are used to design the robot control scheme. The proposed scheme strives to make the results of rehabilitation training close to the natural movement of the human body.

Using t-SNE visualization [20] as shown in Fig. 3, we can observe that generated samples by our GAN model show better overlap with the real samples than the Gaussian components fitting method in [8–10]. Kullback-Leibler divergence (KL) [21] of the generated data distribution and the real data distribution is also calculated, and the results are KL (GAN) = 0.172, and KL (Gaussian components method) = 1.22. Since the Gaussian components fitting model has fewer parameters than GAN and requires manual adjustment, there is certain challenge in representing the high-dimensional and variable human movement. In contrast, GAN can automatically learn the distribution characteristics of human movement through the adversarial training of two deep neural networks with a large number of parameters.

We further use ergodicity to evaluate the difference between the generated movement trajectories by GAN and the real movement trajectories ($p = 0.342$). Ergodicity can be defined as the sum $\varepsilon(\boldsymbol{p}(t))$ of the weighted square distance between the spatial Fourier transform coefficients $\mathrm{c_k}$ of the trajectory $\boldsymbol{p}(t)$ and the Fourier coefficients $\boldsymbol{\phi}_k$ of the target distribution, see in [4].

$$\varepsilon(\boldsymbol{p}(t)) = \sum_{k_1=0}^{K} \cdots \sum_{k_n=0}^{K} \left| \Lambda_k \mathrm{c_k} - \boldsymbol{\phi}_k \right|^2 \tag{16}$$

where the state is n-dimensional at each time step and there are $\boldsymbol{K}+1$ coefficients along each dimension. $\boldsymbol{\Lambda}_k$ is the weight on the harmonics defined in [22].

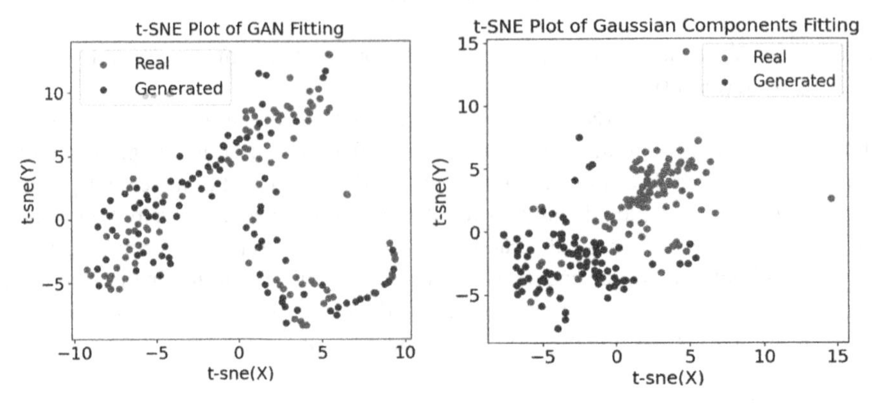

Fig. 3. t-SNE visualization of GAN fitting and the Gaussian components fitting. For high-dimensional data in the dimensionality reduction space, the blue scatter points represent the distribution of the generated data, and the red scatter points represent the distribution of the real data. (Color figure online)

Since $p > 0.05$, there is no significant difference between the generated data and the real data. The result of statistical analysis of ergodicity (Fig. 4) shows that the trajectories generated by GAN are highly close to the real trajectories in terms of movement distribution. The effectiveness of the developed GAN in capturing human movement distribution is of great significance for the subsequent design of robot control scheme.

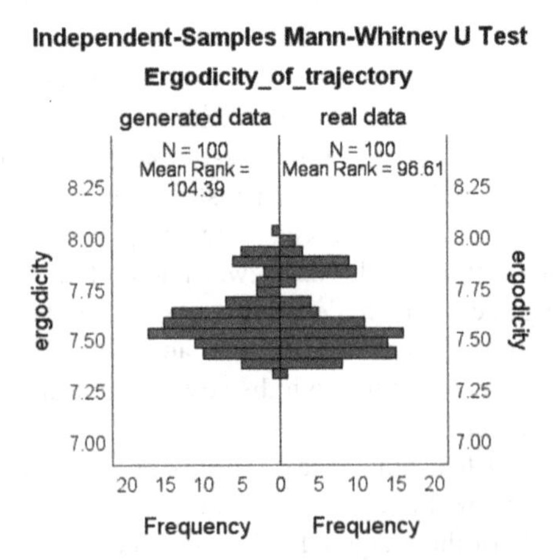

Fig. 4. Ergodicity of the generated trajectories and the real trajectories Mann-Whitney U Test.

4 Conclusion

In this study, we propose a GAN based two-stage control scheme for the CDRR. In the first stage, we collect a dataset of natural human movement, and then the developed GAN model is used to capture the distribution of the data. As for the second stage, the trained generator produces a predicted position of the end-effector at each time step. The predicted position will be used to calculate the robot force later. The results of our GAN model show potential performance of extracting human movement features. Future work will focus on conducting further experiments based on the proposed control scheme, integrating human movement distribution into the control of real CDRR (Fig. 5), recruiting subjects for robot training experiments and evaluation.

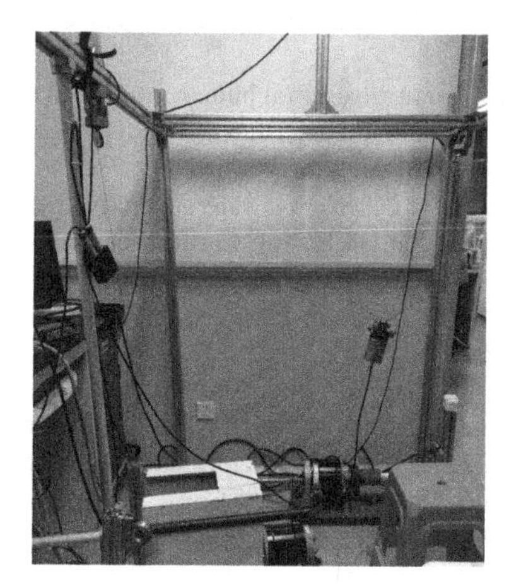

Fig. 5. The real CDRR.

References

1. Stinear, C.M., Lang, C.E., Zeiler, S.: Advances and challenges in stroke rehabilitation. Lancet Neurol. **19**(4), 348–360 (2020)
2. Lo, A.C., Guarino, P.D., Richards, L.G.: Robot-assisted therapy for long-term upper-limb impairment after stroke. N. Engl. J. Med. **362**(19), 1772–1783 (2010)
3. Nordin, N., Xie, S.Q., Wünsche, B.: Assessment of movement quality in robot-assisted upper limb rehabilitation after stroke: a review. J. Neuroeng. Rehabil. **11**(1), 1–23 (2014)
4. Fitzsimons, K., Acosta, A.M., Dewald, J.P.: Ergodicity reveals assistance and learning from physical human-robot interaction. Sci. Robot. **4**(29), eaav6079 (2019)
5. Huang, F.C., Patton, J.L.: Individual patterns of motor deficits evident in movement distribution analysis. In: 2013 IEEE 13th International Conference on Rehabilitation Robotics (ICORR), pp. 1–6. IEEE, Seattle (2013)
6. Wright, Z.A., Fisher, M.E., Huang, F.C.: Data sample size needed for prediction of movement distributions. In: 2014 36th Annual International Conference of the IEEE Engineering in Medicine and Biology Society, pp. 5780–5783. IEEE, Chicago (2014)

7. Huang, F.C., Patton, J.L.: Movement distributions of stroke survivors exhibit distinct patterns that evolve with training. J. Neuroeng. Rehabil. **13**(1), 1–13 (2016)
8. Patton, J.L., Mussa-Ivaldi, F.A.: Robot-assisted adaptive training: custom force fields for teaching movement patterns. IEEE Trans. Biomed. Eng. **51**(4), 636–646 (2004)
9. Wright, Z.A., Lazzaro, E., Thielbar, K.O.: Robot training with vector fields based on stroke survivors' individual movement statistics. IEEE Trans. Neural Syst. Rehabil. Eng. **26**(2), 307–323 (2017)
10. Patton, J.L., Aghamohammadi, N.R., Bittman M.F.: Error Fields: Robotic training forces that forgive occasional movement mistakes. PREPRINT (Version 1) available at Research Square (2022). https://doi.org/10.21203/rs.3.rs-1277924/v1
11. Goodfellow, I., Pouget-Abadie, J., Mirza, M.: Generative adversarial nets. Adv. Neural Inf. Process. Syst. **27** (2014)
12. Wang, Z., Chai, J., Xia, S.: Combining recurrent neural networks and adversarial training for human movement synthesis and control. IEEE Trans. Visual Comput. Graphics **27**(1), 14–28 (2019)
13. Zhao, R., Su, H., Ji, Q.: Bayesian adversarial human movement synthesis. In: Proceedings of the IEEE/CVF Conference on Computer Vision and Pattern Recognition (CVPR), pp. 6225–6234. IEEE, Washington (2020)
14. Wang, J., Yan, S., Dai, B., Lin, D.: Scene-aware generative network for human movement synthesis. In: Proceedings of the IEEE/CVF Conference on Computer Vision and Pattern Recognition (CVPR), pp. 12206–12215. IEEE (2021)
15. Nishimura, Y., Nakamura, Y., Ishiguro, H.: Human interaction behavior modeling using generative adversarial networks. Neural Netw. **132**, 521–531 (2020)
16. Gulrajani, I., Ahmed, F., Arjovsky, M.: Improved training of wasserstein gans. Advances in Neural Information Processing Systems (NIPS), vol. 30. MIT Press, Los Angeles (2017)
17. Zi, B., Duan, B.Y., Du, J.L.: Dynamic modeling and active control of a cable-suspended parallel robot. Mechatronics **18**(1), 1–12 (2008)
18. Li, Y., Ge, S.S.: Human–robot collaboration based on movement intention estimation. IEEE/ASME Trans. Mechatron. **19**(3), 1007–1014 (2013)
19. Maurice, P., Huber, M.E., Hogan, N.: Velocity-curvature patterns limit human–robot physical interaction. IEEE Robot. Autom. Lett. **3**(1), 249–256 (2017)
20. Van der Maaten, L., Hinton, G.: Visualizing data using t-SNE. J. Mach. Learn. Res. **9**(11), 2579–2605 (2008)
21. Kullback, S., Leibler, R.A.: On information and sufficiency. Ann. Math. Stat. **22**(1), 79–86 (1951)
22. Mathew, G., Mezić, I.: Metrics for ergodicity and design of ergodic dynamics for multi-agent systems. Physica D **240**(4–5), 432–442 (2011)

Learning from Human Demonstrations for Redundant Manipulator

Kinematic Analysis and Optimization of a New 2R1T Redundantly Actuated Parallel Manipulator with Two Moving Platforms

Lingmin Xu[1], Xinxue Chai[1,2], Ziying Lin[1], and Ye Ding[1](✉)

[1] State Key Laboratory of Mechanical System and Vibration, School of Mechanical Engineering, Shanghai Jiao Tong University, Shanghai 200240, China
y.ding@sjtu.edu.cn

[2] Faculty of Mechanical Engineering and Automation, Zhejiang Sci-Tech University, Hangzhou 310018, Zhejiang, China

Abstract. Redundantly actuated parallel manipulators (RAPMs) with two rotations and one translation (2R1T) are suitable for some manufacturing applications where high precision and speed are required. In this paper, a new 2R1T RAPM with two output moving platforms, called Var2 RAPM, is proposed. The Var2 RAPM is a (2R$\underline{P}$R-R)-R$\underline{P}$S-U$\underline{P}$S RAPM, which is actuated by four $\underline{P}$ joints (where R denotes a revolute joint, $\underline{P}$ denotes an actuated prismatic joint, S denotes a spherical joint, and U denotes a universal joint). First, the mobility analysis shows that the second moving platform can achieve two rotations and one translation. Then, the inverse kinematics is presented by using the constraints of joints. In addition, one local and two global kinematic performance indices are used to evaluate the motion/force transmissibility of the proposed Var2 RAPM. Finally, with respect to the two global kinematic indices, the architectural parameters of Var2 RAPM are optimized by using the parameter-finiteness normalization method. The Var2 RAPM has the potential for the machining of curved workpieces.

Keywords: Redundantly actuated parallel manipulator · Kinematic analysis · Optimization

1 Introduction

The 3-degrees of freedom (DOFs) parallel manipulators (PMs) with two rotations and one translation are often called 2R1T PMs, which have attracted much interest and attention from both academia and industry [1–5]. Because of the special output characteristics, some 2R1T PMs have been successfully used as parallel modules in the hybrid machines [3–5]. By integrating with a 2-DOF serial gantry or a 2-DOF serial tool head, the hybrid machines can achieve output motion with three translations and two rotations, and thus can be used for machining workpieces with complicated structures.

The work is supported by the National Natural Science Foundation of China (Grants No. 51935010, 52005448), and China Postdoctoral Science Foundation (Grant No. 2021M702123).

H. Liu et al. (Eds.): ICIRA 2022, LNAI 13456, pp. 47–57, 2022.
https://doi.org/10.1007/978-3-031-13822-5_5

Among these selected 2R1T PMs, the famous examples are 3PRS PM [3], 2UPR-SPR PM [4], and 3UPS-UP PM [5], which have been respectively used in the Ecospeed machine center, Exechon robot, and Tricept robot. In this paper, the P, P, R, S, and U denote the actuated prismatic joint, passive prismatic joint, revolute joint, spherical joint, and universal joint, respectively.

Compared with the above-mentioned 2R1T PMs without redundancy actuation, the 2R1T redundantly actuated PMs (RAPMs) have better stiffness, and thus are more suitable for machining applications that require a large load capacity. Normally, the ways to implement the actuation redundancy can be divided into two types: the first one is to add new actuated limbs in the original PM without changing the DOF, and the other is to replace the passive joints in the original limbs with actuated ones [6]. The first way to design a RAPM is better than the other one due to the better force distribution and stiffness, and there is much study on this [7–10]. In this paper, we also focus on this type.

In this paper, a new 2R1T (2RPR-R)-RPS-UPS RAPM is presented, which is actuated by four P joints. Because the designed RAPM looks like a combination of two V-shaped structures, it is called Var2 RAPM here. The main feature of Var2 RAPM is there are two moving platforms in the RAPM. The moving platforms 1 and 2 are connected with each other by using one R joint, which not only reduces the number of kinematic joints but also increases the dexterity of output motion to some extent. Besides the advantage of actuation redundancy, the Var2 RAPM is an overconstrained system, namely, there are some dependent constraints in the structure. Therefore, it can resist more external loads, and thus is more suitable for some machining applications that require high accuracy and stiffness [11].

This paper is organized as follows. In Sect. 2, the structural design of Var2 RAPM is introduced in detail. Section 3 analyzes the mobility by using the screw theory. The inverse kinematics are presented in Sect. 4, respectively. In Sect. 5, the local and global performances of Var2 RAPM are evaluated based on the motion/force transmissibility. Section 6 presents the dimensional optimization of Var2 RAPM with respect to the global indices. The conclusions are summarized in Sect. 7 finally.

2 Structure Description of the Var2 RAPM

The computer-aided design (CAD) model of Var2 RAPM is shown in Fig. 1, which is composed of a fixed base, two moving platforms, and four kinematic limbs. Compared with most existing PMs, the main difference of the designed Var2 RAPM is there are two moving platforms in the mechanism structure, which are called moving platform 1 and moving platform 2 respectively. They are connected with each other by using an R joint on the moving platform 1, and the rotational center is defined as point C. In addition, the moving platforms 1 and 2 are connected to the fixed base by two identical kinematic RPR limbs, and one RPS limb and one UPS limb, respectively. For the first limb RPR (A_1B_1) and second limb RPR (A_2B_2), the rotational axes of the four R joints are parallel to each other, and they are always perpendicular to the P joint. For the third limb RPS (A_3B_3) and fourth limb UPS (A_4B_4), they are connected to the moving platform 2 by two S joints. The first rotational axis of the U joint in the UPS limb is parallel to the axes of R joints in the two R<u>P</u>R limbs, and the second rotational axis of the U joint is always perpendicular to the <u>P</u> joint. For the R joint on the moving platform 1, its rotational direction is always perpendicular to the axes of R joints in the two R<u>P</u>R limbs. It should be noted that the points A_1, A_2, A_3, and A_4 are not in the same plane.

The fixed and moving coordinate frames should be established before analyzing the kinematics of Var2 RAPM. As shown in Fig. 1, the fixed coordinate frame *O-XYZ* is attached to the fixed base, in which the origin *O* is at the midpoint of A_1A_2. The *X*-axis passes through the point A_1, the *Y*-axis is parallel to the A_3A_4, and the *Z*-axis points vertically upward. For the moving coordinate frame *P-uvw*, the origin *P* is at the midpoint of B_3B_4. The *u*-axis is parallel to the B_1B_2, the *v*-axis passes through the point B_3, and the *w*-axis points vertically upward. An end-effector tool is mounted on the moving platform 2, and the end point of tool is defined as *p*. The architectural lengths of the Var2 RAPM are defined as follows: $OA_1 = OA_2 = l_1$, $CB_1 = CB_2 = l_2$, $A_3A_4 = 2l_3$, $PB_3 = PB_4 = l_4$, $CP = h_1$, $Pp = h_3$. In addition, the distance between the A_1A_2 and A_3A_4 is defined as h_2.

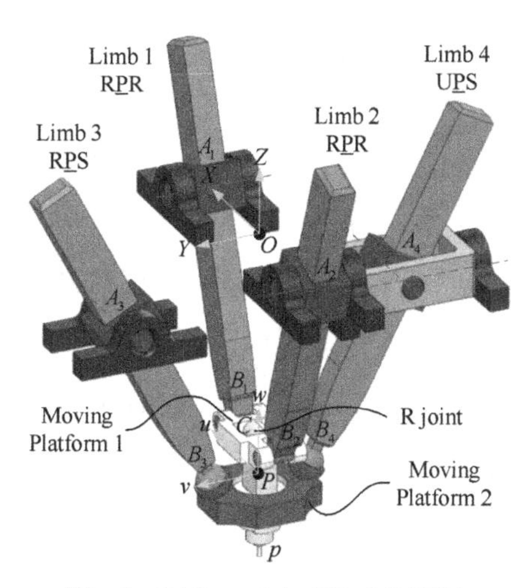

Fig. 1. CAD model of Var2 RAPM.

3 Mobility Analysis

The mobility analysis of the Var2 RAPM is conducted here by using the screw theory [12, 13]. A brief introduction to the used theory is necessary for understanding the derivative process of mobility better. In the screw theory, the unit twist screw of the jth kinematic joint of limb *i*, $\boldsymbol{\$}_{ij}$, is defined as follows:

$$\boldsymbol{\$}_{ij} = \left(\boldsymbol{s}_{ij}; \boldsymbol{s}_{0ij}\right) = \left(\boldsymbol{s}_{ij}; \boldsymbol{r}_{ij} \times \boldsymbol{s}_{ij} + h_{ij}\boldsymbol{s}_{ij}\right) \tag{1}$$

where $\boldsymbol{s}_{ij}$ denotes the unit direction vector, $\boldsymbol{r}_{ij}$ denotes the position vector of any point on the direction axis, and h_{ij} denotes the screw pitch.

Assuming there are n single-DOF kinematic joints in the limb i, $\{\boldsymbol{\$}_{i1}, \boldsymbol{\$}_{i2}, \ldots, \boldsymbol{\$}_{in}\}$, then the 6-$n$ unit constraint wrench screws $\boldsymbol{\$}_{im}^{r} = \left(\boldsymbol{s}_{im}^{r};\ \boldsymbol{s}_{im}^{0r}\right)$ that are reciprocal to these n twist screws can be calculated by using the following relationship:

$$\boldsymbol{\$}_{ij} \circ \boldsymbol{\$}_{im}^{r} = \boldsymbol{s}_{ij} \cdot \boldsymbol{s}_{im}^{0r} + \boldsymbol{s}_{im}^{r} \cdot \boldsymbol{s}_{0ij} = 0 \quad (j{=}1, \ldots, n \ \ m{=}1, \ldots, 6\text{-}n) \tag{2}$$

where $\circ$ denotes the reciprocal product.

The coordinates of joint points in the proposed Var2 RAPM should be defined firstly for the sake of mobility analysis clearly. The position vectors of points A_i and B_i (i = 1, 2, 3, 4) with respect to the frame O-XYZ are defined as $\boldsymbol{OA}_1 = \left(l_1\ 0\ 0\right)^{\mathrm{T}}$, $\boldsymbol{OA}_2 = \left(-l_1\ 0\ 0\right)^{\mathrm{T}}$, $\boldsymbol{OA}_3 = \left(0\ l_3\ -h_2\right)^{\mathrm{T}}$, $\boldsymbol{OA}_4 = \left(0\ -l_3\ -h_2\right)^{\mathrm{T}}$, $\boldsymbol{OB}_1 = \left(x_{B_1}\ 0\ z_{B_1}\right)^{\mathrm{T}}$, $\boldsymbol{OB}_2 = \left(x_{B_2}\ 0\ z_{B_2}\right)^{\mathrm{T}}$, $\boldsymbol{OB}_3 = \left(0\ y_{B_3}\ z_{B_3}\right)^{\mathrm{T}}$, and $\boldsymbol{OB}_4 = \left(x_{B_4}\ y_{B_4}\ z_{B_4}\right)^{\mathrm{T}}$, respectively. The coordinates of points C, P, and p are defined as $\boldsymbol{OC} = \left(x_1\ 0\ z_1\right)^{\mathrm{T}}$, $\boldsymbol{OP} = \left(x_2\ y_2\ z_2\right)^{\mathrm{T}}$, and $\boldsymbol{Op} = \left(x_{tool}\ y_{tool}\ z_{tool}\right)^{\mathrm{T}}$, respectively.

In this paper, the limb 1, limb 2, and moving platform 1 are integrated into an overconstrained substructure, and its twist system in the general configuration can be expressed in the fixed coordinate frame O-XYZ as follows:

$$\begin{cases} \boldsymbol{\$}_{11} = \left(0\ \ 1\ \ 0;\ \ 0\ \ 0\ \ l_1\right) & \boldsymbol{\$}_{21} = \left(0\ \ 1\ \ 0;\ \ 0\ \ 0\ \ -l_1\right) \\ \boldsymbol{\$}_{12} = \left(0\ \ 0\ \ 0;\ \ x_{B_1} - l_1\ \ 0\ \ z_{B_1}\right) & \boldsymbol{\$}_{22} = \left(0\ \ 0\ \ 0;\ \ x_{B_2} + l_1\ \ 0\ \ z_{B_2}\right) \\ \boldsymbol{\$}_{13} = \left(0\ \ 1\ \ 0;\ \ -z_{B_1}\ \ 0\ \ x_{B_1}\right) & \boldsymbol{\$}_{23} = \left(0\ \ 1\ \ 0;\ \ -z_{B_2}\ \ 0\ \ x_{B_2}\right) \\ \boldsymbol{\$}_{m1} = \left(l_{m1}\ \ 0\ \ n_{m1};\ \ 0\ \ z_1 l_{m1} - x_1 n_{m1}\ \ 0\right) & \end{cases} \tag{3}$$

where l_{m1} and n_{m1} denotes the components of the direction vector of the R joint on the moving platform 1.

Using the screw theory, the independent wrench system of this substructure is composed of two constraint wrench screws (CWSs), and can be obtained as follows:

$$\boldsymbol{\$}_{11}^{r} = \left(0\ \ 0\ \ 0;\ \ -n_{m1}\ \ 0\ \ l_{m1}\right),\ \ \boldsymbol{\$}_{12}^{r} = \left(0\ \ 1\ \ 0;\ \ -z_1\ \ 0\ \ x_1\right) \tag{4}$$

where $\boldsymbol{\$}_{11}^{r}$ denotes the constraint couple perpendicular to the rotational axes of the R joints in the R$\underline{\text{P}}$R limbs and moving platform 1 simultaneously. $\boldsymbol{\$}_{12}^{r}$ denotes the constraint force passing through the points C and along the direction of Y-axis.

Using the similar procedure, the wrench systems of the third R$\underline{\text{P}}$S limb can be obtained as follows:

$$\boldsymbol{\$}_{31}^{r} = \left(1\ \ 0\ \ 0;\ \ 0\ \ z_{B_3}\ \ -y_{B_3}\right) \tag{5}$$

where $\boldsymbol{\$}_{31}^{r}$ denotes the constraint force passing through the point B_3 and parallel to the direction of the X-axis.

Since the fourth U$\underline{\text{P}}$S limb is a 6-DOF limb, there are no constraint couples or forces within this limb. According to the Eqs. (4) and (5), the wrench system of the Var2 RAPM is the union of the wrench systems of limbs, and thus can be expressed as follows:

$$\begin{cases} \$^r_{C1} = \begin{pmatrix} 0 & 0 & 0; & -n_{m1} & 0 & l_{m1} \end{pmatrix} & \$^r_{C3} = \begin{pmatrix} 1 & 0 & 0; & 0 & z_{B_3} & -y_{B_3} \end{pmatrix} \\ \$^r_{C2} = \begin{pmatrix} 0 & 1 & 0; & (n_{m1}x_1 - l_{m1}z_1)/l_{m1} & 0 & 0 \end{pmatrix} & \end{cases} \tag{6}$$

Therefore, the twist system of the designed Var2 RAPM can be obtained by using the screw theory as follows:

$$\begin{cases} \$_{pm1} = \begin{pmatrix} l_{m1} & 0 & n_{m1}; & y_{B_3}n_{m1} & z_1 l_{m1} - x_1 n_{m1} & 0 \end{pmatrix} & \$_{pm3} = \begin{pmatrix} 0 & 0 & 0; & 0 & 0 & 1 \end{pmatrix} \\ \$_{pm2} = \begin{pmatrix} 0 & 1 & 0; & -z_{B_3} & 0 & 0 \end{pmatrix} & \end{cases} \tag{7}$$

Equation (7) shows that the Var2 RAPM has 3 output DOFs, namely, 2R1T. They are one rotation around the B_1B_2, one rotation passing through the point B_3 and parallel to the direction of the Y-axis, and one translation along the Z-axis, respectively. Therefore, the proposed Var2 RAPM can be used as a parallel module of a hybrid five-axis machine, and applied for machining the curved workpieces.

4 Inverse Kinematics

The inverse kinematics of Var2 RAPM is an essential step for the performance evaluation and further control, which involves the calculation of the actuated displacements q_i ($i =$ 1, 2, 3, 4) given the position of the center point of moving platform 2 $\begin{pmatrix} x_2 & y_2 & z_2 \end{pmatrix}^{\mathrm{T}}$. By using the Z-X-Y Euler angle, the orientation of the moving frame P-uvw with respect to the fixed frame O-XYZ can be described by a rotation matrix $\boldsymbol{R}$ as:

$$\boldsymbol{R} = \boldsymbol{R}_Y(\theta)\boldsymbol{R}_X(\psi)\boldsymbol{R}_Z(\varphi) = \begin{pmatrix} s\theta s\psi s\varphi + c\theta c\varphi & s\theta s\psi c\varphi - c\theta s\varphi & s\theta c\psi \\ c\psi s\varphi & c\psi c\varphi & -s\psi \\ c\theta s\psi s\varphi - s\theta c\varphi & c\theta s\psi c\varphi + s\theta s\varphi & c\theta c\psi \end{pmatrix} \tag{8}$$

where s and c denote the sine and cosine functions, respectively. The $\boldsymbol{R}_Z(\varphi)$, $\boldsymbol{R}_X(\psi)$, and $\boldsymbol{R}_Y(\theta)$ denote the rotation matrices around the Z-, X-, and Y-axis, respectively.

Before expressing the actuated displacements, the relationship between the rotational angles $\begin{pmatrix} \varphi & \psi & \theta \end{pmatrix}$ and position $\begin{pmatrix} x_2 & y_2 & z_2 \end{pmatrix}^{\mathrm{T}}$ should be determined firstly, which can be finished by using three constraint conditions of the Var2 RAPM. These conditions can be described as follows: (A) the Y-axis component of the position vector $\boldsymbol{OC}$ is equal to zero; (B) the Y-axis component of the position vector $\boldsymbol{OB}_1$ is equal to zero; (C) the X-axis component of the position vector $\boldsymbol{OB}_3$ is equal to zero. Based on the above conditions, the angles $\begin{pmatrix} \varphi & \psi & \theta \end{pmatrix}$ can be obtained as:

$$\psi = \arcsin(y_2/h_1), \quad \varphi = 0, \quad \theta = \arcsin(-x_2/(l_4 s\psi)) = \arcsin(-x_2h_1/(y_2l_4)) \tag{9}$$

To sum up, the expressions of angles ψ and θ with respect to the coordinates $\begin{pmatrix} x_2 & y_2 & z_2 \end{pmatrix}^{\mathrm{T}}$ have been obtained. Based on these results, the positions of points B_i ($i =$ 1, 2, 3, 4) also can be expressed with respect to the coordinates $\begin{pmatrix} x_2 & y_2 & z_2 \end{pmatrix}^{\mathrm{T}}$. Therefore, the actuated displacements q_i ($i = 1, 2, 3, 4$) can be obtained as follows:

$$q_i = \|\boldsymbol{OB}_i - \boldsymbol{OA}_i\|_2 \tag{10}$$

5 Kinematic Performance Evaluation

Kinematic performance evaluation is to calculate and analyze the kinematic performance of the Var2 RAPM within the reachable workspace, which is the basis of the optimization design. Up to data, there are several kinematic indices widely used [14–18], including the condition number [14], manipulability [15] based on the algebra characteristics of the kinematic Jacobian matrix, and the motion/force transmission index [16–18] based on the screw theory. Among these indices, the motion/force transmission index is selected to evaluate the kinematic performance of the proposed 2R1T Var2 RAPM here, which not only avoids the inconsistency problem caused by the Jacobian-based indices [19], but also can measure the power transmission between the inputs and outputs clearly.

Performance evaluation can be divided into the local and global levels. In this paper, the local transmission index (LTI) proposed by Li et al. [17, 18] is used for the motion/force transmissibility of the Var2 RAPM in a certain configuration, which considers the input transmission index (ITI) and output transmission index (OTI) simultaneously. The used index was constructed by virtually separating the original RAPM into multiple 1DOF RAPMs with two actuators, and considering the motion/force transmission characteristics of all the 1DOF RAPMs comprehensively. In the following parts, the LTI procedure of the Var2 RAPM in the workspace will be illustrated, which mainly consists of four steps.

Step 1: Determine the transmission wrench screw (TWS) in each limb of the Var2 RAPM.

According to Ref. [20], the TWS in each limb should be linearly independent of all the constraint wrench screws and passive twist screws. Based on the results in Fig. 1 and Eq. (6), the four unit TWSs are the wrench screws passing through the point A_i and along the directions of A_iB_i.

Step 2: Determine the virtual output twist screws (OTSs) of the Var2 RAPM.

Through locking the two actuators in the Var2 RAPM, the corresponding two TWSs then turn into the new CWSs exerted on the moving platform 2. In this case, the Var2 RAPM is subjected to a new wrench system with five CWSs, and the OTS of moving platform 2 can be derived by using the screw theory. The output motion with instantaneous single DOF is driven by two active actuators. For the Var2 RAPM with four actuators, there are 6 combination cases ($C_4^2 = 6$) of locking strategy in a certain configuration.

The derivative process of OTS in the case 1 where the actuators of limb 1 and 2 are locked is taken as example here. Based on the first new wrench system of moving platform $\boldsymbol{U}^1 = \{\boldsymbol{\$}_{C1}^r, \boldsymbol{\$}_{C2}^r, \boldsymbol{\$}_{C3}^r, \boldsymbol{\$}_{T1}, \boldsymbol{\$}_{T2}\}$, the OTS of the moving platform 2, $\boldsymbol{\$}_O^1$, can be derived by using the screw theory as:

$$\boldsymbol{\$}_O^1 \circ \boldsymbol{U}^1 = 0 \tag{11}$$

Using the similar way, the other five OTSs can be obtained.

Step 3: Calculate the ITI and OTI of the Var2 RAPM.

The ITI can reflect the power transmissibility from the actuator to the limb. For the Var2 RAPM, there are four actuated limbs in the mechanism, and the ITI of each limb is be defined as follows:

$$\lambda_i = \left|\boldsymbol{\$}_{Ai} \circ \boldsymbol{\$}_{Ti}\right| / \left|\boldsymbol{\$}_{Ai} \circ \boldsymbol{\$}_{Ti}\right|_{\max} \quad (i=1, 2, 3, 4) \tag{12}$$

where $\boldsymbol{\$}_{Ai}$ denotes the input twist screw, and it is determined by the $\underline{P}$ joint in each limb of Var2 RAPM.

Different from the ITI, the OTI can evaluate the power transmissibility from the limb to the moving platform. In Ref. [17], the OTI is defined as the average value of output transmission performance of all the single DOF RAPMs. The process of output performance in case 1 is also taken as an example here, which is influenced by the active limbs 3 and 4, and can be expressed as follows:

$$\eta^1 = \min\left\{\frac{\left|\boldsymbol{\$}_O^1 \circ \boldsymbol{\$}_{T3}\right| + \left|\boldsymbol{\$}_O^1 \circ \boldsymbol{\$}_{T4}\right|}{\left|\boldsymbol{\$}_O^1 \circ \boldsymbol{\$}_{T3}\right|_{max} + \left|\boldsymbol{\$}_O^1 \circ \boldsymbol{\$}_{T4}\right|}, \frac{\left|\boldsymbol{\$}_O^1 \circ \boldsymbol{\$}_{T3}\right| + \left|\boldsymbol{\$}_O^1 \circ \boldsymbol{\$}_{T4}\right|}{\left|\boldsymbol{\$}_O^1 \circ \boldsymbol{\$}_{T3}\right| + \left|\boldsymbol{\$}_O^1 \circ \boldsymbol{\$}_{T4}\right|_{max}}\right\} \tag{13}$$

The physical meaning of the above equation is the minimum value of coupled effects from the two active limbs. After obtaining the η^g ($g = 1, 2, 3, 4, 5, 6$) in all the cases, the OTI of Var2 RAPM in a certain configuration can thus be written as follows:

$$\eta = \sum_{g=1}^{6} \eta^g / 6 \tag{14}$$

Step 4: Determine the LTI of the Var2 RAPM. Considering the ITI and OTI in a certain configuration simultaneously, the LTI of the Var2 RAPM can be expressed as:

$$\Gamma = \min\{\lambda_1, \lambda_2, \lambda_3, \lambda_4, \eta\} \tag{15}$$

The procedure for obtaining the LTI of Var2 RAPM has been illustrated above, from which the LTI distributions in the reachable workspace can be calculated given the architecture parameters. In this paper, the ranges of orientations of the Var2 RAPM are defined as follows: $\psi \in [-30°, 30°]$, and $\theta \in [-30°, 30°]$, and the link parameters are set as follows: $l_1 = 300\,\text{mm}$, $l_3 = 300\,\text{mm}$, $l_2 = 100\,\text{mm}$, $l_4 = 120\,\text{mm}$, $h_1 = h_2 = 110\,\text{mm}$, and $h_3 = 150\,\text{mm}$.

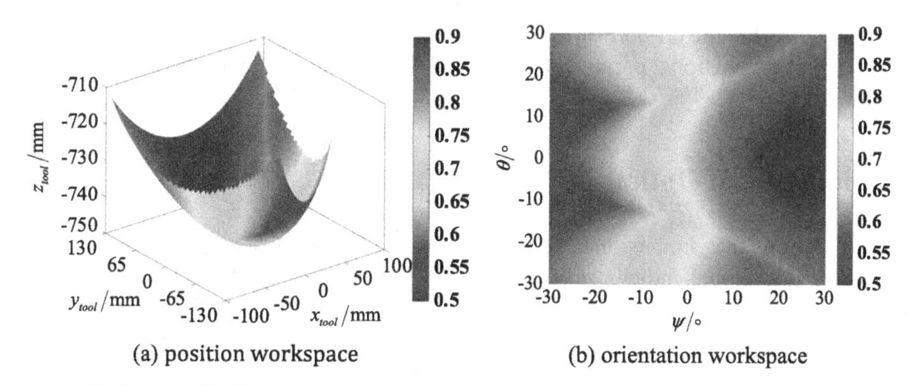

Fig. 2. LTI distributions of Var2 RAPM in the workspace when $z_2 = -600\,\text{mm}$.

For clarity, the LTI distributions of Var2 RAPM in the position and orientation workspaces when $z_2 = -600\,\text{mm}$ are shown in Fig. 2. Figure 2(a) shows the LTI distributions in the position workspace of the end point of tool, from which we can find that motion/force transmissibility in the region where $y_{tool} > 0$ is better than that in the region where $y_{tool} \leq 0$, which is caused by the difference between the R$\underline{P}$S and U$\underline{P}$S limbs. In addition, the distributions are symmetrical about the plane $x_{tool} = 0$, which is consistent with the structural characteristics of the Var2 RAPM. The LTI distributions in the orientation workspace are shown in Fig. 2(b), which are also symmetrical about the angle θ. The general trend in Fig. 2(b) is that the motion/force transmissibility of the mechanism is getting better with the increase of angle ψ.

Only the LTI, which is used to measure the transmissibility of a certain configuration, is not enough to evaluate the overall performance of Var2 RAPM within these link parameters. Two global performance indices are therefore defined here to describe the performances in a set of poses. They are the global transmission index (GTI) and good transmission workspace (GTW), respectively. In this paper, the GTI is defined as the average value of the performance in the workspace, which can be expressed as:

$$\sigma_{\text{GTI}} = \int_S \Gamma dW \Big/ \int_S dW \tag{16}$$

where W denotes the workspace, and S denotes the area of workspace.

The GTW is defined as the proportion of workspace in which the value of LTI is greater than 0.7 [21], and it can be expressed as:

$$\sigma_{\text{GTW}} = \int_{S_G} dW \Big/ \int_S dW \tag{17}$$

where S_G denotes the area of workspace with $\Gamma \geq 0.7$.

Both the σ_{GTI} and σ_{GTW} range from zero to unity, and the bigger value means the overall performance of the Var2 RAPM is in the better level. Based on the selected link parameters, the values of σ_{GTI} and σ_{GTW} of Var2 RAPM in the workspace when $z_2 = -600$ mm can be calculated based on the Eqs. (16) and (17), which are 0.739 and 0.598, respectively. Results show that the average value of overall performance is good in the workspace, but the proportion of workspace with better motion/force transmissibility is fewer. The dimensional parameters of Var2 RAPM should be optimized before actual manufacturing.

6 Dimensional Optimization

In this section, the dimensional parameters of Var2 RAPM are optimized by using the above two global indices, σ_{GTI} and σ_{GTW} simultaneously. The rotational ranges are limited as $\psi \in [-30°, 30°]$ and $\theta \in [-30°, 30°]$, and the operational height are set as $z_2 = -2l_1$. The link l_2, l_4, and l_1 are set as design parameters here, and $h_1 = h_2 = (l_2 + l_4)/2$ here. Using the parameter-finiteness normalization method [22], these design parameters can be expressed as follows:

$$l_1 = l_3,\ D = (l_1 + l_2 + l_4)/3,\ \text{and}\ r_1 = l_2/D,\ r_2 = l_4/D,\ r_3 = l_1/D = l_3/D \tag{18}$$

where D denotes a normalized factor, and r_i (i = 1, 2, 3) denote normalized nondimensional parameters, which should satisfy the following relationships:

$$0 < r_1,\ r_2,\ r_3 < 3 \text{ and } r_1,\ r_2 \leq r_3 \tag{19}$$

Based on the Eqs. (18) and (19), the parameter design space can be obtained, as shown in Fig. 3, including the spatial (Fig. 3(a)) and plan (Figs. 3(b)–(d)) views with all possible points. The relationship between (r_1, r_2, r_3) and (s, t) can be described as follows:

$$r_1 = 3/2 + \sqrt{3}t/2 - s/2,\ r_2 = 3/2 - \sqrt{3}t/2 - s/2,\ r_3 = s \text{ or } s = r_3,\ t = (r_1 - r_2)/\sqrt{3} \tag{20}$$

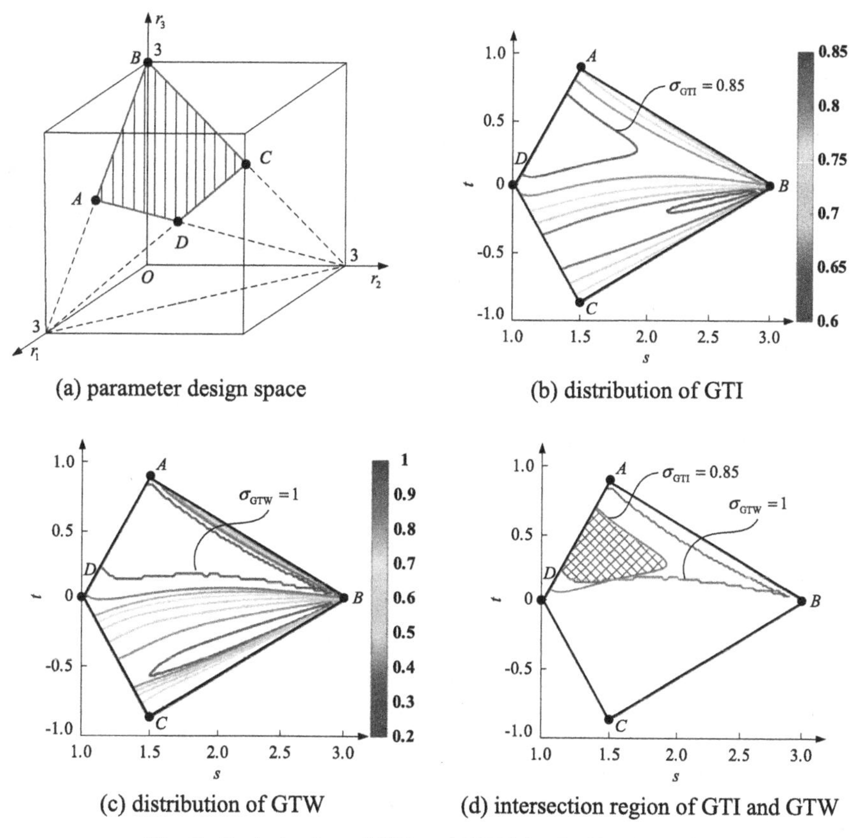

Fig. 3. Optimization of GTI and GTW for the Var2 RAPM.

The distributions of σ_{GTI} and σ_{GTW} in the plan view are shown in Figs. 3(b) and (c), from which one can find that the variation tendencies of two plots are similar. Through intersecting these two results, the optimized (s, t) with better σ_{GTI} and σ_{GTW} can be obtained, as shown in Fig. 3(d). In the intersection region, all the σ_{GTW} are equal to unity, and all the σ_{GTI} are bigger than 0.85. For example, when s = 1.44 and

$t = 0.4$, the normalized nondimensional parameters r_i $(i = 1, 2, 3)$ can be obtained by substituting (s, t) into Eq. (20), which can be calculated as $r_1 = 1.1264$, $r_2 = 0.4336$, and $r_3 = 1.44$. Combining the actual situations, the normalized factor D is set as 300mm, and thus the optimized link parameters of Var2 RAPM can be obtained by using Eq. (18), i.e., $l_1 = l_3 = 432$ mm, $l_2 = 337.92$ mm, $l_4 = 130.08$ mm, and $h_1 = h_2 = 234$ mm. The LTI distributions of the Var2 RAPM with optimized link parameters in the position and orientation workspaces are shown in Fig. 4(a) and (b), respectively, which are also symmetrical about the plane $x_{tool} = 0$ and angle θ. Compared with the results in Sect. 5, the σ_{GTI} and σ_{GTW} with optimized parameters are increased from 0.739 to 0.882 (improved by 19.35%), and from 0.598 to 1 (improved by 67.22%), respectively, which verifies the effectiveness of optimization. For different applications, therefore, the designers or engineers can select suitable optimized parameters in the regions.

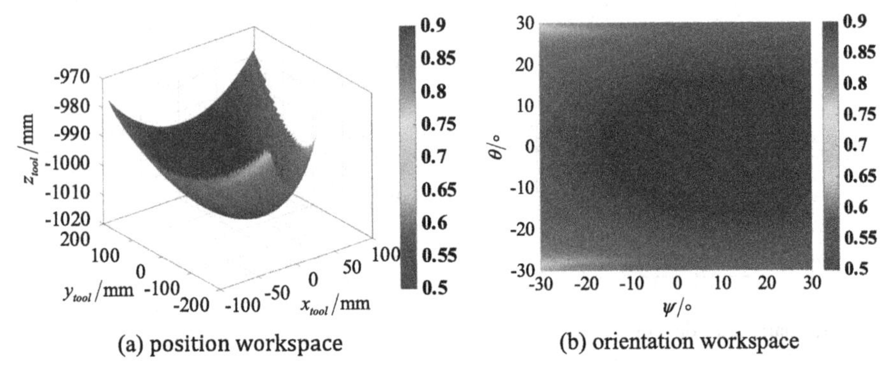

Fig. 4. LTI distributions of the Var2 RAPM with optimized link parameters.

7 Conclusions

A new 2R1T RAPM, Var2 RAPM, is designed in this paper, which is actuated by four P joints. Compared with other existing RAPMs, the main difference is there are two moving platforms in the Var2 RAPM, which not only enlarge the ranges of orientation angles but also increase the dexterity of output motion. The motion/force transmissibility is used to evaluate the kinematic performance of Var2 RAPM, and great overall performances are obtained by optimizing the link parameters by using two global indices. The proposed Var2 RAPM can be used as parallel modules of hybrid five-axis machines, and applied for machining the curved workpieces.

References

1. Liu, J.F., Fan, X.M., Ding, H.F.: Investigation of a novel 2R1T parallel mechanism and construction of its variants. Robotica **39**(10), 1834–1848 (2021)

2. Zhao, C., Chen, Z.M., Li, Y.W., Huang, Z.: Motion characteristics analysis of a novel 2R1T 3-UPU parallel mechanism. ASME J. Mech. Des. **142**(1), 012302 (2020)
3. Wahl, J.: Articulated Tool Head. US Patent No. 6431802B1 (2002)
4. Bi, Z.M., Jin, Y.: Kinematic modeling of Exechon parallel kinematic machine. Robot. Comput. Integr. Manuf. **27**(1), 186–193 (2011)
5. Siciliano, B.: The Tricept robot: inverse kinematics, manipulability analysis and closed-loop direct kinematics algorithm. Robotica **17**(4), 437–445 (1999)
6. Gosselin, C., Schreiber, L.T.: Redundancy in parallel mechanisms: a review. ASME Appl. Mech. Rev. **70**(1), 010802 (2018)
7. Lee, M., Jeong, H., Lee, D.: Design optimization of 3-DOF redundant planar parallel kinematic mechanism based finishing cut stage for improving surface roughness of FDM 3D printed sculptures. Mathematics **9**(9), 961 (2021)
8. Sun, T., Liang, D., Song, Y.M.: Singular-perturbation-based nonlinear hybrid control of redundant parallel robot. IEEE Trans. Industr. Electron. **65**(4), 3326–3336 (2018)
9. Shen, X.D., Xu, L.M., Li, Q.C.: Motion/force constraint indices of redundantly actuated parallel manipulators with over constraints. Mech. Mach. Theory **165**, 104427 (2021)
10. Chai, X., Zhang, N., He, L., Li, Q., Ye, W.: Kinematic sensitivity analysis and dimensional synthesis of a redundantly actuated parallel robot for friction stir welding. Chin. J. Mech. Eng. **33**(1), 1 (2020). https://doi.org/10.1186/s10033-019-0427-6
11. Jin, Y., Kong, X.W., Higgins, C., Price, M.: Kinematic design of a new parallel kinematic machine for aircraft wing assembly. In: Proceedings of the 10th IEEE International Conference on Industrial Informatics, pp. 669–674. Beijing, China (2012)
12. Huang, Z., Li, Q.C.: Type synthesis of symmetrical lower-mobility parallel mechanisms using the constraint-synthesis method. Int. J. Robot. Res. **22**(1), 59–82 (2003)
13. Li, S., Shan, Y., Yu, J., Ke, Y.: Actuation spaces synthesis of lower-mobility parallel mechanisms based on screw theory. Chin. J. Mech. Eng. **34**(1), 1–12 (2021). https://doi.org/10.1186/s10033-021-00546-7
14. Xu, L.M., Chen, Q.H., He, L.Y., Li, Q.C.: Kinematic analysis and design of a novel 3T1R 2-(PRR)^{2}RH hybrid manipulator. Mech. Mach. Theory **112**, 105–122 (2017)
15. Stoughton, R.S., Arai, T.: A modified stewart platform manipulator with improved dexterity. IEEE Trans. Robot. Autom. **9**(2), 166–173 (1993)
16. Chong, Z.H., Xie, F.G., Liu, X.J., Wang, J.S., Niu, H.F.: Design of the parallel mechanism for a hybrid mobile robot in wind turbine blades polishing. Robot. Comput. Integr. Manuf. **61**, 101857 (2020)
17. Li, Q.C., Zhang, N.B., Wang, F.B.: New indices for optimal design of redundantly actuated parallel manipulators. ASME J. Mech. Robot. **9**(1), 011007 (2017)
18. Xu, L.M., Chai, X.X., Li, Q.C., Zhang, L.A., Ye, W.: Design and experimental investigation of a new 2R1T overconstrained parallel kinematic machine with actuation redundancy. ASME J. Mech. Robot. **11**(3), 031016 (2019)
19. Merlet, J.P.: Jacobian, manipulability, condition number, and accuracy of parallel robots. ASME J. Mech. Des. **128**(1), 199–206 (2006)
20. Joshi, S.A., Tsai, L.W.: Jacobian analysis of limited-DOF parallel manipulators. ASME J. Mech. Des. **124**(2), 254–258 (2002)
21. Tao, D.C.: Applied Linkage Synthesis. Addison-Wesley (1964)
22. Liu, X.J., Wang, J.S.: A new methodology for optimal kinematic design of parallel mechanisms. Mech. Mach. Theory **42**(9), 1210–2122 (2007)

Control Design for a Planar 2-DOF Parallel Manipulator: An Active Inference Based Approach

Duanling Li[1], Yixin He[1], Yanzhao Su[2], Xiaomin Zhao[3], Jin Huang[2(✉)], and Liwei Cheng[1]

[1] Beijing University of Posts and Telecommunications, Beijing 100876, China
[2] Tsinghua University, Bejing 100084, China
huangjin@tsinghua.edu.cn
[3] Hefei University of Technology, Hefei 230009, Anhui, China

Abstract. Active inference controller has the advantages of simplicity, high efficiency and low computational complexity. It is based on active inference framework which is prominent in neuroscientific theory of the brain. Although active inference has been successfully applied to neuroscience, its application in robotics are highly limited. In this paper, an active inference controller is adopted to steer a 2-DOF parallel manipulator movement from the initial state to the desired state. Firstly, the active inference controller is introduced. Secondly, Dynamic model of parallel manipulator system with constraints is established by Udwaida-Kalaba equation. Thirdly, apply the active inference controller to the parallel manipulator system. Finally, the simplicity and effectiveness of control effect are verified by numerical simulations.

Keywords: Active inference · Free energy principle · Parallel manipulator · Udwaida-Kalaba equation · Dynamic modeling

1 Introduction

With the development and research of industrial automation control, electromechanical and computer science technology, the technology of robot is rapidly rising and widely applied in machinery manufacturing [1], logistics distribution [2], aerospace industry [3], national defense [4], deep-sea exploration [5], medical industry [6] and so on. In these different types of robots, the industrial robots attract the most attention and play an important and irreplaceable role in complex, harsh and unsuitable for human survival environments. Serial robot has been developed more mature and has the advantages of large motion space, simple structure and low cost. Compared with the serial robot, parallel robot has the characteristics of strong bearing capacity, higher precision and smaller error [7].

Supported by the NSFC Programs (No. 52175019, No. 61872217), Beijing Natural Science Foundation (No. 3212009).

H. Liu et al. (Eds.): ICIRA 2022, LNAI 13456, pp. 58–69, 2022.
https://doi.org/10.1007/978-3-031-13822-5_6

Nowadays, many scenarios don't need robots with higher mobility indeed and these works can be completed by robots with fewer degrees of freedom [8]. Parallel manipulator with fewer degrees of freedom is simple structure, easier control design and inexpensive and have a perfect practical application prospect [9].

The traditional PID control [10] has been difficult to meet the requirements of robot control of high-precision and high-performance. In addition to PID control, common robot control methods include sliding mode control [11], robust control [12], adaptive control [13] and so on. When designing a sliding mode controller, it does not obtain a completely accurate mathematical model of the controlled system and can solve the disturbance caused by the uncertain model. Robust control has prominent suppression effect on external disturbance and uncertainty. However, for the nonlinear complex system of robot manipulator, robust controller needs to linearize the nonlinear system and the design process of controller is complicated and has errors. Adaptive control plays an important role in solving the problems of parameter uncertainty and external interference in robot manipulator. At present, adaptive control has been combined with other control methods by researchers, such as sliding mode adaptive control [14], robust adaptive control [15], fuzzy adaptive control [16] and neural network adaptive control [17], and these combined control methods have better control effect.

At present, the robot control is inspired by active inference which is a prominent brain theory and has been successful in the field of neuroscience [18]. Active inference provides an approach to understand the decisions made by agents. The optimal behavior of agents is based on the principle of free energy minimization. Active inference is related to Bayesian inference. It is necessary for agent to introduce generation model of world and describing the state. Agents make predictions with the help of the above model and realize these predictions through taking actions. Active inference can be applied to robot control and because this method does not need inverse dynamic model, it has aroused the interest of researchers [19]. Léo Pio-Lopez et al. [20] demonstrate the feasibility of active inference in robot control and simulate to control a 7 DOF arm of a PR2 robot by using the active inference framework in Robot Operating System. Guillermo Oliver et al. [21] apply the active inference model to a real humanoid robot for the first time and show the advantage and robustness of this control algorithm in reaching behavior. Corrado Pezzato et al. [22] designed a controller based on active inference which is the first model free and apply it to a real 7-DOF series manipulator. Mohamed Baioumy et al. [23] present an approach for robotic manipulator on the basis of the active inference framework by introducing a temporal parameter and the approach improves the robustness against poor initial parameters and damps oscillations. Active inference has a promising prospect, however, there is no application of scholars applying it to parallel manipulator.

In this paper, we apply an active inference controller to a 2-DOF parallel manipulator successfully and achieve the movement of parallel manipulator from the initial state to the desired state. The rest of paper is organized as follows: In Sect. 2, we introduce the active inference controller, which is composed of three parts: free energy principle, belief update and control actions. In Sect. 3

we adopt the Udwadia-Kalaba equation and complete the dynamics model of 2-DOF parallel manipulator system. In Sect. 4, the performance of the active inference controller is verified via numerical simulations. In Sect. 5, conclusions are drawn, and some works we have to do in the future are proposed.

2 Active Inference Controller

The active inference controller [22] is related to the classical PID controller and when the temporal parameter approaches 0, the controller is similar to a PID controller [23]. The model free controller using active inference for robot control consists of three parts: free energy equation, belief update and control actions. The free energy equation provides a way to calculate the free energy. Based on the minimizing free energy principle, the belief update and control action adopt the gradient descent method to reduce the free energy of the robot system, and finally make each joint of the robot reach the desired position from the initial position.

2.1 Free Energy Equation

Free energy principle is related to Bayesian rules. Given dynamic generative model of agent's world, using Bayesian rules, we can get a posterior probability $p(m|n)$ which is the probability of being in the state m under the condition of sensory input n. Since solving $p(n)$ usually involves difficult integrals, an auxiliary probability distribution $q(m)$ is introduced to approximate the true posterior distribution $p(m|n)$. The Kullback-Leibler divergence between the posterior distribution $p(m|n)$ and auxiliary probability distribution $q(m)$ is calculated and minimize it.

$$D_{KL}(q(m)||p(m|n)) = \int q(m) \ln \frac{q(m)}{p(m|n)} dx = F + \ln p(n) \tag{1}$$

In the above equation, F is the free energy. While minimizing free energy F, $D_{KL}(q(m)||p(m|n))$ can also be minimized, so that the auxiliary probability distribution $q(m)$ is approaches to the true posterior distribution $p(m|n)$. Suppose $q(m)$ is Gaussian distribution whose mean is μ. According to the Laplace approximation, F can be simplified to

$$F \approx -\ln p(\mu, n) \tag{2}$$

μ is referred to as the 'belief' about the true state m. Since Eq. (2) is still a general expression, we must further specify the joint probability $p(\mu, n)$ in order to numerically calculate F. Here, generalized motion is introduced, using increasingly higher order time derivatives, to more accurately describe the belief μ and sensory input n of the system. Using additional theoretical knowledge of the generalized motion, the equation of free energy is further obtained.

$$F = -\ln p(\tilde{\mu}, \tilde{n}) \tag{3}$$

$\tilde{\mu}$ and $\tilde{n}$ are higher order time derivatives about the belief μ and sensory input n, which are $\tilde{\mu} = [\mu, \dot{\mu}, \ddot{\mu}, ...]$ and $\tilde{n} = [n, \dot{n}, \ddot{n}, ...]$ respectively. The sensory input of a certain dynamic order is only related to the beliefs of the same dynamic order. However, the state of a certain dynamic order is only related to the state of the lower order. In addition, the noise of each dynamic order is irrelevant. Since the position and velocity can be measured by the position and velocity sensors, the generalized motion is only up to the second order. Then, $p(\tilde{\mu}, \tilde{n})$ can be written as

$$p(\tilde{\mu}, \tilde{n}) = p(n|\mu)\, p(\dot{\mu}|\mu)\, p(\dot{n}|\dot{\mu})\, p(\ddot{\mu}|\dot{\mu}) \tag{4}$$

n and μ are the sensory observation and the internal belief of angles of robotic joints, $\dot{n}$ and $\dot{\mu}$ is the sensory observation and the internal belief of angular velocities of robotic joints, $\ddot{\mu}$ is the internal belief of angular accelerations of robotic joints. Two functions $r(\mu)$ and $t(\mu)$ need to be introduced to evaluate the free energy. $r(\mu)$ shows the mapping between the sensory input n and the belief μ. $t(\mu)$ is a function about the belief μ, which guides the robot to the desired position μ_d. μ_d represents the ideal angles of robotic joints. $r(\mu)$ and $t(\mu)$ are

$$r(\mu) = \mu \tag{5}$$

$$t(\mu) = \mu_d - \mu \tag{6}$$

For getting the numerical expression of free energy, we also need to establish a generation model about the belief μ, namely

$$\dot{\mu} = t(\mu) + h \tag{7}$$

where h is gaussian noise.

According to the above equation, we can further get

$$\ddot{\mu} = \frac{\partial t}{\partial \mu}\dot{\mu} + \dot{h} = -\dot{\mu} + \dot{h} \tag{8}$$

Substituting Eq. (4), Eq. (5)–(8) into Eq. (3), we can obtain the numerical expression of free energy

$$\begin{aligned} F = {} & \tfrac{1}{2}(n-\mu)^{\mathrm{T}} \textstyle\sum_n^{-1} (n-\mu) + \tfrac{1}{2}(\dot{n}-\dot{\mu})^{\mathrm{T}} \textstyle\sum_{\dot{n}}^{-1} (\dot{n}-\dot{\mu}) \\ & + \tfrac{1}{2}(\dot{\mu}+\mu-\mu_d)^{\mathrm{T}} \textstyle\sum_{\mu}^{-1} (\dot{\mu}+\mu-\mu_d) + \tfrac{1}{2}(\ddot{\mu}+\dot{\mu})^{\mathrm{T}} \textstyle\sum_{\dot{\mu}}^{-1} (\ddot{\mu}+\dot{\mu}) \end{aligned} \tag{9}$$

where

$$\begin{aligned} \textstyle\sum_n = \sigma_n I_n, \sum_{\dot{n}} = \sigma_{\dot{n}} I_n \\ \textstyle\sum_\mu = \sigma_\mu I_n, \sum_{\dot{\mu}} = \sigma_{\dot{\mu}} I_n \end{aligned} \tag{10}$$

σ_n, $\sigma_{\dot{n}}$, σ_μ and $\sigma_{\dot{\mu}}$ are constants, I_n is identity matrix.

2.2 Belief Update

Belief update adopts the gradient descent method which is based on the free energy principle, to minimize F by finding the belief μ. The belief update law can be written as follow

$$\Delta\tilde{\mu} = \frac{d}{dt}\tilde{\mu} - k_\mu \frac{\partial F}{\partial \tilde{\mu}} \tag{11}$$

where k_{μ} is the learning rate of belief update, which can be regarded as a tuning parameter.

Therefore, the generalized motion of the belief μ in the next moment is

$$\tilde{\mu}_{(t+1)} = \tilde{\mu}_{(t)} + \Delta t \cdot \Delta \tilde{\mu}_{(t)} \tag{12}$$

2.3 Control Actions

The control actions τ is a vector which represents the driving torque of each joint of the robot. It can guide the parallel manipulator system from the initial position to the desired position. This process also adopts the gradient descent method based on the free energy principle. Although the free energy equation is not directly related to the control action τ, the control actions τ can modify the sensory input n to minimize free energy. The actions update law is given by

$$\Delta \tau = -k_a \frac{\partial \tilde{n}}{\partial \tau} \frac{\partial F}{\partial \tilde{n}} \tag{13}$$

where k_a is the learning rate of actions update, like k_{μ}, it also can be regarded as a tuning parameter. $\frac{\partial \tilde{n}}{\partial \tau}$ is considered to be the identity matrix.

Therefore, the control actions τ in the next moment is

$$\tau_{(t+1)} = \tau_{(t)} + \Delta t \cdot \Delta \tau_{(t)} \tag{14}$$

3 Dynamic Model of Parallel Manipulator System

It is difficult to establish analytical dynamics model of parallel manipulator by traditional dynamics modeling methods, such as Newton-Euler method, Lagrangian method, and so on. Huang Jin et al. [24] proposed a novel approach for solving the dynamics problem of parallel manipulator by Udwaida-Kalaba equation which is proposed by Udwadia and kalaba [25]. Firstly, the parallel manipulator system is divided into several subsystems, and then these subsystems are integrated through kinematic constraints. Compared with other traditional dynamics modeling methods, Udwadia-Kalaba theory does not need to use additional auxiliary variables to establish motion equations, and the method is more systematic and easier to implement.

In this section, the 2-DOF parallel manipulator system is chosen [26]. The 2-DOF parallel manipulator system can be segmented into three subsystems, with each containing an active link and a passive link. This parallel manipulator system can be seen in Fig. 1. The dynamic modeling of the unconstrained subsystem can be obtained by the traditional Lagrangian approach. By introducing a geometric constraint among the subsystem and using the Udwaida-Kalaba formulation, the three subsystems are combined together and the dynamic model of 2-DOF parallel manipulator system is established (Fig. 3).

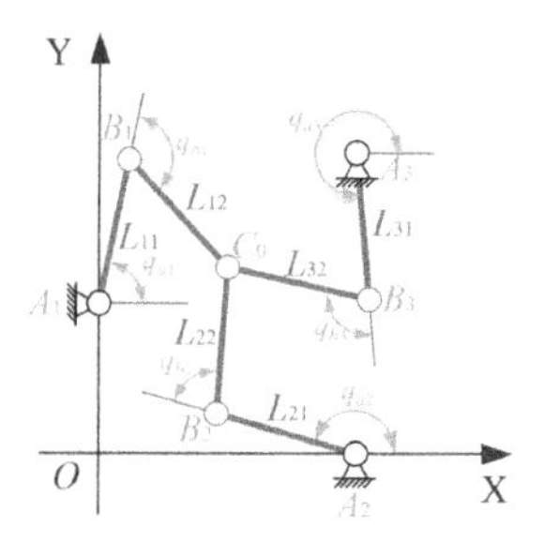

Fig. 1. The initial position of parallel manipulator system

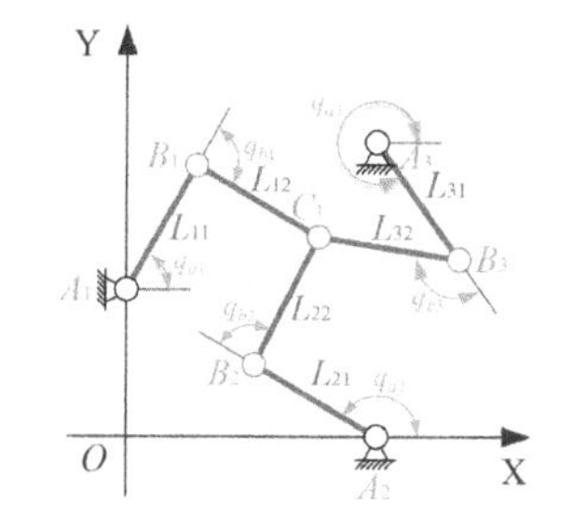

Fig. 2. The desired position of parallel manipulator system

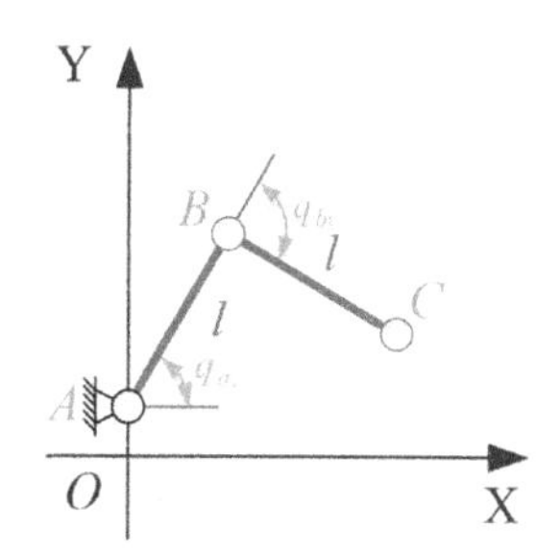

Fig. 3. A subsystem of 2-DOF parallel manipulator system

3.1 Dynamic Model of Subsystem Without Constraint

The subsystem of 2-DOF parallel manipulator system can be seen as a two-link mechanism and the active and passive links of each subsystem are regarded as ideal rigid bodies. The length of both links is l, the masses of the active link and the passive link are m_{ai} and m_{bi}, the mass center of the two links are located at r_{ai} and r_{bi} away from joint A of active link and joint B of passive link, their coordinates in the Cartesian coordinates are (x_{cai}, y_{cai}) and (x_{cbi}, y_{cbi}), and the inertia moments of the two links relative to the center of mass are I_{ai} and I_{bi} respectively. The subsystem of 2-DOF parallel manipulator system is shown in Fig. 2.

Because the parallel manipulator system only moves on the plane, the gravitational potential energy of the link is not considered when calculating the energy of the subsystem. Therefore, the Lagrange function of the subsystem is equal to the sum of the kinetic energy of active link E_{ai} and the kinetic energy of passive link E_{bi}, in which the kinetic energy of each link is composed of rotational kinetic energy and translational kinetic energy. The kinetic energy of active link E_{ai} and the kinetic energy of passive link E_{bi} are given by

$$\begin{cases} E_{ai} = \frac{1}{2} I_{ai} \dot{q}_{ai}^2 + \frac{1}{2} m_{ai} (\dot{x}_{cai}^2 + \dot{y}_{cai}^2) \\ E_{bi} = \frac{1}{2} I_{bi} (\dot{q}_{ai} + \dot{q}_{bi})^2 + \frac{1}{2} m_{bi} (\dot{x}_{cbi}^2 + \dot{y}_{cbi}^2) \end{cases} \tag{15}$$

where

$$\begin{array}{c} \dot{x}_{cai} = -r_{ai} \sin q_{ai} \dot{q}_{ai}, \dot{y}_{cai} = r_{ai} \cos q_{ai} \dot{q}_{ai} \\ \dot{x}_{cbi} = -l \sin q_{ai} \dot{q}_{ai} - r_{bi} \sin(q_{ai} + q_{bi})(\dot{q}_{ai} + \dot{q}_{bi}) \\ \dot{y}_{cbi} = l \cos q_{ai} \dot{q}_{ai} + r_{bi} \cos(q_{ai} + q_{bi})(\dot{q}_{ai} + \dot{q}_{bi}) \end{array} \tag{16}$$

Thus, the Lagrange coefficient of the subsystem is

$$L_i = E_{ai} + E_{bi} \tag{17}$$

Equation (18) is the Euler-Lagrange equation.

$$\begin{cases} \frac{d}{dt} \frac{\partial L_i}{\partial \dot{q}_{ai}} - \frac{\partial L_i}{\partial q_{ai}} = \tau_{ai} \\ \frac{d}{dt} \frac{\partial L_i}{\partial \dot{q}_{bi}} - \frac{\partial L_i}{\partial q_{bi}} = \tau_{bi} \end{cases} \tag{18}$$

The Eq. (18) can be written in matrix forms and the dynamic model of the parallel robot subsystem is obtained.

$$M_i\left(q_i,t\right)\ddot{q}_i + C_i\left(q_i,\dot{q}_i,t\right)\dot{q}_i = \tau_i\left(t\right) \tag{19}$$

where

$$M_i = \begin{bmatrix} \alpha_i+\beta_i+2\gamma_i+m_2l^2 & \beta_i+\gamma_i \\ \beta_i+\gamma_i & \beta_i \end{bmatrix}, C_i = \begin{bmatrix} 2\omega\dot{q}_2 & \omega\dot{q}_2 \\ -\omega\dot{q}_1 & 0 \end{bmatrix}, \tau_i = \begin{bmatrix} \tau_{ai} \\ \tau_{bi} \end{bmatrix} \tag{20}$$

$$\begin{gathered} \alpha_i = I_{ai} + m_{ai}r_{ai}^2, \beta_i = I_{bi} + m_{bi}r_{bi}^2 \\ \gamma_i = r_{bi}lm_{bi}(\sin q_{ai}\sin(q_{ai}+q_{bi}) + \cos q_{ai}\cos(q_{ai}+q_{bi})) \\ \omega_i = r_{bi}lm_{bi}(\sin q_{ai}\cos(q_{ai}+q_{bi}) - \cos q_{ai}\sin(q_{ai}+q_{bi})) \end{gathered} \tag{21}$$

$M_i\left(q_i,t\right)$ is the mass inertia matrix, $C_i\left(q_i,\dot{q}_i,t\right)$ is the Coriolis force and centrifugal force matrix and $\tau_i\left(t\right)$ is the corresponding torque of each joint.

3.2 Constraint Design for Parallel Manipulator System

The three subsystems of the parallel manipulator system are connected at the end. Thus, we can get a geometric constraint between the three subsystems. Differentiate the geometric constraint twice to time and then convert it into the form of Eq. (22).

$$A\left(q,t\right)\ddot{q} = b\left(q,\dot{q},t\right) \tag{22}$$

where

$$\begin{gathered} A = \begin{bmatrix} -ls_{a1}-ls_{ab1} & -ls_{ab1} & ls_{a2}+ls_{ab2} & ls_{ab2} & 0 & 0 \\ lc_{a1}+lc_{ab1} & lc_{ab1} & -lc_{a2}-lc_{ab2} & -lc_{ab2} & 0 & 0 \\ -ls_{a1}-ls_{ab1} & -ls_{ab1} & 0 & 0 & ls_{a3}+ls_{ab3} & ls_{ab3} \\ lc_{a1}+lc_{ab1} & lc_{ab1} & 0 & 0 & -lc_{a3}-lc_{ab3} & -lc_{ab3} \end{bmatrix} \\ b = \begin{bmatrix} lc_{a1}\dot{q}_{a1}^2 + lc_{ab1}(\dot{q}_{a1}+\dot{q}_{b1})^2 - lc_{a2}\dot{q}_{a2}^2 - lc_{ab2}(\dot{q}_{a2}+\dot{q}_{b2})^2 \\ ls_{a1}\dot{q}_{a1}^2 + ls_{ab1}(\dot{q}_{a1}+\dot{q}_{b1})^2 - ls_{a2}\dot{q}_{a2}^2 - ls_{ab2}(\dot{q}_{a2}+\dot{q}_{b2})^2 \\ lc_{a1}\dot{q}_{a1}^2 + lc_{ab1}(\dot{q}_{a1}+\dot{q}_{b1})^2 - lc_{a3}\dot{q}_{a3}^2 - lc_{ab3}(\dot{q}_{a3}+\dot{q}_{b3})^2 \\ ls_{a1}\dot{q}_{a1}^2 + ls_{ab1}(\dot{q}_{a1}+\dot{q}_{b1})^2 - ls_{a3}\dot{q}_{a3}^2 - ls_{ab3}(\dot{q}_{a3}+\dot{q}_{b3})^2 \end{bmatrix} \end{gathered} \tag{23}$$

$$\begin{gathered} s_{a1} = \sin q_{a1}, c_{a1} = \cos q_{a1}, s_{ab1} = \sin\left(q_{a1}+q_{b1}\right), c_{ab1} = \cos\left(q_{a1}+q_{b1}\right) \\ s_{a2} = \sin q_{a2}, c_{a2} = \cos q_{a2}, s_{ab2} = \sin\left(q_{a2}+q_{b2}\right), c_{ab2} = \cos\left(q_{a2}+q_{b2}\right) \\ s_{a3} = \sin q_{a3}, c_{a3} = \cos q_{a3}, s_{ab3} = \sin\left(q_{a3}+q_{b3}\right), c_{ab3} = \cos\left(q_{a3}+q_{b3}\right) \end{gathered} \tag{24}$$

3.3 Dynamic Model of Parallel Manipulator System with Constraints

As shown in Fig. 1, according to the Udwadia-Kalaba equation, the dynamic model of the parallel manipulator system with constraint can be described as:

$$M\left(q,t\right)\ddot{q} = Q\left(q,\dot{q},t\right) + Q^C\left(q,\dot{q},t\right) \tag{25}$$

where

$$\begin{aligned} Q &= \tau - C\dot{q} \\ Q^C &= M^{\frac{1}{2}}\left(AM^{-1/2}\right)^+ \left[b - AM^{-1}Q\right] \end{aligned} \tag{26}$$

Further, the acceleration can be elicited as

$$\ddot{q} = M^{-1}Q + M^{-\frac{1}{2}}\left(AM^{-1/2}\right)^+ \left[b - AM^{-1}Q\right] \tag{27}$$

where

$$M = \begin{bmatrix} M_1 & 0 & 0 \\ 0 & M_2 & 0 \\ 0 & 0 & M_3 \end{bmatrix}, C = \begin{bmatrix} C_1 & 0 & 0 \\ 0 & C_2 & 0 \\ 0 & 0 & C_3 \end{bmatrix}, q = \begin{bmatrix} q_{a1} \\ q_{b1} \\ q_{a2} \\ q_{b2} \\ q_{a3} \\ q_{b3} \end{bmatrix}, \tau = \begin{bmatrix} \tau_{a1} \\ \tau_{b1} \\ \tau_{a2} \\ \tau_{b2} \\ \tau_{a3} \\ \tau_{b3} \end{bmatrix} \tag{28}$$

The parallel manipulator system has three actuators. The relationship between actuator torque u and driving torque τ can be expressed as:

$$u = \left({J_2}^+\right)^{\mathrm{T}} \left({J_1}^{\mathrm{T}}\tau\right) \tag{29}$$

where

$$J_1 = \begin{bmatrix} a_{1x}c_1 & a_{1y}c_1 \\ -b_{1x}c_1 & -b_{1y}c_1 \\ a_{2x}c_2 & a_{2y}c_2 \\ -b_{2x}c_2 & -b_{2y}c_2 \\ a_{3x}c_3 & a_{3y}c_3 \\ -b_{3x}c_3 & -b_{3y}c_3 \end{bmatrix}, J_2 = \begin{bmatrix} a_{1x}c_1 & a_{1y}c_1 \\ a_{2x}c_2 & a_{2y}c_2 \\ a_{3x}c_3 & a_{3y}c_3 \end{bmatrix}, u = \begin{bmatrix} u_1 \\ u_2 \\ u_3 \end{bmatrix} \tag{30}$$

$$\begin{aligned} a_{ix} &= l\cos\left(q_{ai} + q_{bi}\right), a_{iy} = l\sin\left(q_{ai} + q_{bi}\right) \\ b_{ix} &= l\cos q_{ai} + l\cos\left(q_{ai} + q_{bi}\right), b_{iy} = l\sin q_{ai} + l\sin\left(q_{ai} + q_{bi}\right) \\ c_i &= \frac{1}{a_{iy}b_{ix} - a_{ix}b_{iy}} \end{aligned} \tag{31}$$

4 Simulation

In this section, the performance of the active inference controller for the parallel manipulator system are verified by numerical simulations. The parameter values for the parallel manipulator system are $l = 0.2440$ m, $m_{a1} = m_{a2} = m_{a3} = 1.440$ kg, $m_{b1} = m_{b2} = m_{b3} = 0.720$ kg, $r_{a1} = r_{a2} = r_{a3} = 0.0657$ m, $r_{b1} = r_{b2} = r_{b3} = 0.1096$ m. The parameter values for the active inference controller are $\sigma_n = 1$, $\sigma_{\dot{n}} = 1$, $\sigma_\mu = 20$, $\sigma_{\dot{\mu}} = 15$, $k_a = 900$, $k_\mu = 150$.

The initial state and the desired state of parallel manipulator are shown in Fig. 1 and Fig. 2. The initial angles of each joint of parallel manipulator are $q_{a1} = 1.3627$ rad, $q_{b1} = -2.1848$ rad, $q_{a2} = 2.8647$ rad, $q_{b2} = -1.3687$ rad, $q_{a3} = 4.8012$ rad, $q_{b3} = -1.8788$ rad. The desired angles of each joint of parallel manipulator are $q_{a1} = 1.0472$ rad, $q_{b1} = -1.5720$ rad, $q_{a2} = 2.6180$ rad, $q_{b2} = -1.5216$ rad,

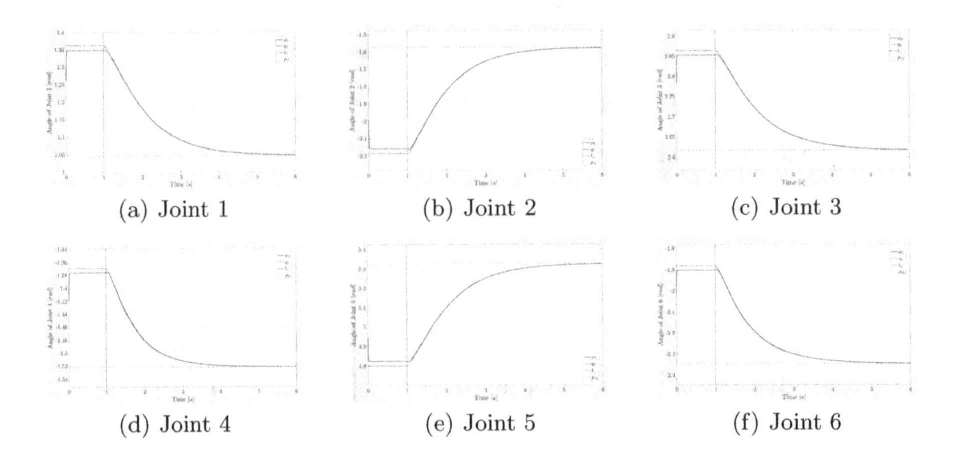

(a) Joint 1 (b) Joint 2 (c) Joint 3

(d) Joint 4 (e) Joint 5 (f) Joint 6

Fig. 4. Angle curve of each joint. (Color figure online)

$q_{a3} = 5.3294$ rad, $q_{b3} = -2.3444$ rad. The initial beliefs about the states of the parallel manipulator are $\mu_{a1} = 1.2627$ rad, $\mu_{b1} = -1.9848$ rad, $\mu_{a2} = 2.7847$ rad, $\mu_{b2} = -1.4187$ rad, $\mu_{a3} = 4.9512$ rad, $\mu_{b3} = -2.0288$ rad. The task is to steer the parallel manipulator to move from the initial state to the desired state with the help of active inference controller.

Through belief update and action update based on the free energy, the belief μ and the control actions τ of the next moment can be obtained. The action τ, namely torque, is brought into the dynamic model of the parallel manipulator system to get the angular acceleration of each joint, and then the angular velocity and angle of each joint are further obtained through integration. In order to generate the sensor data, Gaussian noise is introduced. The calculated angular velocity and angle are combined with Gaussian noise to obtain the sensor data, that is, the sensory input n. Substitute the above calculated belief μ and sensory input n into the free energy equation to obtain the free energy of the next moment. If the loop goes on like this, the parallel manipulator will finally reach the predetermined state.

The performance of active inference controller in simulation is depicted in Figs. 4. Figures 4 show the angle curve of each joint of the parallel manipulator. Blue solid line represents the curve of the belief μ of each joint, black dotted line represents the curve of the sensory input n of each joint, green dotted line represents the desired position of each joint. Red dotted line represents start executing action line, that is, from the first second, the controller starts to output the action τ. Each joint finally reaches the ideal states in 6 s and the curve is smooth and stable without jittering from start executing action. The angle velocity curve of each joint of the parallel manipulator is presented in Figs. 5. Blue solid line represents the curve of the belief $\dot{\mu}$ of each joint, black dotted line represents the curve of the sensory input $\dot{n}$ of each joint. The angular velocity curve which is smooth and stable without jittering presents a trend of rising first

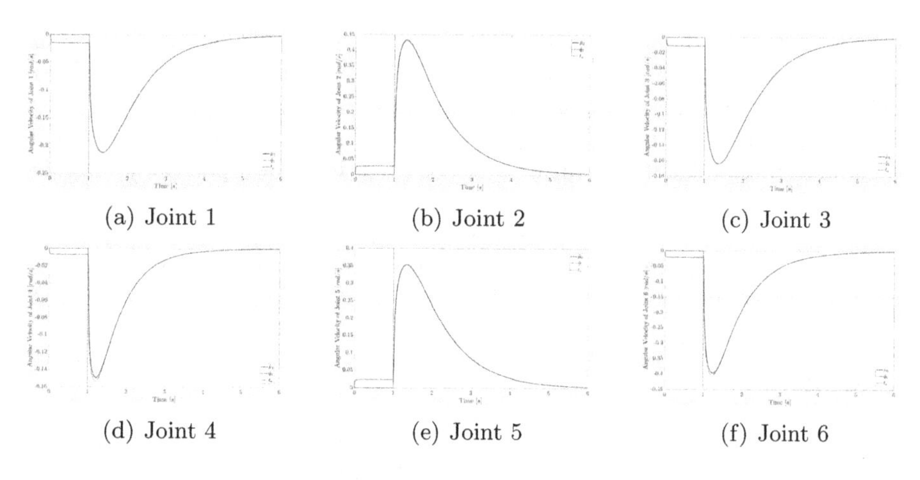

(a) Joint 1 (b) Joint 2 (c) Joint 3

(d) Joint 4 (e) Joint 5 (f) Joint 6

Fig. 5. Angle velocity curve of each joint. (Color figure online)

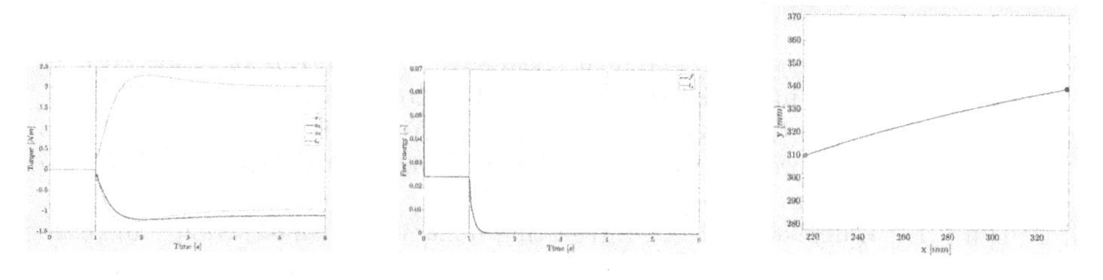

Fig. 6. Actuator torque. (Color figure online) **Fig. 7.** Free energy **Fig. 8.** Trajectory of C

and then falling. At 1.5 s, the angular velocity reaches the maximum value, and then gradually converges to 0 (Figs. 7 and 8).

Figures 6 show the curve of three actuator torques, the free energy of the parallel manipulator system and the actual trajectory of end point C respectively. The curve of actuator torques is smooth and stable overall, except for a little oscillation near the first second, and finally show a convergence trend. The free energy curve shows a step downward trend and finally converges to 0. In 0–1 s, since the active inference controller does not output the action τ, the free energy is always 0.024. The green point and red point represent the starting point and terminal point of end point C individually in Fig. 6. The actual trajectory of end point C can be approximately regarded as a straight line, which conforms to the free energy minimization principle.

5 Conclusion

This paper adopts an active inference controller and apply it to a 2-DOF parallel manipulator system. The parallel manipulator realizes the movement from the initial state to the desired state and the control effect of the active inference controller is perfect from the above simulation results. Compared with other control methods, the active inference controller is simpler, more efficient, computationally inexpensive and has also better adaptability. Active inference controller is sensitive to parameters, such as the learning rate of belief and action update. If the set parameters are not appropriate, more visible oscillations will appear in the simulation results. In the future work, we will introduce parallel manipulator with higher degrees of freedom, and consider more complex scenarios, such as placing some obstacles in the process of movement of robotic manipulator. In addition, we will attempt to find a method to adjust parameters to achieve better control effect.

References

1. Pedersen, M.R., et al.: Robot skills for manufacturing: from concept to industrial deployment. Robot. Comput. Integr. Manuf. **37**, 282–291 (2016)
2. Qi, B.Y., Yang, Q.L., Zhou, Y.Y.: Application of AGV in intelligent logistics system. In: Fifth Asia International Symposium on Mechatronics (AISM 2015). IET (2015)
3. Roveda, L., et al.: EURECA H2020 CleanSky 2: a multi-robot framework to enhance the fourth industrial revolution in the aerospace industry. In: Proceedings of International Conference on Robotics and Automation (ICRA), Workshop Ind. Future, Collaborative, Connected, Cogn. Novel Approaches Stemming from Factory Future Industry 4.0 Initiatives (2017)
4. Dufourd, D., Dalgalarrondo, A.: Integrating human/robot interaction into robot control architectures for defense applications. In: 1th National Conference on Control Architecture of Robots, April 2006
5. Kunz, C., et al.: Deep sea underwater robotic exploration in the ice-covered arctic ocean with AUVs. In: 2008 IEEE/RSJ International Conference on Intelligent Robots and Systems. IEEE (2008)
6. Feng, M., et al.: Development of a medical robot system for minimally invasive surgery. Int. J. Med. Robot. Comput. Assist. Surg. **8**(1), 85–96 (2012)
7. Merlet, J.-P.: Parallel manipulators: state of the art and perspectives. Adv. Robot. **8**(6), 589–596 (1993)
8. Carricato, M., Parenti-Castelli, V.: Singularity-free fully-isotropic translational parallel mechanisms. Int. J. Robot. Res. **21**(2), 161–174 (2002)
9. Qin, Y., et al.: Modelling and analysis of a rigid-compliant parallel mechanism. Robot. Comput. Integr. Manuf. **29**(4), 33–40 (2013)
10. Su, Y.X., Duan, B.Y., Zheng, C.H.: Nonlinear PID control of a six-DOF parallel manipulator. IEE Proc. Control Theory Appl. **151**(1), 95–102 (2004)
11. Kim, N.-I., Lee, C.-W., Chang, P.-H.: Sliding mode control with perturbation estimation: application to motion control of parallel manipulator. Control. Eng. Pract. **6**(11), 1321–1330 (1998)
12. Li, Y., Qingsong, X.: Dynamic modeling and robust control of a 3-PRC translational parallel kinematic machine. Robot. Comput. Integr. Manuf. **25**(3), 630–640 (2009)
13. Cazalilla, J., et al.: Adaptive control of a 3-DOF parallel manipulator considering payload handling and relevant parameter models. Robot. Comput. Integr. Manuf. **30**(5), 468–477 (2014)

14. Bennehar, M., et al.: A novel adaptive terminal sliding mode control for parallel manipulators: design and real-time experiments. In: 2017 IEEE International Conference on Robotics and Automation (ICRA). IEEE (2017)
15. Achili, B., et al.: A robust adaptive control of a parallel robot. Int. J. Control **83**(10), 2107–2119 (2010)
16. Nguyen, V.T., et al.: Finite-time adaptive fuzzy tracking control design for parallel manipulators with unbounded uncertainties. Int. J. Fuzzy Syst. **21**(2), 545–555 (2019)
17. Kamalasadan, S., Ghandakly, A.A.: A neural network parallel adaptive controller for dynamic system control. IEEE Trans. Instrum. Meas. **56**(5), 1786–1796 (2007)
18. Buckley, C.L., et al.: The free energy principle for action and perception: a mathematical review. J. Math. Psychol. **81**, 55–79 (2017)
19. Mercade, A.C.: Robot manipulator control under the Active Inference framework (2018)
20. Pio-Lopez, L., et al.: Active inference and robot control: a case study. J. R. Soc. Interface **13**(122), 20160616 (2016)
21. Oliver, G., Lanillos, P., Cheng, G.: Active inference body perception and action for humanoid robots. arXiv preprint arXiv:1906.03022 (2019)
22. Pezzato, C., Ferrari, R., Corbato, C.H.: A novel adaptive controller for robot manipulators based on active inference. IEEE Robot. Autom. Lett. **5**(2), 2973–2980 (2020)
23. Baioumy, M., et al.: Active inference for integrated state-estimation, control, and learning. In: 2021 IEEE International Conference on Robotics and Automation (ICRA). IEEE (2021)
24. Huang, J., Chen, Y.H., Zhong, Z.: Udwadia-Kalaba approach for parallel manipulator dynamics. J. Dyn. Syst. Meas. Control **135**(6) (2013)
25. Kalaba, R., Udwadia, F.: Analytical dynamics with constraint forces that do work in virtual displacements. Appl. Math. Comput. **121**(2–3), 211–217 (2001)
26. Luo, L.: The dynamic modeling of 2-DOF redundant actuation parallel robot based on U-K theory. Dissertation, Chang'an University

Geometrical Parameter Identification for 6-DOF Parallel Platform

Jian-feng Lin, Chen-kun Qi(✉), Yan Hu, Song-lin Zhou, Xin Liu, and Feng Gao

State Key Laboratory of Mechanical System and Vibration, School of Mechanical Engineering, Shanghai Jiao Tong University, Shanghai 200240, China
chenkqi@sjtu.edu.cn

Abstract. The positioning accuracy of the 6-degree-of-freedom (DOF) parallel platform is affected by the model accuracy. In this paper, the parameter identification of the 6-DOF parallel platform is carried out to obtain the accurate kinematic model. The theoretical inverse kinematic model is established by closed-loop vector method. The rotation centers of each spherical joint and hooker joint are selected as geometric parameters to be identified. The inverse kinematic model with geometric error is established. The geometric parameters are identified by iterative least square method. Simulation results show that the parameter identification method is correct.

Keywords: Parallel platform · Kinematics · Parameter identification · Iterative least squares

1 Introduction

Parallel mechanisms are widely used in industrial and medical fields due to the high stiffness, fast response and good stability [1, 2]. Parallel mechanisms can be divided into less degrees of freedom [3, 4] and six degrees of freedom [5, 6]. However, the actual model of parallel mechanism is often different from the theoretical model. The model errors will lead to the decrease of positioning accuracy and affect the working performance of the parallel mechanism. Therefore, the study of error calibration is of great significance to improve the accuracy of parallel mechanism.

The error sources of parallel mechanisms are mainly divided into geometric error [7, 8] and non-geometric error [9]. Geometric error is caused by machining error in manufacturing process and assembly error in assembly process. Non-geometric error is caused by friction, hysteresis and thermal error. The proportion of non-geometric error does not exceed 10% of the total error source of parallel mechanism. It is important to study the calibration of geometric parameters.

The geometric errors of the parallel mechanism cause some differences between the real kinematics model and the theoretical kinematics model. Kinematics model calibration can effectively solve the defect, and the economic cost is low. The purpose of robot kinematic calibration is to identify the error between the real model parameters and the theoretical model parameters to correct the theoretical kinematic model [10]. The

H. Liu et al. (Eds.): ICIRA 2022, LNAI 13456, pp. 70–76, 2022.
https://doi.org/10.1007/978-3-031-13822-5_7

calibration process can be divided into four steps: error modeling, error measurement, error identification and error compensation.

As for the 2-DOF translational parallel manipulator, two types of kinematic calibration methods are proposed based on different error models [11]. A practical self-calibration method for planar 3-RRR parallel mechanism is proposed in literature [12]. On the premise that the structure of the mechanism remains unchanged, the line ruler is used to measure and record the output pose of the mechanism. A kinematic calibration method for a class of 2UPR&2RPS redundant actuated parallel robot is proposed in reference [13]. The geometric error model of the robot is established based on the error closed-loop vector equation, and the compensable and uncompensable error sources are separated. The error Jacobian matrix of 54 structural parameters for 6PUS parallel mechanism is established in literature [14]. The kinematic parameters are calibrated by separating position parameters and angle parameters [15]. There are few literatures about the calibration of 6-DOF platform.

The structure of this paper is as follows. Section 2 presents the kinematics of 6-DOF parallel platform. Section 3 provides the parameter identification algorithm. In Sect. 4, the simulation is conducted. Section 5 gives the conclusion.

2 Kinematics of 6-DOF Parallel Platform

2.1 6-DOF Parallel Platform

The 6-DOF parallel platform is shown in Fig. 1. It consists of an upper platform, a lower platform and six chains. Each chain is composed of a prismatic joint (P), a hooker joint (U) and a spherical joint (S). The upper platform can achieve the motion of six degrees of freedom.

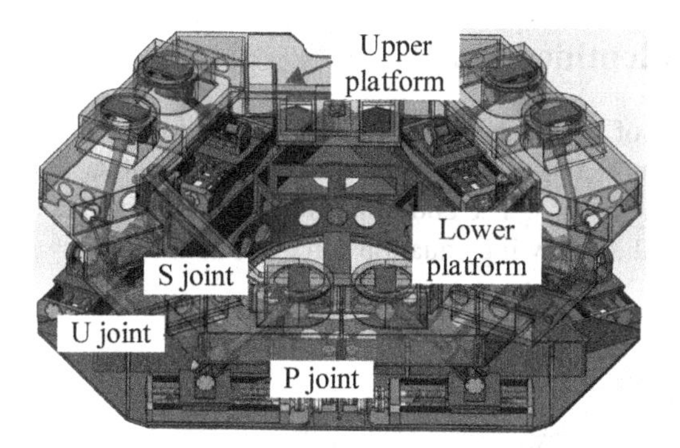

Fig. 1. 6-DOF parallel platform

2.2 Kinematic Model

The diagram of 6-DOF parallel platform is shown in Fig. 2. The Cartesian coordinate system o_1-$x_1y_1z_1$ is established on the center of lower platform. The Cartesian coordinate system o_2-$x_2y_2z_2$ is established on the center of upper platform. The moving coordinate system is o_2-$x_2y_2z_2$, while the fixed coordinate system is o_1-$x_1y_1z_1$. According to the closed-loop vector method, the equation can be obtained as:

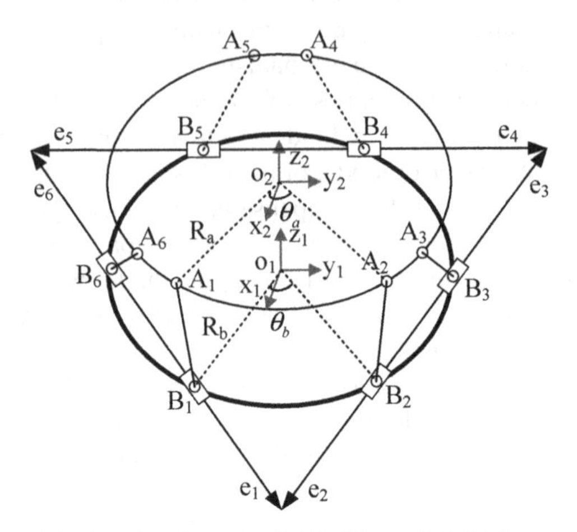

Fig. 2. The diagram of 6-DOF parallel platform

$$\boldsymbol{p} + \boldsymbol{R}\boldsymbol{a}'_i = \boldsymbol{b}_i + q_i\boldsymbol{e}_i + \boldsymbol{l}_i \tag{1}$$

where $\boldsymbol{p}$ is the position matrix, $\boldsymbol{R}$ is the rotation matrix, i is the ith chain, $\boldsymbol{b}_i$ is the coordinates of B_i in the coordinate system o_2-$x_2y_2z_2$, q_i is the driving displacement of the ith chain, $\boldsymbol{e}_i$ is the unit vector of the ith chain.

3 Parameter Identification

The rotation centers of each spherical joint and hooker joint are selected as geometric parameters to be identified. Assuming that the coordinates of A_i in the moving coordinate system are $(x_{A_i}, y_{A_i}, z_{A_i})$. The coordinates of B_i in the fixed coordinate system are $(x_{B_i}, y_{B_i}, z_{B_i})$. Then the following equation can be obtained as:

$$\delta \boldsymbol{q} = \boldsymbol{K}\delta\boldsymbol{\Theta} \tag{2}$$

where $\boldsymbol{K}$ is the coefficient matrix.

$$\delta\boldsymbol{q} = [\delta q_1, \delta q_2, \cdots, \delta q_i]^T, \quad (i = 1, \cdots, 6)$$

$$\delta\boldsymbol{\Theta} = [\delta x_{A_1}, \delta y_{A_1}, \delta z_{A_1}, \delta x_{B_1}, \delta y_{B_1}, \delta z_{B_1}, \cdots, \delta x_{A_i}, \delta y_{A_i}, \delta z_{A_i}, \delta x_{B_i}, \delta y_{B_i}, \delta z_{B_i}]^T$$

Taking the chain 1 for example, it can be obtained as:

$$\begin{bmatrix} {}^{mea}q_{11} - q_{11} \\ {}^{mea}q_{12} - q_{12} \\ \vdots \\ {}^{mea}q_{1j} - q_{1j} \end{bmatrix}_{j\times 1} = \begin{bmatrix} M_{11} & M_{21} & M_{31} & M_{41} & M_{51} & M_{61} \\ M_{12} & M_{22} & M_{32} & M_{42} & M_{52} & M_{62} \\ \vdots & \vdots & \vdots & \vdots & \vdots & \vdots \\ M_{1j} & M_{2j} & M_{3j} & M_{4j} & M_{5j} & M_{6j} \end{bmatrix}_{j\times 6} \cdot \delta\boldsymbol{\Theta}_1 \tag{3}$$

where $\delta\boldsymbol{\Theta}_1 = [\delta x_{A_1}, \delta y_{A_1}, \delta z_{A_1}, \delta x_{B_1}, \delta y_{B_1}, \delta z_{B_1}]^T$, ${}^{mea}q_{1j}$ is the driving displacement obtained by the jth measurement.

According to the least square method, it can be obtained as:

$$\delta\boldsymbol{\Theta}_1 = (\boldsymbol{M}^T \cdot \boldsymbol{M})^{-1} \cdot \boldsymbol{M}^T \cdot \begin{bmatrix} {}^{mea}q_{11} - q_{11} \\ {}^{mea}q_{12} - q_{12} \\ \vdots \\ {}^{mea}q_{1j} - q_{1j} \end{bmatrix}_{j\times 1} \tag{4}$$

where

$$\boldsymbol{M} = \begin{bmatrix} M_{11} & M_{21} & M_{31} & M_{41} & M_{51} & M_{61} \\ M_{12} & M_{22} & M_{32} & M_{42} & M_{52} & M_{62} \\ \vdots & \vdots & \vdots & \vdots & \vdots & \vdots \\ M_{1j} & M_{2j} & M_{3j} & M_{4j} & M_{5j} & M_{6j} \end{bmatrix}_{j\times 6}$$

Because the actual geometric error of the 6-DOF parallel platform is different from the differential component, the least square iterative method is introduced to identify geometric parameters. The identification process is shown in Fig. 3.

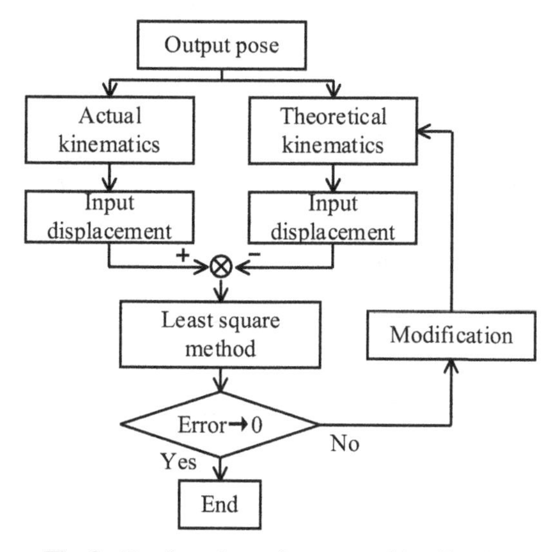

Fig. 3. The flow chart of parameter identification

4 Simulation

Given the output poses of the upper platform, the corresponding driving displacements can be calculated by theoretical inverse kinematics and actual inverse kinematics. The iterative least square algorithm is used to identify geometric errors.

The correctness of parameter identification method is verified. Given different output poses, the input displacements are calculated by the actual inverse kinematics model. The Newton iteration method is used to calculate the output poses of the identified model and the theoretical model. The comparison of errors before and after calibration is shown in Fig. 4. The pose errors after calibration is shown in Fig. 5.

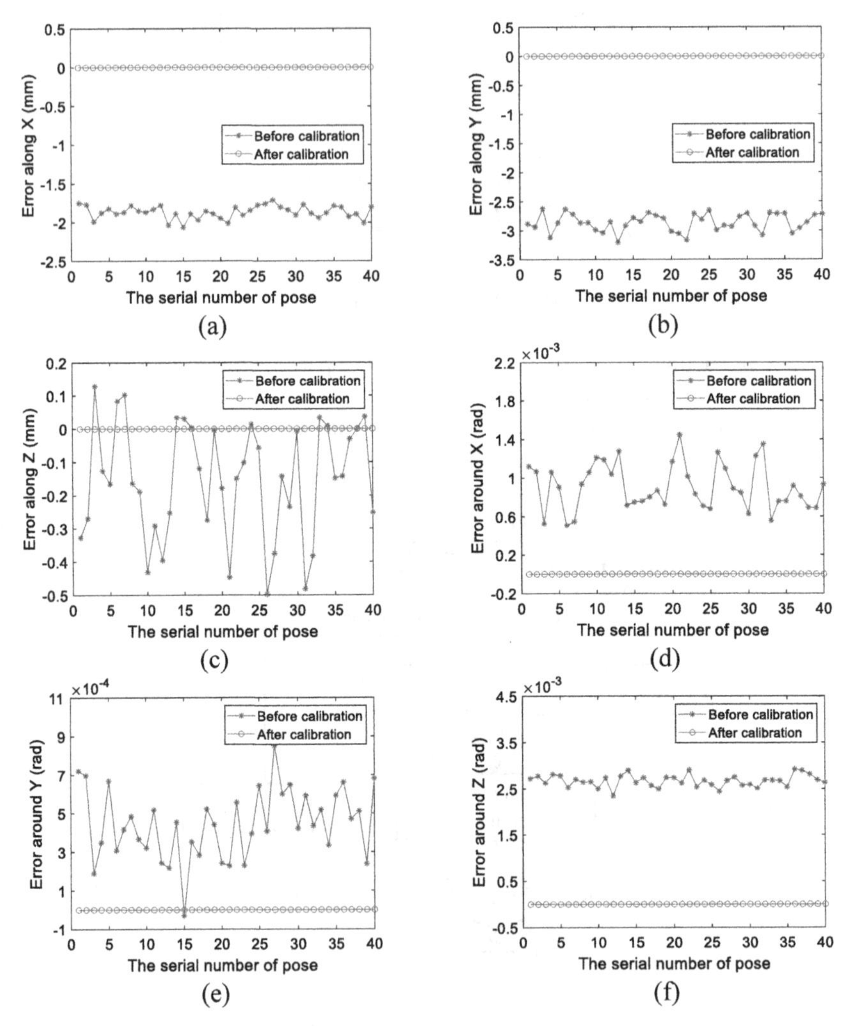

Fig. 4. Comparison of errors before and after calibration

As can be seen from the figure, after calibration, the positioning accuracy of the 6-DOF parallel platform along the X direction is improved from -2.064 mm to 9.659×10^{-10} mm. The positioning accuracy along the Y direction is improved from -3.202 mm to -6.242×10^{-10} mm. The positioning accuracy along the Z direction is improved from -0.4974 mm to 1.75×10^{-10} mm. The positioning accuracy around the X direction is

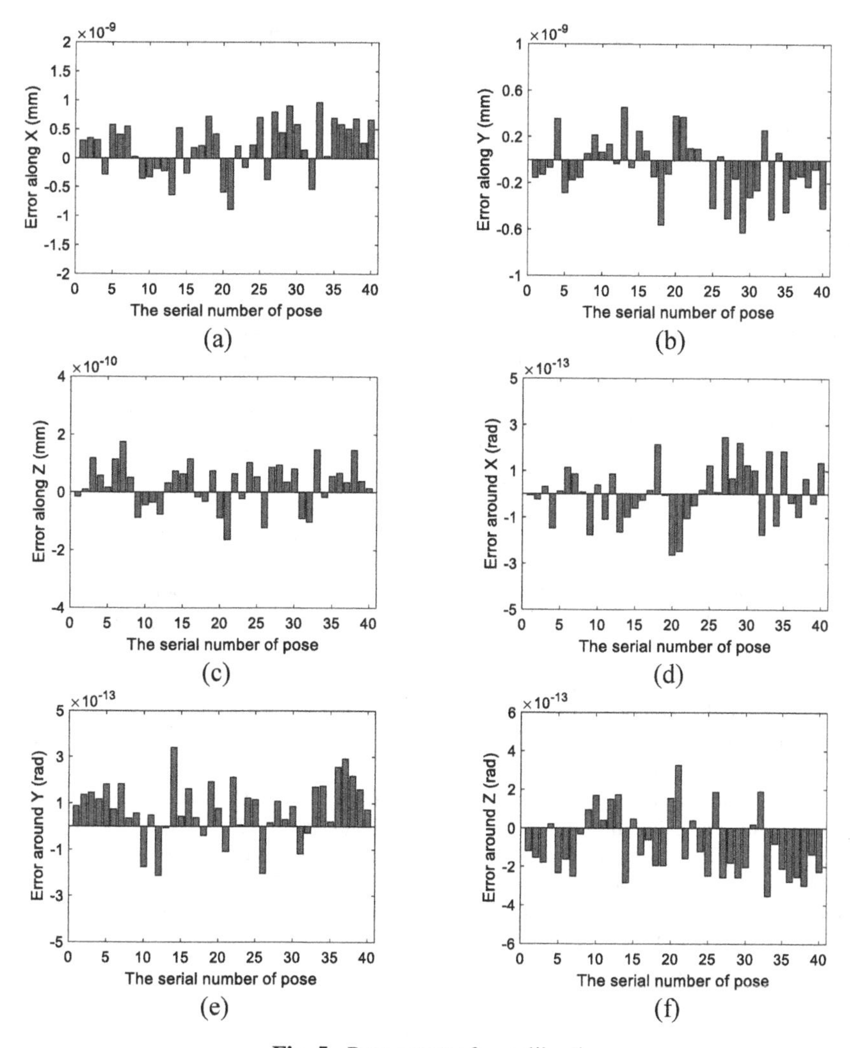

Fig. 5. Pose errors after calibration

improved from 0.00145 rad to -2.614 $\times$ 10^{-13} rad. The positioning accuracy around the Y direction is improved from 0.00085 rad to 3.398 $\times$ 10^{-13} rad. The positioning accuracy around the Z direction is improved from 0.00292 rad to -3.525 $\times$ 10^{-13} rad.

5 Conclusion

In this paper, the kinematics model calibration of the 6-DOF parallel platform is carried out. The theoretical inverse kinematic model is established by closed-loop vector method. The rotation centers of each spherical joint and hooker joint are selected as geometric parameters to be identified. The iterative least square method is introduced to identify the geometric parameters. Simulation results show that the parameter identification method is correct.

Funding. The work is partially supported by a grant Shanghai Natural Science Foundation (Grant No. 19ZR1425500).

References

1. Sun, T., Zhai, Y., Song, Y., Zhang, J.: Kinematic calibration of a 3-DoF rotational parallel manipulator using laser tracker. Robot. Comput. Integr. Manuf. **41**, 78–91 (2016)
2. Dasgupta, B., Mruthyunjaya, T.S.: The Stewart platform manipulator: a review. Mech. Mach. Theory **35**(1), 15–40 (2000)
3. Ma, N., et al.: Design and stiffness analysis of a class of 2-DoF tendon driven parallel kinematics mechanism. Mech Machine Theor **129**, 202–217 (2018)
4. Song, Y., Zhang, J., Lian, B., Sun, T.: Kinematic calibration of a 5-DoF parallel kinematic machine. Precis. Eng. **45**, 242–261 (2016)
5. Feng, J., Gao, F., Zhao, X.: Calibration of a six-DOF parallel manipulator for chromosome dissection. Proc. Inst. Mech. Eng. Part C **226**(4), 1084–1096 (2011)
6. Dong, Y., Gao, F., Yue, Y.: Modeling and prototype experiment of a six-DOF parallel micro-manipulator with nano-scale accuracy. Proc. Inst. Mech. Eng. C J. Mech. Eng. Sci. **229**(14), 2611–2625 (2015)
7. Daney, D.: Kinematic calibration of the Gough Platform. Robotica **21**(6), 677–690 (2003)
8. Renders, J.-M., Rossignol, E., Becquet, M., Hanus, R.: Kinematic calibration and geometrical parameter identification for robots. IEEE Trans. Robot. Autom. **7**(6), 721–732 (1991). https://doi.org/10.1109/70.105381
9. Gong, C., Yuan, J., Ni, J.: Nongeometric error identification and compensation for robotic system by inverse calibration. Int. J. Mach. Tools Manuf **40**(14), 2119–2137 (2000)
10. Renaud, P., Andreff, N., Martinet, P., Gogu, G.: Kinematic calibration of parallel mechanisms: a novel approach using legs observation. IEEE Trans. Robot. **21**(4), 529–538 (2005)
11. Zhang, J., et al.: Kinematic calibration of a 2-DOF translational parallel manipulator. Adv. Robot. **28**(10), 1–8 (2014)
12. Shao, Z.: Self-calibration method of planar flexible 3-RRR parallel manipulator. J. Mech. Eng. **45**(03), 150 (2009). https://doi.org/10.3901/JME.2009.03.150
13. Zhang, J., Jiang, S., Chi, C.: Kinematic calibration of 2UPR&2RPS redundant actuated parallel robot. Chin. J. Mech. Eng. **57**(15), 62–70 (2021)
14. Fan, R., Li, Q., Wang, D.: Calibration of kinematic complete machine of 6PUS parallel mechanism. J. Beijing Univ. Aeronaut. Astronaut. **42**(05), 871–877 (2016)
15. Chen, G., Jia, Q., Li, T., Sun, H.: Calibration method and experiments of robot kinematics parameters based on error model. Robot **34**(6), 680 (2012). https://doi.org/10.3724/SP.J.1218.2012.00680

Multi-robot Cooperation Learning Based on Powell Deep Deterministic Policy Gradient

Zongyuan Li, Chuxi Xiao, Ziyi Liu, and Xian Guo(✉)

College of Artificial Intelligence, Nankai University, China Institute of Robotics and Automatic Information System, Tianjin, China
guoxian@nankai.edu.cn

Abstract. Model-free deep reinforcement learning algorithms have been successfully applied to a range of challenging sequential decision making and control tasks. However, these methods could not perform well in multi-agent environments due to the instability of teammates' strategies. In this paper, a novel reinforcement learning method called Powell Deep Deterministic Policy Gradient (PDDPG) is proposed, which integrates Powell's unconstrained optimization method and deep deterministic policy gradient. Specifically, each agent is regarded as a one-dimensional variable and the process of multi-robot cooperation learning is corresponding to optimal vector searching. A conjugate direction in Powell-method is constructed and is used to update the policies of agents. Finally, the proposed method is validated in a dogfight-like multi-agent environment. The results suggest that the proposed method outperforms much better than independent Deep Deterministic Policy Gradient (IDDPG), revealing a promising way in realizing high-quality independent learning.

Keywords: Multi-agent cooperation · Powell's method · Deep deterministic policy gradient

1 Introduction

Model-free deep reinforcement learning algorithms have been successfully applied to a range of challenging sequential decision making and control tasks. In 2013, DQN [1] first stabilized deep reinforcement learning and started a new era. After that, more and more reinforcement learning methods appeared. Double-DQN [2] reduced the error of estimate value and Dueling-DQN [3] improved sampling efficiency. Actor-Critic base methods like A2C, A3C and SAC showed a great ability of AC structure in a large range of tasks. REINFORCE [4] and DDPG [5] introduced direct policy search method into RL area. TRPO [6] and PPO [7] introduced trust region policy optimization methods and are widely used in many practical problems.

This work is supported by the National Natural Science Foundation of China (62073176). All the authors are with the Institute of Robotics and Automatic Information System, College of Artificial Intelligence, Nankai University, China.

H. Liu et al. (Eds.): ICIRA 2022, LNAI 13456, pp. 77–87, 2022.
https://doi.org/10.1007/978-3-031-13822-5_8

However, traditional single-agent algorithms usually perform unsatisfactorily in multi-agent environments. Independent learning algorithms like independent Q-Learning [8] and IDDPG [5] allocate each agent a unique policy and update policies independently in training, but sometimes get bad results in multi-agent problems because of instability of teammates strategies. CTDE algorithms, known as Centralized Training Decentralized Execution methods, such as COMA [9] and MADDPG [10], suffer from huge input information of state, and in some real-world situations, it is not easy to realize centralized training. Value decompose methods like VDN [11] and QMIX [12] perform well in many tasks, but it is hard for them to realize imitate learning because human-generated data usually do not contain action value.

Considering that the problems of CTDE methods and Value Decomposition methods are fundamental structural problems, it is difficult to realize centralized training in a group of different types robots or accurately describe action value in every environments, we turned our attention to independent learning.

The main problem of independent learning is the instability of teammates, which makes it possible that updating one agent rebounds on the whole team. However, there is an intuitive way to reduce the instability that we can fix some agents' policies and update others, which shares the idea of some unconstrained optimization algorithms.

Inspired by unconstrained optimization algorithms like Univariate Search Technique and Powell's Method, a method called Powell-Method Based DDPG was proposed. We find that updating one agent with others' policies fixed is a good way to decrease instability of teammates strategies, of which Powell's Method first performs one direction search and then generate a conjugate direction in the end of each round of searching. We regard the training process as searching for a team strategy in a strategy space, doing one direction search in each round, generating a conjugate direction to optimize the searching direction and using the directions to avoid being disturbed by other agents.

Our new RL algorithm is evaluated on a dogfight-like multi-agent environment to evaluate its performance. Results show that PDDPG's conjugate direction shortened the process of updating, helping a team form a team strategy, and performs much better than IDDPG.

Compared with the existing work, the main contributions of this paper are:

(1) The problem of multi-robot cooperative optimization is described as a multivariable optimization problem and solved along with the same idea as unconstrained optimization algorithms.
(2) Powell's conjugate direction is constructed in our reinforcement learning algorithm, and agents can be optimized using such direction.
(3) Experiments show that our method can help agents learn to cooperate with their teammates, and our algorithm may be a feasible scheme for realizing high-quality independent learning.

2 Related Work

2.1 Deep Deterministic Policy Gradient

As a well-known Actor-Critic based RL algorithm, Deep Deterministic Policy Gradient plays a nonnegligible role in the area of continuous control problems. It uses an actor to generate deterministic action and adds a gaussian noise to encourage exploration. Not the same as other AC based methods, the critic of DDPG evaluates action value instead of state value, thus the input of our critic is (s, a) instead of (s). We define the deterministic action as $a = \mu(s;\theta)$, and action value is presented as $Q(s,a) = r(s,a) + \gamma E_{s',a'}[Q_\pi(s',a')]$, where θ denote the parameters of policy μ.

Note that the output of the actor is a deterministic action. To encourage exploration, we add a zero-centered gaussian noise ϵ with an adaptive variance, which encourages more exploration while getting a bad result and encourages more exploitation while doing well. In this case, the sampling strategy $\pi(\theta)$ equals to $\mu(\theta) + \epsilon$ and our agent selects action using policy $\pi(\theta)$ instead of $\mu(\theta)$ unless its exploring noise is banned.

To maximize the discounted cumulative return, the critic can be trained by minimizing the Mean Square Loss:

$$Loss = Q(s,a) - (r(s,a) + \gamma Q^{-}(s',a')) \tag{1}$$

and update the actor by backpropagating the gradient

$$\nabla_\theta J(\theta) = E_{s\sim\mu_\theta}[\nabla_a Q_\mu(s,a)|_{a=\mu(s)} \nabla_\theta \mu_\theta(s)] \tag{2}$$

Besides, we update the target nets as:

$$\begin{cases} \mu(\theta^{-}) \leftarrow \tau\mu(\theta) + (\tau - 1)\mu(\theta^{-}) \\ Q(\omega^{-}) \leftarrow \tau Q(\omega) + (\tau - 1)Q(\omega^{-}) \end{cases} \tag{3}$$

where τ is the soft rate for updating target nets.

2.2 Powell's Method

Powell's Method is a method based on Univariate Search Technique.

The Univariate Search Technique is a famous unconstrained optimization method that does a one-dimension search to find the maximum value along with corresponding parameters. For function $f(x_1, x_2, ..., x_n)$, Univariate Search Technique does search in one dimension, fixing parameters of other dimensions, changing direction with regular intervals, and repeating the process for several rounds. It directly explores the optimal solution of the objective function without calculating the derivative of the function but suffers from oscillation when near the optimal solution.

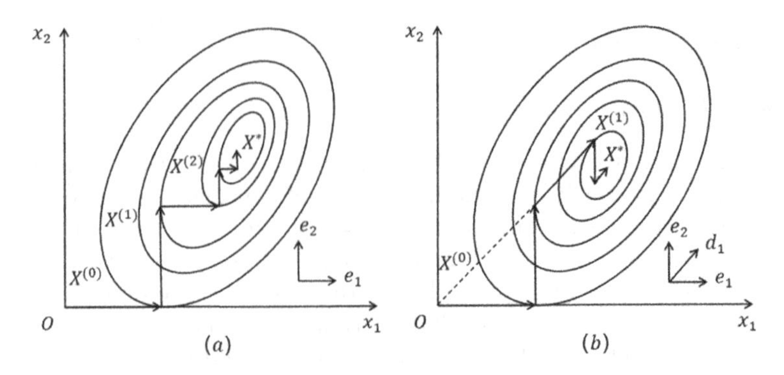

Fig. 1. Univariate Search Technique and Powell's Method with two variables. (a) Start with $X^{(0)} = (x_0^{(0)}, x_1^{(0)})$, two variables are updated one by one in the direction of its axis, and finally find the maximum in $X^* = (x_0^*, x_1^*)$. (b) Start with $X^{(0)} = (x_0^{(0)}, x_1^{(0)})$, two variables are updated according to the direction list $d = [e_0, e_1]$ and conjugate direction d_k. The conjugate direction are generated as $d_k = X_n^{(k)} - X_0^{(k)}$ and the method finally find the maximun in $X^* = (x_0^*, x_1^*)$.

To decrease the oscillation of the Univariate Search Technique and shorten the training process, J.D. Powell introduced the conjugate direction into the Univariate Search Technique algorithm. Powell's Method does the same as the Univariate Search Technique at first, searching for the optimal solution in each direction. At the end of round k, it generates a conjugate direction $d_k = X_n^{(k)} - X_0^{(k)}$ and search for the best strategy in the generated direction. After that, delete the first direction of the direction list and append the new direction to the list. Then, search for new rounds according to the directions of the list and the generated conjugate direction, and update the list for each round. For example, Fig. 1 shows a problem with two variables $X^{(0)} = (x_0^{(0)}, x_1^{(0)})$, two variables are first updated according to the direction of the direction list $d = [e_0, e_1]$, then a conjugate direction d_k will be generated in the end of round k. Variables will be updated according to d_k and the direction list will be updated to $d = [e_1, d_0]$.

Powell's Method inspired us that we can regard one agent as a variable, and do one dimension search for the best team strategy, and a direction means updating a specific combination of some agents with different learning rates or exploring noises, instead of updating only one agent just like Univariate Search Technique does. This idea makes some agents learn faster and explore more while other agents assist them as far as possible. In some sense, our method relieves the instability of teammates and shows the ability in helping agents learn to cooperate with teammates.

3 Method

3.1 PDDPG

Inspired by the idea of the Univariate Search Technique and Powell's Method, we regard an agent as a dimension in unconstrained optimization problem at first, describe the value of optimized function as the discounted cumulative return. In PDDPG, agents' parameters are updated in a specific direction, which means more than one agent may be updated at one time with different learning rates and exploring noise.

Powell-Method Based DDPG is the main contribution of our work. To stabilize teammates' policies, we first build IDDPG structure, with the same replay buffers as DQN used, then, selected a direction and reset learning rate and noise scale for each agent according to it. For every m episodes, it changes a direction to update and repeat this until reaching the maximum number of episodes.

In PDDPG, the direction list is initialized as $d = eye(n)$, where a specific direction can be described as $d[i]$. After using a direction in derection list, the difference of actors before and after updating will be calculated to generate the direction d_k. Just like what Powell did in his method, our conjugate direction is also proportional to the updated quantity. We extract the parameters of all agents' actors before and after m episodes updating, calculating $d_k[j]$ as:

$$d_k[j] = sum(abs(\theta^{new}_{all-agents} - \theta^{old}_{all-agents})) + \delta \tag{4}$$

where δ is a regularization term, usually, $\delta = 1e - 6$. After updating n agents, a $1 \times n$ sized direction vector d_k will be constructed and all agents could be updated according to it. After using d_k, we delete $d[0]$ and append d_k to d to update the direction list. the process could be described as:

$$d[n][j] = d_k[j](j = 0, 1, 2, ..., n - 1) \tag{5}$$

Note that each element of d_k is always much larger than 1, two normalization methods have been tried. The first one normalizes the maximum to 1, while the second normalizes the sum to 1. In practice, we find both the methods are acceptable, but the first one sometimes makes our algorithm too similar to IDDPG and may lead to the same problem of it.

To balance exploration and exploitation, we enlarge the variance of the gaussian noise when the team performs badly and decreases it when performing well in the sampling stage. Variances of the gaussian noise are set as:

$$\begin{cases} \sigma_j = \sigma_0 \times (1.5 - WinRate) \times d[i][j](i < n) \\ \sigma_j = \sigma_0 \times (1.5 - WinRate) \times d_k[j](i = n) \end{cases} \tag{6}$$

at the same time, set learning rates as:

$$\begin{cases} lr_j = lr_0 \times d[i][j](i < n) \\ lr_j = lr_0 \times d_k[j](i = n) \end{cases} \tag{7}$$

where i is direction id, j is agent id, k is round id, $d_k[j]$ represents a component of the round-k conjugate direction. To increase the utilization of collected data, we do 50 times updates by sampling minibatch with 64 transitions in the updating stage. The pseudo codes PDDPG are:

Algorithm 1. PDDPG

```
initialize the environment, n agents, direction list d
save current parameters as θ^{old}_{all-agents}
while episodes ¡ MaxEpisodes do
  reset environment and get initial state s
  while steps ¡ MaxSteps do
    select action a for all agents according to s, generate joint action
    exacute joint action and receive r, s', done
    store transition (s,a,r,s') for each agent
    if done then
      break
    else
      update the state
    end if
  end while
  i = j = (episodes//m)%(n + 1)
  k = (episodes//m)//(n + 1)
  if i < n then
    according to d[i], update agents, set noise variance
    if episodes%m = m − 1 then
      d_k[j] = sum(abs(θ^{new}_{all-agents} − θ^{old}_{all-agents})) + δ
      save current parameters as θ^{old}_{all-agents}
    end if
  else
    according to d_k, update agents, set noise variance
    if episodes%m = m − 1 then
      delete d[0] and append d_k to d
      save current parameters as θ^{old}_{all-agents}
    end if
  end if
end while
```

4 Experiments

Our approach is validated in a two agents cooperating environment.

4.1 Experiments Setup

Overall Introduction. In our environment, there are two teams, one of which is our controlled team, and the other is our opponent. There are two agents in our team and one agent in opponent's. The goal of both teams is to catch its enemy. Figure 3 shows the environment (Fig. 2).

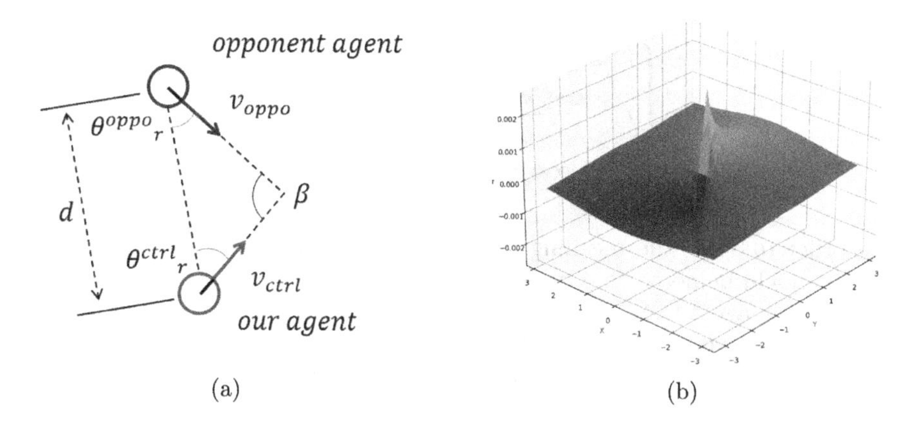

Fig. 2. Diagram of environment and intrinsic reward. (a) Both the agents goals at catching its enemy. If $d < MinDistance$ and $\theta_r < \pi/4$, it could be regarded that one agent has been catched by another. (b) Diagram of Part r_1 of the Intrinsic Reward. The direction of x-axis is the same as the speed direction of opponent, and the origin attaches to the opponent. Entering the area in front of a opponent will greatly punish our agent, while entering the area behind means a lot of positive rewards.

State. The state of the environment are composed of each agent's state $[x, y, \theta]$ where θ is defined as the angle of rotation from x-axis forward counterclockwise. All agents are initialized in fixed location with a direction fixed initial velocity $v_0 = 0.5 \times v_{max}$.

Action. All of our agents select actions using their actor networks, while the opponent is controlled by a model-based method. The action of each agent is a combination of its linear velocity and angular velocity, described as $[v, \omega]$, and the range of v and ω is $[0.1, 1.2]$ and $[-0.8, 0.8]$. The opponent does a one-step search to find the action that makes it closest to our agents. Considering that this is a very aggressive strategy, making it extremely difficult to train our agents and always turning a loss game into a draw for an opponent, we slightly reduced the opponent's max velocities to ensure it is possible for us to win.

Reward. The reward consists of two parts. The first part is environment return, it is a sparse reward and only appears at the end of an episode. For a winning game, all of our agents get a reward scaled at 100, while a punishment scale of -100 will be received if our agents lose the game. Nothing will be received if two teams tied with each other. The second part is an artificial force field liked intrinsic reward, it is a dense reward and used as a guide to encourage our agents to circle behind the opponent. The intrinsic rewards are:

$$\begin{cases} r_{intrinsic} = 0.001 \times (r_1 + r_2) \\ r_1 = \dfrac{-cos(\theta_r^{oppo}) \times |cos(\theta_r^{oppo})|}{\sqrt{d}} \\ r_2 = \dfrac{cos(\theta_r^{ctrl}) \times |cos(\theta_r^{ctrl})|}{\sqrt{d}} \end{cases} \tag{8}$$

In order to show the meaning of the intrinsic reward, we visualized r_1 in Fig. 4. It should be noted that r_1 is a reward that comes from the enemy and r_2 comes from an agent itself. In our practice, we set absolute value of environment return 10 times larger than the absolute value of cumulative intrinsic return, making our agent pay more attention to winning, instead of accumulating useless intrinsic rewards.

Conditions for Victory. When an agent goes beyond the map or is caught by opponents, the team it belongs to will lose the current game. If no agent catches others within $MaxSteps$, or two agents catch each other at the same time, we regard the two sides are tied.

The formula for winning rate is:

$$WinRate = (1 \times n_{win} + 0.5 \times n_{draw} + 0 \times n_{loss})/n \tag{9}$$

4.2 Results

In our experiments, agents were trained in a 2v1 environment using IDDPG, PDDPG and some degraded versions of PDDPG. Each experimental group underwent five repeated trials and the best result of all of them was selected for comparison. In our experiments, we set discount factor $\gamma = 0.99$, default learning rate $lr_0 = 1e-3$, default noise variance $\sigma_0 = 0.15$, and set epoch-interval for changing direction as 100.

The result in Fig. 3(a) shows that 2v1 tasks trained by IDDPG sometimes perform even worse than 1v1 DDPG, which signifies both the agents disturbed teammates and did not find a strategy to defeat the opponent. However, PDDPG successfully find strategies to win and exceeded the winning rate of 90%, winning nearly all games against the aggressive opponent and maintaining stability with a high winning rate. It is worth noting that, degraded PDDPG without conjugate direction can also win 80% games, showing that just fixing some agents can also decrease teammates' instability. We find that the degraded version sometimes perform well in the second half of updating an agent but perform terribly in the first half, due to the imprecise prediction of best actions after fixing too many episodes. Our method relieves the problem by updating both agents with different speeds when using the conjugate direction, at the same time, the conjugate direction avoids the problem of two agents disturbing each other.

To ensure the final result comes from two agents' cooperation instead of the capability of a single agent, we also made an experiment with a series of opponent's difficulties. We set the difficulty as $[0.6, 0.7, 0.8, 0.9, 1.0]$, training a single

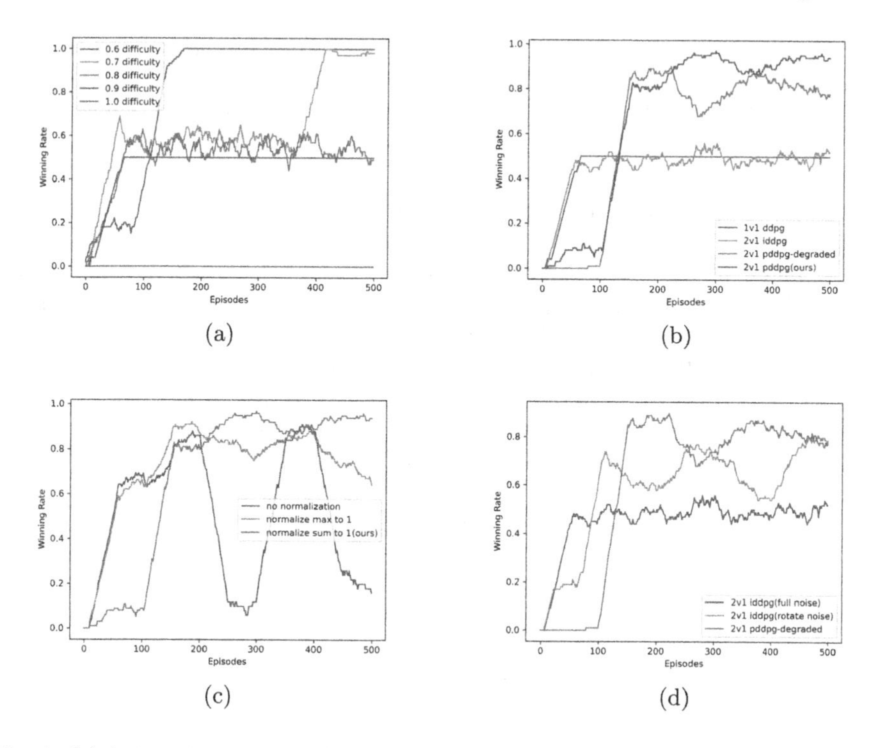

Fig. 3. (a) 2v1 against enemy of 0.9 difficulty. It is obvious that PDDPG outperforms all others methods, get highest winning rate after 250 episode. Difficulty there is a hyper-parameter about opponent's max speed. (b) 1v1 against enemy of a series of difficulties. Opponent with 1.0 difficulty could always beat our agent while opponent with difficulty lower than 0.9 could be defeated by a single agent within 500 episode training. (c) Supplement experiment on different normalization method. Both the methods are acceptable, though normalizing the maximum to 1 makes our algorithm similar to IDDPG. However, normalization is necessary. Direction without normalization will lead to terribly large learning rate and exploring noise and destroy the training process while using the generated conjugate direction. (d) Supplement experiment on different exploring noise scale. Rotating the noise also has effects on improving performance, at the same time, changing agents to train is also important.

agent using DDPG. The difficulty is equal to the opponent's max velocity over ours, resembling the opponent's reaction speed. The result in Fig. 3(b) suggests that an opponent with 1.0 difficulty can always defeat our agents, while an opponent of less than 0.9 difficulty is possible to be defeated by a single agent. So, 0.9 difficulty was selected as a baseline and used as the control group in the experiment mentioned above.

To further study the reason of getting better results, we did two more supplement experiments. The first one is about the normalization methods and the other one is about noise scale. Results are shown in Fig. 3(c) and Fig. 3(d).

In the first supplement experiment, two normalization methods are evaluated, the first method normalize the maximum to 1, reducing other values in the same scale, while the second normalize the sum to 1. In our experiment, we find that both the normalization methods are acceptable, but normalize the sum to 1 present a better result.

In the second supplement experiment, the influence of the scale of exploring noise is studied. We tried IDDPG with rotation of noise. This new kind of IDDPG updates all agents at a same time, but only one agent selects action using policy $\pi(\theta)$ while others using policy $\mu(\theta)$. the result shows that, the rotation of noise also make it possible for agents to learn a team strategy, but there is still a certain gap with PDDPG and its degraded versions in performance.

Also, some trajectories of our agents were collected. Figure 4(a) shows a very clever strategy while Fig. 4(b) shows how to tie with the opponent. These two trajectories come from PDDPG trained policies and IDDPG trained policies.

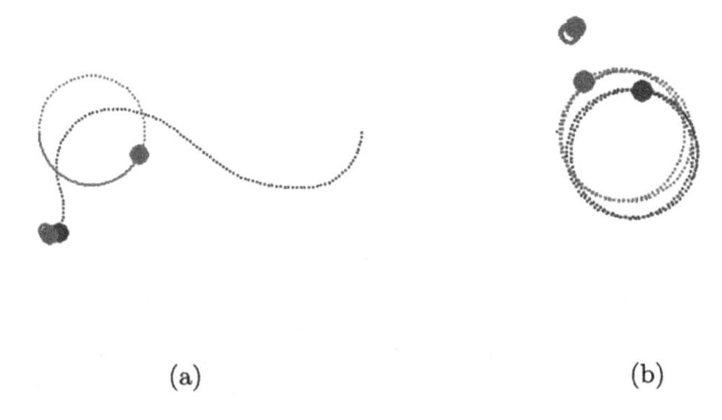

(a) (b)

Fig. 4. Trajectory of a winning game and a drawing game. (a) One agent makes a circular motion with a big radius to lure the opponent into the trap and the other one rotate with a very small radius, catching the opponent before it aims at our agent. (b) Our agents make a circular motion to get out of the opponent and run out of all time steps to avoid the punishment.

5 Conclusions

We proposed Powell's Method Based Deep Deterministic Policy Gradient as a new approach to train agents to cooperate with their teammates. In our experiment, it is proved that our method outperforms IDDPG and the rotate-noise version IDDPG, significantly improve the performance of independent learning methods and proves the effectiveness of Powell's method, especially the effectiveness of Powell's conjugate direction. We also analyzed the influence of normalization tricks for conjugate direction and the influence of the scale of exploring noise, finding that normalization is necessary and the setting of exploration noise also has a great effect on the result.

References

1. Mnih, V., et al.: Playing Atari with deep reinforcement learning. arXiv e-prints arXiv:1312.5602, December 2013
2. van Hasselt, H., Guez, A., Silver, D.: Deep reinforcement learning with double Q-learning. arXiv e-prints arXiv:1509.06461, September 2015
3. Wang, Z., Schaul, T., Hessel, M., van Hasselt, H., Lanctot, M., de Freitas, N.: Dueling network architectures for deep reinforcement learning. arXiv e-prints arXiv:1511.06581, November 2015
4. Williams, R.J.: Simple statistical gradient-following algorithms for connectionist reinforcement learning. Mach. Learn. **8**(3–4), 229–256 (1992)
5. Lillicrap, T.P., et al.: Continuous control with deep reinforcement learning. arXiv e-prints arXiv:1509.02971, September 2015
6. Schulman, J., Levine, S., Moritz, P., Jordan, M.I., Abbeel, P.: Trust region policy optimization. arXiv e-prints arXiv:1502.05477, February 2015
7. Schulman, J., Wolski, F., Dhariwal, P., Radford, A., Klimov, O.: Proximal policy optimization algorithms. arXiv e-prints arXiv:1707.06347, July 2017
8. Tampuu, A., et al.: Multiagent cooperation and competition with deep reinforcement learning. arXiv e-prints arXiv:1511.08779, November 2015
9. Foerster, J., Farquhar, G., Afouras, T., Nardelli, N., Whiteson, S.: Counterfactual multi-agent policy gradients. arXiv e-prints arXiv:1705.08926, May 2017
10. Lowe, R., Wu, Y., Tamar, A., Harb, J., Abbeel, P., Mordatch, I.: Multi-agent actor-critic for mixed cooperative-competitive environments. arXiv e-prints arXiv:1706.02275, June 2017
11. Sunehag, P., et al.: Value-decomposition networks for cooperative multi-agent learning. arXiv e-prints arXiv:1706.05296, June 2017
12. Rashid, T., Samvelyan, M., Schroeder de Witt, C., Farquhar, G., Foerster, J., Whiteson, S.: QMIX: monotonic value function factorisation for deep multi-agent reinforcement learning. arXiv e-prints arXiv:1803.11485, March 2018

Model Construction of Multi-target Grasping Robot Based on Digital Twin

Juntong Yun[1,2,3], Ying Liu[2,4,5](✉), and Xin Liu[3,5]

[1] Key Laboratory of Metallurgical Equipment and Control Technology of Ministry of Education, Wuhan University of Science and Technology, Wuhan 430081, China
[2] Hubei Key Laboratory of Mechanical Transmission and Manufacturing Engineering, Wuhan University of Science and Technology, Wuhan 430081, China
liuying3025@wust.edu.cn
[3] Precision Manufacturing Research Institute, Wuhan University of Science and Technology, Wuhan 430081, China
[4] Research Center for Biomimetic Robot and Intelligent Measurement and Control, Wuhan University of Science and Technology, Wuhan 430081, China
[5] Longzhong Laboratory, Xiangyang 441000, Hubei, China

Abstract. In the unstructured environment, how to grasp the target object accurately and flexibly as human is always a hot spot in the field of robotics research. This paper proposes a multi-object grasping method based on digital twin to solve the problem of multi-object grasping in unstructured environment. The twin model of robot grasping process is constructed to realize multi-object collision-free grasping. Firstly, this paper analyzes the composition of digital twin model based on the five-dimensional digital twin structure model, and constructs the twin model of unknown target grasping process combining with multi-target grasping process of robot. Then, a digital model of the key elements in the multi-target grasping process of the robot is established to realize the interaction between the physical entity and the virtual model, and an evaluation model of the virtual model grasping strategy is established to screen the optimal grasping strategy. Finally, based on the twin model of multi-target grasping process, a multi-target grasping process based on digital twin is constructed to realize the multi-target stable grasping without collision.

Keywords: Robot · Digital twin · Unstructured environment · Multi-object grasping

1 Introduction

Although the grasping operation of objects in the environment is a simple and common daily behavior for humans, it is a great challenge for robots [1]. In the unstructured environment, how the robot can grasp the object accurately and flexibly as human is always a hot spot in the field of robot research [2, 3].

Digital Twin (DT) is to simulate physical objects in virtual space and create a virtual model or a group of physical entities corresponding to the virtual model in Digital form

H. Liu et al. (Eds.): ICIRA 2022, LNAI 13456, pp. 88–97, 2022.
https://doi.org/10.1007/978-3-031-13822-5_9

[4]. Through the test and analysis of the model, we can complete the understanding, analysis and optimization of physical entities [5, 6]. Digital twin has two characteristics: first, it classifies and integrates all kinds of data and metadata in physical objects. Digital twin is a faithful mapping of physical objects. Second, digital twin can truly reflect the change of the whole life cycle of physical products, and constantly accumulate relevant knowledge, and optimize and analyze the current situation together with objects to complete the improvement work. Therefore, from the perspective of digital twin, the digital model of industrial robot is the digital twin of robot physical object [7–9].

Based on the advantages of digital twin technology, this paper proposes a multi-target grasping method for robot based on digital twin technology. Firstly, based on the concept of digital twin 5D structure model and combined with multi-target grasping process, the twin model of unknown target grasping process of robot is constructed. Then the key elements of multi-target grasping process are modeled digitally to realize the interaction between physical entity and virtual model. Then, the evaluation model of virtual model grasping strategy is constructed based on the evaluation index of physical entity grasping state. Finally, based on the twin model and function module of multi-target grasping process, a multi-target grasping process based on digital twin is designed to achieve multi-target stable grasping without collision.

2 Related Work

Although the grasping operation of objects in the environment is a simple and common daily behavior for humans, it is a great challenge for robots. In the unstructured environment, how the robot can grasp the object accurately and flexibly as human is always a hot spot in the field of robot research [10].

Robotic planar gripping is mainly used for 2D planar gripping inspection in industrial scenarios, where the robot arm must approach the target to be gripped from a direction approximately orthogonal to the image [11]. The gripping position is usually represented by the gripping rectangle and the angle of rotation in the plane. The graspable geometry model is defined as the grasping rectangular box in the RGB-D image. The representation gives the position and orientation of the two-finger manipulator before it successfully grasps the object.

In order to further improve the speed and accuracy of grasping pose detection, many researchers carry out in-depth research based on grasping rectangular frame. Lenz et al. [12] uses multiple convolutional networks to represent features in the sliding window detection channel and used the final classification for grab detection, but this method runs at a speed of 13.5 s per frame, with an overall accuracy of about 75% and a 13.5% delay between perception and grab detection. Redmon and Angelova [13] used the same data set as Lenz et al., but took grasping pose detection as a regression problem and solved it using convolutional neural network. Yu et al. [14] also designed a three-level convolutional neural network based on Lenz method to select the grasping position and posture of the robot. Yan et al. [15] first detects the target object in the image by using the method of target detection to obtain the category and position of the object, and then adopted Lenz method to grab the object by using the mechanical arm. This method reduces the possible grasping position of the object by using target detection and improved the decision-making efficiency of grasping.

Andreas et al. [16] used traditional methods to screen the candidate grasp positions and poses, and used CNN network to estimate the grasp quality based on depth images from three angles. In literature, the traditional method is also used to screen candidate grasping positions, and then the depth map is not used to evaluate the grasping quality, but the Point cloud inside the grasping is used, and the 3D neural network Point Net is used to estimate the grasping quality. The 6-DOF Grasp Net detection model proposed by Nvidia [17] improves the screening of candidate grasp poses and the evaluation of grasping quality.

In addition, many researchers have conducted extensive research on object grasping in complex scenes. Dogar et al. proposes a combination of push and grab, which can not only shorten the processing time of the target, but also deal with some objects that are too big or too heavy to be grasped [18]. Lan et al. [19] proposes a robot fetching method suitable for multi-object stacking scenarios. Two important problems in multi-object grasping are considered: how to grasp a specific object and how to ensure the safety in the process of grasping, so that the object will not be damaged because of the grasping sequence.

To sum up, the current under the unstructured environment of multi-objective grab is aimed at grasping pose detection parts and robot grab, a lack of understanding of the types of objects in the scene and the dependency relationship between object and grab parts of judgment, and, more importantly, these methods did not consider fetching strategy and influence on the stability of other objects. Therefore, it is very necessary to study how to generate multi-target grasping strategy to achieve multi-target collision-free and stable grasping of manipulator.

3 Multi-target Grasping Digital Twin Model of Robot

3.1 Digital Twin

Digital Twin is a technical means to create a virtual model of physical entities in a Digital way and simulate, verify, predict and control the whole life-cycle process of physical entities with the help of historical data, real-time data and algorithm model. In short, it is to build a virtual model with physical entities, evolve in a virtual environment, feedback and influence the real world.

The research focus of digital twin technology should be on model and data. Digital twin technology can be divided into 5D structure model, including physical object, virtual model, connection, data and service system, as shown in Fig. 1. According to the digital twin five-dimensional structure model, in order to realize the practical application of digital twin technology, a series of characteristics of physical objects are analyzed first, and a virtual model is constructed according to these characteristics, and the data interaction between virtual and real space is completed through the connecting channel, and the data is analyzed.

Physical system is objective existence, it is composed of a variety of subsystems, and through the cooperation of each subsystem to complete the production and manufacturing operation. Virtual system is the digital mirror image of physical objects. It gathers the basic elements of physical objects together, among which the most important elements are geometry, rules, behavior and physical information. Connection is the

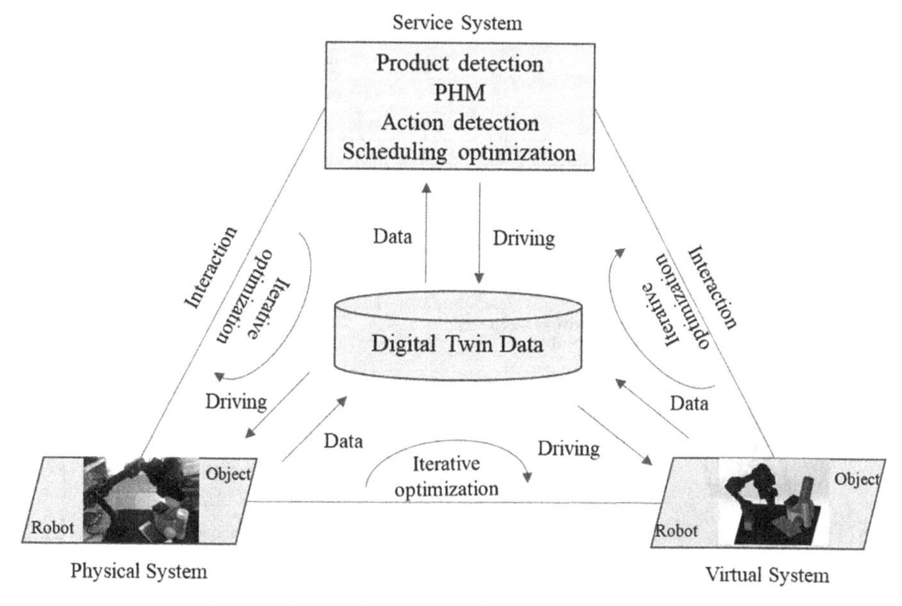

Fig. 1. Schematic diagram of digital twin 5D model.

channel between the physical system and the virtual system. The data can be effectively transmitted in real time, so as to realize real-time interaction and ensure the consistency between the physical system and the virtual system. Data includes all the data of physical objects and virtual models, which are fused with multi-domain data and continuously updated and iterated with the output of real-time data. Service system includes all kinds of information of the whole system. Real-time optimization is made based on physical entities and virtual models to ensure intelligent, accurate and reliable service quality.

3.2 Twin Model of Multi-object Grasping Process for Robot

Digital twin is driven by multi-dimensional virtual model and fused data, and realizes monitoring, simulation, prediction, optimization and other practical functional services and application requirements through virtual and real closed-loop interaction. The construction of digital twin model is the premise to realize the application of digital twin. The robot grasping framework based on digital twin is shown in Fig. 2, including physical entity layer, virtual model layer, twin data layer and application layer in physical space.

The physical entity layer is the base of the robot grasping process based on digital twin, and is the carrier of the robot grasping process digital twin system. The work of other layers needs to rely on the physical entity layer to carry out. Involved in the physical layer contains a robot grasping of the target, the environment, the robot and process all the physical elements combined with hardware resources (with the target, robot, camera, sensor, computer, etc.), software resources (with auxiliary software, application system and network resources, etc.), and the relevant properties of physical objects. Various

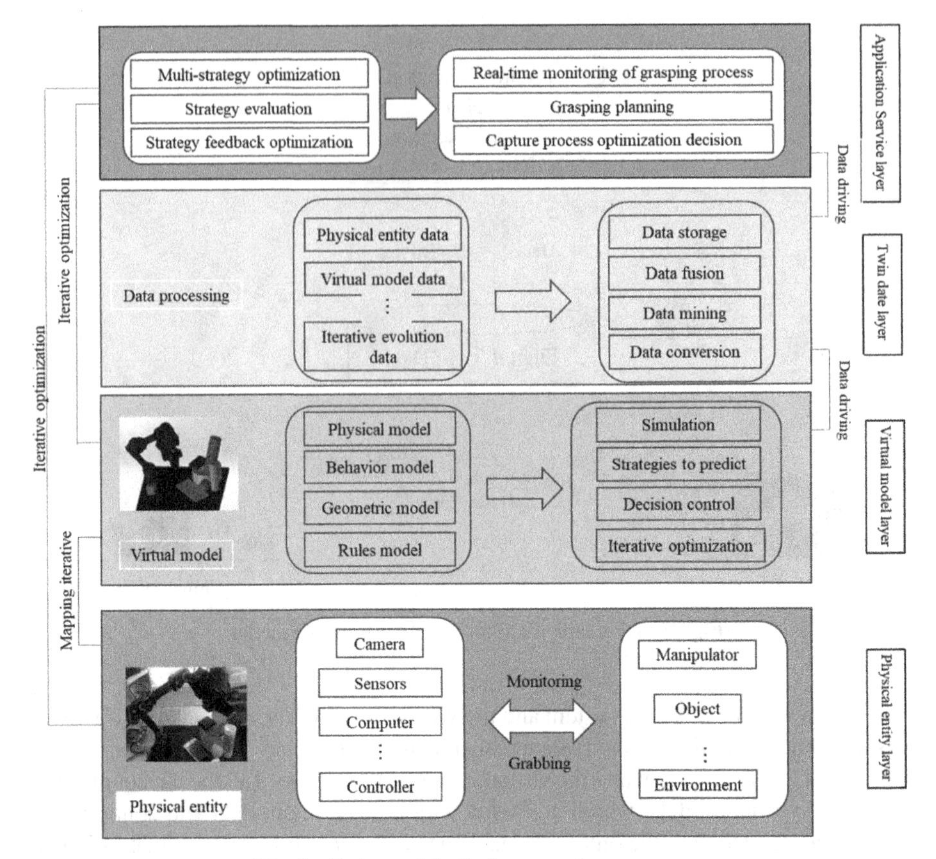

Fig. 2. Twin model of robot grasping process

sensors deployed on the entity element collect relevant data in the grasping process for monitoring and analysis.

Virtual model layer refers to the real mirror image of physical entity layer in the information space. It realizes faithful mapping from four aspects of physics, behavior, geometry and rule through complex digital modeling technology, and its essence is a collection of various models. Driven by virtual model, simulation, diagnosis and prediction, decision control, iterative optimization and other processes of physical entities are realized, and real-time feedback to the physical layer.

Twin data layer is based on the number of twin robot important support, and also a fully multi-level and complex database collection, including the physical entity in the process of fetching data, virtual model data, the application service data and derived data fusion. The state of robot grasping is analyzed by using the obtained data. Related data includes static data and dynamic data. This kind of data requires dynamic monitoring and tracking by multiple types of equipment. In the whole process of robot grasping, the twin data layer stores, fuses, mines and transforms data, and updates data in real time, thus providing more comprehensive and accurate information for the digital twin system.

The application service layer is the top layer of the robot grasping framework based on digital twin. The service layer of robot grasping application based on digital twin includes multi-strategy optimization, strategy evaluation, strategy feedback optimization, etc., and real-time monitoring, grasping planning and decision-making of robot grasping process are carried out to realize robot grasping.

4 Robotic Multi-target Fetching Based on Digital Twin

4.1 Digital Modeling of Key Elements of Robot Multi-target Grasping Process

The digital twin model of robot grasping process is conducive to analyzing the operation mechanism of information space. The construction of the twin model needs to be oriented to different physical entities and diverse data generated by it, and its realization needs to establish a unified logical data model. The specific modeling content includes: physical entity digital modeling, virtual model digital modeling, virtual and real mapping association modeling.

In the process of robot grasping, the key elements involved include: robot, target to be grasped, etc. The robot and the object interact naturally in the physical space, and its state is constantly updated with the dynamic operation of the manufacturing process. Taking the above factors into consideration, this paper adopts formal modeling language to model the key elements of mechanical grasping process, and the model is defined as follows:

$$PS = \{PE, PP\} \tag{1}$$

where, PS represents the physical space of key elements in the process of robot grasping; PE represents the collection of various components of the physical space robot; PP represents a collection of objects in physical space.

The digital twin model needs to be highly modularized, extensible and dynamically adaptable, and the model can be constructed in virtual model by parameterized modeling method. The virtual model not only describes the geometric information and topological relations in the process of robot grasping, but also contains the complete dynamic information description of each physical object. Then, the multi-dimensional attributes of the model are parameterized to realize real-time mapping of robot grasping process. Formal modeling language is used to model the digital Leeson model in the virtual model. The specific definition of the model is as follows:

$$VS = \{VE, VP\} \tag{2}$$

where, VS represents the virtual model of key elements in the process of robot grasping; VE represents the set of robot components in the virtual model; VP represents the collection of objects in the virtual model. Both VE and VP are dynamic sets, and they interact naturally. Set elements and their states in the virtual model are updated synchronously with the dynamic operation of the manufacturing process in the physical space.

Based on the establishment of physical space entity model *PS* and information space virtual model *VS*, the virtual-real mapping association between the two is further established, and the formalized modeling language is adopted to model the virtual-real mapping association relationship. The model definition is as follows:

$$PS \overset{1:1}{\leftrightarrow} VS \tag{3}$$

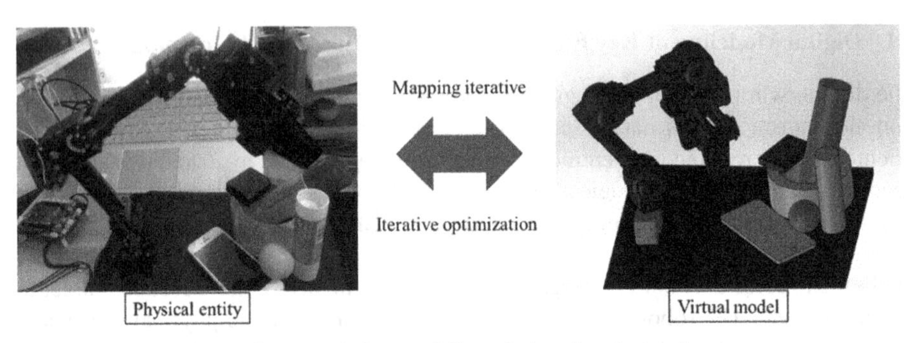

Fig. 3. Association modeling of virtual-real mapping

where, $PS = \{PE, PP\}$, $VS = \{VE, VP\}$, $\overset{1:1}{\leftrightarrow}$ represents the bidirectional real mapping between physical space entity model and virtual model of information space. The bidirectional real mapping between physical entity robot *PE* and virtual model robot *VE*, physical entity targets *PP* and virtual model target *VP* should be synchronized with each other, as shown in Fig. 3.

4.2 Twin Data Perception

There are three main steps to realize the data acquisition process of robot grasping process, as shown in Fig. 4, including static data acquisition, dynamic data perception and data analysis. Firstly, the static data of the virtual model of the robot is obtained by determining the parameters of each part of the robot. The second step is to realize the dynamic perception of data, first register the basic information of all kinds of sensors in the management system, including sensor type, sensor location, sensor key function parameters, etc., using a unified data interface is to realize the data interaction process. For example, by configuring displacement sensors to sense the grasping state of the robot. The third step is the analysis process after data perception, which aims to process the captured real-time data into useful information by using relevant principles, so as to comprehensively judge the overall state of robot grasping, and then use the obtained data to make intelligent decisions.

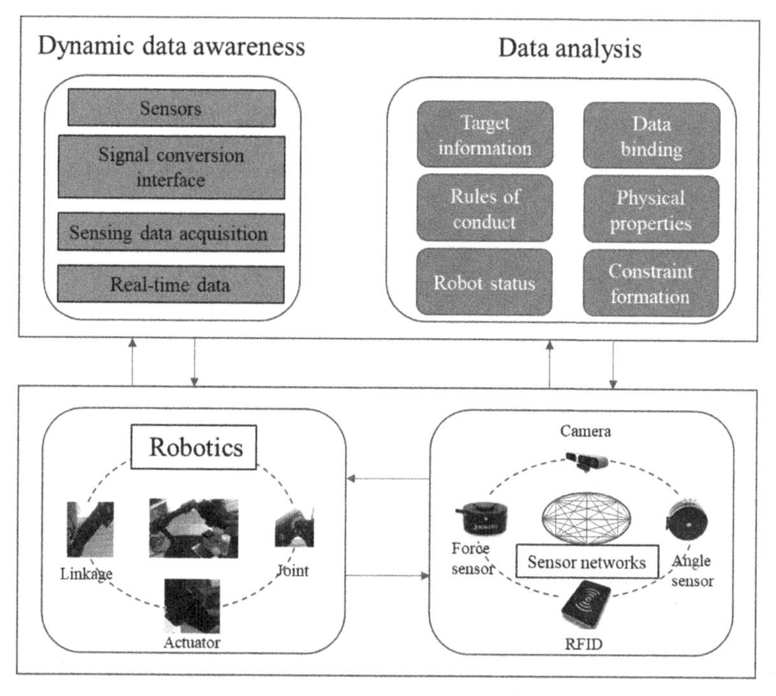

Fig. 4. Data perception of twin models

4.3 Evaluation of Robot Virtual Model Grasping Strategy

In view of the virtual model simulation prediction of grasping strategy, the evaluation model of grasping strategy is established to determine the optimal grasping strategy. The model diagram of grasping strategy evaluation is shown in Fig. 5, in which yellow represents three kinds of objects, and connecting line represents the connection between nodes.

Through the comprehensive evaluation score S of all elements in the robot grasping process in the virtual model, the grasping strategy is evaluated, and the highest score is the optimal grasping strategy.

$$S = (x,y,z) \tag{4}$$

where, x is used to describe the state of the robot in the virtual environment; y represents the state of uncaptured target; z represents the state of the captured target.

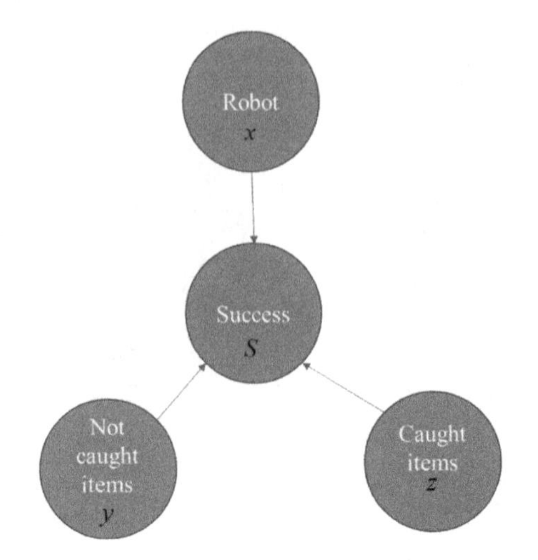

Fig. 5. Grasping strategy evaluation model diagram

5 Conclusion

In this paper, a multi-target grasping method based on digital twin is proposed to achieve the stable grasping without collision. Based on the concept of digital twin 5D structure model and combined with the robot grasping process, a multi-object grasping process twin model was constructed. Digital modeling is carried out for key elements of multi-target grasping process of robot to realize the interaction between physical entity and virtual model. The evaluation model of virtual model grasping strategy is constructed to realize the evaluation and optimization of grasping strategy and generate the optimal grasping strategy. Finally, based on the twin model and function module of multi-target grasping process, a multi-target grasping process based on digital twin is designed to achieve multi-target stable grasping without collision.

References

1. Han, Y., Chu, Z., Zhao, K.: Arget positioning method in binocular vision manipulator control based on improved canny operator. Multimedia Tools Appl. **13**(79), 9599–9614 (2020)
2. Zhukov, A.: Improvement and extension of the capabilities of a manipulator based on the probe of an atomic-force microscope operating in the hybrid mode. Instrum. Exp. Tech. **62**(3), 416–420 (2019)
3. Wang, L., Yan, J., Cao, T.: Manipulator control law design based on Backstepping and ADRC Methods. Lect. Notes Electr. Eng. **705**, 261–269 (2021)
4. Tuegel, E., Ingraffea, A.R., Eason, T.G.: Reengineering aircraft structural life prediction using a digital twin. Int. J. Aerosp. Eng. **2011**, 1687–5966 (2011)
5. Tao, F., Cheng, J., Qi, Q.: Digital twin-driven product design, manufacturing and service with big data. The Int. J. Adv. Manuf. Technol. **94**(9–12), 3563–3576 (2018)
6. Schleich, B., Anwer, N., Mathieu, L.: Shaping the digital twin for design and production engineering. CIRP Ann. **66**(1), 141–144 (2017)

7. Tao, F., Sui, F., Liu, A.: Digital twin-driven product design framework. Int. J. Prod. Res. **2018**, 1–19 (2018)
8. Grieves, M.: Irtually intelligent product systems: digital and physical twins. Journal **2**(5), 99–110 (2016)
9. Zhang, H., Liu, Q.: A digital twin- based approach for designing and decoupling of hollow glass production line. IEEE Access **5**, 26901–26911 (2017)
10. Singh, S., Raval, S., Banerjee, B.: Roof bolt identification in underground coal mines from 3D point cloud data using local point descriptors and artificial neural network. Int. J. Remote Sens. **42**(1), 367–377 (2021)
11. Gupta, M., Muller, J., Sukhatme, G.: Using manipulation primitives for object sorting in cluttered environments. IEEE Trans. Autom. Sci. Eng. **12**(2), 608–614 (2015)
12. Lenz, I., Lee, H., Saxena, A.: Deep learning for detecting robotic grasps. Int. J. Robot. Res. **34**(4/5), 705–724 (2015)
13. Redmon, J., Angelova, A.: Real-time grasp detection using convolutional neural networks. In: 2015 IEEE International Conference on Robotics and Automation, pp. 1316–1322. IEEE, Piscataway, USA (2015)
14. Qiu, Z., Zhang, S.: Fuzzy fast terminal sliding mode vibration control of a two-connected flexible plate using laser sensors. J. Sound Vib. **380**, 51–77 (2016)
15. Yang, C., Peng, G., Li, Y.: Neural networks enhanced adaptive admittance control of optimized robot-environment interaction. IEEE Trans. Cybern. **49**(7), 2568–2579 (2019)
16. Andreas, P., Marcus, G., Kate, S.: Grasp pose detection in point clouds. The Int. J. Robot. Res. **36**(13–14), 1455–1473 (2017)
17. Tran, Q., Young, J.: Design of adaptive kinematic controller using radial basis function neural network for trajectory tracking control of differential-drive mobile robot. Int. J. Fuzzy Logic Intell. Syst. **19**(4), 349–359 (2019)
18. He, W., Dong, Y.: Adaptive fuzzy neural network control for a constrained robot using impedance learning. IEEE Trans. Neural Netw. Learn. Syst. **29**(4), 1174–1186 (2018)
19. Dumlu, A.: Design of a fractional-order adaptive integral sliding mode controller for the trajectory tracking control of robot manipulators. J. Syst. Control Eng. **232**(9), 1212–1229 (2018)

Mechanism Design, Control and Sensing of Soft Robots

A New Geometric Method for Solving the Inverse Kinematics of Two-Segment Continuum Robot

Haoran Wu[1], Jingjun Yu[1], Jie Pan[1], Guanghao Ge[1], and Xu Pei[2](✉)

[1] Robotics Institute, Beihang University, Beijing 100191, China
[2] Department of Mechanical Design, Beihang University, Beijing 100191, China
goingxu@163.com

Abstract. The inverse kinematics (IK) of continuum robot (CR) is an important factor to guarantee motion accuracy. How to construct a concise IK model is very essential for the real-time control of CR. A new geometric algorithm for solving the IK of CR is proposed in this paper. Based on Piecewise Constant Curvature (PCC) model, the kinematics model of CR is constructed and the envelope surface of the single segment is calculated. The IK of CR is obtained by solving the intersection of surfaces. The spatial problem is transformed into a plane using a set of parallel planes, and the intersection is solved using the k-means clustering algorithm. The algorithm's real-time performance and applicability are improved further by increasing the sampling rate and decreasing the range of included angles. A distinct sequence is designed for solving the IK of CR. The efficiency and effectiveness of geometric are validated in comparison to some of the most popular IK algorithms. Finally, the accuracy of the algorithm is further validated by a physical prototype experiment.

Keywords: Continuum robot · Constant curvature model · Inverse kinematics · Geometric method

1 Introduction

Continuum robot (CR) is a new type of bionic robot with a continuous backbone without joints 1. Compared with traditional rigid discrete robots, they can change their shape flexibly according to environmental obstacles and has strong adaptability to small environments and unstructured spaces [2]. CR has been used in exploration [3], medicine [4], rescue [5], aviation manufacturing [6], and other fields. As a kind of hyper redundant robot, modeling and solving the inverse kinematics (IK) of CR proves to be a very interesting and challenging problem. A large proportion of efforts in the area have focused on the establishment and solution of the kinematic model.

The traditional method of solving the IK of CR is the Jacobian method [7–9]. The CR is transformed into a rigid body model and solved by iteration of the Jacobian inverse matrix. However, the calculation of the pseudo-inverse of the Jacobian matrix is large. And non-convergence and singularities are also inevitable through this method

H. Liu et al. (Eds.): ICIRA 2022, LNAI 13456, pp. 101–112, 2022.
https://doi.org/10.1007/978-3-031-13822-5_10

[10, 11]. Neural network [12–14] is also used to solve the IK of CR. However, the excessive calculation may be existed due to the large training set when the robot has many degrees of freedom. Another method is to establish the model according to the geometric relationship, and the inverse kinematics can be solved by numerical iteration, which is also applied in this paper. A geometric method for IK of the hyper-redundant robot is proposed by Yahya [15] and the angles of adjacent links are set to be the same. Williams and Svenja [16] analogized the movement of a snake's tail following the head's trajectory and proposed a follow-the-leader (FTL) heuristic algorithm for the hyper-redundant robot. Given the collision-free tip trajectory, the other joints of the robot follow the end [17, 18]. However, the potential mobile ability of the endpoint is restricted by the FTL algorithm. Sreenivasan [19] applied the geometric properties of a three-dimensional tractive line to solve the IK solution and the CR is discretized into a sufficiently large number of rigid links. Gong [20] regarded the two-segment underwater soft manipulator as arcs with opposite curvature. The reverse constant curvature model is constructed to obtain the IK of CR. However, few types of robots are suitable for this algorithm due to the limitation of the algorithm.

The main contribution of this paper is to propose a geometric method for solving the IK of two-segment CR. The remainder of the paper is constructed as follows: First of all, the kinematics model of CR and the envelope surface of the single-segment are solved based on Piecewise Constant Curvature (PCC) model [21, 22]. In the following section, a geometric method is proposed for solving the inverse kinematics (IK) of CR. And a numerical iterative strategy is applied to improve the real-time performance of the algorithm. Finally, the effectiveness is further verified through physical prototype experiments and numerical comparison with some of the most popular algorithms.

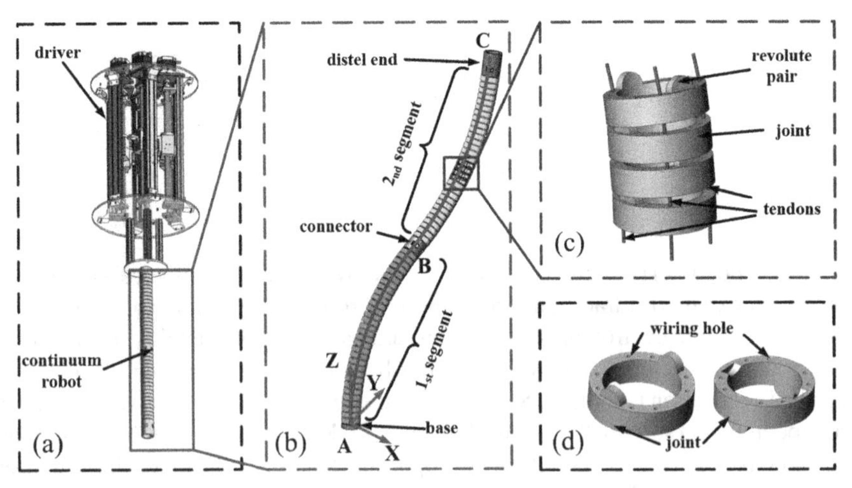

Fig. 1. (a) 3D model of CR (b)Two-segment CR (c) The driving tendons of CR (d) Single joint of CR.

2 Design of the Continuum Robot

The two-segment continuum robot is illustrated in Fig. 1, each segment of the continuum robot is driven by three tendons. The whole continuum robot is driven by six step motors. The wiring hole is distributed evenly around the joint. Each joint is connected by rotating pair and adjacent rotating pairs are vertically distributed. The Z-axis is on the line where the initial position of the CR located, and the Y-axis is determined by the bending plane where the orientation angle of the first segment of the CR is zero, and the coordinate system is established as shown in Fig. 1.

The control framework of the continuum robot is illustrated in Fig. 2. The whole control system is divided into two parts: control hardware and lower computer, upper computer. The whole control system is controlled by an STM32F407 microprocessor. The main functions of the upper computer include real-time kinematics solving and providing a visual interactive interface. The PC communicates with the lower computer via a serial port, sending step-motor movement data instructions to the lower computer. The IMU JY901 is mounted at the end of the continuum robot, measuring the end pose of the continuum robot. The geometric method is applied to solve the IK of the continuum robot according to the pose obtained by MCU. The effectiveness of the IK algorithm is verified through forward kinematics.

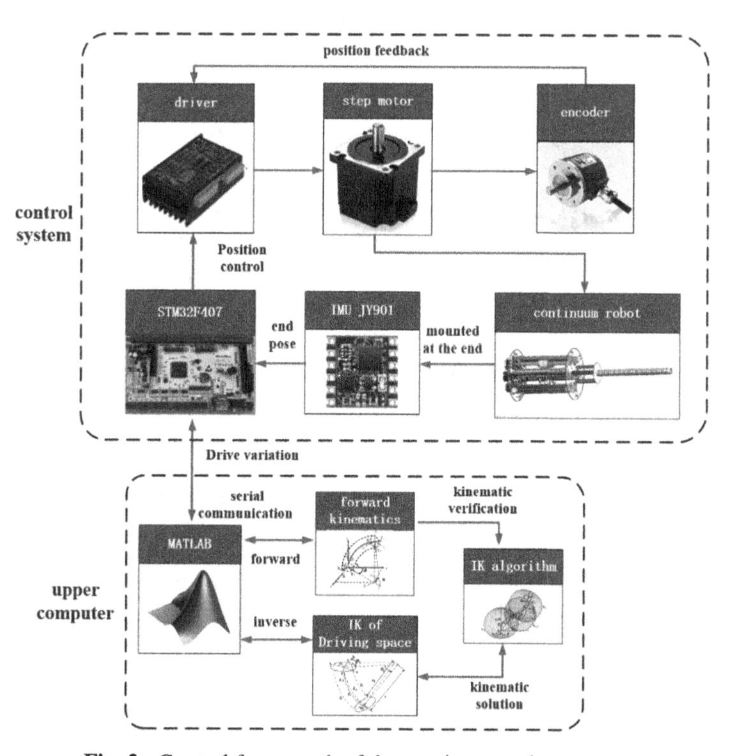

Fig. 2. Control framework of the continuum robot system.

3 Geometric Algorithm for Solving Inverse Kinematics

3.1 Kinematics Model of Continuum Robot

The graphical algorithm is mainly devised for solving the mapping from task space to joint space. To further simplify the calculation, the kinematics model is constructed based on the PCC model. The single-segment of continuum robot is equivalent to an arc, and the length l of the arc is assumed to remain constant. The configuration space of a continuum robot can be described by two parameters: bending angle α and direction angle θ. As illustrated in Fig. 3a, the end pose of CR can be obtained by the D-H method.

According to Eq. (1), the end envelope surface of single-segment CR can be obtained. If the base of CR is known to be at point O, the tip point P must be on the surface illustrated in Fig. 2b below, and vice versa.

$$\begin{aligned} T(k) &= \text{Rot}(z, -\theta)\text{Trans}\left(0, \frac{l(1-\cos\alpha)}{\alpha}, 0\right)\text{Trans}\left(0, 0, \frac{l\sin\alpha}{\alpha}\right)\text{Rot}(x, -\alpha) \\ &= \begin{bmatrix} \cos\theta & -\cos\alpha\sin\theta & -\sin\theta\sin\alpha & -l(1-\cos\alpha)\sin\theta/\alpha \\ -\sin\theta & \cos\alpha\cos\theta & \cos\theta\sin\alpha & l(1-\cos\alpha)\cos\theta/\alpha \\ 0 & -\sin\alpha & \cos\alpha & l\sin\alpha/\alpha \\ 0 & 0 & 0 & 1 \end{bmatrix} \end{aligned} \tag{1}$$

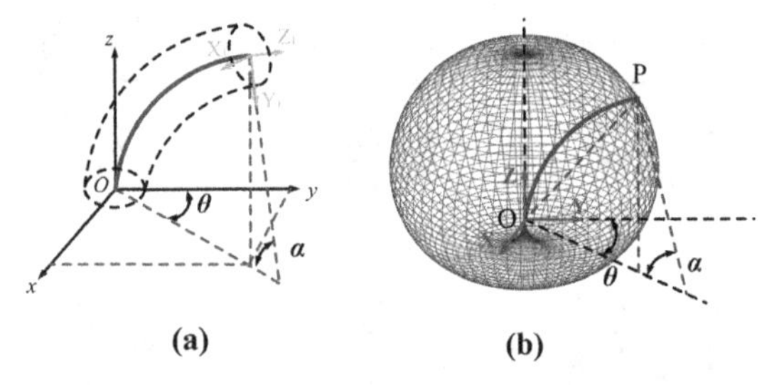

Fig. 3. (a) Kinematics model of CR (b) End envelope surface of single-segment CR

3.2 Inverse Kinematics of Two-Segment Continuum Robot in Space

Surface V_1 and surface V_2 can be constructed based on the final pose and base, and the geometric relationship between the two surfaces can then be judged. Then, auxiliary surface $V_{1\text{-}2}$ can be constructed at the intersection or closest point of two surfaces. The second segment continuum robot must be situated on surface V_2 as well as auxiliary surface $V_{1\text{-}2}$.

As illustrated in Fig. 4a, two surfaces do not have intersections. Find the nearest point B on surface V_1 to surface V_2 and create auxiliary surface $V_{1\text{-}2}$ on that point. Find a point in the auxiliary plane that is tangential to the end pose. The approximate IK can be obtained according to B and C_1.

As shown in Fig. 4b and Fig. 4c, there has the intersection of two surfaces. Make auxiliary surface V_{1-2} on intersecting line S. By traversal, the point C_1 with the same tangential direction and closest to the end pose can be found. Then, the approximate or exact IK can be found.

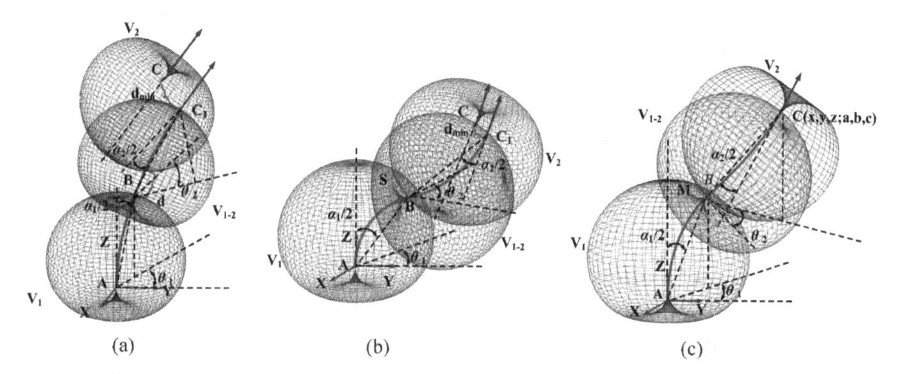

Fig. 4. (a)There is no intersection of two surfaces (b) There is an intersection but tangential non-coincidence (c) There is a tangential coincidence at the intersection.

Find Feasible Solutions

As illustrated in Fig. 5, a set of parallel planes is used to intercept two surfaces. The spacing between parallel planes p_i is h, the distance from point p_i to the plane is d_i, and the points less than 0.5h from the plane are projected onto the current plane. In this way, the spatial problem is transformed into a plane problem.

$$P_i \in p_i \left\{ d_i \leq \frac{h}{2} \right\} \tag{2}$$

As shown in Fig. 6, the k-means clustering method is used to solve the intersections of two curves. The coordinates of each point are $x^{(i)}$. Firstly, find the points where the distance between the two curves is less than a certain threshold. And then the clustering center μ_j is initialized and the classification of each point $c^{(i)}$ is judged.

$$c^{(i)} = \arg\min_j \|x^{(i)} - \mu_j\|^2 \tag{3}$$

The clustering centers μ_j are recalculated according to the attribution of the updated points.

$$\mu_j = \frac{\sum_{i=1}^{m} 1\{c^{(i)} = j\} x^{(i)}}{\sum_{i=1}^{m} 1\{c^{(i)} = j\}} \tag{4}$$

The above steps are repeated until the cost function J is less than a certain threshold. At this moment the clustering center is the intersection of two curves.

$$J(c, \mu) = \sum_{i=1}^{m} \|x^{(i)} - \mu_{c^{(i)}}\|^2 \tag{5}$$

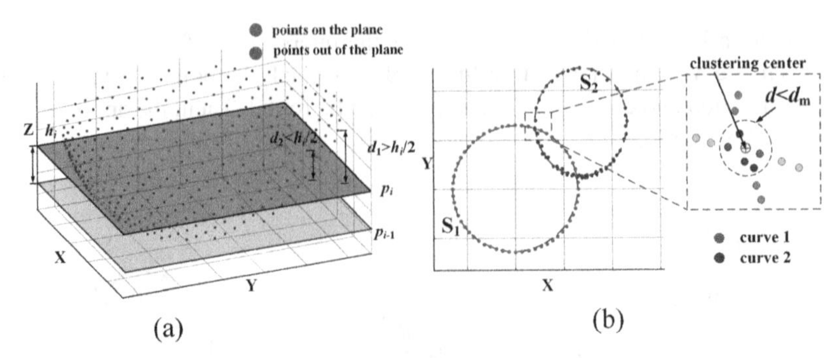

Fig. 5. (a) Project the discrete points onto a plane (b) Schematic diagram of clustering algorithm

Iteration and Exact Solution

Although the exact IK solution can be solved when the sampling rate and accuracy of the clustering algorithm are set high, it will lead to a larger amount of calculation. Therefore, the sampling rate of a surface is reduced to solve the approximate range of the IK solution. Then a more accurate solution is solved with surfaces in the range of approximate solution. A feasible solution with higher precision can be obtained by iteration, and the efficiency and real-time performance of the algorithm can be greatly improved.

As illustrated in Eq. (6), k is the constant that determines the rate of iteration convergence, and ε is the accuracy of the current calculation.

$$\varepsilon_i = k\varepsilon_{i-1} \tag{6}$$

The range of bending angle α and direction angle θ in the next iteration is shown in Eq. (7).

$$\begin{cases} \alpha_i \in (\alpha_s - \varepsilon_i\alpha_o, \alpha_s + \varepsilon_i\alpha_o) \\ \theta_i \in (\theta_s - \varepsilon_i\theta_0, \theta_s + \varepsilon_i\theta_0) \end{cases} \tag{7}$$

Two sections of surface are made within the range of bending angle and direction angle. And the precision of clustering algorithm d_i and spacing between h_i cutting parallel planes is further updated to improve the accuracy of the algorithm.

$$\begin{cases} h_i = \varepsilon_i h_{i-1} \\ d_i = \varepsilon_i d_{i-1} \end{cases} \tag{8}$$

The specific calculation process is as follows. As illustrated in Fig. 6a, the approximate range of IK of CR is calculated. And then make the surface of the next iteration within the solved range. As illustrated in Fig. 6c, another iteration is executed to improve the accuracy of the IK solution. In the specific calculation process, if the requirements for accuracy are not very high, the two iterations can fully meet the requirements. If the distance from the target pose is still greater than a certain threshold after two iterations, it indicates that there is no accurate feasible solution at this pose. The current solution is the approximate IK of CR.

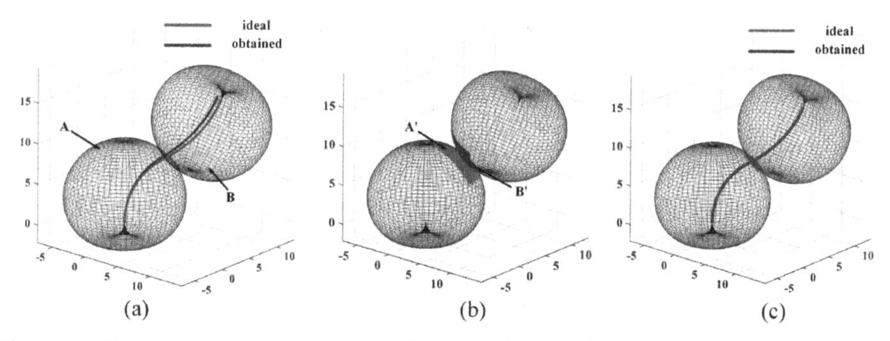

Fig. 6. (a) Obtain approximate solution (b) Make surfaces in the range of approximate solution interval (c) Obtain exact solution.

The general steps for solving the inverse kinematics of CR are illustrated in Fig. 7. Two surfaces are generated based on the orientation of the base and end, and the geometric relationship between the two surfaces is judged. The two surfaces are intercepted by a series of parallel surfaces if there is an intersection between the two surfaces. And k-means clustering algorithm is used to solve the intersection of two surfaces. A feasible solution of CR can be obtained based on the intersection of two surfaces. Another integration is performed until the required accuracy is met.

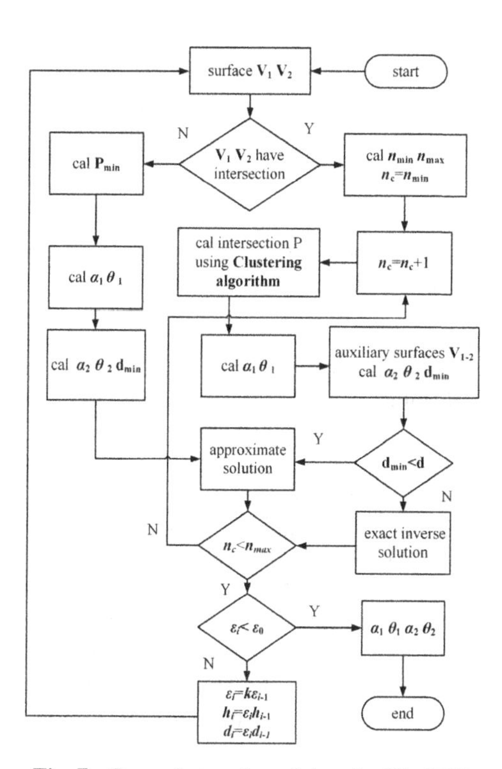

Fig. 7. General steps for solving the IK of CR

4 Comparison and Experiment of Inverse Kinematics Algorithm

4.1 Comparison of Geometric Algorithm

To verify the efficiency and accuracy of the graphical method proposed in the paper, some of the most popular IK methods have been tested against the graphical method. All tests were run on Intel i7-10850 H (2.3 GHZ) processor, 16 GB memory, and MATLAB 2019b equipment. And the convergence threshold of the end is set as 10^{-5} m. Figure 8(a) illustrates the convergence of the geometric method. The number of iterations needed to reach the target against the distance between the target and the end effector is illustrated as follows. For reachable targets, the algorithm has a faster convergence speed and the average number of iterations is 4.14 times. And for unreachable targets, the distance from the target tends to be constant after the third iteration. Thus, the existence of a feasible solution can be inferred by convergence results. The time distribution of the graphical method is shown in Fig. 8(b). The end pose of CR can be calculated by forward kinematics and the IK of the end pose is further solved. The average calculation time of this algorithm is 101 ms, and the calculation time is approximately following a normal distribution.

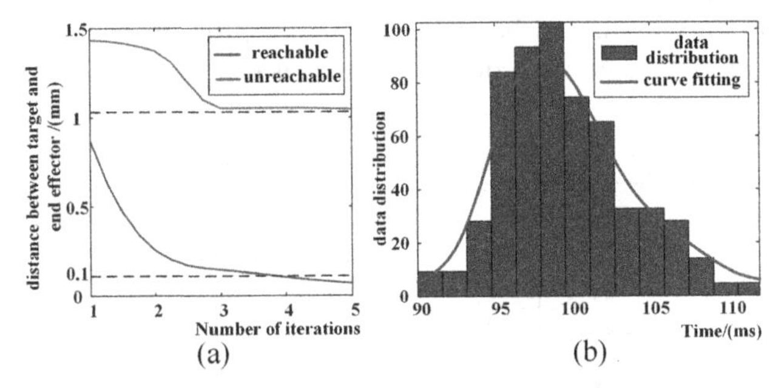

Fig. 8. (a) The number of iterations needed to reach the target against the distance between the target and end effector (b) Time distribution of graphical method (100 runs)

Table 1 depicts the computational efficiency of each method, which means the time required to obtain the IK solution with the requested degree of accuracy. The mean running time of the geometric method is 101 ms, which is 26.5 times faster than the Jacobian method and 2.8 times faster than the FTL method. Due to the iterative strategy of gradually improving the accuracy is used, the geometric method has a faster iterative convergence rate. Simultaneously, the suitable sampling rate of the surface can save the calculation time of a single iteration. The pseudo-inverse of the Jacobian matrix is calculated in each iteration, so the single iteration of the Jacobian algorithm takes a long time. The slower convergence rate of the Jacobian method leads to its poor real-time performance. And there may also be a non-convergence and singular pose when the Jacobian method is used to solve the IK. The single-step iteration time of the FTL algorithm is similar to that of the geometric method. The process of iteration depends largely on the current pose. Therefore, non-convergence may occur in the process of iteration.

Table 1. Average results (100 runs) for solving the IK with different methods

	Graphic	Jacobian	FTL
Number of iterations	4.1	127.8	11.9
MATLAB exe. time/(s)	101.0	2683.3	282.3
Interactions per seconds	41.0	21.0	42.2
Time (per interaction)/(ms)	24.4	47.6	23.7

4.2 Experimental Verification of Algorithm

Since it is difficult to measure the configuration parameters of CR directly, the indirect method is applied to measure and analyze. The change of driving rope length can be calculated according to the parameters of configuration space. And control the CR move to the specified pose. A gyroscope is attached to the end of CR to measure the end pose. The geometric method is applied to calculate the configuration parameters of CR.

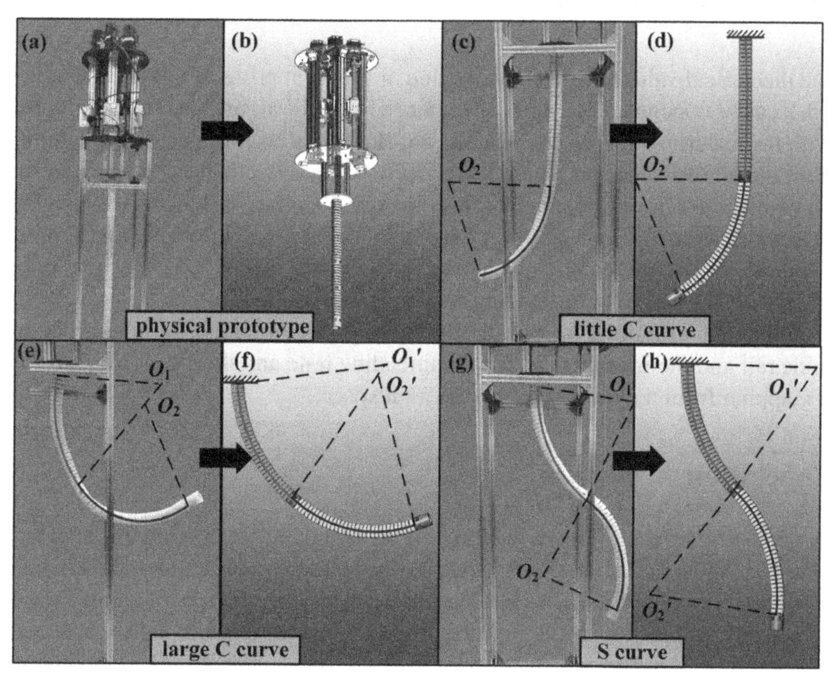

Fig. 9. Experimental verification of IK algorithm (a–b) physical prototype (c–d) little C-curve (e–f) large C-curve (g–h) S-curve

As illustrated in Fig. 9, a set of posture curves for continuum are designed to verify the performance of the algorithm, which includes 'little C-curve', 'large C-curve', and 'S-curve'. Figure 9 c e g are the configurations achieved by physical prototype and Fig. 9 d f h are inverse kinematics obtained by the graphic method. The 'little C-curve' of CR is illustrated in Fig. 9 c-d, the root segment of CR remains stationary and the end segment bends into a specified arc. Figure 9 e–f is a schematic diagram of the 'large C-curve' of CR, the two segments of CR complete the 'small C-curve' motion in the same direction. The 'S-curve' of CR is shown in Fig. 9 g–h, the two segments of CR complete the 'small C-curve' motion in the opposite direction.

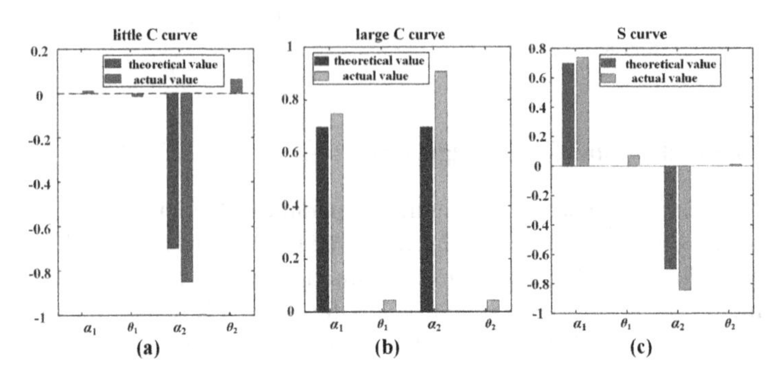

Fig. 10. Actual and calculated values of configuration parameters of continuum robot

The theoretical value and calculated value of different poses of CR are illustrated in Fig. 10. It can be concluded that the pose obtained by the IK algorithm is relatively consistent with the experimental results. And the absolute error of configurational parameters is within the allowable range which further proved the effectiveness of the algorithm. On the whole, the error of bending angle is greater than that of direction angle. This is owing to the deformation caused by gravity cannot be ignored in practical experiments. Therefore, there is a systematic error in the PCC model itself. And this type of error increases with the increase of bending angle. This leads to a larger error in 'large C-curve' compared with other poses. Simultaneously, due to the positioning accuracy of the guide rail and the friction between the threading hole and the alloy rod, errors exist in the mapping from the drive space to the joint space.

5 Conclusion

A new geometric method for solving the inverse kinematics of continuum robot is proposed in this paper. Based on the Piecewise Constant Curvature model, the kinematics model of the continuum robot is constructed and the envelope surface of a single segment is calculated. Combined with the geometric and numerical method, a distinct sequence is proposed in this paper, which is well fitted for two segment continuum robots in space. And the algorithm can also be applied to obtain approximate feasible solutions to unreachable targets. The efficiency and real-time performance of the algorithm are further ameliorated through iterative strategies that gradually increase accuracy. The efficiency and accuracy of the geometric method are validated through numerical comparison and prototype experiments. Compared with the traditional Jacobi algorithm, this algorithm has higher efficiency and no non-convergent or singular solution. As this algorithm is capable of definite geometric meaning and better real-time performance, so it is easy to be accepted and applied. Therefore, the graphical method provides a new scheme for the real-time control of a two-segment continuum robot. For multi-segment continuum robots, further research can be carried out based on the geometric method which treats the two-segment as a subproblem.

Funding. This work was supported in part by the National Natural Science Foundation of China (Grant No. U1813221) and the National Key R&D Program of China (Grant No. 2019YFB1311200).

Appendix

Table 2.

Table 2. Parameters of the physical prototype

Parameters	Value	Parameters	Value
The outer diameter of joint	40 mm	The inner diameter of joint	30 mm
The pitch circle of wiring hole	35 mm	Joint height	15 mm
Diameter of wiring hole	1.5 mm	Diameter of the driven rope	0.8 mm
Length of the root control segment	442.5 mm	Joint number of the root control segment	32
Length of the end control segment	437.5 mm	Joint number of the end control segment	31

References

1. Pritts, M.B., Rahn, C.D.: Design of an artificial muscle continuum robot. In: IEEE International Conference on Robotics and Automation, 2004. Proceedings. ICRA'04, pp. 4742–4746 (2004)
2. Chikhaoui, M.T., Burgner-Kahrs, J.: Control of continuum robots for medical applications: state of the art. In: ACTUATOR 2018; 16th International Conference on New Actuators, pp. 1–11. VDE (2018)
3. Buckingham, R., Graham, A.: Nuclear snake-arm robots. Ind. Robot: an Int. J. **39**, 6–11 (2012)
4. Burgner-Kahrs, J., Rucker, D.C., Choset, H.: Continuum robots for medical applications: a survey. IEEE Trans. Rob. **31**(6), 1261–1280 (2015)
5. Xu, K., Simaan, N.: An investigation of the intrinsic force sensing capabilities of continuum robots. IEEE Trans. Robot. **24**(3), 576–587
6. Buckingham, R., et al.: Snake-arm robots: a new approach to aircraft assembly (No. 2007-01-3870). SAE Technical Paper (2007)

7. Dulęba, I., Opałka, M.: A comparison of Jacobian-based methods of inverse kinematics for serial robot manipulators. Int. J. Appl. Math. Comput. Sci. **23**(2), 373–382 (2013)
8. Rucker, D.C., Webster, R.J.: Computing jacobians and compliance matrices for externally loaded continuum robots. In: 2011 IEEE International Conference on Robotics and Automation, pp. 945–950. IEEE (2011)
9. Cobos-Guzman, S., Palmer, D., Axinte, D.: Kinematic model to control the end-effector of a continuum robot for multi-axis processing. Robotica **35**(1), 224–240 (2017)
10. Machado, J.T., Lopes, A.M.: A fractional perspective on the trajectory control of redundant and hyper-redundant robot manipulators. Appl. Math. Model. **46**, 716–726 (2017)
11. Jiang, H., et al.: A two-level approach for solving the inverse kinematics of an extensible soft arm considering viscoelastic behavior. In: 2017 IEEE International Conference on Robotics and Automation (ICRA), pp. 6127-6133 (2017)
12. Melingui, A.: Neural networks-based approach for inverse kinematic modeling of a compact bionic handling assistant trunk. In: 2014 IEEE 23rd International Symposium on Industrial Electronics (ISIE), pp. 1239–1244 (2014)
13. Thuruthel, T.G., Shih, B., Laschi, C., Tolley, M.T.: Soft robot perception using embedded soft sensors and recurrent neural networks. Sci Robot. **4**(26), eaav1488 (2019)
14. George Thuruthel, T., Ansari, Y., Falotico, E., Laschi, C.: Control strategies for soft robotic manipulators: a survey. Soft Rob. **5**(2), 149–163 (2018)
15. Yahya, S., Moghavvemi, M., Mohamed, H.A.: Geometrical approach of planar hyper-redundant manipulators: inverse kinematics, path planning and workspace. Simul. Model. Pract. Theory **19**(1), 406–422 (2011)
16. William II, R.L., Mayhew IV, J.B.: Obstacle-free control of the hyper-redundant nasa inspection manipulator. In: Proc. of the Fifth National Conf. on Applied Mechanics and Robotics, pp. 12–15 (1997)
17. Cho, C.N., Jung, H., Son, J., Sohn, D.K., Kim, K.G.: An intuitive control algorithm for a snake-like natural orifice transluminal endoscopic surgery platform: a preliminary simulation study. Biomed. Eng. Lett. **6**(1), 39–46 (2016)
18. Xiong, Z., Tao, J., Liu, C.: Inverse kinematics of hyper-redundant snake-arm robots with improved tip following movement. Robot **40**, 37–45 (2018)
19. Sreenivasan, S., Goel, P., Ghosal, A.: A real-time algorithm for simulation of flexible objects and hyper-redundant manipulators. Mech. Mach. Theory **45**(3), 454–466 (2010)
20. Gong, Z., et al.: A soft manipulator for efficient delicate grasping in shallow water: Modeling, control, and real-world experiments. The Int. J. Robot. Res. **40**(1), 449–469 (2021)
21. Jones, B.A., Walker, I.D.: Kinematics for multisection continuum robots. IEEE Trans. Rob. **22**(1), 43–55 (2006)
22. Webster, R.J., III., Jones, B.A.: Design and kinematic modeling of constant curvature continuum robots: a review. The Int. J. Robot. Res. **29**(13), 1661–1683 (2010)

All-in-One End-Effector Design and Implementation for Robotic Dissection of Poultry Meat

Han Liu and Jingjing Ji(✉)

State Key Laboratory of Digital Manufacturing Equipment and Technology, Huazhong University of Science and Technology, Wuhan 430074, Hubei, China
jijingjing@hust.edu.cn

Abstract. The meat dissection is yet a labor-intensive operation in the poultry industry all around the world, and the technologies making it unmanned are emerging. A low-cost robotic poultry meat dissection processing cycle, and its end-effector, are proposed in this work. An all-in-one end-effector capable of multi-configuration for a series of dissection tasks, e.g., off-shackle, cone-fixture, and butterfly-harvesting, is implemented for a single manipulator. The structural operation environment in the processing line is highly utilized as a priori; that way the demand for complicated perception capability of a robot is significantly reduced. As a whole, the hardware and processing proposed is developed with an in-depth adaptation to the industry demand. An experimental platform testing the robotic poultry dissection was settled up, and the processing cycle and hardware proposed was validated. The manipulator-harvested butterfly obtained in our platform demonstrates a smoothness quality on the breast separation surfaces that is clearly comparable with that of the manually handled one. It turns out that the robotic dissection processing proposed is of the satisfactory in terms of the industrial application demand.

Keywords: Robotic dissection · Poultry processing · Low cost · End-effector

1 Introduction

The poultry consumption is keeping increasing worldwide [1]. The annual poultry production in China reached 23.61 million tons in 2020. It continues its growth rate 5%–10% year on year [2]. At present, the meat dissection is yet a labor-intensive operation in the poultry industry all around the world. But the technologies making it unmanned are emerging in the sight vision of the researchers, in response to the booming labor cost. The recent outbreak COVID-19 pandemic [3], together with the reduction of the robot prices, makes the trend of robotization in poultry dissection even accelerated. However, the poultry objects are nature unstructured, which means they highly differ individually with intricate interfaces among bones and tissues. This is in general not the case in a conventional robotic application facing consistent mechanical parts. The traditional manual handling of poultry dissection is indeed a significantly comprehensive task employing

H. Liu et al. (Eds.): ICIRA 2022, LNAI 13456, pp. 113–124, 2022.
https://doi.org/10.1007/978-3-031-13822-5_11

the vision, touch, and intelligent decision-making capability of a human, which brings a great challenge for robot implementation to replace people. Taking advantage of the structural environment prescribed by the processing line in a poultry facility, and of the a priori data of stipulated operation beat, a robotic poultry dissection processing cycle targeting to the industrial applications is put forward in this paper. The readily-for-use point-to-point trajectory planning capability of a serial robotic manipulator is utilized while all other complex operation functionalities being allocated to an elaborately realized end-effector. That way a fully unmanned poultry dissection cycle, including a series of tasks like grasping, emplacing, and manipulating, is established at a low cost using one serial manipulator and its one end-effector.

Some related research work has been reported in recent decade. Zhou *et al.* [4] identified the cutting locations on chicken front halves by studying the anatomic structure of the shoulder. They then developed a 3-DOF automated cutting device for butterfly (wings and breast), capable of size adaptation and deformation compensation with the help of force controlled robot. Based on a comprehensive statistical analysis, the Georgia Tech Research Institute's Food Processing Technology Division successfully predicted the internal shoulder joint location of a chicken with the help of two wing-body attachment points and the keel point, at an error less than 3mm. Hu *et al.* [5] thereby identified the characterizing points then the cutting path by way of geometric deduction. The deviation from nominal cutting path was corrected by force control thus that the shoulder deboning was elaborately conducted. Maithani *et al.* [6] assisted the industrial meat cutting task in conjunction with a physical Human-Robot Interaction strategy. An impedance-control-based system was built in which the cutting trajectory was controlled manually with the assisting force provided by a manipulator. The adaptation of the cutting on various meats was thus realized.

For a human handling of poultry dissection, one holds the meat by one hand, and do the cut by the other. Multiple technologies as aforementioned, *e.g.*, computer vision, force control, and intelligent decision-making, have been employed to make a machine capable of simulating this human handling. However, getting a machine dissector fully humane-like using these sophisticated strategies escalated cost of development, and it is to some extent out of line with the actual industry demand. Taking this into regard, we propose an intuitive tearing-based robotic poultry dissection. Pneumatic gripper is employed here for the grasping and tearing tasks building up the dissection for harvesting. This novel tearing-based processing method poses a challenge on the design and implementation of the end-effector.

For a better grasping and manipulating of objects with irregular shapes, multiple achievements have been presented upon the design of manipulator end-effectors. Monta *et al.* [7] combined a dual-finger with a suction pad so that a suction-drag-pick harvesting manner was formed up. Tomatoes could be this way harvested one by one from a cluster. Ciocarlie *et al.* [8] proposed an enveloping finger grasper capable of holding objects of different shapes or sizes, by additional degree of freedoms (DoF) obtained by multiple tendon-driven joints. Hou *et al.* [9] composed a dual-finger grasper by introducing variable stiffness of two negative pressure-controlled jamming soft fingertips. A remarkable grasp stability was acquired using these soft fingertips in a geometrically closure form. In the case of robotic poultry dissection, the chicken's legs and the wings are individual objects of irregular shape and different postures. There are demands in the industry that to grasp leg or wing firmly in the presence of grease, and to leave the object intact in operation. Unfortunately, upon reviewing the literatures, we were unable to identify any commercially available end-effector that is readily capable for fulfilling these demands. In view of this, a novel all-in-one end-effector and its supporting control system is developed in this paper in order to fulfill the grasp requirement aforementioned. An industry-friendly robotic poultry dissection operation test using the end-effector together with a manipulator has been conducted successfully.

The remainder of this paper is organized as follows. Section 2 describes the manipulator-based poultry dissection and the implementation of the end-effector. The robotic dissection consisting of a series of tasks of grasping, emplacing, and tearing is presented together with the hardware realization, utilizing the a priori known structural operation environment in the industry facility. The all-in-one end-effector design covering the complicated operation cycle, including leg-gripping, cone-fixture, and butterfly-harvesting, is described as well. Its configuration switching component and the clamping components are depicted in detail.

Section 3 presents the experimental validation and results discussion. The end-effector realization and its control system are demonstrated. An experimental platform of the robotic poultry dissection automation line was settled up. And the full processing cycle proposed was validated then recorded there. All achievements are concluded in Sect. 4.

2 End-Effector Design for Robotic Dissection of Poultry Meat

The full robotic processing in dissecting a chicken is conceptually depicted in Fig. 1 with its hardware layout. Before each dissection cycle, the overhead conveyor sends a chicken body (with gut removed) into the prescribed location A. Then the manipulator conducts the dissection cycle as a sequence of operations I-II-III-IV-I listed in Fig. 1(b), at a sequence of operation locations A-B-C-D-A in Fig. 1(a), respectively. The yellow arrows give the trajectory of the manipulator end in a cycle. In one dissection cycle, the manipulator (I) clamps the chicken legs at location A then off-shackle it from the conveyor; (II) rotates the chicken to fix it onto a cone at location B then leave it on the cone; then moves from B to C while an individual equipment lifting the wings to an expanded pose; (III) clamps the two lifted wings at C then tears the butterfly (wings and breast) off the carcass; (IV) carries and releases the harvested butterfly to its expected destination D, then moves back to A for the next arriving chicken and a new dissection cycle.

In order to conduct these sequential operations all in one by a manipulator, a novel end-effector is conceived. Its principle is outlined in Table 1. The timing belt delivers the torque intermittently applied by rotary cylinder ② to grippers ③ and ④ that are attached to the two driven shafts, respectively. That way ③ and ④ could be driven rotating simultaneously between two configurations for either clamping legs or wings, each around its own driven shaft axis. Every time the two grippers rotate to a prescribed configuration, a locking pair driven by linear cylinder (pen cylinder) ① activates to align up and hold the grippers against the load torque by the chicken in operation, locking them from any deviation from the configuration. The full CAD model of an end-effector is demonstrated in Fig. 2(a). An end-effector is composed of the configuration-switching component and the clamping component, as conceptually boxed in dashed red lines. The configuration-switching component includes the timing belt system and the locking pair, driven by a rotary cylinder ② (yellow) and a pen cylinder ① (tan), respectively. The clamping component consists primarily of the two pneumatic grippers ③ and ④. The indented jaw on the grippers is carefully designed thus that stable grasp on multiple portions of a chicken body could be achieved.

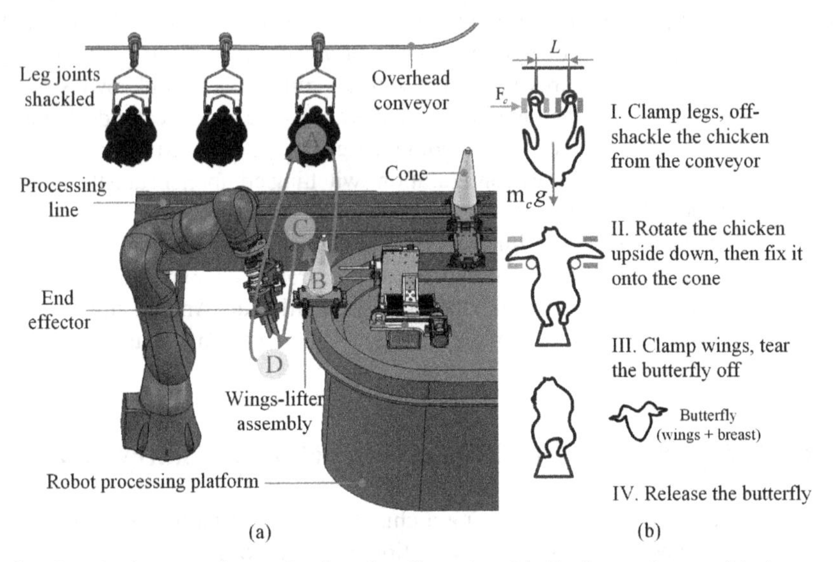

Fig. 1. The robotic processing cycle of poultry dissection. (a): Equipment layout; (b): A sequence of operations I-II-III-IV operations at positions A, B, C, and D.

Table 1. Scheme and outline of the end-effector

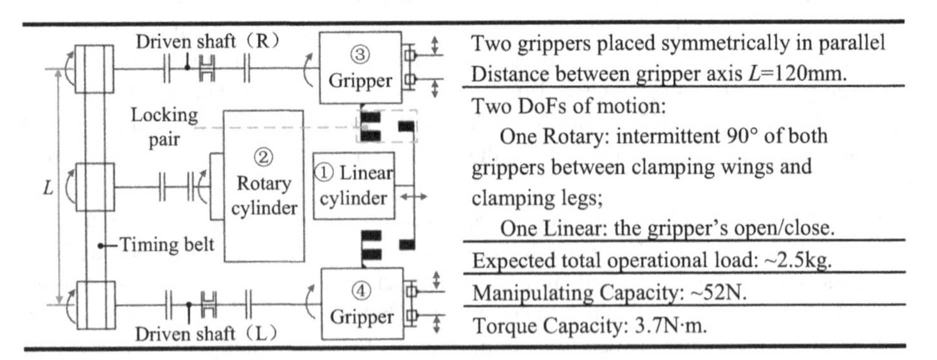

Two grippers placed symmetrically in parallel Distance between gripper axis L=120mm.
Two DoFs of motion: One Rotary: intermittent 90° of both grippers between clamping wings and clamping legs; One Linear: the gripper's open/close.
Expected total operational load: ~2.5kg.
Manipulating Capacity: ~52N.
Torque Capacity: 3.7N·m.

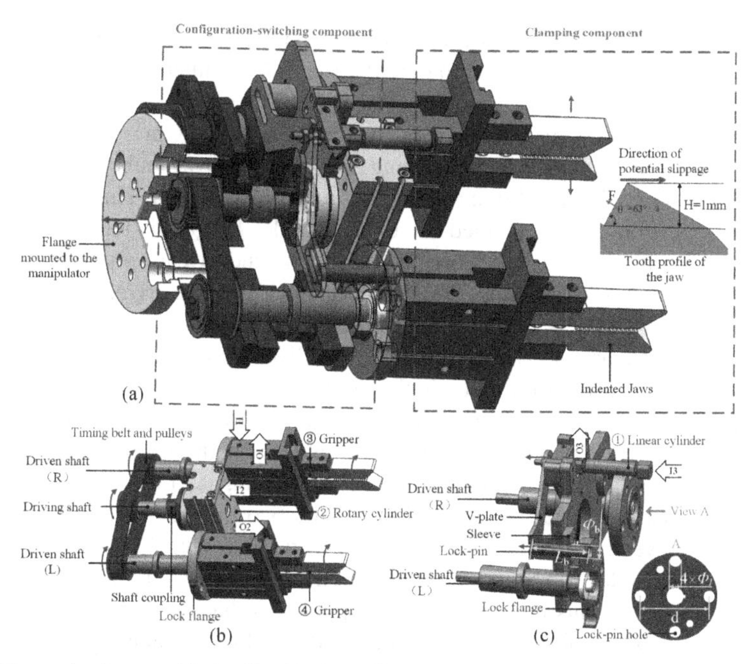

Fig. 2. The end-effector with details. (a): Overall assembly; (b): The timing belt system; (c): The locking pair.

2.1 Configuration-Switching Component

For the rotary cylinder ②, compressed air is fed into Port I2, rotates the output shaft of the cylinder clockwise around the Z-axis (see Fig. 2(a)) before being exhausted from Port O2. The output torque is delivered to the driving shaft via a shaft coupling, rotates the driving pulley clockwise as well, then rotates the two symmetrically located driven pulleys by the timing belt. Each driven pulley is rigidly fixed by a driven shaft to a lock flange to which a gripper is mounted. So that we have a synchronized rotary activity of the output of cylinder ② and both grippers around the Z-axis. The gear ratio of the three pulleys is set to 1:1:1. It is of importance determining the specification of cylinder ②, regarding it being the power driving the grippers' switching between configurations. The overall size of the cylinder lays an intense influence on the principle setup and the structure complexity, in view of the restriction set by the driven base distance L, see Table 1. The data could be obtained as in Eq. (1a) and (1b),

$$T = KI\dot{\omega}, \text{ where } \dot{\omega} = 2\theta/t^2 \tag{1a}$$

$$E_{max} = 0.5I\omega_{max}^2, \text{ where } \omega_{\max} = 2\theta/\mathrm{t} \tag{1b}$$

where T is the required output torque (N·m) for the rotary cylinder in driving its load, *i.e.*, the grippers; $K = 5$ is the load margin factor; I is the moment of inertia (kg·m^2) of the load, namely the grippers and the shaft system of timing belt; $\dot{\omega}$ is the angular acceleration (rad·s^{-2}) stipulated by the rotary angle θ and the transient time t of a configuration switching; E_{max} is the maximum angular kinetic energy of the two grippers and the shaft system of timing belt. In compliance with the manipulation scheme, θ is determined 90°. The transient period t is set to 0.5 s. The moment of inertia of the two grippers and the shaft system is calculated to be $I = 6.4 \times 10^{-4}$ kg·m^2, with a simplifying assumption that the shape of a gripper being approximated by a cube. Substituting all these values into Eq. (1a) yields the required torque $T = 0.08$ N·m. Then $E_{max} = 0.013$ J for two gripper-rotating is obtained. A rotary cylinder HRQ10 is thus selected, with its maximum torque capacity 1.1 N.m > T and kinetic energy capacity 0.04 J > E_{max}.

In the configuration-switching component, the locking pair, driven by a pneumatic linear cylinder (pen cylinder) ①, has the on-off control over the rotary DoF of the clamping component around Z-axis, see Fig. 2(c). Given that air being compressed into the Port O3 and exhausted from Port I3, the piston rod of pen cylinder ① goes linearly in the negative Z-direction. The piston rod and the V-plate then cause a simultaneous movement of the two lock-pins in the same direction. Now both lock-pins align and insert into their corresponding lock-pin holes on the locking flanges to which the grippers are mounted, eliminating the Z-axis rotary DoF of the two grippers. We may regain the rotary DoF by inversing the air flow to withdraw the lock-pins. The V-plate takes its notched shape in order to reduce the part weight. For each configuration switching, a gripper rotates 90°. This requires four lock pin holes that are evenly distributed on the lock flange at 90° apart. The performance of the locking accounts primarily on the stiffness of the pins. And the major load against the pen cylinder ① is just the friction during lock/unlock. The friction may be obtained as in Eq. (2),

$$F_f = m_l g \mu \mathrm{K} \tag{2}$$

where m_l is the effective weight of the moving locking parts (Lock-pin and V-plate); the gravity $g = 9.8$ m·s^{-2}, μ is the friction coefficient; and $K = 5$ again is the friction margin factor. The effective weight could be determined by the 3D model and the associated material property that $m_l = 0.085$ kg. The friction coefficient between aluminum alloy surfaces at dry condition is $\mu = 1.05$. Substituting all these values into Eq. (2) yields the friction $F_f = 4.37$ N. In order to eliminate any unnecessary bend of the piston rod in the presence of some lateral load in operation, a pen cylinder PB10 is selected of which the rod diameter is 4 mm. The piston stroke is determined 20 mm to avoid any collision with other parts.

2.2 Clamping Component and Its Jaw

The clamping component consists of two pneumatic parallel-finger cylinders, or the grippers, and the jaws mounted on the fingers. This component is highlighted by the right dashed red box in Fig. 2(a). Each gripper finger slides with a linear DoF in X or Y-direction. Given that the powering air being fed into Port I1 and flow out from Port O1, the two fingers of a gripper move apart so that the object is released. The two fingers close up to hold the object if the air flow is inversed. The demand that to carry and to manipulate the chicken body firmly and smoothly yields some requirement on the clamping force by the two jaws mounted on the fingers. The clamping force is determined as

$$F_c = m_c g K / 4\mu \tag{3}$$

where $\mu = 0.2$ is the friction factor of chicken skin, m_c measures the weight of the chicken body, the gravity $g = 9.8$ m·s^{-2}, and $K = 4$ is the margin factor of reliability. Substituting these values into Eq. (3) yields the required minimum clamping force of one gripper $F_c = 45.9$ N. A pneumatic gripper HFK32 is selected of which the maximum clamping force reaches 160 N. The capability of HFK32 ensures a secured clamping on the chicken leg even when the body is incorrectly fed to the manipulator by the conveyor with only one leg on the shackle.

A jaw is the interface between the chicken meat and the machine while clamping. The primary concerns during a clamping include eliminating the damage to the tissue and preventing any potential slippage. Some clamping structures that are commonly deployed in medical purpose [10] are referenced in terms of reducing tissue damage. And a peak-to-peak jaw design [11] is employed for prohibiting the slippage. The teeth of jaw are right-angled, see Fig. 2.

The detailed specification of the components that form an end-effector is tabulated in the following Table 2.

Table 2. Specification of an end-effector

Item		Specification
Components for configuration-switching (Rotating the grippers 90 ° intermittently)	Pneumatic rotary cylinder	Model: HRQ10; Angle adjustment range: 0–190°; Repeatable precision: 0.2°; Theoretic moment (@0.5 MPa): 1.1 N·m
	Timing belt and pulleys	Type: S3M; Max tensile load: 780 N; Transmitted power: 0.001–0.9 kW Center-to-center driven base distance (L): 120 mm; Number of pulley teeth: 32; Gear ratio: 1:1:1
Components for locking the grippers	Pneumatic pen cylinder	Model: PB10; Stroke: 20 mm; Thrust (@0.5 Mpa): 23.5 N
	Lock-pin	Material: stainless steel 304; Diameter ($Ø_b$): 10 mm; Length (L_b): 45 mm
	Lock flange	Material: aluminum alloy 6061-t6; Diameter ($Ø_l$): 10 mm; Diameter (d): 70 mm
Components for clamping	Two-finger pneumatic gripper	Model: HFK32; Gripping stroke (both sides): 22 mm; Gripping force: 50–220 N
	3D-printed indented jaws	Material: Resin; Tooth profile: see Fig. 2

3 Experiment and Results Discussion

An experimental platform was settled up in accordance with the robotic dissection hardware configuration and the end-effector specifications conceived. The platform, as demonstrated in Fig. 3, was purposed to validate the poultry dissection processing proposed and the all-in-one end-effector. The platform was comprised of a commercial robot KUKA LBR iiwa 14 R820 to which the end-effector is mounted, a cone, an overhead rail conveyor with a shackle, and the supporting system including an Arduino UNO single chip microcomputer (SCM), four electromagnetic relays, four two-position-five-way solenoid pneumatic valves, and a controlling PC. A horizontal bar located 1.5 m above the ground played the role of overhead conveyor; with a chicken shackle moving freely on it. All relays and pneumatic valves were 24 V DC powered using a regulated supply. The Arduino SCM was employed in controlling the relays' on-off by switching signal level at SCM's ports. And the pneumatic cylinders in the end-effector was controlled by the relays via activating the solenoid pneumatic valves. The communication between the manipulator and the SCM was governed by the PC.

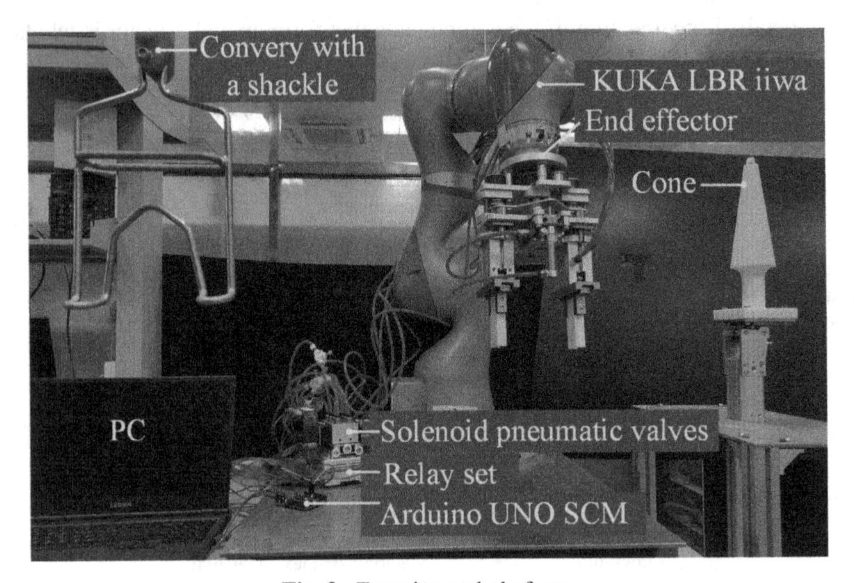

Fig. 3. Experimental platform.

3.1 Motion Control

The tasks that motion control needs to conduct over the end-effector include the clamp-release and the configuration-switching.

For the former one, the motion control on clamp-release involves the control over both the robot and the end-effector. The activity of the robot in a dissection cycle was regulated by manual teaching, and robot's trajectory connecting these clamp-release locations was driven in a point-to-point manner. Every time the manipulator reaches a clamp-release location, it triggers a signal to the Arduino SCM via PC informing that its being ready for an operation. The SCM then switches the voltage applied to the pneumatic valve 1 through relay 1, feeding the air into Port I1 to move both grippers clamping. Note that the controlling air flow is shared by the two grippers, as they are expected always operating shoulder to shoulder. The releasing is similar but in the reversed direction.

For the latter one, the motion control on configuration-switching does not involve the communication between the robot and the SCM. Every time the end-effector completed releasing a butterfly or leg, relay 2 switches pneumatic valve 2 to feed air into Port I2. The pen cylinder rod moves linearly in the positive Z-direction, thereby unlock the rotary DoF of the two grippers around Z-axis. The SCM then triggers pneumatic valve 3 via relay 3 to activate the rotary cylinder to rotate both grippers at 90°, namely the configuration switching. Once the configuration-switching completed, pneumatic valve 2 is switched back again by SCM via relay 2. The pen cylinder's rod then moves back to lock the rotary DoF of the grippers around Z-axis (Fig. 4).

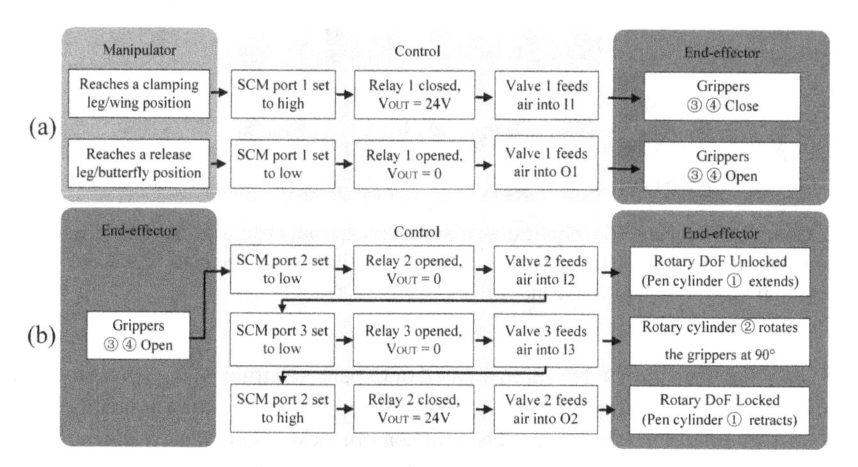

Fig. 4. Scheme of motion control. (a): clamp and release; (b): configuration switching.

3.2 Results and Discussion

Figure 5 illustrates the full operation cycle of the robotic poultry dissection processing proposed. The sequence in a cycle is composed of, off-shackling the chicken body from the overhead conveyor (Fig. 5 i–ii), rotating the chicken by the robot then fixing the chicken onto the cone (Fig. 5 iii–vi), switching the grippers to a wing-clamp configuration (Fig. 5 vii–viii), clamping the chicken at the wings (Fig. 5 ix), and finally tear-harvesting the butterfly off the carcass (Fig. 5 x). The full cycle was conducted 10 times on the platform and a 100% success was achieved. The quality of the manipulator-harvested butterfly is compared with that of a manually handled butterfly in Fig. 6. Based on 10 repeated dissection cycle tests in our experiment, the average time share by each operation step and the average total time cost were calculated and tabulated in Table 3.

Upon reviewing the experiment results as illustrated in Fig. 5, Fig. 6, and Table 3, we reach the discoveries that follow.

(1) This robotic processing methodology takes advantage of the a priori known information of structural environment for determining the prescribed operation location and

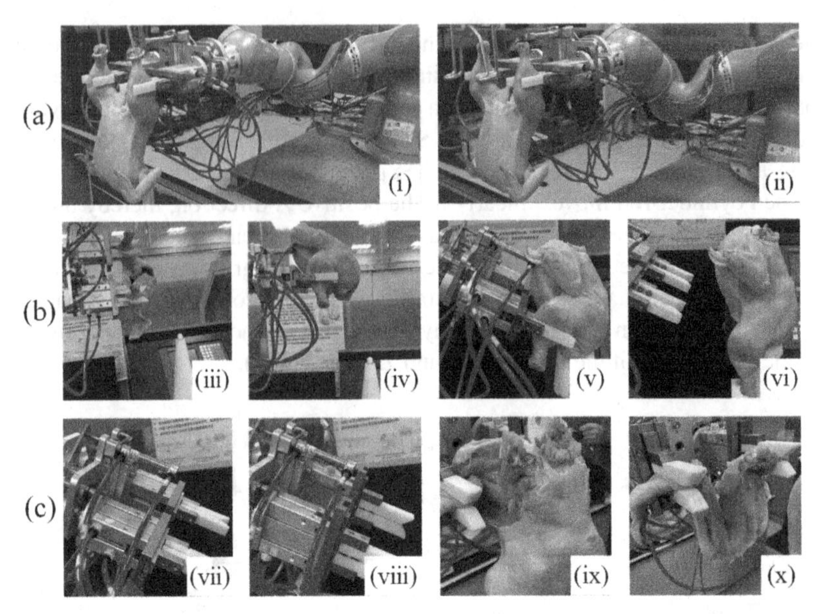

Fig. 5. Complete sequence of a robotic dissection of chicken butterfly. (a): Clamping legs and off-shackling; (b): Rotating the chicken upside down then fixing it onto the cone; (c): Configuration switch, clamping wings and tearing the butterfly off the carcass.

orientation trajectories. No sophisticated or expensive strategy, *e.g.*, machine vision or intelligence algorithm, is necessary in completing the dissection as no more online trajectory planning is required. The chicken object was kept steady during the dissection cycle. No unexpected relative motion of the chicken from either the manipulator or the cone was identified. The completeness of the manipulator-harvested butterfly could put on a par with that of a manually handled one.

(2) Two DoFs are allocated to one end-effector. The firm and stable clamping on either wings or legs could be thus conducted sequentially based on one manipulator. The grippers in the end-effector is capable of enforcing the tearing operation, to get the butterfly off the carcass without unexpected slippage. Nevertheless, the tradeoff for the stable clamping is the obvious indentation marked by the gripper jaw onto the chicken skin. We could thus tell that the air pressure applied is of great significance for a non-damage clamp operation. Comparing with those of the manually dissected chicken, however, the tissue intactness on the separation surfaces and the smoothness on the carcass surface are just satisfied but way to improve [12]. See Fig. 6. The quality deviation suggests a necessity of further optimizing the moving trajectory of the manipulator in tearing, which is yet a linear motion thus far.

(3) As repeated 10 times on our platform, the total time cost for each dissection cycle is averaged around 32 s, in which the total time period occupied by the effective operations, e.g., off-shackle, cone-fixture, clamping and tearing, is averaged 17 s. The remaining 15 s are occupied by the transfer process of the manipulator between operation locations, which accounts for less than half length of a full cycle. The

validation tests were conducted under a laboratory condition. Both end-effector operation and manipulator transferring could be sped up through the subsequent process decomposition.

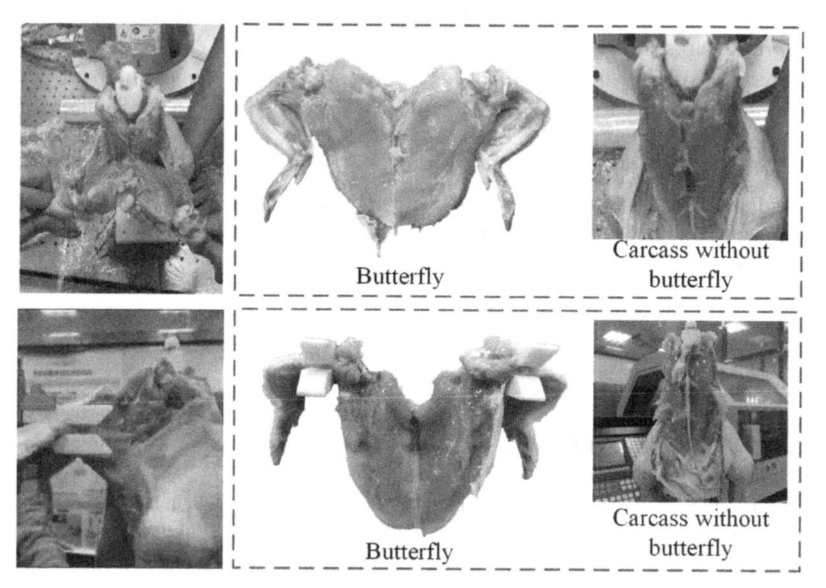

Fig. 6. Results comparison. Upper half: manually handled; bottom half: manipulator-harvested.

Table 3. Time cost in a dissection cycle (in s)

Time cost for	Steps			Sum up
Transferring	From A to B	From B to C and do the configuration-switching	From C to D	15
	9	4	2	
Operations	Cone-Fixture	Clamp and tear the butterfly	Release the butterfly	17
	5	10	2	

4 Conclusion

A novel robotic poultry meat dissection processing is put forward in this paper together with its hardware implementation. An all-in-one multi-purpose end-effector is realized accordingly for the robotic dissection presented. The end-effector is pneumatic powered, capable of off-shackling the chicken from the conveyor, fixing it onto a cone, then clamping, tearing the butterfly off the carcass. The processing designed and the end-effector implemented were validated by the dissection experiments conducted on a test platform. And the non-slippage clamping by the grippers were validated as well, which implies its potential extension to some downstream processing in the poultry industry, promoting the improvement of unmanned poultry lines. The butterfly harvested by the robotic dissection proposed would be later fed to a similar dissection step to get the wings and the breast separated. And the legs would be ripped off the carcass in another step that follows. The robotic dissection processing proposed sets an elementary methodology and an initial possibility for a prospective full-cycle robotic unmanned product-oriented poultry processing line.

Acknowledgements. This work was supported in part by the National Key Research and Development Program of China (Grant No. 2019YFB1311005), the National Natural Science Foundation of China (Grant No. 52175510) and Young Elite Scientists Sponsorship Program by CAST (Grant No. 2019QNRC001).

References

1. Mallick, P., Muduli, K., Biswal, J., Pumwa, J.: Broiler poultry feed cost optimization using linear programming technique. J. Oper. Strateg. Plan. **3**(1), 31–57 (2020)
2. Chen, Y., Feng, K., Jiang, Y., Hu, Z.: Design and research on six degrees of freedom robot evisceration system for poultry. In: the 2020 3rd International Conference on E-Business, Information Management and Computer Science, pp. 382–386. Association for Computing Machinery, Wuhan, China (2020)
3. Noorimotlagh, Z., Jaafarzadeh, N., Martínez, S., et al.: A systematic review of possible airborne transmission of the COVID-19 virus (SARS-CoV-2) in the indoor air environment. Environ. Res. **193**, 110612 (2021)
4. Zhou, D., Holmes, J., Holcombe, W., McMurray, G.: Automation of the bird shoulder joint deboning. In: IEEE/ASME International Conference on Advanced Intelligent Mechatronics, pp. 1–6. Zurich, Switzerland (2007)
5. Hu, A., Bailey, J., Matthews, M., McMurray, G., Daley, W.: Intelligent automation of bird deboning. In: IEEE/ASME International Conference on Advanced Intelligent Mechatronics, pp. 286–291. Kaohsiung, Taiwan (2012)
6. Maithani, H., Ramon, J.A.C., Lequievre, L., Mezouar, Y., Alric, M.: Exoscarne: assistive strategies for an industrial meat cutting system based on physical human-robot interaction. Appl. Sci. **11**(9), 3907 (2021)
7. Monta, M., Kondo, N., Ting, K.: End-effectors for tomato harvesting robot. Artif. Intell. Rev. **12**(1), 11–25 (1998)
8. Ciocarlie, M., et al.: The Velo gripper: a versatile single-actuator design for enveloping, parallel and fingertip grasps. The Int. J. Robot. Res. **33**(5), 753–767 (2014)
9. Hou, T., et al.: Design and experiment of a universal two-fingered hand with soft fingertips based on jamming effect. Mech. Mach. Theory **133**, 706–719 (2019)
10. Wang, J., Li, W., Wang, B., Zhou, Z.: Influence of clamping parameters on the trauma of rabbit small intestine tissue. Clin. Biomech. **70**, 31–39 (2019)
11. Shi, D., Wang, D., Wang, C., Liu, A.: A novel, inexpensive and easy to use tendon clamp for in vitro biomechanical testing. Med. Eng. Phys. **34**(4), 516–520 (2012)
12. Barbut, S.: Pale, soft, and exudative poultry meat—reviewing ways to manage at the processing plant1. Poult. Sci. **88**(7), 1506–1512 (2009)

Motor Imagery Intention Recognition Based on Common Spatial Pattern for Manipulator Grasping

Wenjie Li[1], Jialu Xu[2], Xiaoyu Yan[2], Chengyu Lin[2](✉), and Chenglong Fu[2]

[1] Shenzhen Foreign Languages School, Shenzhen, China
[2] Department of Mechanical and Energy Engineering, Southern University of Science and Technology, Shenzhen, China
11930524@mail.sustech.edu.cn

Abstract. With further development of brain-computer interface (BCI) technology and the BCI medicine field, people with movement disorders can be treated by using artificial limbs or some external devices such as an exoskeleton to achieve sports function rehabilitation. At present, brain-computer interfaces (BCI) based on motor imagination mainly have problems such as accuracy that need to be improved. The key to solving such problems is to extract high-quality EEG signal features and reasonably select classification algorithms. In this paper, the efficient decoding of dichotomy EEG intentions is verified by combining features such as Common Spatial Pattern (CSP) with different classifiers in the PhysioNet public data set. The results show that: CSP is used as the feature, and the support vector machine (SVM) has the best classification effect. The highest classification accuracy of five subjects reaches 87% and the average accuracy reaches 80%. At the same time, the CSP-SVM algorithm was used in the qbrobotics manipulator to conduct a grasping experiment to verify the real-time performance and effectiveness of the algorithm. This provides a new solution for the brain-computer interface to control robotic auxiliary equipment, which is expected to improve the daily life of the disabled.

Keywords: Brain-computer interfaces · Motor imagery · Feature extraction · Common spatial pattern · Classification algorithm

1 Introduction

People with neuromuscular diseases or motor system injuries, such as stroke, and spinal cord injury patients, lack the ability to freely move or limbs controlling, which causes some limitations in daily life. Fortunately, they preserve the relevant nerves that generate motor function similar to healthy individuals. Although the brain regions controlling limb movement have lost some function, the brain nerves that control the limbs still exist [1]. Invasive brain-computer interface (BCI) can monitor and decode the neural activity of the brain such as the sensorimotor cortex, and detect the activity of the brain by implanting bypassing electrodes in the brain. These arrays cover a few square millimeters with a high

H. Liu et al. (Eds.): ICIRA 2022, LNAI 13456, pp. 125–135, 2022.
https://doi.org/10.1007/978-3-031-13822-5_12

signal-to-noise ratio. Many relevant research results have been made [2, 3]. However, these invasive methods face the risk of postoperative complications and infection. So non-invasive BCI is more suitable to assist the disabled improve lost motor ability [4].

At present, many scholars have made fruitful progress in non-invasive BCI. University of Minnesota researchers completed complex grasping tasks by using a non-invasive BCI to control the mechanical arm [5]. The experimenter effectively learned to operate the mechanical arm to grasp and randomly move objects in limited 3D space and was able to maintain their control over multiple objects for 2–3 months. Researchers at the University of South Florida have used a wheelchair-mounted mechanical arm to compensate for the defects of disabled upper limbs, a technology that exceeds the ability of current devices of this type. Expansion of the traditional basic technology is accomplished by controlling this nine-degree of freedom system to improve the daily life of the disabled, which provides scalability for future research [6].

Motor imagery is described as performing motor imagery in the brain rather than truly performing movement, which has naturally occurring characteristics and is so widely used in BCI, moreover, motor imagery can be used not only for the control of artificial limbs by the disabled but also for rehabilitation after stroke [7]. The study of motor imagery is more meaningful and more challenging. However, a brain-computer interface based on motor imagery has some disadvantages such as not high classification accuracy and slow response speed, which make it hard to be applied to practical BCI control.

There are many feature extraction methods for motor imagery, which are broadly divided into three common methods. First, in terms of the feature extraction on the time series and on the frequency, and the mixed feature extraction on the time series and the frequency, the feature extraction on the time series is related to the amplitude of the EEG signal, while the frequency feature is the energy feature of the EEG. Before the feature extraction, the signal should be pre-processed such as re-reference and filtering to improve the quality of the signal. Second, the method of parametric model analysis is used, and this type of method is suitable for online analysis. Third, a spatial domain feature extraction method represented by Common Spatial Patterns (CSP) [2], is mainly aimed at the processing of dichotomy motor imagery EEG. The accuracy of dichotomy motor imagery intention recognition is high, which is recommended by based research [8]. After extracting effective features, accurate classifiers are also required to decode EEG signals. Many effective classifiers have recently emerged, including linear classifiers and nonlinear classifiers. Linear Discriminant Analysis (LDA), regularized LDA, and Support Vector Machines (SVMs) have high efficiency and generalization in motor imagery classification. In particular, SVM can solve both linear and nonlinear problems, is robust to noise, can well solve nonlinear classification and small sample problems, and has a good effect on EEG classification [9, 10]. Therefore, LDA and SVM are widely used in EEG signal classification [11].

In addition to traditional classification algorithms, a lot of research progress on deep learning-based classifiers has also been made [12, 13]. Dose [14] et al. proposed a deep learning model that learns generalized features using convolutional neural networks and classifies using traditional full connected layers with an average accuracy of 68.51%. Chiarell [15] et al. significantly improved BCI performance by using multimodality recordings combined with advanced nonlinear deep learning classification procedures. Wang et al. proposed a classification method based on long and short-term memory, which achieved high accuracy [16]. However, BCI based on deep learning requires a large number of training data to optimize the model, which is not effective in practical applications and is hard to possess a real-time control external device with high response speed, so the traditional classifier is more suitable for real-time BCI control [17].

While remarkable achievements have been made in BCI research based on motor imagery, there is still a need for improved accuracy to achieve highly robust systems. Combining feature vectors with appropriate classification algorithms will improve the accuracy and real-time performance of BCI. As described above, BCI has great application prospects in the field of rehabilitation, but there are still problems of low accuracy and poor stability in motor imagery. In this study, we proposed to organically combine features such as CSP with different classifiers to achieve better classification results, and use qbrobotics mechanical arm grasping experiments to verify the real-time performance and effectiveness of the algorithm.

2 Experiments and Methods

2.1 Experimental Description

Physionet provides a lot of open-source EEG signals. In these experiments, we randomly selected 5 subjects from the EEG motor movement/image dataset. The subjects performed different tasks according to the prompts during the experiment, meanwhile, a 64-channel EEG device was used for data acquisition, the EEG is shown in Fig. 1. Each subject performed the four tasks in each experiment. To realize the recognition of motor imagery, this study selected imagined hands or feet opening or grasping as the experimental data. To unify the data dimension, 4 s of data were selected because each task was performed for about 4 s.

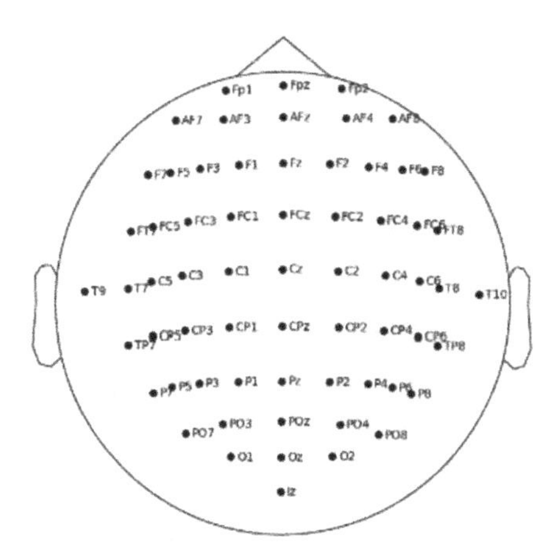

Fig. 1. 64-channel EEG

2.2 Pre-processing

EEG signal is a comprehensive potential signal of the cerebral cortex, which contains a lot of noise and interference. The quality of data can be effectively improved by preprocessing. It is mainly achieved through the following steps:

1) EEG re-reference: The selection of appropriate EEG reference can greatly affect the classification accuracy and sensitivity to artifacts, and the average reference was used to eliminate channel noise in this study.
2) Filter: Before calculating the spatial filter, the EEG signals after re-reference were filtered in the 5–40 Hz band. The data before and after filtering are shown in Fig. 2 and 3. FIR filter was used. The broadband provided better classification results than the narrow band.
3) Electro-oculogram noise removal: Through Independent Component Analysis (ICA), the eye movement components in the EEG were extracted. Eye movement noise often appears on the forehead of the brain, and the quality of the data can be further improved by removing eye movements.

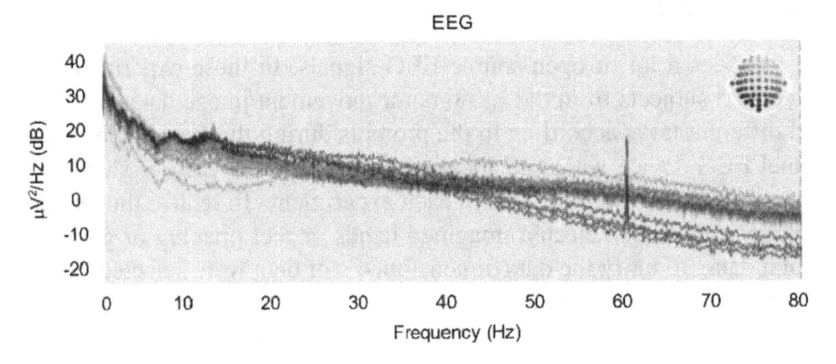

Fig. 2. Channel power spectral density before filtering

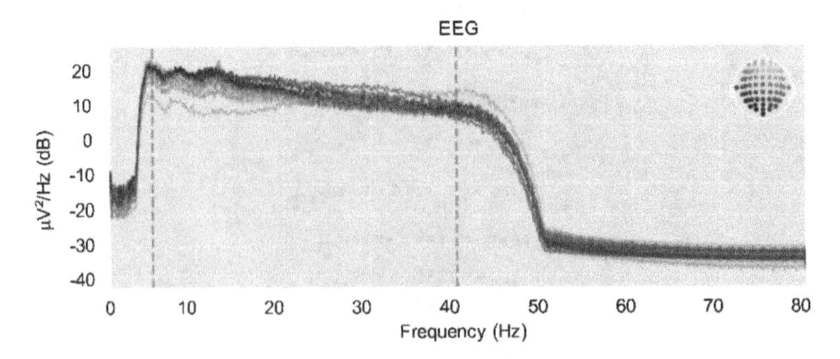

Fig. 3. Channel power spectral density after filtering

2.3 Feature Extraction

The common spatial pattern (CSP) algorithm uses a supervised learning method to create a common space filter. The optimization goal is to maximize one kind of variance and minimize another kind of variance at the same time. It is suitable for feature extraction of the binary classification task.

The process of the common spatial pattern (CSP) algorithm is as follows:

(1) Segment the raw data by category. Data E is binary classified data, E_1 s the first type of EEG data based on motor imagination, E_2 is the second type of EEG data based on motor imagination.

(2) Firstly, the covariance matrix is calculated according to two types of data, and the calculation formula is:

$$C_i = \frac{E_i \cdot E_i^{\mathrm{T}}}{\mathrm{trace}\left(E_i \cdot E_i^{\mathrm{T}}\right)}(i = 1, 2) \tag{1}$$

where trace(E) is the trace of matrix E.

C_1 is the expectation of the spatial covariance matrix of first type EEG data based on motor imagination, C_2 is the expectation of the spatial covariance matrix of second type EEG data based on motor imagination, C_c represents the sum of two covariance matrices, with

$$C_c = C_1 + C_2 \tag{2}$$

(3) Then orthogonal whitening transformation and diagonalization are carried out:

$$C_c = U_c \Lambda_c U_c^T \tag{3}$$

where C_c is a positive definite matrix, U_c is the feature vector, Λ_1 is a diagonal matrix.

Whitening transformation matrix U_c, we can get:

$$P = \frac{1}{\sqrt{\Lambda_c}} \cdot U_c^{\mathrm{T}}. \tag{4}$$

Apply P to $C_1 C_2$, we can get:

$$S_1 = PC_1P^T \tag{5}$$

$$S_2 = PC_2P^T \tag{6}$$

The following conditions are met:

$$S_1 = B\Lambda_1 B^T \tag{7}$$

$$S_2 = B\Lambda_2 B^T \tag{8}$$

$$\Lambda_1 + \Lambda_2 = I \tag{9}$$

where I is an identity matrix.

(4) Compute projection matrix

When S_1 is maximum, S_2 is minimum. The projection matrix is:

$$W = \left(Q^T P\right)^T. \tag{10}$$

(5) The feature matrix is obtained by projection

The feature matrix is obtained by projecting the EEG data of motor imagination:

$$Z_{M\times N} = W_{M\times M} E_{M\times N}. \tag{11}$$

(6) Feature normalization

$$y_i = \log\left(\frac{\mathrm{var}(Z_i)}{\sum_{n=1}^{2m} \mathrm{var}(Z_n)}\right) \tag{12}$$

The data extracted by CSP is shown in Fig. 4.

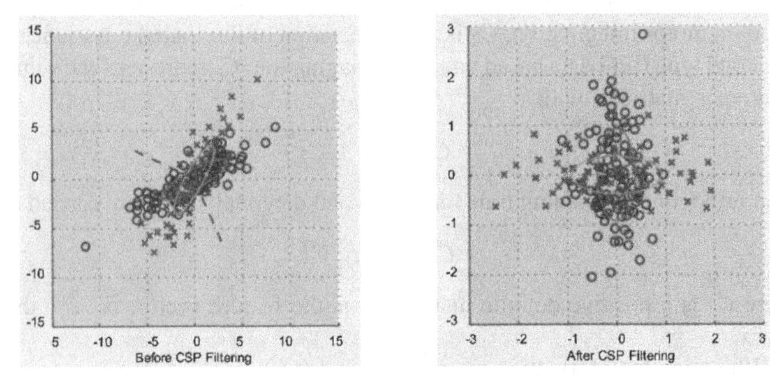

Fig. 4. Data distribution before and after CSP extraction

Source power modulation (SPoC) is also a common feature of EEG signals, similar to CSP. The result of this method is a set of spatial filters W, which directly optimize the power of the target variable using spatial filtering signals.

2.4 Classification Algorithm

Linear Discriminator (LDA) and Support Vector Machine (SVM) are commonly used classifiers in EEG, which are more suitable for real-time BCI systems due to their good generalization and high response rate.

LDA aims to separate different classes of data using hyperplanes. For both classes of problems, the class of the feature vector depends on which side of the hyperplane this vector was located (see Fig. 5). The classification hyperplane was obtained by finding the projection with the largest distance between the two types of means and the smallest variance between the classes. Multiple hyperplanes were used to solve the Class N problem. The strategy commonly used for multiclass BCIs is an OVR strategy that separates each class from all others. This method has low latitude and a small computational load and is suitable for real-time BCI systems. Linear discriminator has the advantage of a high operation rate. This classifier has better generalization capability.

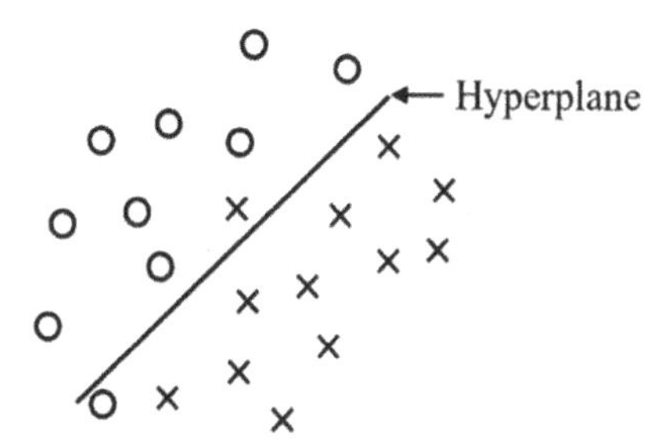

Fig. 5. Dichotomy LDA hyperplane

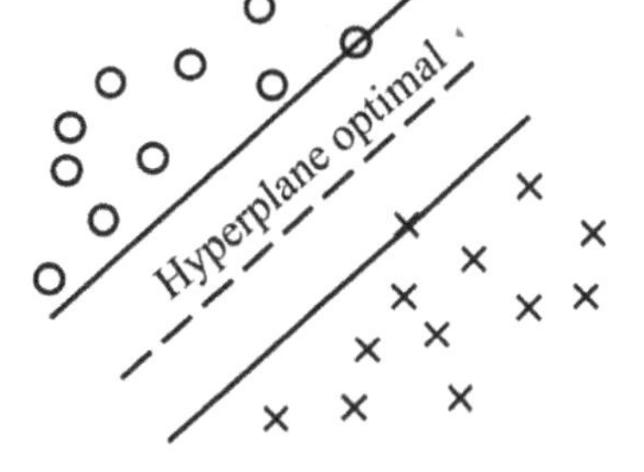

Fig. 6. SVM optimal hyperplane

SVM uses a discriminant hyperplane to distinguish samples of different classes. However, for SVM, the selected hyperplane is the hyperplane that maximizes the edge value as shown in Fig. 6, and it is known that maximizing the edge value can improve the generalization ability. Linear SVM uses linear decision boundaries for classification. Increasing the complexity of the classifier by changing the kernel can create non-linear decision boundaries. The kernel function commonly used in motor imagery classification is Gaussian or Radial Basis Function (RBF) kernel. The RBF kernel-based support vector machine is very effective in BCI classification.

3 Analysis of Experimental Results

The experiment uses the combination of two EEG features (CSP, SPoC) and two classification algorithms (LDA, SVM) to extract and classify the features of motor imagination. Through the combination of feature and classification algorithms, there are four algorithms to evaluate the effect of classification. They are common spatial pattern as the feature, linear discriminator as the classifier (CSP_LDA), common spatial pattern as the feature, support vector machine as the classifier (CSP_SVM), source power modulation as the feature, linear discriminator as the classifier (SPoC_LDA), source power modulation as the feature, and support vector machine as the classifier (SPoC_SVM).

The components of CSP and SPoC are 25 and 20 respectively. SVM uses RBF as kernel function and 5×10 cross-validation results. The average classification results of different algorithms for all subjects are shown in Fig. 7.

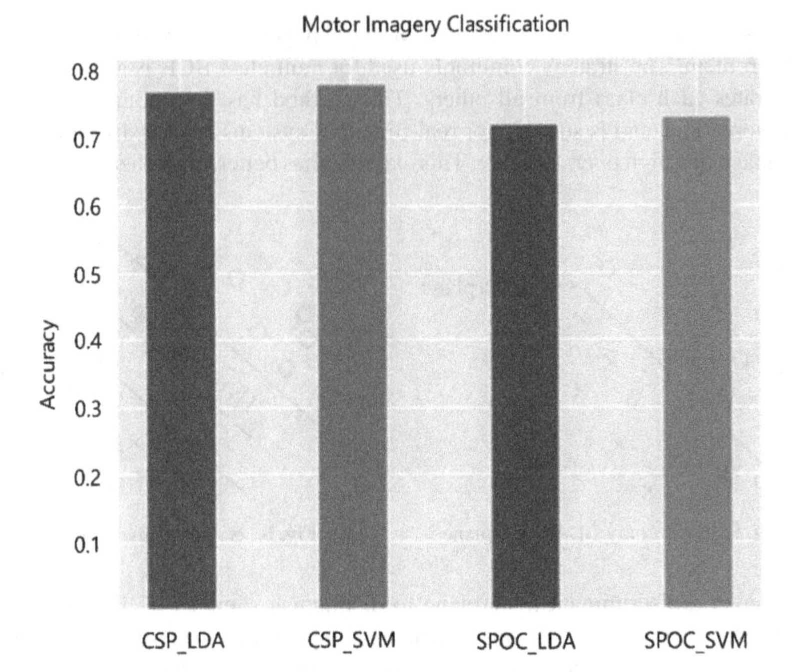

Fig. 7. Classification results with different classifiers

Both CSP and SPoC could extract effective EEG information, while both LDA and SVM had better classification results. The classification results of CSP as feature and support vector machine as the classifier (CSP_SVM) were better.

The specific classification results of the five subjects are shown in Table 1. The proposed feature extraction and classification algorithm gained good results on most subjects.

Table 1. Classification results of 5 subjects

Subject number	CSP_LDA	CSP_SVM	SPOC_LDA	SPOC_SVM	Best accuracy
S1	0.76	0.82	0.73	0.77	0.82
S2	0.86	0.78	0.78	0.77	0.86
S3	0.73	0.80	0.71	0.76	0.80
S4	0.63	0.66	0.67	0.62	0.67
S5	0.87	0.84	0.71	0.72	0.87
Average	**0.77**	**0.78**	**0.72**	**0.73**	**0.80**

4 Validation of Grasping Experiment

Qbrobotics is an anthropomorphic robotic hand based on soft-robotics technology. It is flexible, adaptable, and able to interact with the surrounding environment. The custom-made electronic board inside the qb SoftHand Research is composed of a logic stage for communication and low-level computation, and a power stage for motion control. This, together with a DC motor and its absolute encoder, establishes a simple position and current control feedback regulated by a properly tuned PID controller (shown in Fig. 8).

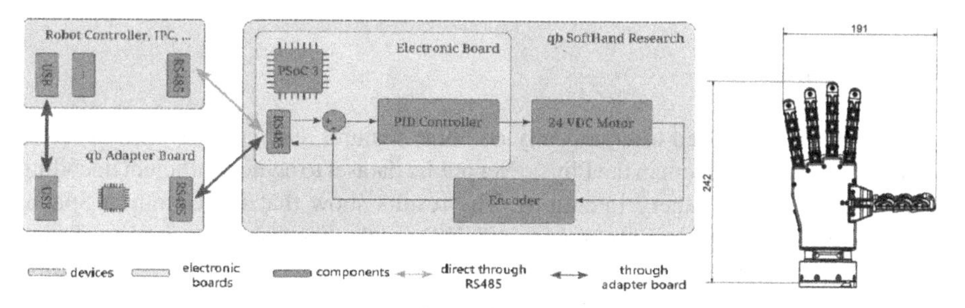

Fig. 8. Qbrobotics hardware simplified scheme

In this study, an application experiment was performed in a practical daily scenario using a qbrobotics mechanical arm. The experimental Scenario is shown in Fig. 9. Users wore 32-channel BrainProducts EEG devices to control the qbrobotics mechanical arm through a motor imagery paradigm. The experiment utilized a sliding window of two seconds to sample the data. Then, by combining the CSP and support vector machine (CSP_SVM), it identified the person's motor imagery intention and imagine bimanual movement to control the grasping or opening of a mechanical arm. The experimental results show that the proposed method in this study can effectively identify the user's motor imagery intention and realize the grasping or opening of the mechanical arm.

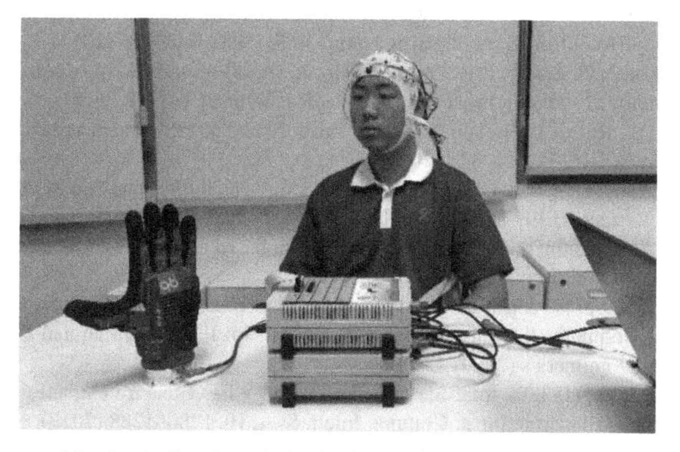

Fig. 9. Verification of qbrobotics mechanical arm grip test

5 Discussion

The robustness of the algorithm remains to be improved due to the small amount of data and the large difference between users. The component selection of CSP has a great impact on feature extraction, and the more components, the more complex the spatial patterns, and the higher requirements for the algorithm. EEG signals are highly correlated with timing and frequency, and in practical applications, appropriate features should be selected according to different task types to improve the robustness of the BCI.

6 Conclusion

In this study, we proposed to organically combine features such as CSP with different classifiers and validate them in the PhysioNet public dataset to achieve efficient decoding of dichotomy motor imagery intentions. The results show that the Common Spatial Pattern (CSP) and support vector machine (SVM) are the best classifiers in the selected data set. The highest classification accuracy reaches 87% and the average classification accuracy reaches 80%. At the same time, the real-time performance and effectiveness of the algorithm in this study were verified in a qbrobotics mechanical arm grasping experiment using CSP-SVM. This study provides a new solution to the application of BCI technology and offers the possibility of improving future life.

Acknowledgements. This work was supported by the National Key R&D Program of China [Grant 2018YFB1305400 and 2018YFC2001601]; National Natural Science Foundation of China [Grant U1913205] and Special fund for the Cultivation of Guangdong College Students' Scientific and Technological Innovation (pdjh2022a0453).

References

1. Hochberg, L.R., Bacher, D., Jarosiewicz, B., et al.: Reach and grasp by people with tetraplegia using a neurally controlled robotic arm. Nature **485**(7398), 372–375 (2013)
2. Wang, L., Liu, X.C., Liang, Z.W., Yang, Z., Hu, X.: Analysis and classification of hybrid BCI based on motor imagery and speech imagery. Measurement **147**, 12 (2019)
3. Grosprêtre, S., Rufno, C., Lebon, F.: Motor imagery and corticospinal excitability: a review. Eur. J. Sport Sci. **16**(3), 317–324 (2016)
4. Zhang, H., Yang, H., Guan, C.: Bayesian learning for spatial filtering in an EEG-based brain–computer interface. IEEE Trans. Neural Netw. Learn. Syst. **24**(7), 1049–1060 (2013)
5. Meng, J., et al.: Noninvasive electroencephalogram based control of a robotic arm for reach and grasp tasks. Sci. Rep. **6**, 38565 (2016)
6. Palankar, M., et al.: Control of a 9-DoF Wheelchair-mounted robotic arm system using a P300 Brain Computer Interface: initial experiments. In: IEEE International Conference on Robotics & Biomimetics (2009)
7. Uyulan, C., Erguzel, T.T.: Analysis of time-frequency EEG feature extraction methods for mental task classification. Int. J. Comput. Intell. Syst. **10**, 1280–1288 (2017)

8. Mahmoudi, M., Shamsi, M.: Multi-class EEG classification of motor imagery signal by finding optimal time segments and features using SNR-based mutual information. Australas. Phys. Eng. Sci. Med. **41**(4), 957–972 (2018). https://doi.org/10.1007/s13246-018-0691-2
9. Chen, M., Tan, X., Li, Z.: An iterative self−training support vector machine algorithm in brain−computer interfaces. Intell. Data Anal. **20**(1), 67–82 (2016)
10. Thomas, K.P., Guan, C., Lau, C.T., Vinod, A.P., Ang, K.K.: A new discriminative common spatial pattern method for motor imagery brain-computer interfaces. IEEE Trans. Biomed. Eng. **56**, 2730–2733 (2009)
11. Garcia, G.N., Ebrahimi, T., Vesin, J.-M.: Support vector eeg classification in the fourier and time-frequency correlation domains. In: Conference Proceedings of the First International IEEE EMBS Conference on Neural Engineering (2003)
12. Li, Y., Zhang, X.R., Zhang, B., Lei, M.Y., Cui, W.G., Guo, Y.Z.: A channel-projection mixed-scale convolutional neural network for motor imagery EEG decoding. IEEE Trans. Neural Syst. Rehabil. Eng. **27**, 1170–1180 (2019)
13. Virgilio, G.C.D., Sossa, A.J.H., Antelis, J.M., Falcón, L.E.: Spiking Neural Networks applied to the classification of motor tasks in EEG signals. Neural Netw. **122**, 130–143 (2020)
14. Dose, H., et al.: An end-to-end deep learning approach to MI-EEG signal classification for BCIs. Expert Syst. Appl. **114**(Dec), 532–542 (2018)
15. Chiarelli, A.M., et al.: Deep learning for hybrid EEG-fNIRS brain-computer interface: application to motor imagery classification. J. Neural Eng. **15**(3), 036028.1–036028.12 (2018
16. Ping, W., et al.: LSTM-based EEG classification in motor imagery tasks. IEEE Trans. Neural Syst. Rehab. Eng. (11), 1–1 (2018)
17. Xu, J., Zheng, H., Wang, J., Li, D., Fang, X.: Recognition of EEG signal motor imagery intention based on deep multi-view feature learning. Sensors. **20**(12), 3496 (2020)

Event-Triggered Secure Tracking Control for Fixed-Wing Aircraft with Measurement Noise Under Sparse Attacks

Guangdeng Chen, Xiao-Meng Li, Wenbin Xiao, and Hongyi Li(✉)

Guangdong Province Key Laboratory of Intelligent Decision and Cooperative Control, School of Automation, Guangdong University of Technology, Guangzhou 510006, China
lihongyi2009@gmail.com

Abstract. This study focuses on the secure tracking control problem of fixed-wing aircraft with multiple output sensors under sparse attacks and measurement noise, where the main challenge is to achieve secure state estimation. An output estimation method based on the Grubbs test is proposed to exclude outliers before estimation to improve the estimated accuracy. Moreover, an event-triggered mechanism is designed to transmit data only when more than half of the measurement data meet the triggering conditions, thereby alleviating the interference of sparse attacks with normal triggering. Based on the event-triggered output estimation, an extended high-gain state observer is designed, which ensures the output prediction accuracy by using the reference output, and estimates the matched model uncertainty as an extended state to decrease the estimation errors. Finally, the tracking controller is designed with the estimated states. The control performance of the proposed method is verified by a numerical simulation.

Keywords: Event-triggered mechanism · Fixed-wing aircraft · Measurement noise · Sparse attacks · Tracking control

1 Introduction

Fixed-wing aircraft have played a vital role in the military and civilian fields, and increasingly stringent standards have been placed on their safety. With the rapid development of electronic countermeasures technology, fixed-wing aircraft need to improve not only their failure tolerance but also their ability to resist malicious attacks. Among the attack types, sensor attacks pose a tremendous threat to flight control systems that rely on sensors to provide accurate flight feedback. Unlike sensor failure signals whose prior knowledge such as differentiable or bounded can be acquired in advance, the sensor attack signal is maliciously manufactured by the adversary, so it is necessary to propose novel control strategies to deal with sensor attacks [1].

H. Liu et al. (Eds.): ICIRA 2022, LNAI 13456, pp. 136–148, 2022.
https://doi.org/10.1007/978-3-031-13822-5_13

Due to the equipment and energy constraints, the adversary cannot attack multiple sensors at the same time, so sensor redundancy strategies are developed to make the attacks sparse [2]. To deal with sparse attacks, many effective secure state estimation methods have been established in recent years. Shoukry et al. [3] designed a multi-modal Luenberger observer based on satisfiability modulo theory to reduce memory usage. In [4], an algorithm based on constrained set partitioning has been proposed to ensure computational efficiency and estimation accuracy. In [5], a multi-observer-based method has been designed for nonlinear systems to ensure robust input-to-state stable estimation. However, the methods in [3–5] are not only computationally intensive, but also restrict that only a subset of sensors can be attacked. When an output is measured by multiple sensors, some less restrictive methods have been proposed. In [6], the output is estimated from the mean value of the safety data, but the computational reduction was not significant enough. The methods proposed in [7] and [8] use the median of all measurements as the output estimation for linear and nonlinear systems, respectively. This estimation method is computationally simple, but the estimated output may contain large measurement noise. Moreover, the methods in [6–8] would encounter huge data processing pressure due to the data to be processed at each sampling moment. The data generated by multiple sensors also consumes considerable communication resources when sensors and controller are connected through a network. Event-triggered control is often used to reduce unnecessary data transmissions [9–11]. In particular, more and more attention has been attracted to sampled-data-based event-triggered (SDET) control, which does not require continuous monitoring of triggering conditions and inherently avoids Zeno behavior [12]. However, when sensors are attacked, it is difficult for the existing triggering strategies to simultaneously maintain control performance and reduce data transmission. Therefore, one of the challenges of this study is to design an effective SDET mechanism and an output estimation method with low computational complexity under sparse attacks.

Since the data from digital sensors and networks are discrete, only discrete output estimates are available even without SDET mechanisms. Therefore, the observer based on continuous output cannot be applied. A method to construct discrete observers has been proposed in [13], but the system needs to be discretized first. In [14], a Luenberger observer has been constructed by maintaining the latest available output, but it is difficult to achieve satisfactory estimation accuracy. An observer without feedback terms that directly corrects the estimation at the sampling moment has been proposed in [15], but the obtained estimated states are not differentiable. Farza et al. [16] proposed a high-gain observer based on output prediction, which not only has high estimation accuracy but also ensures the differentiability of the estimated states. However, this observer is not suitable for SDET output with large output update intervals. Moreover, although high gain observers can suppress the effects of model uncertainty, they are also more sensitive to measurement noise, so the estimation errors cannot be reduced by simply increasing the gain. Therefore, another challenge of this study is to design a suitable high-gain observer for the SDET output, which can reduce the effect of model uncertainty under measurement noise.

Based on the above investigations, in this study, a secure SEDT mechanism is first designed to reduce data transmission, and a data processing method based on the Grubbs test is designed to estimate the system output. Next, an extended high-gain observer based on output prediction is developed to reconstruct the system states, and a tracking controller is designed by the backstepping method. The contributions of this study are described as follows:

(1) A novel triggering strategy is proposed, which decides whether to vote for each measurement data according to the tracking performance, and whether to transmit all data according to the voting result, thereby achieving data transmission reduction and satisfactory tracking performance under sparse attacks.
(2) The accuracy of the output estimation under the SDET mechanism is ensured by introducing the desired output, so that the observer based on the output estimation can be established. Moreover, the matched model uncertainty is estimated as an extended state to improve the estimation performance when the uncertainty is slowly time-varying.
(3) Different from [7,8], this study uses the Grubbs test to exclude outliers before output estimation to improve the estimation accuracy under sparse attacks and measurement noise.

Notation: $\mathbb{N}_+$ is the set of positive integers; sign$(\cdot)$ is the symbolic function; supp(a) represents the number of nonzero elements in vector a; $\mathcal{E}_r$ is an r-dimensional vector whose elements are all one; $\Lambda(f \geq 0)$ is a function such that $\Lambda(f \geq 0) = 1$ if $f \geq 0$ and $\Lambda(f \geq 0) = 0$ otherwise; $\mathbf{Mea}(\bar{\Gamma})$, $\mathbf{Med}(\bar{\Gamma})$, and $\mathbf{S}(\bar{\Gamma})$ represent the mean, median, and standard deviation of $\bar{\Gamma}$, respectively; $G_c(a, b)$ represents the Grubbs critical value, a and b represent the significance level and sample size, respectively; $\bar{\Gamma}\backslash\varsigma$ represents a vector obtained by excluding element ς from $\bar{\Gamma}$; and $\lambda_{\max}(Q)$ and $\lambda_{\min}(Q)$ represent the maximum and minimum eigenvalues of matrix Q, respectively.

2 Problem Formulation

Following [17], the short-period dynamics of a fixed-wing aircraft can be formulated as

$$\begin{aligned}
\dot{x}_1 &= x_2 + \frac{P_\beta}{W} x_1 \\
\dot{x}_2 &= M_\rho u + M_\beta x_1 + M_q x_2 + d_2(x_1, u) \\
y &= x_1
\end{aligned} \tag{1}$$

where x_1 and x_2 represent the angle of attack (AOA) and pitch rate, respectively, $P_\beta = -P_\alpha$, P_α is the lift curve slope at x_1, W is the trimmed airspeed, M_q is the aircraft pitch damping, u is the elevator deflection, and M_ρ and M_β are known parameters. Moreover, model uncertainty is expressed as $d_2(x_1, u) = M_\rho \phi(x_1, u)$, and the unknown nonlinear-in-control effect $\phi(x_1, u)$ is expressed as

$$\phi(x_1, u) = ((1 - C_0)\mathbf{e}^{-\frac{(x_1-\beta_0)^2}{2\sigma^2}} + C_0)(\tanh(u + z_0) + \tanh(u - z_0) + 0.01u)$$

with C_0, β_0, σ and z_0 being unknown parameters.

It is essential to upgrade civilian and military aircraft to resist malfunction and electronic warfare weapons. Therefore, the AOA, a critical state of flight control, is measured by multiple sensors. These sensors are synchronously sampled in time series $\{t_k^s\}_{k=1}^{+\infty}$, and the upper and lower bounds of the sampling interval $\mu_k = t_{k+1}^s - t_k^s$ are $\mu_{\min}$ and $\mu_{\max}$, respectively. Therefore, the output vector of the sensors is

$$\bar{Y}_{a_1}(t_k^s) = [\bar{y}_1, ..., \bar{y}_r]^T = \mathcal{E}_r y(t_k^s) + \upsilon(t_k^s) + a_1(t_k^s)$$

where r is the number of sensors, $\upsilon(t_k^s) = [\upsilon_1(t_k^s), ..., \upsilon_r(t_k^s)]^T$ is the Gaussian white noise vector at t_k^s, and $a_1(t_k^s) = [a_{1,1}(t_k^s), ..., a_{1,r}(t_k^s)]^T$ represents the sensor attack vector or fault vector at t_k^s. If $a_{1,i}(t_k^s) = 0$, it means that the ith sensor is not attacked at t_k^s. The adversary may attack the transmission network of the measurement data, that is, the network output is

$$\bar{Y}_a(t_j^e) = \bar{Y}_{a_1}(t_j^e) + a_2(t_j^e)$$

where t_j^e represents the data transmission instant in the network, and $a_2(t_j^e) = [a_{2,1}(t_j^e), ..., a_{2,r}(t_j^e)]^T$ is the network attack vector. The network will not transmit $a_2(t)$ to the control center until $\bar{Y}_{a_1}(t)$ is received.

The goal of this study is to design an event-triggered secure control strategy for system (5) so that the AOA can track the desired signal $y_r(t)$ with a small error, and the network data transmission is reduced.

The control structure shown in Fig. 1 is constructed to achieve the above goal. The assumptions and lemmas required for this control structure are listed below.

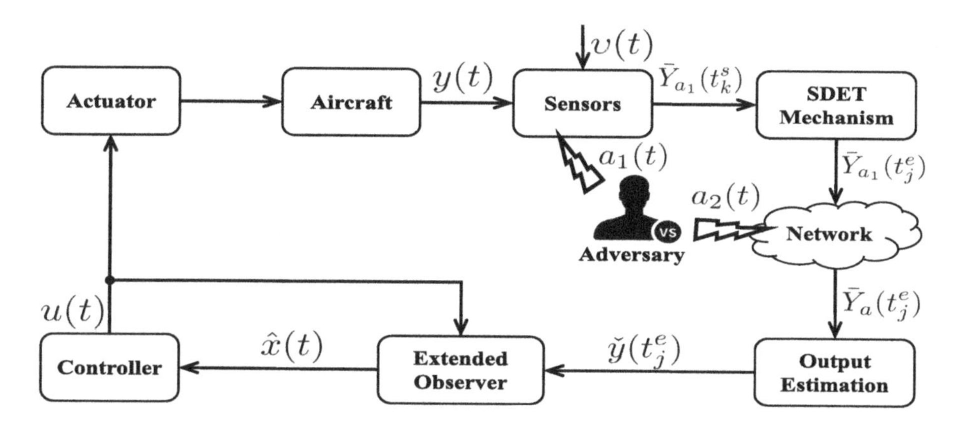

Fig. 1. The structure of an aircraft with the proposed secure control strategy.

Assumption 1 [6]. *a_1 and $a = a_1 + a_2$ satisfy* $\text{supp}(a_1(t_k^s)) < r/2, k \in \mathbb{N}_+$ *and* $\text{supp}(a(t_j^e)) < r/2, j \in \mathbb{N}_+$, *respectively.*

Assumption 2 [18]. *There exist constants $\bar{d}$ and $\bar{v}$ such that $|\dot{d}_2(x_1, u)| \leq \bar{d}$ and $|v_i(t)| \leq \bar{v}$, $i = 1, ..., r$.*

Assumption 3 [19]. *The desired signal and its first to third order derivatives are bounded.*

Lemma 1 [20]. *Consider a function $V(t) \geq 0$ that satisfies*

$$\dot{V}(t) \leq -\varkappa V(t) + \varrho \int_{t_k}^{t} V(\tau) d\tau + \eta, \forall t \in [t_k, t_{k+1})$$

with $k \in \mathbb{N}_+$, $t_0 \geq 0$, where $\varkappa, \varrho > 0$, $\eta \geq 0$, $0 < \mu_{\min} \leq t_{k+1} - t_k \leq \mu_{\max}$ with $\mu_{\max}$ and $\mu_{\max}$ being positive constants. Let $\kappa = (\varkappa - \varrho\mu_{\max}) \mathrm{e}^{-\varkappa\mu_{\max}}$. If $\kappa > 0$,

$$V(t) \leq \mathrm{e}^{-\kappa(t-t_0)} V(t_0) + \eta\mu_{\max} \frac{2 - \mathrm{e}^{-\kappa\mu_{\min}}}{1 - \mathrm{e}^{-\kappa\mu_{\min}}}, \forall t \in [t_0, +\infty)$$

Lemma 2 [21]. *Consider the tracking differentiator*

$$\begin{aligned} \dot{\varphi} &= \varphi_d, \\ \dot{\varphi}_d &= -R^2 \text{sign}(\varphi - \alpha) |\varphi - \alpha|^{b_1} - R^2 \text{sign}\left(\frac{\varphi_d}{R}\right) \left|\frac{\varphi_d}{R}\right|^{b_2} \end{aligned} \tag{2}$$

where R, b_1, and b_2 are positive parameters, b_1 and b_2 satisfy $b_1, b_2 \in (0, 1)$ and $(1 + b_1)(2 - b_2) = 2$. If there exists a time series $\{t_k\}_{k=1}^{+\infty}$ such that $\dot{\alpha}(t)$ and $\ddot{\alpha}(t)$ are bounded on time intervals (t_k, t_{k+1}), $k \in \mathbb{N}_+$, and the right derivatives $\dot{\alpha}(t_k^+)$ and $\ddot{\alpha}(t_k^+)$ and left derivatives $\dot{\alpha}(t_{k+1}^-)$ and $\ddot{\alpha}(t_{k+1}^-)$ are bounded, then for any $l \in (0, \min_{0 \leq k < +\infty}(t_{k+1} - t_k))$, there exists a positive constant σ such that $|\varphi(t) - \alpha(t)| \leq \sigma$, $t \in (t_k + l, t_{k+1})$.

3 Main Results

3.1 Secure SDET Mechanism Design

To reduce network data transmission, a novel triggering mechanism that can attenuate the impact of attacks and noise is designed as

$$\begin{aligned} \Phi_i(t_k^s) &= |\bar{y}_i(t_k^s) - y_r(t_k^s)| - \vartheta_i(t_k^s), t_k^s \in [t_j^e, t_{j+1}^e) \\ t_{j+1}^e &= \inf\left\{ t_k^s > t_j^e | \sum_{i=1}^{r} \Lambda(\Phi_i(t_k^s) \geq 0) > \frac{r}{2} \right\}, t_1^e = 0 \end{aligned} \tag{3}$$

where $\Phi_i(t_k^s) \geq 0$ and $\vartheta_i(t_k^s) > 0$ are the triggering condition and the time-varying threshold of the ith sensor, respectively, and there exists a constant ϵ such that $\vartheta_i(t_k^s) \leq \epsilon$. SDET mechanism (3) ensures that measurement data are transmitted over the network only when more than half of the sensors meet their triggering conditions.

3.2 Output Estimation Algorithm Design

For conciseness, (t_j^e) is omitted below when it does not cause ambiguity. Let $\bar{Y}_a$ be represented as $\bar{Y}_a = [\gamma_1, \ldots, \gamma_r]^T$. To estimate output from $\bar{Y}_a$ under measurement noise and sparse attacks, an output estimation algorithm based on the Grubbs test is designed as follows:

Algorithm 1. Output Estimation Against Sparse Attacks and Noise

Input: $\bar{Y}_a$, a, r;
Output: $\check{y}$, Γ, b;
1: Set $\Gamma := \bar{Y}_a$, $b := \text{supp}(\Gamma)$, $\bar{\gamma} := \arg\max_{\gamma_j \in \Gamma}\{|\gamma_j - \mathbf{Mea}(\Gamma)|\}$, $G(\Gamma) := |\bar{\gamma} - \mathbf{Mea}(\Gamma)|/\mathbf{S}(\Gamma)$;
2: **while** $b > 1 + r/2$ and $G(\Gamma) > G_c(a, b)$ **do**
3: $\Gamma := \Gamma \backslash \bar{\gamma}$, $b := \text{supp}(\Gamma)$, $\bar{\gamma} := \arg\max_{\gamma_j \in \Gamma}\{|\gamma_j - \mathbf{Mea}(\Gamma)|\}$, $G(\Gamma) := |\bar{\gamma} - \mathbf{Mea}(\Gamma)|/\mathbf{S}(\Gamma)$;
4: **end while**
5: **if** $b \leq 1 + r/2$ **then** $\check{y} := \mathbf{Mea}[\Gamma]$;
6: **else** $\check{y} := \mathbf{Med}[\Gamma]$;
7: **end if**

Theorem 1. *Consider system output subjected to attacks and noise, if Assumption 2 holds, then according to Algorithm 1, for all $j \in \mathbb{N}_+$, we have*

$$y(t_j^e) - \omega_M(t_j^e) \leq \check{y}(t_j^e) \leq y(t_j^e) + \upsilon_M(t_j^e), \upsilon_M(t_j^e) = \max\{|\upsilon_i(t_j^e)|, i = 1, ..., r\} \quad (4)$$

Proof. Let $\varpi = [\varpi_1, \ldots, \varpi_b]^T = \Gamma - \mathcal{E}_b y$ with Γ and b being the algorithm outputs. The proof will be carried out in three cases. 1) No data are excluded. We define $\breve{\varpi} = \mathbf{Med}(\varpi)$, then $\check{y} = y + \breve{\varpi}/2$. Under Assumption 1, we can always find two non-attacked elements ϖ_m and ϖ_n in ϖ such that $\varpi_m \leq \breve{\varpi} \leq \varpi_n$. Therefore, we have $y - (|\varpi_m + \varpi_n|)/2 \leq \check{y} \leq y + (|\varpi_m + \varpi_n|)/2$, then inequality (4) holds. 2) The excluded data includes data that is not attacked. Define the excluded non-attacked data as γ_E and let $\varpi_E = \gamma_E - y$. Therefore, $|\varpi_j| \leq |\varpi_E| \leq \upsilon_M$ holds for $j = 1, ..., b$. Obviously, inequality (4) holds. 3) The excluded data are all attacked data. If $b > 1 + r/2$, similar to case 1, we can prove that inequality (4) holds. If $b \leq 1 + r/2$, all elements in $\bar{\Gamma}$ are not attacked, inequality (4) is established. ■

3.3 Extended High-Gain Observer Design

Since the negative impact of noise increases with gain, the high-gain observer cannot rely on increasing gain to suppress model uncertainty. As a result, the model uncertainty will reduce the state estimation accuracy and increase the tracking error, which in turn leads to the performance degradation of the tracking performance-based mechanism (3). To estimate the system states and $d_2(x_1, u)$, system (1) is reformulated as

$$\begin{aligned} \dot{x} &= Ax + Bu + d(x,u) \\ y &= Cx \end{aligned} \quad (5)$$

where $x = [x_1, x_2, x_3]^T$ with $x_3 = d_2$ is the unavailable state vector, $d(x,u) = [0, 0, \dot{d}_2(x,u)]^T$ is the unknown nonlinearity. $A = [P_\beta/W, 1, 0; M_\beta, M_q, 1; 0, 0, 0]$, $B = [0, M_\rho, 0]^T$, and $C = [1, 0, 0]$ are known matrices.

An output-prediction-based high-gain observer is constructed to estimate x in light of discrete data $\breve{y}(t_j^e)$, that is,

$$\dot{\hat{x}}(t) = A\hat{x}(t) + Bu(t) - \delta \Xi^{-1} S \left(C\hat{x}(t) - \chi(t)\right), t \in [t_k^s, t_{k+1}^s) \quad (6)$$

where $\hat{x} = [\hat{x}_1, \hat{x}_2, \hat{x}_3]^T$, δ is the designed gain parameter, Ξ is a diagonal matrix with diagonal elements 1, $1/\delta$, and $1/\delta^2$ in order, the vector $S = [s_1, s_2, s_3]^T$ is designed such that $A - SC$ is Hurwitz, namely, for a third-order identity matrix I_3 and any constant $\rho > 0$, there exists a symmetric positive definite matrix Q such that $Q(A - SC) + (A - SC)^T Q \leq -2\rho I_3$. Moreover, $\chi(t)$ is the prediction of $y(t)$, governed by

$$\dot{\chi}(t) = \hat{x}_2 + \frac{P_\beta}{W}\hat{x}_1, t \in [t_k^s, t_{k+1}^s)$$

If a trigger occurs at the sampling moment t_k^s, $\chi(t_k^s) = \breve{y}(t_k^s)$, otherwise, $\chi(t_k^s) = y_r(t_k^s)$.

We define $f(t) = C\hat{x}(t) - \chi(t)$, for $t \in [t_k^s, t_{k+1}^s)$, its derivative is $\dot{f}(t) = -\delta s_1 f(t)$. Furthermore, integrating its derivative from t_k^s to t ($t < t_{k+1}^s$) can get $f(t) = (C\hat{x}(t_k^s) - \chi(t_k^s))\,\mathrm{e}^{-\delta s_1 (t - t_k^s)}, t \in [t_k^s, t_{k+1}^s)$. Therefore, observer (6) can be expressed as

$$\begin{aligned} \dot{\hat{x}}(t) &= A\hat{x}(t) + Bu(t) - \delta \Xi^{-1} S f(t) \\ f(t) &= (C\hat{x}(t_k^s) - \chi(t_k^s))\,\mathrm{e}^{-\delta s_1 (t - t_k^s)}, t \in [t_k^s, t_{k+1}^s) \end{aligned} \quad (7)$$

The state estimation error is represented as $\tilde{x} = \hat{x} - x$. Let $z = \Xi\tilde{x}$. The performance of the proposed state estimation strategy can be stated as the following theorem.

Theorem 2. *Consider system (1) subject to attack $\bar{a}$ and noise v. Under Assumptions 1 and 1, the triggering mechanism (3), Algorithm 1, and observer (7) can ensure the observation error satisfies*

$$\|\tilde{x}(t)\| \leq \frac{\delta^2\sqrt{\lambda_{\max}(Q)}}{\sqrt{\lambda_{\min}(Q)}}\|\tilde{x}(0)\|\,\mathrm{e}^{-\kappa t} + \frac{\delta^2 \eta \mu_{\max}\left(2 - \mathrm{e}^{-\kappa\mu_{\min}}\right)}{\sqrt{\lambda_{\min}(Q)}\left(1 - \mathrm{e}^{-\kappa\mu_{\min}}\right)}$$

for $\varrho\mu_{\max} < \varkappa$, *where* $\eta = \sqrt{\lambda_{\max}(Q)}\left(\bar{d} + \delta\|S\|(\bar{v} + \epsilon)\right)$, $\kappa = (\varkappa - \varrho\mu_{\max})\,\mathrm{e}^{-\varkappa\mu_{\max}}$, $\varkappa = \delta\rho/\lambda_{\max}(Q)$, *and* $\varrho = \delta^2|P_\beta|\,\|S\|\sqrt{\lambda_{\max}(Q)}/(|W|\sqrt{\lambda_{\min}(Q)})$.

Proof. We define a Lyapunov function as $V_1 = z^T Qz$ and its derivative as

$$\dot{V}_1 = \delta z^T \left(Q(A - SC) + (A - SC)^T Q\right) z + 2z^T Q(\delta S\left(C\bar{z} - f(t)\right) - \Xi d(x, u)) \quad (8)$$

According to Assumption 2, one obtains

$$-2z^T Q\Xi d(x, u) \leq 2\bar{d}\sqrt{\lambda_{\max}(Q)}\sqrt{V_1} \quad (9)$$

According to Assumptions 1 and 2, Theorem 1, and the bound of the triggering threshold, we have

$$Cz(t_k^s) - f(t_k^s) = -x_1(t_k^s) + \chi(t_k^s) \leq \bar{\upsilon} + \epsilon$$

For $t \in [t_k^s, t_{k+1}^s)$, $k \in \mathbb{N}_+$, it follows from $C\dot{z}(t) - \dot{f}(t) = \delta \frac{P_\beta}{W} z_1 + \delta z_2$ that

$$2\delta z^T QS\left(C\bar{z} - f(t)\right) \leq 2\delta \left\|S\right\| \sqrt{\lambda_{\max}(Q)}\sqrt{V_1} \left(\bar{\upsilon} + \epsilon + \frac{\delta |P_\beta|}{|W|\sqrt{\lambda_{\min}(Q)}} \int_{t_k^s}^t \sqrt{V_1} d\tau\right) \quad (10)$$

Substituting (9)–(10) into (8) yields

$$\begin{aligned}\dot{V}_1 \leq & -\frac{2\delta\rho}{\lambda_{\max}(Q)} V_1 + 2\delta \left\|S\right\| \sqrt{\lambda_{\max}(Q)}\sqrt{V_1} \left(\bar{\upsilon} + \epsilon + \frac{\delta |P_\beta|}{|W|\sqrt{\lambda_{\min}(Q)}} \int_{t_k^s}^t \sqrt{V_1} d\tau\right) \\ & + 2\bar{d}\sqrt{\lambda_{\max}(Q)}\sqrt{V_1}, t \in [t_k^s, t_{k+1}^s), k \in \mathbb{N}_+\end{aligned}$$

Since $\dot{V}_1 = 2\sqrt{V_1} d(\sqrt{V_1})/dt$, it follows that

$$\frac{d(\sqrt{V_1})}{dt} \leq -\varkappa\sqrt{V_1} + \varrho \int_{t_k^s}^t \sqrt{V_1} d\tau + \eta, t \in [t_k^s, t_{k+1}^s), k \in \mathbb{N}_+$$

According to Lemma 1, when the appropriate parameters are chosen such that $\kappa > 0$, we have

$$\sqrt{V_1(t)} \leq \sqrt{V_1(0)}\mathbf{e}^{-\kappa t} + \eta\mu_{\max}\frac{2 - \mathbf{e}^{-\kappa\mu_{\min}}}{1 - \mathbf{e}^{-\kappa\mu_{\min}}}, \forall t \in [0, +\infty)$$

Therefore, the observation error satisfies

$$\left\|\tilde{x}(t)\right\| \leq \frac{\delta^2\sqrt{\lambda_{\max}(Q)}}{\sqrt{\lambda_{\min}(Q)}} \left\|\tilde{x}(0)\right\| \mathbf{e}^{-\kappa t} + \frac{\delta^2\eta\mu_{\max}\left(2 - \mathbf{e}^{-\kappa\mu_{\min}}\right)}{\sqrt{\lambda_{\min}(Q)}\left(1 - \mathbf{e}^{-\kappa\mu_{\min}}\right)}, \forall t \in [0, +\infty)$$

The proof is finished. ∎

3.4 Tracking Controller Design

Based on the obtained estimated states, we define the following error surfaces:

$$\begin{aligned} e_1 &= \hat{x}_1 - y_r \\ e_2 &= \hat{x}_2 - \varphi \end{aligned}$$

where φ is the estimate of the virtual control signal α by the tracking differentiator (2). Under Assumption 3, we construct the control laws with positive gains k_1 and k_2 as

$$\alpha = -k_1 e_1 - \frac{1}{2}e_1 - \frac{P_\beta}{W}\hat{x}_1 + \delta s_1 f(t) + \dot{y}_r \tag{11}$$

$$u = \frac{1}{M_\rho}\left(-k_2 e_2 - e_1 - M_\beta \hat{x}_1 - M_q \hat{x}_2 - \hat{x}_3 + \delta^2 s_2 f(t) + \dot{\varphi}_d\right) \tag{12}$$

Theorem 3. *Under Assumptions 1–3, the control strategy composed of the tracking differentiator (2), triggering mechanism (3), Algorithm 1, observer (7), and control signals (11) and (12) can ensure that all closed-loop signals of system (5) are bounded and the system output converges in the vicinity of the desired signal.*

Proof. Select a Lyapunov function as $V_2 = e_1^2/2 + e_2^2/2$. Substituting the control signals (11) and (12) into the derivative of V_2 yields

$$\dot{V}_2 \leq e_1\left(\varphi - \alpha - k_1 e_1 - \frac{1}{2}e_1\right) - k_2 e_2^2$$

Since AOA is bounded, it follows from Assumption 3 and observer (7) that the first and second order derivatives of α satisfy the conditions of Lemma 2. Therefore, according to the boundedness of $\alpha(t_{k+1}^{s+}) - \alpha(t_{k+1}^{s-})$, there exists a constant σ such that $|\varphi(t) - \alpha(t)| \leq \sigma, \forall t \geq 0$. Consequently, we have $\dot{V}_2 \leq -\bar{k}V_2 + \sigma^2/2$ with $\bar{k} = \max\{k_1, k_2\}$, which demonstrates that all signals of the closed-loop system are bounded. Moreover, the tracking error $\tilde{y} = y - y_d$ satisfies

$$\lim_{t\to\infty} |\tilde{y}| \leq \lim_{t\to\infty} (|\tilde{x}_1| + |e_1|) \leq \frac{\sigma^2}{\sqrt{\bar{k}}} + \frac{\delta^2 \eta \mu_{\max}\left(2 - \mathrm{e}^{-\kappa\mu_{\min}}\right)}{\sqrt{\lambda_{\min}(Q)}\left(1 - \mathrm{e}^{-\kappa\mu_{\min}}\right)}$$

Thus, the proposed control strategy can drive the system output to converge in the vicinity of the desired signal. ■

4 Numerical Simulation

A numerical simulation is performed to demonstrate the control performance of the obtained theoretical results. The model parameters are chosen as $P_\beta = 511.538$, $W = 502$, $M_\rho = -0.1756$, $M_\beta = 0.8223$, $M_q = -1.0774$, $C_0 = 0.1$, $\beta_0 = 2.11$, $\sigma = 0.25$, and $z_0 = 0.14$. The initial values are set as $x(0) = [-10; 15]$, $\hat{x}(0) = [-8; 14]$, and $\varphi(0) = \varphi_d(0) = 0$. The triggering threshold $\vartheta_i(t_k^s)$ is designed as a constant 0.025. The significance level a is chosen as 0.05. The control parameters are chosen as $\delta = 2$, $\rho = 100$, $k_1 = k_2 = 3$, $R = 50$, and $b_1 = 0.5$. The system output is measured by six sensors with a sampling period of 0.01 s. The measurement noise v_i is considered as zero-mean Gaussian noise with a variance of 0.0002 and a bound of 0.5. Moreover, the attack signals are considered as

$$a_1 = \begin{cases} [0, \sin(y), 0, 0, 10\sin(10t), 0]^T, & t < 10 \text{ s} \\ [0, 0, \cos(y), 0, 0, t^{0.5}]^T, & t \geq 10 \text{ s} \end{cases}, a_2 = \begin{cases} [0, y, 0, 0, 3 + \sin(t+2), 0]^T, & t < 10 \text{ s} \\ [0, 0, -y, 0, 0, 3 + \sin(2t)]^T, & t \geq 10 \text{ s} \end{cases}$$

To implement an aggressive maneuver flight [17], the desired output is designed as

$$y_r = \frac{171}{\pi}\left(\frac{0.5}{1+\mathbf{e}^{t-8}} + \frac{1}{1+\mathbf{e}^{t-20}} - \mathbf{e}^{-0.2t} - 0.5\right) \text{deg}$$

To show the effect of Grubbs test and uncertainty estimation, the proposed control strategy is investigated from three cases. Case 1 adopts the proposed control strategy. In Case 2, the output is estimated by the method proposed in [7], i.e., $\breve{y}(t_j^e) = \mathbf{Med}[\bar{Y}_a(t_j^e)]$. In Case 3, the output is estimated by the method proposed in [7] and d_2 is not estimated as an extended state.

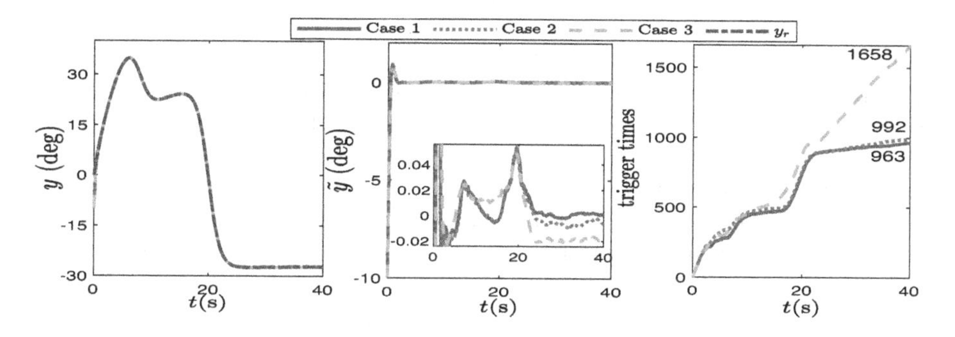

Fig. 2. Output tracking performance and number of triggers.

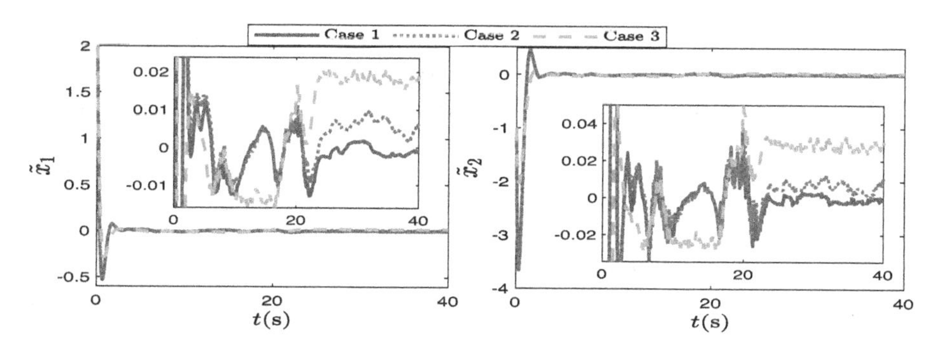

Fig. 3. System state estimation errors.

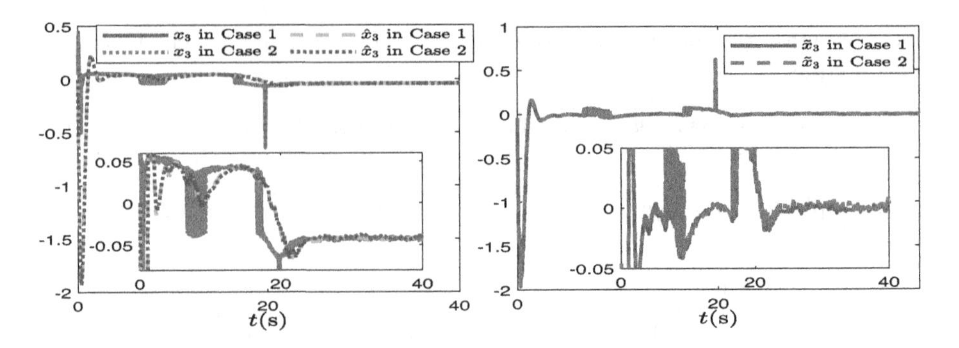

Fig. 4. Estimation effect of the extended state $x_3 = d_2$.

The simulation results are shown in Figs. 2, 3 and 4. It can be seen from Figs. 2, 3 and 3 that the proposed control method can accurately reconstruct the system states, track the desired signal, and reduce network data transmission despite being affected by sparse attacks, measurement noise, and model uncertainty. The simulation results of Case 3 show that although the output does not change significantly after 25 s, due to the model uncertainty, the tracking error is not stable around 0 but around −0.02, which makes the designed triggering mechanism unable to effectively reduce data transmission. Comparing Case 2 and Case 3, it can be seen that the compensation for model uncertainty improves the control performance and reduces the number of triggers. Moreover, by comparing Case 1 and Case 2, it is shown that the proposed output estimation method based on the Grubbs test can further improve the tracking performance and state estimation accuracy.

5 Conclusion

This study proposes an event-triggered secure state estimation and tracking control algorithm for fixed-wing aircraft with measurement noise under sparse attacks. A voting SDET mechanism based on tracking performance is designed to reduce data transmission while ensuring satisfactory tracking performance. The influence of measurement noise on secure output estimation is alleviated by a Grubbs-test-based output estimation algorithm. This study combines the SDET output and the desired output for output prediction to ensure the prediction accuracy, and effectively estimates the model uncertainty by constructing an extended state, so as to ensure that the constructed high-gain observer can accurately reconstruct the system states under measurement noise and model uncertainty. Then, the tracking controller can be designed with the backstepping method. The simulation results illustrate the satisfactory control performance of the proposed method. Future research will investigate how to improve the noise immunity of state estimation and achieve fixed-time stabilization [22,23].

Acknowledgements. This work was partially supported by the National Natural Science Foundation of China under Grant Nos. 62121004, 62141606, 62033003, and 62003098, the Local Innovative and Research Teams Project of Guangdong Special Support Program under Grant No. 2019BT02X353, and the Key Area Research and Development Program of Guangdong Province under Grant No. 2021B0101410005.

References

1. Fawzi, H., Tabuada, P., Diggavi, S.: Secure estimation and control for cyber-physical systems under adversarial attacks. IEEE Trans. Autom. Control **59**(6), 1454–1467 (2014)
2. Petit, J., Stottelaar, B., Feiri, M., Kargl, F.: Remote attacks on automated vehicles sensors: experiments on camera and LiDAR. In: Proceedings of Black Hat Europe, p. 995 (2015)
3. Shoukry, Y., et al.: SMT-based observer design for cyber-physical systems under sensor attacks. ACM Trans. Cyber-Phys. Syst. **2**(1), 1–27 (2018). Art. no. 5
4. An, L., Yang, G.H.: State estimation under sparse sensor attacks: a constrained set partitioning approach. IEEE Trans. Autom. Control **64**(9), 3861–3868 (2019)
5. Yang, T., Murguia, C., Kuijper, M., Nešić, D.: A multi-observer based estimation framework for nonlinear systems under sensor attacks. Automatica **119**, 109043 (2020)
6. Yang, T., Lv, C.: A secure sensor fusion framework for connected and automated vehicles under sensor attacks. IEEE Internet Things J. (early access). https://doi.org/10.1109/JIOT.2021.3101502
7. Ao, W., Song, Y., Wen, C.: Distributed secure state estimation and control for CPSs under sensor attacks. IEEE Trans. Cybern. **50**(1), 259–269 (2020)
8. Chen, G., Yao, D., Li, H., Zhou, Q., Lu, R.: Saturated threshold event-triggered control for multiagent systems under sensor attacks and its application to UAVs. IEEE Trans. Circ. Syst. I Reg. Papers **69**(2), 884–895 (2022)
9. Tabuada, P.: Event-triggered real-time scheduling of stabilizing control tasks. IEEE Trans. Autom. Control **52**(9), 1680–1685 (2007)
10. Yao, D., Li, H., Shi, Y.: Adaptive event-triggered sliding mode control for consensus tracking of nonlinear multi-agent systems with unknown perturbations. IEEE Trans. Cybern. (early access). https://doi.org/10.1109/TCYB.2022.3172127
11. Chen, G., Yao, D., Zhou, Q., Li, H., Lu, R.: Distributed event-triggered formation control of USVs with prescribed performance. J. Syst. Sci. Complex., 1–19 (2021). https://doi.org/10.1007/s11424-021-0150-0
12. Heemels, W.P.M.H., Donkers, M.C.F., Teel, A.R.: Periodic event-triggered control for linear systems. IEEE Trans. Autom. Control **58**(4), 847–861 (2013)
13. Mao, Z., Jiang, B., Shi, P.: Fault-tolerant control for a class of nonlinear sampled-data systems via a Euler approximate observer. Automatica **46**(11), 1852–1859 (2010)
14. Selivanov, A., Fridman, E.: Observer-based input-to-state stabilization of networked control systems with large uncertain delays. Automatica **74**, 63–70 (2016)
15. Etienne, L., Hetel, L., Efimov, D., Petreczky, M.: Observer synthesis under time-varying sampling for Lipschitz nonlinear systems. Automatica **85**, 433–440 (2017)
16. Farza, M., M'Saad, M., Fall, M.L., Pigeon, E., Gehan, O., Busawon, K.: Continuous-discrete time observers for a class of MIMO nonlinear systems. IEEE Trans. Autom. Control **59**(4), 1060–1065 (2014)

17. Young, A., Cao, C., Patel, V., Hovakimyan, N., Lavretsky, E.: Adaptive control design methodology for nonlinear-in-control systems in aircraft applications. J. Guid. Control Dyn. **30**(6), 1770–1782 (2007)
18. Ma, R., Shi, P., Wu, L.: Sparse false injection attacks reconstruction via descriptor sliding mode observers. IEEE Trans. Autom. Control **66**(11), 5369–5376 (2021)
19. Cao, L., Yao, D., Li, H., Meng, W., Lu, R.: Fuzzy-based dynamic event triggering formation control for nonstrict-feedback nonlinear MASs. Fuzzy Sets Syst. (early access). https://doi.org/10.1016/j.fss.2022.03.005
20. Bouraoui, I., Farza, M., Ménard, T., Ben Abdennour, R., M'Saad, M., Mosrati, H.: Observer design for a class of uncertain nonlinear systems with sampled outputs-application to the estimation of kinetic rates in bioreactors. Automatica **55**, 78–87 (2015)
21. Guo, B., Zhao, Z.: Weak convergence of nonlinear high-gain tracking differentiator. IEEE Trans. Autom. Control **58**(4), 1074–1080 (2013)
22. Liu, Y., Li, H., Zuo, Z., Li, X., Lu, R.: An overview of finite/fixed-time control and its application in engineering systems. IEEE/CAA J. Autom. Sinica (early access). https://doi.org/10.1109/JAS.2022.105413
23. Liu, Y., Yao, D., Li, H., Lu, R.: Distributed cooperative compound tracking control for a platoon of vehicles with adaptive NN. IEEE Trans. Cybern. (early access). https://doi.org/10.1109/TCYB.2020.3044883

Backstepping Fuzzy Adaptive Control Based on RBFNN for a Redundant Manipulator

Qinlin Yang[1], Qi Lu[2(✉)], Xiangyun Li[3,4], and Kang Li[2,3,4]

[1] College of Electrical Engineering, Sichuan University, Chengdu 610041, Sichuan, China
[2] Department of Mechanical Engineering, Sichuan University-Pittsburgh Institute, Sichuan University, Chengdu 610041, Sichuan, China
qi.lu@scu.edu.cn
[3] West China Biomedical Big Data Center, West China Hospital, Sichuan University, Chengdu 610041, Sichuan, China
[4] Med-X Center for Informatics, Sichuan University, Chengdu 610041, Sichuan, China

Abstract. Redundant manipulator is a highly nonlinear and strongly coupled system. In practical application, dynamic parameters are difficult to determine due to uncertain loads and external disturbances. These factors will adversely affect the control performance of manipulator. In view of the above problems, this paper proposes a backstepping fuzzy adaptive control algorithm based on the Radial Basis Function Neural Network (RBFNN), which effectively eliminates the influence of the internal uncertainty and external interference on the control of the manipulator. Firstly, the algorithm adopts the backstepping method to design the controller framework. Then, the fuzzy system is used to fit the unknown system dynamics represented by nonlinear function to realize model-free control of the manipulator. The fuzzy constants are optimized by RBFNN to effectively eliminate the control errors caused by unknown parameters and disturbance. Finally, in order to realize RBFNN approximating the optimal fuzzy constant, an adaptive law is designed to obtain the weight value of RBFNN. The stability of the closed-loop system is proved by using Lyapunov stability theorem. Through simulation experiments, the algorithm proposed in this paper can effectively track the target joint angle when the dynamic parameters of the 7-DOF redundant manipulator are uncertain and subject to external torque interference. Compared with fuzzy adaptive control, the tracking error of the algorithm in this paper is smaller, and the performance is better.

Keywords: Redundant manipulator · Fuzzy system · RBF neural network · Adaptive control

---

The work has been financially supported by Natural Science Foundation of China (Grant Nos. 51805449 and 62103291), Sichuan Science and Technology Program (Grant Nos. 2021ZHYZ0019 and 2022YFS0021) and 1·3·5 project for disciplines of excellence, West China Hospital, Sichuan University (Grant Nos. ZYYC21004 and ZYJC21081). All findings and results presented in this paper are by those of the authors and do not represent the funding agencies.

H. Liu et al. (Eds.): ICIRA 2022, LNAI 13456, pp. 149–159, 2022.
https://doi.org/10.1007/978-3-031-13822-5_14

1 Introduction

At present, the control task of manipulator is mainly to locate, move and transport equipment. The efficient manipulator controller can control the manipulator to complete the control task and achieve the control goal, even in the presence of uncertain parameters, external force and torque. The traditional controller design usually needs to derive the accurate mathematical model of the manipulator. However, the redundant manipulator has a complex structure, and it is difficult to determine the dynamic parameters in engineering practice, and it is difficult to obtain an accurate mathematical model. With the deepening, extensive and complex application of manipulators, these reasons promote the development of manipulator control, but also pose great challenges to the traditional controller design. For example, in manipulator-assisted rehabilitation training, the external force on the manipulator will also change due to the different muscle strength of patients [1].

Among the traditional control methods, such as PID [2, 3] control can achieve model-free control of the manipulator. Due to the simple structure of PID control algorithm and obvious intent of mathematical expression, However, PID control has been widely used in the case of disturbance and dynamic uncertainty of the manipulator. It is difficult to achieve high-precision motion control of manipulator at high speed [4]. Most cases are applied to low load and low speed control with poor adaptive ability. Therefore, in the design process of control algorithm of manipulator, intelligent control has received much attention due to various reasons. Such as in terms of intelligent control, fuzzy network [5, 6], deep learning [7] have been applied. In [8], for n DOF manipulator, this paper proposes an adaptive control algorithm based on neural network, in the process of control, eliminate the dynamic time- varying parameter uncertainty and unknown delay, and meet the state and input constraints, by using saturation function and Lyapunov-Krasovskii functional, the influence of actuator saturation and time delay is eliminated, but in the experiment part only about 2 DOF manipulator through simulation experiments, not in a high DOF manipulator for validation. Therefore, it is difficult to realize unmodeled manipulator control, real-time compensation of external interference and effective proof of controller stability at the same time.

To sum up, this paper focuses on the study of manipulator with uncertain dynamics, affected by external disturbance and uncertain load, and proposes a joint space backstepping adaptive fuzzy adaptive control algorithm based on RBF neural network. Compared with previous work, the main contributions of this paper are summarized as follows.

(1) The controller designed in this paper does not need to obtain the fuzzy constants in the fuzzy system according to the expert knowledge, which are approximated by RBF (Radial Basis Function) neural network, which really strengthens the applicability of the controller.
(2) Combining the traditional backstepping controller design with the fuzzy system and RBF neural network in intelligent control framework, the controller not only maintains a concise structure, which is convenient for stability verification, but also can realize the model-free control of nonlinear and strongly coupled systems.

2 System Description and Problem Formulation

2.1 System Description of Redundant Manipulator

We consider an n degree-of-freedom (DOF) redundant manipulator dynamic model in the joint space as follow:

$$\begin{aligned} M(q)\ddot{q} + C(q,\dot{q})\dot{q} + g(q) + d &= \tau \\ y &= q \end{aligned} \tag{1}$$

where $q, \dot{q}, \ddot{q} \in R^n$ are the joint angular displacement, velocity and acceleration vectors, $M(q) \in R^{n \times n}$ is the symmetric and positive definite inertia matrix, $C(q,\dot{q})\dot{q} \in R^n$ is the vector of Coriolis and centripetal torque, $g(q) \in R^n$ denotes the vector of the gravitational torque. $\tau \in R^n$ denotes the vector of the control input torque, $d \in R^n$ denotes the vector of unknown disturbance observer in manipulator dynamics and it can be friction disturbance or other forms of unknown disturbance. $y \in R^n$ denotes the output vector of the manipulator, which are the joint angles. Because the seven- DOF redundant manipulator is used in this paper, then $n = 7$.

In the process of model-free control, the dynamic parameters of the manipulator are unknown, but usually they are bounded. The system has the following properties:

(1) The inertial matrix $M(q)$ is a positive definite symmetric matrix, and lower and upper bounded as $m_1\|\chi\|^2 \le \chi^T M(q)\chi \le m_2\|\chi\|^2$, where m_1, m_2 are known positive constants and $\|*\|$ is the Euclidean 2-norm,$\chi \in R^n$ is an arbitrary vector.
(2) The inertia matrix $M(q)$ and centripetal force and Coriolis moment matrix $C(q,\dot{q})$ satisfy the following relations: $\dot{q}^T(\dot{M} - 2C)\dot{q} = 0$.

We define $x_1 = q$,$x_2 = \dot{q}$. Rewrite the dynamic equation of the manipulator as shown below, and then use the Backstepping controller design method.

$$\begin{cases} \dot{x}_1 = x_2 \\ \dot{x}_2 = M^{-1}(x_1)\tau - M^{-1}(x_1)C(x_1,x_2) - M^{-1}(x_1)g(x_1) - M^{-1}(x_1)d \\ y = x_1 \end{cases} \tag{2}$$

where $C(x_1,x_2)$, $M^{-1}(x_1)$ and $g(x_1)$ are unknown nonlinear smooth functions related to the redundant manipulator of the controlled object, and y_d are defined as the desired joint angles, which are second-order differentiable.

2.2 Problem Formulation

The Control objective of this paper is formulated as when the dynamic parameters of redundant manipulator are uncertain and affected by external interference, a backstepping fuzzy adaptive control algorithm based on RBFNN is designed to adjust the joint angles y of the manipulator and realize the accurate tracking of the target joint angles y_d.

3 Backstepping Fuzzy Adaptive Control Based on RBFNN

3.1 Backstepping Control

The actual joint angle is y, the desired joint angle is y_d. We define the error as

$$z_1 = y - y_d \tag{3}$$

If the estimate of x_2 is defined as virtual control α_1, the error is defined as

$$z_2 = x_2 - \alpha_1 \tag{4}$$

In order to drive z_1 goes to 0 in the control process, α_1 needs to be designed. According to the following formula:

$$\dot{z}_1 = \dot{x}_1 - \dot{y}_d = x_2 - \dot{y}_d = z_2 + \alpha_1 - \dot{y}_d \tag{5}$$

The virtual control is selected as

$$\alpha_1 = -\lambda_1 z_1 + \dot{y}_d \tag{6}$$

for the first subsystem of (3), the Lyapunov function is constructed as

$$V_1 = \frac{1}{2} z_1^T z_1 \tag{7}$$

Then we can obtain that

$$\dot{V}_1 = z_1^T \dot{z}_1 = z_1^T (\dot{y} - \dot{y}_d) = z_1^T (\dot{x}_1 - \dot{y}_d) = z_1^T (x_2 - \dot{y}_d) \tag{8}$$

In order for the first subsystem to be stable, we need z_2 converge to 0.

3.2 Application of Fuzzy System in Backstepping Control

According to (3) and (5), the following equation can be obtained:

$$\dot{z}_2 = \dot{x}_2 - \dot{\alpha}_1 = -M^{-1}Cx_2 - M^{-1}g - M^{-1}d + M^{-1}\tau - \dot{\alpha}_1 \tag{9}$$

For the second subsystem

$$\dot{x}_2 = M^{-1}(x_1)\tau - M^{-1}(x_1)C(x_1, x_2) - M^{-1}(x_1)g(x_1) - M^{-1}(x_1)d \tag{10}$$

construct the Lyapunov function as follow:

$$V_2 = V_1 + \frac{1}{2} z_2^{\mathrm{T}} M z_2 \tag{11}$$

therefore

$$\begin{aligned} \dot{V}_2 &= \dot{V}_1 + \frac{1}{2} z_2^{\mathrm{T}} M \dot{z}_2 + \frac{1}{2} \dot{z}_2^{\mathrm{T}} M z_2 + \frac{1}{2} z_2^{\mathrm{T}} \dot{M} z_2 \\ &= -\lambda_1 z_1^T z_1 + z_1^T z_2 + z_2^{\mathrm{T}}(-C\alpha_1 - g - M\dot{\alpha}_1 + \tau) - z_2^{\mathrm{T}} d \\ &= -\lambda_1 z_1^T z_1 + z_1^T z_2 + z_2^{\mathrm{T}}(f + \tau) - z_2^{\mathrm{T}} d \end{aligned} \tag{12}$$

by satisfying the stability of Lyapunov function, the control law is obtained as follow:

$$\tau = -\lambda_2 z_2 - z_1 - \varphi \tag{13}$$

where φ denotes the fuzzy system output. Where (12), $f = -C\alpha_1 - g - M\dot{\alpha}_1$, substitute the obtained control law (13) into (12), we can obtain:

$$\dot{V}_2 = -\lambda_1 z_1^T z_1 - \lambda_2 z_2^T z_2 + z_2^T(f - \varphi) - z_2^T d \tag{14}$$

according to (14), a fuzzy system output φ should be used to fit f, because $f = -C\alpha_1 - g - M\dot{\alpha}_1$, which contains the inertial matrix M and Coriolis force matrix C of the manipulator. Because the dynamic parameters of the manipulator is uncertain and difficult to measure, model-free control scheme is used. In this paper, the fuzzification method will be used as single-valued fuzzification. The defuzzification method is product inference machine and barycentric average.

If there are N fuzzy rules to form a fuzzy system, the i-th fuzzy rule can be expressed as

$$R^i = \text{IF } x_1 \text{ is } \mu_1^i \text{ and ...and } x_n \text{ is } \mu_n^i, \text{ then y is } \mathrm{B}^i (i = 1, 2, ..., N) \tag{15}$$

where μ_j^i denotes the subordinate function of linguistic variables $x_j (j = 1, 2, ..., n)$. We define the output of the fuzzy system as follow:

$$y = \frac{\sum_{i=1}^{N} \theta_i \prod_{j=1}^{n} \mu_j^i(x_j)}{\sum_{i=1}^{N} \prod_{j=1}^{n} \mu_j^i(x_j)} = \xi(x)\theta \tag{16}$$

where $\xi = [\xi_1(x)\xi_2(x)...\xi_N(x)]^T$, $\xi_i(x) = \frac{\prod_{j=1}^{n} \mu_j^i(x_j)}{\sum_{i=1}^{N} \prod_{j=1}^{n} \mu_j^i(x_j)}$, $\theta = [\theta_1\theta_2...\theta_N]^T$, where θ are fuzzy constants. In (14), f will be obtained through the fitting of the fuzzy system, the research object of this paper is a 7-DOF redundant manipulator, where $f \in R^7$ represents the torque control quantity of the fitted dynamic part on the manipulator of 7-DOF, which can be described as $f(1), f(2)...f(7)$. The fuzzy system is expressed as follows:

$$\varphi_1(x) = \frac{\sum_{i=1}^{N} \theta_{1i} \prod_{j=1}^{n} \mu_j^i(x_j)}{\sum_{i=1}^{N} \prod_{j=1}^{n} \mu_j^i(x_j)} = \xi_1^T(x)\theta_1 \; \ldots \; \varphi_7(x) = \frac{\sum_{i=1}^{N} \theta_{7i} \prod_{j=1}^{n} \mu_j^i(x_j)}{\sum_{i=1}^{N} \prod_{j=1}^{n} \mu_j^i(x_j)} = \xi_7^T(x)\theta_7 \tag{17}$$

we define the fitting result of f as Φ, we have

$$\Phi = [\varphi_1\varphi_2...\varphi_7]^T = \begin{bmatrix} \xi_1^T & \ldots & 0 \\ \ldots & \ldots & \ldots \\ 0 & \ldots & \xi_7^T \end{bmatrix} \begin{bmatrix} \theta_1 \\ \ldots \\ \theta_7 \end{bmatrix} \tag{18}$$

3.3 RBFNN Combined with Fuzzy System

In (18), θ usually unknown and requires expert knowledge to make judgment and selection, because θ is abundant and difficult to obtain, it is difficult to obtain an effective control law. Therefore, this paper uses a simple-structured RBF neural network to obtain fuzzy constants to achieve efficient control of the manipulator. RBFNN is a kind of forward neural network, which consists of three layers of network.

The first layer is the input layer, where $x = [x_1, x_2, ..., x_n]^T$ is the input of the network and n represents the dimension of the input.

The second layer is the hidden layer, whose output is $h(x) = [h_1, h_2, ..., h_m]^T$, and the Gaussian basis function is used as the subordinate function of the input layer introduced as follow:

$$h_j(x) = \exp(-\frac{\|x - c\|^2}{2b^2}) \tag{19}$$

where c is the coordinate vector of the center point of the gaussian basis function of the hidden layer, b is the width of the gaussian basis function of the hidden layer, $j = 1, 2...m$, RBF neural network weight value is given as

$$W = [W_1, W_2, ..., W_m]^T \tag{20}$$

The third layer is the output layer of neural network, and the output as follow

$$y = W^T h(x) = W_1 h_1 + ... + W_m h_m \tag{21}$$

We define the optimal approximation fuzzy constant θ_i in (22), and the RBF neural network is used to approximate θ_i. The input of neural network is $x = [e, \dot{e}]^T$, and the ideal output of neural network as

$$\theta_i = W^{*T} h(x) - \varepsilon_i \tag{22}$$

where W^* is the ideal weight of the RBF neural network, $h(x)$ is the output of the Gaussian function, for a given arbitrary small constant $\varepsilon_i(\varepsilon_i > 0)$, the following inequality holds, $\|\theta_i - W^{*T} h(x)\| \leq \varepsilon_i$ is the approximation error of the network, and the actual output of the neural network is set as

$$\hat{\theta}_i = \hat{W}^T h(x) \tag{23}$$

where $\hat{W}$ is the actual weight of the neural network, we define $\tilde{W} = W^* - \hat{W}, \tilde{\theta}_i = \theta_i - \hat{\theta}_i$, where $\tilde{W}$ is the error between the ideal weight and the actual weight, and $\tilde{\theta}_i$ is the error between the ideal output and the actual output. Therefore, the adaptive law based on RBF neural network is given as

$$\dot{\hat{W}} = \gamma h(x)(z_2^T \xi^T(x))^T - 2k\hat{W} \tag{24}$$

Then, Fig. 1 shows the system control block diagram, from (10)–(24), we can achieve the control law as follows:

$$\tau = -\lambda_2 z_2 - z_1 - \xi \int (\gamma h(z_2^T \xi^T)^T - 2k\hat{W})dt^T h \tag{25}$$

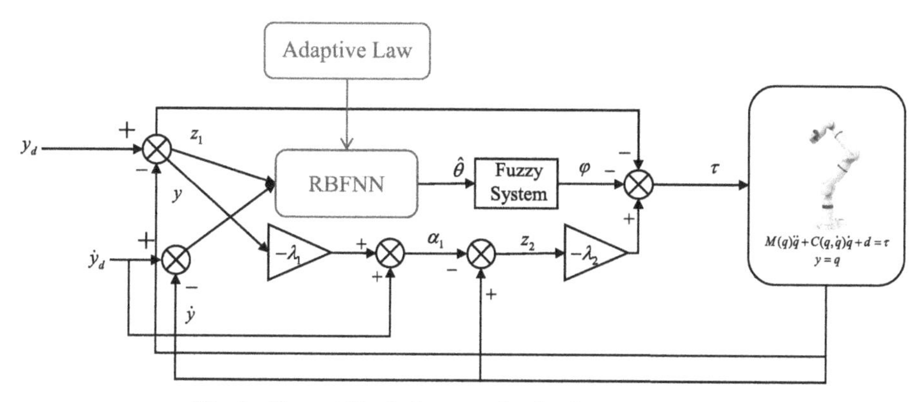

Fig. 1. Control block diagram of redundant manipulator.

3.4 Stability Analysis

For the whole system, we define the Lyapunov function candidate as follow

$$V = \frac{1}{2}z_1^{\mathrm{T}}z_1 + \frac{1}{2}z_2^{\mathrm{T}}Mz_2 + \frac{1}{2\gamma}\tilde{W}^{\mathrm{T}}\tilde{W} = V_2 + +\frac{1}{2\gamma}\tilde{W}^{\mathrm{T}}\tilde{W} \tag{26}$$

where $\gamma > 0$, hence

$$\begin{aligned}\dot{V} =& -\lambda_1 z_1^T z_1 - \lambda_2 z_2^T z_2 + z_2^{\mathrm{T}}\Big[f - \xi(x)\Big(W^{*T}h(x)\Big)\Big] \\ &+ z_2^{\mathrm{T}}\Big[\xi(x)\Big(W^{*T}h(x)\Big) - \xi(x)\Big(\hat{W}^T h(x)\Big)\Big] - z_2^{\mathrm{T}}d - \frac{1}{\gamma}\tilde{W}^{\mathrm{T}}\dot{\hat{W}}\end{aligned} \tag{27}$$

According to the basic inequality $a^2 + b^2 \geq 2a(a, b \in R$, the equality sign holds if and only if $a = b)$, substitute the obtained adaptive law (24) into the above equation, we obtain

$$\begin{aligned}\dot{V} \leq& -\lambda_1 z_1^T z_1 - (\lambda_2 - 1)\lambda_2 z_2^T z_2 + \frac{k}{\gamma}(2W^{*\mathrm{T}}\hat{W} - 2\tilde{W}^{\mathrm{T}}\hat{W}) \\ &+ \frac{1}{2}(\xi(x)\varepsilon)^T(\xi(x)\varepsilon) + \frac{1}{2}d^T d\end{aligned} \tag{28}$$

According to $(\hat{W} - W^*)^{\mathrm{T}}(\hat{W} - W^*) \geq 0$, we obtain $2W^{*\mathrm{T}}\hat{W} - 2\hat{W}^{\mathrm{T}}\hat{W} \leq -\hat{W}^{\mathrm{T}}\hat{W} + W^{*\mathrm{T}}W^*$, plug in the formula above

$$\begin{aligned}\dot{V} \leq& -\lambda_1 z_1^T z_1 - (\lambda_2 - 1)\lambda_2 z_2^T z_2 + \frac{k}{\gamma}(-\hat{W}^{\mathrm{T}}\hat{W} - W^{*\mathrm{T}}W^*) \\ &+ \frac{2k}{\gamma}W^{*\mathrm{T}}W^* + +\frac{1}{2}(\xi(x)\varepsilon)^T(\xi(x)\varepsilon) + \frac{1}{2}d^T d\end{aligned} \tag{29}$$

According to $(W^* + \hat{W})^{\mathrm{T}}(W^* + \hat{W}) \geq 0$, we obtain $-W^{*\mathrm{T}}\hat{W} - \hat{W}^{\mathrm{T}}W^* \leq W^{*\mathrm{T}}W^* + \hat{W}^{\mathrm{T}}\hat{W}$.

then

$$\tilde{W}^{\mathrm{T}}\tilde{W} = (W^{*\mathrm{T}} - \hat{W}^{\mathrm{T}})(W^{*} - \hat{W}) \leq 2W^{*\mathrm{T}}W^{*} + 2\hat{W}^{\mathrm{T}}\hat{W} \tag{30}$$

Therefore $-W^{*\mathrm{T}}W^{*} - \hat{W}^{\mathrm{T}}\hat{W} \leq -\frac{1}{2}\tilde{W}^{\mathrm{T}}\tilde{W}$, we obtain as follow

$$\begin{aligned} \dot{V} \leq & -\frac{2}{2}\lambda_1 z_1^T z_1 - (\lambda_2 - 1)\frac{2}{2}\lambda_2 z_2^T M^{-1} M z_2 \\ & - \frac{k}{2\gamma}\tilde{W}^{\mathrm{T}}\tilde{W} + \frac{2k}{\gamma}W^{*\mathrm{T}}W^{*} + \frac{1}{2}(\xi(x)\varepsilon)^T(\xi(x)\varepsilon) + \frac{1}{2}d^T d \end{aligned} \tag{31}$$

We take $\lambda_2 > 1$, because $M \leq \sigma_0 I$, thus

$$\begin{aligned} \dot{V} \leq & -\frac{2}{2}\lambda_1 z_1^T z_1 - (\lambda_2 - 1)\frac{2}{2\sigma_0}\lambda_2 z_2^T M z_2 - \frac{k}{2\gamma}\tilde{W}^{\mathrm{T}}\tilde{W} \\ & + \frac{2k}{\gamma}W^{*\mathrm{T}}W^{*} + \frac{1}{2}(\xi(x)\varepsilon)^T(\xi(x)\varepsilon) + \frac{1}{2}d^T d \end{aligned} \tag{32}$$

We define $c_0 = \min\{2\lambda_1, 2(\lambda_2 - 1)\frac{1}{\sigma_0}, k\}$, therefore

$$\dot{V} \leq -c_0 V + \frac{2k}{\gamma}W^{*\mathrm{T}}W^{*} + \frac{1}{2}(\xi(x)\varepsilon)^T(\xi(x)\varepsilon) + \frac{1}{2}d^T d \tag{33}$$

Because $\xi_i(x) = \frac{\prod_{j=1}^{n}\mu_j^{\mathrm{i}}(x_j)}{\sum_{i=1}^{N}\prod_{j=1}^{n}\mu_j^{\mathrm{i}}(x_j)}$, thus $|\xi(x)| \leq 1$, then

$$\dot{V} \leq -c_0 V + \frac{2k}{\gamma}W^{*\mathrm{T}}W^{*} + \frac{1}{2}\varepsilon^T\varepsilon + \frac{1}{2}d^T d \tag{34}$$

Since interference $d \in R^n$ is bounded, $D > 0$ exists and $d^T d \leq D$ is satisfied, then

$$\dot{V} \leq -c_0 V + \frac{2k}{\gamma}W^{*\mathrm{T}}W^{*} + \frac{\varepsilon^T\varepsilon}{2} + \frac{D}{2} = -c_0 V + c_{V_{\max}} \tag{35}$$

where $c_{V_{\max}} = \frac{2k}{\gamma}W^{*\mathrm{T}}W^{*} + \frac{\varepsilon^T\varepsilon}{2} + \frac{D}{2}$. Solving (35) and we obtain as

$$\dot{V}(t) \leq V(0)\exp(-c_0 t) + \frac{c_{V_{\max}}}{c_0}(1 - \exp(-c_0 t)) \leq V(0) + \frac{c_{V_{\max}}}{c_0} \tag{36}$$

where $V(0)$ is the initial value of $V(t)$ and defines a compact set as

$$\Omega_0 = \left\{X \,|\, V(X) \leq V(0) + \frac{c_{V_{\max}}}{c_0}\right\} \tag{37}$$

According to $\left\{z_1, z_2, \tilde{W}\right\} \in \Omega_0$, V is bounded, and all signals of the closed-loop system are bounded. z_1 and z_2 are required to converge in order to effectively track the target value of the joint Angle and angular velocity of the manipulator. Unknown disturbance will determine the accuracy of the convergence value.

4 Simulation Experiment and Result Analysis

In this section, the effectiveness of the backstepping fuzzy adaptive control algorithm based on RBF neural network is verified by numerical simulation. The ROKEA redundant manipulator model xMate is selected for simulation. ROKEA has 7 revolute joints and a load capacity of 3kg. Under the action of the controller, the manipulator realizes the tracking of the target joint vector in the joint space. The controller controls the torque of the seven joints. The ROKEA xMate structure is shown in Fig. 2, Table 1 lists the masses of each link.

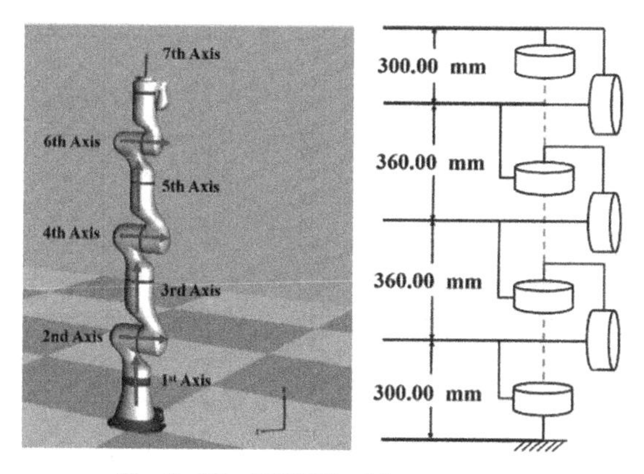

Fig. 2. The ROKEA xMate structure.

Table 1. The masses of each link.

Link (i)	1	2	3	4	5	6	7
Mass/kg	3.44	3.62	3.46	2.38	2.22	2.18	0.51

In the actual task execution, it is necessary to load a gripper at the end of the manipulator for handling tasks, and in order to expand the function of the manipulator, add cameras, sensors with different functions and other equipment on the manipulator link. In this case, it will affect dynamic parameters such as the original mass, center of mass, and moment of inertia of the manipulator affect the control performance of the manipulator. The controller uses RBF fuzzy backstepping control to realize the fitting of the nonlinear system, so as to realize the control of the manipulator without accurate manipulator model information.

Compared with the mass of the link of the original manipulator in Table 1, the mass change of each link is $[+0, +1, +0, +2, +1, +0, +1]$ kg. The added mass of the 7th link can be regarded as the load. The control parameters $\lambda_1 = 80$, $\lambda_2 = 70$, $k = 1.5$, $b = 6$, set the external interference torque as $d = [0, 0.5, 0, 0, 0.5, 0.5, 0]^T$, and $q_d = [0.9, 0.5, 0.5, 0.9, 0.5, 0.5, 0.5]^T \cdot \sin(0.5\pi \cdot t)$ the desired joint angle trajectories are set to The simulation results are as follows.. The actual and desired angles of each joint and the angle errors of each joint are shown in Fig. 3. The 3D trajectory is shown in Fig. 4. It can be observed that the RBFNN combined with fuzzy system has better robustness, smaller control error, compared with the fuzzy adaptive control in [9], smoother trajectory and better control performance under the impact of the uncertainty of dynamic parameters and unknow load of the manipulator.

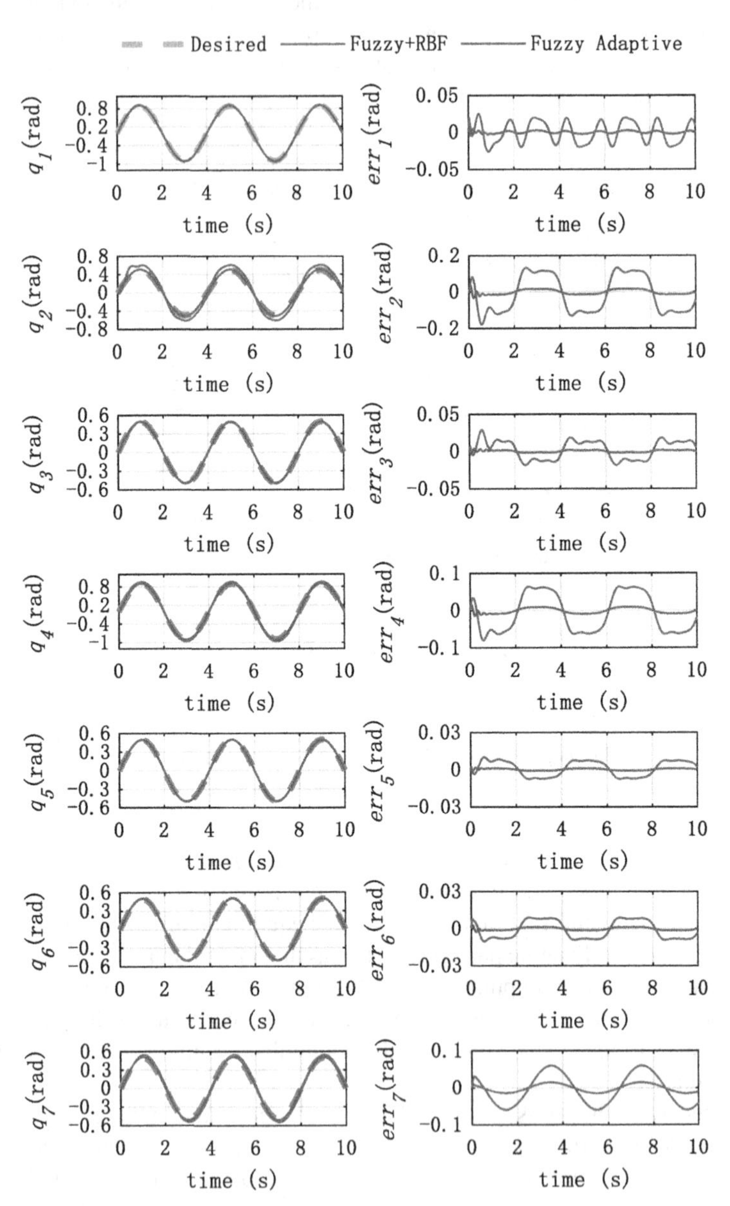

Fig. 3. The motion of the manipulator under uncertain dynamic parameters and external disturbance torque.

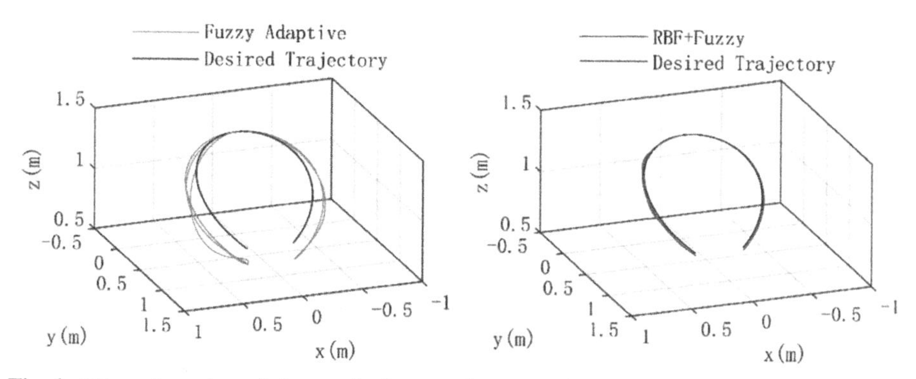

Fig. 4. 3D motion view of the manipulator under uncertain dynamic parameters and external disturbance torque.

References

1. Burgar, C.G., Lum, P.S., Shor, P.C.: Development of robots for rehabilitation therapy: the Palo Alto VA/Stanford experience. J. Rehabil. Res. Dev. **37**(6), 663–674 (2000)
2. Souza, D.A., Batista, J.G., dos Reis, L.L.N., Júnior, A.B.S.: PID controller with novel PSO applied to a joint of a robotic manipulator. J. Braz. Soc. Mech. Sci. Eng. **43**(8), 1–14 (2021). https://doi.org/10.1007/s40430-021-03092-4
3. Li, J.: Robot manipulator adaptive motion control with uncertain model and payload. Beijing Institute of Technology (2016)
4. Ren, Z.: The present situation and development trend of industrial robot. Equip. Manuf. Technol. **3**, 166–168 (2015)
5. Yao, Q.: Adaptive fuzzy neural network control for a space manipulator in the presence of output constraints and input nonlinearities. Adv. Space Res. **67**(6), 1830–1843 (2021)
6. Ghafarian, M., Shirinzadeh, B., Al-Jodah, A., et al.: Adaptive fuzzy sliding mode control for high-precision motion tracking of a multi-DOF micro/nano manipulator. IEEE Robot. Autom. Lett. **5**(3), 4313–4320 (2020)
7. Bui, H.: A deep learning-based autonomous robot manipulator for sorting application. In: IEEE Robotic Computing, IRC 2020 (2020)
8. Song, Z., Sun, K.: Prescribed performance adaptive control for an uncertain robotic manipulator with input compensation updating law. J. Franklin Inst. **358**(16), 8396–8418 (2021)
9. Zhang, Z., Yan, Z.: An adaptive fuzzy recurrent neural network for solving the nonrepetitive motion problem of redundant robot manipulators. IEEE Trans. Fuzzy Syst. **28**(4), 684–691 (2019)

A Flexible Electrostatic Adsorption Suction Cup for Curved Surface

Haotian Guan, Shilun Du, Yaowen Zhang, Wencong Lian, Cong Wang, and Yong Lei(✉)

State Key Laboratory of Fluid Power and Mechatronic Systems, Zhejiang University Hangzhou, Zhejiang, China
ylei@zju.edu.cn

Abstract. As an important electrical pipeline, the GIS (gas insulated metal-enclosed switchgear) pipeline is applied in many scenarios. However, it has a narrow internal place and non-magnetic wall, which makes existing robot adsorption structure difficult to meet the needs of this type of the pipe. This paper combines the flexible electrostatic adsorption electrode and the flexible material to design a suction cup adapted to curved surface and narrow pipeline inner wall. The relationship between electrostatic adsorption force and the applied voltage is studied through designing a coplanar bipolar electrostatic adsorption membranes. And the adsorption force simulation are completed by the Ansoft Maxwell, the structure of the flexible electrostatic suction cup analyzed by Abaqus. After making the suction cup prototypes with different structures, the experiment testing adsorption force determines the final structure and verifies the adsorption capacity of electrostatic suction cup.

Keywords: Pipeline adsorption · Flexible electrostatic adsorption · Concentric electrode

1 Introduction

Pipelines play an important role in petrochemical, power, construction and other fields [1]. However, safety problems such as cracking and embrittlement are prone to occur during usage, which may easily cause certain economic losses, environmental pollution and even casualties. Especially in the field of electricity industry, the GIS (gas insulated metal-enclosed switchgear) pipeline [2] is a kind of important electrical pipeline, which is narrow and non-magnetic. The use of pipeline climbing adsorption robots can effectively inspect the case of pipelines [3].

The research on climbing robots first started in Japan. The world's earliest wall-climbing robot was developed in 1966 by the A. NISHI [4] team of Osaka

This work is supported in part by the National Key Research and Development Project under Grant 2019YFB1310404, and the National Natural Science Foundation of China Grant No. 52072341.

H. Liu et al. (Eds.): ICIRA 2022, LNAI 13456, pp. 160–170, 2022.
https://doi.org/10.1007/978-3-031-13822-5_15

University, Japan, which was a negative pressure adsorption crawler robot using a centrifugal fan. In 1975, the team developed the second-generation machine with a single suction cup structure on the basis of the first-generation machine. Mahmoud [5] and others from the University of Coimbra in Portugal studied a climbing robot using ferromagnetic adsorption. The surface adsorption was realized by designing a central adsorption body and three auxiliary adsorption magnetic wheels at the bottom of the robot. The Chinese University of Hong Kong Tin [6] has developed a tree-climbing robot with a continuous manipulation structure based on the motion characteristics of tree-climbing animals, which can achieve climbing motions on trunks and branches of various diameters and shapes. Inspired by the chemical composition of the feet of climbing insects, Kathryn [7] of the United States has designed a small robot that can walk on smooth walls, but due to the influence of adhesive materials, its performance is poor in a dusty environment. As it can be seen in existing literature, the wall-climbing robot is to realize stable adsorption and moving of inclined surfaces [8,9]. However, the existing wall-climbing robots are difficult to adapt to the inner wall of the pipeline because of the narrow space and the large curvature surface, and also have problem such as inflexible movement and easy winding of cables.

In this work, a flexible electrostatic adsorption suction cup for curved surface is proposed to provide stable support of the pipeline climbing robot. The rest of the paper is organized as follow. Section 2 introduces the objectives and overall design of the suction cup. Section 3 introduces the design of the electrostatic adsorption membranes and the flexible structure of the suction cup with the related simulation verification. In Sect. 4, the optimal parameters of the electrostatic adsorption membrane are determined through experiments, and the adsorption force test of the electrostatic suction cup is carried out. Finally, conclusion and future works are provided in Sect. 5.

2 Overall Design

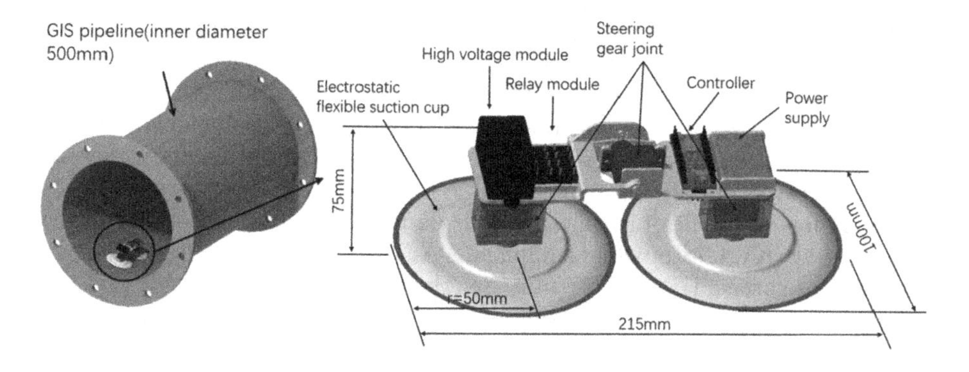

Fig. 1. Overall view of the electrostatic adsorption robot

According to the characteristics of GIS pipeline, the designed pipeline climbing robot needs to achieve stable adsorption on the inner wall of the pipeline and full-area movement, and has a wireless control system. The adsorption capacity can provide absorption force of 300 g, which is the self-weight of robot.

The suction cup can be divided to support structure and control hardware. The hardware system of the suction cup mainly includes the controller, high voltage generator, bluetooth module, steering gear, power supply and other modules. As for hardware selection, light and small hardware equipment is more likely to be selected under the premise of meeting the needs, and overall weight of suction cup is within 300 g. In order to adapt to the narrow space inside the pipeline, the overall size is also designed to be very small. A rotary joint is arranged between two suction cups to achieve the movement of the suction cups on the inner wall of pipelines. The specific overall structure design is shown in Fig. 1.

3 Design of the Flexible Electrostatic Suction Cup

3.1 Suction Cup Structure Design

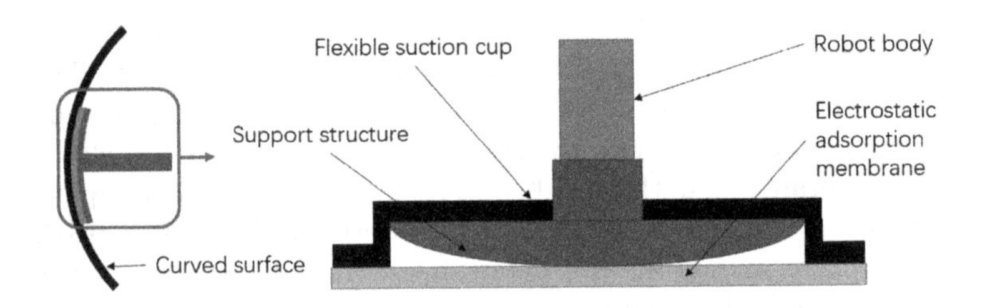

Fig. 2. Schematic diagram of the structure of electrostatic adsorption suction cup

Compared with the rigid suction cup, the flexible suction cup has a certain flexibility and can better adapt to the shape of the curved surface [10]. Based on the structure of the vacuum suction cup, this paper proposes a supporting electrostatic suction cup structure that meets the above requirements, including flexible electrostatic adsorption membrane, rigid support structure and flexible suction cup. The simplified structure diagram is shown in Fig. 2. The core of the structure is to use the spherical bottom-shaped support structure to ensure that the suction cup can adapt to the curved surface, while also making the electrostatic adsorption membranes uniformly stressed to a certain extent, thereby effectively preventing peeling.

The radius of the flexible suction cup is designed to 50 mm according to the size of the 110 KV GIS pipeline equipment. And the radius R corresponding to the spherical surface is set to 500 mm. Set the thickness of the flexible body at different positions of the flexible suction cup to be the same as d, the section radius of the spherical surface at the bottom of the support structure to be δ, the fillet radius at the outer edge of the suction cup as ρ and the height of flexible

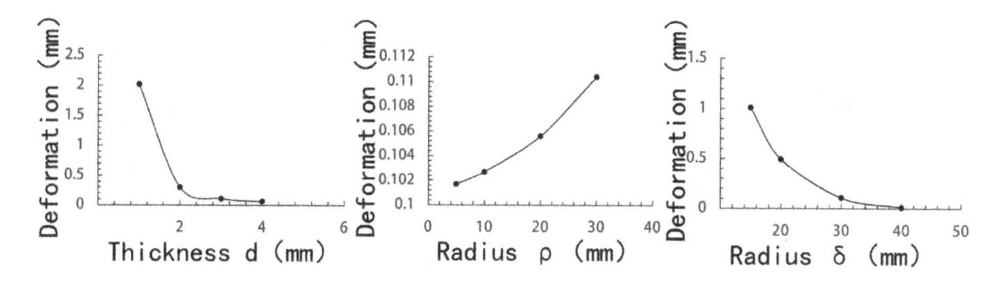

Fig. 3. The relationship between parameters and deformation variables

suction cup (σ) can be calculated by section diameter and flexible body thickness. Respectively set the thickness of the flexible body d = 1/2/3/4 mm, the section radius of the spherical structure at the bottom of the support structure $\delta = 15/20/30/40$ mm, and the radius at the outer edge of the suction cup $\rho = 5/10/20/30$ mm.

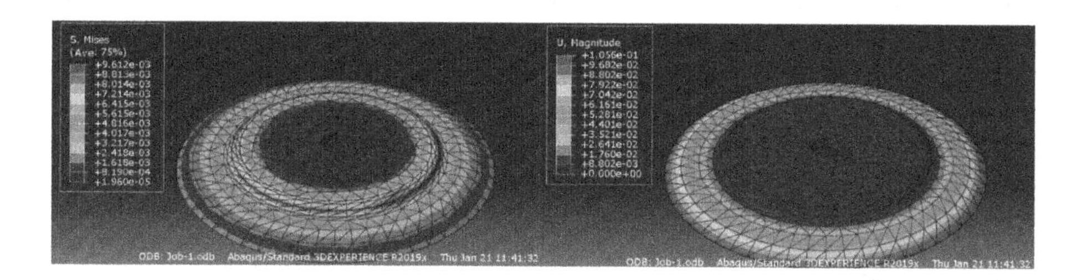

Fig. 4. The flexible suction cup deformation cloud map

Use ABAQUS to simulate and analyze the performance of the flexible suction cup. According to the maximum deformation of the flexible suction cup with different structural parameters obtained by the simulation under the same load, the relationship between each parameter and the deformation variable is drawn as shown in the Fig. 3. It can be seen from the relationship diagram that for the thickness parameters of the suction cup, as the thickness of the suction cup increases, the deformation of the flexible suction cup decreases when the pressure is applied, and the adaptability to the curved surface gradually decreases. Combined with the results of simulation, the final designed flexible suction cup structure parameter is selected as d = 3 mm, $\delta = 30$ mm, $\rho = 20$ mm. The simulation results under this parameter are shown in the Fig. 4.

A load of 500 g is applied to the end of the support rod above the soft suction cup, and this position is regarded as the center of the robot gravity. It can be seen from the Fig. 5 that the maximum deformation is 0.1047 mm, and the displacement of the applied load position is 1.117 mm, which is much smaller than the structural size of the suction cup. The simulation results prove that the flexible suction cup can meet the requirements of the robot.

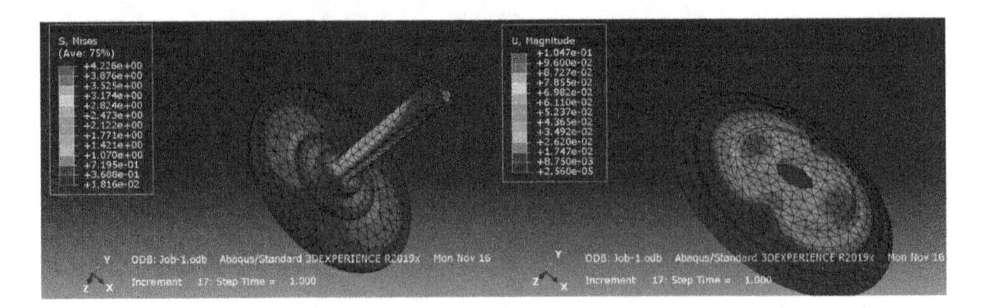

Fig. 5. Suction cup simulation deformation cloud map

3.2 Adsorption Membranes Design

Electrostatic adsorption electrodes usually have unipolar and bipolar types. Bipolar electrostatic adsorption electrode is chosen because it does not require auxiliary electrodes and has the advantages of easy control and fast response speed. But its working space is the edge rather than the center, which leads uneven distribution of electrostatic adsorption force. Therefore, it is necessary to use the Schwarz-Christopher transformation [11] to complete the capacitor edge problem. In the process of determining the electrostatic adsorption force, the electrode can be directly equivalent to the capacitance for calculation. However, the disadvantage of bipolar electrostatic adsorption electrode is that the adsorption capacity is weak. Therefore, it is necessary to increase the number of electrode pairs. The concentric electrode structure is proposed in this paper, as shown in Fig. 6.

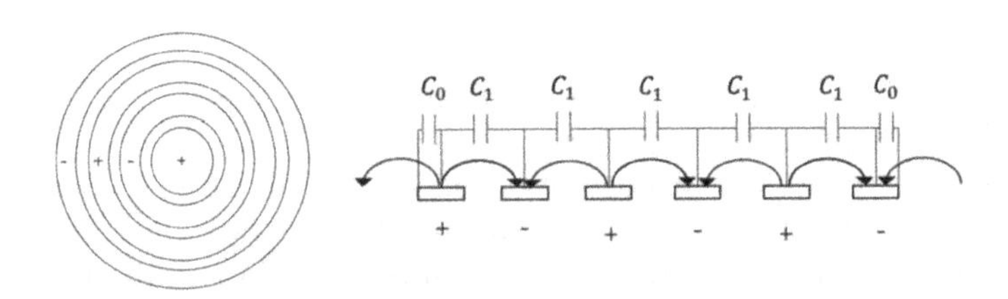

Fig. 6. Schematic diagram of concentric electrode structure and equivalent capacitance of multiple groups of electrodes

The concentric electrode structure can be approximately regarded as the parallel connection of multiple capacitor units from the perspective of cross-sectional structure analysis. The adjacent coplanar electrodes and the half electrodes on the edge respectively constitute two types of capacitors, the capacitances of which are C0 and C1, and multiple capacitors are connected in series. The Schematic diagram of the equivalent capacitance of multiple groups of electrodes is shown in Fig. 6. Since the width of a single electrode is small relative to the entire electrostatic adsorption electrode, it can be ignored for the sake of simplifying the calculation. The adsorption capacitance C is mainly determined by the dielectric coefficient of the material and electrode spacing b, so these influencing factors will be discussed.

The Influence of Applied Voltage U and Electrode Spacing B on the Adsorption Force. The greater the applied voltage U, the greater the electrostatic adsorption force generated by the electrode. But in actual application, if the voltage U becomes excessive, the electrode will be broken down. Therefore, the relationship between the electrode spacing and the breakdown voltage can be drawn according to the Parsing law, which shows that when the electrode spacing b is less than 0.8 mm, there is a rapid sudden change for the breakdown voltage U. Therefore, the curve with electrode spacing greater than 0.8 mm should be selected, and when the size of the electrode spacing b is determined, the maximum voltage allowed to be applied by the electrodes is determined by the Parsing law.

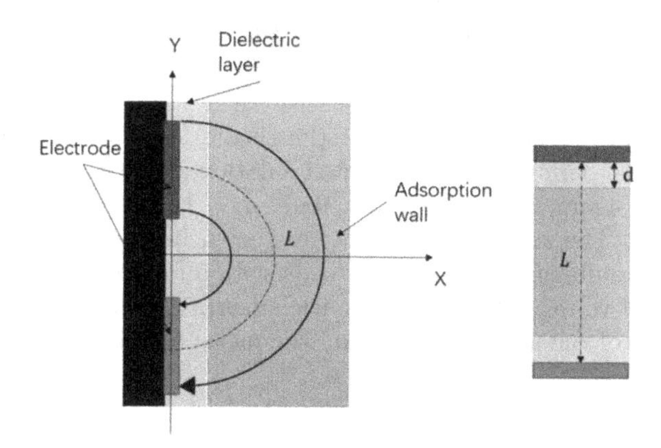

Fig. 7. Equivalent dielectric coefficient calculation model

The Influence of the Dielectric Coefficient ε_r of the Material on the Adsorption Force. In the actual application, as shown in Fig. 7, the size of the air layer relative to the dielectric layer and the wall surface is small, and the relative dielectric coefficient is 1 which can be ignored to simplify the calculation. Simplify the circular electric field from the positive electrode through the dielectric layer, adsorption wall, and dielectric layer to the straight electric field. The dielectric constant of the dielectric layer is ε_{medium}, and the dielectric constant of the wall is ε_{wall}. Assuming that the thickness of the insulating medium layer is d, the dielectric coefficient of the medium filled between the two electrodes can be calculated by volume weighted [10].

In summary, in order to maximize the adsorption force of the electrostatic adsorption electrode in a certain area, this problem can be transformed into a

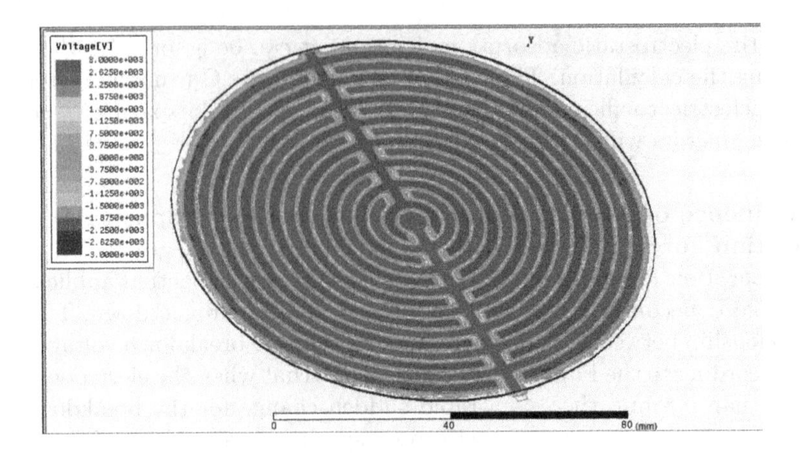

Fig. 8. Adsorption electrode potential distribution

nonlinear optimization problem under constrained conditions. The parameters of electrode can be optimized with Matlab. The theoretical electrostatic adsorption force of the electrode is calculated to be 55.2971 N when the applied voltage is 3 kv.

In order to verify the correctness of the theoretical analysis above, Ansoft Maxwell is used to perform finite element simulation analysis. In this simulation, U = 3170, b = 4 mm, a = 1 mm, k = 0.67, d = 13 um, where k is the void ratio($k = \frac{b}{b+2a}$). The simulation results are shown in the Fig. 8. According to the simulation results of the electrostatic adsorption force obtained by Maxwell, when the control voltage of the electrostatic adsorption film reaches 3000 v, the adsorption force of the adsorption film on the metal aluminum can reach about 56 N.

4 Experiment

4.1 Electrostatic Adsorption Membranes Test

The experiment tests the influence of different parameters of the suction cup structure and voltage changes on the adsorption force. In order to further verify the theoretical analysis above, three kinds of electrostatic adsorption membranes are designed and fabricated, and the corresponding electrostatic adsorption force tests on non-metallic wall are planned under different voltages and adsorption wall conditions. The parameters of each adsorption membranes are shown in the Table 1.

Table 1. Membranes design parameter table

Membranes number	Electrode width (a)	Electrode interval (b)	Void ratio (k)	Insulation thickness
1	1 mm	4 mm	0.67	13 um
2	1 mm	1 mm	0.5	13 um
3	1 mm	1 mm	0.5	25 um

Voltmeter
Force sensor
Adsorption membrane
Voltage regulating module
Adsorption plane
DC power supply
Experiment platform

Fig. 9. Figure of experimental principle and real surface tests of different materials

During the experiment, the voltage applied to the electrostatic adsorption membrane is indirectly controlled by controlling the magnitude of the input voltage, and then the tensile force can be slowly increased by pulling the tension dynamometer until the electrostatic adsorption membrane slides, and the maximum tensile force can be measured. The specific schematic diagram is in Fig. 9.

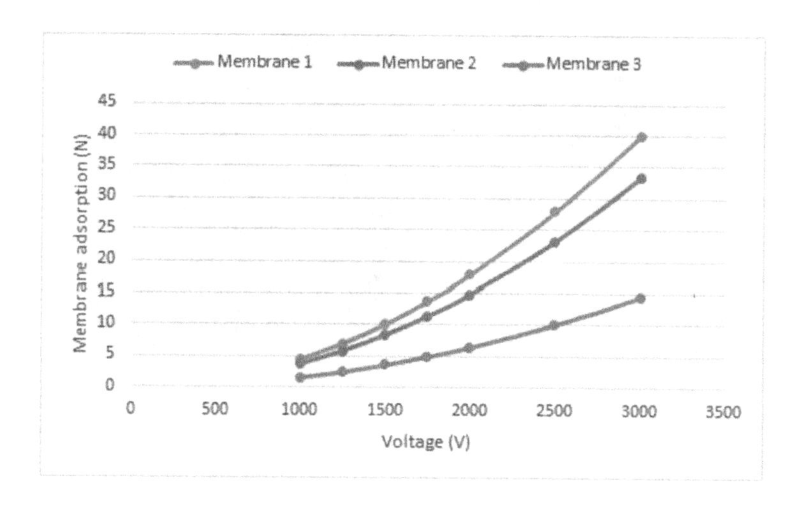

Fig. 10. Experimental results graph

According to the data analysis of the experimental results in Fig. 10, it can be known that the electrostatic adsorption force has a square relationship with the applied voltage, that is, the electrostatic adsorption force can be increased by increasing the voltage amplitude within the maximum allowable applied voltage. The increase of the thickness of the dielectric layer will obviously weaken the adsorption force, so a smaller thickness of the dielectric layer should be selected in the actual application process. The influence coefficient of adsorption force and void ratio is also nearly proportional to the square. The experimental results verify the above analysis and demonstration. The parameters of membrane 1 are used to make a prototype and the working voltage is determined to be 3000 v, which can achieve the design requirement.

4.2 Electrostatic Adsorption Flexible Suction Cup Test

The adsorption force test is carried out using the electrostatic suction cup prototype under the actual condition. A metal arc pipe with an inner diameter of 500 mm is selected as the experimental condition for the adsorption force test. In order to facilitate observation and operation, the pipe wall used is a partial arc-shaped pipe wall, but its structure does not affect the performance test of the suction cup prototype. During the test, the suction cup is attached to the curved surface of the pipe wall with a thrust of 2.5 N to simulate the falling action of the robot suction cup, and the control voltage is connected. The adsorption force test is carried out until the suction cup stably adsorbs (Fig. 11).

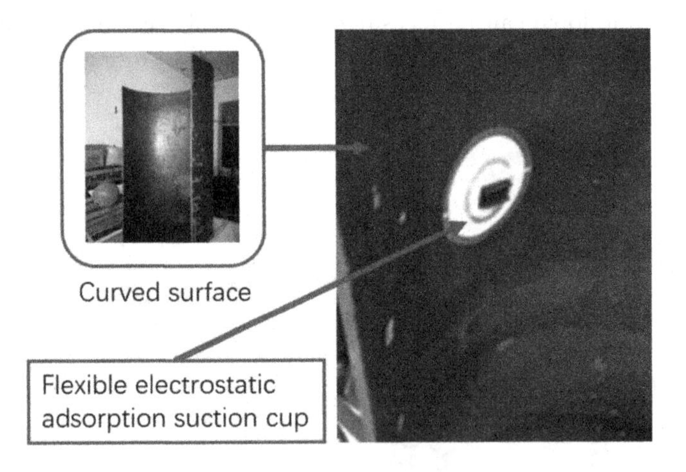

Fig. 11. Electrostatic suction cup adsorption test

The experimental test results are shown in Table 2. According to the experimental results, the graph is shown in Fig. 12. It can be seen that the relationship between the absorption force of the suction cup and the applied voltage is also similar to the square relationship. The results show that the suction cup can better achieve the fitting and adsorption of the arc-shaped pipeline wall. And it can be seen that compared with the electrostatic adsorption membrane, the overall adsorption capacity of the suction cup is reduced, but it can meet the design goals in the previous paragraph.

Table 2. The adsorption force test of the electrostatic flexible suction cup

Applied voltage (V)	1000	1250	1500	1750	2000	2500	3000
Adsorption force (N)	3.509	5.784	9.073	11.405	15.637	25.431	37.254

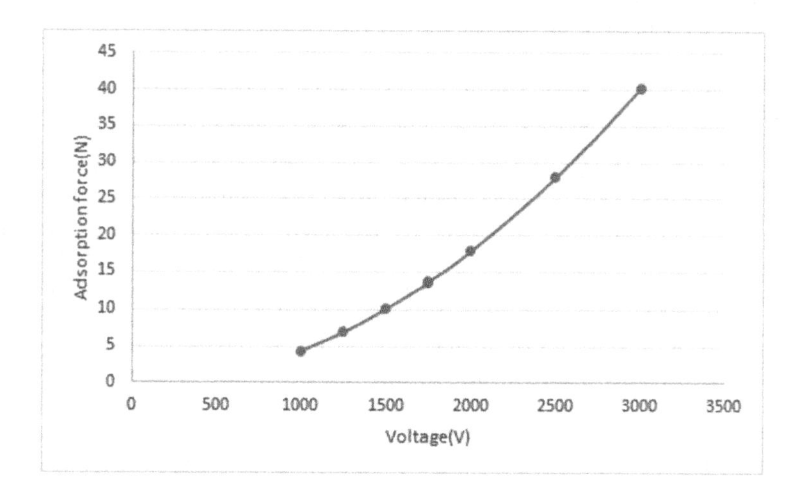

Fig. 12. Electrostatic suction cup adsorption test

5 Conclusion

Aiming at the problem of GIS or other industries pipelines with narrow interior space, curved walls and non-magnetic properties, it is difficult for conventional wall-climbing robots to enter the pipeline for inspection work. In this paper, a flexible suction cup structure based on the adsorption membrane is designed for curved surface. The parameters of the electrostatic adsorption membrane are optimized by theoretical calculation and simulation. The adsorption effect of the suction cup is tested by simulation and experiment, which is verified to provide an adsorption force of 56 N for robot, which is far greater than the self-weight of robot. The performance test of the suction cup prototype is carried out on the curved surface of the pipe wall. The experimental results show that the overall flexible structure can achieve curved surface adaptability with enough adsorption force.

Future work will focus on adapting the electrostatic adsorption membrane and the flexible structure of the electrostatic suction cup to accommodate more complex surface.

References

1. Rudd, S., McArthur, S.D.J., Judd, M.D.: A generic knowledge-based approach to the analysis of partial discharge data. IEEE Trans. Dielectr. Electr. Insul. **17**(1), 149–156 (2010)

2. Reid, A.J., Judd, M.D., Fouracre, R.A., Stewart, B.G., Hepburn, D.M.: Simultaneous measurement of partial discharges using IEC60270 and radio-frequency techniques. IEEE Trans. Dielectr. Electr. Insul. **18**(2), 444–455 (2011)
3. Sawabe, H., Nakajima, M., Tanaka, M., Tanaka, K., Matsuno, F.: Control of an articulated wheeled mobile robot in pipes. Adv. Robot. **33**(20), 1072–1086 (2019)
4. Nishi, A.: Development of wall-climbing robots. Comput. Electr. Eng. **22**(2), 123–149 (1996)
5. Tavakoli, M., Viegas, C., Marques, L., Norberto Pires, J., De Almeida, A.T.: OmniClimbers: omni-directional magnetic wheeled climbing robots for inspection of ferromagnetic structures. Robot. Auton. Syst. **61**(9), 997–1007 (2013)
6. Lam, T.L., Xu, Y.: Biologically inspired tree-climbing robot with continuum maneuvering mechanism. J. Field Robot. **29**(6), 843–860 (2012)
7. Daltorio, K.A., Gorb, S., Peressadko, A., Horchler, A.D., Ritzmann, R.E., Quinn, R.D.: A robot that climbs walls using micro-structured polymer feet. In: Tokhi, M.O., Virk, G.S., Hossain, M.A. (eds.) Climbing and Walking Robots. Springer, Heidelberg (2006). https://doi.org/10.1007/3-540-26415-9_15
8. Li, J., Gao, X., Fan, N., Li, K., Jiang, Z., Jiang, Z.: Adsorption performance of sliding wall-climbing robot. Chin. J. Mech. Eng. (Engl. Ed.) **23**(6), 733–741 (2010)
9. Jian, Z., Dong, S., Tso, S.-K.: Development of a tracked climbing robot. J. Intell. Robot. Syst. **35**(4), 427–443 (2002)
10. Surachai, P.: Development of a wall climbing robot. J. Comput. Sci. **6**(10), 1185 (2010)
11. Chuang, J.M., Gui, Q.Y., Hsiung, C.C.: Numerical computation of Schwarz - Christoffel transformation for simply connected unbounded domain. Comput. Methods Appl. Mech. Eng. **105**(1), 93–109 (1993)

SCRMA: Snake-Like Robot Curriculum Rapid Motor Adaptation

Xiangbei Liu[1,2], Fengwei Sheng[1,2], and Xian Guo[1,2](✉)

[1] College of Artificial Intelligence, Nankai University, Tianjin, China
guoxian@nankai.edu.cn
[2] Institute of Robotics and Automatic Information System, Tianjin, China

Abstract. Controllers for underwater snake-like robots are difficult to design because of their high DOF and complex motions. Additionally, because of the complex underwater environment and insufficient knowledge of hydrodynamics, the traditional control algorithms based on environment or robot modeling cannot work well. In this paper, we propose an SCRMA algorithm, which combines the characteristics of the RMA algorithm for rapid learning and adaptation to the environment, and uses curriculum learning and save&load exploration to accelerate the training speed. Experiments show that the SCRMA algorithm works better than other kinds of reinforcement learning algorithms nowadays.

Keywords: Underwater snake-like robot · Reinforcement learning · RMA

1 Introduction

With the development of society and industry, there is an increasing demand for human beings to exploit marine resources. Currently, there are two main types of underwater work: manual and robotic. The advantages of underwater snake-like robots are low cost, low accident rate, and independence from complex underwater environments. Amongst underwater robots, snake-like robots have a wide range of locomotion gait and can move flexibly through complex underwater environments. This makes them ideal for working in environments with small operating spaces, dense obstacle distribution, or complex water conditions. Therefore, the study of underwater snake-like robots has great practical importance.

There are three main gait generation methods for snake-like robots, namely the curve-based method, dynamic model-based method, and CPG-based method. Hirose implemented multiple gait [1] through a snake-like robot and Ma proposed the Serpentine curve [2]. These methods have the advantage that they

This work is supported by the National Natural Science Foundation of China (62073176). All the authors are with the Institute of Robotics and Automatic Information System, College of Artificial Intelligence, Nankai University, China.

H. Liu et al. (Eds.): ICIRA 2022, LNAI 13456, pp. 171–182, 2022.
https://doi.org/10.1007/978-3-031-13822-5_16

do not require an accurate mechanical model, but the causality between its internal parameters and the motion of the robot can only be analyzed qualitatively and is less able to learn about the environment. The advantage of Dynamic model-based approaches is that their physical significance of model parameters is obvious [3–5], but for the complex high-DOF underwater snake-like robot, it is hard to build an accurate model and generate a gait. The CPG-based approach is built by imitating the nervous system of vertebrates and can generate different gait patterns [6,7]. Ouyang divided the control into three parts, in addition to the CPG, using a cerebellar model articulation controller (CMAC) similar to the cerebellum for fast learning and adapting to perturbations in the environment, and a proportional-derivative controller (PD) for faster convergence [8]. This approach allows online parameter updating and is computationally cheap and fast, but model parameter tuning is complex and has no real physical meaning.

All these methods are based on accurate modeling of the snake-like robot and its environment. However, modeling underwater snake-like robot are difficult due to their complex structure and the modeling of complex underwater situations such as ocean currents is still a difficult subject due to the current lack of knowledge in hydrodynamics. Therefore, the existing methods are not accurate enough to estimate the unknown disturbances, which leaves much room for improvement in the accuracy and speed of the snake-like robot motion control.

Biological snakes clearly do not understand the principles of muscle movement, or complex physics, but they are still more flexible and adaptable than human-made snake-like robot. This is because the biological snake's locomotor ability comes from constantly trying out different locomotor strategies in various environments, learning from the feedback given by the environment. This method is very similar to reinforcement learning algorithms, an AI method that is also widely used to generate snake-like robot controllers.

In recent years, with the rapid development of artificial intelligence, reinforcement learning has become increasingly popular, which has led to the proposal of many reinforcement learning algorithms [9–11]. These algorithms are also widely used for robot control [12–14], which also includes snake-like robots. Sartoretti et al. used the asynchronous advantage actor-critic (A3C) algorithm to achieve online snake-like robot motion on unstructured terrain [15]. Bing et al. designed a controller for gait generation with PPO algorithm to obtain a new energy-efficient slithering gait. But these algorithms are also unable to learn and adapt to the environment quickly, just like the non-reinforcement learning methods in the previous section.

In 2021, Kumar et al. proposed the Rapid Motor Adaptation (RMA) algorithm [16], which enables a land legged robot to learn and adapt quickly to complex environments, allowing the robot to move smoothly in a variety of complex environments, including sand, mud, hiking. Inspired by this algorithm, and incorporating ideas from Curriculum learning [17], we propose the Snake-like robot Curriculum Rapid Motor Adaptation(SCRMA) algorithm to solve the problem of fast learning and adaptation of underwater snake-like robot.

The main contributions are summarized as follows:

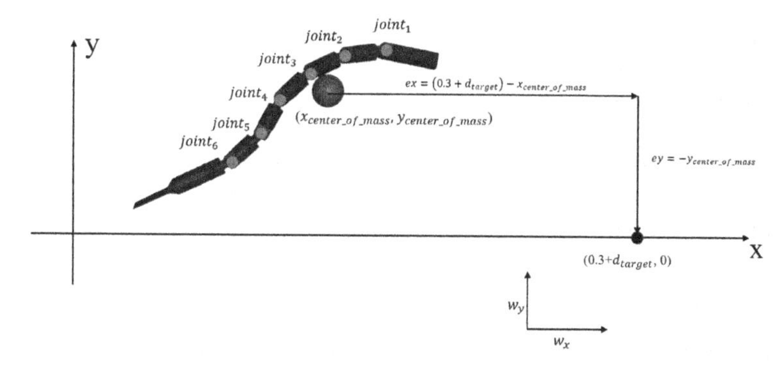

Fig. 1. This figure shows the structure of the underwater snake-like robot. In this figure, these blue dots represent the joints, the grey dot represents the robot's center of mass, and the black dot represents the target point. We use the coordinate of the center of mass to refer to the coordinate of the snake-like robot in this paper. At the beginning of each episode, the target point is generated on the x-axis according to d_{target}, the snake-like robot is placed on the origin, and finally, the wind on the x-axis and y-axis is randomly generated. If the snake-like robot can move to the target point, then we call the robot wins in this episode, otherwise, the robot loses. (Color figure online)

1) SCRMA algorithm is proposed, which enables fast learning and adaptation to underwater ocean currents, enabling the snake-like robot to move to the target point under the action of random currents.
2) Combining with the idea of the Curriculum learning method and proposing the save&load exploration method, which greatly reduces the training time.
3) Compared SCRMA with three other algorithms and demonstrate its high efficiency, accuracy, and robustness.

The remainder of this paper is organized as follows: In Sect. 2, we model the underwater snake-like robot and illustrate the training goal. In Sect. 3, we describe the model in terms of the Markov Decision Process (MDP) and then we propose the SCRMA algorithm. In Sect. 4, we introduce the simulation environment and train 4 agents to demonstrate the superiority of the SCRMA algorithm. In Sect. 5, we summarize the contributions of this paper and explain future research directions.

2 Modeling of Underwater Snake-Like Robots

In this paper, we use the coordinate of the center of mass to refer to the coordinate of the snake-like robot.

As shown in Fig. 1, the underwater snake-like robot used in this paper consists of a snake-head module, five snake-body modules, and a snake-tail module, which are connected by six rotating joints in the middle. The controller controls the motion of the snake-like robot by controlling the joint angle of the six rotating joints. There are winds in the environment both in the x-axis and y-axis directions, namely w_x and w_y in the bottom right corner of Fig. 1. This can be used to represent surface winds, underwater waves, or ocean currents. These forces can blow the snake-like robot in the wind's direction, representing the randomization of the environment. To be consistent with the environmental parameters simulated in Sect. 4, the term 'wind' is used directly below to refer to this fluid action with direction and magnitude.

2.1 Hydrodynamic Model

Since the snake-like robot studied in this paper is located underwater, both gravity and buoyancy forces act on the center of mass and have equal magnitude and opposite directions, thus the effects cancel each other out. This proves that the underwater snake-like robot moves in a two-dimensional plane, and we only need to analyze the kinematics and dynamics model in this two-dimensional plane.

According to Taylor's theory [18], each module of the snake-like robot is assumed to be a long cylinder having a viscosity resistance of each part as formula (1):

$$\begin{cases} f_1 = \rho\pi r l C_f(v_1 + |v_1|v_1)/2 \\ f_2 = \rho\pi l C_d(v_2 + |v_2|v_2) \\ f_3 = \rho\pi l C_d(v_3 + |v_3|v_3) \end{cases} . \tag{1}$$

where ρ is the density of the water, r is the radius of the long cylinder, l is the length and v_1, v_2, v_3 is the velocity relative to the water current.

As the snake-like robot swims through the water, the water exerts a viscosity resistance on each of the snake-like robot's modules separately. As the snake-like robot moves in a Serpentine curve, each of its elements of volume is subjected to a force perpendicular to its body and in the direction opposite to its velocity. The components of these forces along the direction of the snake-like robot's advance constitute the driving force for the snake-like robot's advance.

2.2 Training Goal

The snake-like robot will be initialized at the beginning of each episode. At this time, the coordinates are (0, 0), the angle of each joint is 0, and the snake's head points to the positive direction of the x-axis. Each episode randomly generates a two-dimensional wind as $(w_x, w_y) \in [-wl, wl]^2$ where $wl > 0$ represents the absolute value of the maximum wind speed on each axis.

The training goal of this paper is to train an agent to swim to the target point at an arbitrary direction and size of the wind. Namely, the agent has strong robustness of wind. So we set the target point as $(d_{target} + 0.3, 0)$, where $d_{target} > 0$.

If the snake-like robot can move to a point satisfying $|ex| < 0.01$ and $|ey| < 0.3$ within given time steps, then we call the snake-like robot wins in this episode, otherwise the robot loses.

The reason why the target's x coordinate is $d_{target} + 0.3$ instead of d_{target} is that there is 50% possibility that the x-axis wind is positive so that the agent can win the episode in the first time step if $d_{target} = 0$. Thus there is 50% winning rate at the beginning of the training, so the agent will tend to be conservative in subsequent exploration, so the distance between the target point and the initial position should be at least 0.3, which is the biggest allowable ey.

3 SCRMA

3.1 MDP Representation

To apply the reinforcement learning algorithm, the task is first described as a Markov Decision Process (MDP) [19], including *state* (s_t), *action* (a_t), and *reward* (r_t). To use SCRMA, we also need to describe the environment parameters using *environment code* (e_t).

The *state* needs to be able to fully represent all the information about the robot, ensuring that future decisions are related to the s_t only, so the $s_t \in R^{118}$ is defined to include the following: the angle and angular velocity of the robot's six joints, the pose of the robot's seven modules, as well as the velocity and angular velocity, and the robot's error concerning the target point.

$$\begin{aligned} s_t = \{&\alpha_1, \alpha_2, ..., \alpha_6, \omega_1, \omega_2, ..., \omega_6, \\ &\xi_1, \xi_2, ..., \xi_7, v_1, v_2, ..., v_7, ex, ey\} \end{aligned} \tag{2}$$

The *action* is the target angle for each joint at the next time step, with a size between $-\pi/2$ and $\pi/2$, as illustrated in formula (3).

$$a_t = \{\theta_1, \theta_2, ..., \theta_6\} \in [-\pi/2, \pi/2]^6. \tag{3}$$

The *reward* represents the training goal and determines how good the policy will be. The training goal in this paper is to enable an underwater snake-like robot to swim to the target point at an arbitrary direction and size wind. It is desired to ensure that ey does not exceed 0.3 while reducing ex.

$$r_t = \begin{cases} 1 - ey_t^2 & , if\ agent\ wins \\ [ex_{t-1}^2 - ex_t^2] - ey_t^2 & , otherwise \end{cases}. \tag{4}$$

Namely, punish the ey and encourage the agent to reduce the ex. When finally the snake-like robot wins, give it a little reward equalling to 1.

Environment code represents the force that the environment giving to the robot, including passive force exported by the simulation environment (Mujoco) in Sect. 4, and also including the wind information.

Namely:

$$e_t = \{f_{passive}, w_x, w_y\} \tag{5}$$

where $f_{passive} \in R^{12}$.

3.2 RL-Based Gait Generator

The aim of RL is to find the optimal policy π^* that maximizes the expectation of cumulative return and PPO [10] algorithm is a very efficient RL algorithm.

The PPO algorithm's neural networks include actor network and critic network. The objective function to the update the actor network is

$$L^{PPO}(\theta) = \hat{E}_t[min(r_t(\theta)\hat{A}_t, clip(r_t(\theta), 1-\epsilon, 1+\epsilon)\hat{A}_t)] \tag{6}$$

where $r_t(\theta) = \frac{\pi_\theta(a_t|s_t)}{\pi_{\theta old}(a_t|s_t)}$, and $\hat{A}_t = Q_\pi(s_t, a_t) - V_\pi(s_t)$.

The mean square error (MSE) is used as the loss function of critic network $V\omega$ to update:

$$\begin{aligned} L_{critic}(\omega) &= \frac{1}{T}\sum_{t=0}^{T}(G_t - V_\omega(s_t))^2 \\ &= \frac{1}{T}\sum_{t=0}^{T}((\sum_{t_i=t}^{T-1}\gamma^{t_i-t}r_{t_i} + \gamma^{T-t}V_\omega(s_T)) - V_\omega(s_t))^2. \end{aligned} \tag{7}$$

This method can be used to update Env Factor Encoder (ϕ) too.

$$L_\phi = \frac{1}{T}\sum_{t=0}^{T}(z_t - \hat{z}_t)^2 = \frac{1}{T}\sum_{t=0}^{T}(z_t - \phi((s_{t-5}, a_{t-5}, ..., s_{t-1}, a_{t-1})))^2. \tag{8}$$

Define $s_t = s_0$ and $a_t = a_0$ for $t < 0$ (Fig. 2).

3.3 SCRMA Framework

Human's ability to analyze the environment in real-time when walking comes not only from direct observation from eyes but also from the information we have obtained from past interactions with the environment. For example, we can walk steadily on a city road but the same gait is easy to slip on the ice. However, after a very short time, humans will subconsciously adjust the frequency and amplitude of their legs to adapt to the low friction environment of the ice. SCRMA algorithm is similar to this capability.

SCRMA is divided into two phases. In the first phase we consider (s_t, e_t) as *state*, π and μ as total policy, and train with the PPO [10] algorithm to allow the agent to make optimal decisions in any known environment. In the second phase, we use $z_t = \mu(e_t)$ as the label and $(s_{t-5:t-1}, a_{t-5:t-1})$ as the input vector for supervised learning, which enables the agent to learn the environment parameter from the past information.

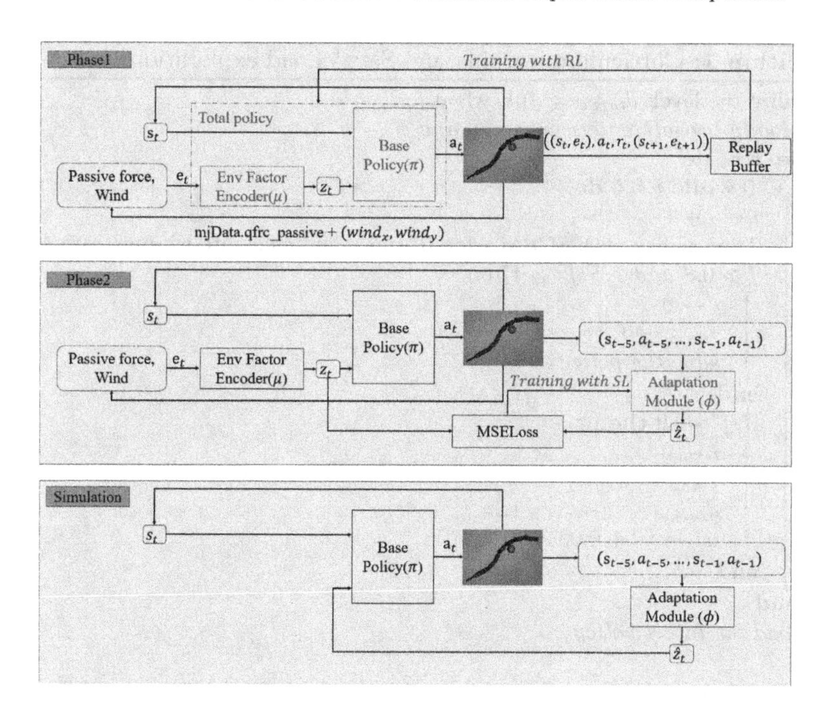

Fig. 2. SCRMA is divided into two phases, and the network marked by the red box in the figure is the network trained in this phase. When doing simulation, only two networks, Base Policy (π) and Adaptation Module (ϕ) are needed. (Color figure online)

3.4 Curriculum Learning and Save&Load Exploration

If we naively train the agent with SCRMA algorithm, it is likely to fail because the target point is too far or the wind is too strong. Therefore, curriculum learning [17] is used to speed up the training. Two parameters directly determine the training difficulty: d_{target} and wl. So set $d_{target_0} = 0.1$ and $wl_0 = 0$ at the beginning of the training, and train the agent in this environment until the agent can win in more than 90% of episodes. Then, boost these two parameters to train the agent for harder courses.

Another problem is policy collapse. Since the RL exploration is random, it is likely that the more training, the worse policy will be. Therefore, this paper proposes the save&load exploration method. When training the agent, every time the policy is updated enough times, the winning rate of the agent is calculated, and if the winning rate is the highest in training history, save the policy, and if the winning rate goes worse, the latest saved policy will be loaded and retrained the agent. This method can effectively prevent the accumulation of randomly worse exploration and accelerate the training speed.

Curriculum learning and Save&Load exploration are shown in Algorithm 1.

Algorithm 1: Curriculum learning and Save&Load exploration

```
1  Initialize the level: d_target = 0.1, wl=0, f_max = 0 ;
2  Randomly initialize networks π, μ and φ ;
3  while True do
4      i ← 0 while i < 5 do
5          i ← i + 1
6          Train policy with SCRMA for 10 times and calculate winning rate f if
           f > 0.8 and f > f_max then
7              i ← 0
8              f_max ← f
9              save the policy
10         end
11         if f > 0.9 then
12             i ← 0
13             f_max ← 0
14             d_target ← d_target * 1.1
15             wl ← wl + 0.005
16         end
17     end
18     load the latest policy
19 end
```

4 Experiment Result

4.1 Simulation Setup

In this paper, we use DeepMind's Mujoco as the simulation environment. There are three parameters describing the hydrodynamic environment in Mujoco: wind, density, and viscosity. We fix density as 3500 and viscosity as 0. In this paper, we set 22 levels of curriculum learning, from the initial $level_1$ with $d_{target} = 0$ and $wl = 0$, to the final $level_{22}$ with $d_{target} = 0.7$ and $wl = 0.1$.

We trained 4 agents with different algorithms. The training algorithm of each agent is shown in the following table (Table 1).

Table 1. Four agents with different training algorithms. Where agent1 is used to demonstrate the effectiveness of the SCRMA algorithm, the others are the control group.

Agent	SCRMA	PPO	Curriculum	Save&load
1	✓	–	✓	✓
2	–	✓	✓	✓
3	–	✓	✓	–
4	–	✓	–	–

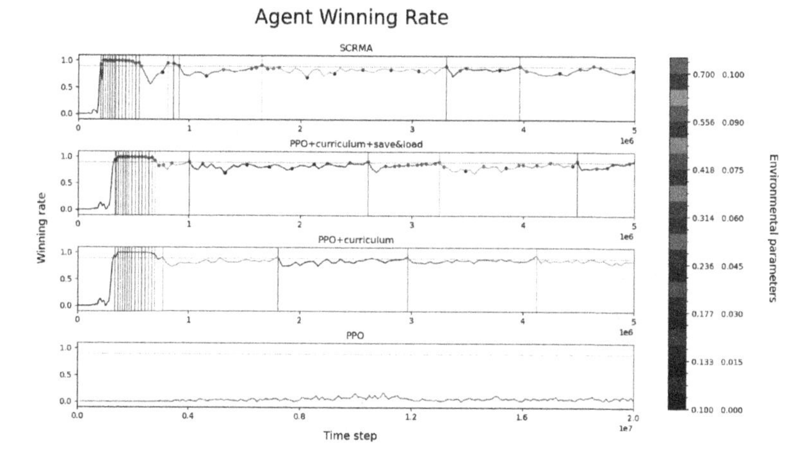

Fig. 3. The first three agents were trained with 5 million time steps, and the last one with only the PPO algorithm was trained with 20 million time steps. Each plot is divided into many segments with different colors, and each color represents a pair of environmental parameters, namely the level of the curriculum. The reflection between color and the environmental parameter is shown in the color bar. In the subplots of the two agents that apply save&load exploration, red dots represent saving models and black dots represent loading models. (Color figure online)

4.2 Winning Rate

Figure 3 shows the winning rate of these four agents while training, and since the win rate of agent4 never exceeded 10% after 5 million time steps, we trained it for another 15 million time steps.

From this fig, it can be seen that even though agent4, which does not use curriculum learning, has the longest training time, its winning rate still does not exceed 20%, which is the worst performance among the four agents. This demonstrates the importance for curriculum learning. We can also see from the figure that after 5 million time steps training, agent3 learned $level_2 0$ in green, agent2 learned $level_2 1$ in blue, and agent1 learned $level_2 2$ in red, which demonstrates the efficiency of SCRMA.

4.3 Error Distribution of Winning Sample

The winning rate, the mean and standard deviation of ey of the four agents are shown in Table 2.

We can find that after 5 million time steps of training, agent1 has the highest winning rate and the mean value is the closest to 0 in these four agents. Agent2 has a higher winning rate than agent3 because agent2 is more efficient in training. Agent4 has trained for 20,000,000 time steps, but its winning rate is the lowest. The standard deviation of all four agents is similar, which is probably due to the similarity of the neural network.

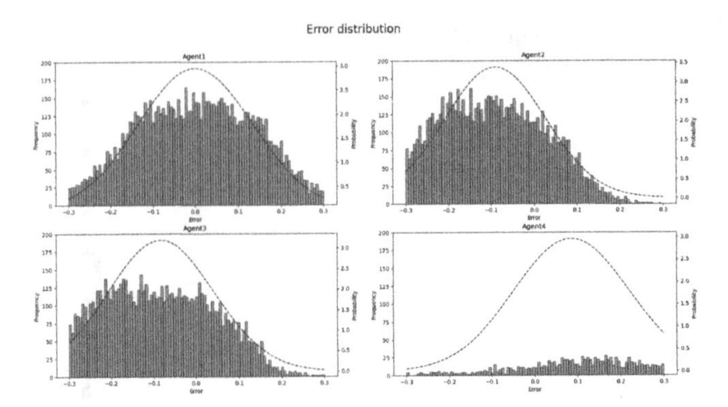

Fig. 4. We count the *ey* of successful samples from 4 agents and draw a normal distribution curve based on the mean and standard deviation of these data. The absolute value of the *ey* does not exceed 0.3 due to the task setting.

Table 2. We compare the performance of the four agents. The first column is the winning rate, and the remaining two columns calculate the mean and standard deviation of the final *ey* of the agent's winning sample.

Agent	Winning rate	Average	Std
1	95.96%	0.00058	0.13507
2	83.42%	−0.08965	0.11791
3	78.37%	−0.08213	0.12494
4	11.30%	0.08594	0.13491

The distribution of *ey* is shown in Fig. 4. Agent1 is closest to the normal distribution and is almost symmetric about the y-axis. This proves the accuracy of SCRMA.

4.4 Wind Histogram of Failed Samples

We also study the winds of the failed samples, as shown in Fig. 5, for analyzing the robustness of different algorithms for wind.

We can find that the failed samples after applying SCRMA are mainly concentrated when the wind in the positive direction of both x and y axes is large. At this time, the *ex* decreases too fast and the *ey* grows too fast, so the agent does not have enough time to eliminate the *ey* and eventually the *ey* becomes too large when the *ex* satisfies the condition.

Agent1, agent2, and agent3 are also sensitive to the wind in the negative direction of x-axis, and it is guessed that the agent does not detect the upwind environment in time and is less robust, so it cannot reduce *ex* to 0 within the given time steps.

No obvious pattern was found in the failed sample of Agent4, which needs to be further study.

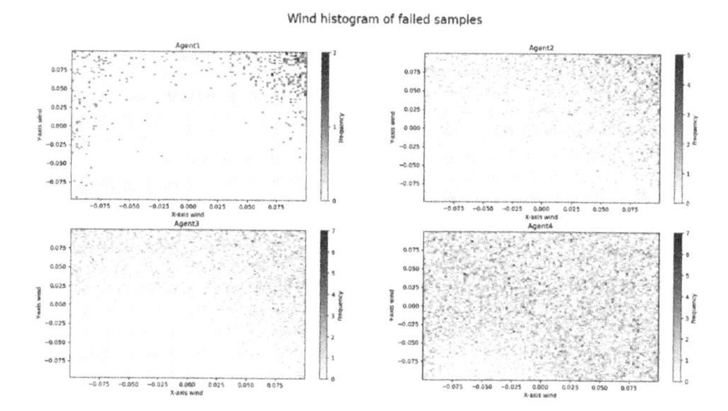

Fig. 5. We used a two-dimensional histogram to calculate the wind in the failed sample, divided into 100 bins. The figures show the ability of 4 agents to adapt to winds of different sizes and directions.

In summary, we can infer that the SCRMA algorithm is designed to significantly improve the winning rate by increasing the sensitivity to wind.

5 Conclusion and Future Work

In this paper, we propose the SCRMA algorithm for training the underwater snake-like robot, so that the robot can complete the task with a high winning rate. The efficiency, accuracy, and robustness of the SCRMA algorithm are demonstrated by comparing it with three other algorithms.

There is still a long way to go. First, the simulation environment can add the randomness of density and viscosity to better simulate the realistic hydrodynamic environment. Second, the target can be generated in the plane rather than on the x-axis, so that the agent can be trained to move in the entire 2D plane. The third is to try to make the agent able to move in the vertical direction to extend its mobility to 3D.

References

1. Hirose, S.: Biologically Inspired Robots: Snakelike Locomotors and Manipulators. Oxford University Press (1993)
2. Ma, S.: Analysis of snake movement forms for realization of snake-like robots. In: Proceedings of the 1999 IEEE International Conference on Robotics and Automation (1999)
3. Matsuno, F., Mogi, K.: Redundancy controllable system and control of snake robots based on kinematic model. In: IEEE Conference on Decision & Control (2000)
4. Mohammadi, A., Rezapour, E., Maggiore, M., Pettersen, K.Y.: Maneuvering control of planar snake robots using virtual holonomic constraints. IEEE Trans. Control Syst. Technol. **24**(3), 884–899 (2016)
5. Ariizumi, R., Matsuno, F.: Dynamic analysis of three snake robot gaits. IEEE Trans. Rob. **33**(5), 1075–1087 (2017)

6. Hasanzadeh, S., Tootoonchi, A.A.: Ground adaptive and optimized locomotion of snake robot moving with a novel gait. Auton. Robot. **28**(4), 457–470 (2010)
7. Tesch, M., Schneider, J.G., Choset, H.: Using response surfaces and expected improvement to optimize snake robot gait parameters. In: IEEE/RSJ International Conference on Intelligent Robots & Systems (2011)
8. Ouyang, W., Liang, W., Li, C., Zheng, H., Ren, Q., Li, P.: Steering motion control of a snake robot via a biomimetic approach. Front. Inf. Technol. Electron. Eng. **20**(1), 32–44 (2019)
9. Minh, V., et al.: Human-level control through deep reinforcement learning. nature **518**, 529–533 (2015)
10. Schulman, J., Wolski, F., Dhariwal, P., Radford, A., Klimov, O.: Proximal policy optimization algorithms (2017)
11. Lillicrap, T.P., et al.: Continuous control with deep reinforcement learning. Computer Science (2015)
12. Okal, B., Kai, O.A.: Learning socially normative robot navigation behaviors with Bayesian inverse reinforcement learning. In: IEEE International Conference on Robotics and Automation (ICRA) (2016)
13. Kretzschmar, H., Spies, M., Sprunk, C., Burgard, W.: Socially compliant mobile robot navigation via inverse reinforcement learning. Int. J. Robot. Res. **35**(11), 1289–1307 (2016)
14. Zhu, Y., et al.: Target-driven visual navigation in indoor scenes using deep reinforcement learning (2016)
15. Sartoretti, G., Shi, Y., Paivine, W., Travers, M., Choset, H.: Distributed learning of decentralized control policies for articulated mobile robots (2018)
16. Kumar, A., Fu, Z., Pathak, D., Malik, J.: RMA: rapid motor adaptation for legged robots (2021)
17. Bengio, Y., Louradour, J., Collobert, R., Weston, J.: Curriculum learning. In: Proceedings of the 26th Annual International Conference on Machine Learning, ICML 2009, Montreal, Quebec, Canada, 14–18 June 2009 (2009)
18. Taylor, G.: Analysis of the swimming of long and narrow animals. Proc. R. Soc. Lond. **214**(1117), 158–183 (1952)
19. Bellman, R.: A Markovian decision process. Indiana Univ. Math. J. **6**(4), 679–684 (1957)

Investigation on the Shape Reconstruction of Cable-Driven Continuum Manipulators Considering Super-Large Deflections and Variable Structures

Yicheng Dai, Xiran Li, Xin Wang, and Han Yuan(✉)

Harbin Institute of Technology Shenzhen, Shenzhen 518055, China
yuanhan@hit.edu.cn

Abstract. The Cable-driven continuum manipulator is a kind of robot with high flexibility and dexterity. It has been attracting extensive attention in the past few years due to its potential applications for complex tasks in confined spaces. However, it is a challenge to accurately obtain the manipulator's shape, especially for super-large deflections, like more than 180°. In this paper, a novel shape reconstruction method based on Bézier curves is proposed for cable-driven continuum manipulators considering super-large defections and variable structures. In addition, a variety of structures of continuum manipulators are considered, where five kinds of manipulators with variable cross sections are manufactured and tested. The feasibility of the proposed method is verified by both simulations and experiments. Results show that the shape reconstruction errors (root-mean-square-error) are all less than 2.35 mm, compared to the 180 mm long manipulators, under various manipulator structures.

Keywords: Continuum manipulators · Shape reconstruction · Bézier curve · Super-large deflection · Variable structures

1 Introduction

In comparison with conventional robots, continuum robots have much more flexibility and compliance with excellent dexterity, and can adjust to the environment without causing any damage to either the environment or the robot themselves [1,2]. Continuum manipulators are the functional imitation and electronic mechanization of soft animals in the biological world, such as the snake, octopus tentacle and elephant trunk. They have great potentials in minimally invasive surgeries [3,4] and maintenance in confined spaces [5].

However, there exists a disadvantage in shape controlling of the manipulator for its virtual joints' rotation cannot be controlled independently. Therefore the backbone deformation cannot be controlled as desired [6]. Reconstruct the shape of the continuum manipulator becomes very important in practical applications especially when there is a need for avoiding obstacles. There are many ways to obtain the shape of manipulators [7]. They are mainly divided into two categories, one is the measuring method, and another is calculation method. The first kind of methods are mainly based on detecting devices, such as the Optitrack [8], which can obtain the position information by tracking the target balls. Some research proposed that the shape of manipulators could be

H. Liu et al. (Eds.): ICIRA 2022, LNAI 13456, pp. 183–194, 2022.
https://doi.org/10.1007/978-3-031-13822-5_17

measured by X-ray [9,10], Ultrasound [11,12], or magnetic resonance imaging (MRI) [13]. An emerging detection method is using fiber Bragg grating sensors [14–18]. It is a promising tool due to its thin, flexible, and weightless nature. And another method is measuring the shape of manipulators by electromagnetic(EM) sensors [19].

The calculation methods are based on geometry, kinematics and statics, including Cosserat rod theory [20], elliptical integrals [21], statics model [22] and modal approach [23]. And there is another shape reconstruction method based on the 3rd-order Bézier curve. But this method can only be used in small deflection angle [6], since unacceptable errors will be observed when the manipulator's deflection angle is larger than 180°.

In this paper, we present a shape reconstruction method based on 4th-order and 5th-order Bézier curve for the cable-driven continuum manipulator. The proposed method is based on the length of the robot and the pose information of the initial point and the distal point, which can be obtained by EM sensors. The performance of this method is veridied by simulations and experiments. Moreover, the adaptability of the proposed method in continuum manipulators with variable structures is validated. And the adapted range of Bézier curve in different order will be given. The main contribution of this paper are summarized as follows.

- A general shape reconstruction method for continuum robots is proposed.
- It is applicable to continuum manipulators with traditional constant structures and variable structures.
- This method is applicable to continuum manipulators in super-large deflections which exceed 180°.
- The application range of different orders of Bézier curves is given.

The organization of this paper is as follows. In Sect. 2, the structure and mathematics model of the Bézier curve will be presented. In Sect. 3, the simulation of shape reconstruction based on Bézier curves are conducted. In Sect. 4, experiments of shape reconstruction on continuum manipulators in different conditions are conducted, and the simulation experiment results will be shown and analyzed. Conclusions will be made in Sect. 5.

2 Shape Reconstruction Method Based on Bézier Curve

In this part, the shape reconstruction method based on Bézier curve will be introduced. A Bézier curve is a parametric curve frequently used to model smooth curves in computer graphics and related fields, which can be formulated as:

$$\mathbf{P}(t) = \sum_{i=0}^{n} \mathbf{P}_i \mathbf{B}_{i,n}(t), t \in [0,1] \tag{1}$$

$$\mathbf{B}_{i,n}(t) = C_n^i t^i (1-t)^{n-i} \tag{2}$$

where $\mathbf{P}_0$ is the start point and $\mathbf{P}_n$ is the end point of the curve; $\mathbf{P}_1 \sim \mathbf{P}_{n-1}$ are control points. The curve starts at $\mathbf{P}_0$ going toward $\mathbf{P}_1 \sim \mathbf{P}_{n-1}$ and arrives at $\mathbf{P}_n$. The first and the last control points of an nth-order Bézier curve can be expressed as:

$$\mathbf{P}_1 = \mathbf{P}_0 + S_{01}\mathbf{H}_0 \tag{3}$$

$$\mathbf{P}_{n-1} = \mathbf{P}_n - S_{n-1,n}\mathbf{H}_n \tag{4}$$

In the equations above, S_{01} and $S_{n-1,n}$ represent the length of $\mathbf{P}_0\mathbf{P}_1$ and $\mathbf{P}_{n-1}\mathbf{P}_n$ respectively. $\mathbf{H}_0$ is the direction vector from $\mathbf{P}_0$ to $\mathbf{P}_1$. $\mathbf{H}_0$ and $\mathbf{P}_0$ are related to the experiment setup. $\mathbf{H}_n$ is the direction vector from $\mathbf{P}_{n-1}$ to $\mathbf{P}_n$. $\mathbf{H}_0$ and $\mathbf{H}_n$ are the tangents at $\mathbf{P}_0$ and $\mathbf{P}_n$, respectively. $\mathbf{H}_n$ and $\mathbf{P}_n$ can be measured by EM sensor or other method.

In the expression of a Bézier curve, there are two unknown parameters S_{01}^3 and S_{23}^3 for the 3rd-order Bézier curve. For the 4th and the 5th-order Bézier curves, except for $\mathbf{P}_1$ and $\mathbf{P}_{n-1}$ there are other control points that can't be expressed by S_{01} and S_{34}. In a planar model, the number of unknown parameters is 4 including coordinates of two points for the 4th-order Bézier curve. It is 6 for the 5th-order Bézier curve including coordinates of three points.

In order to obtain the unknown parameters, an error evaluation function based on the Levenberg Marquardt (LM) algorithm will be established. For a continuum manipulator, there only exists radial bending, in the deflecting process. The manipulator won't be stretched or compressed in the axial direction. And the length of the continuum manipulator is constant. Based on this character, the optimization function can be expressed as:

$$f = \sum_{i=1}^{n} (L_i - L_{bi})^2 \tag{5}$$

$$L_{bi} = \int_{(i-1)/n}^{i/n} \left\| \mathbf{B}'(t) \right\| dt \tag{6}$$

where n is the number of points that are used to estimate the curve length for the nth-order Bézier curve. In this paper, n is set to be equal to the segment number of the robot. Therefore, each joint in the robot will have a corresponding point on the curve. And L_i represents the real length of the backbone between two disks. L_{bi} is the corresponding length of the Bézier curve. Then the coordinates parameters of unknown control points for Bézier curve can be calculated by minimize the objective function f.

3 Simulations of Shape Reconstruction Methods Based on Bézier Curves

In this part, we carry out a set of simulations on a normal continuum manipulator with one section to compare the shape reconstruction performance of different orders of Bézier curves. The reference shape is generated by numerical method. In Fig. 1, except for the coordinate origin, each point represents a disk, and there are 18 disks in the manipulator.

From the simulation results shown in Fig. 1 we can see that, the proposed methods based on the 4th-order and the 5th-order Bézier curves have a much better performance than the method based on the 3rd-order Bézier curve when the manipulator is in a large deflection (larger than 180°, shape 1, 2, 9), while the performance are relatively close when the deflection is small (smaller than 180°).

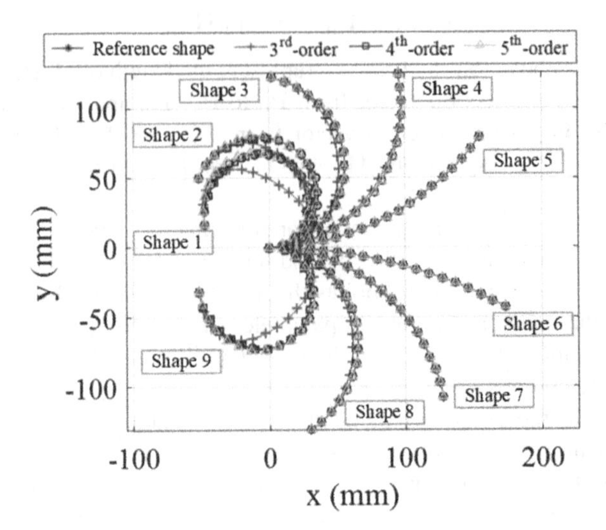

Fig. 1. Simulation results of shapes obtained by different order Bézier curves.

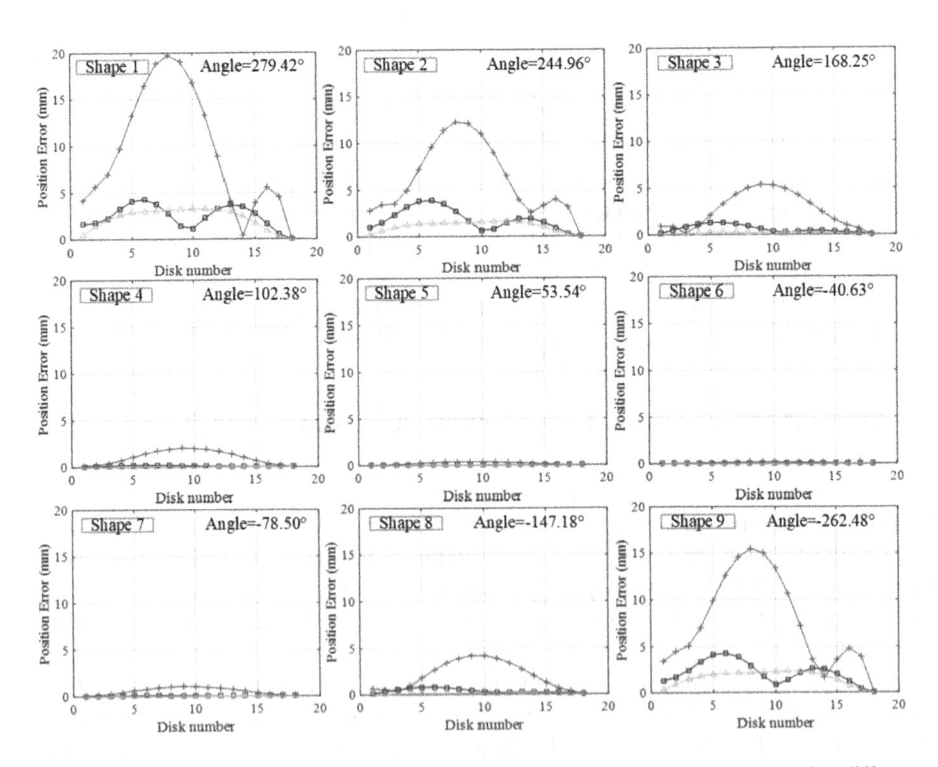

Fig. 2. The position error between reconstructed disks and numerical disks in different deflections.

Results shown in Fig. 2 indicate that the largest position error happens in the middle of the shape, for the start and end point are known while the other control point are calculated. And the position error reduces as the deflection decreases, and ascends as the deflection increases. It should be noted that, in the shape of 1, 2, 3, 8 and 9, there exists a decrease in position error, for there is a cross between the reconstructed shape and the reference shape (Fig. 3).

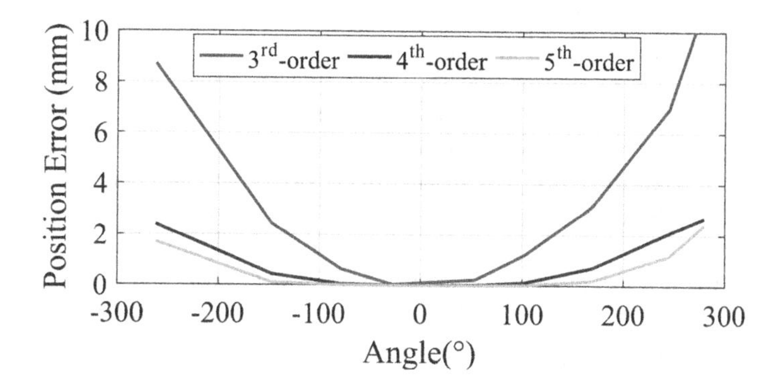

Fig. 3. RMSE of reconstructed shapes in different deflections.

According to the RMSE between the reconstructed shapes and the reference shapes, the proposed methods based on the high order Bézier curves have better performance, especially when the deflection angle is about larger than 120°, where the RMSE of the shape is about 2 mm.

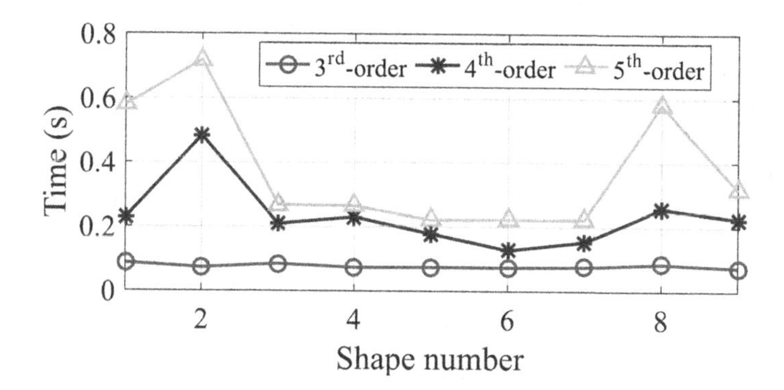

Fig. 4. Time-consuming of different orders Bézier curves.

Except for the shape reconstruction performance, the time-consuming is also important. Results in Fig. 4 shows the time-consuming of different order curves. We can see that the higher order curve needs more time in optimizing calculation. In practical, methods can be chosen by considering both the shape reconstruction performance and time-consuming.

4 Experiment

In this section, experiments will be performed on 5 types of continuum manipulators. They are designed in different ways as shown in Fig. 5. The spacers are fabricated by 3D printer with the thickness of 2 mm while the flexible backbone employs superelastic NiTi backbones with the thickness of 0.2 mm. The driven cables are normal wire ropes, with a diameter of 0.4 mm. In Fig. 5, K_{rate} represents the decreasing rate of interval between two disks, while K_{slope} represents the cable path slope rate relative to the backbone. The 5 prototypes are introduced below.

1. Type-1: the interval between two spacers is constant while the cross-section width of the intermediate flexible backbone as well as the size of spacers keeps constant;
2. Type-2: the interval between two spacers decreases alone the backbone, while the cross-section width of flexible backbone and the size of spacers keeps constant;
3. Type-3: the size of spacers scales down while the interval between two spacers and the cross-section width of flexible backbone keeps constant;
4. Type-4: the cross-section width of flexible backbone scales down while the interval between two spacers and the size of spacers keeps constant;
5. Type-5: both the cross-section width of flexible backbone and the size of spacers decreases while the interval between two spacers keeps constant.

4.1 Experimental Setup

As shown in Fig. 6, the continuum manipulator is fixed on a plastic base installed on the optical flat panel. There are two modes of the experimental device. In mode I, the cable is driven by standard weight, as shown in the upper right corner of Fig. 6. It is used in verifying the application of the proposed method on the 5 prototypes. Making a small change, the device will be altered into mode II, in which the cable is driven by a servo motor. This mode is used in the trajectory following experiment.

The manipulator profile which means the real shape is measured by a camera, through identifying the red markers mounted at each sapcers of the manipulator. The tip position and orientation are measured by a EM sensor. The little needle EM sensor and the data wire are extremely lightweight, which can be ignored.

4.2 Experiments on 5 Types of Continuum Manipulators

The distance root-mean-square error (RMSE) of the center of disks between the camera result and the reconstruction result is used to evaluate the performance. Shape reconstruction based on constant curvature is a common method. In this part, it serves as a comparison method, which is represented in magenta line with circle.

According to the simulation experiment results shown in Fig. 7, the constant curvature method has worst shape reconstruction performance, for it has strict conditions of use. It is not applicable to continuum manipulators with frictions and variable strictures. As for methods based on Bézier curves, they are applicable to continuum manipulator with constant structures or variable structures. As the results show, they all have good

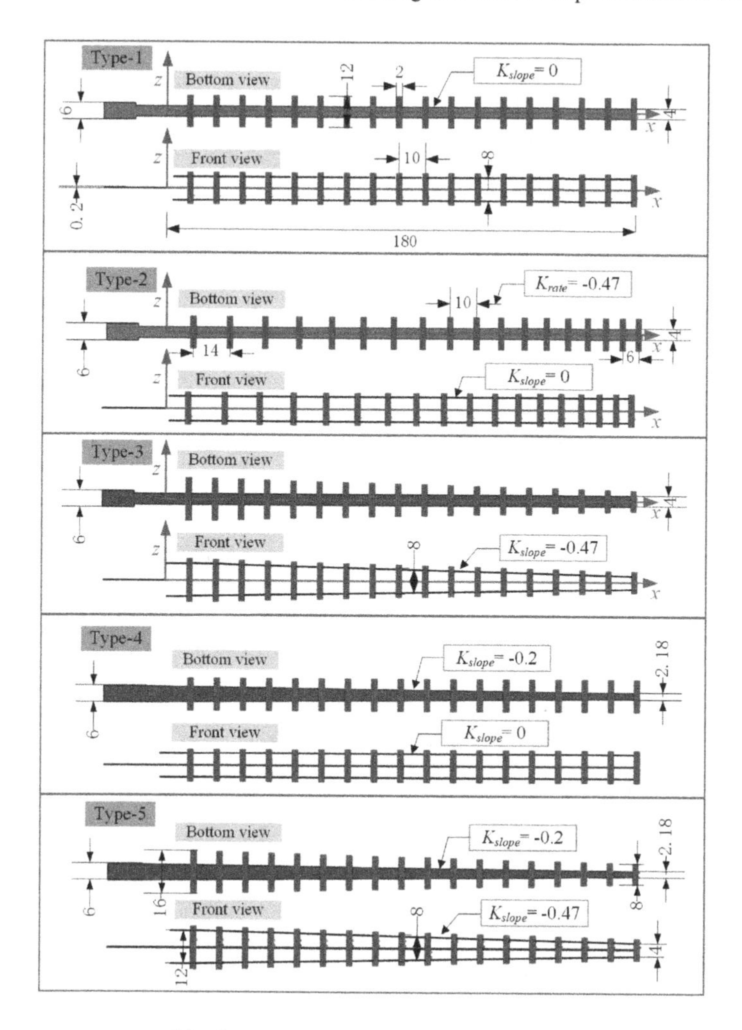

Fig. 5. 5 types of continuum manipulators.

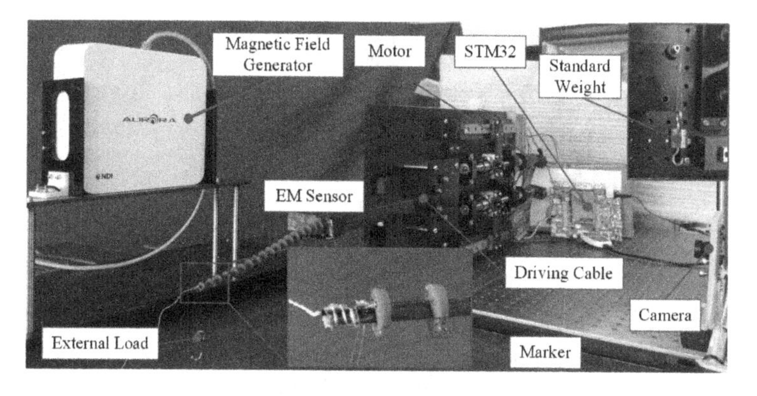

Fig. 6. Experimental setup.

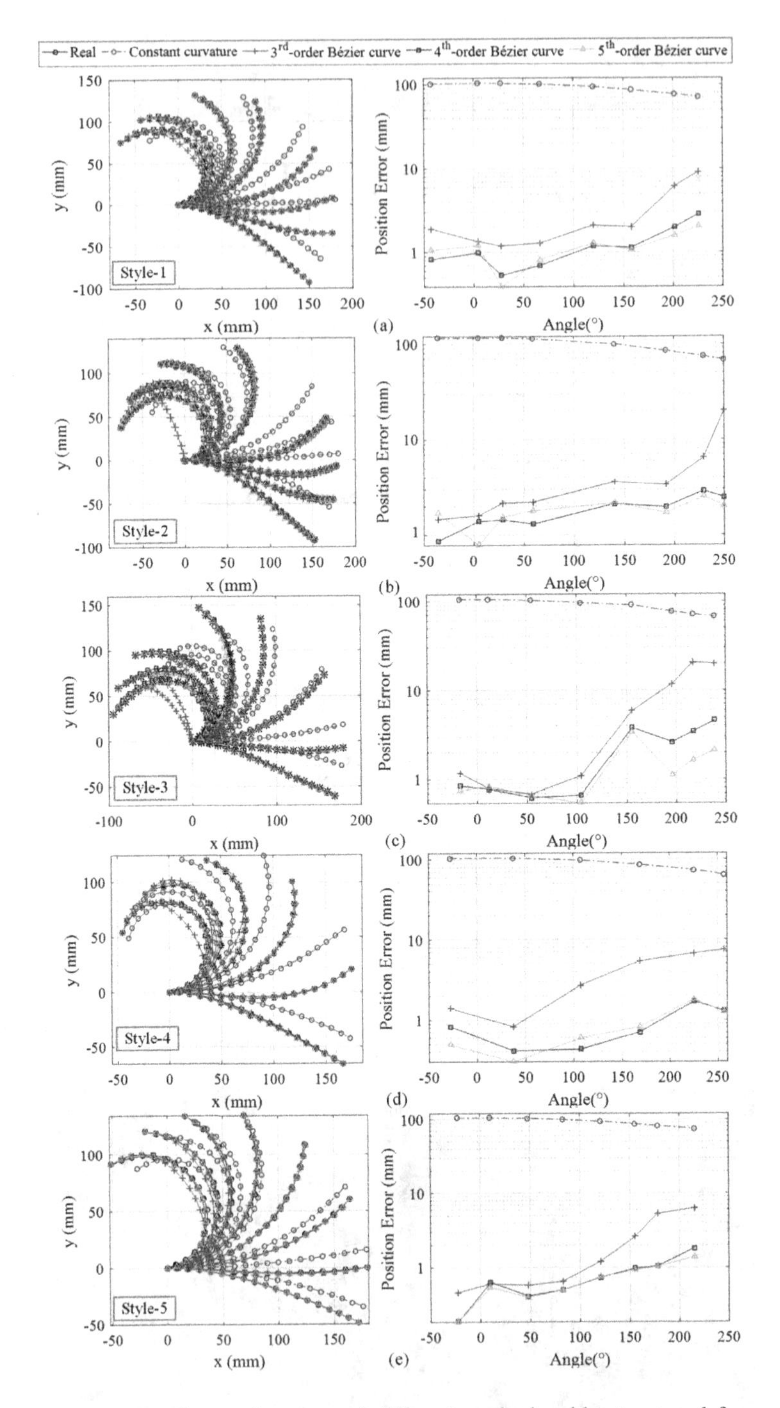

Fig. 7. Shape reconstruction performance of different methods without external force. (In the experiments, 8 curves are generated in (a, b, c, e), with cable force from 0g to 350g, with an interval of 50g; and 6 curves are generated in (d), with cable force from 0g to 500g, with an interval of 100g.).

shape reconstruction performance when the deflection is small. But when the deflection becomes larger, the RMSE of the 3rd-order curve increases quickly while RMSEs of the 4th and 5th-order curves are still small. This results demonstrates that the higher order Bézier curves have a good shape reconstruction performance in super-large deflections.

Table 1. Simulation experiments errors of different shape reconstruction methods.

Category	n^{th}-order	Deflection angle (°)		
		0–100	100–180	180–250
Style-1	3	1.44	2.05	7.69
	4	0.76	1.16	2.41
	5	0.87	1.17	1.81
Style-2	3	1.86	3.58	9.86
	4	1.25	2.11	2.43
	5	1.46	2.16	2.06
Style-3	3	0.89	3.47	17.34
	4	0.75	2.23	3.55
	5	0.75	1.97	1.63
Style-4	3	1.12	2.71	6.52
	4	0.62	0.43	1.23
	5	0.39	0.59	1.29
Style-5	3	0.57	1.21	4.73
	4	0.43	0.73	1.27
	5	0.41	0.74	1.11
Average	3	1.17	2.60	9.23
	4	0.76	1.33	2.18
	5	0.77	1.33	1.58

According to the error analysis in Tablc 1, when the deflection angle of the manipulator is in the range of 0–90°, the average RMSEs of these 5 types of manipulators are 1.17 mm, 0.76 mm and 0.77 mm respectively. When the deflection angle is in the range of 90–180°, the average RMSEs are 2.6 mm, 1.33 mm and 1.33 mm respectively. And in the range of 180–250°, the average RMSEs are 9.23 mm, 2.18 mm and 1.58 mm respectively. It can be concluded that, the effective application range of the 3rd-order Bézier curve is 0–90°, and it is 90–180° for the 4th-order Bézier curve, and 180–250° for the 4th-order Bézier curve.

In this part, experiments on cable-driven continuum manipulators with a wide range of structures have been conducted, and the results show that methods based on the Bézier curves have a broad adaptability in continuum manipulators with different structures.

4.3 Trajectory Following Experiments

To further verify the shape reconstruction performance of higher order Bézier curves, an open-loop experiment is carried out on the third prototype. The one section manipulator is a planar robot with one DOF.

In the experiment, the tip orientation which means the deflection angle of the continuum manipulator that will be controlled. For cable-driven robots, the cable lengths are closely related to the shape of the manipulator. In this experiment, the distal end of the manipulator will be controlled to arrive at the desired deflection angle by controlling driving cables.

Results in Fig. 7 illustrate that, different orders of Bézier curves have different reconstruction shape. Therefore the corresponding lengths are different, and different shape control accuracy will be obtained by using these cable length datas. To reconstruct the shape of a continuum manipulator by using Bézier curves, both the position and the orientation should be known. In this experiment, these informations will be obtained first by an EM sensor in a previous movement, in which the continuum manipulator deflected from 0 to 230°. Shapes and the corresponding lengths will be obtained through different reconstruction methods. Then the lengths will be used to control the manipulator.

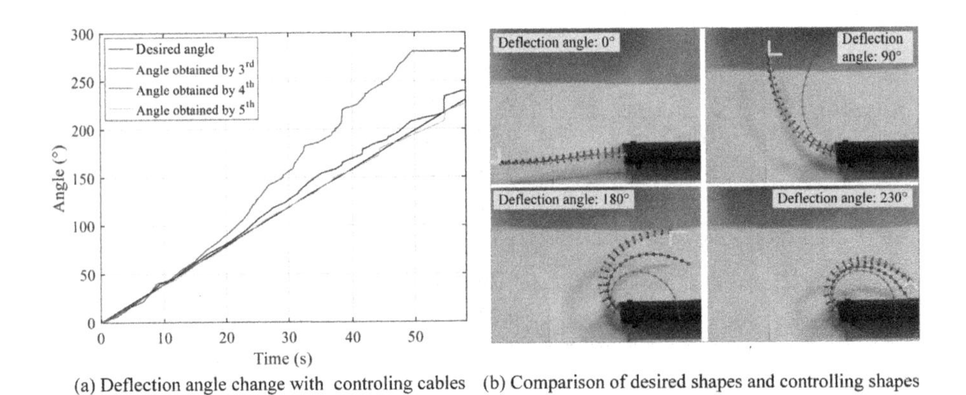

(a) Deflection angle change with controling cables (b) Comparison of desired shapes and controlling shapes

Fig. 8. Trajectory following experiments.

As shown in Fig. 8(a), the control angles obtained by cable lengths of different orders of Bézier curves are pretty close when the deflection is small, and the difference ascends as the deflection angle increases, especially the result obtained by the method based on 3rd-order Bézier curve. Four shape results of the trajectory following experiment are shown in Fig. 8(b). The shapes of ground truce in deflection of 0°, 90°, 180°, and 230° are shown in the pictures, and the shapes obtained by controlling driving cables are shown in colored lines. These results illustrate that the cable lengths obtained by the 3rd-order Bézier curve is invalid in controlling the manipulator to reach the desired deflection. As for the cable length obtained by the 4th-order Bézier curve, the controlling effect is good when the desired deflection is smaller than 180°. The cable length obtained by the 5th-order Bézier curve has the best controlling effect in almost all the deflections. These results indicate that the proposed method has a good shape reconstruction performance in super-large deflections.

5 Conclusion

In this paper, a shape reconstruction method of continuum manipulator based on a high order Bézier curve considering large deflection and variable structures is proposed and verified by simulations and experiments. Five prototypes, with constant structure and variable structures, are fabricated and tested. Some conclusions are made according to the simulation experiment results and analyses. The shape reconstruction method based on the 3rd-order Bézier curve is valid when the deflection angle is smaller than 90°, and the average RMSE is 1.2 mm, while it is 2.6 mm in the range of 90–180°, and invalid when the deflection angle surpassing 180°. The 4th-order Bézier curve is effective when the deflection angle is smaller than 180°, and the average RMSE is 1.33 mm, while it is 2.18 mm in the range of 180–250°. The 5th-order Bézier curve has the best shape reconstruction performance with average RMSE of 0.77 mm in 0–90°, 1.32 mm in 90–180°, and 1.58 mm in 180–250°.

Results show that the shape reconstruction performances of the 4th-order and the 5th-order Bézier curve methods are much better than the 3rd-order Bézier curve method, when the manipulator is in super-large deflections (larger than 180°), and the average RMSEs for the 4th-order and the 5th-order Bézier curve methods are 2.18 mm and 1.58 mm respectively, while it is 9.23 mm for the 3rd-order Bézier curve method. Among these five prototypes, the maximum RMSEs for the 4th-order and the 5th-order Bézier curve methods are 4.95 mm and 2.35 mm, with the corresponding relative errors of 2.75% and 1.31%, compared to the 180 mm long manipulators. They are much smaller than that for the 3rd-order Bézier curve method, which are 20 mm and 11.1%.

We believe in the near future, continuum robots will be used in flexible grasping, like an real elephant trunk. The proposed shape reconstruction method based on the high order Bézier curve, which is applicable in super-large deflections, will play an important role.

Acknowledgements. This work was supported by the National Natural Science Foundation of China (Grants No. 62173114), and the Science and Technology Innovation Committee of Shenzhen (Grants No. JCYJ20210324115812034, JCYJ20190809110415177, GXWD20201230155427003-20200821181254002).

References

1. Xu, K., Simaan, N.: An investigation of the intrinsic force sensing capabilities of continuum robots. IEEE Trans. Rob. **24**(3), 576–587 (2008)
2. Morales Bieze, T., Kruszewski, A., Carrez, B., Duriez, C.: Design, implementation, and control of a deformable manipulator robot based on a compliant spine. Int. J. Robot. Res. **39**(14), 1604–1619 (2020)
3. Fontanelli, G.A., Buonocore, L.R., Ficuciello, F., Villani, L., Siciliano, B.: An external force sensing system for minimally invasive robotic surgery. IEEE/ASME Trans. Mechatron. **39**(14), 1543–1554 (2020)

4. Burgner-Kahrs, J., Rucker, D.C., Choset, H.: Continuum robots for medical applications: a survey. IEEE Trans. Rob. **31**(6), 1261–1280 (2015)
5. Dong, X., et al.: Development of a slender continuum robotic system for on-wing inspection/repair of gas turbine engines. Robot. Comput. Integr. Manuf. **44**, 218–229 (2017)
6. Song, S., Li, Z., Meng, M.Q.-H., Yu, H., Ren, H.: Real-time shape estimation for wire-driven flexible robots with multiple bending sections based on quadratic Bézier curves. IEEE Sens. J. **15**(11), 6326–6334 (2015)
7. Shi, C., et al.: Shape sensing techniques for continuum robots in minimally invasive surgery: a survey. IEEE Trans. Biomed. Eng. **64**(8), 1665–1678 (2016)
8. Song, J., Ma, J., Tang, L., Liu, H., Kong, P.: Deformation measurement of drogue in wind tunnel test based on OptiTrack system. In: 2020 5th International Conference on Automation, Control and Robotics Engineering (CACRE), pp. 565–569. IEEE (2020)
9. Lamecker, H., Wenckebach, T.H., Hege, H.-C.: Atlas-based 3d-shape reconstruction from X-ray images. In: 18th International Conference on Pattern Recognition, ICPR 2006, vol. 1, pp. 371–374 (2006)
10. Lobaton, E.J., Fu, J., Torres, L.G., Alterovitz, R.: Continuous shape estimation of continuum robots using X-ray images. In: 2013 IEEE International Conference on Robotics and Automation, pp. 725–732 (2013)
11. Boctor, E.M., Choti, M.A., Burdette, E.C., Webster, R.J., III.: Three-dimensional ultrasound-guided robotic needle placement: an experimental evaluation. Int. J. Med. Robot. Comput. Assist. Surg. **4**(2), 180–191 (2008). https://doi.org/10.1002/rcs.184
12. Kim, C., Chang, D., Petrisor, D., Chirikjian, G., Han, M., Stoianovici, D.: Ultrasound probe and needle-guide calibration for robotic ultrasound scanning and needle targeting. IEEE Trans. Biomed. Eng. **60**(6), 1728–1734 (2013)
13. Park, Y.-L., et al.: Real-time estimation of 3-D needle shape and deflection for MRI-guided interventions. IEEE/ASME Trans. Mechatron. **15**(6), 906–915 (2010)
14. Henken, K., Van Gerwen, D., Dankelman, J., Van Den Dobbelsteen, J.: Accuracy of needle position measurements using fiber Bragg gratings. Minim. Invasive Ther. Allied Technol. **21**(6), 408–414 (2012)
15. Kim, B., Ha, J., Park, F.C., Dupont, P.E.: Optimizing curvature sensor placement for fast, accurate shape sensing of continuum robots. In: 2014 IEEE International Conference on Robotics and Automation (ICRA), pp. 5374–5379 (2014)
16. Roesthuis, R.J., Janssen, S., Misra, S.: On using an array of fiber Bragg grating sensors for closed-loop control of flexible minimally invasive surgical instruments. In: 2013 IEEE/RSJ International Conference on Intelligent Robots and Systems, pp. 2545–2551 (2013)
17. Ryu, S.C., Dupont, P.E.: FBG-based shape sensing tubes for continuum robots. In: 2014 IEEE International Conference on Robotics and Automation (ICRA), pp. 3531–3537 (2014)
18. Moon, H., et al.: FBG-based polymer-molded shape sensor integrated with minimally invasive surgical robots. In: 2015 IEEE International Conference on Robotics and Automation (ICRA), pp. 1770–1775 (2015)
19. Bajo, A., Simaan, N.: Kinematics-based detection and localization of contacts along multisegment continuum robots. IEEE Trans. Rob. **28**(2), 291–302 (2011)
20. Rucker, D.C., Webster, R.J., III.: Statics and dynamics of continuum robots with general tendon routing and external loading. IEEE Trans. Rob. **27**(6), 1033–1044 (2011)
21. Xu, K., Simaan, N.: Analytic formulation for kinematics, statics, and shape restoration of multibackbone continuum robots via elliptic integrals. J. Mech. Robot. **2**(1), 011006 (2009)
22. Yuan, H., Zhou, L., Xu, W.: A comprehensive static model of cable-driven multi-section continuum robots considering friction effect. Mech. Mach. Theor. **35**, 130–149 (2019)
23. Chirikjian, G.S., Burdick, J.W.: A modal approach to hyper-redundant manipulator kinematics. IEEE Trans. Robot. Autom. **10**(3), 343–354 (1994)

Intelligent Perception and Control of Rehabilitation Systems

An Improved Point-to-Feature Recognition Algorithm for 3D Vision Detection

Jianyong Li[1(✉)], Qimeng Guo[1], Ge Gao[1], Shaoyang Tang[1], Guanbo Min[1], Chengbei Li[1], and Hongnian Yu[2,3]

[1] College of Computer and Communication Engineering, Zhengzhou University of Light Industry, Zhengzhou 450001, China
lijianyong@zzuli.edu.cn

[2] School of Electrical Engineering, Zhengzhou University, Zhengzhou 450001, China

[3] School of Engineering and the Built Environment, Edinburgh Napier University, Edinburgh EH11 4BN, UK

Abstract. Vision-detection-based grasping is one of the research hotspots in the field of automated production. As the grasping scenes become more and more diversified, 3D images are increasingly chosen as the input images for object recognition in complex recognition scenes because they can describe the morphology and pose information of the scene target objects more effectively. With object recognition and pose estimation in 3D vision as the core, this paper proposes an improved pose estimation algorithm based on the PPF feature voting principle for the problems of low recognition rate and poor real-time performance in vision detection systems. The algorithm firstly performs preprocessing measures such as voxel downsampling and normal vector calculation on the original point cloud to optimize the point cloud quality and reduce the interference of irrelevant data. Secondly, an improved point cloud downsampling strategy is proposed in the point cloud preprocessing stage, which can better preserve the surface shape features of the point cloud and avoid introducing a large number of similar surface points. Finally, an improved measure of scene voting ball is proposed in the online recognition stage. The recognition and matching experiments on the public dataset show that the proposed algorithm has an average recognition rate improvement of at least 0.2%.

Keywords: 3D images · Pose estimation · PPF feature voting · Point cloud processing

Supported by National Natural Science Foundation of China (62103376), Robot Perception and Control Support Program for Outstanding Foreign Scientists in Henan Province of China (GZS2019008), China Postdoctoral Science Foundation (2018M632801) and Science & Technology Research Project in Henan Province of China (212102310253).

H. Liu et al. (Eds.): ICIRA 2022, LNAI 13456, pp. 197–209, 2022.
https://doi.org/10.1007/978-3-031-13822-5_18

1 Instruction

The rising cost of labor in China has led to an increase in production costs, and the aging of society has also exacerbated the problem of low productivity in the manufacturing sector due to the lack of a large workforce. Automation in the production field is the first way to solve the high production cost and low productivity of enterprises. At the same time, with the introduction of industrial planning, more and more production departments are using automation to replace workers in tedious operations, especially in the field of parts assembly and industrial sorting. Nowadays, although robots can replace workers to perform some tedious tasks, their ability to perceive the environment and their flexibility to accomplish tasks are greatly reduced compared to humans. A key issue and an important tool to improve the environment perception capability of industrial robots and to enhance the flexibility of industrial robots for production tasks is to use computer vision for target object recognition and pose estimation.

Initially, target object recognition and pose estimation is performed by 2D images obtained from planar RGB cameras. This algorithm should extract the key points based on the obtained scene image and the object model, then perform some specific descriptor operation on the local area based on the key points, then match the descriptors of the scene image with the descriptors of the model image in the feature space, and finally restore the 6D pose of the target object according to the PnP algorithm. Lepetit et al. [1] proposed the SIFI scale-invariant feature transformation algorithm is the most representative algorithm of this type. The SIFI scale-invariant feature transformation algorithm has good stability and will not be affected by the transformation of the viewpoint. Although SIFT is the most robust locally invariant descriptor, the algorithm is designed for grayscale images, and many objects are incorrectly assigned if color features are ignored in the description and matching of objects. To address this problem, Abdel et al. [2] proposed the CSIFT algorithm, which constructs descriptors in color invariant space and was shown to be more robust to color and luminosity changes compared to the traditional SIFT algorithm. Kehl et al. [3] then proposed an SSD target detection algorithm, which is based on deep learning and is able to restore the 6D pose of a target object from a single 2D RGB image.

With the development of depth cameras and their applications, studying the poses of target objects based on 3D images has gradually become a mainstream. Rusu et al. [4] proposed the point feature histogram PFH descriptor, which performs recognition operations by describing local features around points in a 3D point cloud dataset. However, this algorithm is time consuming, and for this problem (Rusu et al. [5]) proposed the FPFH algorithm, which solves the problem of 3D alignment of overlapping point cloud views by modifying the mathematical expressions as well as rigorous analysis, thus reducing the computational time. Tombari et al. [6] proposed the SHOT algorithm, a novel surface feature representation scheme, as well as the addition of the striatal CSHOT descriptor, Tombari et al. [7] has significant recognition rate in complex scenes. Yulan Guo et al. [8] conducted a comparative experiment of 10 descriptors including the above-mentioned local descriptors commonly used in 3D object recognition, 3D shape retrieval and 3D modeling on a publicly available dataset. Rusu et al. [9] extended the local descriptor FPFH from local description to global description, named VFH global descriptor, which is robust to surface noise and images with missing depth information. Aldoma et al. [10] proposed the clustered viewpoint feature histogram CVFH algorithm based on the improvement of the VFH algorithm, which is able to present good results under the influence of partial occlusion and noise. Drost et al. [11] proposed the PPF-based point-to-feature voting algorithm is a representative algorithm based on the Hough voting algorithm. The algorithm describes the model 3D point cloud globally and the scene 3D point cloud locally, and invokes the Hough voting mechanism to make positional assumptions about the target object in the parameter space by voting. The PPF point pair feature algorithm has a high recognition rate under noise, clutter and partial occlusion, and is a relatively advanced algorithm at that time. Choi et al. [12] proposed color point pair features based on PPF point pair features to make full use of color information and applied them to a voting scheme to obtain more effective pose estimation. Drost et al. [13] proposed a novel multimodal, scale and rotation invariant feature based on the PPF point pair feature algorithm combined with intensity information to describe the contour of the object as well as the surface morphology simultaneously, improving the robustness of the algorithm to occlusions and clutter more. Birdal et al. [14] applied weighted Hoff voting and interpolation recovery of the positional parameters in the matching process to improve the accuracy of the final pose estimation. Hinterstoisser et al. [15] invoked a new sampling as well as voting scheme on the basis of the Drost-PPF algorithm, which very significantly improved the anti-interference ability of the algorithm against the background as well as the sensor and the recognition accuracy. The algorithm causes the problem of high recognition accuracy but also high algorithm complexity because of the expansion of the left and right neighborhoods of the four-dimensional descriptors, etc. Therefore, the application scenario of this algorithm is more suitable for those who are willing to give up the cost to pursue accuracy. Wang et al. [16] et al. proposed a new voting scheme based on the algorithm proposed by Hinterstoisser, which includes a series of improvements such as field point index interruption, field point division, and support point removal strategies, resulting in a significant reduction in computation time, and tests on a public dataset showed that the improved algorithm did not sacrifice much recognition performance, but was much faster than the algorithm proposed by Hinterstoisser, the Hs-PPF algorithm.

This paper first starts with the original Drost-PPF, analyzes and introduces the basic principle of its algorithm and its defects, then describes the principle of Hs-PPF algorithm proposed by Hinterstoisser and its defects, and then proposes an improved Hs-PPF algorithm for the defects.

2 Related Work

2.1 Drost-PPF

The earliest PPF point-to-feature voting algorithm was proposed by Drost et al. [11] in 2010, and the overall framework of this algorithm is shown in Fig. 1.

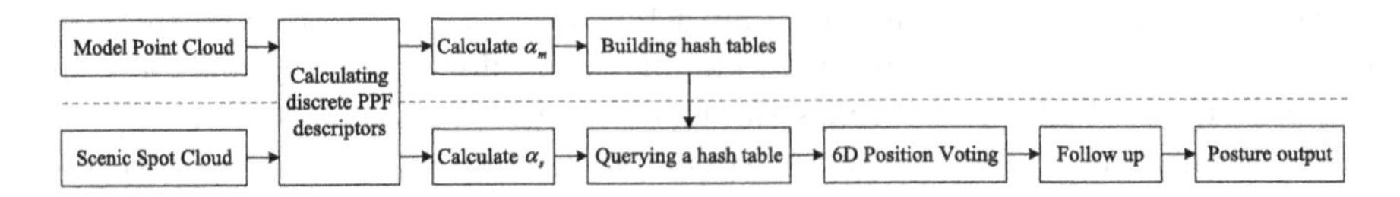

Fig. 1. Algorithm framework diagram of Drost PPF description method.

The basic descriptive unit of the Drost-PPF point pair recognition algorithm is a four-dimensional descriptor that is composed of two points in a point cloud image, as shown in Fig. 2.

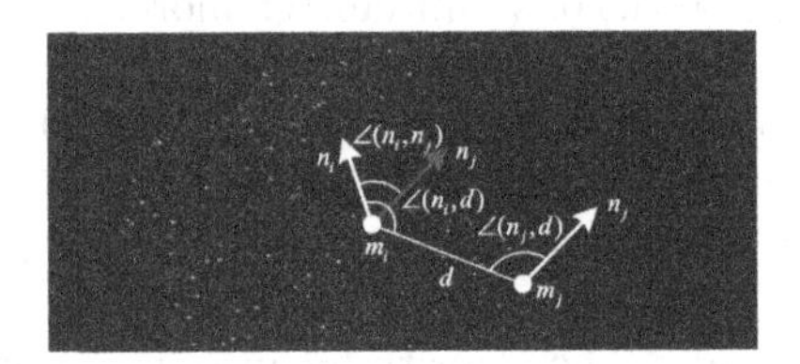

Fig. 2. Schematic diagram of PPF descriptors.

The Stanford rabbit point cloud diagram has two points m_i, m_j, which form a point pair. The specific definition of this descriptor of the point cloud is shown in Eq. (1), n_i and n_j are the surface normal vectors of the plane in which the two points m_i and m_j are located, respectively. $d = m_i - m_j$ is the distance between these two points. $\angle(n_i, d)$ is the angle between the normal n_i and the lines m_i and m_j. $\angle(n_j, d)$ is the angle between the normal n_j and the lines m_i and m_j. $\angle(n_i, n_j)$ is the angle between the normal n_i and the normal n_j.

$$F(m_i, m_j) = (\|d\|_2, \angle(n_i, d), \angle(n_j, d), \angle(n_i, n_j)) \tag{1}$$

The Drost-PPF point pair recognition algorithm is discretized immediately after calculating the continuous descriptors of the point pairs in order to accelerate the matching between the descriptors.

The principle of Drost-PPF positional calculation can be understood in conjunction with Fig. 3.

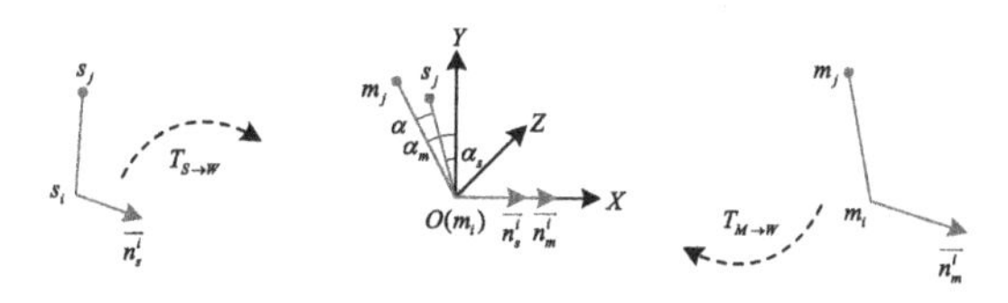

Fig. 3. Schematic diagram of Drost-PPF positional matrix solution.

Where point s_r is called a reference point on the site and point s_j, also located on the site cloud, forms a point pair (s_r, s_j) whose normal vector is $\overrightarrow{n_s^r}$. There are also corresponding point pairs m_r and normal vectors $\overrightarrow{n_m^r}$ on the model point cloud. Move the field point s_r to the origin O of the intermediate reference coordinate system, this step translation transformation can be labeled as T_S, and rotate its normal vector $\overrightarrow{n_s^r}$ to coincide with the X-axis of the intermediate reference coordinate system, the rotation matrix is labeled as R_S, the transformation at this point can be labeled as $T_{S \to W} = T_S \cdot R_S$. Similarly, the reference point m_r of the model point cloud is moved to the origin of the middle reference coordinate system, and the normal vector $\overrightarrow{n_m^r}$ of m_r is rotated to coincide with the X-axis of the reference coordinate system, and the transformation mark of this step is $T_{M \to W} = T_M \cdot R_M$. The vector $\overrightarrow{s_r s_j}$ of field point pairs (s_r, s_j) differs from the Y-axis of the reference coordinate system by α_s. The vector $\overrightarrow{m_r m_j}$ of model point pairs (m_r, m_j) differs from the Y-axis of the reference coordinate system by α_m. To reduce the computational effort, the vector $\overrightarrow{m_r m_j}$ and vector $\overrightarrow{s_r s_j}$ angle difference is calculated as in Eq. (2).

$$\alpha = \alpha_m - \alpha_s \tag{2}$$

The six-degree-of-freedom positional relationship between the model M and the scene object S can be expressed in terms of ${}^M T_S$, the parameter (m_r, α), which will be referred to as the LC coordinates. The expression of the calculation matrix ${}^M T_S$ is shown in Eq. (3), which expresses the transformation relationship from model point pairs to field point pairs.

$${}^M T_S = T_{M \to W} R(a) (T_{S \to W})^{-1} \tag{3}$$

Off-line Training. The offline phase is to perform global description for the model point cloud and store its descriptors in a hash table, which has 3 main steps in this part. In step 1, the model point cloud is downsampled and solved for normal vectors according to the preset dimensions, and then the discrete PPF descriptors of any point pair in the model are calculated.

In step 2, calculate the rotation angle of any point pair about the intermediate coordinate system α_m.

In step 3, a global description of the model is constructed. The indexes of the hash table are discrete four-dimensional descriptors that store the intermediate rotation parameters of the corresponding point pairs.

Online Identification. The online recognition phase can be divided into 3 steps.

In step 1, a local description of the scene is created. The computation process is the same as that for offline training.

In step 2, the model descriptor is used as the Key for querying the global hash table, calculating the difference between the model rotation parameters and the scene rotation parameters, and storing the result in the set. The specific process is shown in Fig. 4.

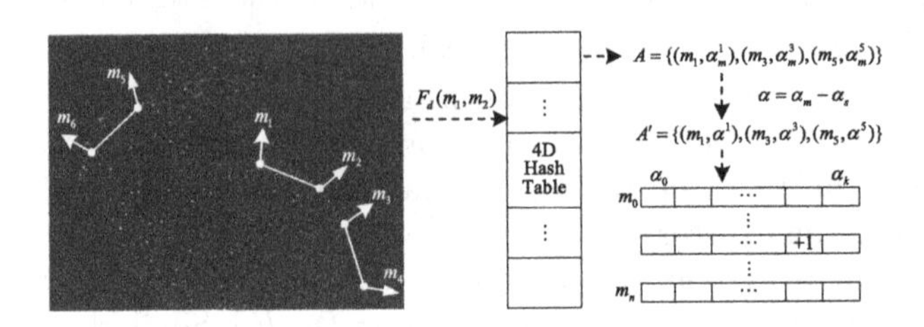

Fig. 4. Drost-PPF's online voting process.

In step 3, a two-bit accumulator is constructed to vote according to Hoff voting, and the bit pose corresponding to the position with the most votes is the optimal bit pose of the reference point. The most accurate poses are calculated based on the positional clustering.

Then all the parameters in the set are voted according to the format of the accumulator, and the pose corresponding to the position with the most votes is the optimal pose of the reference point. Finally, the most accurate pose is calculated according to the bit-pose clustering, based on the optimal pose provided by all reference points.

2.2 Hs-PPF

Although the Drost-PPF algorithm has some advantages over other algorithms, Drost-PPF loses its original advantages when the object is obscured, or when the background clutter and sensor noise are serious. Hs-PPF algorithm improves on Drost-PPF in three places, which is perfect to correct some defects of Drost-PPF.

Improved Downsampling Strategy. When Drost-PPF downsamples the model point cloud and the scene point cloud voxels, the accuracy of downsampling will be lost when the difference between the normal vector of the model point or the scene point and the angle of the normal vector in the voxel lattice in which the point is located is large, leading to a decrease in the recognition rate later. Therefore, when the difference between the two normal vector angles exceeds a certain threshold, the point will be retained, and vice versa, the point will be deleted, and the process is shown in Fig. 5.

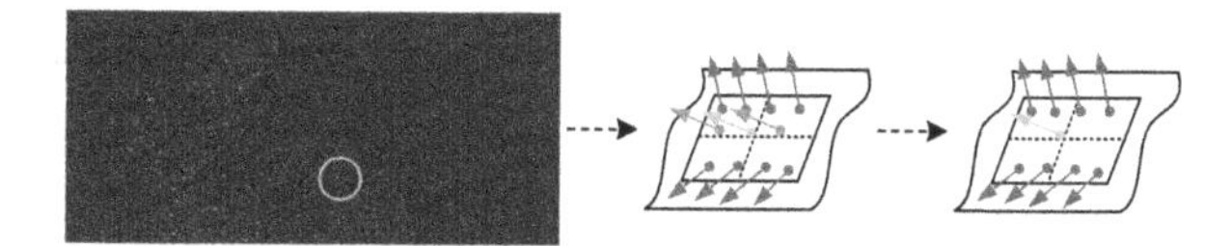

Fig. 5. Drost-PPF downsampling mechanism.

Intelligent Sampling of Point Pairs. As shown in Fig. 6, the minimum enclosing box of the target object is calculated and the lengths of the three sides of the minimum enclosing box are found.

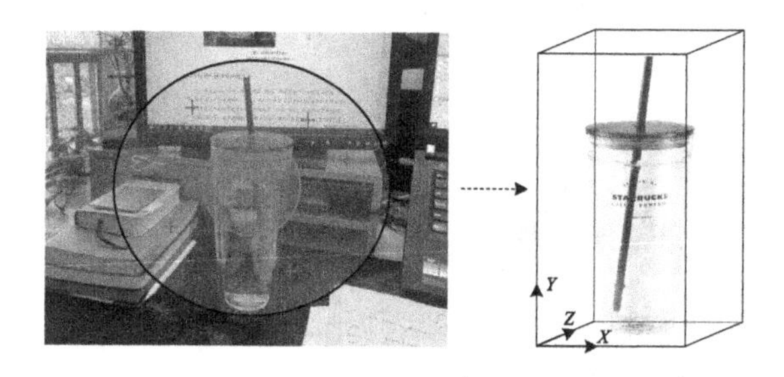

Fig. 6. Voting ball construction strategy.

Set two voting balls, where the large ball radius d_{obj} is the maximum distance between any two points of the target object, and to avoid more background points, define a smaller radius $R_{\text{min}} = \sqrt{d_{\text{min}}^2 + d_{\text{med}}^2}$, where d_{min} is the minimum edge of the three sides of the enclosing box of the target object, and d_{med} is the medium edge of the three sides, R_{min} defined as the minimum radius of the voting ball of the scene.

The reference point s_r of the scene first forms R_{min} point pair with the points inside the ball of radius a and votes to obtain the potential pose of the target object of the scene. The reference point then forms a point pair with the points in the outer circle (rejected by the ball of radius R_{min} and accepted by the ball of radius d_{obj}) to calculate the descriptor matching hash table and votes to detect the maximum value of the two-dimensional accumulator again, and then obtain the six-degree-of-freedom pose of the target object corresponding to this reference point.

Improvements for Sensor Noise. The Drost-PPF algorithm discrete point pairs of descriptors for fast access. However, sensors can change the discretization library due to some irresistible factors, making some descriptors not matched correctly. Therefore, the Hs-PPF algorithm not only stores the PPF descriptors in their original locations, but also expands the four dimensions of the PPF descriptors, as shown in Fig. 7, and the expansion will result in $3^4 - 1$ expanded descriptor.

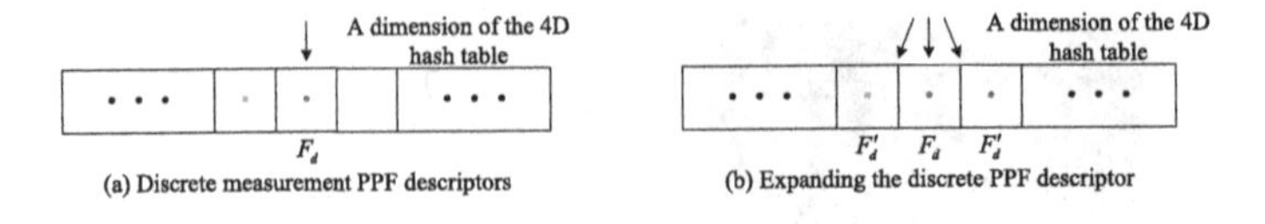

Fig. 7. Diagram of PPF neighborhood expansion strategy.

The rotation angle also faces noise effects, so the rotation angle is also expanded in this way.

As shown in Fig. 8, the neighborhood expansion causes the reference points to have the same descriptors and rotation angles as the point pairs consisting of the same background points, thus causing more false votes in the 2D accumulator and affecting the correct results.

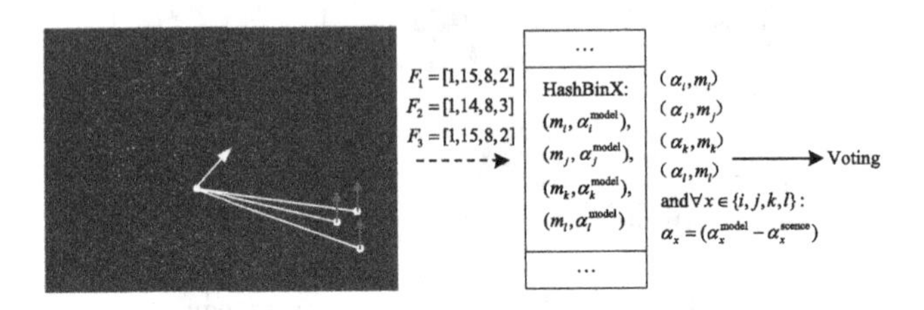

Fig. 8. Schematic diagram of duplicate voting in pairs of adjacent points of the scene.

Therefore, a 32-bit array b is assigned to each discrete descriptor, each corresponding to a discrete angle value and initialized to 0. The first query sets the bit corresponding to the discrete angle of rotation of this descriptor to 1 and allows this descriptor to query the hash table and vote. If the second time is still this descriptor and the discrete value of the corresponding angle is also this bit, querying the hash table and voting is forbidden because this position has already participated and voted before.

3 Improved Pose Estimation Algorithm Based on PPF Feature Voting

3.1 Improved Point Cloud Downsampling Strategy

From the Drost-PPF point pair identification algorithm and the Hs-PPF point pair identification algorithm, it can be seen that the downsampling result of the first algorithm is shown in Fig. 9(b), which removes all non-prime points except prime points; the second algorithm only removes the non-prime points with small differences from the normal vector of prime points, as shown in Fig. 9(c), and the density of non-prime points with similar characteristics will increase the computational effort. Therefore, a new differential downsampling strategy

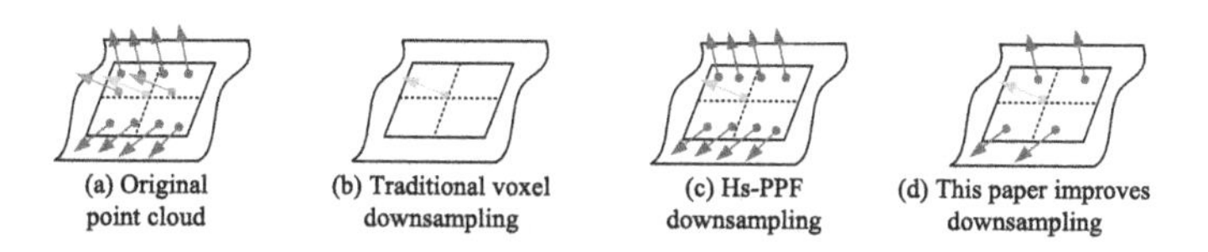

Fig. 9. Comparison diagram of the three downsampling methods.

is proposed. As shown in Fig. 9(d), it is not only to check whether the normal vector of the non-prime point is too different from the prime normal vector as in the Hs-PPF point pair identification algorithm, but also to check whether the distance between the retained non-prime points exceeds a certain threshold. For example, if the normal vector of a non-prime point is different from the normal vector of a prime point, but the distance between the non-prime point and the retained points does not exceed the threshold, the improved algorithm will still choose to delete the non-prime point. This method can avoid retaining too many similar points, while preserving the surface information of the point cloud to the maximum extent.

3.2 Improved Scene Voting Ball Strategy

Hs-PPF algorithm is to take the reference point as the center of the ball, and when the point voting of the inner voting ball is finished calculate the six degrees of freedom bit pose of the target object which is less contaminated by the background points at this time, then the farthest distance is used as the radius to form the outer voting ball, and the points rejected by the inner voting ball but accepted by the outer voting ball form the point pair with the reference point, calculate the descriptor and vote, and again detect the target object two-dimensional at this time The Hs-PPF algorithm point pairs are intelligently sampled with large voting spheres in order not to miss any point of the target object and small voting spheres in order to exclude the interference of background points. However, the six-degree-of-freedom positional voting process performed twice not only increases the complexity during positional clustering, but also may divide the points that are supposed to be target objects inside the large voting sphere and outside the small voting sphere or the points that are supposed to be background points inside the small voting sphere thus leading to incorrect voting. Therefore, it is proposed to give the voting value of the inner voting ball a different number from that of the outer voting ball to strengthen the points of the target object itself while weakening the interference of the background points as much as possible. Both ensure that all field points participate in the voting and weaken the voting influence of the background points.

4 Experimental Results and Analysis

4.1 Experimental Results of Downsampling Variable Distance

The ACCV dataset of [17] was selected as the dataset for this comparison experiment. The dataset is publicly available with over 18,000 real images containing 15 different objects as well as the ground, and each object has its own 3D point cloud model. Because the state-of-the-art method only considers relatively clear objects, the experiment removes the bowls and cups contained in the dataset. This experiment was done to compare the recognition rates of the three algorithms for 13 objects on the dataset. The distance thresholds between non-center-of-mass points are set to 1/2, 2/3, and 3/4 times of the downsampling size, respectively. The larger the threshold value, the fewer points are retained and the lower the recognition rate. From the line Fig. 10, we can see that 0.5 times of the downsampling size is chosen as the distance threshold.

4.2 Experimental Results of the Ballot Value Size of the Scenario Voting

The ACCV dataset was selected as the dataset for this comparison experiment. Three scenarios were used for the scenario voting ball size radius. (1) 0.6 for the small ballot and 0.4 for the large ballot; (2) 0.7 for small ballot ballots and 0.3 for large ballots; (3) 0.8 for small ballots and 0.2 for large ballots.

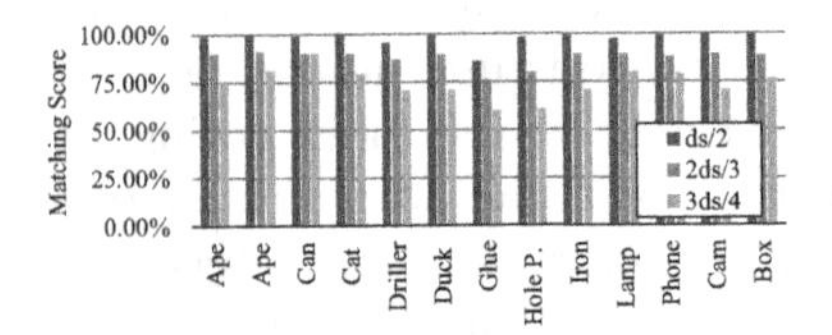

Fig. 10. Comparison diagram of three kinds of downsampling distance thresholds.

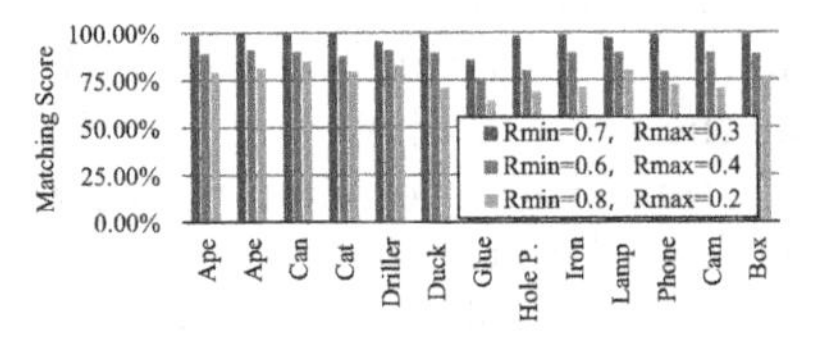

Fig. 11. Scene voting ball different voting ballot value of the three choices of comparison chart.

The experimental results are shown in Fig. 11. After the experiment, it is found that it is better to choose a small voting ball vote value of 0.7 and a large voting ball vote value of 0.3. Comprehensive consideration of the selection of this scheme as a scene voting ball of different voting ball vote value.

4.3 Experimental Results of ACCV Dataset

The ACCV dataset was selected as the dataset for this comparison experiment. From Table 1, it can be seen that this algorithm is higher than the Hs-PPF algorithm in terms of accuracy, and it can also be seen that the accuracy is significantly higher than the original algorithm due to more improvements in the Hs-PPF algorithm.

Table 1. Experimental results of ACCV data set comparison.

Comparison object	Different algorithms		
	This paper algorithm	Hs-PPF	Drost-PPF
Ape	98.9%	98.5%	86.5%
Bench V.	99.8%	99.8%	70.7%
Can	99.1%	98.7%	80.2%
Cat	99.9%	99.9%	85.4%
Driller	95.4%	93.4%	87.3%
Duck	99.1%	98.2%	46.0%
Glue	85.4%	75.4%	57.2%
Hole P.	98.1%	98.1%	77.4%
Iron	98.8%	98.3%	84.9%
Lamp	97.2%	96.0%	93.3%
Phone	98.9%	98.6%	80.7%
Cam	99.5%	99.3%	78.6%
Box	99.2%	98.8%	97.0%

4.4 Experimental Results for the Krull Dataset

The Krull dataset of [18] was selected as the dataset for this comparison experiment. The ACCV dataset is almost free of occlusion, so to demonstrate robustness to occlusion, we tested our method on the Krull occlusion dataset. It is noisier than the previous occlusion dataset Mian and includes more background clutter. It includes over 1200 real depth and color images of 8 objects and their ground truth 3D poses. The results of the comparison experiments using the present algorithm, Hs-PPF algorithm and Drost-PPF algorithm on 8 objects of the dataset are shown in Table 2.

From Table 2, it can be seen that the present algorithm is still higher than the Hs-PPF algorithm in the case of occlusion, and it can also be seen that

Table 2. Experimental results of Krull data set comparison.

Comparison object	Different algorithms		
	This paper algorithm	Hs-PPF	Drost-PPF
Ape	88.9%	81.4%	76.5%
Can	95.1%	94.7%	80.2%
Cat	79.9%	55.2%	45.4%
Driller	87.4%	86.0%	67.3%
Duck	89.1%	79.7%	46.0%
Box	75.4%	65.6%	57.2%
Glue	78.1%	52.1%	47.4%
Hole P.	98.8%	95.5%	84.9%

Table 3. Comparison of the two algorithms.

Comparison object	Different algorithms	
	This paper algorithm takes time/s	Drost-PPF takes time/s
Driller	10	11
Bench V.	12	13
Can	6	7
Box	7	9

the accuracy of the Hs-PPF algorithm is significantly higher than the original algorithm due to more improvements in the Hs-PPF algorithm.

4.5 Time-Consuming Experiments

A comparison of the time consumption of the two algorithms is shown in Table 3. It can be seen that the algorithm outperforms the Hs-PPF point pair recognition algorithm in terms of time performance, and the algorithm will be further optimized subsequently considering the requirements of the real-time performance of the algorithm for the grasping task on the actual production pipeline.

5 Conclusion

In this paper, we propose an improved point pair recognition algorithm based on the Hs-PPF point pair recognition algorithm. By experimenting on two datasets, the ACCV dataset and the occlusion type dataset, we verify that the improved algorithm has higher recognition rate in both normal and occlusion cases, and the improved algorithm also has improved computational speed compared with the original Hs-PPF point pair recognition algorithm, but the algorithm still has a disadvantage in time performance.

References

1. Lepetit, V., et al.: EPnP: an accurate o(n) solution to the PnP problem. Int. J. Comput. Vis. **81**(2), 155–166 (2009). https://doi.org/10.1007/s11263-008-0152-6
2. Abdel-Hakim, A.E., et al.: CSIFT: a SIFT descriptor with color invariant characteristics. In: 2006 IEEE Computer Society Conference on Computer Vision and Pattern Recognition, pp. 1978–1983. IEEE (2006). https://doi.org/10.1109/CVPR.2006.95
3. Kehl, W., et al.: SSD-6D: making RGB-based 3D detection and 6D pose estimation great again. In: 2017 IEEE International Conference on Computer Vision, pp. 1521–1529. IEEE (2017). https://doi.org/10.1109/ICCV.2017.169
4. Rusu, R.B., et al.: Aligning point cloud views using persistent feature histograms. In: 2008 IEEE/RSJ International Conference on Intelligent Robots and Systems, pp. 3384–3391. IEEE (2008). https://doi.org/10.1109/IROS.2008.4650967

5. Rusu, R.B., et al.: Fast point feature histograms (FPFH) for 3D registration. In: 2009 IEEE International Conference on Robotics and Automation, pp. 3212–3217. IEEE (2009). https://doi.org/10.1109/ROBOT.2009.5152473
6. Tombari, F., Salti, S., Di Stefano, L.: Unique signatures of histograms for local surface description. In: Daniilidis, K., Maragos, P., Paragios, N. (eds.) ECCV 2010. LNCS, vol. 6313, pp. 356–369. Springer, Heidelberg (2010). https://doi.org/10.1007/978-3-642-15558-1_26
7. Tombari, F., et al.: A combined texture-shape descriptor for enhanced 3D feature matching. In: 2011 18th IEEE International Conference on Image Processing, pp. 809–812. IEEE (2011). https://doi.org/10.1109/ICIP.2011.6116679
8. Guo, Y., Bennamoun, M., Sohel, F., Lu, M., Wan, J., Kwok, N.M.: A comprehensive performance evaluation of 3D local feature descriptors. Int. J. Comput. Vis. **116**(1), 66–89 (2015). https://doi.org/10.1007/s11263-015-0824-y
9. Rusu, R.B., et al.: Fast 3D recognition and pose using the viewpoint feature histogram. In: 2010 IEEE/RSJ International Conference on Intelligent Robots and Systems, pp. 2155–2162. IEEE (2010). https://doi.org/10.1109/IROS.2010.5651280
10. Aldoma, A., et al.: CAD-model recognition and 6DOF pose estimation using 3D cues. In: 2011 IEEE International Conference on Computer Vision Workshops, pp. 585–592. IEEE (2011). https://doi.org/10.1109/ICCVW.2011.6130296
11. Drost, B., et al.: Efficient and robust 3D object recognition. In: 2010 IEEE Computer Society Conference on Computer Vision and Pattern Recognition, pp. 998–1005. IEEE (2010). https://doi.org/10.1109/CVPR.2010.5540108
12. Choi, C., et al.: 3D pose estimation of daily objects using an RGB-D camera. In: 2012 IEEE/RSJ International Conference on Intelligent Robots and Systems, pp. 3342–3349. IEEE (2012). https://doi.org/10.1109/IROS.2012.6386067
13. Drost, B., et al.: 3D object detection and localization using multimodal point pair features, modeling, processing, visualization & transmission. In: 2012 2nd Joint 3DIM/3DPVT Conference: 3D Imaging, Modeling, Processing, Visualization & Transmission, pp. 9–16. IEEE Computer Society (2012). https://doi.org/10.1109/3DIMPVT.2012.53
14. Birdal, T., et al.: Point pair features based object detection and pose estimation revisited. In: 2015 International Conference on 3D Vision, pp. 527–535. IEEE (2015). https://doi.org/10.1109/3DV.2015.65
15. Hinterstoisser, S., Lepetit, V., Rajkumar, N., Konolige, K.: Going further with point pair features. In: Leibe, B., Matas, J., Sebe, N., Welling, M. (eds.) ECCV 2016. LNCS, vol. 9907, pp. 834–848. Springer, Cham (2016). https://doi.org/10.1007/978-3-319-46487-9_51
16. Wang, G., et al.: An improved 6D pose estimation method based on point pair feature. In: 2020 Chinese Control And Decision Conference, pp. 455–460. IEEE (2020). https://doi.org/10.1109/CCDC49329.2020.9164326
17. Hinterstoisser, S., et al.: Model based training, detection and pose estimation of texture-less 3D objects in heavily cluttered scenes. In: Lee, K.M., Matsushita, Y., Rehg, J.M., Hu, Z. (eds.) ACCV 2012. LNCS, vol. 7724, pp. 548–562. Springer, Heidelberg (2013). https://doi.org/10.1007/978-3-642-37331-2_42
18. Krull, A., et al.: Learning analysis-by-synthesis for 6D pose estimation in RGB-D images. In: 2015 IEEE International Conference on Computer Vision, pp. 954–962. IEEE (2015). https://doi.org/10.1109/ICCV.2015.115

Characteristic Analysis of a Variable Stiffness Actuator Based on a Rocker-Linked Epicyclic Gear Train

Zhisen Li[1], Hailin Huang[1], Peng Xu[1], Yinghao Ning[1], and Bing Li[1,2(✉)]

[1] School of Mechanical Engineering and Automation, Harbin Institute of Technology, Shenzhen 518055, People's Republic of China
libing.sgs@hit.edu.cn

[2] State Key Laboratory of Robotics and System, Harbin Institute of Technology, Harbin 150001, People's Republic of China

Abstract. This paper presents the characteristic analysis and mechanical realization of a novel variable stiffness actuator based on a rocker-linked epicyclic gear train (REGT-VSA). The stiffness adjustment of the actuator works by converting the differential motion of the planetary gear train into the linear motion of the elastic element. The unique design of the rocker-linked epicyclic gear train ensures excellent compactness and easy controllability, which enables the actuator to be qualified for constructing a manipulator toward cooperation applications. However, the output position and stiffness of the actuator may be affected by the mechanism clearance. The paper introduces characteristic analysis of stiffness and clearance, and carries out a series of related simulations. The analysis results can provide guidelines for the high-quality assembly of lever-based VSA.

Keywords: Variable stiffness actuator · Epicyclic gear train · Clearance analysis

1 Introduction

Variable stiffness actuator (VSA) realizes a novel class of actuation systems, characterized by the property that the apparent output stiffness can be changed independently of the output position. Actuators in this category are particularly significant when used on robots that must interact safely with humans and feature properties such as energy conservation, robustness and high dynamics [1, 2]. According to the working principle, VSAs are generally classified into antagonistic-type VSAs and serial-type VSAs [3]. The serial-type VSAs are gaining in popularity because their position and stiffness are independently driven by two motors [4]. There are four methods to regulate the stiffness for the serial-type VSAs, including adjusting the spring preload [5], using the curved or inclined surface of special parts [6, 7], changing the effective length of the leaf spring [8, 9], and changing the effective ratio of the lever mechanism [10]. Compared with other methods, the variable ratio of the lever mechanism permits a wide range of stiffness regulation and low energy consumption [11]. By varying the positions of the pivot point, force point, and spring positioning point, the variable lever mechanism can make the

H. Liu et al. (Eds.): ICIRA 2022, LNAI 13456, pp. 210–221, 2022.
https://doi.org/10.1007/978-3-031-13822-5_19

lever amplification ratio from zero to infinity, so as to achieve a large range of the transmission ratio between the elastic elements' force and the actuator's output force [12]. Moreover, thanks to the vertical relationship between stiffness adjustment and spring deformation, the energy consumption is significantly reduced [13].

The clearance analysis of the regulation mechanism is another concern in the structural design of a lever-based VSA. When regulating the positions of the pivot or the spring module, these movements are realized through the form of sliding or rolling rather than block pairs for the limited structural size, resulting in clearances inevitably. These clearances would directly lead to rotation angle error between the input shell and the output shell of the actuator, thus deteriorating the performance of position accuracy [14]. Moreover, they may have a negative impact on the position of the pivot and spring module, thereby reducing the stiffness control accuracy. Therefore, it is necessary to analyze the influence of clearances in the stiffness adjustment mechanism on the actuator.

This paper presents a novel variable stiffness actuator based on a rocker-linked epicyclic gear train and its characteristic analysis. The design of the rocker-linked epicyclic gear train ensures the compactness of structure and the accuracy of transmission. The simultaneous relocations of the pivot and torsion spring permit a wide-range stiffness adjustment. The structural clearances affecting the output accuracy of stiffness and position are analyzed. Then a series of related simulations are conducted. The analysis results are beneficial for the high-quality assembly of lever-based VSA.

The layout of this paper is as follows: Sect. 2 describes the mechanism concept of the novel actuator. Section 3 analyzes the character of the stiffness and the clearance. Section 4 establishes the mechanical realization of REGT-VSA. Finally, a conclusion is given in Sect. 5.

2 Mechanism Concept

This section introduces the working principle concept of REGT-VSA. The variable stiffness mechanism of REGT-VSA is mainly composed of a lever with a cam follower, a guide, a movable pivot, a movable torsion spring module and the rocker-linked epicyclic gear train, as shown in Fig. 1. The torsion spring module mainly includes two pretensioned torsion springs in opposite directions and a torsion spring bracket for fixing. The headend of each rocker is connected with planet gear by a bearing, while the rearends of the two rockers are respectively connected with the pivot and the torsion spring module. When the sun gear and the ring gear in the epicyclic gear train move at different speeds, the planet gears revolve around the sun gear and drive the rockers to shake. Additionally, the rockers drive the pivot and the torsion spring module synchronously away from or approach the center of the actuator. Thus, the amplification ratio between the spring force and the output force is changed to adjust the stiffness.

The variable stiffness mechanism of REGT-VSA has three different states: the minimum stiffness state, the maximum stiffness state and the working state. When the rocker-linked epicyclic gear train drives the pivot and the torsion spring module to the nearest distance by counterclockwise differential rotation, the actuator is in the minimum stiffness state, as shown in Fig. 1(a). In contrast, if the rocker-linked epicyclic gear train drives the pivot and the torsional spring module to the farthest distance by clockwise differential rotation, the actuator is in the maximum stiffness state, as shown in Fig. 1(b). In the working state, the external force introduced from the output point drives the lever to rotate around the pivot, and the cam follower at the end of the lever also rotates around the pivot to squeeze the torsion spring, thus generating a force to resist the external force, as shown in Fig. 1(c).

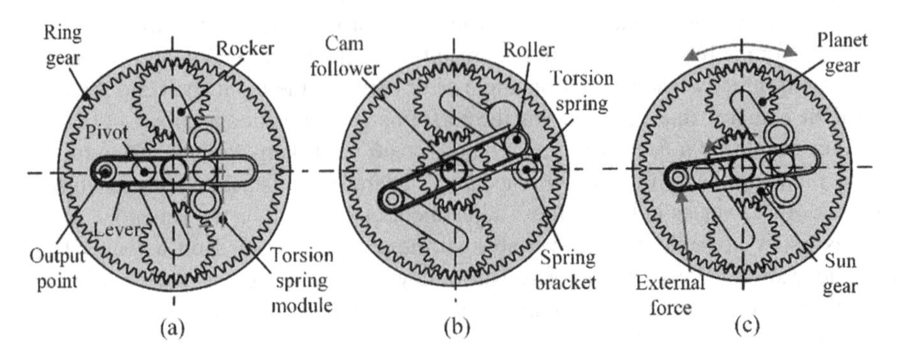

Fig. 1. Principle of the REGT-VSA: (a) minimum stiffness state, (b) maximum stiffness state, and (c) working state.

3 Characteristic Analysis

3.1 Stiffness Analysis

The deflection angle of the actuator is the angle difference between the output shell and the input shell. The deflection causes the lever to rotate an angle around the pivot. Then, the cam follower on the lever squeezes the torsion spring on one side, as shown in Fig. 2(a). The force exerted on the lever by squeezing the cam follower can be expressed as

$$F'_B = 2K_t|PB|\sin\theta/|BH|^2 = F'_A|PA'|/|PB'|, \tag{1}$$

where $|PA'|/|PB'|$ is the lever ratio, F'_A is the release force produced by F'_B at point A', θ is the deflection angle of the lever, K_t is the spring constant of the torsion spring.

The external force exerted on the output point can be expressed as

$$F_\tau = \frac{F'_A}{\cos(\theta-\varphi)} = \frac{2K_s|PB|^2\sin^2\theta}{R\cos(\theta-\varphi)\cos\theta\sin\varphi}, \tag{2}$$

where φ represents the deflection angle between the output shell and the input shell; R is the distance of the output point relative to the actuator axis.

Given that the angle φ and the angle θ are less than 0.175 rad, the difference between radians and their sinusoids is negligible. The relationship between angles θ and φ can be approximately written as

$$\theta = \frac{R\varphi}{R-|PO|}. \tag{3}$$

Due to the geometric relationship $|PB'| = |PB|/\cos\theta$ and $|PA'| = R\sin\varphi/\sin\theta$, the external force exerted on the output point can be expressed as

$$F_\tau = \frac{2RK_s(|PB|_{\min} + l_s)^2}{(R - |PO|_{\min} - l_s)^2}\varphi. \tag{4}$$

where $|PB|_{\min}$ is the shortest distance between the point P and the point B; $|PO|_{\min}$ is the shortest distance between the point P and the point O; and l_s is the distance that the pivot (torsion spring) has moved.

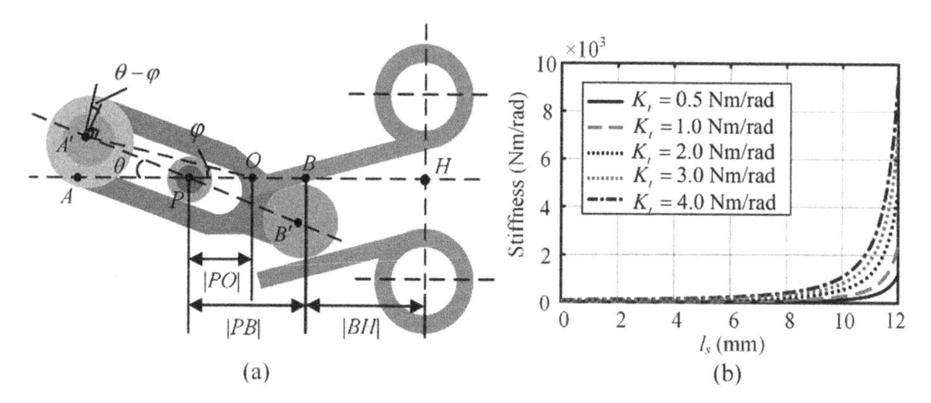

Fig. 2. Stiffness analysis: (a) schematic of the variable ratio lever mechanism, and (b) the relationship curves between stiffness and l_s under different spring coefficients.

The output torque of the actuator is

$$M_\tau = \frac{2R^2K_s(|PB|_{\min} + l_s)^2}{(R - |PO|_{\min} - l_s)^2}\varphi. \tag{5}$$

The stiffness of the actuator can be expressed as

$$K = \frac{\partial(M_\tau)}{\partial(\varphi)} = \frac{2R^2K_t(|PB|_{\min} + l_s)^2}{(R - |PO|_{\min} - l_s)^2|BH|^2}, \tag{6}$$

where $|BH| = |BH|_{\min} + l_s$ is the distance from point B to point H.

According to Eq. (6), if the parameters K_t, $|BH|_{\min}$, $|PB|_{\min}$, $|PO|_{\min}$ and l_s are invariable, the output torque is proportional to the angle φ and the stiffness is constant. Since the above five parameters directly affect the actuator's stiffness, the change of stiffness is mainly the adjustment process of the parameters. Among the parameters, the size of l_s can significantly affect the lever ratio.

By analyzing the mathematical stiffness model, the relationship curves between the stiffness and the value l_s under different spring coefficients are as shown in Fig. 2 (b). It is found that the stiffness increases significantly with the increase of l_s, when the value of l_s exceeds 10 mm. In contrast, when the range of l_s is [0 mm, 10 mm], the change of stiffness is slow. The reason for the phenomenon is that the pivot is close to the output

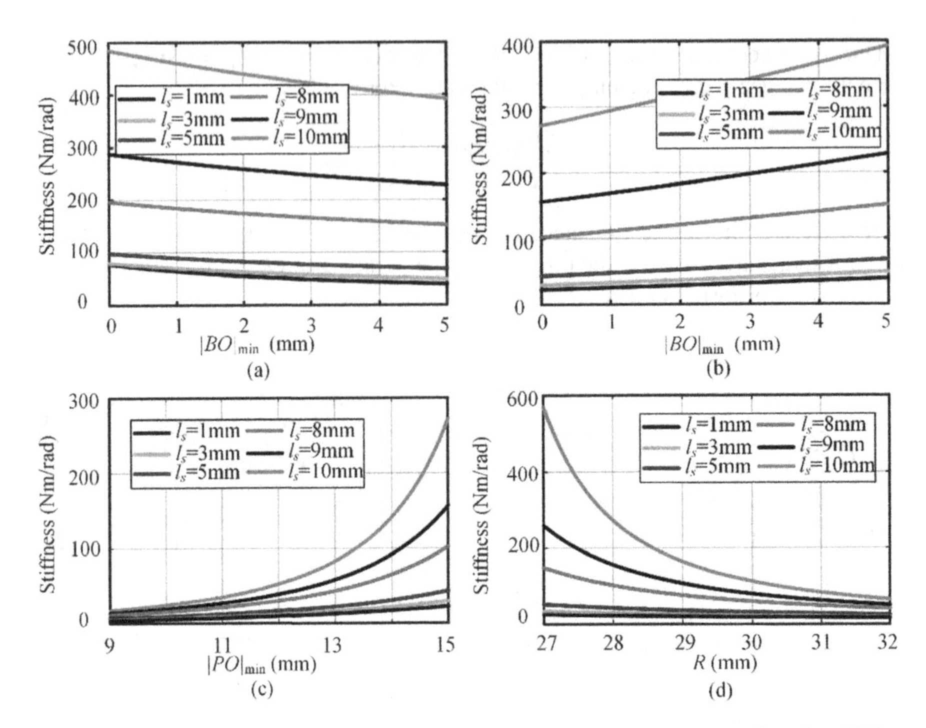

Fig. 3. Curves of stiffness: (a) when $|BO|_{\min}$ changes and $|HO|$ is constant, (b) when $|BO|_{\min}$ changes and $|BH|$ is constant, (c) when $|PO|_{\min}$ is variable, and (d) when R is variable.

point when the value of l_s exceeds 10 mm, and the lever ratio will change rapidly with the further increase of l_s. In addition, the average stiffness is positively correlated with the spring coefficient.

Figure 3 shows the curves of stiffness varying with parameters. From the results, it can be found that when $|HO|$ remains constant, the stiffness decreases with the increase of $|BO|_{\min}$, while the stiffness increases with the increase of $|BO|$ when $|BH|$ remains unchanged. In addition, the stiffness is positively correlated with $|PO|_{\min}$, while inversely correlated with the value of R.

3.2 Clearance Analysis

This section will discuss the influence of clearance on the stiffness and output torque of the actuator. The clearances affecting the actuator's performance mainly include: the clearance between pivot and lever ($\Delta\delta_1$); the clearance of cam follower caused by bearing clearance ($\Delta\delta_2$); the clearances of transmission gears and rocker bearings which would cause elastic elements' linear-motion errors (Δl_s).

Specifically, the clearances at the pivot and the cam follower have a negative impact on the accuracy of angular position. Figure 4 illustrates the schematic of the analysis for clearances only at the pivot or cam follower. As shown in Fig. 4, $\Delta\varphi$ is the error of deflection angle between the output shell and the input shell, and $\Delta\theta$ is the error of the lever's deflection angle.

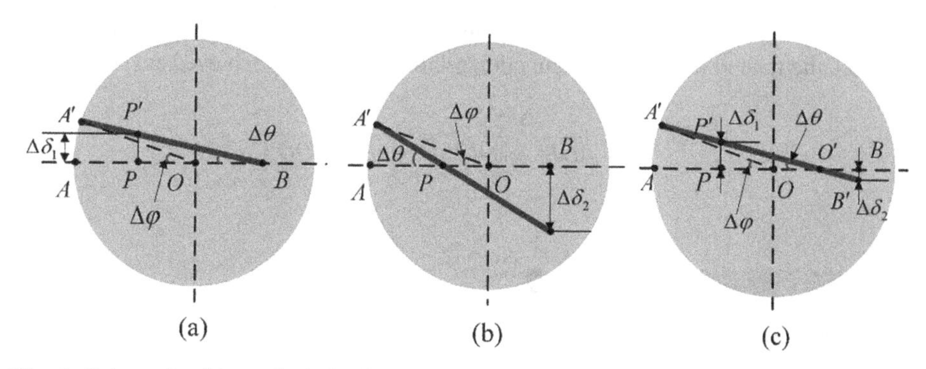

Fig. 4. Schematic of the analysis for clearances: (a) clearance exists only at the pivot, (b) clearance exists only at the cam follower, and (c) clearances exist at both the pivot and the cam follower.

Firstly, when the clearance only exists at the pivot, as shown in Fig. 4(a), the related angular constraints can be expressed as

$$\begin{cases} \frac{\sin(\Delta\varphi-\Delta\theta)}{|OB|} = \frac{\sin\Delta\theta}{R} \\ \tan\Delta\theta = \frac{\Delta\delta_1}{|PB|} \end{cases}. \tag{7}$$

As the angles $\Delta\varphi$ and $\Delta\theta$ are smaller than 5° in this model, the difference between radian and its trigonometric function is negligible. It could be supposed that $\sin\Delta\varphi = \Delta\varphi$, $\tan\Delta\varphi = \Delta\varphi$, $\sin\Delta\theta = \Delta\theta$, $\tan\Delta\theta = \Delta\theta$. According to Eq. (7), the deflection angle error caused by clearance can be obtained

$$\Delta\varphi = \frac{R + |OB|}{R|PB|}\Delta\delta_1. \tag{8}$$

Combining Eq. (8) and Eq. (5), the torque error ΔM caused by the clearance can be obtained as

$$\Delta M = \frac{2RK_s(|PB|_{\min} + l_s)(R + |OB|)}{(R - |PO|_{\min} - l_s)^2}\Delta\delta_1. \tag{9}$$

Secondly, for the occasion where the clearance only exists at the cam follower, as shown in Fig. 4(b), the related angular constraints can be expressed as

$$\begin{cases} \frac{\sin(\Delta\theta-\Delta\varphi)}{|PO|} = \frac{\sin\Delta\theta}{R} \\ \tan\Delta\theta = \frac{\Delta\delta_2}{|PB|} \end{cases}. \tag{10}$$

Therefore, the deflection angle error $\Delta\varphi$ caused by the clearance is

$$\Delta\varphi = \frac{R - |PO|}{R|PB|}\Delta\delta_2. \tag{11}$$

Combining Eq. (11) and Eq. (5), the torque error ΔM caused by the clearance can be obtained as

$$\Delta M = \frac{2RK_s(|PB|_{\min} + l_s)}{(R - |PO|_{\min} - l_s)}\Delta\delta_2. \tag{12}$$

Thirdly, when clearances exist at both the pivot and the cam follower, as shown in Fig. 4(c), the related trigonometric function relationship can be achieved as

$$\tan \Delta\theta = \frac{\Delta\delta_2}{|OB| - |OO'|} = \frac{\Delta\delta_1}{|OP| + |OO'|}. \tag{13}$$

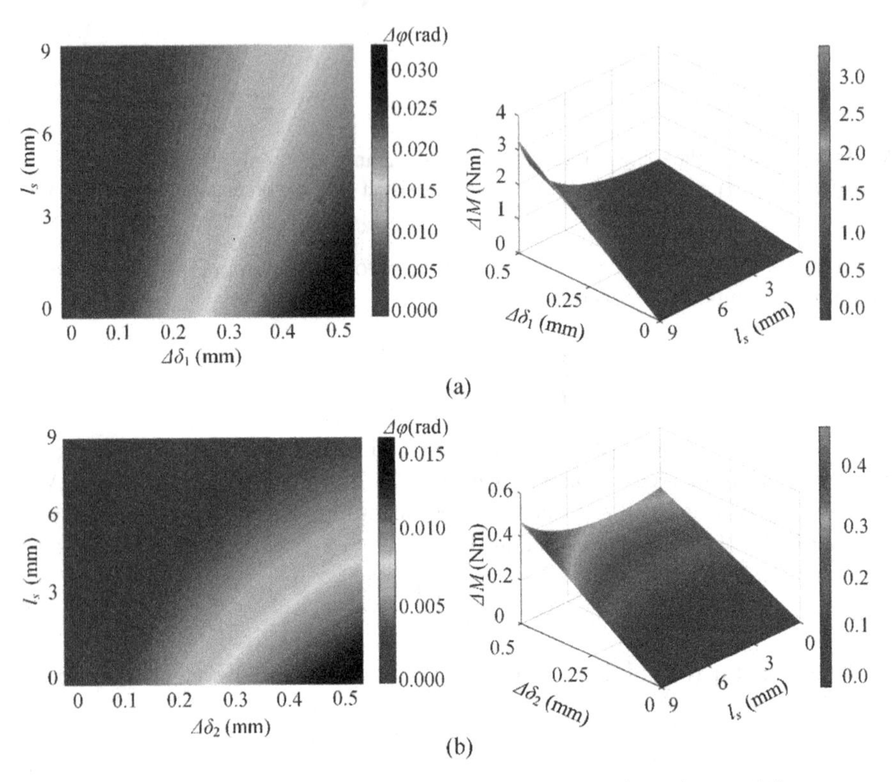

Fig. 5. Simulation results of clearance: (a) at the pivot, and (b) at the cam follower.

It can be derived as

$$\begin{cases} \tan\Delta\theta = \frac{\Delta\delta_1 + \Delta\delta_2}{|OB| + |OP|} \\ |OO'| = \frac{|OB|\Delta\delta_1 - |OP|\Delta\delta_2}{\Delta\delta_1 + \Delta\delta_2} \end{cases}. \tag{14}$$

Besides,

$$\frac{\sin(\Delta\varphi - \Delta\theta)}{|OO'|} = \frac{\sin \Delta\theta}{R}. \tag{15}$$

The deflection angle error caused by the $\Delta\delta_1$ and $\Delta\delta_2$ can be expressed as

$$\Delta\varphi = \frac{R + |OO'|}{R}\Delta\theta, \tag{16}$$

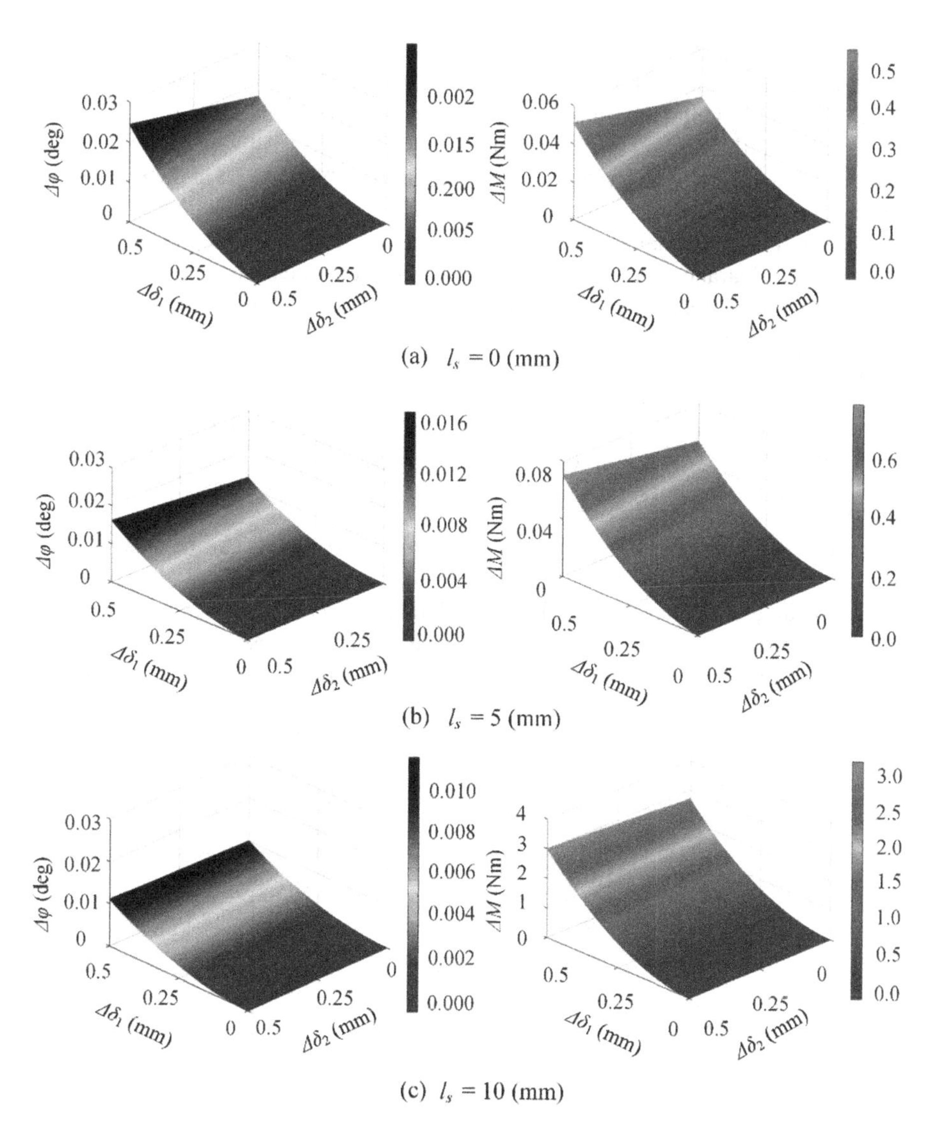

Fig. 6. Simulation results of clearances at both the pivot and the cam follower: (a) when l_s = 0 mm, (b) when l_s = 5 mm, and (c) when l_s = 10 mm.

Combining Eq. (16) and Eq. (5), the torque error ΔM caused by the clearance can be written as

$$\Delta M = \frac{2RK_s(|PB|_{\min} + l_s)^2(\mathrm{R} + |OO'|)}{(R - |PO|_{\min} - l_s)^2}\Delta\theta, \tag{17}$$

where the value of $|OO'|$ and $\Delta\theta$ are decided by Eq. (14).

Figure 5(a) illustrates the mathematical analysis of the clearance at the pivot point, and Fig. 5(b) shows the analysis of the clearance at the cam follower. Figure 6 depicts the simulation results of the clearances at both the pivot and the cam follower. In Figs. 5 and 6, the left subgraphs show the influence of clearance on deflection angle φ, while the right subgraphs depict the influence of clearance on the output torque M.

According to the results, the values of $\Delta\varphi$ and ΔM are positively correlated with the values of clearance, which rise with the increase of $\Delta\delta_1$ or $\Delta\delta_2$. On the other hand, when l_s increases, the deflection angle error $\Delta\varphi$ increases more slowly with the clearance, while the rotation torque deflection error ΔM increases faster with the clearance.

Finally, the clearances of transmission gears and rocker bearings directly affect the parameter l_s, so the effects of Δl_s (the error of l_s) on the accuracy of stiffness can be used for simplifying the investigation of the influence of the above clearances. According to Eq. (6), The stiffness error ΔK corresponding to Δl_s can be expressed as

$$\Delta K = \frac{2R^2 K_t}{|BH|^2}\left(\frac{(|PB|_{\min} + l_s)^2}{(|\mathrm{M}| - l_s)^2} - \frac{(|PB|_{\min} + l_s + \Delta l_s)^2}{(|\mathrm{M}| - l_s - \Delta l_s)^2}\right), \tag{18}$$

where $|\mathrm{M}| = R - |PO|_{\min}$.

Figure 7 illustrates the relationship between ΔK and Δl_s. It can be seen that the error of stiffness increases with the growth of the absolute value of Δl_s. Furthermore, when the parameter l_s increases resulting in a larger value of stiffness, the accuracy of stiffness deteriorates heavily with the certain values of Δl_s. These results demonstrate that the larger the stiffness is, the worse the accuracy of stiffness with certain values of Δl_s.

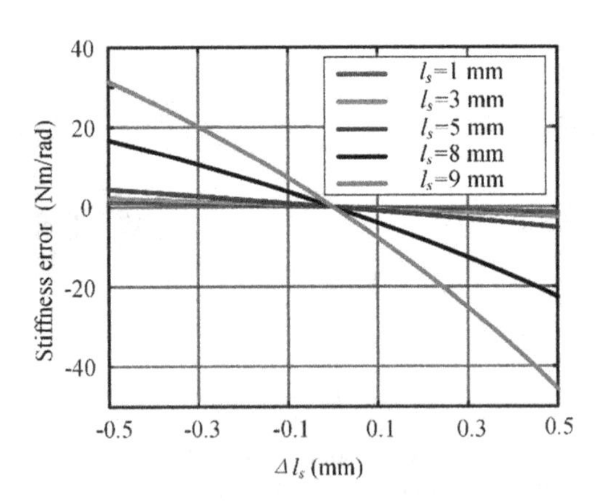

Fig. 7. Variation curves of stiffness error ΔK with Δl_s.

4 Mechanical Realization

The structure of REGT-VSA is shown in Fig. 8(a). The actuator contains a position motor M1 and a stiffness adjusting motor M2. The input torque of the actuator is mainly provided by the motor M1. The output shaft of motor M1 is connected with the spindle through a flat key. The sun gear of the epicyclic gear train is fixed on the spindle and rotates synchronously with the spindle. Two planet gears are placed on the planet carrier, which can self-rotate or revolve around the sun gear. Each planet gear is connected with a rocker through a bearing. The guide bracket is installed on the top of the output shaft. The linear guide is placed on the guide bracket. There are two sliders along the linear guide, supporting the pivot and spring bracket, respectively. Furthermore, each slider is linked with a rocker through a bearing. When the sun gear and ring gear move at different speeds, the planet gear revolves around the sun gear and drives the rocker to shake, thus driving the sliders away from or close to the actuator's center. A cam follower is installed at the end of the lever, which can rotate around its axis and convert sliding friction into rolling friction. The torsion springs are fixed on the spring bracket by bolts. The output shell is connected with the lever through the output point, which can rotate around the input shell through a cross roller bearing. The encoder and the torque sensor are fixed on the end cap of the output shell. The deflection angle between the input shell and the output shell is measured using a pair of gears, namely gear III and gear IV.

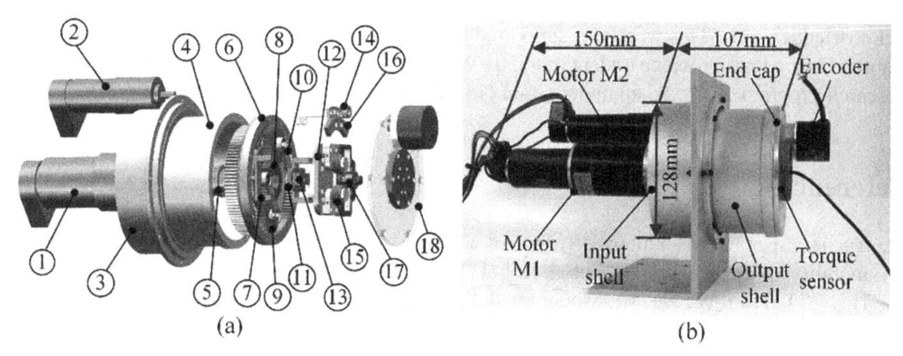

Fig. 8. Mechanical realization of the actuator: (a) Mechanical composition of the actuator, and (b) Physical prototype of the actuator. 1, M1; 2, M2; 3, input shell; 4, output shell; 5, spindle; 6, ring gear; 7, sliding fork; 8, sun gear; 9, planet gear; 10, rocker; 11, guide bracket; 12, slider; 13, linear guide; 14, gear III; 15, torsion spring; 16, gear IV; 17, cam follower; 18, end cap.

To evaluate the performance, a prototype of REGT-VSA is developed, as shown in Fig. 8(b). The maximum diameter of the prototype is 128 mm, the length excluding the driving motors is 107 mm, and the total mass is 4.5 kg. The actual total length of the lever is 42 mm. Two DC motors (M1: Maxon Re40, 150 W; M2: Maxon Re30, 40 W) with gear reducers provide the input power. The reduction ratio of M1 is 230:1, and that of M2 is 66:1. An encoder with a resolution of 500 pulses per revolution is installed at each motor to detect the motors' position. An additional absolute encoder is utilized with a resolution of 16384 pulses per revolution to measure the relative rotation angle between the input shell and the output shell. The spring coefficient of the torsion spring is 1.0 Nm/rad.

5 Conclusions

In this paper, the characteristic analysis of a novel variable stiffness actuator based on a rocker-linked epicyclic gear train was presented. The actuator transforms the differential rotation of the epicyclic gear train into linear motion, which not only improves its compactness but also ensures convenient control of transmission. The relationship between stiffness and its parameters is studied. The structural clearances affecting the accuracy of stiffness (e.g., the clearance at the pivot and the cam follower) are analyzed and a series of simulations are conducted. From the simulation results, it can be concluded that the position error and torque error have an enlarging trend along with the increase of clearances. Besides, the clearances of transmission gears and rocker bearings will affect the stiffness's accuracy. Therefore, reducing the assembly clearances of key components such as lever and torsion spring is meaningful for a better output accuracy of stiffness and position. These results could be guidelines for the manufacture of lever-based VSA. In the future, we will carry out experiments on the proposed VSA to verify the relevant theories.

Acknowledgement. This work was supported by the Key-area Research and Development Program of Guangdong Province under Grant 2019B090915001, and in part by the Key Fundamental Research Program of Shenzhen under Grant No. JCYJ20200109112818703.

References

1. Yu, H., Spenko, M., Dubowsky, S.: An adaptive shared control system for an intelligent mobility aid for the elderly. Auton. Robot. **15**(1), 53–66 (2003)
2. Huang, J., Huo, W., Xu, W., Mohammed, S., Amirat, Y.: Control of upper-limb power-assist exoskeleton using a human-robot interface based on motion intention recognition. IEEE Trans. Autom. Sci. Eng. **12**(4), 1257–1270 (2015)
3. Liu, Y., Liu, X., Yuan, Z.: Design and analysis of spring parallel variable stiffness actuator based on antagonistic principle. Mech. Mach. Theory **140**, 44–58 (2019)
4. Sun, J., Guo, Z., Zhang, Y., et al.: A novel design of serial variable stiffness actuator based on an archimedean spiral relocation mechanism. IEEE-ASME Trans. Mech. **23**(5), 2121–2131 (2018)
5. Van Ham, R., Vanderborght, B., Van Damme, M., et al.: MACCEPA, the mechanically adjustable compliance and controllable equilibrium position actuator: design and implementation in a biped robot. Robot. Auton. Syst. **55**(10), 761–768 (2007)
6. Ning, Y., Xu, W., Huang, H., et al.: Design methodology of a novel variable stiffness actuator based on antagonistic-driven mechanism. Proc. Inst. Mech. Eng. Part C J. Mech. Eng. Sci. **233**(19–20), 6967–6984 (2019)
7. Nam, K.H., Kim, B.S., Song, J.B.: Compliant actuation of parallel-type variable stiffness actuator based on antagonistic actuation. J. Mech. Sci. Technol. **24**(11), 2315–2321 (2010)
8. Liu, H., Zhu, D., Xiao, J.: Conceptual design and parameter optimization of a variable stiffness mechanism for producing constant output forces. Mech. Mach. Theory **154**, 104033 (2020)
9. Tao, Y., Wang, T., Wang, Y., Guo, L., et al.: A new variable stiffness robot joint. Ind. Robot. **42**(4), 371–378 (2015)
10. Sun, J., Guo, Z., Sun, D., et al.: Design, modeling and control of a novel compact, energy-efficient, and rotational serial variable stiffness actuator (SVSA-II). Mech. Mach. Theory **130**, 123–136 (2018)

11. Visser, L.C., Carloni, R., Stramigioli, S.: Energy-efficient variable stiffness actuators. IEEE Trans. Robot. **27**(5), 865–875 (2011)
12. Ning, Y., Huang, H., Xu, W., et al.: Design and implementation of a novel variable stiffness actuator with cam-based relocation mechanism. ASME J. Mech. Robot. **13**(2), 021009 (2021)
13. Jafari, A., Tsagarakis, N.G., Sardellitti, I., et al.: A new actuator with adjustable stiffness based on a variable ratio lever mechanism. IEEE-ASME Trans. Mech. **19**(1), 55–63 (2012)
14. Schrade, S.O., Dätwyler, K., Stücheli, M., et al.: Development of VariLeg, an exoskeleton with variable stiffness actuation: first results and user evaluation from the CYBATHLON 2016. J. Neuroeng. Rehabil. **15**(1), 1–18 (2018)

Facial Gesture Controled Low-Cost Meal Assistance Manipulator System with Real-Time Food Detection

Xiao Huang, LongBin Wang, Xuan Fan, Peng Zhao(✉), and KangLong Ji

Lanzhou University, Lanzhou 730000, China
zhaopeng@lzu.edu.cn

Abstract. In view of the increasing number of people with independent eating disorders in the world, the need to design and develop a low-cost and effective meal assistance manipulator system has become more and more urgent. In this paper, we propose a low-cost meal assistance manipulator system that uses facial gesture based user intent recognition to control the manipulator, and integrates real-time food recognition and tracking to achieve different food grabbing. The system is divided into three stages. The first stage is real-time food detection and tracking based on yolov5 and deepsort, which completes the classification, positioning and tracking of food and pre-selects the target food. The second stage is user intent recognition based on facial gesture. The user selects the target food through the change of facial gesture and controls the feeding timing. The third stage is the grabbing and feeding based on image servo. Finally, four volunteers tested the system to verify the effectiveness of the system. See videos of our experiments here: https://www.bilibili.com/video/BV1Z54y1f78E/

Keywords: Meal assistance manipulator · Real-time food detection · Facial gesture detection

1 Introduction

A significant proportion of the world's population is unable to eat independently, and this phenomenon is becoming increasingly serious due to the aging of the population [1] and the continuing increase in the disabled population [2]. Continuing to hire professional nursing care for them will not only bring financial pressures that many families cannot bear, but it will also bring enormous psychological pressure on them. Therefore, finding suitable alternatives for such elderly and disabled people to solve their feeding problems has become an urgent challenge. Although people have made a lot of attempts in the field of business and research [3–8], this problem has not yet been effectively solved.

This work was supported in part by the Double Tops Construction Fund of Lanzhou University 561120202. School of Information Science & Engineering, Lanzhou University Lanzhou, 730000, China.

H. Liu et al. (Eds.): ICIRA 2022, LNAI 13456, pp. 222–231, 2022.
https://doi.org/10.1007/978-3-031-13822-5_20

How to control a feeding-assist robotic arm according to the user's intention is a major challenge in such research. Since the inability to eat independently largely implies limited hand movement, controlling the robotic arm by a simple button or joystick [3,4] is not applicable to assisted eating robotic arms. In addition, eye-movement-based control methods [5] are often not stable enough, and methods based on EOG [6], EMG [7], and EEG [8] signals require the wearing of expensive and cumbersome hardware devices, and this work also requires the assistance of others, which is contrary to the original intention of this research. Since modern image processing techniques are becoming more sophisticated, Arm-A-Dine [9] has shown us the feasibility of facial expression-based interaction, and both elderly and upper limb disabled groups have the ability to control facial posture. Therefore, this paper adopts the method of recognizing the user's facial posture and mapping the facial posture and eating intention to control the assisted feeding robotic arm.

The type of food is also an important issue; most of the previous studies used a spoon as the end-effector of the assisted-feeding robotic arm, which led to targeting only a single fine-grained food or paste-like food, which does not match the eating habits of ordinary people. In addition, we believe that assisted eating is a topic of elderly assistance, and researchers should try to avoid using expensive hardware devices [10], therefore, how to reduce the cost of the assisted eating robotic arm is also an issue of concern.

To solve this series of problems, this paper designs and implements a low-cost assisted feeding robotic arm controlled by facial pose, which uses a slender gripping claw as the end effectors to simulate the use of chopsticks, and based on yolov5. The advanced food recognition algorithm and deepsort target tracking algorithm enable the robotic arm to obtain different block foods. At the same time, the user can select the food through different facial postures and control the feeding timing of the robotic arm.

The rest of this paper is organized as follows: Sect. 2 is a systematic overview of the proposed assisted feeding robotic arm, Sect. 3 is a description of the food detection and tracking system, Sect. 4 is a description of a facial pose-based user intent detection system, and Sect. 5 is a description of an image-servo-based gripping and feeding system for a robotic arm. Section 6 presents the experiments and results, and Sect. 7 proposes our future research directions.

2 System Overview

The experimental platform used in this paper is shown in Fig. 1, and the whole system consists of a 6-degree-of-freedom desktop robot arm Lzbot and its control system, as well as an image acquisition and visual perception system. The two USB cameras and the interactive screen are connected to the host PC. Camera 1 is responsible for collecting food images and camera 2 is responsible for collecting face images, during the operation of the system, the interaction screen displays the images from the end camera, which ensures a unified view of the robotic arm and the user. The upper computer is responsible for processing the image data,

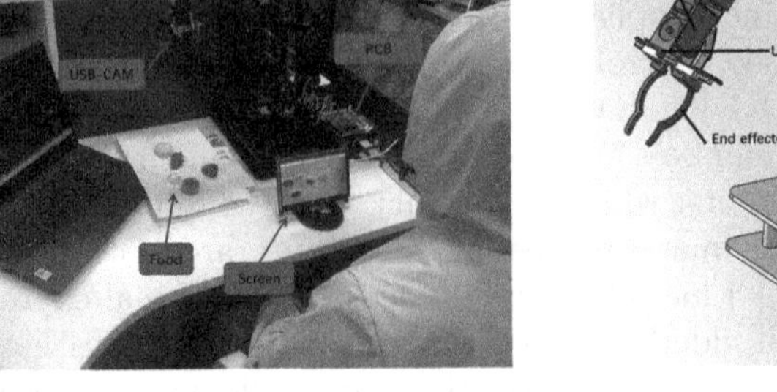

Fig. 1. Hardware structure.

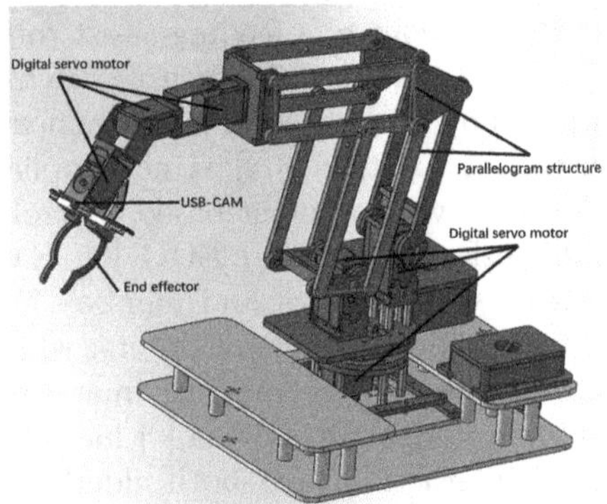

Fig. 2. The manipulator.

obtaining food information and user intentions and sending them to the PCB, which generates the corresponding control signals to each motor.

Lzbot is a low-cost 6-DOF robotic arm developed by our lab. As shown in Fig. 2, the six degrees of freedom ensure sufficient space and flexibility. The 60 kg analog motor provides sufficient power and accuracy. The combined structure design of the duo manipulator and the desktop manipulator uses a parallelogram link to stabilize the direction of the end, which makes the control more convenient. A USB camera is installed at the end, and the camera's field of view is exactly the same as the manipulator's activity space, which can be fully utilized space. We designed the control circuit PCB of the robot arm based on GD32 (main MCU) and packaged the corresponding driver.

The control flow of the proposed assisted feeding robotic arm system in this paper is shown in the Fig. 3. It consists of three main stages: yolov5-based real-time food detection, facial pose-based user intent detection, and image servo-based grasping and feeding. Among them, the food detection and facial pose detection algorithms are encapsulated in the host computer by the ROS method, and communicated using the topic method. The grasping and feeding procedures of the robotic arm are encapsulated on the bottom PCB, and the upper computer and the lower computer communicate via serial ports.

3 Stage 1: Real-Time Food Detection

The current mainstream target detection algorithms are R-CNN series, YOLO series [11] and transformer [12] based on the target detection method. The two-stage target detection method based on R-CNN first extracts the preselected frame based on the image, and then does secondary detection to get the result based on the preselected frame. This will result in high detection accuracy, but slower detection speed. The detection speed of the transformer-based detection method is also slower than that of the YOLO algorithm. Therefore, we choose

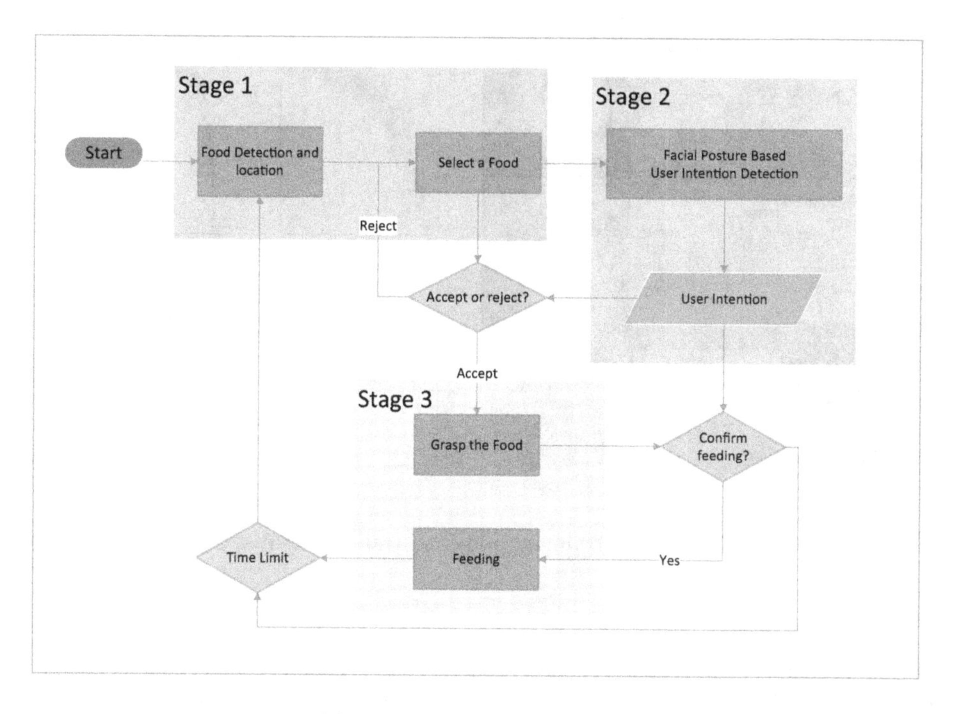

Fig. 3. The control algorithm.

the yolov5 algorithm that combines detection accuracy and speed for real-time food detection.

We collected a dataset of 600 images of mooncakes, buns, and strawberries, and annotated each image. The labeled food images are shown in Fig. 4. We notice that even the smallest strawberry has an object box larger than $32 * 32$ in the image, and according to the definition of small objects [13], small objects do not exist in our dataset. Therefore we remove the Mosaic data augmentation method created for small objects. After a series of data preprocessing and hyperparameter adjustment, we divided the dataset into a training set and a test set in a ratio of 3:1, and trained for 300 rounds under the yolov5s network model. The results are shown in the Fig. 4. It can be seen that our final mAP@0.5 reaches 0.97, and mAP@0.5:0.95 also reaches 0.8, which proves that the model we trained has a very good detection effect on our food dataset.

After detection, we also fused the deepsort [14] target tracking algorithm with yolov5. Based on the results of yolov5 detection, deepsort uses the Hungarian algorithm to determine whether a target in the current frame is the same as a target in the previous frame, and uses Kalman filter predicts the position at the current moment. We track the detected food in real time through the deepsort algorithm, and assign a unique ID to each food, which facilitates subsequent food switching and tracking of the target food.

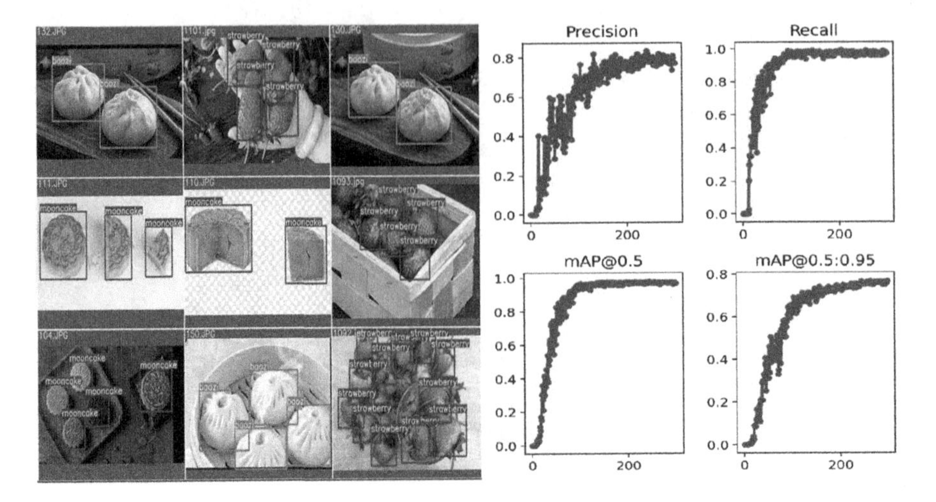

Fig. 4. The training foods and training results.

After the system is running, it will first enter the food detection stage, and the robotic arm will automatically run to a preset posture, so that the optical axis of the end camera is perpendicular to the plane where the food is located, so as to facilitate the observation of the food. At this time, the host computer runs the food detection node, and the detection result will be displayed on the interactive screen. We use green rectangles to select all the foods individually to tell the user all the food options, and randomly pre-select a target food, we use a red rectangle to select it to prompt the user's current food choice, as shown in the Fig. 5.

4 Stage 2: Facial Gesture Detection

After the system has successfully obtained food detection results, we present these results on a display in front of the user. The user perceives the food currently selected by the system and all the food in the field of view through the food detection information on the display. At this time, the user can convey his intentions by swinging some facial gestures, and the system will convert these intentions to obtain the interactive control instructions transmitted by the user.

At the beginning of the algorithm, each component in the system will maintain the initial state. For the user who is seated, we will perform continuous face detection on the user. Only when the user always stays in the system's field of view, it will keep running. When the system successfully detects the user's face, the algorithm will extract the key points of the user's face, and the extracted feature matrix will be used as an important input for the user's gesture judgment. After that, we will set the number of multi-frame combinations in combination with the frame rate of the camera and the duration of the action completed by the target user under the prior knowledge. The imaging rate of the camera used

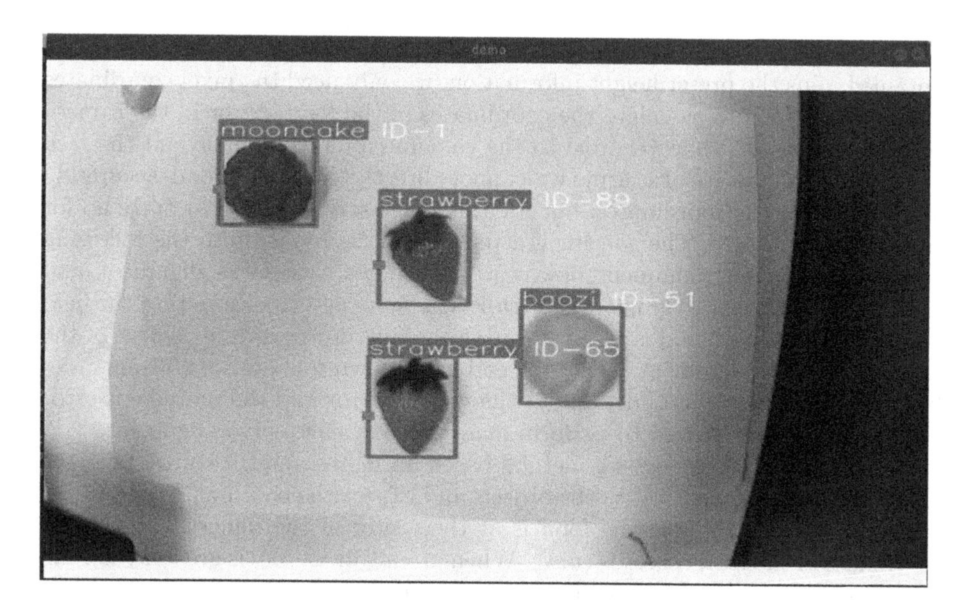

Fig. 5. Real-time detection result.

by the system is fast, and the target user's action lasts for a long time. We often cannot accurately detect the user's gesture within one frame, and the detection result in one frame is usually not the user's true intention. Therefore, for the user's posture discrimination, a unit of multiple frames is usually used. In this system, we select 12 frames as a group for detection.

After the algorithm successfully combines multiple frames, the system will detect in this unit. Using the relevant pose estimation algorithm, we estimate the relevant poses of the user's nodding, shaking his head and mouth, and count and sort the results of each frame in the combination, and get the result with the highest count value as the final result in the combination. We compare this result with preset rules to convert the detection results into the user intent.

Taking the user shaking his head as an example, if the user is dissatisfied with the current grabbing food, he only needs to shake his head slightly, and the algorithm will correspondingly convert the detected shaking gesture into the user's "rejection" intention. The intention will be understood as a control instruction of "switching food" to the system, and the system will switch the grab target after obtaining the instruction, and perform user intention detection again.

5 Stage 3: Image Servo Based Grasping and Feeding

In the third stage, the robotic arm will complete the work of grabbing and feeding the target food. When the user's intention to confirm the food is recognized, the robotic arm will immediately grab the target food. Since the camera at the end has been calibrated in advance and the internal parameters of the camera are obtained, plus the preset height information, we only need the pixel coordinates of the target food to calculate the coordinates of the food center in the camera coordinate system. However, due to the camera calibration error and the lack of precision of the robotic arm, we cannot directly grasp the food accurately after obtaining the coordinates, but use the image servo method to make up for the lack of hardware. The specific grasping action is divided into the following four steps: high-level alignment, downward movement, close servo alignment, and downward movement to grasp. The high-level alignment can ensure that the field of view of the target food is not lost during the first downward movement of the robot arm. After alignment, we let the end move down to a position 50 mm away from the target. At this time, due to the camera error and the movement error of the robot arm, we need to perform more precise image servo alignment.

We adopt an image-serving method based on image point features, take the midpoint of the target food as the object, and expect the coordinates to be (u0, v0), that is, the center of the food is at the center of the image. The current coordinate of the food center is (u, v). When $|u-u0|$ or $|v-v0|$ is greater than the set threshold, let the end approach slowly in the corresponding direction. When both $|u-u0|$ and $|v-v0|$ are smaller than the threshold, it can be considered that the end has reached the position where the grasping can be completed, and the servo alignment is completed.

After alignment, due to camera calibration and mechanical errors have become small enough at close range, it can be directly moved down to grab. After completing the grasping, once the user's intention to confirm eating is detected, the robotic arm will execute to the preset pose for feeding. When it is detected that the user has obtained food or the time limit is exceeded, the robotic arm returns to the first stage to restart the task.

6 Experiments and Result

To validate the system proposed in this paper, a total of four volunteers participated in this experiment. We had each volunteer test 15 feedings, with each feeding consisting of two shaking heads, one nodding, and one mouth opening. We evaluated the system in terms of response time, feeding success rate, and average feeding time. The experimental results are shown in Table 1.

Table 1. The experimental results.

	Average response time	Feeding success rate	Average feeding time
Volunteer 1	2.9 s	100%	33.4 s
Volunteer 2	2.4 s	93.3%	29.6 s
Volunteer 3	2.5 s	93.3%	25.8 s
Volunteer 4	3.6 s	86.7%	28.2 s
All	2.85 s	93.3%	29.25 s

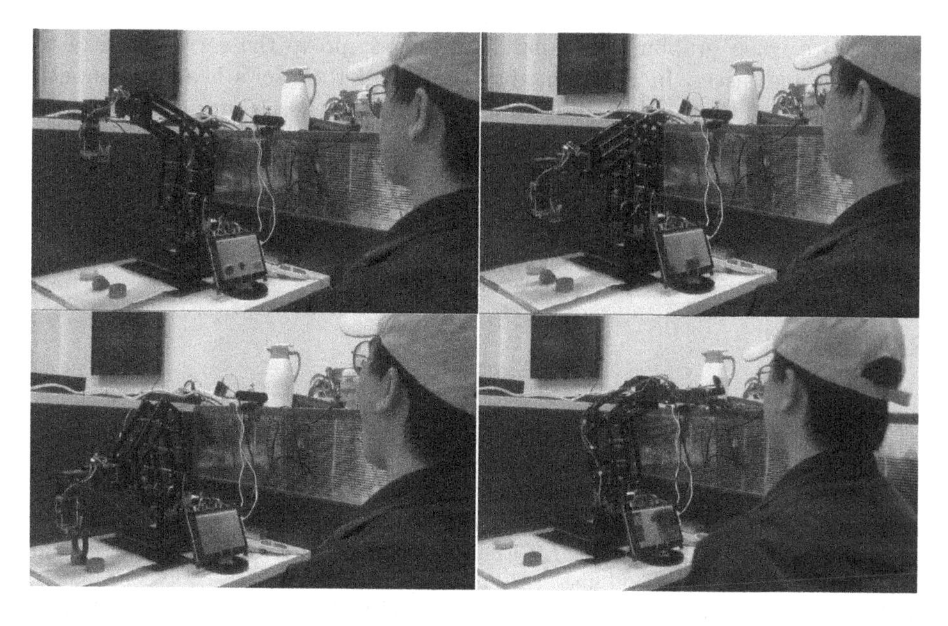

Fig. 6. The experimental process.

The experimental process is shown in the Fig. 6, the system was able to identify the intentions of all volunteers very accurately, but the response time was slightly different due to the individual's movement range, movement speed and other habits. Due to the existence of irregular food such as strawberries in the food, only the central point of the food is considered in the grasping strategy, and eventually the food slips without finding the optimal grasping point. The average command response time of less than three seconds and the average feeding time of around 30s proves that the system is effective and reliable.

7 Conclusions and Future Driction

In order to solve the problem of independent eating disorders of some elderly people and some disabled people, this paper designs and implements an effective feeding assistance robotic arm. The robot is controlled based on user intent recognition of facial gestures, and integrates real-time food detection and tracking based on yolov5 and deepsort. Four volunteers participated in the experiment to verify the feasibility of the system. All volunteers were able to easily control the robotic arm with facial gestures. The final results showed that the average response time for a single command was 2.85 s, and the average feeding time was 29.25 s, which proves the effectiveness of the system.

The future improvement direction of this system is as follows: optimize the relevant target detection algorithm so that it can be deployed on Raspberry Pi or jetson nano, fully use edge computing power, and further reduce the cost of the entire system. Improve the grabbing strategy of the system to have a higher success rate when grabbing irregular food, and improve the servo strategy to make grabbing faster. In addition, the tracking of the user's head and mouth can be added, so that each feeding can go directly to the vicinity of the mouth, while making the system safer and more feasible.

Acknowledgement. This work was supported in part by the Double Tops Construction Fund of Lanzhou University 561120202. 1. School of Information Science & Engineering, Lanzhou University, Lanzhou, 730000, China.

References

1. United Nations survey on ageing. www.un.org/zh/global-issues/ageing. Accessed 9 May 2022
2. WHO. www.who.int/en/news-room/fact-sheets/detail/disability-and-health. Accessed 9 May 2022
3. Ishii, S., Tanaka, S., Hiramatsu, F.: Meal assistance robot for severely handicapped people. In: Proceedings of 1995 IEEE International Conference on Robotics and Automation (ICRA), pp. 1308–1313 (1995)
4. Meal Buddy. www.performancehealth.com/meal-buddy-systems. Accessed 9 May 2022
5. Lopes, P., et al.: Icraft-eye-controlled robotic feeding arm technology. Technical report (2012)
6. Perera, C.J., Naotunna, I., Sandaruwan, C., Kelly-Boxall, N., Gopura, R.C., Lalitharatne, T.D.: Electrooculography signal based control of a meal assistance robot. In: IASTED International Conference on Biomedical Engineering, Innsbruck, Austria, pp. 133–139 (2016)
7. Shima, K., Fukuda, O., Tsuji, T., Otsuka, A., Yoshizumi, M.: EMG-based control for a feeding support robot using a probabilistic neural network. In: 2012 4th IEEE RAS & EMBS International Conference on Biomedical Robotics and Biomechatronics (BioRob), Rome, pp. 1788–1793 (2012)
8. Perera, C.J., Lalitharatne, T.D., Kiguchi, K.: EEG-controlled meal assistance robot with camera-based automatic mouth position tracking and mouth open detection. In: 2017 IEEE International Conference on Robotics and Automation (ICRA), pp. 1760–1765 (2017)
9. Mehta, Y.D., Khot, R.A., Patibanda, R., et al.: Arm-A-Dine: towards understanding the design of playful embodied eating experiences. In: Proceedings of the 2018 Annual Symposium on Computer-Human Interaction in Play, pp. 299–313 (2018)
10. Park, D., Hoshi, Y., Mahajan, H.P., et al.: Active robot-assisted feeding with a general-purpose mobile manipulator: design, evaluation, and lessons learned. Robot. Auton. Syst. **124**, 103344 (2020)
11. Zhao, Z.-Q., Zheng, P., Xu, S.-T., Wu, X.: Object detection with deep learning: a review. IEEE Trans. Neural Netw. Learn. Syst. **30**(11), 3212–3232 (2019). https://doi.org/10.1109/TNNLS.2018.2876865
12. Carion, N., Massa, F., Synnaeve, G., Usunier, N., Kirillov, A., Zagoruyko, S.: End-to-end object detection with transformers. In: Vedaldi, A., Bischof, H., Brox, T., Frahm, J.-M. (eds.) ECCV 2020. LNCS, vol. 12346, pp. 213–229. Springer, Cham (2020). https://doi.org/10.1007/978-3-030-58452-8_13

13. Kisantal, M., Wojna, Z., Murawski, J., et al.: Augmentation for small object detection. arXiv preprint arXiv:1902.07296 (2019)
14. Wojke, N., Bewley, A., Paulus, D.: Simple online and realtime tracking with a deep association metric. In: 2017 IEEE International Conference on Image Processing (ICIP), pp. 3645–3649. IEEE (2017)

Improved Cascade Active Disturbance Rejection Control for Functional Electrical Stimulation Based Wrist Tremor Suppression System Considering the Effect of Output Noise

Changchun Tao[1], Zan Zhang[1](✉), Benyan Huo[1], Yanhong Liu[1], Jianyong Li[2], and Hongnian Yu[1,3]

[1] School of Electrical and Information Engineering, Zhengzhou University, Zhengzhou 450001, China
zanzan@zzu.edu.cn
[2] College of Computer and Communication Engineering, Zhengzhou University of Light Industry, Zhengzhou 450001, China
[3] School of Engineering and the Built Environment, Edinburgh Napier University, Edinburgh EH11 4BN, UK

Abstract. The wrist tremor suppression system designed based on functional electrical stimulation technology has been welcomed by the majority of tremor patients as a non-invasive rehabilitation therapy. Due to the complex physiological structure characteristics of the wrist musculoskeletal system, it is difficult to accurately model it, and the measurement noise at the output port of the wrist tremor suppression system is difficult to avoid. These problems seriously affect the performance of tremor suppression. In order to solve the above problems, an improved linear active disturbance rejection control (LADRC) scheme is proposed based on the cascade strategy. The simulation results show that the proposed improved cascade LADRC can not only meet the requirements of system control performance against model uncertainty, but also attenuate the influence of output noise on the system, so as to effectively improve the suppression performance of tremor.

Keywords: Functional electrical stimulation · Musculoskeletal model · Wrist tremor suppression system · Output noise · Improved cascade LADRC

Supported by National Natural Science Foundation of China (62103376), Robot Perception and Control Support Program for Outstanding Foreign Scientists in Henan Province of China (GZS2019008), China Postdoctoral Science Foundation (2018M632801) and Science & Technology Research Project in Henan Province of China (212102310253).

H. Liu et al. (Eds.): ICIRA 2022, LNAI 13456, pp. 232–244, 2022.
https://doi.org/10.1007/978-3-031-13822-5_21

1 Instruction

Tremor, an involuntary periodic movement, is a common movement disorder that often occurs in the wrist of the upper limb, mainly in the elderly [1]. Due to the surgical risks and obvious side effects of traditional tremor suppression schemes such as invasive therapy and drug therapy, it is difficult to meet the treatment needs of patients with tremor. With the development of science and technology, auxiliary tremor suppression schemes have been widely used as an important means. Among them, functional electrical stimulation (FES) technology plays an important role in the field of assisted tremor suppression because it can inhibit the pathological tremor and accelerate rehabilitation process at the same time [2]. FES technology induces muscle contraction by artificially applying low-level electrical pulse to stimulate the nerve. The electrical pulse signal is transmitted through the surface electrode attached to the skin. The FES signal is used to stimulate the muscles of the sick part of the patient to generate the anti-phase muscle torque relative to tremor movement, so as to achieve the purpose of inhibiting pathological tremor of the wrist. The quality of the inhibition effect depends on the control performance of the designed controller [3].

Considering the problem that it is difficult to accurately model the musculoskeletal system, and the designed controller needs to have engineering practicability, this paper applies linear active disturbance rejection control (LADRC) to the design of wrist tremor suppression control system with functional electrical stimulation. LADRC does not depend on the mathematical model of the system and has strong robustness and anti-interference [4,7,8]. Furthermore, the output noise of the FES based tremor suppression system is also an important issue to be considered. At present, relevant researchers have put forward many different schemes to solve the problem that LADRC performance affected by noise. Literature [12] designed a gain switchable observer, that is, large gain is used to reconstruct the system state, and when the observation error is reduced to a certain value, it is switched to small gain to reduce the influence of high-frequency noise. Literature [13] and Literature [14] deal with the measurement noise by introducing random approximation strategy and fast filter into the high gain observer respectively, so as to maintain the original characteristics of the observer to the greatest extent. Literature [15,16] directly designed an adaptive observer with on-line gain adjustment to solve this problem. Although it has better noise fitting effect than the first two schemes, the observer design is complex and difficult to realize in engineering.

In view of the above research problems, an improved cascade LADRC method for wrist tremor suppression control system affected by output noise is proposed in this paper. Firstly, The filter noise processing method with engineering practicability is improved, and a LADRC method based on cascade control strategy is proposed to effectively compensate the amplitude and phase loss of the filtered output. Then, aiming at the noise problem of the wrist tremor suppression control system, in order to ensure the control performance of the system and the safety and comfort of the subjects, the influence of noise on LADRC is analyzed from the perspective of frequency domain, and an improved cascade

LADRC method is further proposed. Finally, numerical simulation results show that the proposed improved cascade LADRC can not only suppress tremor more effectively, but also output electrical stimulation signals with smaller amplitude, which ensures the safety of subjects and the tremor suppression performance of the system.

2 Modeling of Wrist Musculoskeletal System

Flexor carpi radialis (FCR) and flexor carpi radialis (ECR) drive the radius to realize the flexion and extension motion of the wrist [17]. According to the physiological characteristics of the musculoskeletal system, this paper uses Hammerstein structure to establish a mathematical model of the human wrist system. This structure includes nonlinear module and dynamic linear module, which is not only simple in structure, but also can better characterize the characteristics of muscle system [18]. The structure of musculoskeletal system is shown in Fig. 1, where $u_{\text{fcr}}(t)$ and $u_{\text{ecr}}(t)$ represent the electrical stimulation pulses applied to the radial muscles; $f_{\text{fcr}}(u_{\text{fcr}})$ and $f_{\text{ecr}}(u_{\text{ecr}})$ represent the nonlinear characteristics of muscle recruitment; $G_{\text{fcr}}(s)$ and $G_{\text{ecr}}(s)$ represent linear activation dynamics (LAD) of the wrist muscle model; $\tau(t)$ represents the torque acting on the radius of the wrist; $G_{\text{RBD}}(s)$ represents the wrist skeletal model; $d(t)$ represents tremor signal; $y(t)$ is the wrist position angle.

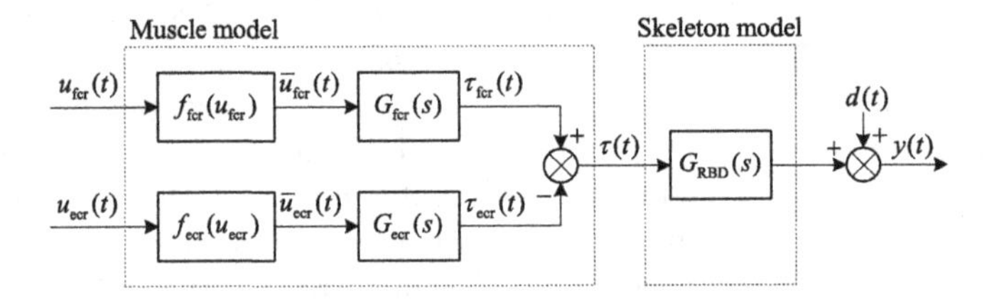

Fig. 1. Hammerstein musculoskeletal system model.

In order to simplify the model structure and the corresponding parameter identification procedure, $u_{\text{fcr}}(t)$ and $u_{\text{ecr}}(t)$ can be compacted to a new control input $u(t)$ to generate radial flexion and extension movement. Therefore, the two-input two-output muscle nonlinear recruitment model can be converted into a single-input single-output model. The nonlinear characteristics of muscle recruitment is represented by $f(u)$, and its mathematical expression is

$$f(u) = r_0 + r_1 u + r_2 u^2 + \cdots + r_i u^i \tag{1}$$

where i is the order of nonlinear model, $r_0, r_1, \cdots, r_i$ are the model parameters to be identified. When $i = 3$, the model has a good fitting effect for the nonlinear characteristics of muscle recruitment [20].

Since different wrist muscles have similar LAD [21], $G_{\text{fcr}}(s) \approx G_{\text{ecr}}(s) = G_{\text{LAD}}(s)$. Therefore, $G_{\text{RBD}}(s)$ and $G_{\text{LAD}}(s)$ can be combined to obtain the equivalent linear musculoskeletal system model $G(s)$. The equivalent linear musculoskeletal model can be expressed as

$$G(s) = \frac{b}{s^n + c_{n-1}s^{n-1} + c_{n-2}s^{n-2} + \cdots + c_1 s + c_0} \quad (2)$$

where b, c_0, c_1, $\cdots$, c_{n-1} are the model parameters to be identified. By the open-loop identification experiment and using the input and output data set, the nonlinear and linear parameters can be obtained. $f(u) = 2.6267 \times 10^{-4} - 6.6848 \times 10^{-5}u - 9.5845 \times 10^{-9}u^2 - 1.4905 \times 10^{-10}u^3$. $G(s) = \frac{3.438}{s^2 + 10.24s + 39.86}$.

3 Design of Improved Cascade LADRC for Tremor Suppression System

3.1 Design of Cascade LADRC

For the noise problem existing at the output port, a low-pass filter is usually added to the output to process the output signal including high-frequency noise, and the processed signal is used as the real output of the system. However, due to the existence of inertial filtering link, the amplitude and phase of the system output will be affected. In order to compensate the changes in amplitude and phase caused by the low-pass filter, a control strategy based on cascaded LADRC (CLADRC) algorithm is proposed.

The CLADRC method has two control loops, outer loop LADRC and inner loop LADRC. The idea of CLADRC is to use the output $u_v(t)$ of the outer loop LADRC to control the output $y_l(t)$. The control value $\bar{u}'(t)$ of the inner loop LADRC is the actual control action acting on the musculoskeletal model. Due to the nonlinear recruitment $f(u)$ existing in the musculoskeletal system, a feed-forward linearization controller $f^{-1}(\bar{u}')$ is designed first, i.e., $f(u) \cdot f^{-1}(\bar{u}') \approx 1$. By adding a first-order inertial filter, the original system is transformed into a composite system not affected by noise, in which the output of the filter is taken as the new state of the composite system. The block diagram of the cascade active disturbance rejection controller of the composite system is shown in Fig. 2.

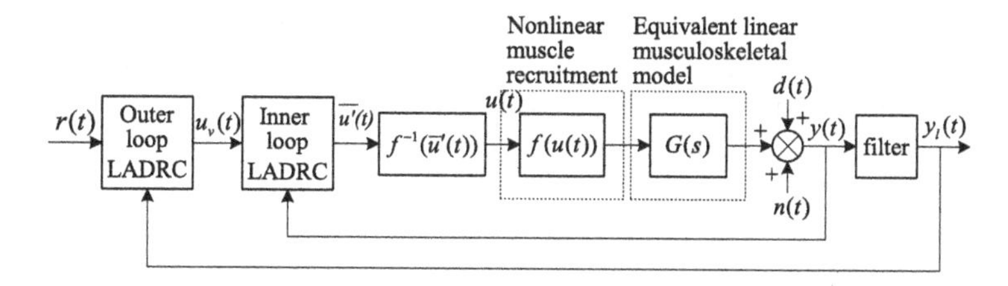

Fig. 2. Structure diagram of CLADRC tremor suppression system.

The CLADRC is designed as

$$\begin{cases} \dot{\hat{x}} = A\hat{x} + Bu + L(y - C\hat{x}) \\ \hat{y} = C\hat{x} \\ u = k_p(u_v - \hat{x}_1) - k_{d_1}\hat{x}_2 - \hat{x}_3/b_0 \\ \dot{\hat{x}}_l = A_l\hat{x}_l + B_l u_v + L_l(y_l - C_l\hat{x}_l) \\ u_v = k_{l_p}(r - \hat{x}_{l_1}) - \hat{x}_{l_2}/b_v \\ \hat{y}_l = C_l\hat{x}_l \end{cases} \tag{3}$$

where $A_l = \begin{bmatrix} 0 & 1 & 0 \\ 0 & 0 & 1 \\ 0 & 0 & 0 \end{bmatrix}$, $B_l = \begin{bmatrix} 0 \\ b_0 \\ 0 \end{bmatrix}$, $C_l = \begin{bmatrix} 1 & 0 & 0 \end{bmatrix}$. L_l and K_l are the observer gain and controller gain of the outer loop LADRC respectively,

$$L_l = \begin{bmatrix} l_{l1} & l_{l2} \end{bmatrix}^T, K_l = k_{lp} \tag{4}$$

For the selection of the order of the inertial filter, we should not only consider its filtering effect, but also consider whether the filter can protect the stability of the original system on the premise of effectively filtering the system noise. Although the higher the order of inertial filter is, the better the filtering effect is. However, the higher the order, the greater the possibility of destroying the stability of the original system. The inertial filter is selected as $1/(Ts+1)$ on considering the above problem in this research. In order to synthesize the noise suppression performance of the filter and its influence on the phase delay, the filter factor $T = 0.2$ is selected. According to the inner loop LADRC parameter tuning method, the pole assignment method is also used to select the parameters of the outer loop LADRC controller. $k_{lp} = \omega_{lc}^2$; $l_{l1} = 2\omega_{lo}$, $l_{l2} = \omega_{lo}^2$. For this system, the filter still adopts the first-order inertia link strategy without changing the value of T. The outer loop LADRC compensation coefficient $b_v = 1.45/T$. For the selection of outer loop controller bandwidth and observer bandwidth, it is necessary to meet the requirements of the system for control performance and control value. The inner loop LADRC parameters are chosen as $\omega_o = \omega_c = 20$, $b_0 = 3.5$ and the outer loop LADRC parameters are chosen as $\omega_{lc} = 8$, $\omega_{lo} = 10$.

In order to verify the tremor suppression performance of the proposed CLADRC method, a comparative simulation test is carried out with LADRC (with only inner loop and without output filter) method. The simulation results of the step input signal are shown in Fig. 3. The controller designed based on CLADRC algorithm can effectively filter the noise at the output port and ensure the tracking and anti-interference performance of the system. However, it can be found that the control value of the CLADRC is large at the step point. When the system enters into the steady state, the control value of the CLADRC is slightly larger than that of the LADRC.

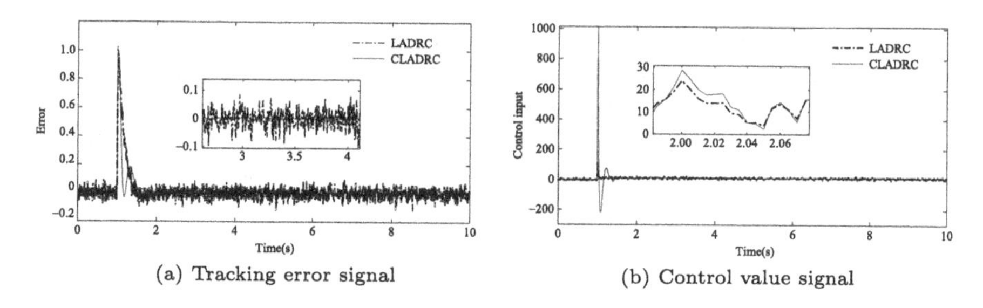

(a) Tracking error signal (b) Control value signal

Fig. 3. Tracking error and control value under step input signal.

The simulation results under sinusoidal signal are shown in Fig. 4. As shown in Fig. 4, CLADRC algorithm can effectively filter the noise at the output port and ensure the tracking and anti-interference performance of the control system. However, there is also the problem of large control value of the system. The control value is obviously larger and changes faster than that of the LADRC algorithm.

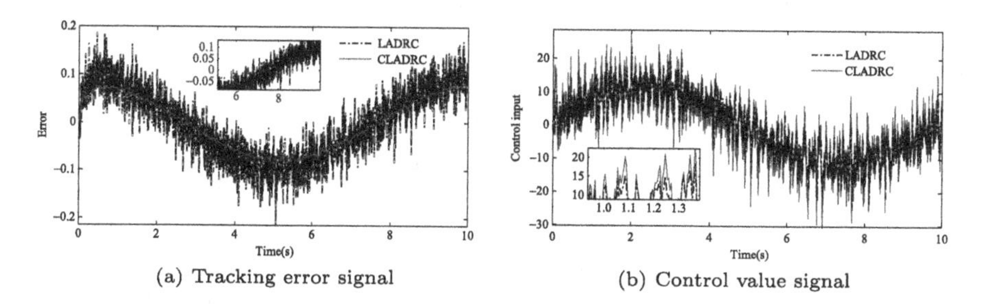

(a) Tracking error signal (b) Control value signal

Fig. 4. Tracking error and control value under sinusoidal input signal.

To sum up, although CLADRC algorithm improves the control performance of the closed-loop control system and eliminates the influence of noise, it will increase the amplitude of the control value. In order to sort out the above problems, this study theoretically deduces the relationship between the inner loop LADRC performance and noise of CLADRC system, so as to analyze it through structural transformation and deduce the existing problems from the perspective of frequency domain.

3.2 Performance Analysis of LADRC System Under the Influence of Noise

The influence of measurement noise on the output of the original system is obvious. For the cascade control system, the influence of noise on the performance of its inner loop LADRC must be analyzed first. In order to explore the relationship between the inner loop LADRC and noise, correlation analysis is carried out from the perspective of frequency domain as follows.

Substitute $l_1 = 3\omega_o$, $l_2 = 3\omega_o^2$, $l_3 = \omega_o^3$ into the linear extended state observer (LESO) to obtain the output ($\hat{x}_1$, $\hat{x}_2$, $\hat{x}_3$) of transfer function of LESO,

$$\hat{x}_1 = \frac{3\omega_o s^2 + 3\omega_o^2 s + \omega_o^3}{(s+\omega_o)^3} y + \frac{b_0 s}{(s+\omega_o)^3} \bar{u} \tag{5}$$

$$\hat{x}_2 = \frac{(3\omega_o^2 s + \omega_o^3)s}{(s+\omega_o)^3} y + \frac{b_0(s+3\omega_o)s}{(s+\omega_o)^3} \bar{u} \tag{6}$$

$$\hat{x}_3 = \frac{\omega_o^3 s^2}{(s+\omega_o)^3} y + \frac{b_0 \omega_o^3}{(s+\omega_o)^3} \bar{u} \tag{7}$$

Substitute $k_p = \omega_c^2$ and $k_d = 2\omega_c$ to calculate the LADRC output control value

$$\bar{u}' = \frac{1}{b_0}[\omega_c^2(r - \hat{x}_1) - 2\omega_c \hat{x}_2 - \hat{x}_3] \tag{8}$$

Because $f(u) \cdot f^{-1}(\bar{u}') \approx 1$, so $\bar{u}' \approx \bar{u}$. Substitute (5)–(7) into (8), then

$$\begin{aligned} \bar{u}' = & \frac{1}{b_0} \frac{(s+\omega_o)^3}{(s+\omega_o)^3 + 2\omega_c s^2 + (\omega_c^2 + 6\omega_o\omega_c)s - \omega_o^3} \\ & \left[\omega_c^2 r - \frac{(3\omega_c^2\omega_o + 6\omega_c\omega_o^2 + \omega_o^3)s^2 + (3\omega_c^2\omega_o^2 + 2\omega_c\omega_o^3)s + \omega_c^2\omega_o^3}{(s+\omega_o)^3} y \right] \end{aligned} \tag{9}$$

The block diagram of the inner-loop second-order LADRC controller with the CLADRC structure shown in Fig. 2 is converted into a two-DOF structure by Eq. (9), and the influence of the output port measurement noise on the performance of the LADRC controller is analyzed. The simplified block diagram is shown in Fig. 5.

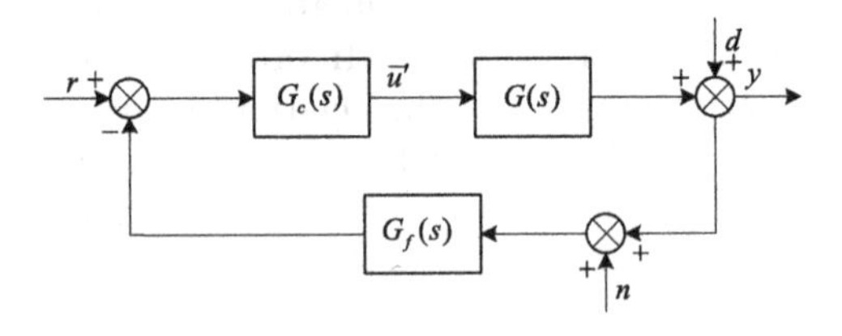

Fig. 5. Two-degree-of-freedom structure of LADRC.

$$G_c(s) = \frac{\omega_c^2(s^3 + 3\omega_o s^2 + 3\omega_o^2 s + \omega_o^3)}{b_0[s^3 + (3\omega_o + 2\omega_c)s^2 + (3\omega_o^2 + 6\omega_o\omega_c + \omega_c^2)s]} \tag{10}$$

$$G_f(s) = \frac{(3\omega_c^2\omega_o + 6\omega_c\omega_o^2 + \omega_o^3)s^2 + (3\omega_c^2\omega_o^2 + 2\omega_c\omega_o^3)s + \omega_c^2\omega_o^3}{\omega_c^2(s^3 + 3\omega_o s^2 + 3\omega_o^2 s + \omega_o^3)} \tag{11}$$

According to Fig. 5, the transfer function from noise to system output is

$$\frac{y}{n} = \frac{G_c(s)G(s)G_f(s)}{1 + G_c(s)G(s)G_f(s)} \tag{12}$$

Substitute (10) and (11) into (12), then we got

$$\frac{y}{n} = \frac{bB(s)}{b_0 A(s)(s^2 + a_1 s + a_0) + bB(s)} \tag{13}$$

where

$$A(s) = s^3 + (3\omega_o + 2\omega_c)s^2 + (3\omega_o^2 + 6\omega_o\omega_c + \omega_c^2)s \tag{14}$$

$$B(s) = (3\omega_c^2\omega_o + 6\omega_c\omega_o^2 + \omega_o^3)s^2 + (3\omega_c^2\omega_o^2 + 2\omega_c\omega_o^3)s + \omega_c^2\omega_o^3 \tag{15}$$

According to the identified model $G(s) = \frac{3.438}{s^2+10.24s+39.86}$, where $b \approx 3.5$, $a_0 = 39.86$, $a_1 = 10.24$. Substitute these model parameters into (13) and do not change the value of b_0. Take different observer bandwidth ω_o and controller bandwidth ω_c, analyze the frequency domain characteristics of the system affected by noise, and obtain the relationship between the noise signal and the LADRC performance.

According to the parameters of the inner loop LADRC obtained by parameter tuning in the previous section($\omega_o = \omega_c = 20$, $b_0 = 3.5$), using the principle of the single variable method to select inner loop LADRC. Firstly, $\omega_o = 20$ remains unchanged, and ω_c is taken as 10, 20, 30, and 40 respectively. The domain characteristics are shown in Fig. 6(a). Then, $\omega_c = 20$ remains unchanged, and ω_o is chosen as 10, 20, 30, and 40 respectively. The frequency domain characteristics of the inner-loop LADRC control system are shown in Fig. 6(b).

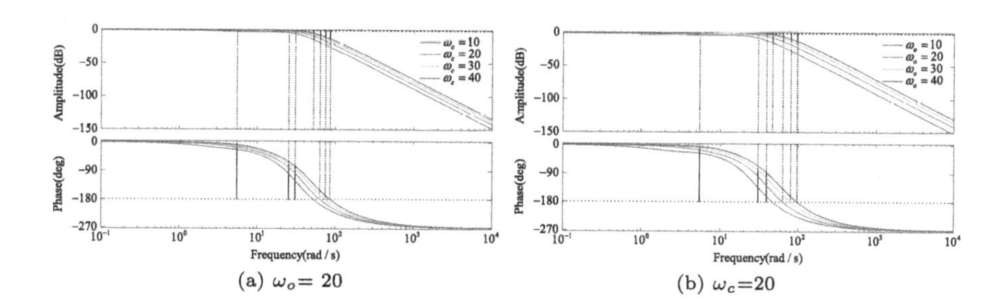

(a) ω_o= 20 (b) ω_c=20

Fig. 6. Frequency domain characteristic curve of the system affected by noise.

As seen from Fig. 6, with the increase of ω_c and ω_o, the gain of the high frequency band of the system and the amplification effect on the noise signal increase, which will limit the anti-noise ability of the closed-loop control system, and the control value will change dramatically. If the value of ω_c and ω_o are reduced, the system tracking and anti-disturbance performance cannot meet the control requirements. From this, we can know the relationship between the control performance of the closed-loop control system and the control value of the system. Therefore, how to design a controller that can meet the control performance requirements and make the system less sensitive to measurement noise is particularly important.

3.3 Design of Improved CLADRC

Aiming at the problems of the cascade control strategy, an improved cascade active disturbance rejection controller is designed to solve the problems of rapid change of control value and rate caused by noise. The block diagram of improved CLADRC (ICLADRC) is shown in the Fig. 7, in which the outer loop LADRC is designed as a first-order active disturbance rejection controller, and the inner loop adopts a mixed data-driven and model information-assisted control strategy to design the controller [25]. The inner loop of the cascade structure is designed as a model-assisted LADRC (MLADRC) algorithm, and the controller based on the MLADRC algorithm improves the original cascade system.

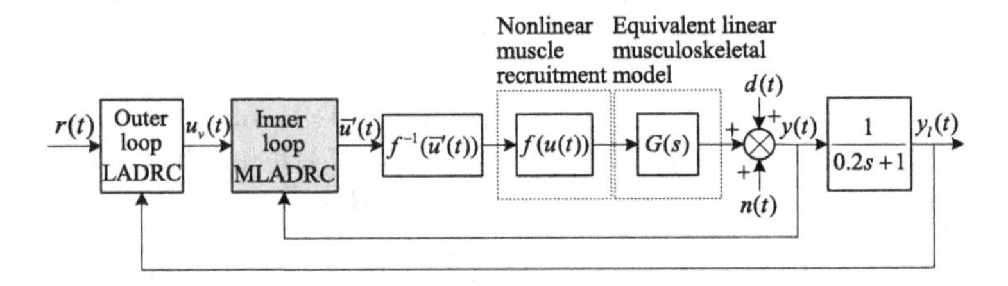

Fig. 7. Structure diagram of ICLADRC.

For the inner loop controller of the ICLADRC control system, the design steps of the controller based on the MLADRC algorithm are as follows:

The equivalent musculoskeletal system is

$$\ddot{y} = -10.24\dot{y} - 39.86y + f' + b_0\bar{u} \tag{16}$$

where, $f' = d + (b - b_0)\bar{u}$ is the total disturbance of the system.

Adding the identified model information of the system to LESO, the state space expression of MLADRC's LESO can be written as

$$\dot{\hat{x}} = A_m\hat{x} + B_m\bar{u} + L_m(y - C\hat{x}) \tag{17}$$

where, $A_m = \begin{bmatrix} 0 & 1 & 0 \\ -a_0 & -a_1 & 1 \\ 0 & 0 & 0 \end{bmatrix}$, $B_m = \begin{bmatrix} 0 \\ b_0 \\ 0 \end{bmatrix}$, $C = \begin{bmatrix} 1\,0\,0 \end{bmatrix}$, and $L_m = \begin{bmatrix} l_{m1}\; l_{m2}\; l_{m3} \end{bmatrix}^{\mathrm{T}}$ is the observer error feedback gain matrix of the MLADRC algorithm to be tuned. Also with the idea of pole placement, l_{m1}, l_{m2}, l_{m3} are designed, and the elements of the observer error feedback gain matrix are obtained as

$$\begin{cases} l_{m1} = 3\omega_o - a_1 \\ l_{m2} = 3\omega_o^2 - 3a_1\omega_o - a_0 + a_1^2 \\ l_{m3} = \omega_o^3 \end{cases} \tag{18}$$

The relationship between the total disturbance of the control system designed based on the MLADRC algorithm and the total disturbance of the original system can be expressed as: $f' = f - a_0\hat{x}_1 - a_1\hat{x}_2$. Then, the control value of the MLADRC can be expressed as

$$\bar{u}' = \frac{1}{b_0}\left[k_p(r - \hat{x}_1) - k_d\hat{x}_2 - (\hat{x}_3 - a_0\hat{x}_1 - a_1\hat{x}_2)\right] \tag{19}$$

It can be seen from the analysis in the previous section that the smaller the observer bandwidth and controller bandwidth, the lower the high-frequency band gain of the system and the smaller the impact of noise on the performance of the controller. Therefore, the sensitivity of the system to noise will be significantly reduced, which will weaken the fluctuation range of the control output of the system. The closed-loop control system based on ICLADRC algorithm can not only meet the requirements of expected performance, but also ensure the safety of experimental subjects. The block diagram of the controller designed based on MLADRC algorithm is shown in Fig. 8.

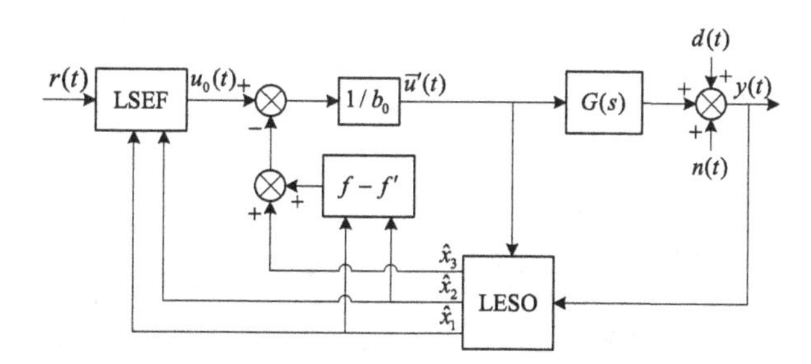

Fig. 8. MLADRC controller structure diagram.

In order to verify its performance, the ICLADRC designed based on MLADRC algorithm is verified by numerical simulation. According to the ICLADRC algorithm, the controller is also designed by the tuning idea. Considering the characteristics of the inner loop controller, the MLADRC parameters of the inner loop are determined after tuning: $\omega_c = 12$, $\omega_o = 10$, $b_0 = 3.5$. The LADRC parameters of the outer loop are: $\omega_{lc} = 7$, $\omega_{lo} = 14$, $b_v = 15$. The simulation results of different controllers under the step signal are shown in Fig. 9.

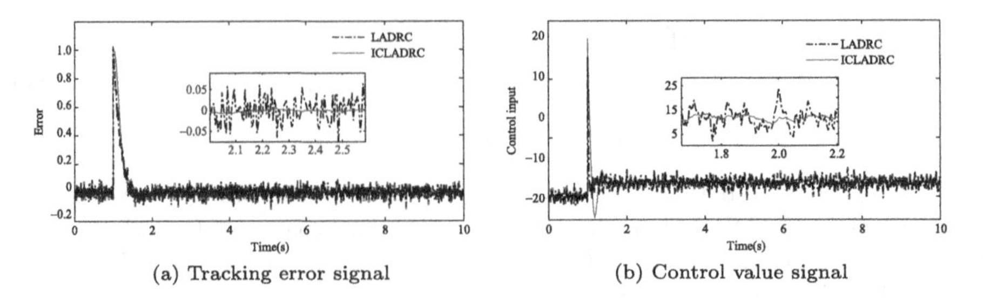

(a) Tracking error signal
(b) Control value signal

Fig. 9. Tracking error and control value under step input signal.

It can be seen from Fig. 9(a) that under the step signal, the ICLADRC control algorithm can effectively filter the noise existing at the output port and the tracking and anti-disturbance performance of the controller is improved as well. As seen from Fig. 9(b), it can be found that the control value of the closed-loop control system designed based on the ICLADRC algorithm is significantly smoother than that of the closed-loop control system designed based on the LADRC algorithm.

The simulation results of different controllers under the sinusoidal input signal are shown in Fig. 10. From Fig. 10(a), the simulation results are the same as the step input signal. By analyzing all the above simulation results, it can be clearly seen that the ICLADRC closed-loop control system based on the MLADRC algorithm is more suitable for the needs of the FES based tremor suppression system.

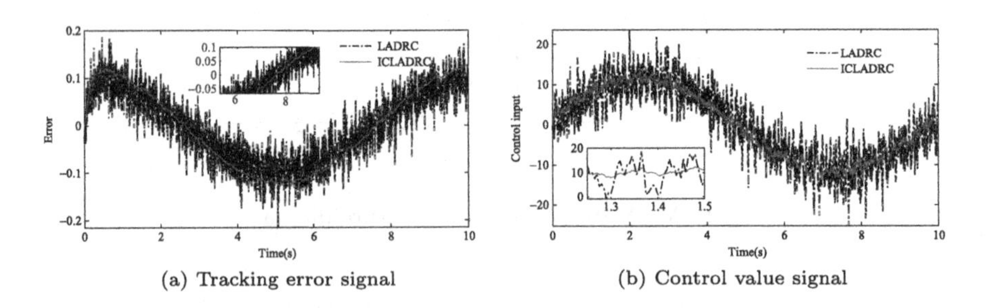

(a) Tracking error signal
(b) Control value signal

Fig. 10. Tracking error and control value under sinusoidal input signal.

4 Conclusion

In this paper, considering the complex physiological and structural characteristics of the wrist musculoskeletal system, the Hammerstein model is used to establish the wrist system. Firstly, considering the influence of the measurement noise existing at the output port, the CLADRC algorithm is designed based on the cascade control strategy, and the influence of noise on the performance of LADRC is analyzed from the perspective of frequency domain. Although

CLADRC algorithm improves the control performance of the closed-loop control system and eliminates the influence of noise, it increases the amplitude of the control value. Then, the controller based on the MLADRC algorithm is designed to improve the original CLADRC system. Simulation results verify that the controller designed based on the ICLADRC algorithm can not only suppress the tremor more effectively, but also make the controller output more appropriate electrical stimulation pulse signal.

References

1. Tousi, B., Cummings, J. (eds.): Neuro-Geriatrics. Springer, Cham (2017). https://doi.org/10.1007/978-3-319-56484-5
2. Wen, Y., Huang, X., Tu, X., Huang, M.: Functional electrical stimulation array electrode system applied to a wrist rehabilitation robot. J. Huazhong Univ. Sci. Technol. (Nat. Sci. Edn.) **41**(S1), 332–334, 342 (2013). https://doi.org/10.13245/j.hust.2013.s1.086
3. Copur, E.H., Freeman, C.T., Chu, B., Laila, D.S.: Repetitive control of electrical stimulation for tremor suppression. IEEE Trans. Control Syst. Technol. **27**(2), 540–552 (2019). https://doi.org/10.1109/TCST.2017.2771327
4. Han, J.: From PID to active disturbance rejection control. IEEE Trans. Ind. Electron. **56**(3), 900–906 (2009). https://doi.org/10.1109/TIE.2008.2011621
5. Li, Y., Zhang, C., Song, J., Li, X., Duan, B.: An active disturbance rejection control strategy for a three-phase isolated matrix rectifier. IEEE Trans. Transp. Electrif. **8**(1), 820–829 (2021). https://doi.org/10.1109/TTE.2021.3100544
6. Chen, Z., Ruan, X., Li, Y.: Dynamic modeling of a cubical robot balancing on its corner. Control Decis. **34**(6), 1203–1210 (2019). https://doi.org/10.13195/j.kzyjc.2017.1559
7. Liang, Q., Wang, C., Pan, J., Wen, Y., Wang, Y.: Parameter identification of b0 and parameter tuning law in linear active disturbance rejection control. Control Decis. **30**(9), 1691–1695 (2015). https://doi.org/10.13195/j.kzyjc.2014.0943
8. Gao, Y., Wu, W., Gao, L.: Linear active disturbance rejection control for high-order nonlinear systems with uncertainty. Control Decis. **35**(2), 483–491 (2020). https://doi.org/10.13195/j.kzyjc.2018.0550
9. Gao, Z.: Scaling and bandwidth-parameterization based controller tuning. In: American Control Conference, pp. 4989–4996. Institute of Electrical and Electronics Engineers Inc., Denver (2003). https://doi.org/10.1109/ACC.2003.1242516
10. Chen, Z., Sun, M., Yang, R.: On the Stability of linear active disturbance rejection control. Acta Automatica Sinica **39**(5), 574–580 (2013). https://doi.org/10.3724/SP.J.1004.2013.00574
11. Gao, Y., Wu, W., Wang, Z.: Cascaded linear active disturbance rejection control for uncertain systems with input constraint and output noise. Acta Automatica Sinica **48**(3), 843–852 (2022). https://doi.org/10.16383/j.aas.c190305
12. Prasov, A.A., Khalil, H.K.: A nonlinear high-gain observer for systems with measurement noise in a feedback control framework. IEEE Trans. Autom. Control **58**(3), 569–580 (2013). https://doi.org/10.1109/TAC.2012.2218063
13. Lee, J., Choi, K., Khalil, H. K.: New implementation of high-gain observers in the presence of measurement noise using stochastic approximation. In: European Control Conference, pp. 1740–1745. Institute of Electrical and Electronics Engineers Inc., Aalborg (2016). https://doi.org/10.1109/ECC.2016.7810542

14. Teel, A. R.: Further variants of the Astolfi/Marconi high-gain observer. In: American Control Conference, pp. 993–998. Institute of Electrical and Electronics Engineers Inc., Boston (2016). https://doi.org/10.1109/ACC.2016.7525044
15. Battilotti, S.: Robust observer design under measurement noise with gain adaptation and saturated estimates. Automatica **81**, 75–86 (2017). https://doi.org/10.1016/j.automatica.2017.02.008
16. Nair, R.R., Behera, L.: Robust adaptive gain higher order sliding mode observer based control-constrained nonlinear model predictive control for spacecraft formation flying. IEEE/CAA J. Automatica Sinica **5**(1), 367–381 (2018). https://doi.org/10.1109/JAS.2016.7510253
17. Bo, A.P.L., Poignet, P., Zhang D., Ang, W. T.: FES-controlled co-contraction strategies for pathological tremor compensation. In: IEEE/RSJ International Conference on Intelligent Robots and Systems, pp. 1633–1638. IEEE Computer Society, Louis (2009). https://doi.org/10.1109/IROS.2009.5354397
18. Liu, Y., Qin, Y., Huo, B., Wu, Z.: Functional electrical stimulation based bicep force control via active disturbance rejection control. In: 5th IEEE International Conference on Advanced Robotics and Mechatronics, pp. 306–311. Institute of Electrical and Electronics Engineers Inc., Shenzhen (2020). https://doi.org/10.1109/ICARM49381.2020.9195304
19. Yang, X., Chen, J., Li, Y., Zhou, Y.: A method of modulation and evaluation for functional electrical stimulation based on muscle activation properties. J. Fuzhou Univ. (Nat. Sci. Edn.) **47**(3), 346–351 (2019)
20. Alibeji, N., Kirsch, N., Farrokhi, S., Sharma, N.: Further results on predictor-based control of neuromuscular electrical stimulation. IEEE Trans. Neural Syst. Rehabil. Eng. **23**(6), 1095–1105 (2015). https://doi.org/10.1109/TNSRE.2015.2418735
21. Colacino, F.M., Emiliano, R., Mace, B.R.: Subject-specific musculoskeletal parameters of wrist flexors and extensors estimated by an EMG-driven musculoskeletal model. Med. Eng. Phys. **34**(5), 531–540 (2012). https://doi.org/10.1016/j.medengphy.2011.08.012
22. Yuan, D., Ma, X., Zeng, Q., Qiu, X.: Research on frequency-band characteristics and parameters configuration of linear active disturbance rejection control for second-order systems. Control Theor. Appl. **30**(12), 1630–1640 (2004). https://doi.org/10.7641/CTA.2013.30424
23. Li, H., Zhu, X.: On parameters tuning and optimization of active disturbance rejection controller. Control Eng. China **11**(5), 419–423 (2004)
24. Lin, F., Sun, H., Zheng, Q., Xia, Y.: Novel extended state observer for uncertain system with measurement noise. Control Theor. Appl. **22**(6), 995–998 (2005)
25. Li, G., Pan, L., Hua, Q., Sun, L., Lee, K.Y.: Water Pump Control: A hybrid data-driven and model-assisted active disturbance rejection approach. Water (Switz.) **11**(5), 1066 (2019). https://doi.org/10.3390/w11051066
26. Lenz, F.A., Jaeger, C.J., Seike, M.S., Lin, Y., Reich, S.G.: Single-neuron analysis of human thalamus in patients with intention tremor and other clinical signs of cerebellar disease. J. Neurophysiol. **87**(4), 2084–2094 (2002). https://doi.org/10.1152/jn.00049.2001

An Anti-sideslip Path Tracking Control Method of Wheeled Mobile Robots

Guoxing Bai, Yu Meng(✉), Qing Gu, Guodong Wang, Guoxin Dong, and Lei Zhou

School of Mechanical Engineering, University of Science and Technology Beijing, Beijing 100083, China
myu@ustb.edu.cn

Abstract. Anti-sideslip has not been paid much attention by most researchers of wheeled mobile robots. And some existing anti-sideslip path tracking control methods based on switching control have problems such as relying on design experience. To enable the wheeled mobile robot to prevent sideslip and track the reference path at the same time, we propose an anti-sideslip path tracking control method based on a time-varying local model. The principle of this method is to make model predictions and rolling optimizations in the robot coordinate system in each control period. The proposed controller is tested by MATLAB simulation. According to the simulation results, the proposed controller can prevent sideslip when the wheeled mobile robot tracks the reference path. Even if the ground adhesion coefficient is low, the maximum lateral speed of the robot is only 0.2159 m/s. While preventing sideslip, the proposed controller is able to keep the displacement error of path tracking within 0.1681 m. Under the same conditions, the maximum absolute value of the displacement error of the proposed controller is at least 55.15% smaller than that of the controller based on the global model.

Keywords: Anti-sideslip · Local coordinate system · Path tracking · Predictive control · Wheeled mobile robot

1 Introduction

As early as 2008, researchers have pointed out that if people want to make wheeled mobile robots have the ability to travel at high speeds, the possibility of sideslip must be taken into account [1]. But up to now, most researchers still focus on low-speed wheeled mobile robots and do not consider sideslip in their research work [2–6].

Of course, several researchers have noticed the effect of sideslip on the path tracking of wheeled mobile robots, and have carried out some research work. In 2015, Song incorporated the sideslip into a kinematics model of a wheeled mobile robot in his thesis [7]. But he did not consider the source of the sideslip from the dynamics, nor did he consider reducing or eliminating the sideslip.

H. Liu et al. (Eds.): ICIRA 2022, LNAI 13456, pp. 245–256, 2022.
https://doi.org/10.1007/978-3-031-13822-5_22

In 2019, we introduced the mature tire mechanics in the vehicle control field into the path tracking of wheeled mobile robots and established a nonlinear model predictive controller based on the dynamics model. It was found that although the control method based on the dynamics model can reduce the sideslip, due to the coupling relationship between the longitudinal speed and the reference path coordinate points, the controller is hard to reduce the sideslip by reducing the longitudinal speed. Thus, we added a yaw rate-based longitudinal speed switch on the basis of the aforementioned nonlinear model predictive controller to reduce the reference longitudinal speed when turning, thereby further reducing the sideslip [8]. However, although the above switch can better prevent sideslip, it also leads to problems such as the reduction of the longitudinal speed due to disturbances such as positioning errors.

In 2020, we proposed an anti-sideslip path tracking control method based on fuzzy control and nonlinear model predictive control. In the path tracking control system, a fuzzy controller is used as a switch to adjust longitudinal speed [9]. This control method has better performance than the control method using the simple switch in [8], but the fuzzy controller still depends on the design experience and is difficult to get the optimal solution.

All in all, considering the above-mentioned problems of switch control such as fuzzy control, we believe that it is necessary to improve the performance of anti-sideslip path tracking in other ways. According to [8], in addition to changing the reference longitudinal speed, another method to improve the longitudinal speed adjustment capability is to weaken the coupling relationship between the longitudinal speed and the reference path coordinate points. In our research on the path tracking control of vehicles, we found that in the local coordinate system of the vehicle body, the longitudinal speed has a strong coupling relationship with the coordinates of the forward direction, but a weak coupling relationship with the coordinates perpendicular to the forward direction, so we proposed a path tracking control method based on a time-varying local model, which can improve the accuracy of path tracking control by actively adjusting the longitudinal speed [10]. To realize anti-sideslip path tracking of wheeled mobile robots, we completed the following work referring to [10]. Firstly, we build a time-varying local prediction model in the robot body coordinate system of each control period. Secondly, we build an anti-sideslip path tracking controller on this basis. The remainder of this paper is organized as follows. Section 2 introduces the dynamics model of the wheeled mobile robot, which is the essential theoretical basis of this paper. Section 3 is the main contribution of this paper, that is, the anti-sideslip path tracking controller based on the time-varying local model. Section 4 presents the MATLAB simulation results. Section 5 is a conclusion.

2 Dynamics Model

The dynamics model is the basis of this paper, we adopt the model in [8]:

$$\begin{bmatrix} \dot{X} \\ \dot{Y} \\ \dot{\theta} \\ \dot{v}_y \end{bmatrix} = \begin{bmatrix} v_x \cos\theta - v_y \sin\theta \\ v_x \sin\theta + v_y \cos\theta \\ \omega \\ \frac{1}{M}(f_{\mathrm{MF}}(\alpha_{\mathrm{r}}) + f_{\mathrm{MF}}(\alpha_{\mathrm{l}})) - v_x\omega \end{bmatrix} \tag{1}$$

where X and Y are the abscissa and ordinate in the coordinate system, θ is the heading angle, v_x and v_y are the longitudinal speed and the lateral speed, ω is the yaw rate, M is the mass, f_{MF} is the Magic Formula for lateral force, α_{r} and α_{l} are the slip angle of the right drive wheel and the left drive wheel.

We adopt the Magic Formula in [11, 12], and considering that the wheeled mobile robot has no camber, we simplify the camber angle:

$$\begin{cases} f_{\mathrm{MF}}(\alpha) = -\mu D \sin(C \arctan(B\alpha - E(B\alpha - \arctan(B\alpha)))) \\ C = a_0 \\ D = a_1 F_z^2 + a_2 F_z \\ B = a_3 \sin(2\arctan(F_z/a_4))/(CD) \\ E = a_5 F_z^2 + a_6 F_z + a_7 \end{cases} \tag{2}$$

where α is the slip angle, μ is the ground adhesion coefficient, F_z is the vertical load of the tire, a_0 to a_7 are the parameters of the magic formula.

Slip angles can be calculated from longitudinal speed and yaw rate:

$$\begin{cases} \alpha_{\mathrm{r}} = \arctan\left(\dfrac{v_y}{v_x + \omega W/2}\right) \\ \alpha_{\mathrm{l}} = \arctan\left(\dfrac{v_y}{v_x - \omega W/2}\right) \end{cases} \tag{3}$$

where W is the width.

Then, we can combine (1) to (3) and write the model in vector form:

$$\dot{\xi} = f(\boldsymbol{\xi}, \boldsymbol{\delta}) \tag{4}$$

where:

$$\begin{cases} \boldsymbol{\xi} = \begin{bmatrix} X & Y & \theta & v_y \end{bmatrix}^{\mathrm{T}} \\ \boldsymbol{\delta} = \begin{bmatrix} v_x & \omega \end{bmatrix}^{\mathrm{T}} \end{cases} \tag{5}$$

3 Controller

The design process of the controller based on nonlinear model predictive control can be divided into two parts, the first part is to design the prediction model, and the second part is to design the optimization objective function.

3.1 Prediction Model in Time-Varying Local Coordinate System

We are going to build the prediction model in the robot body coordinate system. Since the robot body coordinate system changes with time, we call it a time-varying local coordinate system. The process of transforming from the global coordinates to the robot body coordinate system is shown in Fig. 1. The subscript GC represents the global coordinate system.

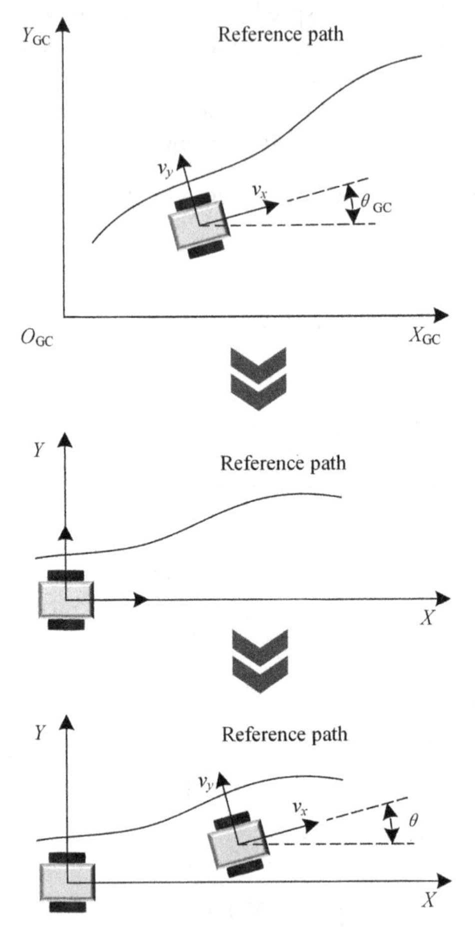

Fig. 1. The process of transforming from the global coordinates to the robot body co-ordinate system.

The dynamics model of the wheeled mobile robot does not change with the coordinate system, but there are some differences in the starting point of the prediction model. In the global coordinate system, the starting point of the prediction model is usually the position and attitude represented by the current state quantities such as the abscissa, ordinate and heading angle of the wheeled mobile robot. Of course, these state quantities are global coordinate information. In the robot body coordinate system in each control period, the starting point of the prediction model is the origin of the coordinate system, namely:

$$\boldsymbol{\xi}(0) = \left[0\;0\;0\;v_y(0)\right]^{\mathrm{T}} \tag{6}$$

If we use numbers to represent the ordinal numbers of the quantities in the prediction model, the prediction model can be written as:

$$\begin{cases} \boldsymbol{\xi}(1) = \boldsymbol{\xi}(0) + Tf(\boldsymbol{\xi}(0), \boldsymbol{\delta}(1)) \\ \vdots \\ \boldsymbol{\xi}(i) = \boldsymbol{\xi}(i-1) + Tf(\boldsymbol{\xi}(i-1), \boldsymbol{\delta}(i)) \\ \vdots \\ \boldsymbol{\xi}(m) = \boldsymbol{\xi}(m-1) + Tf(\boldsymbol{\xi}(m-1), \boldsymbol{\delta}(m)) \\ \vdots \\ \boldsymbol{\xi}(n) = \boldsymbol{\xi}(n-1) + Tf(\boldsymbol{\xi}(n-1), \boldsymbol{\delta}(m)) \end{cases} \tag{7}$$

where T is the control period, m is the control horizon, n is the prediction horizon.

3.2 Optimization Objective Function for Anti-sideslip Path Tracking

Generally, the path planning system provides global reference paths. So, for designing the controller, we also need to transform the reference path into the robot body coordinate system, as shown in Fig. 1.

However, since the reference path may be very long, transforming all paths in each control period is a waste. Thus, we usually only transform a part of the reference path, that is some points on the reference path in front of the robot.

To achieve this, we first find the closest point to the robot on the reference path. Then, based on this point, we take n points along the forward direction of the wheeled mobile robot, and the arc length between two adjacent points is equal to the product of the control period and the given longitudinal speed. In this way, we can obtain a sequence of points on the reference path.

These points can be represented as:

$$\boldsymbol{\xi}_{\mathrm{refGC}} = \left[X_{\mathrm{refGC}}(i)\;Y_{\mathrm{refGC}}(i)\;\theta_{\mathrm{refGC}}(i)\;v_{\mathrm{yref}}(i)\right]^{\mathrm{T}},\; i = 1, 2, \cdots, n \tag{8}$$

where the subscript ref represents reference path.

Since the sideslip is not desired, therefore:

$$v_{\mathrm{yref}}(i) = 0,\; i = 1, 2, \cdots, n \tag{9}$$

After obtaining these sequence points, we need to transform them into the robot body coordinate system.

At each certain moment, the coordinates and heading angle of the wheeled mobile robot in the global coordinate system are given by the positioning system. We can represent this information as:

$$\boldsymbol{\xi}_{\mathrm{GC}}(t) = \left[X_{\mathrm{GC}}(t)\ Y_{\mathrm{GC}}(t)\ \theta_{\mathrm{GC}}(t)\ v_y(t) \right]^{\mathrm{T}} \tag{10}$$

Then, we perform the following coordinate transformations:

$$\begin{cases} X_{\mathrm{ref}}(i) = (X_{\mathrm{refGC}}(i) - X_{\mathrm{GC}}(t))\cos\theta_{\mathrm{GC}}(t) \\ \quad +(Y_{\mathrm{refGC}}(i) - Y_{\mathrm{GC}}(t))\sin\theta_{\mathrm{GC}}(t) \\ Y_{\mathrm{ref}}(i) = (Y_{\mathrm{refGC}}(i) - Y_{\mathrm{GC}}(t))\cos\theta_{\mathrm{GC}}(t)\,,\ i = 1, 2, \cdots, n \\ \quad -(X_{\mathrm{refGC}}(i) - X_{\mathrm{GC}}(t))\sin\theta_{\mathrm{GC}}(t) \\ \quad \theta_{\mathrm{ref}}(i) = \theta_{\mathrm{refGC}}(i) - \theta_{\mathrm{GC}}(t) \end{cases} \tag{11}$$

In this way, the reference path in the time-varying local coordinate system can be obtained:

$$\boldsymbol{\xi}_{\mathrm{ref}} = \left[X_{\mathrm{ref}}(i)\ Y_{\mathrm{ref}}(i)\ \theta_{\mathrm{ref}}(i)\ 0 \right]^{\mathrm{T}},\ i = 1, 2, \cdots, n \tag{12}$$

On the basis of the above, the optimization objective function can be designed as:

$$\begin{aligned} \min_{v_x(i),\omega(i)} J &= \sum_{i=1}^{n} \left\| \boldsymbol{\xi}(i) - \boldsymbol{\xi}_{\mathrm{ref}}(i) \right\|_{\mathbf{Q}}^{2} + \sum_{i=0}^{m-1} \left\| \boldsymbol{\delta}(i+1) - \boldsymbol{\delta}(i) \right\|_{\mathbf{R}}^{2} \\ \text{s.t.}\ & -\Delta v_{x\,\mathrm{lim}} \le \Delta v_x(i) \le \Delta v_{x\,\mathrm{lim}} \\ & -\Delta\omega_{\mathrm{lim}} \le \Delta\omega(i) \le \Delta\omega_{\mathrm{lim}} \end{aligned} \tag{13}$$

where **Q** and **R** are weight matrixes, Δv_x is the velocity increment per control period, $\Delta\omega$ is the yaw rate increment per control period, the subscript lim represents the upper bound of the constraint.

4 Simulation

We tested the proposed controller via MATLAB. The computer model used for simulation is Dell G15 5510, and the processor is Intel(R) Core(TM) i5-10200H CPU @ 2.40 GHz. The control period is set to 0.05 s. The proposed controller is compared with the controller based on the global model, and the controller parameters are shown in Table 1. In the table, the proposed controller is abbreviated as PC, and the compared controllers are abbreviated as GC1 and GC2, in which the parameters of GC1 are exactly the same as those of PC, and the parameters of GC2 are more suitable for global coordinates. The parameters of the wheeled mobile robot are shown in Table 2. The simulation is divided into high-adhesion ground simulation and low-adhesion ground simulation. The adhesion coefficient of high-attachment ground is 0.8, and the adhesion coefficient of low-attachment ground is 0.5. In the simulation, the reference path consists of straight lines and a circular arc with a radius of 2.5 m.

Table 1. Parameters of controllers.

Controller	PC	GC1	GC2
m	1	1	1
n	10	10	10
Q	$\begin{bmatrix} 1 & & & \\ & 10 & & \\ & & 10 & \\ & & & 10 \end{bmatrix}$	$\begin{bmatrix} 1 & & & \\ & 10 & & \\ & & 10 & \\ & & & 10 \end{bmatrix}$	$\begin{bmatrix} 10 & & & \\ & 10 & & \\ & & 10 & \\ & & & 10 \end{bmatrix}$
R	$\begin{bmatrix} 1 & \\ & 1 \end{bmatrix}$	$\begin{bmatrix} 1 & \\ & 1 \end{bmatrix}$	$\begin{bmatrix} 1 & \\ & 1 \end{bmatrix}$

Table 2. Wheeled mobile robot parameters.

Symbol	Value	Symbol	Value
M	250 kg	W	0.5 m
$\Delta v_{x\,\mathrm{lim}}$	0.2 m/s	$\Delta\omega_{\mathrm{lim}}$	0.33 rad/s
a_0	1.505	a_1	-1.147×10^{-4}
a_2	1.229	a_3	1.602×10^{3}
a_4	6.204×10^{3}	a_5	-1.1×10^{-8}
a_6	2.646×10^{-4}	a_7	-1.127

4.1 High-Adhesion Ground Simulation

The results of the high-adhesion road simulation are shown in Fig. 2 to Fig. 6. It can be seen from Fig. 2 that when the ground adhesion coefficient is high, both the proposed controller and the controller based on the global model can control the wheeled mobile robot to track the reference path. Figure 3 shows the displacement error, which is defined in [9]. The maximum absolute value of the displacement error of the proposed controller is 0.1198 m, which is at least 55.15% smaller than that of the controller based on the global model. Figure 4 shows the heading error. The maximum absolute value of the heading error of the proposed controller is 0.1282 rad.

Figure 5 shows the lateral speed of the wheeled mobile robot. The proposed controller can keep the absolute value of the lateral speed of the wheeled mobile robot within 0.1646 m/s. This value is at least 26.97% less than the maximum absolute value of the lateral speed under the control of the controller based on the global model. Figure 6 shows the longitudinal speed of the wheeled mobile robot.

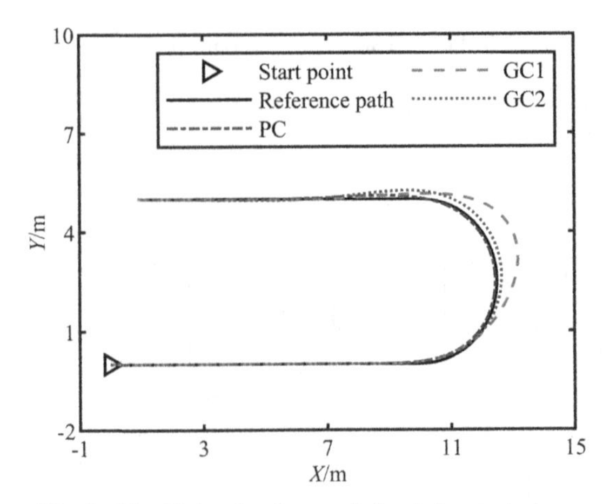

Fig. 2. The high-adhesion road simulation trajectory.

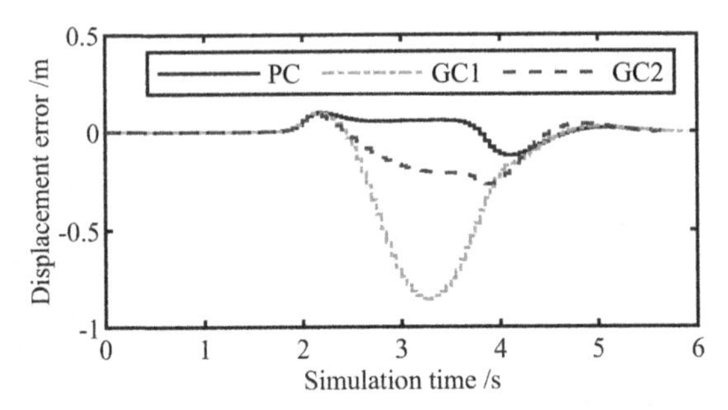

Fig. 3. Displacement error of the high-adhesion road simulation.

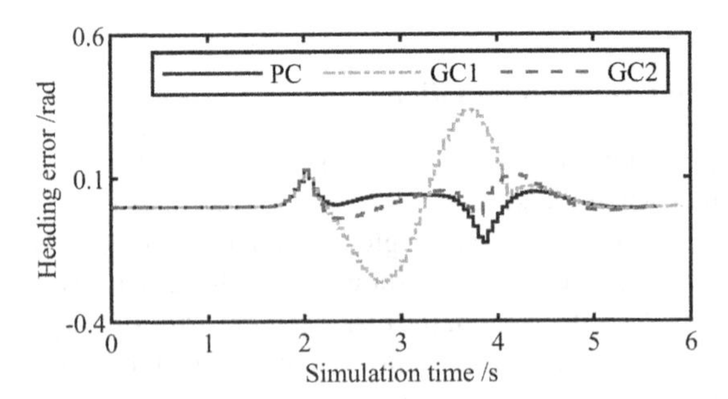

Fig. 4. Heading error of the high-adhesion road simulation.

It can be seen that the proposed controller adjusts the longitudinal speed to a lower value when turning. When the controller based on the global model adopts the same parameters as the proposed controller, the minimum value of longitudinal velocity is lower, but the fluctuation is large, and the displacement error and the heading error are also very large. But when the parameters suitable for the global coordinate system are used, the longitudinal speed reduction value of this controller is very small. Therefore, combined with the displacement error and heading error, it can be considered that the proposed controller has a stronger ability to improve the control accuracy by reducing the longitudinal speed than the controller based on the global model.

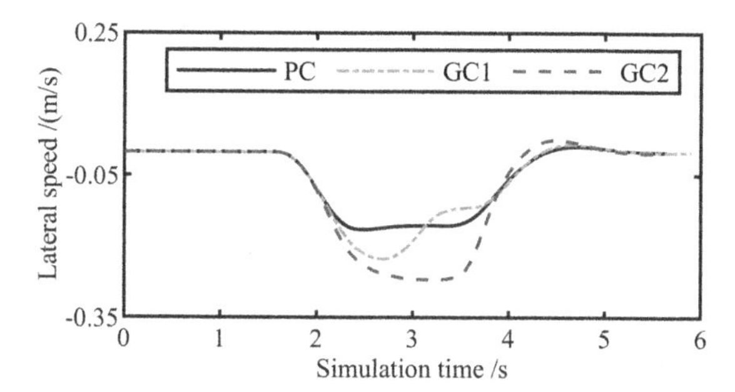

Fig. 5. Lateral speed of the high-adhesion road simulation.

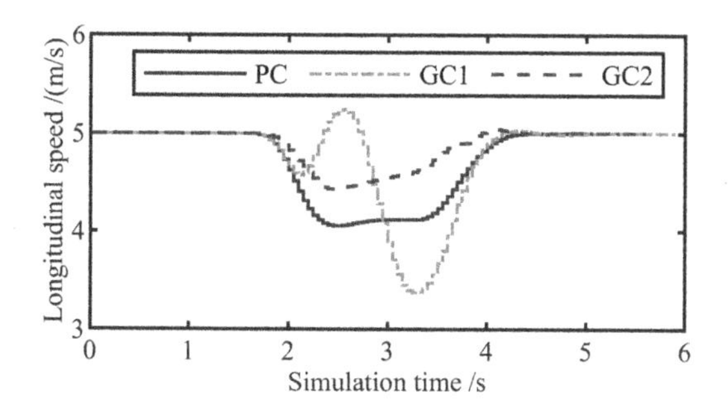

Fig. 6. Longitudinal speed of the high-adhesion road simulation.

4.2 Low-Adhesion Ground Simulation

The results of the low-adhesion road simulation are shown in Fig. 7 to Fig. 11. It can be seen from Fig. 7 that when the ground adhesion coefficient is low, the proposed controller can control the wheeled mobile robot to track the reference path, but the controller based on the global model cannot. Figure 8 shows the displacement error. The maximum absolute value of the displacement error of the proposed controller is 0.1681 m. The displacement error of the controller based on the global model diverges. Figure 9 shows the heading error. The maximum absolute value of the heading error of the proposed controller is 0.1304 rad. The heading error of the controller based on the global model diverges. Figure 10 shows the lateral speed of the wheeled mobile robot. The proposed controller can keep the absolute value of the lateral speed of the wheeled mobile robot within 0.2159 m/s. Figure 11 shows the longitudinal speed of the wheeled mobile robot.

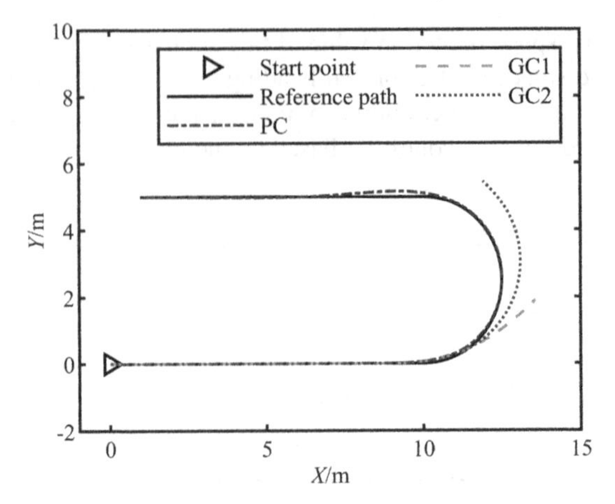

Fig. 7. The low-adhesion road simulation trajectory.

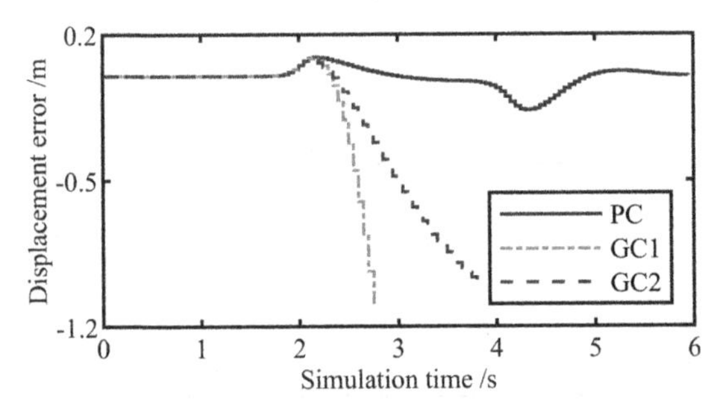

Fig. 8. Displacement error of the low-adhesion road simulation.

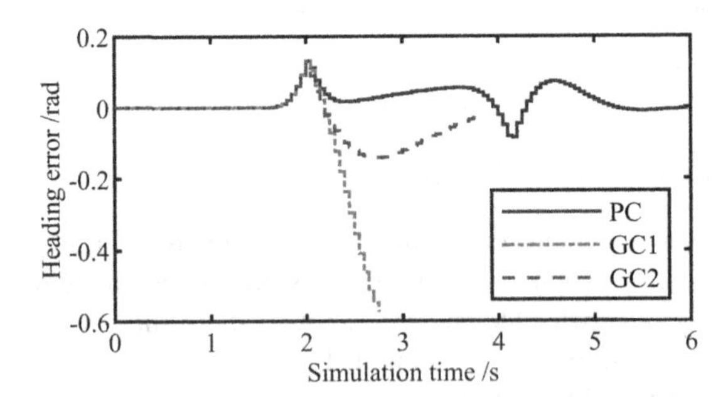

Fig. 9. Heading error of the low-adhesion road simulation.

It can be seen that the proposed controller adjusts the longitudinal speed to a lower value when turning. The longitudinal speed of the controller based on the global model fails to decrease significantly. Combined with the displacement error and heading error, it can be considered that the proposed controller has a stronger ability to improve the control accuracy by reducing the longitudinal speed than the controller based on the global model.

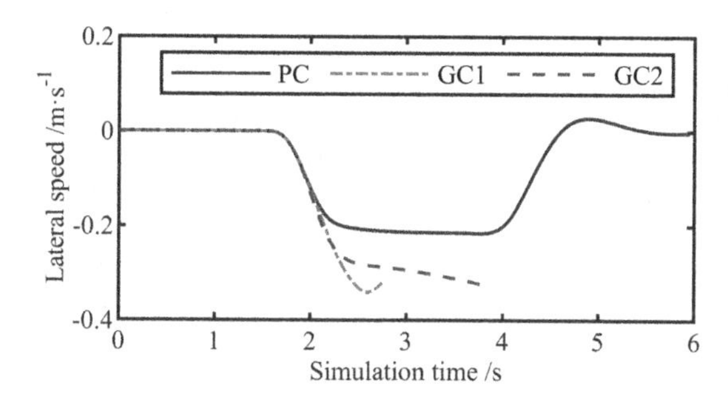

Fig. 10. Lateral speed of the low-adhesion road simulation.

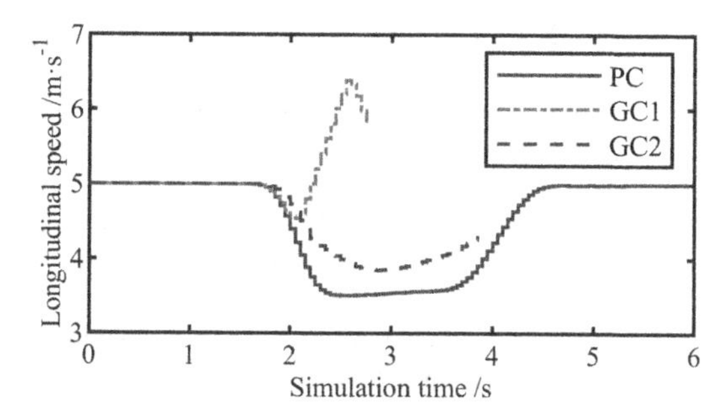

Fig. 11. Longitudinal speed of the low-adhesion road simulation.

5 Conclusion

For the anti-slip path tracking of wheeled mobile robots, we propose a control method based on a time-varying local model. According to the simulation results, we got the following conclusions.

Firstly, our proposed path tracking controller based on the time-varying local model can keep the lateral speed within 0.2159 m/s when controlling the wheeled mobile robot to track the reference path even if the ground adhesion coefficient is low. That is, our controller can prevent sideslip and track the reference path simultaneously.

Secondly, the proposed controller can ensure that the error of the path tracking of the wheeled mobile robot is within a small range while preventing sideslip, and the maximum absolute value of the displacement error is 0.1681 m. At the same time, under the same working conditions, the maximum absolute value of the displacement error of the proposed controller is at least 55.15% smaller than that of the controller based on the global model.

References

1. Sidek, N., Sarkar, N.: Dynamic modeling and control of nonholonomic mobile robot with lateral slip. In: Third International Conference on Systems (ICONS 2008), pp. 35–40. Cancun, Mexico (2008)
2. Yang, H., Wang, S., Zuo, Z., Li, P.: Trajectory tracking for a wheeled mobile robot with an omnidirectional wheel on uneven ground. IET Control Theory A. **14**(7), 921–929 (2020)
3. Ibraheem, G.A.R., Azar, A.T., Ibraheem, I.K., Humaidi, A.J.: A novel design of a neural network-based fractional PID controller for mobile robots using hybridized fruit fly and particle swarm optimization. Complexity **2020**, 3067024 (2020)
4. Khalaji, A.K., Jalalnezhad, M.: Robust forward\backward control of wheeled mobile robots. ISA T. **115**, 32–45 (2021)
5. Zhao, L., Jin, J., Gong, J.: Robust zeroing neural network for fixed-time kinematic control of wheeled mobile robot in noise-polluted environment. Math. Comput. Simulat. **185**, 289–307 (2021)
6. Gao, X., Gao, R., Liang, P., Zhang, Q., Deng, R., Zhu, W.: A hybrid tracking control strategy for nonholonomic wheeled mobile robot incorporating deep reinforcement learning approach. IEEE Access **9**, 15592–15602 (2021)
7. Song, X.G.: Mobile System Modeling of Wheeled Robots and Model Learning-Based Research on Tracking Control. Harbin Institute of Technology, Harbin (2015, in Chinese)
8. Bai, G.X., Liu, L., Meng, Y., Luo, W.D., Gu, Q., Wang, J.P.: Path tracking of wheeled mobile robots based on dynamic prediction model. IEEE Access **7**, 39690–39701 (2019)
9. Bai, G.X., Meng, Y., Liu, L., Luo, W.D., Gu, Q., Li, K.L.: Anti-sideslip path tracking of wheeled mobile robots based on fuzzy model predictive control. Electron. Lett. **56**(10), 490–493 (2020)
10. Bai, G.X., Zhou, L., Meng, Y., Liu, L., Gu, Q., Wang, G.D.: Path tracking of unmanned vehicles based on time-varying local model. Chin. J. Eng. Online published (2022, in Chinese). https://doi.org/10.13374/j.issn2095-9389.2022.03.18.003
11. Ji, X., He, X., Lv, C., Liu, Y., Wu, J.: A vehicle stability control strategy with adaptive neural network sliding mode theory based on system uncertainty approximation. Veh. Syst. Dyn. **56**(6), 923–946 (2018)
12. Ji, X., Yang, K., Na, X., Lv, C., Liu, Y., Liu, Y.: Feedback game-based shared control scheme design for emergency collision avoidance: a fuzzy-linear quadratic regulator approach. J. Dyn. Syst. Meas. Control Trans. ASME **141**(8), 081005 (2019)

Design of a Practical System Based on Muscle Synergy Analysis and FES Rehabilitation for Stroke Patients

Ruikai Cao[1], Yixuan Sheng[1,2], Jia Zeng[2], and Honghai Liu[1(✉)]

[1] State Key Laboratory of Robotics and System, Harbin Institute of Technology, Shenzhen, China
honghai.liu@hit.edu.cn

[2] State Key Laboratory of Mechanical System and Vibration, Shanghai Jiao Tong University, Shanghai, China

Abstract. In the diagnosis and rehabilitation of motor function for stroke patients, the combination of motor function assessment based on Muscle Synergy Analysis (MSA) and rehabilitation using Functional Electrical Stimulation (FES) is a new strategy, which has been validated its feasibility and superiority in clinical rehabilitation. However, it is difficult to be extended to a larger patient population because of low equipment integration, high cost, and complicated operation. This paper designed a hardware and software integrated system for MSA and FES, to achieve functional integration, portability, and simplicity of operation. The hardware system implements multi-channel sEMG acquisition and FES. The software system achieves device control, data processing, and algorithm analysis with a simple and clear user interface. The functions of the system were preliminarily validated by the data of healthy people. This system solves the current problems of equipment function separation and complicated data processing. It realizes the integration of diagnosis and rehabilitation processes, and helps to promote the further development and application of stroke intelligent rehabilitation system.

Keywords: Rehabilitation system · Muscle synergy analysis · Functional electrical stimulation · Stroke

1 Introduction

Humans are facing increasing challenges from stroke. Hemiplegia will appear in post-stroke patients, which may also lead to complications such as cardiovascular weakness, osteoporosis, and muscle atrophy, if effective rehabilitation intervention is not carried out in time. To achieve better recovery results, more effective motor function assessment and rehabilitation methods need to be provided.

Currently, most clinical tests on motor function assessment are at the behavioral and kinematic levels, which cannot reflect the abnormal muscle coordination patterns due to neural deficits. Muscle Synergy Analysis (MSA), however, could

H. Liu et al. (Eds.): ICIRA 2022, LNAI 13456, pp. 257–268, 2022.
https://doi.org/10.1007/978-3-031-13822-5_23

offer insight into both underlying neural strategies for movement and functional outcomes of muscle activity [1]. It's hypothesized to reflect the neural strategy of the central nervous system to control activation of different motor modules. MSA has been used in stroke patients, who often exhibit abnormal muscle synergies in their paretic side compared to healthy people. With the knowledge of neural strategy change after stroke, clinicians are more likely to understand the pathology underlying the patient and make better rehabilitation decisions.

For rehabilitation methods, there is an increasing interest in Functional Electrical Stimulation (FES), a technique that uses short pulses of electrical current to stimulate neuromuscular areas, activating the motor units in the area and causing the muscles to contract [2]. Studies have shown that FES can enhance the recovery of muscle strength in patients after stroke [3]. And activity-dependent interventions may help to enhance the plasticity of the central nervous system and facilitate motor relearning [4], which is consistent with the method of FES.

In recent studies, Junghwan Lim et al. [5] investigated a multi-channel FES rehabilitation method that analyzed the patient's muscle synergy and walking posture. Two stroke patients were involved in the clinical evaluation, and both of them enhanced the muscle synergy similarity with healthy subjects and walking performance. Simona Ferrante et al. [6] also designed a multi-channel FES controller based on healthy muscle synergies. They also synchronized the FES strategy with the patient's volitional gait. The intervention method increased the number of extracted synergies and resemblance with the physiological muscle synergies in two chronic stroke patients.

Although the above studies demonstrate the effectiveness of the MSA + FES approach, the experiment equipment they used was rather complex. At least two separate devices are needed in their experiments to achieve the desired function, one for collecting surface Electromyography (sEMG) data and another for FES stimulation. Moreover, they also face the problem of processing data and analyzing with professional software such as MATLAB or Python. These facts all discourage the clinical practice of this method for a larger population of patients. To solve the above limitations, this paper designed a hardware and software integrated system for muscle synergy analysis and functional electrical stimulation. The main contributions and advantages of this system are summarized as follows.

- Functional integration of sEMG acquisition and FES into one device.
- A user-friendly software to do signal processing and data analysis for non-programmers.
- Compact device size for better portability.
- Lower cost compared to multiple devices.

2 System Overview

2.1 Hardware System

The hardware system implements multi-channel sEMG acquisition and functional electrical stimulation. At the circuit level, it mainly consists of power module, microcontroller module, EMG acquisition module, electrical pulse generation module, switch module, and Bluetooth module, etc. The architecture of the hardware system is shown in Fig. 1a.

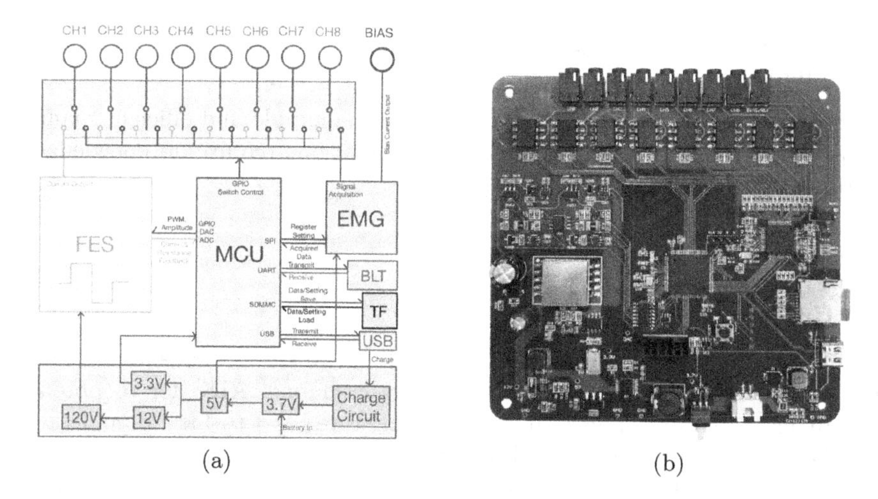

Fig. 1. Hardware system architecture and device photograph. (a) Architecture of the hardware system. (b) Device photograph

The device is powered by a 3.7 V Li battery, different voltage levels are obtained by a series of voltage converter circuits. The microcontroller module is based on STM32H730VIT6 (STMicroelectronics), which is a high-performance MCU with a maximum operating frequency of 480 MHz and a rich set of peripheral functions. The EMG acquisition module is based on ADS1299 (Texas Instruments). The integrated chip solution significantly reduces the size of the PCB compared to discrete components, ensuring miniaturization and portability of the device. The FES module is able to output symmetric/asymmetric biphasic current pulse, with frequency range 20–50 Hz, amplitude 1–100 mA, and pulse width 100–400 µs. The device has 8 channels, plus one for bias current output during EMG acquisition. Each channel can either be configured as EMG acquisition or FES.

The photograph of the device is shown in Fig. 1b, the size of which is about 12 * 12 * 3 cm, making it suitable for portable use.

2.2 Software System

The software system is developed based on WPF (C# + XAML) and mainly contains three interfaces: the EMG acquisition interface, the muscle synergy analysis interface, and the FES interface. The main functions of each interface include:

- EMG Acquisition Interface: Paradigm selection and video demonstration; Channel configuration; Real-time data presentation; Signal filtering; Frequency domain analysis.
- Muscle Synergy Interface: Signal filtering; Data segmentation; Muscle activation calculation; Muscle synergy analysis based on Non-Negative Matrix Factorization (NNMF) algorithm.
- FES Interface: Channel configuration; Current & impedance monitoring.

The software is user-friendly and operates in a simple and efficient manner. All data processing and algorithm analysis operations require no knowledge of programming or professional software usage.

3 System Design

3.1 Hardware Design

The integrated chip solution is chosen for EMG acquisition to ensure the portability of the device. ADS1299 is a low noise, 8-channel ADC for biopotential measurements. In our application, the sampling rate is set to 1 kSPS, and gain is set to 24. The lowest 6 bits and highest 2 bits of the converted data are discarded. The conversion range of the remaining 16 bits of data is 1.43 μV–46.875 mV. A low pass RC filter is placed at each input, the cut-off frequency of which is set to 500 Hz. For differential signals, it's necessary to ensure the common-mode voltage is around the middle of the analog supply voltage. This is achieved by the on-chip common-mode voltage measurement and negative feedback current output functions. The output current is delivered by the bias electrode to the human body.

The electrical stimulator circuit generates biphasic pulse current. The constant current source is realized by negative feedback operational amplifier (op-amp) [7]. The current flows through a sampling resistor, which is connected to the inverting input of the op-amp. Given a reference voltage at the non-inverting input, the op-amp will adjust the conductivity of the MOS in the feedback loop, to maintain a constant voltage across the sampling resistor. A constant level of current is therefore achieved. Further, bipolar current flow is realized using H bridge, which consists of four MOSFETs and is controlled by the PWM signals. By combining the negative feedback op-amp constant current circuit with the H-bridge circuit, a constant current source circuit that can output bipolar pulse wave is realized, as shown in Fig. 2.

Due to the size limitation of the device, this device contains only one electrical pulse generation circuit instead of eight. In order to achieve multi-channel output capability, a scheme of polling between different channels within the same cycle is used. This is achievable because the FES pulse has an extremely short current output time. For example, a symmetric biphasic pulse with a period of 25 ms and pulse width of 400 μs takes up less than 1 ms to output the current. So the pulse generation circuit can switch to other channels during the remaining time of the cycle to deliver the required currents. Figure 3 shows the scheme of polling when 4 channels are turned on.

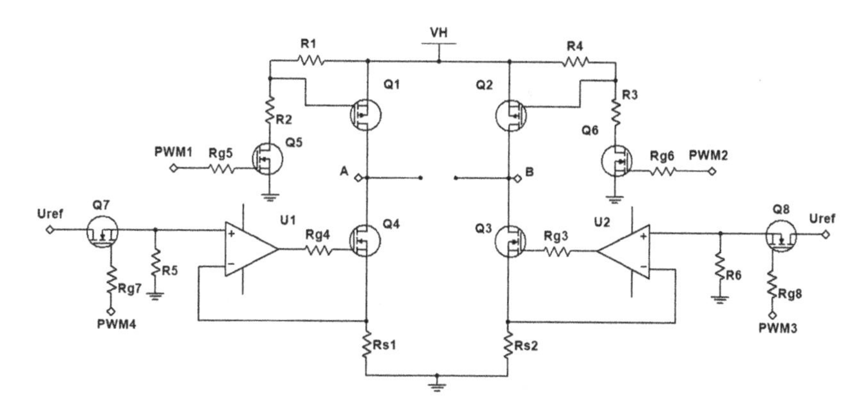

Fig. 2. The schematic of FES waveform generation circuit

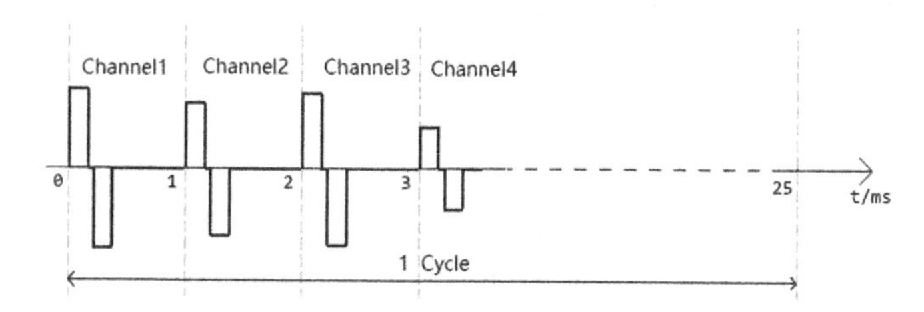

Fig. 3. The polling scheme of multi-channel FES

3.2 Software Design

EMG Acquisition Interface. Figure 4 shows the EMG acquisition interface. In the experiment preparation stage, the clinician selects the experimental paradigm and sets the acquisition parameters. Common parameters include acquisition duration, filtering type, and range. Each channel can be individually configured with the on/off status, muscle name, and Maximum Voluntary Contraction (MVC). On the right side of the interface, real-time data and filtered data are displayed, with the filtered data lagging behind the original one by 1–2 s. The filtering process is done by Python, using butter() and filtfilt() function in scipy. The powerline filter is a 2nd-order Butterworth bandstop filter with cutoff frequency set at 50 ± 2 Hz. The bandpass filter is a 4th-order Butterworth filter.

The software is also able to display the frequency spectrum of the signal, which is analyzed by Short Time Fourier Transform (STFT). Results will show on the right of the time domain view. This function is executed by Python using stft() function in scipy.

Muscle Synergy Analysis Interface. EMG signal pre-processing and muscle synergy analysis are implemented in this interface. The data pre-processing here includes: data filtering, data segmentation, full-wave rectification, low-pass filtering, and muscle activation calculation. The filtering process is the same as

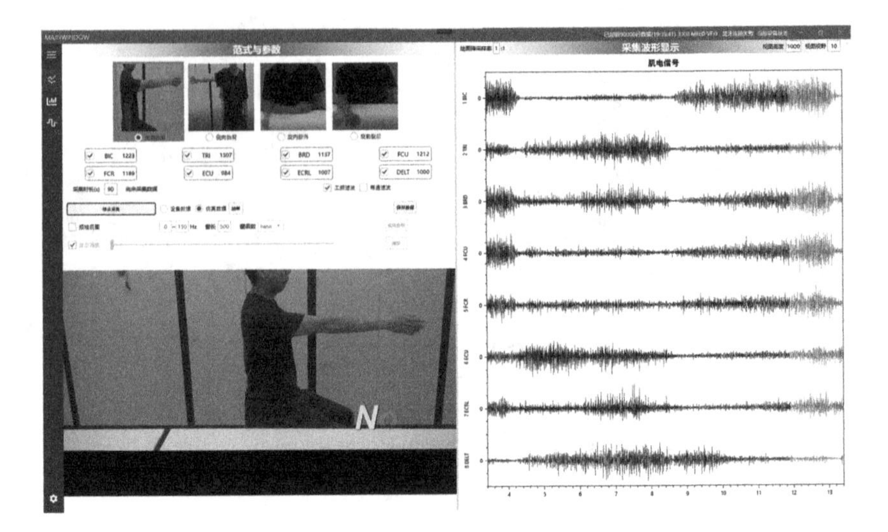

Fig. 4. The EMG acquisition interface

described in the EMG interface. Data segmentation extracts the periodic movement segments in the data, which can be done by manual selection or automatic threshold segmentation [8]. The general steps of threshold segmentation are: data downsampling, data summation of all channels, moving average filtering, median filtering, and threshold selection. Users can specify the parameters of each step to get the best segmentation results. Figure 5 gives one example of the segmentation result.

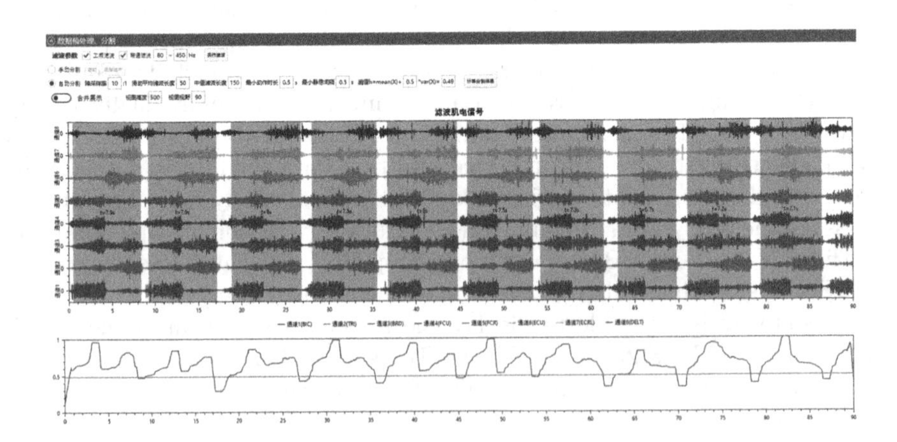

Fig. 5. Data segmentation view

In order to make the EMG signal amplitude reflect muscle force information more accurately, muscle activation model which is formulated in two steps were used [9,10]. First, the EMG signals e(t) are transformed into neural activation u(t):

$$u(t) = \alpha e(t-d) - \beta_1 u(t-1) - \beta_2 u(t-2) \tag{1}$$

where d is the electromechanical delay. α is a scale factor used to account for inter-subject differences in muscle parameters, β_1, β_2 are the coefficients that determine the dynamic characteristics of the second-order system. The above parameters must satisfy the conditions $\beta_1 = \gamma_1 + \gamma_2, \beta_2 = \gamma_1 \times \gamma_2, |\gamma_1|, |\gamma_2| < 1, \alpha - \beta_1 - \beta_2 = 1$.

Secondly, to better represent the nonlinear relationship between EMG amplitude and force, using

$$a(t) = \frac{e^{Au(t)} - 1}{e^A - 1} \tag{2}$$

where a(t) is the muscle activation, A is a coefficient indicating the degree of nonlinearity, taking values in the range of $[-3, 0)$, with -3 indicating exponential correlation and 0 indicating linear correlation.

The muscle activation results described above are used as input to the muscle synergy analysis. There are several algorithms available, including Principal Component Analysis (PCA), Factor Analysis (FA), Independent Component Analysis (ICA), and NNMF, etc. The most commonly used one is NNMF [11], for its non-negativity constraint on data is consistent with the non-negativity of muscle activation level, which makes it easier to be translated into physiological meaning. The algorithm decomposes the matrix V(m by n) into the product of two low-rank matrices W(m by r) and H(r by n), where r is the rank of decomposed matrices. In muscle synergy analysis, m is the number of muscles, n is the length of the temporal sampling data, and r is the number of synergies underlying the movement. W reveals the muscle composition of each synergy and the relative level of activation of each muscle. H reflects the activation timing of each synergy during the execution of the movement.

In practice, however, the actual number of synergies is not known before performing the analysis, so it needs to be estimated by some criteria, such as MSE and VAF [12]. VAF is used here, which is calculated as

$$VAF = (1 - \frac{||V - WH||^2}{||V||^2}) \times 100\% \tag{3}$$

The number of synergies is determined based on empirical rule, as specified in the article [8]: the minimum number of synergies that achieves VAF >85%, with less than a <5% increase upon addition of another synergy.

The execution of the NMF algorithm is implemented by the NMF() function in Python's sklearn library. Results are calculated for each segment. The algorithm will be repeated multiple times to eliminate the incidental results caused by random initialization. The analysis interface is shown in Fig. 9 in the validation section. For each segment, the results are presented in the form of mean $\pm$ std, which represents the average solution and variability due to random initialization. The top-left graph shows the variation of VAF with the number of synergies, the top-right bar graph shows the synergy distribution matrix W, and the bottom curve shows the synergy activation coefficients H.

The order of synergies is random across repeated trials, so similarity matching needs to be performed before averaging. The cosine similarity matching method (calculate the inner product between normalized synergy vectors) provided in the article [12] is implemented here to sort synergies in the results.

Overall results that combine all segments can also be calculated in this interface, where the mean and standard deviation across segments are shown. In some cases, the user needs to manually adjust the order of synergies before combining, if they are found to be inconsistent among different segments.

FES Interface. During rehabilitation, patients are treated with electrical stimulation according to preset parameters. In this Interface, parameters can either be directly loaded from files or manually set by clinicians. After the stimulation starts, the real-time status including progress, current, and impedance will be shown. This information is helpful to check whether the FES current is delivered properly. Figure 6 shows the FES interface.

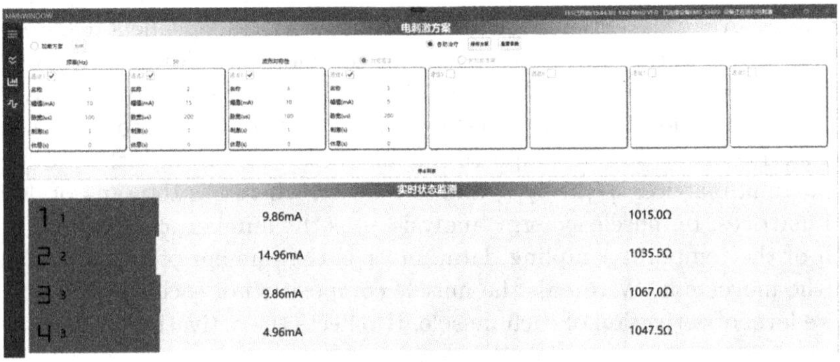

Fig. 6. The FES interface

4 System Validation

4.1 EMG Acquisition Validation

The EMG acquisition function was firstly verified by square and sine signal with 1 Hz and amplitude 1 mV. The results showed that the amplitude error is less than 1%, with std value being 23 μV. Thus the original signal is well preserved. Secondly, human EMG data were acquired according to certain experimental paradigms. The analysis of the time domain amplitude and frequency domain spectrum showed that the data are valid. Figure 7 illustrates the EMG signal of the anterior, mid and posterior head of the deltoid muscles when performing forward arm raise movement.

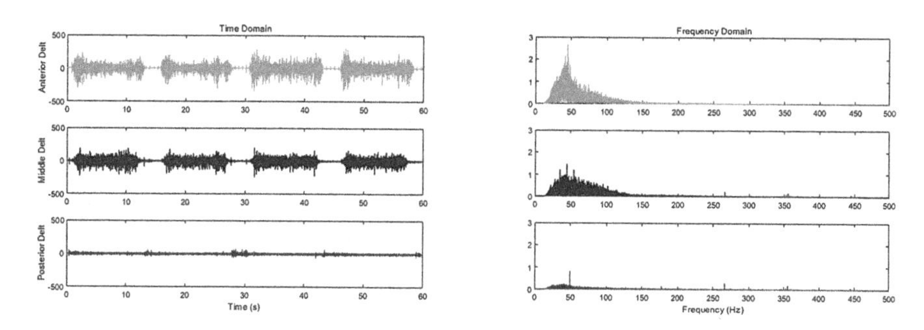

Fig. 7. EMG data acquired from anterior, mid and posterior head of DELT

4.2 FES Validation

The functionality of single-channel FES was validated on a 1 kΩ resistor. Figure 8a and 8b show the symmetric and asymmetric biphasic pulse. The control accuracy of frequency, amplitude, and pulse width were further verified, and the results showed that the control errors of these parameters are negligible within the parameter range. Except for current amplitudes lower than 5 mA, the error may be higher than 5%, which is acceptable because usually the FES amplitude is higher than 5 mA to make the muscle contract. In the multi-channel output functionality test, 4 channels were enabled and each channel was configured with different parameters and connected to a 1 kΩ resistor. The result is shown in Fig. 8c, which proves that the logic of polling and control of parameters are correct.

4.3 Muscle Synergy Analysis Validation

To validate the function of MSA in the software, data from article [13] were used, in which EMG data were collected from 10 healthy subjects during elbow and shoulder joint movements. The EMG of 8 muscles were collected, including Biceps Brachii (BIC), Triceps Brachii (TRI), Brachioradialis (BRD), Flexor Carpi Ulnaris (FCU), Flexor Carpi Radialis (FCR), Extensor Carpi Ulnaris (ECU), Extensor Carpi Radialis Longus (ECRL), and Deltoid (DELT).

The elbow and shoulder joint movements of one subject at the velocity of 30°/s were chosen. Data were loaded using our software and pre-processed, which 80 Hz highpass filtering, data segmentation, full-wave rectification, 2 Hz low pass filtering, and muscle activation calculation. Next, muscle synergy analysis was performed for each movement. The results showed that both elbow and shoulder joint movements could be represented by two synergies with VAFs equal to 96.11% and 95.12%, respectively. The analysis results of elbow joint movement are shown in Fig. 9.

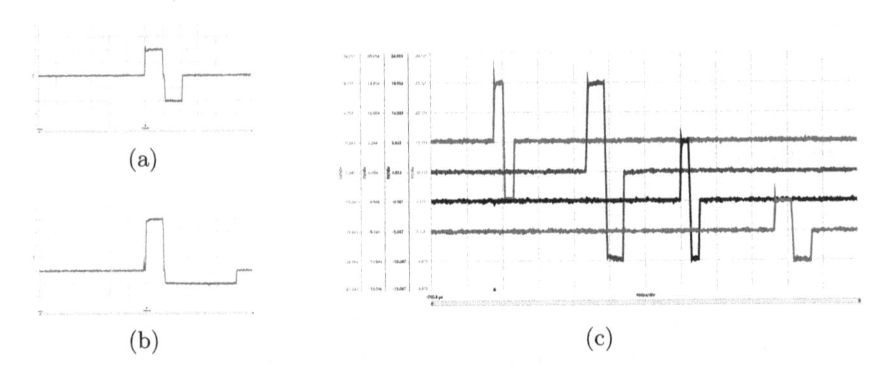

(a)

(b)

(c)

Fig. 8. Real FES waveform. (a) Symmetric waveform. (b) Asymmetric waveform. (c) Four channel polling

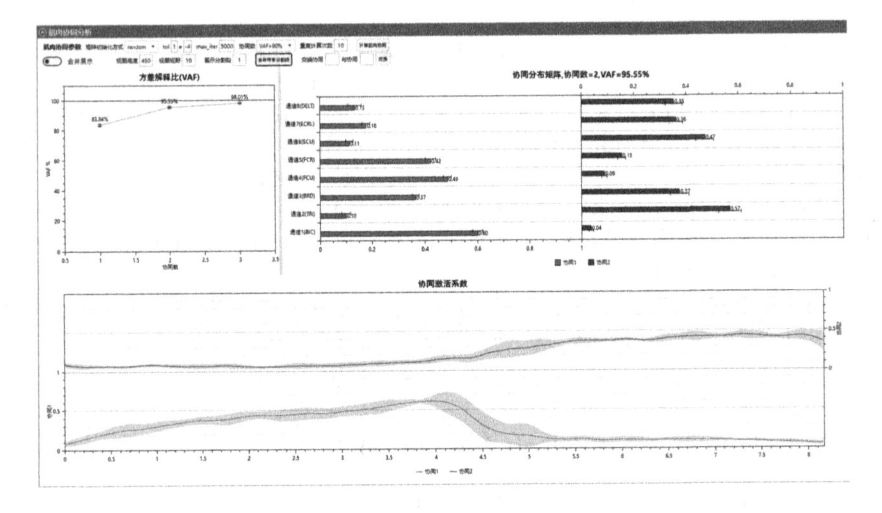

Fig. 9. MSA results of elbow joint movement. The top-left graph shows the variation of VAF with the number of synergies, the top-right bar graph shows the synergy distribution matrix W, and the bottom curve shows the synergy activation coefficients H.

Our results are consistent with the results obtained in article [13], although the relative weights between synergies are not exactly consistent, which is mainly due to individual variability. The results given in the article are the average results of 10 subjects, but only one of them is shown here, which is still within a reasonable range. So the validity of the muscle synergy analysis function of the software can be verified.

5 Conclusion

In this paper, a functionally integrated system for MSA and FES is designed. The hardware of the system integrates 8-channel EMG signal acquisition and FES functions. It has the advantages of small size, portability, and low cost compared to existing complex commercial devices. The software enables flexible device control, convenient data processing, and algorithm analysis. For researchers and clinicians, this means that they can get rid of the difficulties of using professional data analysis software and programming to do data analysis. This will undoubtedly accelerate the progress of patient diagnosis and rehabilitation. In addition, the functions of the currently designed software are extendable. Any new or different data processing requirements can be added to the current software in a similar manner.

Currently, the system has been preliminarily validated in healthy individuals and has not yet been used for stroke patients. Next, we will verify the functions of the system for stroke patients and apply the system to a larger population. We believe that the proposed system will further promote the development and application of stroke intelligent rehabilitation system and thus provide more convenience for the treatment of stroke patients.

References

1. Safavynia, S., Torres-Oviedo, G., Ting, L.: Muscle synergies: implications for clinical evaluation and rehabilitation of movement. Top. Spinal Cord Inj. Rehabil. **17**(1), 16–24 (2011)
2. Zhou, Y., Zeng, J., Li, K., Hargrove, L.J., Liu, H.: sEMG-driven functional electrical stimulation tuning via muscle force. IEEE Trans. Industr. Electron. **68**(10), 10068–10077 (2021)
3. Lynch, C.L., Popovic, M.R.: Functional electrical stimulation. IEEE Control Syst. Mag. **28**(2), 40–50 (2008)
4. Ting, L.H., et al.: Neuromechanical principles underlying movement modularity and their implications for rehabilitation. Neuron **86**(1), 38–54 (2015)
5. Lim, J., et al.: Patient-specific functional electrical stimulation strategy based on muscle synergy and walking posture analysis for gait rehabilitation of stroke patients. J. Int. Med. Res. **49**(5), 03000605211016782 (2021)
6. Ferrante, S., et al.: A personalized multi-channel FES controller based on muscle synergies to support gait rehabilitation after stroke. Front. Neurosci. **10**, 425 (2016)
7. Zhou, Y., Fang, Y., Gui, K., Li, K., Zhang, D., Liu, H.: sEMG bias-driven functional electrical stimulation system for upper-limb stroke rehabilitation. IEEE Sens. J. **18**(16), 6812–6821 (2018)
8. Israely, S., Leisman, G., Machluf, C., Shnitzer, T., Carmeli, E.: Direction modulation of muscle synergies in a hand-reaching task. IEEE Trans. Neural Syst. Rehabil. Eng. **25**(12), 2427–2440 (2017)
9. Manal, K., Buchanan, T.S.: A one-parameter neural activation to muscle activation model: estimating isometric joint moments from electromyograms. J. Biomech. **36**(8), 1197–1202 (2003)

10. Sheng, Y., Liu, J., Zhou, Z., Chen, H., Liu, H.: Musculoskeletal joint angle estimation based on isokinetic motor coordination. IEEE Trans. Med. Robot. Bionics **3**(4), 1011–1019 (2021)
11. Lee, D.D., Seung, H.S.: Learning the parts of objects by non-negative matrix factorization. nature **401**(6755), 788–791 (1999)
12. Cheung, V.C., d'Avella, A., Tresch, M.C., Bizzi, E.: Central and sensory contributions to the activation and organization of muscle synergies during natural motor behaviors. J. Neurosci. **25**(27), 6419–6434 (2005)
13. Sheng, Y., Zeng, J., Liu, J., Liu, H.: Metric-based muscle synergy consistency for upper limb motor functions. IEEE Trans. Instrum. Meas. **71**, 1–11 (2022)

Adaptive Force Field Research on Plane Arbitrary Training Trajectory of Upper Limb Rehabilitation Robot Based on Admittance Control

Gao Lin[1,2,3], Dao-Hui Zhang[1,2(✉)], Ya-Lun Gu[1,2], and Xin-Gang Zhao[1,2]

[1] State Key Laboratory of Robotics, Shenyang Institute of Automation, Chinese Academy of Sciences, Shenyang, China
zhangdaohui@sia.cn

[2] Institutes for Robotics and Intelligent Manufacturing, Chinese Academy of Sciences, Shenyang, China

[3] China Medical University, Shenyang, China

Abstract. In order to further improve the flexibility, safety and training pertinence of the desktop end-traction type upper limb rehabilitation robot, this paper proposes a realization method of the adaptive force field of plane arbitrary training trajectory based on admittance control. In ROS, the flexible drag of the robot end is realized through the admittance control algorithm. For the drag trajectory, the function equation is obtained by fitting, and then the desired pose is adjusted in real time through the end pose to realize the trajectory force field. For the regular trajectory force field, a simpler plane segmentation method is used to achieve. The size of the force field is adaptively adjusted by means of position-force control, and the force is compensated in the tangential direction of the desired position, so as to realize the active rehabilitation training of the adaptive force field. The experimental results show that adding an adaptive force field to any plane training trajectory can improve training compliance, safety and training pertinence, which has an important reference for the application of personalized compliance active training to desktop upper limb rehabilitation robots.

Keywords: End traction · Compliant control · Adaptive force field · Human-computer interaction · Personalized training

1 Introduction

The number of stroke patients in China is increasing day by day, and the current number of patients is about 13 million, ranking first in the world. 44% of stroke patients have upper extremity movement disorders [1]. Clinical medical research has proved that upper limb rehabilitation training can improve the muscle strength of the[1] affected limb, prevent

[1] This work was supported in part by the National Natural Science Foundation of China (61903360, 92048302, U20A20197, U1813214), and LiaoNing Revitalization Talents Program (XLYC1908030).

H. Liu et al. (Eds.): ICIRA 2022, LNAI 13456, pp. 269–280, 2022.
https://doi.org/10.1007/978-3-031-13822-5_24

muscle atrophy, and also stimulate the nervous system and promote the recovery of brain cell function [2]. Traditional rehabilitation therapy is taught by rehabilitation therapists, which undoubtedly increases the workload of therapists. However, most of the existing upper limb rehabilitation robot training methods are passive rehabilitation, which lacks safety and flexibility [3].

The active compliance training method of end-traction upper limb rehabilitation robot has always been a research hotspot at home and abroad. Hogan and Krebs [4] of Massachusetts Institute of Technology realized stable and smooth rehabilitation training of end-traction rehabilitation robot through impedance control strategy combined with robot vision. It has high safety and comfort, and has been clinically applied and has achieved good evaluation, but it has high requirements on hardware. The MIME-III desktop mirror-based rehabilitation robot designed by Burgar [5] at Stanford University maps the motion of the healthy side arm to the paralyzed side arm, so that it can complete the same motion. The robot cannot be back-driven, which may cause injury especially to patients with spasticity. Zhang Zuozu [6] of Nanjing University of Aeronautics and Astronautics proposed a control strategy for active interactive training of upper limb rehabilitation robot based on Lyapunov function, and proved through active training experiments that different intensities of training can be achieved by adjusting admittance parameters. In addition, Le Yuyi [7] of Shanghai University realized the acceleration of the upper limb rehabilitation robot by applying a small interactive force at low speed by mapping the admittance coefficient of the interactive force. Stable force is required.

To sum up, the current upper limb rehabilitation robots mainly have the following two problems. The first problem is that most of the upper limb rehabilitation robots take the robot system as the core, set the predetermined parameters by professionals to precisely control the robot, and complete the movement assistance for people regularly and quantitatively. There is a lack of real-time interaction between humans and robots. The function of making real-time adjustments to subjective movements has low human compliance, resulting in low compliance and safety. The second problem is that for the individual differences of different upper limb hemiplegia patients, it is impossible to carry out targeted trajectory training, and it is impossible to make the training scientific and effective.

In response to the above problems, this study proposes a method for targeted trajectory compliance training for different stroke patients with upper limb hemiplegia. An adaptive force field is added to the obtained trajectory, and the feasibility of the proposed method is verified through adaptive force field experiments, trajectory compensation force experiments and multi-trajectory compliance experiments. The main contributions of this paper are as follows:

- An adaptive force field based on position-interaction force control is proposed, and it is verified by comparative experiments that the proposed adaptive force field can effectively improve the flexibility, safety and human-computer interaction ability of active training.
- The regular training trajectories such as straight lines and circles are extended to arbitrary training trajectories on the plane, which solves the problem of training trajectories caused by individual differences of patients.

2 Construction of Upper Limb Rehabilitation Robot System

2.1 Hardware Composition

As shown in Fig. 1, the system hardware is mainly composed of 6 parts:

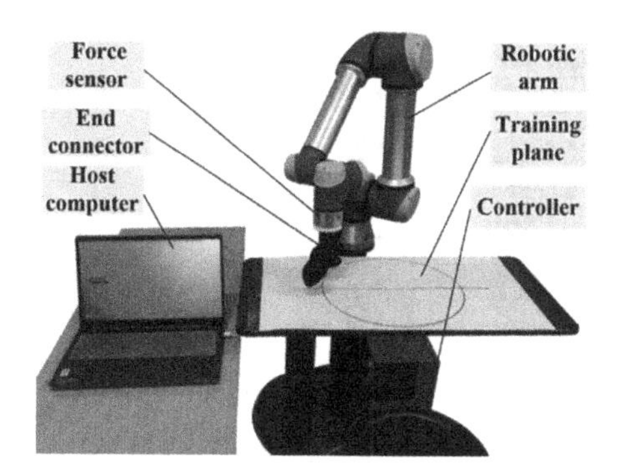

Fig. 1. Hardware composition

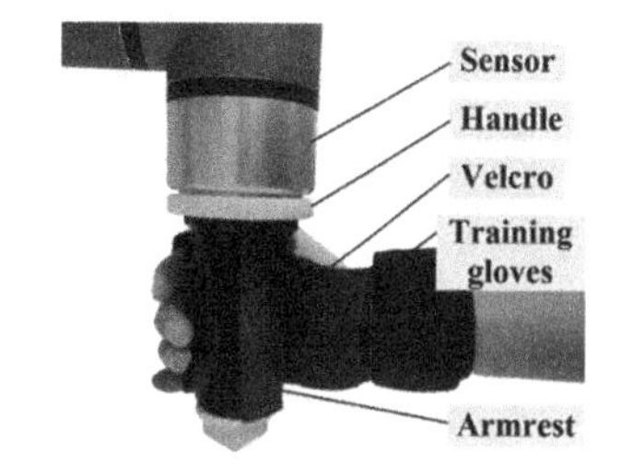

Fig. 2. End connector wearing effect

The upper computer is loaded with the Robot Operating System (ROS) for receiving, processing and sending data. The controller is connected with the host computer through a network cable, and is used to receive user commands from the host computer. The robotic arm used is the ur5e designed to imitate the range of motion of the human arm. It has six rotating joints, a payload of 5 kg and a working radius of 850 mm, which meets the requirements of upper limb rehabilitation training. The model of the six-dimensional force sensor is ATI Delta IP60, the accuracy is 0.01%, the maximum force/torque measurement is 660N and 60 Nm respectively, and the sensor parameters meet the training requirements. The end connector consists of a handle, training glove and hook and loop. The handle is made of 3D printing and can transmit the patient's interactive force information to the sensor. The hook surface of the hook and loop is sewn to the palm surface of the training glove, and the wool surface is fixed to the surface of the handle. After the patient puts on the training gloves, they can be connected with the handle through the hook and loop, so that the upper limb hemiplegia patient can also "grip" the handle tightly. While the equipment is easy to wear, when the robot runs abnormally, it is convenient to separate the human and the machine, which ensures the safety of the training process. Figure 2 shows the wearing effect of the right hand in a relaxed state.

2.2 Software Implementation

This study is based on the Robot Operating System (ROS) to realize a modular and easily scalable upper limb rehabilitation robot compliant interactive rehabilitation system. ROS supports two programming languages, C++ and Python. In order to make the robot get faster and more stable control, this research is based on C++ programming. ROS provides four communication methods: Topic, Service, Action and Parameter Service. The communication architecture of this system is shown in Fig. 3.

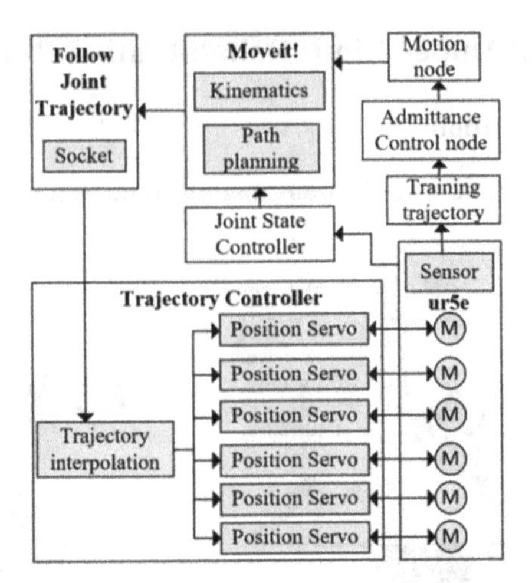

Fig. 3. Control system communication architecture

Subscribe to the force/torque information from the six-dimensional force sensor through topic communication. In the admittance control node, the force/torque is processed by coordinate transformation and admittance controller to obtain the position offset in Cartesian space, and then the position offset is added to the current end position to obtain the next moment the end of the manipulator needs to be position to move to. Then input this position information to the robot motion node, and set the motion time of each end point. The position coordinates are solved by kinematics and path planning in Moveit, and the robot position information described by the position, velocity and acceleration of the six axes in the joint space is obtained. The position information in the joint space is encapsulated into Action type information through follow_joint_trajectory and then communicates with the controller. The trajectory controller sends instructions to the six position servos in the form of trajectory interpolation, and realizes the process of coarse interpolation to fine interpolation of the trajectory and then sent to each position servo. At the same time, the joint state controller is used to monitor the state of the motor in real time, and the joint motion information is fed back to Moveit, so that Moveit can obtain the motion state of the robot in real time.

3 Compliant Control

3.1 Acquisition and Transformation of Force/Torque Information

The force/torque information collected by the six-dimensional force sensor is obtained by subscribing to Wrench.msg through topic communication. The force/torque obtained in the force sensor coordinate system needs to be transformed to the base coordinate system to represent. The transformation of force/moment vector between these two coordinate systems in Wrench.msg is realized by $W^A = F_B^A W^B$. It is carried out in the following two steps.

$$\overrightarrow{f_b} = M^{-1}\overrightarrow{f} \tag{1}$$

$$\overrightarrow{t_b} = M^{-1}\left(\overrightarrow{t} - \overrightarrow{p} \times \overrightarrow{f}\right) \tag{2}$$

$\overrightarrow{f_b}$ and $\overrightarrow{t_b}$ represent the transformed values of the force/torque vector respectively, M^{-1} is the inverse of the force sensor coordinate system relative to the attitude matrix in the base coordinate, and $\overrightarrow{p}$ is the position matrix of the force sensor coordinate system relative to the base coordinate.

3.2 Admittance Controller

In the end-traction rehabilitation robot, the introduction of admittance control can change the compliance of the robot end and improve the human-machine interaction ability between the robot and the patient. The admittance control makes there a dynamic relationship between the interaction force and the position deviation, so as to achieve the purpose of compliance. The relationship between the force and the position is shown in formula (3).

$$M\ddot{x}_e + B\dot{x}_e + Kx_e = F_{ext}, x_e = x_d - x \tag{3}$$

Among them, F_{ext} represents the interaction force, and M, B and K represent the inertia, damping and stiffness properties, respectively. x_e is the difference between the actual pose x_d described in the workpiece or base coordinate system and the desired pose x. $\dot{x}_e$ and $\ddot{x}_e$ are the first and second derivatives of x_e, representing the difference between the velocity and the acceleration.

In the process of rehabilitation training, the movement speed of the end of the robotic arm needs to correspond to the magnitude of the interaction force. That is to say, under the action of constant force, the end of the robotic arm can maintain a certain speed movement without changing the speed, so as to ensure the stability of the rehabilitation training. Therefore, when moving the end of the manipulator on the plane, let the inertial characteristic M = 0, and the simplified formula is shown in formula (4). In addition to movement, it also includes rotation in the z-axis direction. In order to make the rotation direction change immediately with the change of the torque direction during rotation, the inertia characteristic M and damping characteristic B are set equal to 0 when the z-axis rotation of the end of the manipulator is performed. The simplified formula is shown in formula (5).

$$B(\dot{x}_d - \dot{x}) + K(x_d - x) = F_{xy} \tag{4}$$

$$K(x_d - x) = F_{Rz} \tag{5}$$

Among them, F_{xy} and F_{Rz} represent the external force in the direction of the horizontal plane and the moment in the direction of the z-axis, respectively.

4 Realization of Adaptive Force Field for Planar Arbitrary Trajectory

4.1 Addition of Drag Track Adaptive Force Field

For plane arbitrary trajectories, there is no need to consider the transformation of the movement step size in the z-axis direction. Change the desired pose to the current pose

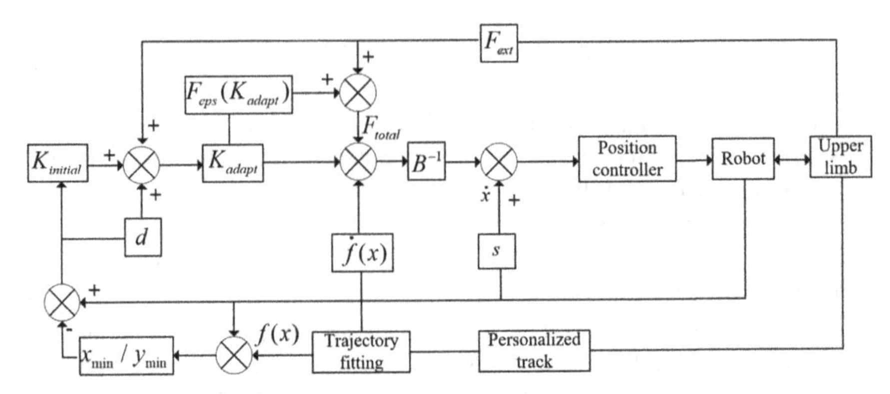

Fig. 4. The control block diagram of the system

with the update of the actual pose of the terminal, and then add the position offset converted by the admittance controller by accumulating the position in the x and y directions to achieve a derivative-based approach. Nano-controlled plane free drag function, expressed as $p^{t+1} = (x^t + \Delta x, y^t + \Delta y, C_z, C_{roll}, C_{pitch}, yaw^t + \Delta\theta_z)$. Then, the drag track is added to the admittance plane by means of position control (Fig. 4).

$$d_{\min} = \sqrt{(x - x_{\min})^2 + (y - y_{\min})^2} \tag{6}$$

where (x_{min},y_{min}) is on the fitted trajectory and the tangent unit direction vector along the closest point is $(\frac{1}{\sqrt{1+\dot{f}(x)}}, \frac{\dot{f}(x)}{\sqrt{1+\dot{f}(x)}})$. Then, the admittance coefficient is adjusted in real time through the offset distance of the end and the interactive force. The specific relationship is as follows:

$$K_{adapt} = K_{initial} + C_0 d + C_1 F_{ext} + C_2 \Delta v \tag{7}$$

$$\Delta v = \frac{p^{t+1} - 2p^t + p^{t-1}}{T} \tag{8}$$

Among them, d is the distance from the current end position to the desired position point, F_{ext} is the magnitude of the interaction force, and Δv is the velocity change. The larger the d, F_{ext}, and Δv, the larger the K_{adapt}. With the change of the force field, in order to make the resistance in the training direction not change with the change of the force field, an additional compensation force needs to be added along the direction of the training trajectory. The compensation force F_{cps} is related to K_{adapt}.

$$K_{adapt} \propto F_{cps} + C_3 \tag{9}$$

Among them, C_3 is a constant. When C_3 is 0, only F_{cps} acts, and the effect is 0 assist (completely active). As C_3 increases, the boosting effect along the trajectory becomes more pronounced, changing the training from fully active to passive training. Take the subject's upper limb to perform elbow joint and shoulder joint forward flexion and extension movement as an example (see Figs. 5, 6).

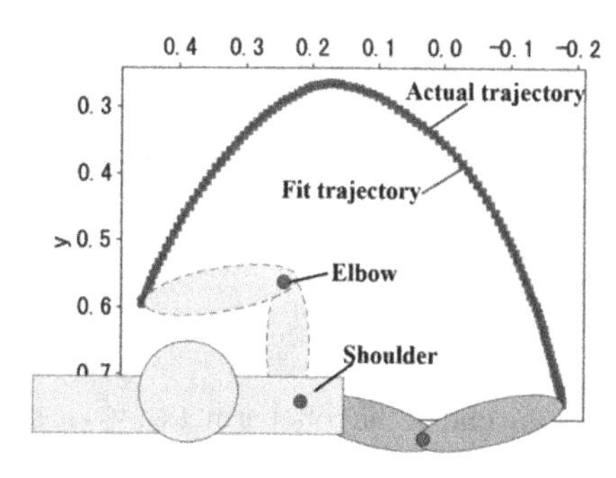

Fig. 5. Drag Track

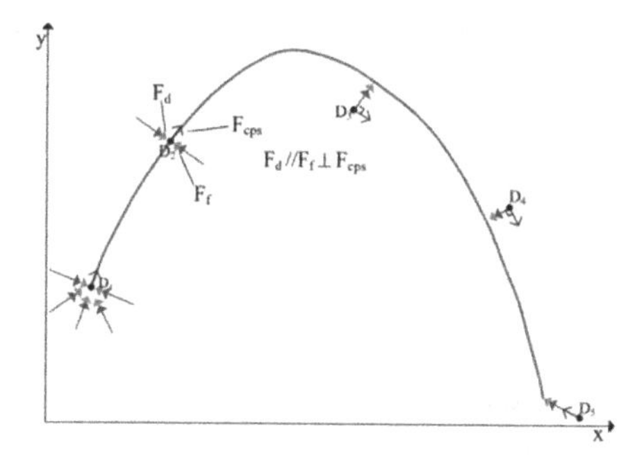

Fig. 6. Stress at different locations

4.2 Adaptive Force Field of Regular Trajectory Based on Plane Segmentation Method

It is difficult to obtain a regular trajectory when physicians or patients drag the end by themselves. Compared with the above-mentioned implementation method of the drag trajectory adaptive force field, a simpler implementation method is proposed here for the regular training trajectory adaptive force field. The plane segmentation method is to divide the training plane into multiple regions, each region receives a different force field direction, and the trajectory function pointed to by the desired position is also different. For example, triangles, squares, polygons, etc. can be pieced together by dividing planes by multiple straight lines, and the regular curve graphics are represented as circles.

4.2.1 Realization of Adaptive Force Field for Straight Lines

The x direction in the desired position is set to a fixed value or f(x) is a straight line equation, so as to realize a straight line training trajectory of an applied force field. Taking x as a certain value as an example, the implementation is as follows.

$$x_l^{t+1} = C + \Delta x \tag{10}$$

$$y_l^{t+1} = y_l^t + \Delta y \tag{11}$$

$$yaw^{t+1} = yaw^t + \Delta\theta_z \tag{12}$$

Among them, C is a constant value, Δx, Δy, Δc are the pose offsets, and the position offset Δx in the x direction is expressed as follows.

$$\Delta x_l^{t+1} = \Delta x_l^t + \frac{T_l\left[F_x - K_{adapt}\Delta x_l^t\right]}{B} \tag{13}$$

where T_l is the time constant obtained by integrating $\dot{x}_e$ with time, K_{adapt} is the adaptive admittance coefficient, and F_x is the component of the interaction force plus the compensation force in the x-axis direction. The linear force field is shown in Fig. 7.

4.2.2 Realization of Circumferential Adaptive Force Field

For the plane circular trajectory, on the basis of the free drag of the plane realized by admittance control, the circular force field is formed by dividing the plane in the plane by formula (14).

$$(x_c, y_c) = (a, b) + r\frac{\vec{d}}{|\vec{d}|}, \vec{d} = (x^t - a, y^t - b) \tag{14}$$

Among them, (x^t, y^t) is the current position of the end of the robot arm, (x_c, y_c) is the corresponding circular trajectory position, (a, b) is the center of the circle, r is the radius of the circular trajectory, and $\vec{d}$ indicates the direction from the center of the circle to the end position of the robot arm vector.

The direction vector of the compensation force F_{cps} is perpendicular to the direction vector from the center of the circle to the end position of the manipulator. So $\vec{n} \times \vec{d} = 0$, and the components of F_{cps} in the x and y directions of the plane can be obtained by the following formula.

$$F_{cx} = F_{cps} \times \frac{b - y}{|\vec{d}|}, F_{cy} = F_{cps} \times \frac{x - a}{|\vec{d}|} \tag{15}$$

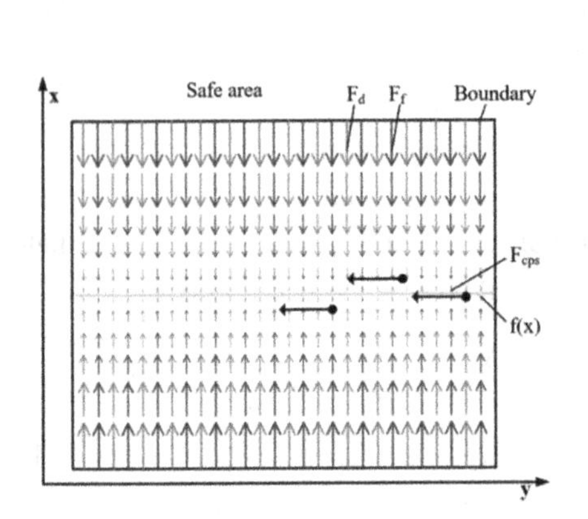

Fig. 7. Linear adaptive force field

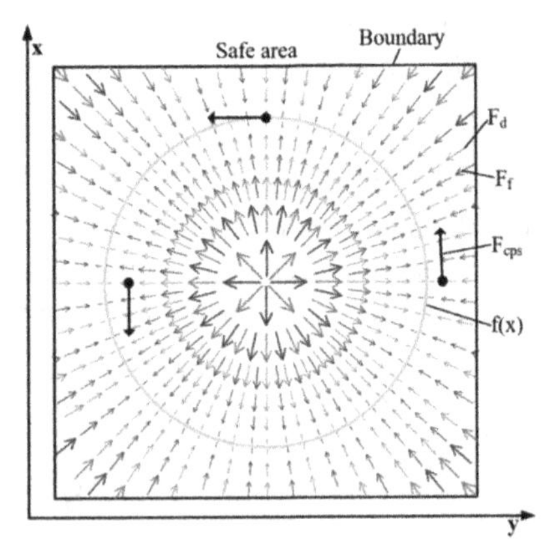

Fig. 8. Circumferential adaptive force field

The position offset in the x-direction and the position of the robot arm end in the x-axis direction at the next moment are shown in Eq. (16–17).

$$\Delta x^{t+1} = \Delta x^t + \frac{T_c(f_{cx} - K_{adapt}\Delta x^t)}{B} \tag{16}$$

$$x_{final}^{t+1} = a + \frac{r(x^t - a)}{|\vec{d}|} + \Delta x^{t+1} \tag{17}$$

Similarly, the force and position in the y-axis direction can be obtained. The circular force field is shown in Fig. 8.

5 Experiment and Analysis

5.1 Experiments with Adaptive Force Fields

The training stiffness is adaptively adjusted by the robot end position and interaction force, see Eq. (7). A total of four groups of experiments were carried out: position control experiment, position-interaction force control experiment, low stiffness experiment (K = 100 N/m) and high stiffness experiment (K = 300 N/m). In the position control experiment, $K_{initial} = 100$ N/m, the position offset coefficient $C_1 = 1000$, when the offset distance d = 20 cm, the stiffness reaches 300 N/m. In the position-force control experiment, $K_{initial} = 100$ N/m, the position offset coefficient $C_1 = 500$, the interaction force coefficient $C_2 = 1.7$, when the offset distance d = 20 cm, the stiffness reaches 300 N/m. 10 reciprocating experiments were carried out respectively, and the recording frequency of the end position and the interactive force was 10 Hz, and the relationship between the interactive force and the offset distance under each control mode was obtained as shown in Fig. 9.

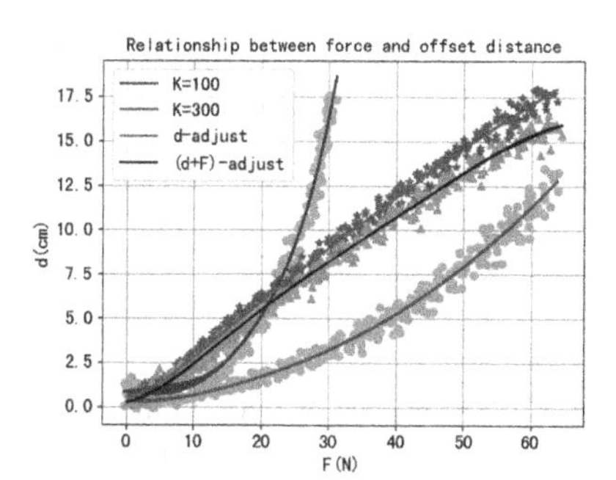

Fig. 9. Relationship between force and offset position under four different force fields

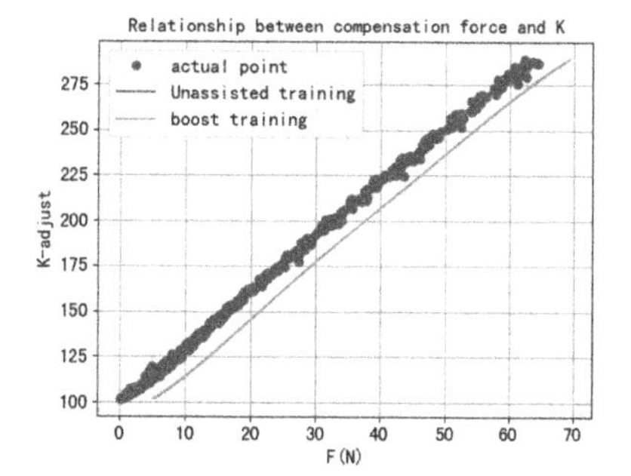

Fig. 10. Relationship between interaction force and stiffness

Experiments show that when the stiffness is fixed, under the action of small interactive force, the end offset is small, that is, the training activity and flexibility are poor. In the low stiffness experiment, with the increase of the interaction force, the offset distance increases rapidly. When the interaction force reaches 30N, the offset distance reaches 17.5 cm, and the safety is relatively poor. In the high stiffness experiment, the force-position curve was always lower than the other three groups of experiments, indicating that the safety is higher at high stiffness, but the flexibility and training activity are greatly limited. For the position control experiment, when the interaction force is small, the compliance and training activity are significantly improved compared with the constant stiffness, but with the increase of the interaction force, the offset still maintains a certain increase, and the safety is not high. For the position-force experiment, when the interaction force is small, the flexibility and training activity are similar to those of the position control. With the increase of the interaction force, the offset distance increases slowly and gradually, which effectively improves the training safety. Based on the above analysis, position-force control has better flexibility, training range of motion and safety.

5.2 Compensation Force Test Experiment

For position-force control, the relationship between the interaction force and K_{adjust} is obtained by curve fitting (see Fig. 8): $F = 0.00012K^2 + 0.28751K - 29.313842$. See formula (9), take this force as the compensation force F_{cps} in the tangential direction of the trajectory (completely active, no traction in the training direction), and C_3 is set to 1N (there is 1N force traction in the training direction). By adding a gradual interference force of ±10N and a mutational interference force of ±10N and 20N in the direction of the vertical training trajectory, the change of the speed in the tangential direction was observed. The results are shown in Fig. 11. The experimental results show that for the gradual force, the velocity remains unchanged, indicating that the compensation force can correctly offset the change in stiffness in the direction of the training trajectory. For the mutation force, under the influence of Δv, the velocity in the training direction produces a small jitter, which is negligible during the training process.

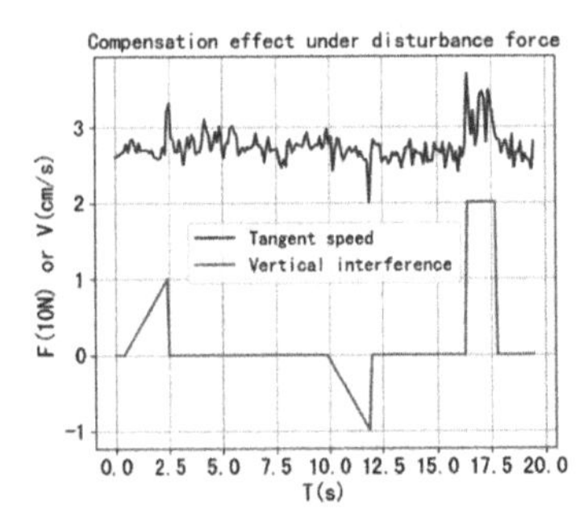

Fig. 11. Influence of disturbance force on velocity in trajectory direction

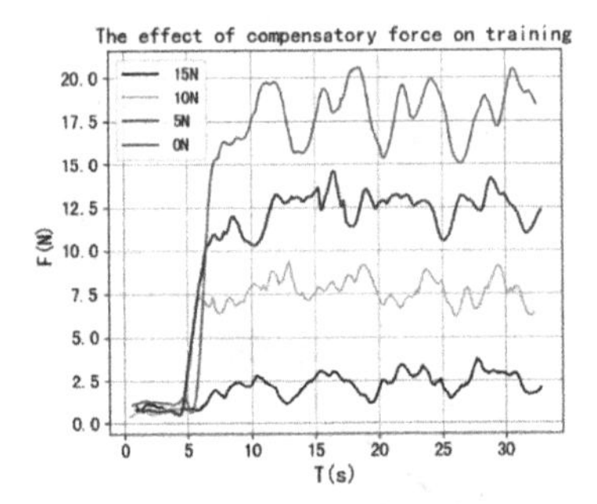

Fig. 12. The change of active training force under different compensation force

On the basis of the compensation force, a constant force was added along the tangential direction of the training trajectory to test the effect of active training under the adaptive force field (yellow trajectory in Fig. 10). Set $C_3 = 0$ (full active), $C_3 = 5$ (training direction assist 5N), $C_3 = 10$ (assist 10N), $C_3 = 15$ (assist 15N, close to passive training). The active training effect of the adaptive force field is shown in Fig. 12. Under the condition of a certain training speed, the magnitude of the assist is inversely proportional to the magnitude of the training force. When the assist is 0, the training force is about 17.5N; when the resistance is 5, the training force is about 12.5N; when the assist is 10, the training force is about 7.5N; when the assist is 15, the training force is about 2.5N. In each mode, the difference in training force is about 5N, which is the same as the difference in power assist. It can be seen from the figure that with the increase of power assist, the deviation and range of training force gradually becomes smaller. When the assist is 0, it is fully active training. At this time, there is no traction to assist the exercise, and the training force deviation is the largest, and the deviation range is [0–5N]. When the assist is 15N, the training force deviation range is reduced to [0–2.5N]. This result shows that the greater the assist, the more conducive to training.

5.3 Visualization and Training Compliance Analysis

Figure 13 corresponds to the visualization effect of artificially applying interference force not along the trajectory direction during the power reciprocating training from left to right, corresponding to the drag trajectory of the elbow joint, the forward flexion and extension of the shoulder joint, the regular trajectory line and the circle. Among them, the red mark ① represents the application of the interference force, and the blue mark ② represents the reset effect of the robot end after the interference force is cancelled. Under the real-time change expectation, there is the effect of the compensation force constant C_3 (5N) and the admittance coefficient B (100), so that the robot end can return to the motion trajectory with a smaller motion slope after deviating from the motion trajectory.

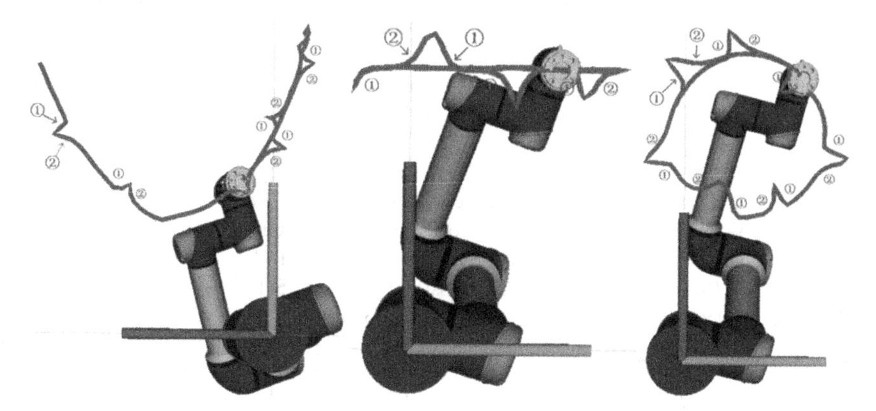

Fig. 13. Visualized training effect of each training track in the plane

6 Conclusion

Combining with the actual application requirements of the current desktop end traction upper limb rehabilitation robot, this paper proposes an implementation method of an adaptive force field based on the admittance control of the arbitrary trajectory of the training plane. Through sufficient comparison experiments with the existing methods in the current literature, the adaptive force field of arbitrary drag trajectories and arbitrary regular trajectories is realized, and finally the test of active training is completed. The experimental results show that the position-interactive force adaptive compliance control method proposed in this paper has greatly improved training compliance, human-computer interaction and safety compared with position control or constant stiffness control. Secondly, the correctness of the compensation force in the stiffness change training is verified through the force interference experiment, and finally the fully active-gradually passive active training test is completed with different assist magnitudes, illustrating the proposed adaptive force field under any trajectory. is feasible and effective. Finally, the training compliance between the plane dragging trajectory and the regular trajectory realized by the plane segmentation method was tested and analyzed, and a solution was provided for the differences in the training trajectory of different upper limb hemiplegia patients.

References

1. Zhou, H., Zhao, J., Li, B.J.: Correlation of peripheral nerve injury to motor function of upper limb in convalescent patients with peripheral paralysis after stroke. Chinese J. Rehabil. Theory Pract. **26**(11), 1333–1338 (2020). https://doi.org/10.3969/j.issn.1006-9771.2020.11.015
2. Gao, P.C., Tang, F.: Effect of intelligent rehabilitation training system on upper limb and hand function of patients with stroke. Chinese J. Rehabil. Theory Pract. **26**(10), 1198–1203 (2020). https://doi.org/10.3969/j.issn.1006-9771.2020.10.013
3. Li, Y.Q., Zeng, Q., Huang, G.Z.: Application of robot-assisted upper limb rehabilitation for stroke (review). Chinese J. Rehabil. Theory Pract. **26**(3), 310–314 (2020). https://doi.org/10.3969/j.issn.1006-9771.2020.03.009
4. Krebs, H.I., Ferraro, M., Buerger, S.P.: Rehabilitation robotics: pilot trial of a spatial extension for MIT-manus. J. Neuroengineering Rehabil. **1**(1), 1–15 (2004). https://doi.org/10.1186/1743-0003-1-5
5. Mihelj, M., Nef, T., Riener, R.: ARMin - Toward a six DoF upper limb rehabilitation robot. In: IEEE/RAS-EMBS International Conference on Biomedical Robotics & Biomechatronics, pp. 1154–1159, BioRob (2022). https://doi.org/10.1109/BIOROB.2006.1639248
6. Zhang, Z.G.,Wo, Q.C.: Adaptive active interactive training control of upper limb rehabilitation robot based on Barrier Lyapunov function. Instrum. Technol. **5**(19), 1–10 (2022). https://kns.cnki.net/kcms/detail/11.2179.TH.20220223.1903.006.html
7. Le, Y.Q., Guo, S.: Research on compliance control method of upper limb rehabilitation robot based on variable admittance control. Ind. Control Comp. **34**(10), 4 (2021). https://doi.org/10.3969/j.issn.1001-182X.2021.10.005

FES-Based Hand Movement Control via Iterative Learning Control with Forgetting Factor

Guangyu Zhao[1], Qingshan Zeng[1], Benyan Huo[1(✉)], Xingang Zhao[2], and Daohui Zhang[2]

[1] School of Electrical Engineering, Zhengzhou University, Zhengzhou 450001, China
huoby@zzu.edu.cn

[2] Shenyang Institute of Automation, Chinese Academy of Sciences, Shenyang 110169, China

Abstract. Functional electrical stimulation (FES) is an effective approach to restore hand movement function for patients with stroke. In this paper, a multi-electrode hand rehabilitation system is presented. Iterative learning control (ILC) with forgetting factor algorithm is employed to achieve an accuracy position control of multi-joint hand movement. A mapping matrix is identified to model the gains from the multi-electrode inputs to the multiple joints of the hand. The convergence conditions of ILC with forgetting factor for the proposed method are analyzed. Finally, experiments on healthy subjects are carried out to verify the performance of the proposed control method.

Keywords: Functional electrical stimulation · Iterative learning control · Electrode arrays

1 Introduction

Stroke is a major cause of paralysis and motor dysfunction, and its high incidence and disability rates pose a huge threat to human health [1,2]. Most patients have different degrees of hand function impairment after a stroke. The hand function is fundamental to activities of daily living (ADLs) [3], which includes communication function, grasping function and manipulation function. Experimental studies have shown that hand training for stroke patients can improve their motor function, reduce neurological loss and improve their daily living ability,

This work is partially supported by the National Natural Science Foundation of China (62103376), China Postdoctoral Science Foundation (2018M632801), Science & Technology Research Project in Henan Province of China (212102310253) and Joint fund of Science & Technology Department of Liaoning Province and State Key Laboratory of Robotics (2021-KF-22-10). & National Natural Science Foundation of China (U1813214).

H. Liu et al. (Eds.): ICIRA 2022, LNAI 13456, pp. 281–292, 2022.
https://doi.org/10.1007/978-3-031-13822-5_25

which is beneficial for patients to return to normal life as soon as possible [4,5]. Therefore, it's necessary to provide a specific and effective hand rehabilitation training approach for patients with stroke.

FES is an effective method for hand rehabilitation. Surface electrodes apply short electrical pulses series to produce muscle contraction and functional movement [6,7]. Many results of researches and experiments have shown that FES has a very positive effect on the rehabilitation of patients' motor functions to improve the range of joint motion and promote neural plasticity [8]. Surface electrodes are used to conduct stimulating electrical currents, which induces muscle contraction and produces joint movements, allowing paralyzed muscles to rebuild or restore function for the purpose of treatment and functional rehabilitation [9].

The complexity of the human musculoskeletal system and the difficulty of muscle selectivity and recruitment imply the complexity of control. The hand, in particular, has a complex structure and many movement patterns [10]. To address the hand movement control problem in hand rehabilitation training, a multi-electrode hand rehabilitation system is presented, in which The multi-channel electrical stimulator and an electrode array are combined with each other to stimulate muscles related to hand joints so that the hand can be stimulated to perform specified movements, and the number of electrodes and the stimulation intensity could be adjusted to be suitable for the experimental subject to carry out the rehabilitation training.

Hand rehabilitation training is a process of constant repetition of movements, which could take the advantages of iterative learning control (ILC) algorithm adequately [11]. Many improved algorithms of ILC have been proposed. ILC was combined with neural networks to improve the trajectory tracking performance for a multi-axis articulated industrial robot [12]. An unified framework for practical ILC was established from the robust adaptive control [13]. A gain optimized P-type ILC was constructed to improve the robustness of the system to internal disturbances [14]. In order to enhance the robustness of the algorithm and reduce the influence of modeling error on the control effect, the forgetting factor [15] is introduced into ILC. By tunning the forgetting factor, the algorithm could abandon the historical information at a desired rate. Therefore, ILC with forgetting factor could adapt to the variation of the controlled object with anti-disturbances ability. In FES-based rehabilitation process, the dynamics of stimulated muscle will vary due to muscle fatigue which exacerbates gradually. To improve the control precision of the hand movement, ILC with forgetting factor is employed in this work.

The remainder of this paper is organized as follows. Section 2 describes the experimental platform, experiment design, hand modeling process, and the structure of ILC with forgetting factor. In Sect. 3, the experimental results are presented and analyzed. Finally, the conclusions are drawn.

2 Method

2.1 Design of the FES-Based Hand Rehabilitation System

Figure 1 shows the hand rehabilitation system designed in this paper. The system is mainly divided into hardware module and software module. The hardware module consists of five parts, which are the upper computer, router, Kinect depth camera, multi-channel electrical stimulator, electrode array and Raspberry Pi controller. Among them, Kinect depth image sensor is connected with the upper computer, and collects angle information of hand joints which is transmitted to the host computer for analysis and processing. The software of the whole rehabilitation system runs on the upper computer. The multi-channel electrical stimulator is connected to a wearable electrode array, which includes 24 independent stimulation channels. According to the control command, the electrical stimulator generates stimulation signal and transmits to the patient's target muscle to induce the contraction of these muscles. The controller, the upper computer and the multi-channel electrical stimulator are connected with the router to establish a wireless data link. The software module includes the implementation of human-computer interaction interface, hand angle information extraction and transmission. The human-computer interaction interface is realized in MATLAB/Simulink. 3Gear's hand tracking server and Kinect are used together to acquire the angle information of hand joints.

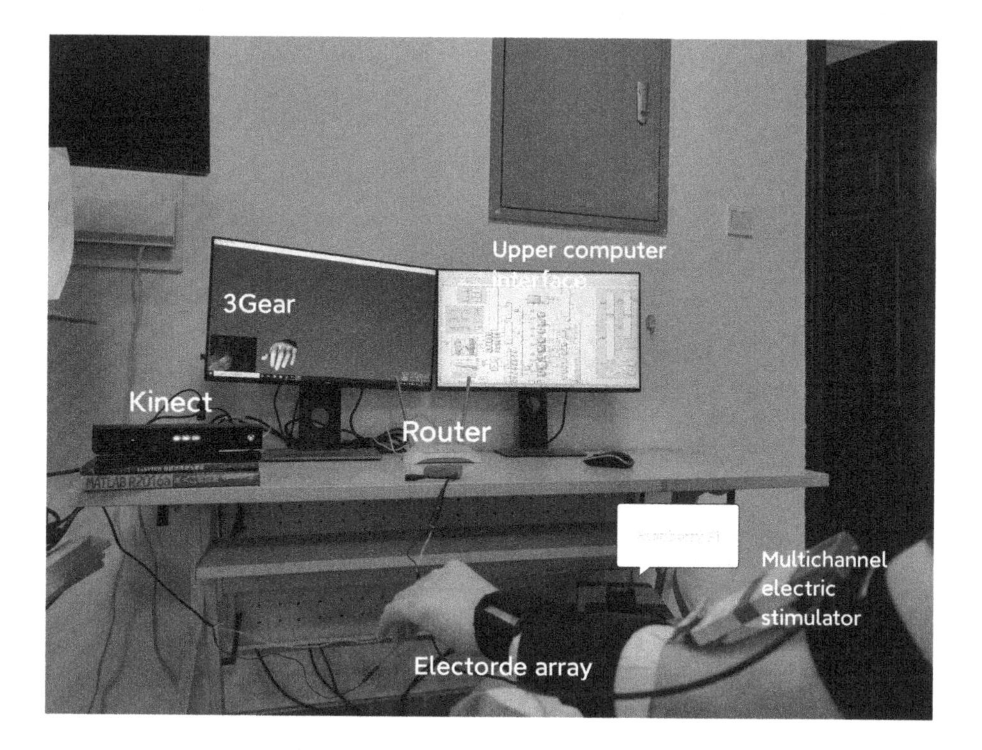

Fig. 1. Experimental platform

In addition, the system is equipped with an emergency stop function, which can stop the system immediately, so as to ensure the safety of patients in the process of rehabilitation.

2.2 Experiment Design

The hand movement control experiments are carried out on the proposed rehabilitation system. The experiment procedure is as follows.

(1) The subject places the forearm on the armrest of the chair and keep it relax.
(2) The electrode array and electrical stimulator are attached to the external extensor area of the subject's forearm, and then the position of the Kinect sensor is adjusted to capture the movement of hand joints continuously.
(3) The maximum stimulation voltage test is carried out to determine the maximum voltage that the subject could withstand to ensure the safety of the experiment.
(4) The initial states and the target position are set up by recording the angle data of hand joints in the initial position and the desired position as shown in Fig. 2.
(5) The mapping matrix from the stimulated inputs to the hand joints responses is identified by applying triangular stimulated signal to each stimulated channel sequentially.
(6) The proposed ILC with forgetting factor algorithm are carried out until the hand movement basically matches the desired position.

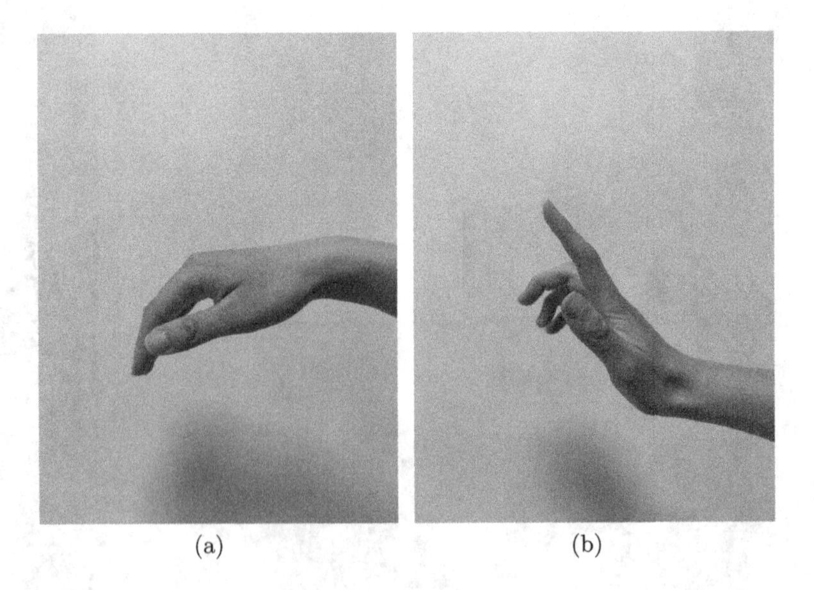

(a) (b)

Fig. 2. Experiment setup. (a) Initial position; (b) desired position.

2.3 Hand Model

During rehabilitation, surface electrodes are usually attached to the external extensor and flexor muscles of the forearm to arouse the movement of the hand. The hand rehabilitation system is a multiple-input, multiple-output system involving multiple muscles, multiple joints and multiple electrodes. If the specific muscle corresponding to the hand joint movement is modelled, the matching relationship between the electrodes and the muscle and how to handle the coupling effect of other electrodes on that muscle needs to be considered, and the movement of the hand will change the relative position of the electrode and the muscle, which will increase the complexity of modeling. In order to reduce the complexity of the hand system model, this paper proposes a new relationship model called linear mapping matrix. This model can directly construct the relationship between joint motion and electrodes without considering the parameters of the actual muscle model. Using electrode arrays for electrical stimulation, the stimulation points can be adjusted in real time according to the situation, which can largely improve the flexibility of the system.

The hand rehabilitation training for stroke patients is mainly based on wrist and finger extension movements. Therefore, only the movements of wrist and fingers caused by forearm muscles are considered in the linear mapping matrix, and their joint angular states are defined as the output of the system. The motion state of the finger joints can be defined as:

$$y_{fi} = y_f(\theta_{i1}, \theta_{i2}). \tag{1}$$

In Eq. (1), $i = 1, 2, \cdots 5$ denotes the five fingers of the hand, respectively, y_f denotes the joint angle vector of the fingers, and θ_{i1}, θ_{i2} denotes the first two joint angle of each finger, respectively. During the temporal electrical stimulation of forearm muscles, the wrist also produces flexion and extension movements in sagittal plane and Frontal plane, so the motion state of the wrist is defined as:

$$y_{wj} = y_w(\theta_{wj}). \tag{2}$$

In Eq. (2), y_w represents the joint angle of the wrist, $j = 1, 2$ represents wrist motion in sagittal plane and Frontal plane, respectively. And θ_{wj} represents the angular value of motion in each plane. Then, the output of the hand rehabilitation system can be expressed as $Y = \left[y_{fi}\; y_{wj}\right]$,where $i = 1, 2, \cdots, 5$, $j = 1, 2$. There are 12 joint angles of hand motion.

The relationship between electrodes and hand joints motion is:

$$y_{ij} = g_{ij} u_j, \tag{3}$$

where i is the number of joints in the hand rehabilitation system, $j = 1, \cdots, N$ represents the number of electrodes, and g_{ij} represents the relationship model between the j-th electrode and the i-th joint. By establishing the relationship model between the joint and the electrode, the corresponding actual muscle model can be represented by this mapping matrix. For the i-th joint, the output

y_i is a linear superposition of multiple electrode stimuli when multiple electrodes are used for stimulation. The relationship model g_{ij} between the joint and all electrodes in the electrode array together characterizes the corresponding actual muscle model.

$$G_i = \left[g_{i1}\ g_{i2} \cdots g_{iN},\right] \tag{4}$$

G is the relationship model matrix of the system. For the hand rehabilitation system, the relationship between the output and input of the system after the simplification can be described as:

$$Y = GU, \tag{5}$$

where Y is the hand joints angle, G is $p \times q$ dimensional linear mapping matrix, p represents the number of joints, q represents the number of electrodes, and $U = \left[u_1\ u_2 \cdots u_N\right]^T$ is the input to each electrode. This hand system model directly describes the relationship between joint motion and electrodes.

When the system was used with different subjects, the G needed to be re-identified due to changes in factors such as electrode array position or muscle state. The identification process is as follows:

Firstly, it is necessary to generate the delta wave electrical stimulation commands. The triangular wave stimulation signal used for the experiment is:

$$u = \begin{cases} \frac{2t}{T} u_{max} & \mathrm{t} < \frac{T}{2} \\ \left(2 - \frac{2t}{T}\right) u_{max} & \frac{T}{2} \leq t < T \end{cases}, \tag{6}$$

where T is the length of the triangular wave stimulation signal, t is the number of samples, and the maximum value of the triangular wave electrical stimulation signal u_{max} is set according to the tolerance threshold voltage of different subjects.

Next, the generated triangular wave stimulation signal u is applied to the 24 electrodes in sequence. At the same time, the angular data y_{ij} of each joint movement of the hand when each electrode applied electrical stimulation was recorded, where i denotes the hand joint and j denotes the electrode to which electrical stimulation was applied.

Finally, after the j-th electrode is applied with electrical stimulation, the angular data y_{ij} of the hand joint movement and the applied electrical stimulation signal u_j are recorded, and the relationship model G_j between this electrode and the hand joint is established according to the simplified modeling method of the hand system proposed above, where $j = 1, \cdots, N$, N is the number of electrodes in the electrode array. Similarly, after all the N electrodes in the electrode array are stimulated in turn, the simplified model matrix G of the whole hand system can be established.

For the hand rehabilitation system designed in this paper with multiple inputs and multiple outputs, and the number of electrodes is larger than the output of the system. In order to realize the cooperative control of multiple electrodes in the electrode array and improve the control accuracy of the system, this paper uses ILC as the framework and add the forgetting factor to the ILC control law. The forgetting factor can increase the robustness of the system and increase the anti-disturbances capability of the system.

2.4 Iterative Learning Control with Forgetting Factor

For the hand rehabilitation system, the desired position of the hand gesture in ILC control can be obtained by collecting angular data via Kinect. With the previous modeling of the hand rehabilitation system and the simplification of the model, the input-output relationship can be expressed as $y_i = G_i u$. Then for the whole system, at the k-th iteration, the input-output relationship can be expressed as:

$$y_k = Gu_k + d \tag{7}$$

$$u_k = G^{-1}(y_k - d), \tag{8}$$

where k denotes the number of iterations, u_k, y_k denote the input and output at the k-th iteration, respectively. And G^{-1} is the inverse of G.

The iterative learning control law used in this system is:

$$u_{k+1} = au_0 + (1-a)u_k + \rho e_k \tag{9}$$

where k denotes the number of iterations, u_k denotes the input and output at the k-th iteration, a is the forgetting factor, u_0 is the initial input to the system, ρ is the learning gain, and $e_k = y_d - y_k$ denotes the tracking error of the system at the k-th iteration.

Let $\rho = \beta G^{-1}$:

$$u_{k+1} = au_0 + (1-a)u_k + \beta G^{-1} e_k \tag{10}$$

where β is the learning gain matrix.

Let $\Delta u_k = u_\infty - u_k, u_{k+1} = u_\infty - u_{k+1}$, u_∞ is the desired output, then from (10) it follows that:

$$\begin{aligned} \Delta \mathrm{u_{k+1}} &= \mathrm{u_\infty} - \mathrm{u_{k+1}} = \Delta \mathrm{u_k} - \mathrm{u_{k+1}} + \mathrm{u_k} \\ &= \Delta \mathrm{u_k} - \mathrm{au_0} + \mathrm{au_k} - \rho \mathrm{e_k} \\ &= \Delta \mathrm{u_k} - \mathrm{au_0} + \mathrm{a}\left(\mathrm{u_\infty} - \Delta \mathrm{u_k}\right) - \rho \mathrm{e_k} \\ &= (1-\mathrm{a})\Delta \mathrm{u_k} - \rho \mathrm{e_k} + \mathrm{a}\left(\mathrm{u_\infty} - \mathrm{u_0}\right) \end{aligned} \tag{11}$$

let $e_k = y_r - y_k = G(u_\infty - u_k)$,

$$\begin{aligned} \Delta \mathrm{u_{k+1}} &= (1-\mathrm{a})\Delta \mathrm{u_k} - \rho \mathrm{G}\left(\mathrm{u_\infty} - \mathrm{u_k}\right) + \mathrm{a}\Delta \mathrm{u_0} \\ &= (1-\mathrm{a}-\beta)\Delta \mathrm{u_k} + \mathrm{a}\Delta \mathrm{u_0} \\ &= \varepsilon \Delta \mathrm{u_k} + \mathrm{a}\Delta \mathrm{u_0} \end{aligned} \tag{12}$$

Therefore, as long as ρ, δ, r, β are chosen appropriately such that $\varepsilon < 1$, then $\Delta u_k \to 0$ as $k \to \infty$, the system can achieve zero tracking error.

3 Result

Three healthy subjects were recruited to participate in the experiment. Before the start of each experiment, each subject was asked to relax their arm muscles, and then the experimental procedure described in Sect. 2.2 was performed. During the experiment, the experimenter should strictly follow the experimental procedure and keep observing the data information in the upper computer interface. If there is any abnormal data or the subject feels discomfort in the arm, the experiment should be stopped immediately. For the same subject, rest time should be reserved between two experiments to avoid muscle fatigue and soreness.

After completing the experiment according to the predetermined steps, the data of all sampling points during the experiment are listed first. Then the error between the joint angle data and the desired joint angle data is calculated separately for all the sampling points in each experiment. The set with the smallest sum of all joint angle errors is considered as the best angle data for this experiment. It is worth mentioning that since the stimulus voltage was gradually increased during the iterations, the best angle data were often generated in the second half of the experiment. The experimental results for the three subjects are shown in Fig. 3.

Except for the initial position angle data, the other curves in the Fig. 3 represent the best angle data selected in different iterations of the experiment. The angle of the hand joint becomes progressively smaller as the hand undergoes extension movements under the effect of electrical stimulation. This data also shows the size of the range of motion of the finger under the action of electrical stimulation.

This experiment is mainly for the flexion and extension of the index finger, so the angle values of the index finger joints corresponding to the horizontal coordinates 9 and 10 are mainly compared. The maximum error between the actual control results and the desired angle data during the iteration of subject 1 was about 10°, and the minimum error was about 1°; the maximum error of subject 2 was about 6°, and the minimum error was about 2°; the maximum error of subject 3 was about 5°, and the minimum error was about 0.1°. The horizontal coordinates 1 and 2 indicate the angles of the upper and lower, left and right directions of the wrist, respectively. It can be seen from the figure that all three subjects had a significant wrist lift during the experiment. The horizontal coordinates 3, 4; 5, 6; 7, 8 indicate the joint angles of the little finger, ring finger, and middle finger, respectively. On the one hand, due to the increase of the stimulation voltage during the experiment, the coupling effect between the fingers causes the angles of the three fingers to be sometimes smaller than the desired angle. On the other hand, during the control process, the hand joint motion data collected by Kinect may have dropped frames due to the influence of light and shadow, thus causing errors in the collected data.

In all sampling points, when the voltage reaches the threshold voltage, 100 sets of joint angle data are selected to represent the overall trend of hand movements during the whole experiment. The difference between the joint angle data

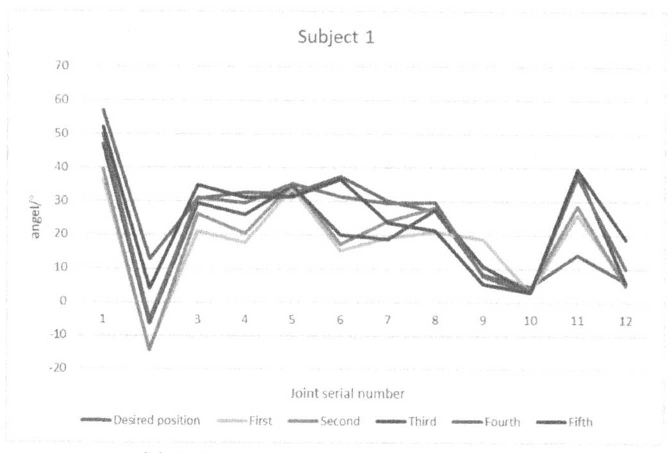

(a) Subject 1: five iterative experiments

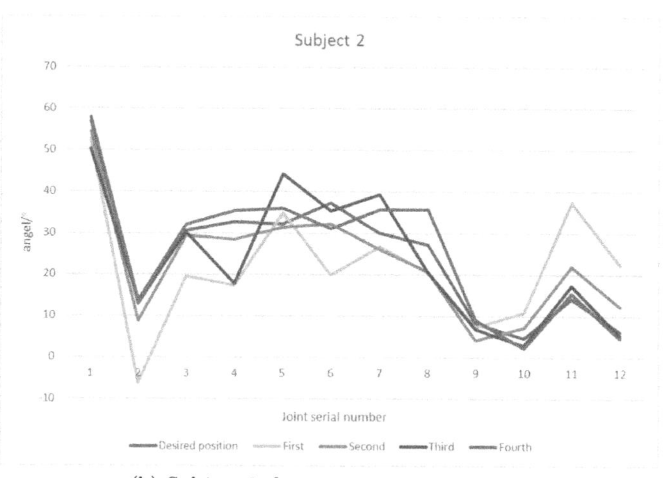

(b) Subject 2: four iterative experiments

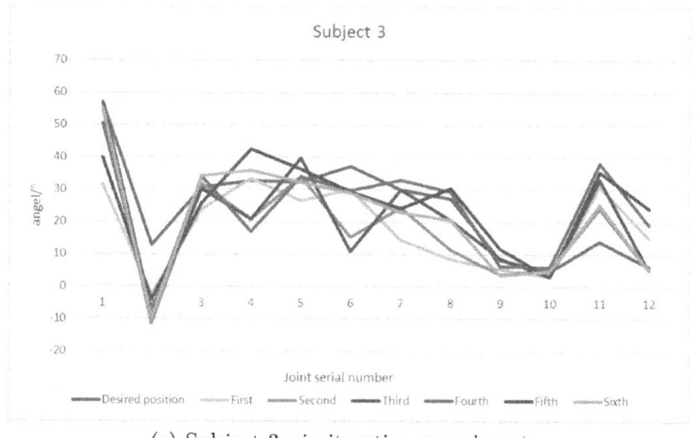

(c) Subject 3: six iterative experiments

Fig. 3. Convergence of joint angles of three subjects during the iterative experiments

and the expected value was calculated for each sampling point separately. The two joint angle errors for each finger were then summed and used to represent the motion of the whole finger. The angular errors of the two directions of the wrist are also summed to represent the motion of the wrist. Finally, the sum of the errors of the 100 sets of joint angle data and the desired joint angle is calculated in each iteration. The results are shown in Fig. 4.

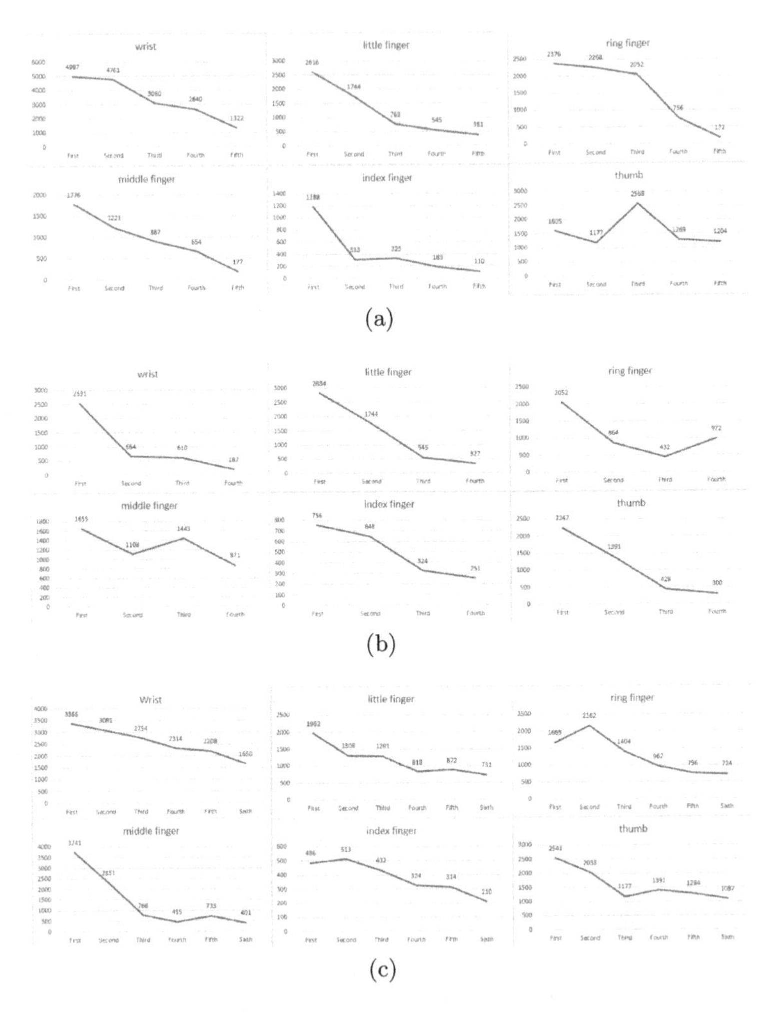

Fig. 4. (a), (b) and (c) are the joint angle errors of the wrist and fingers of the three experimenters respectively

From Fig. 4, it can be concluded that the tracking error of each finger joint shows a decreasing trend as the number of iterations increases, and the iterative learning can also correct the data deviation in time to complete the tracking of the desired angle when the data sampling error occurs. For the three subjects, the maximum joint error averaged about 5° during the experiment, and the minimum joint error was the index finger joint, with the angle converging to about 2°. It can be seen that the hand basically completes the desired gesture movement during the experiment.

4 Conclusions

In this paper, a multi-electrode hand rehabilitation system is designed for hand rehabilitation of stroke patients. To reduce the difficulty of hand modeling caused by multiple joints, muscles and electrodes in the hand rehabilitation system, a simplified modeling method of hand model is proposed to establish a linear mapping matrix. The mapping matrix represents the relationship of system input and output without considering specific muscle parameters, which increases the flexibility of the system. The maximum threshold voltage test balances the safety of the subject and the effectiveness of the electrical stimulation. ILC with forgetting factor is used to control hand joint movements according to the set desired motion. The proposed methods are verified via experiments on healthy subjects. Except for data errors due to finger coupling and Kinect sampling errors, the position error of the desired joint angle is about 5°. In the future, more subjects will be recruited to validate the proposed methods, and the proposed method will be extended to the position tracking control of hand joints movement.

References

1. Langhorne, P.P., Bernhardt, P.J., Kwakkel, G.: Stroke care 2: stroke rehabilitation. Lancet **377**(9778), 1693–1702 (2011)
2. Garratt, E., Mistry, D., Boyle, C., Fellerdale, M., Southcott, V., King, S.: Arming our patients: empowering patients to increase self-directed upper-limb activity at the Oxfordshire stroke rehabilitation unit. Physiotherapy **113**, e50–e51 (2021)
3. Dobkin, B.H.: Strategies for stroke rehabilitation. Lancet Neurol. **3**(9), 528–536 (2004)
4. Farmer, J., Zhao, X., van Praag, H., Wodtke, K., Gage, F.H., Christie, B.R.: Effects of voluntary exercise on synaptic plasticity and gene expression in the dentate gyrus of adult male Sprague–Dawley rats in vivo. Neuroscience **124**(1), 71–79 (2004)
5. Annetta, N.V., et al.: A high definition noninvasive neuromuscular electrical stimulation system for cortical control of combinatorial rotary hand movements in a human with tetraplegia. IEEE Trans. Biomed. Eng. **66**(4), 910–919 (2019). https://doi.org/10.1109/TBME.2018.2864104
6. Bae, D.-Y., Shin, J.-H., Kim, J.-S.: Effects of dorsiflexor functional electrical stimulation compared to an ankle/foot orthosis on stroke-related genu recurvatum gait. J. Phys. Ther. Sci. **31**(11), 865–868 (2019). https://doi.org/10.1589/jpts.31.865
7. Lim, J., et al.: Patient-specific functional electrical stimulation strategy based on muscle synergy and walking posture analysis for gait rehabilitation of stroke patients. J. Int. Med. Res. **49**(5), 425–1385 (2021)

8. Usman, H., Zhou, Y., Metcalfe, B., Zhang, D.: A functional electrical stimulation system of high-density electrodes with auto-calibration for optimal selectivity. IEEE Sens. J. **20**(15), 8833–8843 (2020)
9. RaviChandran, N., Teo, M.Y., Aw, K., McDaid, A.: Design of transcutaneous stimulation electrodes for wearable neuroprostheses. IEEE Trans. Neural Syst. Rehabil. Eng. **28**(7), 1651–1660 (2020). https://doi.org/10.1109/TNSRE.2020.2994900
10. Kutlu, M., Freeman, C.T., Hallewell, E., Hughes, A.M., Laila, D.S.: Upper-limb stroke rehabilitation using electrode-array based functional electrical stimulation with sensing and control innovations. Med. Eng. Phys. **38**(4), 366–379 (2016)
11. Madady, A.: An extended PID type iterative learning control. Int. J. Control Autom. Syst. **11**(3), 470–481 (2013)
12. Chen, S., Wen, J.T.: Industrial robot trajectory tracking control using multi-layer neural networks trained by iterative learning control. Robotics **10**(1), 50 (2021)
13. Chen, W., Li, J., Li, J.: Practical adaptive iterative learning control framework based on robust adaptive approach. Asian J. Control **13**(1), 85–93 (2010)
14. Hussain, I., Ruan, X., Liu, C., Liu, Y.: Linearly monotonic convergence and robustness of p-type gain-optimized iterative learning control for discrete-time singular systems. IEEE Access **9**, 58337–58350 (2021)
15. Wang, H., Dong, J., Wang, Y.: Research on open-closed-loop iterative learning control with variable forgetting factor of mobile robots. Discret. Dyn. Nat. Soc. **2016**, 1–6 (2016)

Actuation, Sensing and Control of Soft Robots

3D Printing of PEDOT:PSS-PU-PAA Hydrogels with Excellent Mechanical and Electrical Performance for EMG Electrodes

Hude Ma[1,2], Jingdan Hou[1,2], Wenhui Xiong[1,2], Zhilin Zhang[1,2], Fucheng Wang[1,2], Jie Cao[1], Peng Jiang[1,4], Hanjun Yang[1,2], Ximei Liu[1,2(✉)], and Jingkun Xu[1,3(✉)]

[1] Jiangxi Key Lab of Flexible Electronics, Flexible Electronics Innovation Institute, Jiangxi Science and Technology Normal University, Nanchang 330013, Jiangxi, People's Republic of China
liuxm@jxstnu.edu.cn, xujingkun1971@yeah.net

[2] School of Pharmacy, Jiangxi Science and Technology Normal University, Nanchang 330013, Jiangxi, People's Republic of China

[3] School of Chemistry and Molecular Engineering, Qingdao University of Science and Technology, Qingdao 266042, Shandong, People's Republic of China

[4] Xi'an Physical Education University, Xi'an 710068, Shanxi, People's Republic of China

Abstract. Bioelectronics has been developed for recording the electrophysiological activity of diagnostic and therapeutic devices. However, current bioelectrodes still imperfectly comply with tissues, which results in high interfacial impedance and even mechanical detachment. Herein, we report a simple yet effective approach to overcome such hurdles by designing a highly conductive, adhesive hydrogel composite based on freeze-dried poly(3,4-ethylenedioxythiophene):poly (styrene sulfonate) (PEDOT:PSS), polyurethane (PU), and poly(acrylic acid) (PAA). With the continuous phase-separation of PEDOT:PSS, PU, and PAA, the resultant composite hydrogels can simultaneously achieve high adhesion (lap-shear strength > 8 kPa), stretchability (fracture strain > 1100%), and electrical conductivity (conductivity > 2 S/m) by overcoming the traditional trade-off between mechanical and electrical properties in conducting polymer hydrogels. Moreover, such hydrogels are readily applicable to advanced manufacturing techniques such as 3D printing. We further fabricated skin electrodes and achieved high quality and high signal-to-noise ratio EMG signal recording of the forearm.

Keywords: Conductive hydrogel · PEDOT:PSS · Poly (acrylic acid) · Adhesion · Continuous phase-separation

1 Introduction

In tandem with recent advances in miniaturized and flexible electronics, the integration of bioelectronic devices in close contact with biological tissues has been intensively

H. Ma, J. Hou, W. Xiong — Contribute equally to this work.

H. Liu et al. (Eds.): ICIRA 2022, LNAI 13456, pp. 295–304, 2022.
https://doi.org/10.1007/978-3-031-13822-5_26

explored in a broad range of diagnostic and therapeutic applications [1]. These bioelectronic devices rely on different forms of bioelectrodes to achieve electrophysiological signal recording or stimulation [2]. However, current bioelectrodes do not always imperfectly comply with tissues, which results in high interfacial impedance and even mechanical detachment [3].

Modification of electrodes using advanced interfacial materials has been employed to narrow the mismatch [4–8]. One effective strategy is to coat advanced materials, inclusive of ionically conductive polymer and electrically conductive carbon materials (e.g., carbon nanotubes and graphene), conducting polymers (CPs) (e.g., poly(3,4-ethylenedioxythiophene):polystyrene sulfonate (PEDOT:PSS) on electrodes due to their superior electrical properties, satisfying biocompatibility, and tissue-like softness [9, 10]. Among such interfacial materials, PEDOT has been at the forefront in terms of its tunable electrical conductivity, flexibility, and stretchability [11, 12]. Fabio Cicoira et al. electropolymerized PEDOT:tetrafluoroborate on platinum-iridium (PtIr) microelectrodes in acetonitrile for deep brain stimulation and recording. The as-prepared PEDOT:tetrafluoroborate coatings exhibited higher charge storage capacity, higher capacitance, and lower charge transfer resistance than bare PtIr microelectrodes [13]. David C. Martin et al. coated amethylamine-functionalized EDOT derivatives (EDOT-NH_2) onto conducting substrates, which showed greatly enhanced adhesion and comparable electroactivity to pristine PEDOT coatings [7]. However, PEDOT-coated metal electrodes still have issues like incompetent mechanical properties and complex synthetic processes, resulting in decreased device performance and lifetime of the electrodes.

Herein, to develop an advanced interfacial material and an effective approach to overcome such hurdles, we adopt the continuous phase-separation strategy (Fig. 1a) to prepare the PEDOT:PSS-PU-PAA hydrogel which exhibits excellent electrical conductivity, adhesion, stretchability, and long-term stability. Benefit from these, we further fabricate EMG electrodes via 3D printing technology (Fig. 1b) and demonstrate their capability for recording high signal-to-noise ratio signals in diverse grip forces.

2 Results and Discussion

2.1 Design Principle and Preparation of PEDOT:PSS-PU-PAA Hydrogel

We prepare PEDOT:PSS-PU-PAA hydrogel with continuous phase-separation strategy to achieve integration properties including conductivity, adhesion, stretchability, and long-term stability (Fig. 1a). PEDOT:PSS, PU, and PAA was select as the conductive [14, 15], mechanical [16], and adhesive phase [17], respectively. PEDOT:PSS was lyophilized and redispersed in a deionized water-DMSO mixture (water:DMSO = 85:15 v/v) to form phase-separated PEDOT:PSS nanofibers, followed by being sheared at high speed with PU and PAA for uniform mixing and to form a continuous phase-separated hydrogel. In addition to the PEDOT:PSS phase separation that increases the electrical conductivity of PEDOT:PSS-PU-PAA hydrogel, the PEDOT:PSS-PU-PAA continuous phase in which PAA dopes PEDOT [18] and a continuous conductive path of PEDOT is formed also contribute to the high electrical conductivity of the hydrogel. The carboxyl groups on PAA can interact with different substrates (e. g., electrostatic interaction and coordination, etc.), which endow the hydrogel with adhesion.

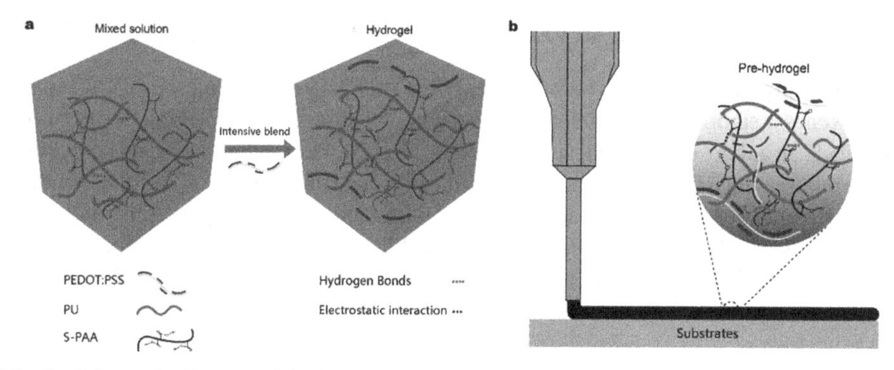

Fig. 1. Schematic diagram of the formation of PEDOT:PSS-PU-PAA hydrogel and the manufacturing strategy developed for 3D printing. (a) Design principle of PEDOT:PSS-PU-PAA hydrogel. (b) 3D printing process of PEDOT:PSS-PU-PAA hydrogels.

2.2 Optimization of PEDOT:PSS-PU-PAA Hydrogels

Adhesion and electrical conductivity are important factors that determine the performance of EMG electrodes. We prepared five ratios of hydrogels with different PAA contents (from 0 wt.% to 75 wt.%) and obtained the best performing material (25%) by comparing their adhesion (lap-shear strength) and electrical conductivity. As it is illustrated (Fig. 2), the adhesion of PEDOT:PSS-PU-PAA hydrogels increases, while the electrical conductivity decreases with the increasement of PAA content. This is due to low connectivity between electrical phases caused by excessive PAA. But the increasement of the carboxy group which bonds with substrates contribute to a higher adhesion [19]. The PEDOT:PSS-PU-PAA hydrogel with 25 wt.% PAA content is adopted in the following experiments.

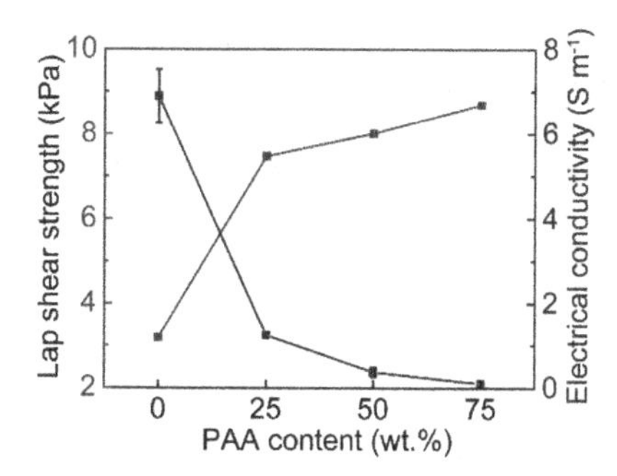

Fig. 2. Lap-shear strength and electrical conductivity of PEDOT:PSS-PU-PAA hydrogels with different PAA contents.

2.3 Mechanical Properties of PEDOT:PSS-PU-PAA Hydrogel

Adhesion and conformal mechanical properties of EMG electrode are crucial for the effectiveness, stability, and fidelity of the signal detection. The PEDOT:PSS-PU-PAA hydrogel can adhere on different substrates (e.g., Au, Al, Pt, polyimide (PI) and polyethylene terephthalate (PET)) (Fig. 3a) owing to the presence of carboxyl groups that bonding with different substrates [20] via hydrogen bonding, electrostatic interaction, and coordination, etc. PEDOT:PSS-PU-PAA hydrogel exhibits optimal adhesion properties with a lap-shear strength of 8.63 kPa on a polyethylene terephthalate film (PET). This is because PET possesses more cross-linking sites than others.

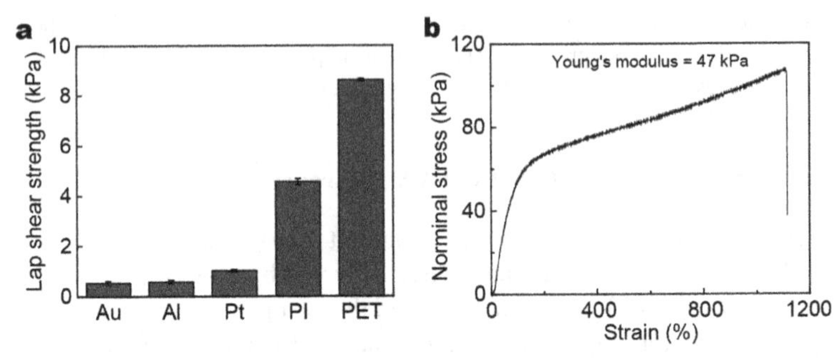

Fig. 3. Mechanical performance of PEDOT:PSS-PU-PAA hydrogels with 25% PAA contents. (a) Lap shear strength of PEDOT:PSS-PAA-PU hydrogels integrated on different substrates (tensile rate: 50 mm min^{-1}). The maximum adhesion was reaching on PET at 8.63 kPa. (b) Representative stress-strain curve of PEDOT:PSS-PU-PAA hydrogel (tensile rate: 50 mm min^{-1}).

Apart from adhesion, EMG electrode need a certain degree of softness to adapt to biological tissues [21], thereby ensuring the efficient connection. The stress-strain curve of the PEDOT:PSS-PU-PAA hydrogel is obtained through a mechanical testing machine (Fig. 3b). The measured Young's modulus is 47 kPa which is similar to many biological tissues (20–100 kPa) [22]. PEDOT:PSS-PU-PAA hydrogel exhibits a large elongation (1100%), which can meet the requirements of bioelectronic interface materials in extreme cases such as large deformation. The mechanical properties of hydrogels primarily depend on the type and density of cross-linked molecules, and the swelling caused by the hydrophilic-hydrophobic balance [23, 24]. Such mechanical performance is attributed to the physical cross-linkage between PEDOT:PSS, PU, and PAA. During the deformation process, these dynamic cross-linkages giving the hydrogel low Young's modulus and high stretchability [22, 25].

2.4 Electrochemical Performance of PEDOT:PSS-PU-PAA Hydrogel

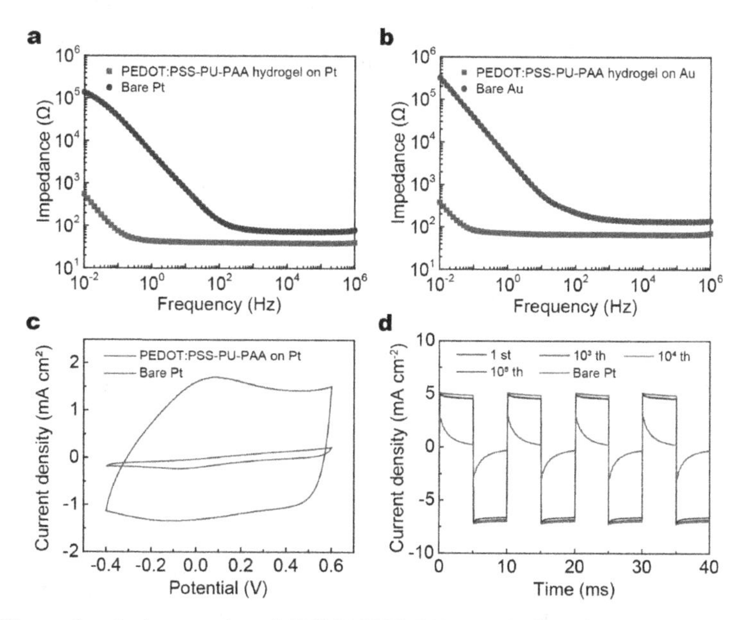

Fig. 4. Electrochemical properties of PEDOT:PSS-PU-PAA hydrogels. (a) and (b) EIS curves of PEDOT:PSS-PU-PAA hydrogel coated on Pt and Au electrodes. (c) CV characterization for PEDOT:PSS-PU-PAA hydrogel coated on a Pt electrode under between –0.4 V and 0.6 V vs. Ag/AgCl. (d) Cyclic electrochemical current pulse injection curves from 1st to 10^5th cycle under between –0.5 V and 0.5 V vs. Ag/AgCl.

In addition to the strong adhesion, the interfacial material should improve the electrochemical properties of electrodes with a maximal extent [26]. To investigate this requirement, we systematically characterize the electrochemical properties of PEDOT:PSS-PU-PAA hydrogel coated electrodes. All coated electrodes show superb electrochemical performance compared to bare electrodes after coating PEDOT:PSS-PU-PAA hydrogels. We print the PEDOT:PSS-PU-PAA hydrogel on Pt and Au electrodes and carry out an electrochemical impedance test. It can be seen that the PEDOT:PSS-PU-PAA hydrogel provides a lower impedance than the bare Pt electrode (Fig. 4a) and Au electrode (Fig. 4b) in the low frequency region. The positive impedance attribute to the continuous phase which results in a good connectivity between electrical phases and thus, good electrical conductivity [11, 27]. In addition, the secondary doping of PEDOT:PSS by PAA increases the carrier concentration, which induces a lower impedance [28, 29].

In addition to low impedance, high efficiency EMG signal recording requires a set of desired electrode features including high charge storage capacity (CSC) and high charge injection capacity (CIC) [2, 30]. In the cyclic voltammetry, PEDOT-PU-PAA hydrogel is doped during the anodic scan and de-doped during the cathodic scan [13]. The PEDOT:PSS-PU-PAA coated Pt electrode shows an enhanced CSC (Fig. 4c) (227 μC cm^{-2}) with respect to the bare Pt (14 μC cm^{-2}). We further adopt long-term biphasic stimulation to demonstrate the stability of the PEDOT:PSS-PU-PAA hydrogel in PBS. The measured CIC (Fig. 4d) of the PEDOT:PSS-PU-PAA hydrogel (527 μC cm^{-2}) is over 3 times greater than the Pt electrode (161 μC cm^{-2}). In addition, it maintains favorable CIC over 10,000 charging and discharging cycles (decreased by 3.8%). These promising electrochemical results imply a particular advantage for the novel EMG coatings.

2.5 EMG Signal Recording of PEDOT:PSS-PU-PAA Hydrogel Based EMG Electrode

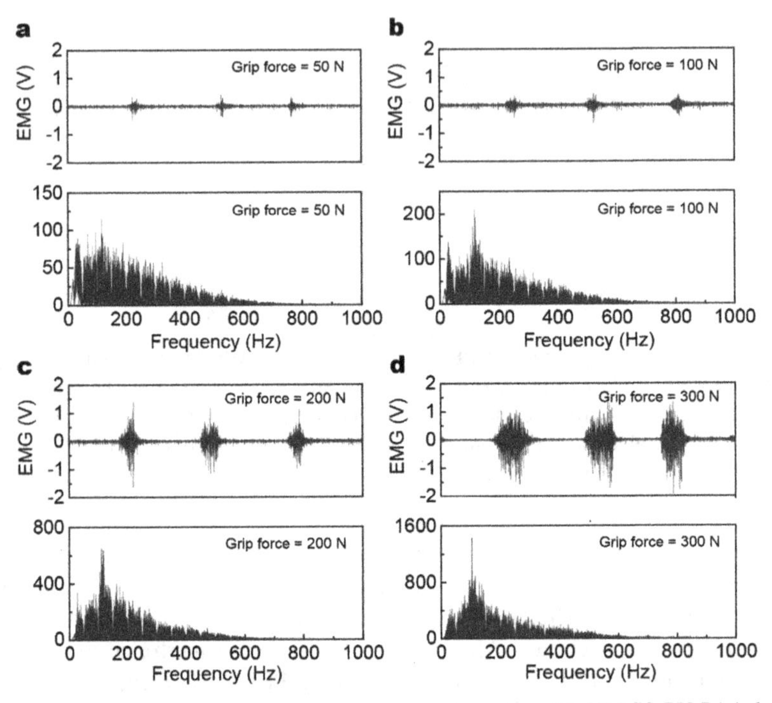

Fig. 5. EMG signals and frequency spectrum transmitted by PEDOT:PSS-PU-PAA hydrogel based electrode.

The adhesive and conductive PEDOT:PSS-PU-PAA hydrogel coating enables the ability to reliably monitor high-quality EMG signal recording [5]. We prepared a EMG electrode by attaching 3D-printed hydrogel to a conductive patch. These electrodes can precisely detect and distinguish EMG signals with different grip force. The signal-to-noise (SNR) ratio and frequency spectrum energy increase with the grip force (the grip forces of 50 N, 10 N, 100 N, and 200 N correspond to the SNR of 13.38 dB, 15.30 dB, 21.10 dB, and 29.98 dB, respectively) (Fig. 5a–5d). With the boost of gaining units, any given unit of EMG signal input can expand the regional response [31]. In this case, the increased amplitude and unchanged noise amplitude lead to the increased SNR [32]. These results imply the potential application of PEDOT:PSS-PU-PAA hydrogel coatings to enhance EMG electrode performance for precise monitoring and distinguishing human limb motion from other daily activities.

3 Conclusion

In this work, we provide a continuous phase-separation strategy to develop an adhesive and conductive hydrogel as well as a fabrication approach for 3D printing to coat hydrogels on different substrates. The long-term stability and superb electrochemical performance of this hydrogel were also demonstrated. Benefiting from these advantages, we further prepare an EMG electrode and achieve high-quality EMG signal recording with high SNR. This paradigm could be anticipated in bioelectronics in terms of the fabrication of bioelectronic devices and the choice of interfacial materials.

4 Materials and Experimental

4.1 Materials

PEDOT:PSS conductive particles were purchased from Agfa-Gevaert N.V.. Dimethyl sulfoxide (DMSO, >99.8%) was acquired from Sigma Aldrich. Polyacrylic acid (PAA, 25 wt.% aqueous solution, M_W = 240,000) was purchased from Hebei Bailingwei Superfine Material Co., LTD. Polyurethane (PU, HydroMed D3) was acquired from AdvanSource Biomaterials. Ethanol (>99.7%) was bought from Sigma Aldrich.

4.2 Preparation of PEDOT:PSS-PU-PAA Hydrogel

PEDOT:PSS conductive pellets (0.7 g) were blended uniformly with 10 ml of mixed solvent (deionized water: DMSO = 85 vol%: 15 vol%) in a syringe, followed by filtering agglomerates. The water-soluble polyurethane HydroMed D3 was dissolved by adding it repeatedly in small quantities to a mixture of ethyl and deionized water (ethyl: deionized water = 95: 5, v/v) and was stirred at room-temperature. The dissolution procedure was performed in a sealed environment to prevent evaporation of the ethanol. PEDOT:PSS-PU solution was prepared by uniformly mixing PEDOT:PSS aqueous dispersion and PU solution in a syringe at a weight content ratio of 1:2. PEDOT:PSS-PU-PAA solutions (PEDOT:PSS-PU-PAA ink) were prepared by uniformly blending PAA solution (25 wt.%) and the above mixture, followed by being centrifuged in a high-speed of 8000 rpm/s for 10 min to remove air bubbles. After printing, the 3D-printed PEDOT:PSS-PU-PAA ink was dried at a room temperature for 24 h, followed by being equilibrated in PBS to be converted into a hydrogel state.

4.3 Mechanical Characterization of PEDOT:PSS-PU-PAA Hydrogel

To test the adhesion performance of PEDOT:PSS-PU-PAA hydrogels, we adhered the hydrogel (0.5 cm × 0.5 cm) to the surface of the substrate and implemented a standard lap-shear test with a universal test machine (ZHIQU Precision Instrument, ZQ-990LB,

China). To measure the mechanical property of PEDOT:PSS-PU-PAA hydrogels, pure-shear tensile test of dumbbell-shaped samples (35 mm × 6 mm × 1 mm) were conducted with the universal test machine. All the tests were performed at a constant tensile speed (50 mm min-1) and three samples were used.

4.4 Electrical and Electrochemical Characterization of PEDOT:PSS-PU-PAA Hydrogel

The electrical conductivity of PEDOT:PSS-PU-PAA hydrogel was measured by standard four-probe method (Keithley 2700 digital multimeter, Keithley). PEDOT:PSS-PU-PAA hydrogel was cut into rectangles (20 mm × 2 mm), and its ends are connected to the instrument via copper wires. For the characterization of the electrochemical performance, the PEDOT:PSS-PU-PAA hydrogel was coated on a platinum wire (1 mm diameter), followed by being soaked in PBS. The hydrogel-modified platinum wire, platinum sheet, Ag/AgCl (saturated KCl), and PBS functioned as the working electrode, counter electrode, reference electrode, and electrolyte, respectively.

4.5 Fabrication of EMG Electrode

A commercial pad which composed of a butylene terephthalate film (PBT) and an electrode layer was cleaned with alcohol and preserved in N_2 before use. PEDOT:PSS-PU-PAA hydrogel was directly printed on the side of butylene terephthalate film on the commercial pad.

4.6 Detection of EMG Signal

The EMG electrode what we have fabricated was used as the working electrode and commercial electrodes were used as the reference and counter electrode. The three electrodes were adhered on the forearm and spaced 5 cm apart. All the electrodes were connected to an electrophysiological test system (HE1LP, Qubit systems). EMG signals were collected from 3 subjects during griping. The grip force was measured using a dynamometer and simultaneously recorded with the HE1LP.

Acknowledgments. This work was supported by the National Natural Science Foundation of China (51963011), Technological Expertise and Academic Leaders Training Program of Jiangxi Province (20194BCJ22013), Training Program of Natural Science Foundation of China Youth Fund (20202ZDB01007), Jiangxi Provincial Double Thousand Talents Plan-Youth Program (JXSQ2019201108), Natural Science Foundation of Jiangxi Province (20202ACB214001), Jiangxi Key Laboratory of Flexible Electronics (20212BCD42004), the Science and Technology Project of the Education Department of Jiangxi Province (GJJ201101), the Science and Technology Project of the Education Department of Jiangxi Province (GJJ211141), Jiangxi Science and Technology Normal University (2020XJZD006), and Jiangxi Science & Technology Normal University for the Provincial Postgraduate Innovation Program grants (YC2021-S754).

Author Contributions. H.M. and X.L. proposed the concept and application of PEDOT:PSS-PU-PAA hydrogels. H.M. and W.X. developed the materials and methods for PEDOT:PSS-PU-PAA hydrogels. H.M., W.X., and J.H. performed the adhesion and conductivity tests. H.M. and

J.H. performed the EMG experiments. H.M. analyzed the EMG data. H.M., J.H., and X.L. prepared Figures. H.M., J.H., W.X., F.W., and X.L. analyzed the data and wrote the manuscript. J.X., P.J., and X.L. provided funding to support experiments. X.L. and J.X. supervised the work and provided critical feedback on the development of design strategy, device fabrication, data interpretation, application, and critical revision.

Conflict of Interest. The authors declare no competing interests.

References

1. Deng, J., et al.: Electrical bioadhesive interface for bioelectronics. Nat Mater **20**(2), 229–236 (2021)
2. Yuk, H., Lu, B., Zhao, X.: Hydrogel bioelectronics. Chem Soc Rev **48**(6), 1642–1667 (2019)
3. Pan, L., et al.: A compliant ionic adhesive electrode with ultralow bioelectronic impedance. Adv Mater **32**(38), 2003723 (2020)
4. Kim, S., Jang, L.K., Jang, M., Lee, S., Hardy, J.G., Lee, J.Y.: Electrically conductive polydopamine-polypyrrole as high-performance biomaterials for cell stimulation in vitro and electrical signal recording in vivo. ACS Appl Mater Interfaces **10**(39), 33032–33042 (2018)
5. Yang, M., et al.: Poly(5-nitroindole) thin film as conductive and adhesive interfacial layer for robust neural interface. Adv Funct Mater **31**(49), 2105857 (2021)
6. Yao, B., et al.: High-stability conducting polymer-based conformal electrodes for bio-/iono-electronics. Mater. Today **53**, 84–97 (2022)
7. Ouyang, L., Wei, B., Kuo, C.C., Pathak, S., Farrell, B., Martin, D.C.: Enhanced PEDOT adhesion on solid substrates with electrografted P(EDOT-NH_2). Sci Adv **3**(3), e1600448 (2017)
8. Inoue, A., Yuk, H., Lu, B., Zhao, X.: Strong adhesion of wet conducting polymers on diverse substrates. Sci. Adv. **6**(12), eaay5394 (2020)
9. Fantino, E., et al.: 3D printing of conductive complex structures with in situ generation of silver nanoparticles. Adv Mater **28**(19), 3712–3717 (2016)
10. Lu, B., et al.: Pure PEDOT:PSS hydrogels. Nat Commun **10**(1), 1043 (2019)
11. Fan, X., et al.: PEDOT:PSS for flexible and stretchable electronics: modifications. Strategies Appl., Adv Sci **6**(19), 1900813 (2019)
12. Donahue, M.J., et al.: Tailoring PEDOT properties for applications in bioelectronics. Mater Sci Eng R Rep **140**, 100546 (2020)
13. Bodart, C., et al.: Electropolymerized Poly(3,4-ethylenedioxythiophene) (PEDOT) coatings for implant-able deep-brain-stimulating microelectrodes. ACS Appl Mater Interfaces **11**(19), 17226–17233 (2019)
14. Yao, B., et al.: Ultrahigh-conductivity polymer hydrogels with arbitrary structures. Adv Mater **29**(28), 1700974 (2017)
15. Zhao, Q., et al.: Robust PEDOT:PSS-based hydrogel for highly efficient interfacial solar water purification. Chem Eng J **442**, 136284 (2022)
16. Seyedin, M.Z., Razal, J.M., Innis, P.C., Wallace, G.G.: Strain-responsive Polyurethane/PEDOT: PSS elastomeric composite fibers with high electrical conductivity. Adv Funct Mater **24**(20), 2957–2966 (2014)
17. Yuk, H., et al.: Dry double-sided tape for adhesion of wet tissues and devices. Nature **575**(7781), 169–174 (2019)
18. Kim, N., et al.: Highly conductive PEDOT:PSS nanofibrils induced by solution-processed crystallization. Adv Mater **26**(14), 2268–2272 (2014)

19. Gao, Y., Wu, K., Suo, Z.: Photodetachable adhesion. Adv Mater **31**(6), 1806948 (2019)
20. Taboada, G.M., et al.: Overcoming the translational barriers of tissue adhesives. Nat Rev Mater **5**(4), 310–329 (2020)
21. Sun, J.Y., Keplinger, C., Whitesides, G.M., Suo, Z.: Ionic skin. Adv Mater **26**(45), 7608–7614 (2014)
22. He, K., et al.: An on-skin electrode with anti-epidermal-surface-lipid function based on a zwitterionic polymer brush. Adv Mater **32**(24), 2001130 (2020)
23. Zhao, X., Chen, X., Yuk, H., Lin, S., Liu, X., Parada, G.: Soft materials by design: unconventional polymer networks give extreme properties. Chem Rev **121**(8), 4309–4372 (2021)
24. Gan, D., et al.: Conductive and tough hydrogels based on biopolymer molecular templates for controlling in situ formation of polypyrrole nanorods. ACS Appl Mater Interfaces **10**(42), 36218–36228 (2018)
25. Zhao, X.: Multi-scale multi-mechanism design of tough hydrogels: building dissipation into stretchy networks. Soft Matter **10**(5), 672–687 (2014)
26. Jeong, J.W., Shin, G., Park, S.I., Yu, K.J., Xu, L., Rogers, J.A.: Soft materials in neuroengineering for hard problems in neuroscience. Neuron **86**(1), 175–186 (2015)
27. Shi, H., Liu, C., Jiang, Q., Xu, J.: Effective Approaches To Improve The Electrical Conductivity Of PEDOT:PSS: a review. Adv Electron Mater **1**(4), 1500017 (2015)
28. Nevrela, J., et al.: Secondary doping in poly(3,4-ethylenedioxythiophene):Poly(4-styrenesulfonate) thin films. J Polym Sci B Pol Phys **53**(16), 1139–1146 (2015)
29. Hu, F., Xue, Y., Xu, J., Lu, B.: PEDOT-based conducting polymer actuators. Front Robot AI **6**, 114 (2019)
30. Yuk, H., et al.: 3D printing of conducting polymers. Nat Commun **11**(1), 1604 (2020)
31. Teki, S., et al.: The right hemisphere supports but does not replace left hemisphere auditory function in patients with persisting aphasia. Brain **136**(6), 1901–1912 (2013)
32. Aceves-Fernandez, M.A., Ramos-Arreguin, J.M., Gorrostieta-Hurtado, E., Pedraza-Ortega, J.C.: Methodology proposal of EMG hand movement classification based on cross recur-rence plots. Comput Math Methods Med **2019**, 6408941 (2019)

Stretchable, Conducting and Large-Range Monitoring PEDOT: PSS-PVA Hydrogel Strain Sensor

Zhilin Zhang[1,2], Hude Ma[1,2], Lina Wang[1,2], Xinyi Guo[2], Ruiqing Yang[2], Shuai Chen[2], and Baoyang Lu[1,2](✉)

[1] Jiangxi Key Lab of Flexible Electronics, Flexible Electronics Innovation Institute, Jiangxi Science and Technology Normal University, Nanchang 330013, Jiangxi, People's Republic of China
luby@jxstnu.edu.cn

[2] School of Pharmacy, Jiangxi Science and Technology Normal University, Nanchang 330013, Jiangxi, People's Republic of China

Abstract. Highly stretchable conducting polymer hydrogel strain sensors are widely used in many wearable electronic devices such as human exercise health monitoring, human-machine interface, and electronic skin. Flexible strain sensors can convert the sensed mechanical tensile deformation into electrical signal output. The structure and detection principle are simple. However, the strain sensors reported so far still face the problems of low stretchability, poor mechanical property and low sensitivity. In this work, we prepared an anisotropic tough conducting polymer hydrogel via a simple ice template-soaking strategy. The ice template can effectively control the growth of ice crystals, thereby forming a honeycomb-like micro-nano structure network; then the frozen hydrogel is immersed in a high concentration of sodium citrate salt solution. During the soaking process, phase separation of PEDOT:PSS and strong aggregation and crystallization of PVA were induced. The prepared conductive polymer hydrogels have excellent tensile properties, mechanical property and stable resistance changes. Conductive polymer hydrogels can be used as wearable strain sensors for detection of minute physiological movements and motion monitoring under large strains.

Keywords: Conducting polymer · Strain sensor · Human motion monitoring

1 Introduction

Conductive polymer hydrogel is a three-dimensional polymer network swollen by a large amount of water [1]. Due to its excellent electrical and mechanical properties, it has attracted extensive attention for its excellent development prospects in wearable electronic devices [2], electronic skins [3, 4], and soft robots [5, 6]. However, conductive hydrogels suffer from the relatively loose and irreversible cross-linking of their polymer

Z. Zhang and H. Ma—These authors contribute equally to this work.

H. Liu et al. (Eds.): ICIRA 2022, LNAI 13456, pp. 305–314, 2022.
https://doi.org/10.1007/978-3-031-13822-5_27

networks [1], resulting in poor inherent fatigue resistance and difficulty in detecting stable electrical signals during continuous mechanical deformation. In addition, most hydrogels rarely simultaneously possess large stretchability, high sensitivity and wide detection range, limiting practical applications.

The pores and microstructures in hydrogels are mostly randomly oriented. According to research, materials with ordered pores and microstructures show unique advantages in terms of physicochemical and mechanical properties. By constructing ice templates, the grown ice crystals act as both pores and self-assembly drivers, so that the precursor solution is arranged around the ice crystals to form a honeycomb-like micro-nano network structure [7]. Liu et al. [8] fabricated a honeycomb porous structure with ordered nanocrystalline domains by a unidirectional ice template-freeze-drying-annealing method, and designed an anisotropic anti-fatigue PVA hydrogel. And it can be used as a universal method to prepare other soft materials.

The Hofmeister effect, also known as the ion-specific effect, refers to the different abilities exhibited by different salt solutions to precipitate proteins from aqueous solutions [9]. For hydrogels, the ion-specific effect refers to the hydration of the hydrophilic functional groups on the hydrophobic chain of different ion pairs. He et al. [10] found that ions have a specific effect on the gelation of PVA by freeze-immersion method, and PVA hydrogels follow the Hofmeister effect in salt solutions. In addition to different types of ions, higher salt solution concentrations also enhance the effect on the mechanical properties of PVA hydrogels. The hydrogels treated with different salts exhibit different mechanical properties, which are due to the specific interactions between ions, water molecules and polymer molecules leading to different degrees of aggregation of polymer chains, and by regulating the aggregation of polymer chains state to adjust its mechanical properties. Furthermore, the Hofmeister effect is universal for hydrophilic polymers, so this strategy can be extended to other systems containing hydrophilic polymer compositions.

We previously prepared PEDOT:PSS-PVA hydrogels by physical-chemical cross-linking [11] and demonstrated the 3D printability of PEDOT:PSS [12], but the resulting hydrogels had poor mechanical strength, which was not sufficient to support their use in the field of strain sensors Applications. In this work, we fabricated a tough, highly stretchable conducting polymer hydrogel using a synergistic ice-template-salting-out strategy. By constructing the ice template, the ice crystals grow vertically, and the polymer chains are arranged around the ice crystals, finally forming a honeycomb-like nanofiber network. It was then soaked in a 1.5 M sodium citrate solution. During the soaking process, the electrostatic interaction of PEDOT:PSS is weakened, and the phase separation of PEDOT:PSS is induced to form a hydrophobic hard region of PEDOT-rich and a hydrophilic soft region of PSS-rich [13]. In addition, PVA also undergoes strong aggregation and crystallization. The prepared conductive polymer hydrogel not only possesses excellent mechanical properties, but also exhibits sensitive electrical signals under different strains as a strain sensor after encapsulation, and the relative resistance change remains stable during 500 loading-unloading cycles. PEDOT:PSS-PVA hydrogels can also monitor various human movements of different intensities and ranges, including large-scale movements such as joint flexion, finger flexion, and microscopic movements such as swallowing. It has great potential in various advanced fields such as wearable electronics, electronic skin, and soft robotics.

2 Experimental Section

2.1 Materials

Poly(vinyl alcohol) (PVA, Mw 146,000–186,000, 99 + % hydrolyzed) was purchased from Sigma-Aldrich. Poly(3,4-ethylenedioxythiophene)-poly(styrenesulfonate) (PEDOT:PSS) (Dry re-dispersible pellets) was purchased from Kaivo, Zhuhai. Dimethyl sulfoxide (DMSO, > 99.8%, GC), anhydrous ethanol (AR, ≥ 99.7%) and trisodium citrate dihydrate (AR, 98%) were obtained by Aladdin, Shanghai.

2.2 Preparation of PEDOT: PSS-PVA Hydrogels

0.7 g of PVA was added to 9.3 g of deionized water, and heated at 90 °C for 1 h to obtain a colorless and transparent 7 wt% PVA solution. Using PEDOT:PSS conductive particles, dissolved in a mixed solvent of deionized water and dimethyl sulfoxide (DMSO) (85 vol%:15 vol%) to obtain a black viscous solution, which was mixed hundreds of times in a syringe and filter until well mixed. The two-phase solution was mixed and filtered according to a certain solid content of PEDOT:PSS until the mixture was uniform. After centrifugation, it was introduced into a mold and immersed in an ethanol bath of –80 °C at a speed of 1 mm/min^{-1}. Immediately after freezing, put into 1.5 M sodium citrate solution and gel for four days. Complete swelling with PBS buffer was then performed before subsequent testing.

2.3 Mechanical Measurement

The mechanical properties of the conductive polymer hydrogel films were determined by a universal tensile testing machine, which could measure the stress-strain curves during single stretching and cyclic stretching. From the stress-strain curve, basic mechanical information such as Young's modulus, tensile strength, elongation at break, stability and durability of tensile samples can be calculated. During the test, the dumbbell heads at both ends of the dumbbell-shaped hydrogel were wrapped with gauze to increase the friction between the sample and the clamps of the tensile machine, and prevent the test effect from being affected by the sample slipping during the test. The air humidity was maintained by a micro-humidifier to avoid water loss of the hydrogel samples and maintain their inherent mechanical properties. The test samples should choose samples with no defects and uniform film formation and good quality, so as not to affect the test accuracy. At least 3 groups of each sample were tested as error analysis.

2.4 Sensing Measurement

In order to explore the strain sensing properties of conductive polymer hydrogels and avoid the loss of hydrogel water during long-term testing to affect its stretchability and sensing properties, the prepared dumbbell-shaped hydrogel films were encapsulated with two transparent upper and lower layers. of high-stretch VHB double-sided tape (3 M, 4905; 0.5 mm). Its high stretchability (>500%), sealing, transparency and viscosity will not affect the inherent properties of the hydrogel, and ensure that the hydrogel does not lose water during long-term testing. The encapsulated hydrogel sample is fixed between the slide fixtures, and commands are sent to the Arduino Nano development board through the Arduino software to generate a pulse signal to control the driver to drive the stepper motor to cyclically stretch the sample to be tested.

2.5 Conductive Polymer Hydrogel Strain Sensor Assembly and Application

The assembly of the strain sensor is to mold the conductive polymer hydrogel into the desired shape, paste the conductive carbon cloth tape at both ends and encapsulate it between two layers of VHB double-sided tape. Pay attention to exhaust gas during packaging to avoid introducing air bubbles, and the size of the VHB can be cropped as needed. The assembled sensor can be directly attached to the skin surface, and related applications are attempted based on the strain sensing performance measured in the early stage of the encapsulated conductive polymer hydrogel. For example, it can monitor human movements under large strains, such as bending and straightening of elbows and fingers; it can also sense the activities of human organs under small strains such as swallowing.

3 Results and Discussion

3.1 Preparation of PEDOT: PSS-PVA Hydrogels

In this study, a simple blend of PEDOT:PSS and PVA was used to prepare conductive polymer hydrogels through an ice-template-soak strategy, which simultaneously exhibited excellent mechanical properties, electrical properties, strain sensitivity, and stability. We poured the blended PEDOT:PSS-PVA composite solution into a special mold, the bottom of the mold is glass, surrounded by customized acrylic material. Immerse the mold vertically into an ethanol bath at –80 °C at a speed of 1 mm/min^{-1}, and take it out after it is completely immersed in the ethanol bath. Soak in the prepared 1.5 M sodium citrate solution and soak for four days to completely gel. Through directional freezing, ice crystals grow from bottom to top, and after freezing, a honeycomb-like nanofiber network is formed. It was then immersed in a 1.5 M sodium citrate salt solution. In a high concentration of sodium citrate salt solution, the electrostatic interaction between PEDOT and PSS in PEDOT:PSS would induce phase separation of PEDOT:PSS. PEDOT-rich hydrophobic hard domains and PSS-rich hydrophilic soft domains are formed. The gelation of PVA during the soaking process is due to the salting-out effect, the PVA molecular chain undergoes chain rearrangement, and ions induce strong aggregation and crystallization of PVA. In such a phase-separated PEDOT:PSS-PVA hydrogel, the hydrophobic PEDOT-rich semi-crystalline domains enable strain sensing with stable electrical performance, while the hydrophilic PVA crystalline domains can guarantee the mechanical stretchability and robustness through multiple molecular interactions including electrostatic, hydrogen bonding, and chain entanglement.

3.2 Mechanical Properties

We prepared a series of PEDOT:PSS-PVA hydrogels with different PEDOT:PSS solid contents, and their stress-strain curves were tested at a rate of 100 mm/min^{-1} (Fig. 1a). The tensile strains at break of 10 wt% PEDOT:PSS-PVA, 15 wt% PEDOT:PSS-PVA, 20 wt% PEDOT:PSS-PVA and 30 wt% PEDOT:PSS-PVA hydrogels were 422% and 279%, 218% and 121% (Fig. 1c), respectively, and the breaking strengths are 1226 kPa, 437 kPa, 397 kPa and 172 kPa (Fig. 1d), respectively. With the increase of PEDOT:PSS solid content, the stress and strain showed a rapid decrease because the excess PEDOT:PSS brought more defects to the mechanical properties of the hydrogel. First, the introduction of PEDOT:PSS greatly reduced the density of the PVA polymer chain, resulting in a decrease in the crosslinking density of PVA, and the mechanical properties of the PEDOT:PSS-PVA hydrogel decreased. Secondly, the decrease in the density of the PVA polymer chain also greatly reduces the number of hydrogen bonds between PVA, so that the mechanical properties of the hydrogel decrease with the increase of the solid content of PEDOT:PSS. Figure 1b shows that the Young's modulus of the PEDOT:PSS-PVA hydrogel also showed a decreasing trend with the increase of the solid content of PEDOT:PSS.

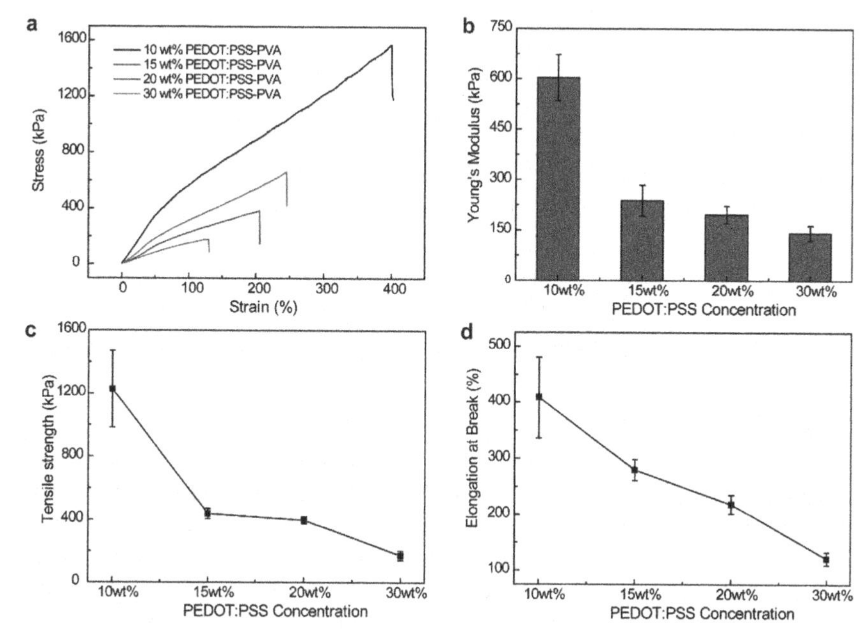

Fig. 1. Mechanical properties of PEDOT:PSS-PVA hydrogels. a) Stress-strain curves of 10 wt% PEDOT:PSS-PVA, 15 wt% PEDOT:PSS-PVA, 20 wt% PEDOT:PSS-PVA, and 30 wt% PEDOT:PSS-PVA hydrogels. b) Young's modulus of PEDOT:PSS-PVA hydrogels with different PEDOT:PSS solid contents. c) Breaking strength of PEDOT:PSS-PVA hydrogels with different PEDOT:PSS solid contents. d) Elongation at break of PEDOT:PSS-PVA hydrogels with different PEDOT:PSS solid contents.

3.3 Sensing Properties

Due to the presence of the conductive polymer PEDOT:PSS, the PEDOT:PSS-PVA hydrogel has been shown to be a strain sensor with high sensitivity and stability. We chose 10 wt% PEDOT:PSS-PVA with better performance as the strain sensing material for subsequent tests. In general, the sensitivity of a strain sensor is evaluated by the gauge factor (GF), $GF = (R - R_0/R_0)/\varepsilon$ [14]. where R is the real-time resistance, R_0 is the initial resistance, and ε is the strain during stretching. The resistance change of the PEDOT:PSS-PVA hydrogel strain sensor can be accurately recorded by LCR in order to study its performance under various strains, such as electrical stability, gauge factor and response time. We encapsulate it with VHB tape and use conductive carbon cloth to connect with LCR electrode clips for testing. By comparing the in situ resistance of PEDOT:PSS-PVA hydrogels (Fig. 2) with different PEDOT:PSS solid contents, we found that its resistance decreased with the increase of PEDOT:PSS solid content, indicating that its electrical conductivity increased with the increase of PEDOT:PSS solid content become better.

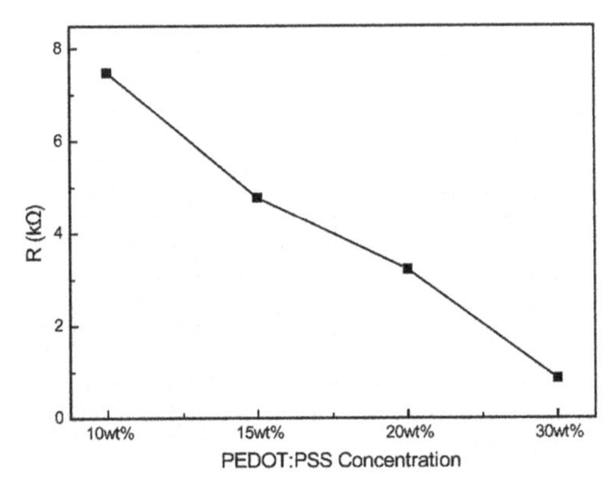

Fig. 2. In situ resistance of PEDOT:PSS-PVA hydrogels with different solid contents of PEDOT:PSS.

As shown in Fig. 3a, the PEDOT:PSS hydrogel strain sensor has good strain sensitivity within 0%–300% strain, and we tested its real-time performance at 50%, 100%, 200%, and 300% strain, respectively. In addition, the sensor has good sensitivity and reproducibility. Subsequent analysis of the resistance change during a single loading-unloading stretching process (Fig. 3b) showed a linearity of 0.98, almost no hysteresis was observed, and a response time of 4.5 s. It is worth noting that Fig. 3d shows the long-term cycling stability test of the PEDOT:PSS-PVA hydrogel strain sensor under 100% strain. The resistance change is stable during 500 cycles, showing excellent durability. At the beginning of the test, the resistance change showed a slight downward trend, which was due to the fact that the polymer chain sliding in the hydrogel did not reach an equilibrium state at the beginning of the stretching, while after cyclic stretching for a period of time, the polymer chains in the hydrogel did not reach equilibrium. The sliding area of the chain is stable, so the resistance change starts to stabilize after a period of time after the test starts. Therefore, PEDOT:PSS-PVA hydrogels with excellent fatigue resistance and electrical properties show great potential in the field of smart wearable devices.

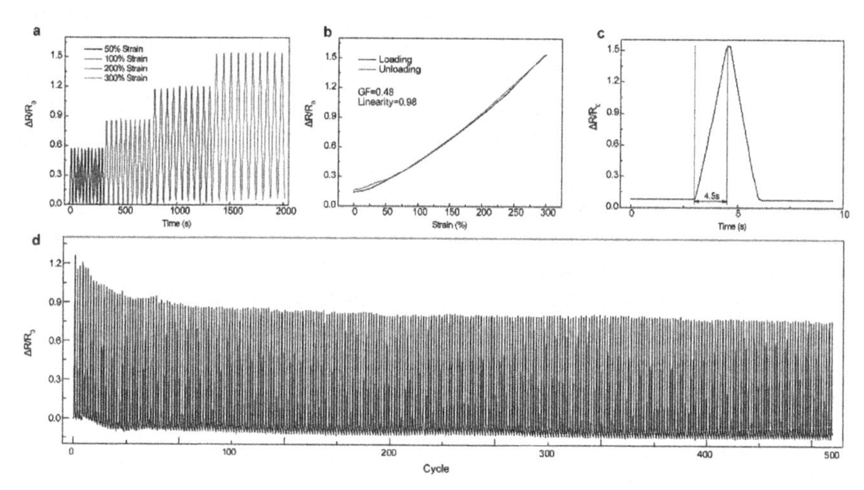

Fig. 3. Sensing performance of PEDOT:PSS-PVA hydrogel strain sensor. a) Resistance response signals under different strains. b) Linearity and sensitivity at 300% strain. c) Response time. d) 500 cycles of stretching versus resistance change.

In order to demonstrate the excellent sensing performance of the directionally frozen PEDOT:PSS-PVA hydrogel strain sensor (Fig. 4), we prepared non-directionally frozen samples to test the sensing performance. Under the condition of keeping the stretching speed and strain the same, there is a clear difference in the sensitivity between the two, which is due to the fact that the cellular nanofiber network formed by directional freezing during stretching is more conducive to the change of the conductive path.

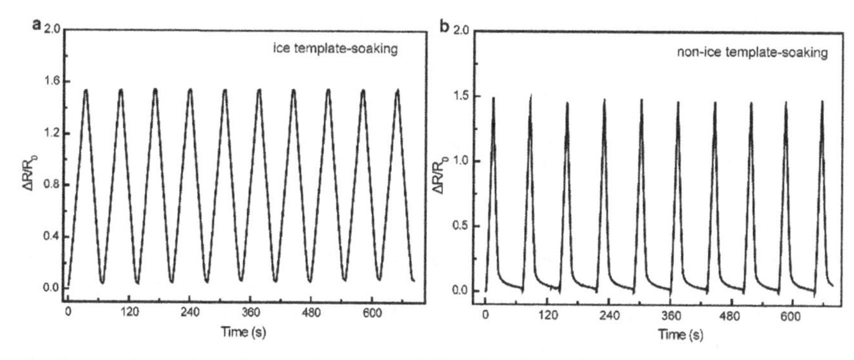

Fig. 4. Comparison of sensing performance of directional freezing and non-directional freezing PEDOT:PSS-PVA hydrogels.

3.4 Human Movement Monitoring

The PEDOT:PSS-PVA hydrogel can be used as a strain sensor to monitor various large-scale human motions, such as finger bending, knee motion, and can also detect some micromotions such as swallowing. We attached the encapsulated PEDOT:PSS-PVA hydrogel strain sensor on the human body surface to detect various motion signals. By attaching the hydrogel strain sensor to the finger to bend at a certain time interval, a series of regular resistance change signals can be monitored (Fig. 5a). When attaching the hydrogel strain sensor to the elbow (Fig. 5b) and knee (Fig. 5d) to bend, electrical signal changes can also be monitored. In addition, the attached larynx can also detect subtle physiological movements during swallowing (Fig. 5c). It is worth noting that the hydrogel strain sensor has good repeatability monitoring and can distinguish large-scale movement from subtle physiological movement by relative resistance changes, which is due to the good electrical stability and good sensitivity of the hydrogel strain sensor. Therefore, we designed conductive polymer hydrogels that are highly stretchable, soft, and good strain-sensitivity, with excellent wearable sensors for monitoring human motion.

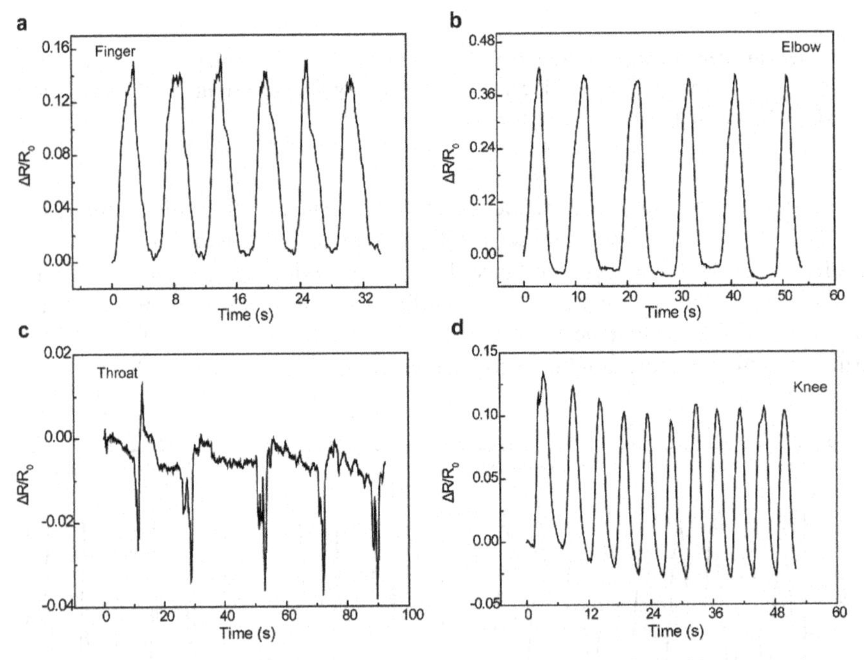

Fig. 5. Application of PEDOT:PSS-PVA hydrogel strain sensor in the field of human motion monitoring. a) Resistance change response when the finger is bent. b) Resistance change response when the elbow is bent. c) Resistance change response of the larynx during swallowing. d) Resistance change response when the leg is bent.

4 Conclusion

In summary, a stretchable, conductive, and strain-sensitive conductive polymer hydrogel strain sensor is fabricated using a facile method of ice-template-immersion synergy. The ice template constructed the honeycomb nanofiber network inside the hydrogel, which was then immersed in sodium citrate solution to make it gel. During the soaking process, PEDOT:PSS phase separation and strong aggregation and crystallization of PVA were induced. In addition, the synergistic interaction of physical crosslinking and hydrogen bonding can effectively dissipate energy, endow the hydrogel with stability and fatigue resistance. Due to its excellent electrical properties, the assembled hydrogel strain sensor exhibits different resistance responses under different strains, and can detect both large-scale human motion and subtle physiological signals, which can be applied to human motion monitoring and has a wide range of applications in wearable electronic devices.

Funding. This work was supported by the National Natural Science Foundation of China (51963011), Technological Expertise and Academic Leaders Training Program of Jiangxi Province (20194BCJ22013), Training Program of Natural Science Foundation of China Youth Fund (20202ZDB01007), Jiangxi Provincial Double Thousand Talents Plan-Youth Program (JXSQ2019201108), Natural Science Foundation of Jiangxi Province (20202ACB214001), Jiangxi Key Laboratory of Flexible Electronics (20212BCD42004), Research Project of State Key Laboratory of Mechanical System and Vibration (MSV202013), the Science and Technology Project of the Education Department of Jiangxi Province (GJJ201101), and Jiangxi Science & Technology Normal University (2020XJZD006). Z. Z. thank Jiangxi Science & Technology Normal University for the Provincial Postgraduate Innovation Program grants (YC2021-S754).

References

1. Cui, C., Shao, C., Meng, L., Yang, J.: High-Strength, self-adhesive, and strain-sensitive chitosan/poly (acrylic acid) double-network nanocomposite hydrogels fabricated by salt-soaking strategy for flexible sensors. ACS Appl. Mater. Interfaces **11**, 39228–39237 (2019)
2. Liu, Q., Chen, J., Li, Y., Shi, G.: High-performance strain sensors with fish-scale-like graphene-sensing layers for full-range detection of human motions. ACS Nano **10**, 7901–7906 (2016)
3. Chou, H.H., Nguyen, A., Chortos, A., To, J.W., Lu, C., Mei, J., et al.: A chameleon-inspired stretchable electronic skin with interactive colour changing controlled by tactile sensing. Nat. Commun. **6**, 8011 (2015)
4. Yuk, H., Lu, B., Zhao, X.: Hydrogel bioelectronics. Chem. Soc. Rev. **48**, 1642–1667 (2019)
5. Miao, L., Song, Y., Ren, Z., Xu, C., Wan, J., Wang, H., et al.: 3D temporary-magnetized soft robotic structures for enhanced energy harvesting. Adv. Mater. **33**, e2102691 (2021)
6. Soft Robotics based on Electroactive Polymers (2021)
7. Shahbazi, M.-A., Ghalkhani, M., Maleki, H.: Directional freeze-casting: A bioinspired method to assemble multifunctional aligned porous structures for advanced applications. Adv. Eng. Mater. **22**, 2000033 (2020)
8. Liang, X., Chen, G., Lin, S., Zhang, J., Wang, L., Zhang, P., et al.: Anisotropically fatigue-resistant hydrogels. Adv. Mater. **33**, e2102011 (2021)
9. Jungwirth, P., Cremer, P.S.: Beyond Hofmeister. Nat. Chem. **6**, 261–263 (2014)
10. Wu, S., Hua, M., Alsaid, Y., Du, Y., Ma, Y., Zhao, Y., et al.: Poly (vinyl alcohol) hydrogels with broad-range tunable mechanical properties via the Hofmeister effect. Adv. Mater. **33**, e2007829 (2021)

11. Zhao, Q., Liu, J., Wu, Z., Xu, X., Ma, H., Hou, J., et al.: Robust PEDOT:PSS-based hydrogel for highly efficient interfacial solar water purification. Chem. Eng. J. **442**, 136284 (2022)
12. Yuk, H., Lu, B., Lin, S., Qu, K., Xu, J., Luo, J., et al.: 3D printing of conducting polymers. Nat. Commun. **11**, 1604 (2020)
13. Lu, B., Yuk, H., Lin, S., Jian, N., Qu, K., Xu, J., et al.: Pure PEDOT:PSS hydrogels. Nature Communications **10** (2019)
14. Wei, J., Xie, J., Zhang, P., Zou, Z., Ping, H., Wang, W., et al.: Bioinspired 3D printable, self-healable, and stretchable hydrogels with multiple conductivities for skin-like wearable strain sensors. ACS Appl. Mater. Interfaces **13**, 2952–2960 (2021)

A Virtual Force Sensor for Robotic Manipulators Based on Dynamic Model

Yanjiang Huang[1,2], Jianhong Ke[1,2], and Xianmin Zhang[1,2(✉)]

[1] School of Mechanical and Automotive Engineering, South China University of Technology, Guangzhou 510640, China
zhangxm@scut.edu.cn

[2] Guangdong Province Key Laboratory of Precision Equipment and Manufacturing Technology, South China University of Technology, Guangzhou 510640, Guangdong, People's Republic of China

Abstract. In human-robot interactions, a force sensor should be equipped in the robot to detect the force or torque to guarantee the safety of operators. However, mounting force sensors will increase the manufacturing cost. In this paper, a virtual force sensor that only requires motion information is designed to deal with this problem. Firstly, parameter identification is performed to obtain the dynamic model of the robot. After that, a virtual force sensor based on generalized momentum is designed to estimate the disturbance force or torque. Experiments are performed in a UR5 robot in the simulation software Coppeliasim, and the results validate that the virtual force sensor can estimate the force or torque in joint space.

Keywords: Parameter identification · Force estimation · Virtual force sensor

1 Introduction

With the advancement of robotics, robot manipulators play a crucial role in manufacturing industry. In human-robot interaction, unforeseen contact can not be totally avoided, resulting in damage to operators and robots. Therefore, estimating the force or torque induced by interaction with the environment is necessary. There are several approaches to estimate the external force or torque. One way is to mount a force or torque sensor in the end-effector of the robot to measure the contact force [1]. Another way is to equip a joint torque sensor in the robot. The KUKA LWR robot [2] and the Franka Emika Panda [3] mount joint torque sensors in each joint. Moreover, artificial skin [4] can be wrapped in the robot to detect force and torque. However, all these methods mentioned above will complicate the structure design and increase manufacturing costs. Therefore, a sensorless method is proposed to deal with this problem. This method calculates the difference between measured torque due to the actual movement of the robot

H. Liu et al. (Eds.): ICIRA 2022, LNAI 13456, pp. 315–324, 2022.
https://doi.org/10.1007/978-3-031-13822-5_28

and the theoretical torque computed with the dynamic model of robot [5]. Reference [6] proposed a nonlinear disturbance observer (NDO) for sensorless torque control. However, NDO requires computing the inverse of the inertia matrix, which might result in numerical instability. Reference [7,8] combined the filtered dynamic equations with the dynamic model of the robot to estimate the force or torque. The only parameter to tune is the cutoff frequency of the filter. The main contribution of this paper is designing a visual force sensor based on the generalized momentum, which can be used online to estimate the external force or torque.

The paper is structured as follows. Section 2 gives a detailed problem formulation of designing a virtual force sensor for robot manipulators. Section 3 shows the experiments and analysis of the virtual force sensor. Moreover, Sect. 4 presents the conclusion and future work.

2 Problem Formulation

2.1 Parameter Identification

Generally, the dynamic model of a serial industrial robot can be written as:

$$\boldsymbol{M}(\boldsymbol{q})\ddot{\boldsymbol{q}} + \boldsymbol{C}(\boldsymbol{q},\dot{\boldsymbol{q}})\dot{\boldsymbol{q}} + \boldsymbol{G}(\boldsymbol{q}) + \boldsymbol{F}_r(\dot{\boldsymbol{q}}) = \boldsymbol{\tau} + \boldsymbol{\tau}_{ext} \tag{1}$$

where $\boldsymbol{q}$, $\dot{\boldsymbol{q}}$ and $\ddot{\boldsymbol{q}}$ are the joint position, joint velocity and joint acceleration, respectively. $\boldsymbol{M}(\boldsymbol{q})$ and $\boldsymbol{C}(\boldsymbol{q},\dot{\boldsymbol{q}})$ are the inertia matrix and the Coriolis matrix. $\boldsymbol{G}(\boldsymbol{q})$ and $\boldsymbol{F}_r(\dot{\boldsymbol{q}})$ are the gravity and the friction torque of joint. $\boldsymbol{\tau}$ is the actual torque of joint, and $\boldsymbol{\tau}_{ext}$ is torque induced by the interaction with environment. If there is no contract or collision with the environment, the $\boldsymbol{\tau}_{ext}$ should be equal to $\boldsymbol{0}$. Typically, the friction torque $\boldsymbol{F}_r(\dot{\boldsymbol{q}})$ can be formulated with Coulomb friction and viscous friction as:

$$\boldsymbol{F}_r(\dot{\boldsymbol{q}}) = \boldsymbol{F}_c sign(\dot{\boldsymbol{q}}) + \boldsymbol{F}_v \dot{\boldsymbol{q}} \tag{2}$$

where $\boldsymbol{F}_c$ and $\boldsymbol{F}_v$ are the Coulomb and viscous coefficient, respectively.

One way to obtain the dynamic model of the robot is using the computer aided design (CAD) technique. Another way that is more accurate is to perform parameter identification in the robot. In this paper, the second approach is adopted to obtain the dynamic model of the robot. In parameter identification, the robot performs the designed trajectory in free space, which means $\boldsymbol{\tau}_{ext} = \boldsymbol{0}$. Moreover, dynamic model (1) should be rewritten into linear regression form as:

$$\boldsymbol{Y}(\boldsymbol{q},\dot{\boldsymbol{q}},\ddot{\boldsymbol{q}})\boldsymbol{\pi} = \boldsymbol{\tau} \tag{3}$$

where $\boldsymbol{Y}(\boldsymbol{q},\dot{\boldsymbol{q}},\ddot{\boldsymbol{q}})$ is the regression matrix, which can be computed with the help of OpenSYMORO [9]. And $\boldsymbol{\pi}$ is the base parameter vector to be identified.

For an accurate solution of the linear regression formulation (3), an excitation trajectory should be well designed to excite the robot. A fifth polynomial trajectory [10] or a finite Fourier series-based trajectory [11] is usually used to stimulate the robot system. This paper adopts the Fourier series-based trajectory to excite the robot system. For the i-th joint, the excitation trajectory in joint space can be defined as:

$$\begin{aligned} q_i(t) &= q_{i0} + \sum_{k=1}^{5}(\frac{a_k}{\omega_f k}\sin(\omega_f kt) - \frac{b_k}{\omega_f k}\cos(\omega_f lk)) \\ \dot{q}_i(t) &= \sum_{k=1}^{5}(a_k \cos(\omega_f kt) + b_k \sin(\omega_f kt)) \\ \ddot{q}_i(t) &= \omega_f \sum_{k=1}^{5}(-a_k k \sin(\omega_f kt) + b_k k \cos(\omega_f kt)) \end{aligned} \tag{4}$$

where ω_f is the base angular frequency, q_{i0} is the initial position of the i-th joint, and a_k, b_k are the coefficients. The coefficients a_k and b_k can be obtained by solving the optimization problem as:

$$\begin{aligned} &\min cond(\boldsymbol{Y}(\boldsymbol{q}, \dot{\boldsymbol{q}}, \ddot{\boldsymbol{q}})) \\ &s.t. : \boldsymbol{q}_{min} \leq \boldsymbol{q} \leq \boldsymbol{q}_{max} \\ &\quad\quad \dot{\boldsymbol{q}}_{min} \leq \dot{\boldsymbol{q}} \leq \dot{\boldsymbol{q}}_{max} \\ &\quad\quad \ddot{\boldsymbol{q}}_{min} \leq \ddot{\boldsymbol{q}} \leq \ddot{\boldsymbol{q}}_{max} \end{aligned} \tag{5}$$

where $cond(\boldsymbol{Y}(\boldsymbol{q}, \dot{\boldsymbol{q}}, \ddot{\boldsymbol{q}}))$ means the condition number of the regression matrix, $\boldsymbol{q}_{min}$, $\boldsymbol{q}_{max}$, $\dot{\boldsymbol{q}}_{min}$, $\dot{\boldsymbol{q}}_{max}$, $\ddot{\boldsymbol{q}}_{min}$ and $\ddot{\boldsymbol{q}}_{max}$ are the limitation of joint positions, velocities and accelerations, respectively.

After performing the excitation trajectory in the robot, the motion information can be obtained, which contains joint position $\boldsymbol{q}$, joint velocity $\dot{\boldsymbol{q}}$, and joint torque $\boldsymbol{\tau}$. Moreover, the joint acceleration $\ddot{\boldsymbol{q}}$ can be computed based on $\dot{\boldsymbol{q}}$. Therefore, the linear regression problem (3) can be solved by least square (LS) [12] as follows:

$$\boldsymbol{\pi} = (\boldsymbol{Y}^T\boldsymbol{Y})^{-1}\boldsymbol{Y}^T\boldsymbol{\tau} \tag{6}$$

It is noted that the joint position $\boldsymbol{q}$, joint velocity $\dot{\boldsymbol{q}}$, joint acceleration $\ddot{\boldsymbol{q}}$, and joint $\boldsymbol{\tau}$ are required for parameter identification. However, the joint acceleration $\ddot{\boldsymbol{q}}$ can not be obtained directly for most robot manipulators. Therefore, the identification process is performed offline.

2.2 Design of Virtual Force Sensor

Once the base parameter set $\boldsymbol{\pi}$ is obtained, it is possible to estimate the external torque $\boldsymbol{\tau}_{ext}$ caused by the environment. In this paper, the virtual force sensor is designed based on the generalized momentum of the robot:

$$\boldsymbol{p} = \boldsymbol{M}\dot{\boldsymbol{q}} \tag{7}$$

Therefore, the time derivative of (7) is:

$$\dot{\boldsymbol{p}} = \boldsymbol{M}\ddot{\boldsymbol{q}} + \dot{\boldsymbol{M}}\dot{\boldsymbol{q}} \tag{8}$$

According to the skew-symmetry of matrix $\dot{\boldsymbol{M}} - 2\boldsymbol{C}$, it follows that:

$$\dot{\boldsymbol{M}} = \boldsymbol{C} + \boldsymbol{C}^T \tag{9}$$

Substitute (1) and (9) into (8):

$$\dot{\boldsymbol{p}} = \boldsymbol{\tau} + \boldsymbol{C}^T\dot{\boldsymbol{q}} - \boldsymbol{G} - \boldsymbol{F}_r + \boldsymbol{\tau}_{ext} \tag{10}$$

From [13], the disturbance estimation can be defined as:

$$\dot{\hat{\boldsymbol{\tau}}}_{ext} = \boldsymbol{K}(\dot{\boldsymbol{p}} - \dot{\hat{\boldsymbol{p}}}) \tag{11}$$

where $\boldsymbol{K}$ is the diagonal gain matrix.

Finally, by integrating both parts of (11), the virtual force sensor based on generalized momentum has the following form:

$$\hat{\boldsymbol{\tau}}_{ext} = \boldsymbol{K}\left(\boldsymbol{p} - \boldsymbol{p}(0) - \int_0^t (\boldsymbol{\tau} + \boldsymbol{C}^T\dot{\boldsymbol{q}} - \boldsymbol{G} - \boldsymbol{F}_r + \hat{\boldsymbol{\tau}}_{ext})ds\right) \tag{12}$$

where $\boldsymbol{p}(0)$ is the initial value of the generalized momentum.

In order to use the virtual force sensor (12) to estimate the force or torque induced by the environment, a numerical approximation method should be adopted. Based on the Newton-Euler iteration method, we have the following equation:

$$RNEA(\boldsymbol{q}, \dot{\boldsymbol{q}}, \ddot{\boldsymbol{q}}) = \boldsymbol{M}(\boldsymbol{q})\ddot{\boldsymbol{q}} + \boldsymbol{C}(\boldsymbol{q}, \dot{\boldsymbol{q}})\dot{\boldsymbol{q}} + \boldsymbol{G}(\boldsymbol{q}) \tag{13}$$

where $RNEA(\boldsymbol{q}, \dot{\boldsymbol{q}}, \ddot{\boldsymbol{q}})$ can be easily obtained after performing parameter identification. Then, numerical procedure can be followed:

$$\begin{aligned} \boldsymbol{G}(\boldsymbol{q}) &= RNEA(\boldsymbol{q}, \boldsymbol{0}, \boldsymbol{0}) \\ \boldsymbol{P} &= RNEA(\boldsymbol{q}, \boldsymbol{0}, \dot{\boldsymbol{q}}) - \boldsymbol{G}(\boldsymbol{q}) \\ \boldsymbol{C}(\boldsymbol{q}, \dot{\boldsymbol{q}})\dot{\boldsymbol{q}} &= RNEA(\boldsymbol{q}, \dot{\boldsymbol{q}}, \boldsymbol{0}) - \boldsymbol{G}(\boldsymbol{q}) \\ \boldsymbol{M}_i(\boldsymbol{q}) &= RNEA(\boldsymbol{q}, \boldsymbol{0}, \boldsymbol{e}_i) - \boldsymbol{G}(\boldsymbol{q}) \end{aligned} \tag{14}$$

where $\boldsymbol{M}_i$ means the i-th column of $\boldsymbol{M}$, and $\boldsymbol{e}_i$ is a normalized vector whose i-th element is 1. Then, the derivative of $\boldsymbol{M}$ can be computed as:

$$\dot{\boldsymbol{M}}(\boldsymbol{q}_t) = (\boldsymbol{M}(\boldsymbol{q}_t) - \boldsymbol{M}(\boldsymbol{q}_{t-\triangle t}))/\triangle t \tag{15}$$

where $\triangle t$ is the sampling period. Then, $\boldsymbol{C}^T$ can be obtained according to (9), (14) and (15). If the sampling period $\triangle t$ is small enough, the trapezoidal rule is sufficient to calculate the integration item. Therefore, all elements in the virtual force sensor (12) can be computed, and the virtual force sensor can be established to estimate the force or torque induced by the environment.

It should be noted that the virtual force sensor based on generalized momentum only need the joint position $\boldsymbol{q}$, joint velocity $\dot{\boldsymbol{q}}$ and joint $\boldsymbol{\tau}$ according to (12), (14) and (15). The joint acceleration $\ddot{\boldsymbol{q}}$ is not necessary for the virtual force sensor. Because $\boldsymbol{q}$, $\dot{\boldsymbol{q}}$, and $\boldsymbol{\tau}$ can be accessed directly when the robot is executing, the virtual force sensor can be used online to estimate the external force or torque.

3 Experiment and Analysis

3.1 Experiment Setup

In this part, two kinds of experiments are performed in a UR5 robot in CoppeliaSim with the release version of 4.2.0 [14], which is a versatile and ideal robotics simulator software. Figure 1 shows the whole simulation framework. CoppeliaSim is the server-side where a UR5 robot can perform the experiment. Furthermore, Visual Studio Code is the client-side to control the robot via C++. The server-side communicates with the client-side via $B0-based\ Remote\ API$ provided by CoppeliaSim. Moreover, CoppeliaSim provides four kinds of physical engines for robot dynamic simulation, and the $Bullet2.83$ physical engine is chosen in our simulation to perform the experiment. The simulation time step is set to 10ms. It is worth mentioning that due to the property of $Bullet2.83$, there is no friction torque $\boldsymbol{F}_r = \boldsymbol{0}$ in the simulation experiment.

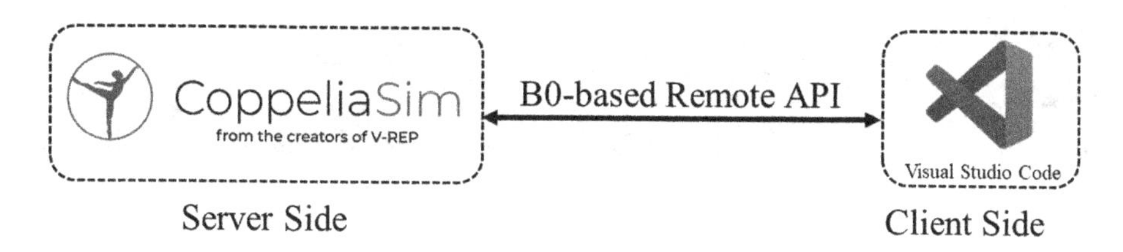

Fig. 1. Simulation framework

3.2 Parameter Identification of UR5

Table 1 is the modified Denavit-Hartenbegr (m-DH) parameters of UR5, which are used to establish the kinematic and dynamic model of UR5. Firstly, a fifth Fourier series-based trajectory is designed to excite the robot system. Here, in our experiment, the base angular frequency is $\omega_f = 0.1\pi$, and the period of trajectory is the 20s. The initial pose $\boldsymbol{q}(0) = [0, -\frac{\pi}{2}, 0, -\frac{\pi}{2}, 0, 0]^T$ is the same with the Table 1. Then the Matlab function $fmincon$ is used to solve the optimization problem (5). Therefore, the trajectory can be used as the input in CoppeliaSim to excite the UR5 robot. Figure 2 is the visualization of the trajectory of the end-effector when UR5 performs the designed trajectory.

As mentioned earlier, the period is the 20 s, and the simulation step time is 10ms. Therefore, 2000 groups of motion information that include joint position $\boldsymbol{q}$, joint velocity $\dot{\boldsymbol{q}}$, and joint torque $\boldsymbol{\tau}$ can be collected. Then, the joint acceleration $\ddot{\boldsymbol{q}}$ can be calculated from joint velocity $\dot{\boldsymbol{q}}$ based on the central difference algorithm. Therefore, the linear regression problem (6) would be solved offline with the collected motion information.

Figure 3 is the comparison between the measured torque and the estimated torque. The measured torque is directly read from CoppeliaSim, and the estimated torque is calculated with (3). It is seen that the estimated torque displays

Table 1. Modified DH parameter of UR5

Link	α (rad)	a (mm)	d (mm)	q (rad)
1	0	0	89.16	$q_1(0)$
2	$\frac{\pi}{2}$	0	0	$q_2(-\frac{\pi}{2})$
3	0	−425	0	$q_3(0)$
4	0	−392.25	109.15	$q_4(-\frac{\pi}{2})$
5	$\frac{\pi}{2}$	0	94.65	$q_5(0)$
6	$-\frac{\pi}{2}$	0	82.30	$q_6(0)$

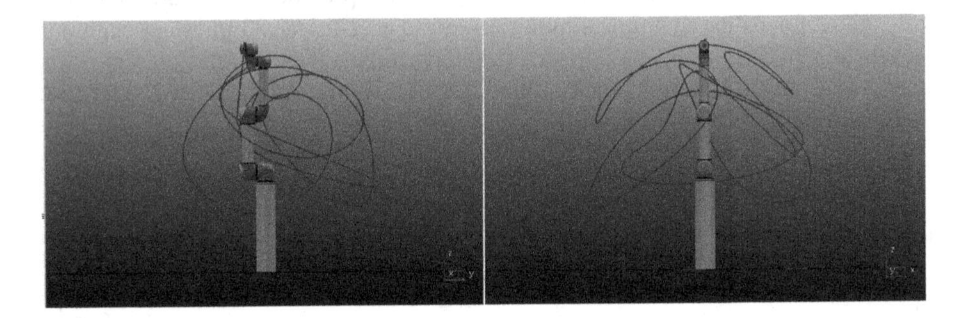

Fig. 2. Visualization of the excitation trajectory of end-effector of UR5.

the same trend as the measured torque except for the last joint, which might be caused by the small measured torque of the last joint. In order to validate the identification result, two different trajectories are used for cross-validation, and the root mean squared error (RMSE) is computed as the evaluation index. Table 2 shows the RMSE of three trajectories. It can be seen that for all trajectories, the same joint has a similar RMSE value, which means the identification result is robust enough.

Table 2. RMSE of three trajectories

	joint1	joint2	joint3	joint4	joint5	joint6
Identification trajectory	0.3172	1.8725	0.8276	0.2081	0.1476	0.1891
Validation trajectory 1	0.2959	1.4059	0.6468	0.2294	0.1717	0.1431
Validation trajectory 2	0.2366	1.5172	0.6852	0.1385	0.0557	0.0994

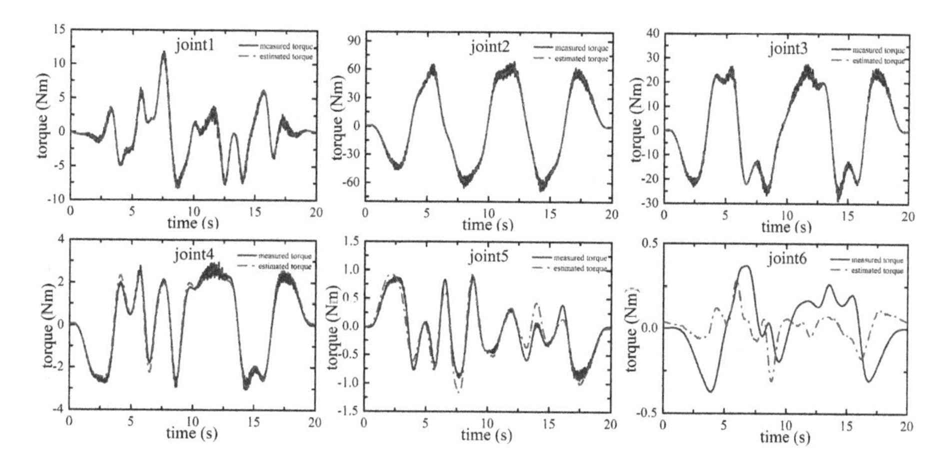

Fig. 3. Comparison between the measured torque and the estimated torque. Blue solid line and the dot dash line donate the measured torque and the estimated torque, respectively. (Color figure online)

3.3 Experiments and Discussion of Virtual Force Sensor of UR5

For the virtual force sensor, the gain matrix $\boldsymbol{K}$ should be tuned in order to obtain a good performance. Because the gain matrix $\boldsymbol{K}$ is a diagonal matrix, the gain element of each joint can be tuned separately. It should be noted that the larger the gain is, the faster the virtual force sensor will converge to estimate the external force or torque. However, a large gain value will amplify the noise in estimation. In this paper, $\boldsymbol{K} = diag(10, 5, 9, 18, 17, 15)$, and the sampling period $\triangle t = 10\,\text{ms}$, which is consistent with the simulation step time.

In the first experiment, the virtual force sensor is used to evaluate constant torque in each joint. For the joint1 to joint3, the disturbance torque is 5 Nm, and for the joint4 to joint6, the disturbance torque is 2 Nm. Figure 4 shows the estimation result of the constant disturbance torque. It can be seen that the virtual force sensor performs an accurate estimation of the disturbance torque. However, the noise amplification occurs in each joint, especially in joint2 to joint4, which might be caused by numerical calculation error.

To further verify the property of the virtual force sensor, a sinusoidal disturbance torque is applied in each joint. Figure 5 presents the evaluation result of the sinusoidal disturbance torque. Similar to the result in Fig. 4, the virtual force sensor can estimate time-varying external torque. Nevertheless, the phase lag phenomenon happens in almost every joint in the estimation. The same phenomenon also occurs in Fig. 4. It is because the virtual force sensor based on generalized momentum can be seen as a first-order low pass filter [7], which will result in phase lag in estimation.

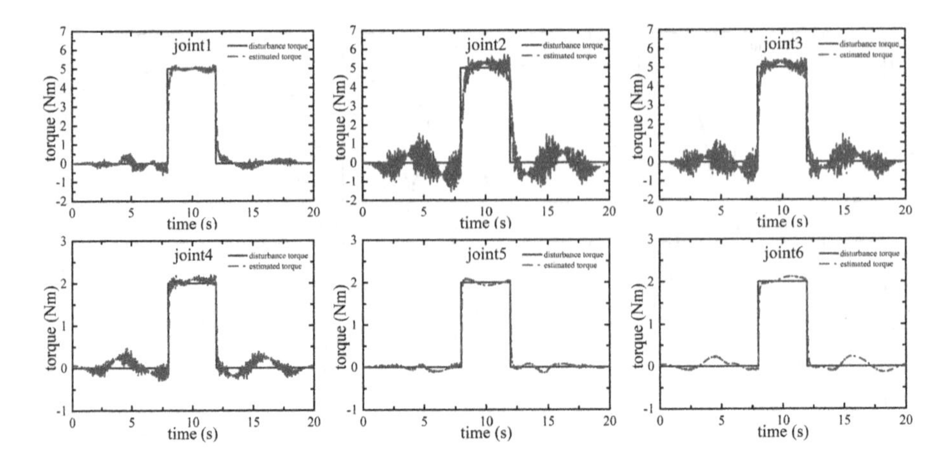

Fig. 4. Comparison between the constant disturbance torque and the estimated torque. Blue solid line and the red dot dash line donate the disturbance torque and the estimated torque, respectively. (Color figure online)

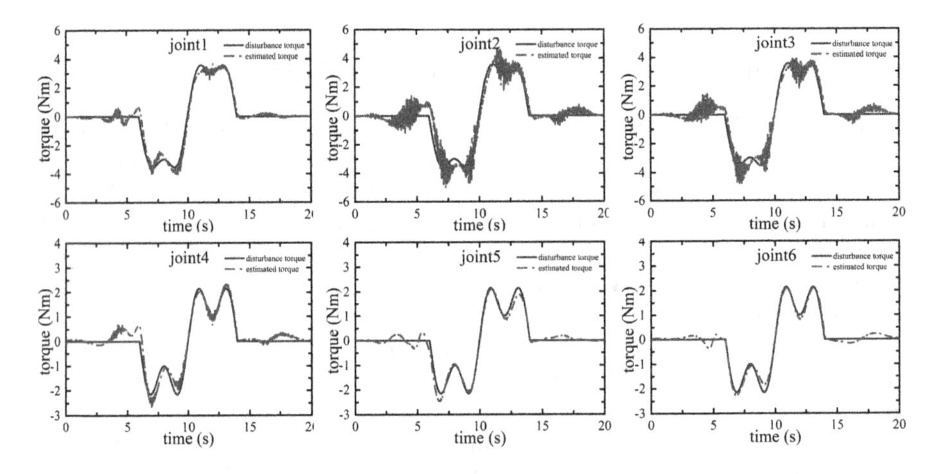

Fig. 5. Comparison between the sinusoidal disturbance torque and the estimated torque. Blue solid line and the red dot dash line donate the disturbance torque and the estimated torque, respectively. (Color figure online)

4 Conclusion and Future Work

This paper designs a virtual force sensor to estimate the disturbance torque in joint space. Firstly, parameter identification is performed offline in a UR5 robot to obtain the dynamic model. After that, a virtual force sensor based on generalized momentum is designed to estimate the disturbance force or torque induced by the interaction with the environment. Experiments are conducted in a UR5 robot in the simulation software CoppeliaSim. The result shows that the virtual force sensor based on the dynamic model and generalized momentum can evaluate both constant force and time-varying force. However, the experiments are performed only in simulation. Moreover, friction is not considered in the estimation method, which simplifies the estimation process. In the future, the virtual force sensor will be validated in a real UR5 robot, and each joint's friction should be considered.

Acknowledgment. This work was supported in part by the National Natural Science Foundation of China under Grant 52075178, in part by the Guangdong Basic and Applied Basic Research Foundation under Grant 2019A1515011154.

References

1. Buondonno, G., De Luca, A.: Combining real and virtual sensors for measuring interaction forces and moments acting on a robot. In: 2016 IEEE/RSJ International Conference on Intelligent Robots and Systems (IROS), pp. 794–800 (2016)
2. Magrini, E., Flacco, F., De Luca, A.: Control of generalized contact motion and force in physical human-robot interaction. In: 2015 IEEE International Conference on Robotics and Automation (ICRA), pp. 2298–2304 (2015)
3. Gaz, C., Cognetti, M., Oliva, A., Robuffo Giordano, P., De Luca, A.: Dynamic identification of the Franka Emika Panda robot with retrieval of feasible parameters using penalty-based optimization. IEEE Robot. Autom. Lett. **4**(4), 4147–4154 (2019)
4. Tiwana, M.I., Redmond, S.J., Lovell, N.H.: A review of tactile sensing technologies with applications in biomedical engineering. Sens. Actuators A **179**, 17–31 (2012)
5. Chen, S., Wang, J., Kazanzides, P.: Integration of a low-cost three-axis sensor for robot force control. In: 2018 2nd IEEE International Conference on Robotic Computing (IRC), pp. 246–249 (2018)
6. Chen, W.-H., Ballance, D.J., Gawthrop, P.J., O'Reilly, J.: A nonlinear disturbance observer for robotic manipulators. IEEE Trans. Ind. Electron. **47**(4), 932–938 (2000)
7. Van Damme, M., et al.: Estimating robot end-effector force from noisy actuator torque measurements. In: 2011 IEEE International Conference on Robotics and Automation, pp. 1108–1113 (2011)
8. Ho, C.N., Song, J.B.: Collision detection algorithm robust to model uncertainty. Int. J. Control Autom. Syst. **11**(4), 776–781 (2013)
9. Khalil, W., Vijayalingam, A., Khomutenko, B., Mukhanov, I., Lemoine, P., Ecorchard, G.: OpenSYMORO: an open-source software package for symbolic modelling of robots. In: 2014 IEEE/ASME International Conference on Advanced Intelligent Mechatronics (2014)
10. Atkeson, C.G., An, C.H., Hollerbach, J.M.: Estimation of inertial parameters of manipulator loads and links. Int. J. Robot. Res. **5**(3), 101–119 (1986)
11. Swevers, J., Verdonck, W., De Schutter, J.: Dynamic model identification for industrial robots. IEEE Control Syst. Mag. **27**(5), 58–71 (2007)
12. Gautier, M., Khalil, W.: Direct calculation of minimum set of inertial parameters of serial robots. IEEE Trans. Robot. Autom. **6**(3), 368–373 (1990)

13. De Luca, A., Mattone, R.: Actuator failure detection and isolation using generalized momenta. In: 2003 IEEE International Conference on Robotics and Automation (Cat. No.03CH37422), vol. 1, pp. 634–639 (2003)
14. Rohmer, E., Singh, S.P.N., Freese, M.: CoppeliaSim (formerly V-REP): a versatile and scalable robot simulation framework. In: Proceedings of the International Conference on Intelligent Robots and Systems (IROS) (2013). www.coppeliarobotics.com

A Variable Stiffness Soft Actuator with a Center Skeleton and Pin-Socket Jamming Layers

Rong Bian[1], Ningbin Zhang[1], Xinyu Yang[1], Zijian Qin[1], and Guoying Gu[1,2](✉)

[1] State Key Laboratory of Mechanical System and Vibration, School of Mechanical Engineering, Shanghai Jiao Tong University, Shanghai 200240, China
guguoying@sjtu.edu.cn

[2] Meta Robotics Institute, Shanghai Jiao Tong University, Shanghai 200240, China

Abstract. Soft actuators have benefits of compliance and adaptability compared to their rigid counterparts. However, most soft actuators have finite enough stiffness to exert large forces on objects. This article proposes a tendon-driven soft actuator with a center skeleton and pin-socket jamming layers, allowing variable stiffness under different input air pressures. Two tendons drive the actuator to bend. The center skeleton reduces the required layers (1.6 mm total thickness) for a wide range of variable stiffness. The pin-socket design restricts the unexpected movement of layers and broadens its maximum angle when unjammed. Then, we develop a prototype and experimentally characterize its variable stiffness. Finally, we make a gripper with three proposed actuators to demonstrate the validity of the development.

Keywords: Soft actuator · Variable stiffness · Layer jamming · Gripper

1 Introduction

Soft actuators have demonstrated tremendous potential in compliance and adaptability [1–5]. Compared with their rigid counterparts, soft actuators have the advantage to adapt to the environments passively, especially in unstructured environments [6, 7]. However, soft actuators normally exert limited forces and hardly resist external loads due to their body of low stiffness [8, 9], restricting their further application.

To enhance soft actuators' stiffness and output forces, scientists raise many variable stiffness mechanisms. Tunable stiffness mechanisms make soft actuators flexible and rigid in different conditions. The technology can be categorized by stimuli modes. For example, the stiffness can be promoted by (1) heat, such as low-melting-point alloy hardened by phase change and shape memory materials [10, 11]; (2) magnetism, such as magnetorheological fluid and elastomers whose yield stress increase through the activation of electro-magnet [12, 13]; (3) pressure, for example, a jamming structure which can be hardened by friction forces between parts (particles, fibers, layers, etc.) inside when pressure changes [14–17].

H. Liu et al. (Eds.): ICIRA 2022, LNAI 13456, pp. 325–332, 2022.
https://doi.org/10.1007/978-3-031-13822-5_29

Layer jamming is attracting increasing attention for its simpleness and stability [16–18]. Sealed with outer membranes, stacks of flexible layers are compressed by an external stimulus (such as a vacuum) in a closed system. Friction forces between layers hinder their slides and deformation, thus solidifying the whole structure. Based on the principle, scientists proposed variable stiffness flexure joints [17], soft grippers [19], shoes [20], orthoses [21], manipulator minimally invasive surgery [22], and variable compliance kangaroo tail [23].

There are some insights into a few weaknesses restricting the performance of soft actuators with layer jamming. More layers should be integrated into the actuator structure for higher stiffness when jammed [17, 24], reducing flexibility when unjammed and adding extra weight. Lucas Gerez [17] theoretically and experimentally verified that the maximum load applied to the middle of layer jamming structure is nearly proportional to the number of jamming layers. On the other hand, as actuators move and layers slide, some unexpected random behaviors happen and damage jamming performance [25]. The layers sometimes slide into unexpected orientation randomly when unjammed, which is detailed in Section Fabrication.

In this study, we design a novel actuator with a center skeleton and pin-socket layers which are driven by tendons and hardened by layer jamming. Shortened and elongated tendons driven by a motor can drive the bending of the center skeleton. The center skeleton increases compliance when unjammed and can be hardened by the jamming layers when jammed under a vacuum. The pin-shaped and socket-shaped layers are assembled in a pair so that their movements are restricted by each other, avoiding random unexpected behavior when unjammed. Experiments are conducted to characterize the variable stiffness of the actuator. We make a gripper with the proposed actuators and test its grasp performance.

The rest of the article is organized as follows: Sect. 2 exhibits the materials, Sect. 3 presents experimental methods, and Sect. 4 shows results.

2 Design and Fabrication

2.1 Design

The CAD model of the actuator is shown in Fig. 1. It is comprised of a center skeleton, flexible pin-socket layers, fixtures, two tendons, a motor, and an airtight membrane.

The center skeleton consists of a beam and a series of branches as shown in Fig. 2(a). The skeleton made up of TPU can easily bend. The flexible pin-socket layers are used for jamming. The arrangement is demonstrated in Fig. 2(b). A pin-shaped layer and a socket-shaped layer overlap and are pasted to each side of the skeleton. The pin's head is penetrated through the socket's hole restricting the moving direction, which is named a pin-socket pair. The membrane encases the center skeleton and layers. Because the membrane is soft and airtight, it can be pressed on the layers and create positive pressure between layers when a vacuum is applied inside.

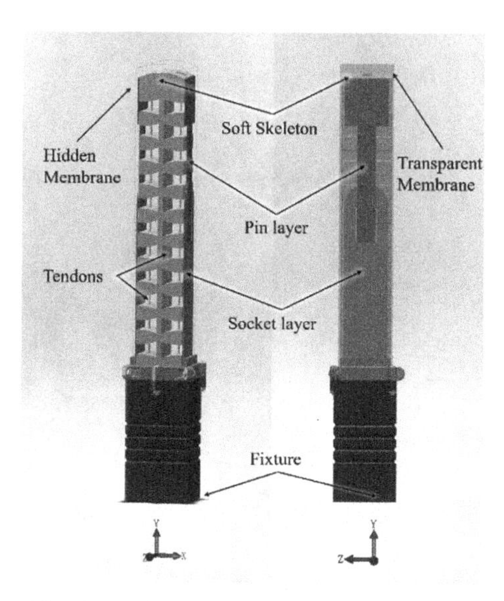

Fig. 1. CAD model of the proposed actuator.

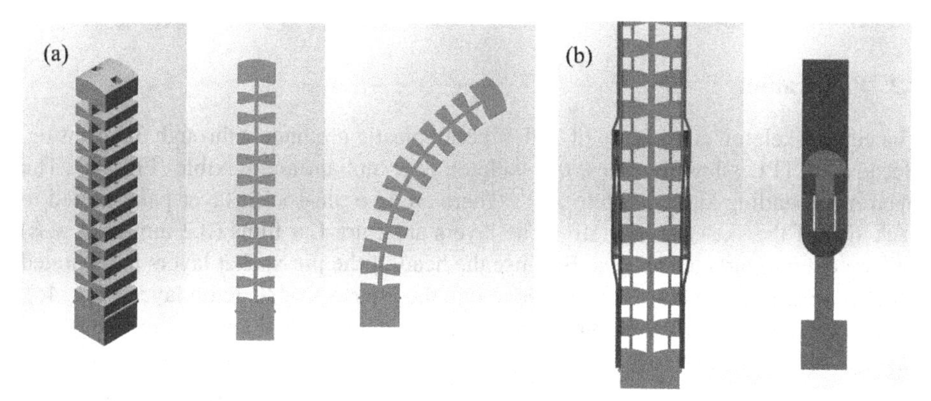

Fig. 2. (a) Axis view of center skeleton, side view of center skeleton and bending skeleton. (b) Arrangement of layers and pin-socket pair.

The bending is activated by tendons and a motor. As shown in Fig. 3, two tendons drive the skeleton with one end fixed to the top of the skeleton and one end fixed to the motor horn. The motor is glued on the fixtures which seal the open mouth of the membrane to ensure airtightness.

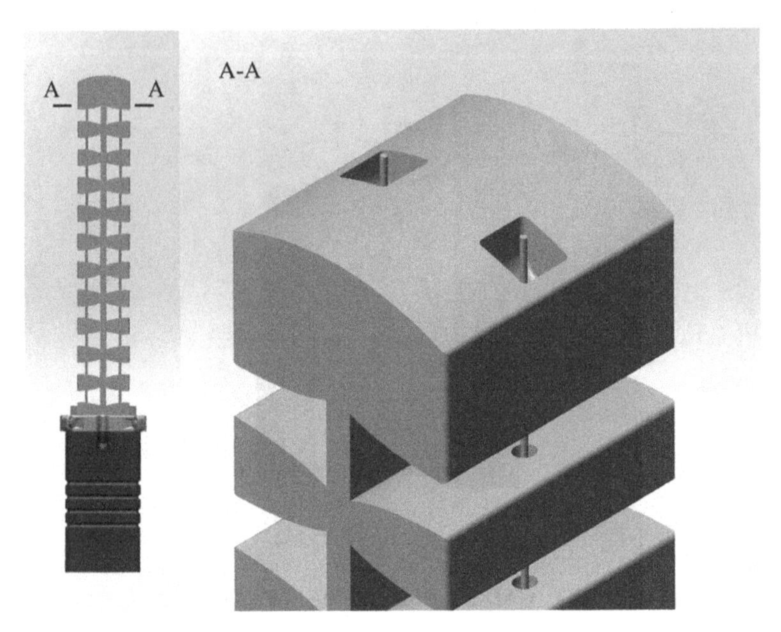

Fig. 3. Layout of tendons

2.2 Fabrication

The center skeleton is made up of TPU (thermoplastic urethanes) through 3D printing. Because of TPU's low modules, the skeleton is compliant and flexible (Fig. 4a). The maximum bending angle is up to 270°. There are two pin-socket layer pairs pasted to each side of the skeleton (Fig. 4b). The layers are Dura-Lar films (0.2 mm thickness) and are cut into targeted shapes. Because the head of the pin-socket layers is separated from the skeleton, the layers do not move into the intervals of skeleton layers (Fig. 4c), which is an unexpected random movement seen in conventional layers (Fig. 4d).

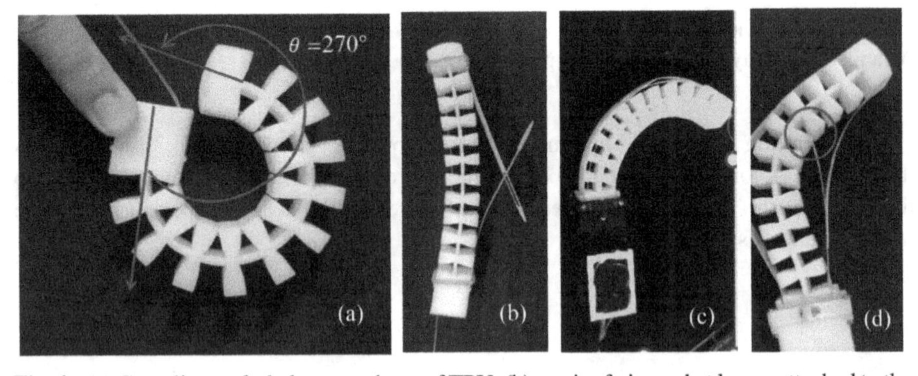

Fig. 4. (a) Compliant soft skeleton made up of TPU, (b) a pair of pin-socket layers attached to the skeleton, (c) the bending actuator and its moving pin-socket layers, and (d) an unexpected random movement of layers without the pin-socket design.

A latex membrane with 0.5 mm thickness encases the skeleton. The open side of the membrane is connected to the fixture tightly. The fixtures are 100%-infill-density resin to ensure airtightness and immobilize the motor (MKS-HV75K) inside. An air tube is glued to the bottom of the fixtures to provide the vacuum.

3 Experimental Methods

Experiments are conducted to assess the variable stiffness and actuator's compliance. We characterize the bending stiffness of the actuator under different pressures. We build a platform to actuate the actuator and measure its deformation as shown in Fig. 5a. Moreover, to exhibit its performance, we make a three-finger gripper with the actuators and test its grasp ability as shown in Fig. 5b.

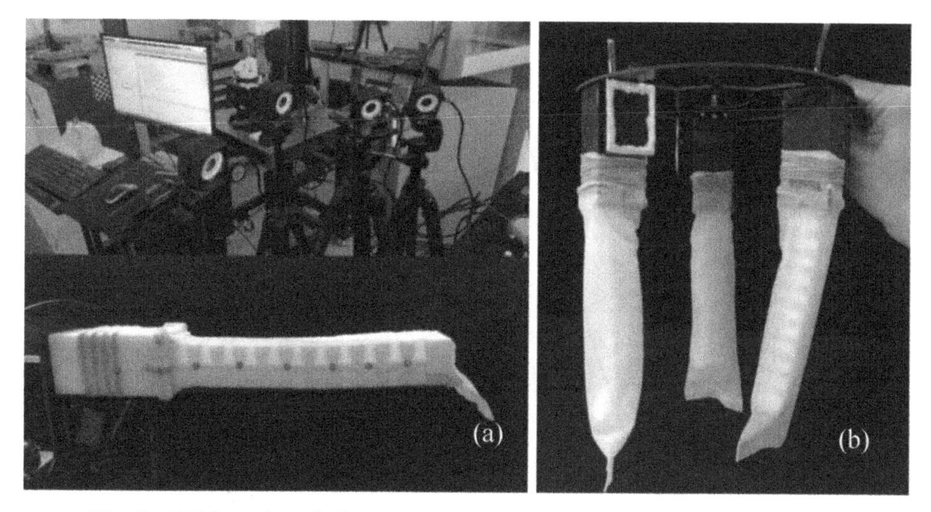

Fig. 5. (a) Measuring platform. (b) gripper composed of the proposed actuators.

To characterize the relationship between the deformation of the actuator and the force it bears, the platform includes a visual measurement system (Motive cameras) and a holder that fixes the actuator. Motive cameras detect the position of markers pre-pasted to the actuator and increasing weights are put on the tip of the actuator.

The air pressure in the actuator is controlled by the opening time of the valve with the feedback of the pressure sensor. The air is extracted by an air pump (370-C, Willitz Inc.) and the pressure is synchronously measured by the thin-film pressure sensor (XGZP6874). The valve (X-Valve, Parker Inc.) closes when the pressure reaches its target.

Dexterous grasp experiments are conducted to verify the gripper is dexterous. In the dexterous grasp experiments, daily living objects with different shapes and rigidity are passed on to the gripper. The grip fingers bend, contact the objects and apply forces to fix their relative positions. The grasp experiment is repeated three times and the holding time is longer than 3 s.

4 Results

Figure 6 shows the results of deformation-force curves under different vacuum pressure. The different vacuum pressure (5 kPa, 10 kPa, 15 kPa, 20 kPa) is applied to the actuator. It is noticed that the jammed actuator's force-displacement curve is approximately linear under low force and gets nonlinear when the force increases. The actuator under 20 kPa vacuum pressure exhibit the largest linear region, which means larger forces it can withstand.

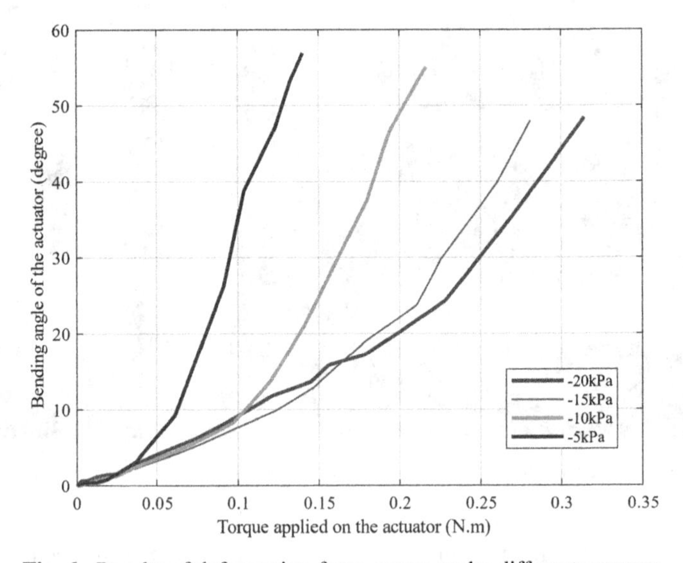

Fig. 6. Results of deformation-force curves under different pressure.

Figure 7 shows the results of grasp experiments of the three-finger gripper made up of the variable stiffness actuators. Objects with different shapes and rigidity are grasped successfully by the compliant gripper due to the actuator's compliance.

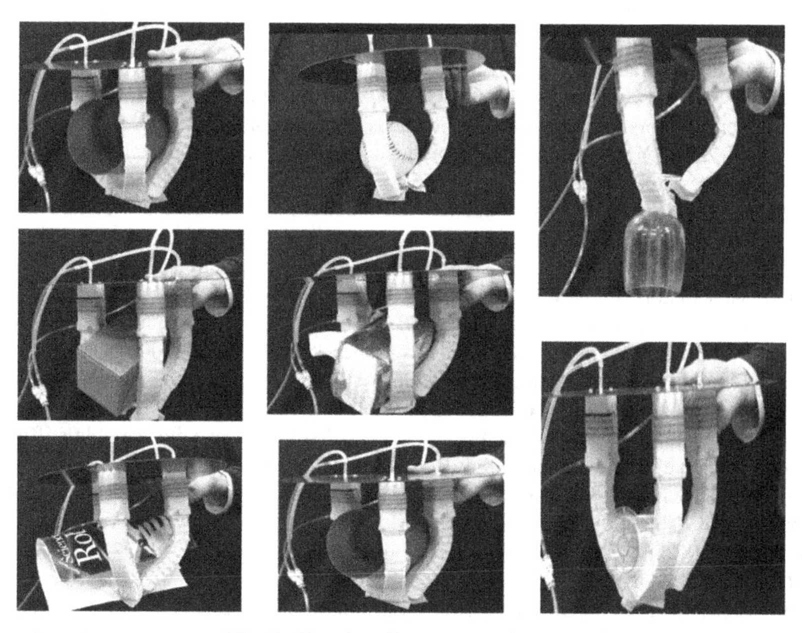

Fig. 7. Results of grasp experiments.

5 Conclusions

This study proposes a soft variable stiffness actuator driven by tendons based on layer jamming. We detail the design and fabrication of the actuator. The soft center skeleton increases the actuator's compliance and can be hardened by the pin-socket layers under vacuum pressure. The pin-socket design restricts the layer's unexpected random behaviors and broadens the maximum bending angle of the actuator. Experiments are conducted to assess the actuator bending stiffness and potential application.

Although the proposed actuator exhibits variable stiffness and compliance, its movement is limited to one direction. Regarding future directions, we may design a new skeleton with multiple degrees of freedom and a matched variable stiffness structure.

Acknowledgements. This work was supported in part by the the State Key Laboratory of Mechanical Transmissions (SKLMT-ZDKFKT-202004) and the Science and Technology Commission of Shanghai Municipality (20550712100).

References

1. Leddy, M.T., Dollar, A.M.: Preliminary design and evaluation of a single-actuator anthropomorphic prosthetic hand with multiple distinct grasp types. In: Proceedings of the 7th IEEE International Conference on Biomedical Robotics and Biomechatronics (Biorob), pp. 1062–1069. IEEE, Enschede (2018)
2. Zhang, Y.-F., et al.: Fast-response, stiffness-tunable soft actuator by hybrid multimaterial 3D printing. Adv. Funct. Mater. **29**(15), 1806698 (2019)

3. Gu, G., et al.: A soft neuroprosthetic hand providing simultaneous myoelectric control and tactile feedback. Nat. Biomed. Eng. (2021)
4. Zhang, N., Ge, L., Xu, H., Zhu, X., Gu, G.: 3D printed, modularized rigid-flexible integrated soft finger actuators for anthropomorphic hands. Sens. Actuators, A **312**, 112090 (2020)
5. Zhang, N., Zhao, Y., Gu, G., Zhu, X.: Synergistic control of soft robotic hands for human-like grasp postures. Sci. China-Technol. Sci. **65**(3), 553–568 (2022)
6. Zou, M., et al.: Progresses in tensile, torsional, and multifunctional soft actuators. Adv. Func. Mater. **31**(39), 2007437 (2021)
7. Rupert, L., Saunders, B.O., Killpack, M.D.: Performance metrics for fluidic soft robot rotational actuators. Front. Robot. Ai **8**, 249 (2021)
8. Dou, W., Zhong, G., Cao, J., Shi, Z., Peng, B., Jiang, L.: Soft robotic manipulators: Designs, actuation, stiffness tuning, and sensing. Adv. Mater. Technol. **6**(9), 2100018 (2021)
9. Long, F., et al.: Latest advances in development of smart phase change material for soft actuators. Adv. Eng. Mater. **24**(3), 2100863 (2022)
10. Wang, W., Ahn, S.-H.: Shape memory alloy-based soft gripper with variable stiffness for compliant and effective grasping. Soft Rob. **4**(4), 379–389 (2017)
11. Shintake, J., Schubert, B., Rosset, S., Shea, H., Floreano, D.: Variable stiffness actuator for soft robotics using dielectric elastomer and low-melting-point alloy. In: Proceedings of the International Conference on Intelligent Robots and Systems, pp.1097–1102, IEEE, Hamburg (2015)
12. Pettersson, A., Davis, S., Gray, J.O., Dodd, T.J., Ohlsson, T.: Design of a magnetorheological robot gripper for handling of delicate food products with varying shapes. J. Food Eng. **98**(3), 332–338 (2010)
13. Diez, A.G., Tubio, C.R., Etxebarria, J.G., Lanceros-Mendez, S.: Magnetorheological Elastomer-Based Materials and Devices: State of the Art and Future Perspectives. Adv. Eng. Mater. **23**(6), 2100240 (2021)
14. Behringer, R.P., Chakraborty, B.: The physics of jamming for granular materials: a review. Rep. Prog. Phys. **82**(1), 012601 (2019)
15. Brancadoro, M.: Fiber jamming transition as a stiffening mechanism for soft robotics. Soft Robot. **6**(7), 663–674 (2019)
16. Choi, W.H., Kim, S., Lee, D., Shin, D.: Soft, multi-DoF, variable stiffness mechanism using layer jamming for wearable robots. IEEE Robot. Autom. Lett. **4**(3), 2539–2546 (2019)
17. Gerez, L., Gao, G., Liarokapis, M.: Laminar jamming flexure joints for the development of variable stiffness robot grippers and hands. In: Proceedings of the 2020 IEEE/RSJ International Conference on Intelligent Robots and Systems (IROS). IEEE, Las Vegas (2020)
18. Narang, Y.S., Vlassak, J.J., Howe, R.D.: Mechanically versatile soft machines through laminar jamming. Adv. Func. Mater. **28**(17), 1707136 (2018)
19. Wang, T., Zhang, J., Li, Y., Hong, J., Wang, M.Y.: Electrostatic layer jamming variable stiffness for soft robotics. IEEE/ASME Trans. Mechatron. **24**(2), 424–433 (2019)
20. Arleo, L., Dalvit, M., Sacchi, M., Cianchetti, M.: Layer jamming for variable stiffness shoes. IEEE Robot. Autom. Lett. **7**(2), 4181–4187 (2022)
21. Ibrahimi, M., Paternò, L., Ricotti, L., Menciassi, A.: A layer jamming actuator for tunable stiffness and shape-changing devices. Soft Rob. **8**(1), 85–96 (2020)
22. Kim, Y.-J., Cheng, S., Kim, S., Iagnemma, K.: A novel layer jamming mechanism with tunable stiffness capability for minimally invasive surgery. IEEE Trans. Rob. **29**(4), 1031–1042 (2013)
23. Santiago, J.L.C., Godage, I.S., Gonthina, P., Walker, I.D.: Soft robots and kangaroo tails: Modulating compliance in continuum structures through mechanical layer jamming. Soft Rob. **3**(2), 54–63 (2016)
24. Caruso, F., Mantriota, G., Afferrante, L., Reina, G.: A theoretical model for multi-layer jamming systems. Mech. Mach. Theory **172**, 104788 (2022)
25. Gao, Y., Huang, X., Mann, I.S.: A novel variable stiffness compliant robotic gripper based on layer jamming. J. Mech. Robot. **12**(5), 051013 (2020)

A Parameter Optimization Method of 3D Printing Soft Materials for Soft Robots

Anqi Guo[1,2], Wei Zhang[1,2], Yin Zhang[1,2], Lining Sun[1,2], and Guoqing Jin[1,2](✉)

[1] School of Mechanical and Electrical Engineering, Suzhou University, Suzhou, China
gqjin@suda.edu.cn

[2] Robotics and Microsystem Research Centre, Soochow University, Suzhou, China

Abstract. With the increasingly complex structure design of soft robot, 3D printing of soft materials is one of the main research directions of soft robot manufacturing, which will give it new opportunities. Firstly, a soft material 3D printing platform compatible with embedded printing and external printing was built in this paper. Secondly, through the analysis and research of printing materials, pressure-flow rate simulation results of the different needles and the parameter experiment of printing influencing factors, the best combinations of printing parameters of functional magnetic material embedded printing and single component silica gel external printing were obtained respectively. Finally, based on this, bionic gecko magnetic drive soft robots and soft grippers were manufactured, and relevant performance experiments were carried out to verify the feasibility of integrated manufacturing of fully flexible soft robots using soft material 3D printing technology, highlighting the significant advantages of direct manufacturing and flexible model changing of 3D printing.

Keywords: 3D printing · Soft materials · Soft robot

1 Introduction

Soft robot is a new type of robot with bionic design based on soft creatures in nature [1–4], which has demonstrated great potential in areas as diverse as industrial production, disaster rescue, medical services and military exploration compared with traditional rigid robots because of its better safety and environmental adaptability [5–7]. The progress of soft material 3D printing technology provides a new opportunity for the integrated manufacturing of soft robots [8]. Furthermore, embedded 3D printing technology is an emerging technology, that can manufacture complex internal structures [9]. In the manufacturing process, complex cavity structures can be freely formed in the matrix, which solves the problem that low elastic modulus materials need additional support materials during the 3D printing process [10–13]. In recent years, researchers at home and abroad made some excellent achievements in manufacturing soft robots using 3D printing technology. In 2016, Truby, Ryan L's research group used embedded 3D printing to manufacture the full soft robot 'Octobot' [14]. The Federal University of Technology in Zurich, Switzerland, studied a bionic elephant trunk soft driver with a multi hardness

H. Liu et al. (Eds.): ICIRA 2022, LNAI 13456, pp. 333–345, 2022.
https://doi.org/10.1007/978-3-031-13822-5_30

gradient based on 3D printing [15]. JoM.P.Geraedts et al. used 3D printing technology to produce human-computer interaction soft hands [16]. P. Ei et al. produced a glove-like electronic skin combining a strain sensor and pressure sensor [17]. The embedded printing flexible sensor manufactured by RyanL. Truby et al. could realize real-time feedback of tactile and temperature sense, which further promoted the development of intelligent robot sensing and wearable medical detection equipment [18]. Eliot FGomez et al. used 3D printing technology to build a modular robot with complete self-healing ability under multiple injuries [19]. Harvard University used embedded 3D printing technology to produce high cell density in vitro tissues with vascular network and specific organ functions [20].

In summary, at the manufacturing level, the emphasis is mostly on the structural performance design and new material research of soft robots, while the experimental research on the optimization of soft material 3D printing parameters is relatively few [21]. In this paper, based on a comprehensive platform suitable for external printing of soft materials and embedded 3D printing, the performance of experimental soft materials was analyzed, followed by motion simulation analysis of needle flow modules with different shapes. Then, the printing parameters of external printing of soft materials and embedded printing of magnetic functional materials were optimized. Examples such as soft grippers and bionic gecko magnetic drive soft robots were manufactured to verify the feasibility of integrated manufacturing soft robots based on soft material external printing and embedded 3D printing.

2 Printing Platform

Based on the relatively stable extrusion system, the gantry 3D printing platform that can support heavier extrusion devices was refitted in this paper, so that a DIW soft material 3D printer with a working area of 250 * 250 * 300 (mm^3) was built based on the extrusion principle of a syringe (as shown in Fig. 1). By configuring different needle extrusion

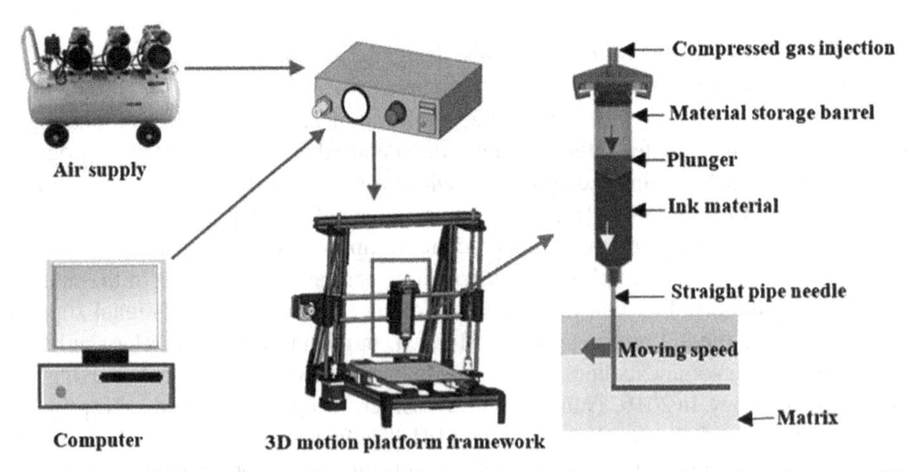

Fig. 1. 3D printing platform. Different needle extrusion devices are configured on the gantry 3D printing platform to realize external printing and embedded 3D printing of soft materials.

devices, external printing and embedded 3D printing of soft materials can be carried out on the platform.

3 Soft Material Analysis

In this paper, single component silica gel DC737 was used for external printing. When the embedded printing experiment was carried out, the mixed slurry of magnetic particle NdFeB and two-component silica gel Ecoflex 00–30 was used as the printing material. First of all, magnetic mixtures of different proportions were prepared to observe the curing conditions. The same plate was used to ensure that the thickness of the sheet was equal and the magnetic force was measured (as shown in Fig. 2(a)). The scatter box diagram of the test results is shown in Fig. 2(b). Considering the magnetic and mechanical properties (as shown in Fig. 2(c, d)) of magnetic pastes with different mixing ratios, the mixing ratio of Ecoflex 00–30 and NdFeB was 3:1,which was selected to be the mixing ratio of subsequent experiments.

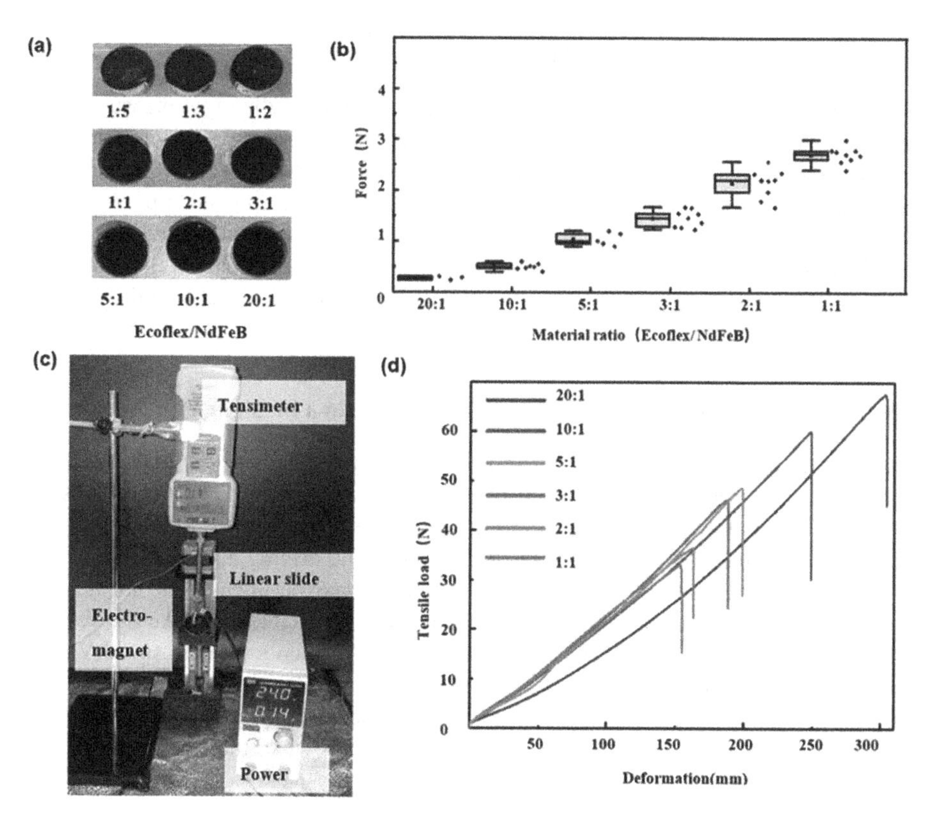

Fig. 2. (a) Magnetic slurries with different mixing ratios (b) Drawing box diagram of magnetic slurry with different mixing proportion (c) Magnetic measuring device (d) Relationship between tensile load and deformation of tensile specimens with different mixing ratios

4 Modeling and Simulation of Soft Material Printing

4.1 Simulation of Internal Flow Field of Different Needles

In this paper, the two common types of needles (conical needle and straight pipe needle) were analyzed by the finite element method, the pressure-velocity distribution of two different types of needles under the same air pressure (40 psi) was simulated, and the logarithmic coordinate comparison diagram of the velocity of the central plane of two needles under fixed air pressure was drawn in Fig. 3(a). As the inner diameter of the conical needle decreased gradually, the flow velocity of the slurry increased exponentially, and reached the maximum value at the outlet of the needle. The discharge was smooth and not easy to block so that only a small pressure was required to ensure high-precision molding. In contrast, the flow rate and pressure in the straight pipe needle were almost the same along the length of needle. Therefore, conical needle was preferred for external printing, while straight pipe needle was selected for embedded printing to avoid serious stirring effect of the needle in the base material.

4.2 Simulation of the Internal Flow Field Under Different Pressures

When the inner diameter of the two needles fixed at 0.6 mm, the input pressure P was used as a parameterized simulation scan to simulate the pressure and flow rate U of the needle (as shown in Fig. 3(b)). The same point between the two needles is that the flow rate on the central axis raises with the increase of the driving extrusion air pressure. The difference is that the straight pipe needle produces a large pressure at the sudden change, and the flow rate from the sudden change to the outlet pressure is almost the same; The velocity of conical needle increases exponentially from inlet to outlet, and the pressure decreases exponentially to atmospheric pressure.

4.3 Flow Field Simulation of Functional Materials in a Straight Pipe Needle

Based on the above material research and selection, embedded printing magnetic functional materials were simulated, it can be seen in Fig. 3(c) that the extrusion pressure P directly affects the final flow rate, and the internal flow rate of the slender needle is smaller. The simulation was carried out by inputting the driving air pressure P to provide guiding data for subsequent parameter optimization experiments. The needle outer diameter d_0 and needle moving speed V_0 were used as variables to make the velocity shear rate nephogram, as shown in Fig. 3(d). When the needle moving speed V_0 was controlled at 1.5 mm/s and 3 mm/s, the maximum flow rate was slightly greater than the needle moving speed. When the printing speed increased, the maximum shear rate increased, the range of high shear rate increased slightly, while the range of high flow rate was almost unchanged.

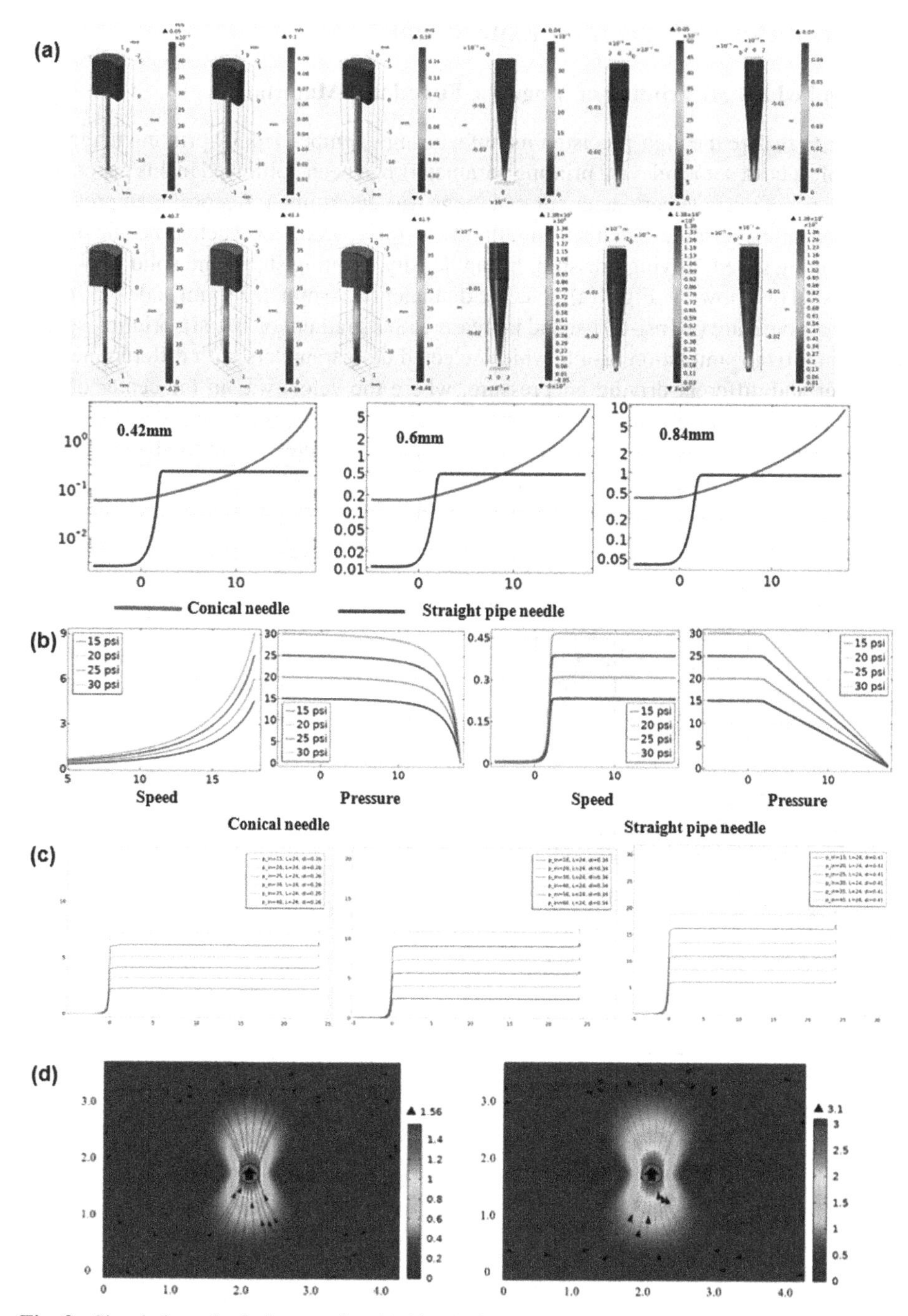

Fig. 3. Simulation of printing needles (a) Simulation analysis diagram of pressure and flow rate of straight pipe and conical needle under 40 psi (b) Simulation analysis diagram of needle (the inner diameter is 0.6 mm) (c) Simulation analysis diagram of magnetic slurry in straight pipe needle (d) Flow velocity simulation diagram of needle under different printing speeds

5 Parameter Optimization Experiments

5.1 Embedded 3D Printing of Magnetic Functional Materials

In order to realize the high-precision manufacturing of embedded 3D printing magnetic drive soft robots, some relevant printing parameters had been optimized in this paper, and finally a set of optimal parameter combinations was determined. According to previous experimental material research and simulation results, parameters such as needle diameter, printing speed, driving pressure, filling density, printing shape and solid thickness were tested (as shown in Fig. 4(a)). Needle diameter (0.26 mm, 0.33 mm and 0.41 mm), driving air pressure (10 psi-60 psi and increase with a gradient of 10 psi), printing speed according to the simulation results which selected the parameters with different needle diameter and different driving air pressure, where the velocity U in the center of the needle is equal to the needle moving speed v. The line was printed and the line width was measured to obtain an optimized combination of parameters. Printing shapes (circle, square, triangle and pentagram) and filling density (20%, 40%, 60%, 70%, 80%, 90% and 100%) were selected by measuring the diffusivity of different shapes with the same

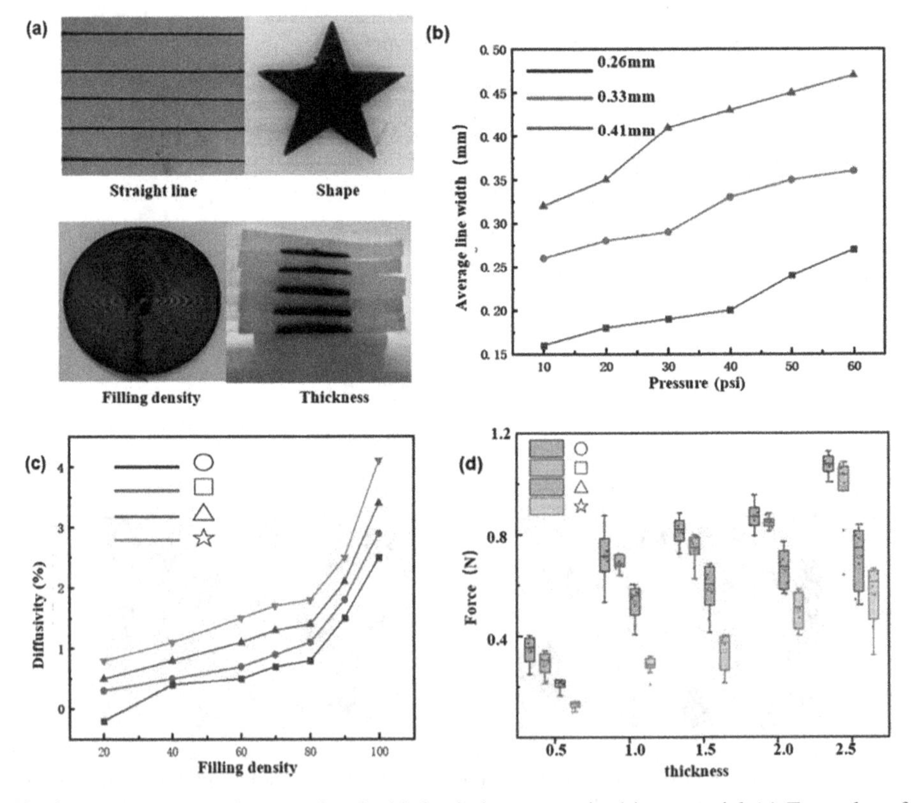

Fig. 4. Parameter experiments of embedded printing magnetic drive material (a) Examples of parameter experiment (b) Experimental results of monofilament printing (c) Results of diffusivity with different filling rates (d) Magnetic force of different thickness

printing area printed with different filling densities. Print thickness (0.5 mm, 1.0 mm, 1.5 mm, 2.0 mm and 2.5 mm) were selected by measuring the magnetic force of different shapes of printed products with different thickness.

Through the parameter optimization experiment, the average linewidth of embedded printing lines with different needle diameters and different combinations of air pressure and flow rate could be obtained, as shown in Fig. 4(b), from which a more stable and accurate parameter combination was obtained. At the same time, the measurement shows that under a single factor, the circular print with the same area has the best magnetism and low diffusivity, followed by square, triangle and star; The Fig. 4(c) shows that the diffusivity raised with the increase of filling density, and when the filling density was greater than 80%, the growth rate of diffusivity increased sharply. The magnetic force grown with the increase of thickness t, and when the thickness $t > 1$ mm, the growth rate of it decreased gradually, and reached the peak when $t = 2.5$ mm (as shown in Fig. 4(d)). To achieve the purpose of manufacturing a magnetic drive soft robot with high precision, the following printing parameter combinations were used in subsequent experiments: needle diameter 0.41 mm, driving air pressure 30 psi, needle moving speed 6 mm/s, filling density 80%, printing shape circular and printing thickness 1.5 mm.

5.2 Optimization of External Printing Parameters of Soft Materials

Firstly, a monofilament printing experiment was carried out. Taking the air pressure P and moving speed v_0 as cyclic variables (as shown in Table 1), the influence of printing parameters on the actual linewidth c was studied through cyclic experiments, and the best printing parameters have been obtained (Fig. 5(a)). When the air pressure was too small or the speed was too high, discontinuous deposition and poor printing stability easily occurred; in contrary, the deposition volume exceeded the limit of the layer height when the air pressure was too high or the speed was too small, there were obvious marks scratched by the needle on the surface of the printed monofilament (Fig. 5(c) and (d)), the deposited wire was gourd shaped, and the line width was uneven and difficult to control. Based on this, the optimal parameter combination was that when the needle diameter was 0.6 mm, the air pressure was 22.5 psi and the printing speed was 14.5 mm/s.

Then, a single line wall printing experiment was carried out to study the bonding between layers during printing (Fig. 5(b)), in order to obtain the appropriate parameters of the minimum printing spacing, to print more complex structures. The single line wall was formed by continuous vertical stacking of extruded silica gel. The square outline with a side length of 20 mm was printed, and the number of layers was 10. After curing, the steady-state area was analyzed to qualitatively evaluate the influence of printing floor height ton printing a single line wall (as shown in Fig. 5(e)). Finally, $t = 0.5$ mm was selected for subsequent experiments.

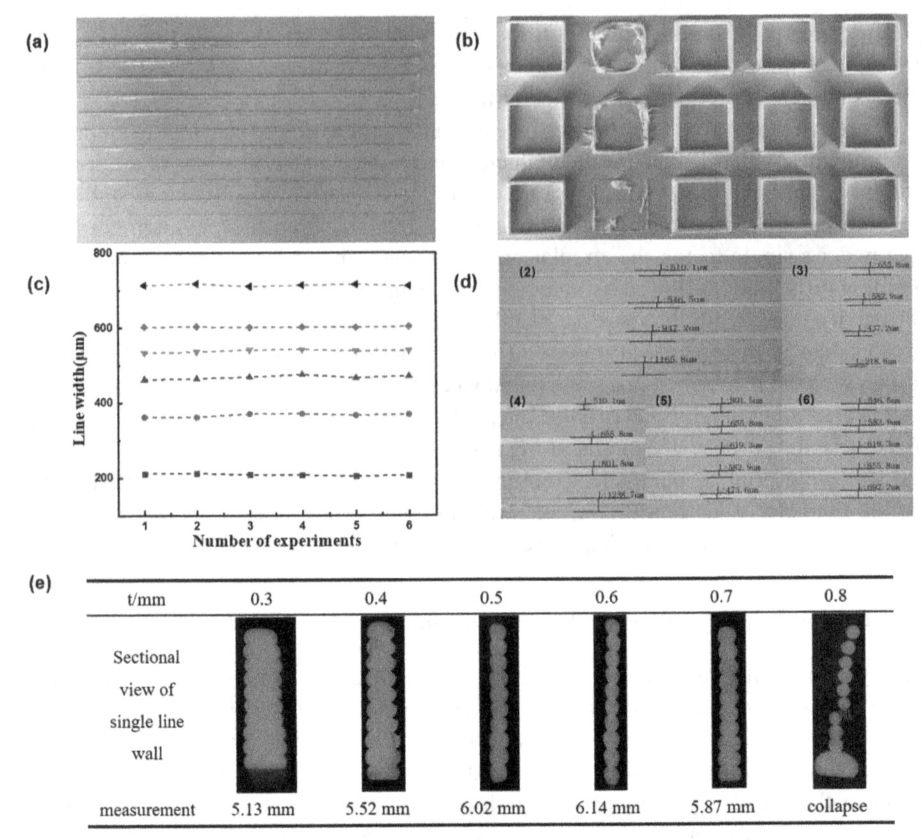

Fig. 5. Optimization of external printing parameters of soft materials (a) Single line printing experiment (b) the single line wall printing experiment (c) Results of the first single line experiment (d) Results of the second to sixth single line experiments (e) Results of single line wall printing

Table 1. Experiment of single wire printing

Number	Invariant factors	Changing factors
1	v0 = 16 mm/s	P = 14–44 psi
2	P = 29 psi	v0 = 11/13/15/17 mm/s
3	v0 = 14 mm/s	P = 14/17/20/23 psi
4	P = 23 psi	v0 = 12/13/14/15 mm/s
5	v0 = 14.5 mm/s	P = 21/22/23/24/25 psi
6	P = 22.5 psi	v0 = 14.3–14.7 mm/s

6 Soft Robot Manufacturing and Experiment

6.1 Embedded 3D Printing Magnetic Drive Gecko Soft Robots

In this paper, magnetic gecko soft robots were poured with a mold, and magnetic paste was embedded printed in four soles. First, a gait experiment of the printed bionic gecko soft robots was carried out, and five different vertical crawling modes were studied on vertical paper, as shown in the Table 2 below, in which red represents the movement of the foot, tested 40 times, and the average step length and success rate were recorded. The experiment showed that the success rate of single foot crawling was greater than 85%, but the average step length was approximately 60% of the bipedal movement. Rollover easily occurred during ipsilateral bipedal crawling, and the success rate was low. In contrast, the alternating movement of the front and rear feet was more successful.

Table 2. Study on different gait of bionic gecko

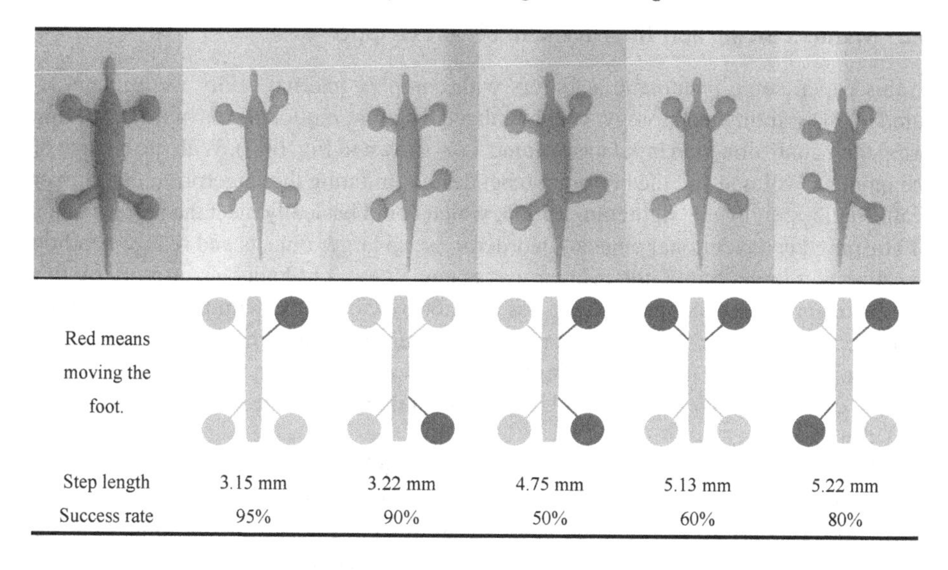

Red means moving the foot.					
Step length	3.15 mm	3.22 mm	4.75 mm	5.13 mm	5.22 mm
Success rate	95%	90%	50%	60%	80%

Then, vertical plane crawling experiments on six different materials with the same thickness were carried out. The same number of rubidium magnets was used on the back to attract, and the single step distance, maximum height and success rate of vertical crawling in several cases were recorded. The experiment showed that with material roughness increasing, the one-step stride of the bionic gecko gradually decreased, the maximum crawling height decreased together, but the success rate would increase (as shown in Table 3).

Table 3. Study on bionic gecko working in vertical plane with different materials

Material	Glass	Aluminum	Paper	PVC	PC
Friction coefficient	0.5	1.4	1.8	2.4	3.5
Step length	5.86 mm	5.77 mm	5.58 mm	5.43 mm	5.22 mm
Success rate	30%	45%	55%	70%	80%

6.2 Manufacturing and Improvement of Soft Gripper

In this paper, some pneumatic actuators with complex internal cavity were manufactured by 3D printing technology, which realized different motion forms of a 3D printing gripper by controlling the input gas volume V(as shown in Fig. 6(a)). With the change of the input gas volume and the bottom ratchet design imitating the fingerprint, the gripper realized the opening and tightening action, which could basically meet the requirements of grasping unconventional objects. In order to grasp larger objects and realize bending grasping at a larger angle, the gripper was extended to eight knuckles or more by filling the linear array of knuckle structures (Fig. 6(c)). Compared with pouring molding, high-precision 3D printing soft material manufacturing is direct manufacturing, which eliminates the process of mold making and demolding. Moreover, it is relatively convenient to change the size of the model, and the manufacturing time consumption is short (Fig. 6(b)).

The soft gripper designed and manufactured in the Fig. 6(c) could successfully grasp large objects, on the other hand, in order to further improve the working performance of the soft grippers so that it could grasp small objects, a magnetic working area was added at the bottom of the soft gripper. First, the embedded printing technology was used to inject magnetic slurry into the bottom surface of the gripper, and then the soft material external printing technology was used to manufacture the rest of the gripper on the bottom surface after it was cured. Through the joint action of pneumatic and magnetic drive parts, the grasping of small cylinder (4 mm), steel and plastic balls (5 mm) was realized.

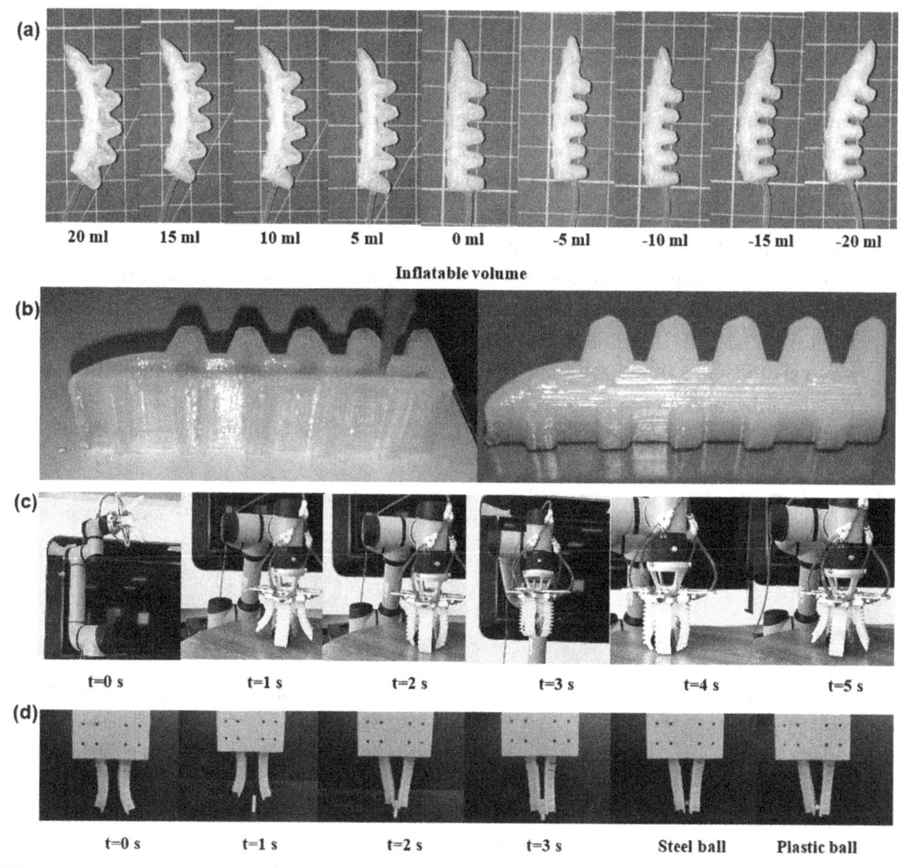

Fig. 6. (a) different motion forms of a 3D printing gripper by controlling the input gas volume V (b) External printing gripper process and results (c) Grasping experiment of eight knuckles grippers (d) Grasping experiment of eight knuckles grippers with magnetic working area

7 Conclusions

Firstly, a soft material 3D printing platform compatible with embedded and external printing was introduced in this paper. Through material research, motion simulation and parameter experiments, the best parameter combinations of functional magnetic material embedded printing and external printing were finally obtained. Based on this, a bionic gecko magnetic drive soft robot and soft grippers were manufactured and relevant performance experiments were carried out. The main conclusions are as follows:

(1) Comparing the simulation results of internal air pressure and flow rate between conical needle and straight pipe needle, the discharge of conical needle is smooth, and only a small extrusion pressure can ensure high-precision molding in the printing process, which is not easy to cause congestion.

(2) The performance of materials for embedded and external printing was studied. Then, based on the parameter simulation results, more stable and accurate parameter combinations were obtained for subsequent experiments through parameter experiments, which provided an effective guarantee for the high-precision integrated manufacturing of soft robots.
(3) A bionic gecko magnetic drive soft robot and soft grippers were designed and printed. Crawling experiments of bionic magnetic drive geckos with different gaits and different material planes were carried out. The deformation of the soft gripper under different inflation volumes was studied, and the corresponding improvement was made for grasping larger and smaller objects. The feasibility of integrated manufacturing of fully flexible soft robots using soft material 3D printing technology was verified, and the significant advantages of 3D printing, direct manufacturing and flexible model changing were highlighted.

Acknowledgment. Research supported by National Natural Science Foundation of China (61773274).

References

1. Rus, D., Tolley, M.T.: Design, fabrication and control of soft robots. Nature **521**(7553), 467–475 (2015)
2. Florian, H., Melanie, B., Martin, K.: Sustainability in soft robotics: Becoming sustainable, the new frontier in soft robotics. Adv. Mater. **33**(19), 2170143 (2021)
3. Wang, J., Gao, D., Lee, P.S.: Artificial muscles: Recent progress in artificial muscles for interactive soft robotics. Adv. Mater. **33**(19), 2170144 (2021)
4. Won, P., Ko, S.H., Majidi, C., Feinberg, A.W., WebsterWood, V.A.: Biohybrid actuators for soft robotics: Challenges in scaling up. Actuators **9**(4), 96–96 (2020)
5. Gul, J.Z., et al.: 3D printing for soft robotics - a review. Sci. Technol. Adv. Mater. **19**(1), 243–262 (2018)
6. Gu, G., Zou, J., Zhao, R., Zhao, X., Zhu, X.: Soft wall-climbing robots. Sci. Robot. **3**(25), eaat2874 (2018)
7. Wallin, T.J., Pikul, J., Shepherd, R.F.: 3D printing of soft robotic systems. Nat. Rev. Mater. **3**(6), 84–100 (2018)
8. Truby, R.L., Lewis, J.A.: Printing soft matter in three dimensions. Nature **540**(7633), 371–378 (2016)
9. Lee, H.-K., et al.: Normal and shear force measurement using a flexible polymer tactile sensor with embedded multiple capacitors. J. Microelectromech. Syst. **17**(4), 934–942 (2008)
10. Grosskopf, A.K., et al.: Viscoplastic matrix materials for embedded 3D printing. ACS Appl. Mater. Interfaces **10**(27), 23353–23361 (2018)
11. Sachyani Keneth, E., Kamyshny, A., Totaro, M., Beccai, L., Magdassi, S.: 3D printing materials for soft robotics. Adv. Mater. **33**(19), e2003387 (2020)
12. Zhang, B., Kowsari, K., Serjouei, A., Dunn, M.L., Ge, Q.: Reprocessable thermosets for sustainable three-dimensional printing. Nat. Commun. **9**(1), 1831 (2018)
13. Wang, J., Liu, Y., Fan, Z., Wang, W., Wang, B., Guo, Z.: Ink-based 3D printing technologies for graphene-based materials: A review. Adv. Compos. Hybrid Mater. **2**(1), 1–33 (2019). https://doi.org/10.1007/s42114-018-0067-9

14. Wehner, M., Truby, R.L., Fitzgerald, D.J., et al.: An integrated design and fabrication strategy for entirely soft, autonomous robots. Nature **536**(7617), 451–455 (2016)
15. Hinton, T.J., Hudson, A., Pusch, K., et al.: 3D printing PDMS elastomer in a hydrophilic support bath via freeform reversible embedding. ACS Biomater. Sci. Eng. **2**(10), 1781–1786 (2016)
16. Noor, N., Shapira, A., Edri, R., et al.: 3D printing of personalized thick and perfusable cardiac patches and hearts. Adv. Sci. **6**, 1900344 (2019)
17. Wei, P., Yang, X., Cao, Z., et al.: Flexible and stretchable electronic skin with high durability and shock resistance via embedded 3D printing technology for human activity monitoring and personal healthcare. Adv. Mater. Technol. **4**(9), 1900315 (2019)
18. Luo, G., Yu, Y., Yuan, Y., et al.: Freeform, reconfigurable embedded printing of all-aqueous 3D architectures. Adv. Mater. **31**(49), 1904631 (2019)
19. Gomez, E.F., et al.: 3D-printed self-healing elastomers for modular soft robotics. ACS Appl. Mater. Interfaces **13**, 28870–28877 (2021)
20. SkylarScott, M.A., et al.: Biomanufacturing of organ-specific tissues with high cellular density and embedded vascular channels. Sci. Adv. **5**(9), eaaw2459 (2019)
21. Plott, J., Shih, A.: The extrusion-based additive manufacturing of moisture-cured silicone elastomer with minimal void for pneumatic actuators. Addit. Manuf. **17**, 1–14 (2017)

A Novel Soft Wrist Joint with Variable Stiffness

Gang Yang[1], Bing Li[1,2], Yang Zhang[1], Dayu Pan[1], and Hailin Huang[1](✉)

[1] School of Mechanical Engineering and Automation, Harbin Institute of Technology, Shenzhen 518055, People's Republic of China
huanghailin@hit.edu.cn

[2] State Key Laboratory of Robotics and System, Harbin Institute of Technology, Harbin 150001, People's Republic of China

Abstract. This paper presents a novel soft wrist joint made up of two soft variable stiffness bending joints and one soft torsion joint. The soft variable stiffness bending joint can achieve bending motion with variable stiffness and the motion of the soft bending actuator does not influence the terminal position of the soft bending joint, which provides a prerequisite for subsequent accuracy control. The torque of the soft torsion actuator is generated from four uniformly distributed spiral chambers and fiber-reinforced ropes, which can enhance torsion. The soft wrist joints are obtained by assembling them in series, providing a new solution for the soft robotic arm.

Keywords: Soft variable stiffness actuator · Soft torsion joint · Soft wrist joint

1 Introduction

Soft actuators have drawn extensive concern in recent years for their excellent performance and multiple application scenarios like the grasp of irregular objects [1, 2] multifunctional operation under complex work conditions [3–5], and medical devices [6–9]. In contrast to conventional actuators in robot arms which have a better load capacity, soft actuators process unique advantages in lightweight and safety reliability. This paper provides the design of a novel soft wrist joint with two soft variable stiffness bending joints and a soft torsion joint in detail, and lays a technical foundation for further research in the complete soft robotic arm.

For most of the available bending joints with variable stiffness, jamming [10, 11] and embedding variable stiffness objects like SMA [12] or SMP [13] are adopted frequently. The stiffness can be significantly improved by these means; however, the length of the soft actuators usually changes which makes it against accuracy control. In this paper, a novel variable stiffness joint is presented whose stretching does not influence terminal position which provides a prerequisite for its accuracy control.

As it provides different motions by varying angles, the fiber-reinforced soft actuator is used as a torsion joint in the soft wrist joint. [14, 15] Besides, in this paper, spiral chambers are adopted to enhance torsion.

In this paper, we focus on two types of soft actuators, study the performances of each joint, and combine them in series to form a soft wrist joint.

H. Liu et al. (Eds.): ICIRA 2022, LNAI 13456, pp. 346–356, 2022.
https://doi.org/10.1007/978-3-031-13822-5_31

2 Design and Fabrication

2.1 Design of the Wrist Joint

This section introduces the design of the pneumatic wrist joint which is consist of five parts including the base, bending joint 1, link 1, bending joint 2, and torsion joint, as shown in Fig. 1(a), which are bolted to each other. The detailed design of the torsion joint is shown in Fig. 1(b), and the torsion joint is made up of a soft torsion actuator, pneumatic fitting, joint connection, and spiral chambers. Four spiral chambers are distributed uniformly, guaranteeing good torsion when inflated. From Fig. 1(c), it can be seen that the bending joint is composed of an upper platform, soft bending actuator, variable stiffness part, and lower platform. The upper platform is connected to link 1 and the lower platform is connected to the base. The variable stiffness part shown in Fig. 1(d) and (e), consists of a soft variable stiffness balloon, pneumatic fittings, revolute pair, and a pair of half shafts. The stiffness of the bending joint increases when inflated, as the soft variable stiffness balloon swells to increase the friction between the half shafts and revolute pair.

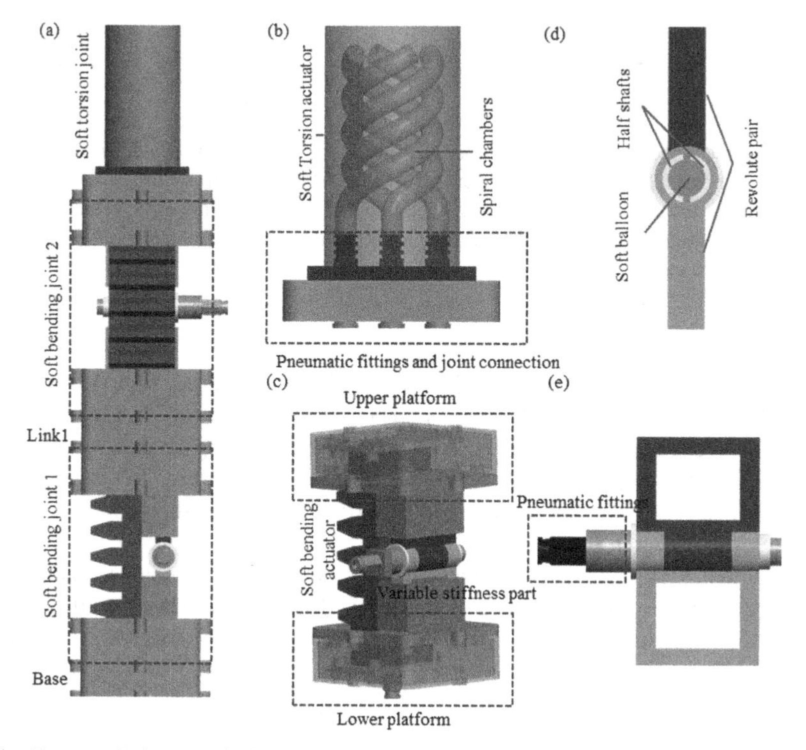

Fig. 1. Concept design: (a) design of the wrist joint, (b) detailed design of the torsion joint (c) detailed design of the variable stiffness bending joint, and (d)-(e) two views of the variable stiffness part.

2.2 Fabrication of Soft Actuator

The wrist joint has two different soft actuators, a soft bending actuator, and a soft torsion actuator. A soft bending actuator is excepted to possess both good deformation and load capacity. As a result, the Dragon Skin™ 30 rubber is chosen, and the fabrication can be divided into two parts. In the first step shown in Fig. 2, reasonable molds and proper silicone rubber treatment are required to get a reliable soft actuator without two sides and underside shown in Fig. 2(e). Besides, we can see a special sealing design with serval raised rings.

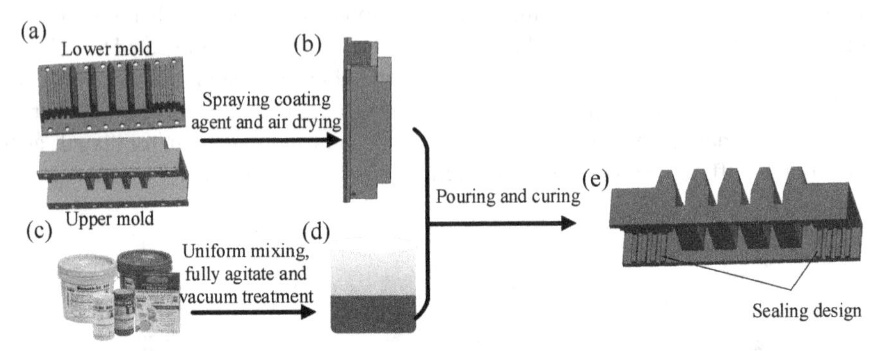

Fig. 2. The first step of the fabrication process. (a) Prepare corresponding molds including upper mold and lower mold, spray release agent evenly on the surface, and air drying. (b) The assembled mold combining lower and upper mold. (c) The Dragon Skin™ 30 rubber. (d) Rubber solution after corresponding treatment. (d) Pour rubber solution into the assembled mold and cure, the excepted structure can be obtained.

Then combining the molds of step 2 shown in Fig. 3(b), pouring silicone rubber, and curing, we can get a soft bending actuator without two sides shown in Fig. 3(d). Then, stick it to both end caps shown in Fig. 3(e), and the soft bending actuator is obtained as shown in Fig. 3(f). The whole soft bending actuator shown in Fig. 3(i) is obtained by sticking it to both end caps shown in Fig. 3(e), a reinforced layer shown in Fig. 3(f), and a bottom layer shown in Fig. 3(g).

Similar to the fabrication of the soft bending actuator, the soft torsion actuator adopts the method of silicone rubber casting to fabricate. However, the torsion actuator adopts Ecoflex™ 00–30 to fabricate for its good torsion effect. The fabrication process is shown in Fig. 4.

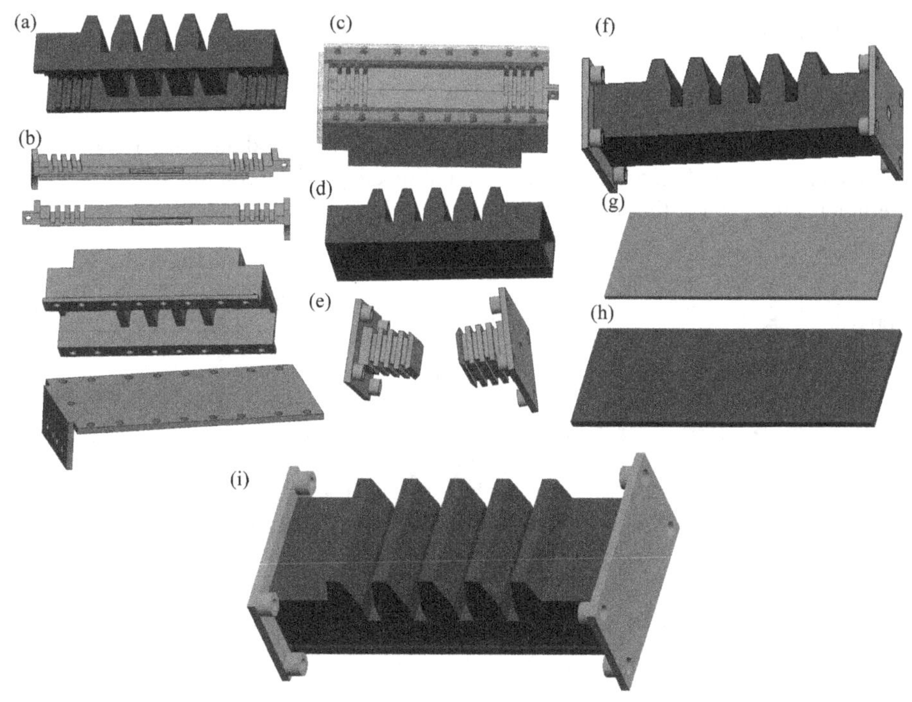

Fig. 3. Illustrations of the following fabrication process. (a) The structure we get from the first step. (b) Molds of the second step. (c) The assembled mold of step 2. (d) Soft bending actuator without two sides. (e) Two end caps. (f) The soft bending actuator without the reinforced layer (g) and the bottom layer (h). (i) Completed soft bending actuator.

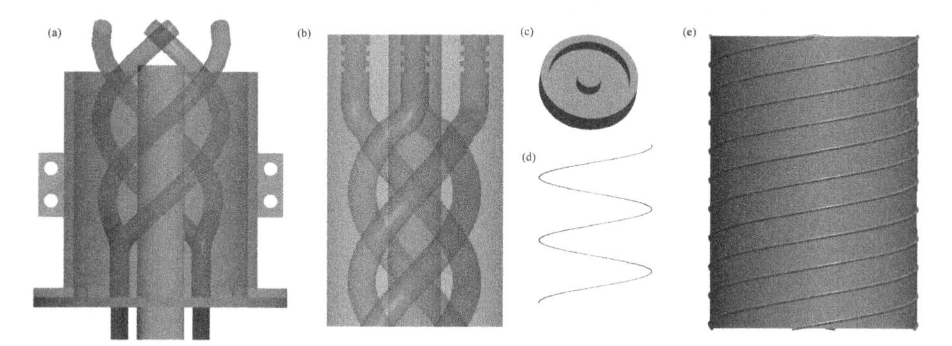

Fig. 4. Illustrations of the fabrication process of soft torsion actuator. (a) The assembled mold of step 1. (b) The structure can be obtained from step 1. (c) Mold of step 2. (d) Fiber-reinforced rope. (d) The whole soft torsion actuator.

3 Analysis and Simulation of Soft Actuator

3.1 Soft Bending Actuator

For soft bending actuators, Abaqus simulation is used for aided analysis, the result of which can be seen in Fig. 5(a). Besides, simulation adapts the Yeoh model and material coefficients refer to the literature [16].

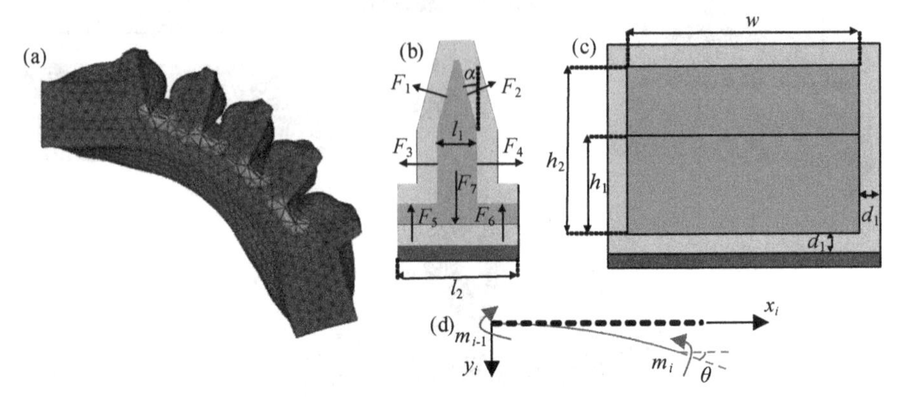

Fig. 5. Illustrations of force and size of the soft bending actuator. (a) Simulation result of soft bending actuator. (b), (c) Two views of one unit of soft bending actuator. (d) Force model in a unit.

The forces of each unit can be seen in Fig. 5(b), and the detailed size of each unit can be seen in Fig. 5(b) and Fig. 5(c). CBCM (Chain Beam Constraint Model) is adopted to analyze the statics of the soft bending actuator. And the model of one unit is shown in Fig. 5(d). Based on the aforementioned analysis, we have

$$F_1 = F_2 = \frac{Pw(h_2 - h_1)}{\cos\alpha} \tag{1}$$

$$F_3 = F_4 = Ph_1 w \tag{2}$$

$$F_5 = F_6 = \frac{P(l_2 - l_1)w}{2} \tag{3}$$

$$F_7 = Pl_2 w \tag{4}$$

$$m_i - m'_{i-1} = (F_1 + F_2)\cos\alpha\left(d_1 + \frac{h_1 + h_2}{2}\right) + (F_3 + F_4)(d_1 + 0.5h_1) \tag{5}$$

$$\theta_i = \frac{(m_i - m'_{i-1})l_2}{EI} \tag{6}$$

where $\theta_i (i = 1, 2, 3, 4)$ represents the bending angle of unit i corresponding to unit i-1, α represents the top wedged angle, P represents input pressure, EI represents the stiffness of the beam cross-section, and h_2, h_1, l_2, l_1, d_1 and w represent the corresponding length shown in Fig. 5(b) and Fig. 5(c).

The bending angle of soft bending actuator θ can be expressed as the sum of the bending angle of each unit $\theta_i (i = 1, 2, 3, 4)$, that is

$$\theta = \theta_1 + \theta_2 + \theta_3 + \theta_4 \tag{7}$$

3.2 Soft Torsion Actuator

For soft torsion actuators, the result of Abaqus simulation can be seen in Fig. 6(a). Torque comes from two sources, one of which is from spiral chamber T_1 shown in Fig. 6(b), and the other of which is from the fiber rope shown in Fig. 6(c) and Fig. 6(d).

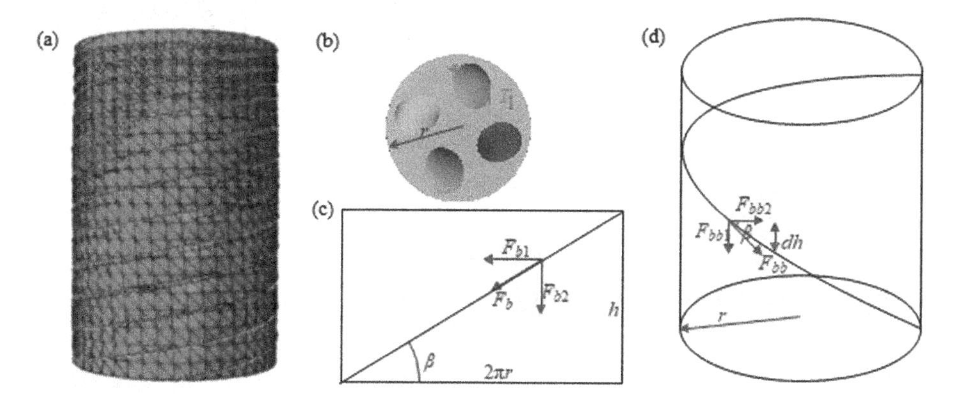

Fig. 6. Illustrations of force diagram of soft torsion actuator. (a) Simulation result of soft torsion actuator. (b) Horizontal cross-section of spiral chambers and torque from spiral chambers. (c) Force diagram of one unit. (d) Force diagram of a differential element.

In each unit, the force meets the equation

$$F_{b1} = F_b \cos\beta \tag{8}$$

$$F_{b2} = F_b \sin\beta \tag{9}$$

$$\tan\beta = h/2\pi r \tag{10}$$

$$L_b = \sqrt{h^2 + (2\pi r)^2} \tag{11}$$

$$q = F_b/L_b \tag{12}$$

where F_b represents the force of fiber in one unit on the soft torsion actuator, β represents the angle of the helical fiber, h represents the height of one unit, r is the radius of the soft torsion actuator, L_b represents the length of fiber restraint rope in one unit, and q is the distribution of the force.

For each differential element, we have the equation as

$$F_{bb} = \frac{q dh_0}{\sin\beta} \tag{13}$$

$$F_{bb1} = F_{bb}\cos\beta \tag{14}$$

where F_{bb} represents the force of fiber in one differential element on a soft torsion actuator and F_{bb1} represents the component of the circumferential tangential force F_{bb}. As a result, the torque T_2 caused by fiber in one unit is produced which by be expressed as

$$T_2 = \int_0^h F_{bb1} dh_0 \tag{15}$$

Therefore, torsion occurs under the combination of T_1 and T_2.

4 Experiment Validation and Discussion

4.1 Experiment Validation

After the analysis and simulation, it is necessary to verify by experiment.

For the soft bending actuator and torsion actuator, the bending and torsion angles are changed by varying input pressure shown in Fig. 7.

What is more, to explore the influence of the variable stiffness part on bending angle, bending angles are obtained under the combined pressure of a soft bending actuator and variable stiffness balloon, which are shown in Fig. 8.

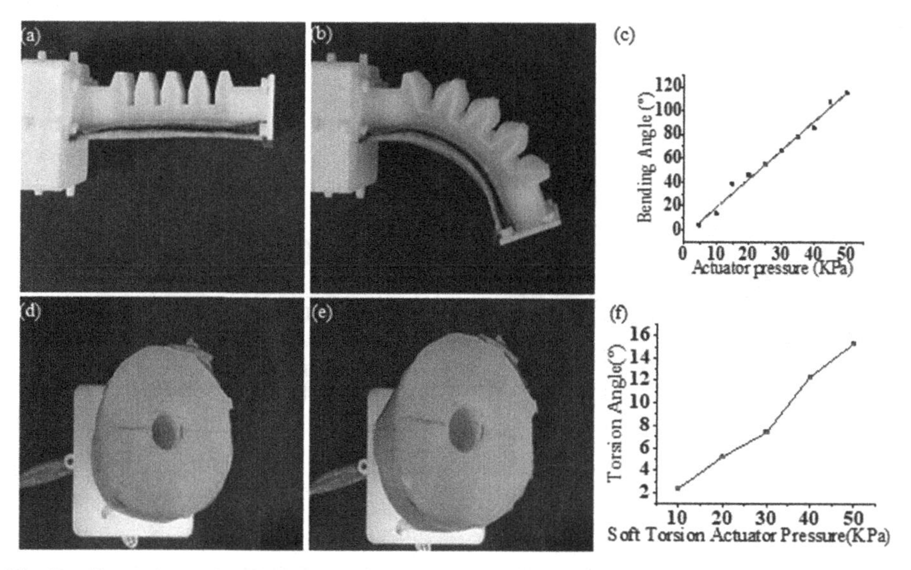

Fig. 7. Illustrations of the relationship between input pressure and bending or torsion angle. (a) The initial state of the soft bending actuator. (b) Bending state of the soft bending actuator at 30 kPa. (c) The relationship between the pressure and bending angle of the soft bending actuator. (d) The initial state of the soft torsion actuator. (b) Torsion state of the soft torsion actuator at 40 kPa. (c) The relationship between the pressure and torsion angle of the soft torsion actuator.

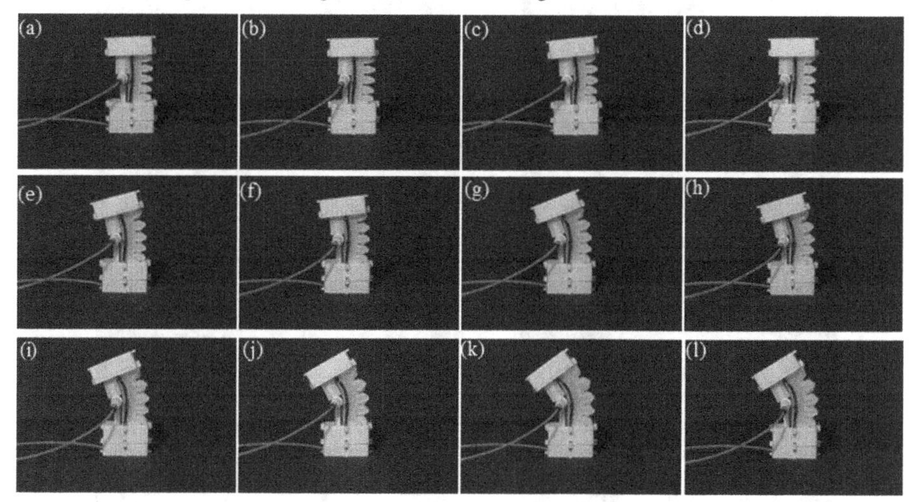

Fig. 8. The experiment of the bending angle under the combined pressure of a soft bending actuator and soft variable stiffness balloon. The pressure of the soft bending actuator is 10 kPa in (a-b), 15 kPa in (c-d), 20 kPa in (e-f), 25 kPa in (g-h), 30 kPa in (i-j), and 35 kPa in (k-l). The pressure of the soft variable stiffness balloon is 0 in (a, c, e, g, i, and l), 120 kPa in (b and d), and 150 kPa in (f, h, j, and l).

The physical graph of the soft variable stiffness balloon and a pair of half shafts is shown in Fig. 9(a), and the effect of variable stiffness can be seen in Fig. 9(b).

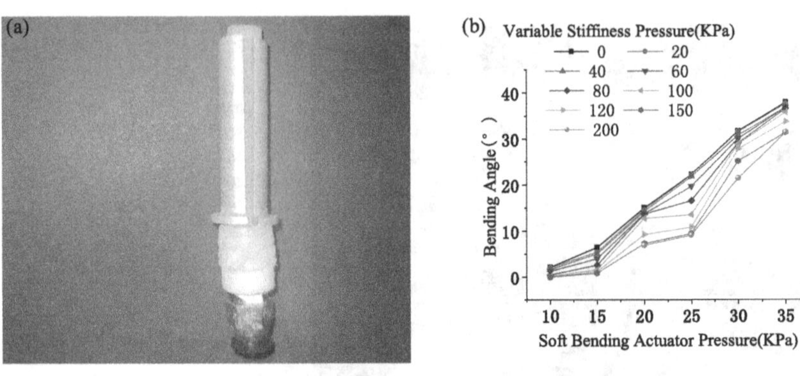

Fig. 9. (a) Illustrations of the soft variable stiffness balloon and a pair of half shafts. (b) The effect of variable stiffness.

Assemble two soft variable stiffness bending joints and one soft torsion in series, the soft wrist joint is obtained shown in Fig. 10.

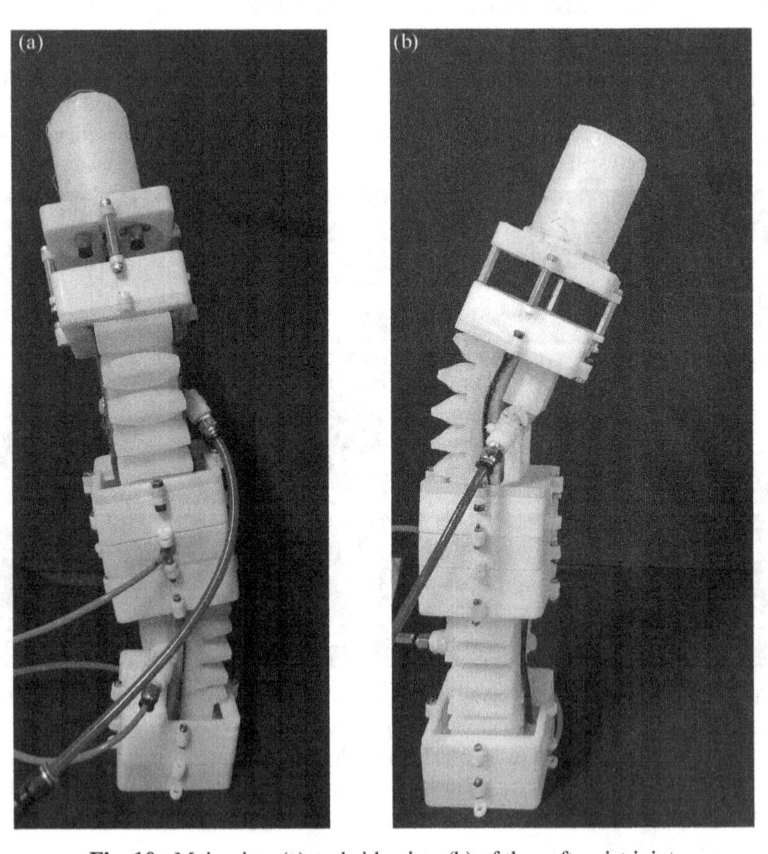

Fig. 10. Main view (a) and side view (b) of the soft wrist joint.

4.2 Discussion

According to the theoretical analysis and experiment validation, we have.

1) When the pressure of variable stiffness is constant, the higher the input pressure of the soft bending actuator is, the higher the bending angle will be.
2) When the input pressure of the soft bending actuator remains unchanged, the higher the pressure of the input variable stiffness balloon is, the smaller the corresponding bending angle will be.
3) When the input pressure of the soft bending actuator is small, corresponding to the pressure of 10 kPa 15 kPa, the variable stiffness part also has an obvious effect under a small pressure.
4) When the pressure input of the soft bending actuator is small, corresponding to the pressure of 30 and 35 kPa in Fig. 9, the effect of small variable stiffness pressure on changing the bending is not obvious. Only when the pressure of the variable stiffness is larger, such as 150 kPa 200 kPa, the effect of the variable stiffness is obvious, which is a new challenge for the sealing capability of soft variable stiffness balloons.
5) Contrast the bending of the soft bending actuator, we will find that the bending angle with the variable stiffness part is far less, which means the variable stiffness will greatly limit the bending effect. As a result, how to alleviate the inhibitory effect will be a big challenge.

5 Conclusion

In this paper, we presented a novel wrist joint consisting of two soft variable stiffness bending joints and one torsion joint, and the detailed fabrication processes of two soft actuators were explained. Then finite element analyses and corresponding experiments of two soft actuators were carried out\n. What is more, some potential reasons were discussed to explain the difference between reality and simulation.

Besides, there is still much room for improvement for further study. For example, the workspace of the soft bending actuator and torsion actuator can be improved, and there is no closed-loop control, which are our key contents of follow-up work.

Acknowledgment. This work was supported in part by the National Natural Science Foundation of China under Grants 51835002 and 52075113, in part by Guangdong Science and Technology Research Council under Grant No. 2020B1515120064, in part by Guangdong Basic and Applied Basic Research Foundation under Grant 2019A1515011897.

References

1. Chen, C., Sun, J., Wang, L., et al.: Pneumatic bionic hand with rigid-flexible coupling structure. **15**(4), 1358 (2022)
2. Wang, R., Huang, H., Xu, R., et al.: Design of a novel simulated "soft" mechanical grasper. Mech. Mach. Theory **158**, 104240 (2021)

3. Novel Design of a Soft Lightweight Pneumatic Continuum Robot Arm with Decoupled Variable Stiffness and Positioning. **5**(1), 54–70 (2018)
4. Naclerio, N.D., Karsai, A., Murray-Cooper, M., et al.: Controlling subterranean forces enables a fast, steerable, burrowing soft robot. **6**(55), eabe2922 (2021)
5. Yang, C., Geng, S., Walker, I., et al.: Geometric constraint-based modeling and analysis of a novel continuum robot with Shape Memory Alloy initiated variable stiffness. Int. J. Robotics Res. **39**(14), 1620–1634 (2020)
6. Cacucciolo, V., Shintake, J., Kuwajima, Y., et al.: Stretchable pumps for soft machines. Nature **572**(7770), 516-519 (2019)
7. Chen, G., Pham, M.T., Redarce, T.: Sensor-based guidance control of a continuum robot for a semi-autonomous colonoscopy. Robot. Auton. Syst. **57**(6), 712–722 (2009)
8. Kim, D.H., Macdonald, B., Mcdaid, A., et al.: User perceptions of soft robot arms and fingers for healthcare. In: Proceedings of the 25th IEEE International Symposium on Robot and Human Interactive Communication (IEEE RO-MAN), Columbia Univ, Teachers Coll, New York City, NY, F Aug 26–31, (2016)
9. Polygerinos, P., Wang, Z., Galloway, K.C., et al.: Soft robotic glove for combined assistance and at-home rehabilitation. Robot. Auton. Syst. **73**, 135–143 (2015)
10. Bamotra, A., Walia, P., Prituja, A.V., et al.: Layer-jamming suction grippers with variable stiffness. J. Mechanisms Robotics-Trans. Asme **11**(3) (2019)
11. Xiangxiang, W., Linyuan, W., Bin, F., et al.: Layer jamming-based soft robotic hand with variable stiffness for compliant and effective grasping. Cogn Comput Syst (UK) **2**(2), 44–49 (2020)
12. Li, J.F., Zu, L., Zhong, G.L., et al.: Stiffness characteristics of soft finger with embedded SMA fibers. Compos. Struct. **160**, 758–764 (2017)
13. Yang, Y., Chen, Y.H., Li, Y.T., et al.: Bioinspired robotic fingers based on pneumatic actuator and 3D printing of smart material. Soft Rob. **4**(2), 147–162 (2017)
14. Bishop-Moser, J., Kota, S.: Design and modeling of generalized fiber-reinforced pneumatic soft actuators. IEEE Trans. Rob. **31**(3), 536–545 (2015)
15. Connolly, F., Polygerinos, P., Walsh, C.J., et al.: Mechanical programming of soft actuators by varying fiber angle. Soft Rob. **2**(1), 26–32 (2015)
16. Marechal, L., Balland, P., Lindenroth, L., et al.: Toward a common framework and database of materials for soft robotics. Soft Rob. **8**(3), 284–297 (2021)

Rangefinder-Based Obstacle Avoidance Algorithm for Human-Robot Co-carrying

Xiong Guo[1,2,3], Xinbo Yu[1,3], and Wei He[1,2(✉)]

[1] Institute of Artificial Intelligence, University of Science and Technology Beijing, Beijing, China
weihe@ieee.org

[2] School of Automation and Electrical Engineering, University of Science and Technology Beijing, Beijing, China

[3] Shunde Graduate School, University of Science and Technology Beijing, Foshan, China

Abstract. The mobile robot (follower) in the human-robot co-carrying task needs to follow the human partner (leader) online, and also needs to avoid obstacles. Most existing local reactive obstacle avoidance algorithms cannot be applicable. This paper proposes a rangefinder-based obstacle avoidance algorithm for omnidirectional mobile robots in human-robot co-carrying. The proposed obstacle avoidance algorithm is implemented using a rangefinder, and obstacle avoidance can be achieved combined with controller design on our designed co-carrying system. Finally, simulation results are carried out to show the effectiveness of our proposed obstacle avoidance algorithm.

Keywords: Human-robot collaboration · Real-time obstacle avoidance · Mobile robots · Cooperative tasks

1 Introduction

Recently, intelligent robots which perform tasks in dynamic and uncertain environments have become an important development trend [1,2]. Due to the uncertainty of the dynamic environment, it is quite challenging for the robot to respond

This work was financially supported by the Beijing Natural Science Foundation under Grant JQ20026, in part by the National Natural Science Foundation of China under Grant No. 62003032, 62061160371, 62073031, in part by the China Postdoctoral Science Foundation under Grant 2020TQ0031 and Grant 2021M690358, in part by Beijing Top Discipline for Artificial Intelligent Science and Engineering, University of Science and Technology Beijing, in part by the Technological Innovation Foundation of Shunde Graduate School of University of Science and Technology Beijing under Grant BK20BE013, in part by Guangdong Basic and Applied Basic Research Foundation (2020B1515120071), in part by the Interdisciplinary Research Project for Young Teachers of USTB (Fundamental Research Funds for the Central Universities) (FRF-IDRY-20-019).

H. Liu et al. (Eds.): ICIRA 2022, LNAI 13456, pp. 357–368, 2022.
https://doi.org/10.1007/978-3-031-13822-5_32

to many emergencies in time, which puts forward higher requirements for the real-time performance and safety of the robot.

This paper is devoted to the study of omnidirectional mobile robots and human partners to complete co-carrying tasks. Mobile robots have a wider range of activities, more emergencies to be dealt with, and more severe environmental conditions to be applied in [3]. The co-carrying system in this paper adopts a leader-follower architecture, the leader is used for guiding the system to move towards the destination, and the follower will consider the obstacle avoidance when following the leader's motion.

There have been some previous studies on co-carrying works [4–7]. [8] employed distributed "robot helpers" (D_R Helpers) to help humans carry objects. The authors proposed adaptive dual caster action to strengthen the maneuverability of the co-carrying system, and a passivity-based controller was designed to keep the stability of human-robot interaction. A path planning method (*Riesenfeld Spline*) was applied into D_R Helpers to generate path for mobile robots in [9,10].

In [11], a hybrid framework was proposed for co-carrying tasks, and an observer was designed for estimating the control of the human partner. In addition, an adaptive impedance-based control strategy with *Neural Networks* was proposed. In [12,13], a fault-tolerant control and a data fusion method were proposed. The active disturbance rejection control was used to keep the end effector in a stable pose in [14]. Inspired by fault-tolerant systems and anti-jamming systems, we designed a mechanism with motion redundancy for co-carrying, which can tolerate the motion error to fluctuate within a certain range.

Different from the work mentioned in the above literature, our co-carrying system considers real-time obstacle avoidance. *Potential field method* is a method of avoiding obstacles through virtual forces, which brings some inspiration to our obstacle avoidance algorithm.

In [15–19], *Artificial Potential Field Approach* based algorithms were proposed to perform local obstacle avoidance, and they were applied to an independent mobile robot with no constraints. In addition, a vision-based obstacle avoidance method was proposed in [20–22], which has important implications for our work.

This paper mainly proposes a rangefinder-based obstacle avoidance algorithm for omnidirectional mobile robots with our designed mechanism in a co-carrying system. The remainder is formed as: Our designed co-carrying system is introduced in Sect. 2. The obstacle avoidance algorithm is described in Sect. 3. A mobile robot controller is presented in Sect. 4. In Sect. 5, simulations and results are shown.

2 Co-carrying System

A co-carrying system is shown in Fig. 1, an omnidirectional mobile robot with a mechanism is designed to perform co-carrying an object with a human partner. The mechanism here can measure the partner's motion intention and tolerate the motion error within a certain range.

We designed a mechanism for object loading as shown in Fig. 2. When performing co-carrying, the slider will be fixed with the object and follow the object's movement, while the displacement sensor measures the distance of the slider. The disc on the load base can rotate freely and record the current angle.

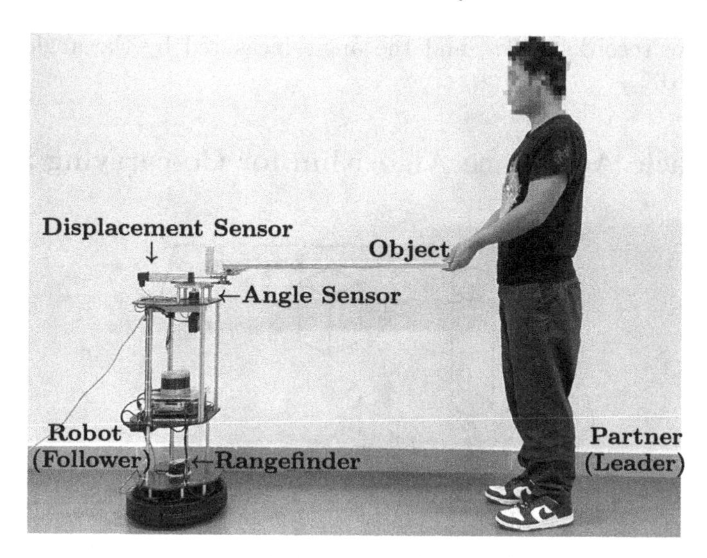

Fig. 1. Schematic diagram of the co-carrying system.

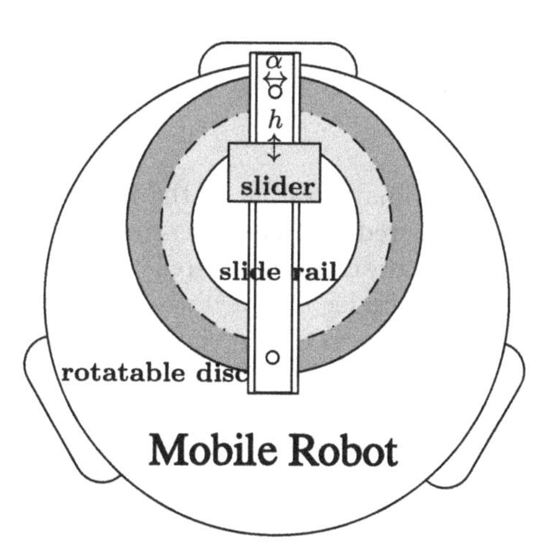

Fig. 2. Top view of the designed mechanism.

Due to the disc can rotate freely and the slider can move on the slide rail in a certain range, our co-carrying system owns the ability to buffer errors in robot motion. In other words, our system has a certain fault tolerance due to the redundancy in the mechanism.

The displacement sensor can measure the displacement of the slider on the slide rail, we record it as h, and the angle measured by the angle sensor is recorded as α.

3 Obstacle Avoidance Algorithm for Co-carrying System

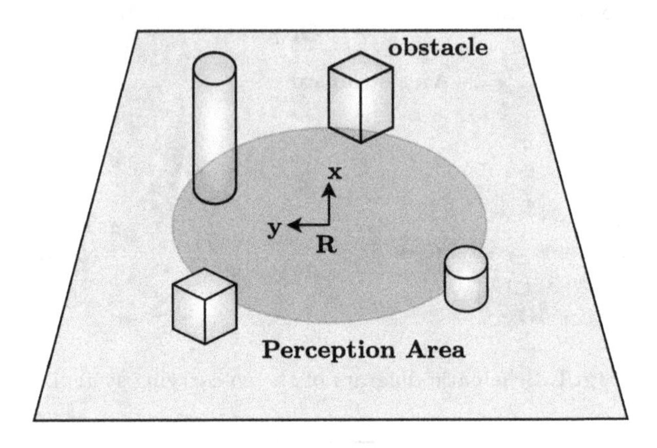

Fig. 3. The diagram of perception area (red), and obstacles (yellow). (Color figure online)

As shown in Fig. 3, the coordinate system of the rangefinder and the robot is aligned on the X-Y plane, the red area is the sensing area of the rangefinder we set, and we only focus on the obstacles in the red area.

The raw data of the rangefinder is a vector containing n sampled distance values. We design a piecewise function to preprocess the rangefinder raw data:

$$\mathbf{S}(i) = \begin{cases} \mathbf{S}_{\text{raw}}(i) & \mathbf{S}_{\text{raw}}(i) < R_{\text{p}} \\ R_{\text{p}} & \mathbf{S}_{\text{raw}}(i) \geq R_{\text{p}} \end{cases} \tag{1}$$

where $\mathbf{S}_{\text{raw}} \in \mathbb{R}^n$ is the raw data from rangefinder, R_{p} is a distance threshold (radius of the perception area), $\mathbf{S} \in \mathbb{R}^n$ denotes the output distance array, $i = 1, 2, ..., n$.

After raw data processing, we convert S to two-dimensional Cartesian space:

$$\mathbf{S}_{\text{x}}, \mathbf{S}_{\text{y}} = f_{\text{cts}}(\mathbf{S}) \tag{2}$$

where f_{cts} denotes the convert function from $\mathbf{S}$ to $\mathbf{S}_{\text{x}}$, $\mathbf{S}_{\text{y}}$. $\mathbf{S}_{\text{x}}$ and $\mathbf{S}_{\text{y}}$ are two coordinate value arrays on x-axis, y-axis, respectively.

The linear velocity for avoiding obstacles can be obtained via

$$v_{\mathrm{y}} = \eta \sum_{i=1}^{n} \mathbf{S}_{\mathrm{y}}(i) \tag{3}$$

where η denotes obstacle avoidance factor, which is used to adjust obstacle avoidance sensitivity, and n is the length of the array $\mathbf{S}_{\mathrm{y}}$.

4 Mobile Robot Control

First, we define error $\mathbf{e(k)}$ as:

$$\mathbf{e(k)} = [h(k), v_{\mathrm{y}}(k), \alpha(k)]^{\mathrm{T}} \tag{4}$$

where $k = 1, 2, ...$ denotes time kT, T is the period of system data updating.

To keep the slider at a desired position of the slide rail stably, we designed an improved position PID controller to ensure the stable operation of the mechanism.

Considering a traditional positional PID algorithm, the relationship between robot control input $\mathbf{u(k)}$ and error $\mathbf{e(k)}$ is

$$\mathbf{u(k)} = \mathbf{K_P}[\mathbf{e(k)}] + \mathbf{K_I}[\sum_{i=0}^{k} \mathbf{e(i)}] + \mathbf{K_D}[\mathbf{e(k)} - \mathbf{e(k-1)}] \tag{5}$$

where $\mathbf{K_P} = \mathbf{diag}(k_{\mathrm{Px}} \quad k_{\mathrm{Py}} \quad k_{\mathrm{Pz}})$, $\mathbf{K_I} = \mathbf{diag}(k_{\mathrm{Ix}} \quad k_{\mathrm{Iy}} \quad k_{\mathrm{Iz}})$, $\mathbf{diag}()$ denotes diagonal matrix, $\mathbf{K_D} = \mathbf{diag}(k_{\mathrm{Dx}} \quad k_{\mathrm{Dy}} \quad k_{\mathrm{Dz}})$.

Robot control input at time kT is defined as $\mathbf{u(k)} = [v_{\mathrm{cx}}(k), v_{\mathrm{cy}}(k), \omega_{\mathrm{cz}}(k)]^{\mathrm{T}}$, where $v_{\mathrm{cx}}(k)$, $v_{\mathrm{cy}}(k)$, $\omega_{\mathrm{cz}}(k)$ denote the command linear velocity applied to mobile robot on x-axis and y-axis, angular velocity respectively. We can calculate them by the following equations:

$$v_{\mathrm{cx}}(k) = \left(k_{\mathrm{Px}}, 0, 0\right)[\mathbf{e(k)}] + \left(k_{\mathrm{Ix}}, 0, 0\right) M_{vx}(\mathbf{e}) + \left(k_{\mathrm{Dx}}, 0, 0\right)[\mathbf{e(k)} - \mathbf{e(k-1)}] \tag{6}$$

$$v_{\mathrm{cy}}(k) = \left(0, k_{\mathrm{Py}}, 0\right)[\mathbf{e(k)}] + \left(0, k_{\mathrm{Iy}}, 0\right) M_{vy}(\mathbf{e}) + \left(0, k_{\mathrm{Dy}}, 0\right)[\mathbf{e(k)} - \mathbf{e(k-1)}] \tag{7}$$

$$\omega_{\mathrm{cz}}(k) = \left(0, 0, k_{\mathrm{Pz}}\right)[\mathbf{e(k)}] + \left(0, 0, k_{\mathrm{Iz}}\right) M_{wz}(\mathbf{e}) + \left(0, 0, k_{\mathrm{Dz}}\right)[\mathbf{e(k)} - \mathbf{e(k-1)}] \tag{8}$$

where $M_{vx}(\mathbf{e}) = \sum_{i=0}^{k} P_{vx}[\mathbf{e(i)}]$, $M_{vy}(\mathbf{e}) = \sum_{i=0}^{k} P_{vy}[\mathbf{e(i)}]$, $M_{wz}(\mathbf{e}) = \sum_{i=0}^{k} P_{wz}[\mathbf{e(i)}]$. Among them

$$P_{vx}[\mathbf{e(i)}] = \begin{cases} (0,0,0)^{\mathrm{T}}, & \text{if} \quad (|h(i)| \leq \epsilon_v) \\ [h(i), 0, 0]^{\mathrm{T}}, & \text{otherwise} \end{cases} \tag{9}$$

$$P_{vy}[\mathbf{e}(\mathbf{i})] = \begin{cases} (0,0,0)^{\mathrm{T}}, & \text{if} \quad (|v_{\mathrm{y}}(i)| \le \epsilon_v) \\ [0, v_{\mathrm{y}}(i), 0]^{\mathrm{T}}, & \text{otherwise} \end{cases} \tag{10}$$

$$P_{wz}[\mathbf{e}(\mathbf{i})] = \begin{cases} (0,0,0)^{\mathrm{T}}, & \text{if} \quad (|\alpha(i)| \le \epsilon_\omega) \\ [0, 0, \alpha(i)]^{\mathrm{T}}, & \text{otherwise} \end{cases} \tag{11}$$

where ϵ_v is a positive constant, which is a threshold to control the mobile robot linear velocity error. ϵ_ω is a threshold to ensure that angular velocity error keeping in a certain range.

5 Simulations

In this part, an Open Dynamics Engine based simulation environment and an omnidirectional mobile robot with our designed mechanism are built in Gazebo, mainly for testing the performance of co-carrying and obstacle avoidance. The test content of the simulation is three-fold:

1) Obstacle avoidance performance.
2) Partner's motion following error.
3) The velocity smoothness of the object.

Evaluation metrics:

1) The co-carrying system can complete the task without collisions in the obstacle scenario we designed.
2) During the co-carrying process, the position error fluctuates within the allowable range, which means our co-carrying system has certain fault tolerance.
3) The velocity of the object cannot be abruptly changed, and it is smooth enough, which ensures stability and safety.

5.1 Simulation Setup

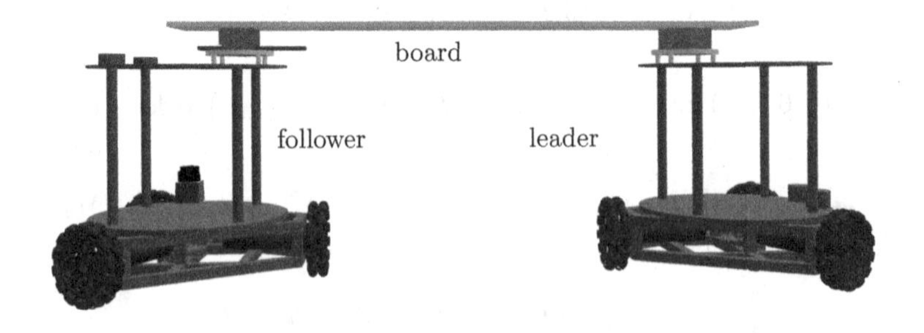

Fig. 4. Co-carrying system in Gazebo.

Gazebo is a well-designed simulator for robot simulation, which offers the ability to accurately and efficiently simulate populations of robots in complex indoor environments, so we use Gazebo as the simulation tool. The human-robot co-carrying system we designed in the simulation environment is shown in Fig. 4.

The simulation diagram is shown in Fig. 5. In Fig. 5, the right one is an omnidirectional mobile robot that can be controlled by a human (leader), we use it to simulate the partner's motion. The robot on the left is an omnidirectional mobile robot (follower), which cooperates with the partner to complete co-carrying tasks in simulations. A rangefinder is installed on the robot center to obtain the surrounding environment information. The Blue dotted rectangle in Fig. 5 is a board used to simulate the object, which is connected to two rotatable points on two omnidirectional mobile robots. "Follow" means the follower following the leader's motion, "Avoid" means the robot performing obstacle avoidance.

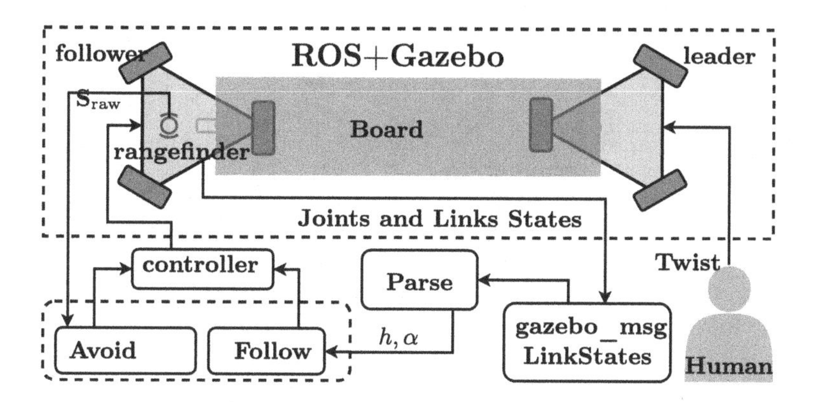

Fig. 5. Simulation diagram. Where yellow points represent two rotatable points, which are used to connect the two ends of the object. The grey rectangle represents a board. The Blue circle is a rangefinder. Twist, LinkStates are two ROS message types. (Color figure online)

We designed two scenarios to test the performance of our co-carrying system in Gazebo as follows:

Scenario 1. This scenario contains some irregularly placed obstacles, we manually control the leader to guide the robot's moving. The follower must ensure that it does not hit obstacles during several tests and the objects it carries do not fall.

Scenario 2. There are no obstacles in this scenario, which is used to test the co-carrying system's performance when it is backward. We use the keyboard to control the leader to move backward, while the follower needs to follow the leader's motion and ensure that the board moves smoothly.

5.2 Simulation Results

For the performance show of obstacle avoidance and following, we show the trajectory of the co-carrying system during the entire task execution process, including the trajectory of the follower, the trajectory of the leader, and the partial trajectory of the board. To more clearly show the performance of the follower following, we connect the center point of the slider with the leader's grasping point.

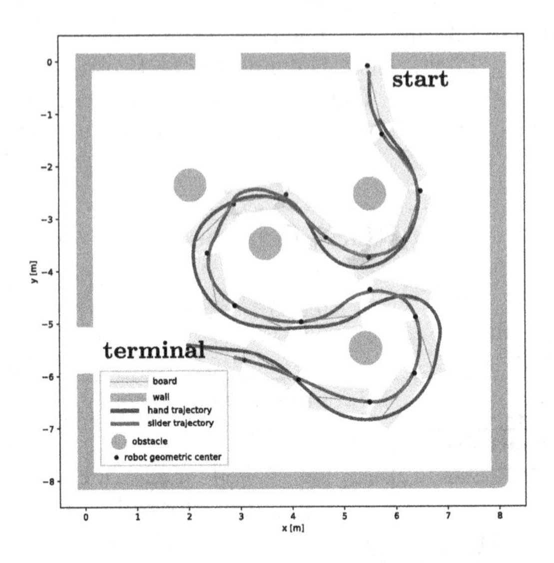

Fig. 6. Trajectories of the co-carrying system while performing obstacle avoidance.

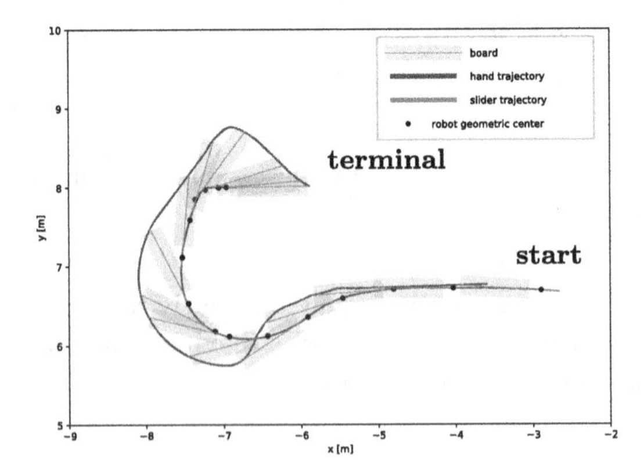

Fig. 7. Trajectories of co-carrying system while performing moving backward.

Co-carrying Performance. Figure 6 shows trajectories in the process of performing obstacle avoidance, the length of the board is 1.2 m and the width is 0.4 m. Figure 7 shows trajectories in the process of performing moving backward.

Position Error. Figure 8 corresponds to Fig. 6, they are from the same test.

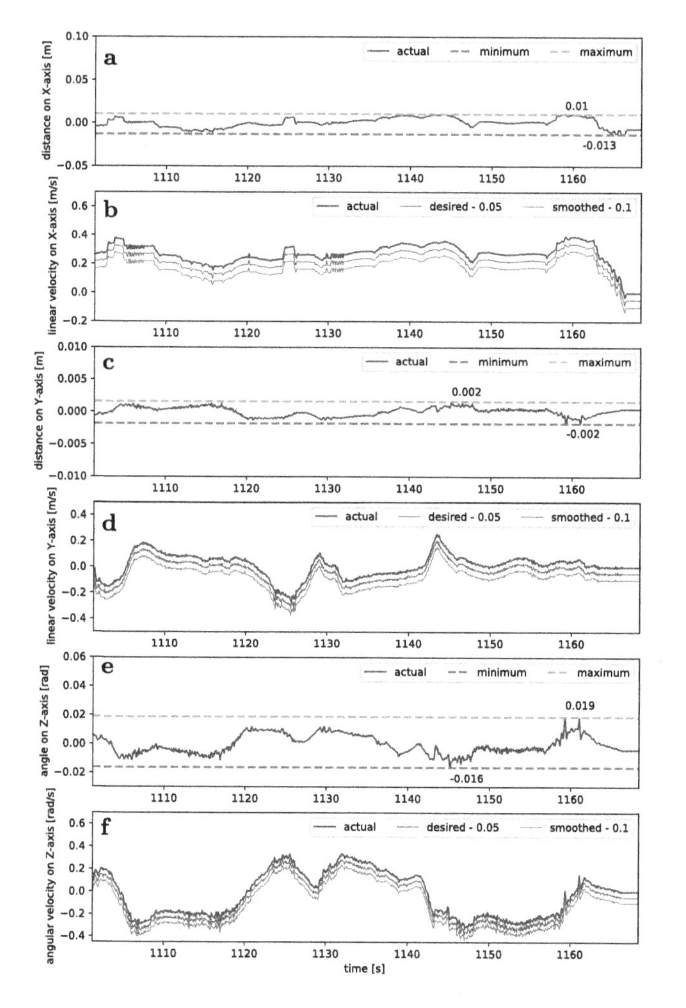

Fig. 8. States of the co-carrying system while performing obstacle avoidance. The desired values of a, c, and e are 0, 0, 0, respectively, and these three subfigures show the following position errors. Subfigures b, d and f show the velocity values of the follower.

In general, when the co-carrying system performs the co-carrying tasks, the follower's following error is within an allowable range (within the tolerance of the system), which can ensure that the board is carried safely. For the relationship of the following error with velocity, it can be seen in Fig. 8. Obviously, when the linear velocity increases sharply, the follower's following error will increase, but it will not exceed the allowable range of the system.

Velocity Smoothness. Figure 9 shows the real-time velocity of the board during being carried. From the perspective of linear velocity and angular velocity, it is smooth, which means the board is moving smoothly and stably. The enlarged parts are to show the smoothness of the velocity more clearly.

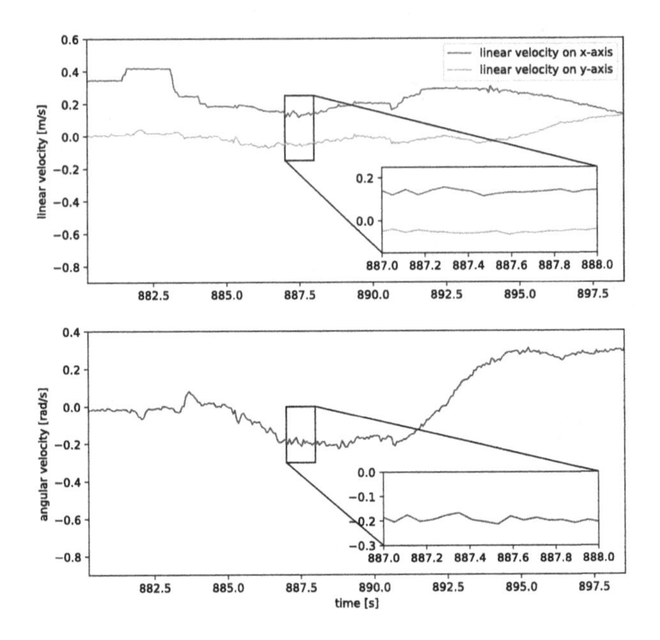

Fig. 9. Real-time board velocity when the co-carrying system performs obstacle avoidance.

6 Conclusion

This paper proposed a rangefinder-based obstacle avoidance algorithm for the omnidirectional mobile robot in human-robot co-carrying. An improved positional PID controller was employed to drive the mobile robot. In addition, a tolerance mechanism was designed to improve the robustness of the co-carrying system. Finally, the effectiveness of the obstacle avoidance algorithm was verified in simulations.

References

1. He, W., Li, Z., Chen, C.L.P.: A survey of human-centered intelligent robots: issues and challenges. IEEE/CAA J. Automatica Sinica **4**(4), 602–609 (2017)

2. Kong, L., He, W., Chen, W., Zhang, H., Wang, Y.: Dynamic movement primitives based robot skills learning. Mach. Intell. Res. (2022)
3. Truong, X.T., Ngo, T.D.: Toward socially aware robot navigation in dynamic and crowded environments: a proactive social motion model. IEEE Trans. Autom. Sci. Eng. **14**(4), 1743–1760 (2017)
4. Fu, L., Zhao, J.: Robot compliant catching by Maxwell model based Cartesian admittance control. Assem. Autom. **41**(2), 133–143 (2021)
5. Li, Z., Li, X., Li, Q., Su, H., Kan, Z., He, W.: Human-in-the-loop control of soft exosuits using impedance learning on different terrains. IEEE Trans. Robot. (2022)
6. Liu, Z., Han, Z., Zhao, Z., He, W.: Modeling and adaptive control for a spatial flexible spacecraft with unknown actuator failures. Sci. China Inf. Sci. **64**(5), 1–16 (2021). https://doi.org/10.1007/s11432-020-3109-x
7. He, W., Mu, X., Zhang, L., Zou, Y.: Modeling and trajectory tracking control for flapping-wing micro aerial vehicles. IEEE/CAA J. Automatica Sinica **8**(1), 148–156 (2020)
8. Hirata, Y., Kosuge, K.: Distributed robot helpers handling a single object in cooperation with a human. In: Proceedings 2000 IEEE International Conference on Robotics and Automation, ICRA, Millennium Conference, Symposia Proceedings (Cat. No.00CH37065), vol. 1, pp. 458–463 (2000)
9. Hirata, Y., Takagi, T., Kosuge, K., Asama, H., Kaetsu, H., Kawabata, K.: Map-based control of distributed robot helpers for transporting an object in cooperation with a human. In: Proceedings 2001 IEEE International Conference on Robotics and Automation (ICRA) (Cat. No.01CH37164), vol. 3, pp. 3010–3015 (2001)
10. Hirata, Y., Takagi, T., Kosuge, K., Asama, H., Kaetsu, H., Kawabata, K.: Motion control of multiple DR helpers transporting a single object in cooperation with a human based on map information. In: Proceedings 2002 IEEE International Conference on Robotics and Automation (Cat. No. 02CH37292), vol. 1, pp. 995–1000 (2002)
11. Yu, X., He, W., Li, Q., Li, Y., Li, B.: Human-robot co-carrying using visual and force sensing. IEEE Trans. Industr. Electron. **68**(9), 8657–8666 (2021)
12. Lin, H., Zhao, B., Liu, D., Alippi, C.: Data-based fault tolerant control for affine nonlinear systems through particle swarm optimized neural networks. IEEE/CAA J. Automatica Sinica **7**(4), 954–964 (2020)
13. Zhou, T., Chen, M., Yang, C., Nie, Z.: Data fusion using Bayesian theory and reinforcement learning method. Sci. China Inf. Sci. **63**(7), 1–3 (2020)
14. Cheng, C., et al.: Stability control for end effect of mobile manipulator in uneven terrain based on active disturbance rejection control. Assem. Autom. **41**(3), 369–383 (2021)
15. Khatib, O.: Real-time obstacle avoidance for manipulators and mobile robots. In: Proceedings of the 1985 IEEE International Conference on Robotics and Automation, vol. 2, pp. 500–505 (1985)
16. Tang, L., Dian, S., Gu, G., Zhou, K., Wang, S., Feng, X.: A novel potential field method for obstacle avoidance and path planning of mobile robot. In: 2010 3rd International Conference on Computer Science and Information Technology, vol. 9, pp. 633–637 (2010)
17. Zhou, L., Li, W.: Adaptive artificial potential field approach for obstacle avoidance path planning. In: 2014 7th International Symposium on Computational Intelligence and Design, vol. 2, pp. 429–432 (2014)
18. Li, X., Fang, Y., Fu, W.: Obstacle avoidance algorithm for multi-UAV flocking based on artificial potential field and Dubins path planning. In: 2019 IEEE International Conference on Unmanned Systems (ICUS), pp. 593–598 (2019)

19. Liu, Y., Ding, P., Wang, T.: Autonomous obstacle avoidance control for multi-UAVs based on multi-beam sonars. In: Global Oceans 2020: Singapore - U.S. Gulf Coast, pp. 1–5 (2020)
20. Fu, Q., Wang, J., Gong, L., Wang, J., He, W.: Obstacle avoidance of flapping-wing air vehicles based on optical flow and fuzzy control. Trans. Nanjing Univ. Aeronaut. Astronaut. **38**(2), 206 (2021)
21. Fu, Q., Wang, X., Zou, Y., He, W.: A miniature video stabilization system for flapping-wing aerial vehicles. Guidance Navig. Control **2**(01), 2250001 (2022)
22. Huang, H., He, W., Wang, J., Zhang, L., Fu, Q.: An all servo-driven bird-like flapping-wing aerial robot capable of autonomous flight. IEEE/ASME Trans. Mechatron. (2022). IEEE

Design and Analysis of a Novel Magnetic Adhesion Robot with Passive Suspension

Hao Xu, Youcheng Han, Weizhong Guo(✉), Mingda He, and Yinghui Li

State Key Laboratory of Mechanical System and Vibration, School of Mechanical Engineering, Shanghai Jiao Tong University, Shanghai 200240, China
wzguo@sjtu.edu.cn

Abstract. The steel lining of large facilities is an important structure that experiences extreme environments and requires periodical inspection after manufacture. However, due to the complexity of their internal environments (crisscross welds, curved surfaces, etc.), high demands are placed on stable adhesion and curvature adaptability. This paper presents a novel wheeled magnetic adhesion robot with passive suspension named NuBot, which is mainly applied in nuclear power containment. Based on the kinematic model of the 3-DOF independent suspension, a comprehensive optimization model is established, and global optimal dimensions are properly chosen from performance atlases. Then, a safety adhesion analysis considering non-slip and non-overturning condition is conducted to verify the magnetic force meet the safety demand. Experiments show that the robot can achieve precise locomotion on both strong and weak magnetic walls with different inclination angles, and can stably cross the 5 mm weld seam. Besides, its maximum payload capacity reaches 3.6 kg. Results show that NuBot has good comprehensive capabilities of surface-adaptability, adhesion stability, and payload. Besides, the robot can be applied in more ferromagnetic environments with more applications and the design method offers guidance for similar wheeled robots with passive suspension.

Keywords: Wheeled robot · Passive suspension · Magnetic adhesion · Steel lining inspection

1 Introduction

Currently, the nuclear power industry widely adopts steel lining construction as the inner wall of security containment, which is the last barrier against accidental release of radioactive material [1]. Considering the steel lining consists of thousands of welds for hundreds of curved steel plates, it is quite important to inspect the welds periodically. Traditionally, the welds are mainly inspected by the workers standing on the scaffold, with disadvantages of long-period time, great danger, and high cost. Hence, it is significant to develop a robot platform to inspect the welds. Existing research mainly focuses on the surface adaptation and adhesion method:

The surface-adaptive locomotion mechanism to move on the steel lining and adapt to the surface. In previous research, the climbing robots include: a) the crawler robots.

H. Liu et al. (Eds.): ICIRA 2022, LNAI 13456, pp. 369–380, 2022.
https://doi.org/10.1007/978-3-031-13822-5_33

Sitti [2] adopted active compliant joints and Fan [3] adopted the linkage-spring system to connect two track modules. However, the active joint increases the number of actuators, and the hard metal track scratches the steel lining plating; b) the legged robots. Guan [4] imitated the inchworm and proposed a biped climbing robot; Seo [5] presented a robot with eight footpads and one active *degree of freedom* (DOF). However, the discrete gait of the legged locomotion slows the speed; c) the wheeled robots with no more than three wheels [6–8] can always keep each wheel in contact with the surface. However, more wheels mean stronger adhesion force and a larger driving force. Considering the unevenness of most surfaces, an adaptive suspension or compliant mechanism is needed: Eto [9] developed a passive rocker arm suspension and spherical wheels, and Guo [10] proposed a robot with an electromagnetically driven compliant beam to adapt to the surface. In conclusion, the wheeled robot with passive suspension is a feasible option, which offers sufficient DOFs to ensure all wheels adapt to the wall without increasing actuators.

The stable and practical adhesive approach to attach to the steel lining. The adhesion method is the direct factor to determine the adhesion stability and payload. Generally, magnetic adhesion [6–8, 11–13] works well on the ferromagnetic surface, which is an effective choice for the steel lining. In contrast, pneumatic adhesion [4, 14] can be employed for more occasions. However, the required pump limits the size and weight, and the adhesion force will decrease on the rough surface. Mechanical adhesion [15–17] imitates the structure of insects, hooking tiny grooves on rough walls. But they cannot adapt to the steep and smooth walls. Considering environmental constraints, NuBot is expected to be endowed with comprehensive capabilities of surface adaptability, adhesion stability, surface harmlessness, and lightweight. Based on the above analysis and demands, magnetic adhesion is the optimal option for the steel lining.

This paper proposes a magnetic adhesion robot called NuBot to inspect the steel lining weld in nuclear power containment. As aforementioned, the magnetic adhesion wheeled robot with passive suspension is the optimal option in the special environment. Hence, this research will mainly focus on the kinematic modeling and parameter optimization of the robot. And experiments are conducted to verify its locomotion.

2 Overall Design and Modeling of the Robot

Figure 1(a) and (b) show the structure of NuBot, including two passive suspensions, six magnetic adhesion wheels, the image capture system, and a control system. When inspecting the weld seam inside the containment, the technician controls NuBot to move and monitor the images captured by the camera. During locomotion, the 3-DOF suspension ensures that all wheels are in contact with the surface of steel lining, and adapt to the surrounding environment like the support structure.

The suspension can be seen as a five-bar mechanism, and its kinematic model can be established. As shown in Fig. 2, l_4, l_5, l_6 are regarded as three inputs, and the relative motions (two translations and one rotation) of the suspension to the steel lining are considered as outputs. Hence, *B*, *C*, and *E* are three independent point positions that are chosen to install three wheels, as a reflection of three outputs.

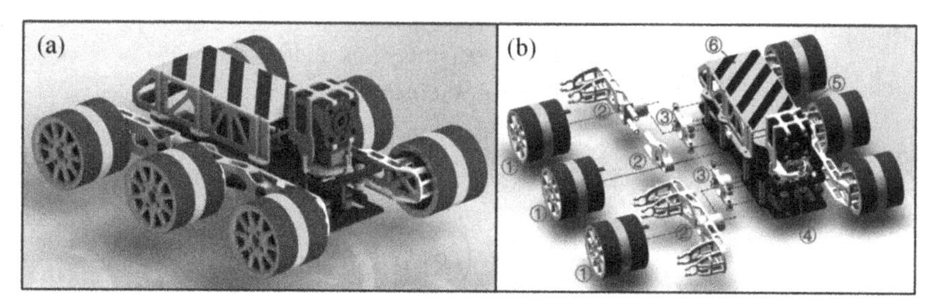

Fig. 1. (a) Axonometric view of the NuBot; (b) Exploded view of the NuBot, ① magnetic adhesion wheel, ② suspension, ③ support, ④ chassis, ⑤ camera, ⑥ shell.

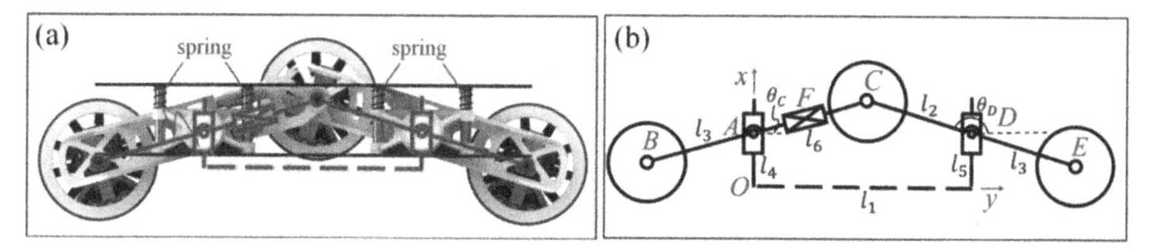

Fig. 2. Sketch of (a) the suspension structure and (b) the mechanism.

The positions of B, C, and E are all taken as the outputs, so the position kinematics of the suspension can be defined by a loop-closure equation given as

$$\begin{aligned} \boldsymbol{r}_{OA} + \boldsymbol{r}_{AC} &= \boldsymbol{r}_{OD} + \boldsymbol{r}_{DC} \\ \boldsymbol{r}_{OB} &= \boldsymbol{r}_{OA} + \boldsymbol{r}_{AB} \\ \boldsymbol{r}_{OE} &= \boldsymbol{r}_{OD} + \boldsymbol{r}_{DE} \end{aligned} \tag{1}$$

where $\boldsymbol{r}_{OA}$ is a vector from O to A, and other vectors have the same naming rule.

Next, three differential kinematic equations can be derived to denote the influence of outputs on the inputs. According to the mechanism theory, the linkage between C and F has three DOFs, while AB and DE are 2-DOF linkages, due to the redundancy of the θ_C and θ_D. The inverse Jacobian matrix of C can be obtained as $\dot{\boldsymbol{q}} = \left[\dot{l}_4, \dot{l}_5, \dot{l}_6\right] = \boldsymbol{J}_{IK}^{C}\boldsymbol{v}_C$, where $\boldsymbol{q}$ denotes the input matrix, $\boldsymbol{v}_C$ is the velocity of point C, and $\boldsymbol{J}_{IK}^{C}$ is the inverse Jacobian matrix expressed by

$$\boldsymbol{J}_{IK}^{C} = \begin{bmatrix} -\tan\theta_C & 1 - \frac{x_C}{\cos^2\theta_C} \\ \pm\frac{l_1 - x_C}{\sqrt{l_2^2 - (l_1 - x_C)^2}} & 1 & 0 \\ \frac{1}{\cos\theta_C} & 0 & \frac{x_C\sin\theta_C}{\cos^2\theta_C} \end{bmatrix}, \boldsymbol{v}_C = \begin{bmatrix} \dot{x}_C \\ \dot{y}_C \\ \dot{\theta}_C \end{bmatrix} \tag{2}$$

The inverse Jacobian matrices of linkages AB and DE can be obtained from pseudo-inverse of Jacobian matrices as $\boldsymbol{v}_B = \left[\dot{x}_B, \dot{y}_B\right]^T = \boldsymbol{J}_{FK}^{B}\dot{\boldsymbol{q}}$, and $\boldsymbol{v}_E = \left[\dot{x}_E, \dot{y}_E\right]^T = \boldsymbol{J}_{FK}^{E}\dot{\boldsymbol{q}}$, where $\boldsymbol{v}_B$ and $\boldsymbol{v}_E$ are the velocities of wheels B and E, and $\boldsymbol{J}_{FK}^{B}$ and $\boldsymbol{J}_{FK}^{E}$ are their Jacobian matrices obtained by

$$\boldsymbol{J}_{FK}^{B} = \begin{bmatrix} \kappa_1\sin\theta_D & -\kappa_1\sin\theta_D & \kappa_1\cos(\theta_C - \theta_D) \\ 1 - \kappa_2\sin\theta_D & \kappa_2\sin\theta_D & -\kappa_2\cos(\theta_C - \theta_D) \end{bmatrix} \tag{3}$$

$$J_{FK}^{E} = \begin{bmatrix} \kappa_4 \sin\theta_D & -\kappa_4 \sin\theta_D & \kappa_3 \sin\theta_D \\ \kappa_4 \cos\theta_D & 1 + \kappa_4 \cos\theta_D & -\kappa_3 \cos\theta_D \end{bmatrix} \tag{4}$$

where

$$\kappa_1 = \frac{l_3 \sin\theta_C}{l_6 \sin(\theta_C - \theta_D)} \qquad \kappa_2 = \frac{l_3 \cos\theta_C}{l_6 \sin(\theta_C - \theta_D)}$$
$$\kappa_3 = \frac{l_3}{l_2 \sin(\theta_C - \theta_D)} \; \kappa_4 = \left(\cos\theta_D + \frac{\sin\theta_D}{\tan(\theta_C - \theta_D)}\right)\frac{l_3}{l_2}$$

Hence

$$J_{IK}^{B} = \left(J_{FK}^{B}\right)^{\dagger}, J_{IK}^{E} = \left(J_{FK}^{E}\right)^{\dagger} \tag{5}$$

where J_{IK}^{B} and J_{IK}^{E} represent the inverse Jacobian matrices of wheels B and E, and the symbol $\dagger$ denotes pseudo-inverse.

3 Optimization of the Suspension

To analyze the relations between the suspension parameters and performances of NuBot, we utilize the performance atlas [18] for its visualization and global optimization. To establish the optimization model, the design variables are chosen to be l_1, l_2, l_3, while l_4, l_5, l_5 are prismatic joint variables. Besides, the translational and rotational ranges of all joints are chosen to be the constraint conditions. Finally, three criteria (i.e. objective functions) are employed to give a comprehensive evaluation [19]:

(1) The *workspace area index* (WAI) is utilized to describe the motion range that determines the surface adaptability formed by the wheels to traverse the surface. The wheels should reach more areas to satisfy different situations, and can be expressed as

$$\eta_W = \int_S dS = \int_{x_{min}}^{x_{max}} dx \int_{y_{min}}^{y_{max}} dy \tag{6}$$

where x and y are the integral variables of the three outputs workspace.

(2) The *global payload index* (GPI) is used to evaluate the extremum payload force that the suspension can bear. The force equation is $F = G\tau$, where F is the output force at wheel centers while τ is the input force at prismatic joints A, D, and F, and G_{IK} is the inverse force Jacobian matrix. With $G = J_{IK}^{-T}$, the GPIs can be given by

$$\eta_{Fmax} = \frac{\int_S \sqrt{\max(|\lambda_{Fi}|)} dS}{\int_S dS}, \eta_{Fmin} = \frac{\int_S \sqrt{\min(|\lambda_{Fi}|)} dS}{\int_S dS} \tag{7}$$

where λ_{Fi} are the eigenvalues of the matrix $G^T G$.

(3) The *global stiffness index* (GSI) is used to evaluate the extreme linear displacement capacity. The suspension should possess larger mechanism stiffness to produce linear displacement in the input joints, so the robot can adapt to the surface with

the displacement of the input joints. The deformation equation is: $\boldsymbol{P} = \boldsymbol{CF}$, where $\boldsymbol{P}$ is the linear displacement of the three prismatic joints, $\boldsymbol{C} = \boldsymbol{J}_{FK}\boldsymbol{J}_{FK}^{T}$ are the compliance matrix. So the GSIs can be given by

$$\eta_{Pmax} = \frac{\int_S \sqrt{\max(|\lambda_{Pi}|)}dS}{\int_S dS}, \eta_{Pmin} = \frac{\int_S \sqrt{\min(|\lambda_{Pi}|)}dS}{\int_S dS} \tag{8}$$

where λ_{Pi} are the eigenvalues of the matrix $\boldsymbol{C}^T\boldsymbol{C}$.

There are three optimization parameters in the suspension model, including l_1, l_2 and l_3. The mean of the parameters is obtained by $(l_1 + l_2 + l_3)/3 = T$. And we can obtain the normalized equation by $t_1 + t_2 + t_3 = 3$, where $t_i = l_i/T(i = 1, 2, 3)$. Thus, the optimization parameters are converted to $t_i(i = 1, 2, 3)$.

Considering the constraint $l_2 < l_1$, the constraint equations can be given by

$$\begin{cases} t_2 < t_1 \\ 0 < t_1, t_3 < 3 \\ 0 < t_2 < 1.5 \end{cases} \tag{9}$$

Thus, the parameter design space of the suspension can be shown in Fig. 3.

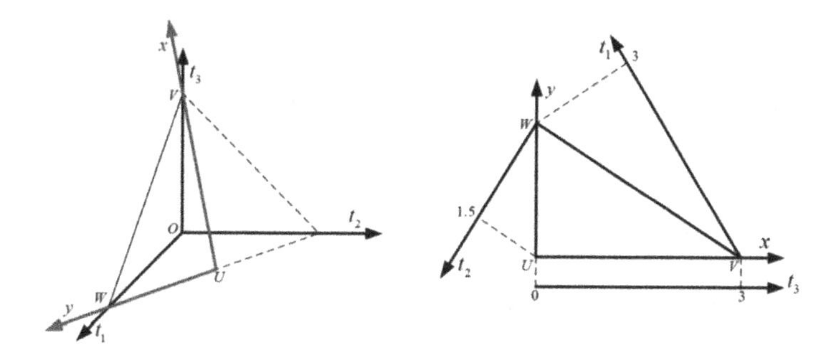

Fig. 3. Parameter design space of the suspension.

The mapping function between frame $O - t_1t_2t_3$ and frame $U - xyz$ is

$$\begin{cases} x = (3 - t_1 + t_2)/\sqrt{2} \\ y = (3 - t_1 - t_2)/\sqrt{6} \end{cases} \tag{10}$$

Consequently, the ASI, GPI and GSI performance atlases of the suspension from wheels *B*, *C*, and *E* can be obtained. Herein we take the GPIs (denoted as η_{Fmax}, η_{Fmin}) and GSIs (denoted as η_{Pmax}, η_{Pmin}) atlases of wheel *B* as an example. As shown in Fig. 4(a), they show different distribution principles with the change of design variables. Considering a larger payload capacity of the suspension, the solution domain is finally assigned as $\Omega_{\text{GPI}} = \{(t_1, t_2, t_3)|\eta_{Fmin}^B \geq 0.2, \eta_{Fmin}^C \geq 0.7, \eta_{Fmin}^E \geq 0.5\}$. Likewise, due to better adaptability to the complex surface, the solution domain is assigned as $\Omega_{\text{GSI}} = \{(t_1, t_2, t_3)|\eta_{Pmin}^B \geq 0.5, \eta_{Pmin}^C \geq 0.9, \eta_{Pmin}^E \geq 0.3\}$.

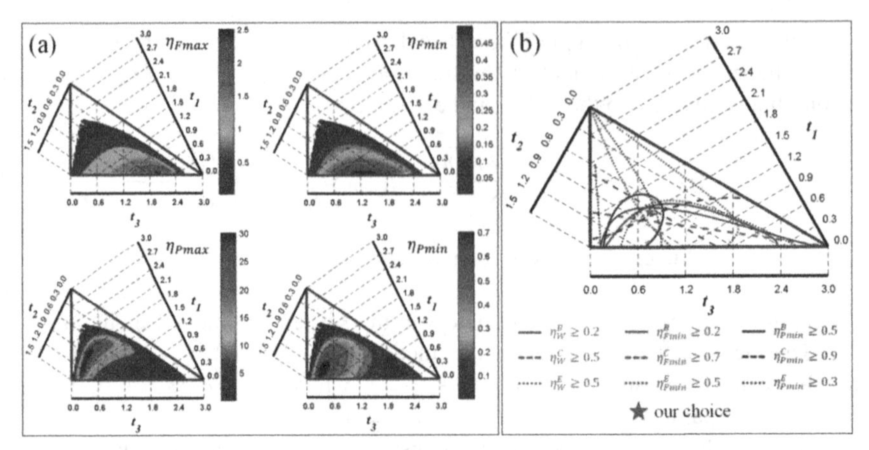

Fig. 4. The performance atlases of wheel B: (a) η_{Fmax}, (b) η_{Fmin}, (c) η_{Pmax}, (d) η_{Pmin}. The global solution domain and the optimum solution of suspension.

The above results provide us with reliable evidence to compare these parameters to satisfy the optimum comprehensive performances of the suspension. Therefore, we can conclude that the optimum domain solution of the suspension (denoted as Ω_{susp}) is the intersection set of Ω_{WAI}, Ω_{GPI}, and Ω_{GSI}. Further, if we assign more strict demands to each sub-domain (Ω_{WAI}, Ω_{GPI}, or Ω_{GSI}), the size of Ω_{susp} will be smaller and smaller and tends to be a unique value. In this case, the global optimum solution is finally identified. Besides, it can also be assigned directly in Ω_{susp} according to the designer's initiative (Fig. 4(b)). Following a series of trade-offs among WAI, GPI, and GSI, the red star represents our final identification: the global optimum non-dimensional solution with $t_1 = 1.5$ and $t_2, t_3 = 0.75$, that is, $l_1 : l_2 : l_3 = 2 : 1 : 1$.

4 Safety Adhesion Analysis

Safety is the primary factor to prevent the robot from falling from the surface. Considering the magnetic force decreases when NuBot crosses the weak-magnetic weld, the experimental magnetic forces under four typical situations in Fig. 5(a) are obtained by slowly pulling a single wheel until the wheel is separated from the wall. In Fig. 5(b), the largest magnetic force is 46.75 N in case 1; the smallest force is only 31.26 N in case 3. When moving on the steel lining, there is no protection for the robot from falling. Therefore, the magnet must provide sufficient adhesion force, which means non-slip and non-overturning conditions must be satisfied.

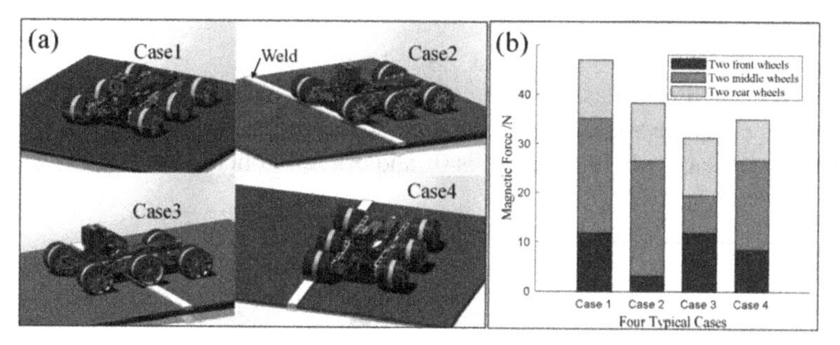

Fig. 5. (a) Four typical robot states on the wall. Case 1: all wheels on the wall; Case 2: two front or rear wheels on the weld; Case 3: two middle wheels on the weld; Case 4: three left or right wheels on the weld. (b) The experimental magnetic forces of the wheels under different cases.

4.1 Non-slip Condition

Firstly, the adhesion force should be large enough to prevent slippage:

$$\begin{cases} \mu_s F_N \geq Gsin\varphi \\ F_N = F_M/k - Gcos(\pi - \varphi) \end{cases} \tag{11}$$

where μ_s is the sliding friction coefficient between the wheel and the wall. G is the gravity of NuBot. φ is the inclination of the wall, which ranges from 90° to 180° on the inner containment. F_N is the positive pressure on the wall, F_M is the total magnetic adhesion force, and k is the safety factor.

By substituting the maximum and minimum magnetic forces into Eq. 11, we can obtain in Fig. 6 that the magnetic force to meet the non-slip condition reaches the maximum when the inclination is 116.6°. Among the four cases, the robot in case 1 is the safest, with a safety factor of 2.133; and case 3 has a minimum safety factor of 1.425.

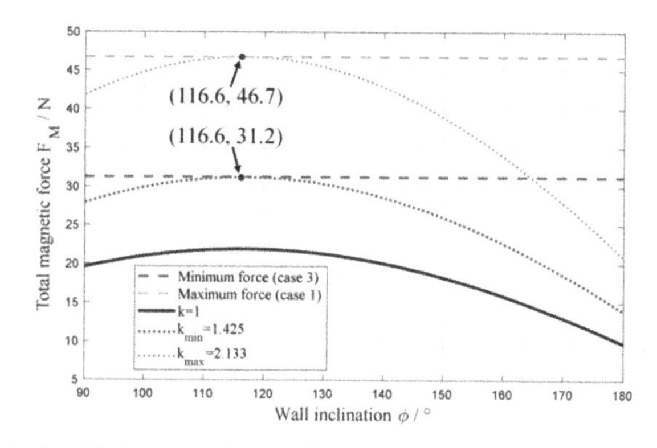

Fig. 6. Relationship between the required magnetic force and surface inclination.

4.2 Non-overturning Condition

Overturning is another occasion to affect the adhesion safety of NuBot. The pitch of NuBot is defined as the angle between its forward direction and the horizon. Herein we only consider the case where the pitch is 0° and 90°, and obtain

$$\begin{cases} 2l_{lr}F_{lr} \geq -kGcos\left[\varphi + \text{atan}\left(\frac{h}{l_{lr}}\right)\right]\sqrt{h^2 + {l_{lr}}^2} \\ l_{fr}(2F_{fr} + F_m) \geq -kGcos\left[\varphi + \text{atan}\left(\frac{h}{l_{fr}}\right)\right]\sqrt{h^2 + {l_{fr}}^2} \end{cases} \tag{12}$$

where l_{lr} and l_{fr} denote half of the wheelbase of the adjacent left and right wheels, and the front and rear wheels; h is the height of the centroid; F_{lr}, F_{fr}, and F_m refers to the magnetic force from the left or right three wheels, the two front or rear wheels, and the two middle wheels, respectively. And the left parts of the inequalities represent the magnetic moment under the two pitches.

Based on the experimental magnetic force, the minimum moment of the force when the pitch is 0° and 90° can be obtained as 1.74 N·m in case 4 and 2.97 N·m in case 2, respectively. One can see from Fig. 7 that the magnetic moment to meet the non-overturning condition first increases and then decreases with the increase of inclination angle. And the minimum safety factor k is 2.2 and 2.9 when the pitch is 0° and 90°, respectively. Thus, it can be concluded that the robot can meet the non-overturning conditions when on steel lining.

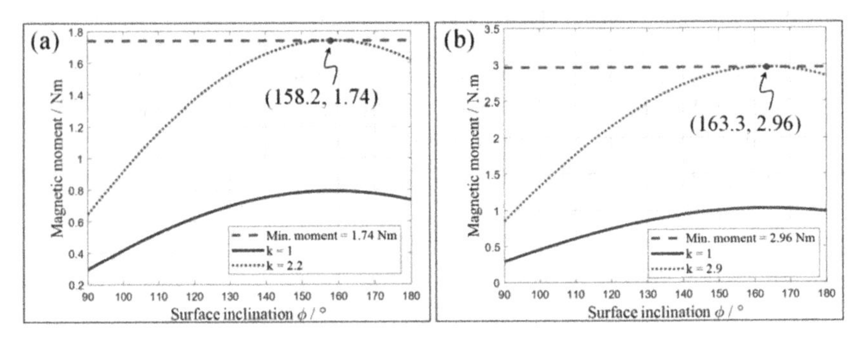

Fig. 7. Relationship between the required magnetic moment and surface inclination with the pitch angle of (a) 0° and (b) 90°.

5 Prototype and Experiment

5.1 Prototype Design

So far, the design parameters and specific structure of NuBot have been obtained. After topology optimization of each part (Fig. 8(a)), the prototype with six magnetic wheels, two pairs of suspensions, and a platform is developed, as shown in Fig. 8(b).

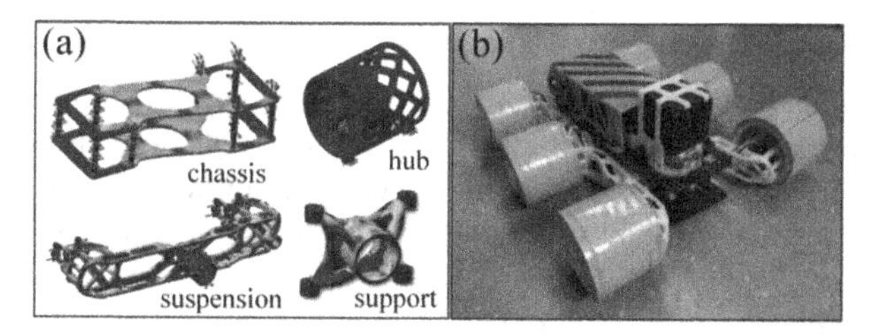

Fig. 8. (a) Topology optimization of NuBot parts; (b) The NuBot prototype.

5.2 Locomotion on Different Types of Walls

Figure 9(a) shows the locomotion on a strong magnetic curved wall with a maximum speed of 0.5 m/s, which is around 1.92 times the robot length. Figure 9(b) shows the locomotion on a weak magnetic blackboard with a minimum turning radius of 0.2 m. Locomotion experiments on the blackboard with an inclination angle of 180° are conducted in Fig. 9(c), which validate the adhesion stability of NuBot.

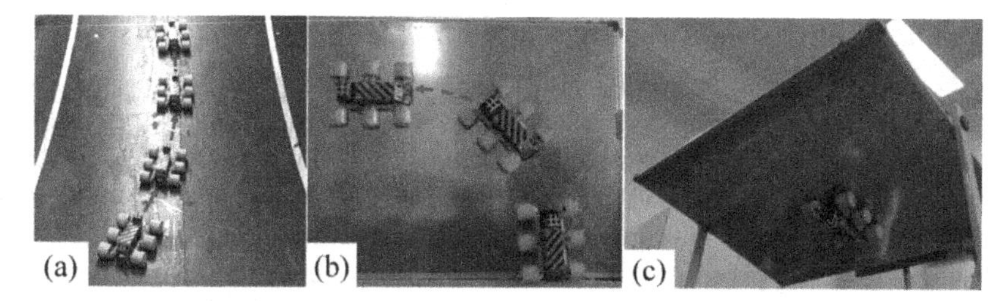

Fig. 9. Locomotion on (a) strong magnetic wall; (b) vertical blackboard; (c) inverted blackboard.

Experiments on the vertical blackboard are conducted to verify the precise movement. The desired trace can be obtained through the given wheel speed on both sides. And the actual trace is obtained by accumulating the mileage from the wheel encoders. In Fig. 10, the robot turns around a circle at a constant speed. The actual trace coincides well with the desired one, and the trace error tends to be a minor value.

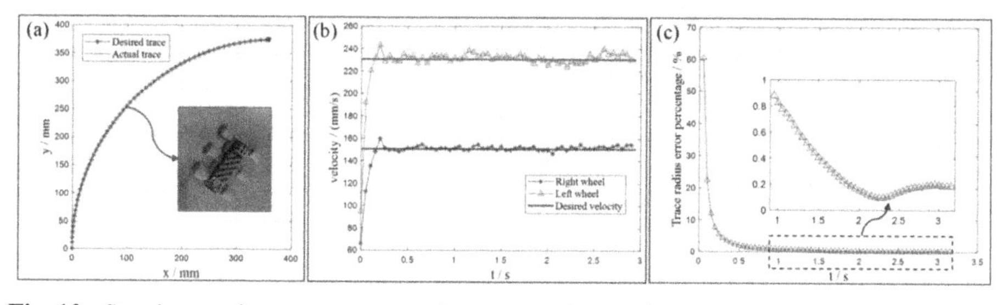

Fig. 10. Steering motion accuracy experiments. (a) Comparison between the actual trace and desired trace (the desired radius is 375 mm); (b) Change of left and right wheel speed (desired wheel speed: left 230 mm/s and right 150 mm/s); (c) Error percentage of trace radius.

5.3 Validation of Weld-crossing Capacity

Crossing the welds may cause instability of the robot body, which in turn affects the adhesion stability. Figure 11 shows the weld-crossing experiments with 3D printed welds with two typical heights of 3 mm and 5 mm (the maximum weld height). Results show that the robot can still keep stable during the process of crossing welds.

Fig. 11. The robot crosses the weld with a height of (a)–(c): 3 mm; (d)–(e): 5 mm.

5.4 Validation of Payload Capacity

Payload capacity is a significant index of the climbing robot, and Fig. 12 shows experiments of payload capacity under various walls. The maximum load of the robot on the blackboard is 0.6 kg (Fig. 12(a)), and the wheels will slip when the load increases. Besides, the load on the curved wall reaches 0.8 kg (Fig. 12(b)). In either case, the robot can load enough weight of the equipment for weld inspection.

Besides, extreme payload tests are conducted to verify the payload capacity (Fig. 12(c)-(f)). NuBot in Fig. 12(c) possesses the maximum payload of 3.6 kg. Notably, the magnetic force in Fig. 12(f) is not the smallest, but the payload is the worst, which indicates the asymmetry of magnetic force distribution reduces the payload capacity.

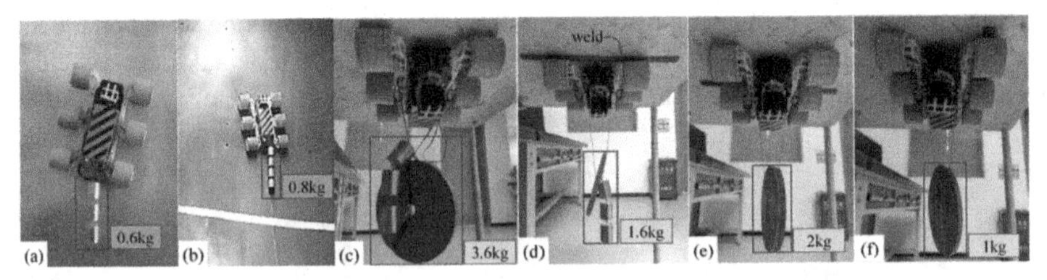

Fig. 12. Payload capacity experiments. Locomotion on the (a) vertical blackboard, (b) curved wall; Static adhesion to the inverted wall with: (c) all wheels on the wall, (d) two front, (e) two middle, (f) two lateral wheels crossing the weld.

6 Conclusion

This paper proposes a magnetic adhesion robot with the passive suspension and six wheels named NuBot, which is used to inspect the steel lining weld of the ferromagnetic wall (nuclear power containment is chosen as an application example here).

Based on the kinematic model, the suspension is reasonably optimized utilizing the performance atlas method considering the performance indices of WAI, GPI, and GSI. And optimal parameters are properly chosen from the performance atlases perceptually. Then, the structures of the robot, especially the magnet unit inside the wheel can be designed with the assistance of engineering software. In addition, the electronic system with a PI controller is constructed to accomplish a higher control accuracy.

Experiments show that NuBot can move with a maximum speed of 0.5 m/s—1.92 times the robot length per second. And the actual locomotion trace has a small error compared to the desired one. Besides, the robot can stably cross the welds and move on the walls with various inclination angles. In addition, the robot reaches a maximum load of 3.6 kg, which is 3.67 times its mass. In conclusion, NuBot meets the comprehensive requirements of size, weight, locomotion, payload, and adhesion stability.

Acknowledgments. This work was supported in part by Shanghai Nuclear Star Nuclear Power Technology Co., Ltd., National Natural Science Foundation of China (Grant No. 51735009), and State Key Lab of Mechanical System and Vibration Project (Grant No. MSVZD202008).

References

1. Sun, H.H., Cheng, P.D., Miu, H.X., Zhang, W.Z., Zhu, X.G., Weng, M.H.: Third Generation Nuclear Power Technology AP1000, 2nd edn. China Electric Power Press, Beijing (2010)
2. Seo, T., Sitti, M.: Tank-like module-based climbing robot using passive compliant joints. IEEE/ASME Trans. Mechatron. **18**(1), 397–408 (2012)
3. Gao, F., Fan, J., Zhang, L., Jiang, J., He, S.: Magnetic crawler climbing detection robot basing on metal magnetic memory testing technology. Robot. Auton. Syst. **125**, 103439 (2020)
4. Guan, Y., et al.: A modular biped wall-climbing robot with high mobility and manipulating function. IEEE/ASME Trans. Mechatron. **18**(6), 1787–1798 (2012)
5. Liu, Y., Kim, H., Seo, T.: AnyClimb: a new wall-climbing robotic platform for various curvatures. IEEE/ASME Trans. Mechatron. **21**(4), 1812–1821 (2016)
6. Song, W., Jiang, H., Wang, T., Ji, D., Zhu, S.: Design of permanent magnetic wheel-type adhesion-locomotion system for water-jetting wall-climbing robot. Adv. Mech. Eng. **10**(7), 1687814018787378 (2018)
7. Tâche, F., Fischer, W., Caprari, G., Siegwart, R., Moser, R., Mondada, F.: Magnebike: a magnetic wheeled robot with high mobility for inspecting complex-shaped structures. J. Field Robot. **26**(5), 453–476 (2009)
8. Tavakoli, M., Marques, L., de Almeida, A.T.: Omniclimber: an omnidirectional light weight climbing robot with flexibility to adapt to non-flat surfaces. In: 2012 IEEE/RSJ International Conference on Intelligent Robots and Systems, Vilamoura-Algarve, Portugal, pp. 280–285 (2012)
9. Eto, H., Asada, H.H.: Development of a wheeled wall-climbing robot with a shape-adaptive magnetic adhesion mechanism. In: 2020 IEEE International Conference on Robotics and Automation (ICRA), Paris, France, pp. 9329–9335 (2020)
10. Guo, J., Lee, K.M., Zhu, D., Yi, X., Wang, Y.: Large-deformation analysis and experimental validation of a flexure-based mobile sensor node. IEEE/ASME Trans. Mechatron. **17**(4), 606–616 (2011)
11. Fan, J., Xu, T., Fang, Q., Zhao, J., Zhu, Y.: A novel style design of a permanent-magnetic adsorption mechanism for a wall-climbing robot. J. Mech. Robot. **12**(3), 035001 (2020)

12. Lee, G., Wu, G., Kim, J., Seo, T.: High-payload climbing and transitioning by compliant locomotion with magnetic adhesion. Robot. Auton. Syst. **60**(10), 1308–1316 (2012)
13. Liu, G., Liu, Y., Wang, X., Wu, X., Mei, T.: Design and experiment of a bioinspired wall-climbing robot using spiny grippers. In: 2016 IEEE International Conference on Mechatronics and Automation, Harbin, China, pp. 665–670 (2016)
14. Lee, G., Kim, H., Seo, K., Kim, J., Kim, H.S.: MultiTrack: a multi-linked track robot with suction adhesion for climbing and transition. Robot. Auton. Syst. **72**, 207–216 (2015)
15. Asbeck, A.T., Kim, S., Cutkosky, M.R., Provancher, W.R., Lanzetta, M.: Scaling hard vertical surfaces with compliant microspine arrays. Int. J. Robot. Res. **25**(12), 1165–1179 (2006)
16. Birkmeyer, P., Gillies, A.G., Fearing, R.S.: Dynamic climbing of near-vertical smooth surfaces. In: 2012 IEEE/RSJ International Conference on Intelligent Robots and Systems, Vilamoura-Algarve, Portugal, pp. 286–292 (2012)
17. Carpenter, K., Wiltsie, N., Parness, A.: Rotary microspine rough surface mobility. IEEE/ASME Trans. Mechatron. **21**(5), 2378–2390 (2015)
18. Liu, X.J., Wang, J.: A new methodology for optimal kinematic design of parallel mechanisms. Mech. Mach. Theory **42**(9), 1210–1224 (2007)
19. Han, Y., Guo, W., Peng, Z., He, M., Gao, F., Yang, J.: Dimensional synthesis of the reconfigurable legged mobile lander with multi-mode and complex mechanism topology. Mech. Mach. Theory **155**, 104097 (2021)

Machine Vision for Intelligent Robotics

Pose Estimation of 3D Objects Based on Point Pair Feature and Weighted Voting

Sen Lin[1], Wentao Li[2,3](✉), and Yuning Wang[4]

[1] School of Automation and Electrical Engineering, Shenyang Ligong University, Shenyang 110159, China

[2] State Key Laboratory of Robotics, Shenyang Institute of Automation, Chinese Academy of Sciences, Shenyang 110016, China

liwentao@sia.cn

[3] Institutes for Robotics and Intelligent Manufacturing, Chinese Academy of Sciences, Shenyang 110169, China

[4] Unit 32681 of the People's Liberation Army, Tieling 112609, China

Abstract. 3D object pose estimation is an important part of machine vision and robot grasping technology. In order to further improve the accuracy of 3D pose estimation, we propose a novel method based on the point pair feature and weighted voting (WVPPF). Firstly, according to the angle characteristics of the point pair features, the corresponding weight is added to the vote of point pair matching results, and several initial poses are obtained. Then, we exploit initial pose verification to calculate the coincidence between the model and the scene point clouds after the initial pose transformation. Finally, the pose with the highest coincidence is selected as the result. The experiments show that WVPPF can estimate pose effectively for a 60%–70% occlusion rate, and the average accuracy is 17.84% higher than the point pair feature algorithm. At the same time, the WVPPF has good applicability in self-collected environments.

Keyword: 3D object · Point pair feature · Weighted voting · Pose verification · Pose estimation

1 Introduction

With the rapid development of 3D imaging technology, the acquisition of 3D information has become more convenient and fast, and is widely used in machine vision, intelligent robot grasping, target recognition and other fields [1]. As an essential part of robot applications, the 3D object pose estimation methods have attracted increasing attention [2].

The methods of the 3D object pose estimation can be divided into two directions: based on template matching and 3D point cloud. Among the template matching methods, the most representative one is the Linemod algorithm. However, the sparse sampling of the template affects the accuracy of the initial pose estimation [3], and there exists the problem of a high mismatch rate in the case of target occlusion [4]. Based on the Linemod

H. Liu et al. (Eds.): ICIRA 2022, LNAI 13456, pp. 383–394, 2022.
https://doi.org/10.1007/978-3-031-13822-5_34

algorithm, A. Tejani et al. [5] propose the latent-class hough forests (LCHF) framework to eliminate background clutter and foreground occlusion in the scene to improve the accuracy of pose estimation. The 3D point cloud methods are the focus of research at present and have rich research results [6, 7]. The iterative closest point (ICP) is the most classical algorithm, but is now mostly used for the optimization calculation after obtaining the initial pose [8]. Aiming at the problem that there are multiple target point clouds in the scene, and they are missing or occluded, the voting algorithm based on point pair feature (PPF) proposed by B. Drost et al. [9] is one of the effective pose estimation methods. However, there are still some problems with the PPF algorithm. Because three elements of the point pair features in the overall four elements are related to the angle of the normal vector, the characteristics of the point pair features in the approximate plane region are very similar, easy to mismatch the non-corresponding point pairs. This kind of mismatching may increase the voting number of the wrong pose and affect the accuracy of the final results.

In order to reduce the vote proportion of mismatched poses, the WVPPF algorithm based on 3D object pose estimation is proposed in this paper. For the point pairs with similar features, the mismatched pose is given a lower voting weight to reduce the total voting number of the unreliable ones, improving the accuracy of pose candidates filtering and clustering. In addition, the initial pose verification is added after pose clustering, and the final pose result is selected by calculating the coincidence of the model and scene point clouds.

2 Principle of the PPF Algorithm

As shown in Fig. 1, the point pair feature descriptor is a four-dimensional vector composed of the distance between any two points on the point cloud and its normal vector.

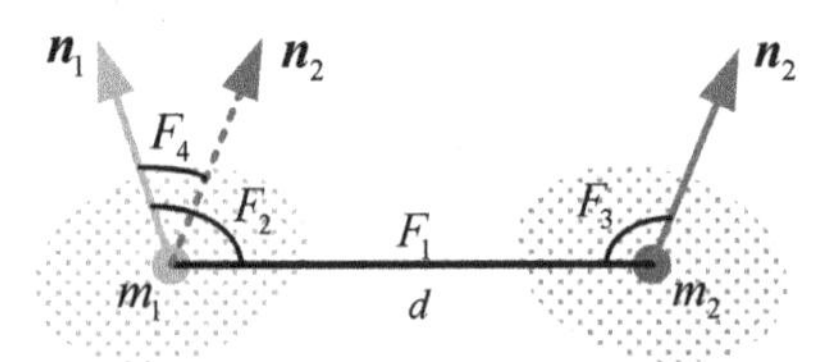

Fig. 1. Point pair feature

Supposing that there is a point pair (m_1, m_2) in the space, whose characteristics can be expressed as follows:

$$\boldsymbol{F}(m_1, m_2) = (\|d\|_2, \angle(\boldsymbol{n}_1, d), \angle(\boldsymbol{n}_2, d), \angle(\boldsymbol{n}_1, \boldsymbol{n}_2)) \tag{1}$$

where $F_1 = ||d||_2$ is the distance between two points. $\boldsymbol{n}_1$, $\boldsymbol{n}_2$ is the normal vector corresponding to the point m_1, m_2. $F_2 = \angle(\boldsymbol{n}_1, d)$, $F_3 = \angle(\boldsymbol{n}_2, d)$ denotes the angle between two normal vectors and the line between two points, $F_4 = \angle(\boldsymbol{n}_1, \boldsymbol{n}_2)$ is the angle between two normal vectors. The point pair feature is asymmetric, which is $F(m_1, m_2) \neq F(m_2, m_1)$. The point pairs with the same features can be stored in the same location of the hash table, after obtaining the point pair features of the model.

Supposing that the point pair (s_r, s_i) in the scene has the same characteristic as the point pair (m_r, m_i) in the model, then they are matched. The transformation matrix between the two points is as follows:

$$s_i = T_{s \to g}^{-1} R_x(\alpha) T_{m \to g} m_i \tag{2}$$

where $R_x(\alpha)$ denotes the rotation angle α, at which the point m_i rotates around the positive half axis of the x-axis to s_i. $T_{s \to g}$ and $T_{m \to g}$ represent the rotation and translation matrices of the normal vectors to the x-axis of the local reference frame, respectively.

As shown in Fig. 2, after matching the scene pairs with the model pairs and obtaining the rotation angles α_i $(i = 1, 2, \cdots, n)$, we add one to the corresponding angle in the accumulator.

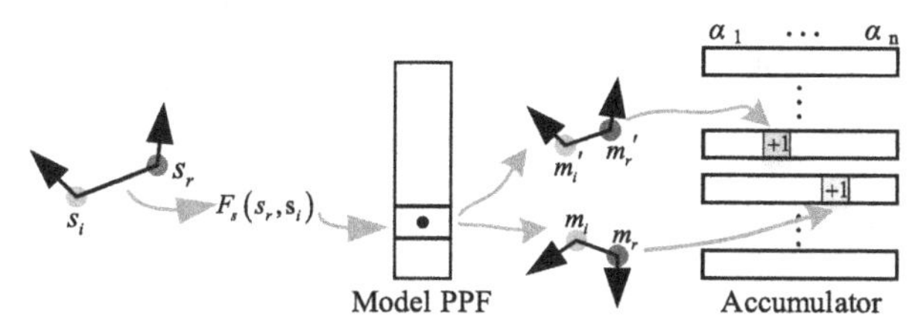

Fig. 2. Pose voting process

After voting, the poses whose votes are greater than a certain threshold are selected as candidates and rearranged in descending order according to the number of votes. The threshold is set to 90% of the highest number of votes. The pose with the highest votes is selected as the first cluster center, and the differences between other poses, the cluster center in translation and rotation are compared in turn. The difference measure is shown in Eq. (3):

$$\arccos\left[\frac{trace\left(R_c^T R_j\right) - 1}{2}\right] \leq \varepsilon, \ \left\| t_j - t_c \right\|_2 \leq \sigma \tag{3}$$

where R_c and t_c are the rotation matrix and translation vector of the cluster center, respectively. R_j and t_j $(j = 1, 2, \cdots, m$, m is the number of remaining poses) are the rotation matrix and translation vector of other poses. $trace(\cdot)$ represents the trace of matrix, ε is the threshold of the rotation angle, and σ is the distance threshold of the translation vector. The values are set as $\pi/15$ and $0.1D_M$ (D_M is the maximum diameter of the target object).

If the difference between the poses is less than the setting threshold, we can say they belong to the same category. Otherwise, the pose is taken as the new clustering center, and the remaining poses will be compared with the difference until all the poses are planned into the corresponding category.

3 Improve the Algorithm

3.1 Weighted Voting

As shown in Fig. 2, the vote of PPF for each group of matching results is the same (equals to 1). However, there will be a large number of mismatching votes when there are lots of similar point pairs. So we propose a weighted voting method.

The normal vector characteristics of the point pair features on different planes are shown in Fig. 3. As shown in Fig. 3(a), the normal vectors of point pairs on the approximate plane are nearly parallel, with the included angle of normal vectors close to 0° or 180°. Suppose the distance between two points is also equal. In that case, the similar point pair features will lead to mismatch with non-corresponding point pairs, so considering that the discrimination of point pairs with the included angle of normal vectors close to 0° or 180° is low. As shown in Fig. 3(b), at the position with larger curvature of the point cloud, the normal vector angle and the relative angle of the point pairs are varied. The feature of the point pair is more obvious conducive to the correct matching of the point pair. Therefore, the discrimination of the point pair features whose normal vector angle is close to 90° is higher.

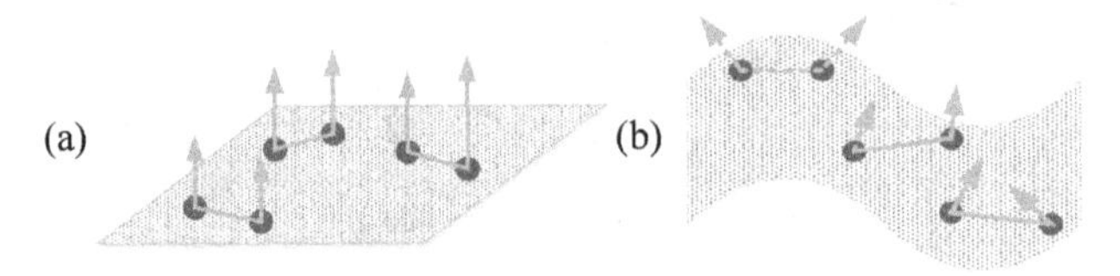

Fig. 3. The normal vectors of point pairs on different planes. (a) is the point pair features on approximate plane, and (b) is the point pair features on curved surface.

Therefore, combined with the influence of the angle of the normal vector on the feature discrimination, the lower voting weight is assigned to the point pair whose normal vector is close to parallel. Similarly, the higher voting weight is assigned to the point pair whose normal vector is close to vertical. The improved weighted voting method is shown in Eq. (4):

$$v = 1 - \lambda|\cos\theta| \tag{4}$$

where v is the voting value is limited to $[0,\ 1]$, θ is the angle between the normal vectors of the point pair. λ is the weighted parameter, the selection of the λ is explained in the experiment section.

If the normal vectors of the point pairs are close to parallel, then $|\cos\theta|$ is close to 1, and v is close to $1 - \lambda$. If the normal vectors of the point pairs are close to vertical, $|\cos\theta|$ is close to 0, and v is close to 1. In this way, the voting proportion of the point pairs with obvious discrimination can be increased, so as to eliminate part of the wrong poses before the pose clustering.

3.2 Initial Pose Verification

In order to select the pose result accurately, the initial pose verification is added on the basis of PPF. The improved algorithm selects the final pose by calculating the coincidence between the transformed model point cloud and the scene point cloud. Judging whether the model point (x_m, y_m, z_m) and the scene point (x_s, y_s, z_s) are coincident according to Eq. (5) and Eq. (6):

$$d = \sqrt{(x_s - x_m)^2 + (y_s - y_m)^2 + (z_s - z_m)^2} < r \tag{5}$$

$$r = D_M \cdot samplingrate \tag{6}$$

where d is the distance between the scene and the model point after pose transformation, r is the sampling step, obtained by multiplying the maximum diameter of the object and the sampling rate. When the distance d is less than the threshold of r, the model point is considered to coincide with the scene point. The calculation of coincidence is shown in Eq. (7):

$$\delta = \frac{P_C}{P_M} \tag{7}$$

where δ represents the coincidence between the model and the scene point clouds, P_C represents the number of model points overlapped with the point cloud in the scene, and P_M is for the total number of model points.

If the result of a pose is correct, the model point cloud transformed by the pose will have a high coincidence with the corresponding scene point cloud, so the pose with the highest coincidence is saved as the final result. The overall flowchart of WVPPF is shown in Fig. 4.

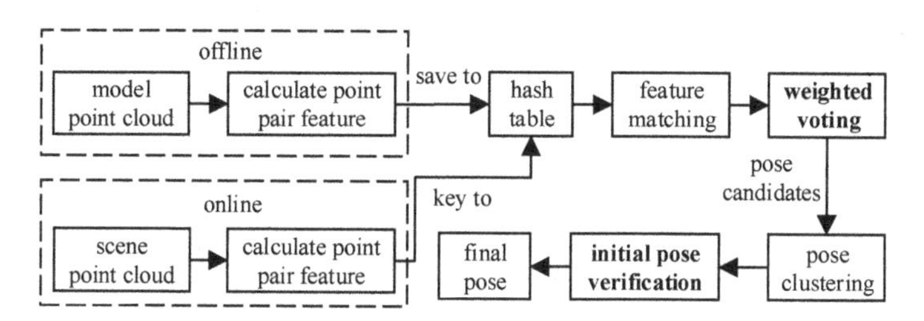

Fig. 4. Flowchart of WVPPF

4 Experimental Results and Analysis

4.1 Select Appropriate λ

The experiments were carried out on Intel i5-4210U, CPU with 8 GB memory. The asian conference on computer vision (ACCV) dataset [10], Mian dataset [11] and self-collected dataset were used to test the accuracy and practicability of the algorithm.

In order to select the appropriate λ for the algorithm, four objects shown in Fig. 5 are selected from the ACCV dataset for experiments. The number of clusters of pose and corresponding rotation matrix and translation vector error when $\lambda = 0,\ 0.1,\ \cdots,\ 1$ are recorded. The calculation methods of the two errors are as follows:

$$e_{\boldsymbol{R}} = ||\boldsymbol{R}_g - \boldsymbol{R}_t||_F \tag{8}$$

$$e_{\boldsymbol{T}} = ||\boldsymbol{T}_g - \boldsymbol{T}_t||_2 \tag{9}$$

where $\boldsymbol{R}_g$ and $\boldsymbol{T}_g$ are the ground-truth recorded in the dataset, $\boldsymbol{R}_t$ and $\boldsymbol{T}_t$ are the pose estimation results.

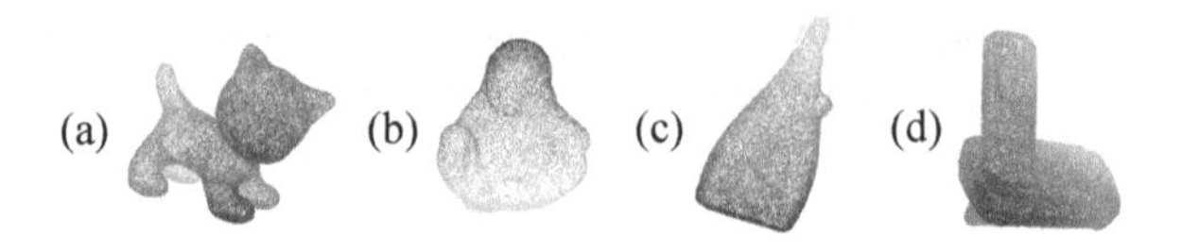

Fig. 5. Experimental objects: (a) Cat, (b) Ape, (c) Glue, (d) Phone

The number of pose clustering results with different λ is shown in Fig. 6. With the increase of λ, the number of pose clustering decreased gradually. After $\lambda = 0.8$, the cluster numbers of ape, phone and glue increased, significantly on the category of glue, While the cluster number of cat continued to decrease. There are two main reasons for the decrease in the number of pose clustering results: (1) The weighted voting reduces the number of mismatched pose votes. When using the threshold to select the pose candidates, part of the wrong poses can be eliminated. (2) In pose clustering, most of the selected pose candidates are close to the correct pose with little differences among them. Therefore, there are more pose candidates belonging to the same category, and the overall number of pose clustering results will be reduced. As shown in Fig. 6, the number of clusters of ape, glue, and phone reaches the lowest value at the same time. Note that the number of cat clusters is at a low level even though it is not the lowest.

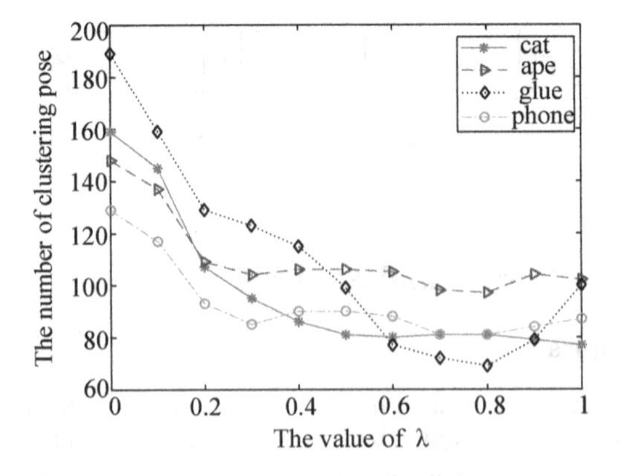

Fig. 6. Influence of λ on the number of pose clustering results

The effect of the value of λ on pose estimation error is shown in Fig. 7(a) and (b). When λ is around 0.8, the errors of rotation matrix and translation vector are in a relatively low and stable position, indicating a small pose estimation error in this range. Combined with the influence on the number of pose clusters, the proposed algorithm is considered to be better when λ is set to 0.8.

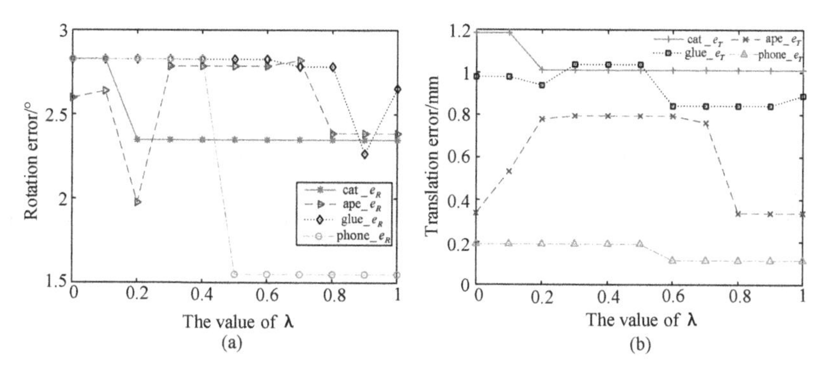

Fig. 7. Influence of λ on pose estimation errors. (a) is the influence on rotation error, and (b) is the influence on translation error

4.2 Experiments and Analysis on Mian Dataset

The 3D object models selected from Mian dataset are shown in Fig. 8(a), and the three experimental scenes selected are shown in Fig. 8(b). The occlusion rate of each object in different scenes is shown in Table 1. The pose estimation errors of PPF, Fast-PPF [12] and WVPPF are compared in this experiment. The pose estimation error is calculated as shown in Eq. (10):

$$error = \underset{m \in M}{avg} \ ||(\boldsymbol{R}_g m + \boldsymbol{t}_g) - (\boldsymbol{R}_e m + \boldsymbol{t}_e)||_2 \tag{10}$$

where M is the model point cloud, m is the point on the model point cloud. $\boldsymbol{R}_g$ and $\boldsymbol{t}_g$ are the ground-truth recorded in the data set. $\boldsymbol{R}_e$ and $\boldsymbol{t}_e$ are the poses estimated by PPF, Fast-PPF, and WVPPF.

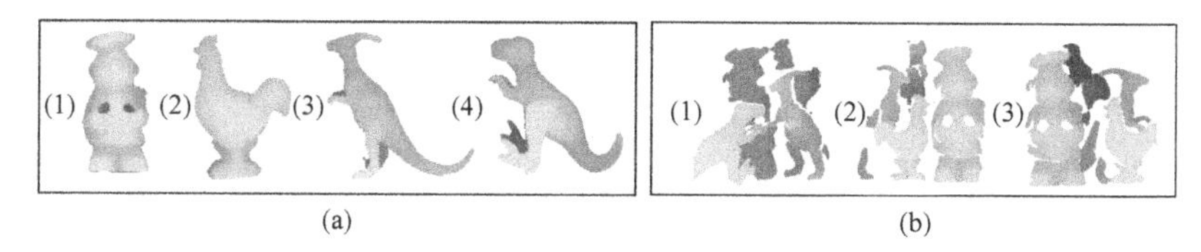

Fig. 8. Four 3D object models and three experimental scenes. In (a), (1) is chef, (2) is chicken, (3) is parasaurolophus, and (4) is t-rex. In (b), (1) is scene 1, (2) is scene 2, and (3) is scene 3.

The pose estimation results of each algorithm in scene 1, 2 and 3 are shown in Fig. 9, 10 and 11. The red point cloud represents the position of the 3D model transformed by the results. The pose estimation error values of the three experimental groups are recorded in Table 2, 3 and 4. The experimental results in scene 1 show that the three algorithms have relatively accurate pose estimation results for chef, parasalolophus, and t-rex, but not for chicken. According to Table 1, the occlusion rate of chicken in scene 1 is 85%, which is too high, leading to the low estimation accuracy of PPF, Fast-PPF, and WVPPF for chicken. However, the accuracy of the WVPPF algorithm is higher than the other two algorithms, which is 12.17% higher than PPF. The occlusion rate of chef in scene 1 is 77.2%, which also belongs to the object with high occlusion in this scene. As can be seen from Fig. 9(1), when estimating the pose of chef by PPF, the pose is probably correct, but the head and foot of the model are reversed. While the pose of chef estimated by WVPPF is more accurate than Fast-PPF. Because the curvature of the head and foot of the chef model changes little, most of the point pairs generated in these regions have nearly parallel normal vectors, and the feature discrimination is not obvious, leading to the matching error for the chef model. WVPPF can reduce the voting proportion of low discrimination pairs to reduce the possibility of the 3D model getting the wrong pose in the scene. For chef model, the accuracy of WVPPF is 94.85% higher than that of PPF and 17.23% higher than that of Fast-PPF. According to Fig. 9(3) and (4), the occlusion rate of parasaurolophus and t-rex in the scene 1 is relatively low, so the pose estimation results are relatively successful.

Table 1. Occlusion rate of objects in each scene

Object	Occlusion rate/%		
	Scene 1	Scene 2	Scene 3
Chef	77.2	62.8	62.3
Chicken	85.0	62.5	65.6
Parasaurolophus	67.8	84.1	81.3
T-rex	69.3	84.0	87.4

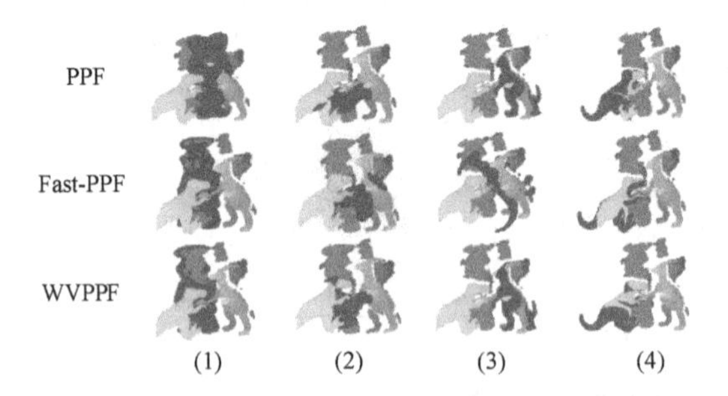

Fig. 9. Pose estimation results in scene 1: (1) Chef, (2) Chicken, (3) Parasaurolophus, (4) T-rex

The pose estimation effects of each 3D model in scene 2 are shown in Fig. 10(a). The occlusion rates of parasaurolophus and t-rex are high, so the pose estimation results of

Table 2. Pose estimation error in scene 1

Error/(mm)	Chef	Chicken	Parasaurolophus	T-rex
PPF	151.32	167.09	**17.16**	6.84
Fast-PPF	9.4	**132.48**	83.48	**4.61**
WVPPF	**7.78**	146.75	**17.16**	6.71

the three algorithms are not accurate. Both of them have the phenomenon of mismatching with the foreground objects. However, the accuracy of WVPPF is higher than that of PPF, the accuracy of parasaurolophus and t-rex is improved by 10.08% and 2.05%, respectively, and the estimation error is lower than that of Fast-PPF. In scene 2, the occlusion rates of chef and chicken are low, and they are both foreground objects. PPF and WVPPF can effectively estimate the pose of these two objects. As shown in Fig. 10(1), after the process of the PPF algorithm, the chef model has a certain tilt relative to the target position in the scene, while the pose estimation result of the WVPPF algorithm is closer to the target position in the scene, and the accuracy is 42.27% higher than PPF. The chicken has enough point cloud information, but Fast-PPF is not practical. It can be seen from Fig. 10(2) that PPF and WVPPF have a similar effect on the pose estimation of chicken. However, by observing the details, the WVPPF algorithm is closer to the scene target in the outline, and its pose estimation accuracy is 37.86% higher than PPF.

The pose estimation effects of each 3D model in scene 3 are shown in Fig. 10(b). Similar to the experimental results in scene 1 and scene 2, for chef and chicken with low occlusion rates, the three algorithms can be successfully applied, while for parasaurolophus and t-rex with high occlusion rates, there is much missing information in the point cloud, which leading to unsatisfactory pose estimation results. However, compared with PPF, the accuracy of chef, chicken, parasaurolophus and t-rex is improved by 0.71%, 24.32%, 17.35% and 3.07%, respectively.

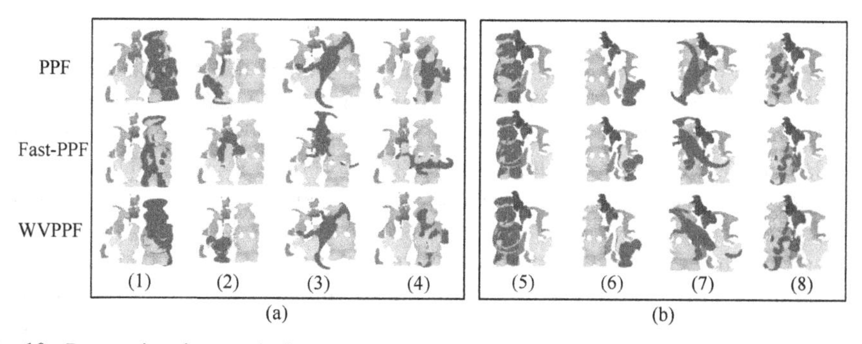

Fig. 10. Pose estimation results in scene 2 and scene 3. (a) is the result of scene 2, and (b) is the result of scene 3. In (a), (1) is chef, (2) is chicken, (3) is parasaurolophus, and (4) is t-rex. In (b), (5) is chef, (6) is chicken, (7) is parasaurolophus, and (8) is t-rex.

Table 3. Pose estimation error in scene 2

Error/(mm)	Chef	Chicken	Parasaurolophus	T-rex
PPF	12.16	7.00	184.61	163.89
Fast-PPF	10.91	126.54	198.40	178.55
WVPPF	**7.02**	**4.35**	**166.01**	**160.53**

Table 4. Pose estimation error in scene 3

Error/(mm)	Chef	Chicken	Parasaurolophus	T-rex
PPF	8.44	5.92	170.04	168.48
Fast-PPF	**8.38**	8.08	178.35	**163.31**
WVPPF	**8.38**	**4.48**	**140.53**	**163.31**

According to the pose estimation results of three scenes, we can see the application of the three algorithms will be affected when the occlusion rate of the target in the scene exceeds 80%, but the average accuracy of WVPPF is 8.94% higher than PPF. When the occlusion rate is between 60% and 70%, the accuracy of the WVPPF algorithm is 17.84% higher than that of PPF. In general, when the occlusion rate is less than 70%, the improved WVPPF algorithm can effectively estimate the pose of the objects in the multi-target occlusion environment, and can reduce the feature mismatching of similar point pairs.

4.3 Experiments and Analysis on Self-collected Dataset

The practical application of the WVPPF algorithm is evaluated by self-collected data set. The structured light scanning platform is used to collect the real object model and the point cloud data of aliasing scenes. As shown in Fig. 11(a), the correct pose of the glass box can be obtained in both scenes. There is a certain error in the second scene of Fig. 11(a)(1), which does not coincide with the glass box in the scene, but most main part of the target still coincides with the glass box. This won't affect the subsequent operations.

As shown in Fig. 11(b), the mouse can also get the correct pose in both real scenes. The red outline of the mouse is below the blue scene in the third scene of Fig. 11(b)(2), because the mouse is blocked by the upper objects during the acquisition, the estimated mouse pose (red point cloud) is displayed under the blue point cloud. At the same time, it is verified that a certain degree of occlusion does not affect the pose estimation ability of the WVPPF algorithm.

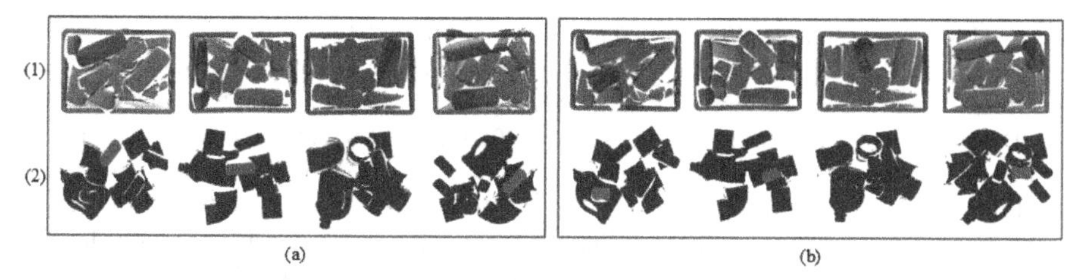

Fig. 11. (a) is the pose estimation result of the glass box in different scenes, and (b) is the pose estimation result of the mouse in different scenes. (1), (2) indicates two different scenes.

5 Conclusion

In order to solve the problem of generating a large number of mismatched poses for similar features, a weighted voting method based on the PPF algorithm is proposed. Our proposed method can effectively fulfill the needs of robot pose estimation in the mixed occlusion environment combined with the initial pose verification to estimate the pose of 3D objects. The proposed method reduces the voting proportion of the wrong pose efficiently. The coincidence of the model and scene point clouds is verified to ensure the accuracy of the final pose results. Experimental results on public datasets show that the proposed WVPPF can achieve or outperform the state-of-the-art methods. Through the experiment of self-collected dataset, the practical application of WVPPF is verified in the actual mixed occlusion environment, resulting in accurate pose estimation.

Funding. Supported by the Major Program of the National Natural Science Foundation of China (Grant No. 61991413).

References

1. Tejani, A., Kouskouridas, R., Doumanoglou, A., Tang, D., Kim, T.K.: Latent–class hough forests for 6 DoF object pose estimation. IEEE Trans. Pattern Anal. Mach. Intell. **40**(1), 119–132 (2018)
2. Li, M.Y., Hashimoto, K.: Accurate object pose estimation using depth only. Sensors **18**(4), 1045 (2018)
3. Liu, S., Cheng, H., Huang, S., Jin, K., Ye, H.: Fast event-inpainting based on lightweight generative adversarial nets. Optoelectron. Lett. **17**(8), 507–512 (2021). https://doi.org/10.1007/s11801-021-0201-8
4. Pan, W., Zhu, F., Hao, Y.M., Zhang, L.M.: Pose measurement method of three-dimensional object based on multi-sensor. Acta Optica Sinica **39**(2), 0212007 (2019)
5. Tejani, A., Tang, D., Kouskouridas, R., Kim, T.-K.: Latent-class hough forests for 3D object detection and pose estimation. In: Fleet, D., Pajdla, T., Schiele, B., Tuytelaars, T. (eds.) ECCV 2014. LNCS, vol. 8694, pp. 462–477. Springer, Cham (2014). https://doi.org/10.1007/978-3-319-10599-4_30
6. Chen, T.J., Qin, W., Zou, D.W.: A method for object recognition and pose estimation based on the semantic segmentation. Electron. Technol. **49**(1), 36–40 (2020)
7. Li, S.F., Shi, Z.L., Zhuang, C.G.: Deep learning-based 6D object pose estimation method from point clouds. Comput. Eng. **47**(8), 216–223 (2021)

8. Yu, H.S., Fu, Q., Sun, J., Wu, S.L., Chen, Y.M.: Improved 3D-NDT point cloud registration algorithm for indoor mobile robot. Chin. J. Sci. Instrum. **40**(9), 151–161 (2019)
9. Drost, B., Ulrich, M., Navab, N., Ilic, S.: Model globally, match locally: Efficient and robust 3D object recognition. In: Proceedings of the 2010 IEEE Computer Society Conference on Computer Vision and Pattern Recognition, pp. 998–1005. IEEE, San Francisco (2010)
10. Hinterstoisser, S., Lepetit, V., Ilic, S., Holzer, S., Bradski, G., Konolige, K., Navab, N.: Model based training, detection and pose estimation of texture-less 3D objects in heavily cluttered scenes. In: Lee, K.M., Matsushita, Y., Rehg, J.M., Hu, Z. (eds.) ACCV 2012. LNCS, vol. 7724, pp. 548–562. Springer, Heidelberg (2013). https://doi.org/10.1007/978-3-642-37331-2_42
11. Choi, C., Christensen, H.I.: 3D pose estimation of daily objects using an RGB-D camera. In: Proceedings of the 2012 IEEE/RSJ International Conference on Intelligent Robots and Systems, pp. 3342–3349. IEEE, Algarve, Portugal (2012)
12. Li, M.Y., Koichi, H.: Fast and robust pose estimation algorithm for bin picking using point pair feature. In: Proceedings of the 24th International Conference on Pattern Recognition, pp. 1604–1609. IEEE, Beijing, China (2018)

Weakly-Supervised Medical Image Segmentation Based on Multi-task Learning

Xuanhua Xie[1,2,3], Huijie Fan[2,3](✉), Zhencheng Yu[2,3], Haijun Bai[1], and Yandong Tang[2,3]

[1] College of Information Engineering, Shenyang University of Chemical Technology, Shenyang 110142, China
[2] State Key Laboratory of Robotics, Shenyang Institute of Automation, Chinese Academy of Sciences, Shenyang 110016, China
{xiexuanhua,fanhuijie,yuzhencheng,ytang}@sia.cn
[3] Institutes for Robotics and Intelligent Manufacturing, Chinese Academy of Sciences, Shenyang 110016, China

Abstract. Medical image segmentation plays an important role in the diagnosis and treatment of diseases. Using fully supervised deep learning methods for medical image segmentation requires large medical image datasets with pixel-level annotations, but medical image annotation is time-consuming and labor-intensive. On this basis, we study a medical image segmentation algorithm based on image-level annotation. Existing weakly supervised semantic segmentation algorithms based on image-level annotations rely on class activation maps (CAM) and saliency maps generated by preprocessing as pseudo-labels to supervise the network. However, the CAMs generated by many existing methods are often over-activated or under-activated, while most of the saliency map generated by preprocessing is also rough and cannot be updated online, which will affect the segmentation effect. In response to this situation, we study a weakly supervised medical image segmentation algorithm based on multi-task learning, which uses graph convolution to correct for improper activation, and uses similarity learning to update pseudo-labels online to improve segmentation performance. We conducted extensive experiments on ISIC2017 skin disease images to validate the proposed method and experiments show that our method achieves a Dice evaluation metric of 68.38% on this dataset, show that our approach outperforms state-of-the-art methods.

Keywords: Weakly supervised semantic segmentation · Multi-task learning · Graph convolution

This work is supported by the National Natural Science Foundation of China (61873259, U20A20200, 61821005), and the Youth Innovation Promotion Association of Chinese Academy of Sciences (2019203).

H. Liu et al. (Eds.): ICIRA 2022, LNAI 13456, pp. 395–404, 2022.
https://doi.org/10.1007/978-3-031-13822-5_35

1 Introduction

Accurate and robust segmentation of organs or lesions from medical images plays an important role in many clinical applications, such as diagnosis and treatment planning. In recent years, due to the maturation of deep complete Convolution Neural Networks (CNN), semantic segmentation has made great progress [11]. The success of these models [2,25,26] comes from a large number of training datasets with pixel-level labels. However, because medical images are highly professional and complex, pixel-level labeling of a large number of medical images for disease diagnosis will be time-consuming and labor-intensive. In order to reduce the labeling burden, we hope to use weaker labels to train the semantic segmentation network and try to obtain segmentation performance equivalent to fully supervised methods. Various types of weak labels have been studied for this purpose, including bounding boxes [8,14,15], scribbles [12,16], points [1], and image-level labels [5,6,9,19,22,24], the labeling strength and cost of the four labeling types decrease sequentially. In this work, we focus on a method with the lowest annotation cost, which is weakly supervised semantic segmentation with image-level annotations.

To the best of our knowledge, most advanced approaches for weakly supervised semantic segmentation follow a two-step process, first generating a Class Activation Map (CAM) [27] using the classification network as the pseudo-label needed for the segmentation network, and then training the segmentation network using the pseudo-label. The CAM is an effective method to localize objects by image classification labels. Although the generated CAM can activate the most discriminative parts of the object, the existing methods usually cannot activate the image correctly, which is easy to cause over-activation or under-activation. Saliency methods have been introduced into weakly supervised semantic segmentation by many research methods [7,13,17,18] because they can identify the most obvious objects and regions in an image, and have achieved good results. Therefore, using CAM to locate objects in each class and using saliency maps to select background regions has become a very popular practice in the field of computer vision. For example, [19] uses CAM to find object regions and employs saliency maps to find background regions to train segmentation models; [21] uses saliency maps to refine CAM generated by classification networks to produce more accurate object regions.

In this paper, we also use a combination of CAM and saliency maps to generate the pseudo-labels and train the segmentation network. However, firstly, we found the CAM generated by the common methods is likely to cause over-activation or under-activation of some areas; Secondly, the saliency map generated by the pre-trained model is relatively rough, which is easy to cause unclear boundaries and wrong background information, and in most cases, the saliency map cannot be updated online. Which will affect segmentation performance. To solve these problems, we propose a weakly supervised segmentation network based on multi-task learning, hoping to improve the effectiveness of the segmentation network. In brief, First, in order to alleviate the degree of inappropriate CAM activation, we introduce a graph-based global reasoning unit in the image classification branch to mine the classification information of images and obtain the correct CAM as possible. Second, we use the global context block in the saliency detection branch and semantic segmentation branch to capture the similarity of feature vectors at any two locations to capture the semantic relationship of spatial locations, then the features learned from the global context block are integrated to optimize the two branches and optimizes the original coarse saliency map through iterative learning. Finally, we use a loss function to constrain the learning of the entire network and optimize the network parameters for better segmentation results.

In summary, our proposed weakly supervised semantic segmentation framework can achieve better segmentation results using only image-level ground-truth labels, on which our method can fully utilize the high correlation between segmentation maps and saliency maps, saliency maps are updated online to generate better pseudo-labels through an iterative learning method. The remainder of this paper is organized as follows. Section 2 presents a deep learning network diagram for semantic segmentation of weakly supervised medical images and points out the method used in this paper. To illustrate the effectiveness of the proposed method, experiments performed on various environments are presented in Sect. 3. Finally, Sect. 4 concludes the paper.

2 The Proposed Approach

Overview: In this paper, we focus on the problem of weakly supervised semantic segmentation based on image-level labels. The proposed network framework is shown in Fig. 1. Given a set of training images with image-level labels, it is first fed to a shared backbone network to extract image features. Next, the feature is transferred to three task-specific branches: the image classification branch, the saliency detection branch, and the semantic segmentation branch to predict the category probability, the dense saliency map, and the dense semantic segmentation map, respectively. Finally, the loss function is used to constrain the whole network learning, and to optimize the pseudo-label using an iterative learning approach in order to achieve better segmentation results. Our approach used in this paper is carefully described in the section below.

2.1 Image Classification Generates Better CAMs

The quality of the CAM seriously affects the segmentation effect. An excellent CAM is not easy to cause improper activation, so as to provide the correct foreground range and help the network to perform image segmentation. In this case, in order to generate a better CAM, Inspired by [3], we introduce a graph-based global reasoning unit in the image classification branch to mine image information at a deeper level, generate the correct CAM, and reduce the over-activated and under-activated areas. The architecture diagram of the graph-based global reasoning unit is shown in Fig. 2. This unit uses graph convolution to achieve global information reasoning through feature mapping and reverse

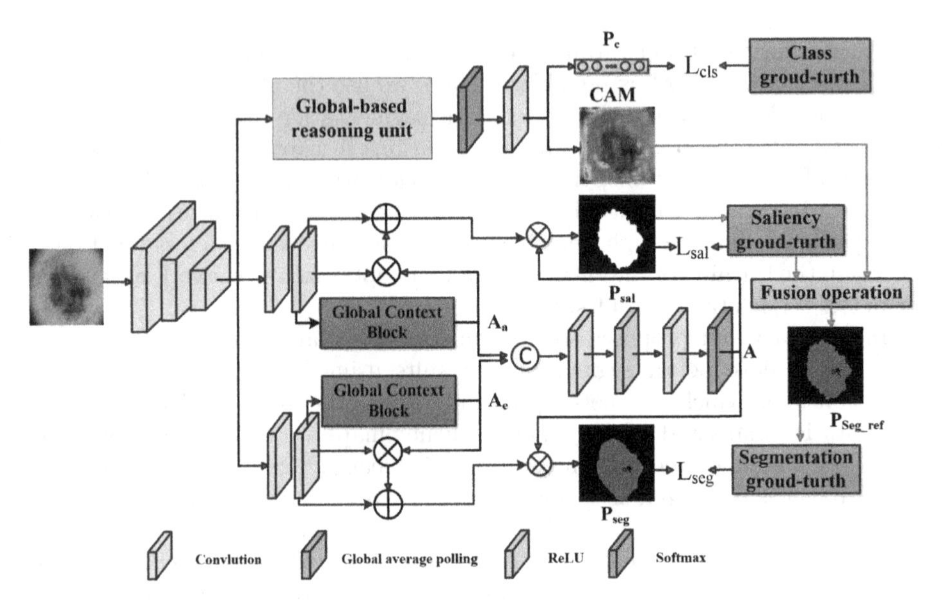

Fig. 1. Network diagram of the WSSS method proposed in this paper. First, the skin disease images are fed into the ResNet38 backbone network to extract features, and then the features are transferred to three specific task branches to predict class probabilities, dense saliency map, and dense semantic segmentation maps, respectively. A graph-based global reasoning unit is introduced in the classification branch to generate better CAM, and the global context block is used in the saliency detection and image segmentation branches to obtain two branch attention features respectively, and integrate them into global attention features. The feature further optimizes two branches. Finally, the loss function is used to constrain the learning of the entire network and update the pseudo-labels using an iterative learning method. The black line in the network diagram represents network training, and the green line represents label update (Color figure online)

mapping so that the network can generate a better CAM. Briefly, for a particular input $x \in R^{(H \times W \times C)}$, where c is the dimension, the coordinate space $\Omega \in R^{(H \times W \times C)}$ is obtained after a 1×1 convolution. The coordinate space of the original features is mapped into the interaction space using the projection equation $V = F(x) \in R^{(N \times C)}$ and a linear union between the nodes is obtained as follows, and the process can be defined as:

$$V_i = b_i X = \sum_{\forall j} b_{ij} x_j \tag{1}$$

where $B = [b_1, ..., b_N] \in R^{N \times W \times H}$, $X_j \in R^{1 \times C}$ N is the number of graph nodes. After the mapping in the previous step, the long-distance relational reasoning becomes the relational reasoning between the nodes in the corresponding graph, and then the graph convolution is introduced to calculate the correlation of each node in the interactive space, and the process can be defined as:

$$Z = GVA_g = ((I - A_g)V)W_g \tag{2}$$

A_g represents the adjacency matrix of the N nodes, W_g represents the parameters of the graph convolution, G represents an $\mathrm{N} \times \mathrm{N}$ adjacency matrix, and I represents the graph node.

Finally, the node feature$Z \in R^{N \times C}$ is mapped back to the original coordinate space by the inverse projection function $Y = \mathrm{g}(Z) \in R^{W \times H \times C}$to obtain the following features, and the process can be defined as:

$$y_i = d_i Z = \sum_{\forall j} d_{ij} z_j \tag{3}$$

where $D = [d_1, ..., d_N] = B^T$

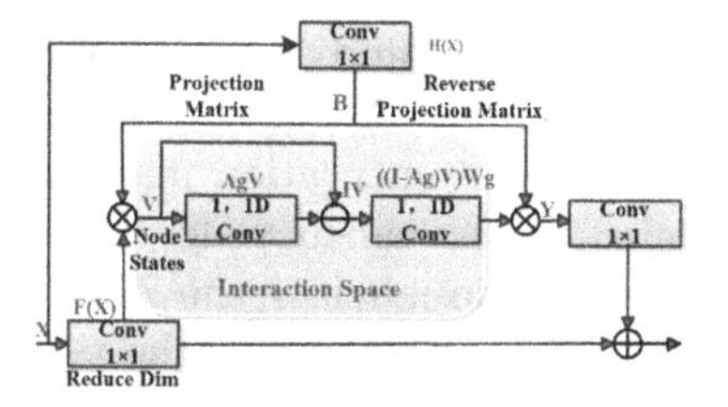

Fig. 2. Graph-based global reasoning unit

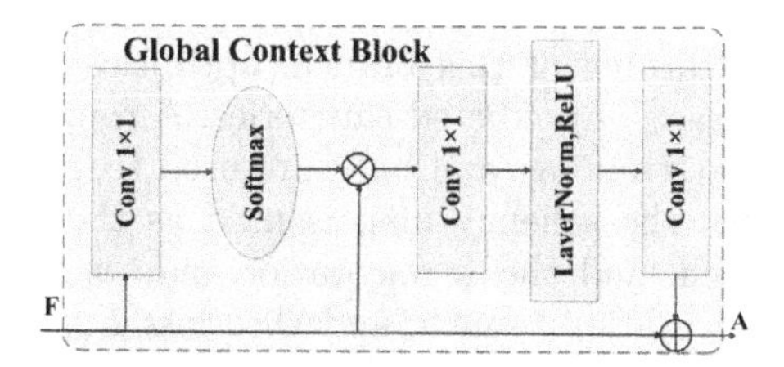

Fig. 3. Global context block

2.2 Cross-Task Similarity Learning

We note that the saliency maps used in many existing weakly supervised semantic segmentation algorithms cannot be updated online, they simply use pre-training to generate a rough saliency map as a pseudo-label input into the network, which will lead to poor segmentation results. In response to this problem, we study a cross-task similarity learning method, which exploits both saliency detection and semantic segmentation are features of dense pixel prediction, to build a network to learn the context information and semantic relations in the spatial dimension of these densely predicted feature maps. As shown in Fig. 3, we use the global context block to capture the similarity of the feature vectors of any two locations to capture the semantic relationship of spatial locations. Given a feature map F, through this block, we can get a global context informational attention map A. More specifically, the global context block goes through three steps: (1) global attention pooling: use 1×1 convolution and softmax function to obtain relevant attention weights and then perform attention pooling to obtain global context features; (2) feature transformation: Feature transformation through two 1×1 convolutions to capture inter-channel dependencies; (3)

Feature aggregation: Use addition to aggregate global contextual features onto features at each location.

After obtaining the attention features A_a and A_e of a single branch, in order to learn the pixel similarity more accurately, we integrate the two attention features through two convolutional layers and a softmax layer to obtain the global attention feature A. The global attention feature can simultaneously learn the edge information of the saliency detection feature and the semantic information of the semantic segmentation feature, then use the complementary characteristics of the two to extract a better global attention feature, and use this feature to optimize the saliency detection branch and semantic segmentation branch to generate better prediction maps.

2.3 Network Training

The method in this paper uses image-level labels as the ground-truth labels of the image classification branch, uses multi-label soft margin loss L_{cls} to constrain the learning of this branch, optimizes network parameters, and improves the accuracy of predicted categories. A rough saliency map and CAM are obtained by pre-training, and both are used to generate an initial segmentation map, in which the saliency map is used as the pseudo-label of the saliency detection branch, and the segmentation map is used as the pseudo-label of the image segmentation branch, and Dice loss L_{sal} and L_{seg} is used to constrain the two branches. study. The overall learning objective function of the network is the sum of the losses of the three tasks.:

$$L_{all} = L_{cls} + L_{sal} + L_{seg} \tag{4}$$

We use iterative learning to update the pseudo-labels. After each network training, the prediction map generated by the saliency detection branch and semantic segmentation is replaced by the pseudo-labels of a new round of training to realize the online update of pseudo-labels to achieve a better segmentation effect. In the network diagram shown in Fig. 1, the black line represents network training, and the green line represents label update.

3 Experiments

We validate our method on the ISIC2017 skin disease dataset [4], which contains 3 common skin diseases and backgrounds. The official dataset contains 2000 training images, 150 validation images, and 600 testing images. Note that we only train our network with image-level classification ground truth labels. We use the common metrics (Dice Coefficient) for medical image segmentation to validate our results. In our experiments, the model is validated in PyTorch. We use ResNet38 [20] as the backbone of the network. In addition, we use data augmentation methods such as random horizontal flips and random crops to enlarge the dataset.

3.1 Comparison with State-of-the-Art Experiments

We show the visualization results of our WSSS method and other WSSS methods in Fig. 4, and the results of the test images in Table 1, respectively. Compared with state-of-the-art methods for WSSS [10,23], our method achieves better results, and Dice assessment metrics reached 68.38%. Our method only utilizes image-level ground-truth labels and uses preprocessing to compute a coarse saliency map to train the network. Compared with other methods that use saliency map supervision, our method outperforms 3.02% and 6.58% respectively.

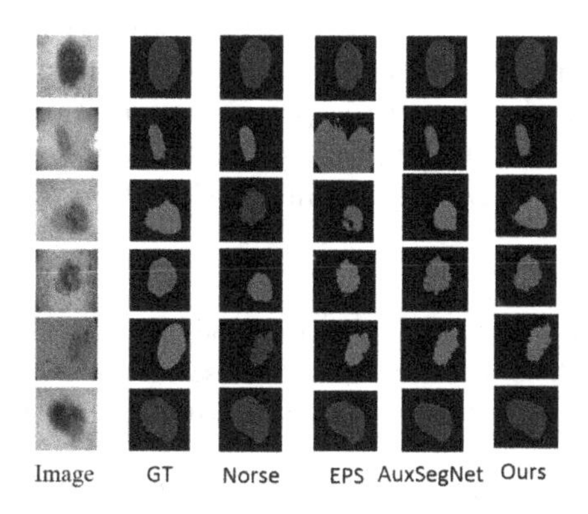

Fig. 4. Visualization results of our WSSS method and state-of-the-art WSSS methods.GT represents the ground-truth label of the image

Table 1. Comparison with state-of-the-art experiments, I represents use image-level label, and S represents participation in network learning using saliency map

Method	Backbone	Sup	Accuracy (%)	Dice (%)	Jaccard index (%)
Norse [23]	VGG16	I+S	84.14	65.36	56.11
EPS [10]	ResNet38	I+S	81.09	61.80	52.39
Ours	ResNet38	I+S	86.76	68.38	58.13

3.2 Ablation Experiments

We add three modules to the backbone network, namely a graph-based global reasoning unit, and two global context blocks. The graph-based global reasoning unit helps to generate better CAMs, which can alleviate incorrect activation to a certain extent, and the global context block can learn the feature similarity of two locations in the image, highlighting many to capture more effective information, we also integrate the features learned by the two global context blocks into one global feature and use this feature to optimize the initial features of the saliency detection branch and semantic segmentation.

To verify the effectiveness of these modules, we conduct ablation experiments as shown in Table 2. First, we did an experiment based on Baseline, and the measured Dice result was 60.04%. Second, applying the global context block to the semantic segmentation branch improves the Dice result by 1.22%. Then the global context block is applied to the saliency detection branch and the semantic segmentation branch at the same time, and the features of the two branches are integrated into a global feature and the two branches are optimized. After experiments, it can be found that the Dice result is 62.19%. Finally, a graph-based global reasoning unit is added, and the segmentation performance is the best, with a Dice result of 63.83%. This proves that our added modules all contribute to the improvement of image segmentation performance. We use iteratively update pseudo-labels to improve segmentation performance. In the experiments, we performed 6 iterations to obtain better results. Table 3 shows the segmentation effect of Dice, experiments show that the fifth iteration gives the best results.

Table 2. Ablation experiment, where B is Baseline, GC (seg) represents the segmentation branch using the global context block, GC (sal) represents the saliency detection branch using the global context block, and GR is a graph-based global reasoning unit

Config	Dice (%)
B	60.04
B+GC(seg)	61.26
B+GC(seg)+GC(sal)	62.19
B+GC(seg)+GC(sal)+GR	63.38

Table 3. Number of iterations. We performed a total of five iterations and achieved the best results at the fifth time

Iteration number	1	2	3	4	5	6
Dice (%)	63.38	65.48	67.21	68.37	68.38	68.33

4 Conclusion

In this work, we propose to use multi-task learning for weakly supervised semantic segmentation of medical images, and this work achieves better segmentation results using only image-level ground-truth labels. More specifically, we use image classification and saliency detection as auxiliary tasks to regulate the learning of the semantic segmentation task. We add a graph-based global reasoning unit to the image classification task to learn the global reasoning of the image to generate better CAM. We utilize a global context block to learn feature similarities at different locations in an image, integrating the feature similarities of the two branches to improve saliency prediction, providing improved pseudo-labels for saliency detection. The use of better CAM and improved pseudo-label produces more accurate segmentation predictions. Thus the segmentation performance is gradually improved using an iterative learning approach and the effectiveness of our method is validated on the ISIC2017 skin disease dataset, and in future work, we will further improve the network to improve segmentation performance.

References

1. Bearman, A., Russakovsky, O., Ferrari, V., Fei-Fei, L.: What's the point: semantic segmentation with point supervision. In: Leibe, B., Matas, J., Sebe, N., Welling, M. (eds.) ECCV 2016. LNCS, vol. 9911, pp. 549–565. Springer, Cham (2016). https://doi.org/10.1007/978-3-319-46478-7_34
2. Chen, L.C., Papandreou, G., Schroff, F., Adam, H.: Rethinking atrous convolution for semantic image segmentation (2017)
3. Chen, Y., Rohrbach, M., Yan, Z., Yan, S., Feng, J., Kalantidis, Y.: Graph-based global reasoning networks. IEEE (2020)
4. Codella, N., et al.: Skin lesion analysis toward melanoma detection: a challenge at the 2017 international symposium on biomedical imaging (ISBI), hosted by the international skin imaging collaboration (ISIC) (2017)
5. Fan, J., Zhang, Z., Song, C., Tan, T.: Learning integral objects with intra-class discriminator for weakly-supervised semantic segmentation. In: 2020 IEEE/CVF Conference on Computer Vision and Pattern Recognition (CVPR) (2020)
6. Fan, R., Hou, Q., Cheng, M.-M., Yu, G., Martin, R.R., Hu, S.-M.: Associating inter-image salient instances for weakly supervised semantic segmentation. In: Ferrari, V., Hebert, M., Sminchisescu, C., Weiss, Y. (eds.) ECCV 2018. LNCS, vol. 11213, pp. 371–388. Springer, Cham (2018). https://doi.org/10.1007/978-3-030-01240-3_23
7. Huang, Z., Wang, X., Wang, J., Liu, W., Wang, J.: Weakly-supervised semantic segmentation network with deep seeded region growing. In: 2018 IEEE/CVF Conference on Computer Vision and Pattern Recognition (CVPR) (2018)
8. Khoreva, A., Benenson, R., Hosang, J., Hein, M., Schiele, B.: Simple does it: weakly supervised instance and semantic segmentation. IEEE Computer Society (2016)
9. Lee, J., Kim, E., Lee, S., Lee, J., Yoon, S.: FickleNet: weakly and semi-supervised semantic image segmentation using stochastic inference. IEEE (2019)
10. Lee, S., Lee, M., Lee, J., Shim, H.: Railroad is not a train: saliency as pseudo-pixel supervision for weakly supervised semantic segmentation (2021)
11. Long, J., Shelhamer, E., Darrell, T.: Fully convolutional networks for semantic segmentation. IEEE Trans. Pattern Anal. Mach. Intell. **39**(4), 640–651 (2015)
12. Meng, T., Perazzi, F., Djelouah, A., Ayed, I.B., Boykov, Y.: On regularized losses for weakly-supervised CNN segmentation (2018)
13. Oh, S.J., Benenson, R., Khoreva, A., Akata, Z., Schiele, B.: Exploiting saliency for object segmentation from image level labels. IEEE (2017)
14. Papandreou, G., Chen, L.C., Murphy, K.P., Yuille, A.L.: Weakly-and semi-supervised learning of a deep convolutional network for semantic image segmentation. In: IEEE International Conference on Computer Vision (2016)

15. Song, C., Huang, Y., Ouyang, W., Wang, L.: Box-driven class-wise region masking and filling rate guided loss for weakly supervised semantic segmentation. IEEE (2019)
16. Vernaza, P., Chandraker, M.: Learning random-walk label propagation for weakly-supervised semantic segmentation (2018)
17. Wang, X., You, S., Li, X., Ma, H.: Weakly-supervised semantic segmentation by iteratively mining common object features. IEEE (2018)
18. Wei, Y., Feng, J., Liang, X., Cheng, M.M., Zhao, Y., Yan, S.: Object region mining with adversarial erasing: a simple classification to semantic segmentation approach. IEEE (2017)
19. Wei, Y., Xiao, H., Shi, H., Jie, Z., Fe Ng, J., Huang, T.S.: Revisiting dilated convolution: a simple approach for weakly- and semi- supervised semantic segmentation. IEEE (2018)
20. Wu, Z., Shen, C., Hengel, A.: Wider or deeper: Revisiting the resnet model for visual recognition. Pattern Recogn. (2016)
21. Xiao, H., Feng, J., Wei, Y., Zhang, M.: Self-explanatory deep salient object detection (2017)
22. Yang, G.Y., Li, X.L., Martin, R.R., Hu, S.M.: Sampling equivariant self-attention networks for object detection in aerial images. arXiv e-prints (2021)
23. Yao, Y., et al.: Non-salient region object mining for weakly supervised semantic segmentation (2021)
24. Zhang, B., Xiao, J., Wei, Y., Sun, M., Huang, K.: Reliability does matter: an end-to-end weakly supervised semantic segmentation approach. In: Proceedings of the AAAI Conference on Artificial Intelligence, vol. 34, no. 7, pp. 12765–12772 (2020)
25. Zhao, H., Shi, J., Qi, X., Wang, X., Jia, J.: Pyramid scene parsing network. IEEE Computer Society (2016)
26. Zhao, H., et al.: PSANet: point-wise spatial attention network for scene parsing. In: European Conference on Computer Vision (2018)
27. Zhou, B., Khosla, A., Lapedriza, A., Oliva, A., Torralba, A.: Learning deep features for discriminative localization. IEEE Computer Society (2016)

A Deep Multi-task Generative Adversarial Network for Face Completion

Qiang Wang[1,2,3], Huijie Fan[2,3](✉), and Yandong Tang[2,3]

[1] The Key Laboratory of Manufacturing Industrial Integrated, Shenyang University, Shenyang, China
wangqiang@sia.cn

[2] The State Key Laboratory of Robotics, Shenyang Institute of Automation, Chinese Academy of Sciences, Shenyang 110016, China
{fanhuijie,ytang}@sia.cn

[3] Institutes for Robotics and Intelligent Manufacturing, Chinese Academy of Sciences, Shenyang 110016, China

Abstract. Face completion is a challenging task that requires a known mask as prior information to restore the missing content of a corrupted image. In contrast to well-studied face completion methods, we present a Deep Multi-task Generative Adversarial Network (DMGAN) for simultaneous missing region detection and completion in face imagery tasks. Specifically, our model first learns rich hierarchical representations, which are critical for missing region detection and completion, automatically. With these hierarchical representations, we then design two complementary sub-networks: (1) DetectionNet, which is built upon a fully convolutional neural net and detects the location and geometry information of the missing region in a coarse-to-fine manner, and (2) CompletionNet, which is designed with a skip connection architecture and predicts the missing region with multi-scale and multi-level features. Additionally, we train two context discriminators to ensure the consistency of the generated image. In contrast to existing models, our model can generate realistic face completion results without any prior information about the missing region, which allows our model to produce missing regions with arbitrary shapes and locations. Extensive quantitative and qualitative experiments on benchmark datasets demonstrate that the proposed model generates higher quality results compared to state-of-the-art methods.

Keywords: Mutli-task · Generative adversarial network · Region detection · Face completion

1 Introduction

Image inpainting [2, 14] aims at reconstructing the missing or masked regions of an image. It is one of the most fundamental operations in image processing [5], image editing [7, 17], and other-low level computer visions. The purpose of image inpainting

Q. Wang—This work is supported by the Natural Foundation of Liaoning Province of China (Grant 2021-KF-12-07).

H. Liu et al. (Eds.): ICIRA 2022, LNAI 13456, pp. 405–416, 2022.
https://doi.org/10.1007/978-3-031-13822-5_36

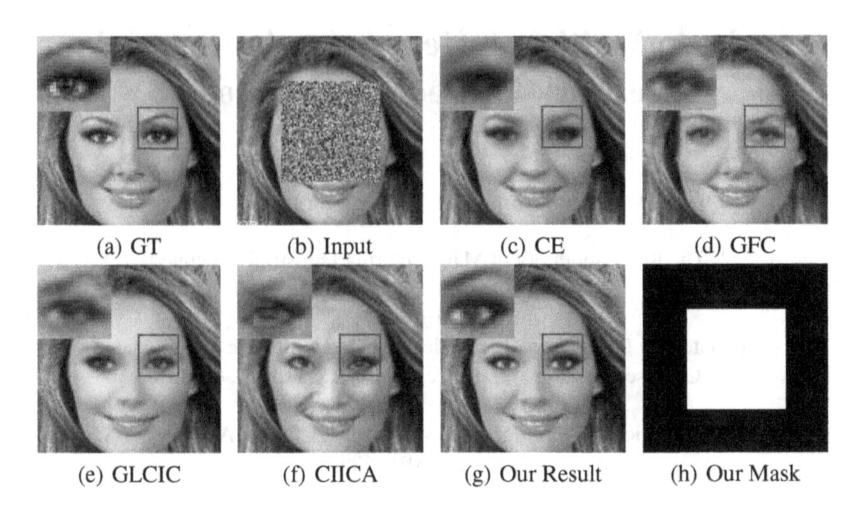

Fig. 1. Qualitative illustration of the face completion task. Given an face image (128 × 128) (a) with a missing hole (64×64) (b), our model can detect the missing region (h) and generate sharper and more coherent content (g) compared to Context Encoder (CE) [19] (c), Generative Face Completion (GFC) [24] (d), Globally and Locally Consistent Image Completion (GLCIC) [10] (e) and Generative Image Inpainting with Contextual Attention (GIICA) [26] (f).

is to generate semantically plausible and context-aware content. The generated content can either be as accurate as the original or maintain coherence well with the known context such that the recovered image appears to be realistic and natural-looking. Early methods [1,17,21–23] based on texture synthesis techniques address the hole-filling problem via searching similar textures from known regions and synthesizing the content of missing regions in a coarse-to-fine manner. These methods using low-level visual features perform well for background completion, but they cannot address the situation whereby an object (e.g., face) is partly missing in an image. More recently, the powerful capabilities of Generative Adversarial Networks (GANs) [4,16,18,25,27] have achieved excellent performances in texture synthesis [6] and image restoration [19,26]. Many models [19,25] based on GANs have been proposed to solve the problem of face completion. Figure 1 illustrates the results of recent research. Because GANs have a powerful capability to learn high-level features of input images, these GAN-based models can effectively obtain a semantical understanding of input and generate the global structure of the missing region of a face image. Although these GAN-based models could achieve good performance in face completion tasks through the use of different network topologies and training procedures, they have two major limitations:

- **requiring prior information of the missing region** Existing GAN-based methods for face completion require the location and geometry information of the missing region. These methods either fix the location and shapes of the missing region [19] or use a binary mask as input [10]. The corrupted image contains this information, which can be effectively obtained by the network itself without any prior.

- **lack of spatial information** To achieve a good semantical understanding, the pooling operation needs to be performed to obtain high-level features of the input image [19,24]. However, this strategy loses a substantial amount of spatial information, which is critical for structural reconstruction and detail recovery. Single-level features cannot provide sufficient spatial information for face completion. Simply combining multi-level features with skip connection [20] also cannot address this problem completely in face completion.

To overcome these limitations, we design a Deep Multi-task Generative Adversarial Network (DMGAN), which can simultaneously detect and complete the missing region in face imagery effectively. The proposed DMGAN includes two parallel networks, DetectionNet and CompletionNet, to complete missing region detection and completion tasks, respectively. One network is fully convolutional and is used to learn a mask that illustrates the missing regions where should be completed, while the other network integrates multi-level features to predict the missing region in a coarse-to-fine manner. Other than directly fusing the same-scale features with skip connections [8,20], we introduce a novel connection structure that can transfer all the high-level features to the shallower layers in our model. We also design a corresponding loss function to ensure that the model can generate sufficient spatial information and high-frequency details at each scale and thus restore the face image gradually.

The main contributions of this paper are summarized as follows:

- We propose an end-to-end Deep Multi-task Generative Adversarial Network (DMGAN) for simultaneous missing region detection and completion for face imagery.
- The proposed model can directly obtain the geometric information (such as location, sizes, and shapes) of a missing region, which is always given as a prior for face completion in previous methods.
- By constructing a hierarchical structure from different scale feature maps of the input image, the proposed model can learn texture information under different granularities and complete the missing regions in a coarse-to-fine manner.

2 Proposed Method

We proposed a deep multi-task GAN that can simultaneously detect missing regions in a corrupted face image and complete them. The proposed model consists of three subnets, i.e., DetectionNet, CompletionNet, and DiscriminationNet, as shown in Fig. 2. In this section, we first discuss the motivation and architecture of our model and then present the details of the training procedure.

2.1 DetectionNet

An overview of the DetectionNet architecture can be seen in Fig. 2, we build the DetectionNet upon the recent successes of fully convolutional nets for image segmentation and transfer learning. DetectionNet uses the hierarchical features extracted to learn an

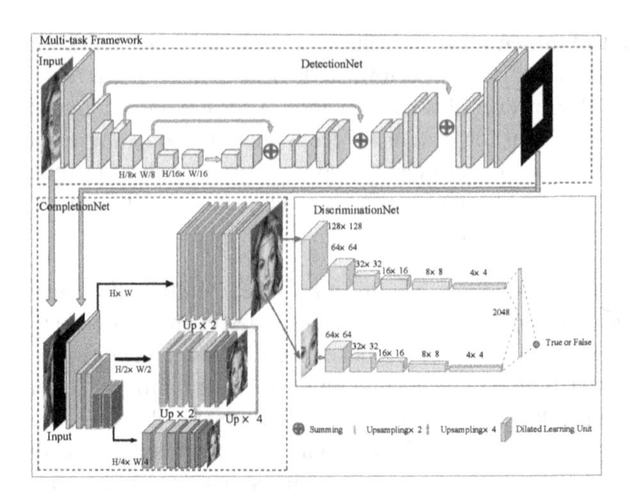

Fig. 2. The architecture of DMGAN: CompletionNet (upper left), DetectionNet (bottom), GlobalNet (bottom left) and Discriminators (upper right)

image mask. We combine the output of deep blocks with that of shallow blocks. Moreover, DetectionNet has a wider perspective and can learn the mask in a complex corrupted image.

We first upsample the output of the last block with $\frac{1}{2}$ stride deconvolutional layers; then, we sum the upsample features with the output of the previous block for fine detection. We initialize this layer with a random initialization and allow it to be jointly optimized with all the other layers. The remainder can be done in the same manner. The final output of DetectionNet is upsampled back to the size of the original input image. DetectionNet is an end-to-end model that detects the missing region in a coarse-to-fine manner. The entire DetectionNet framework is a cascade of convolutional nets with a similar structure. Finally, we use a sigmoid function to transfer the learned feature map into a binary mask that represents the missing region of the input image.

2.2 CompletionNet

An accurate missing region completion method should be performed from two aspects: 1) understand the image content from multiple perspectives and 2) combine the low-level and high-level features for content recovery. Hence, we design a novel network with short connection structure to complete the face image. As shown in Fig. 2, CompletionNet has three parts: (1) Feature Learning, (2) Short Connections and (3) Image Completion. CompletionNet utilizes multi-level features and transfers high-level features to low-level layers with a short connection structure. The low-level layers refine the coarse features shared from high levels and supply sufficient spatial information for local detail recovery.

Feature Learning. Feature Learning contains three convolutional blocks. Each block, except for the last block, has a convolutional layer (along with batch normalization and LReLU activation) and a strided layer (along with batch normalization and LReLU activation). The convolutional layers act as a feature extractor or semantic attribute extractor, and the strided layers reduce the resolution to $\frac{1}{2}$ of an input. In this paper, we design three blocks to represent the input into hierarchical features, which is used for completion. The shallower blocks, which capture more local appearance information, cannot provide much semantic information. The deeper blocks, with more layers, can capture more semantic context well but fails to obtain contexture details due to the coarse features. In the following step, we take advantage of these features extracted from different blocks to learn sufficient representation for detection and completion.

At each level, the first convolutional layer transforms the output features of Feature Learning into appropriate dimensions. Then, the sufficient convolutional layers are followed to learn more features for generating the completed image in the current level. The output of the feature learning is connected to two different layers: (1) a convolutional layer for predicting a face image at the current level and (2) deconvolutional layers to upsample the features for the shallow level. Note that we perform the feature learning process at the coarse resolution and generate upsampling features for the high resolution with deconvolutional layers.

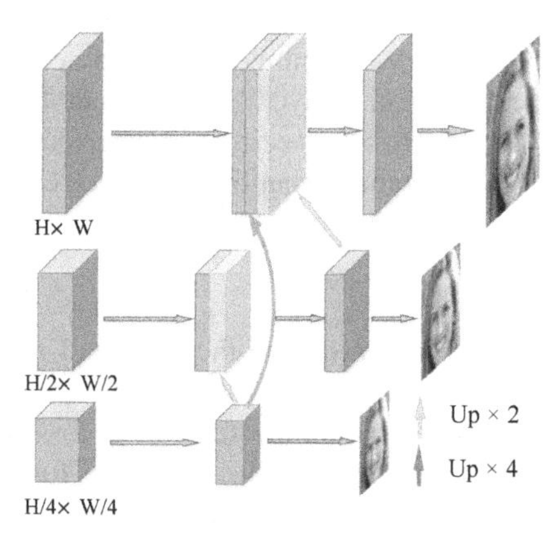

Fig. 3. Illustration of short connections

Short Connections. We adopt a short connection mechanism (Fig. 3) in each level except for the finest level. Each level has deconvolution layers for directly connecting the shallow levels. The shallow levels combine the upsampling features shared from deep levels with the learned features at the current level together to predict the completed image in the corresponding scales.

The motivation of this architecture is that the deep layers are capable of learning semantical content of a missing region but at the expense of a loss of high-frequency details, while the shallow layers focus on low-level features but do achieve a good semantic understanding. This phenomenon inspired us to utilize multi-scale features extracted from GlobalNet and transfer the high-level features to low-level layers for learning more valuable content. The shallow layers can learn additional rich features that contain high-frequency details as well as achieve a good semantical understanding from deep layers to predict the missing content in a coarse-to-fine manner.

Image Completion. In this step, we combine the features shared from deep levels and learned features of the current level into a representation with a concatenation layer. Using the combined representation, our model predicts the completed image at the current level. The reason for completing the face image at every level is that, using the coarse completion results in a higher level, the low-level layers can learn both additional high-frequency details and spatial information to refine the results of the high levels.

2.3 DiscriminationNet

DiscriminationNet is used to discern whether an input image is real. Inspired by [10], we proposed DiscriminationNet, which trains two contexture discriminators to ensure the consistency of the completed image. One discriminator takes the entire completed image as input, while the other discriminator takes a 64 × 64-pixel image patch random selected from the completed image as input. The outputs of the two discriminators are all 1024-dimensional vectors. We combine these two vectors into a single 2048-dimensional vector with a concatenation layer. Using a single fully connected layer and a sigmoid function, the 2048-dimensional vector is transferred into a value in the range of [0, 1], which represents the probability that the input image is real or completed.

2.4 Loss Function

Let x be an input image, and let $\hat{x}, y$ denote the predicted image and ground truth, respectively. To train our multi-task network, two loss functions are jointly used: the pixel loss (Eq. 2) for training stability and the adversarial loss (Eq. 4) to improve how realistic the results are. Using the mixture of these two loss functions can achieve high performance on face completion tasks. The entire training process is performed via backpropagation.

Detection Loss. DetectionNet has a single output. The aim of DetectionNet is to learn a region mask $\hat{M_x}$ to illustrate which regions should be completed (regions to be completed have a value of 1 and 0 otherwise). M_x denotes the ground truth of the image mask, and θ_d is the set of DetectionNet parameters to be optimized. The detection loss function is defined as follows:

$$\ell_{\text{det}}(\hat{M_x}, M_x; \theta_d) = \frac{1}{N} \sum_{i=1}^{N} \|(\hat{M_x} - M_x)\|^2, \tag{1}$$

where N is the total number of training samples in each batch.

Completion Loss. In contrast to DetectionNet, CompletionNet has multiple outputs. The aim of CompletionNet is to learn a mapping function $G(x)$ for generating high-quality completed images $\hat{x}$. Let y_s and $\hat{x}_s$ be the ground truth and completed image of x at level s, respectively. We define $s = 0$ as the finest level of our model. θ_c denotes the set of network parameters of CompletionNet to be optimized. $\hat{M}_x$ is the output of DetectionNet. The completion loss is a joint loss function considering the predicted image mask $\hat{M}_x$.

We employ the pixel loss in the pixel space as follows:

$$\ell_{\mathrm{p}}(\hat{x}_s, y_s;) = \frac{1}{N}\sum_{i=1}^{N} \rho(\hat{x}_s - y_s), \tag{2}$$

where $\rho(x) = \sqrt{x^2 + \varepsilon^2}$ is the Charbonnier penalty function [11] (a differentiable variant of ℓ_2), and ε is set as $1e^{-3}$ in this paper. N is the total number of training samples in each batch. The completion Loss ℓ_{com} is defined by

$$\ell_{\mathrm{com}}(x;\theta_c) = \sum_{s=1}^{S} \ell_{\mathrm{p}}(\hat{x}_s, y_s;) + M_x \odot \{\ell_{\mathrm{p}}(\hat{x}_0, y_0)\}, \tag{3}$$

where $\odot$ is pixel-wise multiplication and S is the number of levels. In this paper, we set $S = 2$ in all experiments.

Adversarial Loss. In our model, we use DiscriminationNet D for our CompletionNet, where D is only defined on the finest level. DiscriminationNet takes the masked region or entire image as input to ensure the local and global consistency of the generated image. The adversarial loss of D is defined as

$$\ell_{adv} = \max_{D} \log D_{\varphi}(y_0) + \log(1 - D_{\varphi}(\hat{x}_0)), \tag{4}$$

where y_0 and $\hat{x}_0$ represent the ground truth and output of the finest level. In the training phase, DiscriminationNet is jointly updated with the generative network at each iteration. For CompletionNet and DiscriminationNet, the optimization is

$$\ell_{adv} = \max_{D} \log D_{\varphi}(y_0, M_d) + \log(1 - D_{\varphi}(\hat{x}_0, M_d)), \tag{5}$$

where M_d is a 64×64-pixel image patch randomly selected from $\hat{x_0}$ as input. By combining these loss functions, the objective loss of our completion model becomes

$$\ell = \ell_{\mathrm{com}} + \lambda_{adv}\ell_{\mathrm{adv}}, \tag{6}$$

where λ_{adv} is the pre-defined weight of the adversarial loss. Note that here ℓ_{adv} consists of two context discriminators.

3 Experiments

In this section, we present our empirical comparisons with several state-of-the-art image inpainting approaches. The experiments are conducted on three public datasets: Celeb-Faces Attribute Dataset (CelebA) [15], Labeled Faces in the Wild Home (LFW) [9] and Cross-Age Celebrity Dataset (CACD) [3]. The comparisons in terms of qualitative and quantitative results are presented in this section.

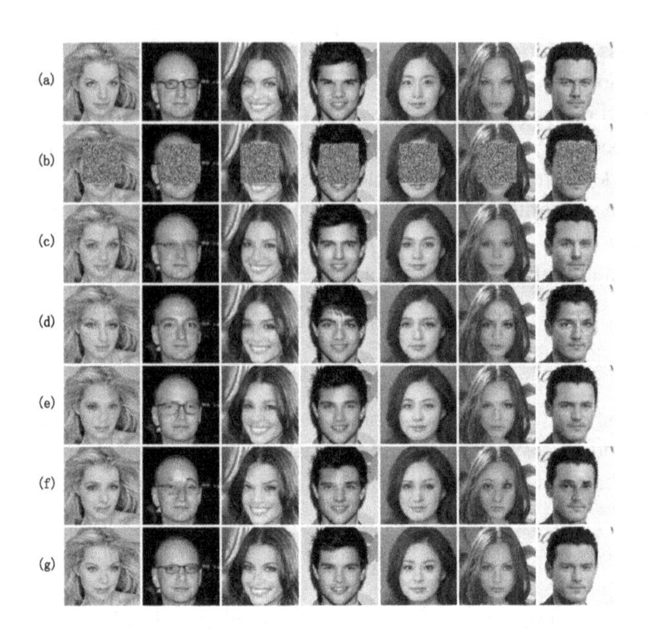

Fig. 4. Visual comparisons of CelebFaces [15] with center missing region. From top to bottom: original image (a), input image (b), CE [19] (c), GFC [24] (d), GLCIC [10] (e), GIICA [26] (f) and our results (g).

3.1 Completion Results

We compared our proposed model with the following methods: CE [19], GFC [12], GLCIC [10], GIICA [26] and PConv [13].

Center Region Completion. We first compare our model with the CE [19], GFC [12], GLCIC [10] and GIICA [26] method on 128×128-pixel images with fixed 64×64-pixel regions missing in the center of the image. We conduct extensive experiments using three datasets: (CelebA) [15], Labeled Faces in the Wild Home (LFW) [9] and Cross-Age Celebrity Dataset (CACD) [3]. From the results in Fig. 4, we can notice that:

- Our model achieves the best performance among the state-of-the-art models without any prior information about the missing region.
- The CE, GFC and GLCIC models have obvious artifacts and blurry in the generated regions since these models only use single-level features to predict the missing regions. Although the GIICA with attention mechanism performs better than others in term of avoiding artifacts and blurry, this model has obvious unnatural content (e.g. eyes or brows) in the predicted content. The generated content with our model has a clean facial contour and sufficient high-frequency details which make the completed face image look natural and realistic. This also validates that our model with short connection can leverage the high-frequency recovery and the scale-space problem, which generates more close results to the originals.

- We present a special case in Fig. 4 (e.g., wearing glasses). The results show that our model is capability of restoring the masked eye and glasses information by filling in realistic eyes and glasses.

Table 1. Quantitative comparison results of CE [19], GFC [24], GLCIC [10], GIICA [26] and Ours on the CelebA data [3].

Model	CelebA PSNR/SSIM	LFW PSNR/SSIM	CACD PSNR/SSIM
CE	21.30/0.719	19.84/0.706	19.25/0.697
GFC	21.51/0.824	–	–
GLCIC	22.73/0.847	20.43/0.761	19.89/0.747
GIICA	22.15/0.816	19.73/0.723	19.41/0.703
Ours	**23.63/0.886**	**21.6/0.786**	**20.44/0.761**

We evaluated our model with two commonly used image quality metrics: Structural Similarity Index (SSIM) and Peak-Signal-to-Noise Ratio (PSNR). Specifically, the SSIM measures the holistic similarity between the completed face image and the original image, and the PSNR measures the difference between two images at the pixel level. For a fair comparison, we retrained the compared methods with the same experimental setting. We cropped the input images to 128×128 pixels first and fixed 64×64-pixel corresponding masked regions in the center of the inputs. As shown by the results in Table 1, our method performs substantially better than all reference methods: CE [19], GFC [12], GLCIC [10] and GIICA [26]. The quantitative results demonstrate that our model (in terms of PSNR/SSIM) achieves better performance on the three public datasets.

3.2 Model Analysis

Short Connections. We empirically find short connections to be crucial for face completion. As shown in Table 2, the same model without short connections performs much more poorly than the model with short connections. Specifically, we observe around 0.3 dB performance drop across all three datasets. We conclude that the model with short connections can well leverage the multi-level features and improve the complete accuracy of our model.

Dilated Convolutions. To demonstrate the effect of the dilated convolutional layer, we retrain our model with the same number of dilated convolutional layers and general convolutional layers on the three datasets. Furthermore, for a fair comparison, we use the same parameter setting and iterations, which is essential to compare the results fairly. As shown in Table 2, dilated convolutional layers can generate content with a much larger receptive field, which is important for face completion. By using the same number of parameters and amount of computational resources, the dilated convolutional layer can learn more realistic content.

Table 2. Quantitative evaluationof the proposed DMGAN and its variants.

Model	CelebA PSNR/SSIM	LFW PSNR/SSIM	CACD PSNR/SSIM
w/o short connection	23.28/0.859	21.37/0.767	20.25/0.743
w/o dilated convolutions	23.57/0.872	21.51/0.781	20.40/0.752
DMGAN	**23.63/0.886**	**21.60/0.786**	**20.44/0.761**

4 Conclusion

We proposed a multi-task generative face completion framework to detect and recover missing regions automatically without any additional prior information about the missing regions. The comparison results show that our model can detect missing regions accurately and generate state-of-the-art results in semantic inpainting and high-frequency detail recovery. Our DetectionNet model based on a fully convolutional network can learn and assign every pixel in the input image a semantic label (1 or 0) to illustrate whether the pixel is corrupted. Based on the results of DetectionNet, CompletionNet with its short connection structure transfers the high-level features to low-level layers and generates realistic information for face completion. This strategy of fusing semantic understanding and shallow features can generate sufficient spatial information in a coarse-to-fine manner. Our model can be extended to low-level vision applications such as denoising, rain removal and other image restoration problems. Although there remain cases in which our approach may produce discontinuities or artifacts, we have achieved state-of-the-art results in semantic inpainting using the proposed multi-task GAN. In future work, we will focus on the image inpainting on natural scenes or structures.

References

1. Barnes, C., Shechtman, E., Finkelstein, A., Dan, B.G.: PatchMatch: a randomized correspondence algorithm for structural image editing. ACM Trans. Graph. **28**(3), 1–11 (2009)
2. Bertalmio, M., Vese, L., Sapiro, G., Osher, S.: Simultaneous structure and texture image inpainting. IEEE Trans. Image Process. **12**(8), 882–9 (2003)
3. Chen, B.-C., Chen, C.-S., Hsu, W.H.: Cross-age reference coding for age-invariant face recognition and retrieval. In: Fleet, D., Pajdla, T., Schiele, B., Tuytelaars, T. (eds.) ECCV 2014. LNCS, vol. 8694, pp. 768–783. Springer, Cham (2014). https://doi.org/10.1007/978-3-319-10599-4_49
4. Dosovitskiy, A., Springenberg, J.T., Brox, T.: Learning to generate chairs with convolutional neural networks. In: Proceedings of the IEEE Conference on Computer Vision and Pattern Recognition, pp. 1538–1546 (2015)
5. Gatys, L.A., Ecker, A.S., Bethge, M.: Texture synthesis using convolutional neural networks. In: Proceedings of the International Conference on Neural Information Processing Systems, pp. 262–270 (2015)

6. Gregor, K., Danihelka, I., Graves, A., Rezende, D.J., Wierstra, D.: DRAW: a recurrent neural network for image generation. Computer Science, pp. 1462–1471 (2015)
7. Hays, J., Efros, A.A.: Scene completion using millions of photographs. In: ACM SIGGRAPH, p. 4 (2007)
8. Hou, Q., Cheng, M.M., Hu, X., Borji, A., Tu, Z., Torr, P.: Deeply supervised salient object detection with short connections. IEEE Trans. Pattern Anal. Mach. Intell. (99), 3203–3212 (2016)
9. Huang, G.B., Ramesh, M., Berg, T., Learned-Miller, E.: Labeled faces in the wild: a database for studying face recognition in unconstrained environments. Technical report 07–49, University of Massachusetts, Amherst, October 2007
10. Iizuka, S., Simo-Serra, E., Ishikawa, H.: Globally and locally consistent image completion. ACM Trans. Graph. (ToG) **36**(4), 107 (2017)
11. Lai, W.S., Huang, J.B., Ahuja, N., Yang, M.H.: Deep Laplacian pyramid networks for fast and accurate super-resolution. In: Proceedings of the IEEE Conference on Computer Vision and Pattern Recognition, pp. 624–632 (2017)
12. Li, Y., Liu, S., Yang, J., Yang, M.H.: Generative face completion. In: Proceedings of the IEEE Conference on Computer Vision and Pattern Recognition, pp. 3911–3919 (2017)
13. Liu, G., Reda, F.A., Shih, K.J., Wang, T.-C., Tao, A., Catanzaro, B.: Image inpainting for irregular holes using partial convolutions. In: Ferrari, V., Hebert, M., Sminchisescu, C., Weiss, Y. (eds.) ECCV 2018. LNCS, vol. 11215, pp. 89–105. Springer, Cham (2018). https://doi.org/10.1007/978-3-030-01252-6_6
14. Liu, Y., Caselles, V.: Exemplar-based image inpainting using multiscale graph cuts. IEEE Trans. Image Process. **22**(5), 1699–1711 (2013)
15. Liu, Z., Luo, P., Wang, X., Tang, X.: Deep learning face attributes in the wild. In: Proceedings of the IEEE International Conference on Computer Vision, pp. 3730–3738 (2015)
16. Lucas, A., Lopez-Tapiad, S., Molinae, R., Katsaggelos, A.K.: Generative adversarial networks and perceptual losses for video super-resolution. IEEE Trans. Image Process. (99), 1–1 (2018)
17. Lundbæk, K., Malmros, R., Mogensen, E.F.: Image completion using global optimization. In: IEEE Computer Society Conference on Computer Vision and Pattern Recognition, pp. 442–452 (2006)
18. Noh, H., Hong, S., Han, B.: Learning deconvolution network for semantic segmentation. In: Proceedings of the IEEE International Conference on Computer Vision, pp. 1520–1528 (2015)
19. Pathak, D., Krähenbühl, P., Donahue, J., Darrell, T., Efros, A.A.: Context encoders: feature learning by inpainting. In: Proceedings of the IEEE Conference on Computer Vision and Pattern Recognition, pp. 2536–2544 (2016)
20. Wang, C., Xu, C., Wang, C., Tao, D.: Perceptual adversarial networks for image-to-image transformation. IEEE Trans. Image Process. **27**(8), 4066–4079 (2018)
21. Wexler, Y., Shechtman, E., Irani, M.: Space-time video completion. In: Proceedings of the IEEE Conference on Computer Vision and Pattern Recognition, pp. 120–127 (2004)
22. Wilczkowiak, M., Brostow, G.J., Tordoff, B., Cipolla, R.: Hole filling through photomontage. In: Proceedings of the British Machine Vision Conference 2005, Oxford, UK, September, pp. 492–501 (2005)
23. Xu, Z., Sun, J.: Image inpainting by patch propagation using patch sparsity. IEEE Trans. Image Process. **19**(5), 1153 (2010)
24. Yang, C., Lu, X., Lin, Z., Shechtman, E., Wang, O., Li, H.: High-resolution image inpainting using multi-scale neural patch synthesis. In: Proceedings of the IEEE Conference on Computer Vision and Pattern Recognition, pp. 6721–6729, July 2017

25. Yeh*, R.A., Chen*, C., Lim, T.Y., Schwing, A.G., Hasegawa-Johnson, M., Do, M.N.: Semantic image inpainting with deep generative models. In: Proceedings of the IEEE Conference on Computer Vision and Pattern Recognition, pp. 5485–5493 (2017). * equal contribution
26. Yu, J., Lin, Z., Yang, J., Shen, X., Lu, X., Huang, T.S.: Generative image inpainting with contextual attention. In: Proceedings of the IEEE Conference on Computer Vision and Pattern Recognition, pp. 5505–5514, June 2018
27. Zeiler, M.D., Fergus, R.: Visualizing and understanding convolutional networks. In: Fleet, D., Pajdla, T., Schiele, B., Tuytelaars, T. (eds.) ECCV 2014. LNCS, vol. 8689, pp. 818–833. Springer, Cham (2014). https://doi.org/10.1007/978-3-319-10590-1_53

Transformer Based Feature Pyramid Network for Transparent Objects Grasp

Jiawei Zhang[1,2], Houde Liu[1,2](✉), and Chongkun Xia[1,2]

[1] Tsinghua University, Hai Dian, Beijing, China
zjw20@mails.tsinghua.edu.cn, ai-robot@sz.tsinghua.edu.cn
[2] AI and Robot Laboratory, Tsinghua Shenzhen Graduate School, Shen Zhen, China

Abstract. Transparent objects like glass bottles and plastic cups are common in daily life, while few works show good performance on grasping transparent objects due to their unique optic properties. Besides the difficulties of this task, there is no dataset for transparent objects grasp. To address this problem, we propose an efficient dataset construction pipeline to label grasp pose for transparent objects. With Blender physics engines, our pipeline could generate numerous photo-realistic images and label grasp poses in a short time. We also propose TTG-Net - a transformer-based feature pyramid network for generating planar grasp pose, which utilizes features pyramid network with residual module to extract features and use transformer encoder to refine features for better global information. TTG-Net is fully trained on the virtual dataset generated by our pipeline and it shows 80.4% validation accuracy on the virtual dataset. To prove the effectiveness of TTG-Net on real-world data, we also test TTG-Net with photos randomly captured in our lab. TTG-Net shows 73.4% accuracy on real-world benchmark which shows remarkable sim2real generalization. We also evaluate other main-stream methods on our dataset, TTG-Net shows better generalization ability.

Keywords: Transparent objects · Transformer · Synthetic dataset

1 Introduction

Grasping is an important topic in the robotic field, as ascending research is being done to make the robot more intelligent for grasping objects fastly and accurately. Current approaches have shown satisfying performance on public robotic grasping datasets, such as Cornell [1] and Jacquard grasping dataset [2]. However, these datasets are focused on opaque objects and few works show good performance on grasping transparent objects.

Transparent objects are very common in daily life, such as glass bottles, plastic cups, and many others made with transparent materials. These objects are fragile and not easy to spot, once broken they may hurt people. So it's necessary

Supported by the Shenzhen Science Fund for Distinguished Young Scholars (RCJC20210706091946001) and the Guangdong Special Branch Plan for Young Talent with Scientific and Technological Innovation (2019TQ05Z111).

H. Liu et al. (Eds.): ICIRA 2022, LNAI 13456, pp. 417–429, 2022.
https://doi.org/10.1007/978-3-031-13822-5_37

Fig. 1. Example images rendered by Blender. We import various HDRI environments to provide various lighting conditions.

for robots to detect or grasp transparent objects, as the deployment of home robots and industrial robots is rapidly intensifying. However, it's challenging for existing vision methods to regress stable grasping pose for transparent objects due to its unique vision properties. Transparent objects' appearance changes as backgrounds changes, making it challenging for vision approaches like CNNs to extract stable features for regressing grasping poses.

To address this problem of grasping transparent objects, we propose TTG-Net - an attention-based feature pyramid network for generating planar grasping pose. TTG-Net introduces a transformer into the feature extraction pipeline to provide a global receptive field which is vital for grasping pose generation since we want TTG-Net to focus more on the whole transparent object rather than its diverse appearance. To train TTG-Net, we also construct a large-scale virtual dataset consisting of over 2000 images. We train and evaluate TTG-Net on our dataset and achieved satisfying accuracy of 80.4%, even our dataset consists of some images difficult to recognize even for human. We also conduct a real-world test benchmark with 64 images of transparent objects, TTG-Net achieved 75.0% accuracy on the real-world dataset which shows its impressive generality on sim2real. T We evaluate other grasping pose generation methods on our dataset, and TTG-Net significantly outperforms these methods especially on real-world benchmarks.

In summary, the contributions of our work are threefold. First, we propose a transformer-based network for generating planar grasping pose for transparent objects. Our method could exploit the global field to focus more on the whole transparent objects, despite the diverse appearance inherited from backgrounds. We propose the first large grasping dataset for transparent objects with over 2000 images. Using our method, a large-scale dataset with sufficient backgrounds for transparent objects can be constructed in a short time, without costly hand-labeling. We conduct both virtual valuation and read-world test benchmark with TTG-Net and our dataset. To evaluate the generality of our method, we also compared TTG-Net with other grasping network and TTG-Net shows impressive performance on sim2real transfer.

2 Related Work

2.1 Robotic Grasping

Learning-based approaches have been proven effective in this field. Pinto et al. [3] use an AlexNet-liked backbone cascaded by fully connected layers to regress grasp poses. Kumra et al. [4] use ResNet-50 as the backbone and then use a shallow fully connected network directly regress the grasp configuration for the object of interest. Guo et al. [5] introduced a hybrid deep architecture combining the visual and tactile sensing for robotic grasp detection. Morrison et al. [6] propose GG-CNN to predict the quality and pose of grasps at every pixel. Zhou et al. [7] use a detection-based method that regresses grasp rectangles from predefined oriented rectangles, called oriented anchor boxes and classifies the rectangles into graspable and non-graspable. Cao et al. [8] use depth images as input and generate grasp poses by a residual squeeze-and-excitation network with a multi-scale spatial pyramid module. Ainetter et al. [9] propose an end-to-end network for joint grasp detection and dense, pixel-wise semantic segmentation. They also introduce a novel grasp refinement module that can further improve the overall grasp accuracy by combining results from grasp detection and segmentation.

2.2 Transparent Dataset

TransCut [10] is a dataset for transparent objects segmentation, which only contains 49 images. It has limited backgrounds and objects so the diversity is limited. With POV-ray, TOM-Net [11] constructed a large-scale dataset for transparent objects matting, which contains 178k synthetic images. It also contains 876 real images for tests but all these images have no annotation so a quantitative test benchmark is not available. And their backgrounds are RGB images from the COCO dataset, which means its diversity in light intensity and direction is limited. So some images in TOM-Net are not natural. Xie et al. [12] proposed a large scale for segmenting transparent objects, naming Trans10K which contains 10428 real-world images. The objects in Trans10K only have 2 classes: transparent objects and stuff. But Trans10K is labeled manually which is time-consuming and high-cost. In Tran2Seg [13], they refined the annotation of Trans10K and propose Trans10K-V2, which contains 11 fine-grained categories. However, Tran2Seg is also hand-labeled. In [14], polarization images are used for transparent objects segmentation, with an attention-fusion-based Mask-RCNN. Their method shows robust performance even for print-out spoofs. But this method must use a polarization camera to capture images.

In [15], they use optimum thresholding to rule out region proposals for R-CNN to detect transparent objects from RGB images. But their training dataset and test surroundings are limited. ClearGrasp [16] proposed a large-scale dataset for depth completion of transparent objects, using Blender equipped with the ray-tracing Blender Cycles. This dataset consists of 9 CAD models and over 18000 images. They also employed 33 HDRI lighting environments and 65 textures for the ground plane. Based on this dataset, they designed ClearGrasp to infer surface normal, masks, and occlusion boundaries of transparent objects, and then use this information and the raw depth input to complete the depth of transparent objects with global optimization. However, the speed of ClearGrasp is far from satisfying, since it contains three deep neural networks. KeyPose [17] built a large-scale dataset, with 48k stereo and RGBD images, for keypoints detection of transparent objects. However, their data collection pipeline is relatively complex. Both the opaque object and transparent object need to be placed at the same location manually, with a marker as help. Moreover, plane textures need change manually. Using their pipeline, collecting data for a new object needs several hours, under human's careful assistance.

3 Approach

3.1 Overview

Given an image of transparent objects, the grasping network should output an optimal grasp representation for grasping. To grasp an object stably and accurately with a planar gripper, the robot needs to know the coordinates of the grasp center denoted by (x, y), as well gripper rotation θ and width w. With these four items, we can define grasp representation as a four-dimensional vector:

$$g = (x, y, \theta, w) \tag{1}$$

After adding a fifth item h, we can define a new grasp representation as follows:

$$g = (x, y, \theta, w, h) \tag{2}$$

which can be viewed as a rotated grasp bounding box. Grasp bounding box is more convenient for visualization and IOU calculation. Many works, such as [3,5], and [7] use grasp bounding box for grasping representation. An example of this representation is shown in Fig. 1. Transparent objects image can be denoted by a tensor:

$$\mathcal{I} = \mathbb{R}^{3\times H\times W} \tag{3}$$

We refer to set of all the grasps in Images $\mathcal{I}$ as the grasp map, which can be denoted as:

$$\mathbf{G} = (\Theta, \mathbf{W}, \mathbf{Q}) \in \mathbb{R}^{3\times H\times W} \tag{4}$$

where $\mathbf{\Theta}$, $\mathbf{W}$, $\mathbf{Q}$ are each $\in \mathbb{R}^{1\times H\times W}$ and represents grasp angle, width and quality score respectively regressed for each pixel of image $\mathbf{I}$. The height of grasp bounding box is calculated by multiplying the width by a factor a. In this work, we define a as 0.5.

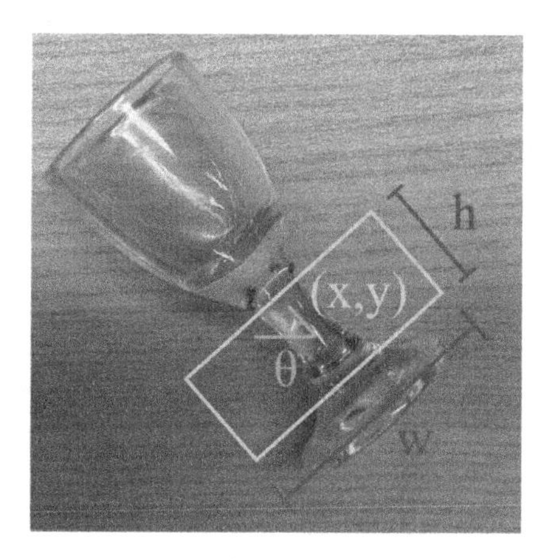

Fig. 2. A grasp bounding box, defined by five items. The yellow labels demark the bounding box and coordinates of grasp center. The green labels show the width and height of grasp bounding box. Orientation is labeled by orange labels (Color figure online)

3.2 Dataset

As transparent objects changes with surroundings, it's necessary to provide various surroundings for the richness of the dataset, since the richness can guarantee the generalization of a deep neural network trained on the dataset. However, it's cumbersome to change surroundings when building a dataset for transparent objects in the real world. In previous work, like KeyPose, they print ten diverse background textures on paper and place them beneath transparent objects.

In trans10k, they collect about 10000 images and annotate them manually. Both these two methods are time-consuming and laborious. To build a dataset efficiently, we use the Blender physics engine to generate images for transparent objects. With ray-tracing rendering engine Cycles built in Blender, we can generate numerous photo-realistic images in a short time.

Our dataset consists of 4 CAD models scanned from real-world transparent objects, including a small handless winecup, one-off plastic cup, shot glass, and ordinary glass cup. We import these models into Blender and adjust their material properties to make them transparent. We employ 8 HDRI environments, which can provide various textures and lighting conditions, including bright living room, dim bar, sunny road, etc. To make transparent objects' appearance change with environments, we orbit the Blender camera around transparent objects and render an image every 20°. Then, 160 high-quality images will be rendered for each CAD model. To avoid annotating all images manually, we make each CAD model synchronous rotate with Blender camera. So each transparent object keeps the same pose in all images. Then we just need to annotate the grasp poses of each object only in one image. After the initial annotation, the raw dataset will be extended through data augmentation, including crop, scaling, rotation, and translation. In the end, the grasping pose of the transparent object in each image is different.

3.3 Network Architecture

The overall pipeline of TTG-Net is shown in Fig. 3. Given an image of transparent objects, TTG-Net generates three images: grasp quality, grasp angle, and grasp width.

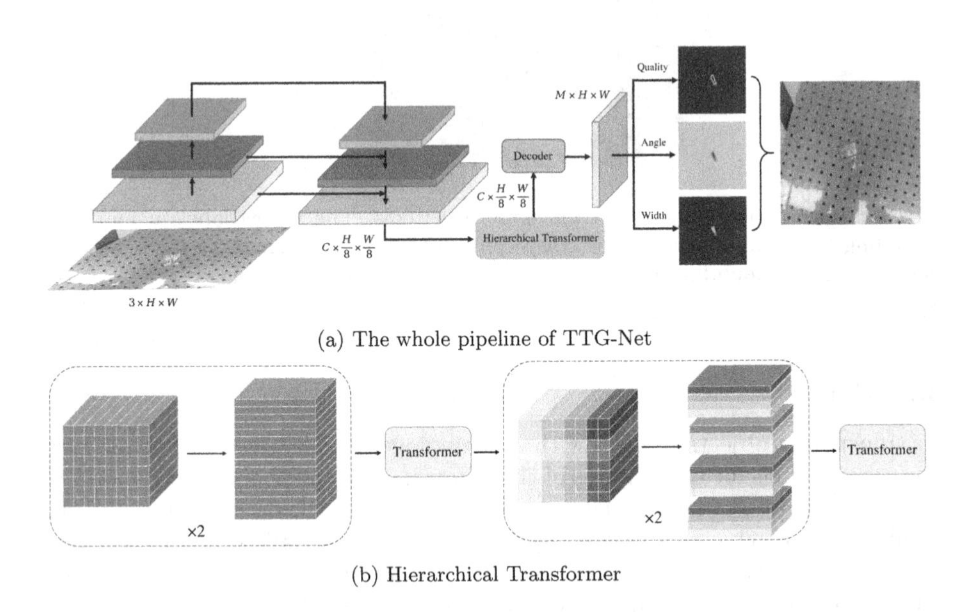

(a) The whole pipeline of TTG-Net

(b) Hierarchical Transformer

Fig. 3. Framework of TTG-Net.

Backbone. To extract features as different scales, we use FPN (Feature Pyramid Network) as the backbone, which is based on ResNet50 [18]. After passing the image through the backbones, the shape of the image is downsampled into a feature map of $(C, \frac{H}{8}, \frac{W}{8})$. Then the feature map is fed into a hierarchical transformer for self-attention.

Hierarchical Transformer. Transformer [19] has been successfully applied in multiple vision tasks [20–24]. The transformer can learn global information from images and provide a global receptive field. So we use transformer to refine the feature outputs from the backbone, in order to make the network pay more attention to the entire shape of transparent objects. Instead of using a traditional transformer whose path size is fixed, we use a hierarchical transformer with multi-scale path size, which pays attention to both non-local and local information of the feature map. The hierarchical transformer consists of two transformer blocks, with different sizes of patches. To make TTG-Net find the transparent objects, we want to enable the transformer to learn global information from the feature map. So the path size is set as $(\frac{H}{8}, \frac{W}{8})$, same as the size of the feature map, which enables the transformer to learn global information from the whole feature map. Then the feature map is reshaped into one flatten patches $x_p \in \mathbb{R}^{(H \times W/64) \times C}$. The transformer takes in the flatten patch and outputs attention map of $(C, \frac{H}{8}, \frac{W}{8})$. After finding the transparent objects, TTG-Net is supposed to generate grasp poses from local information. To enable transformer learn local information from attention map, the size of patches is set as $(\frac{H}{32}, \frac{W}{32})$. So attention map is split into 4×4 paths and fed into a transformer for local self-attention computation. As illustrated in Fig. 2(a), attention computation is performed only for features of the same color.

Decoder. The decoder up-samples the attention map by transposed convolution layers and outputs a feature map of (M, H, W). Then the feature map is fed into three shallow convolution networks to generate three images: Grasp Quality $\mathbf{Q}_I$, Grasp Angle $\mathbf{\Theta}_I$, and grasp width $\mathbf{W}_I$. Pixels in $\mathbf{Q}_I$ range from 0 to 1, which denotes the grasp quality. The angle of each grasp pose is in the range $[-\frac{\pi}{2}, \frac{\pi}{2}]$. Since the antipodal grasp is symmetrical around $\pm\frac{\pi}{2}$, we use two components $sin(2\Phi_T)$ and $cos(2\Phi_T)$ to provide distinct values which are symmetrical at $\pm\frac{\pi}{2}$. The values of pixels in $\mathbf{W}_I$ is in the range [0, 1]. During training, the value of width is scale by $\frac{1}{100}$.

3.4 Loss Function

We define the training loss function as follows:

$$Loss(G, \hat{G}) = \sum_{i}^{N} \sum_{m \in \{q,\theta,w\}} Smooth_{L1}(G_i^m - \hat{G}_i^m) \tag{5}$$

where $Smooth_{L1}$ is:

$$Smooth_{L1} = \begin{cases} (\sigma x)^2/2, & if|x| < 1; \\ |x| - 0.5/\sigma^2, & otherwise. \end{cases} \tag{6}$$

where N denotes the number of grasp candidates, q, θ, w is the grasp quality, the orientation angle and grasp width.

4 Experiment

In this section, we evaluate TTG-Net's ability to estimate grasp pose for transparent objects, on both synthetic and real-world benchmarks. We also select several other grasp pose estimation methods to evaluate our dataset. We also evaluate TTG-Net on two public grasping datasets, Cornell and Jacquard.

4.1 Implementation Details

Although transformer has been proved effective in many computer vision field, their training requires large-scale dataset and carefully designed optimizer and learning rate schedulers. After combining CNN backbone with transformer module, we found transformer become very easy to train, just like convolutional neural networks. We Implement TTG-Net with PyTorch 1.8. TTG-Net is trained on an Nvidia RTX3090 GPU. For loss optimization, we use Adam optimizer where the initial learning rate is set to 0.001 and batch size is set to 24.

4.2 Experimental Results on Synthetic Dataset

TTG-Net is traned on synthetic dataset and evaluate on synthetic dataset. The training dataset contains 1948 images and the evaluation dataset contains 235 images. Transparent objects have different appearance in training and evaluation dataset. Figure 4 shows detection outputs on evaluation dataset. The result shows TTG-Net can effectively estimate grasp pose for transparent objects, in bright or dim environment.

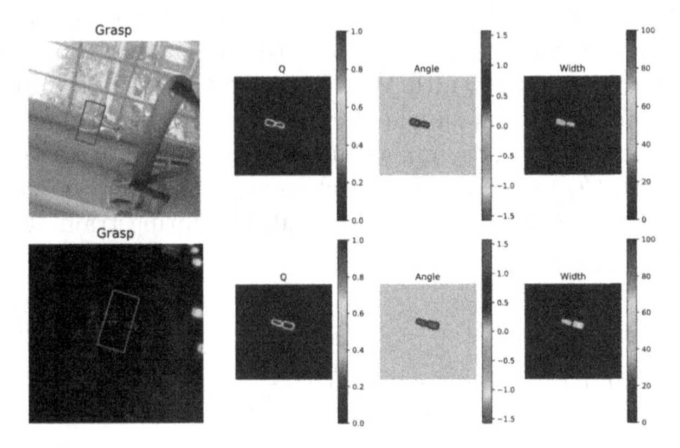

Fig. 4. Experimental results on synthetic dataset

Table 1. Comparison of TTG-Net to GR-ConvNet

Author	Method	Synthetic accuracy (%)	Real-world accuracy (%)
Kumra [4]	GR-ConvNet	74.8	65.6
Our	TTG-Net	80.4	73.4

4.3 Experimental Results on Real-World Dataset

Since TTG-Net is trained and evaluated on the synthetic dataset, we need to evaluate its performance on real-world images of transparent objects. We randomly take 64 photos of transparent objects contained in the synthetic dataset and evaluate TTG-Net on these real-world photos. And the experiment results show that TTG-Net can effectively predict grasp poses for transparent objects in most photos. The selected detection result is shown in Fig. 5. To prove the sim2real ability of TTG-Net, we also do a real-world benchmark with GR-ConvNet trained on the synthetic dataset. Compared to GR-ConvNet, TTG-Net achieves higher accuracy on the real-world benchmark, as shown in Table 1. In Fig. 6, TTG-Net shows better performance on predicting grasp pose for multiple transparent objects, compared to GR-ConvNet.

4.4 Experimental Results on Cornell and Jacquard

We evaluate TTG-Net on two public dataset, Cornell and Jacquard. Cornell and Jacquard are built for grasping opaque objects. Compared with other state-of-art methods, TTG-Net achieves an accuracy of 97.2% in Cornell and 92.3% in Jacquard, which is a satisfied result.

4.5 Ablation Study

We present ablation experiment result in Table 2. In this part, we mainly discuss effectiveness of FPN and hierarchical transformer. Compared to ResNet50, FPN based on ResNet shows higher accuracy both on synthetic and real-world dataset. Furthermore, FPN with hierarchical transformer outperforms FPN. We change numbers of patches to show how hierarchical transformer works. Compared to transformer with fixed size of path, transformer works better when using different scale patches (Table 3).

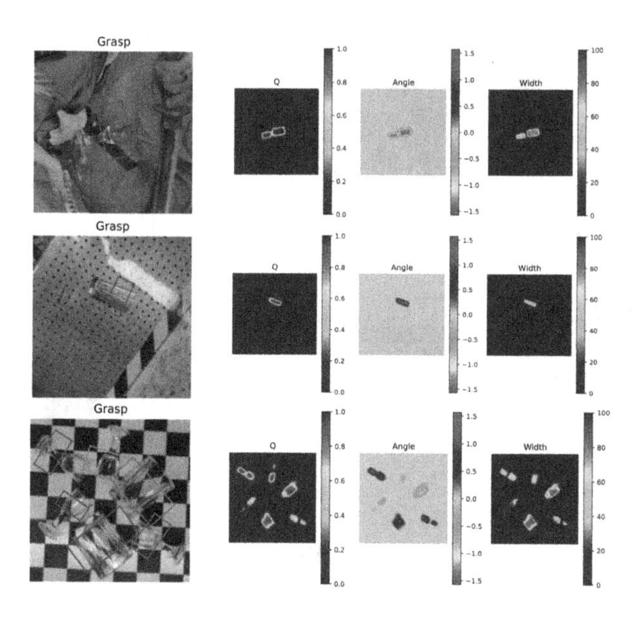

Fig. 5. The real-world benchmark results of TTG-Net

Table 2. Evaluation results on Cornell and Jacquard

Author	Method	Cornell accuracy (%)	Jacquard accuracy (%)
Zhang [25]	ROI-GD	93.6	93.6
Zhou [7]	FCGN	96.6	92.8
Song [26]	RPN	96.2	93.2
Kumra [4]	GR-ConvNet	96.6	91.8
Ainetter [9]	-	98.2	92.9
Cao [8]	RSEN	96.4	94.8
Our	TTG-Net	97.2	92.3

4.6 Evaluation on Robotic Arm

We evaluate TTG-Net on a Franka Panda robotic arm with an Intel RealSense D435 camera. The camera is mounted orthogonal to the working table to restrict workspace to 2d plane. The external parameter of camera and robotic arm is calibrated by easy_handeye. The height of the grab position is fixed, which is set to 1.5 cm above working table. We evaluate our approach with 4 transparent objects which is randomly placed on 10 different positions. Our approach shows a 92.5% grasp accuracy, as shown in Table 4. Since the texture of worktable is simple, the environment of grasp experiment is far more challenging than real-world test dataset. So the grasp accuracy is higher than test accuracy.

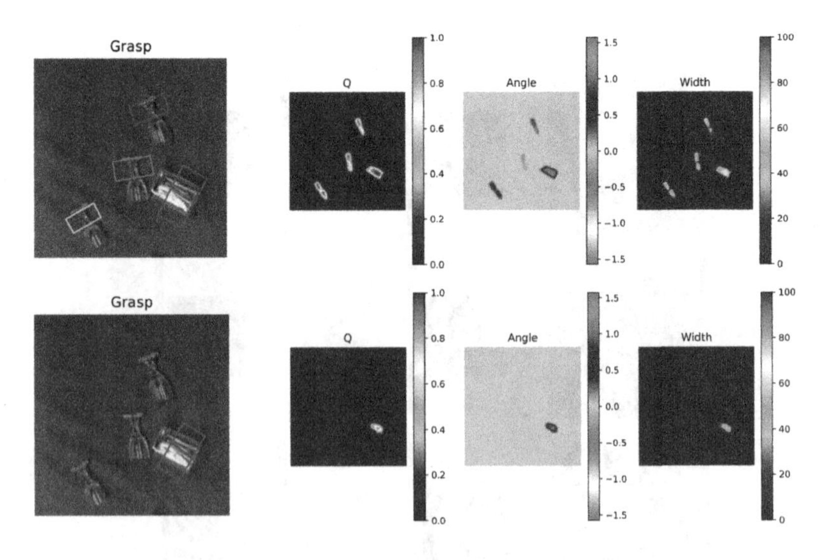

Fig. 6. Comparison of TTG-Net to GR-ConvNet on real-world benchmark

Table 3. Ablation study for different backbone and different setting of hierarchical transformer

Method	Synthetic dataset accuracy (%)	Real-world dataset accuracy (%)
ResNet50	78.6	65.6
PFN	79.1	67.2
PFN + HT(1,1,1,1)	79.8	71.9
FPN + HT(4,4,4,4)	80.0	71.9
FPN + HT(1,1,4,4)	80.4	73.4

Table 4. The result of real-world grasping experiment

Objects	Success rate
Handless winecup	10/10
One-off plastic cup	9/10
Shot glass	9/10
Ordinary glass cup	9/10
Average	92.5%

5 Conclusion

In this work, we propose a synthetic dataset for transparent objects grasp. We also propose TTG-Net, a feature pyramid network with a hierarchical transformer. TTG-Net is trained on our synthetic dataset and we evaluate it on a real-world benchmark. The result shows TTG-Net can predict accurate grasp pose for transparent objects in the real world, with the satisfying ability of sim2real TTG-Net also shows good performance on two public datasets, Cornell and Jacquard. In the future, we would introduce opaque objects into our dataset and extend our work for predicting grasp poses on both transparent and opaque objects.

References

1. Yun, J., Moseson, S., Saxena, A.: Efficient grasping from RGBD images: learning using a new rectangle representation. In: 2011 IEEE International Conference on Robotics and Automation, pp. 3304–3311. IEEE (2011)
2. Depierre, A., Dellandréa, E., Chen, L.: Jacquard: a large scale dataset for robotic grasp detection. In: 2018 IEEE/RSJ International Conference on Intelligent Robots and Systems (IROS), pp. 3511–3516. IEEE (2018)

3. Redmon, J., Angelova, A.: Real-time grasp detection using convolutional neural networks. In: 2015 IEEE International Conference on Robotics and Automation (ICRA), pp. 1316–1322. IEEE (2015)
4. Kumra, S., Kanan, C.: Robotic grasp detection using deep convolutional neural networks. In: 2017 IEEE/RSJ International Conference on Intelligent Robots and Systems (IROS), pp. 769–776. IEEE (2017)
5. Guo, D., Sun, F., Liu, H., Kong, T., Fang, B., Xi, N.: A hybrid deep architecture for robotic grasp detection. In: 2017 IEEE International Conference on Robotics and Automation (ICRA), pp. 1609–1614. IEEE (2017)
6. Morrison, D., Corke, P., Leitner, J.: Closing the loop for robotic grasping: a real-time, generative grasp synthesis approach. arXiv preprint arXiv:1804.05172 (2018)
7. Zhou, X., Lan, X., Zhang, H., Tian, Z., Zhang, Y., Zheng, N.: Fully convolutional grasp detection network with oriented anchor box. In: 2018 IEEE/RSJ International Conference on Intelligent Robots and Systems (IROS), pp. 7223–7230. IEEE (2018)
8. Cao, H., Chen, G., Li, Z., Lin, J., Knoll, A.: Residual squeeze-and-excitation network with multi-scale spatial pyramid module for fast robotic grasping detection. In: 2021 IEEE International Conference on Robotics and Automation (ICRA), pp. 13445–13451. IEEE (2021)
9. Ainetter, S., Fraundorfer, F.: End-to-end trainable deep neural network for robotic grasp detection and semantic segmentation from RGB. In: 2021 IEEE International Conference on Robotics and Automation (ICRA), pages 13452–13458. IEEE (2021)
10. Xu, Y., Nagahara, H., Shimada, A., Taniguchi, R.: Transcut: transparent object segmentation from a light-field image. In Proceedings of the IEEE International Conference on Computer Vision, pp. 3442–3450 (2015)
11. Chen, G., Han, K., Wong, K.-Y.K.: Tom-net: Learning transparent object matting from a single image. In: Proceedings of the IEEE Conference on Computer Vision and Pattern Recognition, pp. 9233–9241 (2018)
12. Xie, E., Wang, W., Wang, W., Ding, M., Shen, C., Luo, P.: Segmenting transparent objects in the wild. In: Vedaldi, A., Bischof, H., Brox, T., Frahm, J.-M. (eds.) ECCV 2020. LNCS, vol. 12358, pp. 696–711. Springer, Cham (2020). https://doi.org/10.1007/978-3-030-58601-0_41
13. Xie, E., et al.: Trans2seg: transparent object segmentation with transformer (2021)
14. Kalra, A., Taamazyan, V., Rao, S.K., Venkataraman, K., Raskar, R., Kadambi, A.: Deep polarization cues for transparent object segmentation. In: Proceedings of the IEEE/CVF Conference on Computer Vision and Pattern Recognition, pp. 8602–8611 (2020)
15. Lai, P.-J., Fuh, C.-S.: Transparent object detection using regions with convolutional neural network. In: IPPR Conference on Computer Vision, Graphics, and Image Processing, vol. 2 (2015)
16. Sajjan, S., et al.: Clear grasp: 3D shape estimation of transparent objects for manipulation. In: 2020 IEEE International Conference on Robotics and Automation (ICRA), pp. 3634–3642. IEEE (2020)
17. Liu, X., Jonschkowski, R., Angelova, A., Konolige, K.: Keypose: multi-view 3D labeling and keypoint estimation for transparent objects. In: Proceedings of the IEEE/CVF Conference on Computer Vision and Pattern Recognition, pp. 11602–11610 (2020)
18. Targ, S., Almeida, D., Lyman, K.: Resnet in resnet: generalizing residual architectures. arXiv preprint arXiv:1603.08029 (2016)
19. Vaswani, A., et al.: Attention is all you need. Advances in Neural Information Processing Systems, vol. 30 (2017)

20. Dosovitskiy, A., et al.: An image is worth 16x16 words: transformers for image recognition at scale. arXiv preprint arXiv:2010.11929 (2020)
21. Wang, W., et al.: Pyramid vision transformer: a versatile backbone for dense prediction without convolutions. In: Proceedings of the IEEE/CVF International Conference on Computer Vision, pp. 568–578 (2021)
22. Zhu, X., Su, W., Lu, L., Li, B., Wang, X., Dai, J.: Deformable DETR: deformable transformers for end-to-end object detection. arXiv preprint arXiv:2010.04159 (2020)
23. Zheng, S., et al.: Rethinking semantic segmentation from a sequence-to-sequence perspective with transformers. In: Proceedings of the IEEE/CVF Conference on Computer Vision and Pattern Recognition, pp. 6881–6890 (2021)
24. Wang, Y., et al.: End-to-end video instance segmentation with transformers. In: Proceedings of the IEEE/CVF Conference on Computer Vision and Pattern Recognition, pp. 8741–8750 (2021)
25. Zhang, H., Lan, X., Bai, S., Zhou, X., Tian, Z., Zheng, N.: Roi-based robotic grasp detection for object overlapping scenes. In: 2019 IEEE/RSJ International Conference on Intelligent Robots and Systems (IROS), pp. 4768–4775. IEEE (2019)
26. Song, Y., Gao, L., Li, X., Shen, W.: A novel robotic grasp detection method based on region proposal networks. Robot. Comput.-Integr. Manuf. **65**, 101963 (2020)

Building Deeper with U-Attention Net for Underwater Image Enhancement

Chi Ma[1], Hui Hu[1(✉)], Yuenai Chen[1], Le Yang[2], and Anxu Bu[2]

[1] The School of Computer Science and Engineering, Huizhou University, 516007 Huizhou, China
{machi,Huhui}@hzu.edu.cn

[2] The School of Computer Science and Software Engineering, University of Science and Technology Liaoning, 114051 Anshan, China
{yangle,buanxu}@ustl.edu.cn

Abstract. The images captured in the underwater scene suffer from color casts due to the scattering and absorption of light. These problems severe interfere many vision tasks in the underwater scene. In this paper, we propose a Deeper U-Attention Net for underwater image enhancement. Different from most existing underwater image enhancement methods, we adequately exploit underlying complementary information of different scales, which can enrich the feature representation from multiply perspectives. Specifically, we design a novel module with self-attention module to enhance the features by the features itself. Then, we design U-Attention block to extract the features at a certain level. At last, we build deeper two level U-structure net with the proposed U-attention block at multiply scale. This architecture enables us to build a very deep network, which can extract the multi-scale features for underwater image enhancement. Experimental results show our method has a better performance on public datasets than most state-of-the-art methods.

Keywords: Underwater image enhancement · U-attention · Multi-scale features · U-Net

1 Introduction

Recently, underwater image enhancement [2,4] has drawn much academic attention in image processing and underwater vision. The aim of the underwater image is to restore the missing information due to wavelength- and distance-dependent attenuation [24]. As the light propagates in the water, it is inevitable to occur the phenomenon of scattering and absorption, which leads to various degrees of color deviations. Therefore, eliminating the color casts and color deviations caused by scattering and absorption is critical in understanding the underwater world.

There are a variety of factors for underwater image degradation, such as scattering, suspended particles, light attenuation, and other optical properties of the underwater environment. Considering all the factors in a unified model is not easy in the image processing. Previous methods want to enhance the underwater image through estimating

H. Liu et al. (Eds.): ICIRA 2022, LNAI 13456, pp. 430–439, 2022.
https://doi.org/10.1007/978-3-031-13822-5_38

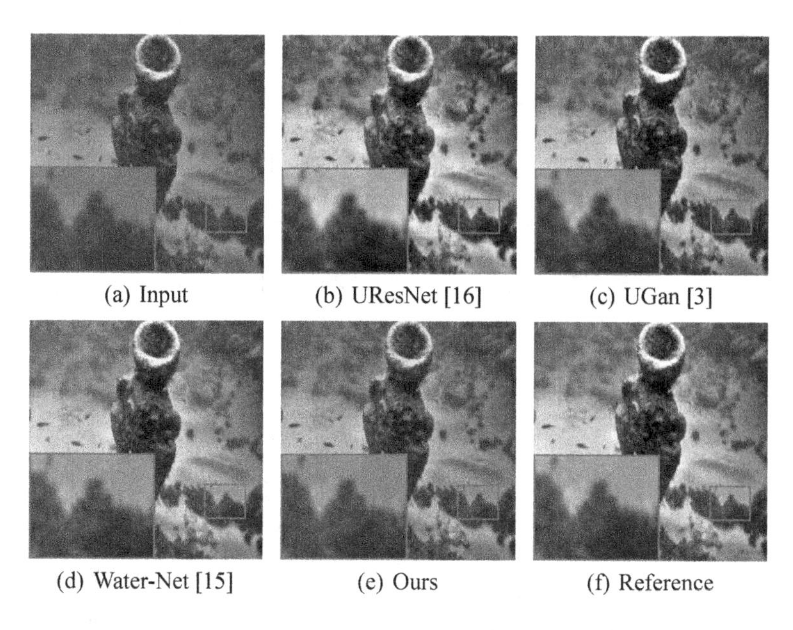

Fig. 1. Qualitative illustration of the underwater image enhancement. Our model can generate sharper and more natural color contents (e) when comparing with UResNet [16] (b), UGan [3] (c) Water-Net [15] (d).

light transmission and background light [1,14,22]. However, these methods can only enhance the underwater images in some cases. They are server limited by their inherent problems: 1) the medium transmission is also affected by multiple factors, estimating the medium transmission accurately is nontrivial; 2) these methods are depended on too much prior information, such as the depth or water-quality, which are not always available.

Recently, CNNs and Gan have shown impressive performance on feature extraction and image generation, since they can use the discriminative features to complete many vision tasks. Hou et al. [8] complete the transmission learning and image enhancement with a residual leaning strategy. Sun et al. [20] introduce an encoder-decoder structure into underwater image enhancement for feature extraction, which has been proved effective in the high-level vision task [19]. As the deeper layers can learn more feature information, Liu et al. in [16] use the VDSR proposed in [11] to extract the features of underwater images, and achieve remarkable performance in the enhanced images as shown in Fig. 1 (b). Although these methods have improved the enhanced performance in the underwater image, there is still much room to design more powerful structures to extract more robust features for underwater image enhancement.

With the development of GAN [5,21,25], more and more methods hope to extract discriminative features [10,17,23], which can be used to generate a pleasing image. In [6], they propose a dense block to extract multi-scale features for underwater image enhancement. This method first validates the multi-scale features can boost the performance of image enhancement. In [3], Fabbri proposed a novel model, named UGAN, to

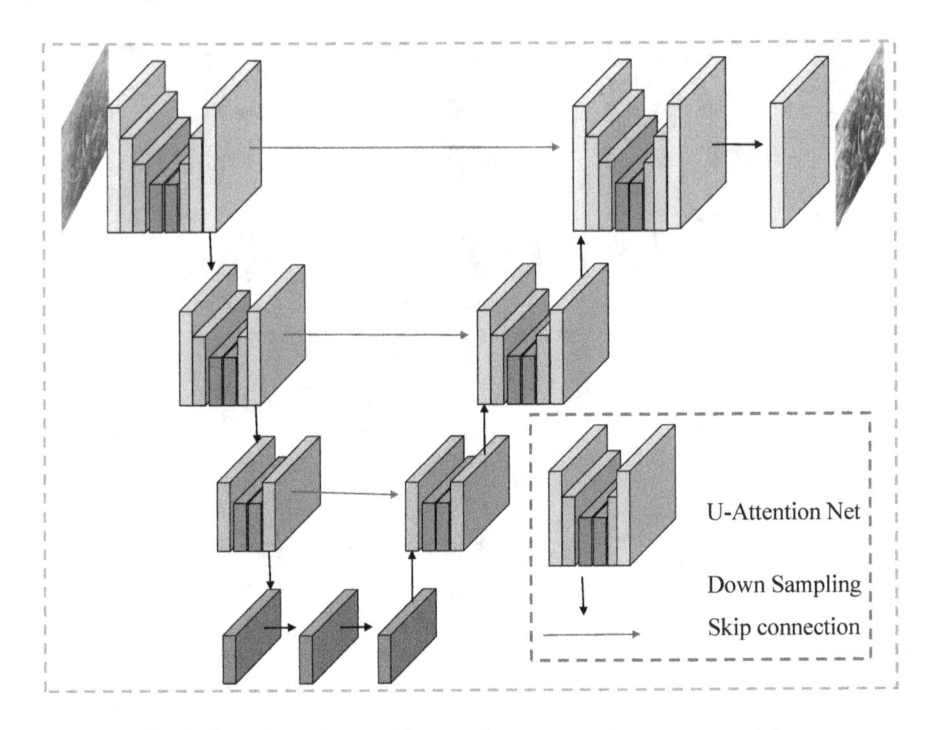

Fig. 2. Overview of our architecture for across-scale attention module.

achieve the underwater image enhancement. This model improves the quality of visual underwater scenes with Generative Adversarial Networks. The result is shown in Fig. 1 (c).In order to eliminate the color cast, Li et al. [13] propose a supervised color migration method, which can correct color distortion with a weakly supervised model. [15] proposes an underwater image enhancement network, called Water-Net, and trains on this benchmark as a baseline, as shown in Fig. 1 (d). In order to explore the information of multi-color space, [14] proposes a multi-color space encoder network, which explores the relationships of multi-color space and incorporates the characteristics to improve the performance of underwater image enhancement. Although these GAN-based method achieve pleasing enhancement results, these GAN-based methods overly focus on the semantic meaning features, which leads the generated images to contain more mesh effect, as shown in Fig. 1 (c).

Inspired by the analysis above, in this paper, we propose a U-attention Net for underwater image enhancement. Specifically, the multi-scale features are extracted and exploited by the U-Attention architecture, which illustrates the features of different scale and models the features well for image enhancement. In the U-Attention, we introduce the pixel-wise attention into U-net structure, which can well capture the features from the spatial-wise attention. Second, we design another U-structure net to explore more deep multi-scale features, which represents the features on different scales and capture more semantic meaning. Finally, we complete the underwater image enhancement with the features extracted from our model, then a clean image can be obtained.

After training the entire model with deep supervision, we evaluate our model on several public benchmarks datasets and real underwater images. The experimental results verify the effectiveness and efficiency of the proposed model.

The contributions of this paper are summarized as follows:

- We propose a novel U-Attention net, which introduce a attention module into the U-Structure formation. The proposed U-attention module can well capture the pixel-level features ate every scale.
- We build a deeper U-Structure net with the proposed U-Attention module, which is able to extract intra-stage multi-scale features to complete the underwater image enhancement effectively.
- Various comparisons on benchmark datasets justify the effectiveness of the proposed model in comparison to several state-of-the-art underwater image enhancement models.

2 Proposed Method

In this section, we first describe the motivation and structure of the proposed U-Attention model and then present the details of the entire structure of our deep network with the U-Attention module for underwater image enhancement.

2.1 Our Network

An accurate underwater enhancement method should have two aspects: 1) extracting the image features from multi-scale perspective and 2) enriching the useful features with the self-attention module form pixel-level. To achieve these two aspects, we proposed a novel U-structure network as shown in Fig. 2, which contain a two levels module to extract the features progressively. The first level U-structure is a U-Attention model, which introduce a pixel-wise attention module into a U-net. The proposed U-Attention module can capture the features from multiple scale with the attention module. The second level U-structure acquires another level semantic meaning by stacking multiple U-Attention module together. At last, we take the extracted features to complete the task of underwater image enhancement.

U-Attention Model. Feature representation from multiple scale is critical in image enhancement since the multiple scale features can understand the image from multiple perspectives. We design a novel U-Attention net that introduce the attention module to capture the pixel-wise attention in the multi-scale features, as shown in Fig. 3 (a). Our attention module is designed on ResNet [7], which is proposed for image recognition. The core component of the attention module is to enhance the features representation through re-calibrating the pixel-wise features responses adaptively, as shown in Fig. 3 (a). The residual learning block first generates an attention map to illustrate the different weights of the pixel-wise features. Then, we re-calibrate the features responses with the learned attentive vector. We formulate the pixel-wise attention learning as follows:

$$F_{ca} = F_{input} + Conv(F_{input} * CA(Conv(F_{input})), \qquad (1)$$

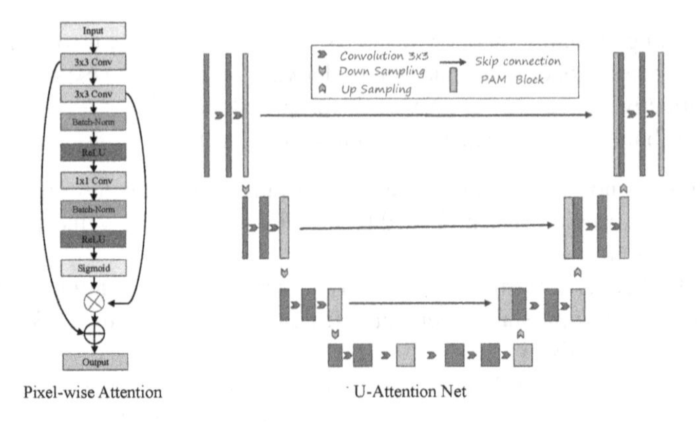

Fig. 3. Overview of our U-Attention module.

where F_{input}, F_{ca} denote the input and features learned by residual learning module. Conv and CA denote the essential convolution operations and pixel-wise attention models with a residual learning strategy.

The residual learning block takes advantage of the pixel-wise relationship to enhance the features representation well. This attention mechanism empower the feature from a pixel-wise level, which can make our model focus on the pixel-level changes, as shown in Fig. 3(a). We denote X as the features of the current layer. Then, we generate the pixel-level attention by a serial of convolutional operators followed by an activation function. The last response M_p characterizes the spatial correlations for each point in the feature X. The final weighted feature F_{pam} can be obtained by:

$$F_{pam} = X * Gaussian(M_p), \tag{2}$$

where X, F_{pam} denotes the input and features reweighted by pixel-attention model. $Gaussian(M_p$ is final the attention map], which is obtained by gating the response map M_p by a Gaussian function.

Based on the discussion mentioned above, we propose our U-Attention model to extract the features from multiply scale by embedding the pixel-wise attention module in to the U-structure formation. The structure of U-Attention is shown in Fig. 3 (c). Where L is the depth of the U-attention, which denotes the number of downsampling. C_{in}, C_{out} denote the number of feature channels of input and output, and M denotes the channel number of the internal layers. Therefore, our U-Attention model likes symmetric encoder-decoder structure with the depth of L which takes the features $F_{in} \in W \times H \times C_{in}$ as input and learns to extract and encoder the multi-scale features $F_{out} \in W \times H \times C_{out}$. The benefits of our U-Attention net are two-fold: 1) the integration of the residual learning and attention mechanism can well guide the net to explore the pixel-wise relationships of the features. 2) The proposed U-Attention model with attention module can well capture the multi-scale information to the maximum degree.

Two Level U-structure Net. The proposed U-Attention model can well capture the multi-scale features with the attention mechanism. However, it may be incapable to

Table 1. Quantitative comparison results in terms of PSNR and SSIM on testing set of the five datesets.

Model	UD [18] PSNR/SSIM	UI [18] PSNR/SSIM	US [18] PSNR/SSIM	UIEBD [15] PSNR/SSIM	UGAN [3] PSNR/SSIM
FUnIE-GAN [9]	21.17/0.878	22.21/0.766	25.48/0.830	19.82/0.834	22.40/0.759
UresNet [16]	20.99/0.873	23.07/0.814	26.13/0.874/	19.31/0.831	22.83/0.796
UGAN [3]	21.12/0.866	24.18/0.831	25.27/0.843	22.78/0.829	24.18/0.822
Water-Net [15]	20.80/0.864	22.50/0.817	22.65/0.820	23.21 /0.868	22.12/0.806
Ours	**21.22/0.881**	**24.37/0.834**	**26.19/0.877**	**23.39/0.876**	**24.52/0.824**

fuse the multi-scale features well. However, stacking multiple U-structure for different tasks have been explored for a while. Then, we propose a two level U-structure, which stacks enough U-Attention module to build a nested U-structure net. The proposed two level U-structure Net as shown in Fig. 2. The top level is a big U-attention net, which can extract the input image features form multi-scale level. The final outputs of the top U-attention net are then filled into another U-attention net which has a relative small size. Therefore, a nested U-structure Net is bulilt for a two-level features extractor. Theoretically, we can build an arbitrary level U-structure net. But the building structure with too many levels will be complicated to be implemented and employed in real applications. In this paper, we design two level U-structure net to completer the underwater image enhancement.

As shown in Fig. 2, in this paper, we empirically design three blocks to reduce the resolution three times, which is enough for underwater image enhancement. The proposed two level U-structure Net has two merits. First, the multi-scale features captured by U-attention help enrich the features. Second, it makes the network more robust by repeated use of up-and down-sampling operations in the encoder-decoder, which can help the U-attention net to capture more robust features for underwater image enhancement.

2.2 Training

Let x be an input underwater image and y be the corresponding ground truth image that has the same size as the input. The aim of our proposed enhancement network is to learn a mapping function $F(x)$ to obtain the enhanced underwater image $\hat{x}$, and make them close to the images in the natural scene.

In the training phase, a pixel loss [12] is used for supervised learning. The pixel loss is defined by:

$$\ell_{\text{pix}} = \rho(\hat{x} - y), \tag{3}$$

where $\rho(x) = \sqrt{x^2 + \varepsilon^2}$ is the pixel penalty function, x, y is the enhanced image and clear image. ε is set as $1e^{-3}$. In principle, we can add more constrain to every scale and enhance the image in multi-resolution. We only enhance the underwater images with the original resolution simply.

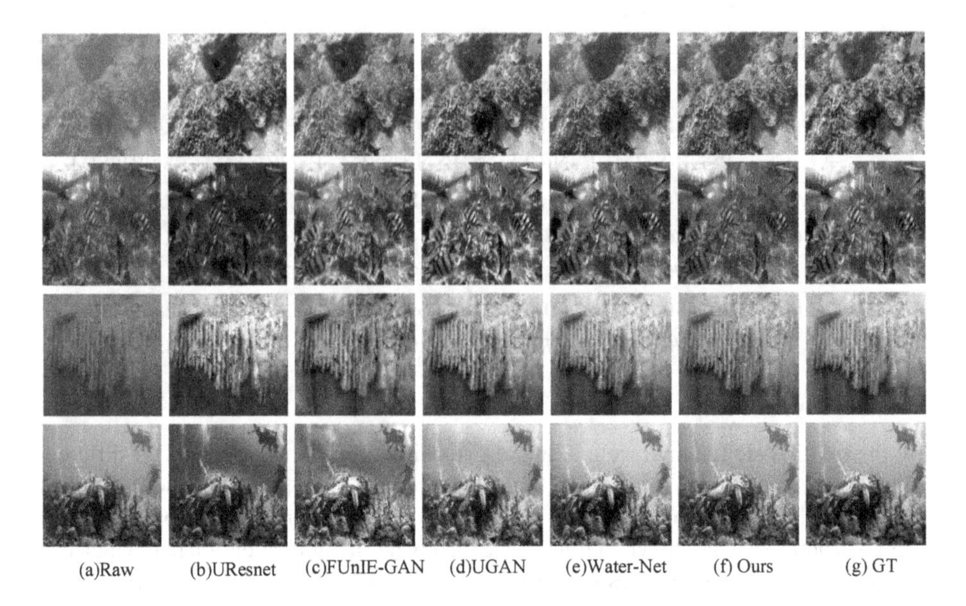

Fig. 4. Visual comparisons on UIEBD datasets [15]. (a) Input (b) UresNet [16], (c) FUnIE-GAN [9], (d) UGAN [3], (e)Water-Net [15], (f) Ours and (g) Ground Truth. We retrain these models and select the best model for comparison.

3 Experiments

In this section, we present our empirical comparisons with several state-of-the-art image deraining models based on deep learning, including FUnIE-GAN [9], UresNet [16], UGAN [3], and Water-Net [15]. The experiments were conducted on the five public datasets: Underwater Dark [18], Underwater ImageNet [18], Underwater Scenes [18], UIEBD [15] and UGAN [3]. The details are as follows. We slightly adapt the training approach: we set the momentum parameter to 0.9 and the weight decay to 0.0001. We also choose a smaller learning rate, which is initialized as 0.001 in the training set.

3.1 Comparison with the State-of-the-Arts

For fair comparison, we retrain the compared model and select the best performance model for comparison. From the results in Fig. 4, we note that our model performs better than other compared enhanced methods in all the datasets. Since our model can fully take advantage of the multi-scale features to represent the image in underwater scene, which can eliminate the color cast well. Such observation implies it is beneficial to use multi-scale features in the underwater image enhancement. Water-net also can well present the perceptual features, which focuses on semantic meaning and enhance the underwater images well. This observation validates that semantic features are very useful in improving the enhancement of the underwater image. UGAN performs worse than our model since it only uses one level u-structure, which can not explore the multi-scale features completely. Our U-attention net tends to focus on multi-scale feature integration while the two level u-structure can extract the features from multiply scale, which can eliminate obvious colored cast in the enhanced images, see the fourth column in Fig. 4. UresNet [16] completes the image enhancement with a very deep net, which takes the multi-level features to improve the performance and achieve competitive performance. However, this model ignore the multi-scale features in the enahncement, which lead to obvious artifact in the enhanced images. Our model learns the features from multiply perspectives with the proposed U-Attention, which takes full advantage of multi-scale features to enhance the image representation. From the results in Fig. 4, we notice that:

- Exploiting the multi-scale features with the proposed U-attention net can well empower the features representation, which is beneficial for the image enhancement.
- Our model can well explore the internal regulation of underwater images from multi-scale features with the two level structure, which has a better performance compared to increasing the depth of the network simply.
- Our model achieves the best performance with the features empowered by the U-Attention model.

In general, our model empowered the features representation through our two level U-structure net, which can well learn and fuse the multi-scale features. This learning strategy of features ensures that the enhanced image is more plausible.

3.2 Quantitative Comparison

We also evaluate our model via two quantitative criteria: Structural Similarity Index (SSIM) and Peak Signal to Noise Ratio (PSNR). The quantitative comparison results are presented in Table 1. From the quantitative comparison, our model achieves a substantial improvement in terms of SSIM and PSNR, compared with the state-of-the-art models. This indicates that the proposed two level U-Structure net provides useful feature representation, which can well boost the underwater enhancement performance. Compared to the best performance model, our model can improve PSNR value from 23.21 to 23.39 in UIEBD.

4 Conclusion

We achieve state-of-the-art performance in underwater image enhancement using the proposed two level u-structure net. The comparison results show that our model with U-Attention is powerful in exploiting the correlation of different scale features. The proposed U-Attention, which models the multi-scale features from high-level to low-level, provides new insight for the research on features representation. The two level u-structure net can well fuse the features from multiply scale. This is substantially useful to represent the image information form multi-perspectives. In the future, we plan to investigate the advantage of the multi-level and multi-scale features to model the underwater image to complete the image enhancement completely.

Acknowledgements. This work is supported by the Foundation Of Guangdong Educational Committee (Grant No. 2021ZDJS082, 2019KQNCX148).

References

1. Berman, D., Levy, D., Avidan, S., Treibitz, T.: Underwater single image color restoration using haze-lines and a new quantitative dataset. IEEE Trans. Pattern Anal. Mach. Intell. **43**, 2822–2837 (2020)
2. Drews, P., Nascimento, E.R., Botelho, S., Campos, M.: Underwater depth estimation and image restoration based on single images. IEEE Comput. Graphics Appl. **36**(2), 24–35 (2016)
3. Fabbri, C., Islam, M.J., Sattar, J.: Enhancing underwater imagery using generative adversarial networks. In: 2018 IEEE International Conference on Robotics and Automation (ICRA), pp. 7159–7165. IEEE (2018)
4. Galdran, A., Pardo, D., Picón, A., Alvarez-Gila, A.: Automatic red-channel underwater image restoration. J. Vis. Commun. Image Represent. **26**, 132–145 (2015)
5. Goodfellow, I.J., et al.: Generative adversarial nets. In: Proceedings of IEEE International Conference on Neural Information Processing Systems, pp. 2672–2680 (2014)
6. Guo, Y., Li, H., Zhuang, P.: Underwater image enhancement using a multiscale dense generative adversarial network. IEEE J. Oceanic Eng. **45**(3), 862–870 (2019)
7. He, K., Zhang, X., Ren, S., Sun, J.: Deep residual learning for image recognition. In: Proceedings of the IEEE Conference on Computer Vision and Pattern Recognition, pp. 770–778 (2016)
8. Hou, M., Liu, R., Fan, X., Luo, Z.: Joint residual learning for underwater image enhancement. In: 2018 25th IEEE International Conference on Image Processing (ICIP) (2018)
9. Islam, M.J., Xia, Y., Sattar, J.: Fast underwater image enhancement for improved visual perception. IEEE Robot. Autom. Lett. **5**(2), 3227–3234 (2020)
10. Jamadandi, A., Mudenagudi, U.: Exemplar-based underwater image enhancement augmented by wavelet corrected transforms. In: Proceedings of the IEEE/CVF Conference on Computer Vision and Pattern Recognition Workshops, pp. 11–17 (2019)
11. Kim, J., Lee, J.K., Lee, K.M.: Accurate image super-resolution using very deep convolutional networks. In: Proceedings of the IEEE/CVF Conference on Computer Vision and Pattern Recognition (CVPR), pp. 1646–1654 (2016)
12. Lai, W.S., Huang, J.B., Ahuja, N., Yang, M.H.: Deep Laplacian pyramid networks for fast and accurate super-resolution. In: Proceedings of the IEEE Conference on Computer Vision and Pattern Recognition, pp. 5835–5843. IEEE (2017)
13. Li, C., Guo, J., Guo, C.: Emerging from water: underwater image color correction based on weakly supervised color transfer. IEEE Sig. Process. Lett. **PP**(99), 1–1 (2017)
14. Li, C., Anwar, S., Hou, J., Cong, R., Guo, C., Ren, W.: Underwater image enhancement via medium transmission-guided multi-color space embedding. IEEE Trans. Image Process. **30**, 4985–5000 (2021)
15. Li, C., et al.: An underwater image enhancement benchmark dataset and beyond. IEEE Trans. Image Process. **29**, 4376–4389 (2019)
16. Liu, P., Wang, G., Qi, H., Zhang, C., Zheng, H., Yu, Z.: Underwater image enhancement with a deep residual framework. IEEE Access **7**, 94614–94629 (2019). https://doi.org/10.1109/ACCESS.2019.2928976
17. Liu, X., Gao, Z., Chen, B.M.: MLFCGAN: multilevel feature fusion-based conditional GAN for underwater image color correction. IEEE Geosci. Remote Sens. Lett. **17**(9), 1488–1492 (2019)
18. Peng, Y.T., Cosman, P.C.: Underwater image restoration based on image blurriness and light absorption. IEEE Trans. Image Process. **26**(4), 1579–1594 (2017)

19. Ronneberger, O., Fischer, P., Brox, T.: U-Net: convolutional networks for biomedical image segmentation. In: Navab, N., Hornegger, J., Wells, W.M., Frangi, A.F. (eds.) MICCAI 2015. LNCS, vol. 9351, pp. 234–241. Springer, Cham (2015). https://doi.org/10.1007/978-3-319-24574-4_28
20. Sun, X., Liu, L., Li, Q., Dong, J., Lima, E., Yin, R.: Deep pixel-to-pixel network for underwater image enhancement and restoration. IET Image Proc. **13**(3), 469–474 (2019)
21. Uplavikar, P.M., Wu, Z., Wang, Z.: All-in-one underwater image enhancement using domain-adversarial learning. In: CVPR Workshops, pp. 1–8 (2019)
22. Wang, K., Shen, L., Lin, Y., Li, M., Zhao, Q.: Joint iterative color correction and dehazing for underwater image enhancement. IEEE Robot. Autom. Lett. **6**(3), 5121–5128 (2021)
23. Yang, M., Hu, K., Du, Y., Wei, Z., Sheng, Z., Hu, J.: Underwater image enhancement based on conditional generative adversarial network. Sig. Process. Image Commun. **81**, 115723 (2020)
24. Zhou, Y., Wu, Q., Yan, K., Feng, L., Xiang, W.: Underwater image restoration using color-line model. IEEE Trans. Circuits Syst. Video Technol. **29**(3), 907–911 (2018)
25. Zhou, Y., Yan, K.: Domain adaptive adversarial learning based on physics model feedback for underwater image enhancement. arXiv preprint arXiv:2002.09315 (2020)

A Self-attention Network for Face Detection Based on Unmanned Aerial Vehicles

Shunfu Hua[1,2,3], Huijie Fan[2,3](✉), Naida Ding[2,3], Wei Li[1], and Yandong Tang[2,3]

[1] College of Information Engineering, Shenyang University of Technology, Shenyang 110870, China
[2] State Key Laboratory of Robotics, Shenyang Institute of Automation, Chinese Academy of Sciences, Shenyang 110016, China
{huashunfu,fanhuijie,dingnaida,ytang}@sia.cn
[3] Institutes for Robotics and Intelligent Manufacturing, Chinese Academy of Sciences, Shenyang 110169, China

Abstract. Face detection based on Unmanned Aerial Vehicles (UAVs) faces following challenges: (1) scale variation. When the UAVs fly in the air, the size of faces is different owing to the distance, which increases the difficulty of face detection. (2) lack of specialized face detection datasets. It results in a sharp drop in the accuracy of algorithm. To address these two issues, we make full advantage of existing open benchmarks to train our model. However, the gap is too huge when we adapt face detectors from the ground to the air. Therefore, we propose a novel network called Face Self-attention Network (FSN) to achieve high performance. Our method conducts extensive experiments on the standard WIDER FACE benchmark. The experimental results demonstrate that FSN can detect multi-scale faces accurately.

Keywords: Face detection · UAVs · Scale variation · Self-attention

1 Introduction

Through the extensive use of face detection in daily life, we are not hard to find that it is a fundamental and important step towards computer vision [15] with many applications, such as tracking, face recognition [4], and facial expression analysis [22]. Hand-crafted features [17] is a main way of early works, but it is not effective in fact. With the development and research of deep learning, recent years have witnessed the tremendous progress made on face detection. Meanwhile, Unmanned Aerial Vehicle is a newly-developing industry because of

This work is supported by the National Natural Science Foundation of China (61873259, U20A20200,61821005), and the Youth Innovation Promotion Association of Chinese Academy of Sciences (2019203).

H. Liu et al. (Eds.): ICIRA 2022, LNAI 13456, pp. 440–449, 2022.
https://doi.org/10.1007/978-3-031-13822-5_39

advanced technology. It is widely used in many fields including military science, fire rescue and public security. We naturally utilize drones to detect faces owing to their convenience and flexibility in specialized tasks. In fact, the main purpose to realize the UAV-based face detection is face recognition, which can help us track and rescue. Thus the face detection is the first step and is meaningful for future work. The trouble is that faces drones detect in the air are totally different from that we see on the ground. It is clearly that the range is broader and the face is smaller. Scale variation problem becomes more and more obvious. Moreover, there are no specialized datasets about face detection based on UAV. We train and test our model on the public face detection benchmark WIDER FACE, then directly apply the face detection algorithm to the drone. However, model detection performance degrades significantly due to the lack of dedicated training data and ground-to-air disparity. We propose the novel network that is called Face Self-attention Network (FSN). Specifically, it follows the similar setting as RetinaNet [9] that achieves state-of-art performance on COCO general object detection [10]. The difference is that we add the self-attention mechanism [16], which is introduced to strengthen the weight of crucial features. Compared with ScaleFace [21], our method achieves 82.0% AP on the hard set of WIDER FACE [20] (threshold of 0.5) with ResNet50 [6], surpassing by 4.8%.

2 Related Work

2.1 Face Detection

A lot of efforts have been put into research owning to the practicality of face detection. Viola-Jones [17] attempts to train a cascade model by Adaboost with Haar-like feature and achieve a good result, which is known as the pioneering work of the face detection. Due to continuous research, the performance of face detection is constantly improving but there is no essential leap until the emergence of deep learning. For current face detection, there are two mainstream methods according to whether following the proposal and refine strategy. One is single-stage method, such as Single-shot multi-box detector(SSD) [11], YOLO [13], RetinaNet. And the other is two-stage method includes R-CNN [5], Fast R-CNN [19], Faster R-CNN [14]. One-stage approaches pay more attention to real-time availability, thus our framework employs a one-stage approach.

2.2 Object Detection Based on UAV

Object detection [11,23] is to locate the object first and then identify the category of the object, which is a very popular research direction in the field of computer vision [15]. At present, the general object detection methods are mainly YOLO, Faster-RCNN, R-FCN [3]. Their difference is whether and where the region proposal is attached. We combine object detection and drones to not only take pictures, but also detect and identify objects. However, objects photographed by drones in the air from a top-down perspective are smaller than what we see standing on the ground. As face detection is a special case of object detection, we can train multiple detectors for different scales like Hu et al. [8], but the method is very time-consuming. SSD and S3FD [23]both design scale-aware anchors, but their extracted features are utilized respectively to predict, which are short of semantics for high-resolution features. Inspired by FPN [8]and self-attention, we propose a novel framework of FSN to address multi-scale problem.

3 Face Self-attention Network

Compared with ordinary face detection, the differences are the range is wider and faces are smaller for the face detection on UAV, which increases the difficulty of face detection. Therefore, we aim to learn more semantic and detailed features at different layers to detect multi-scale faces. In this paper, we utilize a single-stage face detection framework, the overall architecture is shown in Fig. 1.

3.1 Base Framework

In the first stage, we use ResNet50 with a bottom-up pathway (from C2 to C5) as the backbone to extract features. C6 is calculated through a $3*3$ convolution on C5. As the increasing number of the layers, the semantic information is enhanced but the spatial resolution that is beneficial to localization of faces is impaired. Therefore, we utilize Feature Pyramid Network (FPN) to combine visual meaningful features with visually weak but semantically strong features. More specifically, the top-down process is performed by upsampling and the lateral connection is designed to merge the different feature maps. It is beneficial for us to detect faces at a different scale. In the second stage, we employ self-attention mechanism to enhance the role of key features. K, Q, V, represent three identical features and the attention score is calculated by K, Q, V to determine how much attention is devoted to a feature. Then we concatenate and normalize the features obtained by self-attention mechanism. Finally our prediction results include face location and value of confidence. And we use our loss function to compute these two kinds of losses.

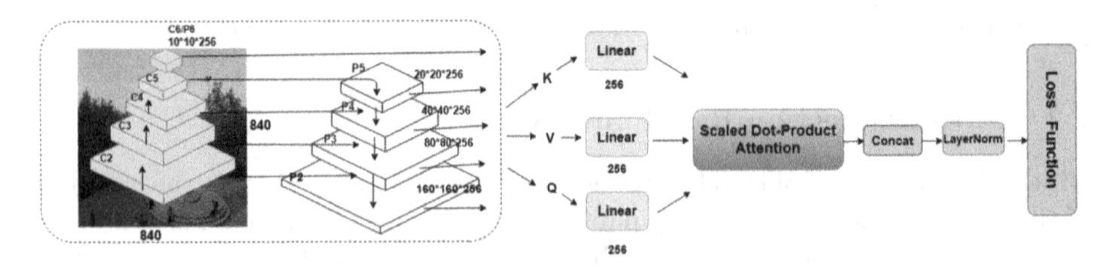

Fig. 1. The network architecture of proposed FSN

3.2 Self-attention Mechanism

Considering that the faces seen by drones are generally small, we adopt a self-attention mechanism to improve the accuracy of face detection. Self-attention is the core mechanism of transformer [16]. We did not employ encoder and decoder because they have a high requirement for computer configuration, but directly use self-attention. After FPN process, we get three different features, assuming feature1, feature2, feature3. Take feature1 as an example, the method uses Q, K, V that are three identical features, they both represent feature1 and their dimensions are 256. First we pass Q, K, V through a linear layer initialized by xavier, and then calculate the weight of each feature through the Scaled Dot-Production Attention module. Finally, the above features obtained are concatenated, and then normalized. Feature2, Feature3 also follow the steps above. Mathematically, we detail our Scaled-Dot Production Attention module as following formula:

$$Attention(Q, K, V) = Softmax(\frac{QK^T}{\sqrt{d_k}})V \tag{1}$$

Q and the transposition of K is multiplied to get the attention score. In order to prevent the result from being too large, we divide it by a scaling factor d_k that is the dimension of Q and K, and then use the Softmax operation to normalize the result to a probability distribution. After this, we multiply the matrix V to get the sum of weights.

3.3 Training

In this section, we mainly introduce four parts: loss function, anchor setting, data augmentation and implementation details.

Loss Function. In order to better train model parameters, we use an efficient loss function that consists of localization loss (loc) and the confidence loss (conf). The key idea is that we select all positive samples and the top n negative samples with large loss by hard negative mining, which avoids the dominant loss value of negative samples. Confidence loss is calculated for both positive and negative samples selected, while localization loss only calculates the loss of positive samples. To this end, our loss function are as follows:

$$L(x, c, l, g) = \frac{1}{N}(L_{conf}(x, c) + \alpha \mathrm{L}_{loc}(x, l, g)) \tag{2}$$

where

$$L_{conf}(x, c) = -\sum_{i \in Pos}^{N} x_{ij}^{p} \log(\hat{c}_i^p) - \sum_{i \in Neg} \log(\hat{c}_i^0) \tag{3}$$

$$L_{loc}(x, l, g) = \sum_{i \in Pos}^{N} \sum_{m \in cx, cy, w, h} x_{ij}^{k} smooth_{L1}(l_i^m - \hat{g}_j^m) \tag{4}$$

where c, l and g represent the class confidences, the predicted boxes and the ground truth boxes respectively. In Eq. 3, we let $x_{ij}^{p} \geq 1$ and have an indicator that is used to match the i-th default box to the j-th ground truth box of category p, namely $x_{ij}^{p} = \{1,0\}$. The localization loss is the smooth-L1 loss weighted by α that is set to 1 by cross validation. It calculates the difference of center point and length and width between the parameters of l and g. N represents the sum of positive and negative samples, namely the number of matched default boxes. When we encounter the special case of $N = 0$, we set the entire loss value to 0.

Anchor Setting. In this paper, there are five feature pyramid levels from P2 to P6. For these five detector layers, we utilize scale-specific anchors. Considering that the features of tiny faces will disappear with the increase of the convolutional layers, we employ P2 to detect small faces. With the input image size at $840 * 840$, the range of anchor is from 16 to 406 pixel on the feature pyramid levels. There are totally 102,300 anchors, of which P2 accounts for 75%. In addition, we set the anchor scale step at $2^{1/3}$ and the aspect ratio of our anchor is set as 1:1. During training, if the values of IoU are larger than 0.5, they are ground-truth boxes otherwise are matched to the background. Meanwhile, we use standard hard negative mining to ensure that the ratio between the negative and positive samples is 3:1 owing to the extreme imbalance between the positive and negative training examples.

Data Augmentation. Since we observe that the tiny faces occupy 20% of the datasets of the WiderFace training set, which is limited and can not be well positioned to meet the need for training of network. We present a strategy of data augmentation that includes cutting, mirror, padding and flip. For cropping, we randomly cut square patches from original images and resize these patches into 840*840 to generate larger training faces. Specifically, the random size we crop is between [0.3,1] of the short edge of the original image. In addition, when we are in the situation where the face is on the crop boundary, we can keep the overlapped part of the ground-truth box if its center is within the sampled patch.

Implementation Details. We conduct experiments on Ubuntu18.04 and employ PyTorch framework. The size of input image is $840 * 840$. We use ResNet50 as our backbone and train it with a batchsize of 4 on GeForce RTX 3090 (24 GB). The optimizer we use is Stochastic Gradient Descent (SGD). Meanwhile, the momentum is set to 0.9 and weight decay of 5e−4. We train our model for 100 epochs and the initial learning rate is set to 1e−3.

4 Experimental Results

In this section, we first introduce the dataset and drone, then compare FSN with other six face detection algorithms. Finally, we prove the effectiveness of self-attention through ablation study experiments.

4.1 About WIDER FACE Benchmark and UAV

WIDER FACE Benchmark. We implement our method on the WiderFace datasets, which consists of 32,203 images with 393,703 annotated faces. Widerface has variable faces in scale, pose, occlusion, expression and illumination. It is split into three subsets: 40% for training, 10% for validation, 50% for testing. According to the detection difficulty, each subset has three levels: Easy, Medium and Hard. Mean average precision is our evaluation metric to compare several methods for face detection. We train our network on the training set of WiderFace and perform evaluations on the validation set.

UAV. The drone we use is Mavic Air2 that is a portable folding aircraft designed by China's DJI in 2020, as shown in Fig. 2. It has a maximum flight speed of 68 km/h and a maximum flight time of about 34 min. Moreover, Mavic Air2 features a brand new camera design that can capture 48-megapixel photos. Meanwhile, DJI Fly that is an app specially equipped with drones allows users to connect their mobile phones for a more intuitive and natural control experience. When UAV is shooting, the len rotates freely, so the captured faces will show various angles. This way increases the difficulty of the experiment and ensures the randomness. In order not to make the face pixels too small, the drone flies at a height of 5 m.

Fig. 2. Live shooting of Mavic Air2

4.2 Evaluation on WIDER FACE Datasets

As shown in Table 1, we conduct experiments on validation set of WiderFace and compare the proposed FSN with other six state-of-the-art face detection algorithms, including MSCNN [1], SSH [12], ScaleFace [21], HR [7], FAN [18], SRN [2]. The data shows that our method is able to achieve significant improvement over the state-of-the-art approaches like ScaleFace and HR using ResNet50, which validates the effectiveness of our algorithm in addressing sale-variant issues to some extent.

Table 1. Comparison of FSN with state-of-art algorithms on the validation set of WIDER FACE (mAP).

Algorithms	MSCNN	SSH	ScaleFace	HR	FAN	SRN	FSN (ours)
Backbone	VGG16	VGG16	ResNet50	ResNet50	ResNet50	ResNet50	ResNet50
Easy	91.6	93.1	86.8	90.7	95.3	96.4	93.9
Medium	90.3	92.1	86.7	89.0	94.2	95.3	92.5
Hard	80.2	84.5	77.2	80.2	88.8	90.2	82.0

4.3 Ablation Studies on Self-attention Mechanism

In order to evaluate the contribution of self-attention mechanism, we conduct comparative experiments with RetinaNet. We first plot the PR curves for our method FSN and our method not using self-attention (RetinaNet) respectively. From the Fig. 3, we can see that the performance of FSN is better.

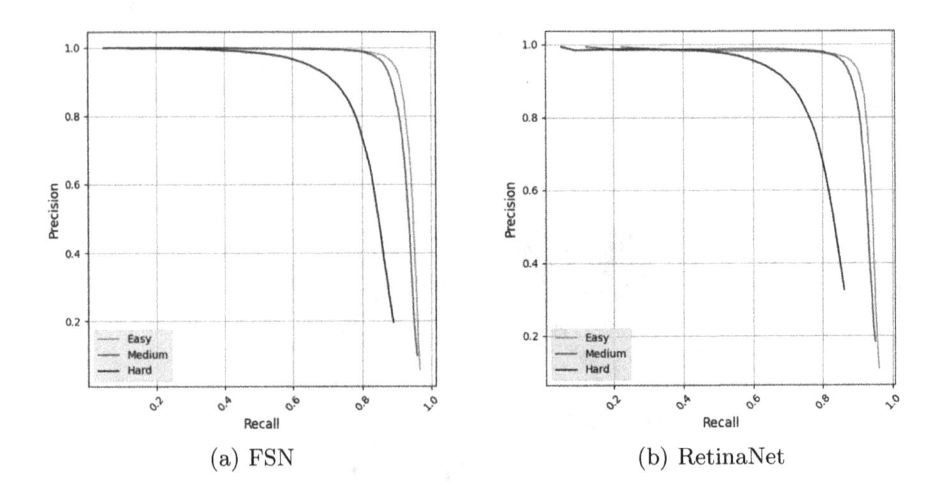

(a) FSN (b) RetinaNet

Fig. 3. Precision-recall curves on WIDER FACE validation set

For better comparison, we use RetinaNet as our comparative method and employ the anchor setting strategy and data augmentation discussed above. As shown in Table 2, our method outperforms the plain RetinaNet by a large margin on hard set. Meanwhile, compared with our method not using self-attention mechanism, FSN obtains improvement 1.5%, 1.3% and 2.4% in easy, medium and hard set respectively.

Table 2. Ablation experiments of FSN on the WIDER FACE validation set.

BaseNet	Anchor assign	Data augmentation	Self-attention	AP (easy)	AP (medium)	AP (hard)
RetinaNet				92.6	91.2	63.4
Ours (RetinaNet)	✓	✓		92.4	91.2	79.6
Ours (FSN)	✓	✓	✓	**93.9**	**92.5**	**82.0**

We compare our method FSN and RetinaNet on images that the drone takes at a height of 5 m. The size of image is 3840 * 2160 and faces in the picture is about 5 to 50 pixels. Since the faces detected by the algorithm can not be seen clearly, we cut out the person from the original and experimental images respectively, then place them in the top left corner of the picture and point out them with red arrow. Pictures (a), (c) employ the method FSN, while Pictures (b), (d) utilize the method RetinaNet. As can be seen in the Fig. 4, we count the number of faces detected by the two methods and draw a conclusion that the method using FSN is slightly more accurate than the method using RetinaNet.

(a) detected faces : 21(FSN)

(b) detected faces : 20(RetinaNet)

(c) detected faces : 11(FSN)

(d) detected faces : 10(RetinaNet)

Fig. 4. Comparison results of images photoed by UAV (Color figure online)

5 Conclusion

In this paper, we propose a novel framework of FSN to target multi-scale problem of face detection based on UAV. Specifically, FPN makes shallow layers have more semantic information and deep layers have more spatial resolution, which is beneficial to detect faces of different sizes. In order to solve the problem that faces are generally smaller for drones, we add the self-attention to make the best of key features of faces. Experimental results on challenging WIDERFACE benchmark validate the effectiveness of our proposed method.

References

1. Cai, Z., Fan, Q., Feris, R.S., Vasconcelos, N.: A unified multi-scale deep convolutional neural network for fast object detection. In: Leibe, B., Matas, J., Sebe, N., Welling, M. (eds.) ECCV 2016. LNCS, vol. 9908, pp. 354–370. Springer, Cham (2016). https://doi.org/10.1007/978-3-319-46493-0_22
2. Chi, C., Zhang, S., Xing, J., Lei, Z., Li, S.Z., Zou, X.: Selective refinement network for high performance face detection. In: Proceedings of the AAAI Conference on Artificial Intelligence, vol. 33, pp. 8231–8238 (2019)
3. Dai, J., Li, Y., He, K., Sun, J.: R-FCN: object detection via region-based fully convolutional networks. In: Advances in Neural Information Processing Systems, vol. 29 (2016)
4. Deng, J., Guo, J., Xue, N., Zafeiriou, S.: ArcFace: additive angular margin loss for deep face recognition. In: Proceedings of the IEEE/CVF Conference on Computer Vision and Pattern Recognition, pp. 4690–4699 (2019)
5. Girshick, R., Donahue, J., Darrell, T., Malik, J.: Rich feature hierarchies for accurate object detection and semantic segmentation. In: Proceedings of the IEEE Conference on Computer Vision and Pattern Recognition, pp. 580–587 (2014)
6. He, K., Zhang, X., Ren, S., Sun, J.: Deep residual learning for image recognition. In: Proceedings of the IEEE Conference on Computer Vision and Pattern Recognition, pp. 770–778 (2016)
7. Hu, P., Ramanan, D.: Finding tiny faces. In: Proceedings of the IEEE Conference on Computer Vision and Pattern Recognition, pp. 951–959 (2017)
8. Lin, T.Y., Dollár, P., Girshick, R., He, K., Hariharan, B., Belongie, S.: Feature pyramid networks for object detection. In: Proceedings of the IEEE Conference on Computer Vision and Pattern Recognition, pp. 2117–2125 (2017)
9. Lin, T.Y., Goyal, P., Girshick, R., He, K., Dollár, P.: Focal loss for dense object detection. In: Proceedings of the IEEE International Conference on Computer Vision, pp. 2980–2988 (2017)
10. Lin, T.-Y., et al.: Microsoft COCO: common objects in context. In: Fleet, D., Pajdla, T., Schiele, B., Tuytelaars, T. (eds.) ECCV 2014. LNCS, vol. 8693, pp. 740–755. Springer, Cham (2014). https://doi.org/10.1007/978-3-319-10602-1_48
11. Liu, W., et al.: SSD: single shot multibox detector. In: Leibe, B., Matas, J., Sebe, N., Welling, M. (eds.) ECCV 2016. LNCS, vol. 9905, pp. 21–37. Springer, Cham (2016). https://doi.org/10.1007/978-3-319-46448-0_2
12. Najibi, M., Samangouei, P., Chellappa, R., Davis, L.S.: SSH: single stage headless face detector. In: Proceedings of the IEEE International Conference on Computer Vision, pp. 4875–4884 (2017)
13. Redmon, J., Divvala, S., Girshick, R., Farhadi, A.: You only look once: unified, real-time object detection. In: Proceedings of the IEEE Conference on Computer Vision and Pattern Recognition, pp. 779–788 (2016)
14. Ren, S., He, K., Girshick, R., Sun, J.: Faster R-CNN: towards real-time object detection with region proposal networks. In: Advances in Neural Information Processing Systems, vol. 28 (2015)

15. Torralba, A., Murphy, K.P., Freeman, W.T., Rubin, M.A.: Context-based vision system for place and object recognition. In: IEEE International Conference on Computer Vision, vol. 2, p. 273. IEEE Computer Society (2003)
16. Vaswani, A., et al..: Attention is all you need. In: Advances in Neural Information Processing Systems, vol. 30 (2017)
17. Viola, P., Jones, M.J.: Robust real-time face detection. Int. J. Comput. Vis. **57**(2), 137–154 (2004)
18. Wang, J., Yuan, Y., Yu, G.: Face attention network: an effective face detector for the occluded faces. arXiv preprint arXiv:1711.07246 (2017)
19. Wang, X., Shrivastava, A., Gupta, A.: A-fast-RCNN: hard positive generation via adversary for object detection. In: Proceedings of the IEEE Conference on Computer Vision and Pattern Recognition, pp. 2606–2615 (2017)
20. Yang, S., Luo, P., Loy, C.C., Tang, X.: Wider face: a face detection benchmark. In: Proceedings of the IEEE Conference on Computer Vision and Pattern Recognition, pp. 5525–5533 (2016)
21. Yang, S., Xiong, Y., Loy, C.C., Tang, X.: Face detection through scale-friendly deep convolutional networks. arXiv preprint arXiv:1706.02863 (2017)
22. Zhang, F., Zhang, T., Mao, Q., Xu, C.: Joint pose and expression modeling for facial expression recognition. In: Proceedings of the IEEE Conference on Computer Vision and Pattern Recognition, pp. 3359–3368 (2018)
23. Zhang, S., Zhu, X., Lei, Z., Shi, H., Wang, X., Li, S.Z.: S3FD: single shot scale-invariant face detector. In: Proceedings of the IEEE International Conference on Computer Vision, pp. 192–201 (2017)

Video Abnormal Behavior Detection Based on Human Skeletal Information and GRU

Yibo Li(✉) and Zixun Zhang

Shenyang Aerospace University, Shenyang, Liaoning, China
Liyibo_sau@163.com

Abstract. Video abnormal behavior detection is an important research direction in the field of computer vision and has been widely used in surveillance video. This paper proposes an abnormal behavior detection model based on human skeletal structure and recurrent neural network, which uses dynamic skeletal features for abnormal behavior detection. The model in this paper extracts key points from multiple frames through the human pose estimation algorithm, obtains human skeleton information, and inputs it into the established autoencoder recurrent neural network to detect abnormal human behavior from video sequences. Compared with traditional appearance-based models, our method has better anomaly detection performance under multi-scene and multi-view.

Keywords: Abnormal behavior detection · GRU · Autoencoder · Skeletal information

1 Introduction

As an important part of the field of computer vision, abnormal behavior detection has developed rapidly in recent years and is widely used in surveillance video. Abnormal behavior detection refers to the appearance of inappropriate actions in the video target scene. Due to the low frequency, unpredictability and variety of abnormal behaviors, semi-supervised [1] or unsupervised [2] algorithms are usually used for detection. An efficient solution is to learn normal patterns in normal training video sequences in an unsupervised setting, and then detect irregular events in videos based on this.

Most of the current abnormal behavior detection methods rely on appearance and motion features, treat a single human body as a target, and extract the centroid, shape and motion features as features to describe the target behavior [3]. Designing behavioral features requires a lot of manpower, and in some scenarios, important information is lost due to noise, making it difficult to generalize detection algorithms to different scenarios. With the development of deep learning, abnormal behavior detection began to use deep neural networks to learn deep features from images. By designing the network according to the rules of feature extraction, it has better universality for abnormal behavior detection in different scenarios. Three-dimensional convolutional neural networks [4], two-stream convolutional neural networks [5] and recurrent neural networks [6] are proposed to extract abnormal behavior features. However, the semantic difference between visual

H. Liu et al. (Eds.): ICIRA 2022, LNAI 13456, pp. 450–458, 2022.
https://doi.org/10.1007/978-3-031-13822-5_40

features and abnormal events leads to the lack of interpretability of the results, and this shortcoming is magnified with the deepening of the network layer, which affects the design of the network structure by researchers.

The human skeleton information features can obtain the position and state information of the key parts of the human body to describe the human behavior through the pose estimation algorithm. Through human body detection and bone key point detection, a human skeleton sequence with behavioral characteristics is formed, which can accurately identify human behavior [7]. Compared with image features, skeletal features are more compact, stronger in structure and rich in semantics [8]. They are more specific for behavior and motion descriptions, and have strong robustness in complex backgrounds.

In this paper, a model is proposed to represent the abnormal behavior characteristics of the human body. The skeleton information is extracted by the human pose algorithm, and the auto-encoder composed of a bunch of recurrent neural networks processes the video frames in an unsupervised manner to obtain the temporal and spatial information in the data. Through end-to-end training, the model extracts the skeletal feature rules of the normal pattern of the data, and effectively detects abnormal events. In order to verify the effectiveness of the model in this paper, the model is applied to the actual data set, and the results show that the method in this paper has high detection accuracy under the condition of real-time detection.

2 Related Work

2.1 Recurrent Neural Network

Recurrent neural network can store information and process time series data, has the ability to memorize input information, and can reflect the relationship of time series data. However, the traditional recurrent network is prone to the problem of gradient disappearance. To solve this problem, the RNN is extended to the Long Short Term Memory (LSTM) [9], the memory unit is used to replace the neuron, and the input gate and output gate are added., Forgetting gate, where the input gate determines the current input information that needs to be retained, the output gate determines the information that needs to be output to the hidden layer at the next moment, and the forgetting gate determines the information at the previous moment that needs to be discarded. The model realizes the overall update of the system state and The result output has better performance for learning features of long sequence data. Gated Recurrent Unit (GRU) [10] reduces LSTM model gating signals to 2, keeping update gate and reset gate. Compared with LSTM, GRU has a simpler structure and faster running speed. The structure is shown in Fig. 1, and the calculation process is shown in formulas (1)–(4).

$$\mathrm{r}_t = \sigma(W_x x_t + U_r h_{t-1}) \tag{1}$$

$$z_t = \sigma(W_z x_t + U_z h_{t-1}) \tag{2}$$

$$\tilde{h}_t = \tanh(W_h x_t + U_h r_h \cdot h_{t-1}) \tag{3}$$

$$h_t = (1 - z_t)h_{t-1} + z_t\tilde{h}_t \tag{4}$$

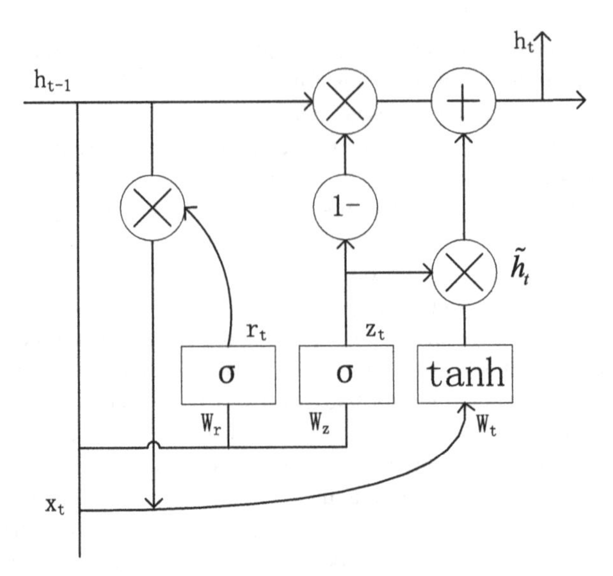

Fig. 1. GRU structure

2.2 Human Pose Estimation Algorithm

The Alphapose human pose algorithm [11] adopts the top-down detection method, and realizes the pose estimation through the regional pose estimation framework. The algorithm framework includes spatial symmetry network, pose non-maximum suppression and pose-guided sample generator, which detects the coordinates (x, y) of 17 skeleton points of the human body in the image, and connects the skeleton key points to form a skeleton through a graph algorithm. shown in Fig. 2.

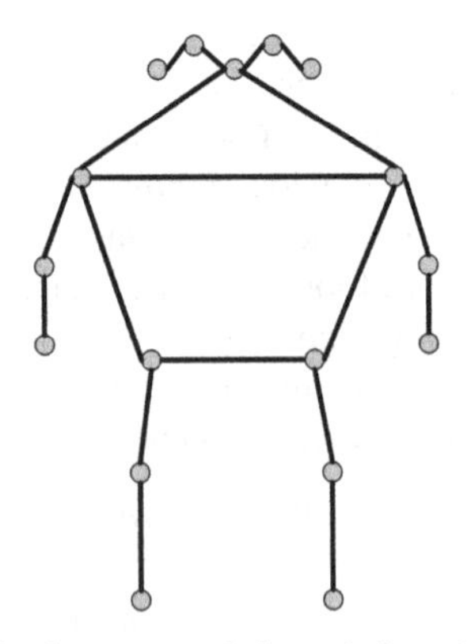

Fig. 2. Human skeleton infomation

3 Video Abnormal Behavior Detection Model

When abnormal behavior occurs in video surveillance, the pedestrian feature of the current frame has a significant change with the pedestrian feature of the past frame, and the abnormal behavior can be detected by comparing the feature changes. This paper proposes an end-to-end detection model consisting of a recurrent neural network and an autoencoder. The model is trained in normal scenarios with the goal of minimizing the error between input video features and output video features. For the trained model, the reconstruction error of detecting normal frame features is small, while the reconstruction error exceeds the threshold when detecting abnormal frames to detect abnormal events [12] (Fig. 3).

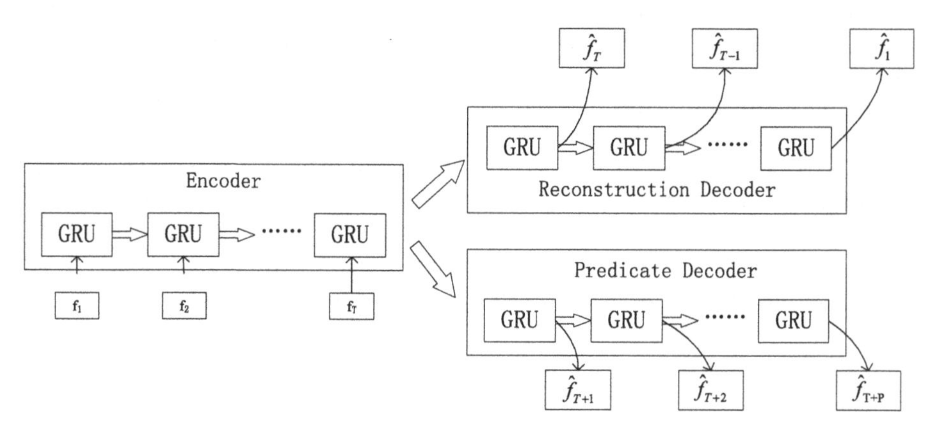

Fig. 3. Model structure

3.1 Human Skeletal Features

Through the human pose estimation algorithm, the coordinate point information of each target bone in the video can be obtained at a certain moment. For any time interval step t, the bone coordinate information of target i in the video frame is (Xi t,Yi t), and X and Y are the set of x and y coordinates of all bone points of a single target. Human skeleton information is obtained by a human pose estimation algorithm.

The model in this paper decomposes the abnormal behavior of human target into two parts: global and local. In the overall dimension, the detection model studies the movement of skeleton information composed of all skeleton positions, and the displacement of the skeleton relative to the center of the video can describe abnormal behavior. In the local dimension, all skeletal information is detected relative to adjacent frame information to detect abnormal behavior. Through the description of two dimensions, the problem of inaccurate detection caused by the size of the human body displayed in the image information can be effectively solved.

3.2 Model Design

An autoencoder consists of two stages: encoding and decoding. The autoencoder learns only behaviors detected in training videos that contain normal behaviors. The reconstruction error of the autoencoder is expected to be higher when we provide input objects with abnormal behavior. Autoencoders often provide better reconstructions of normal behavior. Due to the temporal nature of abnormal behavior, the autoencoder in this paper is built on a recurrent neural network. In the encoding stage, the model obtains the skeleton features of the pedestrians in the video. In the decoding stage, the decoder is divided into reconstructing past frames and predicting skeletal features of future frames. The model is trained in an unsupervised manner by minimizing the reconstruction error of the decoded result of the original input. The autoencoder learns to represent objects detected in training videos that contain only normal behavior. The reconstruction error of the autoencoder is expected to be higher when we provide input objects with abnormal behavior.

The model structure in this paper includes 1 GRU encoder and 2 GRU decoders. The GRU encoder reads in a series of video frame data in chronological order, including all pedestrian skeleton feature data in the video image. 2 GRU decoders need to complete two tasks. First, Reconstruct the feature data input by the encoder, and complete the reconstruction of the past few frames of images. Second, predict the pedestrian skeleton features in the next few frames through the encoder input. When a trained model is used and an abnormal event sequence is input, there will be a large error between the decoder prediction data and the reconstructed data. Based on these data, the model can detect abnormal behaviors that do not conform to normal motion patterns.

3.3 Loss Function

This model defines the loss function in both global and local aspects. When abnormal behaviors occur, pedestrians may undergo dramatic changes in local posture or overall motion. When the pedestrian faints in the video, the pedestrian's local posture will change greatly from the normal behavior pattern, and when the pedestrian is riding, the local posture change is different from the pedestrian's movement in the overall image.

$$L_g = \frac{1}{T}\sum_{t=a}^{a+T}\left\|\hat{X}_t^g - X_t^g\right\| + \frac{1}{P}\sum_{t=a+T}^{a+T+P}\left\|\hat{X}_t^g - X_t^g\right\| \quad (5)$$

$$L_l = \frac{1}{T}\sum_{t=a}^{a+T}\left\|\hat{X}_t^l - X_t^l\right\| + \frac{1}{P}\sum_{t=a+T}^{a+T+P}\left\|\hat{X}_t^l - X_t^l\right\| \quad (6)$$

$$Loss = \lambda_g L_g + \lambda_l L_l \quad (7)$$

The total loss function of the model is shown in formula (7), Lg and Ll represent the global loss and local loss respectively, λ is the weight of the corresponding loss, and the training process minimizes the Loss by optimizing the network parameters.

4 Experiment

4.1 ShanghaiTech Data Set

The ShanghaiTech University campus dataset is one of the largest datasets for anomalous event detection [13]. Unlike other datasets, it contains 13 different scenes with different lighting conditions and camera angles. There are 330 training videos and 107 testing videos in total. Most of the abnormal events in the dataset are related to pedestrians, which is in line with the detection goal of this model. There are 316154 frames in the entire dataset, and each video frame has a resolution of 480 × 856 pixels.

4.2 Evaluation Index

The experimental samples are divided into four categories, in which the true example TP (True Position) indicates that the positive sample is correctly classified as a positive sample by the model; the false positive example FP (False Position) indicates that the negative sample is misclassified as a positive sample by the model; False negative example FN (False Negative) means that positive samples are misclassified as negative samples by the model; TN (True Negative) means that negative samples are correctly classified as negative samples by the model. The ROC curve is established with the true positive rate (TPR) and the false positive rate (FPR) as the vertical and horizontal coordinates, and the area under the ROC curve, AUC [12], can describe the performance of the model.

4.3 Result

During the experiment, the batch was set to 256, the number of model iterations (epoch) was set to 15, the learning rate was 0.001, and the Adam algorithm was used to optimize the model. All models are trained by GeForce RTX 3060 GPU, Python version is 3.6, and the deep learning framework is Tensorflow 2.0 (Fig. 4, 5 and 6).

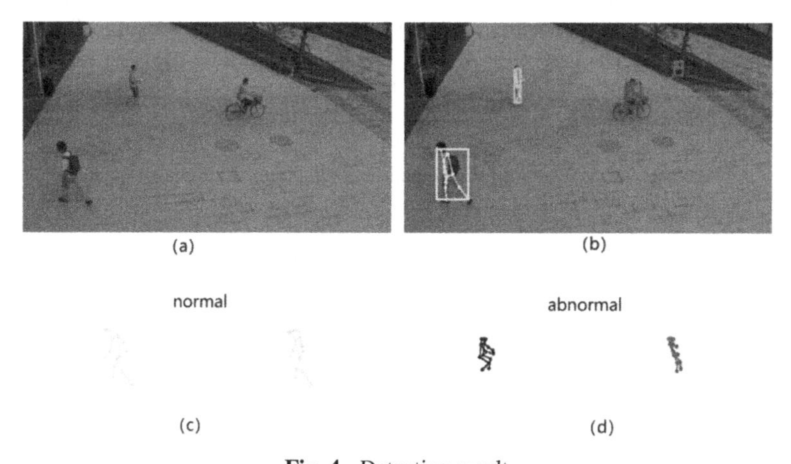

Fig. 4. Detection result

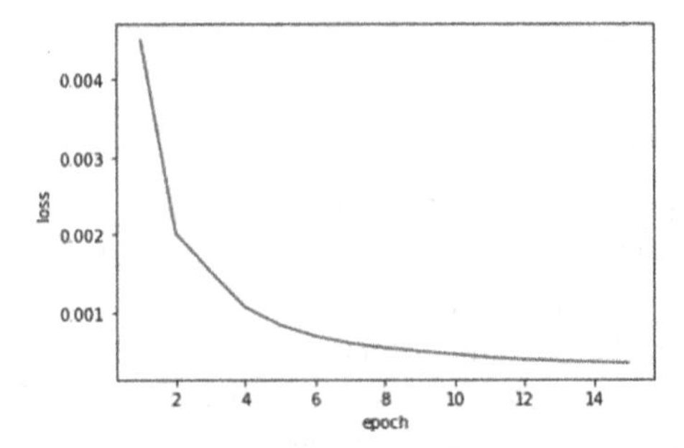

Fig. 5. Loss function

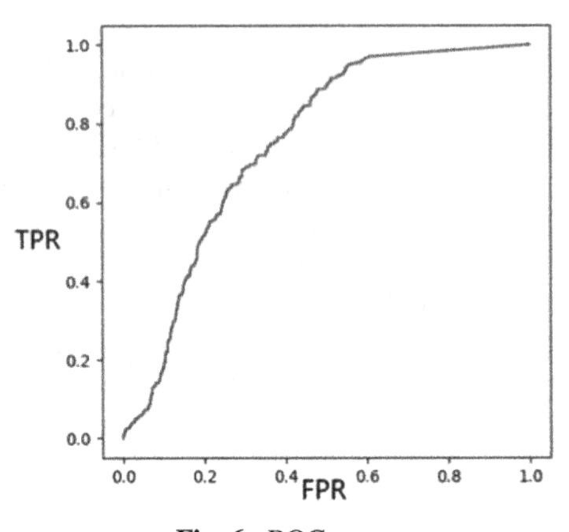

Fig. 6. ROC curve

4.4 Comparison

The model in this paper is compared with methods based on appearance features, including the ConvAE [14] and U-Net [15] algorithms. The model performance is measured by using the area AUC corresponding to the ROC curve. The detection results of each model are shown in Table 1. The detection accuracy of the model in this paper reaches 74.9%, which is higher than the detection accuracy based on appearance feature detection.

Table 1. Comparison of the detection accuracy of the three models.

	ROC AUC
ConvAE	60.9%
U-Net	72.8%
Ours	74.9%

In order to compare the performance of different recurrent neural network units for video abnormal behavior detection, this paper uses traditional RNN, LSTM and GRU three units for horizontal comparison (Table 2).

Table 2. Comparison of accuracy rates of three recurrent neural networks

	ROC AUC
Ours(RNN)	63.6%
Ours(LSTM)	70.8%
Ours(GRU)	74.9%

5 Conclusion

In this paper, a recurrent neural network autoencoder video anomalous behavior detection model based on human skeletal information is proposed, which utilizes human pose estimation algorithms and recurrent neural networks to extract the features of events in videos from spatial and temporal dimensions. The model learns normal event features semi-supervised by an autoencoder, and is verified by experiments with the trained model. The experimental results show that it has high abnormal behavior detection accuracy in multiple scenarios.

References

1. Chalapathy, R., Menon, A.K., Chawla, S.: Robust, deep and inductive anomaly detection. In: Ceci, M., Hollmén, J., Todorovski, L., Vens, C., Džeroski, S. (eds.) ECML PKDD 2017. LNCS (LNAI), vol. 10534, pp. 36–51. Springer, Cham (2017). https://doi.org/10.1007/978-3-319-71249-9_3
2. Ravanbakhsh, M., Nabi, M., Sangineto, E., et al.: Abnormal event detection in videos using generative adversarial nets. In: Proceedings of the 2017 IEEE International Conference on Image Processing (ICIP), pp. 1577–1581. IEEE, Beijing (2017)
3. Li, W., Mahadevan, V., Vasconcelos, N.: Anomaly detection and localization in crowded scenes. IEEE Trans. Pattern Anal. Mach. Intell. **36**(1), 18–32 (2013)
4. Tran, D., Bourdev, L., Fergus, R., et al.: Learning spatiotemporal features with 3D convolutional networks. In: Proceedings of the IEEE International Conference on Computer Vision, pp. 4489–4497. IEEE, Santiago (2015)
5. Feichtenhofer, C., Pinz, A., Zisserman, A.: Convolutional two-stream network fusion for videoaction recognition. In: Proceedings of the IEEE Conference on Computer Vision and Pattern Recognition, pp. 1933–1941. IEEE, Las Vegas (2016)
6. Yue-Hei Ng, J., Hausknecht, M., Vijayanarasimhan, S., et al.: Beyond short snippets: Deep networks for video classification. In: Proceedings of the IEEE Conference on Computer Vision and Pattern Recognition, pp. 4694–4702. IEEE, Boston (2015)
7. Tian, X.M., Fan, J.Y.: Joints kinetic and relational features for action recognition. Signal Process.: The Official Publication of the European Association for Signal Processing (EURASIP) **142**, 412–422 (2018)

8. Morais, R., Le, V., Tran, T., et al.: Learning regularity in skeleton trajectories for anomaly detection in videos. In: Proceedings of the IEEE Conference on Computer Vision and Pattern Recognition, pp. 11996–12004. IEEE, Long Beach (2019)
9. Cho, K., van Merrienboer, B., Bahdanau, D., Bengio, Y.: On the properties of neural machine translation: Encoder-decoder approaches. CoRR, 2014, abs/1409. 1259
10. Hochreiter, S., Schmidhuber, J.: Long short-term memory. Neural Comput. **9**(8), 1735–1780 (1997)
11. Cao, Z., Simon, T., Wei, S.E., et al.: Realtime multi-person 2D pose estimation using part affinity fields. In: Proceedings of the 2017 IEEE Conference on Computer Vision and Pattern Recognition (CVPR), pp. 1302–1310. IEEE, Honolulu (2017)
12. Chong, Y.S., Tay, Y.H.: Abnormal event detection in videos using spatiotemporal autoencoder. In: Cong, F., Leung, A., Wei, Q. (eds.) ISNN 2017. LNCS, vol. 10262, pp. 189–196. Springer, Cham (2017). https://doi.org/10.1007/978-3-319-59081-3_23
13. Luo, W., Liu, W., Gao, S.: A revisit of sparse coding based anomaly detection in stacked rnn framework. In: Proceedings of the IEEE International Conference on Computer Vision, pp. 341–349. IEEE, Venice (2017)
14. Hasan, M., Choi, J., Neumann, J., et al.: Learning temporal regularity in video sequences. In: Proceedings of the IEEE Conference on Computer Vision and Pattern Recognition, pp. 733–742. IEEE, Las Vegas (2016)
15. Liu, W., Luo, W., Lian, D., et al.: Future frame prediction for anomaly detection-a new baseline. In: Proceedings of the IEEE Conference on Computer Vision and Pattern Recognition, pp. 6536–6545. IEEE, Salt Lake City (2018)

A Flexible Hand-Eye and Tool Offset Calibration Approach Using the Least Square Method

Jintao Chen[1,2], Benliang Zhu[1,2(✉)], and Xianmin Zhang[1,2]

[1] School of Mechanical and Automotive Engineering, South China University of Technology, Guangzhou 510640, China
meblzhu@scut.edu.cn

[2] Guangdong Province Key Laboratory of Precision Equipment and Manufacturing Technology, South China University of Technology, Guangzhou, Guangdong 510640, People's Republic of China

Abstract. Hand-eye calibration is the basis of machine vision. The calibration method determines the motion accuracy of the manipulator. In the presented method, the point coordinate in camera frame and robot frame are separately obtained. By stacking the formula, the hand-eye transformation matrix can be calculated using the least square method. Otherwise, a tool offset calibration method is presented and some guidance also of conducting the tool offset calibration experiment is given. Finally, the validity of the proposed method is demonstrated by and experimental studies.

Keywords: Hand-eye calibration · Tool offset calibration · The least square method

1 Introduction

Basically, Robot movement guiding by vision is built on the foundation of well-behaved hand-eye calibration. Hand-eye calibration can be divided into two different categories according to whether the camera is rigidly fixed on the robot (eye-in-hand) or fixed in a place completely irrelevant to the movement of robot (eye-to-hand). In both categories of hand-eye calibration aims at solving for coordinate transformation from camera frame to robot base frame. With the coordinate transformation, when we have the designated point coordinate in camera frame we can easily transform it to coordinate in robot base frame.

The tradition of hand-eye calibration devoted into solve the $AX = XB$ problem [2,4,10,11] where X is the unknown transformation from hand to camera, and A and B are the homogeneous transformations of the eye and hand motions, respectively.

Based on $AX = XB$ problem, $AX = YB$ [3] method was developed. The solution of equation $AX = XB$ can provide an analytical solution of both categories of hand-eye calibration mentioned before. There is a method to conduct

H. Liu et al. (Eds.): ICIRA 2022, LNAI 13456, pp. 459–470, 2022.
https://doi.org/10.1007/978-3-031-13822-5_41

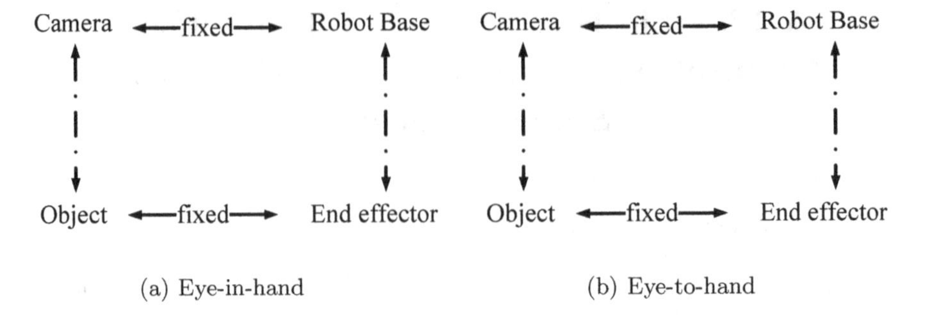

Fig. 1. System diagram of two different categories of eye-hand calibration

the hand-eye calibration based on active vision [5,6] which is sightly different from traditional hand-eye calibration because what it calibrates is transformation matrix between robot and laser pointer instead of camera.

Transformation from camera to object can be solved by EPNP algorithm [7]. Transformation from robot base to end of manipulator can be obtained in the robot control system generally. The condition for equation $AX = XB$ to be established is that exist two pairs of frames which position relatively fixed. Take eye-to-hand as an example, theoretically it needs more than two frames of object to solve the transformation between camera frame and robot base frame. Between different frames, camera must stay still relative to robot base and object must stay still relative to robot base which means that camera cannot be moved during the whole sampling process and object should be rigidly attached to end of manipulator. Analogous to eye-to-hand situation, it also requires camera be rigidly attached to end of manipulator in the eye-to-hand situation. Therefore, to sum up, in both categories of eye-hand calibration it is necessary to design a fixture to connect end of manipulator with camera or object. For eye-in-hand situation it is easily achieved because the fixture is designed when the whole system was designed. However, there is generally a clamp or sucker fixed on the end of manipulator of the robot in eye-to-hand situation. In that case, calibration board cannot be attached to the end of manipulator, so it is necessary to develop a new calibration method which works in this situation. Besides, the result of $AX = XB$ method is the transformation between robot base frame and end of manipulator frame. In general application, we will install a tool at the end of manipulator, so we need one more matrix to conduct the result to tool frame.

The contributions of this paper can be summarized as follows: First, a simple and effective method for hand-eye calibration is developed. It makes up for the shortcomings of traditional calibration methods. Secondly, a tool offset calibration method is presented, and its geometrical interpretation is briefly introduced. Finally, the feasibility and effectiveness of the proposed method are verified by experiment.

The rest of this paper is organized as follows: In Sect. 2, the hand-eye and tool offset calibration system model is established. In Sect. 3, we conducted a series of experiments to validate the effectiveness of the proposed method.

2 Calibration System Model

2.1 Hand-Eye Calibration

Implementation. The way to implement hand-eye calibration method this paper presented can be summarized as followed.

First, capture a frame of image from the camera and compute the corner coordinate in camera frame. Then, move the end of the manipulator to tip the corner one by one to acquire the corresponding coordinate in robot frame. By stacking the coordinate, this problem can be solved by the least square method.

Mathematical Model. Suppose the steps mentioned above are implemented. There are n sets of corresponding point in robot frame and camera frame separately.

$$P_w = \{P_{wi}\} \ , \ P_c = \{P_{ci}\} \tag{1}$$

where $i = 1, \cdots, n$. Let $P_{ci}, P_{wi} \in \mathbb{R}^3$ be the coordinate of the point in camera coordinate system and robot base coordinate system. Given P_{ci}, P_{wi} can be easily computed by a transformation of coordinates: [8]

$$P_{wi} = RP_{ci} + t \tag{2}$$

where $R \in SO(3)$ and $t \in \mathbb{R}^3$. Then, we can calculate $\bar{P_w} = (\sum_{i=1}^{n} P_{wi})/n$, $\bar{P_c} = (\sum_{i=1}^{n} P_{ci})/n$. We can derive that from 2

$$\bar{P_w} = R\bar{P_c} + t \tag{3}$$

(2)–(3):

$$P_{wi} - \bar{P_w} = R(P_{ci} - \bar{P_c}) \tag{4}$$

which has the form as $A = RB$, where $A = \left\{P_{wi} - \bar{P_w}\right\}$, $B = \left\{P_{ci} - \bar{P_c}\right\}$. A linear algebraic theorem shows that $rank\,(B) = 3$ in order for a solution to exit.

In the situation of $rank\,(B) = 3$, we can derive a solution from the least square method [1]:

$$R = C(C^TC)^{-\frac{1}{2}} \tag{5}$$

where

$$C = AB^T \tag{6}$$

By substituting (5) to (3), it can be derived as follows

$$t = \bar{P_{wi}} - C(C^TC)^{\frac{1}{2}}\bar{P_{ci}} \tag{7}$$

Appending a 1 to the coordinate to yield a vector in $\mathbb{R}^4$

$$P_w^* = \left[P_w^T \ 1\right]^T, \ P_c^* = \left[P_c^T \ 1\right]^T \tag{8}$$

Then, we can rewrite (2) as

$$P_w^* = TP_c^* \tag{9}$$

where

$$T = \begin{bmatrix} R & t \\ 0 & 1 \end{bmatrix} = \begin{bmatrix} C(C^TC)^{\frac{1}{2}} & \bar{P_{wi}} - C(C^TC)^{\frac{1}{2}}\bar{P_{ci}} \\ 0 & 1 \end{bmatrix} \tag{10}$$

In the situation of $rank\,(B) = 2$ which means all the point are coplanar but bot collinear. Equation 4 is equivalent to

$$\begin{Bmatrix} P_{w1} - P_{w2} = R(P_{c1} - P_{c2}) \\ P_{w1} - P_{w3} = R(P_{c1} - P_{c3}) \end{Bmatrix} \tag{11}$$

Since R is orthogonal matrix, it keeps the inner product between the vectors unchanged. It follows

$$(P_{w1} - P_{w2}) \times (P_{w1} - P_{w3}) = R\,(P_{c1} - P_{c2}) \times (P_{c1} - P_{c3}) \tag{12}$$

Adding (12) into (11), R in (11) can be solved.

Remark:

a) In this case, we assume that transformation is linear and nonlinear distortion of the camera was reliably corrected.
b) All sets of the point cannot be collinear, otherwise $rank\,(B) = 1$.
c) It can be used in both categories we mentioned in Sect. 1.

2.2 Tool Offset Calibration

Notice the result of hand-eye calibration mentioned above is the transformation matrix between robot base and the end of the manipulator. However, there is usually a tool installed on the end of the manipulator, and it was used to tip the corner on the chessboard in our method. It means that the corner coordinate in camera frame do not correspond with the statistics acquired in the robot system. Therefore, it still needs one more step to compute the tool offset between tip of the tool and end of manipulator.

Implementation. Suppose P_{base}^{tip} be the tip of the tool coordinate in robot base frame and P_{end}^{tip} be the tip of the tool coordinate in the end of manipulator frame. Similar to Eq. 2, it follows

$$P_{base}^{tip} = RP_{end}^{tip} + t \tag{13}$$

Same as $AX = XB$ method, we assumed that we can access the transformation matrix(R and t) between end of manipulator frame and robot base frame.

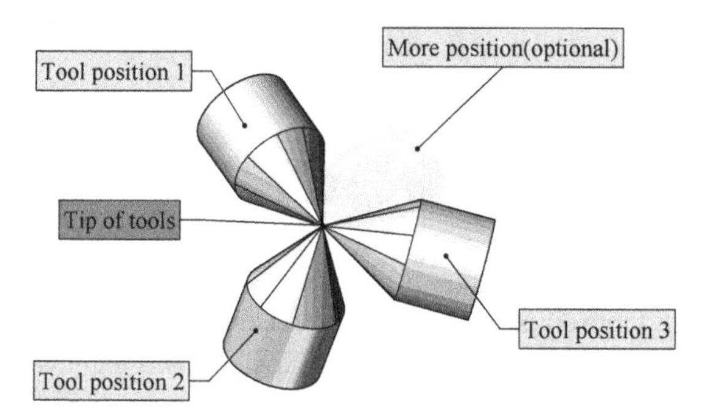

Fig. 2. Tool offset calibration schematic

There are many ways to obtain the offset. For example, if the offset is known in the computer assist design(CAD) software, it can easily be accessed. Here a simple method to calibrate the tool is presented.

As Fig. 2 shows, we can set a fixed point, then tip the point with the tip of the tool with different posture. The coordinate of tip of the tool in end of manipulator frame and coordinate of fixed point in robot base frame is invariable in different posture. Each posture can establish an equation. By stacking the formula, the hand-eye transformation matrix can be calculated using the least square method.

Mathematical Model. We denote the fixed point coordinate in end of the manipulator and in robot base frame by x and y respectively. Suppose there is n frame of posture conducted, it follows

$$y = R_i x + t_i \tag{14}$$

where $i = 1, \cdots, n, R \in SO(3)$ and $t \in \mathbb{R}^3$.

Then, stack the formula as follows

$$A_R x = b_t \tag{15}$$

where

$$A_R = \begin{Bmatrix} R_1 - R_2 \\ \vdots \\ R_1 - R_n \end{Bmatrix} \tag{16}$$

$$b_t = -\begin{Bmatrix} t_1 - t_2 \\ \vdots \\ t_1 - t_n \end{Bmatrix} \tag{17}$$

According to knowledge of linear algebra, x have unique solution when $rank$ $(A_R) = 3$

Lemma 1. *Given R_1, $R_2 \in SO(3)$ and $R_1 \neq R_2$. Exist a unique solution $x = \log\left(R_1^T R_2\right)$ satisfies $(R_1 - R_2)x = 0$*

Proof. Rewriting $(R_1 - R_2)x = 0$ as $x = R_1^T R_2 x$. $R_1^T R_2 \in SO(3)$. Given $w = [w_1, w_2, w_3]^T$ where $w_1, w_2, w_3 \in \mathbb{R}^3$, we denote

$$[w]_\times = \begin{bmatrix} 0 & -w_3 & w_2 \\ w_3 & 0 & -w_1 \\ -w_2 & w_1 & 0 \end{bmatrix} \tag{18}$$

Let $\log(R_1^T R_2)$ denoted by $[r]_\times$ where r satisfies $R_1^T R_2 r = r$. The eigenvalues of $R_1^T R_2$ are 1, $e^{j\theta}$, $e^{-j\theta}$ where $\theta = \|r\|_2$. Noticed the eigenvalues is different from each other, so eigenvector r corresponding to a unique eigenvalue 1.

Lemma 2. *rank* $(A_R) = 2$ *if and only if* $\log(R_1^T R_2) = k_1 log(R_1^T R_3) = \cdots = k_{n-1} log(R_1^T R_n)$ *where* $k_i \in \mathbb{R}$

Proof. Now we consider situation of n=3. For the situation of n > 3, it is similar. Let $\log(R_1^T R_2)$ and $\log(R_1^T R_3)$ denoted by r_1 and r_2. We prove the forward direction first. Firstly, from Lemma 1 it can be derived that $rank(A_R) \geq 2$ Secondly, If it satisfies that $r_1 = k_1 r_2$ r_1, it follows that $(R_1 - R_2)r_1 = (R_1 - R_3)k_1 r_2 = \mathbf{0}$ which prove $rank(A_R) = 2$. It can be derided that $R_1^T R_2 r_1 = k_1 R_1^T R_3 r_2$.

To prove the reverse direction, suppose that it exists x_1, x_2 satisfies $(R_1 - R_2)x_1 = 0$ and $(R_1 - R_3)x_2 = 0$ respectively. We can derive from Lemma 1 $x_1 = R_1^T R_2$, $x_2 = R_1^T R_3$. Meanwhile, $rank(A_R) = rank(R_1 - R_2) = rank(R_1 - R_3) = \cdots = 2$ is equivalent to that it exists x satisfies both $(R_1 - R_2)x = 0$ and $(R_1 - R_3)x = 0$. That mean x_1 and x_2 is linear correlated. So far, Lemma 2 is proved.

Lemma 3. *A sufficient condition proves rank* $(A_R) = 2$ *is that* $\log(R_1) = k_1 log(R_2) = k_2 log(R_3) = \cdots = k_{n-1} log(R_n)$ *where* $k_i \in \mathbb{R}$.

Proof. Let $\log(R_i)$ be denoted by $[w_i]_\times$. Applying Rodrigues formula, 15 can be rewritten as

$$A_R = \begin{Bmatrix} a_1 [w_1]_\times + b_1 [w_1]_\times^2 - a_2 [w_2]_\times - b_2 [w_2]_\times^2 \\ \vdots \\ a_1 [w_1]_\times + b_1 [w_1]_\times^2 - a_{n-1} [w_{n-1}]_\times - b_{n-1} [w_{n-1}]_\times^2 \end{Bmatrix} \tag{19}$$

where $a_i = sin \|w_i\| / \|w_i\|$ and $b_i = (1 - cos \|w_i\|) / \|w_i\|^2$.

It can derive from the given condition as follows

$$[w_i]_\times [w_j]_\times = [w_j]_\times [w_i]_\times \tag{20}$$

where $i,\ j = 1, \ldots, n$.

It follows that $a_1 [w_1]_\times + b_1 [w_1]_\times^2 - a_i [w_i]_\times - b_i [w_i]_\times^2 = (\sqrt{b_1} [w_1] + \sqrt{b_i} [w_i]) (\sqrt{b_1} [w_1] - \sqrt{b_i} [w_i]) + (a_1 [w_1]_\times - a_i [w_i]_\times)$. Therefore, it exists a general solution $x = w_i$ to $Rx = 0$.

Some robot systems use Euler angle or roll-yaw-pitch angle instead the logarithm of rotation matrix to present rotation [8]. Rotation angle and the order of rotation axis codetermine the rotation matrix. Take XYZ Euler angle as an example, XYZ Euler angle means the frame rotate around X-axis of the base frame first, then Y-axis, and Z-axis at last.

Given $\alpha,\ \beta,\ \gamma \in (-\pi, \pi)$ It can be present as follows

$$R = R_Z\left(\gamma\right) R_Y\left(\beta\right) R_X(\alpha) \tag{21}$$

where

$$R_X\left(\alpha\right) = \begin{bmatrix} 1 & 0 & 0 \\ 0 & cos\alpha & -sin\alpha \\ 0 & sin\alpha & cos\alpha \end{bmatrix} \tag{22}$$

$$R_Y\left(\beta\right) = \begin{bmatrix} cos\beta & 0 & sin\beta \\ 0 & 1 & 0 \\ -sin\beta & 0 & cos\beta \end{bmatrix} \tag{23}$$

$$R_Z\left(\gamma\right) = \begin{bmatrix} cos\gamma & -sin\gamma & 0 \\ sin\gamma & cos\gamma & 0 \\ 0 & 0 & 1 \end{bmatrix} \tag{24}$$

Here we briefly introduce the relation between XYZ Euler angle and logarithm of rotation matrix. Given $G_X = [1,\ 0,\ 0]^T$, $G_Y = [0,\ 1,\ 0]^T$ and $G_Z = [0,\ 0,\ 1]^T$, $R_X(\alpha)$, $R_Y\left(\beta\right)$, $R_Z\left(\gamma\right)$ can rewrite as

$$R_X\left(\alpha\right) = e^{[\alpha G_X]_\times} \tag{25}$$

$$R_Y\left(\beta\right) = e^{[\beta G_Y]_\times} \tag{26}$$

$$R_Z\left(\gamma\right) = e^{[\gamma G_Z]_\times} \tag{27}$$

Equation 16 can be rewritten as

$$A_R = \left\{ \begin{matrix} e^{[\gamma_1 G_Z]_\times} e^{[\beta_1 G_Y]_\times} e^{[\alpha_1 G_X]_\times} - e^{[\gamma_2 G_Z]_\times} e^{[\beta_2 G_Y]_\times} e^{[\alpha_2 G_X]_\times} \\ \vdots \\ e^{[\gamma_1 G_Z]_\times} e^{[\beta_1 G_Y]_\times} e^{[\alpha_1 G_X]_\times} - e^{[\gamma_n G_Z]_\times} e^{[\beta_n G_Y]_\times} e^{[\alpha_n G_X]_\times} \end{matrix} \right\} \tag{28}$$

In the case that robot system offers XYZ Euler angle to present rotation, 2 seems not intuitive enough to guide our experiment. Here we present some intuitive guidance.

Lemma 4. *Supposed there is only one parameter of $\alpha,\ \beta,\ \gamma$ was changed in Eq. 16 between different frames, $rank\left(R\right) = 2$.*

Proof. Now we consider situation of only β is changed. The other situation can be proved similarly. Given constant α_0, γ_0 and denote variable β in different frames by $\beta_i (i = 1, \cdots, n)$. According to Rodrigues formula, $e^{[\beta_1 G_Y]_\times} - e^{[\beta_i G_Y]_\times} = a_1\left[\beta_1 G_Y\right]_\times + b_1\left[\beta_1 G_Y\right]_\times^2 - a_i\left[\beta_i G_Y\right]_\times - b_i\left[\beta_i G_Y\right]_\times^2$ where $a_i = sin\left\|w_i\right\| / \left\|w_i\right\|$

and $b_i = (1 - cos\,\|w_i\|)/\,\|w_i\|^2$ According to Lemma 1, n should be greater than 2. In this situation, 15 is equivalent to

$$\begin{Bmatrix} e^{[\beta_1 G_Y]_\times} - e^{[\beta_2 G_Y]_\times} \\ \vdots \\ e^{[\beta_1 G_Y]_\times} - e^{[\beta_n G_Y]_\times} \end{Bmatrix} x' = -e^{[\gamma_0 G_Z]_\times} \begin{Bmatrix} t_1 - t_2 \\ \vdots \\ t_1 - t_n \end{Bmatrix} \tag{29}$$

where $x' = e^{[\alpha_0 G_X]_\times} x$. Applying the easily established identity $[\beta_i G_Y]\,[\beta_1 G_Y] = [\beta_1 G_Y]\,[\beta_i G_Y]$, it follows that the solution that satisfies $(e^{[\beta_1 G_Y]_\times} - e^{[\beta_i G_Y]_\times})x = 0$ is $x = G_Y$. To sum up, it proves that $rank \begin{Bmatrix} e^{[\beta_1 G_Y]_\times} - e^{[\beta_2 G_Y]_\times} \\ \vdots \\ e^{[\beta_1 G_Y]_\times} - e^{[\beta_n G_Y]_\times} \end{Bmatrix} = 2.$

Normally robot system offers the function that move the manipulator purely rotate around a particular axis. According to Lemma 4, we can learn that we should rotate around more than one axis during tool calibration. Notice that Lemma 4 is not a sufficiency condition to prove that it exists unique solution to Eq. 15. It still needs to use Lemma 2 to validate at last, but normally it exists a unique solution when we acquire more than two frames and rotate around more than one axis.

3 Experiment

As shown in Fig. 4, the experiment system is composed of a 6 DOF YASKAWA industrial robot MH3F, a camera with a 12 mm lens, resolution of 5496 × 3672, a chessboard consists of 12 × 8 corners (distance between two adjacent corners is 20 mm) printed by a laser printer and attached on a reliable plane. Image processing which contains threshold, corner detection and intrinsic and extrinsic matrix acquiring is completed based on OpenCV in Visual Studio. Statistics processing is completed in MATLAB.

3.1 Manipulator Movement Accuracy Verification

The hand-eye calibration method we present is directly influenced by the manipulator movement accuracy because translation and rotation information is obtained in the robot system. Thus, it is necessary to check the manipulator movement accuracy. We use Leica laser tracker AT901-B as standard. The result is shown in Table 1. It shows that the movement accuracy is quite reliable. We compare the statistics acquire from the robot system with the laser tracker then compute the error. In terms of translation, the experiment is carried out as followed:

a) Fix the laser tracking point on the end of manipulator.
b) Move the end of manipulator along a particular axis in a certain distance at a time then record the coordinate in robot system and laser tracker.
c) Verify the absolute distance error of the statistics acquired respectively from between laser tracker and robot system.

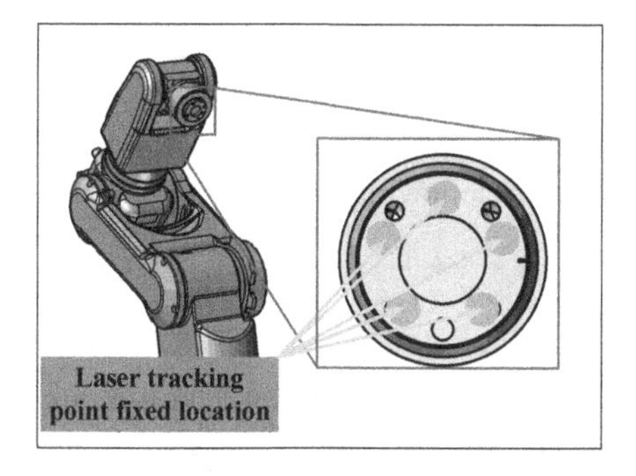

Fig. 3. Rotation verification scheme (Color figure online)

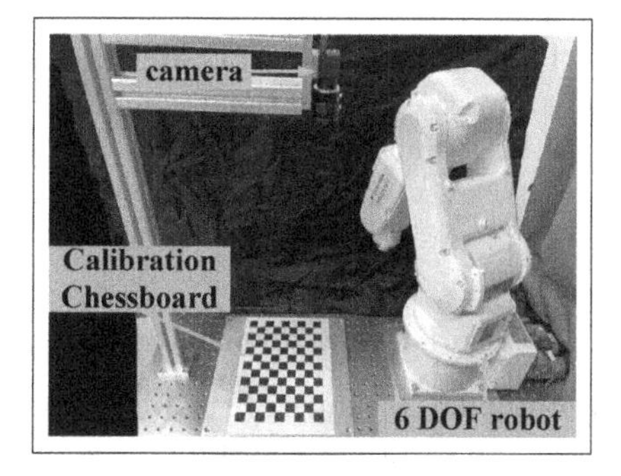

Fig. 4. Experiment platform

In terms of rotation, the experiment is carried out as followed:

a) Establish a reference plane on the home position of the robot. The specific approach to establish a plane is to sample couples of points on different location of the end of the manipulator as shown green points in Fig. 3. Then, with more than three points, the plane can be generated automatically by laser tracker system.
b) Rotate the end of manipulator around a particular axis in a certain angle at a time then record the posture in robot system and laser tracker. Establish a plane in this posture in laser tracker.
c) Verify the absolute angle error of the statistics acquired respectively from between laser tracker and robot system.

Table 1. Manipulator movement accuracy

Axis	x	y	z	Rx	Ry	Rz
Relative error	0.27%	0.07%	0.04%	0.307%	0.16%	0.47%

3.2 Calibration Result

The design size of tool and the calibration result is shown in Table 2.

Table 2. Tool calibration result

Axis	Calibration result	Design size of tool
x	−0.323 mm	0 mm
y	0.263 mm	0 mm
z	29.161 mm	29.1 mm

The result of the tool calibration clearly validate the method we presented since the error is relatively small. It proves that our algorithm is effective and feasible. In fact, the error may not only cause by motion error of manipulator but also manufacturing and assemble error. In the following experiment, we use the calibration result as tool offset.

3.3 Hand-Eye Calibration

We sampled 3 frames of the object which means we have 288 points' coordinate in camera coordinate system so as robot base coordinate system. As a contrast, $AX = XB$ hand-eye calibration method [10] is conducted. Avoiding to lost generality, the same algorithm to detect the corner of the checkerboard and compute the extrinsic matrix is applied, and the same laser printer was used to print checkerboard. 8 frames is acquired to solve the $AX = XB$ problem and here is the result:

Table 3. Hand-eye calibration result using $AX = XB$ method

Parameter	Result
$R_{AX=XB}$	$\begin{bmatrix} -0.0086 & 0.9944 & 0.0038 \\ 0.9940 & 0.0076 & -0.0261 \\ -0.2379 & 0.0042 & -1.0013 \end{bmatrix}$
$t_{AX=XB}$	$\begin{bmatrix} 294.836 & 4.521 & 295.294 \end{bmatrix}^T$

Use this result as a standard to judge our method in the following experiment. Also, a standard to judge the error is necessary. Given that $T_A, T_B \in SE(3)$. Define a metric in $SE(3)$ [9]

$$d(T_A, T_B) = \left\| log\left(R_A^T R_B\right) \right\|^2 + \left\| t_A - t_b \right\|^2 \tag{30}$$

where $T_A = \begin{bmatrix} R_A & t_A \\ 0 & 1 \end{bmatrix}$, $T_B = \begin{bmatrix} R_B & t_B \\ 0 & 1 \end{bmatrix}$

The experiment procedure is organized as followed:

a) Reliably fix the calibration board to the world coordinate system to make sure it cannot be moved easily because we need to move the end of manipulator of the robot to tip the corners on the checkerboard.

b) Acquire a photo from the camera. Then, solve the coordinate transformation from calibration board to camera by EPNP algorithm.
c) Move the end of manipulator of the robot to tip each corner on the board.
d) Repeat step a to step c and make sure the plane of the checkerboard should not be coplanar. At least 4 points be conducted, and these 4 points should not be coplanar.
e) Acquire the tool offset and compensate it in the coordinate

Apply our method to all the point data and compute the distance between the $AX = XB$ method and our method. Here is the result:

Table 4. Hand-eye calibration result using our method

Parameter	Result
R	$\begin{bmatrix} -0.0082 & 0.9999 & 0.0057 \\ 0.9996 & 0.0084 & -0.0276 \\ -0.0277 & 0.0055 & -0.9996 \end{bmatrix}$
t	$\begin{bmatrix} 293.996 & 4.348 & 295.113 \end{bmatrix}^T$
distance	0.7699

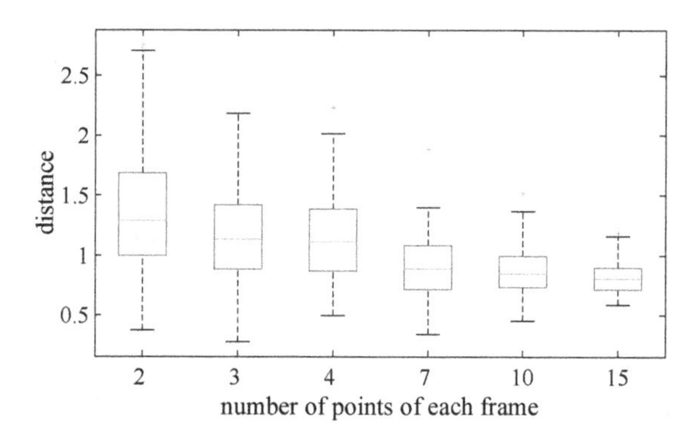

Fig. 5. Error distribution of different number of point sampled

The distance result above shows that our method is effective. However, if it needs to tip so many corners of the checkerboard in each frame, it will be very time-consuming. In this situation, it is necessary to simulate the real hand-eye calibration scenario which just randomly sample a few of corners of each frame to compute the result. Therefore, after acquiring all the statistics, we conducted an experiment to test how does the result perform when only randomly sample few points per frame in the statistics. Figure 5 shows the result that the error decline along with number of samples.

Acknowledgment. This research was supported by the National Natural Science Foundation of China (Grant No. 51975216) and the Guangdong Basic and Applied Basic Research Foundation (Grant No. 2021B1515020053).

References

1. Nadas, A.: Least squares and maximum likelihood estimation of rigid motion. IBM Research Report RC9645 (1978)
2. Fassi, I., Legnani, G.: Hand to sensor calibration: a geometrical interpretation of the matrix Equation AX=XB. J. Robot. Syst. **22**(9), 497–506 (2005). https://doi.org/10.1002/rob.20082
3. Ha, J., Kang, D., Park, F.C.: A stochastic global optimization algorithm for the two-frame sensor calibration problem. IEEE Trans. Industr. Electron. **63**(4), 2434–2446 (2016). https://doi.org/10.1109/TIE.2015.2505690
4. Heller, J., Havlena, M., Pajdla, T.: Globally optimal hand-eye calibration using branch-and-bound. IEEE Trans. Pattern Anal. Mach. Intell. **38**(5), 1027–1033 (2016). https://doi.org/10.1109/TPAMI.2015.2469299
5. Huang, C., Chen, D., Tang, X.: Robotic hand-eye calibration based on active vision. Proceedings - 2015 8th International Symposium on Computational Intelligence and Design, ISCID 2015, vol. 1, no. 3, pp. 55–59 (2016). https://doi.org/10.1109/ISCID.2015.246
6. Izaguirre, A., Pu, P., Summers, J.: A new development in camera calibration calibrating a pair of mobile cameras. In: Proceedings - IEEE International Conference on Robotics and Automation, pp. 74–79 (1985). https://doi.org/10.1109/ROBOT.1985.1087321
7. Lepetit, V., Moreno-Noguer, F., Fua, P.: EPnP: an accurate O(n) solution to the PnP problem. Int. J. Comput. Vision **81**(2), 155–166 (2009). https://doi.org/10.1007/s11263-008-0152-6
8. Murray, R.M., Li, Z., Shankar Sastry, S.: A Mathematical Introduction to Robotic Manipulation. CRC Press (2017). https://doi.org/10.1201/9781315136370
9. Park, F.C., Murray, A.P., McCarthy, J.M.: Designing Mechanisms for Workspace Fit, pp. 295–306 (1993). https://doi.org/10.1007/978-94-015-8192-9_27
10. Park, F.C., Martin, B.J.: Robot sensor calibration: solving AX = XB on the Euclidean group. IEEE Trans. Robot. Autom. **10**(5), 717–721 (1994). https://doi.org/10.1109/70.326576
11. Shiu, Y.C., Ahmad, S.: Calibration of wrist-mounted robotic sensors by solving homogeneous transform equations of the form AX = XB. IEEE Trans. Robot. Autom. **5**(1), 16–29 (1989). https://doi.org/10.1109/70.88014

Weakly Supervised Nucleus Segmentation Using Point Annotations via Edge Residue Assisted Network

Wei Zhang[1,2,3], Xiai Chen[1](✉), Shuangxi Du[4], Huijie Fan[1], and Yandong Tang[1]

[1] State Key Laboratory of Robotics, Shenyang Institute of Automation, Chinese Academy of Sciences, Shenyang 110016, China
chenxiai@sia.cn

[2] Institutes for Robotics and Intelligent Manufacturing, Chinese Academy of Sciences, Shenyang 110169, China

[3] University of Chinese Academy of Sciences, Beijing 100049, China

[4] School of Automation and Electrical Engineering, Shenyang Li Gong University, Shenyang 110159, China

Abstract. Cervical cell nucleus segmentation can facilitate computer-assisted cancer diagnostics. Due to the obtaining manual annotations difficulty, weak supervision is more excellent strategy with only point annotations for this task than fully supervised one. We propose a novel weakly supervised learning model by sparse point annotations. The training phase has two major stages for the fully convolutional networks (FCN) training. In the first stage, coarse mask labels generation part obtains initial coarse nucleus regions using point annotations with a self-supervised learning manner. For refining the output nucleus masks, we retrain ERN with an additional constraint by our proposed edge residue map at the second stage. The two parts are trained jointly to improve the performance of the whole framework. As experimental results demonstrated, our model is able to resolve the confusion between foreground and background of cervical cell nucleus image with weakly-supervised point annotations. Moreover, our method can achieves competitive performance compared with fully supervised segmentation network based on pixel-wise annotations.

Keywords: Nucleus segmentation · Weakly-supervised learning · Point annotation supervision · Cervical cell image segmentation

1 Introduction

In recent years, with the development of deep learning, computer vision field and applied become more wildly in range of industrial application. For example, the sizable well-annotated datasets of medical images obtain a significant trend in healthcare using deep learning technology [1–3] for medical image analysis. However, a medical dataset annotating is expensive, expert-oriented, and time-consuming. The appearance is further exacerbated by a flood of new raw medical images without labels which generated on a

H. Liu et al. (Eds.): ICIRA 2022, LNAI 13456, pp. 471–479, 2022.
https://doi.org/10.1007/978-3-031-13822-5_42

daily basis in clinical practice. Therefore, weakly supervised learning utilize a limited number of labeled data information, as well as a lot of raw unlabeled data, is gaining traction.

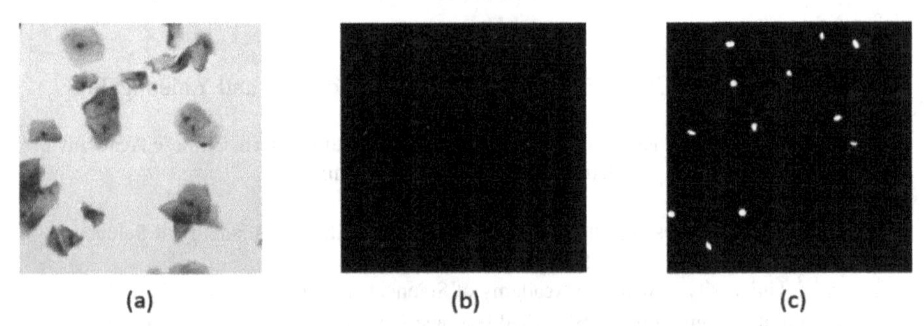

Fig. 1. Different supervision settings for cervical cell nucleus image segmentation. (a) is a nucleus image randomly selected for training, (b) is the full supervision learning where pixel-wise annotations are provided (white and black stands are for foreground and background respectively) and (c) is a weakly supervision learning with only sparse point annotations (white points) are available. It is obvious that full supervision requires more human intervention and annotation effort than weakly supervisions do. (Color figure online)

Weakly Supervised Learning for Medical Image Segmentation. Several weakly supervised learning methods have been proposed for medical image segmentation. Data augmentation is a useful method in weakly supervised learning. Goodfellow [4] proposed the use of GAN for data augmentation, which solves the problem similar to the original image. Besides, transfer learning is also widely used and proved effectiveness in the medical field. Pre-trained model transfer and domain adaptation are most important part to transfer learning. A jointly segmentation and optimized image synthesis framework is proposed by Wang et al. [5] to spleen CT images segmentation with CycleGAN [6].

Weakly Supervised Learning with Point Annotations. G. Zeyu et al. [7] design a model to detect and classify renal cell carcinoma by formulating a hybrid loss. For digital histopathology image segmentation, [8] proposed a weakly supervised framework using point annotations by oversegmentation and transfer learning to recognize complex semantics. A weakly supervised approach is layout towards the goal of neuron segmentation from primate brain images using a peak-shape probability map [9].

For nuclei segmentation, although several deep learning-based methods have been proposed, the cervical cell images nucleus segmentation remained challenging limited with training data nucleus masks. Cervical cell image nucleus is smaller than others, such as pathological images. To segment cervical cell image nucleus is more difficult especially if the label information is inadequate. Alternatively, annotate the nucleus with points is much easier than perform the annotation of full mask (see Fig. 1).

In this work, we propose an edge-aware learning strategy integrating the advantage of edge attention via weakly supervised nucleus segmentation using only point annotation.

In order to address the limitations, our main idea for the fully convolutional networks (FCN) training consists of two aspects: (1) In the first stage, coarse mask labels generation part obtains the initial coarse nucleus regions for FCN using point annotations with a self-supervised learning manner for all training datasets. The Voronoi distance maps are used for this stage. (2) To refine the output masks, we retrain ERN with an additional constraint by our proposed edge residue map at the second stage. In inference part, our whole model can perform well with refined trained FCN directly. After that, the coarse mask and the refined edges of them can all be integrated into the whole learning model via two strategy designed. Our model trained with weakly-supervised data get competitive performance compared with the fully supervised one on CCEDD datasets as our experimental results shown.

2 Method

As shown in Fig. 2, our approach is described in detail. The network has two major stages for the fully convolutional networks (FCN) training. In the first stage, coarse mask labels generation part obtains the initial coarse nucleus regions using point annotations with a self-supervised learning manner for all training datasets. The Voronoi distance maps are used for this stage. To refine the output masks, we retrain ERN with an additional constraint by our proposed edge residue map at the second stage. In inference part, our whole model can perform well with refined trained FCN directly.

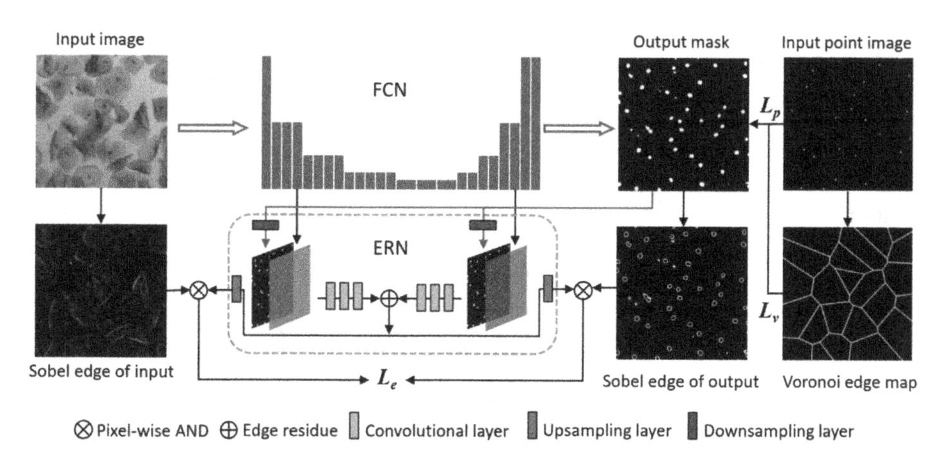

Fig. 2. Overview of our proposed method. (a) FCN is trained with the initial mask by input point labels and its Voronoi edge maps using L_p and L_v loss by self-supervised learning manner. (b) with the output masks of FCN adding to different feature level after pooling layer via edge residue operation, ERN is retrained for edge residue refinement to FCN with loss L_e.

2.1 Coarse Mask Estimation via Self-supervised Learning

This section aims to obtain coarse mask segmentation at the first FCN (f) training stage. We train our FCN networks with a U-Netlike architecture. Although the point annotation information provides some necessary positive pixels, they are too sparse. For this reason, we first transform the point annotations into positive sample maps ($\boldsymbol{P_m}$) with a larger number of annotations as shown in Fig. 3. The cell position likelihood map is used as positive training data where the value decreases gradually by a Gaussian distribution from the point annotation peak. An annotated point of nucleus is expanded to a region map. Then utilize the Voronoi distance map ($\boldsymbol{V_m}$) generated by the point annotation for negative samples to train. Finally we train the FCN model to obtain a coarse segmentation of masks with self-supervised learning strategy.

Self-supervised Learning. For self-supervised learning, the polarization loss employed to restrict the backpropagation of model weights ($\boldsymbol{W}$) of the FCN. The Heaviside function is used to rectify the binary mask of output segmentation map to for realizing the self-supervised strategy. After that, the coarse segmentation mask is generated through FCN model.

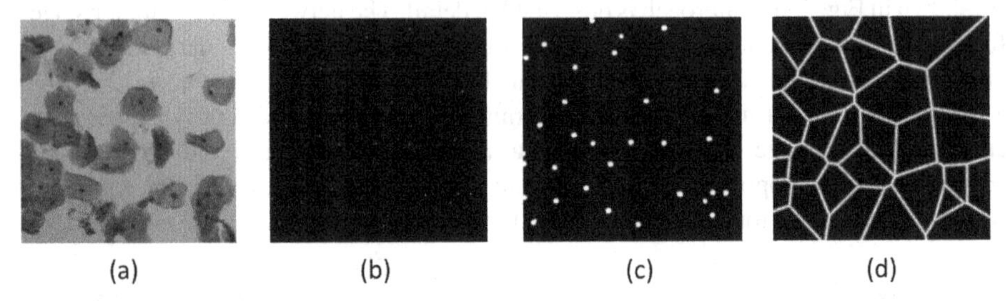

Fig. 3. Sparse annotation maps for course mask estimation via Self-supervised learning.

Among the model training, two sparse losses L_p and L_v are calculated to guide model weights updated. For an input image I, the losses of the model are listed following equations:

$$Lp(\boldsymbol{W}) = \|ReLU(\boldsymbol{P_m}) \cdot (f(\boldsymbol{I}) - \boldsymbol{P_m})\|_F^2 \tag{1}$$

$$Lv(\boldsymbol{W}) = \|\mathrm{ReLU}(\boldsymbol{V_m}) \cdot (f(I) - 0)\|_F^2 \tag{2}$$

where $\cdot$ denote the pixel-wise product, and for extracting mask exactly with the sparse loss, the ReLU operation is used here. With this setting, Lp and Lv can exclusively calculate on the assured positive pixels and negative pixels respectively.

2.2 Edge Residue Map Generation

Segmentation errors are mostly resulted from the loss of spatial resolution recording to some existing segmentation methods [10, 11]. For example, downsampling operations of masks and pooling size for small RoI, which predicting masks in a coarse level of features. For the reason that to predict nuclei edge accurately is more challenging at a low spatial resolution. In this section, inspired by [11], we proposed Edge Residue Map,

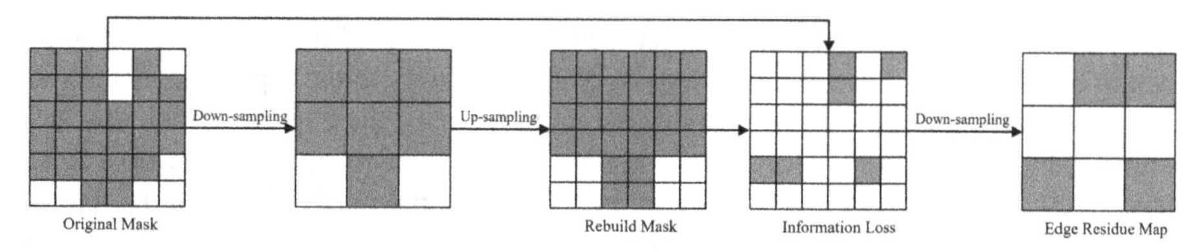

Fig. 4. Illustration on edge residue map definition with mask information loss.

which is lost at mask information by the reduced spatial resolution. We observe that a large portion of the errors are indeed located in these regions. With this method, the contours of nuclei are refined in a contour-sensitive manner.

Edge Residue Map Definition. We simulate the mask information loss at operations of downsampling in Fig. 4. Note that, in a subsequent upsampling step, the mask cannot be rebuild as a result of information is lost in edge map regions. Specifically, we first use an original mask of a nuclei instance at first scale level, expressed with M_s. Grids denote the resolution and differ with an element of 2, where $s \in [0, S]$ is in the range from the finest scale level to coarsest scale level. Downsampling and upsampling operation of nearest neighbors with $2\times$ is denoted by F_{down} and F_{up} respectively. The edge residue map at scale level s is obtained as

$$Es = R_{down}(Ms - 1 \oplus F_{up}(F_{down}(Ms - 1))) \tag{3}$$

where R_{down} is $2 \times$ downsampling with the logical 'or' operation in each neighborhood of 2×2. The symbol $\oplus$ represent the logical 'exclusive or' operation. Thus for a pixel (x, y), if the obtained rebuild mask in the finer scale level and the original mask Ms-1 in the course scale level are different, $Es\ (x, y) = 1$. In particular, pixels in those regions are frequently around nuclei instance boundaries. Therefore we construct the edge residue maps to redress wrong predictions of coarse masks.

Edge Residue Map Generation. For our nucleus segmentation task, the boundaries of the nucleus are inaccurate after rough segmentation. We propose an enhanced boundary-sensitive method to constrain the refined nucleus mask edges.

Mask Feature Inpainting. With the sequential down-sampling and up-sampling operations in the network structure, the boundary information of low-level features lost gradually. We proposed mask feature inpainting operation on the feature maps of different feature layers with the rough nucleus mask results. The purpose is to get the most responsive features of this layer preparing for subsequent edge residue map acquisition. Three convolutional layers are used for enhancement followed by the inpainting operation.

Edge Residue Operation. To get the powerful edge region of FCN outputting mask, we proposed the edge residue operation to obtain the residue regions contain the edge information for mask results. The specific method is the logical 'exclusive or' operation shown in Fig. 1 with $\oplus$ symbol.

Edge Residue Map. This step aims to get the mask edge loss of information from low-level to high-level features. We first employ Sobel operators to the input image and the output mask for detecting edges. For original input images, the edge map includes lots of noisy edges shown in Fig. 1. After that we get intermediate results of each Sobel edge respectively with pixel-wise AND operation. Then the final mask edge residue map is generated by using the cross entropy loss L_e. With mask edge residue map received, the output nucleus mask is refined by retraining ERN exactly. Some visual results are shown in Fig. 5.

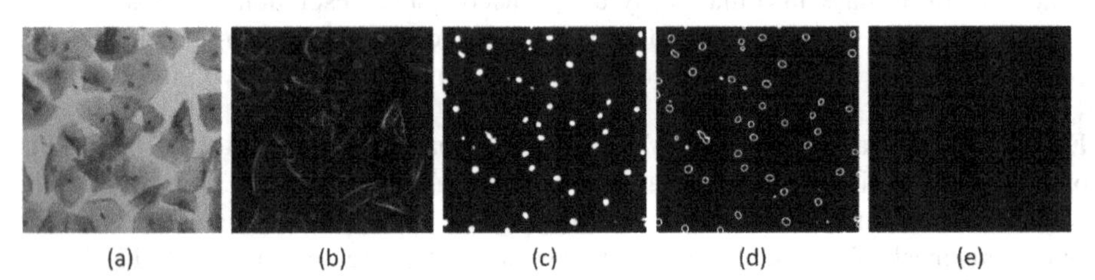

Fig. 5. Visualization results for edge residue map generation process results. From left to right: (a) Input cervical cell image, (b) Sobel edge of input image, (c) The predict mask of output, (d) Sobel edge of output mask, and (e) Edge residue map for mask edge loss information.

3 Experiments

We evaluate our method on the public benchmark named CCEDD [14], and compare with weakly supervised method [9]. We first descript the datasets and four evaluation metrics below, then presenting our results and ablation study to prove the effectiveness of our model.

3.1 Dataset Description

Cervical cell edge detection dataset (CCEDD) [14]. CCEDD dataset was captured digital images with SmartV350D lens, a Nikon ELIPSE Ci slide scanner and a 3-megapixel digital camera collected from Liaoning Cancer Hospital & Institute. CCEDD dataset includes 686 Thinprep Cytologic Test (TCT) cervical cell images with 2,048 × 1,536 pixels. We can get the nucleus mask from cell edge annotation of CCEDD. The point annotation for the training set can be generated by calculating each nucleus mask central point.

3.2 Evaluation Metrics

We compare our method with weakly supervised method [9] and conduct ablation study. For evaluation we take Aggregated Jaccard Index (AJI) [12] and Dice coefficient [13] as object-level metrics. IoU and F1 score are measured as pixel-level metrics.

3.3 Implementation Details

Data augmentation is adopted due to the small size of datasets, such as flipping, rotation, scale, random crop, and affine transformation. The network is initialized with pretrained model parameters. The Adam optimizer is adopted for updating weights. In our weakly supervised network, we train our model for 300 epochs (including 200 epochs for FCN and 100 epochs for ERN) with a learning rate of 1e−4. In fully supervised model training, we also take 200 epochs with a learning rate of 1e−4 on baseline model.

3.4 Results and Comparison

We present our model performance and results for other methods with four evaluation metrics with considerable margins in Table 1. Our method is outperformed significantly compared with method [9] in all metrics on CCEDD datasets. Furthermore, to illustrate our upper limit method, a fully supervised model is trained. Compared with the fully supervised model, our method achieves a result competitively. All comparative scores show the efficacy of our designed model in fusing coarse mask and edge residue refinement process, as well as the effectiveness with learned representation.

Table 2 lists each iteration results for illustrating ablation study of two network stage (FCN and ERN). From Fig. 6, visualization results for the comparison of different iterations are shown. In coarse mask generation FCN stage, the iterations accuracies are gradually increased. The ERN stage shows that after the mask edge residue refinement, the nucleus edges can be better segmented as out model converged. After that, our segmentation model further improved the effectiveness.

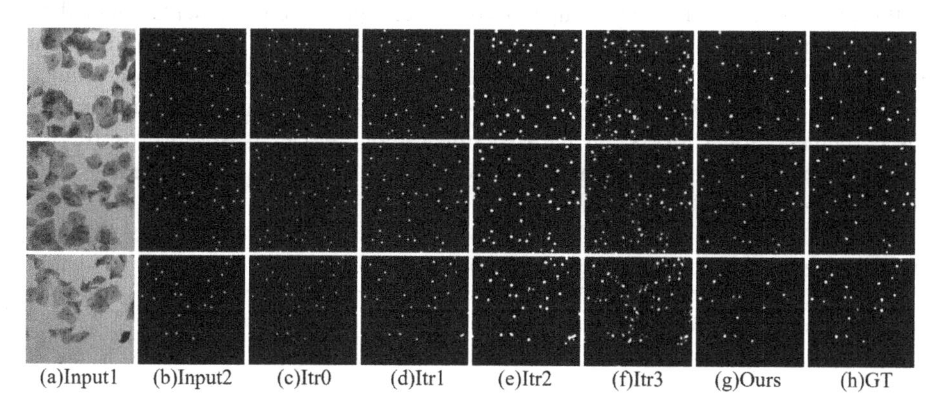

Fig. 6. Visualization results for the comparison of different iterations. From left to right: (a) (b) are input images of the test datasets and the point labels, (c) to (f) are outputs of different iterations with our methods, (g) are our results, and (h) are groundtruth labels.

Table 1. Comparison on the test set of CCEDD.

Method	IoU	F1	Dice	AJI
Fully	0.6512	0.8003	0.7482	0.5417
[9]	–	0.7601	0.7025	0.5147
Baseline	0.6196	0.7785	0.7098	0.5203
FCN	0.6326	0.7808	0.7176	0.5284
FCN + ERN	**0.6473**	**0.7891**	**0.7306**	**0.5367**

Table 2. Comparison of different iterations on the test set of CCEDD.

Iteration	IoU	F1	Dice	AJI
FCN itr1	0.2824	0.4067	0.3478	0.3351
FCN itr2	0.4657	0.6435	0.5523	0.4125
FCN itr3	0.5946	0.7658	0.6984	0.5039
FCN itr4	0.6326	0.7808	0.7176	0.5284
ERN itr1	**0.6473**	**0.7891**	**0.7306**	**0.5367**

4 Conclusion

In this paper we present a weakly supervised cervical cell nucleus segmentation method using only point annotations. We generate the Voronoi distance maps from the point labels and take benefit of the edge residue of different feature level to refine our trained model. Our method can achieves competitive performance compared with fully supervised segmentation network. In the experiments, our model is able to resolve the confusion between foreground and background of cervical cell nucleus image with weakly-supervised point annotations.

Acknowledgements. This work is supported by the National Natural Science Foundation of China (61873259, U20A20200, 61821005), and the Youth Innovation Promotion Association of Chinese Academy of Sciences (2019203).

References

1. Chen, J., Lu, Y., Yu, Q., et al.: Transunet: Transformers make strong encoders for medical image segmentation. arXiv preprint arXiv:2102.04306 (2021)
2. Wang, G., Li, W., Zuluaga, M.A., et al.: Interactive medical image segmentation using deep learning with image-specific fine tuning. IEEE Trans. Med. Imaging **37**(7), 1562–1573 (2018)
3. Zhang, J., Xie, Y., Wu, Q., et al.: Medical image classification using synergic deep learning. Med. Image Anal. **54**, 10–19 (2019)

4. Huang, C., Han, H., Yao, Q., Zhu, S., Zhou, S.K.: 3D U-Net: a 3D universal u-net for multi-domain medical image segmentation. In: Shen, D., et al. (eds.) MICCAI 2019. LNCS, vol. 11765, pp. 291–299. Springer, Cham (2019). https://doi.org/10.1007/978-3-030-32245-8_33
5. Wang, Z., Zou, N., Shen, D., et al.: Non-local U-Nets for biomedical image segmentation. In: AAAI Conference on Artificial Intelligence 2020, vol. 34, no. 04, 6315–6322 (2020)
6. He, K., Zhang, X., Ren, S., et al.: Spatial pyramid pooling in deep convolutional networks for visual recognition. IEEE Trans. Pattern Anal. Mach. Intell. **37**(9), 1904–1916 (2015)
7. Gao, Z., Puttapirat, P., Shi, J., Li, C.: Renal cell carcinoma detection and subtyping with minimal point-based annotation in whole-slide images. In: Martel, A.L., et al. (eds.) MICCAI 2020. LNCS, vol. 12265, pp. 439–448. Springer, Cham (2020). https://doi.org/10.1007/978-3-030-59722-1_42
8. Chen, Z., Chen, Z., Liu, J., et al.: Weakly supervised histopathology image segmentation with sparse point annotations. IEEE J. Biomed. Health Inform. **25**(5), 1673–1685 (2020)
9. Dong, M., et al.: Towards neuron segmentation from macaque brain images: a weakly supervised approach. In: Martel, A.L., et al. (eds.) MICCAI 2020. LNCS, vol. 12265, pp. 194–203. Springer, Cham (2020). https://doi.org/10.1007/978-3-030-59722-1_19
10. He, K., Gkioxari, G., Dollar, P., Girshick, R.: Mask R-CNN. IEEE Trans. Pattern Anal. Mach. Intell. **42**(2), 386–397 (2020)
11. Ke, L., Danelljan, M., Li, X., et al.: Mask Transfiner for High-Quality Instance Segmentation. arXiv preprint arXiv:2111.13673 (2021)
12. Kumar, N., Verma, R., Sharma, S., et al.: A dataset and a technique for generalized nuclear segmentation for computational pathology. IEEE Trans. Med. Imaging **36**(7), 1550–1560 (2017)
13. Sirinukunwattana, K., Snead, D.R.J., Rajpoot, N.M.: A stochastic polygons model for glandular structures in colon histology images. IEEE Trans. Med. Imaging **34**(11), 2366–2378 (2015)
14. Liu, J., Fan, H., Wang, Q., et al.: Local label point correction for edge detection of overlapping cervical cells. Front. Neuroinform. **2022**(16), 895290 (2022)

An Analysis of Low-Rank Decomposition Selection for Deep Convolutional Neural Networks

Baichen Liu[1,2,3], Huidi Jia[1,2,3], Zhi Han[1,2(✉)], Xi'ai Chen[1,2], and Yandong Tang[1,2]

[1] State Key Laboratory of Robotics, Shenyang Institute of Automation, Chinese Academy of Sciences, Shenyang 110016, Liaoning, China
hanzhi@sia.cn

[2] Institutes for Robotics and Intelligent Manufacturing, Chinese Academy of Sciences, Shenyang 110169, Liaoning, China

[3] University of Chinese Academy of Sciences, Beijing 100049, China

Abstract. Deep convolutional neural networks have achieved state of the art results in many image classification tasks. However, the large amount of parameters of the network limit its deployment to storage space limited situations. Low-rank decomposition methods are effective to compress the network, such as Canonical Polyadic decomposition and Tucker decomposition. However, most low-rank decomposition based approaches cannot achieve a satisfactory balance between the classification accuracy and compression ratio of the network. In this paper, we analyze the advantages and disadvantages of Canonical Polyadic and Tucker decomposition and give a selection guidance to take full advantage of both. And we recommend to use Tucker decomposition for shallow layers and Canonical Polyadic decomposition for deep layers of a deep convolutional neural network. The experiment results show that our approach achieves the best trade-off between accuracy and parameter compression ratio, which validates our point of view.

Keywords: Low-rank decomposition · Canonical Polyadic decomposition · Tucker decomposition · Deep convolutional neural networks · Image classification

1 Introduction

Deep convolutional neural networks (DCNNs) have achieved great improvement in many computer vision tasks, such as image classification [5,12,14,20], object segmentation [2,16] and object detection [1,23]. Many DCNNs have a large amount of parameters, which limit its deployment to some mobile devices where

---

This work was supported in part by the National Natural Science Foundation of China under Grant 61903358, 61773367, 61821005, and the Youth Innovation Promotion Association of the Chinese Academy of Sciences under Grant 2022196, Y202051.

H. Liu et al. (Eds.): ICIRA 2022, LNAI 13456, pp. 480–490, 2022.
https://doi.org/10.1007/978-3-031-13822-5_43

the storage space is limited. To overcome this problem, many DCNN compression methods aim to reduce the number of parameters of DCNN to reduce the storage space and accelerate inference speed, including pruning methods, low-rank decomposition methods, quantization methods, knowledge distillation methods and compact design methods.

Low-rank decomposition based methods are effective due to the redundancy in weights of convolutional and fully connected layers. The filters of a convolutional layer can be regarded as a 4-order tensor. Some low-rank tensor decomposition method are applied to compress the parameters of convolutional layers. Canonical Polyadic (CP) decomposition [13] is proposed to compress the number of parameters and their approach obtains 8.5× CPU speedup on AlexNet [12]. [9] proposes a one-shot whole network compression scheme which and compress the filters of convolutional layers via Tucker decomposition. Tensor train (TT) [19] and tensor ring (TR) [22] decomposition can achieve more than 1000× compression ratios because of their unique mathematical property. However, low-rank decomposition based methods always cannot make a satisfactory balance between network performance and compression. When the compression ratio is extremely large, these approaches have significant accuracy drop even after fine-tuning.

We analyze some difficulties of low-rank decomposition based methods. Despite achieving high compression ratio, the rank calculation of CP decomposition is an NP-hard problem [4,10]. It is difficult to set an appropriate CP rank and solve the decomposition. After the decomposition, some important information is lost. It is more easier to estimate the rank of Tucker decomposition. However, the factor matrices obtained by Tucker decomposition are always unbalanced, which affects the performance of the network. To take full use of these two approaches, in this paper, we analyze the advantages and disadvantages of CP and Tucker decomposition and find a balance between these two approaches. We give a guidance of the decomposition selection indicating which decomposition approach to use for each layer. In a DCNN, features of shallow layers help capture global information, which is also important for further feature extraction in deep layers. Tucker decomposition retains more information of the original features due to the core tensor. Therefore, in shallow layers we use Tucker decomposition to compress the layer. The number of channels of a shallow convolutional layer is small, and Tucker decomposition obtains relatively balanced factor matrices. We use a global analytic solution of variational Bayesian matrix factorization (VBMF) [18] to search the rank of Tucker decomposition of each layer. In deep layers where the number of channels of a convolutional layer is large, Tucker decomposition suffers from the unbalanced problem. At this case, we use CP decomposition to avoid the unbalanced problem and achieve higher compression ratio. CP decomposition is solved by alternating least squares (ALS) [10].

To sum up, the main contributions of this paper are presented as follows:

1) We propose to alternatively use CP or Tucker decomposition on the weights of each convolutional layer of DCNNs according to the advantages of both,

which achieves a better trade-off between network performance and parameter compression ratio.

2) We give a decomposition selection guidance of the whole network for many popular DCNNs, including ResNet20, ResNet32 and VGG16.

2 Method

2.1 Preliminary Notations

A d-order tensor is a d dimensional multi-way array. Scalars, vectors and matrices are 0-order, 1-order and 2-order tensors, respectively. In this paper, we denote scalars, vectors and matrices by lowercase letters $(x, y, z, \cdots)$, bold lowercase letters $(\mathbf{x}, \mathbf{y}, \mathbf{z}, \cdots)$ and uppercase letters $(X, Y, Z, \cdots)$, respectively. An N-order tensor (N ≥ 3) is denoted as $\mathcal{X} \in \mathbb{R}^{I_1 \times I_2 \times \cdots \times I_N}$ and each element is denoted as $x_{i_1, i_2, \cdots, i_N}$. The mode-n matrix form of a tensor $\mathcal{X} \in \mathbb{R}^{I_1 \times I_2 \times \cdots \times I_N}$ is the operation of reshaping the tensor into a matrix $X_{(n)} \in \mathbb{R}^{I_n \times (I_1 \cdots I_{n-1} I_{n+1} \cdots I_N)}$.

2.2 CP Decomposition Layers

For an N-order tensor $\mathcal{A} \in R^{I_1 \times I_2 \times \cdots \times I_N}$, its CP decomposition is defined as:

$$\mathcal{A} = \sum_{r=1}^{R} A_{:,r}^{(1)} \circ A_{:,r}^{(2)} \circ \cdots \circ A_{:,r}^{(N)}, \tag{1}$$

where the minimal possible R is called the canonical rank of CP decomposition and $A_{:,r}^{(n)} (n = 1, 2, \cdots, N, r = 1, 2, \cdots, R)$ denote the rth column of the nth factor matrix.

For a convolutional layer with weights $\mathcal{W} \in \mathbb{R}^{d \times d \times C_{in} \times C_{out}}$, the rank-R CP decomposition of $\mathcal{W}$ can be formulated as:

$$\mathcal{W} = \sum_{r=1}^{R} W_{:,r}^{(1)} \circ W_{:,r}^{(2)} \circ W_{:,r}^{(3)} \circ W_{:,r}^{(4)}, \tag{2}$$

where $W^{(n)} (n = 1, 2, 3, 4)$ denote the four factor matrices of $\mathcal{W}$ in size $(d \times R)$, $(d \times R)$, $(C_{in} \times R)$ and $(C_{out} \times R)$ respectively. As the rank calculation of CP decomposition is an NP-hard problem, we set the rank artificially and approximately solve Eq. (2) by ALS [10].

The convolution operation of an ordinary convolutional layer can be formulated as

$$Y = X \otimes \mathcal{W}, \tag{3}$$

where $\otimes$ denotes convolution operation, Y, X and $\mathcal{W}$ denote the output, input and weights of the convolutional layer, respectively. After CP decomposition, the ordinary convolutional layer is replaced by four sequential sub-layers with weights $\mathcal{W}^1 \in \mathbb{R}^{1 \times 1 \times C_{in} \times R}$, $\mathcal{W}^2 \in \mathbb{R}^{d \times 1 \times R \times R}$, $\mathcal{W}^3 \in \mathbb{R}^{1 \times d \times R \times R}$ and $\mathcal{W}^4 \in$

$\mathbb{R}^{1\times 1\times R\times C_{out}}$, respectively. Then the convolution operation is executed on the four sub-layers sequentially, formulated as:

$$Y = (((X \circledast \mathcal{W}^1) \circledast \mathcal{W}^2) \circledast \mathcal{W}^3) \circledast \mathcal{W}^4. \tag{4}$$

Equations (3) and (4) are equivalent.

2.3 Tucker Decomposition Layers

For a tensor $\mathcal{A} \in \mathbb{R}^{I_1\times I_2\times\cdots\times I_N}$, its Tucker decomposition is defined as:

$$\begin{aligned}\mathcal{A} &= \sum_{s_1=1}^{S_1} \times \sum_{s_N}^{S_N} g_{s_1 s_2 \ldots s_N}(u_{s_1}^{(1)} \circ u_{s_2}^{(2)} \circ \cdots \circ u_{s_N}^{(N)}) \\ &= \mathcal{G}\times_1 U^{(1)} \times_2 U^{(2)} \cdots \times_N U^{(N)}.\end{aligned} \tag{5}$$

where $(S_1, S_2, \cdots, S_N)$ denotes the rank of Tucker decomposition, $\mathcal{G} \in \mathbb{R}^{S_1\times S_2\times\cdots\times S_N}$ denotes the core tensor and $U^{(n)} = [u_1^{(n)}, u_2^{(n)}, \cdots, u_N^{(n)}] \in \mathbb{R}^{I_n\times \mathrm{S}_n}(n = 1, 2, \cdots, N)$ denote factor matrices.

For a convolutional layer with weights $\mathcal{W} \in \mathbb{R}^{d\times d\times C_{in}\times C_{out}}$, it can be decomposed via Tucker decomposition as:

$$\mathcal{W}_{i,j,s,t} = \sum_{r_1=1}^{R_1}\sum_{r_2=1}^{R_2}\sum_{r_3=1}^{R_3}\sum_{r_4=1}^{R_4} \mathcal{G}'_{r_1,r_2,r_3,r_4} U_{i,r_1}^{(1)} U_{j,r_2}^{(2)} U_{s,r_3}^{(3)} U_{t,r_4}^{(4)}, \tag{6}$$

where $\mathcal{G}$ is the core tensor of size $R_1 \times R_2 \times R_3 \times R_4$ and $U^{(1)}$, $U^{(2)}$, $U^{(3)}$ and $U^{(4)}$ are factor matrices of sizes $d \times R_1$, $d \times R_2$, $C_{in} \times R_3$ and $C_{out} \times R_4$, respectively.

As the kernel size $d \times d$ is always 3×3 which is too small, we only compress the number of input and output channels. We estimate the rank of input and output channels (R_3, R_4) by VBMF [18]. The global analytic VBMF is promising because it can automatically find rank and provide theoretical condition for perfect rank recovery. Then Eq. (6) is changed to a variant form of Tucker decomposition called Tucker-2 decomposition as:

$$\mathcal{W}_{i,j,s,t} = \sum_{r_3=1}^{R_3}\sum_{r_4=1}^{R_4} \mathcal{G}_{i,j,r_3,r_4} U_{s,r_3}^{(3)} U_{t,r_4}^{(4)}. \tag{7}$$

After Tucker decomposition, an ordinary convolution operation in Eq. (3) is replaced by three sequential sub-layers with weights $\mathcal{W}^1 \in \mathbb{R}^{1\times 1\times C_{in}\times R_3}$, $\mathcal{W}^2 \in \mathbb{R}^{d\times d\times R_3\times R_4}$ and $\mathcal{W}^3 \in \mathbb{R}^{1\times 1\times R_4\times C_{out}}$, respectively. Then the convolution operation is executed on the three sub-layers sequentially, formulated as:

$$Y = ((X \circledast \mathcal{W}^1) \circledast \mathcal{W}^2) \circledast \mathcal{W}^3. \tag{8}$$

2.4 Parameter Analysis

We analyze the number of parameters compressed by CP and Tucker decomposition. The number of parameters of a convolutional kernel $\mathcal{W} \in \mathbb{R}^{d \times d \times C_{in} \times C_{out}}$ denoted as P is calculated as:

$$P = d \times d \times C_{in} \times C_{out}. \tag{9}$$

After rank-R CP decomposition, the number of parameters denoted as P_{cp} is calculated by summing the number of parameters of the four factor matrices as:

$$P_{cp} = C_{in} \times R + d \times R + d \times R + R \times C_{out} = R(C_{in} + C_{out} + 2d) \tag{10}$$

The compression ratio of CP decomposition on a convolutional layer denoted as R_{cp} is defined as:

$$R_{cp} = \frac{P}{P_{cp}} = \frac{d^2 C_{in} C_{out}}{R(C_{in} + C_{out} + 2d)}. \tag{11}$$

After Tucker decomposition, the number of parameters denoted as P_t is calculated by summing the number of parameters of the three sub-layers as:

$$P_t = C_{in} \times R_3 + d^2 \times R_3 \times R_4 + R_4 \times C_{out} \tag{12}$$

The compression ratio of Tucker decomposition on a convolutional layer denoted as R_t is defined as:

$$R_t = \frac{P}{P_t} = \frac{d^2 C_{in} C_{out}}{C_{in} R_3 + d^2 R_3 R_4 + R_4 C_{out}}. \tag{13}$$

When CP rank R and Tucker rank (R_3, R_4) are set similarly, Tucker decomposition has more parameters due to the core tensor part $d^2 \times R_3 \times R_4$. In actual use, CP rank is always set smaller than Tucker rank, which makes CP decomposition achieves much higher compression ratio than Tucker decomposition.

2.5 Guidance of Decomposition Selection

The decomposition selection of each layer influences the trade-off between the performance of the network and the compression ratio of parameters. For a convolutional kernel $\mathcal{W} \in \mathbb{R}^{d \times d \times C_{in} \times C_{out}}$, we discuss how C_{in} and C_{out} influence the decomposition selection.

For relatively deeper layers where C_{in} and C_{out} is relatively large, the Tucker rank (R_3, R_4) estimated by VBMF is about $1/4$ of (C_{in}, C_{out}), which leads to unbalanced factor matrices. At this case, CP decomposition is a better choice to avoid unbalanced factor matrices. We use CP decomposition for deeper layers. The CP rank can be set less than $\frac{1}{5} min(C_{in}, C_{out})$, which achieves higher compression ratio than Tucker decomposition.

For relatively shallow layers where C_{in} and C_{out} is relatively small, the Tucker rank (R_3, R_4) estimated by VBMF is about $\frac{1}{3} min(C_{in}, C_{out})$ to $\frac{1}{2} min(C_{in}, C_{out})$.

As C_{in} and C_{out} is small in shallow layers, the number of parameter of the shallow layers is relatively small. The selection of CP or Tucker decomposition has little impact on the compression ratio of the whole network. At this case, the performance of the network has a higher priority. Compared with CP decomposition, Tucker decomposition retains more information of the original features due to the core tensor. Therefore, we use Tucker decomposition for shallow layers.

Fig. 1. Some examples in CIFAR10.

3 Experiment

We validate our approach on CIFAR10 dataset [11]. CIFAR10 dataset consists of color natural images with 32 × 32 pixels in 10 classes, respectively. It contains 50k images for training and 10k images for testing. Figure 1 shows some examples in CIFAR10. In the experiment, we follow the data augmentation provided in [15] for training: 4 pixels of value 0 are padded on each side of the 32×32 image, and a 32×32 random crop is sampled from the padded image or its horizontal flip. For testing, we only evaluate the original 32×32 images. All the following experiments are conducted on a single GPU of Nvidia GeForce RTX 3090.

3.1 Ablation Study

We analyze and discuss the effectiveness of decomposition selection on ResNet20 [5]. The first layer of ResNet20 is a 3×3 convolutional layer. Then a stack of 3×3 convolutional layers with feature map sizes $\{32, 16, 8\}$ is used sequentially, with 6 layers for each feature map size. Different from ResNet20 in the original paper [5], we discard the global average pooling after the stack of convolutional layers and use a 10-way fully connected layer directly. Note that we use batch normalization [8] after each convolutional layer and before the ReLU activation. We train baseline ResNet20 with an SGD optimizer for 200 epochs, starting with a learning rate of 0.1, divided by 10 at 50 and 150 epochs. The baseline ResNet20 achieves 92.93% classification accuracy on the test set, which is about 1% higher than [5].

Then we conduct decomposition on each convolutional layer of baseline ResNet20. We use VBMF to estimate the rank of Tucker decomposition. For CP decomposition, we set the rank to the nearest integer smaller than $\frac{1}{3}max(C_{in}, C_{out})$. As ResNet20 is a relatively compact network, the performance of the network degrades severely after decomposition and fine-tuning is necessary. We find that the sub-layers of the original convolutional layers are more sensitive to the learning rate. Therefore, we recommend a 0.01 or less initial learning rate for fine-tune. We fine-tune the compressed network for 150 epochs with an initial learning rate 0.01, divided by 10 at 50 and 100 epochs.

As shown in Table 1, We report Top-1 test accuracy and the ratio of the number of parameters to baseline ResNet20 of each compressed network. As for decomposition mode, ‘tcc’ denotes that using Tucker decomposition for the first stage of the baseline ResNet20 and CP decomposition for the middle and last stage. Our point of view is that using Tucker decomposition for shallow layers and CP decomposition for deep layers. Therefore, we recommend ‘tcc’ or ‘ttc’ decomposition for ResNet20 to achieve a better trade-off between accuracy and compression ratio. We evaluate compressed network with pure CP decomposition ‘ccc’ or Tucker decomposition ‘ttt’ for comparison, which conducts CP decomposition or Tucker decomposition on all three stages. In addition, we switch CP and Tucker decomposition for deep and shallow layers by setting decomposition as ‘cct’ and ‘ctt’.

Table 1. Top-1 test accuracy on CIFAR10 and compression ratio of compressed networks. 'R20' denotes the baseline ResNet20.

Decomposition mode	Top-1 accuracy (%)	Number of params. (%)
R20	92.93	100
ttt	92.09	15.0
ttc	**91.84**	**9.7**
tcc	89.95	8.6
cct	89.16	13.1
ctt	91.79	14.2
ccc	86.78	7.8

The network with pure Tucker decomposition achieves 6.7× compression ratio and has 0.84% accuracy decline compared with baseline ResNet20. The network with pure CP decomposition achieves the largest compression ratio 12.8× and has a serious decline in accuracy (6.15%). Our recommended 'ttc' achieves 10.3× compression ratio with only 1.09% accuracy drop and 'tcc' achieves even higher compression ratio 11.6× while 2.98% accuracy drop. If we switch the decomposition mode from 'ttc' to 'ctt' or from 'tcc' to 'cct', the compression ratio becomes smaller while the Top-1 accuracy drops. It validates that it is a better choice to use Tucker decomposition for shallow layers and CP decomposition for deep layers.

We further analyze the rank setting of CP decomposition. CP decomposition is used on all three stages of baseline ResNet20. We change the rank ratio in $\frac{1}{3}max(C_{in}, C_{out})$ from $\frac{1}{10}$ to $\frac{1}{2}$. Table 2 reports Top-1 test accuracy and the ratio of number of parameters to baseline ResNet20 while using different CP rank ratios. For a better trade-off between accuracy and compression ratio, we recommend a ratio of $\frac{1}{5}$. And we use this rank ratio for CP decomposition for all the following experiments.

Table 2. Influence of the rank ratio on CP decomposition. 'R20' denotes the baseline ResNet20.

Rank ratio	Top-1 accuracy (%)	Number of params. (%)
R20	92.31	100
1/2	88.68	12.0
1/3	**86.78**	**7.8**
1/5	86.66	4.6
1/10	82.59	2.2

3.2 Comparison with Recent Approaches on More Networks

We further compare our approach with some recent effective low-rank based network compression approaches [3,6,7,17,21] on more DCNNs in Table 3, including ResNet20, ResNet32 and VGG16. For ResNet32, the first three stages are compressed by Tucker decomposition and the last two stages are compressed by CP decomposition. For VGG16, convolutional layers with the number of channels less than or equal to 256 are decomposed by Tucker decomposition, and convolutional layers with the number of channels equal to 512 are decomposed by CP decomposition.

Table 3. Comparison with low-rank based network compression approaches on CIFAR10. 'BL' denotes the Top-1 test accuracy of the baseline network. 'Top-1' denotes the Top-1 test accuracy of the compressed network. 'Params.' reports the parameter ratio to the baseline network.

Model	Approach	Top-1/BL (%)	Params. (%)
ResNet20	SSS [6]	90.85/92.53	83.4
	Hinge [17]	91.84/92.54	44.6
	RSTR [3]	88.3/90.4	16.7
	LREL [7]	90.8/91.8	59.3
	ADMM-TT [21]	91.03/91.25	14.7
	ours	**91.84**/92.93	**9.7**
ResNet32	RSTR [3]	88.1/92.5	6.6
	LREL [7]	90.7/92.9	32.6
	ADMM-TT [21]	91.96/92.49	17.2
	ours	**92.11**/93.47	11.2
VGG16	LREL [7]	93.5/93.5	50.2
	Hinge [17]	93.59/94.02	19.9
	SSS [6]	91.2/93.5	32.8
	ours	93.12/93.7	**8.4**

The training and fine-tuning procedure of ResNet32 is the same as ResNet20. The training procedure of VGG16 follows the original paper [20]. During fine-tuning, the initial learning rate is set 0.01 to accommodate the sensitivity of the decomposition layers.

We report Top-1 test accuracy together with its corresponding baseline and the parameter ratio to the baseline network. Our approach achieves the best trade-off between Top-1 test accuracy and parameter compression ratio. Our approach achieves the best Top-1 test accuracy and highest parameter compression ratio on ResNet20, the best Top-1 test accuracy on ResNet32 and the highest parameter compression ratio on VGG16.

4 Conclusion

In this paper, we analyze the low-rank decomposition selection criterion of convolutional layers between CP and Tucker decomposition for deep convolutional neural networks. In deep layers of a DCNN, CP decomposition can mitigate the unbalanced problem of Tucker decomposition and achieve relatively larger compression ratio. In shallow layers of a DCNN, Tucker decomposition retains more information of the original features due to the core tensor. Therefore, we recommend to use Tucker decomposition for shallow layers and CP decomposition for deep layers. The experiment results validate our point of view. For future work, we will take more low-rank decomposition methods into consideration and validate on more datasets.

References

1. Ashraf, M.W., Sultani, W., Shah, M.: Dogfight: detecting drones from drones videos. In: Proceedings of the IEEE/CVF Conference on Computer Vision and Pattern Recognition, pp. 7067–7076 (2021)
2. Aygun, M., et al.: 4D panoptic lidar segmentation. In: Proceedings of the IEEE/CVF Conference on Computer Vision and Pattern Recognition, pp. 5527–5537 (2021)
3. Cheng, Z., Li, B., Fan, Y., Bao, Y.: A novel rank selection scheme in tensor ring decomposition based on reinforcement learning for deep neural networks. In: ICASSP 2020–2020 IEEE International Conference on Acoustics, Speech and Signal Processing (ICASSP), pp. 3292–3296. IEEE (2020)
4. Hastad, J.: Tensor rank is np-complete. In: International Colloquium on Automata, Languages, and Programming, pp. 451–460 (1989)
5. He, K., Zhang, X., Ren, S., Sun, J.: Deep residual learning for image recognition. In: IEEE Conference on Computer Vision and Pattern Recognition, pp. 770–778 (2016)
6. Huang, Z., Wang, N.: Data-driven sparse structure selection for deep neural networks. In: Proceedings of the European Conference on Computer Vision (ECCV), pp. 304–320 (2018)
7. Idelbayev, Y., Carreira-Perpinán, M.A.: Low-rank compression of neural nets: learning the rank of each layer. In: IEEE Conference on Computer Vision and Pattern Recognition, pp. 8049–8059 (2020)
8. Ioffe, S., Szegedy, C.: Batch normalization: accelerating deep network training by reducing internal covariate shift. arXiv preprint arXiv:1502.03167 (2015)
9. Kim, Y.D., Park, E., Yoo, S., Choi, T., Yang, L., Shin, D.: Compression of deep convolutional neural networks for fast and low power mobile applications (2016)
10. Kolda, T.G., Bader, B.W.: Tensor decompositions and applications. SIAM Rev. **51**(3), 455–500 (2009)
11. Krizhevsky, A., Hinton, G., et al.: Learning multiple layers of features from tiny images (2009)
12. Krizhevsky, A., Sutskever, I., Hinton, G.E.: ImageNet classification with deep convolutional neural networks, pp. 1097–1105 (2012)
13. Lebedev, V., Ganin, Y., Rakhuba, M., Oseledets, I., Lempitsky, V.: Speeding-up convolutional neural networks using fine-tuned CP-decomposition. In: ICLR 2015: International Conference on Learning Representations 2015 (2015)

14. Lecun, Y., Bottou, L., Bengio, Y., Haffner, P.: Gradient-based learning applied to document recognition. Proc. IEEE **86**(11), 2278–2323 (1998)
15. Lee, C.Y., Xie, S., Gallagher, P., Zhang, Z., Tu, Z.: Deeply-supervised nets. In: Artificial Intelligence and Statistics, pp. 562–570 (2015)
16. Li, X., et al.: PointFlow: flowing semantics through points for aerial image segmentation. In: Proceedings of the IEEE/CVF Conference on Computer Vision and Pattern Recognition, pp. 4217–4226 (2021)
17. Li, Y., Gu, S., Mayer, C., Gool, L.V., Timofte, R.: Group sparsity: the hinge between filter pruning and decomposition for network compression. In: IEEE Conference on Computer Vision and Pattern Recognition, pp. 8018–8027 (2020)
18. Nakajima, S., Tomioka, R., Sugiyama, M., Babacan, S.: Perfect dimensionality recovery by variational Bayesian PCA. In: Advances in Neural Information Processing Systems, vol. 25 (2012)
19. Oseledets, I.V.: Tensor-train decomposition. SIAM J. Sci. Comput. **33**(5), 2295–2317 (2011)
20. Simonyan, K., Zisserman, A.: Very deep convolutional networks for large-scale image recognition. In: ICLR 2015: International Conference on Learning Representations 2015 (2015)
21. Yin, M., Sui, Y., Liao, S., Yuan, B.: Towards efficient tensor decomposition-based DNN model compression with optimization framework. In: Proceedings of the IEEE/CVF Conference on Computer Vision and Pattern Recognition, pp. 10674–10683 (2021)
22. Zhao, Q., Zhou, G., Xie, S., Zhang, L., Cichocki, A.: Tensor ring decomposition. arXiv preprint arXiv:1606.05535 (2016)
23. Zhong, Y., Wang, J., Wang, L., Peng, J., Wang, Y.X., Zhang, L.: DAP: detection-aware pre-training with weak supervision. In: Proceedings of the IEEE/CVF Conference on Computer Vision and Pattern Recognition, pp. 4537–4546 (2021)

Accurate Crop Positioning Based on Attitude Correction of Weeding Robot

Tianjian Wang, Hui Zhou, Jiading Zhou, Pengbo Wang(✉), and Feng Huang

School of Mechanical and Electric Engineering, Jiangsu Provincial Key Laboratory of Advanced Robotics, Soochow University, Suzhou 215123, China
pbwang@suda.edu.cn

Abstract. In order to obtain accurate hoeing distance and reduce the influence of external factors such as vibration and robot platform tilt, this paper proposes a crop positioning point correction algorithm based on attitude information fusion. The algorithm uses the attitude information to calculate the mapping relationship between the image position under the tilt of the target crop camera and the actual ground position, which corrects the crop positioning point. The experimental results show that compared with the uncorrected positioning point, the average error of visual measurement of hoeing distance is reduced by about 58% and 73%, respectively. This algorithm can effectively improve the measurement accuracy of hoeing distance and reduce the possibility of seedling injury.

Keyword: Machine vision · Interplant hoeing · Intelligent weeding robot

1 Introduction

In recent years, organic agriculture without chemicals has attracted more and more attention. In order to ensure the yield increase and meet the market demand, strict field management is needed. The weeding operation is the aspect that consumes the most time and labor cost in agricultural field management [1]. In order to improve production efficiency, the research and development of intelligent automated weeding robot is extremely urgent. The weeding robot based on machine vision uses the visual system to obtain seedling information and calculate the required distance between the blade and the seedling, which control the blade to complete the weeding operation. However, the visual hoeing distance is affected by external factors such as light, shadow, weed density, vibration, and inclination of the carrier platform, resulting in large positioning error of the weeding knife positioning data. In order to reduce the error and improve the stability of the system, it is necessary to use a variety of sensors to collect the variation of each external influencing factor, which combine the information [2–4] and models.

In 2012, Perez et al. [5] in Spain developed an automatic inter-plant weeding machine based on RTK-GPS. The system integrates the positioning information and mileage information in order to improve the frequency and accuracy of crop positioning information update. Amazone Werke [6, 7] developed a weeding prototype based on GPS and tilt sensor fusion technology, which corrected the lateral offset of the machine relative to

H. Liu et al. (Eds.): ICIRA 2022, LNAI 13456, pp. 491–501, 2022.
https://doi.org/10.1007/978-3-031-13822-5_44

the crop to control the lateral movement of the hydraulic lateral frame. The experiment results showed that the fusion of GPS and other sensors could improve the stability and accuracy of the system. R. D. Tillett et al. [8] of Silsoe Institute, UK, used Kalman filter and adaptive step climbing algorithm to fuse mileage information to iteratively optimize positioning information and improve the robustness of positioning algorithm. Zuoxi Zhao et al. [9] applied MEMS gyroscope and accelerometer fusion technology to the horizontal control system of leveling machine. Ziwen Chen et al. [10] analyzed the principle of hoe positioning data correction and visual lag compensation, and used the method of mileage information fusion machine vision to optimize the hoe positioning data. Dynamic experiments show that the distance of the optimized hoe seedling is reduced by 25% under different speeds.

From the above research progress, the research scholars in related fields have made many important research for the positioning technology of intelligent weeding machine based on multi-sensor fusion. However, for the positioning method of small weeding robot, due to the uneven concave and convex between ridges and slopes, the horizontal inclination of the vehicle body often occurs. At this time, the mapping relationship between the camera field of view and the position of the target crop in the image is changed, resulting in inaccurate measurement of the knife-seedling distance and seedling spacing.

In this paper, a method of accurate crop positioning based on attitude correction of weeding robot is proposed. After obtaining the position information of the target crop, the intelligent weeding machine vision system obtains camera tilt angle by IMU. According to the geometric relationship, the mapping relationship between the image position information under camera tilt and the world position information is solved and the hoeing distance are corrected.

2 Material and Method

2.1 Mobile Robot Platform and Vision System

In response to the unstructured and complex environment, our team designed a mobile robot platform, as shown in Fig. 1. The mobile robot platform is mainly composed of walking units on both sides, the middle body frame and the front suspension device. It uses four wheels to drive, and realizes the platform steering by differential speed. Two groups of weeding units are assembled below the beam. In order to adapt to the change of field ridge spacing, the wheel spacing adjustment device is designed for mobile platform. The visual system mainly includes the camera and the industrial computer. The camera type RealsenseD455 is installed in the center of the mobile robot platform about 70 cm from the ground and parallel to the ground.

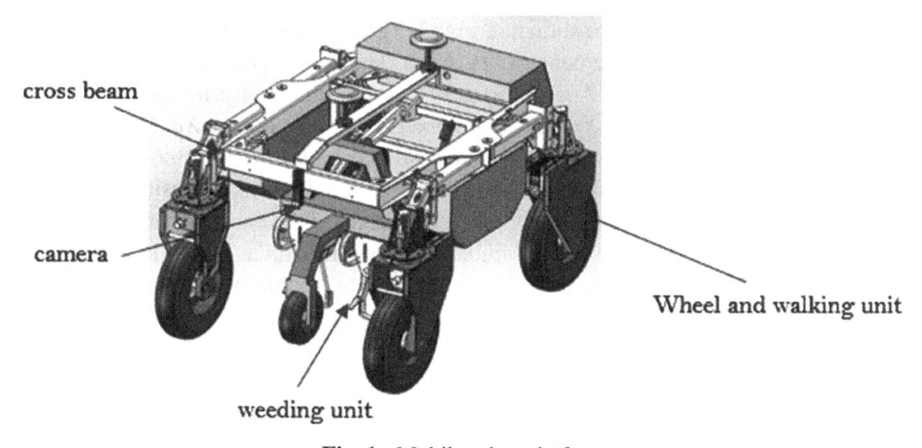

Fig. 1. Mobile robot platform

2.2 Vision System Calibration

In the actual weeding operation, the uneven ridge slope leads to the horizontal tilt of the car. In order to accurately extract the actual position of the crop corresponding to the ground, it is necessary to establish the conversion relationship between image coordinate system and world coordinate system. The IMU and camera external matrix can be obtained by Kalibr calibration tool, and the transformation relationship between image coordinate system and world coordinate system is described in detail.

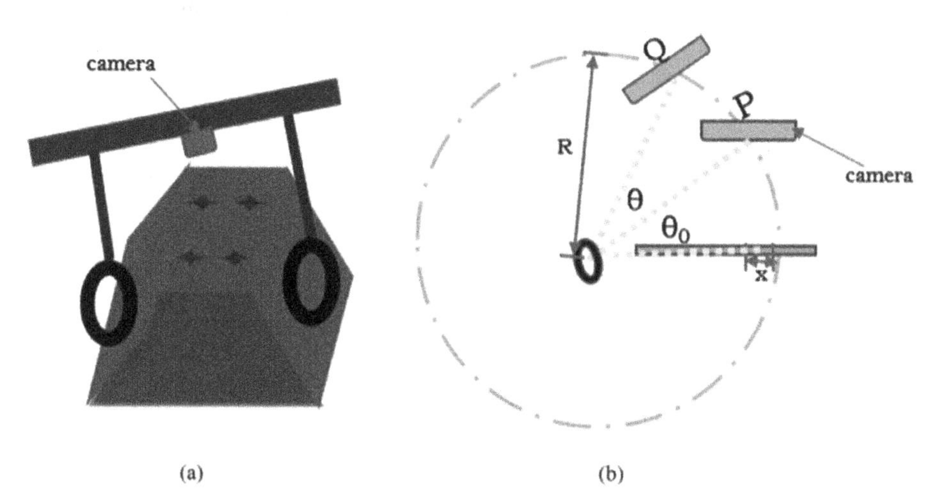

Fig. 2. (a) Robot platform tilting (b) Motion trajectory of camera

When the car body tilts, as shown in Fig. 2(a)(b), the motion trajectory of the camera is a circular trajectory with the tire in the inclined direction as the center and the distance between the tire center and the camera as the radius. The change of the specific field of view and the imaging model are shown in Fig. 3. The camera internal matrix: the pixel size *dx*, *dy*, the pixel coordinate center (*PPx*,*PPy*). The above data can be calibrated by the camera internal reference. According to the camera imaging principle, f is set as the focal length of the camera, *Xp* and *Yp* are pixel coordinate points, *Xc*, *Yc* and *Zc* are camera coordinate points, *Xw*, *Yw* and *Zw* are world coordinate points, the origin of the world coordinate system is the intersection point O of the $Z-axis$ extension line of the camera coordinate system and the ground, z is the height of the camera from the ground (which can be measured by the depth camera), and the horizontal tilt angle is θ (which can be calculated by the IMU attitude). According to the geometric relationship

$$\frac{X_P}{f} = \frac{X_w \cos\theta}{Xw \sin\theta + z} \tag{1}$$

$$\frac{Y_P}{f} = \frac{Y_w}{X_w \sin\theta + z} \tag{2}$$

from the Eq. (1)

$$X_w = \frac{zX_p}{f\cos\theta - X_P \sin\theta} \tag{3}$$

substitute Eq. (3) into Eq. (2)

$$Y_w = \frac{Y_P z\cos\theta}{f\cos\theta - X_P \sin\theta} \tag{4}$$

the conversion relation between pixel coordinates and image coordinates is

$$\begin{cases} X_P = (u - ppx)\mathrm{d}x \\ Y_P = (v - ppy)\mathrm{d}y \end{cases} \tag{5}$$

substitute Eq. (5) into Eq. (3), Eq. (4)

$$X_w = \frac{z(u - ppx)\mathrm{d}x}{f\cos\theta - (u - ppx)dx\sin\theta} \tag{6}$$

$$Yw = \frac{z(v - ppy)\mathrm{d}y\cos\theta}{f\cos\theta - (u - ppx)dx\sin\theta} \tag{7}$$

since the camera pixel sizes dx and *dy* are approximately equal,*dx* and *dy* are reduced to simplify the calculation of Equation (6) and Eq. (7),

$$fx = \frac{f}{dx} \tag{8}$$

$$fy = \frac{f}{dy} \tag{9}$$

$$X_w = \frac{z(u - ppx)}{fx\cos\theta - (u - ppx)\sin\theta} \tag{10}$$

$$Yw = \frac{z(v - ppy)\cos\theta}{fy\cos\theta - (u - ppx)\frac{dx}{dy}\sin\theta} \tag{11}$$

by Eq. (10) and Eq. (11) complete the transformation of the image coordinate system and the world coordinate system of the weeding robot visual system, the actual position of each target crop on the ground can be calculated when the camera is horizontally inclined.

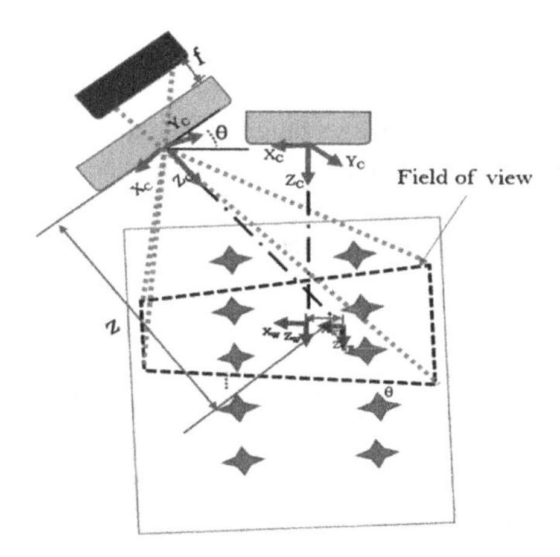

Fig. 3. Camera imaging principle

2.3 Correction of Knife and Seedling Spacing

Accurate positioning information of weeding cutter is the key for small and medium-sized weeding robots based on machine vision to complete weeding and seedling avoidance in the field. In practical application, the positioning information of weeding cutter

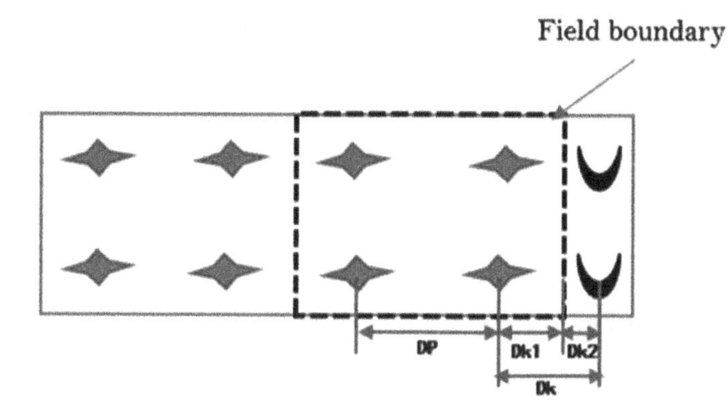

Fig. 4. *DK* and *DP*

refers to the distance between the weeding cutter and the target crop and the distance between the two crops before and after the same ridge. As shown in Fig. 4, it is referred to as the distance between the blade and the seed *Dk* and the distance between the seeding and the seeding *DP*. In the following, the IMU horizontal attitude information is used to optimize the *Dk* and *DP*.

The target detection model based on deep learning is used to infer the image position information between the two adjacent seedlings. The image position information is converted into the positioning point in the world coordinate system through Eqs. (10) and (11). The absolute value of the ordinate difference between the two seedlings is taken, namely, the corrected *DP*.

After the camera is tilted, the boundary below the field of view is changed, as shown in Fig. 2. The field of view changes from a rectangular area to a trapezoidal area, resulting in inaccurate measurement of blade spacing. Therefore, it is necessary to restore the visual field boundary point of the world coordinate system $O_W X_W Y_W Z_W$ when the camera is parallel. As shown in Fig. 2 (b), the distance between the tire and the camera is set the camera is R, and z is the distance from the depth camera to the soil. The initial inclination angle is θ_0. When the camera moves along the circular trajectory, the origin of the camera coordinate system and the origin of the world coordinate system simultaneously move along the negative direction of *Xaxis* and *Zaxis*, that is, from P state to Q state. There is no *Yaxis* component in the migration direction of the two camera coordinate systems. Therefore, after the camera tilts, it is only necessary to calculate the camera level. Height equation is $\mathrm{h} = z cos\theta + R[\sin\theta_0 - sin(\theta_0 + \theta)]$ under the field of view boundary point in the world coordinate system $O_{WR} X_{WR} Y_{WR} Z_{WR}$ position, that is

$$Y_w = \frac{(Y_p - PP_Y)h}{f/\mathrm{dy}} \tag{12}$$

The pixel ordinate $YP = 800$ of the boundary point under the camera field of view is substituted into the formula (12) to obtain the transverse coordinate value of the boundary world coordinate system under the field of view, which is denoted as Y_{w_edge}. After obtaining the horizontal boundary Y_{w_edge}, the boundary under the field of view is widened due to the tilt of the camera. There will be new crops that cannot be seen by the camera under the original field of view. As shown in Fig. 5, the target point greater than the threshold will be removed by using the threshold of the boundary Y_{w_edge} under the horizontal field of view.

After the above threshold judgment, the knife-seedling distance acquisition steps are as follows. Firstly, the target crop image position information can be calculated by image analysis to obtain the corresponding pixel coordinates. The crop position pixel coordinates can be calculated by (10) and (11) formulas to obtain the position under the world coordinate system $O_{WR} X_{WR} Y_{WR} Z_{WR}$, where the ordinate is denoted as Y_{w_plant}. Then, the modified knife-seedling distance is denoted as Y_D by the distance between the boundary point Y_{w_edge} and the ordinate Y_{w_plant} near the boundary target crop positioning point under the field of view.

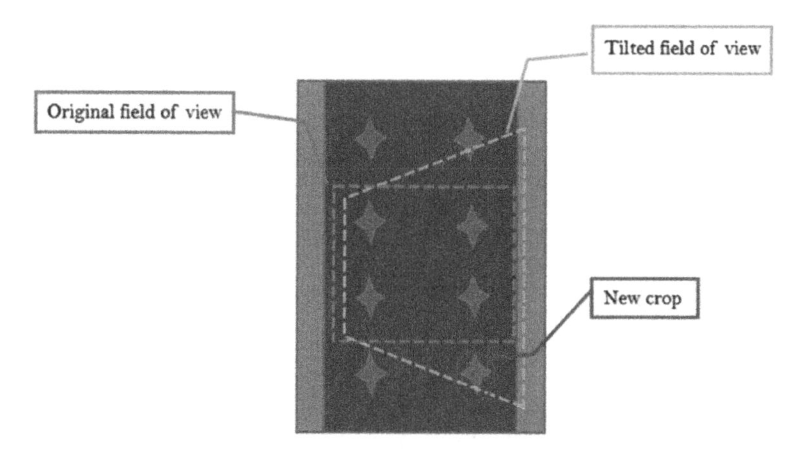

Fig. 5. New crops

3 Result and Discussion

After completing the algorithm, the host computer is mounted on the robot mobile platform, and the experimental site is located in the experimental field of Suzhou University. Due to seasonal reasons, the fake seedlings purchased are used to replace the actual target crops, with a total of two rows, with a row spacing of about 55 cm and a plant spacing of about 40 cm.

The mobile platform of the start-up robot moves along the crop line, and the mobile platform is randomly stopped. The distance between the current crop center and the boundary of the field of view and the distance between the adjacent two seedlings are measured at the level of the camera with a tape ruler as the real value. As shown in Fig. 6 and Fig. 7, the DK and DP of visual detection at 0°, 3°, 7°, 11° and 15° are saved respectively. The real value of the DK and DP, the uncorrected visual measurement of the DK and DP, and the corrected visual measurement of the DK and DP are listed. 10 groups of experiments were carried out and calculated the average error at different angles. Because the measurement method is manual tape measurement, there is a certain measurement error. In order to avoid the error value, the average value is taken after 3 times of measurement.

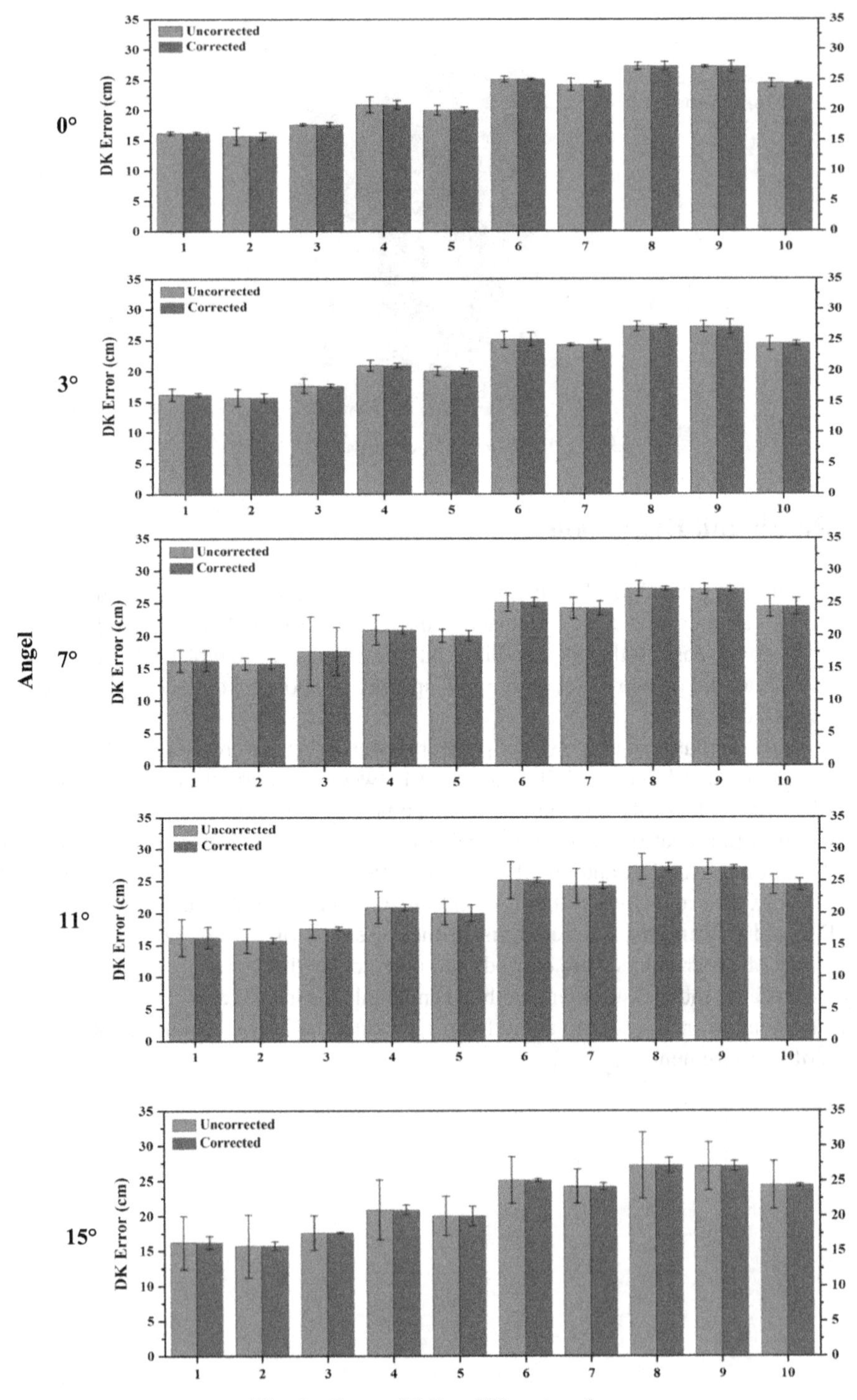

Fig. 6. Error of *DK* at different angles

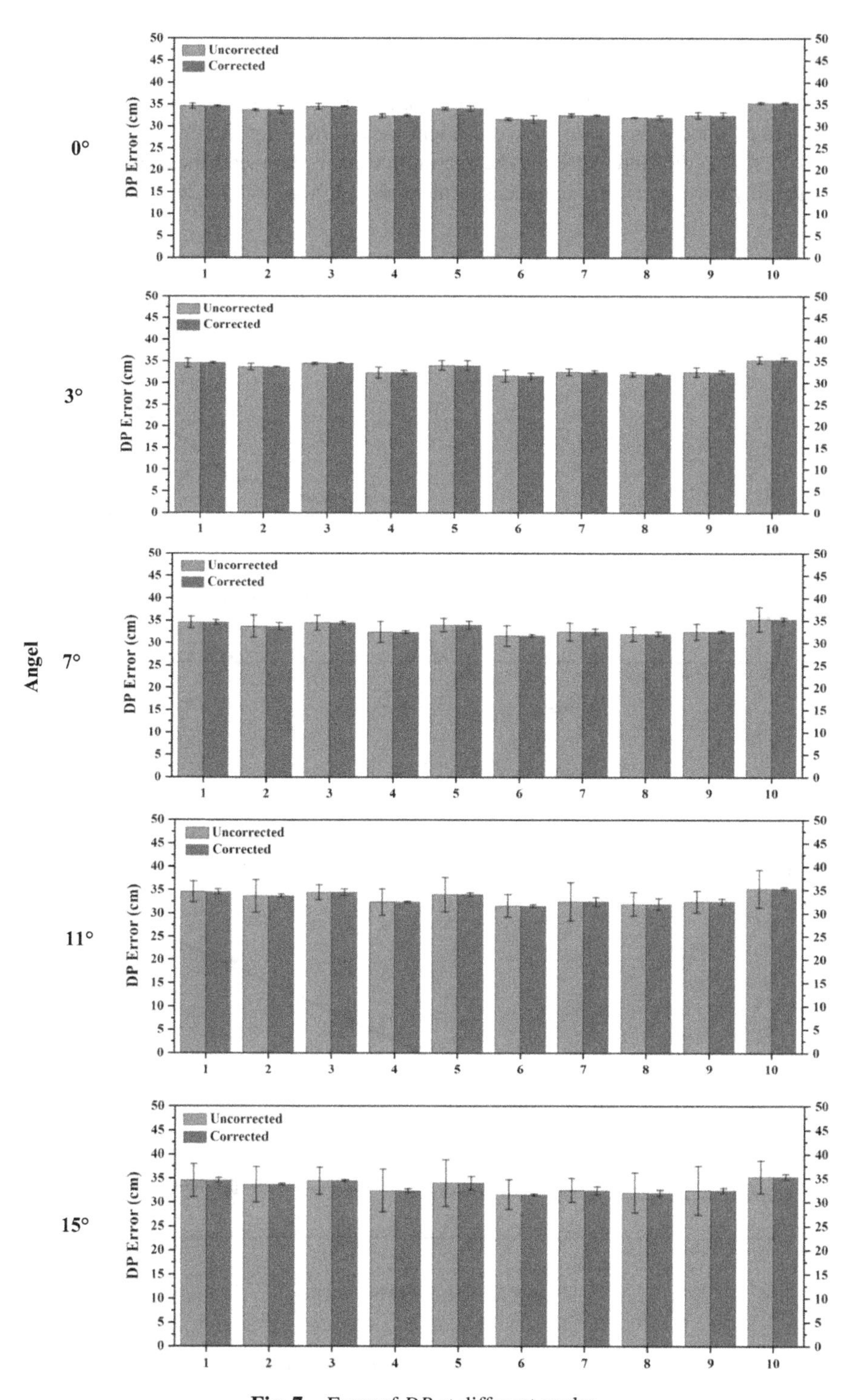

Fig. 7. Error of *DP* at different angles

The errors of DK and DP measured by visual measurement and those measured by manual measurement were statistically analyzed. As shown in Fig. 8 and Fig. 9, the average error increases with the deflection angle. The maximum errors of DK and DP measured by uncorrected visual measurement were 4.7 cm and 5.1 cm, and the average errors were 1.6 cm and 1.9 cm. The maximum errors between the DK and the DP measured by the corrected visual measurement were 1.7 cm and 1.4 cm, and the average

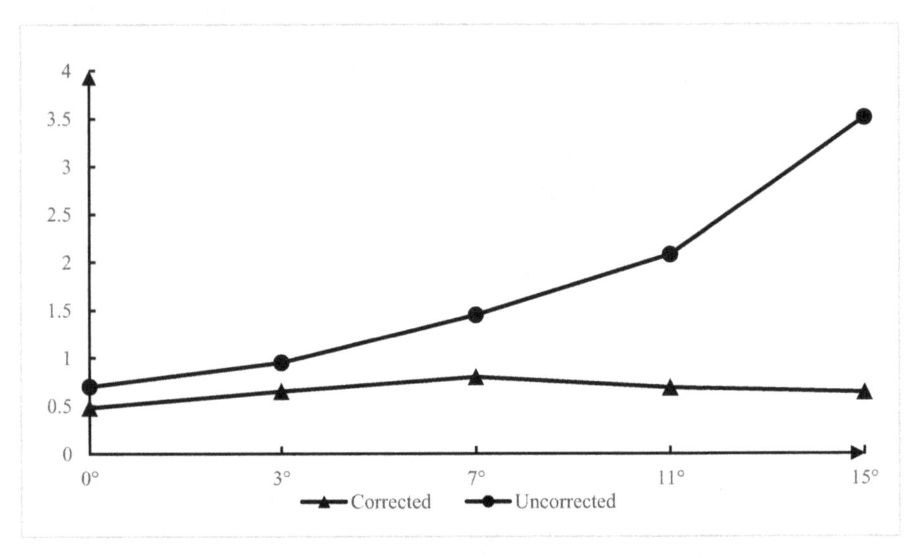

Fig. 8. Average error of *DK* at different angles

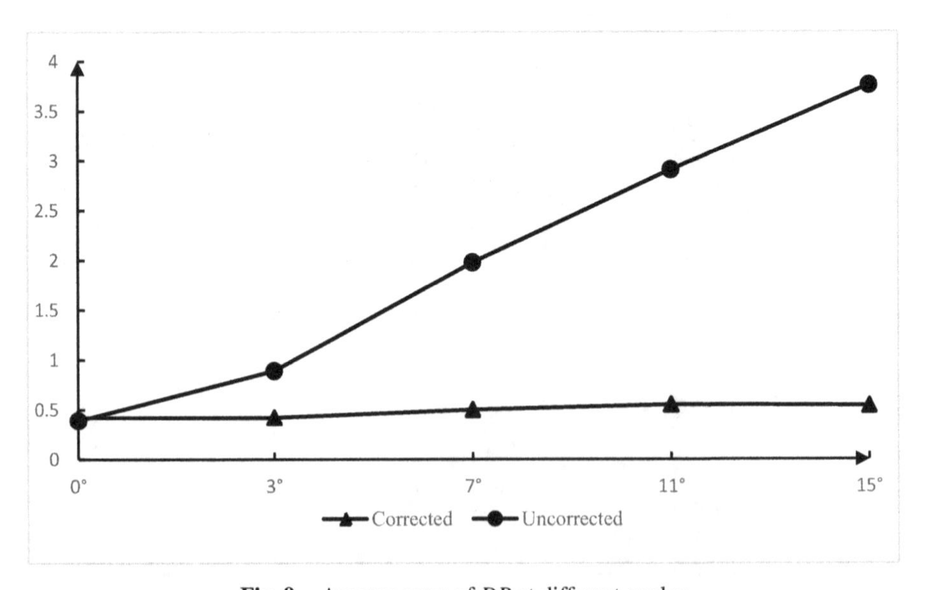

Fig. 9. Average error of *DP* at different angles

errors were 0.7 cm and 0.5 cm, respectively, which were about 58% and 73% lower than uncorrected algorithm.

4 Conclusion

In this paper, aiming at the intelligent weeding robot based on machine vision, a correction algorithm for the positioning data of weeding cutters between anti-tilting plants based on IMU attitude information fusion is proposed. By analyzing the motion state of the horizontal tilt of the car body, the boundary threshold under the restored horizontal field of view is used to effectively filter the new positioning point of the current ridge. For the effective positioning point, the mathematical modeling method is used to correct the positioning data of the weeding knife by calculating the mapping relationship between the image position of the target crop and the actual ground position under the horizontal tilt of the camera. The experiment results show that the maximum errors of the corrected visual measurement *DK* and *DP* are 1.7 cm and 1.4 cm, respectively. Compared with the positioning method without IMU attitude information fusion, the average errors of *DK* and *DP* are reduced by about 58% and 73%, respectively, which effectively improves the accuracy of the visual measurement hoe blade positioning data.

Funding. This research is supported by Modern Agricultural Machinery Equipment and Technology Demonstration and Promotion Project in Jiangsu Province (NJ2020–28).

References

1. Cirujeda, A., Melander, B., Rasmussen, K.: Relationship between speed, soil movement into the cereal row and intra-row weed control efficacy by weed harrowing. Weed Res. **43**(4), 285–296 (2010)
2. Santangelo, E., Assirelli, A., Liberati, P.: Evaluation of sensors for popular cutting detection to be used in intra-row weed control machine. Comput. Electron. Agric. **115**(1), 161–170 (2015)
3. Slaughter, C., Giles, K., Downey, D.: Autonomous robotic weed control systems: a review. Comput. Electron. Agric. **61**(1), 63–78 (2008)
4. Pullen, D., Cowell, A.: An Evaluation of the performance of mechanical weeding mechanisms for use in high speed inter-row weeding of arable crops. J. Agric. Eng. Res. **67**(1), 27–35 (1997)
5. Pérez-Ruiz, M., Slaughter, C., Gliever, J.: Automatic GPS-based intra-row weed knife control system for transplanted row crops. Comput. Electron. Agric. **80**(1), 41–49 (2012)
6. Nørremark, M., Griepentrog, W., Nielsen, J.: Evaluation of an autonomous GPS-based system for intra-row weed control by assessing the tilled area. Precision Agric. **13**(2), 149–162 (2012)
7. Lian, H., Xiwen, L., Shan, Z.: Plant recognition and localization for intra-row mechanical weeding device based on machine vision. Trans. Chin. Soc. Agri. Eng. **29**(10), 12–18 (2013)
8. Tillett, D., Hague, T., Grundy, C.: Mechanical within-row weed control for transplanted crops using computer vision. Biosys. Eng. **99**(2), 171–178 (2008)
9. Zuoxi, Z., Xiwen, L., Qing, L.: Leveling control system of laser-controlled land leveler for field based on MEMS inertial sensor fusion. Trans. Chin. Soc. Agri. Eng. **24**(6), 119–124 (2008)
10. Ziwen, C., Chunlong, Z., Nan, L.: Study review and analysis of high performance intra-row weeding robot. Trans. Chin. Soc. Agri. Eng. **31**(5), 1–8 (2015)

Low-Rank Tensor Tucker Decomposition for Hyperspectral Images Super-Resolution

Huidi Jia[1,2,3], Siyu Guo[1,2,3], Zhenyu Li[1,4], Xi'ai Chen[1,2(✉)], Zhi Han[1,2], and Yandong Tang[1,2]

[1] State Key Laboratory of Robotics, Shenyang Institute of Automation, Chinese Academy of Sciences, Shenyang 110016, Liaoning, China
chenxiai@sia.cn

[2] Institutes for Robotics and Intelligent Manufacturing, Chinese Academy of Sciences, Shenyang 110169, Liaoning, China

[3] University of Chinese Academy of Sciences, Beijing 100049, China

[4] Shenyang University of Technology, Shenyang 110870, China

Abstract. Super-resolution is an important way to improve the spatial resolution of Hyperspectral images (HSIs). In this paper, we propose a super-resolution method based on low-rank tensor Tucker Decomposition and weighted 3D total variation (TV) for HSIs. Global tensor Tucker decomposition and weighted 3D TV regularization are combined to exploit the prior knowledge of data low-rank information and local smoothness. Meanwhile, we use log-sum norm for tensor Tucker Decomposition to approximate the low-rank tensor. Extensive experiments show that our method outperforms some state-of-the-art methods on public HSI dataset.

Keywords: Low-rank tensor · Tucker decomposition · Hyperspectral images · Super-resolution

1 Introduction

Hyperspectral images (HSIs) have high spectral resolution and can show subtle differences in spectral dimensions of objects. Therefore, HSIs have been applied in remote sensing [1,4], object classification [17], product detection and other fields [7]. However, due to the limitations of imaging equipment, data memory and transmission, higher spectral resolution inevitably sacrifices some spatial resolution. Compared to the HSIs, multispectral images (MSIs) and RGB images

Supported in part by the National Natural Science Foundation of China under Grant 61903358, Grant 61873259 and Grant 61821005, in part by the Youth Innovation Promotion Association of the Chinese Academy of Sciences under Grant 2022196 and Grant Y202051, National Science Foundation of Liaoning Province under Grant 2021-BS-023.

H. Liu et al. (Eds.): ICIRA 2022, LNAI 13456, pp. 502–512, 2022.
https://doi.org/10.1007/978-3-031-13822-5_45

have high spatial resolution and low spectral resolution. Based on the complementary characteristics of the two types of data, it has become the trend of HSI super-resolution in recent years to fuse low-resolution HSI (LR-HSI) and high-resolution MSI (HR-MSI) of the same scene to obtain high resolution hyperspectral image (HR-HSI).

In recent years, low-rank method [3,19] has been widely used in computer vision tasks such as denoising, completion and detraining, and achieved great results. HSI naturally has a good low-rank characteristic due to the strong similarity between the spectral segments. Dong et al. [6] expressed the super-resolution problem of HSI as a joint estimation of hyperspectral dictionary and sparse coding, using the prior knowledge of spatial spectral sparsity of HSIs. Wang et al. [16] propose a low-rank representation model based on locality and structure regularized for HSI classification by taking full advantage of the local and global similarity of HSI data. Dian et al. [5] utilize low tensor-train rank to constrain the 4-D tensors composed of nonlocal similar cubes to achieve HSI super-resolution.

Tensors are higher-order generalizations of matrices, which can keep the structural integrity of high order data well [11–13,21]. HSI is naturally a 3-order tensor structure. Tucker Decomposition [15] is one of the most popular tensor decomposition method and achieve satisfactory results in many fields. Different from CP Decomposition [2], where the estimation of rank is a NP-hard problem [9], Tucker Decomposition decomposes tensors into the product of a core tensor with multiple factor matrices. The low-rank constraint can be achieved by minimizing the rank of the core tensor and factor matrices of Tucker decomposition. In our model, we use the log-sum norm rather than the common low-rank norm such as the nuclear norm to constrain the factor matrices of the Tucker decomposition. Many studies [8,18] show that the relaxation of the log-sum norm as the nuclear norm is very reasonable and easy to calculate.

Unfortunately, low-rank regularization is not effective enough to exploit some underlying local structures of tensors for completion. Thus, to make up for the limitation of low-rank constraint, we impose the total variation (TV) regularization to our model. There are also previous methods to combine low-rank regular terms with TV regular terms [14]. Considering that HSI is different from common RGB image, its smoothness in spectral segment direction is also significant. We use 3D TV in our model to maintain local smoothness of hyperspectral image spatial domain and spectral domain at the same time, and weight the TV in the three directions.

In this paper, we propose a hyperspectral images super-resolution model based on low-rank tensor Tucker decomposition and weighted 3D TV for low-resolution hyperspectral image and high-resolution multispectral image fusion. We combine tensor low-rank regularization with weighted 3D TV to recover the main structural information of HSIs while maintaining the edge and local smoothness of data. We use the log-sum norm to perform low-rank estimates on the factor matrices of Tucker factorization rather than the traditional nonconvex norm. Extensive experiments show that our method outperforms some state-of-the-art methods on public HSI dataset.

2 Notation and Preliminaries

A tensor is a multidimensional array with order greater than or equal to three. In this paper, scalars and vectors are regarded as tensors of order zero and order one, they are written in lowercase letters $(a, b, c, \cdots)$ and bold lowercase letters $(\mathbf{a}, \mathbf{b}, \mathbf{c}, \cdots)$, respectively. Matrixes are regarded as tensors of order two and are written in uppercase letters $(A, B, C, \cdots)$. Tensors of order three and above are denoted as $\mathcal{A} \in \mathbb{R}^{I_1 \times I_2 \times \cdots \times I_N}$. and the elements therein are denoted as $a_{i_1, i_2, \cdots, i_N}$. The Frobenius norm of tensor $\mathcal{A}$ is defined as:

$$\|\mathcal{A}\|_F = \sqrt{\langle \mathcal{A}, \mathcal{A} \rangle}, \tag{1}$$

where $\langle \mathcal{A}, \mathcal{A} \rangle$ denotes the inner product of two tensors of the same size, which is defined as:

$$\langle \mathcal{A}, \mathcal{B} \rangle := \sum_{i_1, i_2, \ldots, i_K} a_{i_1, i_2, \ldots, i_K} \cdot b_{i_1, i_2, \ldots, i_K}. \tag{2}$$

A tensor $\mathcal{A} \in \mathbb{R}^{I_1 \times I_2 \times \cdots \times I_N}$ and a matrix $B \in \mathbb{R}^{J \times I_n}$'s mode-n product is:

$$\mathcal{C} = \mathcal{A} \times_n B \in \mathbb{R}^{I_1 \times \cdots I_{n-1} \times J \times I_{n+1} \cdots \times I_n}, \tag{3}$$

each element in $\mathcal{C}$ operates as follows

$$c_{i_1, \cdots, i_{n-1}, j, i_{n+1}, \cdots N} = \sum_{i_n = 1}^{I_n} (a_{i_1, \cdots, a_N} b_{j, i_n}). \tag{4}$$

For an N-order tensor $\mathcal{A} \in \mathbb{R}^{I_1 \times I_2 \times \cdots \times I_N}$, its Tucker decomposition can be written as:

$$\begin{aligned} \mathcal{A} &= \mathcal{G} \times_1 U^{(1)} \times_2 U^{(2)} \cdots \times_N U^{(N)} \\ &= \sum_{s_1 = 1}^{S_1} \times \sum_{s_N}^{S_N} g_{s_1 s_2 \ldots s_N} (u_{s_1}^{(1)} \circ u_{s_2}^{(2)} \circ \cdots \circ u_{s_N}^{(N)}), \end{aligned} \tag{5}$$

where $\mathcal{G} \in \mathbb{R}^{S_1 \times S_2 \times \cdots \times S_N}$ represents the core tensor by Tucker decomposition of the tensor and $\left\{U^{(n)}\right\}_{n=1}^{N} \in \mathbb{R}^{I_n \times \mathrm{S}_n}$ represent the decomposition factors . g and u denote the elements of $\mathcal{G}$ and $U^{(n)}$, respectively.

3 Method

3.1 Proposed Model

We introduce the proposed the HSIs super-resolution method based on low-rank tensor Tucker decomposition and weighted 3D TV in this section. We obtain the desired HR-HSI by fusing LR-HSI and HR-MSI and combining appropriate regularization terms. We represent the desired HR-HSI as $\mathcal{X} \in \mathbb{R}^{W \times H \times S}$, where W, H and S denote the size of width, height and spectral direction. The corresponding LR-HSI and HR-MSI are denoted as $\mathcal{Y} \in \mathbb{R}^{w \times h \times S}$ and $\mathcal{Z} \in \mathbb{R}^{W \times H \times s}$, respectively, where w<W, h<H and s<S. We can formulate our super-resolution model as:

$$\min_{\mathcal{X}} \lambda_1 \|\mathcal{Y} - DS\mathcal{X}\|_F^2 + \lambda_2 \|\mathcal{Z} - \mathcal{X} \times_3 R\|_F^2 + \lambda P(\mathcal{X}), \tag{6}$$

where DS denote the blur and downsampling operations, R denote a transformation matrix mapping the HR-HSI image $\mathcal{Z}$ to its corresponding HR-MSI, and λ_2 and λ_2 represent the weighted coefficients. Based on this Eq. 6, the rationality and validity of the regularization $P(\mathcal{X})$ becomes the key to accurately recover HR-HSI $\mathcal{X}$. Based on the previous discussion, we introduce low-rank tensor Tucker decomposition and weighted 3D TV into the proposed Hyperspectral images super-resolution model.

Low-Rank Tensor Tucker Decomposition. Hyperspectral image naturally has good low-rank characteristic because of the similarity between many spectral segments. We use tensor Tucker Decomposition to exploit the low-rank characteristic. Tucker decomposition decomposes the target tensor $\mathcal{X}$ into the product of N matrixes and a core tensor,

$$\begin{aligned} &\min_{\mathcal{X}} \frac{\lambda_4}{N} \sum_{n=1}^{N} \left\| U^{(n)} \right\|_* + \lambda_5 \|\mathcal{G}\|_F^2 \\ &s.t. \mathcal{X} = \mathcal{G} \times_1 U^{(1)} \times_2 U^{(2)} \cdots \times_N U^{(N)} \end{aligned} \tag{7}$$

where $\mathcal{G}$ and $U^{(n)}$ denote the core tensor and decomposition factors of different directions obtained by the Tucker decomposition of $\mathcal{X}$, respectively. The λ_4 and λ_5 represent the corresponding regularization coefficients of $\mathcal{G}$ and $U^{(n)}$. N is the order of tensor $\mathcal{X}$. The $\|\cdot\|_*$ denotes the low-rank constraint.

Many studies have shown that the log-sum norm is very reasonable and easy to calculate for relaxation of low-rank kernel norm. In our model, we use log-sum norm as a low-rank constraint for decomposition factors of the Tucker factorization. For a given matrix A, log-sum norm is defined as:

$$P_{ls}^*(A) = \sum\nolimits_m (\log(\sigma_m(A) + \varepsilon) - \log(\varepsilon))/(-\log(\varepsilon)), \tag{8}$$

where ε is positive constants of small values and $\sigma_m(A)$ is the mth singular value of matrix A. Based on log-sum nrom, the Eq. 7 can be rewritten as:

$$\begin{aligned} &\min_{\mathcal{X}} \frac{\lambda_4}{N} \sum_{n=1}^{N} \left\| U^{(n)} \right\|_{LS} + \lambda_5 \|\mathcal{G}\|_F^2 \\ &s.t. \mathcal{X} = \mathcal{G} \times_1 U^{(1)} \times_2 U^{(2)} \cdots \times_N U^{(N)} \end{aligned} \tag{9}$$

Weighted 3D TV. Total Variation (TV) recognized as effective in maintaining the local smoothness of an image. Since the target data HSI X has three dimensions of length, width and spectral segment, we choose a more appropriate 3D TV and weighted the three directions respectively. The weighted 3D TV is defined as:

$$3DTV(\mathcal{A}) = \sum_{ijk} w_1 \left|x_{ijk} - x_{ij,k-1}\right| + w_2 \left|x_{ijk} - x_{i,j-1,k}\right| + w_3 \left|x_{ijk} - x_{i-1,j,k}\right| \tag{10}$$

where i, j, k denote the location of pixel in tensor $\mathcal{A}$, and the w_1, w_2 and w_2 denote the weight of three directions respectively.

After adding low-rank regularization and weighted 3D TV regularization terms successively, our super-resolution model is as follows:

$$\begin{aligned} &\min_{X} \lambda_1 \left\|\mathcal{Y} - DS\mathcal{X}\right\|_F^2 + \lambda_2 \left\|\mathcal{Z} - \mathcal{X}\times_3 R\right\|_F^2 + \lambda_3 TV\left(\mathcal{X}\right) \\ &\quad + \frac{\lambda_4}{N}\sum_{n=1}^{N}\left\|U^{(n)}\right\|_{LS} + \lambda_5 \left\|\mathcal{G}\right\|_F^2 \\ &s.t. \mathcal{X} = \mathcal{G}\times_1 U^{(1)}\times_2 U^{(2)}\cdots\times_N U^{(N)} \end{aligned} \tag{11}$$

3.2 Optimization

We use alternating direction method of multipliers (ADMM) algorithm to solve Eq. 11. First, we introduce some auxiliary variables to split the interdependencies of the terms in Eq. 11, then we can obtain:

$$\begin{aligned} &\min_{X} \lambda_1 \left\|\mathcal{Y} - DS\mathcal{X}\right\|_F^2 + \lambda_2 \left\|\mathcal{Z} - \mathcal{X}\times_3 R\right\|_F^2 + \lambda_3 TV\left(\mathcal{X}\right) \\ &\quad + \frac{\lambda_4}{N}\sum_{n=1}^{N}\left\|U^{(n)}\right\|_{LS} + \lambda_5 \left\|\mathcal{G}\right\|_F^2 \\ &s.t. \mathcal{F} = G_w\left(x\right), \mathcal{T} = \mathcal{X}, V^{(n)} = U^{(n)}, \mathcal{T} = \mathcal{G}\times_1 V^{(1)}\times_2 V^{(2)}\cdots\times_N V^{(N)} \end{aligned} \tag{12}$$

The Lagrangian function of the cost function Eq. 12 can be written as:

$$\begin{aligned} &\mathcal{L}\left(\mathcal{X}, \mathcal{F}, U^{(n)}, \mathcal{G}, \mathcal{T}, \mathcal{W}_1, \cdots, \mathcal{W}_4\right) \\ &= \lambda_1 \left\|\mathcal{Y} - DS\mathcal{X}\right\|_F^2 + \lambda_2 \left\|\mathcal{Z} - \mathcal{X}\times_3 R\right\|_F^2 + \frac{w_1}{2}\left\|\mathcal{X} - \mathcal{T}\right\|_F^2 + \left\langle \mathcal{X} - \mathcal{T}, \mathcal{W}_1\right\rangle \\ &+ \lambda_3 \left\|\mathcal{F}\right\|_1 + \frac{w_2}{2}\left\|\mathcal{F} - G_w\left(x\right)\right\|_F^2 + \left\langle \mathcal{F} - G_w\left(x\right), \mathcal{W}_2\right\rangle \\ &+ \frac{\lambda_4}{N}\sum_{n=1}^{N}\left\|U^{(n)}\right\|_{LS} + \frac{w_3}{2}\sum_{n=1}^{N}\left(\left\|U^{(n)} - V^{(n)}\right\|_F^2 + \left\langle U^{(n)} - V^{(n)}, \mathcal{W}_3\right\rangle\right) \\ &+ \lambda_5 \left\|\mathcal{G}\right\|_F^2 + \frac{w_4}{2}\left\|\mathcal{T} - \mathcal{G}\times_1 V^{(1)}\times_2 V^{(2)}\cdots\times_N V^{(N)}\right\|_F^2 \\ &+ \left\langle \mathcal{T} - \mathcal{G}\times_1 V^{(1)}\times_2 V^{(2)}\cdots\times_N V^{(N)}\right\rangle \end{aligned} \tag{13}$$

where $\mathcal{W}_1, \cdots, \mathcal{W}_4$ are the Lagrangian multipliers, and $w_1, \cdots, w_4$ are the positive penalty parameters. Then, we divide Eq. 13 into 6 subproblems and solve them respectively.

Update $\mathcal{X}$. According to ADMM, we fix the other variables in Eq. 13 and extract the terms containing $\mathcal{X}$. We can obtain the objective solution formula for $\mathcal{X}$:

$$\begin{aligned} \min_{\mathcal{X}} \lambda_1 \left\| \mathcal{Y} - DS\mathcal{X} \right\|_F^2 + \lambda_2 \left\| \mathcal{Z} - \mathcal{X} \times_3 R \right\|_F^2 + \frac{w_1}{2} \left\| \mathcal{X} - \mathcal{T} \right\|_F^2 + \langle \mathcal{X} - \mathcal{T}, \mathcal{W}_1 \rangle \\ + \frac{w_2}{2} \left\| \mathcal{F} - G_w\left(x\right) \right\|_F^2 + \langle \mathcal{F} - G_w\left(x\right), \mathcal{W}_2 \rangle \end{aligned} \tag{14}$$

The partial derivative of Eq. 14 with respect to $\mathcal{X}$ is as follows:

$$\begin{aligned} &2\lambda_1 (DS)^T DS\mathcal{X} + w_1\mathcal{X} + w_2 G_w\prime G_w\left(x\right) + 2\lambda_2\mathcal{X} \\ &\quad = 2\lambda_1 (DS)^T \mathcal{Y} + w_1\mathcal{T} - \mathcal{W}_1 + w_2 G_w\prime\mathcal{F} + \mathcal{W}_2 G_w\prime + 2\lambda_2 \left(\mathcal{Z} \times_3 R^T\right), \end{aligned} \tag{15}$$

where $(DS)^T$ denotes the transposes of DS.

Update $\mathcal{F}$. We update $\mathcal{F}$ by minimizing the follow function:

$$\min_{\mathcal{F}} \lambda_3 \|\mathcal{F}\|_1 + \frac{w_2}{2} \left\| \mathcal{F} - G_w\left(x\right) \right\|_F^2 + \langle \mathcal{F} - G_w\left(x\right), \mathcal{W}_2 \rangle \tag{16}$$

The solution of $\mathcal{F}$ is:

$$\mathcal{F} = fold_n \left[soft_{\lambda_3 \alpha_n / w_2} \left(G_w\left(x_{(n)}\right) - \frac{\mathcal{W}_{2(n)}}{w_2} \right) \right] \tag{17}$$

where $soft$ denotes the soft-thresholding operator, and $fold_n$ denote the matrix expansion in mode n direction.

Update $\mathcal{T}$. We update $\mathcal{T}$ by minimizing:

$$\begin{aligned} \min_{\mathcal{T}} \frac{w_1}{2} \left\| \mathcal{X} - \mathcal{T} \right\|_F^2 + \langle \mathcal{X} - \mathcal{T}, \mathcal{W}_1 \rangle + \frac{w_4}{2} \left\| \mathcal{T} - \mathcal{G} \times_1 V^{(1)} \times_2 V^{(2)} \cdots \times_N V^{(N)} \right\|_F^2 \\ + \left\langle \mathcal{T} - \mathcal{G} \times_1 V^{(1)} \times_2 V^{(2)} \cdots \times_N V^{(N)}, \mathcal{W}_4 \right\rangle \end{aligned} \tag{18}$$

The solution of Eq. 18 is:

$$\mathcal{T} = \frac{w_4 C + w_1 \mathcal{X} + \mathcal{W}_1 - \mathcal{W}_4}{w_1 + w_4}, \tag{19}$$

where $C = \mathcal{G} \times_1 V^{(1)} \times_2 V^{(2)} \cdots \times_N V^{(N)}$.

Update $U^{(n)}$. We update $U^{(n)}$ by minimizing:

$$\min_{U^{(n)}} \frac{\lambda_4}{N} \sum_{n=1}^{N} \left\| U^{(n)} \right\|_{LS} + \frac{w_3}{2} \sum_{n=1}^{N} \left\| U^{(n)} - V^{(n)} \right\|_F^2 + \sum_{n=1}^{N} \left\langle U^{(n)} - V^{(n)}, \mathcal{W}_3 \right\rangle. \tag{20}$$

The Eq. 20 is equivalent to the following:

$$\min_{U^{(n)}} \alpha_k \left\| U^{(n)} \right\|_{LS} + \frac{1}{2} \left\| U^{(n)} - V^{(n)} + \frac{\mathcal{W}_3}{w_3} \right\|_F^2, \alpha_k = \frac{\lambda_4}{N w_3}. \tag{21}$$

The solution of $U^{(n)}$ is:

$$U^{(n)} = fold\left(V_1 \sum\nolimits_{\alpha_k} V_2^T \right), \tag{22}$$

where $V_1 diag\left(\sigma_1, \sigma_2, \cdots \sigma_n\right)$ is the singular value decomposition of matrix $V^{(n)} + \frac{\mathcal{W}_3}{w_3}$, and $\sum_{\alpha_k}$ denotes $diag\left(D_{\alpha_k,\varepsilon}\left(\sigma_1\right), D_{\alpha_k,\varepsilon}\left(\sigma_2\right), \cdots D_{\alpha_k,\varepsilon}\left(\sigma_n\right)\right)$.

Update $V^{(n)}$. We update $V^{(n)}$ by minimizing:

$$\min_{V^{(n)}} \frac{w_3}{2} \sum_{n=1}^{N} \left(\left\| U^{(n)} - V^{(n)} \right\|_F^2 + \left\langle U^{(n)} - V^{(n)}, \mathcal{W}_3 \right\rangle \right)$$
$$+ \frac{w_4}{2} \left\| \mathcal{T} - \mathcal{G} \times_1 V^{(1)} \times_2 V^{(2)} \cdots \times_N V^{(N)} \right\|_F^2 + \left\langle \mathcal{T} - \mathcal{G} \times_1 V^{(1)} \times_2 V^{(2)} \cdots \times_N V^{(N)}, \mathcal{W}_4 \right\rangle \tag{23}$$

The solution of $V^{(n)}$ is:

$$V^{(n)} = \left(-\mathcal{W}_3 + w_3 U^{(n)} + \left(\mathcal{W}_4 + w_4 \mathcal{T}\right) V^{(-n)} G_{(n)}^T \right) \left(w_4 I + w_4 G_n V^{(-n)T} V^{(-n)} G_{(n)}^T \right)^{-1} \tag{24}$$

Update $\mathcal{G}$. We update $\mathcal{G}$ by:

$$\begin{aligned} \min_{\mathcal{G}} \lambda_5 \left\| \mathcal{G} \right\|_F^2 + \frac{w_4}{2} \left\| \mathcal{T} - \mathcal{G} \times_1 V^{(1)} \times_2 V^{(2)} \cdots \times_N V^{(N)} \right\|_F^2 \\ + \left\langle \mathcal{T} - \mathcal{G} \times_1 V^{(1)} \times_2 V^{(2)} \cdots \times_N V^{(N)}, \mathcal{W}_4 \right\rangle. \end{aligned} \tag{25}$$

The solution of $\mathcal{G}$ is:

$$vec\left(\mathcal{G}\right) = \left[V^{(-n)T} V^{(-n)} \otimes w_4 V^{(n)T} V^{(n)} + \lambda_5 I \right]^{-1} vec\left(V^{(n)T} \left(\mathcal{W}_4 + w_4 \mathcal{T}\right) V^{(-n)} \right), \tag{26}$$

where $vec()$ denotes the vectorization operation.

4 Experiments

4.1 Experimental Settings

In this section, we evaluate the proposed method qualitatively and quantitatively on a publicly available hyperspectral dataset CAVE [20]. We obtain LR images by blurring and downsampling the original HR images and use HR images as ground truth. In our experiments, we chose the popular Gaussian kernel (size = 7 × 7, mean = 0, standard deviation = 2) as the blurring kernel. We applied our method to the LR images and evaluated whether our method can successfully recover the original high-resolution images.

4.2 Quantitative Comparison

We use three quantitative picture quality indices to evaluate the quality of reconstructed image including the peak signal to noise ratio (PSNR) and the spectral angle mapper (SAM). The PSNR (dB) evaluates the similarity between the reconstructed HR image and the original HR image based on mean squared error. Additionally, we also use the SAM (in rad) calculates the average angle between spectrum vectors of the target HSI and the reference one across all spatial position.

Table 1. Super-resolution reconstruction quantitative results (PSNR and SAM) by different methods on CAVE database

CAVE database								
Methods	CSU		NLSTF		LTTR		Ours	
Indices	PSNR	SAM	PSNR	SAM	PSNR	SAM	PSNR	SAM
Balloons	42.1584	4.6532	43.1864	2.7211	50.0220	2.2487	**50.9525**	**1.3364**
Cd	32.3485	6.0593	37.0761	5.0165	41.1704	3.4161	**43.7338**	**2.8364**
Peppers	42.8925	7.7348	39.3428	5.3400	47.9659	4.0383	**48.7912**	**1.6518**
Jelly	37.5831	7.0011	32.6401	6.4434	41.8461	3.9774	**42.3775**	**2.0880**
Paints	37.3632	6.3599	32.8509	4.7587	42.6135	3.2111	**42.8550**	**2.1355**
Photos	43.2328	9.7382	39.6761	8.6356	45.9969	7.0373	**47.9906**	**2.9348**
Stuffed	36.7863	13.5792	38.3248	5.7997	47.2448	4.5514	**47.4246**	**2.3493**

Table 1 shows part of the quantitative results of the CAVE database. We compare the proposed method with state-of-art methods for HSI super-resolution, including coupled spectral unmixing (CSU) [10], nonlocal sparse tensor factorization method (NLSTF) [4], and low tensor train rank method [5,19]. The proposed method outperforms other competing ones in terms of the PSNR and the SAM (in rad) indices. Lower SAM value indicates that the proposed method can recover spectral information more accurately.

4.3 Visual Quality Comparison

In order to compare the performance of our visually with other methods, we chose a band of four scenes from the database of CAVE presented in Fig. 1. From the detail part in Fig. 1 that we can find that our method can suppress the noise in the smooth region well while retaining a sharper edge. This is due to the reasonable combination of low-rank regularization and 3D TV regularization terms.

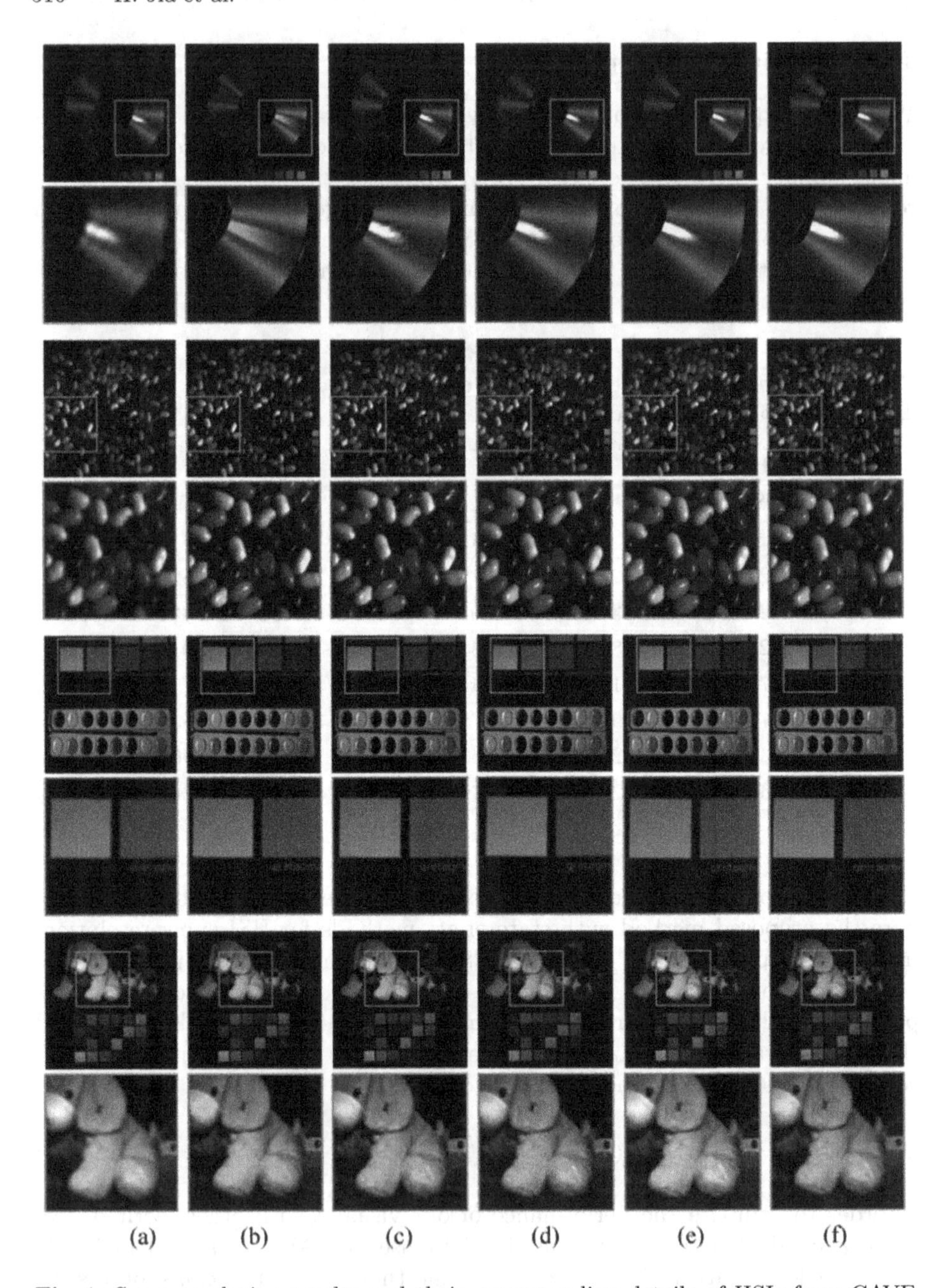

Fig. 1. Super-resolution results and their corresponding details of HSIs from CAVE database: (a) LR, (b) CSU, (c) NLSTF, (d) LTTR, (e) Ours, (f) GT.

5 Conclusion

In this paper, we proposed a new super-resolution method for HSIs based on low-rank tensor Tucker Decomposition and weighted 3D TV. The tensor Tucker Decomposition fully utilized the low-rank characteristic to reconstruct the main structural information of HSI. Meanwhile, We use different weights for 3D TV in each direction according to the characteristics of HSI image, which well maintains the local smoothness of data in spatial domain and spectral domain. Extensive experiments show that the proposed method outperforms some state-of-the-art methods on public HSI dataset.

References

1. Bioucas-Dias, J.M., Plaza, A., Camps-Valls, G., Scheunders, P., Nasrabadi, N., Chanussot, J.: Hyperspectral remote sensing data analysis and future challenges. IEEE Geosci. Remote Sens. Mag. **1**(2), 6–36 (2013). https://doi.org/10.1109/MGRS.2013.2244672
2. Carroll, J.D., Chang, J.J.: Analysis of individual differences in multidimensional scaling via an n-way generalization of "Eckart-Young" decomposition. Psychometrika **35**(3), 283–319 (1970). https://doi.org/10.1007/BF02310791
3. Chen, W., et al.: Tensor low-rank reconstruction for semantic segmentation. In: Vedaldi, A., Bischof, H., Brox, T., Frahm, J.-M. (eds.) ECCV 2020. LNCS, vol. 12362, pp. 52–69. Springer, Cham (2020). https://doi.org/10.1007/978-3-030-58520-4_4
4. Dian, R., Fang, L., Li, S.: Hyperspectral image super-resolution via non-local sparse tensor factorization. In: 2017 IEEE Conference on Computer Vision and Pattern Recognition (CVPR), pp. 3862–3871 (2017). https://doi.org/10.1109/CVPR.2017.411
5. Dian, R., Li, S., Fang, L.: Learning a low tensor-train rank representation for hyperspectral image super-resolution. IEEE Trans. Neural Networks and Learning Systems **30**(9), 2672–2683 (2019). https://doi.org/10.1109/TNNLS.2018.2885616
6. Dong, W., et al.: Hyperspectral image super-resolution via non-negative structured sparse representation. IEEE Trans. Image Process. **25**(5), 2337–2352 (2016). https://doi.org/10.1109/TIP.2016.2542360
7. Fang, L., He, N., Li, S., Plaza, A.J., Plaza, J.: A new spatial–spectral feature extraction method for hyperspectral images using local covariance matrix representation. IEEE Trans. Geosci. Remote Sens. **56**(6), 3534–3546 (2018). https://doi.org/10.1109/TGRS.2018.2801387
8. Gu, S., Zhang, L., Zuo, W., Feng, X.: Weighted nuclear norm minimization with application to image denoising. In: 2014 IEEE Conference on Computer Vision and Pattern Recognition, pp. 2862–2869 (2014). https://doi.org/10.1109/CVPR.2014.366
9. Kolda, T.G., Bader, B.W.: Tensor decompositions and applications. SIAM Rev. **51**(3), 455–500 (2009). https://doi.org/10.1137/07070111X
10. Lanaras, C., Baltsavias, E., Schindler, K.: Hyperspectral super-resolution by coupled spectral unmixing. In: 2015 IEEE International Conference on Computer Vision (ICCV), pp. 3586–3594 (2015). https://doi.org/10.1109/ICCV.2015.409

11. Li, S., Dian, R., Fang, L., Bioucas-Dias, J.M.: Fusing hyperspectral and multispectral images via coupled sparse tensor factorization. IEEE Trans. Image Process. **27**(8), 4118–4130 (2018). https://doi.org/10.1109/TIP.2018.2836307
12. Liu, J., Musialski, P., Wonka, P., Ye, J.: Tensor completion for estimating missing values in visual data. IEEE Trans. Pattern Anal. Mach. Intell. **35**(1), 208–220 (2013). https://doi.org/10.1109/TPAMI.2012.39
13. Prevost, C., Usevich, K., Comon, P., Brie, D.: Coupled tensor low-rank multilinear approximation for hyperspectral super-resolution. In: Proceedings IEEE International Conference on Acoustics, Speech Signal Process, vol. 2019-May, pp. 5536–5540 (2019). https://doi.org/10.1109/ICASSP.2019.8683619
14. Shi, F., Cheng, J., Wang, L., Yap, P.T., Shen, D.: LRTV: MR image super-resolution with low-rank and total variation regularizations. IEEE Trans. Med. Imaging **34**(12), 2459–2466 (2015). https://doi.org/10.1109/TMI.2015.2437894
15. Tucker, L.R.: Some mathematical notes on three-mode factor analysis. Psychometrika **31**(3), 279–311 (1966). https://doi.org/10.1007/BF02289464
16. Wang, Q., He, X., Li, X.: Locality and structure regularized low rank representation for hyperspectral image classification. IEEE Trans. Geosci. Remote Sens. **57**(2), 911–923 (2018)
17. Wang, Q., Lin, J., Yuan, Y.: Salient band selection for hyperspectral image classification via manifold ranking. IEEE Trans. Neural Netw. Learn. Syst. **27**(6), 1279–1289 (2016). https://doi.org/10.1109/TNNLS.2015.2477537
18. Xie, Q., Zhao, Q., Meng, D., Xu, Z.: Kronecker-basis-representation based tensor sparsity and its applications to tensor recovery. IEEE Trans. Pattern Anal. Mach. Intell. **40**(8), 1888–1902 (2018). https://doi.org/10.1109/TPAMI.2017.2734888
19. Xu, Y., Hao, R., Yin, W., Su, Z.: Inverse Probl. Imaging. https://doi.org/10.3934/ipi.2015.9.601
20. Yasuma, F., Mitsunaga, T., Iso, D., Nayar, S.K.: Generalized assorted pixel camera: postcapture control of resolution, dynamic range, and spectrum. IEEE Trans. Image Process. **19**(9), 2241–2253 (2010). https://doi.org/10.1109/TIP.2010.2046811
21. Yin, X.X., Hadjiloucas, S., Chen, J.H., Zhang, Y., Wu, J.L., Su, M.Y.: Tensor based multichannel reconstruction for breast tumours identification from DCE-MRIs. PLoS ONE **12**(3), e0172111 (2017). https://doi.org/10.1371/journal.pone.0172111

Hybrid Deep Convolutional Network for Face Alignment and Head Pose Estimation

Zhiyong Wang[1,2], Jingjing Liu[1,2], and Honghai Liu[2](✉)

[1] The State Key Laboratory of Mechanical System and Vibration, Shanghai Jiao Tong University, Shanghai, China
yzwang_sjtu@sjtu.edu.cn

[2] The State Key Laboratory of Robotics and Systems, Harbin Institute of Technology Shenzhen, Shenzhen, China
honghai.liu@icloud.com

Abstract. Face alignment has been an important focus of vision research because it is the most fundamental step in face analysis, reconstruction, and applications of emotion and attention. However, face alignment still suffers from some problems, such as lack of stability and poor performance in practical applications due to occlusion, illumination, and high training costs. This paper proposes a Dual-Task Hybrid Deep Convolutional Network (DHDCN) to estimate head pose and facial landmark locations simultaneously. By connecting the multi-level features, the local features and global features can be effectively fused. Features common to both tasks are learned in the initial stages of the network, and later stages will train the two tasks independently. Although the results have some gaps compared to the state-of-the-art results, it also demonstrates the feasibility and potential of learning both tasks simultaneously.

Keywords: Face alignment · Head pose · Multi-level feature

1 Introduction

Face alignment, also known as facial feature point positioning, is a well-known field of vision research because it is the most basic and part of face analysis and the most direct description tool for facial features and states [1–3]. This problem has always been highly concerned by the vision community, because it plays an important role in face editing, fatigue detection, expression tracking, head pose estimation, etc. [4–6]. Over the past few years, technology has advanced a lot, and many face alignment algorithms have achieved good results when faces are complete and clear [7,8]. However, still stable face key point detection still faces many challenges in practical applications. When there are occlusions, complex backgrounds, large-angle head movements, and poor lighting conditions,

Supported by the National Natural Science Foundation of China (Grant Nos. 61733011).

H. Liu et al. (Eds.): ICIRA 2022, LNAI 13456, pp. 513–522, 2022.
https://doi.org/10.1007/978-3-031-13822-5_46

the accuracy will drop sharply, resulting in false identification and localization. Head pose estimation is an important issue that needs to be considered in practical applications of face alignment, because there is no guarantee that the face can always face the visual device in an unconstrained environment in a natural state. At the same time, head pose estimation itself also plays an important role in human gaze estimation and emotion analysis, because head pose is not only an important way to express intention, but also the basic response of attention [9,10]. Many applications, such as 3D gaze estimation under free head motion, rely heavily on the accuracy of head pose estimation. However, due to the performance fluctuation of face alignment, the accuracy of head pose estimation is not stable in practical applications [11,12]. The extreme head pose in turn increases the difficulty of face alignment task. Therefore, there is an inherent relationship between the two tasks that restricts and promotes each other.

In an application that requires multiple features, each feature is calculated using an independent network or model, which often results in a serious waste of computing resources, and also affects the overall speed, resulting in poor real-time performance. Integrating multiple tasks into a network and simplifying the network structure is an effective way to solve this problem. Multi-task learning can train multiple interrelated tasks at the same time, and the effect is better than single-task learning [13,14]. Therefore, there are many studies on the task of simultaneously estimating the location and visibility of facial landmarks, head pose, gender, etc. [15]. Although their results have made some progress in model accuracy and simplification, there are also problems such as long training time and poor accuracy on some tasks. In practical applications, including many data sets, it is difficult to mark head poses by artificial or sensor devices, and most of them are calculated by using face key points through face models. Therefore, the location of key points is the basis for head pose estimation. As long as the key points are accurate, the estimation of the head pose is more reliable. However, the large-scale irregular head pose has a decisive effect on the spatial distribution of the facial feature points, and the incorrect pose corresponds to the presentation of different faces in the image. Therefore, the determination of the head pose and the accurate estimation of the feature points can play a promoting role.

Obviously, if the face feature points and head pose can be learned at the same time, the accuracy of the two tasks can theoretically be improved at the same time, and the stability in practical applications can be increased. However, the premise of most face feature localization is that the face needs to be normalized first, that is, to eliminate the rotation of the face, get a frontal face. However, when performing head pose estimation, the original image is not normalized. Therefore, to perform the learning of these two tasks at the same time, it is necessary to make the facial feature points insensitive to the extreme head posture, that is, it can be stably positioned without normalizing.

Based on the above principles, we propose a hybrid deep convolutional network for learning the feature point location and head pose at the same time. By connecting the multi-level features, the local features and global features can be effectively fused, so that feature point localization and head pose estimation

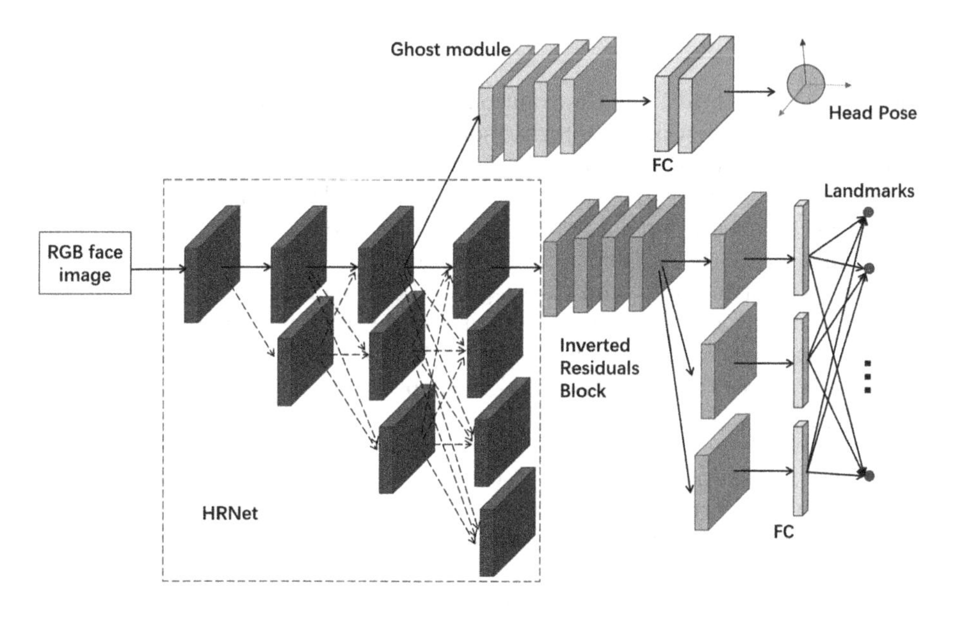

Fig. 1. The structure of the DHDCN.

can be achieved. Two tasks are carried out simultaneously. Features common to both tasks are learned in the initial stages of the network, and later stages will train the two tasks independently.

In summary, the 2 main contributions of this work are as follows:

1) A framework for collaborative learning of associative tasks is proposed, which performs facial feature points and head pose estimation at the same time, and reduces the sensitivity of facial feature points to severe head poses through the interaction of the two task losses. 2) A multi-level feature hybrid network is proposed. The serial part learns common basic features, and the parallel part learns the unique features of each task, which simplifies the calling of multiple modules in face analysis and improves real-time and accuracy.

The rest of this paper is organized as follows: Sect. 2 describes the network and training strategy proposed in this paper in detail. Section 3 provides experimental details and compares the proposed method with the state-of-the-art. Finally, we conclude this work.

2 Method

We propose a Dual-Task Hybrid Deep Convolutional Network (DHDCN) to accurately estimate head pose and facial landmark locations simultaneously. Drawing on the PFLD network structure [16] and combining HRNet [17] and Ghost block [18], the network can effectively integrate multi-level and multi-scale features. The entire structure of our method is shown in Fig. 1. We introduce the proposed method in four parts: network design, training method, and implementation details.

2.1 Network

The Dual-Task Hybrid Deep Convolutional Network (DHDCN) consists of three parts, of which the first part is a high-resolution network (HRNet) [17] containing 4 stages, and the second part is the branch for head pose estimation taking the input the features from HRNet stage3 as input, and the third part is the branch for locating face feature points taking the features from HRNet stage4 as input. The branch network for head pose estimation We employ a GhostNet which contains 3 Ghost modules, and then regress the three Euler angles through two fully connected layers. The branch network of feature point localization uses the Inverted Residuals module of MobileNetV2 [19], and then transforms the output features into three different scales and then splices them to regress the position of the point.

1) HRNet. HRNet is a high-resolution network [17] capable of maintaining high-resolution representations throughout. It starts from the high-resolution sub-network as the first stage, maintains the high-resolution horizontally unchanged, and gradually increases the high-resolution to low-resolution sub-network vertically, forming and connecting more stages. As the network continues to deepen, the scale of parallel sub-networks becomes smaller and smaller, and the low-resolution expression becomes more and more sufficient. In the whole process, each sub-network exchanges information with all the other sub-networks in parallel to perform multi-scale repeated fusion, thus ensuring that the network can always maintain high resolution. The following Fig. 2 shows the schematic structure of the 2-level sub-network interacting with high-resolution network information.
2) Inverted Residuals module. The structure of MobileNetV2 is shown in Fig. 3. The Inverted Residuals module is an operation that first increases the dimension and then reduces the dimension. "Expansion" is a the convolution layer with kernel size of 1×1, in order to increase the number of channels to obtain more features. Then it selects ReLU6 as the activation function of the first two modules, which is presented as Eq. (1). In the "projection" stage, the Linear module is used to reduce the number of channels, which avoids the relatively large instantaneous loss of low-dimensional information caused by ReLU.

$$y = ReLU6(x) = \min(\max(x,\ 6), 6) \tag{1}$$

3) Ghost module. The feature maps gathered from the traditional convolutional neural network usually have a strong similarity which leading to the a very serious waste of computing resources. So we adopted the Ghost bottleneck was employed as the basic convolution bottleneck. The core idea of this Ghost module is to perform linear operations (LO) on the convolutional feature maps to get more feature maps, and then connects the original convolutional feature maps with the newly generated maps. The principle of the Ghost module is shown in Fig. 4. The every feature map obtained from the typical

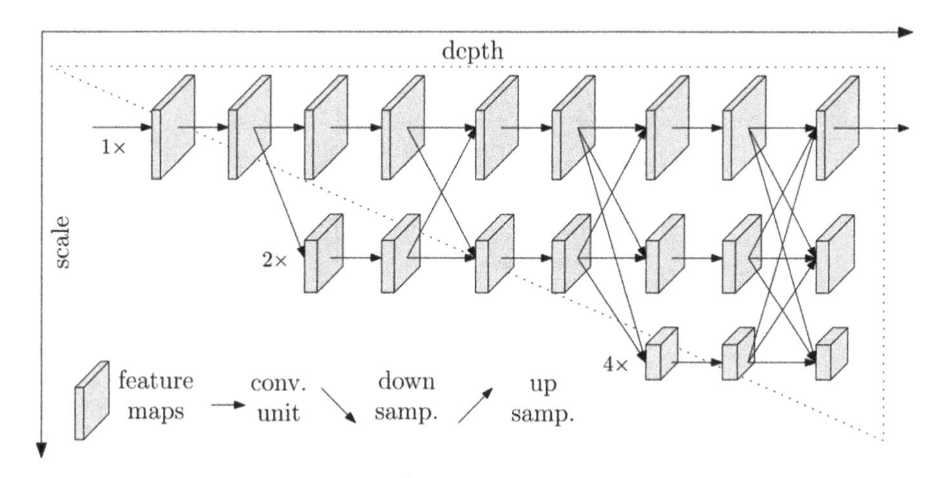

Fig. 2. The structure of HRNet.

convolutional network will be sent to several simple linear operations to correspondingly generate new feature maps. Two types of feature maps will be directly stacked as the output feature map of the module. The Ghost bottleneck consists of two stacked Ghost modules and the batch normalization and ReLU activation are employed between two Ghost modules for better training and nonlinear mapping. The input and output is connected with a shortcut just like the residual block.

2.2 Multi-level and Multi-scale Feature Fusion

Both the head pose estimation and the face feature point estimation need to extract the face pose information from the high-resolution features. Therefore, our network uses HRNet to extract the high-resolution features required by the two tasks. HRnet has the characteristics of multi-scale repeated fusion. By exchanging units across parallel sub-network, the sub-network repeatedly receive information from other parallel sub-network, while ensuring that the network always maintains high resolution. Specifically, the sub-network of a higher level is down-sampled through a convolution with a stride of 2 and a kernel size of 3×3, and communicates with the sub-network of the next level. The lower-level sub-network communicates with the higher-level sub-network through up-sampling. Because head pose estimation relies more on global features, and face feature point estimation expects to obtain smaller local features, we input the features output from the third stage into the head pose estimation branch, and the feature input from the fourth stage. To the facial feature point location branch. In the face feature estimation branch, the feature extraction of multiple Inverted Residuals modules is performed, and the output features are further extracted into multi-scale features through convolutional layers with different

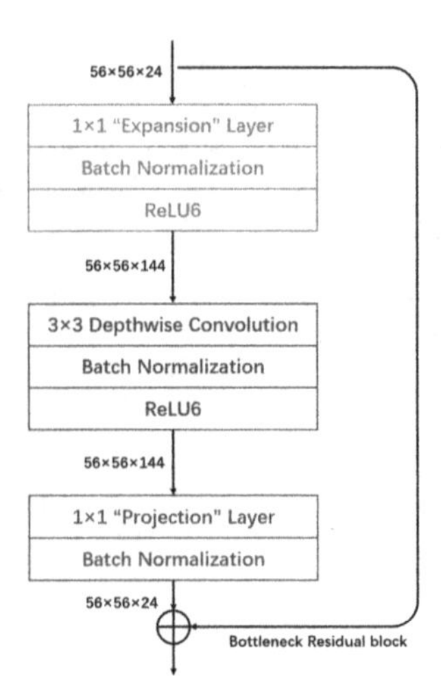

Fig. 3. The structure of MobileNetV2.

kernels (with different sizes). Then it is transformed into 1-dimensional data for splicing, so that for the estimation of features, there is more local information of scale.

2.3 Implementation Details

The input to the network is a 3-channel raw RGB image (without face alignment) of size 112×112. We use HRnet with 4 stages and 4 parallel sub-network as backbone network. Four ghost blocks are used in the head pose estimation branch, and five Inverted Residuals blocks are used in the feature point estimation branch. The initial learning rate is 0.0001. The Adam optimizer was used with the weight decay is 1e−6 and momentum of 0.9. The batch size for both training and testing is 256, and training is performed for 500 epochs. The network is trained and tested on the Window 10 platform, the graphics card is NVIDIA RTX 1080ti, and the PyTorch1.7.0 neural network framework is used.

We use L1 loss for the loss of head pose estimation and L2 loss (MSE) for the loss of feature point estimation. The overall loss follows the loss design in PFLD. For data with a relatively large sample size (such as frontal faces, that is, when the Euler angles of the head pose are relatively small), a small weight is given, and the gradient is When back-propagating, the contribution to model training is smaller; for data with a relatively small sample size (large Euler angles on the head, poor lighting, extreme expressions, blurred pictures, occlusion, and makeup), a relatively small amount is given. Large weights, so that when the gradient is back-propagated, the contribution to the model training is larger. The design of the loss function of the model very cleverly solves the problem of

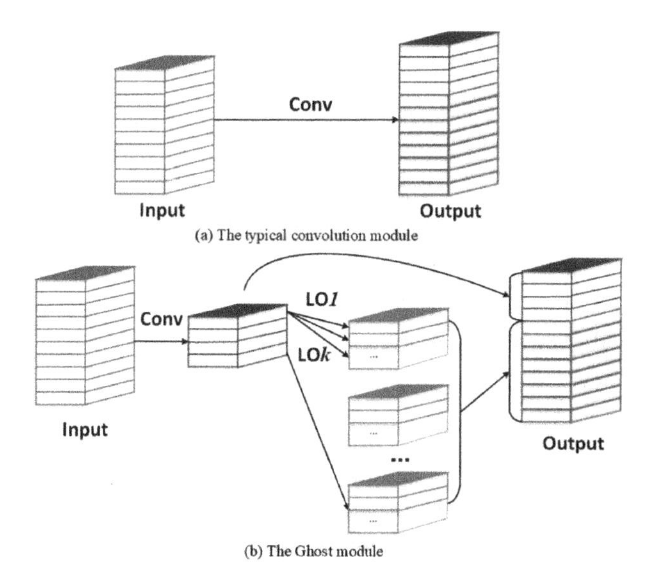

Fig. 4. The principle of the Ghost module. Conv means the convolutional neural network; LO1 means the first linear operation.

unbalanced training samples in various situations. We add the head angle to it, and the loss is calculated as Eq. (2), Eq. (3), and Eq. (5).

$$L_p = \sum_{i}^{n} \sum_{(j=pitch,yaw,roll)} |p_{ij} - p_{ij}^{*}| \tag{2}$$

where p_{ij} and p_{ij}^{*} represent the predicted angle and the ground truth of head pose.

$$L_f = \sum_{i}^{n} \sum_{(j=1)}^{98} ((x_{ij} - x_{ij}^{*})^2 + (y_{ij} - y_{ij}^{*})^2) \tag{3}$$

where x_{ij}, y_{ij}, x_{ij}^{*}, and y_{ij}^{*} represent the predicted location and the ground truth of the feature point.

$$L = w_1 * w_2 * L_f + L_p \tag{4}$$

where w_1 represents the weight for image attribute which includes expression, illumination, make up, occlusion, and blur. The w_2 represents the weight for head pose angle.

3 Experiment and Result

We conducted the experiment to verify the effectiveness of the proposed framework. Comparisons with state-of-the-arts were carried out on the WFLW data set [2]. The performance for head pose was evaluated with the measure of mean

Table 1. Performance comparison of the proposed method and the state-of-the-art methods on WFLW.

Method	NEM (%) ↓	AUC @0.1 ↑	Failure rate (%) ↓	Pose (MAE) ↓
LAB [2]	5.27	0.532	7.56	–
SAN [20]	5.22	0.535	6.32	–
SDFL* [7]	4.25	0.576	2.72	
HIH [21]	4.18	0.597	2.84	–
SLPT [8]	4.14	0.595	2.76	–
Ours method	4.82	0.542	5.36	15.03°

absolute error (MAE). The landmark location error was measured with the normalized mean error (NME), the cumulative error distribution (CED) curve, the area under the curve (AUC), and the failure rate (FR). The NME for the inter-pupil (or inter-ocular) distance is computed as follows:

$$NME = \frac{1}{N} \sum_{(i=1)}^{N} \frac{\|P_i - P_i^*\|_2}{d} \tag{5}$$

where N represents the number of the landmarks; P_i and P_i^* represent the predicted and ground-truth location, respectively; d represent the distance between outer eye corners (inter-ocular).

3.1 Data Set

Wider Facial Landmarks in-the-wild (WFLW) is a data set of faces in natural state that contains 62 life scenes, such as queues, dancing, meetings, family gatherings, etc. It contains a total of 10,000 faces, including 7,500 in the training set and 2,500 in the test machine, and each face is marked with 98 feature points. The advantage of this data set is that the additional attributes of face images in the annotated images include occlusion, pose, makeup, lighting, blur and expression. The way of labeling is, if there is occlusion, big pose, makeup, extreme lighting, blurry picture, exaggerated expression, mark it as 1, otherwise mark it as 0.

3.2 Experiment Result

The results of the proposed method and the other state of the art methods on the WFLW are show in following Table 1. The performance of the proposed method is similar to that of the existing state-of-the-art face alignment methods. Our method achieves NME of 4.90%, Failure rate of 13.2%, and the AUC of 0.595 with the failure threshold at 0.1.

4 Conclusion

Face alignment and head pose estimation have received much attention in the field of face analysis. Aiming at these two related tasks, this paper proposes a Dual-Task Hybrid Deep Convolutional Network (DHDCN). The goal is to reduce the sensitivity of facial landmarks to severe head poses and learn the rough head pose at the same time. In this paper, by referring to the structure of PFLD and combining HRNet, MobileNetV2 and Ghost block [18], the network can effectively integrate multi-level and multi-scale features. After performing the facial pose estimation task, the error of facial feature point localization have some gaps compared to the state-of-the-art face alignment methods. The limitations are summarized as followings: first, the head pose annotation in the WFLW data set comes from the estimation of the face model, which has a large error and is not suitable for accurate estimation of the head pose; second, in practical applications, the accuracy of face detection seriously affects the face alignment and head pose estimation tasks; thirdly, the loss for head pose estimation lack of more efficient designs. In future work, the author will continue to optimize the network and reduce the amount of model parameters on other data set, and study the application of this method in special scenarios, which have no head pose ground truth.

References

1. Kowalski, M., Naruniec, J., Trzcinski, T.: Deep alignment network: a convolutional neural network for robust face alignment. In: Proceedings of the IEEE Conference on Computer Vision and Pattern Recognition Workshops, pp. 88–97 (2017)
2. Wu, W., Qian, C., Yang, S., et al.: Look at boundary: a boundary-aware face alignment algorithm. In: Proceedings of the IEEE Conference on Computer Vision and Pattern Recognition, pp. 2129–2138 (2018)
3. Li, W., et al.: Structured landmark detection via topology-adapting deep graph learning. In: Vedaldi, A., Bischof, H., Brox, T., Frahm, J.-M. (eds.) ECCV 2020. LNCS, vol. 12354, pp. 266–283. Springer, Cham (2020). https://doi.org/10.1007/978-3-030-58545-7_16
4. Li, S., Deng, W., Du, J.P.: Reliable crowdsourcing and deep locality-preserving learning for expression recognition in the wild. In: Proceedings of the IEEE Conference on Computer Vision and Pattern Recognition, pp. 2852–2861 (2017)
5. Wang, Z., Liu, J., et al.: Early screening of autism in toddlers via response-to-instructions protocol. IEEE Trans. Cybern. **52**, 3914–3924 (2022)
6. Wang, Z., Liu, J., He, K., et al.: Screening early children with autism spectrum disorder via response-to-name protocol. IEEE Trans. Ind. Inform. **PP**(99), 1–1 (2019)
7. Lin, C., Zhu, B., Wang, Q., et al.: Structure-coherent deep feature learning for robust face alignment. IEEE Trans. Image Process. **30**, 5313–5326 (2021)
8. Xia, J., Huang, W., Zhang, J., et al.: Sparse local patch transformer for robust face alignment and landmarks inherent relation learning. arXiv preprint arXiv:2203.06541 (2022)

9. Liu, J., Wang, Z., Qin, H., et al.: Free-head pose estimation under low-resolution scenarios. In: 2020 IEEE International Conference on Systems, Man, and Cybernetics (SMC), pp. 2277–2283. IEEE (2020)
10. Ruiz, N., Chong, E., Rehg, J.M.: Fine-grained head pose estimation without keypoints. In: Proceedings of the IEEE Conference on Computer Vision and Pattern Recognition Workshops, pp. 2074–2083 (2018)
11. Albiero, V., Chen, X., Yin, X., et al.: img2pose: face alignment and detection via 6dof, face pose estimation. In: Proceedings of the IEEE/CVF Conference on Computer Vision and Pattern Recognition, pp. 7617–7627 (2021)
12. Yang, T.Y., Chen, Y.T., Lin, Y.Y., et al.: FSA-Net: learning fine-grained structure aggregation for head pose estimation from a single image. In: Proceedings of the IEEE/CVF Conference on Computer Vision and Pattern Recognition, 1087–1096 (2019)
13. Valle, R., Buenaposada, J.M., Baumela, L.: Multi-task head pose estimation in-the-wild. IEEE Trans. Pattern Anal. Mach. Intell. **43**(8), 2874–2881 (2020)
14. Gupta, A., Thakkar, K., Gandhi, V., et al.: Nose, eyes and ears: head pose estimation by locating facial keypoints. In: ICASSP 2019–2019 IEEE International Conference on Acoustics, Speech and Signal Processing (ICASSP), pp. 1977–1981. IEEE (2019)
15. Kumar, A., Alavi, A., Chellappa, R.: Kepler: keypoint and pose estimation of unconstrained faces by learning efficient H-CNN regressors. In: 2017 12th IEEE International Conference on Automatic Face & Gesture Recognition (fg 2017), pp. 258–265. IEEE (2017)
16. Guo, X., Li, S., Yu, J., et al.: PFLD: a practical facial landmark detector. arXiv preprint arXiv:1902.10859 (2019)
17. Wang, J., Sun, K., Cheng, T., et al.: Deep high-resolution representation learning for visual recognition. IEEE Trans. Pattern Anal. Mach. Intell. **43**(10), 3349–3364 (2020)
18. Han, K., Wang, Y., Tian, Q., et al.: Ghostnet: more features from cheap operations. In: Proceedings of the IEEE/CVF Conference on Computer Vision and Pattern Recognition, pp. 1580–1589 (2020)
19. Sandler, M., Howard, A., Zhu, M., et al.: Mobilenetv 2: inverted residuals and linear bottlenecks. In: Proceedings of the IEEE Conference on Computer Vision and Pattern Recognition, pp. 4510–4520 (2018)
20. Dong, X., Yan, Y., Ouyang, W., et al.: Style aggregated network for facial landmark detection. In: Proceedings of the IEEE Conference on Computer Vision and Pattern Recognition, pp. 379–388 (2018)
21. Lan, X., Hu, Q., Cheng, J.: Revisting quantization error in face alignment. In: Proceedings of the IEEE/CVF International Conference on Computer Vision, pp. 1521–1530 (2021)

Research on Multi-model Fusion Algorithm for Image Dehazing Based on Attention Mechanism

Tong Cui[1], Meng Zhang[1(✉)], Silin Ge[2], and Xuhao Chen[2]

[1] College of Artificial Intelligence, Shenyang Aerospace University, Shenyang 110136, China
mengzhang@sau.edu.cn

[2] College of Automation, Shenyang Aerospace University, Shenyang 110136, China

Abstract. In recent years, the researchers of image dehazing mainly focused on deep learning algorithms. However, due to the defective network structure, and inadequate feature extraction, the deep learning algorithm still has many problems to be solved. In this paper, we fuse the physical models including haze imaging model with absorption compensation, multiple scattering imaging model and multi-scale retinex imaging model with convolutional neural network to construct the image dehazing network. Multiple scattering haze imaging model is used to describe the haze imaging process in a more consistent way with the physical imaging mechanism. And the multi-scale retinex imaging model ensures the color fidelity. In the network structure, multi-scale feature extraction module can improve network performance in terms of feature reuse. In the attention feature extraction module, the back-propagating of the important front features is used to enhance features. This method can effectively make up for the deficiency that autocorrelation features cannot share the deep-level information, which is also effective for features replenishment. The results of the comparative experiment demonstrate that our network outperforms state-of-the-art dehazing methods.

Keywords: Image dehazing · Multi-model fusion · Absorption compensation · Attention mechanism

1 Introduction

Haze is caused by absorption and scattering effects of floating atmospheric particles such as dust, mist, and fumes, always undermines image structures, fades image colors and degrades image contrasts. As a noise signal, haze always undermines image structures, fades image colors and degrades image contrasts. To improve the quality of image and upgrade the performances of various high-level computer vision algorithms under the hazy conditions, researchers have to

Supported by Liaoning Education Department General Project Foundation (LJKZ0231); Huaian Natural Science Research Plan Project Foundation (HAB202083).

H. Liu et al. (Eds.): ICIRA 2022, LNAI 13456, pp. 523–535, 2022.
https://doi.org/10.1007/978-3-031-13822-5_47

remove the haze noises from original data firstly. Comparing with other tasks in the area of image restoration, for example, deblurring and denoising which can be attributed to a simple global homogeneous problem, deraining and desnowing that can be resolved with the low rank property [1], haze removal is an especially challenging issue due to the global inhomogeneity of haze. After the well-known haze imaging model [2] was introduced to restore hazy images by Narasimhan and Nayar [3], many physically valid single image dehazing methods have been proposed. But most of these methods usually generate lots of over-saturations especially on close objects. This classical model is expressed as follows:

$$I(x) = t(x)J(x) + (1 - t(x))A \tag{1}$$

where I(x) is the observed intensity of a hazy image at pixel t(x), J(x) indicates the clear scene radiance of a haze-free image, A is an unknown global variable which represents the constant atmospheric light intensity of the hazy image I, the transmission t describes the proportion of scene radiance that gets to the lens through the propagation medium along the cone of vision.

The existing hazy removing methods can be generally split into statistic priors [4,5] boundary constraint [6,7], geometrical cluster distribution [8,9] and deep learning [10–12].The famous Dark Channel Prior (DCP) [4] is based on the statistics that, at the pixels in most non-sky patches of the haze-free outdoor images, at least one of the RGB channels has very low intensity. However, when the scene objects are inherently similar to the atmospheric light or no shadow is cast on them, the dark channel prior will fail. Bayesian statistical prior [8] jointly estimated the scene albedo and depth from a single hazy image.To get accurate transmission, Cui et al. [7] employed different boundary constraint conditions based on DCP in bright and non-bright regions respectively. Fattal [9] proposed the geometrical cluster distribution called color-line fitting method.

Actually, haze is not only lead by the atmospheric scattering but also by the atmospheric absorption. More recently, some single image dehazing methods based on convolutional neural networks (CNNs) have been proposed. The authors attempt to learn a united transmission directly from the training data without any physical model. These deep learning methods can evade the problem that if atmospheric absorption is one cause of haze. Cai et al. [10] built an end-to-end CNN network for transmission estimation and proposed a novel nonlinear activation function BReLU. The above deep-learning algorithms estimate transmission and atmospheric light separately, which is easy to cause error accumulation and even magnification in the two independent estimation processes. Li [20] introduced VGG [16] features and Ll regularization gradient prior into conditional Generative adversarial Network (cGAN) [17] to estimate clear haze-free images. Ren et al. [17] designed a dehazing network with encoding and decoding, combined three derived images to learn confidence graph, and generated dehazing results by fusion of confidence graph. Different from the above algorithm, Zhang et al. [18] proposed the DCPDN which jointly learns transition, atmospheric light and haze-free images to perceive the relationship between them. Deng et al. [19] proposed a deep multi-model fusion network based on the attention mechanism, and the four hierarchical models introduced are not associated with haze imaging. However, due to the defective network structure, inadequate feature extraction, deep learning algorithm still has many problems to be solved. Firstly, the extraction of hazy related feature in existing deep learning dehazing network is not accurate and efficient, which seriously hinders the dahazing performance of the algorithm, resulting in poor effects and color deviation. Secondly, existing deep learning image dehazing algorithms do not carry out network structure design for dense haze, so they cannot effectively process dense hazy images.

According to the physical characteristics of hazy related feature such as atmospheric light and transmission, in this paper we propose targeted "customized" feature extraction strategies for the purpose of improving network dehazing effect and eliminating color difference, rather than simply using shallow network structure or blindly implying network structures designed for other tasks. The overall scheme of this paper adopts the deep learning method based on the fusion of physical model and convolutional neural network, in which the physical model fuses three haze-related imaging models, including haze imaging model with absorption compensation, multiple scattering imaging model and multi-scale retinex imaging model. The haze imaging model with absorption compensation is used to make up for the problem that the classical atmospheric scattering model does not consider atmospheric absorption. The multiple scattering model can accurately express the hazy imaging process in accordance with the physical imaging mechanism. The multi-scale retinex imaging model with component ratio adjustment factor can ensure the strong color fidelity.

2 Related Work

2.1 Absorption Compensation Haze Imaging Model

In the atmospheric environment, the observed intensity of a hazy image is simultaneously determined by the irradiation intensity of atmospheric light and the reflection intensity of the scene target. Therefore, atmospheric absorption attenuation also occurs in the propagation process of atmospheric light to the lens. Absorption compensation haze imaging (ACHI) model which is one of our previous works [7] can be expressed as:

$$I(x) = t_s(x)t_a(x)J(x) + (1 - t_s(x)t_a(x))A \tag{2}$$

where t_s is the scattering transmission and t_a is the supplementary absorption transmission. According to the luminance saturation ratio (LSR) prior and the Beer-Lambert rule describing the attenuation of visible light, the absorption transmission t_a can be expressed as:

$$t_a(x) = e^{-\frac{1}{\ln \max\limits_{y \in \Omega} I_v(y)}(\ln I_v(x) - \ln I_s(x))} \tag{3}$$

where v and s represent brightness and saturation respectively, Ω is the search window for maximum filtering.

2.2 Multiple Scattering Haze Imaging Model

The classical atmospheric scattering model is a simplified single scattering model which cannot express the imaging process of dense haze environment. The multiple scattering haze imaging (MSHI) model takes into account the multiple scattering of light in the propagation medium. Therefore, MSHI model is more consistent with the entire imaging process of the scene reflection light spreading to the camera lens through the scattering of high concentration aerosol, and can better describe the physical imaging mechanism of real haze images. The MSHI model considers that the absorption attenuation and multiple scattering are two independent processes. Multiple scattering is the superposition of multiple single scattering and occurs after single scattering and atmospheric absorption. After repeated scattering, the scene irradiance and atmospheric light reach the camera lens at different angles, forming fuzzy dispersion spots on the imaging plane, namely haze. The spatial redistribution of scene irradiance caused by multiple scattering can be described by Atmospheric Point Spread Function (APSF), so the MSHI model can be expressed as:

$$I(x) = [t(x)J(x) + (1 - t(x))A] \bigotimes h_{APSF} \tag{4}$$

where h_{APSF} is the multi-scattering convolution template; $\bigotimes$ stands for convolution.

In addition to being able to describe the haze imaging process in a more consistent way with the physical imaging mechanism, Eq. (4) is selected as the MSHI model because h_APSF, as a convolution template, is very easy to be embedded into the convolutional neural network. This work intends to improve the solving method of the Radiative Transfer Equation (RTE) proposed by Narasimhan and Nayar et al. to estimate the multiple scattering convolution template h_APSF. Each parameter can be obtained from the distribution of APSE in the image domain. It is easy to generate training set and more efficient than Monte Carlo method.

2.3 Multi-scale Retinex Model

Researchers have devoted a lot of efforts to image dehazing based on retinex theory. Although it has been replaced by the image dehazing algorithm based on physical model, multi-scale retinex has been proved to have significant advantages in color fidelity. In this paper, multi-scale retinex model containing component ratio adjustment factor is integrated into the overall network structure. The employed multi-scale retinex model is able to effectively solve the serious color deviation problem lead by current deep learning algorithm while eliminating hazy noise.

Retinex theory considers that the real world is colorless, and the perception of color is resulted from the interaction between light and matter. The algorithm represents the surface illuminance change of the object by calculating the weighted average value of the central pixel point and the surrounding pixel (the image is convolved by the center surround function).The algorithm removes the surface illuminance change of the object and only retains the essential reflection attribute of the object. The model is as follows:

$$I(x) = R(x)L(x) \tag{5}$$

where I is the image with haze; R is the irradiated light component, i.e., haze-free image; L is the reflected light component, which needs to be eliminated.

We plan to introduce the MSRCR (Multi-Scale Retinex with Color Restore) algorithm proposed by Rehman et al., which introduces a component ratio adjustment factor to reduce the impact of color distortion. The model can be expressed as:

$$R_M SRCR(x) = C(x)R(x) \tag{6}$$

where $R_M SRCR$ is the multi-scale retinex algorithm output image; C is the component ratio adjustment factor, which can be expressed as:

$$C(x) = G\{\log[\alpha I_c(x)] - \log[\sum_{c=1}^{3} I_c(x)]\} \tag{7}$$

where, G represents a gaussian function; $C \in R, G, B$ represents the image three-channel; α is a proportional constant.

3 Our Approach

3.1 Overall Network Structure

This work integrates haze imaging model with absorption compensation, multiple scattering imaging model and multi-scale retinex imaging model with convolutional neural network to construct the image dehazing model. The overall network structure is shown in Fig. 1.

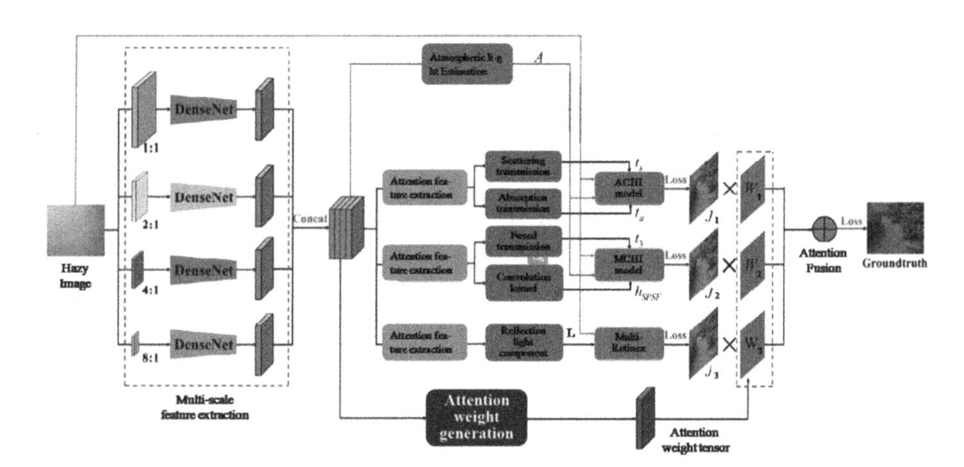

Fig. 1. Overall network structure of multi-model fusion algorithm for image dehazing based on attention mechanism.

Firstly, the multi-scale DenseNet network is used to extract the features of the hazy image. Multi-scale convolutional neural network can help the network adapt to the input of four scales, and can also enable the network to acquire receptive fields of different scales and capture feature information of different scales, which can greatly improve the overall network performance. For image dehazing network, feature ex-traction using multi-scale network structure is also helpful to eliminate halo noise. DenseNet is committed to improving network performance from the perspective of feature reuse. Each layer can get gradient directly from the loss function and obtain the feature information of each node in front, so as to improve the transmission efficiency of feature and gradient in the network. The main cause of haze is multiple scattering of scene irradiance, so DenseNet, which can reuse feature information, will help to extract haze related features.

Then the network will use the attention mechanism to integrate and optimize the cascaded multi-scale features. Attention feature extraction module can use the multi-scale features of complementary between different layers, weighted fuse the shallow feature and the feature of high frequency information carried by the deep-level se-mantic information, and ensure the efficient spread of information. The deep-level semantic information is of great significance for the haze removal. The high frequency information will help to recover the target boundary. The network will also utilize the attention mechanism to organically integrate the dehazing results of the three imaging models. The attention weight matrix can guarantee to give smaller weight to the area where the dehazing result differs greatly from that of ground-truth, and give larger weight to the area where the dehazing result is close to that of ground-truth, so as to realize information complementarity of multiple dehazing results. For each pixel, the network can choose the best one among the three results to transmit to the final result. In addition, according to the physical feature of haze related variables such as atmospheric light, scattering transmission, absorption transmission, APSF convolution kernel and reflected light component, targeted feature extraction strategies and loss functions will be designed.

3.2 Multi-scale Feature Extraction

Multi-scale convolutional neural network can make the network adapt to the input at any scale, and because of the different down-sampling step size, the network can acquire different receptive fields and capture information at different scales. Images of different scales are sent to the network to extract features of different scales for fusion, which will greatly improve the performance of the whole network. It is more suitable for feature extraction of hazy image transmission. Figure 2 shows the proposed network structure of multi-scale feature extraction. In this part of the network, we imply the dense-net to extract the features in the different scales. For the different scales, the decreasing kernel sizes of dense-net are in charge of extracting the haze related features. The kernel sizes of dense-net are 9×9, 7×7, 5×5, and 3×3 which serve on the 1:1, 2:1, 4:1, and 8:1 scale respectively. Then the feature results are concatted as an integral feature tensor which is sent to the attention feature extraction module.

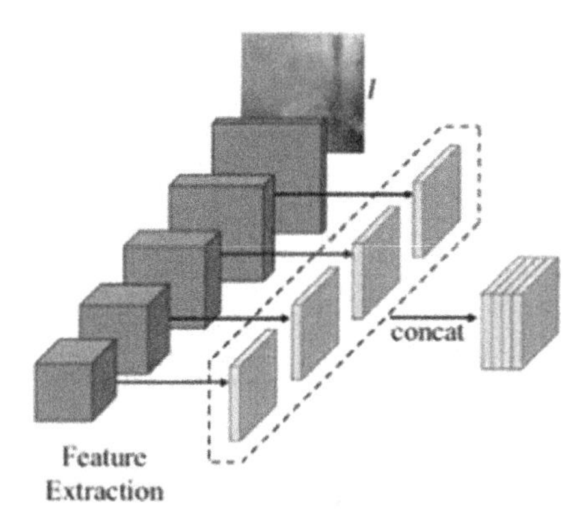

Fig. 2. The network structure of multi-scale feature extraction.

Dense-net can improve the transmission efficiency of information and gradients in the network. Each layer can take gradients directly from the loss function and get input signals directly, so that deeper networks can be trained. This network structure also has regularization effect. While other networks focus on improving network performance in terms of depth and width, Dense-Net focuses on improving network performance in terms of feature reuse. Dense haze images are generated because of multiple scattering of scene irradiance, and dense-Net is able to reuse feature information, so the imaging mechanism of dense haze is consistent and suitable for extracting features of dense haze and underwater images. Figure 3 displays the dense-net structure of multi-scale feature extraction.

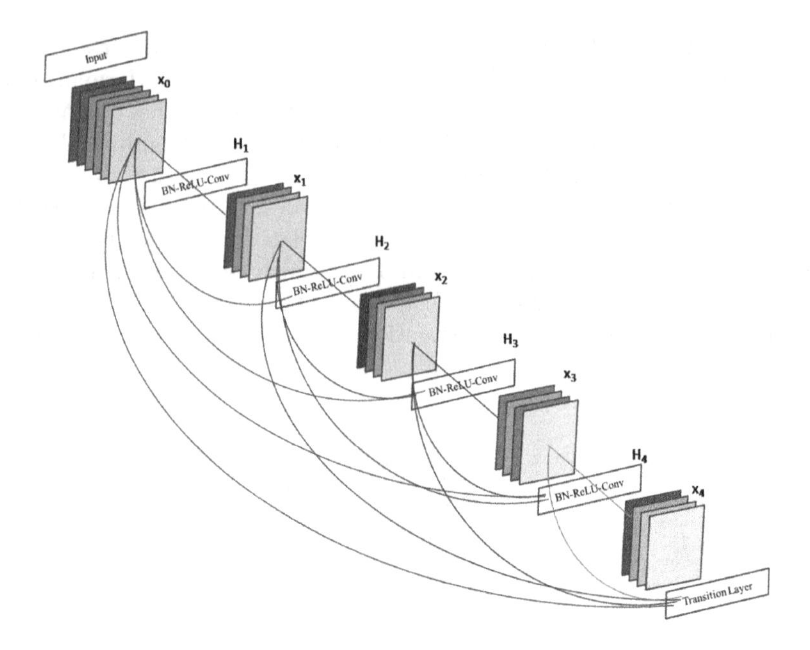

Fig. 3. Dense-net structure of multi-scale feature extraction.

3.3 Attention Feature Extraction

In the attention feature extraction module, we use the back-propagating of the important front features (multi-scale feature extraction result)to enhance subsequent module features. This method can effectively make up for the deficiency that autocorrelation features cannot sharing the deep-level information, which is also effective for features fusion and replenishment. We use the front features of hazy images to calculate the weight matrices of hazy features in space domain and frequency domain, so as to expand the expression of deep-level hazy feature information. The attention module structure is shown in Fig. 4, and the specific model can be expressed as follows:

$$W_d^i = C_d(F_o^{i-1}) \tag{8}$$

where C_d is the deep-level feature correlation enhancing module FDCM; $F(i-1)_o$ is the important front features, the output feature of the previous multi-scale feature ex-traction module; W_d^i represents the deep-level feature correlation weight.

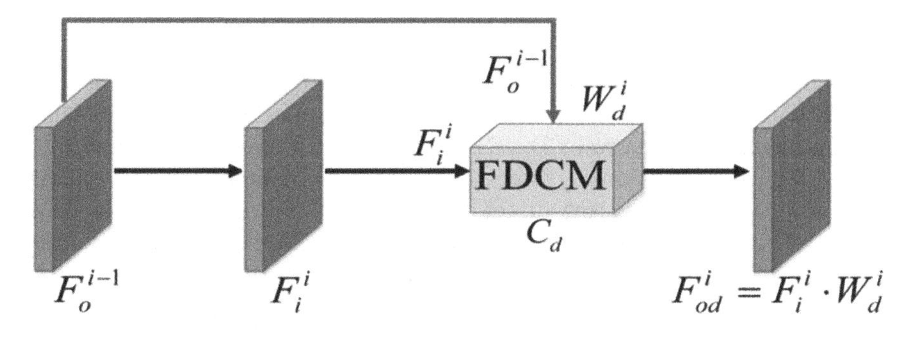

Fig. 4. Themodulestructure of attention mechanism.

The output of deep-level enhanced feature is:

$$F_{od}^i = F_i^i \dot{W}_d^i = F_i^i * C_d(F_o^{i-1}) \tag{9}$$

The overall enhanced features can be further expressed as follows:

$$F_o^i = F_i^i(1 + W_s^i + W_d^i) = F_i^i(1 + C_s(F_i^i) + C_d(F_o^{i-1})) \tag{10}$$

Through the enhancing module C_d, the detail information and texture features of the image are better maintained and transmitted. The features are supplemented and strengthened at the deep-level. The enhanced features are helpful for the model to deeply understand the original texture details and color information of the image.

3.4 Loss Function

In the training, we use attention weights to blend the results of three model firstly. Then optimizing the weights and biases of the proposed network by minimizing the L2 loss on the training set:

$$F_o^i = F_i^i(1 + W_s^i + W_d^i) = F_i^i(1 + C_s(F_i^i) + C_d(F_o^{i-1})) \tag{11}$$

where n is number of training images, R is the blended dehazed result by our network, J is the corresponding ground truth.

4 Our Approach

For the proposed algorithm is a fusion model of physical imaging models and deep learning networks, we compare our dehazing network against not only other dehazing networks including DehazeNet [10], AOD-Net [11], and DA [27], but also one classical method DCP [4]. Among the compared methods, DCP focused on hand crafted features for haze removal, while others are based on convolutional neural networks (CNNs). We retrain the original (released) implementations of these methods or directly report their results on the public datasets.

4.1 Experiment Setup

Train Dataset: Our dehazing model is trained on outdoor datasets, including RESIDE dataset [56], NH-HAZE dataset [57] for homogeneous haze and non-homogeneous haze respectively. For RESIDE dataset, we randomly select 10000 samples from outdoor training subset in RESIDE for training and 300 samples from synthetic objective testing subset in RESIDE for testing. For the NH-HAZE dataset, we synthetic 1000 samples augmented from 50 original high-resolution samples by randomly cropping, flipping, and rotating.

Train Details: Our dehazing model is implemented with PyTorch library and trained on one NVIDIA GeForce RTX 2080 GPU. Our model is trained for 40 epochs on the RESIDE dataset, 20 epochs on the NH-HAZE dataset.

4.2 Experiment Results and Analysis

Fig. 5. Thecomparation results of the proposed method with other four methods.

Figure 5 visually compare the dehazed maps on real-world hazy images and synthetic hazy images. DCP suffers from the halo noises in some images. AOD removes few haze, and there is still a large amount of haze remain in the results. DehazeNet suffers from color distortions in most of the test images. This is particularly evident in the third and fourth images of Fig. 5. Our results are better than that of DA in the dense haze area. Our method can more effectively remove haze while producing realistic colors than these compared state-of-the-art methods.

We use the peak signal-to-noise ratio (PSNR), structural similarity index measure (SSIM) to evaluate the dehazing methods on the dehazed images. Table 1 reports PSNR and SSIM scores of compared dehazing methods. Among all the compared methods, DehazeNet always has the largest PSNR scores, but there are the most color deviation noises in its results. In SSIM scores, our method performs much better than other compared methods in the most experimental images. This demonstrates that the proposed method can preserve the texture structure of hazy image well. The red numbers in the Table 1 express the best metrics value for each image among the compared methods.

Table 1. Quantitative comparisons between our network and compared methods.

Images	Method	DCP	AOD	DehazeNet	DA	Ours
1	PSNR	15.34	12.34	22.19	13.65	21.38
	SSIM	0.787	0.802	0.873	0.769	0.921
2	PSNR	13.65	14.44	19.24	13.78	18.57
	SSIM	0.536	0.753	0.825	0.564	0.838
3	PSNR	14.69	13.34	33.59	21.37	23.67
	SSIM	0.761	0.549	0.976	0.897	0.913
4	PSNR	14.20	14.52	26.37	14.67	22.79
	SSIM	0.532	0.587	0.913	0.631	0.945

5 Conclusion

In this paper, we fuse the physical models including haze imaging model with absorption compensation, multiple scattering imaging model and multi-scale retinex imaging model with convolutional neural network to construct the image dehazing network. In the proposed fusion model, we take the atmospheric absorption attenuation into account. Multiple scattering haze imaging model is able to describe the haze imaging process in a more consistent way with the physical imaging mechanism, especially for the dense haze. And the multi-scale retinex imaging model is helpful to ensure the color fidelity. In the network structure, multi-scale feature extraction module can improve network performance in terms of feature reuse and is consistent and suitable for extracting features of dense haze. In the attention feature extraction module, we use the back-propagating of the important front features to enhance subsequent module features. This method can effectively make up for the deficiency that autocorrelation features cannot sharing the deep-level information, which is also effective for features fusion and replenishment. The results of the comparative experimental demonstrate that our network outperforms state-of-the-art dehazing methods.

References

1. Berman, D., Avidan, S.: Non-local image dehazing. In: Proceedings of the IEEE Conference on Computer Vision and Pattern Recognition, pp. 1674–1682 (2016)
2. McCartney, E.J.: Optics of the Atmosphere: Scattering by Molecules and Particles, pp. 421–432. John Wiley and Sons Inc., New York (1976)
3. Nayar, S.K., Narasimhan, S.G.: Vision in bad weather. In: Proceedings of the Seventh IEEE International Conference on Computer Vision, vol. 2, pp. 820–827 (1999)
4. He, K., Sun, J., Tang, X.: Single image haze removal using dark channel prior. IEEE Trans. Pattern Anal. Mach. Intell. **33**(12), 2341–2353 (2011)
5. Kim, J.-H., Jang, W.-D., Sim, J.-Y., Kim, C.-S.: Optimized contrast enhancement for real-time image and video dehazing. J. Vis. Commun. Image Repres. **24**(3), 410–425 (2013)
6. Meng, G., Wang, Y., Duan, J., Xiang, S., Pan, C.: Efficient image dehazing with boundary constraint and contextual regularization. In: Proceedings of the IEEE international Conference on Computer Vision, pp. 617–C624 (2013)
7. Cui, T., Tian, J., Wang, E., Tang, Y.: Single image dehazing by latent region segmentation based transmission estimation and weighted L1-norm regularization. IET Image Process. **11**(2), 145–154 (2016)
8. Tan, R.T.: Visibility in bad weather from a single image. In: 2008 IEEE Conference on Computer Vision and Pattern Recognition, pp. 1–8 (2008)
9. Fattal, R.: Dehazing using color-lines. ACM Trans. Graphics (TOG) **34**(1), 13 (2014)
10. Cai, B., Xu, X., Jia, K., Qing, C., Tao, D.: Dehazenet: an end-to-end system for single image haze removal. IEEE Trans. Image Process. **25**(11), 5187–5198 (2016)
11. Li, B., Peng, X., Wang, Z., Xu, J., Feng, D.: An all-in-one network for dehazing and beyond. arXiv preprint arXiv:1707.06543 (2017)
12. Ren, W., Liu, S., Zhang, H., Pan, J., Cao, X., Yang, M.-H.: Single image dehazing via multi-scale convolutional neural networks. In: Leibe, B., Matas, J., Sebe, N., Welling, M. (eds.) ECCV 2016. LNCS, vol. 9906, pp. 154–169. Springer, Cham (2016). https://doi.org/10.1007/978-3-319-46475-6_10
13. He, K., Sun, J., Tang, X.: Guided image filtering. IEEE Trans. Pattern Anal. Mach. Intell. **35**(6), 1397–1409 (2013)
14. Li, R., Pan, J., Li, Z., Tang, J.: Single image dehazing via conditional generative adversarial network. In: Proceedings of the IEEE Conference on Computer Vision and Pattern Recognition, pp. 8202–8211 (2018)
15. Simonyan, K., Zisserman, A.: Very deep convolutional networks for large-scale image recognition. arXiv preprint arXiv:1409.1556 (2014)
16. Isola, P., Zhu, J.-Y., Zhou, T., Efros, A.A.: Image-to-image translation with conditional adversarial networks. In: CVPR, pp. 1125–1134 (2017)
17. Ren, W., et al.: Gated fusion network for single image dehazing. In: CVPR, pp. 3253–3261 (2018)
18. Zhang, H., Patel, V.M.: Densely connected pyramid dehazing network. In: CVPR, pp. 3194–3203 (2018)
19. Deng, Z., et al.: Deep multi-model fusion for single-Image dehazing. In: Proceedings of the IEEE International Conference on Computer Vision, pp. 2453–2462 (2019)
20. Li, R., Pan, J., He, M., et al.: Task-oriented network for image Dehazing. IEEE Trans. Image Process. **29**, 6523–6534 (2020)

21. Zhang, J., Tao, D.: Famed-net: a fast and accurate multi-scale end-to-end dehazing network. IEEE Trans. Image Process. **29**, 72–84 (2020)
22. Zhang, Y., Ding, L., Sharma, G.: Hazerd: an outdoor scene dataset and benchmark for Single image dehazing. In: 2017 IEEE International Conference on Image Processing (ICIP), pp. 3205–3209. IEEE (2017)
23. Li, B., et al.: RESIDE: a benchmark for single image dehazing. ArXiv e-prints (2017)
24. Ancuti, C., Ancuti, C.O., Vleeschouwer, C.D.: D-hazy: a dataset to evaluate quantitatively dehazing algorithms. In: 2016 IEEE International Conference on Image Processing (ICIP), pp. 2226–2230. IEEE (2016)
25. Zhang, Y., Ding, L., Sharma, G.: Hazerd: an outdoor scene dataset and benchmark for single image dehazing. In: 2017 IEEE International Conference on Image Processing (ICIP), pp. 3205–3209. IEEE (2017)
26. Shao, Y., Li, L., Ren, W., Gao, C., Sang, N.: Domain adaptation for image dehazing. In: Proceedings of the IEEE Conference on Computer Vision and Pattern Recognition (CVPR), pp. 8202–8211 (2020)

Study on FTO of Permanent Magnet Synchronous Motor for Electric Aircraft Steering Gear

Meng Zhang, Tong Cui(✉), Haoya Zhang, and Nan Gao

College of Artificial Intelligence, Shenyang Aerospace University, Shenyang 110136, China
ct61ct61@126.com

Abstract. As the core part of steering gear in electric aircraft, the failure of motor will threaten flight safety. Aiming at the fault problem of permanent magnet synchronous motor in the steering gear of electric aircraft, a fault-tolerant control (FTC) method based on three-phasefour-switch is proposed. The “Five-stage” SVPWM modulation method is designed to improve the problem of small flux freedom existing in the “Three-stage” SVPWM method, improve the degree of freedom and reduce the switching loss. Further combined with the sliding mode control idea, an improved sliding mode speed controller was designed to replace the traditional PI controller with fixed parameters. Thus, the improved sliding mode FTC strategy of electric aircraft permanent magnet synchronous motor based on “Five-stage” three-phasefour-switch was obtained to improve the robustness and rapidness of the system. The simulation results show that the proposed method can guarantee the stable operation of the steering gear and the flight safety of the electric aircraft even when the system inverter is faulty.

Keywords: Fault-tolerant control (FTC) · Three-phasefour-switch · Permanent magnet synchronous motor · Sliding mode control

1 Introduction

In recent years, the multi-electric or all-electricaircraft has become a hot spot in aerospace and aviation research, and one of the key technologies of electric aircraft is the design of electric steering gear. Motor is the core component of electric steering gear, and its performance directly affects the performance and reliability of electric steering gear [1, 2]. With the development of power electronics technology and permanent magnet motor, permanent magnet synchronous motor has been widely used in rocket, aircraft, missile and other aircraft fields, but also because of its high efficiency, high reliability and other advantages become the first choice of electric steering gear drive motor [3, 4].

Fault tolerance refers to that when one or more components fail in a system, the system isolates the faulty component from the original system and then takes corresponding measures to maintain the original function or continue to operate stably under acceptable performance indicators [5]. Inverter break is one of the most common faults of electric aircraft steering gear. At present, the commonly used inverter FTCschemes

H. Liu et al. (Eds.): ICIRA 2022, LNAI 13456, pp. 536–546, 2022.
https://doi.org/10.1007/978-3-031-13822-5_48

include dual-system backup, bridge arm redundancy, three-phase four bridge arm redundancy, three-phasefour-switch FTC and so on [6, 7]. Thedual-system backup uses two sets of inverters, which provides high reliability but complex system structure, high volume and cost, and low device utilization. Bridge arm redundancy is to realize FTC by isolating the faulty bridge arm from the load and switching the output phase to the auxiliary bridge arm when a certain bridge arm fails. This method is applied to more switches, and the drive and auxiliary device complex. Andin the method of three-phase four-arm bridge, a new bridge arm is needed, and the design process is complicated and the cost is high.

In order to reduce the cost of FTC, the FTC method of three-phasefour-switch has been proposed and attracted more and more researchers' attention. Its topology can be used as the fault-tolerant topology of traditional three-phase six-switch inverter, and it has the advantages of fewer switching devices, less driving circuit, simple circuit design, low cost and small switching loss. Moreover, the method does not require high capacity of switching device, and the more important advantage is that it is easy to realize vector control. In [8], the control principle of three-phasefour-switch is deeply analyzed, and it is proved that this method can be used for FTC of motor. In [9], non-orthogonal coordinate transformation is introduced on the basis of three-phasefour-switch to achieve FTC and suppress the torque ripple of permanent magnet synchronous motor to a certain extent.

To sum up, in order to ensure the safe and stable operation of electric aircraft under fault conditions, three-phasefour-switch FTC is introduced into the motor control of steering gear in this paper. According to the principle of three-phasefour-switch FTC, sectors are divided, turn-on time is calculated and SVPWM control strategy is generated. At the same time, an improved sliding mode speed controller is designed and combined with the three-phasefour-switch to obtain the improved sliding mode FTC strategy of the electric aircraft permanent magnet synchronous motor based on the "Five-stage" three-phasefour-switch. Finally, aFTC simulation model is built in MATLAB/Simulink environment, and the effectiveness of the proposed method is verified by simulation experiments.

2 Permanent Magnet Synchronous Motor Model

The mathematical model of stator voltage equation of surface mount permanent magnet synchronous motor in rotating coordinate system is:

$$\begin{cases} u_d = Ri_d + \dfrac{d}{dt}\psi_d - \omega_e\psi_q \\ u_q = Ri_q + \dfrac{d}{dt}\psi_q + \omega_e\psi_d \end{cases} \tag{1}$$

The stator flux equation is:

$$\begin{cases} \psi_d = L_d i_d + \psi_f \\ \psi_q = L_q i_q \end{cases} \tag{2}$$

The electromagnetic torque equation is:

$$T_e = \frac{3}{2}p_n i_q\left[i_d\left(L_d - L_q\right) + \psi_f\right] \tag{3}$$

where, the cross-axis component of the stator voltageare u_d, u_q. The cross-axis component of the stator currentare i_d, i_q. The stator resistance is *R*. The cross-axis component of the stator flux is ψ_d, ψ_q. The electrical angular velocity is ω_e. The quadrature - direct axis inductance componentare L_d, L_q. The permanent magnet flux is ψ_f.

3 Three-Phasefour-Switch FTC

3.1 Three-Phasefour-Switch FT Inverter Topology

The general topology of three-phasefour-switch inverter is AC-DC-AC. When the inverter is out of phase, the three-phase winding of the motor can still be connected with the main circuit, and its topological structure is shown in Fig. 1.

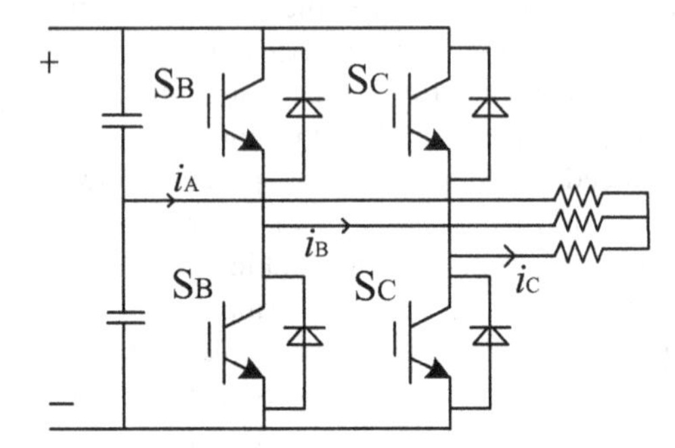

Fig. 1. Topological structure of phase four switching inverter

When the inverter a-phase bridge arm fails, A large current will be generated, leading to the fast disconnection of the fuse, and the system will enter the FTC, and no longer send pulse signals to the two switches of the faulty phase. At the same time, open the bidirectional thyristor and send a signal to make it conductive, and then open the neutral point, complete the FTC conversion in the case of failure, to achieve the normal operation of the motor. When the B and C phase bridge arms fail, the FTC method is similar to that of A phase, and the fault-tolerant state is also entered.

According to literature [9], the voltage formula of each phase of three-phase four-switch FTC stator winding is as follows:

$$\begin{cases} U_a = U_{dc}(1 - S_B - S_C)/3 \\ U_b = U_{dc}(S_B - S_C)/3 \\ u_s = \frac{2}{3}(u_A + u_B e^{j\frac{2\pi}{3}} + u_C e^{j\frac{4\pi}{3}}) = U_a + jU_\beta \end{cases} \tag{4}$$

where U_a, U_b is the voltage component in the two-phase stationary coordinate system. U_{dc} is the DC bus voltage. u_A, u_B, u_C is the three-phase voltage of the inverter. u_s is the space voltage vector. $S_i(i = B/C)$ represents the switching state of a certain phase bridge arm power switch device. If the upper bridge arm of B/C phase is in the on-off state and the lower bridge arm is in the off-off state, then $S_B/S_c = 1$; if the upper bridge arm of B/C phase is in the off-off state and the lower bridge arm is in the on-off state, then $S_B/S_c = 0$.

$$s_i = \begin{cases} 1, \textit{upper bridge arm enabled and lower bridge arm closed} \\ 0, \textit{lower bridge arm enabled and upper bridge arm closed} \end{cases} \quad (5)$$

where, i represents phase B and phase C.

When phase A is in fault-tolerant state, the voltage vector distribution of the three-phasefour-phase switch is shown in Fig. 2.

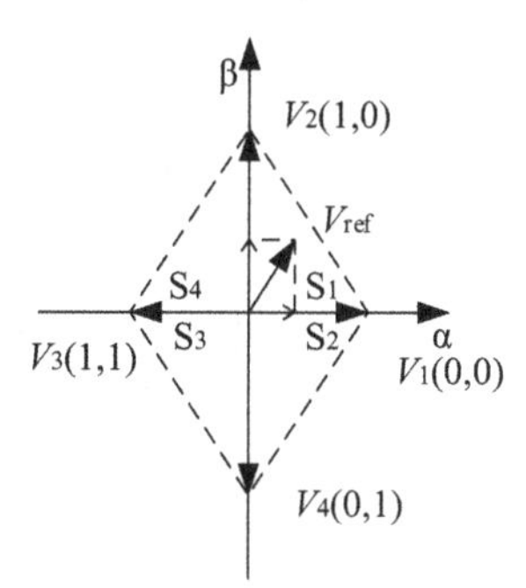

Fig. 2. Three-phasefour-switch voltage vector distribution diagram

As can be seen from Fig. 2, the plane is divided into four parts by four spatial voltage vectors, which are called sectors and are represented by S_1, S_2, S_3 and S_4.

3.2 Conduction Time Calculation

According to the principle of nearest voltage vector, reference voltage U_{ref} can be synthesized by two adjacent effective voltage vectors and zero voltage vectors. Taking the reference voltage vector falling into sector S_1 as an example, based on the principle of vs-second balance:

$$U_{\mathrm{ref}}T_s = U_1T_1 + U_2T_2 + U_0T_0 \quad (6)$$

where T_s is the sampling period.The time of synthesizing effective voltage vectors in a sampling period is expressed by T_1, T_2 and T_0 respectively. The operation time must meet the following relationship:

$$T_1 + T_2 + T_0 = T_s \quad (7)$$

According to Eq. (6), (7) and the voltage vector change relationship, it can be obtained:

$$\begin{cases} T_1 = \frac{3T_s|U_\alpha|}{U_{dc}} \\ T_2 = \frac{3T_s|U_\beta|}{U_{dc}} \\ T_0 = T_s - T_1 - T_2 \end{cases} \quad (8)$$

When $T_1 + T_2 > T_s$, the overmodulation strategy is adopted, which satisfies the following relationship:

$$\begin{cases} T_1 = \frac{T_1}{T_1+T_2} T_s \\ T_2 = \frac{T_2}{T_1+T_2} T_s \end{cases} \quad (9)$$

3.3 "Five-Stage" SVPWM Modulation Strategy

Three-phasefour-switch adopts "Five-stage" SVPWM control mode as shown in Fig. 3. In the first sector, the switching state of the inverter is (00) → (10) → (11) → (10) → (00). In the second sector of inverter switch state change process for (00) → (10) → (11) → (10) → (00); In the third sector, the switching state of the inverter changes as (11) → (01) → (00) → (01) → (11) → (01) → (00), and in the fourth sector, the switching state of the inverter changes as (00) → (01) → (11) → (01) → (00), which satisfies the design rules. The SVPWM waveform of the "Five-stage" three-phasefour-switch can be obtained by comparing the modulation wave and the triangle wave.

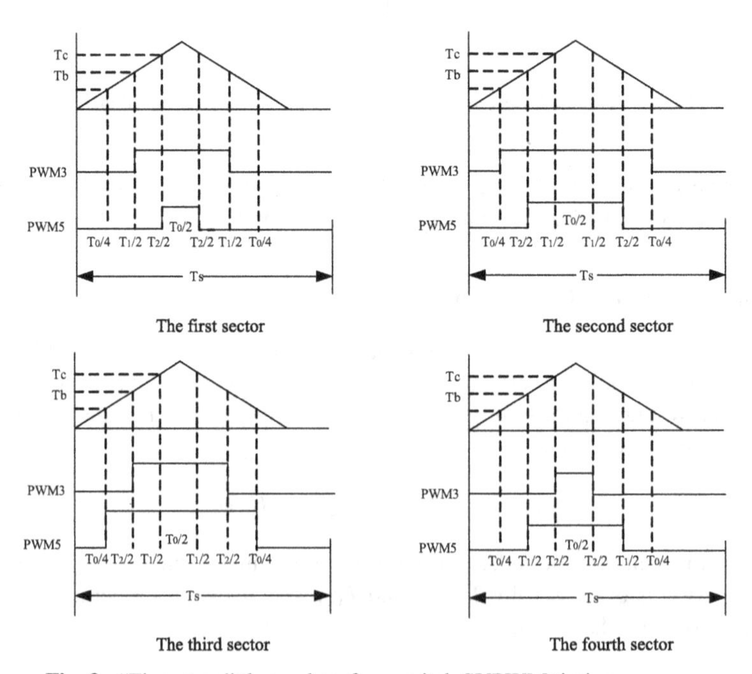

Fig. 3. "Five-stage" three-phasefour-switch SVPWM timing sequence

The four-switch vector synthesis diagram of "Five-stage" SVPWM is shown in Fig. 4.

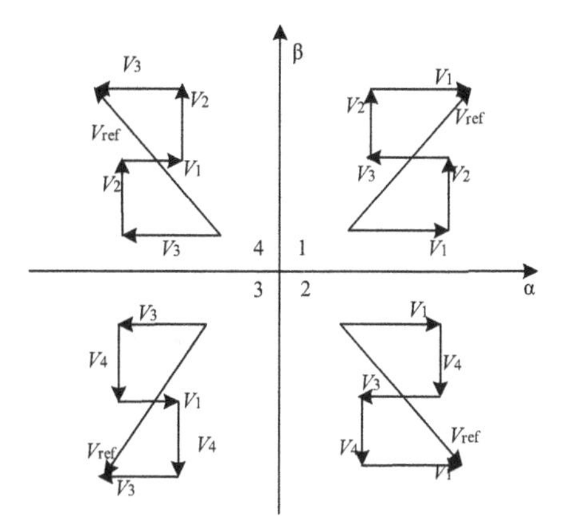

Fig. 4. "Five-stage" three-phasefour-switch SVPWM vector synthesis diagram

Figure 4 shows the "Five-stage" SVPWM vector synthesis process in four sectors. It can be seen from the figure that in the "Five-stage" three-phasefour-switch SVPWM modulation method, the action time of the two adjacent vectors used to synthesize the reference vector is divided into two sections to generate PWM waves in a symmetric way. In this case, the difference between the target vector and the actual composite vector is small, and the difference between the actual flux amplitude and the expected amplitude is also small, and the difference between the actual flux phase and the expected phase is also small, so that the motor torque has a low ripple.

4 Design of Sliding Mode Controller (SMC)

Combined with the sliding mode control idea, an improved sliding mode speed loop controller is designed to replace the traditional PI controller, and an improved sliding mode FTC strategy based on "Five-stage" three-phasefour-switch for electric aircraft permanent magnet synchronous motor is proposed. The vector control principle based on three-phase four-switch is shown in Fig. 5. The SMC part in the figure is the sliding mode controller added to the speed ring.

In this paper, a new approach law and weighted integral gain are introduced to make the sliding mode surface function and integral result approach to zero synchronously in the sliding mode stage, so chattering of the system can be suppressed effectively.

The new exponential approach law is designed as follows:

$$\begin{cases} \dot{s} = -\varepsilon \mathrm{sgn}(s) - qs - K_w|\rho|\mathrm{sgn}(s), \\ \varepsilon > 0, q > 0, K_w > 0 \\ \rho = \int_0^t \left(K_f \rho + s\right) dt \end{cases} \tag{10}$$

where, $-\varepsilon \mathrm{sgn}(s)$ is the constant velocity approaching term.sgn(s) is the sign function. S is the sliding surface. $-qs$ is the exponential approaching term. K_w, K_f are the system parameter. ρ is the integral function.

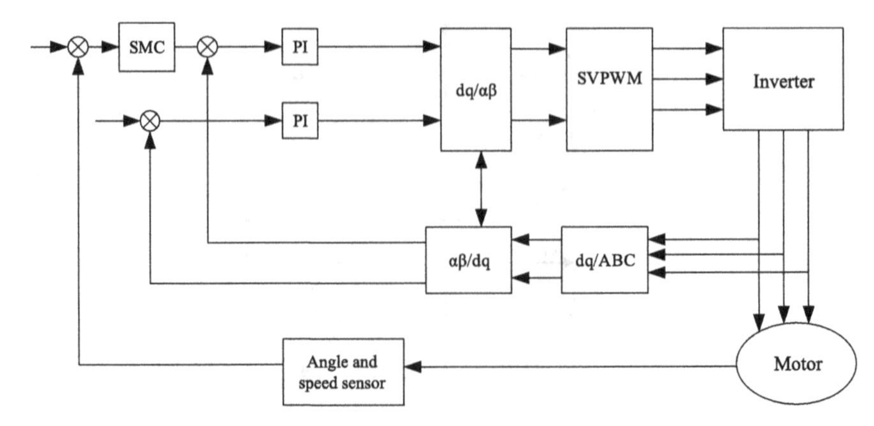

Fig. 5. Three phase four switch vector control principle diagram

State variables of PMSM system:

$$\begin{cases} x_1 = \omega_{ref} - \omega_m \\ x_2 = \dot{x}_1 = -\dot{\omega}_m \end{cases} \tag{11}$$

where, ω_{ref} is the actual speed. ω_m is the given speed. Define $u = \dot{i}_q$, $D=\frac{3p\psi_f}{2J}$, J is the moment of inertia. According to Eq. (1)–(3), it can be obtained after transformation:

$$\begin{bmatrix} \dot{x}_1 \\ \dot{x}_2 \end{bmatrix} = \begin{bmatrix} 0 & 1 \\ 0 & 0 \end{bmatrix} \begin{bmatrix} x_1 \\ x_2 \end{bmatrix} + \begin{bmatrix} 0 \\ -D \end{bmatrix} u \tag{12}$$

The sliding mode surface function is defined as:

$$s = cx_1 + x_2 \tag{13}$$

where, c is the parameter to be designed. $c > 0$, according to the principle of sliding mode control and the principle of stability, the form of sliding mode controller can be deduced as follows:

$$u = \frac{1}{D}[cx_2 + \varepsilon sgn(s) + qs + K_w|\rho|sgn(s)] \tag{14}$$

The integration term is included in the designed approach rate, which can not only eliminate the steady-state error of the system, but also effectively reduce chattering phenomenon and improve the control quality of the control system.

5 Simulation Experiment

In order to verify the feasibility and effectiveness of the above designed controller, a simulation model of permanent magnet synchronous motor vector control system was established in SIMULINK. Simulation conditions are set as follows: DC side voltage is 311 V, PWM switching frequency is 10 kHz, sampling period is 10 μs, variable step size ode23tb algorithm is adopted, relative error is 0.0001, simulation time is 0.4 s. Parameters in the sliding mode controller are selected as follows: $c = 74.311$, $\varepsilon = 251.473$, $q = 105.929$, $K_w = 4651.654$, $K_f = 2.975$, $K_p = 38.192$, $K_i = 7606.97$. Parameters of permanent magnet synchronous motor are shown in Table 1.

Table 1. Permanent magnet synchronous motor parameters.

Parameter	The values	Unit
Polar logarithm (p_n)	4	—
Stator inductance (L_d)	5.25	mH
Stator resistance (R)	0.985	Ω
Moment of inertia (J)	0.003	$kg \cdot m^2$
Damping coefficient (B)	0.008	$N \cdot m \cdot s$
Magnetic chain (ψ_f)	0.1827	Wb

5.1 Fault Simulation of Electric Aircraft

The block diagram of the vector control system of permanent magnet synchronous motor under the condition of inverter failure is shown in Fig. 6. The ideal switch is added into the simulation model to imitate the inverter failure.

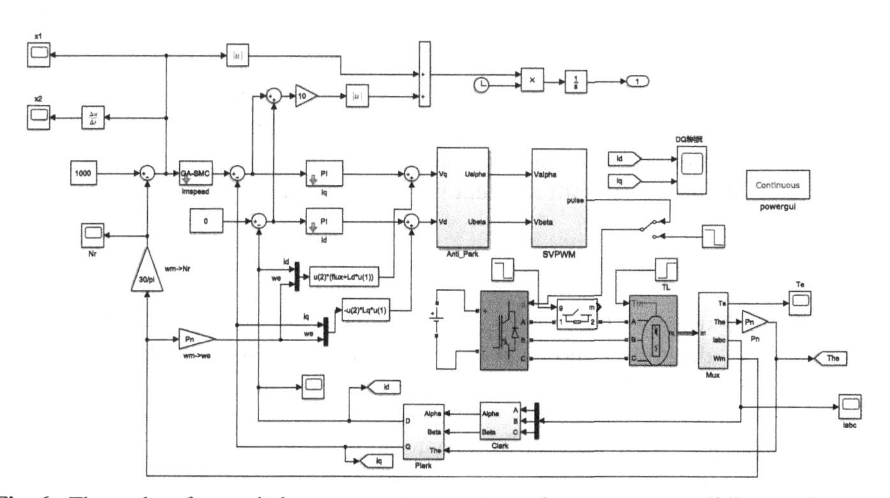

Fig. 6. Three-phasefour-switch permanent magnet synchronous motor sliding mode vector control SIMULINK simulation block diagram

Set the simulation conditions: given the speed of 700 r/min, the initial load torque is 3 N·M, and the failure of a-phase bridge arm is simulated at 0.2 s. The simulation waveform of the speed, torque and current is shown in Fig. 7.

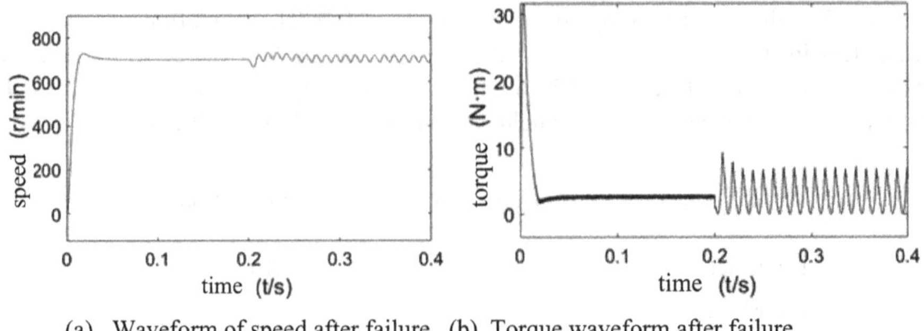

(a) Waveform of speed after failure (b) Torque waveform after failure

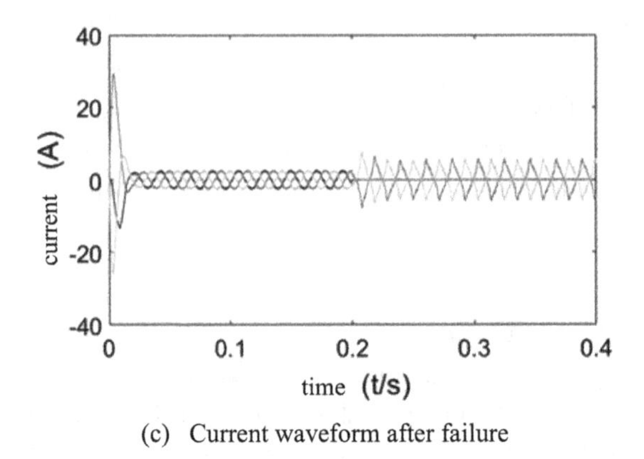

(c) Current waveform after failure

Fig. 7. The waveform of speed, torque and circuit changes in the case of failure

As can be seen from Fig. 7, the motor changes from normal working state to fault state at 0.2 s, and the rotational speed, torque and current fluctuate greatly after 0.2 s, and the precise control of rotational speed cannot be realized. If the aircraft failure is not handled safely in time, an aircraft accident will occur in serious cases.

5.2 Simulation Results Under FTC

Three-phasefour-switch FTC of permanent magnet synchronous motor vector system, simulation block diagram is shown in Fig. 8. Given speed 700 r/min, the initial load torque is 3 N·M. After FTC, the simulation waveform of speed, torque and current is shown in Fig. 9.

As can be seen from Fig. 9, when the system A phase fails, under FTC, the speed can reach A stable state quickly and without overshoot. Compared with the motor failure, the torque fluctuates less and the three-phase current waveform is relatively flat.

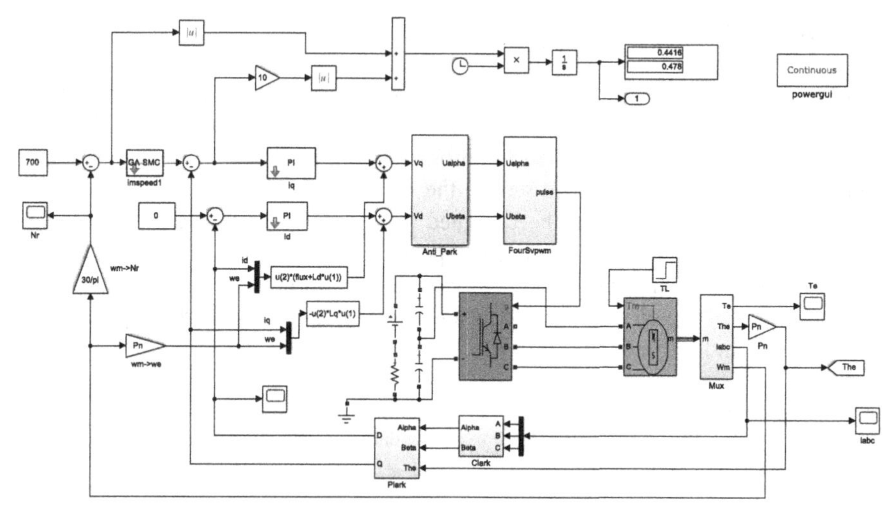

Fig. 8. FTC simulation diagram

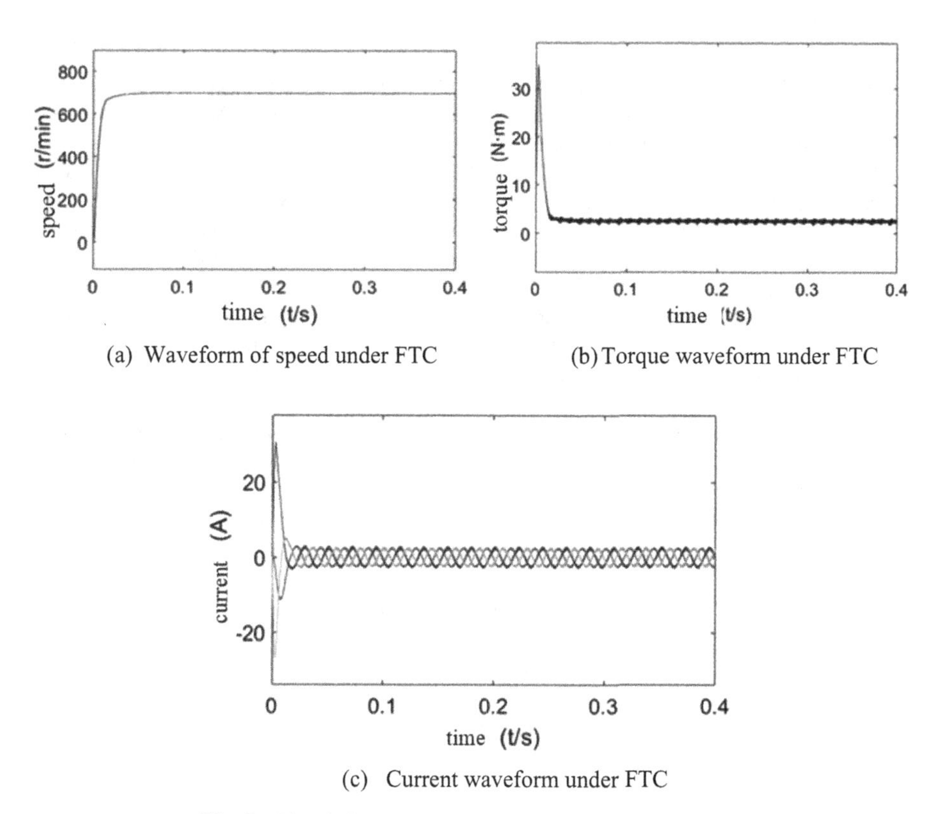

(a) Waveform of speed under FTC

(b) Torque waveform under FTC

(c) Current waveform under FTC

Fig. 9. Simulation results of three-phasefour-switch FTC

6 Conclusion

In this paper, a "Five-stage" SVPWM FTC strategy is proposed for the fault of electric aircraft steering gear. At the same time, a vector control speed loop sliding mode controller is designed. Based on the traditional sliding mode control exponential law, weighted integral gain is added to reduce the system pulsation and suppress the phenomenon of shaking. Based on the designed sliding mode controller, the simulation models of fault and FTC are constructed respectively. The simulation results show that the FTC strategy of "three-phasefour-switch + sliding mode" can ensure the safe and stable operation of the electric aircraft when faults occur in the working process, and ensure the safe and stable flight of the electric aircraft.

Acknowledgement. This work is supported by general project of Education Department of Liaoning Province (Grant No. LJKZ0230), Research foundation of Shenyang Aerospace University (Grant No. 19YB70).

References

1. Suti, A., Di Rito, G., Galatolo, R.: Fault-tolerant control of a three-phase permanent magnet synchronous motor for lightweight UAV propellers via central point drive. Actuators Multidiscip. Digit. Publish. Inst. **10**(10), 253 (2021)
2. Zhao, L., Ge, B., et al.: Fault-tolerant control for reducing harmonic distortion of dual three-phase permanent magnet synchronous motor. Energies **15**(11), 3887 (2022)
3. Li, R., Wu, Z., Li, X.: Review on fault diagnosis and active fault tolerant control of permanent magnet synchronous motor drive system. J. Appl. Sci. Eng. **24**(2), 185–205 (2021)
4. Ullah, K., Guzinski, J., Mirza, A.F.: Critical review on robust speed control techniques for permanent magnet synchronous motor (PMSM) speed regulation. Energies **15**(3), 1235 (2022)
5. Wang, W., Zeng, X., Hua, W., et al.: Phase-shifting fault-tolerant control of permanent-magnet linear motors with single-phase current sensor. IEEE Trans. Industr. Electron. **69**(3), 2414–2425 (2021)
6. Xu, Y., Yan, H., Zou, J., et al.: Zero voltage vector sampling method for PMSM three-phase current reconstruction using single current sensor. IEEE Trans. Power Electron. **32**(5), 3797–3807 (2017)
7. Shanthi, R., Kalyani, S., Devie, P.M.: Design and performance analysis of adaptive neuro-fuzzy controller for speed control of permanent magnet synchronous motor drive. Soft Comput. **25**(2), 1519–1533 (2021)
8. AnQ, T., Sun, X.T., et al.: Control strategy of fault-tolerant three-phase four-switch inverter. Proc. CSEE **30**(3), 14–20 (2010)
9. Chong, Z., Zeng, Z., Zhao, R.: Comprehensive analysis and reduction of torque ripples in three-phase four-switch inverter-fed PMSM drives using space vector pulse-width modulation. IEEE Trans. Power Electron. **32**(7), 5411–5424 (2016)

A Classification Method for Acute Ischemic Stroke Patients and Healthy Controls Based on qEEG

Xiangyu Pan, Hui Chang, and Honghai Liu(✉)

State Key Laboratory of Robotics and System, Harbin Institute of Technology, Shenzhen, China
honghai.liu@icloud.com

Abstract. Stroke has been one of the diseases with high incidence in the world, which is the main reason for adult deformity. It's significant to diagnose stroke quickly and accurately. Currently the main diagnostic method of acute ischemic stroke is also Computed Tomography. Meanwhile, electroencephalogram (EEG) is an electrophysiological manifestation that directly reflects brain activity. Through the analysis of EEG, a large amount of physiological and pathological information can be found. Using qEEG as an indicator for the diagnosis of stroke patients has become a novel and prospective method. This paper proposed a classification method for stroke patients and healthy controls using Quadratic Discriminant Analysis. Four simple features of task-EEG which are RPR of Beta, Delta, DAR and DTABR were obtained by Welch's method. The classification results showed the certain potential for qEEG as an diagnosis method.

Keywords: Stroke · Electroencephalogram (EEG) · Classification · Quadratic discriminant analysis

1 Introduction

Stroke is a disease caused by the disturbance of cerebral blood circulation, including the hemorrhage caused by the rupture of cerebral blood vessels, also known as cerebral hemorrhage. It also includes diseases caused by ischemia caused by cerebrovascular stenosis, also known as cerebral infarction [1]. According to relevant data, there are about 2.5 million new cases of stroke patients in China every year, including 1.5 million deaths due to stroke. Stroke has become one of the most fatal diseases in the world [2].

In addition, stroke has a high disability rate and recurrence rate, which is even worse for ordinary patients and their families. They need to bear huge pain and medical expenses. At the same time, the middle-aged and the elderly account for a large proportion of stroke patients, and stroke often has the characteristics of rapid onset and rapid deterioration. Therefore, the expected goal of stroke

H. Liu et al. (Eds.): ICIRA 2022, LNAI 13456, pp. 547–555, 2022.
https://doi.org/10.1007/978-3-031-13822-5_49

diagnosis is to make a correct diagnosis for stroke patients and enable them to receive effective treatment as soon as possible, so as to prevent the deterioration of their condition [3]. At present, CT and related scales are still the main diagnosis approach with limited efficiency and subjectivity [4, 5].

Electroencephalogram and EEG signals are generated by spontaneous and rhythmic electrical activities carried out by cell groups in the cerebral cortex, which can be continuously recorded and collected by electrodes. EEG signal is usually accompanied by a lot of noise. The purpose of extracting EEG signal is to get the hidden and weak features from its complex and changeable environment for subsequent analysis and research. EEG itself carries a lot of rich pathological information, but also contains a lot of physiological information. EEG can be displayed in the form of graphics for doctors to obtain important and accurate information and understand the patient's condition so that a correct diagnosis can be made [6]. Moreover, the cost of EEG detection is relatively low. If we can quickly realize the effective classification of the stroke disease, it is of great significance to improve the accuracy and efficiency of stroke diagnosis and treatment.

This study has proposed a classification method for AIS patients and healthy controls based on task-EEG [7]. The conditions of different qEEG [8,9] parameters from patients and healthy controls are analysed and compared. Four frequency domain features are obtained for classification and the machine learning method of QDA [10] is applied as the classifier.

2 Patients and Experimental Diagram

There were 25 subjects recruited that were both of 7 healthy controls and 18 patients with AIS, which were confirmed by CT-scan and correlated scales. Healthy patients refers to those who have never had a stroke before, while patients with acute ischemic stroke are those who suffered a stroke due to blockage in the blood vessels of the brain within 72 h after stroke attack. The related clinical conditions of part of patients are shown in Table 1.

One of the symptoms after stroke is limb dysfunction which is also the most obvious and serious symptom. All the patients recruited had a certain degree of motor dysfunction. In order to obtain the task-EEG from all the subjects, the experiments based on isokinetic movement [11] was launched. The experimental design is shown in Fig. 1.

Participants were at rest during baseline trials with the eyes open. ArmMotus, M2 rehabilitation robot (Fourier Intelligence Ltd., CN) was used during isokinetic movement trials. In the isokinetic push-pull paradigm, participants performed an isokinetic alternating push-pull task with their forearm on a gutter support and their hand holding a control stick. The robot provided visual and auditory feedback when the user moved the stick to the desired location.

Table 1. The clinical conditions of patients

Basic info					Assessment	
ID	Gender	Age	StrokeType	StrokeOnset	Bruunstorm	FMA
1	M	72	Ischemic	Apr.13,2020	4/4/3	20
2	F	70	Ischemic	Apr.14,2020	5/5/5	47
3	M	63	Ischemic	May.02,2020	5/5/5	50
4	M	63	Ischemic	May.01,2020	4/5/5	37
5	M	75	Ischemic	Aug.15,2020	4/3/5	19
6	M	64	Ischemic	Sept.17,2020	4/1/5	15
7	F	83	Ischemic	Apr.20,2020	5/5/5	45
8	F	72	Ischemic	Sept.24,2020	4/5/5	38
9	F	76	Ischemic	Jun.7,2020	5/4/5	51
10	M	67	Ischemic	Oct.30,2020	4/3/5	40

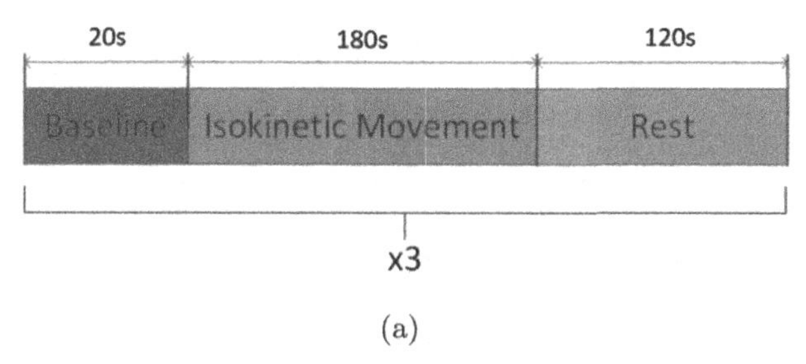

(a)

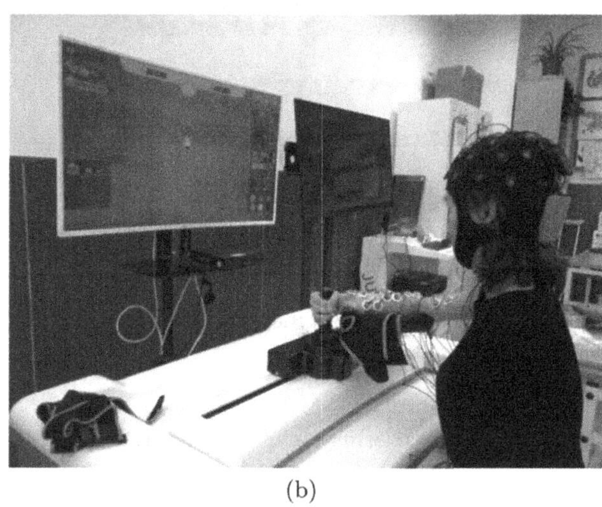

(b)

Fig. 1. Isokinetic movement diagram. (a) Isokinetic task includes 3 consecutive sessions of Baseline session, Isokinetic movement session and Rest session. (b) illustration of a patient performing the isokinetic movement on the rehabilitation robot

3 Methods

In this research, The EEG signals of 32 channels were obtained using the EEG machine from NE Enbio32 System and the recordings were sampled using frequency sampling 500 Hz. The electrodes were placed in accordance with international 10–20 system configuration. The whole data analysis process contains four parts as shown in Fig. 2. After the acquisition of the EEG signal, the second step is the pre-processing of the EEG signals acquired by the device since there would be many noises and artifacts in the signal. The pre-treatment mainly includes the ICA exclusion operation and the filtering operation with 1 Hz high pass filter and 45 Hz low pass filter. The pre-processing operation was realized using EEGLAB of Matlab. The third part is the feature extraction of the signal. Four qEEG characters were applied in this research as features for the classification which are RPR of Delta, Beta, DAR and DTABR. The power spectral of the EEG data of different channels were calculated using Welch method [12] as shown in the following equations:

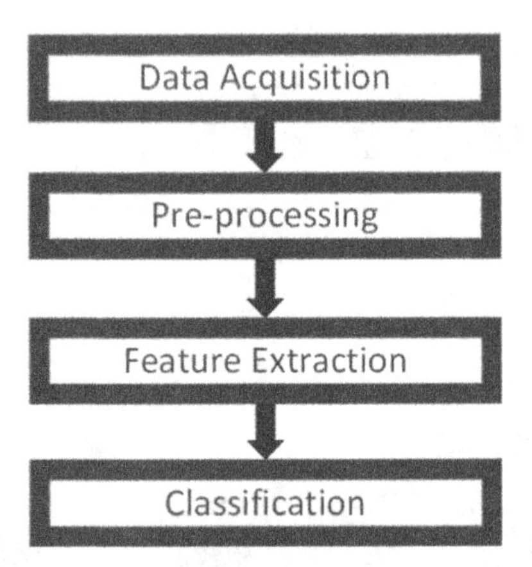

Fig. 2. Data analysis process

$$x_i(n) = x(n + iM - M), 0 \leq n \leq M, 1 \leq i \leq L \tag{1}$$

where x(n) is a series of data which is segmented into L sections and every section has M data points.

$$I_i(\omega) = \frac{1}{U}\left|\sum_{n=0}^{M-1} x_i(n)w(n)e^{-j\omega n}\right|^2, i = 1, 2..., M-1 \tag{2}$$

where w(n) is a window function and U is Normalization factor which is calculated as below:

$$U = \frac{1}{M}\sum_{n=0}^{M-1} w^2(n) \tag{3}$$

The periodic graphs of each segment is treated uncorrelated and the Power Spectral is estimated:

$$P(e^{j\omega}) = \frac{1}{L}\sum_{i=1}^{L} I_i(\omega) \tag{4}$$

Every epoch of the EEG data were segmented into 6 parts with 256 data points in every part to increase the frequency resolution. The calculation of Power Spectral of EEG was realized using Matlab toolbox "psd = pwelch (x)" based on Welch method. Then the relative power ratio of specific frequency band(Delta: 1–4 Hz, Theta: 4–8 Hz, Alpha: 8–13 Hz, Beta: 13–30 Hz) was extracted by calculating the ratio of power spectral of specific frequency band to that of all frequency bands. The DAR and DTABR feature were obtained:

$$DAR = \frac{RPR\ of\ Derta}{RPR\ of\ Alpha} \tag{5}$$

$$DTABR = \frac{RPR\ of\ Derta + Theta}{RPR\ of\ Alpha + Beta} \tag{6}$$

Quadratic discriminant analysis (QDA) is closely related to linear discriminant analysis (LDA), where it is assumed that the measurements from each class are normally distributed [13]. Unlike LDA however, in QDA there is no assumption that the covariance of each of the classes is identical. When the normality assumption is true, the best possible test for the hypothesis that a given measurement is from a given class is the likelihood ratio test. Suppose there are only two groups, with means μ_0, μ_1 and covariance matrices σ_0, σ_1 corresponding to $y = 0$ and $y = 1$ respectively. Then the likelihood ratio is given by:

$$LR = \frac{\sqrt{2\pi\left|\sum_1\right|}^{-1} exp(-\frac{1}{2}(x-\mu_1)^T \sum_1^{-1}(x-\mu_1))}{\sqrt{2\pi\left|\sum_0\right|}^{-1} exp(-\frac{1}{2}(x-\mu_0)^T \sum_0^{-1}(x-\mu_0))} < t \tag{7}$$

The QDA algorithm is applied in the research for the classification of the features using Matlab toolbox "[class, err]=classify(sample, training, group, 'quadratic')".

4 Results and Analysis

In this study, EEG data of 32 channels were obtained but only 10 channels were used for analysis. Since the research was based on task-EEG and the EEG signals were acquired during subjects' isokinetic movements, the channels located in the motor functional areas of the brain are particularly important. On this condition, EEG data from channels of 'F3', 'Fz', 'F4', 'FC5', 'FC1', 'FC2', 'FC6', 'C3', 'Cz', 'C4' were choosed for further analysis. The comparison of average RPR of different frequency bands and channels between stroke patients and healthy controls were analyzed which is shown in Fig. 3. It can be seen that there were differences in RPR values between patients with AIS and normal people.

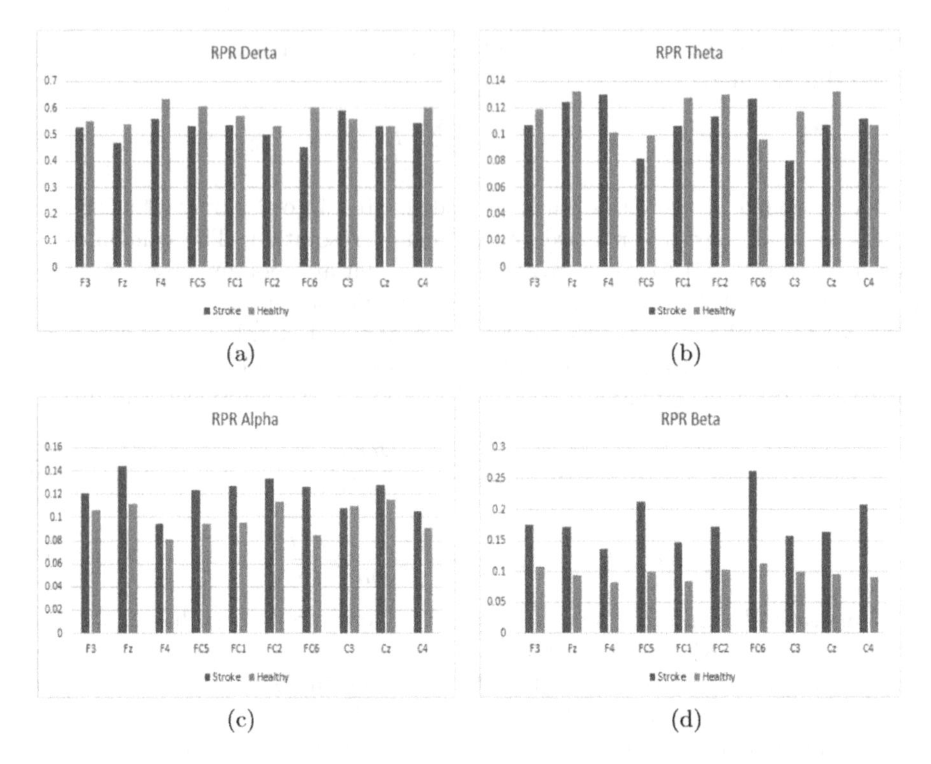

Fig. 3. RPR of different frequency bands in 10 channels from stroke patients and healthy controls. (a) RPR of Delta band. (b) RPR of Theta band. (c) RPR of Alpha band. (d) RPR of Beta band.

Unlike the pattern in resting-state EEG in which the RPR of delta and Theta band should be higher in stroke patients than healthy controls [14], it tends that in the pattern of task-EEG of this study, the RPR of Derta and Theta band were both higher in healthy controls than that in stroke patients. Meanwhile, RPR of Alpha and Beta band showed the opposite trend which were both higher in stroke patients than that in healthy controls. The comparison of DAR and DTABR between AIS patients and healthy controls were also analyzed. Figure 4 shows average DAR and DTABR of stroke patients and healthy controls in different channels. The results showed that the average DAR and DTABR are higher from Stroke patients than healthy controls in almost all the 10 channels. Meanwhile, the trend of the two features changing with the channel is also very similar.

In this study, the four features analyzed above were used for the classification. Before entering the features into the classifier for further training and classification, the four features need to be normalized to the same range (0 to 1), which was realized using the normalization tool in matlab. The 10 fold verification set was constructed using 167 samples of EEG data from different healthy controls and 125 samples of EEG data from different AIS patients. After the pre-

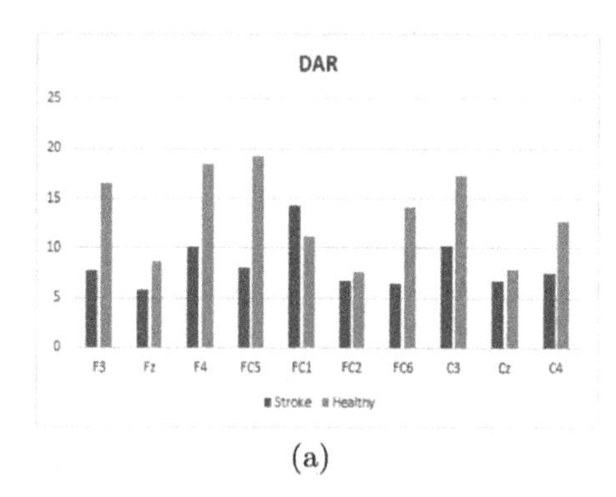

(a)

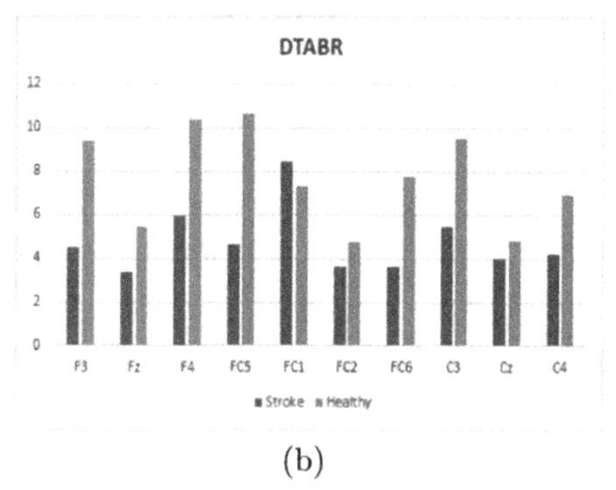

(b)

Fig. 4. DAR and DTABR in 10 channels from stroke patients and healthy controls. (a) DAR (b) DTABR

processing and feature extraction, 10 folds of data were imported into the QDA classifier for recognition. The performance of the classifier is shown in Table 2.

Table 2. The result of 10 folds cross validation

Cross validation	Accuracy (%)	Specificity (%)	Sensitivity (%)
Fold 1	63.33	41.18	92.31
Fold 2	89.21	76.47	1
Fold 3	60.71	43.75	83.33
Fold 4	66.67	58.82	76.92
Fold 5	72.42	75	69.23
Fold 6	65.52	52.94	83.33
Fold 7	63.33	47.06	84.62
Fold 8	60.34	58.82	58.33
Fold 9	71.43	56.25	91.67
Fold 10	70	58.82	84.62
Ave	68.296	56.911	72.536

From the table, it can be seen that the accuracy of the classification results were between 60% and 90% in all the 10 folds and the average accuracy is around 68.3%. Meanwhile, The sensitivity of the classifier was much more higher than the specificity which means that the ability of the classifier for recognizing the healthy controls was stronger than the ability for recognizing the stroke patients.

Three test sets were constructed from different patients and healthy controls each of which contained 120 samples. The classification results of the three test sets were shown in Table 3. The results were consistent with the results of cross validation. From the table, it can be seen that the average accuracy of these three test sets was around 63.17%.

Table 3. The classification result of 3 test sets

Test sets	Accuracy (%)	Specificity (%)	Sensitivity (%)
Set 1	66.67	33.33	84.62
Set 2	63.21	37.50	1
Set 3	59.62	36.26	91.67

5 Conclusion

The paper has proposed a classification method for AIS patients and normal people based on four qEEG features which are RPR of Delta, Beta, DAR and DTABR. The task-EEG data was collected under the specific experimental diagram. The Quadratic Discriminant Analysis algorithm was introduced as the classifier and the average classification accuracy on the test set was 63.17%. The results showed that the classifier had a strong ability of recognizing the healthy controls but with a weak ability of recognizing the stroke patients. This research can provide a reference for the future use of qEEG as a method of clinical diagnosis for stroke disease.

References

1. Whiteford, H.A., et al.: Global burden of disease attributable to mental and substance use disorders: findings from the global burden of disease study 2010. Lancet **382**(9904), 1575–1586 (2013)
2. Johnson, C.O., et al.: Global, regional, and national burden of stroke, 1990–2016: a systematic analysis for the global burden of disease study 2016. Lancet Neurol. **18**(5), 439–458 (2019)
3. Huang, C.-Y., et al.: Improving the utility of the brunnstrom recovery stages in patients with stroke: validation and quantification. Medicine **95**(31), e4508 (2016)
4. Signe Brunnstrom, S.: Motor testing procedures in hemiplegia: based on sequential recovery stages. Phys. Ther. **46**(4), 357–375 (1966)
5. Fugl-Meyer, A.R., Jääskö, L., Leyman, I., Olsson, S., Steglind, S.: The post-stroke hemiplegic patient. 1. a method for evaluation of physical performance. Scandinavian J. Rehabi. Med. **7**(1), 13–31 (1975)
6. Weeren, A.J.T.M., van Putten, M.J.A.M., Sheorajpanday, R.V.A., Nagels, G., De Deyn, P.P.: Reproducibility and clinical relevance of quantitative EEG parameters in cerebral ischemia: a basic approach. Clinical Neurophysiol. **120**, 845–855 (2009)
7. Minfen, S., Zhiwei, L.: Classification of mental task EEG signals using wavelet packet entropy and SVM. In: 8th International Conference on Electronic Measurement and Instruments. IEEE (2007)
8. Finnigan, S., van Putten, M.J.: EEG in Ischaemic stroke: quantitative EEG can uniquely inform (sub-) acute prognoses and clinical management. Clin. Neurophysiol. **124**, 10–19 (2013)
9. Foreman, B., Claassen, J.: Quantitative EEG for the detection of brain ischemia. In: Vincent, J.L. (ed.) Annual Update in Intensive Care and Emergency Medicine 2012. Annual Update in Intensive Care and Emergency Medicine, vol 2012. Springer, Heidelberg (2012). https://doi.org/10.1007/978-3-642-25716-2_67

10. Nair, G.J., Sita, J.: Feature extraction and classification of EEG signals for mapping motor area of the brain. In: International Conference on Control Communication and Computing (ICCC), pp. 463–468. IEEE (2013)
11. Chen, K., Ye, F., Ding, J.S., et al.: Investigation of corticomuscular functional coupling during hand movements using vine copula. Brain Sci. **12**, 754 (2022)
12. Welch, P.: The use of fast fourier transform for the estimation of power spectra: a method based on time averaging over short, modified periodograms. IEEE Trans. Audio Electroacoust. **15**(2), 70–73 (1967)
13. Tharwat, A.: Linear vs. quadratic discriminant analysis classifier: a tutorial. Int. J. Appl. Pattern Recogn. **3**(2), 145–180 (2016)
14. Badri, C., Rahma, O.N., Wijaya, S.K.: Electroencephalogram analysis with extreme learning machine as a supporting tool for classifying acute ischemic stroke severity. In: International Seminar on Sensors, Instrumentation, Measurement and Metrology (ISSIMM), pp. 180–186. IEEE (2017)

Micro or Nano Robotics and Its Application

Non-destructive Two-Dimensional Motion Measurement of Cardiomyocytes Based on Hough Transform

Si Tang[1,2,3,4], Jialin Shi[1,2], Huiyao Shi[1,2,3], Kaixuan Wang[1,2,3], Chanmin Su[1,2], and Lianqing Liu[1,2(✉)]

[1] The State Key Laboratory of Robotics, Shenyang Institute of Automation, Chinese Academy of Sciences, Shenyang 110016, China
lqliu@sia.cn

[2] Institutes for Robotics and Intelligent Manufacturing, Chinese Academy of Sciences, Shenyang 110169, China

[3] University of Chinese Academy of Sciences, Beijing 100049, China

[4] Institute for Stem Cell and Regeneration, Chinese Academy of Sciences, Beijing 100101, China

Abstract. Cardiomyocytes, as one of the few biological cells capable of autonomous beating in vitro, have attracted more and more attention in the fields of biology and robotics. In order to better understand the internal control mechanism of cardiomyocytes as a driver, scholars at home and abroad have proposed many measurement methods for the motion parameters of cardiomyocytes in the past 10 years, and the most common one is two-dimensional measurement. However, existing 2D cardiomyocyte measurement methods are always limited by substrate materials or cannot maintain long-term nontoxic measurements. Here, we proposed in-plane beating measurements of cardiomyocytes based on the microsphere Hough transform under a bright field. Through off-line processing, accurate tracking of the cardiomyocyte beating cycle can be achieved. Due to the simplicity, non-toxicity, and efficacy of the proposed protocol, this method will enable more rapid and accurate detection of cardiomyocyte beating. After further optimizing the data acquisition and data processing methods, we believe that this method will provide a more efficient and useful idea for real-time cardiomyocyte motion measurement.

Keywords: Cardiomyocytes · Hough transform · Measurement

1 Introduction

The state and mechanical properties of living cells are largely influenced by the substrate on which they are grown [1]. Cardiomyocytes, as one of the few cells that can achieve autonomous and regular beating in vitro, have attracted more and more attention from researchers in the field of robotics because they can move without being restricted

H. Liu et al. (Eds.): ICIRA 2022, LNAI 13456, pp. 559–567, 2022.
https://doi.org/10.1007/978-3-031-13822-5_50

by external stimuli. For cardiomyocytes, mainly its contractile force or stress, contractile strain, beating rate, beating rhythm. In recent decades, many scholars have studied related detection methods, such as micro/nanopillar array [2], Traction force microscopy (TFM) [3], Video-based cell motion detection [4], and so on. But they all have their shortcomings. Micropillar/nanopillar arrays can map multidirectional forces with subcellular spatial resolution; however, the limited area of attachment contact between cells and pillars leads to localized stress concentrations on the cell membrane, which may affect cell morphology and physiological properties through mechanotransduction [5]. TFM is to embed nano-scale fluorescent microbeads into the substrate as a target to mark the movement of cells on the substrate [6]. However, the quenching phenomenon of fluorescent dots cannot be avoided under long-term recording, which makes it impossible to take long-term measurements. At the same time, due to the toxicity of fluorescence itself, the impact on cells is unavoidable. Video-based cell motion detection essentially uses optics to reflect cardiomyocytes' shape or continuous motion. However, when cardiomyocytes cover the entire substrate, their highly confluent, thin, and flat shape undoubtedly makes this measurement method extremely difficult [7]. Therefore, in order to obtain a more prehensive movement of cardiomyocytes, a long-term, universal, and non-cytotoxic detection method is urgently needed.

In this paper, we proposed an in-plane two-dimensional measurement method based on the microsphere Hough transform, in which we can achieve cardiomyocytes beating measurements for up to 4 h. And because the microspheres themselves are not cytotoxic, therefore, the stimulation of the production of cells by the measurement method can be minimized. In the whole experiment, the measurement method based on the microsphere Hough transform can always maintain its effectiveness. This facile approach will provide a new method for future studies of cardiomyocyte beating parameters.

2 Materials and Method

2.1 Cell Culture

Cardiomyocytes were obtained from neonatal SD rats (Liaoning ChangSheng Biology Co., Ltd.). The hearts were directly removed from the neonatal rats and placed in Hank's Balanced Salt Solution (HBSS, pH 7.4, Ca^{2+} and Mg^{2+} free, Sigma), washed three times, and then placed in a solution containing trypsin (0.02%, w/v, Gibco), type II Collagenase (0.05%, w/v, Worthington) and pancreatin (0.02%, w/v, Gibco) in HBSS. Digest 4 times on a constant temperature shaker, and mix with an equal volume of DMEM/F-12 medium (with 20% v/v FBS, Gibco) solution after digestion. The mixed solution was centrifuged at 1000 rpm for 10 min and then centrifuged pellet was collected and thawed with DMEM/F-12 solution (with 20% (v/v) FBS, Gibco, and 0.1 mM BrdU, Sigma). The mixed cell suspension was placed in a petri dish, and the cardiomyocytes and fibroblasts were separated by the differential adherence method in a 37 °C incubator. After culturing for 1.5 h, the medium was aspirated and replaced with a new medium.

2.2 Hough Transform

Hough Transform was first proposed by Paul Hough in 1962, which is mainly an algorithm to form a closed graph by connecting edge pixels [8]. Hough transform is divided into ordinary Hough transform (HT) and Circular Hough Transform (CHT). No matter what kind of Hough transform, the fundamental idea is to map the original image space to the parameter space. The points of most conditions are connected to draw the final curve. Since this method involves voting, it is also highly fault-tolerant and adaptable.

Specifically for CHT, its biggest constraint is the circle equation.

$$(x - x_0)^2 + (y - y_0)^2 = r^2 \tag{1}$$

(x_i, y_i) is the coordinate of the center of the circle, and r is the radius value corresponding to the circle. The mapping relationship between Cartesian coordinates and polar coordinates is as follows:

$$\begin{cases} x = x_0 + rcos\theta \\ y = y_0 + rsin\theta \end{cases} \tag{2}$$

In the parameter space, each foreground point in the image space is used as the center of the circle. At this time, the r is unknown, and the results of each calculation are accumulated. The peak point in the parameter space is the point with the most votes, that is, through the point where the center of the circle is calculated by CHT. However, because of the unknown r, the calculation of CHT will take a relatively long time. For this reason, an interval is usually set manually to reduce the calculation time.

Under white light conditions, although the background noise is greatly reduced, the effect of direct CHT tracking is not ideal. Therefore, before the CHT tracks the movement of the microspheres, the collected images need to be preprocessed first. In this paper, binarization and median filtering will be used to reduce the influence of background noise

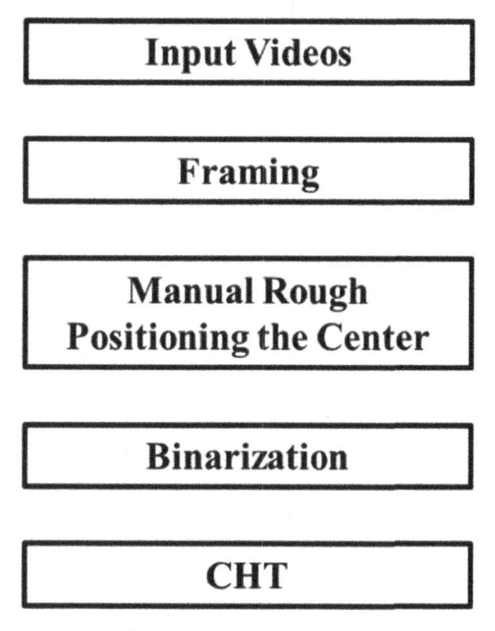

Fig. 1. Algorithm flowchart

on the actual measurement signal. The specific process of identification is as follows (Fig. 1).

3 Results

The system presented in this paper is shown in Fig. 2. The petri dish containing the adherent cardiomyocytes and culture medium was placed on an inverted fluorescence microscope (Ti-2E, Nikon, Inc., Japan), and the light source was the white light source of the microscope. In the absence of any external stimulation, the solution of SO_2 microspheres with a diameter of 10 μm is evenly sprinkled on the cells, and the membrane proteins on the cell membrane bind to the microspheres. This CHT ensures the microspheres and synchronized motion of cardiomyocytes. Since it takes time for the microspheres to bind to the proteins on the cell membrane, so the experiment needs to wait for the stable synchronous movement of cells and microspheres on cardiomyocytes before CMOS starts recording (E3ispm2000kpa, touptek). In order to ensure that the motion of cardiomyocytes can be accurately recorded in each frame, after referring to the previous work [9], we choose 100 frames per second as the sampling bandwidth.

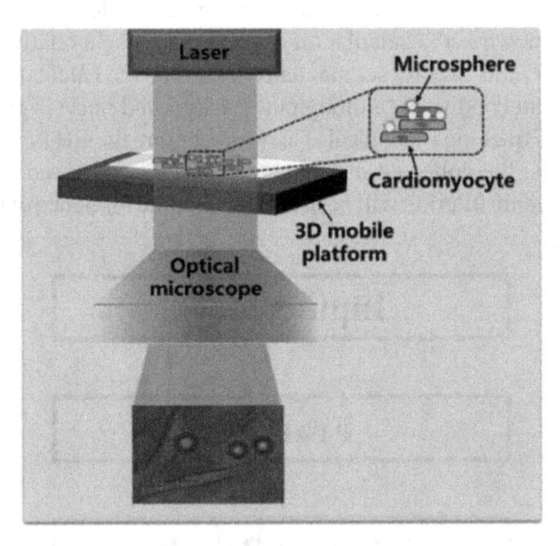

Fig. 2. Scheme of the experimental system

3.1 Cardiomyocytes Beating Cycle Measurement

The cardiomyocytes cultured in vitro produce beating in about 24 h, but the beating is weak at this time, which is not suitable for measurement in the case of microspheres. In this paper, the cardiomyocytes were cultured for 72 h. At this time, the cardiomyocytes showed a large, long-term, and stable beating state. In this state, the influence of the weight of the microspheres on the beating of the cells can also be minimized. Figure 3 shows the scene of the cardiomyocytes cluster and microspheres placed on the cardiomyocytes cluster under white light.

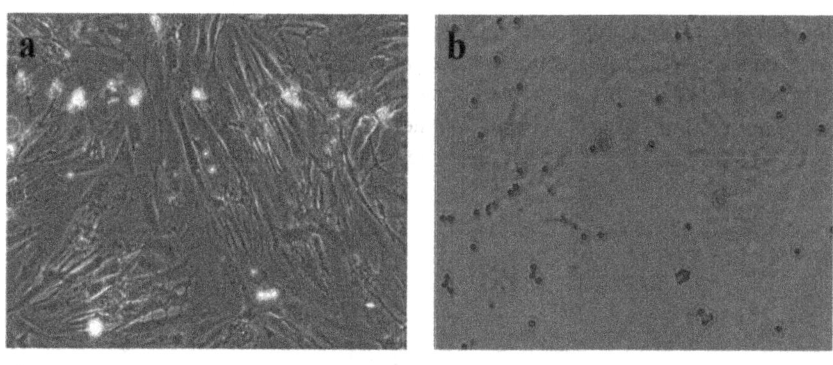

Fig. 3. a: Cardiomyocytes cluster under white light; b: Microspheres and cardiomyocytes cluster co-beat

3.2 Cardiomyocytes Beating Curve Analysis

Under the condition that the room temperature is kept at 25 °C, the SO_2 microsphere solution diluted with the medium is evenly sprinkled in the medium with a Pasteur tube. When the microsphere produces a regular beat, start recording data. The isolated cardiomyocytes can maintain a similar beating frequency for about three and a half hours after the experiment. Between 3.5 h to 4 h, the beat amplitude decreases. Figure 4 shows the in-plane beating of cardiomyocytes after leaving the incubator for 1.5 h.

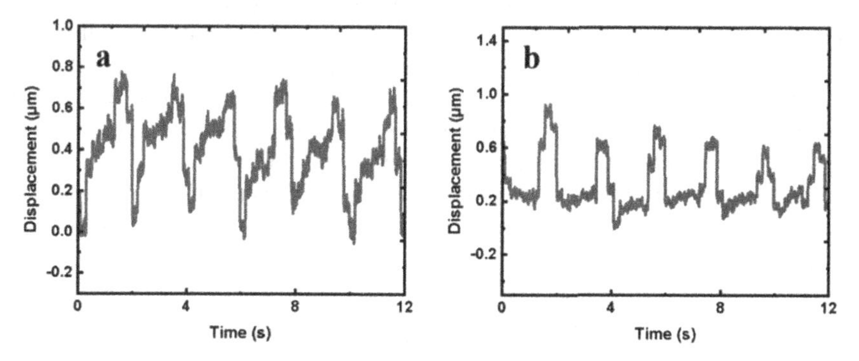

Fig. 4. Beating curve of cardiomyocytes. a: x-direction beat curve; b: y-direction beat curve

It can be seen from Fig. 4 that the beating curves of cardiomyocytes in different directions of motion are not consistent, which may be because the experimental measurement is not a single cardiomyocyte, but the cardiomyocyte cluster composed of multiple cardiomyocytes. In the whole cardiomyocyte clusters, the wave beating state, beating amplitude, and beating rhythm of each cardiomyocyte are not completely consistent. Therefore, when multiple beating states are superimposed, the simple beating initially caused by sarcomere contraction will show different motion states in different directions.

In fact, the beating conditions of cardiomyocytes in different positions also show great differences, which are mainly reflected in the beating frequency and beating amplitude. In order to prove this, we selected another cluster of cardiomyocytes in different positions of the same culture dish and obtained the beating curve of the cardiomyocytes in the same way as above. The beating curve of the cardiomyocytes is shown in Fig. 5.

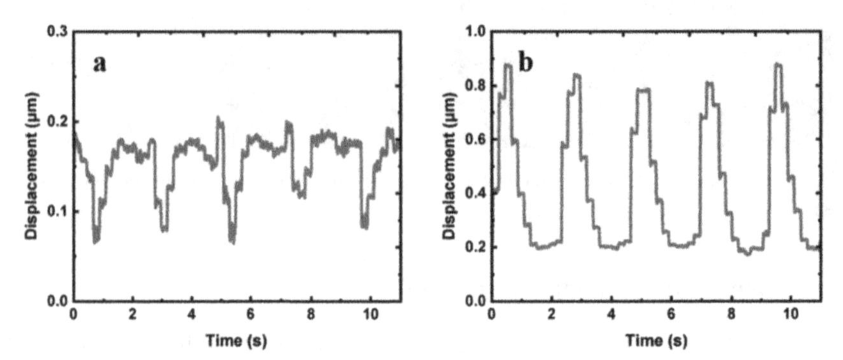

Fig. 5. Another beating curve of cardiomyocytes. a: x-direction beat curve; b: y-direction beat curve

In Fig. 5, the y-direction curve has an obvious stepped sawtooth. All data were obtained under the same conditions, so there was no shortage of sampling points. Combined with the case of focusing on cardiomyocytes, it can be seen that because the cardiomyocytes leave the incubator for a long time at room temperature, and the beating of different cardiomyocytes affects each other, the point where the microspheres are located forms a state of multiple force containment. Under the influence of the joint motion of cardiomyocytes with different forces and different beating frequencies, this kind of curve similar to an insufficient sampling rate is generated.

In the above two experiments, microspheres with a diameter of 10 μm were used. Due to the weight of the ball itself, there may be some stimulation to the cardiomyocytes. At

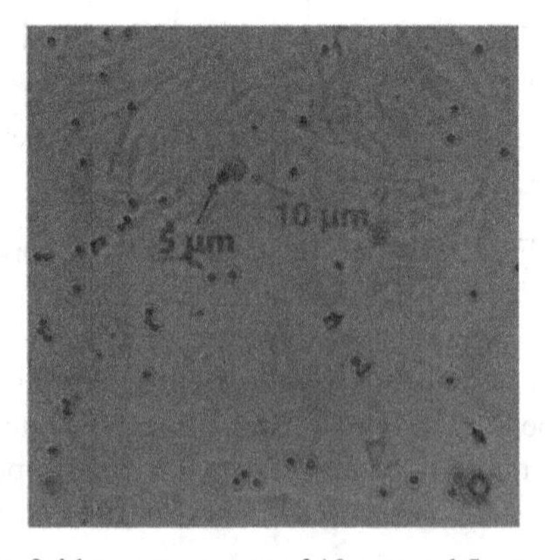

Fig. 6. Bright field measurements of 10 μm and 5 μm microspheres

the same time, when the size of the microspheres is further reduced, the microspheres can stay in a more detailed position. In other words, in the case of measurability, the smaller the size of the microspheres, the more points and information that can be measured. Figure 6 is the measurement effect diagram of the microsphere diameter of 10 μm and the microsphere diameter of 5 μm, respectively.

3.3 Widely Non-cytotoxic Measurement

In the measurement scenario based on the CHT of the microspheres, the only contact of the cardiomyocytes that may have an effect is the microspheres themselves. The microspheres used in the experiment are SO_2 microspheres, which can exist stably in the salt solution and will not react with the culture medium and buffer. The SO_2 at the time of purchase is uniformly dispersed in buffer, so the measurement tool does not affect the state of motion of the cardiomyocytes. Therefore, even at any time, myocardial cells can be synchronously detected by microspheres. However, the status of cardiomyocytes itself will be affected by the growth environment, culture medium environment, whether there are bacteria around, etc. When cardiomyocytes are measured in a bacterial environment at room temperature for a long time, their movement status will inevitably change. Figure 7 shows the XY beating curve of the same cardiomyocyte at an interval of 2 h. As can be seen from the figure, although the beating of the cardiomyocytes just placed into the microspheres was regular, the beating amplitude and frequency were not consistent. After 2 h, the beating curve of the cardiomyocytes showed a common regular movement. At the same time, the beating frequency of heart muscle cells also showed a significant decrease. This matches what we see under a light microscope.

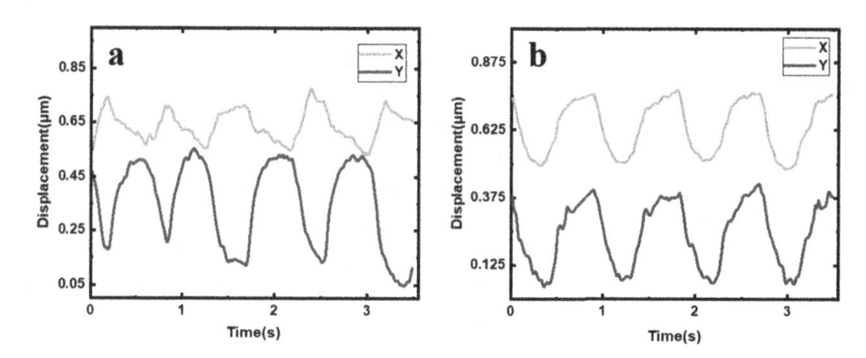

Fig. 7. Beating curve of the same myocardial cell at different time

In the method based on microsphere measurement, the measurement range is no longer limited to a small range. In other words, the measurement range of our proposed measurement method is limited to the imaging range of the microscope. The measurement range of 50X is about 400 μm * 250 μm. When the magnification is reduced, the measurement range can be further expanded. At present, the most accurate measurement method of cardiomyocytes is AFM, and the measurement range of AFM is only 100 μm * 100 μm. In the case of nanoscale measurements, the measurement range of our proposed method is much larger than that of AFM.

4 Discussion

Through experiments, we preliminarily demonstrated the effectiveness of our proposed method for cardiomyocyte measurement. Under our scheme, the motion calibration of cardiomyocytes can be achieved quickly and accurately without the need for high-precision systems, cytotoxic targets, and complex image recognition methods. But in the whole measurement, we do not consider the effect of the ball's weight on the beating of the cardiomyocytes. Cardiomyocytes have been shown to be sensitive to basal or external stimuli in previous studies [10]. As a mechanical stimulus, the existence of the microsphere itself and the stimulation of the cardiomyocytes by the force generated when they move together with the cardiomyocytes are currently unknown.

In addition, in this paper, the measured objects are clustered cardiomyocytes. When the cardiomyocytes grow in clusters, the movement form and state of the cardiomyocytes are significantly different from that of individual cardiomyocytes. Previously, whether it was a one-dimensional AFM measurement or a 2D micro-pillar measurement, the object was a single cardiomyocyte, and the motion, mechanics, and mathematical modeling based on this were all references to the single cardiomyocyte. This paper focuses on clustered cardiomyocytes and uses microspheres to reflect the overall beating pattern and beating frequency of cardiomyocyte clusters. Therefore, the measurement results proposed in the text will inevitably deviate from the previous single-cell results.

5 Conclusion

To summarize, we describe a new method for tracking the beating of cardiomyocytes. Under ordinary white light, we perfectly reproduced the multicellular motion of cardiomyocytes in the medium by tracking the motion of the microspheres. The cell beating rhythm measured by the microsphere method is consistent with the previous research [11], the frequency is around 2 Hz, which proves the validity of the microsphere measurement. At the same time, the motion curve in the 12 s period is drawn. The method proposed in this paper is only a first step in the case of white light. By further optimizing the tracking algorithm, this method will be used to realize the simultaneous online measurement of cardiomyocytes' force and beating in the future, which provides a useful tool for the drug screening of cardiomyocytes in vitro. A quicker measurement method that is easier to operate.

Funding. Supported by the National Natural Science Foundation of China (Grants 61925307, 61927805, 62127811, and 61903359), the Key Research Program of Frontier Sciences, CAS (Grant No. YJKYYQ20210050), 2021 Liaoning Provincial Natural Science Foundation (Grant No. 2021-BS-151), Strategic Priority Research Program of the Chinese Academy of Sciences (XDA16021100).

References

1. Stevens, M.M., George, J.H.: Exploring and engineering the cell surface interface. Science **310**(5751), 1135–1138 (2005)
2. Fu, J., Wang, Y., Yang, M.T., Desai, R.A., Yu, X., et al.: Mechanical regulation of cell function with geometrically modulated elastomeric substrates. Nat. Methods **7**(9), 733–736 (2010)
3. Dou, W., Malhi, M., Zhao, Q., Wang, L., Huang, Z., et al.: Micro engineered platforms for characterizing the contractile function of in vitro cardiac models. Microsyst. Nanoeng. **8**(1) (2022)
4. Hansen, K.J., Favreau, J.T., Gershlak, J.R., Laflamme, M.A., Albrecht, D.R., et al.: Optical method to quantify mechanical contraction and calcium transients of human pluripotent stem Cell-Derived cardiomyocytes. Tissue Eng. Part C Methods **23**(8), 445–454 (2017)
5. Callaghan, N.I., Hadipour-Lakmehsari, S., Lee, S., Gramolini, A.O., Simmons, C.A.: Modeling cardiac complexity: advancements in myocardial models and analytical techniques for physiological investigation and therapeutic development in vitro. APL Bioeng. **3**(1), 11501 (2019)
6. Wang, J.H., Lin, J.: Cell traction force and measurement methods. Biomech. Model. Mechanobiol. **6**(6), 361–371 (2007)
7. Ribeiro, A.J.S., Schwab, O., Mandegar, M.A., Ang, Y., Conklin, B.R., et al.: Multi-Imaging method to assay the contractile mechanical output of micropatterned human iPSC-Derived cardiac myocytes. Circ. Res. **120**(10), 1572–1583 (2017)
8. Illingworth, J., Kittler, J.: A survey of the Hough transform. Comput. Vis. Graph. Image Process. **44**, 87–116 (1988)
9. Ahamadzadeh, E., Jaferzadeh, K., Park, S., Son, S., Moon, I.: Automated analysis of human cardiomyocytes dynamics with holographic image-based tracking for cardiotoxicity screening. Biosens. Bioelectron. **195**, 113570 (2022)
10. Siddique, A., Shanmugasundaram, A., Kim, J.Y., Roshanzadeh, A., Kim, E., et al.: The effect of topographical and mechanical stimulation on the structural and functional anisotropy of cardiomyocytes grown on a circular PDMS diaphragm. Biosens. Bioelectron. **204**, 114017 (2022)
11. Fu, F., Shang, L., Chen, Z., Yu, Y., Zhao, Y.: Bioinspired living structural color hydrogels. Sci. Robot. **3**(16) (2018)

Bubble Based Micromanipulators in Microfluidics Systems: A Mini-review

Yuting Zhou[1,2,3], Liguo Dai[4], Niandong Jiao[1,2(✉)], and Lianqing Liu[1,2]

[1] State Key Laboratory of Robotics, Shenyang Institute of Automation, Chinese Academy of Sciences, Shenyang 110016, China
ndjiao@sia.cn

[2] Institutes for Robotics and Intelligent Manufacturing, Chinese Academy of Sciences, Shenyang 110016, China

[3] University of Chinese Academy of Sciences, Beijing 100049, China

[4] College of Mechanical and Electrical Engineering, Zhengzhou University of Light Industry, Zhengzhou 450002, China

Abstract. Bubbles in liquid have the advantages of controllability, compressibility and biocompatibility, so they are introduced into microfluidic system to drive the fluid and operate micro-objects including cells. In recent years, the acoustic and optothermal bubbles are the two most widely used and efficient bubbles in microfluidic devices. Therefore, the aim of this study is to review recent advances in acoustic bubble-based micromanipulators and optothermal bubble-based micromanipulators in microfluidic systems. The principles and applications of fluid control and micro-object operation of these two kinds of bubble-based manipulators are introduced and the prospects and challenges are discussed.

Keywords: Acoustic bubble · Optothermal bubble · Fluid control · Micro-objects manipulation

1 Introduction

In the microfluidic devices, the Reynolds number of fluid is much smaller than that of macro fluid, which makes the driving of microfluidic more difficult, as well as the operation of micro-objects. Based on the principle of bubble induced acoustic microstreaming, liquid flow can be promoted and controlled by microbubbles, and micro-objects can be operated under the action of Strokes drag force and Bjerknes force in different microstreaming patterns. In addition, the volume change of optothermal bubbles in the process of growth and disappearance will trigger the flow field, which can realize the control of the fluid. Under the combined action of Marangoni effect and balance between surface tension and pressure force, microparticles can be captured on the surface of optothermal bubbles and move with the movement of bubbles, and the microstructures can be pushed to the specified position and adjusted to expected three-dimensional (3D) gesture. In this review, we introduce and discuss recent advances in bubble-based micromanipulators in microfluidic systems. In Sect. 2.1 and Sect. 2.2, we introduce the

H. Liu et al. (Eds.): ICIRA 2022, LNAI 13456, pp. 568–580, 2022.
https://doi.org/10.1007/978-3-031-13822-5_51

working principle, flow control and micro-object manipulation of the acoustic bubble-based manipulator and optothermal bubble-based manipulator, respectively. Finally, we summarize the current limitations of these two bubble-based manipulators and discuss the future development.

2 Results and Discussion

2.1 Acoustic Bubble-Based Manipulator

Working Principle. Acoustic oscillating bubbles include inertial and non-inertial cavitations [1]. Inertial cavitation means that the bubble expands and shrinks sharply or even collapses when the oscillation amplitude is high enough and above a certain threshold. Most oscillating bubbles in microfluids are stable, non-inertial cavitation bubbles. Oscillating bubbles have their natural frequencies, and when they have the same frequency as the exciting acoustic field, the bubbles will reach a maximum vibration amplitude. In microfluidic devices, most acoustic bubbles are trapped in the tube or microchannel and play a role. The resonance frequency f_0 of the bubble in the tube or microchannel is

$$f_0 = \frac{1}{2\pi}\sqrt{\frac{\kappa P_0}{\rho L_0 L_B}} \quad (1)$$

where κ is the polytropic index, P_0 is the constant far-field pressure, ρ is the density of the fluid around the bubble, and L_0 and L_B are the lengths of the liquid in the tube and the bubble column respectively. The microbubbles resonate under the action of acoustic wave, the contact surface between microbubbles and surrounding liquid will vibrate, which will trigger microstreaming and change the flow direction and speed of microfluids with different acoustic excitation frequency [2]. Ahmed et al. found that using ultrasound to oscillate bubbles trapped in a "horse-shoe" structure inside microtubules can affect fluid flow (Fig. 1a) [3]. With the different size and shape of bubbles, the flow microstreaming excited by acoustic bubbles is also different. Therefore, acoustic microbubbles are used to fabricate micromixer, micropump and microvalve.

In addition to manipulating microfluidics, acoustic bubbles can also manipulate particles and micro-objects such as cells. The particle close to the acoustic bubble is affected by two forces at the same time (Fig. 1b) [4]. One is the viscous resistance caused by the streamline direction of the micro flow field, that is, Stokes drag force f_d

$$f_d = 6\pi\eta R_P v_P \quad (2)$$

where η is the dynamic viscosity, R_P is the radius of the particle, and v_P is the flow velocity of solvent relative to the particles, which is also depend on R_P. The other is Bjerknes force f_R caused by the scattering effect of the bubble on the incident acoustic wave, also known as the second radiation force, for a rigid spherical particle

$$f_R = 4\pi\rho\frac{\rho - \rho_P}{\rho + 2\rho_P}\frac{R_b R_P}{d^5}\omega^2\varepsilon^2 \quad (3)$$

where ρ is the density of the fluid of the fluid around the bubble, ρ_P is the density of particles, R_b is the instantaneous radii of a bubble, d is the distance between the bubble and particle centres, ω is the angular frequency and ε is the bubble amplitude. Under different excitation frequencies, the flow field patterns around acoustic bubbles are different. By changing the magnitude of these two forces, a variety of motion patterns of particles can be realized, including capture, rotation and revolution.

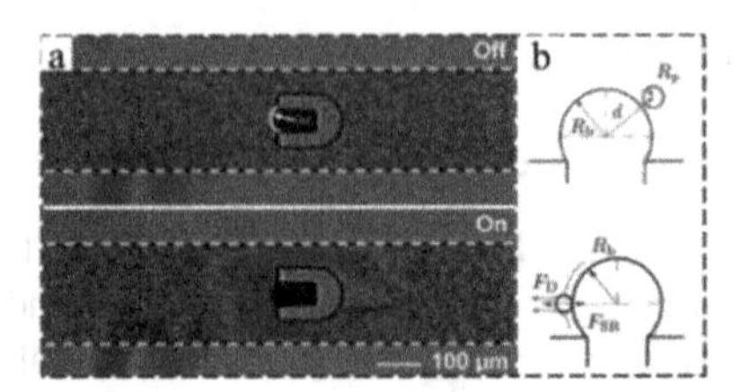

Fig. 1. (a) The acoustic streaming pattern around a bubble located in the horse-shoe structure of the microfluidic pipe. Adapted from Ahmed et al. [3] with permission from Royal Society of Chemistry, Copyright 2009. (b) The geometry and force analysis diagram of the particle trapped on the surface of the microbubble. Adapted from Chen et al. [4] with permission from Royal Society of Chemistry, Copyright 2016.

Flow Control. Cavitation microstreaming generated by acoustic bubbles can enhance fluid mixing, and the induced flow field and mixing intensity can be changed by changing the number [3], position and shape [5], arrangement and mixing area [6] of bubbles in the microfluidic pipe. Orbay et al. [7] introduced an acoustic-bubble based micromixer where the uniform mixing of two high viscosity liquids with low Reynolds number can be realized. Bertin et al. prepared an armoured microbubbles (AWBs) to overcome the problem of short service life of microbubbles [6]. Recently, Conde et al. [8] proposed a hybrid approach combining a substrate with a slab, in which the substrate contained mixing chamber and microfluidic channel, and the slab contained the structure required to capture bubbles (Fig. 2a). This method reduces the acoustic energy lost on the substrate and further improves the mixing efficiency. Micropump based on acoustic oscillation microbubbles have attracted extensive attention because of their simple structure and high specificity [9]. Patel and colleagues proposed a bubble-based micropump [10]. The bubbles in lateral cavity acoustic transducers (LCATs) produce oscillating flow velocity due to the oscillating movement of the air/liquid interface. This structure can separate particles or cells of different sizes. Recently, Gao et al. [11] placed opposite microcavity in the main microchannel to realize efficient bidirectional micropump (Fig. 2b). The flow direction of fluid and particles can be changed by actuating specific microbubbles of opposite directions and different sizes with different acoustic field driving frequencies. As for microvalve, acoustic bubbles can be used to adjust the chemical concentration as a chemical switch [12]. Ahmed et al. utilized the acoustic oscillating bubbles to generate digital chemical waveforms and analog waveforms, and the shape, frequency, amplitude and duty cycle can be easily modulated at the same time. It's a powerful tool for the investigation and characterization of the dynamic properties of many biochemical processes. (Fig. 2c) [13]. Similarly, Liu et al. [14] proposed a tunable chemical gradient generator based on the gas permeability of polydimethylsiloxane (PDMS). The adjustable chemical gradient was realized by changing the excitation combination of bubbles in different channels. In general, the microstreaming generated by acoustic oscillation bubbles can be driven far beyond Stokes boundary layer, and can produce effective fluid control even when dealing with high viscosity fluids. In addition, such acoustic bubble fluid controllers are generally economical and low power consumption.

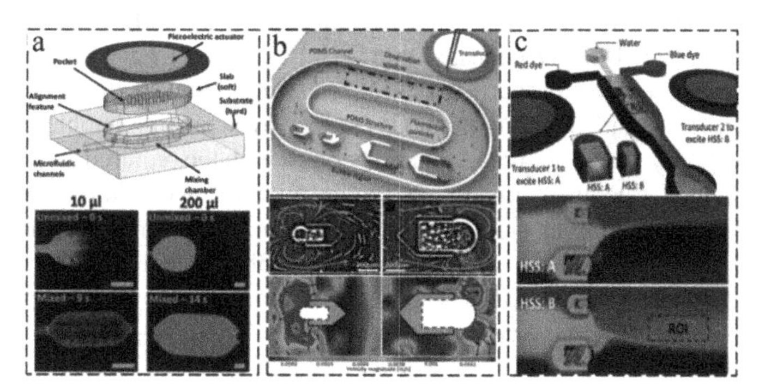

Fig. 2. (a) Up: configuration of the hybrid micromixer. Down: micrographs showing before and after mixing. Adapted from Conde et al. [8] with permission under the terms of CC BY 3.0 license, Copyright 2016. (b) Up: configuration of the acoustic-based bidirectional micropump device. Down: acoustic bubble induced microstreaming flow and the velocity profile given by particle image velocimetry (PIV). Adapted from Gao et al. [11] with permission from Springer Nature, Copyright 2020. (c) Up: schematic of the experimental setup for chemical switching. Down: results demonstrating switching between the blue and red dyes. Adapted from Ahmed et al. [13] with permission under the terms of ACS Author Choice License, Copyright 2014. (Color figure online)

Micro-object Manipulation. The microstreaming caused by oscillating bubbles (inertial and non-inertial) excited by acoustic field can make cell lysis, perforation and deformation. Low intensity ultrasound leads to stable cavitation of microbubbles, which is conducive to the cell formation of small pores and endocytosis. High intensity ultrasound can lead to inertial cavitation and bubble collapse, resulting in cell membrane perforation. In 2019, Meng et al. [15] utilized acoustic microbubbles oscillated with almost the same amplitude and resonance frequency to ensure efficient and uniform acoustic perforation. Recently, Liu et al. [16] realized efficient cell lysis on lab-on-a-chip (LOC) device by using acoustic driven oscillating bubble array. Subsequently, they fabricated a microfluidic device where the cells can be captured and paired with the oscillating bubble surface under acoustic excitation, further realizing the fusion of homotypic or heterotypic cell membranes [17]. Recently, they designed a traveling surface acoustic wave (TSAW) device composed of TSAW chip and PDMS channel to explore single-cell ultrasonic perforation using targeted microbubbles (TMB) in non-cavitation (Fig. 3a) [18]. TSAW was applied to accurately manipulate the movement of TMBs attached to MDA-MB-231 cells, resulting in ultrasonic deformation and reversible perforation at the single cell level.

In microfluidic devices, microbubbles can capture, rotate and operate cells under acoustic excitation through the joint action of Stokes force and Bjerknes force. In these examples, bubbles are usually spontaneously generated due to the hydrophobicity of the structure and trapped in microchannels or microcavities. Ahmed et al. and Ozcelik et al. proposed a technique of acoustofluidic rotational manipulation (ARM) based on the oscillating bubbles trapped within sidewall microcavities of the microfluidic channel to realize the rotation of nematodes, single cells and organisms [19, 20]. Läubli et al. used similar methods to rotate plant cells to achieve high-resolution 3D reconstruction [21]. Tang et al. then realized the 3D reconstruction of HeLa cells [22]. Peng [23] manufactured a microfluidic device where the particles can be focused, captured, extracted and enriched in a non-contact, unmarked and continuous manner by activating bubbles at a specific frequency. Recently, Gao et al. [24] captured bubbles with the required size and position in the microcavity structure of monolayer PDMS microchannels. In the low-frequency acoustic field, the oscillating motion of bubbles induced microfluidic vortices, which were used to break blood clots in blood samples (Fig. 3b). Then, they proposed a tumor-on-a-chip platform (ABSTRACT platform) based on acoustic oscillating bubbles to capture and rotate CTCs to the desired position [25].

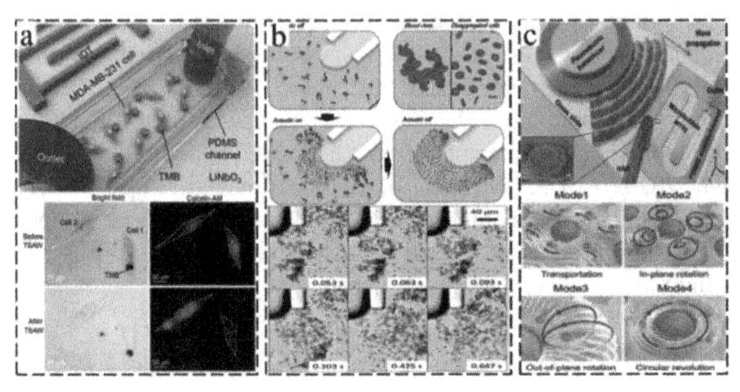

Fig. 3. (a) Up: schematic of the experimental device. Down: bright field and calcein-AM images of the sonoporation achieved before and after the TSAW treatment. Adapted from Liu et al. [18] with permission under the terms of CC BY 4.0 license, Copyright 2022. (b) Up: working mechanism of acoustic bubble-based thrombolysis. Down: blood clots disaggregated to single cells by the acoustically excited bubble. Adapted from Gao et al. [24] with permission from Royal Society of Chemistry, Copyright 2021. (c) Up: schematic of the acoustofluidic multimodal manipulation device. Down: diverse modes of acoustic microstreaming and sample manipulation. Adapted from Zhang et al. [26] with permission from Royal Society of Chemistry, Copyright 2021.

Acoustic bubbles in microfluidic devices can also realize particle and cell sorting. Rogers et al. [27] selectively captured particles by acoustically excited oscillating bubbles. According to the size and resonance frequency of the bubbles, the interaction between Stokes force and Bjekness force was different, resulting in particle attraction or repulsion. Recently, Meng et al. [28] effectively separate two cells with the same size distribution through surface acoustic waves and targeted microbubbles. By specifically adhering targeted microbubbles to MDA-MB-231 cells, the acoustic sensitivity of the cells can be significantly improved, so that MDA-MB-231 cells can be separated from MCF-7 cells. Acoustically excited oscillating bubbles can also realize the transmission and multiple manipulation of particles. Xie et al. [29] controlled the trajectory of particles in a LOC device by using oscillating bubbles trapped in "horseshoe" structure. Once the bubble was excited and oscillated, the particles would be captured by adjacent bubbles. After the transducer was shut down again, the particles immediately returned to the laminar driving state. Recently, Zhang et al. [26] proposed a multifunctional bubble-based acoustofluidic device. By changing the applied acoustic field frequency, the arranged bubbles trapped in the microcavity at the bottom of the chip can be flexibly switched between four different oscillatory motions, so as to generate corresponding microstreaming modes in the microchannel and realize the stable transportation, trapping, 3D rotation and circular revolution of micro-objects (Fig. 3c). In general, due to the different oscillation states and corresponding microfluidic modes of acoustic bubbles, they can achieve a variety of controllable multimodal operations on micro-objects, and has a wide range of applications in the biological field.

2.2 Optothermal Bubble-Based Manipulator

Working Principle. The optothermal effect is usually used to convert light energy into heat energy, and microbubbles are generated at the interface between heat absorbing materials (including metal, amorphous silicon, indium tin oxide or their combination) and liquid [30]. The temperature decreased along the radial direction because of convective cooling along the top and bottom surfaces. This temperature gradient causes a corresponding convective flow which forms a clockwise flow pattern near the bubble–liquid interface (Fig. 4a) [31]. This microscale circulation is caused by the Marangoni effect. The velocity field between liquid layers caused by the thermal Marangoni effect can be described as [32]

$$\eta\left(\frac{\partial \mu}{\partial \mathbf{n}}\right) = \gamma_{\mathrm{T}}\left(\frac{\partial \mathrm{T}}{\partial \mathbf{t}}\right) \tag{4}$$

where η is the dynamic viscosity, μ is the tangential component of the fluid velocity vector at the interface between liquid and air, and **n** and **t** are the unit vectors of the normal and tangential directions of the interface, respectively. $\gamma_{\mathrm{T}} = \partial\gamma/\partial\mathrm{T}$ is the derivative of surface tension γ with respect to temperature.

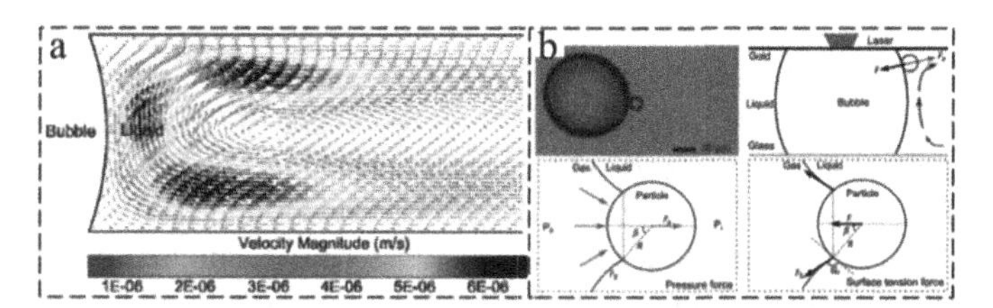

Fig. 4. (a) Convective flow pattern around an optothermal bubble. (b) The experimental image and force analysis image when a particle is trapped on the bubble surface. Adapted from Zhao et al. [31] with permission from Royal Society of Chemistry, Copyright 2014.

Under the action of Marangoni convection, the particles will be brought to the bubble, and will finally be captured on the bubble surface The force balance on the particle located at the bubble–liquid interface can be described as (Fig. 4b) [31]

$$\frac{F}{F_P} = \frac{R_B}{R} \frac{\sin(|\theta_C - \beta|)}{\sin\beta} \tag{5}$$

where F is the combined force of the surface tension F_S, F_P is the pressure force, R_B is the radius of the bubble and R is the radius of the particle, θ_C is the contact angle among the particle and the liquid-gas interface, and β is the half central angle. The balance between these two forces keeps the particles tightly trapped on the bubble surface, and makes it possible for optothermal bubbles to manipulate other micromodules.

Flow Control. The liquid jet produced by cavitation bubble collapse can cause different fluid mixing. In 2007, Hellman et al. utilized laser to produce cavitation bubbles in the microfluidic channel solidified with silicone resin, which realizes the mixing of water and naphthol green dye. For micropump, Dijkink et al. first produced a pump based on the principle of cavitation bubbles generated by laser pulse focusing to drive the flow of microfluidic fluid [33]. As for the microvalve, the bubble can play the role of a valve in the microfluidic system using the characteristic that the bubble volume is affected by temperature. In 2011, Zhang et al. [34] used laser to irradiate the chromium pad immersed in the liquid to produce controllable bubbles, and control the flow direction of the fluid through the growth process of bubbles, which could realize the open and closed state of laminar flow as a microvalve (Fig. 5a). Subsequently, Jian et al. [35] used laser-induced bubbles to selectively control the flow direction in microfluidic chips. The direction and speed of the pumping flow can be adjusted by changing the spot position and laser power. In order to reduce the complexity of optical system, Hyun et al. [36] deposited titanium film on the roughened end-face of silica optical fiber to make optic fiber microheater, and embedded it into microfluidic system to form an optofluidic microvalve-on-a-chip system. The high-speed bubble can grow into a vapor plug, which successfully blocks the liquid flow in the microchannel as a valve. Compared with the acoustic bubble fluid controller, the advantage of the optothermal bubble fluid controller is that it can generate bubbles at almost any position of the microchannel to realize the fluid control of specific parts. However, its disadvantage is that a laser beam can only generate one optothermal bubble, which has a relatively low efficiency; and the equipment is more expensive and complex.

Micro-object Manipulation. When the laser pulse is focused on the buffer interface of the cell solution, local cavitation bubbles will be generated. The hydrodynamic force induced by bubble expansion and subsequent collapse will cause the rupture of cell membrane. Rau et al. analyzed the operation mechanism and hydrodynamics behind the pulsed laser microbeam-induced cell lysis [37]. They found that cavitation bubble expansion was the primary agent of cell lysis. Cavitation bubble collapse, jet formation, and subsequent radial outflow of fluid can also result in the lysis of cells in the central region. Li et al. [38] designed conversion structures to capture a single cell, and then a single cavitation bubble was generated next to a single captured cell by laser. The jet generated by the collapse of the cavitation bubble was directed to the cell, resulting in strong deformation and lysis. Hu et al. utilized the laser to intermittently irradiate the chip to produce bubbles to move cells [39]. With the movement of the laser, the bubbles generated by the last laser irradiation will quickly separate from the substrate and cause optothermal capillary flow. This micro flow will drive the cells to move a certain distance, and the continuous generation and floating bubbles can continuously move the operated cell. Thus, the damage of optothermal bubble temperature to cells is greatly reduced. Fan et al. used the oscillation of bubble volume in the opening and closing cycle of laser pulse to generate shear stress on the cell membrane near the bubble, so as to realize cell perforation [40, 41]. Subsequently, they integrated a single-cell analysis platform for cell perforation, cell lysis and cell manipulation (Fig. 5b) [42]. In microfluidic chips, optothermal bubbles can screen small objects. For example, Wu et al. reported a high speed and high purity pulsed laser triggered fluorescence activated cell sorter (PLACS) (Fig. 5c) [43]. They used the pulsed laser controlled by the detector to generate explosive vapor microbubbles in the outer channel and trigger a high-speed jet in the sample channel, which can make the specific cells deviate from the original path and enter the collection channel to realize the cell sorting function.

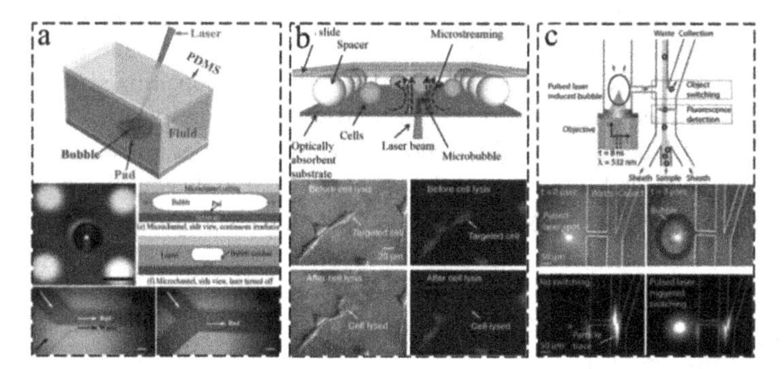

Fig. 5. (a) Up: illustration of bubble generation on the chromium pad. Down: experimental image of bubble growth, an illustration of the bubble behavior and the laser-induced bubble acts as a microvalve. Adapted from Zhang et al. [34] with permission from Royal Society of Chemistry, Copyright 2011. (b) Up: 3D structure of the microfluidic chamber where the cells can be cultured and lysed. Down: brightfield images and fluorescent images before and after the adherent single-cell lysis. Adapted from Fan et al. [42] with permission under the terms of CC BY 4.0 license, Copyright 2017. (c) Up: schematic of the cell sorter. Down: time-resolved images of the laser-induced cavitation bubble and fluorescent particle switching in PLACS. Adapted from Wu et al. [43] with permission from Royal Society of Chemistry, Copyright 2012.

Optothermal bubbles can also manipulate microstructure. Hu et al. utilized optothermal bubbles to realize the movement and arrangement of multiple triangular SU-8 structures, and assembled the hydrogel structures with cells inside into a specific array. Researchers have further developed a system that can generate and move multiple optothermal bubbles at the same time (Fig. 6a) [44]. Cooperation between optothermal bubbles improves the efficiency of moving and operating micromodules. Therefore, optothermal bubble microrobot has a wide application potential in biomedicine and tissue engineering. However, the process of biological tissue reconstruction is complex.

Due to the limitations of the generation and movement of optothermal bubbles in the two-dimensional (2D) plane of the solid-liquid interface, optothermal bubbles can only move the hydrogel structure, but cannot adjust the 3D posture of a single hydrogel structure, which limits the real application in the construction of biological tissue engineering. Fukuda et al. utilized a needle tube to blow bubbles into the solution, and the flow field driven by the bubbles floating in the fluid made the hollow hydrogel module with cell structure float up (Fig. 6b) [45, 46]. They used another needle tube to string up the micromodules. After repeating this action, they string up a series of hydrogel modules with different shapes, and again used the generated bubbles to adjust the posture of the micromodules to make them arranged neatly. Although this method realized the 3D operation of the micromodule, it needs two actuators to control the injection and collection at the same time, which has complex mechanical structure and high control difficulty. Moreover, the collected micromodules must have holes in the center to string to the collector. In order to solve the above problems, Dai et al. creatively proposed a method to utilize the optothermal bubbles on the 2D surface to realize the 3D arrangement and assembly of the hydrogel microstructure (Fig. 6c) [47]. By controlling the position and size of the optothermal bubbles generated at the bottom of the micromodule, they can lift, flip and flexibly adjust the 3D posture of the micromodule. They assembled the hollow ring structure into a 3D vascular structure and nested two ring modules of different sizes. Then, Ge of the same research group utilized the optothermal bubble operation method proposed by Dai to assemble the cell loaded microstructure into a peritoneal tissue structure highly similar to biological peritoneum, which verified the biocompatibility and feasibility of this method [48]. On this basis, Dai et al. further utilized the 3D operation ability and cluster cooperation ability of bubble microrobots to

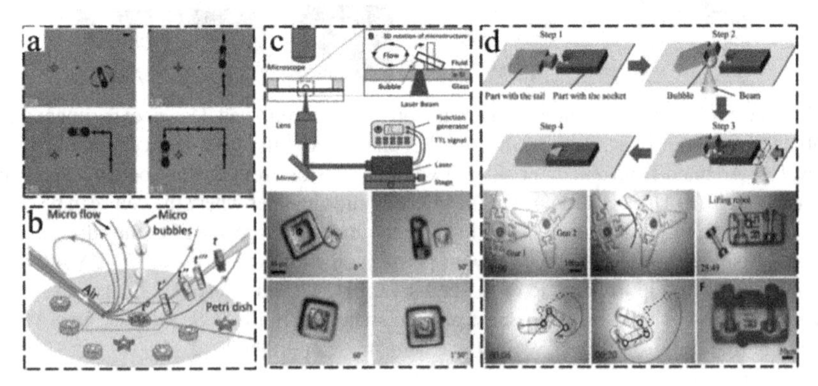

Fig. 6. (a) Micromanipulation of a microstructure using a pair of optothermal bubbles. Adapted from Rahman et al. [44] with permission under the terms of CC BY 4.0 license, Copyright 2017. (b) Schematic of the pickup of micromodules based on microbubble injection. Adapted from Wang et al. [46] with permission from American Chemical Society, Copyright 2017. (c) Up: schematic of the system setup to generate optothermal bubbles. Down: 3D manipulation and assembly of square ring microstructures. Adapted from Dai et al. [47] with permission from John Wiley and Sons, Copyright 2019. (d) Up: schematic of the integrated assembly process of two microparts by optothermal bubbles. Down: the assembled gears, snake-shaped structure and car. Adapted from Dai et al. [49] with permission from American Chemical Society, Copyright 2020.

realize the assembly of multiple micromodules with different interfaces, including mortise and tenon, gear, chain and car, which provided a new solution for the manufacturing, assembly and driving of micro/nanostructure (Fig. 6d) [49]. In general, the optothermal bubbles can realize diversified operation and different forms of motion states of a single micro-object more flexibly and controllably, and plays an important role in the field of precision microoperation and modular assembly.

3 Summary and Outlook

In this mini-review, the basic theory and wide applications of acoustic bubbles and optothermal bubbles manipulators in microfluidic devices were discussed. Bubbles are a double-edged sword in the microfluidic system. Free and uncontrolled bubbles can cause damage, but bubbles under the control of the energy field can become a favorable tool in microfluidic system. Bubble-based manipulators still have great development space in microfluidics. We will further discuss its challenges and future development.

For the acoustic bubble-based manipulator, it has the characteristics of simple equipment, easy implementation, low power consumption and low cost. However, under the long-time oscillation of the gas-liquid interface, the gas in the bubble may dissolve, resulting in changes in the size and shape of the bubble. The resonance frequency of the bubble can be affected, which may lead to the failure of the operation function. Therefore, how to extend the service life of bubbles and improve the effectiveness and stability of the bubble operation is an urgent problem to be solved in future development. For the optothermal bubble-based manipulator, it has high flexibility and controllability. However, the high temperature generated in the laser irradiation area may cause damage to cells and biomolecules, which limits the relevant biological applications. The latest breakthrough of optothermal bubbles is to realize the 3D manipulation and assembly of microstructure on 2D plane [47, 49], which is difficult to realize without relying on other operation technology such as microgripper/micropipette. Moreover, feedback control can be easily combined with the laser-induced optothermal bubbles to realize the automatic operation and the cooperative control of bubbles, which is expected to realize an operation factory.

In conclusion, the action range of the manipulator based on acoustic bubble and optothermal bubble is limited to the 2D plane. Although they can be combined with other operators such as microgripper [50] and magnetic microrobot [51] to complete operation tasks in 3D space, the mechanical structure becomes more complex and is only suitable for an open chip. The combination by superimposing multiple physical fields may further improve the control performance of bubbles. Researchers have combined acoustic-magneto [51], light-acoustic [52] to further expand the application range of bubble-based manipulator, and we look forward to the combination of multiple physical fields to provide more possibilities. The bubble manipulator microfluidics has great application value in biomedicine, microfluidics, clinical application and chemistry, which needs more scholars to study. We hope that in the future, bubble-based manipulators in microfluidic systems can integrate capture, transport and functional applications of micro-objects, just like an automatic assembly line.

Acknowledgement. This work is supported by the National Natural Science Foundation of China (Grant Nos. 91748212, U1613220, 91848201), and the CAS/SAFEA International Partnership Program for Creative Research Teams.

References

1. Hashmi, A.: Oscillating bubbles: a versatile tool for lab on a chip applications. Lab Chip **12**(21), 4216–4227 (2012)
2. Patel, M.V.: Lateral cavity acoustic transducer as an on-chip cell/particle microfluidic switch. Lab Chip **12**(1), 139–145 (2012)
3. Ahmed, D.: A millisecond micromixer via single-bubble-based acoustic streaming. Lab Chip **9**(18), 2738–2741 (2009)
4. Chen, Y.: Onset of particle trapping and release via acoustic bubbles. Lab Chip **16**(16), 3024–3032 (2016)
5. Ozcelik, A.: An acoustofluidic micromixer via bubble inception and cavitation from microchannel sidewalls. Anal. Chem. **86**(10), 5083–5088 (2014)
6. Bertin, N.: Bubble-based acoustic micropropulsors: active surfaces and mixers. Lab Chip **17**(8), 1515–1528 (2017)
7. Orbay, S.: Mixing high-viscosity fluids via acoustically driven bubbles. J. Micromech. Microeng. **27**(1), 015008 (2017)
8. Conde, A.J.: Versatile hybrid acoustic micromixer with demonstration of circulating cell-free DNA extraction from sub-ml plasma samples. Lab Chip **20**(4), 741–748 (2020)
9. Tovar, A.R.: Lateral cavity acoustic transducer. Lab Chip **9**(1), 41–43 (2009)
10. Patel, M.V.: Cavity-induced microstreaming for simultaneous on-chip pumping and size-based separation of cells and particles. Lab Chip **14**(19), 3860–3872 (2014)
11. Gao, Y., Wu, M., Lin, Y., Zhao, W., Xu, J.: Acoustic bubble-based bidirectional micropump. Microfluid. Nanofluid. **24**(4), 1 (2020). https://doi.org/10.1007/s10404-020-02334-6
12. Ahmed, D.: Tunable, pulsatile chemical gradient generation via acoustically driven oscillating bubbles. Lab Chip **13**(3), 328–331 (2013)
13. Ahmed, D.: Acoustofluidic chemical waveform generator and switch. Anal. Chem. **86**(23), 11803–11810 (2014)
14. Liu, B.: A concentration gradients tunable generator with adjustable position of the acoustically oscillating bubbles. Micromachines **11**(9), 827 (2020)
15. Meng, L.: Sonoporation of cells by a parallel stable cavitation microbubble array. Adv. Sci. (Weinh.) **6**(17), 1900557 (2019)
16. Liu, X.: Cell lysis based on an oscillating microbubble array. Micromach. (Basel) **11**(3), 288 (2020)
17. Liu, X.: Rapid cell pairing and fusion based on oscillating bubbles within an acoustofluidic device. Lab Chip **22**(5), 921–927 (2022)
18. Liu, X.: Non-cavitation targeted microbubble-mediated single-cell sonoporation. Micromach. (Basel) **13**(1), 113 (2022)
19. Ozcelik, A.: acoustofluidic rotational manipulation of cells and organisms using oscillating solid structures. Small **12**(37), 5120–5125 (2016)
20. Ahmed, D.: Rotational manipulation of single cells and organisms using acoustic waves. Nat. Commun. **7**, 11085 (2016)
21. Läubli, N.F.: 3D manipulation and imaging of plant cells using acoustically activated microbubbles. Small Methods **3**(3), 1800527 (2019)

22. Tang, Q., Liang, F., Huang, L., Zhao, P., Wang, W.: On-chip simultaneous rotation of large-scale cells by acoustically oscillating bubble array. Biomed. Microdevice **22**(1), 1–11 (2020). https://doi.org/10.1007/s10544-020-0470-1
23. Peng, T.: Trapping stable bubbles in hydrophobic microchannel for continuous ultrasonic microparticle manipulation. Sens. Actuators A Phys. **331**, 113045 (2021)
24. Gao, Y.: Study of ultrasound thrombolysis using acoustic bubbles in a microfluidic device. Lab Chip **21**(19), 3707–3714 (2021)
25. Gao, Y.: Acoustic bubble for spheroid trapping, rotation, and culture: a tumor-on-a-chip platform (ABSTRACT platform). Lab Chip **22**(4), 805–813 (2022)
26. Zhang, W.: Versatile acoustic manipulation of micro-objects using mode-switchable oscillating bubbles: transportation, trapping, rotation, and revolution. Lab Chip **21**(24), 4760–4771 (2021)
27. Rogers, P.: Selective particle trapping using an oscillating microbubble. Lab Chip **11**(21), 3710–3715 (2011)
28. Meng, L.: Microbubble enhanced acoustic tweezers for size-independent cell sorting. Appl. Phys. Lett. **116**(7), 073701 (2020)
29. Xie, Y.L.: Acoustofluidic relay sequential trapping and transporting of microparticles via acoustically excited oscillating bubbles. JALA **19**(2), 137–143 (2014)
30. Ohta, A.T.: Optically actuated thermocapillary movement of gas bubbles on an absorbing substrate. Appl. Phys. Lett. **91**, nihpa130823 (2007)
31. Zhao, C.: Theory and experiment on particle trapping and manipulation via optothermally generated bubbles. Lab Chip **14**(2), 384–391 (2014)
32. Higuera, F.J.: Steady thermocapillary-buoyant flow in an unbounded liquid layer heated nonuniformly from above. Phys. Fluids **12**(9), 2186–2197 (2000)
33. Dijkink, R.: Laser-induced cavitation based micropump. Lab Chip **8**(10), 1676–1681 (2008)
34. Zhang, K.: Laser-induced thermal bubbles for microfluidic applications. Lab Chip **11**(7), 1389–1395 (2011)
35. Jian, A.Q.: Microfluidic flow direction control using continuous-wave laser. Sens. Actuators A Phys. **188**, 329–334 (2012)
36. Kim, H.-T.: Optofluidic microvalve-on-a-chip with a surface plasmon-enhanced fiber optic microheater. Biomicrofluidics **8**(5), 054126 (2014)
37. Rau, K.R.: Pulsed laser microbeam-induced cell lysis: time-resolved imaging and analysis of hydrodynamic effects. Biophys. J. **91**(1), 317–329 (2006)
38. Li, Z.G.: Single cell membrane poration by bubble-induced microjets in a microfluidic chip. Lab Chip **13**(6), 1144–1150 (2013)
39. Hu, W.: An opto-thermocapillary cell micromanipulator. Lab Chip **13**(12), 2285–2291 (2013)
40. Fan, Q.: Laser-induced microbubble poration of localized single cells. Lab Chip **14**(9), 1572–1578 (2014)
41. Fan, Q.: Efficient single-cell poration by microsecond laser pulses. Lab Chip **15**(2), 581–588 (2015)
42. Fan, Q.: Localized single-cell lysis and manipulation using optothermally-induced bubbles. Micromach. (Basel) **8**(4), 121 (2017)
43. Wu, T.H.: Pulsed laser triggered high speed microfluidic fluorescence activated cell sorter. Lab Chip **12**(7), 1378–1383 (2012)
44. Rahman, M.A.: Cooperative micromanipulation using the independent actuation of fifty microrobots in parallel. Sci. Rep. **7**(1), 3278 (2017)
45. Zheng, Z.: 3D construction of shape-controllable tissues through self-bonding of multicellular microcapsules. ACS Appl. Mater. Interfaces **11**(26), 22950–22961 (2019)
46. Wang, H.: Assembly of RGD-modified hydrogel micromodules into permeable three-dimensional hollow microtissues mimicking in vivo tissue structures. ACS Appl Mater Interfaces **9**(48), 41669–41679 (2017)

47. Dai, L.: 2D to 3D manipulation and assembly of microstructures using optothermally generated surface bubble microrobots. Small **15**(45), e1902815 (2019)
48. Ge, Z.: Bubble-based microrobots enable digital assembly of heterogeneous microtissue modules. Biofabrication **14**(2), 025023 (2022)
49. Dai, L.: Integrated assembly and flexible movement of microparts using multifunctional bubble microrobots. ACS Appl. Mater. Interfaces **12**(51), 57587–57597 (2020)
50. Zhou, Y.: Soft-contact acoustic microgripper based on a controllable gas-liquid interface for biomicromanipulations. Small **17**(49), e2104579 (2021)
51. Giltinan, J.: Programmable assembly of heterogeneous microparts by an untethered mobile capillary microgripper. Lab Chip **16**(22), 4445–4457 (2016)
52. Xie, Y.: Probing cell deformability via acoustically actuated bubbles. Small **12**(7), 902–910 (2016)

Fast Locomotion of Microrobot Swarms with Ultrasonic Stimuli in Large Scale

Cong Zhao, Xiaolong Lu(✉), Ying Wei, Huan Ou, and Jinhui Bao

State Key Laboratory of Mechanics and Control of Mechanical Structures, Nanjing University of Aeronautics and Astronautics, Nanjing 210016, People's Republic of China
long_8446110@nuaa.edu.cn

Abstract. In recent years, the acitve motion of micro-nano machines has received wide attention due to its potential applications in biomedicine, biosensing, pollution control and other fields. However, the rapid manipulation of microrobot swarms in a large range is still an unmet challenge. With merits of simple structure, rapid response, and biocompatibility, ultrasonic power has been emerging as one of the most suitable methods to accomplish locomotion for microrobots. Here we demonstrate a fast moving strategy of microrobot swarms with ultrasonic stimuli in a relatively large scale. Both aggregation and dispersion patterns are conducted in a manipulation reservoir with 6 mm in diameter. The aggregation of microrobot swarms is achieved by acoustic radiation force caused by acoustic pressure gradient, which drives the microrobots migrate towards the nearest sound pressure node. The dispersion behavior of the microrobots is driven by the acoustofluidics, which is caused by the localized traveling wave motion of the vibrating substrate under ultrasonic stimuli. Both modes of swarming are completed in several seconds, indicating the great promise for rapidly steering microrobot assembly in diverse potential applications.

Keywords: Ultrasonic · Microrobots · Swarming · Acoustic pressure · Streaming

1 Introduction

The group behavior is a common phenomenon in the biological world. From large colonies of bees and ants to small clusters of cells, they gathered by the thousands and collaborate on a variety of different tasks [1, 2]. Recently, this common phenomenon in biology has inspired researchers to study swarming of synthetic microrobots and achieved remarkable results [3]. However, the practical application of microrobot swarms is still restricted by various factors such as movement speed, control range and biocompatibility.

This work was supported by the National Natural Science Foundation of China (No. 51975278), Research Fund of State Key Laboratory of Mechanics and Control of Mechanical Structures (Nanjing University of Aeronautics and Astronautics) (Grant No. MCMS-I-0321G01) and Fundamental Research Funds for the Central Universities (No. xcxjh20210103).

H. Liu et al. (Eds.): ICIRA 2022, LNAI 13456, pp. 581–589, 2022.
https://doi.org/10.1007/978-3-031-13822-5_52

The aggregation and dispersion are two common behaviors of microrobot swarms, which have great application value in biomedicine, biosensing, pollution control, and other fields. It is the most attractive strategy to control the motion of microrobot swarms by using an external field on a large area, which demonstrates a combination of high productivity and facile operability. Researchers have been able to control some behaviors of different types of synthetic microrobot swarms through external fields such as magnetic field, electric field, ultrasonic field and light [4–7]. However, the swarm behavior controlled by the electric field and magnetic field is limited by material properties and thus unable to control non-magnetized or non-electrified objects. In addition, the bulky external equipment also limits its application in a narrow space. The light-controlled swarm behavior has the limitation of slowly moving speed thus difficult to respond quickly. In contrast, the ultrasonic field has the advantages of simple structure, rapid response, immunity from electromagnetic interference, and the controlled object is not limited by materials [8], which makes it more competitive in diverse applications. Recently, exciting research in the field of acoustic manipulation has confirmed that it is possible to complete specific tasks through ultrasonic manipulation of microrobot swarms on a large scale. Lu et al. have successfully achieved accurate manipulation of single particles or cells through ultrasound [9, 10], and aggregation or dispersion of micro-nano motor swarms used a hybrid sonoelectrode [11]. In addition, the aggregation and separation of micromotors powered by ultrasound [7] has been applied to enhance the signals of biomarker fluorescence detection [12]. In spite of this, it is still strongly desired to investigate an efficient and economical strategy for achieving rapid response of microrobot swarms in a relatively large space.

Here, we demonstrate a fast swarming of microrobots with ultrasonic stimuli in a relatively large scale. We design two swarming modes of microrobots in a manipulation reservoir with 6 mm in diameter. Firstly, the basic structure of manipulation platform for fast swarming movements and the manufacturing method of microrobots are proposed. Then the principle of the two modes of fast locomotion is analyzed by numerically calculating the acoustic pressure and acoustic streaming inside droplets with the ultrasonic stimuli, respectively. At last, vibration characteristics of the manipulation platform for microrobots are measured and the effects of swarming movements in droplets are evaluated with moving trajectories of microrobots under ultrasonic stimuli. This kind of fast swarming movement for microrobots in large areas controlled by a simple device holds great promise in diverse applications.

2 Structure and Working Principle

The fast swarming movement of microrobots is completed on an acoustic manipulation platform (AMP), which is composed of a piezoelectric transducer, a glass substrate, and an insulating layer that limits the shape of droplets. Figure 1 shows the structure and functional schematic diagram of the AMP. Silicone glass with the size of 25 mm × 15 mm × 0.5 mm (length × width × thickness) is selected as the substrate. A ring-shaped piezoelectric transducer (PZT-8, Wuxi Haiying Group, China) with a size of Φ20 mm × Φ10 mm × 1 mm (external diameter × internal diameter × thickness) is pasted on the back of the substrate. In addition, a polymer insulating layer (Kapton tape) with 0.1 mm thickness is covered on the front of the substrate. One Φ6 mm circular cell is cut out from the polymer layer to serve as the reservoir for droplets dispersed with microrobots, limiting the position and boundary of droplet samples to be exactly over the center of the AMP. When AMP is working, 20 μL of droplets dispersed with microrobots are dropped into the reservoir. By adjusting the frequency of ultrasonic excitation, two kinds of swarming motion of microrobots can be achieved respectively.

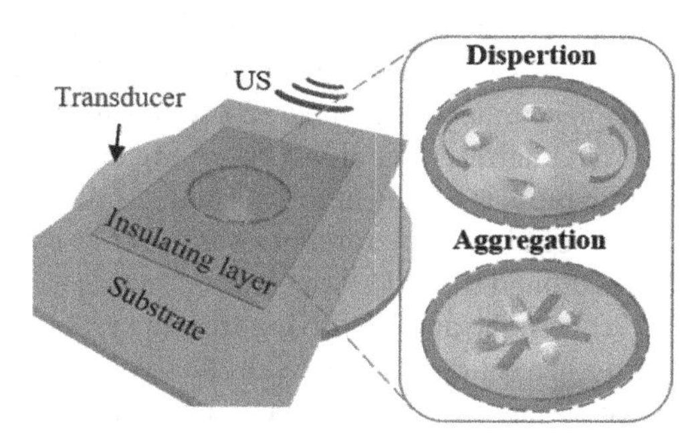

Fig. 1. Structure and functional schematic diagram of the AMP

The AMP has two working modes, one low-frequency mode is used to realize the aggregation of micro-nano robots, and the other high-frequency mode is used to realize the dispersion of microrobots. Figure 2 shows the working principle of two different working modes. When working at low-frequency mode, the substrate resonates under ultrasonic excitation to produce an acoustic pressure gradient toward the droplet center, which enables the microrobots to migrate toward the center.(See Fig. 2(a)) Being different from low frequency mode, when working at high-frequency mode, a kind of droplets resonance with much larger vibration on the surface of the liquid film than on the substrate occurs, which generate strong acoustic streaming on a large scale for the dispersion of the microrobots.(See Fig. 2(b)).

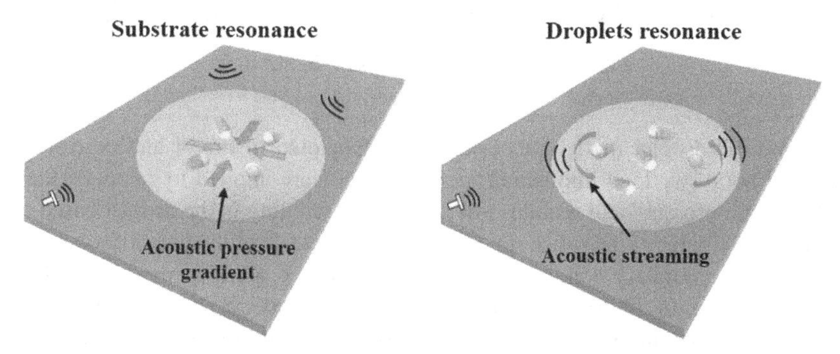

(a) Aggregation in low frequency mode (b) Dispersion in high frequency mode

Fig. 2. Working principle of AMP in two working modes

3 Simulation and Analyses

To better explain the working mechanism of fast swarming movement of microrobots in two working modes, we numerically calculated the acoustic field including the pressure and acoustic streaming with COMSOL Multiphysics software based on the perturbation theory [13] in two working modes, respectively.

To realize the proposed two working principles, we calculated the vibration mode of the AMP at 10–100 kHz and selected the vibration mode as shown in Fig. 3 to analyze the acoustic pressure and acoustic streaming respectively. Calculated vibration frequency corresponding to low frequency mode is around 22.4 kHz which performances a local longitudinal vibration in circular reservoir area. (See in Fig. 3(a)) Such vibration mode is easy to produce acoustic pressure gradient to the center of the droplet, so we choose this mode to realize the aggregation of microrobot swarms. The vibration corresponding to high frequency mode is around 91.2 kHz which performances a local traveling wave movement. (See in Fig. 3(b)) At the applied driving frequency, the vibration transmitted into the droplet will be further enlarged once the droplet volume satisfies the coherent resonance conditions. This enlarged vibration is a major cause of acoustic streaming in the droplet, thus we choose this mode to realize the dispersion of microrobot swarms.

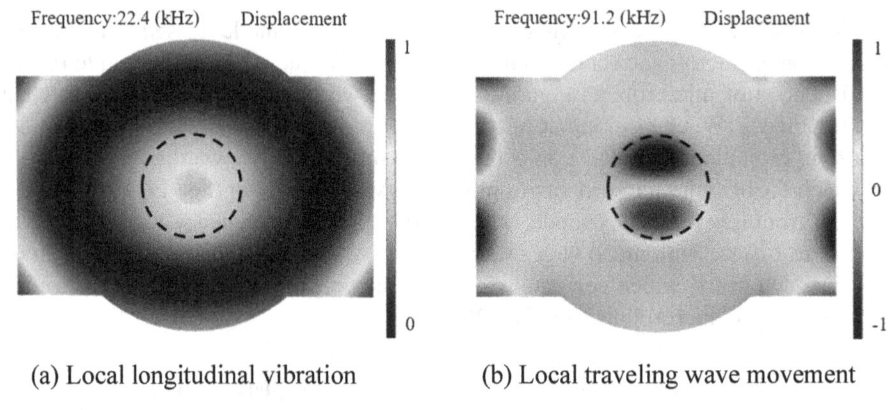

(a) Local longitudinal vibration (b) Local traveling wave movement

Fig. 3. Two vibration modes of the substrate

Then a 3D droplet model was established as shown in Fig. 4 and corresponding simplified boundary load implement methods [13] were applied to calculate the first-order acoustic pressure and second-order acoustic streaming in the droplet based on the two vibration modes of the substrate, respectively. Because of the two different working principles, the boundary conditions imposed by the two working modes are different in the simulation. In the first low-frequency working mode, according to the vibration modes of the substrate, we apply the same boundary condition of velocity along the Z direction on the substrate that vibrates at the bottom of the droplet. In the second high-frequency working mode, the enlarged traveling wave movements with the estimated magnitude (2 μm) are applied as tangential varied velocity boundaries to the outside of the droplet model, while the bottom side movement is neglected compared to the vibration of the droplet at resonance. Then we extracted and analyzed the simulation results based on perturbation theory on y-z cross-sections (See Fig. 4(a)) and x-y cross-sections (See Fig. 4(b)) respectively.

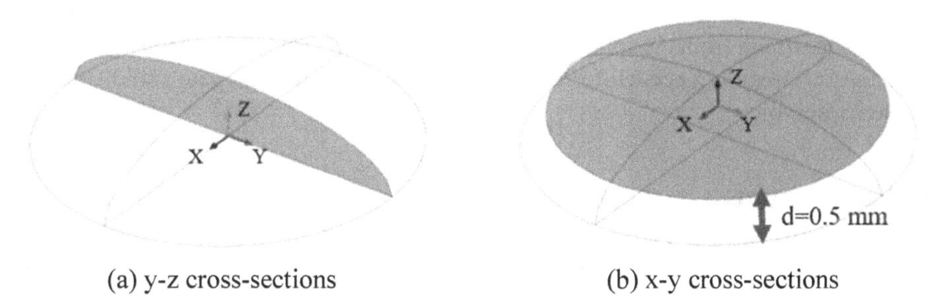

(a) y-z cross-sections (b) x-y cross-sections

Fig. 4. 3D droplet model

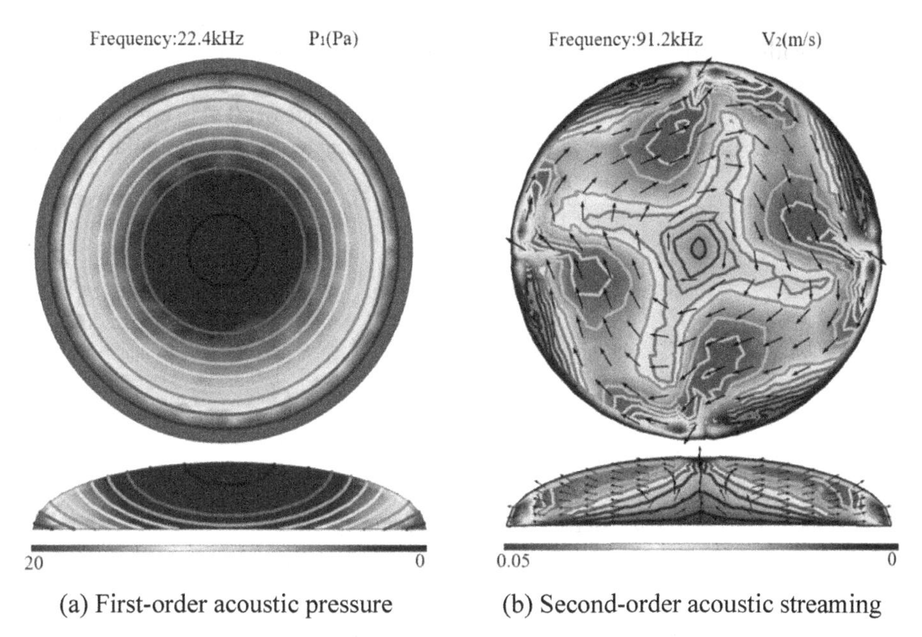

(a) First-order acoustic pressure (b) Second-order acoustic streaming

Fig. 5. First-order acoustic pressure in low frequency mode and second-order acoustic streaming in high frequency mode

Figure 5(a) shows the simulation results of the first-order acoustic pressure generated by the substrate vibration. It is obvious that the acoustic pressure node is generated in the droplet center, which indicates that the microrobot individuals will converge to the droplet center driven by the sound pressure gradient in low frequency mode. Figure 4(b) shows the simulation results of the second-order acoustic streaming generated by the enlarged tangential varied velocity boundaries to the outside of the droplet model. It can be clearly seen from the x-y cross-sections that the acoustic streaming rotates at a clockwise direction around the center of the droplet. The maximum flow rate is up to 50 mm/s, demonstrating a strong acoustic streaming at the vicinity of the outer thin film boundary. Apart from the liquid film, the intensity of acoustic streaming gradually decreases towards the inner middle center, but still effectively rotates at a fast speed to form a notable vortex flow. In general, numerical calculations theoretically verified that the acoustic stimuli from the solid vibration of AMP could generate the acoustic pressure and acoustic streaming needed for clusters of microrobots to aggregate and disperse in droplets reservoir.

4 Experiment

To determine the appropriate vibration mode, the vibration characteristics of the AMP were measured by a 3D laser Doppler vibrometer (PSV-500-3D, Polytec GmbH, Waldbronn, Germany). The results measured under a sweeping frequency are shown in Fig. 6(a). Even though there are many resonant points with significant peak vibration velocity at different frequencies, we select only two frequencies close to the simulation result and with similar modes as working frequencies. Vibration modes at two working frequencies are shown in Fig. 6(b). In low-frequency mode, the working area vibrates longitudinally with a maximum amplitude of 2 μm at a voltage of 20 V. In high-frequency mode, a special vibration mode of local traveling wave appears in the circular working

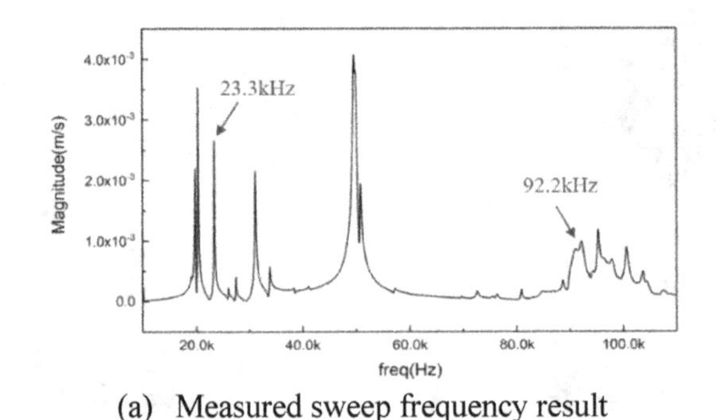

(a) Measured sweep frequency result

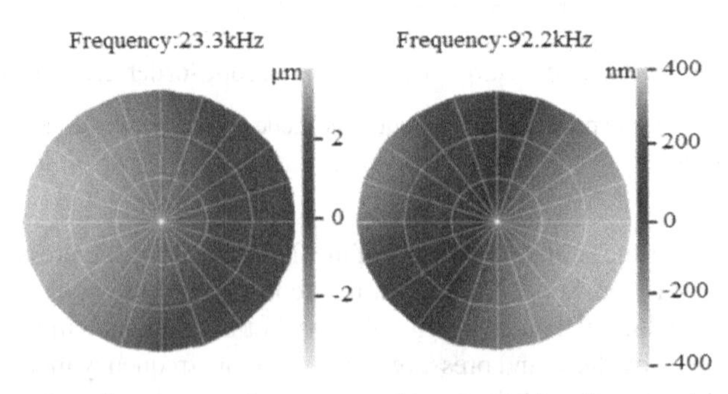

(b) Vibration modes at two selected working frequencies

Fig. 6. Vibration patterns of the working area at two working frequencies

area: A nodal line is generated in the circular area. The amplitude on both sides of the nodal line is equal while the direction is opposite. At the same time, the nodal line rotates clockwise periodically around the center of the circle. The deformation of the reservoir measured at fixed frequency is basically consistent with the local vibration mode obtained by simulation.

In order to verify that the microrobot swarms can achieve rapid aggregation and dispersion under these two working modes, an experimental system was built up as displayed in Fig. 7. The electrical signal generated from the signal generator and amplified by the power amplifier is connected to the piezoelectric transducer for exciting the AMP. An oscilloscope is used to observe the signal applied to the AMP in real-time. The AMP is placed on the stage of an optical microscope, through which the motion of microrobot can be observed.

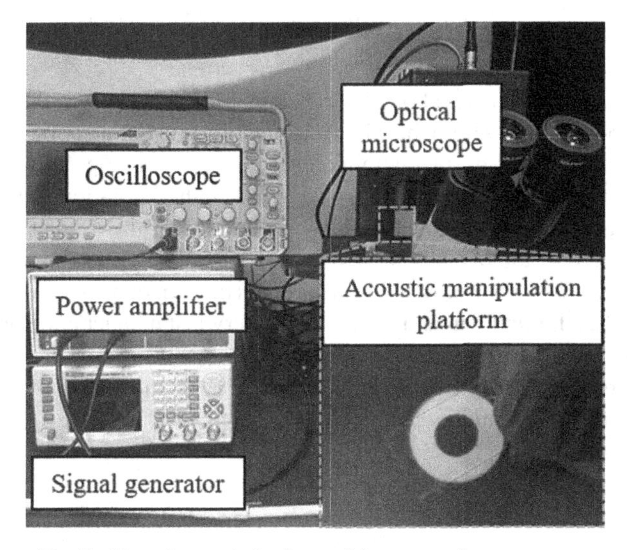

Fig. 7. Experimental platform of fast swarming movement

We used the Mg-based microrobots as experimental subjects to perform on-demand swarming. Its fabrication process is shown in Fig. 8. The Mg-based micromotors were prepared using commercially available magnesium microparticles (catalog #FMW20, TangShan WeiHao Magnesium Powder Co.; average size, 20 ± 5 μm) as the core. The Mg microparticles were initially washed with acetone to eliminate impurities. After drying under a N_2 current, the Mg microparticles were dispersed onto glass slides (10 mg of Mg microparticles per glass slide) and coated with Au layer using a Kurt J. Lesker PVD 75 Proline SP system. The deposition was performed at room temperature with a dc power of 100 W using 12 rpm rotation speed, leading to uniform coatings over the Mg microparticles, while leaving a small opening (~2 μm) at the contact point of the microparticle to the glass slide.

Finally, the actual moving trajectories of the microrobot clusters at both working frequencies was observed and photographed by a CCD camera as show in Fig. 9. When

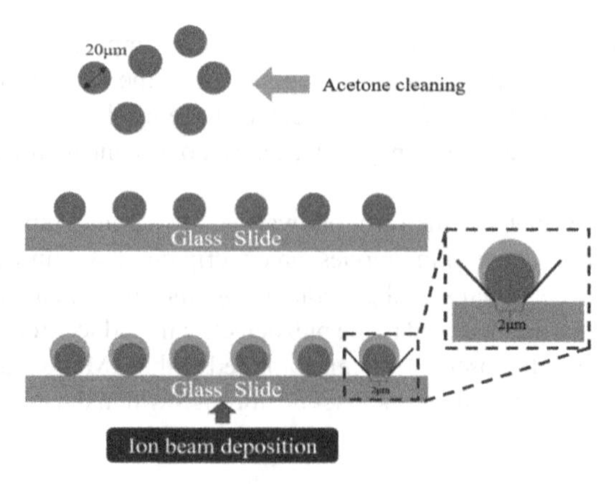

Fig. 8. Fabrication process of Mg-based microrobots

a low-frequency excitation signal of 23.3 kHz and 20 V was given to the AMP, the Mg-based microrobot swarms within the whole droplet of 6 mm diameter began to aggregate rapidly to the center of the droplet within only 1 s as shown in Fig. 9(a). The dispersion of microrobot swarms is a more violent process. When the excitation signal of 92.2 kHz and 20 V was applied, we observed the strong vibration of the droplet with bare eyes. At this working mode, the Mg-based microrobot swarms are dispersed within the whole droplet in less than 2 s. (See Fig. 9(b)) In the process of dispersion, the microrobots floats rapidly and makes the centrifugal movement in a very fast manner compared with other reported microrobot dispersion methods. The experimental results are consistent with the simulation results which verify the reliability of the fast movement of the micro-robot swarms with ultrasonic stimuli.

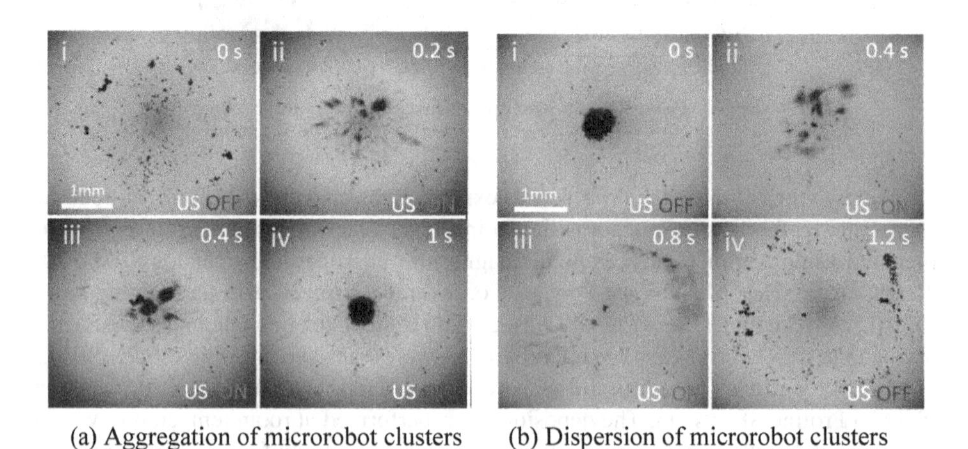

(a) Aggregation of microrobot clusters (b) Dispersion of microrobot clusters

Fig. 9. Fast locomotion of the 20 μm Mg-based microrobots in two modes

In addition, other configurations and sizes of micro nano robots have also been used to carry out the same experiments. Experiments in current stage show that this fast locomotion is also applicable to other types of microrobots.

5 Conclusion

In general, we developed an attractive strategy to achieve two modes of fast locomotion for microrobot swarms on a relatively large scale (6 mm). This method is featured with simple structure, rapid response, and large scope of action compared to other methods. The proposed manipulation strategy of microrobot swarms has a wide range of application prospects, especially for the future development of biosensing, biomedicine, and pollution control, etc.

References

1. Vicsek, T., Zafeiris, A.: Collective motion. Phys. Rep. **517**(3–4), 71–140 (2012)
2. Elgeti, J., Winkler, R.G., Gompper, G.: Physics of microswimmers–single particle motion and collective behavior: a review. Rep. Prog. Phys. **78**(5), 056601 (2015)
3. Wang, Q., Zhang, L.: External power-driven microrobotic swarm: from fundamental understanding to imaging-guided delivery. ACS Nano **15**(1), 149–174 (2021)
4. Yigit, B., Alapan, Y., Sitti, M.: Cohesive self-organization of mobile microrobotic swarms. Soft Matter **16**(8), 1996–2004 (2020)
5. Yan, J., Han, M., Zhang, J., et al.: Reconfiguring active particles by electrostatic imbalance. Nat. Mater. **15**(10), 1095–1099 (2016)
6. Hong, Y., Diaz, M., Ubaldo, M., et al.: Light-driven titanium-dioxide-based reversible microfireworks and micromotor/micropump systems. Adv. Func. Mater. **20**(10), 1568–1576 (2010)
7. Xu, T., Soto, F., Gao, W., et al.: Reversible swarming and separation of self-propelled chemically powered nanomotors under acoustic fields. J. Am. Chem. Soc. **137**(6), 2163–2166 (2015)
8. Lu, X., Soto, F., Li, J., et al.: Topographical manipulation of microparticles and cells with acoustic microstreaming. ACS Appl. Mater. Interfaces **9**(44), 38870–38876 (2017)
9. Lu, X., Zhao, K., Liu, W., et al.: A human microrobot interface based on acoustic manipulation. ACS Nano **13**(10), 11443–11452 (2019)
10. Wei, Y., Lu, X., Shen, H., et al.: An acousto-microrobotic interface with vision-feedback control. Adv. Mater. Technol. **6**, 2100470 (2021)
11. Lu, X., Wei, Y., Ou, H., et al.: Universal control for micromotor swarms with a hybrid sonoelectrode. Small **17**(44), e2104516 (2021)
12. Sun, Y., Luo, Y., Xu, T., et al.: Acoustic aggregation-induced separation for enhanced fluorescence detection of Alzheimer's biomarker. Talanta **233**, 122517 (2021)
13. Lu, X., Zhao, K., Peng, H., et al.: Local enhanced microstreaming for controllable high-speed acoustic rotary microsystems. Phys. Rev. Appl. **11**(4), 044064 (2019)

A Novel Acoustic Manipulation Chip with V-shaped Reflector for Effective Aggregation of Micro-objects

Huan Ou, Xiaolong Lu(✉), Ying Wei, Cong Zhao, and Jinhui Bao

State Key Laboratory of Mechanics and Control of Mechanical Structures, Nanjing University of Aeronautics and Astronautics, Nanjing 210016, People's Republic of China
long_8446110@nuaa.edu.cn

Abstract. Micro-objects to aggregate at a specific location becomes a necessity when the concentration of micro-objects needs to be increased in the local area, such as in lab-on-a-chip devices. However, efficient aggregation is still a big challenge when detecting low concentrations of specimens on manipulation chips. Here, we present an acoustic manipulation chip containing artificial reflector to enhance the aggregation effect. To integrate with the traditional acoustic manipulation chips, V-shaped reflector is introduced into the microchannel as a solid vibration isolator to isolate acoustic waves in a liquid environment. When the V-shaped reflector is immersed into the sample contained in the microchannel to a certain depth, the activation of the piezoelectric transducer will produce acoustic oscillation of the substrate and localized acoustic streaming around the tip of the reflector. Under the localized acoustic streaming, micro-objects in the working area of the acoustic manipulation chip gather beneath reflector. Experimental results illustrate the proposed acoustic chip coupled with the V-shaped reflector offers an effective aggregation strategy for micro-objects, which has broad application prospects in the field of micro-manipulation.

Keywords: Micro-manipulation · Acoustic waves · Reflector · Micro-object aggregation

1 Introduction

With fast development of micro-nanofabrication technologies in last century, tremendous efforts have been devoted to invent intelligent micro/nano systems, including micro laboratories, wearable biosensors, and lab-on-chips [1, 2]. One of the major challenges encountered in this area is miniaturization and active manipulations in order to translate lab-proven concepts into portable, autonomous, and sensitive devices. To overcome such

This work was supported by the National Natural Science Foundation of China (No. 51975278), Research Fund of State Key Laboratory of Mechanics and Control of Mechanical Structures (Nanjing University of Aeronautics and Astronautics) (Grant No.MCMS-I-0321G01) and Fundamental Research Funds for the Central Universities (No. xcxjh20210103).

H. Liu et al. (Eds.): ICIRA 2022, LNAI 13456, pp. 590–599, 2022.
https://doi.org/10.1007/978-3-031-13822-5_53

issues, diverse methods based on different power source have been developed to achieve on-demand micro/nano manipulations [3–5].

In general, manipulations of most micro/nano objects are powered by light, electric, acoustic, magnetic, or chemical energy, among which the acoustic field shows absolute superiority in biocompatibility [6, 7]. On the acoustic manipulation platform, various microstructures are added into microchannels thus complex tasks such as dynamic assembly, directional transportation, and aggregation of micro-objects can be achieved by utilizing localized microstreaming generated near the microstructures at specific acoustic excitation frequencies [8–10]. Conventional acoustic manipulation methods control the motion of microobjects by adjusting the acoustic radiation force through tailoring the distribution of sound pressure nodes or generating locally enhanced microstreaming around the microstructures [11]. However, the combination of reflector and microchannel to achieve rapid particle aggregation still requires to be explored [12].

In this paper, we use acoustic manipulation technology and the resultant local enhanced microstreaming to perform effective aggregation of micro objects. The acoustic manipulation chip contains a new V-shaped reflector in its micro-channel to realize the aggregation of micro/nano objects at low actuation frequency. A simple geometric structure was constructed to explore the influence of artificial reflector on acoustic streaming field. At last, the characteristics of local acoustic streaming field influenced by V-shaped reflector with different size parameters are experimentally discussed.

2 Experimental Section

2.1 Structural Design

In this study, based on the basic structure and functional requirements of the acoustic manipulation chip, an acoustic manipulation chip with two inlets and a micro-object manipulation area is designed. The acoustic manipulation chip limites the flow area of the liquid on the substrate by the microchannel to realize the functional partition of the acoustic manipulation chip. Using laser cutting method to process microchannels can not only get rid of the dependence of mold processing method on high-precision templates, but also process deep closed boundary microchannels and even cut through polymethyl methacrylate (PMMA) plates to process open boundary microchannels. Complex physical and chemical reactions on PMMA surface during laser cutting may also affect its surface properties. For example, ultra-short pulse femtosecond cutting can make PMMA surface from hydrophobic to hydrophilic, enhance the capillary effect of liquid samples in the microchannel, and contribute to the automatic sampling detection of acoustic manipulation chip.

The schematic diagram of device structure is shown in Fig. 1. This acoustic manipulation chip includes a glass substrate, a PMMA microfluidic channel bonded to top of the substrate, a piezoelectric transducer adhered to bottom of the substrate, and a V-shaped reflector fixed above the microchannel. V-shaped reflector was selected as the main configuration of microstructure for acoustic control of micro/nano objects, which was used to form a large-scale continuous acoustic streaming field around the microstructure on the manipulation plane. The size of the substrate is 25 mm in length and 15 mm in width, the diameter of the main working area of the microchannel is 4 mm, and the diameter of the V-shaped reflector is 260 μm. The solution containing micro/nano objects was introduced into the microchannel from both sides. In the middle of the working area, the

V-shaped reflector was immersed into the liquid. Figure 1(a) is the schematic structure and Fig. 1(b) is the acoustic manipulation chip prototype.

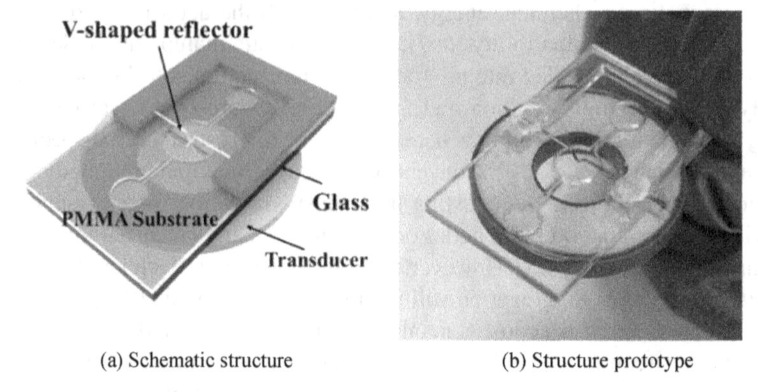

(a) Schematic structure (b) Structure prototype

Fig. 1. Configuration schematic of the acoustic manipulation chip

2.2 Vibration Measurement

The natural frequency of the acoustic manipulation chip and its corresponding vibration mode can be measured using an impedance analyzer and the 3D laser Doppler vibrometer. In the frequency range of 0–80 kHz, the highest velocity exists at 21.1 kHz indicate the resonance of the device. In addition, a 3D laser Doppler vibrometer was used to measure the vibration mode of the working area at 21.1 kHz, as shown in Fig. 2.

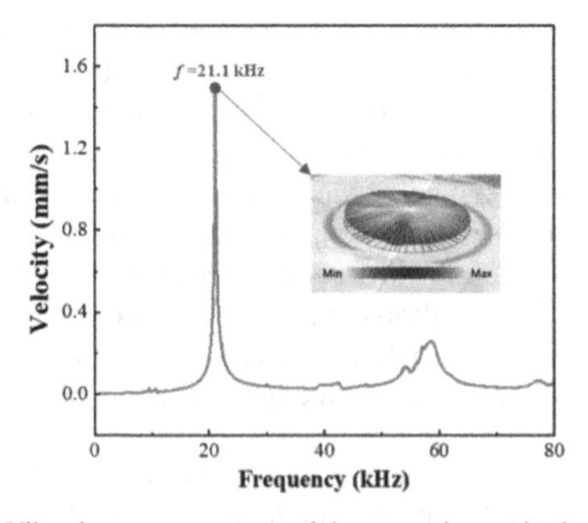

Fig. 2. Vibration measurement of the acoustic manipulation chip

When the driving voltage amplitude is 20 V_{PP}, the out-of-plane longitudinal vibration occured in the working area, and the vibration velocity amplitude at 21.1 kHz is about 1.5 mm/s. Therefore, this frequency is preliminarily selected as the working frequency of the acoustic manipulation chip. Transparent material acoustic manipulation chip needs to paste reflective material to enhance the reflection of the laser on its surface during vibration measurement, so as to ensure the good measurement signal. In addition, the vibration surface of the acoustic manipulation chip needs to be parallel to the measurement laser head, and the chip clamping method is non-horizontal. Therefore, no liquid is injected into the microfluidic channel during vibration measurement. There may be a small deviation between the resonant frequency of the acoustic manipulation chip obtained by vibration measurement and the actual resonant frequency.

2.3 Construction of the Experiment Platform

The driving experimental platform of the acoustic manipulation chip is shown in Fig. 3. The sinusoidal alternating current signal generated by the signal generator is amplified by the power amplifier and applied to the piezoelectric ceramic transducer of the acoustic manipulation chip. Due to the inversed piezoelectric effect, the ultrasonic transducer converts the electrical signal into ultrasonic vibration, which stimulates the overall resonance of the acoustic manipulation chip, and the micro-object in the manipulation area is controlled to move. The acoustic manipulation chip is placed on the optical microscope platform to magnify the manipulation area of micro-objects. Due to the local strengthening effect of the reflector on the acoustic streaming field, micro/nano objects in the working solution can be observed under the microscope to gather under the reflector, which significantly improves the gathering speed and accuracy of micro/nano objects. In experiments, the applied driving signal originally produced by the signal generator with a frequency of 21.3 kHz and a voltage of 2 V_{PP} would then be amplified to 20 V_{PP} by the power amplifier.

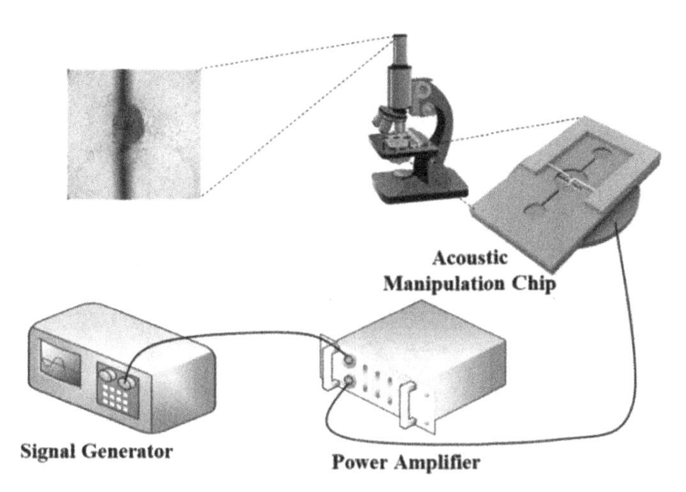

Fig. 3. Schematic diagram and basic experimental setup

3 Results and Discussion

3.1 Influence of Sample Volume on Aggregation

In the experiment, considering the frequency response curve and vibration mode, the sample was added to the acoustic manipulation chip, and reflector was used to manipulate the aggregation of microparticles in the working area of the chip to determine the actual working frequency of the chip in this mode. The experimental results show that the microparticles in the whole working area aggregate towards the bottom of reflector under the ultrasonic excitation of 21.3 kHz. The discrepancy of 0.2 kHz between the measured vibration results and the actual working frequency is the deviation of the natural frequency of the acoustic manipulation chip caused by the addition of the sample.

At the driving voltage of 20 V_{PP} and the driving frequency of 21.3 kHz, polystyrene particles with a diameter of 10 μm were used for experiments in the bright field to confirm the working principle. Since the sample volume directly affects the depth of V-shaped reflector immersed in the liquid, we first explored the influence of sample volume on the aggregation effect. The sample volumes of 20 μL, 24 μL, 28 μL and 32 μL were taken for experiments. When the sample volume was 20 μL, the end of V-shaped reflector was immersed in liquid, and the final aggregation effect was obtained when the experimental time was 4 min, as shown in Fig. 4(a). At 24 μL, the immersion depth was about 200 μm, and the final aggregation effect was obtained when the experimental time was 1 min, as shown in Fig. 4(b). When the sample volumes were 28 μL and 32 μL, the final aggregation effects were the same as that of 24 μL, but the time were 2 min and 3 min respectively. Considering the aggregation efficiency, we choose 24 μL as the sample volume for the subsequent experiment.

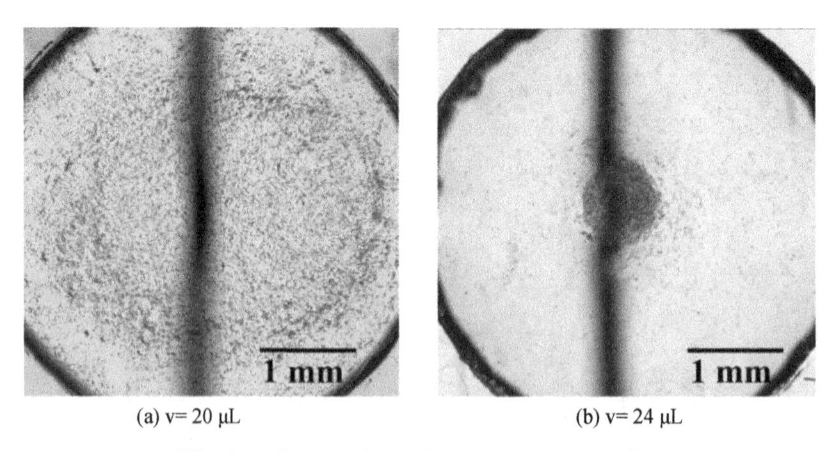

(a) v= 20 μL (b) v= 24 μL

Fig. 4. Influence of sample volume on aggregation

3.2 Influence of Reflectors' Angle on the Aggregation Speed

We can control the speed at which particles aggregate by changing the angle of the reflector. In order to quantify the aggregation behavior and explain the influence of the reflector with different sizes on aggregation, we conducted three groups of experiments (reflectors of 45°, 60°, and 90° respectively) at the driving frequency of 21.3 kHz with the voltage of 20 V_{PP} and measured the moving speed of microparticles. The general process of aggregation was similar to Fig. 5, but the aggregation speed is different. Experimental data are shown in Fig. 6. It can be seen from the figure that the smaller the angle of the reflector, the faster the speed of aggregation to the reflector. When the angle of the reflector is 45°, the aggregation speed is the fastest, which has an average value of 20 mm/s.

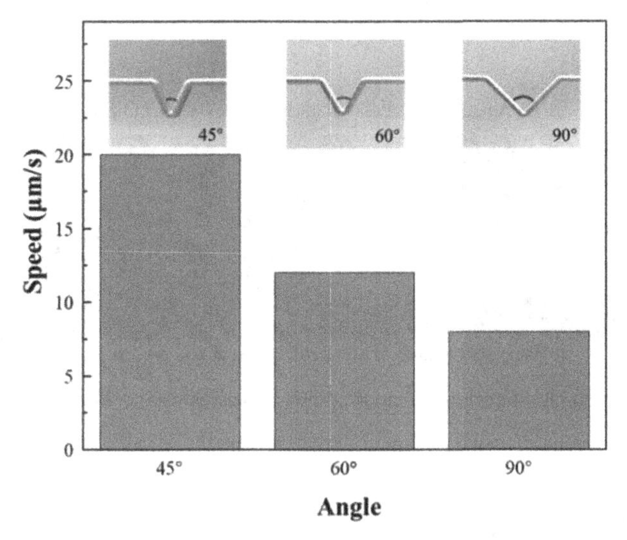

Fig. 5. Influence of reflectors' shapes on the transient collection

3.3 Aggregation Process of Microparticles

Figure 6 shows the aggregation process of microparticles with V-shaped reflector angle of 45° and sample volume of 24 μL. Figure 6(a) i shows that the microparticles are dispersed in the working solution at the beginning of the experiment. Figure 6(a) ii and iii depicts the aggregation of micro-particles in the working area to the middle reflector in the working area during the aggregation process. Figure 6(a) iv depicts the aggregation of a 10 μm diameter particle under the reflector when the aggregation is completed clusters. To clearly capture aggregation trajectory of microparticles, we also utilized fluorescent polystyrene particles with a diameter of 10 μm in the fluorescence field for experiments. The yellow circle in the figure indicates the position of the reflector. The sequence and merged diagrams of the experimental process are shown in Fig. 6(b), illustrating the dynamic aggregation process of micro/nano objects. Figure 6(b) i, ii and iii show the states of the dispersion of microparticles in the working solution, the aggregation of the microparticles towards the reflector and the completion of the cluster by aggregation, respectively. Furthermore, Fig. 6(b) iv also provides the merged microscopic images of the aggregation process. It can be seen from the diagram that the vast majority of micro/nanoparticles in the working area move toward the reflector, which further verifies the aggregation effect of micro/nanoparticles caused by the reflector.

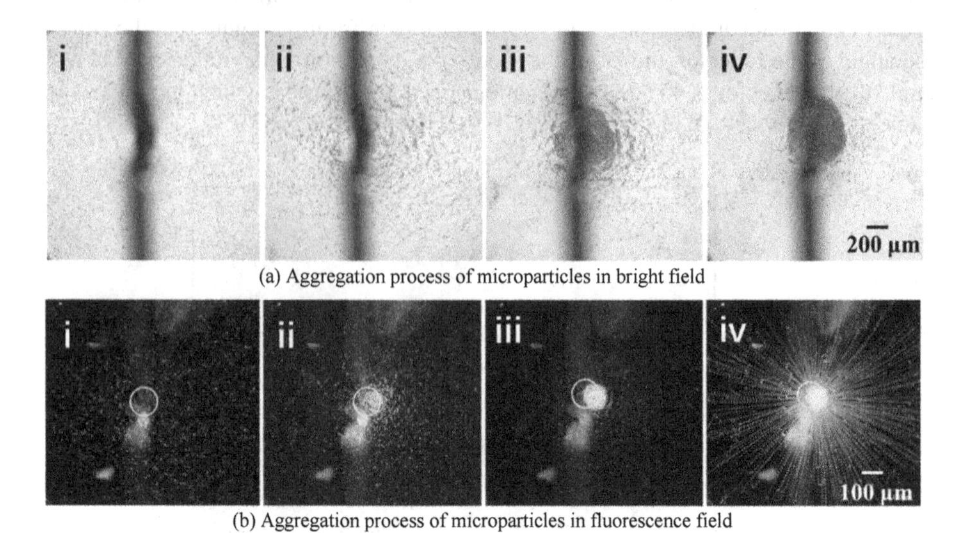

(a) Aggregation process of microparticles in bright field

(b) Aggregation process of microparticles in fluorescence field

Fig. 6. Sequential microscopy images of microparticles were collected by ultrasound

3.4 Simulation Analysis

The acoustic module of multi-physical field coupling analysis software Comsol is used to analyze the acoustic streaming field distribution caused by microstructure. The reflector with an angle of 45° and immersion depth of 200 μm, which corresponds to the fastest aggregation speed in Fig. 5, was selected for simulation calculation. Calculated time-averaged second-order streaming velocity marked with the streamlines (red lines) and flow direction (white arrows) around reflector is shown in Fig. 7. The size parameters of the calculation model are as follows: the diameter of the working area is 4 mm, the depth of the working liquid is 1 mm, the V-shaped reflector material is stainless steel, the included angle is 45°, and the immersed depth of the reflector is 200 μm.

When the driving signal was set as 21.3 kHz and 20 V_{PP}, the simulation results can be obtained: the acoustic streaming field points directly below the reflector with a maximum velocity of 45 mm/s in the centripetal direction, which is consistent with the experimental results. From the flow diagram of the vertical section, it can be seen that the streaming field below the reflector is relatively weak, so the micro/nano objects can be accurately collected below the reflector. It can be seen from the above that the simulation results match the experimental results perfectly, which provides strong evidence for the effective action of the proposed reflector.

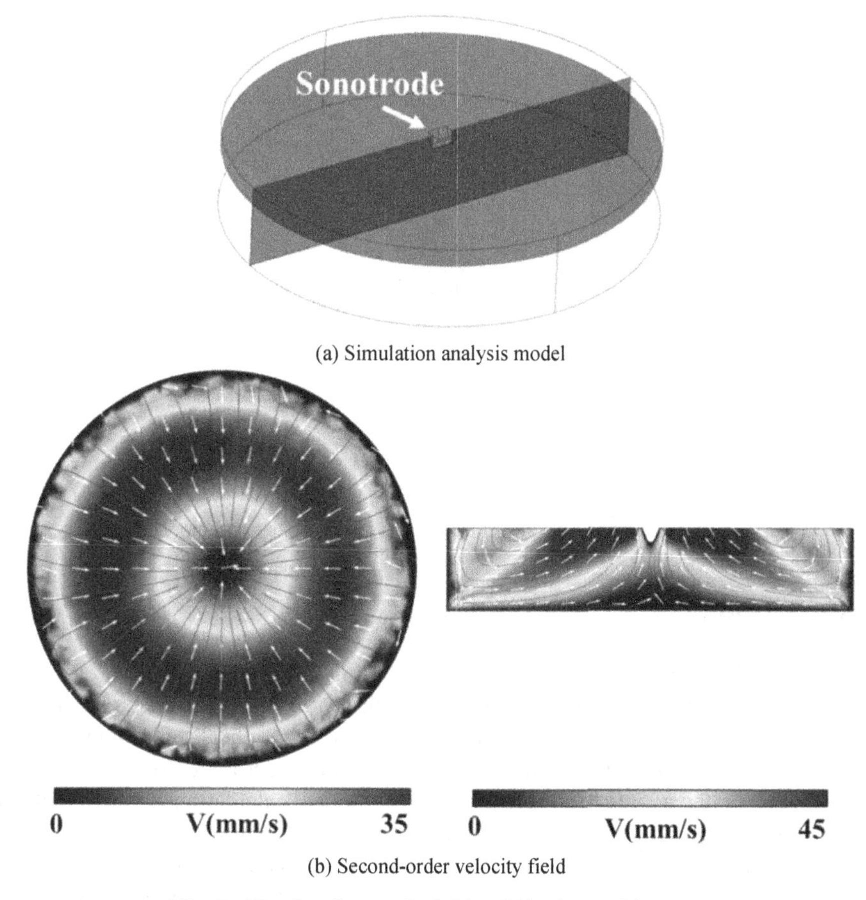

(a) Simulation analysis model

(b) Second-order velocity field

Fig. 7. Simulated acoustic fields within the working area

3.5 Aggregation Effect of Micro/nanomotors

Taking a typical tubular micro/nanomotors as the research object, we verify whether the acoustic manipulation chip containing V-shaped reflector can achieve efficient controlled aggregation in the micro/nanomotors cluster. When an excitation signal of 21.3 kHz is applied on the acoustic manipulation chip, the V-shaped reflector acts as an acoustic barrier device to control the tubular micro/nanomotors to aggregate towards the bottom of the reflector. The aggregation process is similar to Fig. 6, and the effect after aggregation is shown in Fig. 8. Experiments show that, similar to the controlled aggregation of 10 μm polystyrene microspheres, tubular micro/nanomotors can also achieve a large-scale aggregation movement in the acoustic streaming field regulated by V-shaped reflector. The controlled aggregation of micro/nanomotors has broad application prospects in micro/nano manufacturing, pollution degradation, biochemical detection and many other fields [13–15].

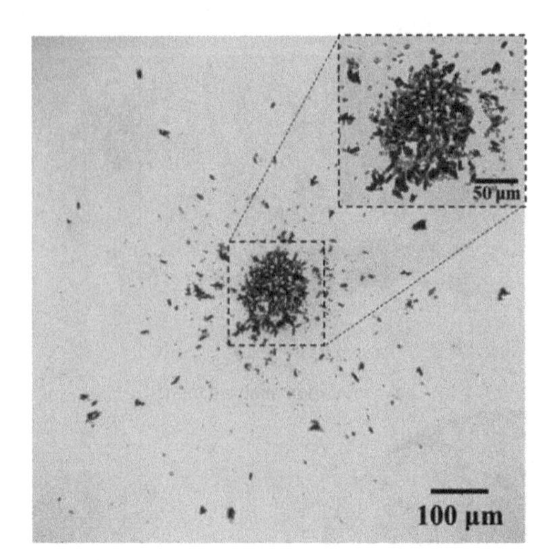

Fig. 8. Aggregation of micromotors beneath the reflector

4 Conclusion

To sum up, we propose a new acoustic manipulation chip with V-shaped reflector to achieve precise aggregation of micro/nano objects. The V-shaped reflector generates local acoustic streaming field in the microchannel, which gathers micro-nano objects in the whole working reservoir. Three different sizes of reflector are proposed to verify the size effect on the aggregation velocity. Both numerical calculation and experimental results show that the proposed reflector has a significant effect on the aggregation of micro/nano objects. At present, how to combine the acoustic manipulation chip with micro/nanomotors to realize the active capture and enrichment of low-concentration objects and realize the integration of electrochemical detection equipment is still a difficult problem to be addressed. Therefore, the proposed method paves a new avenue to perform on-demand enrichment for ultra-sensitive detection based on micro/nano motors, which is a successful step towards the miniaturization and integration of health detection equipment required by space life science.

References

1. Ronkainen, N.J., Halsall, H.B., Heineman, W.R.: Electrochemical biosensors. Chem. Soc. Rev. **39**(5), 1747–1763 (2010)
2. Xu, D., Hu, J., Pan, X., et al.: Enzyme-powered liquid metal nanobots endowed with multiple biomedical functions. ACS Nano (2021)
3. Yu, J., Zhang, Y., Ye, Y., et al.: Microneedle-array patches loaded with hypoxia-sensitive vesicles provide fast glucose-responsive insulin delivery. Proc. Natl. Acad. Sci. USA **112**(27), 8260–8265 (2015)
4. Bernard, I., Doinikov, A.A., Marmottant, P., et al.: Controlled rotation and translation of spherical particles or living cells by surface acoustic waves. Lab. Chip. **17**(14), 2470–2480 (2017)

5. Gao, W., Emaminejad, S., Nyein, H.Y.Y., et al.: Fully integrated wearable sensor arrays for multiplexed in situ perspiration analysis. Nature **529**(7587), 509–514 (2016)
6. Wu, Z., Lin, X., Si, T., et al.: Recent progress on bioinspired self-propelled micro/nanomotors via controlled molecular self-assembly. Small **12**(23), 3080–3093 (2016)
7. Ahmed, D., Ozcelik, A., Bojanala, N., et al.: Rotational manipulation of single cells and organisms using acoustic waves. Nat. Commun. **7**, 11085 (2016)
8. Lu, X., Wei, Y., Ou, H., et al.: Universal Control for Micromotor Swarms with a Hybrid Sonoelectrode. Small, e2104516 (2021)
9. Zhang, P., Bachman, H., Ozcelik, A., et al.: Acoustic microfluidics. Annu. Rev. Anal. Chem. (Palo Alto Calif) **13**(1), 17–43 (2020)
10. Sun, Y., Luo, Y., Xu, T., et al.: Acoustic aggregation-induced separation for enhanced fluorescence detection of Alzheimer's biomarker. Talanta **233**, 122517 (2021)
11. Tian, Z., Shen, C., Li, J., et al.: Programmable acoustic metasurfaces. Adv. Funct. Mater. **29**(13) (2019)
12. Xu, T., Luo, Y., Liu, C., et al.: Integrated ultrasonic aggregation-induced enrichment with raman enhancement for ultrasensitive and rapid biosensing. Anal. Chem. **92**(11), 7816–7821 (2020)
13. Maria-Hormigos, R., Jurado-Sanchez, B., Escarpa, A.: Labs-on-a-chip meet self-propelled micromotors. Lab. Chip **16**(13), 2397–2407 (2016)
14. Hannah, S., Addington, E., Alcorn, D., et al.: Rapid antibiotic susceptibility testing using low-cost, commercially available screen-printed electrodes. Biosens. Bioelectron. **145**, 111696 (2019)
15. Khezri, B., Sheng Moo, J.G., Song, P., et al.: Detecting the complex motion of self-propelled micromotors in microchannels by electrochemistry. RSC Adv. **6**(102), 99977–99982 (2016)

Biosignal Acquisition and Analysis

An Adaptive Robust Student's t-Based Kalman Filter Based on Multi-sensor Fusion

Dapeng Wang, Hai Zhang(✉), and Hongliang Huang

School of Automation Science and Electrical Engineering, Beihang University, Beijing 100191, China
zhanghai@buaa.edu.cn

Abstract. In practical applications, Kalman filter and its variants such as UKF may suffer from the time-varying measurement noise and process-error. Especially, when the process-error is heavy-tailed probability distribution, the Gaussian assumption would be no longer as accurately as expected. Aiming at the time-varying measurement noise and the situation with heavy-tailed process-error problems, in this paper, a new algorithm is proposed based on Student's t-distribution and multi-sensor information fusion. The robustness of the proposed algorithm is guaranteed by the timely estimation of the measurement noise, and the adaptiveness is realized by replacing the Gaussian by the Student's t-distribution. The Kullback-Leible Divergence (KLD) is used as the criterion for distinguishing the Gaussian distribution from the Student's t-distribution. Finally, a challenging target tracking example is presented and the simulation results show that the proposed algorithm achieves a higher accuracy than the other algorithms.

Keywords: Adaptive Kalman filter · Nonlinear filtering · Robust estimation · Target tracking · Student's t-distribution

1 Introduction

State estimation is one of the fundamental problems in many fields such as robotics, navigation, radar applications and so on. Kalman filter and its variants are a serious of promising algorithms which have been being applied in those fields [1]. The nonlinear filtering problem, the uncertainty in process noise and the abnormal observations/measurements are the major challenges of applying the Kalman filters [2].

To deal with the nonlinear problems, the Extend Kalman Filter (EKF), the unscented Kalman filter (UKF), the cubature Kalman filter (CKF), and the particle filter (PF) are the most widely used filters. The EKF is suitable for those scenarios where the non-linearity of the systems is low, because it only linearizes the state equation and/or measurement equation with first order Taylor series expansion [3]. To achieve better performance, the UKF and the CKF take advantage of the deterministic sampling method.

Most of the filters above make the assumption that the probability density function (PDF) of process noise and/or the measurement noise is Gaussian. In some applications, this assumption may work. Unfortunately, for the applications that require higher

H. Liu et al. (Eds.): ICIRA 2022, LNAI 13456, pp. 603–613, 2022.
https://doi.org/10.1007/978-3-031-13822-5_54

accuracy, this won't be satisfied any further. A Gaussian-Student's t mixture (GSTM) algorithm based on variational Bayesian (VB) was proposed to deal with the scenario with Non-stationary Heavy-tailed problems [4]. Inspired by the similar work, in this paper, we use the Student's t-distribution to approximate the heavy-tailed process-error.

As our previous work mentioned, more and more applications could afford two or more redundant sensors or measurement systems with the development of hardware technology [5]. So the redundant measurement noise covariance estimation (RMNCE) method is also adopted in this paper to estimate the measurement noise variance. It is suitable to be deployed on an embed hardware system because the computation complexity of the RMNCE is lightweight [6].

In this paper, a RMNCE and Student's t-based method is proposed to improve the accuracy and the robustness of the UKF. The contributions of the paper are twofold.

Firstly, a dynamic PDF switch framework based on RMNCE is proposed to estimate the variance of measurement noise timely. The frame could be extent to other type of PDF with slightly modification.

Secondly, we use Kullback–Leibler (KLD) divergence between the true posterior PDF and the Gaussian PDF estimated by RMNCE to determine whether the Student's t-distribution or Gaussian will be used for the next iteration. To avoid the expensive computation requirement, we use the same sliding window size in RMNCE to calculate the KLD. Meanwhile, a generalized factor is used to tackle the numerical problem that the sum of the probability would be greater than 1.

The remainder of this paper is organized as follows. The RMNCE -based UKF and the radar tracking dynamic model are presented in Sect. 2. The method we proposed is demonstrated in Sect. 3. In Sect. 4, simulation results are shown and analyzed. The Sect. 5 gives the conclusion.

2 The Dynamic Model of Radar Tracking and the RMNCE-Based UKF

2.1 Dynamic Model of Radar for Target Tracking

For simplicity and without loss of generality, we make the assumption that there are two radars in the system, and that only 2-D ($x - y$ plane) problem is concerned. The two radars are located at the same position for simplicity. Then the state vector of the target at time k can be represented in the Cartesian coordinates as [7]

$$X_k = \left[x_k,\ y_k,\ \dot{x}_k,\ \dot{y}_k, \ddot{x}_k,\ \ddot{y}_k\right] \tag{1}$$

where $x_k,\ y_k$ are the target position, $\dot{x}_k,\ \dot{y}_k$ represent the target velocity, $\ddot{x}_k,\ \ddot{y}_k$ stand for the acceleration of the target.

The dynamic equation for the target movement is as [7]

$$\boldsymbol{X}_k = f(\boldsymbol{X}_{k-1}) + \Gamma_{k-1}\boldsymbol{W}_{k-1},\quad \Gamma_{k-1} = \begin{bmatrix} T^2/2 & 0 \\ 0 & T^2/2 \\ T & 0 \\ 0 & T \\ 1 & 0 \\ 0 & 1 \end{bmatrix} \tag{2}$$

where T denotes the sampling period, Γ_{k-1} is the system noise-driven matrix with compatible dimension, $\boldsymbol{W}_{k-1}$ represents the process noise with compatible dimension. $f(\boldsymbol{X}_{k-1})$ can be given as

$$f(\boldsymbol{X}_{k-1}) = \begin{bmatrix} 1\,0\,T\,0 & T^2/2 & 0 \\ 0\,1\,0\,T & 0 & T^2/2 \\ 0\,0\,1\,0 & T & 0 \\ 0\,0\,0\,1 & 0 & T \\ 0\,0\,0\,0 & 1 & 0 \\ 0\,0\,0\,0 & 0 & 1 \end{bmatrix} \boldsymbol{X}_{k-1} \tag{3}$$

The measurement equations of these two radars can be described as

$$Z_k^i = \begin{bmatrix} r_k^i \\ \varphi_k^i \end{bmatrix} = \begin{bmatrix} \sqrt{x_k^2 + y_k^2} \\ \arctan\left(\frac{y_k}{x_k}\right) \end{bmatrix} + V_k^i,\ i = 1,\ 2 \tag{4}$$

where r_k^i and φ_k^i denote the slant range and azimuth angle in polar coordinates measured by the ith radar, respectively. The measurement noises V_k^i, $i = 1,\ 2$ are independent.

Meanwhile, the statistical properties of processes noise and the measurement noise can be summarized as

$$\begin{cases} E\left[\mathbf{W}_k \mathbf{W}_j^T\right] = \mathbf{Q}_k \delta_{k,j} \\ E\left[\mathbf{V}_k^i \mathbf{V}_j^{iT}\right] = \mathbf{R}_k^i \delta_{k,j} \\ E\left[\mathbf{W}_k \mathbf{V}_j^{iT}\right] = 0 \end{cases} \tag{5}$$

where $\delta_{k,j}$ denotes the Kronecker delta function.

It should be pointed out that the assumption of $\mathbf{W}_{k-1}$ may not be Gaussian in this model.

2.2 The RMNCE-Based UKF

Among the multi-sensor fusion methods, we adopt the RMNCE because not only that it is easily to deployed but also that the estimation of measurement noise variance is isolated from the state estimation.

Assuming $\mathbf{Z}_1(k)$ and $\mathbf{Z}_2(k)$ are independent measurement of $\mathbf{Z}(k)$ from two observations, the model then can be shown as:

$$\begin{cases} \mathbf{Z}_1(k) = \mathbf{Z}_T(k) + S_1(k) + \mathbf{V}_1(k) \\ \mathbf{Z}_2(k) = \mathbf{Z}_T(k) + S_2(k) + \mathbf{V}_2(k) \end{cases} \tag{6}$$

where $\mathbf{Z}_T(k)$ is the ground truth of the $\mathbf{Z}(k)$, $S_1(k)$ and $S_1(k)$ are the steady items of the observation errors, $\mathbf{V}_1(k)$ and $\mathbf{V}_2(k)$ are zero-mean white noise at time epoch k.

If the sample period is short enough, i.e.,

$$S_i(k) - S_i(k-1) \approx 0,\ i = 1,\ 2 \tag{7}$$

The first-order-self-difference $\Delta\mathbf{Z}_1(k)$, $\Delta\mathbf{Z}_2(k)$ and the second-order-mutual-difference $\Delta\mathbf{Z}_{12}$ are defined as

$$\begin{cases} \Delta\mathbf{Z}_1(k) = \mathbf{Z}_1(k) - \mathbf{Z}_1(k-1) \\ \Delta\mathbf{Z}_2(k) = \mathbf{Z}_2(k) - \mathbf{Z}_2(k-1) \\ \Delta\mathbf{Z}_{12}(k) = \Delta\mathbf{Z}_1(k) - \Delta\mathbf{Z}_2(k) \end{cases} \tag{8}$$

The variance of the measurement noise $\mathbf{V}_1(k)$ and $\mathbf{V}_2(k)$ can be estimated by

$$\begin{cases} \sigma_1^2(k) = \mathbf{R}_1 = \frac{E\left[\Delta\mathbf{Z}_{12}(k)\Delta\mathbf{Z}_{12}(k)^T\right] + E\left[\Delta\mathbf{Z}_1(k)\Delta\mathbf{Z}_1(k)^T\right] - E\left[\Delta\mathbf{Z}_2(k)\Delta\mathbf{Z}_2(k)^T\right]}{4} \\ \sigma_2^2(k) = \mathbf{R}_2 = \frac{E\left[\Delta\mathbf{Z}_{12}(k)\Delta\mathbf{Z}_{12}(k)^T\right] - E\left[\Delta\mathbf{Z}_1(k)\Delta\mathbf{Z}_1(k)^T\right] + E\left[\Delta\mathbf{Z}_2(k)\Delta\mathbf{Z}_2(k)^T\right]}{4} \end{cases} \tag{9}$$

The proof of (9) can be found in our previous work [5]. Based on RMNCE, we utilize the UKF to deal with the nonlinear problem.

The main steps in UKF are as follows [8].

Step 1: Initialization.

$$\begin{cases} \hat{\mathbf{X}}_0 = E[\mathbf{X}_0] \\ \mathbf{P}_0 = E\left[\left(\mathbf{X}_0 - \hat{\mathbf{X}}_0\right)\left(\mathbf{X}_0 - \hat{\mathbf{X}}_0\right)^T\right] \end{cases} \tag{10}$$

where $\hat{\mathbf{X}}_0$ is the initial state, $\mathbf{P}_0$ is the initial estimation of error covariance.

Step 2: Sigma points generation.

$$\begin{cases} \chi_{k-1}^{(0)} = \hat{\mathbf{X}}_{k-1} \\ \chi_{k-1}^{(i)} = \hat{\mathbf{X}}_{k-1} + \sqrt{(n+\lambda)\mathbf{P}_{k-1}},\ i = 1,\ 2,\ \ldots,\ n \\ \chi_{k-1}^{(i)} = \hat{\mathbf{X}}_{k-1} - \sqrt{(n+\lambda)\mathbf{P}_{k-1}},\ i = n+1,\ n+2,\ \ldots,\ 2n \end{cases} \tag{11}$$

where n is the dimension of the state; $\lambda = \alpha^2(n+\kappa) - n$ is the fine tuning composite scaling parameter, α is set to $0 \le \alpha \le 1$ and a good default setting on κ is $\kappa = 0$ [5].

Step 3: State prediction.

$$\begin{cases} \chi_{k/k-1}^{(i)} = f\left(\chi_{k-1}^{(i)}, k-1\right),\ i = 0,\ 1,\ 2,\ \ldots,\ 2n \\ \hat{\mathbf{X}}_{k/k-1} = \sum_{i=0}^{2n} \omega_i^{(m)} \chi_{k/k-1}^{(i)} \\ \mathbf{P}_{\mathbf{XX}} = \sum_{i=0}^{2n} \omega_i^{(c)} \left(\chi_{k/k-1}^{(i)} - \hat{\mathbf{X}}_{k/k-1}\right)\left(\chi_{k/k-1}^{(i)} - \hat{\mathbf{X}}_{k/k-1}\right)^T \\ \mathbf{P}_{k/k-1} = \mathbf{P}_{\mathbf{XX}} + \Gamma_{k-1} Q_{k-1} \Gamma_{k-1}^T \end{cases} \tag{12}$$

where $\hat{\mathbf{X}}_{k/k-1}$ is the predicted state mean; and $\mathbf{P}_{\mathbf{XX}}$ is the predicted state covariance. $\omega_i^{(m)}$ and $\omega_i^{(c)}$ are weights, which are calculated by

$$\omega_0^{(m)} = \frac{\lambda}{n+\lambda}$$

$$\omega_0^{(c)} = \frac{\lambda}{n+\lambda} + \left(1 - \alpha^2 + \beta\right)$$
$$\omega_i^{(m)} = \omega_i^{(c)} = \frac{\lambda}{2(n+\lambda)}, \quad i = 1,\ 2,\ \ldots,\ 2n \tag{13}$$

where $\beta \geq 0$ is used to incorporate the higher order information of the distribution, for Gaussian distribution $\beta = 2$. So if the PDF is not Gaussian, the parameter β should be tuned for a better estimation accuracy.

Step 4: Measurement prediction.

$$\gamma_{k/k-1}^{(i)} = h(\chi_{k/k-1}^{(i)}, k),\ i = 0,\ 1,\ 2,\ \ldots,\ 2n$$
$$\hat{\mathbf{Z}}_{k/k-1} = \sum_{i=0}^{2n} \omega_i^{(m)} \gamma_{k/k-1}^{(i)}. \tag{14}$$

Step 5: Kalman gain calculation.

$$\mathbf{P}_{\mathbf{XZ}} = \sum_{i=0}^{2n} \omega_i^{(c)} \left(\chi_{k/k-1}^{(i)} - \hat{\mathbf{X}}_{k/k-1}\right)\left(\gamma_{k/k-1}^{(i)} - \hat{\mathbf{Z}}_{k/k-1}\right)^T$$
$$\mathbf{P}_{\mathbf{ZZ}} = \sum_{i=0}^{2n} \omega_i^{(c)} \left(\gamma_{k/k-1}^{(i)} - \hat{\mathbf{Z}}_{k/k-1}\right)\left(\gamma_{k/k-1}^{(i)} - \hat{\mathbf{Z}}_{k/k-1}\right)^T + \mathbf{R}(k)$$
$$\mathbf{K}_k = \mathbf{P}_{\mathbf{XZ}} \mathbf{P}_{\mathbf{ZZ}}^{-1}. \tag{15}$$

where $\mathbf{R}(k)$ is estimated by RMNCE, and $\mathbf{K}_k$ is the Kalman gain.

Step 6: State estimation and error covariance update.

$$\hat{\mathbf{X}}_k = \hat{\mathbf{X}}_{k/k-1} + \mathbf{K}_k\left(\mathbf{Z}_k - \hat{\mathbf{Z}}_{k/k-1}\right)$$
$$\mathbf{P}_k = \mathbf{P}_{k/k-1} - \mathbf{K}_k \mathbf{P}_{\mathbf{ZZ}} \mathbf{K}_k^T. \tag{16}$$

Step 7: Iterate from steps 2 to 6 until the application is completed.

In practical applications, the process-error and the measurement noise covariance are not only time-varying but also non-Gaussian usually. So if the corresponding covariance matrix Q_{k-1} and $\mathbf{R}(k)$ could be estimated timely, better estimation of the state could be obtained. In our proposed algorithm, we estimate the $\mathbf{R}(k)$ by the RMNCE, then adjust the Q_{k-1} based on the Student's t filter.

3 The Proposed Method

3.1 The Dynamic PDF Switch Frame Base on KLD

In this paper, we employ the KLD to yield an indicator to check whether the posterior PDF of process-error is Gaussian or Student's t-distribution. The original KLD is given by [9]

$$\mathrm{KLD}(P \| Q) = \sum P(x) \log \frac{P(x)}{Q(x)} \tag{17}$$

$$\mathrm{KLD}(P\|Q) = \int P(x)\log\frac{P(x)}{Q(x)}dx \tag{18}$$

where $P(x)$ and $Q(x)$ are two different distributions of random variable x, (17) is the discrete version, (18) is for continues application.

The reason why we take the Student's t-distribution is that many scholars argued that the non-Gaussian heavy-tailed process noise could be approximated by Student's t-distribution with modification [10]. If the result obtained by the KLD indicates that it's better to approximate the posterior PDF by a Student's t-distribution, then the jointly predicted PDF of state and measurement are assumed to be:

$$p(\mathbf{X}_k, \mathbf{Z}_k|\mathbf{Z}_{1:k-1}) = St\left(\begin{bmatrix}\mathbf{X}_k\\ \mathbf{Z}_k\end{bmatrix}; \begin{bmatrix}\hat{\mathbf{X}}_{k|k-1}\\ \mathbf{H}_k\hat{\mathbf{x}}_{k|k-1}\end{bmatrix}, \begin{bmatrix}\mathbf{P}_{k|k-1} & \mathbf{P}_{k|k-1}\mathbf{H}_k{}^T\\ \mathbf{H}_k\mathbf{P}_{k|k-1} & \mathbf{P}_{\mathbf{ZZ}}\end{bmatrix}, \eta\right) \tag{19}$$

where $\mathbf{H}_k = \frac{\partial h}{\partial \mathbf{X}}\Big|_{\mathbf{X}=\hat{\mathbf{X}}_{k/k}}$, and η is the degrees of freedom (dof) parameter.

According to the Bayesian rule and (19), $p(\mathbf{X}_k, \mathbf{Z}_{1:k})$ could be updated as:

$$p(\mathbf{X}_k, \mathbf{Z}_{1:k}) = St\left(\mathbf{X}_k; \hat{\mathbf{X}}'_{k|k}, \mathbf{P}'_{k|k}, \eta'\right) \tag{20}$$

where η' is the dof of the posterior PDF $p(\mathbf{X}_k, \mathbf{Z}_{1:k})$. $\hat{\mathbf{X}}'_{k|k}$ $\mathbf{P}'_{k|k}$, and η' are given by [10]

$$\eta' = \eta + m \tag{21}$$

$$\hat{\mathbf{X}}'_{k|k} = \hat{\mathbf{X}}_{k|k-1} + \mathbf{P}_{k|k-1}\mathbf{H}_k^T\mathbf{P}_{\mathbf{ZZ}}^{-1}\left(\mathbf{Z}_k - \mathbf{H}_k\hat{\mathbf{X}}_{k|k-1}\right) \tag{22}$$

$$\mathbf{P}'_{k|k} = \frac{\eta + \Delta_k^2}{\eta + m}\left(\mathbf{P}_{k|k-1} - \mathbf{P}_{k|k-1}\mathbf{H}_k^T\mathbf{P}_{\mathbf{ZZ}}^{-1}\mathbf{H}_k\mathbf{P}_{k|k-1}\right) \tag{23}$$

$$\Delta_k = \sqrt{\left(\mathbf{Z}_k - \mathbf{H}_k\hat{\mathbf{X}}_{k|k-1}\right)^T\mathbf{P}_{\mathbf{ZZ}k}^{-1}\left(\mathbf{Z}_k - \mathbf{H}_k\hat{\mathbf{X}}_{k|k-1}\right)} \tag{24}$$

To get more accurate estimation, we further adopt the iterative method to generate the $\mathbf{P}_{k|k-1}$ by

$$\mathbf{P}_{k|k-1}^{(i+1)} = \frac{\tau\mathbf{P}_{k|k-1}/\mathrm{E}^{(i+1)}[\xi_k] + \mathbf{D}_k^{(i)}}{\tau + 1} \tag{25}$$

where i is the *ith* iteration, $\tau \in [2, 6]$, in this paper, $\tau = 5$. $\mathrm{E}^{(i+1)}[\xi_k]$ and $\mathbf{D}_k^{(i)}$ please refer to [2].

The mean state vector $\hat{\mathbf{X}}_{k|k}$ and scale matrix $\mathbf{P}_{k|k}$ are obtained as [2]

$$\hat{\mathbf{X}}_{k|k} = \hat{\mathbf{X}}'_{k|k} \qquad \frac{\eta}{\eta - 2}\mathbf{P}_{k|k} = \frac{\eta'}{\eta' - 2}\mathbf{P}'_{k|k} \tag{26}$$

In this paper, we use a sliding window to calculate the KLD between the posterior PDF of process-error and Gaussian. Taking the numerical error into consideration, we use a normalizing factor to guarantee the numerical stability. The normalizing factor is as follows:

$$factor = 1/\sum_{i=k}^{k-width} \mathcal{N}(X(i)) \tag{27}$$

where *width* is the sliding window width, $\mathcal{N}$ represents the Gaussian distribution.

So the final discrete KLD used by this paper is given by:

$$\mathrm{KLD}(P\|\mathcal{N}) = factor * \sum_{i=k}^{k-width} P(X_i)\log\frac{P(X_i)}{\mathcal{N}(X_i)} \tag{28}$$

where P and $\mathcal{N}$ represents the posterior PDF of process-error and the Gaussian distribution respectively, *factor* is given by (27).

3.2 The Proposed New Student's t-Based UKF Method

To reduce the computational burden while maintaining the accuracy of the estimation, the proposed algorithm make use of the first and second moments of the statistics in the sliding window.

If the $\boldsymbol{W}_{k-1}$ is a zero-mean vector with Gaussian distribution, then

$$p(\boldsymbol{W}) = \mathcal{N}(0, Q) \tag{29}$$

Equation (2) then can be expressed as

$$p(\Gamma_{k-1}^T\Gamma_{k-1})^{-1}\Gamma_{k-1}^T(\boldsymbol{X}_k - f(\boldsymbol{X}_{k-1})) = p(\boldsymbol{W}_{k-1}) = \mathcal{N}(0, Q_{k-1}) \tag{30}$$

where $f(\boldsymbol{X}_{k-1})$ is the state function.

The proposed adaptive and robust Student's t UKF (ARS-UKF) algorithm is given hereinafter.

Table 1. The pseudocode of the ARS-UKF algorithm.

Algorithm 1	**ARS-UKF**
Input:	observations from 2 sensors, $Z_k = \left[Z_{1,k},\ Z_{2,k} \right]$.
Output:	position estimations $X_k = \left[x_k,\ y_k \right]$.
Initiation	initiate the filter with $\hat{X}_0, \hat{P}_0, \hat{R}_0^1$ and $\hat{R}_0^2$.
Step1:	state and observation prediction based on UKF through (11) to (15).
Step2:	measurement noise estimation based on RMNCE through (9).
Step3:	Student's t detection based on KDL through (28).
Step4:	if: KLD indicates that the process-error distribution is Gaussian go to Step 5. else: the distribution of process-error maybe Student's t. go to Step 6.
Step5:	state update through (16). Covariance calculation via (26).
Step6:	state update through (19) to (25). Covariance calculation via (26).
Step7:	for the next iteration, Step 1 to Step 6 are executed repeatedly.

4 Simulation

4.1 Simulation Conditions

The initial state of the target is set to $X_0 = \left[1000\,\text{m}, 5000\,\text{m}, 0\,\text{m/s}, 0\,\text{m/s}, 0\,\text{m/s}^2, 0\,\text{m/s}^2\right]$. To make the simulation much closer to the practical situation, the outlier corrupted measurement and process noises are generated according to [5] and [10]

$$\mathbf{w}_k \sim \begin{cases} \mathcal{N}(\mathbf{0}, \mathbf{Q}) & w.p.\, 0.95 \\ \mathcal{N}(\mathbf{0}, 100\mathbf{Q}) & w.p.0.05 \end{cases} \text{t} \in ([0,1/3{*}\text{ST}] \tag{31}$$

$$\mathbf{v}_k^1 = \begin{cases} \mathcal{N}(\mathbf{0}, \mathbf{R}) & \text{t} \in ([0,1/3{*}\text{ST}] \cup [2/3{*}\text{ST,ST}]) \\ \mathcal{N}(\mathbf{0}, 100\mathbf{R}) & \text{t} \in ([1/3{*}\text{ST},2/3{*}\text{ST}]) \end{cases} \tag{32}$$

$$\mathbf{v}_k^2 \sim \mathcal{N}(\mathbf{0}, 10\mathbf{R}) \tag{33}$$

where ST stands for the entire simulation time, **Q** and **R** are nominal process and measurement noise covariance matrices

$$\mathbf{Q} = diag[0.001,\ 0.001]\,\mathbf{R} = diag\left[1^2 \times 100,\ 1^2 \times 0.01\right] \tag{34}$$

Equation (31) indicates that $\mathbf{w}_k$ is most of time generated from a Gaussian with nominal covariance. The process noise values are drawn from Gaussian with severely increased covariance about 5%. This assumption is close to practical application [2]. Equation (32) indicates that the performance of one senor in this simulation is degraded during the middle 1/3 simulation period. Equation (33) shows that the performance of the second sensor used in simulation is worse than the first one, but it keeps unchanged during the whole simulation time.

In our simulation, the standard Kalman filter (KF), the fixed version of the Robust Student's t-distribution Filter (RSKT-F) [2], the sigma version of the Robust Student's t-distribution Filter(RSKT-S) [2] and our proposed method are tested. The sample period in simulation is $T = 0.1$ s.

From 100 times Monte Carlo simulations, the root mean square error (RMSE) of position for each algorithm is obtained as [11]

$$RMSE_k = \sqrt{\frac{1}{N}\sum_{i=1}^{N}\left[\left(\hat{x}_k^i - x_k\right)^2 + \left(\hat{y}_k^i - y_k\right)^2\right]} \tag{35}$$

where N is the total times of simulation, $\left(\hat{x}_k^i, \hat{y}_k^i\right)$ represents the filtering position of the target at time instant k in the ith simulation, and $\left(x_k, y_k\right)$ is the ground truth.

4.2 Results

The positioning and the velocity errors of the algorithms are shown in Fig. 1.

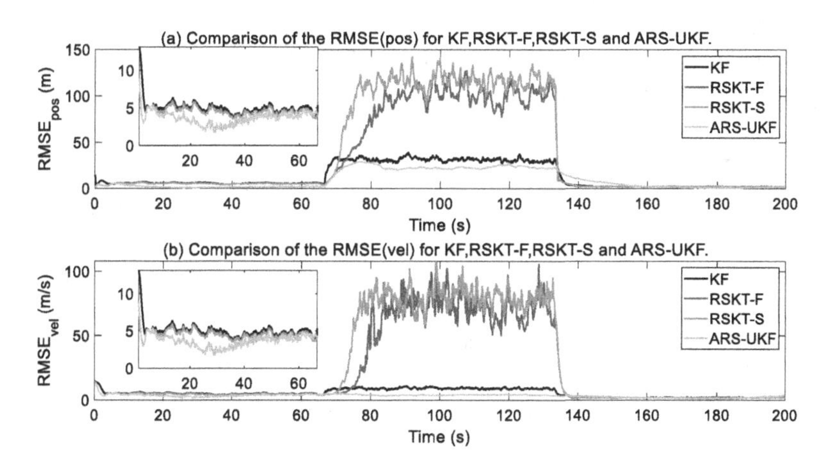

Fig. 1. Positioning and velocity errors of different algorithms.

To illustrate the performance of these algorithms more clearly, the mean of the position errors during the epochs [0, TS*1/3] and [TS*1/3, TS*2/3] are shown in Fig. 2.

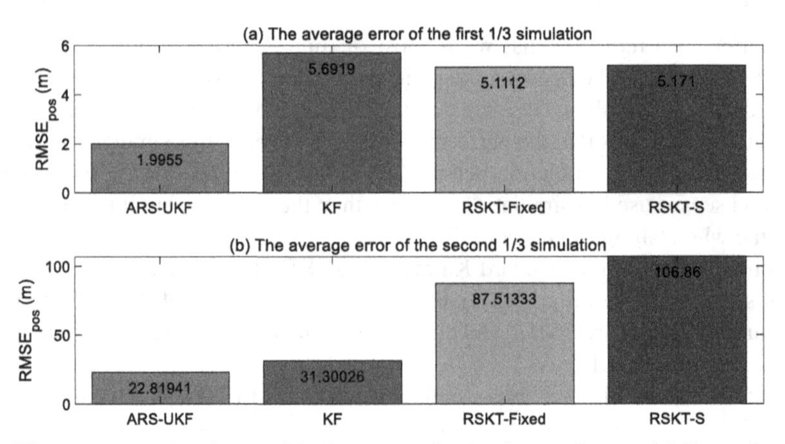

Fig. 2. The average estimation positioning errors in the first and second 1/3 epochs in the simulation.

It can be seen from the above two figures that in the first 1/3 period, the KF could not revisit the process-error which is heavy-tailed, so KF gives the greatest positioning errors. The RSKT-F and RSKT-S achieve better estimation results because that the two algorithms can deal with the heavy-tailed noise. Ours get the smallest estimation error in that the measurement noise variance is also estimated besides taking the Student's t-distribution filter into consideration.

In the second period, the KF performs better than the RSKT-F and RSKT-S, this is because that the latter two algorithms do not take the measurement noise variance change into consideration for a relative long period. While our proposed method outperforms the other algorithms, this is because we not only take the advantage of RSKF-S to deal with the heavy-tailed process-error, but also utilize the RMNCE to estimate the measurement noise variance online in real-time.

5 Conclusions

In this paper, the ARS-UKF algorithm is proposed and simulated. The RMNCE method based filter is first employed with Student's t-distribution together. The results show that positioning estimation accuracy of ARS-UKF method is about 63% better than that of RSKT-F in the first 1/3 simulation (1.9995 m vs 5.1112 m). As for the second 1/3 simulation, the positioning estimation accuracy of ARS-UKF method is about 30% better than that of RSKT-F (22.8194 m vs 31.30026 m). Future work will focus on reducing the estimation errors in the transition period around the 2/3 TS in the simulation.

Acknowledgements. This work is supported by the National Key Research and Development Program of China (2017YFC0821102, 2016YFB0502004).

References

1. Bar-Shalom, X.: Estimation with Applications to Tracking and Navigation. Wiley, New York (2001)

2. Huang, Y., Zhang, Y., Wu, Z., Li, N., Chambers, J.: A novel robust student's t-based Kalman filter. IEEE Trans. Aerosp. Electron. Syst. 1 (2017). https://doi.org/10.1109/TAES.2017.2651684
3. Wang, J., Zhang, T., Jin, B., Zhu, Y., Tong, J.: Student's t-based robust Kalman filter for A SINS/USBL integration navigation strategy. IEEE Sens. J. 1 (2020). https://doi.org/10.1109/JSEN.2020.2970766
4. Jia, G., Huang, Y., Bai, M., Zhang, Y.: A novel robust Kalman filter with non-stationary heavy-tailed measurement noise. IFAC-PapersOnLine **53**, 368–373 (2020). https://doi.org/10.1016/j.ifacol.2020.12.188
5. Wang, D., Zhang, H., Ge, B.: Adaptive unscented Kalman filter for target tacking with time-varying noise covariance based on multi-sensor information fusion. Sensors **21**, 5808 (2021)
6. Li, Z., Zhang, H., Zhou, Q., Che, H.: An adaptive low-cost INS/GNSS tightly-coupled integration architecture based on redundant measurement noise covariance estimation. Sensors **17**, 2032 (2017). https://doi.org/10.3390/s17092032
7. Sun, S., Deng, Z.: Multi-sensor optimal information fusion Kalman filter. Automatica **40**, 1017–1023 (2004)
8. Julier, S., Uhlmann, J.: A General Method for Approximating Nonlinear Transformations of Probability Distributions. http://citeseerx.ist.psu.edu/viewdoc/summary?doi=10.1.1.46.6718. Accessed 04 May 2022
9. Bishop, C.M. (ed.): Pattern Recognition and Machine Learning. ISS, Springer, New York (2006). https://doi.org/10.1007/978-0-387-45528-0
10. Roth, M., Ozkan, E., Gustafsson, F.: A student's t filter for heavy tailed process and measurement noise. In: Acoustics, Speech, and Signal Processing 1988. ICASSP 1988, pp. 5770–5774 (1988)
11. Zhao, Y., Liu, J., Zhou, K., Xu, Q.: An improved TCN-based network combining RLS for bearings-only target tracking. In: Liu, X.-J., Nie, Z., Yu, J., Xie, F., Song, R. (eds.) ICIRA 2021. LNCS (LNAI), vol. 13014, pp. 133–141. Springer, Cham (2021). https://doi.org/10.1007/978-3-030-89098-8_13

Dynamic Hand Gesture Recognition for Numeral Handwritten via A-Mode Ultrasound

Donghan Liu, Dinghuang Zhang, and Honghai Liu(✉)

University of Portsmouth, Portsmouth, UK
honghai.liu@port.ac.uk

Abstract. In recent years, due to the defects of weak sEMG signal, insensitive to fine finger movement and serious impression by noise, researchers consider the need to use A-mode ultrasound (AUS) for gesture decoding. However, the current A-mode ultrasonic gesture recognition algorithm is still relatively basic, which can recognize the recognition function of discrete gestures. However, due to the lack of time information, A-mode ultrasound still lacks an algorithm to recognize the dynamic gesture process. Therefore, we design and experiment a deep learning algorithm model applied to AUS signal, which is a deep learning framework based on LSTM. Due to the principle of LSTM, the model sets a certain number of frames as the whole action process, and constructs the connection of each frame in the whole process, so the time correlation (time characteristic) of AUS signal is constructed. Then, the features from AUS signal are sent to the complete full connection layer to output the classification results. And because AUS signal lacks data set of dynamic gestures, we designed and tested handwritten digits 0–9 as an example of dynamic gestures. Experimental results show that this algorithm can realize the dynamic gesture classification of AUS signal and solve the defect of AUS signal lacking time information. In addition, compared with the experimental action of traditional methods, it gives the practical significance of dynamic gesture in life, which is closer to life.

Keywords: A-mode ultrasound · LSTM · Dynamic hand gesture recognition · Handwritten numeral

1 Introduction

Hand is one of the most important limbs of human beings, which undertakes most of the daily activities and tasks. There are many physically disabled people in the world who are suffering from the inconvenience of life, so it is an urgent research topic to study the use of dexterous hands to improve the quality of life of the disabled. In recent years, with the emergence of computers, the HMI environment is more and more friendly to people, people can pass information to the machine through the mouse and keyboard, and monitor their work through the display.

H. Liu et al. (Eds.): ICIRA 2022, LNAI 13456, pp. 614–625, 2022.
https://doi.org/10.1007/978-3-031-13822-5_55

With the rapidly development of HMI, HMI methods based on gesture recognition have gained widespread attention. Gesture recognition is widely used in many fields, such as smart home control [1], game operation [2], virtual reality (VR)/augmented reality (AR) [3], etc. Gesture recognition is also the process of gesture decoding, so the quality of the result is strongly affected by the original signal. Choosing an appropriate signal is the top priority. At present, many human biological signals have been used for motion decoding, including electromyography (EMG) [4,5], electroencephalography(EEG) [6], and Electrooculography (EOG) [7], etc. In addition, Near-infrared Spectroscopy (NIRS) [8], Force Myography(FMG) [9], Inertial Measurement Unit (IMU) [10], and Ultrasonic(US) sensing [11], etc. have been used in wearable sensing solutions.

Among the non-invasive bioelectric signals of the upper limb, sEMG signal is the easiest and simplest physiological signal. The principle of sEMG signal is that bioelectric signal will be generated when muscle receives the transmission of nerve stimulation pulse. The surface EMG signal can be obtained by guiding and amplifying the wearable device. It is the sum of the active potential difference of multiple motor units. Through the processing and feature extraction of these potential difference waveforms, the functional change characteristics of neuromuscular system can be obtained. In the feature extraction step of gesture recognition based on sEMG signal, the commonly used features can be divided into three categories: time domain features, frequency domain features and time-frequency features [12]. Different feature combinations will have a great impact on the recognition results, so it is very important to select appropriate features according to different tasks. At present, in the research of gesture recognition, scholars have made great progress in using sEMG by using advanced methods. However, surface EMG signals have some defects. The characteristics of sEMG signal principle determine that the signal it produces is very weak and vulnerable to noise interference. Therefore, some research teams try to develop AUS signal to improve the decoding effect of hand gestures [13].

AUS signal is a kind of high-frequency ultrasound, which can detect muscle morphology by receiving echo signal. It detects the echo of sound wave based on the relationship between time and amplitude of sound wave, and has high axial resolution and high positioning accuracy [14]. Compared with the sEMG signal, which can only obtain the surface potential difference, AUS signal can obtain the subtle changes of deep muscles, which has great advantages in recognizing fine gestures. At the same time, the image of AUS signal is less affected by muscle fatigue and has better stability. Compared with B-mode ultrasound, AUS signal equipment is smaller and cheaper, which can meet the application of gesture recognition in portable scenes. A method to obtain forearm muscle information through multiple single element AUS signal sensors [15] was proposed to classify six movements including five fingers bending and resting state, with an average recognition accuracy of 96%. Yang et al. [16] used AUS signal to realize multi-modal synchronization and decoding of proportional wrist/hand kinematics, passed offline data test and online data test, and obtained better performance than sEMG. Although AUS signal can obtain deep muscle characteristics [17], the principle of AUS signal is equivalent to scanning muscles according to a certain frequency, so there is a lack of temporal information. At present, gesture recognition using AUS has made some progress in traditional machine learning, but the traditional feature extraction and classification algorithms have limited ability to extract features from signals, lack of appropriate strategies to identify the types of dynamic actions, and do not solve the problem of lack of time information in AUS signal.

In this paper, we designed and carried out an experiment of handwritten numeral action recognition. Previous studies in [20], they only classify some simple actions, which lack the practical significance of real life, and do not really classify the whole process. We first propose a deep learning framework algorithm for dynamic gesture recognition that can be used on AUS signal, which can be used to solve the defect of AUS signal lacking time information and endow AUS signal with time correlation. The results show that our proposed technology perfectly solves the above problems.

The rest of the paper is organized as follows: Sect. 2 introduces Data acquisition and pre-processing. Section 3 describes methods and experiments. Section 4 describes the results and analysis of different groups. The last section gives a conclusion of this study.

2 Data Acquisition and Pre-processing

The signal acquisition equipment used in this paper is a four channel A-type ultrasonic signal acquisition instrument designed and manufactured by Hangzhou ELONXI company. As shown in Fig. 1, the four type A ultrasonic probes are equipped with elastic straps to fix the probe to the surface of the forearm. The communication mode between the equipment and the computer is Ethernet. In the acquisition process, ultrasonic coupling agent shall be used between the probe and the skin to reduce the influence of air and external environment, so as to ensure the signal quality. The ultrasonic equipment has four channels, which correspond to four ultrasonic probes one by one. The sampling frequency of ultrasonic equipment is 20 MHz, and the working frequency of ultrasonic probe is 2.25 MHz. The number of sampling points per channel is 1000. The software's timer repeats at a frequency 20 Hz, which means it collects 20 frames of data from the ultrasonic device per second. Each frame contains 4000 points, forming a $4 * 1000$ matrix. In order to eliminate the influence of invalid information from the surface and deep layer of the skin, in the process of signal processing, 20 points of the start and end data of each frame of each channel are deleted, and only the middle 960 points are retained. Therefore, the actual matrix is $4 * 960$. In the process of ultrasonic propagation in human tissue, diffusion attenuation, scattering attenuation and absorption attenuation will occur, and its sound intensity will attenuate with the increase of propagation distance. Therefore, we first amplify the echo signal according to the propagation distance through time-varying gain compensation. Generally speaking, the attenuation of ultrasonic wave in the process of human tissue propagation can be considered

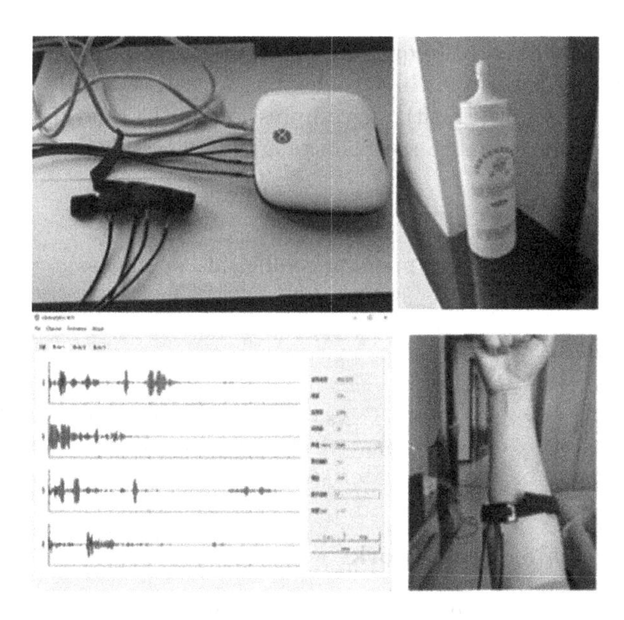

Fig. 1. Four channel ultrasonic equipment wearing and human-computer interaction interface

to conform to the exponential law. Therefore, we design an exponential variable gain amplifier to amplify the echo signal. Assuming $x_{raw}(t)$ is the original ultrasonic signal, the signal $x_{amp}(t)$ after time-varying gain compensation is

$$x_{amp}(t) = x_{raw}(t)e^{\alpha f d}(t) \tag{1}$$

Among them α is the sound attenuation coefficient, f is the center frequency of ultrasonic signal, d is the propagation distance of ultrasonic signal. For human skeletal muscle, the attenuation coefficient is slightly less than 1 dB/(cm ΔMHz). Convert dB to true value, $\alpha \approx 0.058/(\text{cm } \Delta\text{MHz})$. Here we take $\alpha = 0.05/(\text{cm}\Delta\text{MHz})$.

The echo signal after time-varying gain compensation contains obvious low-frequency and high-frequency noise. In order to enhance meaningful signal components, we use Gaussian band-pass filter to filter out noise. Specifically, we first perform Fourier transform on the echo signal, and then multiply it by a Gaussian filter function, as shown below

$$G(f) = \frac{1}{\sqrt{2\pi}\sigma}e - \frac{f^2}{2\sigma^2} \tag{2}$$

where f represents frequency. σ is the standard deviation of Gaussian function, which is calculated from the relative bandwidth of band-pass filter, as shown below

$$\sigma = \frac{bandwidth}{100\%}\Delta\frac{f_c}{(2\sqrt{2lg2})^2} \tag{3}$$

where f_c is the filter center frequency. Then, we get the filtered signal through the inverse Fourier transform. In this chapter, we choose bandwidth as 100%.

Finally, we carry out logarithmic compression on the echo signal to compress the amplitude of the signal to a certain range for easy calculation, as follows

$$x_{logcompression}(t) = log_{1+c}(1 + c\Delta x_{envelope}(t)) \tag{4}$$

where c is the logarithmic compression coefficient, we choose $c = 0.4$ [22].

3 Methodology

In this paper, we design a new strategy using RNN for feature extraction and dynamic hand gestures classification of AUS signal. The specific The details of the experiment will be shown in later sections.

3.1 Deep Learning Framework

Deep learning network is an algorithm for learning data representation. It has the advantages of strong learning ability, strong adaptability, data-driven, portability and so on. It can more fully extract the characteristics of the signal. At present, CNN is the most widely used. It can well extract hierarchical and spatial features [18], so it is used for image recognition, image segmentation, target tracking and other needs. From the principle of AUS signal acquisition [19], we can conclude that AUS signal is taken as a unit by each frame, so each picture unit is to take pictures of muscle depth and signal strength. For the needs of image recognition and image segmentation, researchers first used CNN as a deep learning framework to identify the gesture recognition needs of AUS signals.

In [20], Jia Zeng et al. Verified the superiority of AUS signal in using CNN to identify static motion and wrist motion. However, using CNN can only recognize discrete gestures, analyze and compare each frame separately, and can not build the relationship between frames. Therefore, it can not solve the defect of AUS signal lacking time information, nor can it really recognize the whole process of dynamic gestures. Therefore, researchers propose to combine sEMG signal with time information with AUS signal with high signal strength and deep test depth. However, the sensors with two signals in the same arm are complex, and wearing multiple devices is easy to cause muscle fatigue and inconvenient in life. In order to solve the above problems, we need to introduce a new deep learning framework to construct a dynamic gesture recognition strategy, and take advantage of the high strength of AUS signal to solve the defect of lack of time information of AUS signal.

LSTM is long short term memory [21], which is a variant of recurrent neural network (RNN), mainly to solve the gradient disappearance and gradient explosion problems during the training of long sequences. The main module of this neural network structure is LSTM module, and the LSTM block includes forget gate, input gate, and output gate. The forget gate, which is the key to achieving

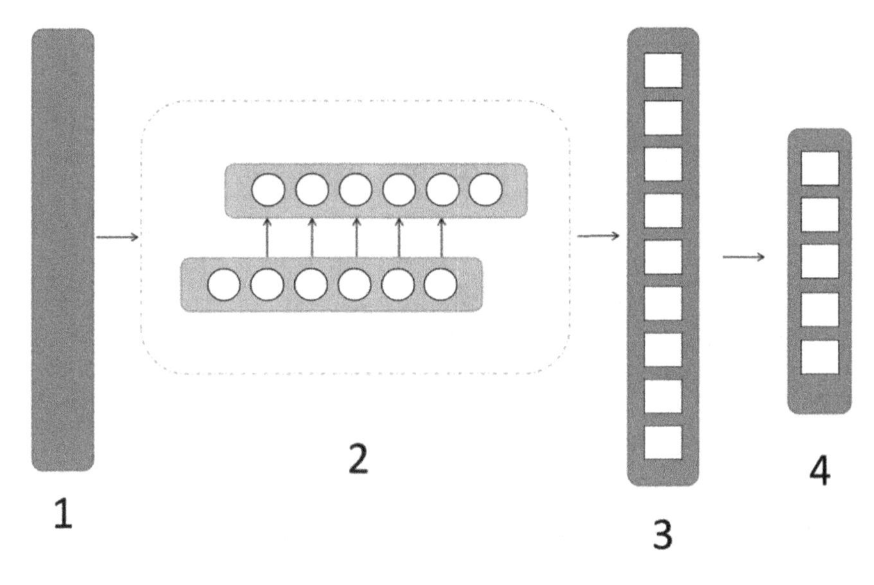

Fig. 2. Recurrent neural network architecture: 1-Input data, 2-Double layer LSTM network, 3-Fully-connected layer of Predictor 4-Fully-connected layer without activation-function - result vector

the long-term memory task, is the most important part of each LSTM block [23]. The LSTM network can capture the temporal dependence of time series and has achieved a series of results in time series prediction [24]. Cyclic neural network has memory, shared parameters and complete Turing, so it has certain advantages in learning the nonlinear characteristics of sequences. Recurrent neural network is applied in natural language processing, such as speech recognition, language modeling, machine translation and other fields. It is also used in all kinds of time series prediction, so it meets our core needs. LSTM can set a certain number of frames as a whole, and can make good use of the temporary characteristics of sequences [25]. Through the gesture recognition strategy with LSTM as the core, we can give AUS signal the lack of time information, which we call time correlation here. Also, with the help of time correlation, we can accurately recognize the whole process of dynamic gesture, which is a problem that has not been solved in previous research.

3.2 Dynamic Gesture Recognition Strategy

Time Correlation (Inter Frame Correlation). Inter Frame Correlation(IFC) mainly means that there must be some connection between adjacent frames to form a whole logical relationship. It is generally used for continuous speech recognition or character emotion recognition. By constructing the inter frame correlation between adjacent frames of AUS signal and setting a certain number of frames as a whole action process, we can not only solve the problem of lack of temporal information in AUS signal, but also classify the whole

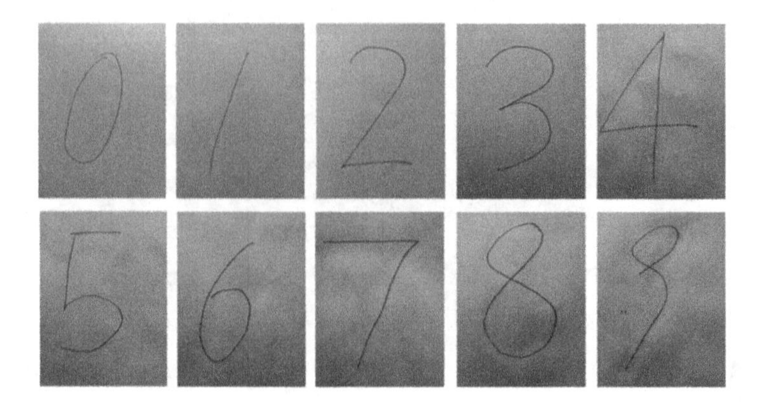

Fig. 3. Handwritten Arabic numerals 0–9

action process as much as possible through this strategy. In this way, we can successfully classify continuous gestures.

Dynamic Gesture Recognition Algorithm Structure. The structure of the whole algorithm is shown in Fig. 2. The LSTM part is a double-layer LSTM neural network with 100 hidden layers. The output result is connected with the full connection layer with 128 hidden layers, and finally sent to the second full connection layer with the set number of categories. The dropout part sets the parameter to 0.5.

After prepossessing the raw signal, we can get a three-dimensional matrix with the dimension of $total\ frames * channels * sampling\ points$, which is called the original data matrix. Because the core of the algorithm is the deep learning framework of LSTM, we need to recast the dimension of the original data matrix into a four-dimensional matrix of $action\ times * frames\ per\ action * channels * sampling\ points$. This matrix is called dynamic action matrix. Through the change of matrix dimension and the use of LSTM, we realize that different from CNN, we take each frame as the object of feature extraction. We take the whole process of action as a whole and realize IFC in the process of a single action. The realization of IFC makes up for the lack of time information of AUS signal, so that it can recognize dynamic actions.

4 Dataset

In this paper, the experimental data set of dynamic gesture recognition is composed of AUS signal. In this experiment, we designed handwritten numeral as the representative of dynamic gesture, because the process of handwritten numeral not only includes the combined movement of hand and wrist, but also reflects the recognition of in-hand manipulation. This handwritten numeral includes ten basic Arabic numerals from numeral 0 to numeral 9, which is shown in Fig. 3.

Fig. 4. Handwritten numeral experiment operation

Through the realization of handwritten basic digit recognition, we can realize the process of recognizing any combination of handwritten digits. The signal acquisition equipment used in this paper is a four channel A-mode ultrasonic signal acquisition instrument designed and manufactured by Hangzhou ELONXI company. The operation demonstration is shown in Fig. 4. The sampling frequency of AUS signal 20 Hz, and the AUS signal has 1000 sampling points per frame. There were four subjects in this experiment. They were all right-handed and had no neuromuscular and joint diseases. This procedure was accorded with the declaration of Helsinki. In this data set, we set the writing time of each number as two seconds. The conclusion drawn from many experiments can not only avoid the meaningless signal of writing for a long time, but also make the writing process blurred because of writing too fast. Each number needs to be written a total of 120 times, and 20 times is set as a group. The experimenters carry out activities according to the prompts of the software. The interval between each writing was 2 s, and 10 min of rest was set between each three groups to avoid the influence of muscle fatigue on the AUS signal experimental data. Only one number is collected for each experiment, and the whole experiment needs to be repeated 10 times. In this experiment, the first 80 out of 120 trials for each subject were selected as the training set, the data from the last 20 trials were used as the test set. And the last data were used as the validate set.

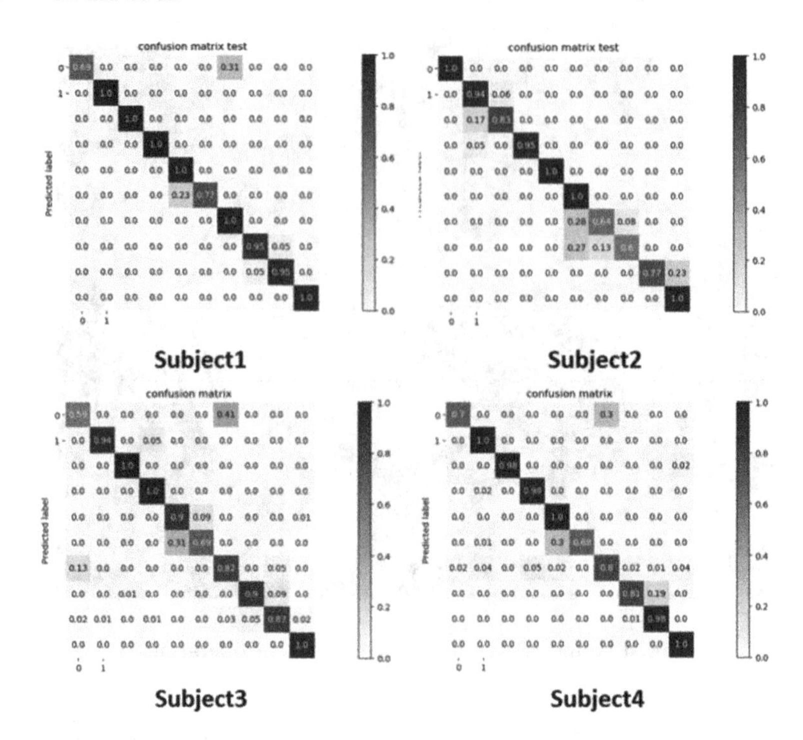

Fig. 5. Handwritten numeral 0–9 personal confusion matrix

5 Result and Discussion

The conclusion of this handwritten numeral action recognition experiment will be analyzed from two aspects. The first is to compare the personal data of the tester with each other, and the second is to use the mixed model that takes all the data of the tester as the training set. By analyzing the data of each tester in Fig. 5, we can draw a conclusion. The overall algorithm classification results are satisfactory. For the writing process with obvious features, the recognition rate can be close to 100%. For gestures with complex features, such as 0 and 6, due to the similar handwriting process, the recognition rate of writing 0 and 6 has a certain error, but the recognition rate can also reach more than 64%. Based on the above four experimental data, we can find that the model is universal. Therefore, it can be proved that the algorithm we designed can meet our needs.

The results of handwritten numeral recognition experiment are shown in Fig. 6. The confusion matrix shows the recognition rate of AUS signals in the whole writing process of handwriting 0 to 9 respectively under the model trained by dynamic gesture recognition algorithm with the mixed owner's data set as the training set. Among them, we can conclude that, similar to the conclusion of individual data analysis, except that the process of writing 0 and 6 is very similar, which affects the recognition rate, the recognition rate of other handwritten numeral processes can reach more than 80%. After converting the specific per-

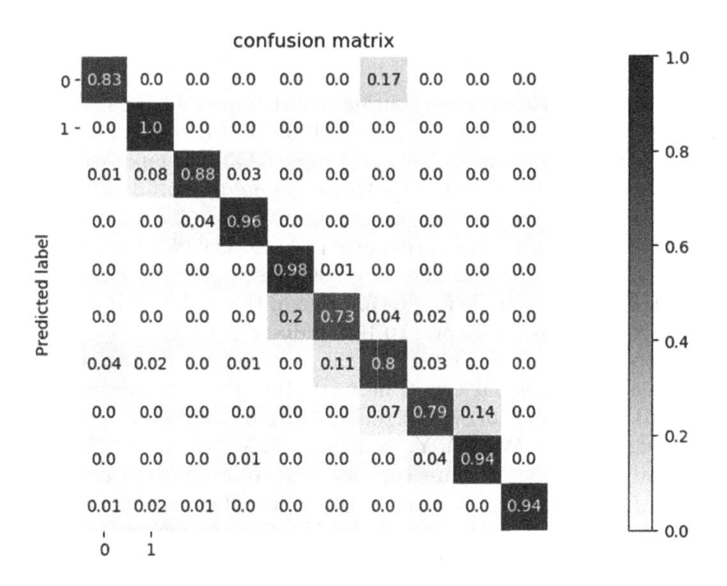

Fig. 6. Confusion matrix of handwritten numeral 0–9 using combination model

sonal data into general model data, we can still accurately classify the dynamic gestures. Therefore, we can conclude that the designed algorithm can meet the needs of dynamic gesture recognition and make up for the lack of time information of AUS signal.

6 Conclusion and Outlook

In this paper, we propose a deep learning algorithm for AUS signal. The algorithm can be used to recognize the classification of dynamic gestures collected by AUS signal. The whole algorithm is a deep learning framework with LSTM network as the core. By using this model, we set 40 frames as the whole process of dynamic action, and correlate the progress of each frame in this process to meet the needs of classifying dynamic action. This algorithm makes up for the lack of time information of AUS signal, gives new dimension information to AUS signal, makes each frame have time correlation, and makes it possible for AUS signal to recognize dynamic actions. What's more, the actions tested in our experiment give the practical application of dynamic action recognition in life for the first time, making the whole research closer to real life.

In future work, we intend to further study the situation of dynamic gesture recognition using AUS signal, including building more data sets of gesture and writing process to make up for the current lack of data sets. In addition, we plan to deepen the current deep learning framework to further improve the accuracy of continuous gesture recognition. Finally, we plan to study the algorithm of online dynamic gesture recognition for real life.

References

1. Zhao, H., Zhao, C.: Gesture recognition smart home. J. Phys. Conf. Ser. **1570**, 012045 (2020). https://doi.org/10.1088/1742-6596/1570/1/012045. (7 pp.)
2. Khalaf, A.S., Alharthi, S.A., Dolgov, I., Toups, Z.O.: A comparative study of hand gesture recognition devices in the context of game design. In: 14th ACM International Conference on Interactive Surfaces and Spaces (ISS), Conference Proceedings, pp. 397–402 (2019). https://doi.org/10.1145/3343055.3360758
3. Wen, F., et al.: Machine learning glove using self-powered conductive superhydrophobic triboelectric textile for gesture recognition in VR/AR applications. Adv. Sci. **7**(14) (2020). https://doi.org/10.1002/advs.202000261
4. Chen, H., Zhang, Y., Li, G., Fang, Y., Liu, H.: Surface electromyography feature extraction via convolutional neural network. Int. J. Mach. Learn. Cybern. **11**(1), 185–196 (2019). https://doi.org/10.1007/s13042-019-00966-x
5. Hu, Y., Wong, Y., Wei, W., Du, Y., Kankanhalli, M., Geng, W.: A novel attention-based hybrid CNN-RNN architecture for sEMG-based gesture recognition. Plos One **13**(10) (2018). https://doi.org/10.1371/journal.pone.0206049
6. Supratak, A., Dong, H., Wu, C., Guo, Y.: DeepSleepNet: a model for automatic sleep stage scoring based on raw single-channel EEG. IEEE Trans. Neural Syst. Rehabil. Eng. **25**(11), 1998–2008 (2017). https://doi.org/10.1109/tnsre.2017.2721116
7. Faust, O., Hagiwara, Y., Hong, T.J., Lih, O.S., Acharya, U.R.: Deep learning for healthcare applications based on physiological signals: a review. Comput. Methods Programs Biomed. **161**, 1–13 (2018). https://doi.org/10.1016/j.cmpb.2018.04.005
8. Ferrari, M., Muthalib, M., Quaresima, V.: The use of near infrared spectroscopy in understanding skeletal muscle physiology: recent developments. Philos. Trans. R. Soc. a-Math. Phys. Eng. Sci. **369**(1955), 4577–4590 (2011). https://doi.org/10.1098/rsta.2011.0230
9. Xiao, Z.G., Menon, C.: A review of force myography research and development. Sensors **19**(20) (2019). https://doi.org/10.3390/s19204557
10. Siddiqui, N., Chan, R.H.M.: Multimodal hand gesture recognition using single IMU and acoustic measurements at wrist. Plos One **15**(1), 12 (2020). https://doi.org/10.1371/journal.pone.0227039
11. Yang, X., Zhou, D., Zhou, Y., et al.: Towards zero re-training for long-term hand gesture recognition via ultrasound sensing. Inst. Electr. Electron. Eng. Inc. (4) (2019). https://doi.org/10.1109/JBHI.2018.2867539
12. Phukpattaranont, P., Thongpanja, S., Anam, K., Al-Jumaily, A., Limsakul, C.: Evaluation of feature extraction techniques and classifiers for finger movement recognition using surface electromyography signal. Med. Biol. Eng. Comput. **56**(12), 2259–2271 (2018). https://doi.org/10.1007/s11517-018-1857-5
13. Xia, W., Zhou, Y., Yang, X.C., He, K.S., Liu, H.H.: Toward portable hybrid surface electromyography/a-mode ultrasound sensing for human-machine interface. IEEE Sens. J. **19**(13), 5219–5228 (2019). https://doi.org/10.1109/Jsen.2019.2903532
14. Dong, J., Zhang, Y., Zhang, H., Jia, Z., Zhang, S., Wang, X.: Comparison of axial length, anterior chamber depth and intraocular lens power between IOLMaster and ultrasound in normal, long and short eyes. Plos One **13**(3) (2018). https://doi.org/10.1371/journal.pone.0194273
15. Li, Y.F., He, K.S., Sun, X.L., Liu, H.H.: Human-machine interface based on multi-channel single-element ultrasound transducers: a preliminary study. In: 18th IEEE International Conference on e Health Networking, Applications and Services (Healthcom), Conference Proceedings, New York, pp. 360–365. IEEE (2016)

16. Yang, X., Zhou, Y., Liu, H.: Wearable ultrasound-based decoding of simultaneous wrist/hand kinematics. IEEE Trans. Ind. Electron. **68**(9), 8667–8675 (2021). https://doi.org/10.1109/tie.2020.3020037
17. He, J., Luo, H., Jia, J., Yeow, J.T.W., Jiang, N.: Wrist and finger gesture recognition with single-element ultrasound signals: a comparison with single-channel surface electromyogram. IEEE Trans. Biomed. Eng. **66**(5), 1277–1284 (2019). https://doi.org/10.1109/tbme.2018.2872593
18. Allard, U.C., et al.: A convolutional neural network for robotic arm guidance using sEMG based frequency-features. In: 2016 IEEE/Rsj International Conference on Intelligent Robots and Systems (IROS 2016), pp. 2464–2470 (2016)
19. Sun, X., Yang, X., Zhu, X., et al.: Dual-frequency ultrasound transducers for the detection of morphological changes of deep-layered muscles. IEEE Sens. J. **18**(4), 1373–1383 (2017)
20. Zeng, J., Zhou, Y., Yang, Y., et al.: Feature fusion of sEMG and ultrasound signals in hand gesture recognition. In: 2020 IEEE International Conference on Systems, Man, and Cybernetics (SMC). IEEE (2020)
21. Yu, Y., Si, X., Hu, C., et al.: A review of recurrent neural networks: LSTM cells and network architectures. Neural Comput. **31**(7), 1235–1270 (2019)
22. Li, Y., He, K., Sun, X., et al. Human-machine interface based on multi-channel single-element ultrasound transducers: a preliminary study. In: 2016 IEEE 18th International Conference on e-Health Networking, Applications and Services (Healthcom). [S.l.]: [s.n.], pp. 1–6 (2016)
23. Zhu, G.M., et al.: Redundancy and attention in convolutional LSTM for gesture recognition. IEEE Trans. Neural Netw. Learn. Syst. **31**(4), 1323–1335 (2020). https://doi.org/10.1109/Tnnls.2019.2919764
24. Tong, R.Z., Zhang, Y., Chen, H.F., Liu, H.H.: Learn the temporal-spatial feature of sEMG via dual-flow network. Int. J. Humanoid Robot. **16**(4) (2019). https://doi.org/10.1142/S0219843619410044. DOI Artn 1941004
25. Ma, C.F., Lin, C., Samuel, O.W., Xu, L.S., Li, G.L.: Continuous estimation of upper limb joint angle from sEMG signals based on SCA-LSTM deep learning approach. Biomed. Signal Process. Control **61** (2020). https://doi.org/10.1016/j.bspc.2020.102024. DOI ARTN 102024

Brain Network Connectivity Analysis of Different ADHD Groups Based on CNN-LSTM Classification Model

Yuchao He[1,3], Cheng Wang[1,2], Xin Wang[1,2], Mingxing Zhu[4], Shixiong Chen[1](✉), and Guanglin Li[1]

[1] CAS Key Laboratory of Human-Machine Intelligence-Synergy Systems, Shenzhen Institutes of Advanced Technology, Chinese Academy of Sciences, Shenzhen 518055, Guangdong, China
sx.chen@siat.ac.cn

[2] Shenzhen College of Advanced Technology, University of Chinese Academy of Sciences, Shenzhen 518055, Guangdong, China

[3] College of Engineering, Southern University of Science and Technology, Shenzhen 518055, Guangdong, China

[4] School of Electronics and Information Engineering, Harbin Institute of Technology, Shenzhen 518055, China

Abstract. Attention deficit hyperactivity disorder (ADHD), as a common disease of adolescents, is characterized by the inability to concentrate and moderate impulsive behavior. Since the clinical level mostly depends on the doctor's psychological and environmental analysis of the patient, there is no objective classification standard. ADHD is closely related to the signal connection in the brain and the study of its brain connection mode is of great significance. In this study, the CNN-LSTM network model was applied to process open-source EEG data to achieve high-precision classification. The model was also used to visualize the features that contributed the most, and generate high-precision feature gradient data. The results showed that the traditional processing of original data was different from that of gradient data and the latter was more reliable. The strongest connections in both ADHD and ADD patients were short-range, whereas the healthy group had long-range connections between the occipital lobe and left anterior temporal regions. This study preliminarily achieved the research purpose of finding differences among three groups of people through the features of brain network connectivity.

Keywords: ADHD · Brain network connectivity · CNN-LSTM model

This work was supported in part by the National Natural Science Foundation of China (#81927804, #62101538), Shenzhen Governmental Basic Research Grant (#JCYJ20180507182241622), Science and Technology Planning Project of Shenzhen (#JSGG20210713091808027, #JSGG20211029095801002), China Postdoctoral Science Foundation (2022M710968), SIAT Innovation Program for Excellent Young Researchers (E1G027), CAS President's International Fellowship Initiative Project (2022VEA0012). #The first two authors have equal contributions to this work.

H. Liu et al. (Eds.): ICIRA 2022, LNAI 13456, pp. 626–635, 2022.
https://doi.org/10.1007/978-3-031-13822-5_56

1 Introduction

Attention deficit hyperactivity disorder (ADHD) is a common neurodevelopmental disorder in children and adolescents. It is mainly manifested in the inability to concentrate, regulate activities, and moderate impulsive behavior [1]. At present, the prevalence of ADHD in the world is about 7.2% [2]. As the pathogenesis of ADHD is not clear, it has gradually become a global public health problem [3]. As one of the subtypes of ADHD (ADHD-I), attention deficit disorder (ADD) is often impaired in intelligence dimension and attention switching ability. The main diagnostic criteria for ADHD are the "Diagnostic and Statistical Manual of Mental Disorders 5th (DSM-5)" developed by the American Psychiatric Association. Doctor makes judgments based on the patient's daily psychological behavior and social activities, and there is no objective and quantifiable classification index [4].

Since traditional diagnosis does not have quantitative classification indicators, and EEG signals, as a nonlinear signal with high temporal resolution, can provide a large amount of objective data related to cognitive activities, deep learning algorithms are used to automatically extract EEG-related features and analyze them. The commonly used deep models for processing EEG signals mainly include convolutional neural network (CNN) [5] and EEGNET [6]. This paper uses CNN and Long Short-Term Memory Model (LSTM) series model CNN-LSTM for classification.

When the human brain completes a task, it does not rely solely on a single neuron or brain region, but through the mutual coordination of multiple brain regions, processing and transmitting information from each brain region to complete the task together. Each brain area can collect, process and process information on the external environment through the autonomous activities of neurons. Brain functional connectivity can well describe the cooperative working state of various brain functional areas. ADHD can cause different degrees of brain damage to patients, so it is of great significance to conduct brain functional connectivity analysis.

Modern network theory is increasingly used in neuroscience to study the brains of healthy and diseased subjects, and the necessity of objectively understanding the network composed of functional connections of different brain regions plays a vital role in neuroscience. In resting EEG, the basic mechanisms of normal and pathological functional brain tissues were revealed by using graph theory and network science tools.

Cerebral connectivity map can analyze the dynamic interaction between brain regions, which can be divided into functional connectivity and effective connectivity [7]. Functional connectivity studies the statistical correlation between brain region dynamics, while effective connectivity represents the causal interaction between activated brain regions and describes the directional effect of one neuronal system on another in the brain [8].

In this paper, CNN-LSTM model was used to realize EEG signal classification processing. The feature that contributes the most to the classification process of the output model is the gradient data [9]. The gradient data was used to draw internal connection diagrams in different brain regions. By comparing the differences in brain area connections in different groups of people, the connection rules of brain areas are obtained.

2 Methods

2.1 Data Process

This study used open-source EEG data, which was approved by the local ethics committee and obtained informed consent from the participants or their legal guardians. All experimental data frameworks can be found in [10]. The data set contains the data of 144 children. The diagnosis was made through the standard guidelines of psychiatrists and psychologists. There were 48 ADHD subjects, 52 ADD subjects, and the other 44 participants were healthy control children. There was no significant difference in the degree of inattention between ADD and ADHD patients, but the hyperactivity and impulsivity in ADHD patients were significantly higher than those in add patients. All the children were no significant difference in age and intelligence quotient (IQ) [11].

2.2 Research Procedure

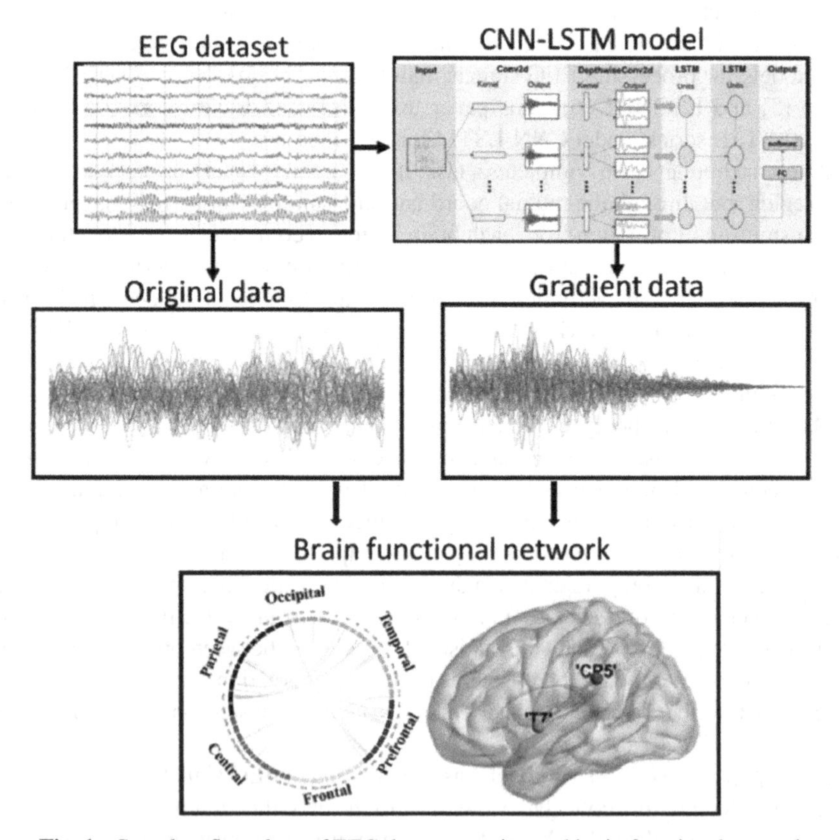

Fig. 1. Complete flow chart of EEG data processing and brain functional network

As shown in Fig. 1, the research procedure was divided into four parts. Firstly, the EEG dataset was input into the CNN-LSTM neural network model to realize accurate classification. Then through visual operation, the input features which contribute the most to the classification process are used and output as gradient data. Thirdly, the original data obtained by traditional processing were used as the control group to draw the brain function network connection diagram. Finally, the differences in brain network connectivity of the three groups of population gradient data and raw data were compared.

2.3 CNN-LSTM Model

In this study, a model in which convolution neural network (CNN) and long short-term memory (LSTM) network were connected in series was designed to learn EEG architecture. As is shown in Fig. 1, The network starts with a temporal convolution (Conv2d part) to learn time-domain filters, then uses a depthwise convolution (DepthwiseConv2d part), connected to each feature map of Conv2d outputs individually, to learn spatial filters. In the recurrent neural network part, two LSTM layers will process the sequence of a temporal summary feature map, which learns how to classify the abstract features of time series.

The input data of the model was channels and time points, which are 56 and 385 in this dataset. In the training stage, the input batch size was 256, the learning rate was 0.001, and training iterations were 300 epochs. The Adam optimizer and categorical cross-entropy loss function were used in the model compile.

The gradient data comes from the dense layer output before the softmax activation function. The gradient data is obtained by generating the network loss parameters, then calculating the gradient of the loss data and the input data, and selecting the maximum channel gradient.

2.4 Brain Connectivity Principle

The essence of the construction of attention function brain network is to abstract the very complex brain into concrete nodes and edges, and then use graph theory to describe the internal connections of brain activity when people pay attention [12]. The brain network construction mainly includes three aspects: the definition of network nodes, the measurement of edges between nodes, and the selection of brain network thresholds.

The nodes of the brain function network are represented by scalp electrodes that collect EEG signal data. The coordinates of each brain area node are represented by the coordinates of all channels, and the frontal lobe, parietal lobe and other brain areas are color-coded to construct a brain area 3D node map that matches the human brain.

The connection edges between nodes are calculated by various phase synchronization features, including coherence coefficient, Pearson correlation coefficient, phase lag index, etc. [13]. Phase lag index (PLI) can represent phase synchronization and is an asymmetry index of phase difference distribution calculated from the instantaneous phases of two time series [14].

Since the sensitivity of PLI to bulk conduction and noise, and discontinuities in detecting changes in phase synchronization are hindered, this experiment uses the weighted phase lag index (WPLI) for calculation. WPLI weights each phase difference by the magnitude of the phase delay, which improves the sensitivity of detecting phase synchronization [1]. Its calculation formula is as follows:

$$WPLI = \left| \frac{\langle |\sin(\Delta\phi(tk))| \mathrm{sgn}(\sin(\Delta\phi(tk))) \rangle}{\sin(\Delta\phi(tk))} \right|$$

where k represents the number of time points in each trial, and $\Delta\varphi$ represents the phase difference between the two time series at time point k. The *sgn* symbol represents the symbolic function, the < > symbol represents the mean value, and the | | symbol represents the absolute value. The instantaneous phase difference is realized by the Hilbert transform.

The clustering coefficient represents the degree of clustering of the network. In order to achieve a more reasonable brain function network construction, this experiment set parameters such as connection edge threshold and node connection strength to make the overall clustering coefficient of the network reach 10%.

3 Results

3.1 CNN-LSTM Classification

5-fold cross-validation was used to measure model performance.in 5-fold classification method, the CNN-LSTM model exceeded 98% in the three group (HC, 98.32%; ADD,98.28%; ADHD, 98.47%).Further analyzing model performance, it can get 99.17% of specificity, 98.36% of sensitivity, 99.66% of AUC score and 98.34% of F1 score. The model has better achieved the classification effect for the ADHD dataset, so it will be more convincing to analyze the gradient data used by the model for brain network.

3.2 Brain Functional Connectivity

In this part, two-dimensional and three-dimensional brain functional connectivity maps were displayed to compare the connectivity differences between the original data and gradient data. As shown in Fig. 2, using the original signal analysis, the brain region connections of healthy people were mainly concentrated between the occipital lobe, temporal lobe, and central lobe. ADD patients and ADHD patients contained too much redundant information, so it was difficult to significantly distinguish brain region connections. Through gradient data analysis, the brain connection activities of healthy people were mainly concentrated between the occipital lobe, temporal lobe, prefrontal lobe, and middle lobe, and those of ADD patients are concentrated between the parietal lobe, middle lobe, and frontal lobe. The prefrontal lobe activities of ADHD patients were slightly higher than those of ADD patients. However, due to too much output data, it was difficult to eliminate the interference of redundant data and could not accurately connect the brain control interval.

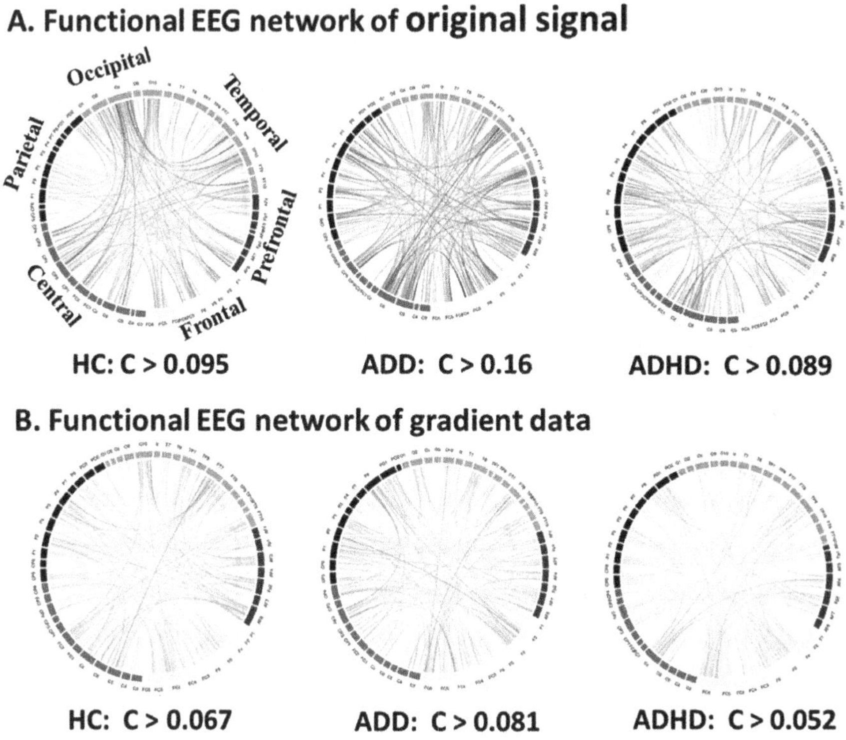

A. Functional EEG network of original signal

Occipital Temporal Parietal Prefrontal Central Frontal

HC: C > 0.095 ADD: C > 0.16 ADHD: C > 0.089

B. Functional EEG network of gradient data

HC: C > 0.067 ADD: C > 0.081 ADHD: C > 0.052

Fig. 2. Brain function connection diagram between original signal (A) and gradient data (B) under low threshold output selection. The outermost circular regions represent different brain regions, in which black represents the parietal lobe, orange represents the occipital lobe, gray represents the temporal lobe, blue represents the prefrontal lobe, yellow represents the frontal lobe, red represents the central region, and red curve connection represents the connection of brain regions. (Color figure online)

In order to reduce the influence of relevant redundant information and increase the signal output threshold, the brain connection diagram of Fig. 3 was generated. According to the analysis of the original data, the brain connections of healthy people were concentrated between parietal lobe, occipital lobe, temporal lobe and prefrontal lobe; ADD patients were connected between parietal lobe, temporal lobe, frontal lobe and middle lobe, and ADHD patients were mainly concentrated between middle lobe, temporal lobe and prefrontal lobe. By analyzing the gradient data, the brain connections of healthy people were concentrated in the occipital lobe, prefrontal lobe and middle lobe; ADD patients are concentrated in the occipital lobe, temporal lobe, prefrontal lobe and middle lobe; and ADHD patients were mainly concentrated in the occipital lobe, frontal lobe, central region and prefrontal lobe.

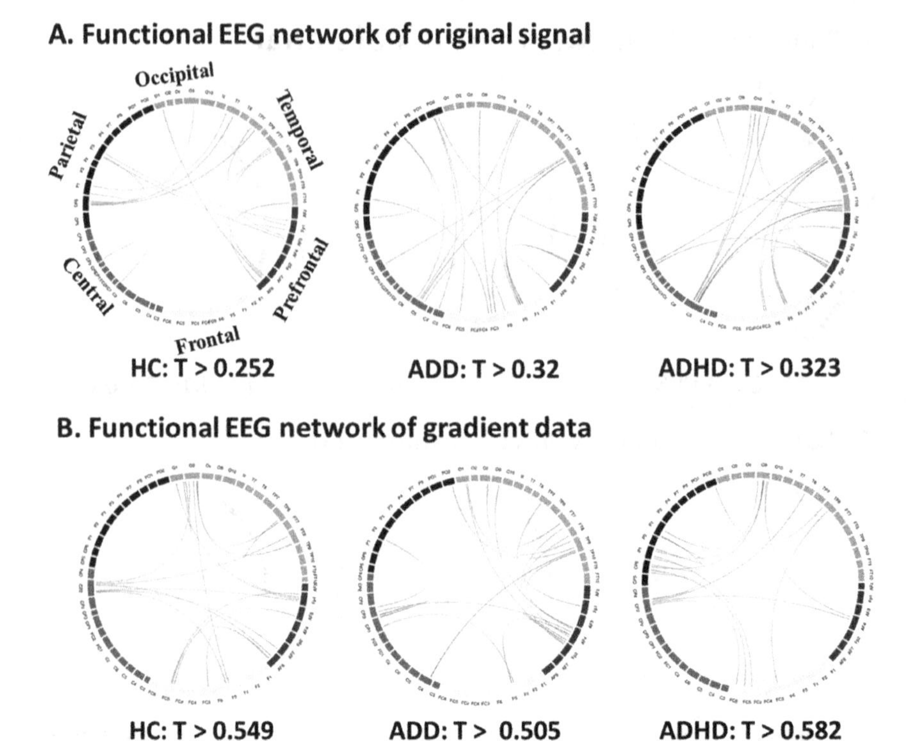

Fig. 3. Brain function connection diagram of original signal and gradient data under high threshold output selection. The outermost circular regions represent different brain regions, in which black represents the parietal lobe, orange represents the occipital lobe, gray represents the temporal lobe, blue represents the prefrontal lobe, yellow represents the frontal lobe, red represents the central region, and red curve connection represents the connection of brain regions. (Color figure online)

The strongest connection diagram of original data and gradient data was different. Figure 4(A) showed the strongest connection of the original signal. The strongest connection of healthy people was between the temporal lobe and occipital lobe. Patients with ADD had connection between the prefrontal lobe and temporal lobe, and patients with ADHD had connection between the frontal lobe and parietal lobe. The gradient data shown in Fig. 4(B) showed the strongest connection. The healthy population had a long-term connection between the temporal lobe and the occipital lobe, and the patients with ADD and ADHD had a short-term connection. The patients with add exist between the frontal lobe and the parietal lobe, and the patients with ADHD have the strongest connection within the parietal lobe.

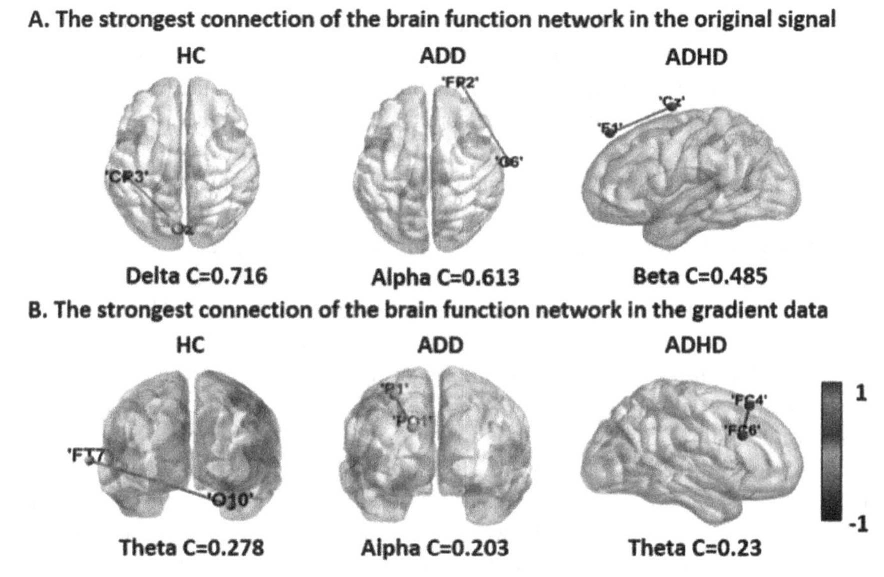

Fig. 4. Sagittal, axial and coronal views of the strongest connections between original signal and gradient data. The color of brain segment represents the strength of corresponding connection. The intensity range is represented by color changes, with red indicating the maximum intensity and blue indicating the minimum intensity (Color figure online)

4 Discussion

In this study, we compared the original signal with gradient data at the level of brain functional connectivity map based on the CNN-LSTM model. Multiple groups of information level brain interval connections were displayed through 2D images, and the connection area and intensity were displayed in the strongest connections through 3D images.

First, the classification accuracy of the CNN-LSTM network model was more than 98%, which was higher than the data classification performance of common EEG. Therefore, visual analysis of its internal feature selection had strong reference significance in finding the features that contributed the most to deep learning classification, and constructing gradient data for image-level analysis and expression.. Because of its classification contribution, the gradient data could better describe the brain activities that played a role in the classification of three types of people.

Then, the multi-signal brain network connection and the strongest brain network connection were displayed through two functional brain network connection modes (2D and 3D) to compare and determine the difference in brain connection. After selecting the connection signals output by two different thresholds, we further compared the original signal with the gradient data. Under the condition of setting a low threshold, it was difficult to distinguish the original signal significantly because the connection was doped with too much redundant information. The gradient data determined the brain interval connection of healthy people, but it was also scattered for patients with ADD and ADHD, which was difficult to determine effectively. To further study the brain connection classification process, we raised the connection data output threshold and determined the brain connection difference between the original signal and gradient data through high-intensity connection. After reducing the signal input, the brain connection was more significant. The two connection areas were similar, but there were still some differences.

Using the numerical analysis of each group, the strongest connection of functional brain network was found, which was conducive to displaying the dominant connection mode. Through the color between connections indicated the intensity, and the length of line segments indicated the long connection and short connection, it could analyze the strongest connection diagram of the brain region of the original signal. The three groups were dominated by the long connection. For the gradient data, there was a constant connection between the temporal lobe and the occipital lobe in the healthy population, while the patients with ADD and ADHD showed that the connection within the brain region was the main connection.

According to different subjects, the display of brain connection would still show some differences. Therefore, there were still some accidental factors in the experimental results under a low amount of data. However, if the amount of data was further expanded, it would help to accurately determine the brain connection and improve the accuracy and universality of the model.

5 Conclusion

Through using gradient data functional brain network analysis, the brain connections of ADD patients were concentrated in the central and prefrontal lobes, and those of ADHD patients were concentrated in the central and occipital lobes. By the strongest connection analysis, the healthy population was mainly the long-distance connection from the frontal lobe, temporal lobe to occipital lobe, and the frequency band was concentrated in θ band. The strongest connection between ADD patients and ADHD patients was short-distance connection, in which ADD patients were connected in the parietal lobe and α band, and ADHD patients were connected between the right anterior midbrain area and θ band. The results of this study could be used as an auxiliary method to help doctors achieve efficient diagnosis and classification in clinical application.

References

1. Thapar, A., Cooper, M.: Attention deficit hyperactivity disorder. Lancet **387**, 1240–1250 (2016)
2. Thomas, R., Sanders, S., Doust, J., et al.: Prevalence of attention–deficit/hyperactivity disorder: a systematic review and meta–analysis. Pediatrics **135**, e994-1001 (2015)
3. Sherman, E.M., Slick, D.J., Connolly, M.B., Eyrl, K.L.: ADHD, neurological correlates and health-related quality of life in severe pediatric epilepsy. Epilepsia **48**, 1083–1091 (2007)
4. American Psychiatric Association. Diagnostic and Statistical Manual of Mental Disorders. 5th ed. Arlington, VA: American Psychiatric Association (2013)
5. Zou, L., et al.: 3D CNN based automatic diagnosis of attention deficit hyperactivity disorder using functional and structural MRI. IEEE Access **5**, 23626–23636 (2017)

6. Lawhern, V.J., Solon, A.J., Waytowich, N.R., Gordon, S.M., Hung, C.P., Lance, B.J.: EEGNet: a compact convolutional neural network for EEG-based brain-computer interfaces. J. Neural Eng. **15**(5), 056013 (2018)
7. M˘alîia, M.-D., et al.: Functional mapping and effective connectivity of the human
8. Allen, E., Damaraju, E., Eichele, T., Wu, L., Calhoun, V.D.: EEG signatures of dynamic functional network connectivity states. Brain Topogr. **31**(1), 101–116 (2018)
9. Simonyan, K., Vedaldi, A., Zisserman, A.J.: Deep inside convolutional networks: visualising image classification models and saliency maps (2013)
10. Vahid, A., Bluschke, A., Roessner, V., Stober, S., Beste, C.: Deep learning based on event-related EEG differentiates children with ADHD from healthy controls. J. Clin. Med, **8**(7) (2019)
11. Döpfner, M., Görtz-Dorten, A., Lehmkuhl, G.: Diagnostik-System für Psychische Störungen im Kindes- und Jugendalter nach ICD-10 und DSM-IV, DISYPS-II; Huber: Bern, Switzerland (2008)
12. Lu, R., Yu, W., Lu, J., et al.: Synchronization on complex networks of network. IEEE Trans. Neural Netw. Learn. Syst. **15**(11), 2110–2118 (2014)
13. Power, J.D., Fair, D.A., Schlaggar, B.L., et al.: The development of human functional brain networks. Neuron **67**(5), 735–748 (2010)
14. Hardmeier, M., Hatz, F., Bousleiman, H., Schindler, C., Stam, C.J., Fuhr, P.: Reproducibility of functional connectivity and graph measures based on the phase lag index (PLI) and weighted phase lag index (wPLI) derived from high resolution EEG. PLoS ONE, **9**, e10864 (2014)
15. Vinck, M., Oostenveld, R., van Wingerden, M., et al.: An improved index of phasesynchronization for electrophysiological data in the presence of volume-conduction, noise and sample-size bias. Neuroimage **55**, 1548–1565 (2011)

Heterogeneous sEMG Sensing for Stroke Motor Restoration Assessment

Hongyu Yang[1], Hui Chang[1], Jia Zeng[2], Ruikai Cao[1], Yifan Liu[1], and Honghai Liu[1](✉)

[1] State Key Laboratory of Robotics and System, Harbin Institute of Technology, Shenzhen, China
honghai.liu@icloud.com

[2] State Key Laboratory of Mechanical System and Vibration, Shanghai Jiao Tong University, Shanghai, China

Abstract. Stroke is a common global health-care problem that has had serious negative impact on the life quality of the patients. Motor impairment after stroke typically affects one side of the body. The main focus of stroke rehabilitation is the recovery of the affected neuromuscular functions and the achievement of independent body control. This paper proposes a heterogeneous sEMG sensing system for motor restoration assessment after stroke. The 136 sEMG nodes have been heterogeneously distributed to sense both dexterous and lump motion. The 128 sEMG nodes are clustered to acquire high density sEMG data acquisition; the other 8 are sparsely arranged for sEMG data acquisition in lump motion. Preliminary experiments demonstrate that the system outperforms existing similar systems and shows potential for evaluating motor restoration after stroke.

Keywords: Heterogeneous sensing nodes · Surface electromyography (sEMG) · Stroke · Motor restoration assessment

1 Introduction

Stroke is a global disease with high mortality and high disability caused by motor cortical damage [1]. According to the statistics, there were approximately 13.68 million new increased stroke patients worldwide a year and about 70% of survivors had different degrees of upper limb and hand movement dysfunction [2]. The recovery of patients' motor function mainly depends on rehabilitation training. However, due to individual difference of patients, it is critical to generate personalized rehabilitation program according to the different motor impairment levels.

Thus far, medical scale methods have been widely used in clinics to evaluate motor functions of stroke patients, including Brunnstrom approach [3,4], Fugl-Meyer (FM) assessment scale [5], Motor Assessment Scale (MAS) [6], and Wolf Motor Function Test (WMF) [7]. Among them, FM assessment is currently

H. Liu et al. (Eds.): ICIRA 2022, LNAI 13456, pp. 636–644, 2022.
https://doi.org/10.1007/978-3-031-13822-5_57

recognized and the most widely used clinical evaluation method [8]. The FM scale was developed based on Brunnstrom approach, where patients completed a series of motions following instructions, and scored by physicians according to the completion degree of the movements [9]. Although the medical scales have good reliability and validity, the results relied on subjective scores from physicians and difficult to reflect minor functional changes due to rough evaluation. Some poststroke patients may have the same evaluation score but with different movement performance. Thus, it would be helpful to quantitatively evaluation with respect to the specific motion of poststroke patients that can reveal the patients' deficits and provide advice on the rehabilitation process.

There still exist problems: Most existing commercial sEMG systems, such as the SAGA (TMSI, Netherland), the Trigno Research+(DELSYS, USA), the EMG-USB2 (OT, Italy), can be divided into two classes: raw sEMG acquisition system and high-density sEMG acquisition system. Human forearm muscles are dense and small, while big arms muscles and lower limbs muscles are large and relatively sparse. The raw sEMG acquisition system cannot fully collect users' small muscles' information, while the high-density sEMG acquisition system cannot fully collect large muscles' information, which brings obstacles to the quantitative evaluation of stroke restoration effect. The heterogeneous sEMG acquisition system integrating two kinds of sEMG signals is desired to solve the problem.

This study proposes an attempt of a heterogeneous sEMG sensing system for stroke motor restoration assessment. The heterogeneous sEMG from subjects provide a basis for objective and quantitative assessment of the motor restoration effect of stroke patients.

2 Heterogeneous sEMG Sensing Integration Solution

This section presents the overall design of the heterogeneous sEMG sensing system. As shown in Fig. 1, the overall architecture is composed of six parts: high-density sEMG acquisition module, raw sEMG acquisition module, microcontroller unit, power supply module, ethernet module and graphical user interface. Every eight high-density sEMG acquisition modules are connected in a daisy chain, and both kinds of acquisition modules communicate with the microcontroller through SPI. The collected analog signal is amplified, filtered, and digitally processed, and then sent to the microcontroller unit for further processing, and then transmitted to the graphical user interface through ethernet.

2.1 Raw sEMG Acquisition Module

The raw sEMG signals are always accompanied with various kinds of noises, such as physiological noise, ambient noise (50 or 60 Hz, power line radiation), electrochemical noise from the skin-electrode interface and so on. So this study makes use of instrumentation amplifier to suppress common mode noise, band pass filter based on operational amplifiers to extract valid frequency band of sEMG signal

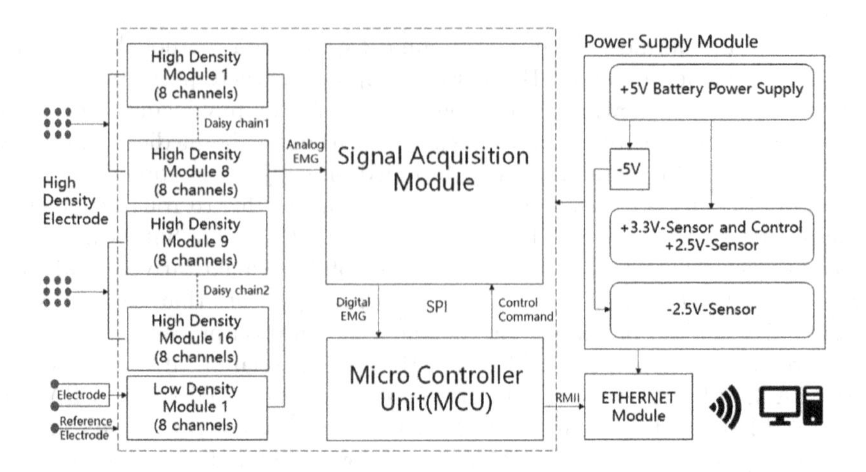

Fig. 1. The architecture of the heterogeneous system

(20–500 Hz) and remove baseline noise as well as movement artifact, and comb filter to supress 50 Hz power line noise and multiples thereof. After all those filtering processes, the signals are digitized by the analog-to-digital converter (ADC) in MCU.

In this study, we chose the ADC integrated chip ADS1299 (Texas Instruments) as the main component of raw acquisition module. As shown in Fig. 2, the raw sEMG signals are amplified, sampled and low-pass filtered in ADS1299 (the sampling rate is 1 kHz). Then the digitized sEMG signals are transmitted into MCU (STM32H743VIT6, STMicroelectronics) through the serial peripheral interface (SPI). The comb filter (center frequency, 50 Hz and multiples thereof) and high-pass filter (cut-off frequency, 20 Hz) towards the signals are realized in MCU to remove the baseline noise and movement artifact.

The low-pass filter is a on-chip digital third-order sinc filter whose Z-domain transfer function is (1). The Z-domain transfer function of the comb filter is (2). The high-pass filter is a six-order Butterworth filter defined by (3) in Z-domain.

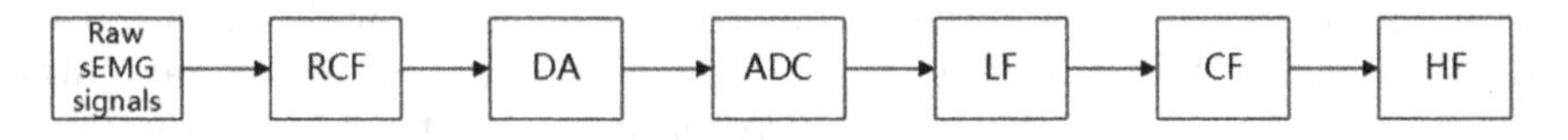

Fig. 2. The sEMG signal processing flow. RCF is the RC filter. DA, ADC and LF are the differential amplification, analog-to-digital conversion and low-pass filter within ADS1299. CF and HF are the comb filter and high-pass filter respectively in MCU.

$$H(Z)_{sinc} = \left| \frac{1 - Z^{-N}}{1 - Z^{-1}} \right|^3 \tag{1}$$

where N is the decimation ratio of the filter.

$$H(Z)_{comb} = \frac{i-1}{i} - \frac{1}{i}Z^{-T} - \frac{1}{i}Z^{-2T} - \cdots - \frac{1}{i}Z^{-iT} \quad (2)$$

where T = fs/fn, fs is the sampling frequency (1 kHz) and fn is the power line noise (50 Hz); i is the filter order, which determines the length of the previous signals being used to estimate noise.

$$H(Z)_{butterworth} = \frac{\sum_{k=0}^{M} b_k Z^{-k}}{1 - \sum_{l=1}^{N} a_l Z^{-l}} \quad (3)$$

where a and b are coefficients; N and M are determined by the filter order, meaning the length of the length of the previous signals to be used.

2.2 High-Density sEMG Acquisition Module

Signal, noise processing and chip selection are all equivalent to the raw sEMG acquisition module. The high-density sEMG acquisition module communicates with the main control module through SPI, and adopts a single-ended input method to lead out the negative poles of multiple high-density EMG acquisition modules, and aggregate them to the high-density electrode interface for connection with the high-density electrodes. The positive electrode is connected to the right-leg driven circuit as shown in Fig. 3a and the lead electrode is connected to the skin so as to avoid saturation of the instrumentation amplifier that amplifies the electrical signal due to the collected bioelectrical signal, and is connected by superimposing the sEMG data of all channels on the skin to suppress common mode interference. But this way of use also brings some problems. A more serious problem is that if the design of the peripheral circuit does not match the user's skin impedance well, this part of the circuit will not only fail to suppress common mode interference, but will instead introduce excess noise, thereby affecting the signal quality. Therefore, another circuit is designed to solve the situation that the user's skin impedance cannot be well matched. As shown in Fig. 3b, another circuit (reference voltage circuit) uses a voltage follower to generate 0V that is not disturbed by the outside world and connect it to the skin to avoid collecting the bioelectrical signal of saturating the instrumentation amplifier used to amplify the electrical signal.

3 System Evaluation and Experiments

Single-channel signal test, multi-channel signal test and sEMG gesture recognition experiments were carried out respectively. It should be noted that all experimental procedures in this study were approved by the School Ethics Committee of Harbin Institute of Technology, and all subjects obtained written informed consent and permission to publish photos for scientific research and educational purposes.

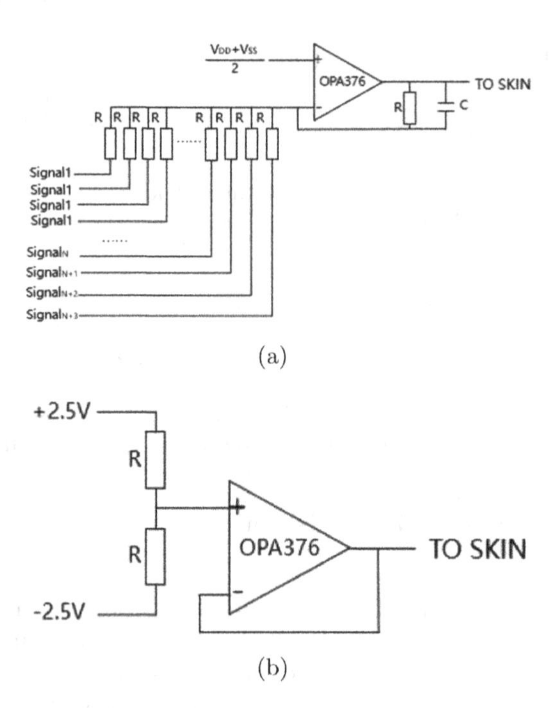

Fig. 3. (a) Right-Leg driven circuit (b) Voltage reference circuit

3.1 Single-Channel Signal Evaluation

In this part, two commercial sEMG sensors DataLOGMWX8 (Biometrics Ltd, UK) and Tringo Wireless (Delsys Inc, USA) were involved to be compared with the heterogeneous sensing system with respect to the time-frequency domain characteristics. A healthy subject was asked to hold on hand close and hand open for 5 s in turn at a moderate level of effort according to the instructions on a computer screen, and the sEMG was recorded at the same time. The protocol was repeated on the same subject for testing different devices. And the electrodes/sensors were placed on the same belly position of flexor carpi ulnaris (FCU) after cleaning with alcohol. As shown in Fig. 4a, the heterogeneous system could well detect sEMG signals 20 Hz and 350 Hz and showed better performance of the suppression 50 Hz and multiples thereof than the two commercial devices.

In order to evaluate the signal quality of the system, signal-noise ratio (SNR) is compared among the heterogeneous system and the commercial devices mentioned before. A healthy subject without nerve and limb disease took part in the experiment. During one trial, the subject was asked to hold on the grip force sensor and keep the grip force at the level of 50% maximum volunteer contraction (MVC) with the vision feedback from computer for 10 s and then relax for 10 s, meanwhile the sEMG signals were detected with the electrodes/sensors placed on the same subject for testing different devices. The SNR was calculated and the result is shown in Fig. 4b. It indicates that the signal quality of the heterogeneous system is better than the two commercial devices.

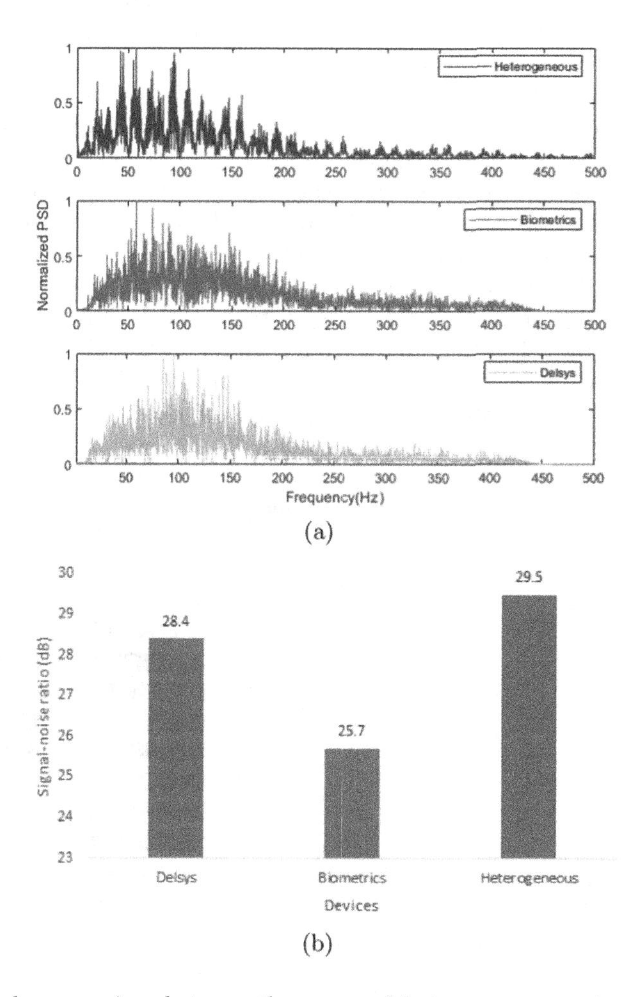

Fig. 4. Signals comparison between the proposed heterogeneous system and commercial sEMG acquisition system in: (a) Nomalized PSD of the acquired sEMG signals (b) Signal-moise ratio

3.2 Multi-channel Signal Evaluation

For the acquisition system with many channels, in addition to testing the signal quality of a single channel, there are some problems that restrict the performance of the system. Therefore, the communication speed and reliability test and the test of the acquisition delay between channels are carried out in this part. After preliminary calculation, we can get that it needs a transmission rate of at least 21 Mbps to meet the basic needs of communication. Considering the communication rate of 25 Mbps, the system design is carried out according to 20 margin, so the communication speed needs to reach 30 Mbps, so according to the requirements of this experiment, the system needs to be able to send data at a rate of 30 Mbps, and in Data integrity is guaranteed during transmission. In this part of the test, in order to simulate the actual application of the system, 512 bytes of data were randomly obtained from the upper computer and transmitted to the upper computer through the upper computer and the data was saved to the upper computer. In the sending process of each cycle, the machine need to send the saved data to assign value and it need to register and send data continuously. After testing, the results show that the average data transmission speed of the system is 43.6 Mbps, the minimum transmission rate is greater than 43 Mbps as shown in Fig. 5a and the data accuracy rate is 100%.

The test method of inter-channel delay is to collect the input signal from different channels, and select the same reference signal of eight channels of the same sampling chip in the experiment for measurement; in another experiment, the test method is to select multiple sampling chips' channels to measure the same reference signal. Two experiments were recorded in the acquired signal. Through the analysis of the two experiments, Fig. 5b shows that almost all channels' waveforms are analyzed under the sampling frequency of 1 KHz and no time delay is found.

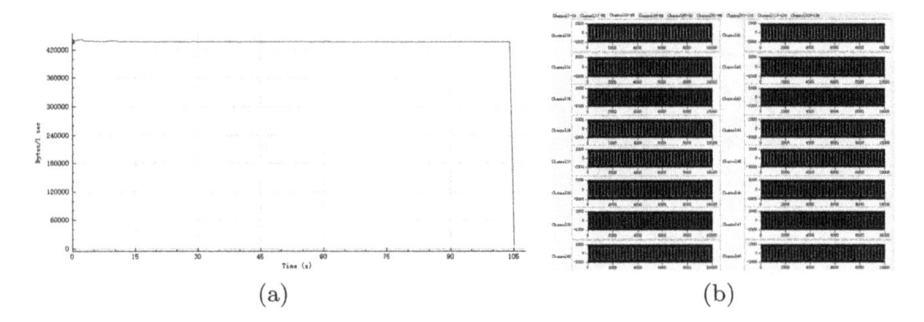

Fig. 5. (a) Data transmission speed of the system (b) Test signal multichannel waveform

3.3 Gesture Recognition Experiments

The experimental process was conducted as follows: 128-channel high-density sEMG electrodes were placed on the forearm and 20 types of gesture movements were identified and classified, of which 20 types of movements are shown in Fig. 5a below. The subjects repeated the above twenty gestures three times in turn, each gesture was held for 5 s, and there was a 5 s rest period after completing a gesture to ensure that the muscles were not fatigued when a certain gesture was completed. The data of the first 1 s and the last 1 s of the 5 s sEMG data collected by each action were discarded, and the rest data was windowed according to the data window length of 200 ms and the window step length of 100 ms. Each channel of the data was activated by sorting and extracting four kinds of time-domain features: absolute mean (MAV), slope sign change (SSC), zero-crossing points (ZC), waveform length (WL), and finally put into LDA classification training in the machine. The experiment result is shown in Fig. 5b and the recognition rate for 20 types of gestures is as high as 90% (Fig. 6).

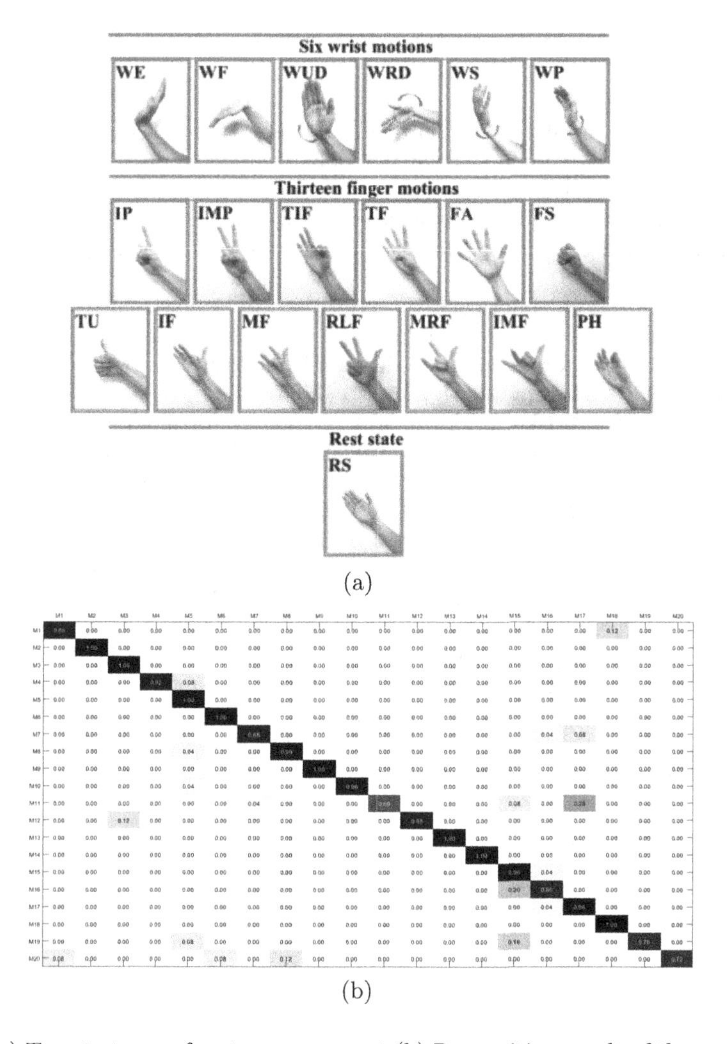

Fig. 6. (a) Twenty types of gesture movement (b) Recognition result of the experiment

4 Conclusion

This paper presents a heterogeneous sEMG sensing system for motor restoration assessment after stroke. The designed system has advantages of multiple channels, portable size and real-time capability. On one hand, the system can be used as raw sEMG acquisition device and high-density sEMG acquisition device respectively; on the other hand, heterogeneous acquisition and analysis of two kinds of sEMG signals has the potential to quantitatively evaluate the rehabilitation effect after stroke.

References

1. Whiteford, H.A., et al.: Global burden of disease attributable to mental and substance use disorders: findings from the global burden of disease study 2010. The lLancet **382**(9904), 1575–1586 (2013)
2. Johnson, C.O., et al.: Global, regional, and national burden of stroke, 1990–2016: a systematic analysis for the global burden of disease study 2016. Lancet Neuro. **18**(5), 439–458 (2019)
3. Huang, C.-Yu., et al.: Improving the utility of the brunnstrom recovery stages in patients with stroke: validation and quantification. Medicine **95**(31), 2016
4. Brunnstrom, S.: Motor testing procedures in hemiplegia: based on sequential recovery stages. Phys. Therapy **46**(4), 357–375 (1966)
5. Fugl-Meyer, A.R., Jääskö, L., Leyman, I., Olsson, S., Steglind, S.: The post-stroke hemiplegic patient. 1. a method for evaluation of physical performance. Scandinavian J. Rehabilitation Med. **7**(1), 13–31 (1975)
6. Carr, J.H., Shepherd, R.B., Nordholm, L., Lynne, D.: Investigation of a new motor assessment scale for stroke patients. Phys. Therapy **65**(2), 175–180 (1985)
7. Wolf, S.L., McJunkin, J.P., Swanson, M.L., Weiss, P.S.: Pilot normative database for the wolf motor function test. Arch. Phys. Med. Rehabilitation **87**(3), 443–445 (2006)
8. Sanford, J., Moreland, J., Swanson, L.R., Stratford, P.W., Gowland, C.: Reliability of the fugl-meyer assessment for testing motor performance in patients following stroke. Phys. Therapy **73**(7), 447–454 (1993)
9. Sullivan, K.J., et al.: Fugl-meyer assessment of sensorimotor function after stroke: standardized training procedure for clinical practice and clinical trials. Stroke **42**(2), 427–432 (2011)

Age-Related Differences in MVEP and SSMVEP-Based BCI Performance

Xin Zhang[1], Yi Jiang[2], Wensheng Hou[1], Jiayuan He[2], and Ning Jiang[2(✉)]

[1] Bioengineering College, Chongqing University, Chongqing 400044, China
zx2929108zx@cqu.edu.cn
[2] West China Hospital, Sichuan University, Chengdu 610041, China
jiangning21@wchscu.cn

Abstract. With the aggravation of the aging society, the proportion of senior is gradually increasing. The brain structure size is changing with age. Thus, a certain of researchers focus on the differences in EEG responses or brain computer interface (BCI) performance among different age groups. Current study illustrated the differences in the transient response and steady state response to the motion checkerboard paradigm in younger group (age ranges from 22 to 30) and senior group (age ranges from 60 to 75) for the first time. Three algorithms were utilized to test the performance of the four-targets steady state motion visual evoked potential (SSMVEP) based BCI. Results showed that the SSMVEP could be clearly elicited in both groups. And two strong transient motion related components i.e., P1 and N2 were found in the temporal waveform. The latency of P1 in senior group was significant longer than that in younger group. And the amplitudes of P1 and N2 in senior group were significantly higher than that in younger group. For the performance of identifying SSMVEP, the accuracies in senior group were lower than that in younger group in all three data lengths. And extended canonical correlation analysis (extended CCA)-based method achieved the highest accuracy (86.39% $\pm$ 16.37% in senior subjects and 93.96% $\pm$ 5.68% in younger subjects) compared with CCA-based method and task-related component analysis-based method in both groups. These findings may be helpful for researchers designing algorithms to achieve high classification performance especially for senior subjects.

Keywords: Age · Brain-computer interface · Steady state motion visual evoked potential · Motion visual evoked potential · Electroencephalogram

1 Introduction

Brain-computer interface (BCI) could acquire and analysis brain signals in real time and provide a non-muscular channel for people communicating with computer or controlling external device [1]. Various spontaneous Electroencephalogram (EEG) signals or

This work was supported by the China Postdoctoral Science Foundation under Grant 2021M700605.

H. Liu et al. (Eds.): ICIRA 2022, LNAI 13456, pp. 645–655, 2022.
https://doi.org/10.1007/978-3-031-13822-5_58

induced EEG signals are selected as the BCI paradigms, such as the event-related desynchronization/synchronization (ERD/ERS) paradigm [2], the P300 paradigm, the steady-state visual evoked potential (SSVEP) paradigm [3] and so on. Particularly, SSVEP is evoked by periodic flicker with a stationary distinct spectrum in EEG recording mainly at the occipital cortical area. SSVEP-based BCI has the advantage of high information transfer rate (ITR) and no need for subjects training. Thus, it is widely used in spelling and brain-controlled robot.

Up to now, BCI applications mainly focus on assisting paralyzed people to participate in daily life activities. Although the paralyzed people can be found among all age groups, senior still account for a large proportion with the aggravation of the aging society. Researcher has shown that there were changes in the brain with age in processing speed, working memory, inhibitory functions, brain structure size, and white matter integrity [4]. Thus, several studies have test the BCI performance in different age groups. Reduced lateralization was found in older subjects. This resulted in a lower classification accuracy in sensory-motor rhythms-based BCI. And the accuracy dropped from 82.3% to 66.4% [5].

For the SSVEP-based BCI, the classification accuracy in older subjects was also lower than the younger subjects. But the decrease of the accuracy was not too much. For example, a BCI spelling performance from young adults (between the ages of 19 and 27) and older adults (between the ages of 54 and 76) was compared. And the results showed a significant difference on the ITR between these two groups [6]. While the statics analysis on the difference in the classification accuracy between these two groups (average accuracies were 98.49% in ten younger subjects and 91.13% in ten older subjects) was not reported. In addition, one study reported the SSVEP-based BCI performance using the medium frequency range (around 15 Hz) and the high-frequency range (above 30 Hz) in 86 subjects age from 18 to 55. Results showed that the age had no significant effect on the classification accuracy [7].

The differences in the performance of SSVEP-based BCI reflect the differences in the brain response from primary visual cortex to the light flash. Actually, there is another pathway in the visual system, i.e., the dorsal pathway, which mainly deal with motion. The perception of motion is one of the fundamental tasks of the higher visual systems. Motion visual evoked potential (mVEP) [8] and steady-state motion visual evoked potential (SSMVEP) have been used to investigate human motion processing. In addition, periodic motion, such as rotation, spiral and radial motion, have been proved to be able to elicit SSMVEP [9]. And SSMVEP-based BCI maintains low-adaptation characteristic and less visual discomfort [10]. As far as we know, there has been no comparison of mVEP and SSMVEP-based BCI performance among different age groups.

In current study, the differences in the brain response to the motion checkerboard paradigm between younger subjects and senior subjects were reported. The feasibility of the checkerboard eliciting the SSMVEP in senior subjects was demonstrated. The transient response i.e., mVEP and the steady state response i.e., SSMVEP were illustrated. And the amplitude and the latency of the transient motion related components were compared between these two age groups. Furthermore, three algorithms were utilized to identify the target stimulus in four checkerboard stimuli. The identification performance between these two groups was compared.

2 Methods

2.1 Participants

Two different age groups of healthy volunteer subjects i.e., younger group and senior group participated in this study. Each group contained eighteen subjects. The Younger Group consisted of nine male and nine female, aging range from 22 to 30 years old (26.17 ± 2.68). The Senior Group consisted of four male and 14 female, aging range from 60 to 75 years old (67.50 ± 4.20). All the 36 volunteer participants were naïve to BCI. The study was approved by the ethical committee of West China Hospital of Sichuan University (20211447). Written informed consent forms were obtained from the participants before the experiment.

2.2 The Checkerboard Paradigm

Checkerboard motion paradigm was chosen as the stimulus in this study. As shown in Fig. 1(A), the checkerboard paradigm consisted of multiple concentric rings. Each ring was divided into white and black lattices with equal sizes and numbers. Thus, the total areas of the bright and dark regions in each ring were always identical.

The multiple rings in the checkerboard contracted as the phase of the sinusoid signal of the targeted frequency changed from 0 degrees to 180 degrees and expanded as the phase changed from 180 degrees to 0 degrees as shown in Fig. 1(B). The red circle is a diagram of the movement of one of the rings. The program presenting the stimulus was developed with MATLAB using the Psychophysics Toolbox. As shown in our previous study [11, 12], the periodic motions of the checkerboard could elicit SSMVEP in the occipital area when the subject gazed at the checkerboard. Due to the checkerboard paradigm did not produce the light flickering, it had the low-adaptation characteristic and less visual discomfort for participants.

2.3 Experimental Design

During the experiment, the participant was seated in a comfortable chair and was briefed on the tasks to be performed. Four targets with motion frequencies 5 Hz, 6 Hz, 4.615 Hz, and 6.667 Hz were displayed on the four corners of the screen in the current study.

Figure 1(C) illustrated the trial sequence in the experiment. Each trial consisted of three phases: cue phase, stimulus phase and relaxation phase. Each trial started with the cue phase (from −2 to 0 s), where one of the four cue letters ('1', '2', '3', '4') would appear at the screen, at one corner of the screen. It indicated the target stimulus for the current trial, at which participant would then engage his or her gaze during the stimulus phase. The stimulus phase would start at 0 s and last 5 s. In this phase, the four checkerboard stimuli appeared on the screen for 5 s, during which the stimuli were modulated at the four frequencies stated above. The participants were asked to gaze at the target appearing in the same position as the letter shown in the cue phase. This was followed by the 2 s long relaxation phase, during which the participant could relax the gaze. And the text 'rest' in Chinese was appearing at the screen.

The experiment consisted of four runs. In each run, each of the four targets was repeated five times in a randomized order i.e., a total of 20 experimental trials. Thus, a total of 80 experimental trials for each participant.

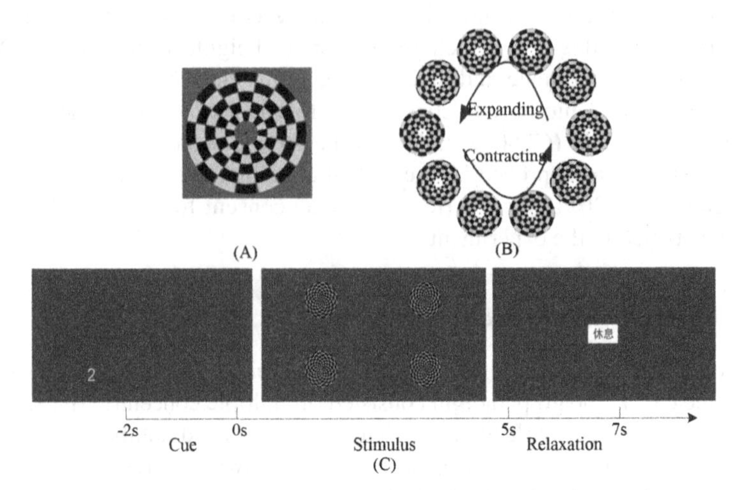

Fig. 1. Experimental design. (A) the checkerboard stimulus (B) the motion process of the checkerboard (C) the trial sequence in the experiment (Color figure online)

2.4 EEG Data Recording

EEG signals were recorded using a 32-channel wireless g.Nautilus EEG system (g.tec, Austria). Electrodes were placed at Fp1, Fp2, AF3, AF4, F7, F3, Fz, F4, F8, FC5, FC1, FC2, FC4, T7, C3, Cz, C4, T8, CP5, CP1, CP2, CP6, P7, P3, Pz, P4, P8, PO7, PO3, PO4, PO8, and Oz according to the extended 10/20 system. The reference electrode was located on the right earlobe, and the ground electrode was located on the forehead. A hardware notch filter at 50 Hz was used, and signals were digitally sampled at 500 Hz. All EEG data, event timestamps (the beginning and the end of each trial) and true labels were recorded for subsequent processing.

2.5 Signal Processing

In this study, canonical correlation analysis (CCA), task-related component analysis (TRCA), and extended CCA-based method (extended CCA) were investigated target stimulus detection and classification. The classification accuracy was computed to evaluate the target identification performance in different age groups. The classification accuracy is defined as the percentage of the correct predictions out of all predictions. EEG data from electrodes P3, Pz, P4, PO7, PO3, PO4, PO8, and Oz were selected for analysis. And EEG data in the stimulus phase were band-pass filtered from 3 Hz to 40 Hz with the Butterworth filter. Then the three above mentioned methods were utilized to do classification. The details of the target identification methods are described below.

CCA-Based Target Identification. CCA is a statistical way to measure the underlying correlation between two multidimensional variables, which has been widely used to detect the frequency of SSVEPs [13]. It is a training free method. Single trial test data are denoted as $X \in \mathbb{R}^{N_c \times N_s}$. Here, N_c is the number of channels, N_s is the number of sampling points in each trial. In the current study, multi-channel EEG data in the occipital region and sine-cosine reference signals were calculated by the following formula

$$\rho(x, y) = \frac{E[w_x^T X Y_f{}^T w_y]}{\sqrt{E[w_x^T X X^T w_y E[w_x^T Y_f Y_f{}^T w_y]]}} \tag{1}$$

where ρ is the CCA correlation coefficient and Y_f is the reference signal.

Sine-cosine reference signals are as follows:

$$Y_f = \begin{Bmatrix} \sin(2 \times \pi \times f \times t) \\ \cos(2 \times \pi \times f \times t) \end{Bmatrix}, \tag{2}$$

where f is the motion frequency.

The target on which the participant focused on could be identified by taking the maximum CCA coefficient.

TRCA-Based Target Identification. Task related component analysis is an approach to extract task-related components from a linear weighted sum of multiple time series. As it had the ability to maximize inter-block covariance and to remove task-unrelated artifacts, TRCA was successfully used as a spatial filter to remove background EEG activities in SSVEP-based BCIs [14]. The spatial filter can be achieved as follow:

$$\omega = \underset{\omega}{\operatorname{argmax}} \frac{\omega^T S \omega}{\omega^T Q \omega} \tag{3}$$

With the help of the Rayleigh-Ritz theorem, the eigenvector of the matrix $Q^{-1}S$ provides the optimal coefficient vector of the objective function in (3).

As there were four individual training data corresponding to four checkerboard stimuli, four different spatial filters could be obtained. Then an ensemble spatial filter $W = [\omega_1 \omega_2 \omega_3 \omega_4]$ was obtained. Through spatial filtering $X^T W$, the test data X were expected to be optimized to achieve maximum performance.

The correlation coefficient was selected as the feature. The Pearson's correlation analysis between the single-trial test signal X and averaging multiple training trials as $\overline{\chi}_i = \frac{1}{N_t} \sum_{h=1}^{N_t} \chi_{ih}$ across trials for i-th stimulus was calculated as $r_i = \rho\left(X^T W, \left(\overline{\chi}_i\right)^T W\right)$. Here, ρ is Pearson's correlation, i indicates the stimulus index, h indicates the index of training trials, and N_t is the number of training trials.

The target on which the participant focused on could be also identified by taking the maximum coefficient as $Target = \max(r_i), i = 1, 2, 3, 4$.

Extended CCA-Based Target Identification Extended CCA-based method [15] incorporated individual template signals obtained by averaging multiple training trials as $\overline{\chi}_i$. The following three weight vectors are utilized as spatial filters to enhance the SNR of SSMVEPs: (1) $W_X(X, \overline{\chi}_i)$ between test signal X and individual template $\overline{\chi}_i$, (2) $W_X(X, Y_f)$ between test signal X and the sine-cosine reference signals Y_f, (3) $W_{\overline{\chi}_i}(\overline{\chi}_i, Y_f)$ between the individual template $\overline{\chi}_i$ and the sine-cosine reference signals Y_f. Then a correlation vector $\hat{r}_i$ is defined as follows.

$$\hat{r}_i = \begin{bmatrix} \rho_{i,1} \\ \rho_{i,2} \\ \rho_{i,3} \\ \rho_{i,4} \end{bmatrix} = \begin{bmatrix} \rho(X, Y_f) \\ \rho(X^T W_X(X, \overline{\chi}_i), \overline{\chi}_i{}^T W_X(X, \overline{\chi}_i)) \\ \rho(X^T W_X(X, Y_f), \overline{\chi}_i{}^T W_X(X, Y_f)) \\ \rho(X^T W_{\overline{\chi}_i}(\overline{\chi}_i, Y_f), \overline{\chi}_i{}^T W_{\overline{\chi}_i}(\overline{\chi}_i, Y_f)) \end{bmatrix} \quad (4)$$

Finally, the following weighted correlation coefficient is used as the feature in target identification.

$$r_i = \sum_{l=1}^{4} sign(\rho_{i,l}) \cdot \rho_{i,l}{}^2 \quad (5)$$

where $sign()$ is used to retain discriminative information from negative correlation coefficients between the test signals and individual templates.

The target on which the participant focused on could be also identified by taking the maximum coefficient.

Besides, a four-fold cross-validation scheme was performed. One run's data were retained as the validation data for testing, and the other three runs' data were used as training data. There was no overlapping part in both training and test subsets. The cross-validation process was then repeated four times, with each of the four runs' data used exactly once as the validation data.

3 Results

3.1 Spectral-Temporal Characteristics of Induced EEG in Younger Subjects and Senior Subjects

To compare the spectral characteristics of SSMVEP induced by the checkerboard stimulus in different age groups, the spectra of the EEG data in stimulus phase (0 to 5 s) were calculated. Figure 2 illustrated the spectra of EEG data averaged from all the trials and all the subjects in the same age group from Oz. As shown in Fig. 2(B), for all stimulus frequencies, a clean peak at the motion frequency was clearly situated at the corresponding frequency when younger subjects each spectrum frequency stared at each of the four checkerboard targets. And the second harmonic of the motion frequencies were either unexpectedly weak or completely absent, which was different from previously reported SSVEP spectra [16]. For the senior group, as shown in Fig. 2(A), the similar results were obtained. The clear peak at the motion frequency occurred and no second harmonic of the motion frequencies were found in the spectra. Furthermore, the averaged amplitudes of the peak at the motion frequencies between younger subjects and senior subjects were highly similar.

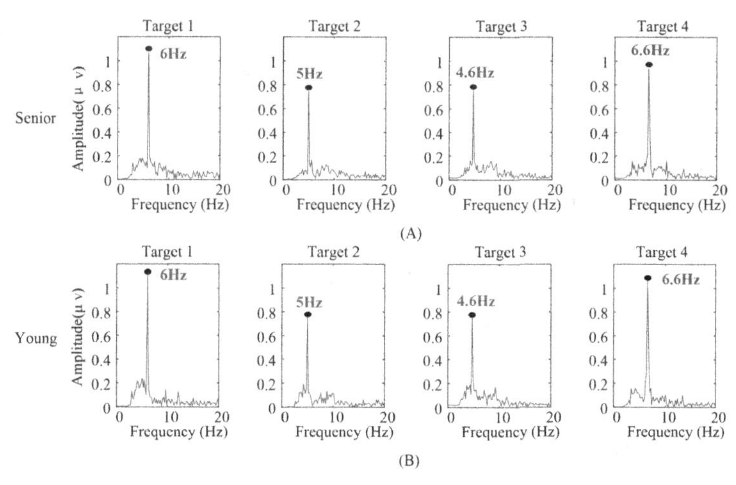

Fig. 2. The spectra of EEG data averaged from all the trials and all the subjects in the same age group at electrode Oz. (A) senior group (B) younger group

In addition to the above steady-state rhythmic EEG components, the transient EEG components immediately following the onset of the visual stimuli were also analyzed. EEG data epochs during the beginning of the stimulus phase (0 to 0.5 s) were averaged over all trials registered from each of the subjects. Figure 3 illustrated the grand average waveforms for all senior subjects (Fig. 3(A)) and all younger subjects (Fig. 3(B)) at electrode PO4. 0 s referred to the moment when the checkerboard occurred on the screen. For both groups, the grand average visually-evoked potentials in response to target stimuli has a clear a sharp positive deflection (P1) with a latency 100–180 ms, followed by a sharp negative deflection (N2) with a latency of 160–230 ms post-motion onset. The amplitudes (mean $\pm$ std) of P1 and N2 were 5.28 ± 2.44 μV and -6.58 ± 2.83 μV in younger group, whereas the amplitudes (mean $\pm$ std) of those two components were 8.79 ± 5.25 μV and -8.05 ± 4.31 μV in senior group. A pair-wised *t*-test revealed

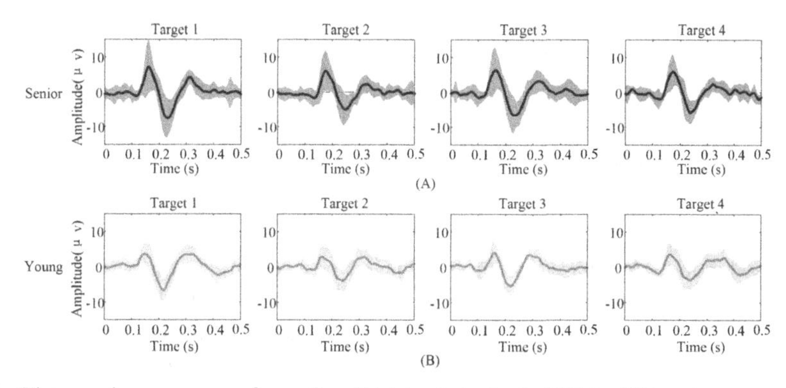

Fig. 3. The grand average waveforms for all trials at electrode PO4 in different age groups. (A) senior group (B) younger group

senior group has a significantly larger the amplitudes of P1 ($t = -5.14, p < 0.001$) and N2 ($t = 2.42, p = 0.017$) than the younger group. In addition, the average latency of P1 in senior group was 6 ms larger than that in younger group. And the latency of P1 in younger group was significant shorter than that in senior group ($t = -2.03, p = 0.044$). There was no significance difference on the latency of N2 ($t = -0.21, p = 0.873$) between these two groups.

3.2 Classification Accuracies in Young Subjects and Senior Subjects

To compare the BCI system performance between younger subjects and senior subjects, three algorithms were utilized to detect and classify the target stimulus. Figure 4 illustrated the classification accuracies with different data lengths in different age groups. As expected, with the increase the EEG epoch length, the classification accuracies increased monotonically in both groups, regardless of algorithm used. And the average accuracies in younger group were higher than that in senior group utilizing CCA-based method (Fig. 4(A)), TRCA-based method (Fig. 4(B)) and extended CCA-based method (Fig. 4(C)). A mixed-effect model of ANOVA was used to quantify the differences on the accuracies. The method with three levels (1: CCA, 2: TRCA, 3: extended CCA), data length with three levels (1: 1 s, 2: 3 s, 3: 5 s), and group with two levels (1: younger, 2: senior) were three fixed factor and subject was a random factor, nested within Group. The analysis results showed that the fix factors method and data length both had a significant effect on the accuracies ($F = 51.35, p < 0.001$; $F = 342.35, p < 0.001$), as expected. However, the fix factor group had no significant effect on the accuracies ($F = 1.6, p = 0.214$).

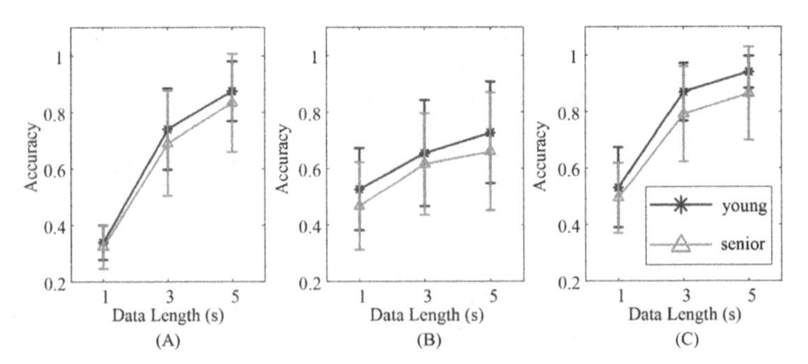

Fig. 4. The classification accuracies utilizing different algorithms in different age groups. (A) CCA-based method (B) TRCA-based method (C) extended CCA-based method

The extended CCA-based method achieved best identification performance in all dada lengths in both groups. The average accuracy achieved 93.96% ± 5.68% within 5 s data length in younger group. And the average accuracy achieved 86.39% ± 16.37% within 5 s data length in senior group. Compared with CCA-based method, the TRCA-based method only achieved superior performance with 1 s data length in both groups.

To further compare the relative identification performance of the checkerboard stimulus targets in different age groups, the confusion matrices of the identification accuracy utilizing extended CCA-based method with 5 s data length for all participants in the same age group were calculated as shown in Fig. 5. The color scale revealed the average classification accuracies and the diagonals labeled the correct classification accuracies among all the participants. We observed that the influence of the four stimulus frequencies on the classification accuracies was not too much. And the target 2 (6 Hz) resulted in the lowest classification accuracy in all cases, which is consistent with the fact its spectral peak is least pronounced among the four stimulation frequencies.

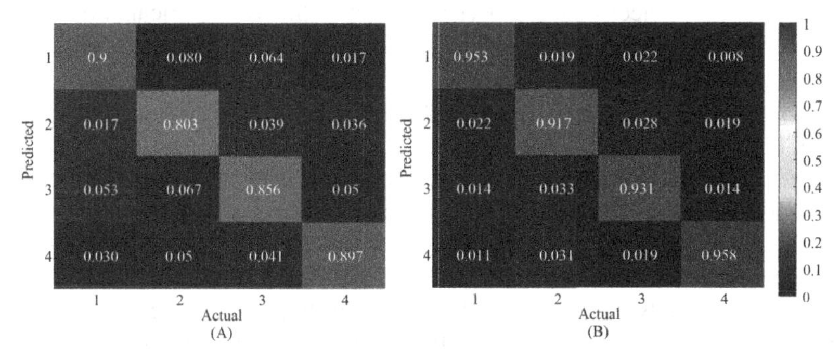

Fig. 5. The confusion matrices of the classification accuracy. (A) senior group (B) younger group (1: target 1, 2: target 2, 3: target 3, and 4: target 4)

4 Discussion and Conclusion

To the best of our knowledge, this is the first study reported the differences in both transient response and steady-state response to the motion checkerboard visual stimuli between younger subjects and senior subjects. The SSMVEP at the motion frequency is clearly elicited in both groups. And two strong transient motion related components i.e., P1 and N2 in mVEP also occur in both groups. The spectral-temporal characteristics of induced EEG are similar with the results in one study [10], which utilized the oscillating Newton's rings as the motion stimuli and recruited younger subjects performing the experiment. It's worth noting that there is a significant difference in some characteristics of mVEP between younger group and senior group. The latency of P1 in senior group is significant longer than that in younger group. The primary reason for this prolongation in senior may be due to ageing of neural pathways and processes [17]. And this prolongation in senior is in agreement with the findings in most studies in spite of utilizing different stimulus.

Interestingly, the significantly larger amplitude in P1 and N2 in the senior group is unexpected. One possible reason is that the senior subject paid more attention on the checkerboard during the stimulus phase than younger subjects. Many senior subjects stated that they were involuntarily counting the times of the movement in the checkerboard when they stared at the stimulus, while no participants in the younger group reported to do so. Previous studies found that the amplitudes of mVEP components can be modulated by attention [18]. 'Counting' may help subjects more concentrate on the target stimulus. In the literature, larger amplitudes of VEP in older females as compared with younger females was reported, in which the authors suggested that this is due to a heightened sensitivity of the older female visual system to patterned stimuli [19]. Furthermore, the characteristics of mVEP is influenced by the type of the visual stimuli. The reason of the larger amplitude in mVEP still need further researched. While the characteristic of the mVEP in senior subjects may provide valuable signal modality in designing more appropriate BCIs for targeted at senior population.

Moreover, the steady state response i.e., SSMVEP provides an even more robust way of determining the subject's choice of visual target. While no significant difference was found, the average accuracy achieved 86.39% in senior subjects, lower than the accuracy in younger subjects (93.96%). The difference in accuracies in these two age groups was similar with the study that reported the difference in the performance of the SSVEP-based BCI between younger subjects and senior subjects [6]. Consistent with this lack of difference, the BCI performance between the two groups were not significant. This might because that the EEG data from channel O1, O2, and POz were not utilized in the classification. Limited by the EEG device, where the positions of the electrodes are fixed and no electrodes were located at O1, O2, and POz.

In the current study, three popular SSVEP algorithms were investigated. In recent literature, many studies utilized TRCA-based method to process SSVEP with short duration (<1 s) and reported enhanced performance [14]. Current study also achieved improved accuracy with 1 s data length compared with the more traditional CCA-based method. However, for the longer data lengths, current study found no difference in performance between CCA-based and TRCA-based methods. Furthermore, extended CCA-based method achieved the highest accuracy in three data lengths.

Overall, current study reports the differences in spectral-temporal characteristics of the EEG induced by the motion checkerboard paradigm in younger group and senior group. The results demonstrated that the SSMVEP-based BCI is feasible for senior population. These findings may be helpful for researchers designing more appropriate BCI systems for senior subjects.

References

1. Wolpaw, J.R., Birbaumer, N., McFarland, D.J., Pfurtscheller, G., Vaughan, T.M.: Brain-computer interfaces for communication and control. Clin. Neurophysiol. **113**, 767–791 (2002)
2. Wang, Z., et al.: BCI monitor enhances electroencephalographic and cerebral hemodynamic activations during motor training. IEEE Trans. Neural Syst. Rehabil. Eng. **27**, 780–787 (2019)
3. Vialatte, F.-B., Maurice, M., Dauwels, J., Cichocki, A.: Steady-state visually evoked potentials: focus on essential paradigms and future perspectives. Prog. Neurobiol. **90**, 418–438 (2010)
4. Park, D.C., Reuter-Lorenz, P.: The adaptive brain: aging and neurocognitive scaffolding (2009)
5. Chen, M.L., Fu, D., Boger, J., Jiang, N.: Age-related changes in vibro-tactile EEG response and its implications in BCI applications: a comparison between older and younger populations. IEEE Trans. Neural Syst. Rehabil. Eng. **27**, 603–610 (2019)
6. Volosyak, I., Gembler, F., Stawicki, P.: Age-related differences in SSVEP-based BCI performance. Neurocomputing **250**, 57–64 (2017)

7. Volosyak, I., Valbuena, D., Lüth, T., Malechka, T., Gräser, A.: BCI demographics II: how many (and what kinds of) people can use a high-frequency SSVEP BCI? IEEE Trans. Neural Syst. Rehabil. Eng. **19**, 232–239 (2011)
8. Heinrich, S.P.: A primer on motion visual evoked potentials (2007)
9. Yan, W., Xu, G., Xie, J., Li, M., Dan, Z.: Four novel motion paradigms based on steady-state motion visual evoked potential. IEEE Trans. Biomed. Eng. **65**, 1696–1704 (2018)
10. Xie, J., Xu, G., Wang, J., Zhang, F., Zhang, Y.: Steady-state motion visual evoked potentials produced by oscillating Newton's rings: Implications for brain-computer interfaces. PLoS ONE **7**, e39707 (2012)
11. Zhang, X., et al.: A convolutional neural network for the detection of asynchronous steady state motion visual evoked potential. IEEE Trans. Neural Syst. Rehabil. Eng. **27**, 1303–1311 (2019)
12. Zhang, X., Xu, G., Xie, J., Zhang, X.: Brain response to luminance-based and motion-based stimulation using inter-modulation frequencies. PLoS ONE **12**, e0188073 (2017)
13. Lin, Z., Zhang, C., Wu, W., Gao, X.: Frequency recognition based on canonical correlation analysis for SSVEP-based BCIs. IEEE Trans. Biomed. Eng. **53**, 2610–2614 (2006)
14. Nakanishi, M., Wang, Y., Chen, X., Wang, Y.T., Gao, X., Jung, T.P.: Enhancing detection of SSVEPs for a high-speed brain speller using task-related component analysis. IEEE Trans. Biomed. Eng. **65**, 104–112 (2018)
15. Nakanishi, M., Wang, Y., Wang, Y.T., Mitsukura, Y., Jung, T.P.: A high-speed brain speller using steady-state visual evoked potentials. Int. J. Neural Syst. **24**, 1450019 (2014)
16. Srihari Mukesh, T.M., Jaganathan, V., Reddy, M.R.: A novel multiple frequency stimulation method for steady state VEP based brain computer interfaces. Physiol. Meas. **27**, 61–71 (2006)
17. Mitchell, K.W., Howe, J.W., Spencer, S.R.: Visual evoked potentials in the older population: Age and gender effects. Clin. Phys. Physiol. Meas. **8**, 317–324 (1987)
18. Torriente, I., Valdes-Sosa, M., Ramirez, D., Bobes, M.A.: Visual evoked potentials related to motion-onset are modulated by attention. Vis. Res. **39**, 4122–4139 (1999)
19. La Marche, J.A., Dobson, W.R., Cohn, N.B., Dustman, R.E.: Amplitudes of visually evoked potentials to patterned stimuli: age and sex comparisons. Electroencephalogr. Clin. Neurophysiol. Evoked Potentials **65**, 81–85 (1986)

HD-tDCS Applied on DLPFC Cortex for Sustained Attention Enhancement: A Preliminary EEG Study

Jiajing Zhao[2,3], Wenyu Li[4], and Lin Yao[1,2,3](✉)

[1] Department of Neurobiology, Affiliated Mental Health Center Hangzhou Seventh People's Hospital, Zhejiang University School of Medicine, Hangzhou 310000, Zhejiang, China
lin.yao@zju.edu.cn
[2] The MOE Frontier Science Center for Brain Science and Brain-machine Integration, Zhejiang University, Hangzhou 310000, Zhejiang, China
[3] The College of Computer Science and Technology, Zhejiang University, Hangzhou 310000, Zhejiang, China
[4] Shanghai Shuyao Information Technology Co., Ltd, Shanghai, China

Abstract. This study investigated the effect of the high-definition transcranial direct current stimulation (HD-tDCS) on sustained attention reaction. The HD-tDCS was applied to the right dorsolateral prefrontal cortex (DLPFC) and the EEG signal was recorded simultaneously. We found that HD-tDCS could be used to enhance the reaction time and the N2 amplitude in response to the flanker stimuli changed significantly after the HD-tDCS stimulation. Moreover, the behavior enhancement was in accordance with the neural response. The HD-tDCS on DLPFC provides a way to enhance the subject's attention to both the target task and the target task accompanied by distractors. The investigated approach may find application for digital medication in patients with attention problems, such as attention deficit hyperactivity disorder (ADHD).

Keywords: HD-tDCS · EEG · Reaction · ERP

1 Introduction

Transcranial direct current stimulation (tDCS), as a component of the Brain-Computer Interference (BCI) based neuromodulation system, has great benefit to modulate human cognitive and behavioral function [1]. Several studies showed that exerting tDCS on the prefrontal cortex enhances cognitive abilities, which were measured by the behavioral difference between conditions before and after stimulation [2,3]. However, it was hard to confirm whether the prefrontal cortex was effectively stimulated because of the current diffuse problem to a huge area of the brain's surface caused by conventional tDCS [4]. By contrast, high-definition tDCS (HD-tDCS) can provide a way to solve the imprecision problem of conventional tDCS [4].

H. Liu et al. (Eds.): ICIRA 2022, LNAI 13456, pp. 656–665, 2022.
https://doi.org/10.1007/978-3-031-13822-5_59

Compared to the huge sponge electrodes of conventional tDCS, HD-tDCS used a 4×1 montage electrodes (1 cm diameter each), with one anode and four cathodes, to release weak current so that HD-tDCS could increase the precision of the target location in the brain. Brain current field modeling has proven that HD-tDCS has a higher focality on target than conventional tDCS [4,5]. HD-tDCS illustrated its positive effects on modulating cognition and behavior [4,6].

Electroencephalography (EEG) was a non-invasive tool to reflect the activity of the brain. Besides EEG itself was used to explore brain function by neuroscience researchers, and have been combined with other technologies to gain more information, such as TMS [7], tDCS [8] and tACS [9]. Most of studies focused on conventional tDCS combined with EEG [8,10–12], but little was known about the HD-tDCS effect on EEG dynamics, and only a few studies [13] had combined HD-tDCS. Moreover, the electrical stimulation on the dorsolateral prefrontal cortex (DLPFC) improves or modulates the cognition function [2,4,6,14–16], and the reaction time of the Flanker task response was enhanced, regardless of the conventional or high definition way [6,15,16]. However, the neural response to the Flanker stimuli after the HD-tDCS was less investigated, especially in the modified flanker task, in which non-response stimuli were included to evaluate the inhibitory control ability.

In this study, we focused on the variation of behavioral data and event-related potentials (ERP) before, in and after stimulation. Behavioral data were collected mainly from the visual Flanker task. We anticipated that the reaction time would be shorter and accuracy would be higher after administration of the stimulation. Moreover, we expected ERP variation due to the HD-tDCS effect.

2 Materials and Methods

2.1 Participants

Four healthy individuals (aged 22–24 years, 2 women, 2 men, right-handed) participated in this study. None of the participants stayed up late or reported the use of any medicine several days before the experiment. All the participants were clearly informed of the safety of the brain stimulation and the tingling sensation during activation of HD-tDCS. According to the recording by the experimenters, all the participants completed experiments following instructions without any abnormal situations. The study was approved by the Ethics Committee of Zhejiang University, Hangzhou, China.

2.2 Study Overview

The study adopted a pre-dur-post test design. After the setup of EEG recording and HD-tDCS device, participants completed a resting state recording part (see Sect. 2.4) at first. Then they finished the visual Flanker task (see Sect. 2.5), which was repeated over two blocks before the activation of HD-tDCS device to assess the normal reaction level. During the activation of the HD-tDCS, participants were also asked to finish two blocks' visual Flanker task so that we can analyze the change in reaction performance during the brain stimulation period (20 min, 1 mA). Given the unpleasant feeling at the beginning of transcranial direct current stimulation, participants started Flanker task 4 min after the start of stimulation. When brain stimulation ended, participants completed another two-blocks Flanker task to assess the effect of HD-tDCS. At the end of the experiment, participants finished a resting state recording again to collect more information to analyze how HD-tDCS changed brain connections. After each block test, there will be a short break of 2–4 min. During the placement of EEG recording and HD-tDCS device, participants practiced the visual Flanker task to be familiar with this experiment. This practice design will reduce the training effect due to the first-time participation. The whole procedure of the study was shown in Fig. 1.

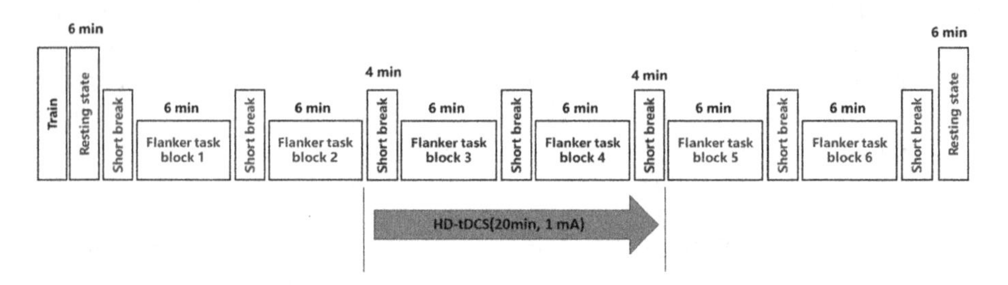

Fig. 1. Experiment procedure with the duration of each part. HD-tDCS was applied before Flanker task block 3. Due to the unpleasant feeling when the current ramped up or ramped down at the beginning and end of stimulation, short breaks before block 3 and after block 4 should last for about 4 min to avoid side effects on performance. EEG recording parts were colored in red and blue. The parts with current stimulation were colored in red. (Color figure online)

2.3 HD-tDCS and EEG

HD-tDCS and EEG electrodes were placed (see Fig. 2) according to the international 10–20 system. The electrode cap with stationary electrode holders was worn under instruction. Electrodes would be placed before the experiment and detached after the experiment in a jiffy to avoid oxidation.

HD-tDCS was administered using five small conductive Ag/AgCl ring-shaped electrodes (4×1 montage, four cathodes and one anode) and the current was controlled by a wireless system (NeuStim NSS14, Neuracle, Changzhou, China). The anode was targeted at F4 to the right dorsolateral prefrontal cortex, and 4 cathodes were targeted at AF4, F2, F6 and FC4. The current lasted for 20 min with 1 mA (firstly ramped up from 0 to 1 mA for 30 s and ramped down to zero for 30 s in the end).

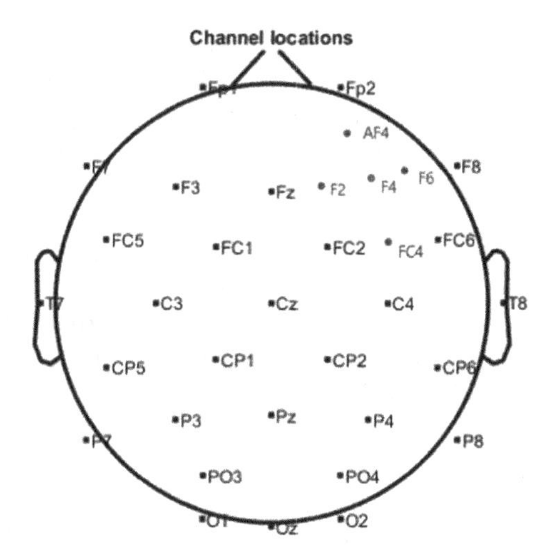

Fig. 2. Channel locations. Black points with names represented 29 EEG recording electrodes. The red point represented the anode of HD-tDCS and blue points represented the cathodes. (Color figure online)

EEG was recorded wirelessly by using 32 electrodes with A1 and A2 as references which were placed at the left and right mastoid(NeuSen W32, Neuracle, Changzhou, China). Because F4 was occupied by HD-tDCS, the EEG data in F4 were not recorded. The raw EEG signals were sampled at 1000 Hz. Then the raw EEG data were filtered by 0.5 Hz high-pass filter 40 Hz low-pass filter and down-sampled 500 Hz using EEGLAB toolbox. The obvious artifacts were removed manually at first and the eye movement components were removed through ICA method [17,18].

2.4 Resting State

Resting-state parts were designed to assess the effects of HD-tDCS on the resting-state brain dynamics, which might modulate brain connections [19]. In this part, participants closed their eyes for 3 min and then opened their eyes and stared at the fixation cross in the center of the screen for another 3 min.

2.5 Visual Flanker Task

The visual flanker task was used to evaluate participants' reaction and response inhibition levels. Participants were seated in front of a 27 in. computer screen and were instructed to react (pressed keys by right hand's finger) as quickly as possible according to the symbol in the center surrounded by other distraction symbols, which appeared for 200 ms. Screen appearing 5 symbols each time combined with left and right arrows and the cross. When the central symbol was a left or right arrow, the participants pressed the corresponding direction keys. When the central symbol was the cross, participants held the response and did nothing. The other symbols were one kind of symbol. The following six situations appeared randomly. "<<<<<" and ">>>>>" were categorized into congruent situation. "<<><<" and ">><>>" were categorized into incongruent situation. "<< + <<" and ">> + >>" were categorized into inhibition situation. These categories were analyzed later. Each block consisted of 250 such trials and there was an average of 1.2 s between each trial. The procedure was conducted with the software PsychoPy [20].

2.6 ERP Analysis

ERP was used to evaluate the effect of HD-tDCS on EEG. ERP is calculated through superposition and average of the event-related EEG and can formulate specific components according to the category of events. The usual visual ERP components are P1, N1, P2, N2 and P3. The N2 and P3 components are also called N200 and P300.

In this study, the ERP data were calculated in the epochs which were extracted from 100 ms before visual Flanker stimulation and 600 ms after that. The 100 ms data before the stimulus were used as the baseline. The ERPs of all the channels were plotted by the EEGLAB toolbox. We selected the obvious channels' ERP (PO3) for the following analysis.

3 Results

We calculate average reaction time and accuracy among the above three categories. The average reaction time (RT) excludes wrong answers' RT and RT below 0.25 s or above 0.6 s. The RT below 0.25 s might have connections with touching by mistakes and few RTs above 0.6 s were considered as outliers.

3.1 Behavior Response to Flanker Task

HD-tDCS could help the brain to react faster to the visual flanker task. Figure 3 showed an average RT of 6 blocks for every participant. In general, it showed a downward trend, i.e., the RT became faster after stimulation. By comparison, the incongruent average RT outcomes had a more obvious downward trend than the congruent one, which might attribute to the difficulty level in the incongruent condition (incongruent Flanker task has a higher RT response than that in congruent condition), which was in accordance with results in other research [21]. The RT of the first block was higher than the second one, which would attribute to the training effect that could not be avoided. When the participants were more and more familiar with the visual Flanker task, normally they improved their performance as time went on. But the fatigue effect would make participants react slower as time went on. In the future study, we will design the control experiments, such as HD-tDCS in sham-stimulation mode, to evaluate how much the training effect and fatigue effect do on the Flanker task performance with HD-tDCS. If the outcomes of RT in the control experiments are lower in the first two blocks and higher gradually in the remaining blocks, it can verify that we can omit the training effect and fatigue effect in the HD-tDCS study.

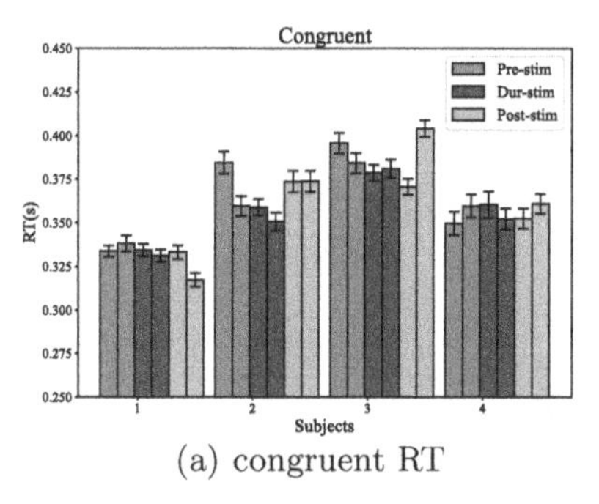

(a) congruent RT

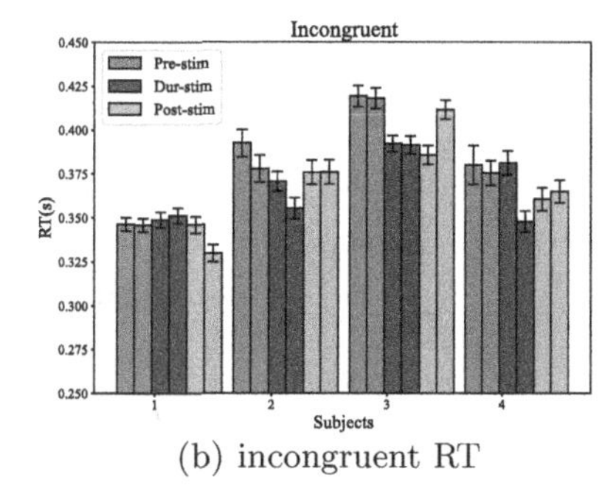

(b) incongruent RT

Fig. 3. Reaction Time of all subjects. Blocks before, during and after stimulation, were colored by red, blue and green. Error bar used the standard error as the parameter. (a)RT of congruent condition. It showed a weak trend that HD-tDCS brought improvement in reacting performance. (b)RT of incongruent condition. It showed a strong trend that HD-tDCS brought improvement in reacting performance. (Color figure online)

The accuracy of the Flanker task was an important reference index. Accuracy in the congruent, incongruent and restrain conditions with respect to the total accuracy was shown in Fig 4. However, the participants we recruited were all students who had high accuracy (most of blocks had more than 90% accuracy) so that the accuracy as a reference index could be hard to assess the HD-tDCS effect. The accuracy of the elders or other individuals with cognitive impairment might be a valuable reference.

3.2 Neural Response to Flanker Task

In this study, most of the participants showed two typical components: N200 and P300. The average of all the subjects' ERP in the PO3 channel was shown in Fig. 5. N200 appeared in about 260 ms and P300 appeared in about 350 ms.

N200 showed a strong regularity effect during and after HD-tDCS, with a higher voltage value than that in the pre-stim blocks (blocks before brain stimulation), i.e., the blocks with HD-tDCS effects would have lower amplitudes of voltage. The ERP of all the participants had this pattern, although it was not so obvious in the restrain condition. The N200 and P300 components might be a valuable marker for investigating the HD-tDCS effect. However, we just collected four individuals' data, we would do more experiments to prove the claim.

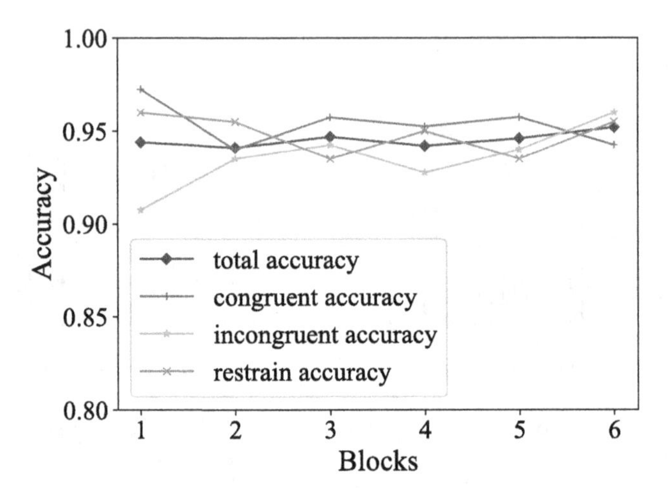

Fig. 4. Accuracy in the congruent, incongruent and restrain conditions besides the total accuracy. The data were averaged by all the participants. Most of the data were more than 90% accurate.

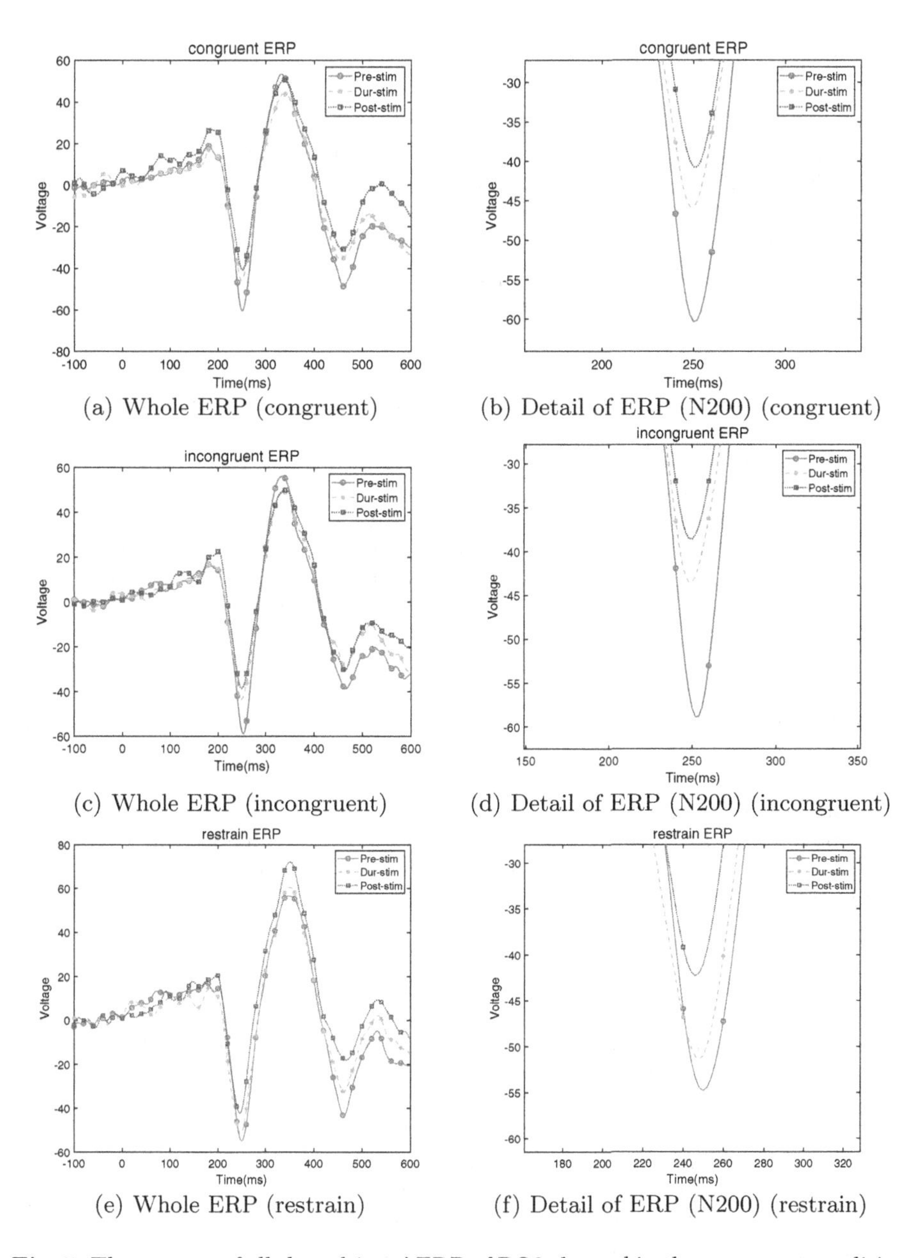

(a) Whole ERP (congruent)

(b) Detail of ERP (N200) (congruent)

(c) Whole ERP (incongruent)

(d) Detail of ERP (N200) (incongruent)

(e) Whole ERP (restrain)

(f) Detail of ERP (N200) (restrain)

Fig. 5. The average of all the subjects' ERP of PO3 channel in the congruent condition. (a) presented the whole ERP between -100 ms and 600 ms in the congruent condition. There were two obvious components: N200 and P300. (b) showed the details of curves of N200 in the congruent condition. This part had strong connections between time and voltage that the blocks during and after stimulation made the voltage higher. However, time $\times$ blocks of the P300 component looked randomly. (c) presented the whole ERP between -100 ms and 600 ms in the incongruent condition. (d) showed the details of curves of N200 in the incongruent condition. (e) and (f) showed the ERP in restrain condition although it was not so obvious compared to the other conditions.

4 Conclusions

In this study, we illustrated that HD-tDCS targeted the right DLPFC cortex could improve sustained attention within a short time period. RT in the incongruent Flanker task condition was slower than that in the congruent condition but had a more obvious decline trend through HD-tDCS modulation, which was partly supported by the ERP difference in N200 and P300 components. However, more subjects were needed to further investigate these preliminary findings and a sham group should be included for a better understanding of the stimulation effect.

Acknowledgement. We thank all volunteers for their participation in the study. This work was partly supported by grants from the National Key R&D Program of China (2018YFA0701400), Key R&D Program of Zhejiang (no. 2022C03011), Chuanqi Research and Development Center of Zhejiang University, the Starry Night Science Fund of Zhejiang University Shanghai Institute for Advanced Study (SN-ZJU-SIAS-002), the Fundamental Research Funds for the Central Universities, Project for Hangzhou Medical Disciplines of Excellence, Key Project for Hangzhou Medical Disciplines, Research Project of State Key Laboratory of Mechanical System and Vibration MSV202115.

References

1. Nitsche, M.A., et al.: Transcranial direct current stimulation: state of the art 2008. Brain Stimul. **1**(3), 206–223 (2008). https://doi.org/10.1016/j.brs.2008.06.004
2. Dubreuil-Vall, L., Chau, P., Ruffini, G., Widge, A.S., Camprodon, J.A.: tDCS to the left DLPFC modulates cognitive and physiological correlates of executive function in a state-dependent manner. Brain Stimul. **12**(6), 1456–1463 (2019). https://doi.org/10.1016/j.brs.2019.06.006
3. Breitling, C., et al.: Improving interference control in ADHD patients with transcranial direct current stimulation (tDCS). Front. Cell. Neurosci. **10**, 72 (2016)
4. Hogeveen, J., Grafman, J., Aboseria, M., David, A., Bikson, M., Hauner, K.: Effects of high-definition and conventional tDCS on response inhibition. Brain Stimul. **9**(5), 720–729 (2016). https://doi.org/10.1016/j.brs.2016.04.015
5. To, W.T., Hart, J., De Ridder, D., Vanneste, S.: Considering the influence of stimulation parameters on the effect of conventional and high-definition transcranial direct current stimulation. Expert Rev. Med. Devices **13**(4), 391–404 (2016). https://doi.org/10.1586/17434440.2016.1153968
6. Gbadeyan, O., Steinhauser, M., McMahon, K., Meinzer, M.: Safety, tolerability, blinding efficacy and behavioural effects of a novel mri-compatible. High-definition tDCS set-up. Brain Stimul. **9**(4), 545–552 (2016). https://doi.org/10.1016/j.brs.2016.03.018
7. Gordon, P.C., et al.: Modulation of cortical responses by transcranial direct current stimulation of dorsolateral prefrontal cortex: a resting-state EEG and TMS-EEG study. Brain Stimul. **11**(5), 1024–1032 (2018). https://doi.org/10.1016/j.brs.2018.06.004
8. Hill, A.T., Rogasch, N.C., Fitzgerald, P.B., Hoy, K.E.: Effects of single versus dual-site High-Definition transcranial direct current stimulation (HD-tDCS) on cortical

reactivity and working memory performance in healthy subjects. Brain Stimul. **11**(5), 1033–1043 (2018). https://doi.org/10.1016/j.brs.2018.06.005

9. Reinhart, R.M.G., Nguyen, J.A.: Working memory revived in older adults by synchronizing rhythmic brain circuits. Nat. Neurosci. **22**(5), 820–827 (2019). https://doi.org/10.1038/s41593-019-0371-x
10. Verveer, I., Hill, A.T., Franken, I.H.A., Yücel, M., van Dongen, J.D.M., Segrave, R.: Modulation of control: Can HD-tDCS targeting the dACC reduce impulsivity? Brain Res. **1756**, 147282 (2021). https://doi.org/10.1016/j.brainres.2021.147282
11. Jindal, U., Sood, M., Chowdhury, S.R., Das, A., Kondziella, D., Dutta, A.: Corticospinal excitability changes to anodal tDCS elucidated with NIRS-EEG joint-imaging: An ischemic stroke study. In: 2015 37th Annual International Conference of the IEEE Engineering in Medicine and Biology Society (EMBC), pp. 3399–3402 (2015). https://doi.org/10.1109/EMBC.2015.7319122
12. Boonstra, T.W., Nikolin, S., Meisener, A.C., Martin, D.M., Loo, C.K.: Change in mean frequency of resting-state electroencephalography after transcranial direct current stimulation. Front. Hum. Neurosci. **10**, 270 (2016)
13. Angulo-Sherman, I.N., Rodriguez-Ugarte, M., Sciacca, N., Ianez, E., Azorin, J.M.: Effect of tDCS stimulation of motor cortex and cerebellum on EEG classification of motor imagery and sensorimotor band power. J. Neuroeng. Rehabil. **14**, 31 (2017). https://doi.org/10.1186/s12984-017-0242-1
14. Gbadeyan, O., McMahon, K., Steinhauser, M., Meinzer, M.: Stimulation of dorsolateral prefrontal cortex enhances adaptive cognitive control: a high-definition transcranial direct current stimulation study. J. Neurosci. **36**(50), 12530–12536 (2016). https://doi.org/10.1523/JNEUROSCI.2450-16.2016
15. Gbadeyan, O., Steinhauser, M., Hunold, A., Martin, A.K., Haueisen, J., Meinzer, M.: Modulation of adaptive cognitive control by prefrontal high-definition transcranial direct current stimulation in older adults. J. Gerontol. Ser. B **74**(7), 1174–1183 (2019). https://doi.org/10.1093/geronb/gbz048
16. Zmigrod, S., Zmigrod, L., Hommel, B.: Transcranial direct current stimulation (tDCS) over the right dorsolateral prefrontal cortex affects stimulus conflict but not response conflict. Neuroscience **322**, 320–325 (2016). https://doi.org/10.1016/j.neuroscience.2016.02.046
17. Comon, P.: Independent component analysis, a new concept? Signal Process. **36**(3), 287–314 (1994). https://doi.org/10.1016/0165-1684(94)90029-9
18. Makeig, S., Bell, A., Jung, T.P., Sejnowski, T.J.: Independent Component Analysis of Electroencephalographic Data. In: Advances in Neural Information Processing Systems, vol. 8. MIT Press (1995)
19. Keeser, D., et al.: Prefrontal transcranial direct current stimulation changes connectivity of resting-state networks during fMRI. J. Neurosci. **31**(43), 15284–15293 (2011). https://doi.org/10.1523/JNEUROSCI.0542-11.2011
20. Peirce, J., et al.: PsychoPy2: experiments in behavior made easy. Behav. Res. Methods **51**(1), 195–203 (2019). https://doi.org/10.3758/s13428-018-01193-y
21. Huebner, R., Lehle, C.: Strategies of flanker coprocessing in single and dual tasks. J. Exper. Psychol.-Hum. Percept. Perform. **33**(1), 103–123 (2007). https://doi.org/10.1037/0096-1523.33.1.103

Reconstructing Specific Neural Components for SSVEP Identification

Lijie Wang, Jinbiao Liu, Tao Tang, Linqing Feng, and Yina Wei(✉)

Research Center for Human-Machine Augmented Intelligence, Research Institute of Artificial Intelligence, Zhejiang Lab, Hangzhou 311100, Zhejiang, China
weiyina039@zhejianglab.com

Abstract. Brain-computer interfaces (BCIs) enable the brain to communicate directly with external devices. Steady-state visual evoked potential (SSVEP) is periodic response evoked by a continuous visual stimulus at a specific frequency in the occipital region of the brain. SSVEP-based BCI has been widely studied because of its high information transfer rate (ITR). However, SSVEP-BCI usually requires a long training period, which prevents its application. Therefore, it is necessary and meaningful to develop a decoding algorithm that can maintain high accuracy with less training time. This study proposed a point-position equivalent reconstruction (PPER) method to reconstruct stable periodic signals. The combination of PPER and ensemble task-related component analysis (eTRCA) algorithm can identify SSVEP with one training sample, whereas the classical methods need at least two training samples. Compared with eTRCA which achieved the highest average ITR of 172.93 ± 4.67 bits/min with two training samples, PPER-eTRCA can achieve the highest average ITR of 172.94 ± 3.75 bits/min with only one training sample. When the time window is 1 s, the accuracy of PPER-eTRCA with one training sample is significantly higher than the accuracy of eTRCA with two training samples (87.45% ± 1.19% vs. 85.32% ± 1.42%). The proposed method provides a feasible solution to reduce the training time while maintaining high performance, which will further facilitate the establishment of a more user-friendly SSVEP-BCI system.

Keywords: Brain-computer interfaces (BCI) · Steady-state visual evoked potential (SSVEP) · Point-position equivalent reconstruction (PPER) · Ensemble task-related component analysis (eTRCA)

1 Introduction

With the development of computer technology and machine learning, the brain-computer interface (BCI) has attracted increasing attention in recent years. It establishes a direct pathway between the central nervous system and the external environment [1, 2]. Steady-state visual evoked potential (SSVEP) is a type of electroencephalography (EEG) signal

Supported by National Natural Science Foundation of China (12101570), Research Initiation Project of Zhejiang Lab (2021KE0PI03, 2021KI0PI02), Key Research Project of Zhejiang Lab (2022KI0AC01), and China Postdoctoral Science Foundation (2021M702974).

H. Liu et al. (Eds.): ICIRA 2022, LNAI 13456, pp. 666–676, 2022.
https://doi.org/10.1007/978-3-031-13822-5_60

evoked by a repetitive visual stimulus, which consists of the fundamental frequency of the stimulus and its harmonics [3]. Because of its simple experiment design and stable performance, SSVEP-BCI has become one of the commonly used paradigms in EEG-based BCI [4].

The decoding algorithms of SSVEP-BCI were usually divided into three categories: training-free methods, subject-independent training methods, and subject-specific training methods [5]. Typical training-free algorithms include power spectral density analysis [6], minimum energy combination [7], and canonical correlation analysis (CCA) [8], etc. Subject-independent training methods obtain system parameters or classifiers suitable for current users through the data of other subjects, such as least absolute shrinkage and selection operator [9] and transfer template-based canonical correlation analysis algorithm. Subject-specific training methods can use the data of the present subject to train the classification model, so they usually show better decoding performance. Recently, the extended CCA (eCCA) [10] and the ensemble task-related component analysis (eTRCA) [11, 12] have become two classical subject-specific training algorithms. However, these subject-specific training methods generally achieve high recognition performance at the expense of a long training time [5, 13]. Excessive visual stimulation will cause users visual fatigue. Therefore, it is very important to develop algorithms that can maintain high accuracy with fewer training samples.

Based on the time-locked and phase-locked features of SSVEP [4], this study proposed a point-position equivalent reconstruction (PPER) method in order to reconstruct specific SSVEP components from EEG. We simulated a mixing signal and reconstructed specific frequency components from it. The results show that the signal reconstructed by the PPER method is close to the target signal. Then, PPER was combined with eTRCA to recognize SSVEP. The results confirm that the classification performance of PPER-eTRCA with one training sample can equal to or exceed the performance of eTRCA with two training samples.

2 Method

2.1 Point-Position Equivalent Reconstruction

Typically, an SSVEP signal measured from one electrode can be defined as a linearly weighted sum of the evoked periodic response and background noise [14]. As SSVEP is time-locked and phase-locked, data points with the same position in different sinusoid cycles can be defined as response equivalent points. Based on this feature, we proposed a PPER method that aimed to recover stable oscillation components from EEG.

A mixing signal is denoted as $L \in R^{N_p}$, where N_p denotes the number of sampling points. It is assumed that the component we expect to reconstruct from L is related to the frequency f_k. We define a sequential sequence:

$$P = \{1, 2, \cdots, N_p\} \quad (1)$$

where the elements of P indicates the original position orders of the sampling points in L.

According to the frequency f_k, P is divided into some cycles. The number of sinusoidal cycles contained in P can be calculated as follows:

$$Q_c = \left\lceil N_P * \frac{f_k}{F_s} \right\rceil \tag{2}$$

where F_s is the sampling rate, $\lceil \ \rceil$ symbol means to take the minimum integer not less than the parameter in the symbol. The sequence points in a complete cycle can be defined as:

$$Q_s = \left\lfloor \frac{f_k}{F_s} \right\rfloor \tag{3}$$

where symbol $\lfloor \ \rfloor$ means to take the maximum integer not greater than the parameter in the symbol. EEG data is sampled discretely, so not every cycle start can be found exactly. The start point of the *m-th* cycle can be approximately described as:

$$F_m = <(m-1) * \frac{f_k}{F_s}> + 1 \tag{4}$$

where symbol $< >$ indicates rounding the variable in symbol to the nearest integer.

After the cycle division and the start point calculation are completed, each point can be assigned an order number within the cycle to which it belongs. Ideally, the response should be equivalent at points with the same order number. Therefore, it is reasonable to randomly shuffle all the points with the same order. For example, for order u, all points in P with this order are picked out and then shuffled. Next, these rearranged points are put back into the vacated positions in P one by one. From order 1 to order Q_s, sequence points in the same order are rearranged in a shuffled order, so we can obtain a new sequence about the order of sampling points.

The process of shuffling and rearrangement of all points in P are repeated in order to obtain more sorting sequences. The total number of sequences that can be generated is:

$$I = \prod\nolimits_{u=1}^{Q_s} A_u \tag{5}$$

where A_u represents the number of all permutations of sequence points at order u. The last cycle may be incomplete. If there are n_l sampling points in the last cycle, the number of full permutations at order u should be:

$$A_u = \begin{cases} \begin{pmatrix} Q_c \\ Q_c \end{pmatrix}, u \le n_l \\ \begin{pmatrix} Q_c - 1 \\ Q_c - 1 \end{pmatrix}, u > n_l \end{cases} \tag{6}$$

where $\begin{pmatrix} b \\ a \end{pmatrix}$ represents the number of all permutations of b elements arbitrarily taken from a different elements, and n_l is the sampling points in a cycle. Therefore, when the sampling rate is large or the sampling time is very long, I will approach infinity.

Although there are many options, we only need to find r sequences with large distances including P. These r sequences are defined as point-position equivalent sequences.

The signal L is transformed separately using the equivalent sequence. Such transformation will introduce a lot of high-frequency noise to the newly generated signal. After low-pass filtering or band-pass filtering, all newly generated signals are averaged. The final averaged signal is what we expect to reconstruct.

2.2 SSVEP Recognition Algorithm

TRCA was originally proposed in 2013 by Tanaka et al. to analyze neuroimaging data and NIRS signals [11]. In 2018, Nakanishi et al. introduced TRCA into SSVEP recognition and developed ensemble TRCA (eTRCA) [12]. Since then, eTRCA has become the mainstream algorithm for SSVEP recognition. In this study, we combined the proposed PPER method with eTRCA for SSVEP recognition. The PPER-eTRCA method is designed to reduce the user's training time while maintaining high accuracy and ITR of SSVEP-BCI.

SSVEP data can be divided into training data and test data when it is classified offline. Training data is generally used to train suitable models or parameters for the classification of test data. We define X_k to represent the training dataset of N_t trials evoked by *k-th* stimulus. The multi-channel EEG data recorded for *i-th* trial is denoted as $X_k^i \in R^{N_c * N_p}$, where N_c represents the number of channels, N_p represents the number of sampling points, and $k = 1, 2, \cdots, N_f$. N_f denotes the number of target stimuli. The frequency of the stimulus that induces X_k^i is f_k.

From target 1 to N_f, we use the PPER method to get the point-position equivalent sequences of each stimulus frequency in turn. At this point, the stimulus frequency is the reconstruction target frequency in PPER. $ES_k = \{P, q_k^1, q_k^2, \cdots, q_k^{r-1}\}$ represents the set of point-position equivalent sequences for *k-th* stimulus, where q_k^n indicates the *n-th* newly generated equivalent sequence. The point-position equivalent sequences of the *k-th* stimulus are used to perform equivalent transformation on the data of each trial in X_k. For example, each channel data of X_k^i is equivalently transformed by the same equivalent sequence, so a new trial data is generated. All data generated by equivalent transformation are taken as training data, so there are $(N_t * r)$ training samples for each stimulus. The equivalent transformation data set of the *k-th* stimulus is denoted as M_k.

We conducted task-related component analysis on equivalent transformation data, the core of which was to recover task-related component as far as possible by maximizing the covariance between trials. The covariance between all trials can be calculated as follows:

$$S_k = \sum_{\substack{i,j=1 \\ i \neq j}}^{N_t * r} Cov\left(M_k^i, M_k^j\right) \tag{7}$$

where M_k^i presents the *i-th* trial in M_k. The variance between all trials is another important calculation item, which can be described as:

$$Q_k = \sum\nolimits_{i=1}^{N_t * r} Cov(M_k^i, M_k^i) \tag{8}$$

The above covariance maximization problem can be converted to solve a suitable mapping vector:

$$\widehat{w}_k = \underset{w}{argmax} \frac{w_k^T S_k w_k}{w_k^T Q_k w_k} \tag{9}$$

where w_k is a spatial filter that represents the coefficient vector to weight the multi-channel data. To find a finite solution, the mapping of variances can ideally be constrained as:

$$w_k^T Q_k w_k = 1 \tag{10}$$

Applying the Rayleigh-Ritz theorem to Eq. (10), we can know that the eigenvectors of $Q_k^{-1} S_k$ are suitable mapping vectors, where Q_k^{-1} represents the inverse of Q_k. By performing the above analysis on the data of each stimulus, an ensemble spatial filter can be constructed as follows:

$$W = [w_1, w_2, \cdots, w_{N_f}] \tag{11}$$

For a single test trial X, we transform it based on the equivalent sequences of all the stimuli that have been generated. For example, we use ES_k to transform X, so r new trials are obtained, denoted as RX_k. Then the correlation between each trial in RX_k and the corresponding template after spatial filtering can be measured:

$$r_k^n = corr((RX_k^n)^T W, \overline{M}_k^T W) \tag{12}$$

where (RX_k^n) represents *n-th* trial in RX_k, $\overline{M}_k$ denotes the template of *k-th* stimulus, and $corr(a, b)$ indicates Pearson's correlation coefficient between a and b. It is well known that SSVEP consists of the fundamental frequency and its harmonics, we apply the filter bank technique into the SSVEP recognition. Then, for a new test trial generated through equivalent transformation, its correlation with the *k-th* stimulus can be expressed as:

$$\rho_k^n = \sum\nolimits_{h=1}^{N_h} a(h).(r_k^n(h)) \tag{13}$$

where N_h is the number of sub-band. If the target frequency of X is f_k, then the newly generated trial in RX_k should retain the main SSVEP components. Ideally, the correlation between any trial of RX_k and template $\overline{M}_k$ is similar. Thus, the correlation between the present single test trial X and the *k-th* stimulus was eventually described as:

$$c\rho_k = \frac{1}{r} \sum\nolimits_{n=1}^{r} \rho_k^n \tag{14}$$

Then, X is transformed with other equivalent sequences in turn to calculate the integral correlation between the generated data and the corresponding template after spatial filtering. The final target to be identified should be:

$$\tau = \underset{k}{argmax}(c\rho_k) \tag{15}$$

where τ is the final identified target.

2.3 Performance Evaluation

In the field of neurophysiological signals, phase-locking value (PLV) is a frequently used indicator, which quantifies the consistency of phase difference between two signals. We calculated PLV between the reconstructed signal and the target signal as follows:

$$PLV_t = \frac{1}{T}\left|\sum\nolimits_{t=1}^{T} e^{i(\varnothing(t)-\theta(t))}\right| \tag{16}$$

where T represents the number of sampling points in the selected time window, $\varnothing(t)$ indicates the phase of the reconstructed signal $\varnothing$ at time t, and $\theta(t)$ indicates the phase of the target signal θ at time t.

Accuracy and information transfer rate (ITR) are two commonly used indicators to evaluate the performance of the SSVEP recognition algorithm. ITR [15] is defined as:

$$\mathrm{ITR} = \left(log_2 M + Alog_2 A + (1-A)log_2\left(\frac{1-A}{M-1}\right)\right)\frac{60}{T_s} \tag{17}$$

where M is the number of target stimulus, A is the accuracy, and T_s is the average selected time. The specifically selected time T_s is defined as:

$$T_s = T_g + T_t \tag{18}$$

where T_g presents the time window length of gazing stimulus, and T_t represents the time for subject to transfer their gaze sight on cue. T_t in this study is 0.5 s.

3 Data Sources

3.1 Simulated Mixing Signal

To simulate the wide-band oscillation characteristics of an EEG, we simulated mixing signals with multiple frequencies:

$$y(t) = y_1(t) + y_2(t) + y_3(t) + \varepsilon(t) \tag{19}$$

where $y(t)$ denotes the synthetic multi-frequency signal, $\varepsilon(t)$ is a white noise signal. In addition, $y_1(t)$, $y_2(t)$ and $y_3(t)$ are three different components:

$$y_1(t) = \sin(2\pi * fr_1 * \mathrm{t}) + 0.75 * \sin(4\pi * fr_1 * \mathrm{t}) \tag{20}$$

$$y_2(t) = \sin(2\pi * fr_2 * \mathrm{t} + 0.5\pi) \tag{21}$$

$$y_3(t) = \sin(2\pi * fr_3 * \mathrm{t} + \pi) \tag{22}$$

where fr_1, fr_2 and fr_3 represents three frequencies, and fr_1 is our target reconstruction frequency.

The signal $y(t)$ is filtered by a bandpass filter. Then we use the PPER method to generate r point-position equivalent sequences for the frequency fr_1. The r sequences are used to transform the signal $y(t)$. Thus, we can obtain r new signals with high-frequency noise. Similarly, these r new signals also need to be bandpass filtered. Next, these r signals are averaged as the constructed signal.

3.2 SSVEP Data

The SSVEP data used in this study comes from a public standard dataset, proposed by Wang et al. in 2017[15]. Thirty-five healthy subjects with normal or corrected-to-normal vision participated in the experiments, eight of whom had previously experienced SSVEP-BCI.

The experiment consists of 40 goals to enable offline spelling. The stimulus squares are displayed on the LCD monitor in the form of 5 rows and 8 columns. A target character is embedded in the middle of each stimulus square. The stimulation interface was coded by the joint frequency and phase modulation (JFPM) approach. It is assumed that (p_x, p_y) denotes the row and column index of the stimulus square. Thus, the frequency here should be:

$$f\left(p_x, p_y\right) = f_0 + \Delta f \times \left[\left(p_y - 1\right) \times 5 + \left(p_x - 1\right)\right] \tag{23}$$

$$\theta\left(p_x, p_y\right) = \theta_0 + \Delta\theta \times \left[\left(p_y - 1\right) \times 5 + \left(p_x - 1\right)\right] \tag{24}$$

where f_0 and θ_0 are 8 Hz and 0 rad respectively, Δf and $\Delta\theta$ are 0.2 Hz and 0.5 π rad.

Each subject has performed six-block experiments, each of which contained 40 trials. Target cues appear randomly at the beginning of each trial. The stimulus square where the target is located turns red for 0.5 s, aiming to guide the subjects to shift their gaze to the target. All stimuli flicker for 5 s. Next, the screen turns black for 0.5 s. At this time, the trial is completed.

The EEG electrodes are placed in a 10–20 system, and the sampling rate is 1k Hz. The collected data is filtered with a bandwidth of 0.15–200 Hz and notched at 50 Hz by hardware system. The collected data is down-sampled to 250 Hz and stored in the dataset.

4 Results

4.1 Reconstructing the Target Component from the Simulated Signal

We simulated a mixed signal $y(t)$ (Fig. 1a, blue) with three different frequencies (fr_1, fr_2 and fr_3 are 8, 10 and 12 Hz respectively) including the target component $y_1(t)$ (Fig. 1a, black) with a frequency of 8 Hz. The sampling rate of $y(t)$ is 1k Hz, and its duration time is 5 s. To reconstruct the target component, we applied the proposed point-position equivalent reconstruction (PPER) method. The number of equivalent transformations (r) is set to 20.

We found that the waveform of the reconstructed signal (Fig. 1a, middle panel, red) almost coincides with the target signal (Fig. 1a, middle panel, black), except at some points. Meanwhile, the frequency domain of the reconstructed signal only contains the target frequency and its harmonics (Fig. 1b, red). Compared with the final reconstructed signal, the single transformation signals not only contain two target frequencies but also contain some extremely weak noise (Fig. 1b, green). Although the single transformation signals introduce some noise, the overlapping line of multiple single transformation signals (Fig. 1a, bottom panel, green) is relatively similar to the target signal (Fig. 1a, bottom panel, black).

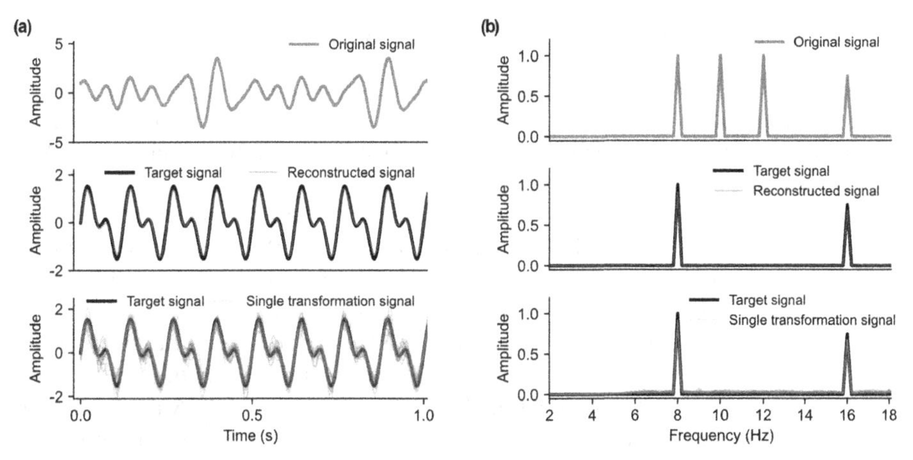

Fig. 1. Comparison between the target signal and the signal generated in reconstruction process. (a) Comparison of reconstruction results in time domain. (b) Comparison of reconstruction results in frequency domain. Original signal (blue) indicates the mixing signal $y(t)$. Target signal (black) indicates $y_1(t)$ in Eq. (20). Reconstruceted signal (red) indicates the final reconstructed signal obtained by PPER method. Single transformation signal (green) indicates the signal produced by a single equivalent transformation during PPER reconstruction. (Color figure online)

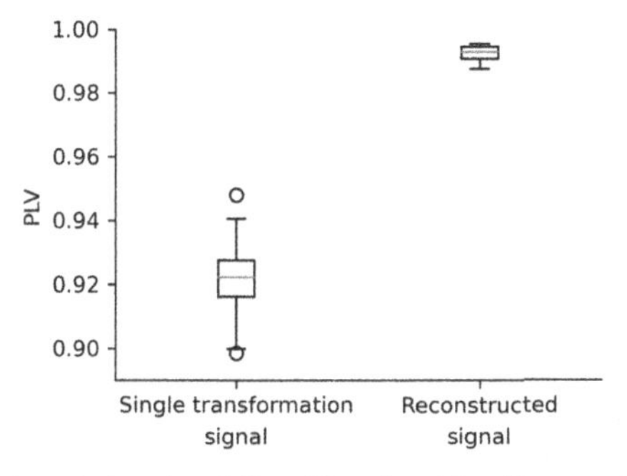

Fig. 2. PLV between the target signal and the other signal in all reconstruction process. The tick values of horizontal x axis indicate exactly what the other signal is.

The reconstruction process was performed 20 times for target signals with different frequencies. During each reconstruction, we analyzed the PLV values between the target signal and all signals generated during the reconstruction (Fig. 2). The mean PLV between the target signal and the reconstructed signal is about 0.9925, indicating that their phase difference is almost the same at different moments. In addition, the mean PLV between the target signal and the single transformation signal is about 0.9214. The result indicates the single transformation signal also has a great similarity with the target signal, which further proves that it is reasonable to take generated test trial (RX_k^n) as the equivalent test signal in the PPER-eTRCA method.

4.2 SSVEP Identification with One Training Sample

The SSVEP classification test in this study was performed using the standard dataset. Since the proposed PPER method can successfully reconstruct the signal equivalent to the target component, we used the combination of PPER and eTRCA to classify SSVEP with one training sample in this study.

The data of the 9 channels in the occipital region were selected for classification, which are Pz, PO5, PO3, POz, PO4, PO6, O1, Oz, and O2. Based on previous studies, the delay time of the visual pathway is 0.14 s. The number of filter bank sub-bands N_h is set to 5, and the leave-one cross-validation strategy is used in the analysis.

We compared the performance of PPER-eTRCA and eTRCA at different time window. We focus on whether the PPER-eTRCA algorithm can maintain the accuracy with less training samples, and traditional methods represented by eTRCA require at least two training samples to train the classification model. Therefore, we set the training sample volume for PPE-eTRCA and eTRCA to 1 and 2, respectively. As the time window increases, the accuracy of the two algorithms increases gradually. When the time window varies from 0.9 s to 1.0 s, the accuracy of PPER-eTRCA is significantly higher than that of eTRCA. The accuracy of PPER-eTRCA is 87.45% 1.19% at 1 s, while the accuracy of eTRCA is 85.32% $\pm$ 1.42%. Among the result of PPER-eTRCA, the accuracy of 23 subjects exceeded 90% (Fig. 3).

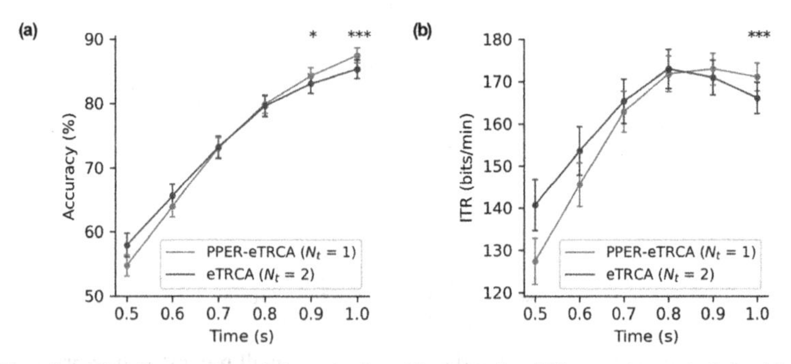

Fig. 3. The classification results of the two algorithms at the different time window. (a) Average accuracy across all subjects. (b) Average ITR across all subjects. The legend N_t denotes the training sample volume of the front algorithm. The asterisks mark denotes the significant difference at the same time window between PPER-eTRCA (red) and eTRCA (blue) (* $P < 0.05$, *** $P < 0.005$). (Color figure online)

The maximum ITR of PPER-eTRCA is 172.94 $\pm$ 3.75 bits/min at 0.9s, while the maximum ITR of eTRCA is 172.93 $\pm$ 4.67 bits/min at 0.8 s. When the time window is 0.9 s, the accuracy across subjects for PPER-eTRCA is about 84.26% $\pm$ 1.30%, which is significantly higher than eTRCA ($p = 0.0003$, two-sample t-test). Moreover, the maximum ITR of PPER-eTRCA can reach 172.94 bits/min with the training time of only 76 s, while the eTRCA method requires the subjects to train for 144s (which does not include the rest time between the two training blocks) to achieve the highest ITR of 172.93 bits/min. When the time window is shorter than 0.6s, PPER-eTRCA did not perform well. One of the reasons may be that when the time window is too short and the sampling rate is relatively low, the number of sample points is too small to do sufficient equivalent transformations. Another reason may be that the PPER-eTRCA is sensitive to signal quality. For several subjects with poor signal quality, the performance of the PPER-eTRCA is much lower than the average level.

In the previous training classification algorithms, there are two samples at least for SSVEP recognition. Our results illustrate that the PPER-eTRCA method is feasible for most people to classify SSVEP with one training sample. In order to achieve the highest ITR, the eTRCA method requires the subjects to train for 144 s, while PPER-eTRCA can be achieved by only 76 s training time, which includes time for gazing stimuli, shifting sight and resting. In conclusion, the PPER method is effective to reconstruct the target component from a broad-band signal when there are enough data points. The proposed PPER-eTRCA algorithm can maintain high accuracy while the training time is greatly shortened.

5 Conclusion

SSVEP-BCI is a possible direction for BCI application. It is important to develop a high-accuracy classification algorithm with a small training sample. In this study, we proposed a PPER method to reconstruct specific components from mixing signals. We first used the simulation signal to verify the effectiveness of the PPER method for recovering periodic oscillations. Then, we developed a PPER-eTRCA algorithm for SSVEP recognition. The results showed that the classification accuracy of PPER-eTRCA with one training sample can reach 87.45% when the time window is 1 s. This result exceeds the accuracy of eTRCA with two training samples in the same time window. To achieve the same highest ITR, eTRCA takes 144 s training time, while the PPER-eTRCA algorithm only needs 76 s. It significantly reduces the user's training time compared to previous algorithms. In summary, the proposed PPER method is effective to reconstruct the target component from a mixed signal, and is easy to extend. The combination of PPER and eTRCA achieved high accuracy with less training time in SSVEP-BCI. This study provides a feasible solution for improving the practical application effect of SSVEP-BCI.

References

1. He, B., Yuan, H., Meng, J., Gao, S.: Brain–computer interfaces. In: Neural Engineering, pp. 131–183 (2020)
2. Wolpaw, J.R.: Brain–computer interfaces. Handb. Clin. Neurol. **110**, 67–74 (2013)
3. Regan, D.: Some characteristics of average steady-state and transient responses evoked by modulated light. Electroencephalogr. Clin. Neurophysiol. **20**(3), 238–248 (1966)
4. Vialatte, F.-B., Maurice, M., Dauwels, J., Cichocki, A.: Steady-state visually evoked potentials: focus on essential paradigms and future perspectives. Prog. Neurobiol. **90**(4), 418–438 (2010)
5. Zerafa, R., Camilleri, T., Falzon, O., Camilleri, K.P.: To train or not to train? A survey on training of feature extraction methods for SSVEP-based BCIs. J. Neural Eng. **15**(5), 051001 (2018)

6. Muller-Putz, G.R., Pfurtscheller, G.: Control of an electrical prosthesis with an SSVEP-based BCI. IEEE Trans. Biomed. Eng. **55**(1), 361–364 (2007)
7. Friman, O., Volosyak, I., Graser, A.: Multiple channel detection of steady-state visual evoked potentials for brain-computer interfaces. IEEE Trans. Biomed. Eng. **54**(4), 742–750 (2007)
8. Bin, G., Gao, X., Yan, Z., Hong, B., Gao, S.: An online multi-channel SSVEP-based brain–computer interface using a canonical correlation analysis method. J. Neural Eng. **6**(4), 046002 (2009)
9. Zhang, Y., Jin, J., Qing, X., Wang, B., Wang, X.: LASSO based stimulus frequency recognition model for SSVEP BCIs. Biomed. Signal Process. Control **7**(2), 104–111 (2012)
10. Chen, X., Wang, Y., Nakanishi, M., Gao, X., Jung, T.P., Gao, S.: High-speed spelling with a noninvasive brain-computer interface. Proc. Natl. Acad. Sci. U. S. A. **112**(44), E6058-6067 (2015)
11. Tanaka, H., Katura, T., Sato, H.: Task-related component analysis for functional neuroimaging and application to near-infrared spectroscopy data. Neuroimage **64**, 308–327 (2013)
12. Nakanishi, M., Wang, Y., Chen, X., Wang, Y.T., Gao, X., Jung, T.P.: Enhancing detection of ssveps for a high-speed brain speller using task-related component analysis. IEEE Trans. Biomed. Eng. **65**(1), 104–112 (2018)
13. Wong, C.M., et al.: Learning across multi-stimulus enhances target recognition methods in SSVEP-based BCIs. J. Neural Eng. **17**(1), 016026 (2020)
14. Cecotti, H.: A self-paced and calibration-less SSVEP-based brain–computer interface speller. IEEE Trans. Neural Syst. Rehabil. Eng. **18**(2), 127–133 (2010)
15. Wang, Y., Chen, X., Gao, X., Gao, S.: A benchmark dataset for SSVEP-based brain-computer interfaces. IEEE Trans. Neural Syst. Rehabil. Eng. **25**(10), 1746–1752 (2017)

Modeling and Recognition of Movement-Inducing Fatigue State Based on ECG Signal

Jingjing Liu[1,2], Jia Zeng[1], Zhiyong Wang[1,2], and Honghai Liu[2](✉)

[1] Shanghai Jiao Tong University, Shanghai 200240, China
lily121@sjtu.edu.cn

[2] Harbin Institute of Technology Shenzhen, Shenzhen 518055, Guangdong, China
honghai.liu@icloud.com

Abstract. Fatigue monitoring is significant during movement process to avoid body injury cased by excessive exercise. To address this issue, we developed an automated framework to recognize human fatigue states based on electrocardiogram (ECG) collected by a smart wearable device. After preprocessing on the raw ECG data, both machine learning solution and deep learning solution were introduced to recognize the fatigue states. Specifically, a set of hand-crafted features were designed which are fed into different machine learning models for comparison. For the deep learning solution, the residual mechanism was employed to build a deep neural network for fatigue classification. The proposed methods were evaluated on data collected from subjects after running exercise and achieved an accuracy of 89.54%.

Keywords: Fatigue analysis · ECG · Machine/deep learning

1 Introduction

Fatigue is a complex state which can be manifested in the form of reduced mental or physical performance [1]. It is an important factor to maintain human safety in realistic scenes such as driving and working. Thus reliable and in-time fatigue analysis is necessary to prevent human beings from possible injuries subsequently.

It is noted that the fatigue-related analysis has a large range considering different realistic applications. In recent years, the driving fatigue and mental fatigue are mostly explored in this area. The drivers' fatigue recognition has been a research hotspot which can play an important role in avoiding accidents. Different methods are employed in this process such as analyzing the videos of drivers' behavior and the vehicle operational performance. Compared with the

This work was supported by the National Natural Science Foundation of China (Grant No. 61733011), Shanghai Key Clinical Disciplines Project and Guangdong Science and Technology Research Council (Grant No. 2020B1515120064).

H. Liu et al. (Eds.): ICIRA 2022, LNAI 13456, pp. 677–685, 2022.
https://doi.org/10.1007/978-3-031-13822-5_61

above indicators, human physiological signals can provide a more reliable and relevant index for modeling the human fatigue states. Specifically, Some researches studied the relationship between the ECG indicators and driving fatigue. In [2], the time or frequency domain parameters of ECG were analyzed to build the relationship with the driving fatigue using traditional statistical approaches. Except from using single ECG indicator, the combination with electromyography (EMG) was explored in [3]. The complexity of EMG, complexity of ECG, and sample entropy (SampEn) of ECG are extracted and the accuracy of the drivers' fatigue recognition model is up to 91%. As for the mental fatigue which are related with overwork or medical issues such as weakness and ageing problems, the collected sensing modalities including HR [4,5], ECG [6], EEG [7], EMG [8] are be used to model the fatigue level in a subjective and automatic way. Yang et al. [6] used eight heart rate variability (HRV) indicators to build fatigue state classifiers. However, the above mentioned approaches mainly assess the human fatigue from the view of mental aspect which the assessment of fatigue in terms of physical performance is barely explored.

There are a few studies in the area of movement-inducing fatigue related analysis using other sensors. Inertial sensors were introduced to analyze the gait parameters to distinguish non-fatigued and fatigued walking trials [9]. In [10], 10 infrared cameras were used to recorded three-dimensional full-body kinematic data and the characteristics of individual movement patterns were observed despite of the fatigue-related changes. Different from the two measures mentioned above, in this work we developed models based on collected ECG signals to classify movement-inducing fatigue states with the recent development of wearable ECGs, which suggests the potential of monitoring and managing human health in daily life. The participants were told to do certain exercises and give self-report fatigue levels after movements. After preprocessing and model construction, we present a recognition framework to map the collected ECG signal to three fatigue levels. Specifically, both machine learning based solutions and deep learning solutions are explored for the modeling.

2 Methodology

In this paper, we aim to develop a fatigue state recognition framework which can automatically output the fatigue level given the input of ECG signal. First a preliminary study was conducted to collect data from 18 participants. Then the preprocessing steps followed by machine learning based solutions and deep learning solutions were introduced. Next the experimental results were analyzed and finally conclusions were given considering the future work.

2.1 Data Collection

In this phase, 18 participants (aged from 23 to 29 years old, 15 males and 3 females) were recruited and data were collected for two trails of running exercise. For each participant, some personal information were collected at first including the gender, age, height, weight, number of exercises per week. Then the ECG data were collected for 5 min when he is in a neutral state. Next in the first trial the participant was told to continuously run for 3 min on the same treadmill. The running speed was set 2.4 m/s and 1.8 m/s for males and females respectively. In the second trial, experiments settings were same as the first trail expect for the running speed (2.8 m/s for the female and 2.2 m/s for the male). After finishing each trial, the ECG data were collected for 5 min and the participants were directed to make self-report about their fatigue states which are set to three levels: No fatigue, Mild fatigue and Moderate fatigue.

As for the ECG collection device, a public available product [11] was used in our collection stage.

2.2 Preprocessing

The original data as well as the ECG signals suffer from noises due to various reasons such as the equipment inducing movement artifacts [12]. To eliminate this, the heart rate variability (HRV) extracted from ECG is considered a more reliable indicator for subsequent analysis. HRV measures the change in the time interval between heartbeats which can be derived from R-R interval (RRI) detection. Given the R-R interval (RRI) detection, the heart rate (HR) could also be computed. During this process, those RRI outliers are removed according to [13] and linearly interpolating is applied for the removed RRIs.

2.3 Machine Learning Solution

In this subsection, we aim to design a set of hand-crafted features and different machine learning methods are employed to map the features to the fatigue level.

As shown in Table 1, we designed 16 features in three domains (personal related characteristics, HRV related features and HR related features).

Given the data of n participants, the above features ($f_i, i = 1, ..., 16$) are computed by applying a time window of length w and step of s. Then the feature matrix $F \in R^{m \times p}$ can be obtained, where m is the number of samples and p is the dimension value of feature. It is noted that the normalization is operated on $F \in R^{m \times p}$ subsequently as shown in Eq. 1.

$$F_s = \frac{F - F_{mean}}{F_{std}} \tag{1}$$

where F_{mean} and F_{std} are matrixes that calculate the mean value and standard deviation value for all samples in F.

Given the normalized feature matrix F_s, our different machine learning algorithms, support vector machine (SVM) [14], Multilayer Perception (MLP) [15], Decision Tree (DT) [16], Random Forest (RF) [17], Adaboost [18], and Gaussian naive Bayes (GNB) [19], were used to build classifiers that automatically recognize the fatigue states. And the parameters for these models are set empirically as shown in Table 2.

Table 1. Hand-crafted features.

Feature name	Explanation
Personal related	
gender	gender = 1(female) gender = 0(male)
age	/years old
height	/cm
weight	/kg
number of exercises per week	/times
HRV Related	
SDNN	standard deviation of NN intervals
TP	total power
LF/HF	low/high frequency power
HR Related, heart rate sequence: $hr_i, i = 1, ..., t$, mean heart rete at the neutral state: $hr_0, hr_i^{'} = (hr_i - hr_0)/hr_0$	
mean(hr)	$f_9 = \sum_{i=1}^{t} hr_i/t$
std(hr)	$f_{10} = \sum_{i=1}^{t}(hr_i - f_9)^2/t)$
max(hr)	$f_{11} = max(hr_1, hr_2, ..., hr_t,)$
min(hr)	$f_{12} = min(hr_1, hr_2, ..., hr_t,)$
mean($hr^{'}$)	$f_{13} = \sum_{i=1}^{t} hr_i^{'}/t$
std($hr^{'}$)	$f_{14} = \sum_{i=1}^{t}(hr_i^{'} - f_{13})^2/t)$
max($hr^{'}$)	$f_{15} = max(hr_1^{'}, hr_2^{'}, ..., hr_t^{'},)$
min($hr^{'}$)	$f_{16} = min(hr_1^{'}, hr_2^{'}, ..., hr_t^{'},)$

Table 2. Parameter setting in machining learning models.

Model	Parameters
SVM	GridSearch: kernel = ('rbf', 'linear', 'poly', 'sigmoid') C = (0.1, 0.5, 1, 2, 5)
MLP	solver = 'adam', alpha = 1e−5, max_iter=1000, hidden_layer_sizes = (100, 50, 30)
DT	criterion ='gini'
RF	n_estimators = 200, max_depth = 6, criterion = 'gini', class_weight = 'balanced'
Adaboost	max_depth = 2, min_samples_split = 20, min_samples_leaf=5, n_estimators = 300, learning_rate = 0.8
GNB	-----

2.4 Deep Learning Solution

In the machine learning solution, features are hand-crafted which suggest the possible information loss. With the recent development in deep learning area, the convolutional neural network is introduced to capture more comprehensive information given the ECG signal as input.

As shown in Fig. 1, we design a convolutional neural network for the fatigue state recognition inspired by the residual mechanism in [20]. The designed network named ResECG can have a better performance by adding the depth of the network through designing sub modules with short connections and stacking multiple sub modules. First, the ECG signal is fed into a stem module to extract basic and low dimensional features which incorporates one convolutional (conv) layer, one batch normalization (bn) layer and one Relu layer. Then the data flow into the first ResBlock. In this block, there are two branches while one branch comprises one conv layer, one bn layer, one Relu layer and one conv layer. Since the convolutional operations in the conv layers change the size of features, pooling operation is applied in another branch to generate feature of the same size. Features from two branches in the first ResBlock are added and fed into subsequent two ResBlocks which have similar architectures. After these three blocks, the generated features have high-dimensional semantic information. In the last sub module, the AdaptiveAveragePooling is employed to eliminate the time dimension by using average pooling. In addition, the AdaptiveAveragePooling layer could fit to different input sizes of ECG signal under different window lengths, and can slow down the over fitting effect. Finally, the feature is mapped to vector of size n through a 1×1 convolution (n represents the number of categories to be classified) and the confidence score of each category is calculated through softmax. There are 340227 parameters in this model.

$$xn_i = \frac{(x_i - min(x))}{max(x) - min(x)} \tag{2}$$

3 Experiments and Results

3.1 Experiments

The collected data suffered from quality issues and after preprocessing only the data of 16 participants were used to train the model. To evaluate the recognition models, we performed 8-fold cross-validation. For each set of validation trial, two participants are divided into test subset while other participants' data are used for training. It is noted that there remains a problem of imbalanced class distribution in the collected data. As shown in Fig. 2, data of no fatigue occupies a larger proportion than the data of mild or moderate fatigue. Considering this issue, the data distribution in the train subset and test subset remains the same.

The recognition performance was evaluated by comparing it to human self-reported fatigue levels. In Table 3, we report the recognition accuracy using different machine learning methods. Among these methods, SVM achieved the best results of 82.6% accuracy. As it can be observed, the performance varies for different Cv settings. Possible reasons could be that there are significant differences between the data of the train subset and the test subset in specific split settings. It can be improved by training a more reliable and precise model based on data of more subjects.

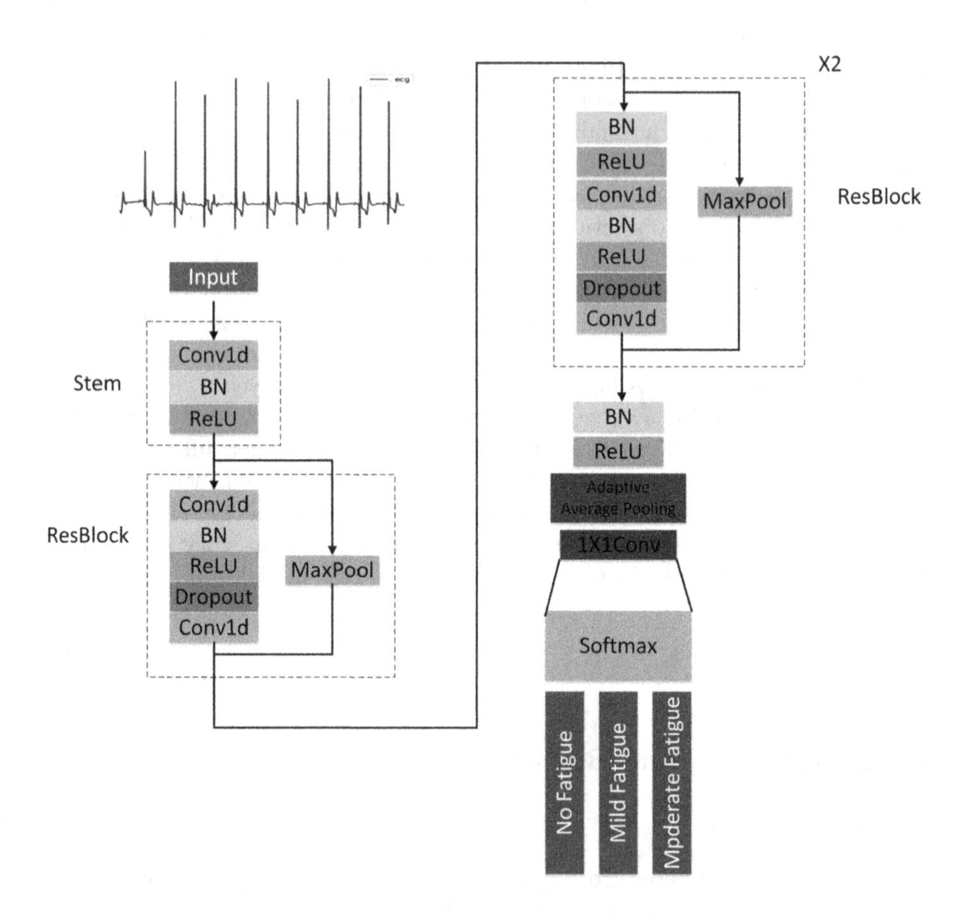

Fig. 1. The architecture of ResECG.

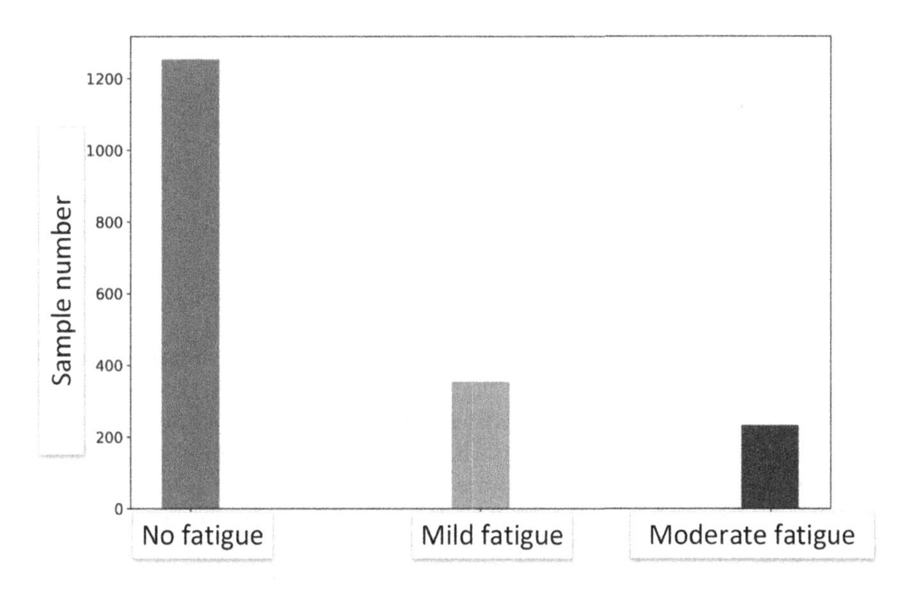

Fig. 2. Data distribution of the collected data.

Table 3. Recognition accuracy (%) for 8-fold Cv using different machine learning methods.

	Cv1	Cv2	Cv3	Cv4	Cv5	Cv6	Cv7	Cv8	Mean
SVM	88.4	86.4	59.7	100	84.2	80.1	91.7	70.3	**82.6**
MLP	56.94	75.3	33.8	100	74.5	76.4	64.8	33.8	64.4
DT	50	48.1	39.8	100	65.7	71.3	55.5	25	56.9
RF	67.6	71.6	19.9	98.1	81.5	72.7	76.8	25.9	64.3
Adaboost	68.5	53.1	29.6	66.7	74.1	75	50	28.7	55.7
GNB	61.1	83.3	57.8	100	89.8	31	57.4	72.7	69.1

As for the deep learning solution, we investigate the effects of using different length of window and the flipping operation for data augmentation. As shown in Table 4, the proposed ResECG network could achieve the best recognition accuracy of 89.54%, with an improvement of 6.94% than the machine learning methods. It is reasonable because the deep learning method could derive more comprehensive information than hand-crafted features. Specifically, the flipping augmentation could bring an improvement of about 2% and the window length of 4096ms is more suitable.

Table 4. Recognition accuracy (%) for different settings using ResECG.

Window length (/ms)	Flipping operation	Accuracy (%)
2048	N	85.24
2048	Y	87.97
4096	N	87.74
4096	Y	89.54

Although promising results could be achieved using our proposed framework, there are still some limitations to be clarified. On the one hand, the performances on different split settings for cross-subject validation are not stable enough. The major reason is the small sample size in this study which limits its application in realistic scenes. If given more subjects, the model could learn to generate more comprehensive features and adjust to the specificities of different subjects. On the other hand, despite of better results achieved by deep learning method, the features cannot be interpreted like machine learning methods. The relevant interpretation of feature sets in the machine learning methods remains to be explored. In the future, we will incorporate more subjects to improve current recognition accuracy and the feature interpretation will also be introduced to generate a interpretable solution.

4 Conclusion

In this paper we aim to develop an automatic framework to evaluate the movement-inducing fatigue based on ECG signal. The framework entails a pipeline from data collection, data preprocessing, machine learning solution and deep learning solution. Both solutions were evaluated on the collected data with significant results. The recognition accuracy could achieve 89.54% under a 8-folder cross subject validation setting. However, this pilot study is limited by the relatively small sample size. In the future, we will develop a more reliable and precise method by enlarging the dataset size and extensive feature interpretation will also be explored.

References

1. Lal, S.K.L., Craig, A.: A critical review of the psychophysiology of driver fatigue. Biol. Psychol. **55**(3), 173–194 (2001)
2. Patel, M., et al.: Applying neural network analysis on heart rate variability data to assess driver fatigue. Expert Syst. Appl. **38**(6), 7235–7242 (2011)
3. Wang, L., Wang, H., Jiang, X.: A new method to detect driver fatigue based on EMG and ECG collected by portable non-contact sensors. Promet-Traffic Transp. **29**(5), 479–488 (2017)

4. Barrios, L., et al.: Recognizing digital biomarkers for fatigue assessment in patients with multiple sclerosis. In: Proceedings of the 12th EAI International Conference on Pervasive Computing Technologies for Healthcare; PervasiveHealth. New York, NY: EAI (2018)
5. Maman, Z.S., et al.: A data-driven approach to modeling physical fatigue in the workplace using wearable sensors. Appli. Ergon. **65**, 515–529 (2017)
6. Huang, S., et al.: Detection of mental fatigue state with wearable ECG devices. Int. J. Med. Inf. **119**, 39–46 (2018)
7. Shen, K.-Q., et al.: EEG-based mental fatigue measurement using multi-class support vector machines with confidence estimate. Clin. Neurophysiol. **119**(7), 1524–1533 (2008)
8. Papakostas, M., et al.: Physical fatigue detection through EMG wearables and subjective user reports: a machine learning approach towards adaptive rehabilitation. In: Proceedings of the 12th ACM International Conference on PErvasive Technologies Related to Assistive Environments (2019)
9. Guaitolini, M., et al.: Sport-induced fatigue detection in gait parameters using inertial sensors and support vector machines. In: 2020 8th IEEE RAS/EMBS International Conference for Biomedical Robotics and Biomechatronics (BioRob). IEEE (2020)
10. Burdack, J., et al.: Fatigue-related and timescale-dependent changes in individual movement patterns identified using support vector machine. Front. Psychol., 2273 (2020)
11. Shi, H., et al.: A mobile intelligent ECG monitoring system based on IOS. In: 2017 International Conference on Sensing, Diagnostics, Prognostics, and Control (SDPC). IEEE (2017)
12. Patel, S.I., et al.: Equipment-related electrocardiographic artifacts: causes, characteristics, consequences, and correction. J. Am. Soc. Anesthesiologists **108**(1), 138–148 (2008)
13. Tanaka, H., Monahan, K.D., Seals, D.R.: Age-predicted maximal heart rate revisited. J. Am. Coll. Cardiol. **37**(1), 153–156 (2001)
14. Suykens, J.AK.: Support vector machines: a nonlinear modelling and control perspective. Eur. J. Control 7(2-3), 311–327 (2001)
15. Ramchoun, H., et al.: Multilayer perceptron: architecture optimization and training (2016)
16. Quinlan, J.R.: Learning decision tree classifiers. ACM Comput. Surv. (CSUR) **28**(1), 71–72 (1996)
17. Segal, M.R.: Machine learning benchmarks and random forest regression (2004)
18. Hastie, T., et al.: Multi-class adaboost. Stat. Interface **2**(3), 349–360 (2009)
19. Zhang, H.: The optimality of naive Bayes. Aa 1.2 (2004): 3
20. He, K., et al.: Deep residual learning for image recognition. In: Proceedings of the IEEE Conference on Computer Vision and Pattern Recognition (2016)

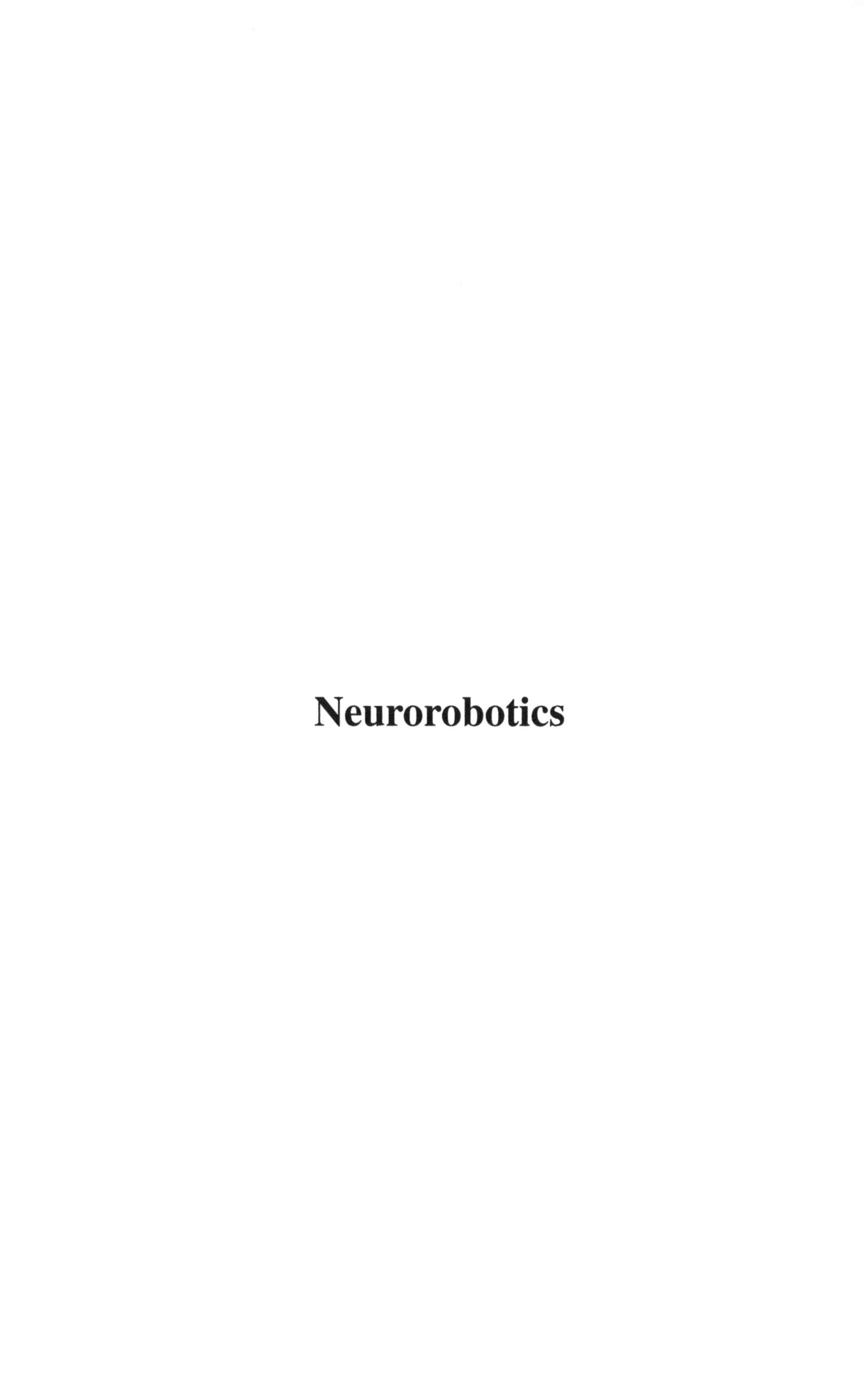

Neurorobotics

Structural Design and Control of a Multi-degree-of-freedom Modular Bionic Arm Prosthesis

Yingxiao Tan[1,2,3], Yue Zheng[1,2,3], Xiangxin Li[1,3](✉), and Guanglin Li[1,3](✉)

[1] Shenzhen Institutes of Advanced Technology (SIAT), Chinese Academy of Sciences, Shenzhen, China
{lixx,gl.li}@siat.ac.cn

[2] Shenzhen College of Advanced Technology, University of Chinese Academy of Sciences, Shenzhen, China

[3] CAS Key Laboratory of Human-Machine Intelligence-Synergy Systems, Shenzhen, China

Abstract. Due to various reasons of natural disasters, car accidents, diseases and so on, different levels of amputations such as hand, wrist and shoulder disarticulation have been caused. A modular structural design of upper limb prosthesis that consists of hands, wrists, elbows, shoulders joints is vital to restore the lost motor functions of amputees and still remains a challenge. This paper designs a modular bionic arm prosthesis with five-degree-of-freedom according to the characteristics of weight, size and range of motion of a natural upper limb. By simulating and analyzing the kinematics of the arm prosthesis, results showed that the range of motion of the prosthesis is relatively wide and can meet the use of daily life. And based on the 3D printing technology, a whole arm prosthesis was printed and assembled modularly. Additionally, a control test of the modular arm prosthesis was conducted. The results showed that the designed prosthesis was operated successfully by the surface electromyography based pattern recognition control. The work of this study provides an effective modular bionic arm prosthesis structure that can restore different motor functions for patients with different levels of amputations.

Keywords: Arm prosthesis · Modular bionic prosthesis · Multiple-degree-of-freedom

1 Introduction

Amputation is commonly caused by trauma, peripheral vascular diseases, tumors, infections, and congenital anomalies [1, 2]. According to statistics, about 2.26 million amputees are in China now, including 0.45 million upper limb amputees, and the number is increasing year by year. Upper limb amputation usually leads to the function loss, which severely limits the activities of daily living, brings social discrimination, and affects the mental health of patients.

Figure 1 shows a schematic diagram of the upper extremity anatomy and the dissection sites. The human arm has three main joints: the shoulder joint, the elbow joint and the

H. Liu et al. (Eds.): ICIRA 2022, LNAI 13456, pp. 689–698, 2022.
https://doi.org/10.1007/978-3-031-13822-5_62

wrist joint. Depending on the connection and function of the bones, the upper limb has six amputation sites: forequarter, shoulder disarticulation, transhumeral, elbow disarticulation, transradial and wrist disarticulation [3, 4]. The above amputation sites can be classified into three types of amputation: forearm amputation, upper arm amputation and shoulder amputation. In order to satisfy the requirements of different amputation types, a modular bionic arm prosthesis need to be designed to reconstruct the corresponding arm function [5].

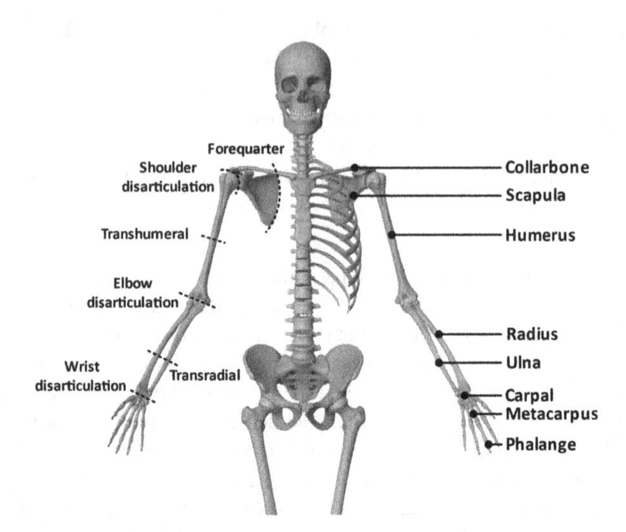

Fig. 1. Anatomy and detachment of the upper extremities.

At present, commercial prostheses mainly include passive prostheses [6], self-powered prostheses [7] and myoelectric prostheses [8]. However, the abandonment rate of existing prostheses is high [9], mainly due to the prosthetic weight [10], control method [11], the prosthetic structure and the price [4]. Compared with human limb function, there is still a large gap in commercial prostheses.

In the current research, the upper limb prostheses, including the LUKE arm designed by the University of Utah [12] and the modular prosthesis designed by Johns Hopkins University [13] perform good bionic performance. But their common problems are large-weight and high price. For example, the LUKE arm [14] weighs 9.8 lb and sells for $100,000. The wearable dexterous prosthetic system [15] designed by Li Zhijun's team has good control accuracy, but the arm span reaches 1m, which reduces the bionic performance. Besides, the current upper limb prosthesis research mainly focuses on increasing the prosthesis's flexibility and augmenting its sensory feedback information. However, the structure of the arm prosthesis, which could be suited for amputees with different amputation types is also worth exploring.

Therefore, in order to meet the needs of different arm amputation types for prosthesis, this paper designs a modular bionic arm prosthesis with five-degree-of-freedom. The modular bionic arm prosthesis could be separated into four modules, which can be applied to different types of amputations through the combination of different modules.

For example, module I, modules combined from module I to III and modules combined from module I to module IV could be adapted to forearm amputation, upper arm amputation, and shoulder amputation, respectively. The designed arm prosthesis also mimics the size, weight, and motion of the natural upper limb. The motions it can complete include flexion and extension of the wrist, elbow and shoulder, and internal and external rotation of the wrist and shoulder.

This paper first describes the structure design of a modular bionic arm prosthesis with five-degree-of-freedom. The range of motion of the prosthesis is simulated and analyzed. Then, a full arm prosthesis was fabricated by 3D printing technology and assembled modularly. Finally, a control test of the arm prosthesis was conducted by the surface electromyography (sEMG) based pattern recognition control.

2 Structural Design

According to the three amputation types mentioned above, this paper designs a modular bionic arm prosthesis with five-degree-of-freedom. According to the different combination of the four modules of the prosthesis, the modules could be used as forearm prosthesis, upper arm prosthesis and full arm prosthesis, respectively.

2.1 Forearm Prosthesis

The forearm prosthesis is suitable for the use of transradial amputees and wrist disarticulation amputees. Module I is used as the forearm prosthesis and its structure is dis-played in Fig. 2. The power transmission of the module I is composed of the Maxon motor I, the reducer I and the turbine worm, which can achieve the flexion and ex-tension function of the wrist joint. The worm gear and worm are used to realize the self-locking function of wrist flexion and extension. The spindle and lock in the module I are responsible for assembling and fixing with the socket and the upper-level module.

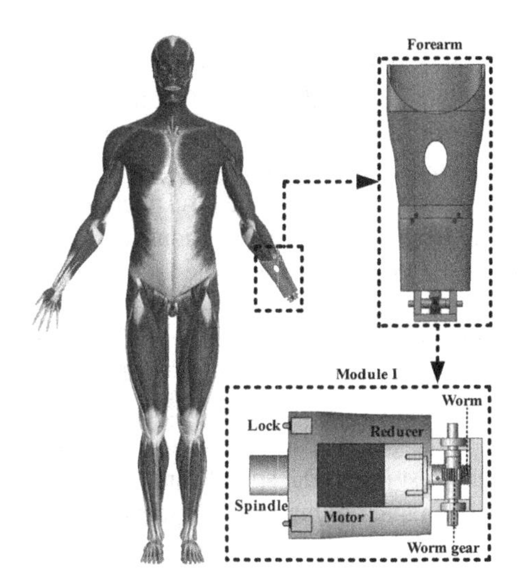

Fig. 2. The structural design of the forearm prosthesis.

2.2 Upper Arm Prosthesis

Figure 3 shows the structural design of the upper arm prosthesis, which is suitable for elbow disarticulation and transhumeral. The upper arm prosthesis is mainly composed of module I, module II and module III, which together constitute three-degree-of-freedom, realize the functions of flexion and extension of the wrist, internal and external rotation of the wrist, and flexion and extension of the elbow.

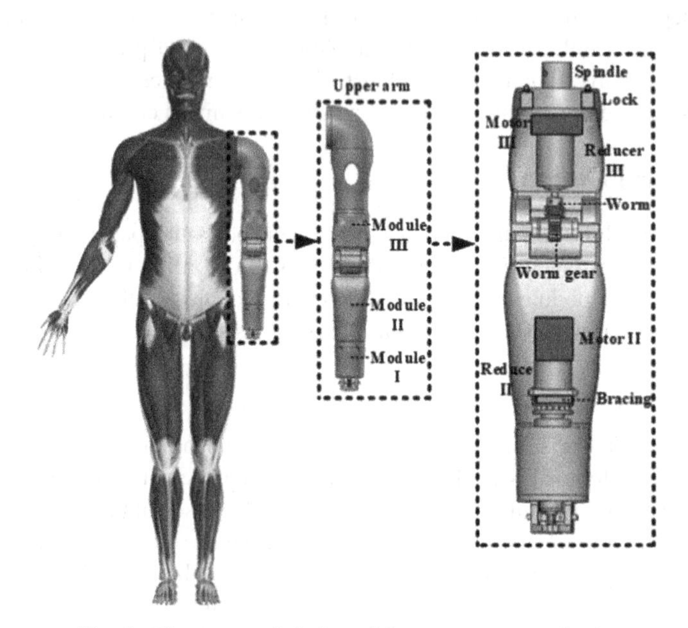

Fig. 3. The structural design of the upper arm prosthesis.

On the basis of the forearm prosthesis, through the splicing of the module I and the module II, the internal and external rotation function of the wrist joint is realized, and the power transmission is composed of the Maxon motor II and the reducer II. In the internal and external rotation of the wrist, self-locking is realized by the lock of the module I, and deep groove ball bearings, plane thrust bearings and bracing components are used at the same time. Bracing has three functions: (1) Connect to the module I; (2) Gravity resist and reduce the axial pressure of the module I and the end equipment on the reducer; (3) Connect to the module II.

The assembly of the module II and the module III forms the flexion and extension function of the elbow joint. The power transmission of elbow joint flexion and extension is composed of Maxon motor III, reducer III and turbine worm, and the joint self-locking is realized through the worm gear and worm, which is similar to the flexion and extension structure of the wrist joint. The spindle and lock in the upper arm module III are responsible for assembling and fixing with the socket or the upper-level module.

2.3 Full Arm Prosthesis

Figure 4 shows the structural design of the full arm prosthesis, which is suitable for two amputation sites, shoulder disarticulation and forequarter amputation. The upper arm prosthesis is mainly composed of module I, module II, module III and module IV, which together constitute five-degree-of-freedom, namely flexion and extension of the wrist join, internal and external rotation of the wrist joint, flexion and extension of elbow joint, internal and external rotation of shoulder joint and flexion and extension of shoulder joint.

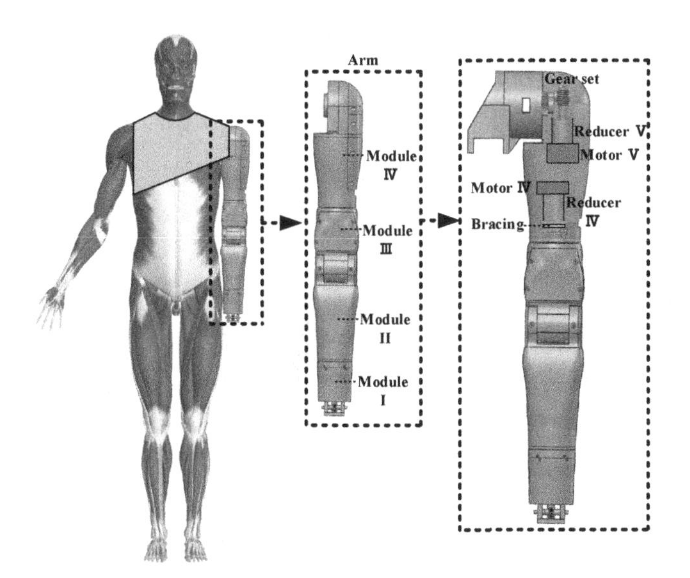

Fig. 4. The structural design of the full arm prosthesis.

On the basis of the upper arm prosthesis, the internal and external rotation of the shoulder joint is realized through the splicing of the module III and the module IV. The power transmission is composed of the Maxon motor IV and the reducer IV. The realization of internal and external rotation is similar to the wrist joint.

The assembly of the module IV and the superior structure (socket, etc.) forms the function of shoulder's flexion and extension. The structure utilized in flexion and extension of the shoulder is similar to the other joints' flexion and extension, both of which realize self-locking through worm gears. The difference is that another gear set is added to the power transmission of shoulder's flexion and extension.

3 Motion Simulation and Control

3.1 Prosthesis Modeling

Figure 5 shows the establishment of the coordinate system of each joint of the full arm prosthesis, coordinate system 0 is the base coordinate system, and coordinate system 6 is the coordinate system of the terminal device. The full arm prosthesis modeling uses modified Denavit–Hartenberg (M-DH) parameters. The specific parameters are shown in Table 1. There are four main parameters in M-DH, namely link length a_i, link twist α_i, link offset d_i and joint angle θ_i.

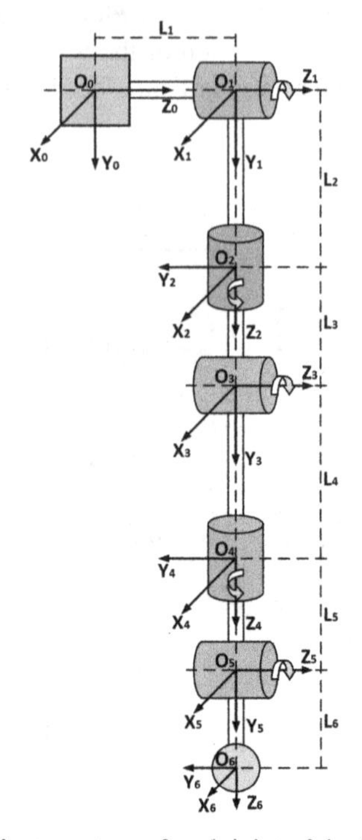

Fig. 5. The coordinate system of each joint of the full arm prosthesis.

Table 1. The M-DH parameters of the full arm prosthesis

i	θ_i	d_i	a_{i-1}	α_{i-1}
1	θ_1	L_1	0	0
2	θ_2	$L_2 + L_3$	0	$-\pi/2$
3	θ_3	0	0	$\pi/2$
4	θ_4	$L_4 + L_5$	0	$-\pi/2$
5	θ_5	0	0	$\pi/2$
6	θ_6	L_6	0	$-\pi/2$

3.2 Motion Simulation

Figure 6 shows the motion simulation of the full arm prosthesis, where (a) is the motion simulation of the reachable workspace of the full arm prosthesis, and (b) is the path planning of the full arm prosthesis. Table 2 shows the comparison between the range of motion of the arm prosthesis designed in this paper and the corresponding human arm. It can be seen that the range of motion of the arm prosthesis is relatively wide. Combined with the simulation result displayed in Fig. 6 and Table 2, it is shown that the function of the full arm prosthesis meets the needs of daily life (Grab, drink, etc.) and can help amputees to restore part of the arm function.

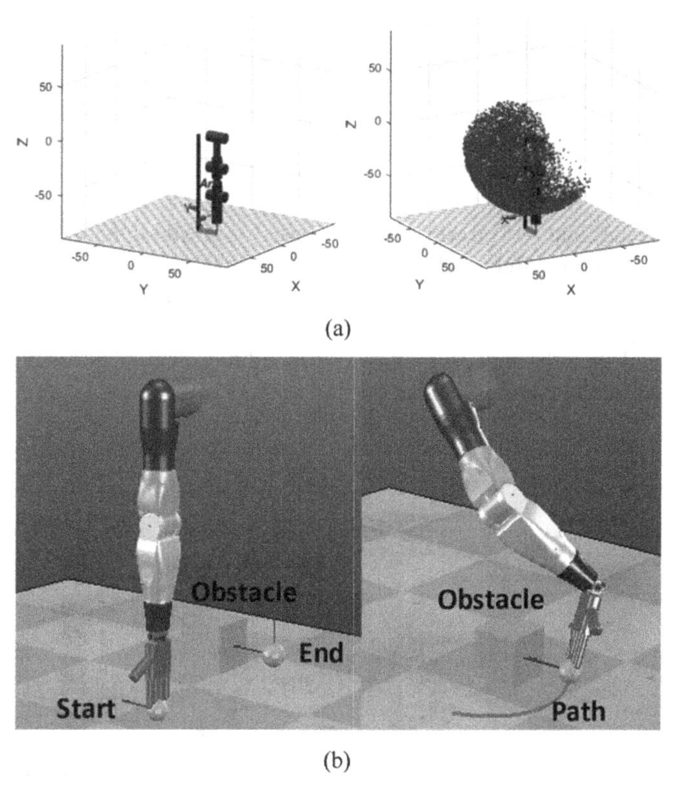

Fig. 6. Full arm prosthesis motion range and kinematics simulation. (a) Modeling and reachable workspace. (b) Kinematics and path planning.

Table 2. The comparison of the motion range between the arm prosthesis and human arm

Name		Human arm [16]	Full arm prosthesis
Wrist joint	Flexion (deg)	50–60	90
	Extension (deg)	35–60	30
	Internal rotation (deg)	80–90	180
	External rotation (deg)	80–90	180

(continued)

Table 2. (*continued*)

Name		Human arm [16]	Full arm prosthesis
Elbow joint	Flexion (deg)	135–150	135
	Extension (deg)	0–10	45
Shoulder joint	Flexion (deg)	70–90	90
	Extension (deg)	40	45
	Internal rotation (deg)	70	180
	External rotation (deg)	70	180

3.3 Motion Control Test

Figure 7 shows the control of an arm prosthesis based on sEMG signals. The control of the arm prosthesis system mainly includes the EMG acquisition unit, the signal processing and the motion control unit. In the arm prosthesis motion control test, the sEMG signal [17] is used. By decoding the sEMG signals, the motion information of the arm joints is obtained, and then the motion of the corresponding joints of the arm prosthesis is controlled. The actual control flow is:

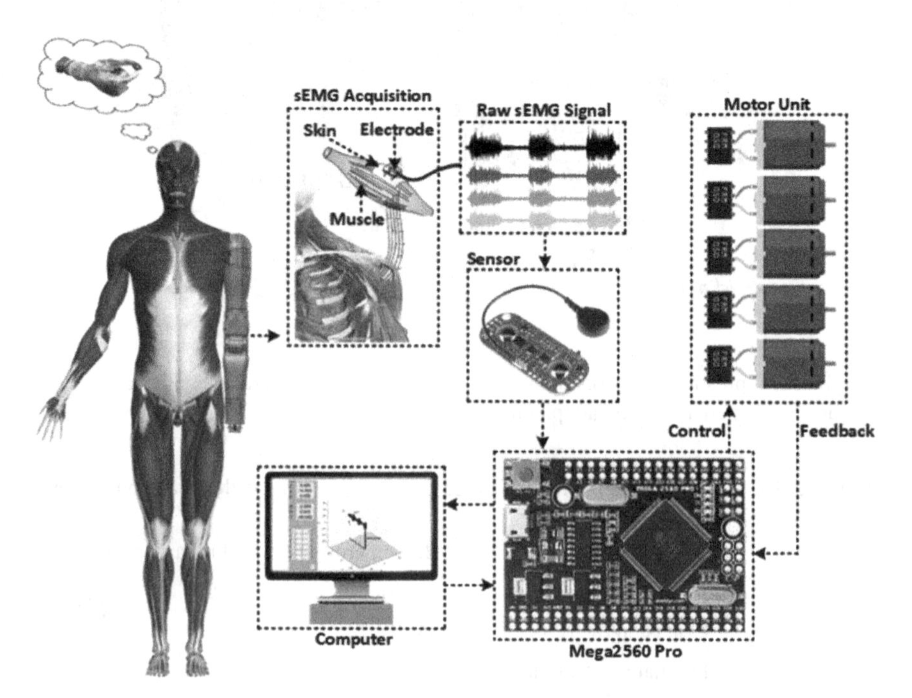

Fig. 7. A modular bionic arm prosthesis system controlled by the sEMG.

1) First, the sEMG signal is acquired through the EMG acquisition sensor MyoWare, and is transmitted to the computer through the Mega2560 Pro.
2) Next, the sEMG signal is decoded by the computer, and the computer will transmits the control signal to the Mega2560 Pro.
3) Then, the Mega2560 Pro will control the motion of the motor unit according to the control signal.
4) Last, the motion state (angle, etc.) of the motor unit will be fed back to the Mega2560 Pro to form a closed-loop control.

The arm prosthesis designed in this article has a total length of 53cm and a weight of about 3kg (including control circuit, battery, etc.). Figure 8 shows the control test of the arm prosthesis, which uses the sEMG signal to control the flexion and extension of the elbow joint. By collecting the sEMG signal of the two channels of the biceps and triceps, the motion intention of the elbow joint is obtained, and the arm prosthesis is controlled to complete the corresponding actions, realizing human-machine coordinated control. In the experimental test, the end device of the arm prosthesis connection is a prosthetic hand independently developed by the laboratory.

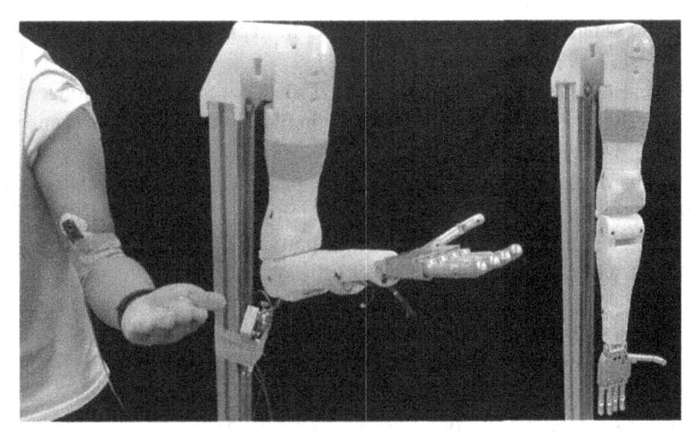

Fig. 8. Elbow flexion and extension control test of arm prosthesis.

4 Conclusion

This paper designs a modular arm prosthesis with five-degree-of-freedom to meet the needs of different arm amputation types for the prosthesis. By simulating and analyzing the kinematics of the arm prosthesis, results showed that the range of motion of the prosthesis is relatively wide and can meet the use of daily life. The arm prosthesis is fabricated by 3D printing technology. Besides, a control test of the modular arm prosthesis was conducted by the sEMG based pattern recognition control. The work of this study provides an effective modular bionic arm prosthesis structure that can restore part of limb functions for amputees with different amputation types and improve their activities of daily living.

Acknowledgments. This work was supported in part by the Key-Area Research and Development Program of Guangdong Province (#2020B0909020004), the National Key Research & Development Program of China (2020YFC2007901, 2020YFC2007905), and the Shenzhen Science and Technology Program (#CJGJZD20200617103002006).

References

1. Magee, R.: Amputation through the ages: the oldest major surgical operation. Aust. N. Z. J. Surg. **68**(9), 675–678 (1998)
2. Olaolorun, D.A.: Amputations in general practice. Niger. Postgrad. Med. J. **8**(3), 133–135 (2001)
3. Cordella, F., et al.: Literature review on needs of upper limb prosthesis users. Front. Neurosci. **10**, 209 (2016)
4. Brack, R., Amalu, E.H.: A review of technology, materials and R&D challenges of upper limb prosthesis for improved user suitability. J. Orthop. **23**, 88–96 (2021)
5. Van Der Niet Otr, O., Reinders-Messelink, H.A., Bongers, R.M., Bouwsema, H., Van Der Sluis, C.K.: The i-LIMB hand and the DMC plus hand compared: a case report. Prosthet. Orthot. Int. **34**(2), 216–220 (2010)
6. Trent, L., et al.: A narrative review: current upper limb prosthetic options and design. Disabil. Rehabili. Assist. Technol. **15**, 604–613 (2019)
7. Huinink, L.H., Bouwsema, H., Plettenburg, D.H., Van der Sluis, C.K., Bongers, R.M.: Learning to use a body-powered prosthesis: changes in functionality and kinematics. J. Neuroeng. Rehabil. **13**(1), 1–12 (2016)
8. Hussain, S., Shams, S., Khan, S. J.: Impact of medical advancement: prosthesis. In: Computer Architecture in Industrial, Biomechanical and Biomedical Engineering, p. 9 (2019)
9. Zhuang, K.Z., et al.: Shared human–robot proportional control of a dexterous myoelectric prosthesis. Nat. Mach. Intell. **1**(9), 400–411 (2019)
10. Biddiss, E.A., Chau, T.T.: Upper limb prosthesis use and abandonment: a survey of the last 25 years. Prosthet. Orthot. Int. **31**(3), 236–257 (2007)
11. Plagenhoef, S., Evans, F.G., Abdelnour, T.: Anatomical data for analyzing human motion. Res. Q. Exerc. Sport **54**(2), 169–178 (1983)
12. Bloomer, C., Kontson, K.L.: Comparison of DEKA Arm and body-powered upper limb prosthesis joint kinematics. Arch. Rehabil. Res. Clin. Transl. **2**(3), 100057 (2020)
13. Bridges, M., et al.: Revolutionizing prosthetics 2009: Dexterous control of an upper-limb neuroprosthesis. Johns Hopkins APL Tech. Digest (Appl. Phys. Lab.) **28**(3), 210–211 (2010)
14. Resnik, L., Klinger, S.L., Etter, K.: The DEKA Arm: Its features, functionality, and evolution during the Veterans Affairs Study to optimize the DEKA Arm. Prosthet. Orthot. Int. **38**(6), 492–504 (2014)
15. Meng, Q., Li, Z., Li, J.: An intelligent upper limb prosthesis with crossmodal integration and recognition. In: 2021 6th IEEE International Conference on Advanced Robotics and Mechatronics (ICARM), pp. 747–752. IEEE (2021)
16. Perry, J.C., Rosen, J., Burns, S.: Upper-limb powered exoskeleton design. IEEE/ASME Trans. Mechatron. **12**(4), 408–417 (2007)
17. Gupta, R., Agarwal, R.: Single muscle surface EMGs locomotion identification module for prosthesis control. Neurophysiology **51**(3), 191–208 (2019)

sEMG-Based Estimation of Human Arm Endpoint Stiffness Using Long Short-Term Memory Neural Networks and Autoencoders

Yanan Ma, Quan Liu, Haojie Liu, and Wei Meng(✉)

School of Information Engineering, Wuhan University of Technology, Wuhan 430070, China
{ma_yanan,quanliu,lhj1989,weimeng}@whut.edu.cn

Abstract. Human upper limb impedance parameters are important in the smooth contact between stroke patients and the upper limb rehabilitation robot. Surface electromyography (sEMG) reflects the activation state of muscle and the movement intention of human body. It can be used to estimate the dynamic parameters of human body. In this study, we propose an estimation model combining long short-term memory (LSTM) neural network and autoencoders to estimate the endpoint stiffness of human arm from sEMG and elbow angle. The sEMG signal is a time varying nonlinear signal. Extracting key features is critical for fitting models. As an unsupervised neural network, autoencoders can select the proper features of sEMG for the estimation. LSTM neural network has good performance in dealing with time series problems. Through a 4-layer LSTM neural network, the mapping relationship between sEMG features and endpoint stiffness is constructed. To prove the superiority of the proposed model, the correlation coefficient between theoretical stiffness calculated by Cartesian impedance model and estimated stiffness and root mean square error (RMSE) is used as the evaluation standard. Compared with two other common models by experiments, the proposed model has better performance on root mean square error and correlation coefficient. The root mean square error and correlation coefficient of proposed model are 0.9621 and 1.732.

Keywords: Endpoint stiffness estimation · Surface electromyography · Long short-term memory neural networks · Autoencoders

1 Introduction

Stroke is a disorder of cerebral blood circulation. According to the pathogenesis, it is mainly divided into hemorrhagic stroke and ischemic stroke [1]. Stroke usually causes varying degrees of sensory and motor dysfunction. Among to stroke survivors, varying degrees of upper extremity motors function impairment is present in approximately 85% of patients. The neural remodeling theory states that patients can achieve better functional recovery from a rehabilitative training modality of motor relearning proactively and actively [2].

Surface electromyography (sEMG) reflects the activation state of muscle and the movement intention of human body, which has the advantage of noninvasive and is

H. Liu et al. (Eds.): ICIRA 2022, LNAI 13456, pp. 699–710, 2022.
https://doi.org/10.1007/978-3-031-13822-5_63

directly related to the expected behavior of human body. It is widely used in recognition of human motion information [3–5]. For motion recognition based on sEMG, the mapping between sEMG and motion information can be constructed by skeletal muscle model, neural network and high-order polynomial fitting [6–8]. The recognized motion information can be divided into discrete classification (such as motion state classification, gesture recognition, etc.) and continuous prediction (such as force prediction, joint angle prediction, joint torque, etc.) [9–11]. Compared with discrete classification, continuous motion information has more advantages in realizing the flexible control of the robot and improving the effect of human-robot interaction [12].

In previous studies, a Hill-based muscle model is a classical method to estimate human dynamic information. In this model, sEMG features are extracted to represent muscle activation. The Hill-type model was applied to calculate muscle contractility combined with physiological information [13]. According to human kinematics model and the size of muscle and bone, the human arm dynamics information can be calculated. Tao et al. proposed a method to predict muscle strength by combining Hill model and sEMG signal of upper limb muscle group [14]. Based on a musculoskeletal model derived from the Hill-type model, Liu et al. incorporate sEMG signal with the joint angle by using a regression method to estimate the joint torque [15]. However, the structure of the constructed skeletal models is complex. It is difficult to measure the physiological parameters of the human body during moving, limiting the use of skeletal muscle models.

Compared with musculoskeletal model with complex physiological parameters, it is concise and intelligible to use the regression method to construct the mapping relationship between sEMG and human arm endpoint stiffness. Traditional regression methods often use polynomial regression models. Hosoda proposes a polynomial regression model to construct the mapping relationship between sEMG of four muscles and the elbow joint torque [16]. In recent years, more and more researchers build regression models through neural networks. Jail et al. map sEMG signal to the motor torque used to control the movement of upper limb rehabilitation equipment through an artificial neural network (ANN) [17]. Chai et al. propose an efficient closed-loop model with sEMG signal as input for estimating joint angles and angular velocities of human upper limbs with joint damping. The model is mainly divided into two parts including a LSTM network and a Zero Neural Network (ZNN) with discrete-time algorithm [18]. In this model, the LSTM network part is an open-loop model, which is used to construct the mapping relationship between sEMG signal and joint motion intention. The ZNN network part is a closed-loop model. As a supplement to the LSTM network, it is used to reduce the prediction error of the open-loop model. In the above methods, the sEMG signal usually needs to be processed, which needs to be windowed to reduce the influence of nonlinearity and time variation [19]. The choice of window size is achieved empirically. It is important to extract reasonable sEMG features for the input of neural networks to estimate human dynamics parameters.

In this study, a neural network model integrating LSTM and autoencoders is proposed to estimate the endpoint stiffness of the human arm. First, we analyze the interactive force information through the torque sensor of the robot end effector, and collect the displacement, velocity and acceleration during the interaction process through the robot kinematics information. sEMG collection system provided the elbow angle and multi-channel sEMG signals. Then, we preprocess the sEMG signal and extract preliminary features (MAV feature) as the input set of autoencoders. After that, the autoencoders automatically extracts valuable features through unsupervised learning which will serve as the input of the LSTM neural network. Finally, the mapping of sEMG signal and elbow angle to human arm endpoint stiffness is constructed through LSTM neural network. The theoretical true value used to train and evaluate the network is calculated by the Cartesian impedance model. The innovation of this study is that we combine the advantages of autoencoders and LSTM to propose a new network model to improve the accuracy of estimating human arm endpoint stiffness from sEMG signals and elbow angle.

2 Materials and Methods

2.1 Data Acquisition

In this study, some healthy testers (two male and one female) have been invited to perform the required trajectory with the robot. We use the signal acquisition equipment (*Delsys, Trigno™, USA*) to record four channel sEMG signals of human arm, which are biceps humeri, triceps humeri, flexor carpi radialis and extensor carpi radialis. In addition, the elbow angle is measured by angle sensor (Delsys, Trigno™, USA). When collecting EMG signals, it is necessary to ensure that the skin area attached to the sensor is smooth and clean. The position of each sensor is shown in Fig. 1.

The theoretical true value of the endpoint stiffness is calculated by the Cartesian impedance model. The position of the end effector is recorded through the manipulator. The interactive force is recorded through the six-dimensional torque sensor of the end effector.

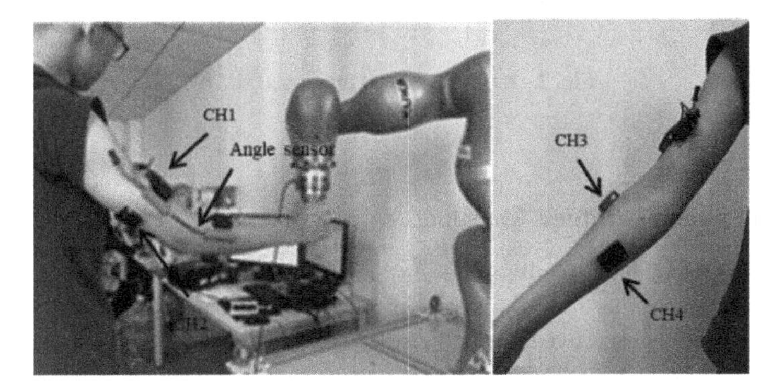

Fig. 1. Data acquisition

2.2 sEMG Processing

The original sEMG signals are usually mixed with noise. The power density spectrum of valid sEMG is concentrated in 20 Hz–500 Hz. Therefore, the original sEMG signals should be filtered through a 20 Hz to 500 Hz band-pass filter. After preprocessing sEMG signals, the amplitude variation can be more clearly observed by full-wave rectification. This process can be expressed as Eq. (1).

$$sEMG_r(n) = |sEMG(n)| \tag{1}$$

where $sEMG(n)$ is the n-th amplitude of the sEMG signals after preprocessing; $sEMG_r(n)$ is the n-th amplitude sample of the sEMG signals obtained after full-wave rectification. Then, we used Eq. (2) to extract the mean absolute value feature.

$$sEMG_s(n) = \frac{1}{N} \sum_{i=nN-N+1}^{nN} sEMG_r(r) \tag{2}$$

where N is the number of times that signals need to be sub-sampled; $sEMG_s(n)$ is the sEMG signals after. Through the above steps, and the preprocessed sEMG signals of four selected muscles are illustrated in Fig. 2.

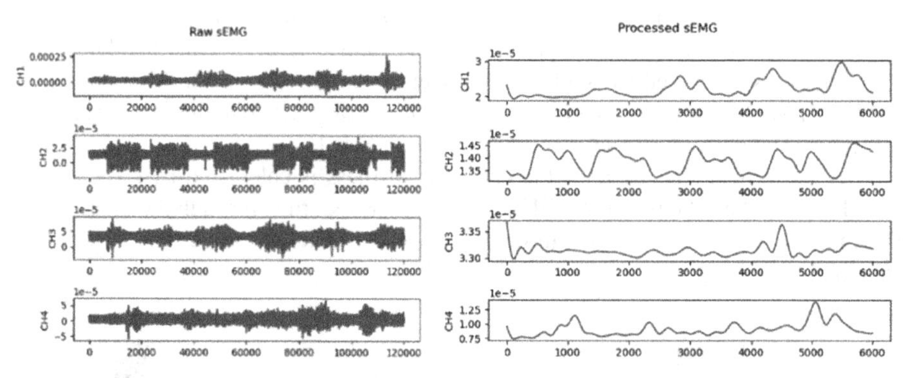

Fig. 2. Raw sEMG and processed sEMG

2.3 Arm Endpoint Stiffness Modeling

Since it is impractical to install torque sensors in all interactive environments, the end stiffness of human upper limbs is inconvenience to be measure directly. The Cartesian impedance model is used to calculate the stiffness, which can be expressed as Eq. (3).

$$M_e\ddot{L} + B_e\dot{L} + K_e(L - L_u) = F \tag{3}$$

where $M_e = diag(M, M, M)$, $B_e = diag(B_x, B_y, B_Z)$, $K_e = diag(K_x, K_y, K_z) \in R^{3*3}$ represented the inertia, damping, stiffness matrices respectively; L, L_u, $F \in R^{3*3}$ represented the real position of endpoint, desire position of the endpoint and force vectors respectively. Equation (3) can be expressed as Eq. (4) –Eq. (8).

$$Ax = b \tag{4}$$

$$x = [M \quad B_x \quad K_x \quad B_y \quad K_y \quad B_z \quad K_z]^T \tag{5}$$

$$b = [\,F_x\;F_y\;F_z\,]^T \tag{6}$$

$$A = \begin{bmatrix} \ddot{X}\;\dot{X}\;X\;0\;0\;0\;0 \\ \ddot{Y}\;0\;\;0\;\dot{Y}\;Y\;0\;0 \\ \ddot{Z}\;0\;\;0\;0\;0\;\dot{Z}\;Z \end{bmatrix} \tag{7}$$

$$x = A^{-1}b \tag{8}$$

According to Eq. (4)–Eq. (8), the endpoint stiffness can be obtained through the values of F and L. In Eq. (8), the traditional unconstrained methods such as the least square method are used to solve the impedance characteristic parameters of the upper limb, which cannot guarantee the non-negative conditions of each variable in the stiffness matrix. There is a certain deviation between the calculated data and the actual situation. The gradient descent method is introduced to solve this problem, which can be expressed as

$$min\|Ax - b\|_2^2 \tag{9}$$

where x is restricted to nonnegative. Figure 3 presents the gradient descent method to solve the optimal solution of Eq. (9), which is used to project the points that do not belong to the constraint range to the point closest to the optimal solution within the constraint range through Euclidean projection. The negative numbers in the impedance characteristic parameters M, B and K are mapped to random numbers between 0 and 1. x is constructed with the processed M, B and K. The iterative operation is continued until the conditions are met. Finally, the impedance characteristic parameters of the end of the human upper limb are obtained.

2.4 Feature Extraction Based on Autoencoders

It is necessary to select suitable features for fitting the model. For this problem, we choose an autoencoder with unsupervised learning capability, which can automatically select representative features from the input data. For this problem, a neural network with unsupervised learning capability can be used for feature extraction. In the study, autoencoders is selected. The structure of the autoencoder can be roughly divided into three layers (see Fig. 5). The layers have the following relationships: the output layer of the autoencoder has the same number of neurons as the output layer. Autoencoder is a neural network that uses backpropagation to make the output value equal to the input value, which an autoencoder consists of two parts: an encoder and a decoder. These two parts can be represented by Eq. (10)–Eq. (11)

$$h_1 = \sigma_e(W_1x + b_1) \tag{10}$$

$$y = W_2h_1 + b_2 \tag{11}$$

where h_1 is the encoding value; x is the raw data; y is the decoding value, W_1, b_1, W_2 and b_2 are the weights value and bias value; σ_e is a nonlinear activation function.

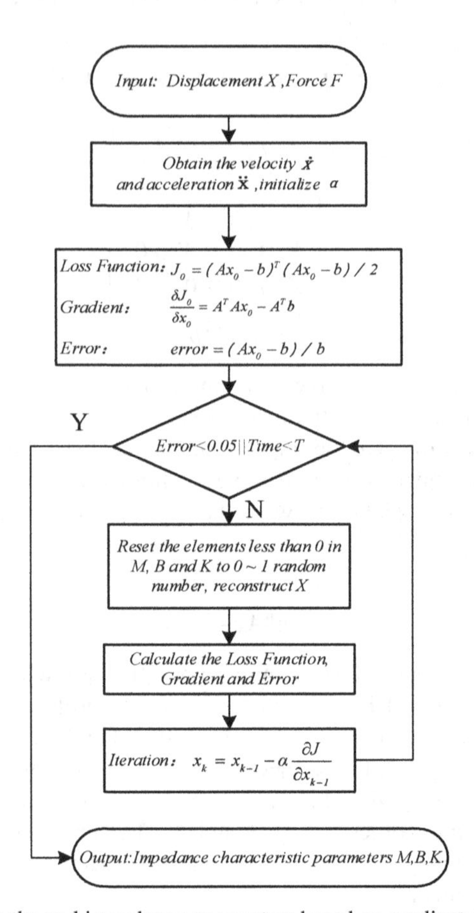

Fig. 3. Obtain the end impedance parameters based on gradient decline method

2.5 Fitting Endpoint Stiffness Based on LSTM Model

To estimate the stiffness from the input data, we construct the relationship between the endpoint stiffness and the output data from the autoencoder model through a LSTM neural network.

In this study, the feature of sEMG signals and the theoretical true value of the endpoint stiffness are supposed as follows:

$$\begin{cases} EMG_i = \left[EMG_{i,1}, EMG_{i,2}, \cdots, EMG_{i,t}\right] \; t = 1, \cdots, j \\ K_i = \left[K_{i,x}, K_{i,y}, K_{i,z}\right] \end{cases} \tag{12}$$

$$K_i = G(EMG_{i,1}, EMG_{i,2}, \cdots, EMG_{i,t}, Angle_i) \tag{13}$$

where EMG_i is the feature of EMG in i-th time, t is the t-th characteristics. K_i is the endpoint stiffness. $Angle_i$ is the angle of elbow. G represents the relationship between EMG_i and K_i. The internal state in the LSTM network is stored in a special unit, which is different from the unit in a standard RNN. The structure of LSTM unit is demonstrated in

Fig. 4. The core structure that distinguishes the RNN unit is three special gate structures. For the forget gate, the process is as follows, where δ is the sigmoid processing; W_F and V_F are weight matrix and projection of input.

$$F_i = \delta(W_F h_{i-1} + V_F a_i) \tag{14}$$

The output data of the cell at the last time enters the forget gate together with the input at the current time. After being processed by the forget gate, it outputs F_i, whose value are [0, 1], which is positively correlated with the degree of importance.

For the input gate, we need to determine which information to use to update the unit state. The process is as follows:

$$I_i = \delta(W_I h_{i-1} + V_I a_i) \tag{15}$$

$$\hat{C}_i = tanh(W_C h_{i-1} + V_C a_i) \tag{16}$$

$$C_i = F_i C_{i-1} + I_i \hat{C}_i \tag{17}$$

where W_I is the weight matrix in the input gate; V_I is the projection matrix in the input gate; $\hat{C}_i$ can be seen as information brought by new input; W_C and V_C can be seen as weights of input respectively; Old unit information C_{i-1} is updated to the new cell information C_i.The final output of the LSTM unit is obtained as follows:

$$O_i = \delta(W_O h_{i-1} + V_O a_i) \tag{18}$$

$$h_i = O_i * tanh(C_i) \tag{19}$$

where W_O is the weight matrix in the output gate; V_O is the projection in the output gate. The dimension of the input layer is 5, which represents the sEMG signals and elbow angle. The dimension of the output layer is 3, which represents the stiffness of human arm endpoint in Cartesian coordinate system.

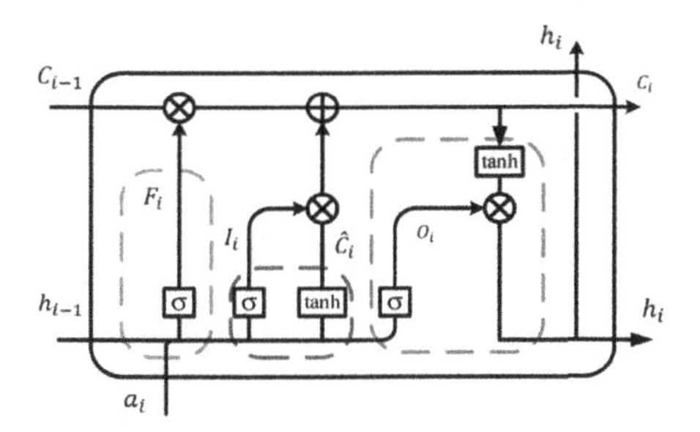

Fig. 4. The structure of LSTM unit.

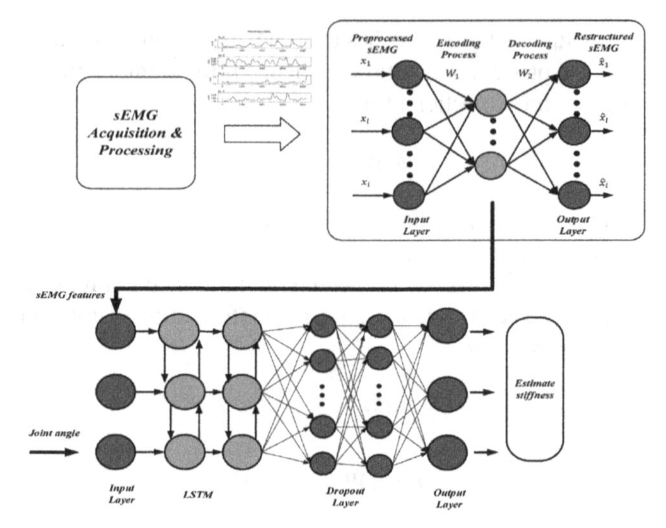

Fig. 5. The structure of whole estimated model

2.6 Performance Evaluation and Statistical Analysis

To compare and verify the effect of the constructed model, two evaluation criteria are introduced to evaluate the proposed model:

$$RMSE = \sqrt{\frac{1}{N}\sum_{k=1}^{N}(x_k - \hat{x}_k)^2} \tag{20}$$

$$R^2 = 1 - \frac{\sum_{k=1}^{n}(x_k - \hat{x}_k)^2}{\sum_{k=1}^{n}(x_k - \overline{x}_k)^2} \tag{21}$$

where *RMSE* is the root mean square error; R^2 is the coefficient of determination; x_k is the theoretical value; $\hat{x}_k$ is the estimated value; N is the total number of samples.

3 Experiments and Results Discussion

3.1 Measurement of Human Arm Endpoint Stiffness

In this study, experimental participants are asked to drag the end effector of the mechanical arm to make circular motion. The default trajectory of the mechanical arm is a circle with a diameter of 20 cm. Set fixed impedance parameters for the mechanical arm to ensure that the tester can drag the end effector to deviate from the predetermined trajectory. Each person collects six groups, and each group makes six circular motions.

First, we analyzed the endpoint stiffness of the human arm by the Cartesian impedance model. According to Eq. (3), the end point stiffness $\{K_x, K_y, K_z\}$ can be calculated by the position $\{L, L_u\}$ and force signal $\{F_x, F_y, F_z\}$. Because we set the appropriate impedance, the motion trajectory can be offset within a certain range according to the wishes of the tester. Displacement is divided into desired value and real value. Figure 6 show the trajectory of the experience, from which it can be seen that the trajectory is a circle with a diameter of 20 cm and allowed to change in a certain range. According to Eq. (9), we use gradient descent method to deal with outliers. Figure 6 show the result of the endpoint stiffness, from which we can see that the stiffness is restricted to nonnegative.

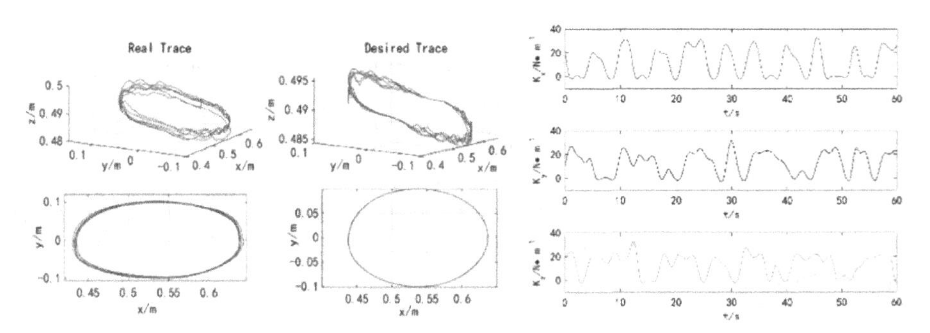

Fig. 6. The trajectory of end effector and the result of the endpoint stiffness

3.2 Network Structure and Results

For each tester, we take six groups of experimental data, which were divided into a training set with four experimental data and a testing set with two experimental data. The sampling frequency of force signal and displacement signal is 20 Hz. The sampling frequency of EMG signal is 2000 Hz. When extracting *MAV* features, the smaller the window length is, the more original information is retained. Here, we use the window length of 20. The input is a 4-channel signal. Therefore, the dimension of the input data was 20, the number of input layer and output layer neurons of the autoencoder was set to 20. To realize the function of feature extraction, the number of neurons in the hidden layer of the autoencoder is less than that in the input layer. The more cell in the hidden layer, the input and output signals of the autoencoder may be more similar. But it may be not the best for the feature extraction. Therefore, seven autoencoders were tested with 8 to 20 hidden layer cell. We increase two cell in each group. The values of *RMSE* for different autoencoders are show in Fig. 7, which connect with the same LSTM network. According to the result, the structure with 16 hidden layer cell was determined.

Processed by the autoencoder, sEMG can be expressed by the value of hidden layers h_1, which is used as the input data of the LSTM network to estimate the endpoint stiffness.In the LSTM network, h_1 and angle of the elbow are the input data to estimate more accurate endpoint stiffness. The number of LSTM layers should not be too many. The increase of the number of layers will lead to the exponential growth of time overhead and memory overhead. Then, the gradient between layers will gradually disappear. Therefore, the number of LSTM layers is four. To avoid over-fitting, we added a dropout layer.

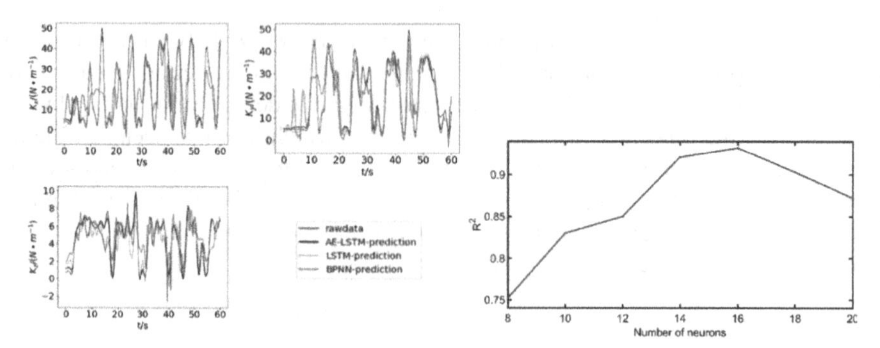

Fig. 7. The comparison of different models and the R^2 for different autoencoders

The Model_1 is our proposed model. The Model_2 just uses the LSTM to estimate. The Model_3 uses the BPNN to estimate. To demonstrate the effectiveness and superiority of the proposed model, R^2 and *RMSE* are calculated according to Eq. (20) –Eq. (21). The Fig. 7 indicates the result of different models. Table 1 lists the experimental results for all participants.

3.3 Results Discussion

Compared the predicted results of different model in Fig. 7, the estimation performance is improved, which appeared that the predicted value was closer to the actual value by the AE-LSTM model. AE-LSTM model was closer than LSTM models, especially the value of K_z during 27–32 s. This is because that autoencoder compresses the sEMG to extract the most representative information in the data, and reduces the dimension of the input information without losing important features. Compared with the predicted value of BPNN, the predicted value of LSTM will not deviate seriously from the theoretical value. This is because LSTM uses the information of the previous time slice to calculate the content of the current time slice, while the output of the hidden node of the traditional model only depends on the input characteristics of the current time slice. Considering the performance of models (Model_1, Model_2 and Model_3) in estimating endpoint stiffness, coefficient of determination derived by model Mode1_1 improve by 6% and 11% compared to those derived by models Model_2, and Model_3.

Table 1. Result of different models & testers

Model name&&Tester	R^2	*RMSE*
Model_1&&Tester1	0.9621	1.732
Model_1&&Tester2	0.9349	2.106
Model_1&&Tester3	0.9486	1.923

(*continued*)

Table 1. (*continued*)

Model name&&Tester	R^2	*RMSE*
Model_2&&Tester1	0.9131	1.979
Model_2&&Tester2	0.8935	3.479
Model_2&&Tester3	0.8762	3.746
Model_3&&Tester1	0.8511	4.124
Model_3&&Tester2	0.8741	3.871
Model_3&&Tester3	0.8544	3.785

4 Conclusion

Human arm endpoint stiffness is of great significance for realizing the robot compliance control in rehabilitation training. The proposed model estimates human endpoint stiffness from sEMG signals and elbow angle. The autoencoder and LSTM were used to train the estimating model. The proposed model was evaluated through experiments by comparing it to two other models. The experimental results show that the addition of the autoencoder makes the estimated stiffness value closer to the theoretical value. The autoencoder extracts effective features from sEMG signal. In future work, we will use the estimated stiffness value to adjust the control of the rehabilitation robot.

Acknowledgments. This project is supported by National Natural Science Foundation of China (Grant 52075398), Application Foundation Frontier Project of Wuhan Science and Technology Program (2020020601012220), and Research Project of Wuhan University of Technology Chongqing Research Institute (YF2021-17).

References

1. Chen, X.: Successive development of ischemic stroke and hemorrhagic stroke in a patient with essential thrombocythemia: a case report. J. Int. Med. Res., 1377–1380 (2021)
2. Fan, C.Y.: Research progress on the mechanism of early rehabilitation training promoting nerve remodeling in stroke. Chin. J. Rehabil. Med., 1377–1380 (2020)
3. Xu, L.: Gesture recognition using dual-stream CNN based on fusion of sEMG energy kernel phase portrait and IMU amplitude image. Biomed. Signal Process. Control **73**, 103364 (2022)
4. Karnam, N.K.: Classification of sEMG signals of hand gestures based on energy features. Biomed. Signal Process. Control **70**, 102948 (2021)
5. Cheng, J.: Position-independent gesture recognition using sEMG signals via canonical correlation analysis. Comput. Biol. Med. **103**, 44–54 (2018)
6. Baumgartner, C.: Using sEMG to identify seizure semiology of motor seizures. Seizure **86**, 52–59 (2021)
7. Shen, X.: Modeling and sensor less force control of novel tendon-sheath artificial muscle based on hill muscle model. Mechatronics **62**, 102243 (2019)
8. Asefi, M.: Dynamic modeling of SEMG–force relation in the presence of muscle fatigue during isometric contractions. Biomed. Signal Process. Control **28**, 41–49 (2016)

9. Zhang, Q.: Simultaneous estimation of joint angle and interaction force towards sEMG-driven human-robot interaction during constrained tasks. Neurocomputing (2021)
10. Nurhanim, K.: Joint torque estimation model of surface electromyography (sEMG) based on swarm intelligence algorithm for robotic assistive device. Procedia Comput. Sci. **42**, 175–182 (2014)
11. Aung, Y.M.: sEMG based ANN for shoulder angle prediction. Procedia Eng. **41**, 1009–1015 (2012)
12. Huang, Y.: Joint torque estimation for the human arm from sEMG using backpropagation neural networks and autoencoders. Biomed. Signal Process. Control **62**, 102051 (2020)
13. Ao, D.: Movement performance of human–robot cooperation control based on EMG-driven hill-type and proportional models for an ankle power-assist exoskeleton robot. Neural Syst. Rehabil. Eng. **25**, 1125–1134 (2017)
14. Tao, Q.: Upper limb muscle strength prediction based on motion capture and sEMG data. In: 25th International Conference on Automation and Computing (ICAC), pp. 1–5 (2019)
15. Liu, H.: Human-robot cooperative control based on sEMG for the upper limb exoskeleton robot. Robot. Auton. Syst. **125**, 103350 (2019)
16. Hosoda, R.: Human elbow joint torque estimation during dynamic movements with moment arm compensation method. IFAC Proc. **47**, 12305–12310 (2014)
17. Jali, M.H., et al.: Joint torque estimation model of sEMG signal for arm rehabilitation device using artificial neural network techniques. In: Sulaiman, H.A., Othman, M.A., Othman, M.F.I., Rahim, Y.A., Pee, N.C. (eds.) Advanced Computer and Communication Engineering Technology. LNEE, vol. 315, pp. 671–682. Springer, Cham (2015). https://doi.org/10.1007/978-3-319-07674-4_63
18. Chai, Y.: A novel method based on long short term memory network and discrete-time zeroing neural algorithm for upper-limb continuous estimation using sEMG signals. Biomed. Signal Process. Control **67**, 102416 (2020)
19. Guo, S.: Comparison of sEMG-based feature extraction and motion classification methods for upper-limb movement. Sensors **15**, 9022–9038 (2015)

Error Related Potential Classification Using a 2-D Convolutional Neural Network

Yuxiang Gao[1], Tangfei Tao[1,2](✉), and Yaguang Jia[1]

[1] School of Mechanical Engineering, Xi'an Jiaotong University, Xi'an 710049, China
taotangfei@mail.xjtu.edu.cn

[2] Key Laboratory of Education Ministry for Modern Design and Rotor-Bearing System, Xi'an, China

Abstract. An error-related potentials (ErrP) is generated in the brain when human's expectations are inconsistent with actual results. The decoding of ErrP can improve the performance of brain-computer systems (BCI). In this paper, we propose an effective ErrP classification method using the proposed attention-based convolutional neural network (AT-CNN). Every 1D EEG signal is transformed into a 2D grayscale image as an input data for the model. In addition, we introduced label smoothing to mitigate the impact of label mismatching data. We evaluate and compare our method using the Monitoring Error-Related Potential dataset. The accuracy of our proposed method is 83.42%, the sensitivity is 69.02%, the specificity is 88.48% and these results outperform the state-of-the-art methods.

Keywords: Brain-computer interface (BCI) · Error-related potential (ErrP) · 2D grayscale image

1 Introduction

A Brain-Computer interface (BCI) is a system that does not rely on peripheral nerves and can convert EEG signals into external control instructions. It can help patients with limb movement difficulties to communicate with the outside world, as well as help stroke patients to restore motor function [1].

Electroencephalography (EEG) is a feasible and well-studied biological signal in brain computer interface (BCI) research and is widely used in BCI systems. It can measure weak potentials in the brain through electrodes placed on the scalp, with the advantages of being easy to capture, high time resolution, low cost and non-invasion [2].

However, because EEG signals are non-stationary and non-linear, decoders tend to misidentify the subject's intentions, which greatly limits the practical application of EEG-based BCIs [3].

An Error Related Potential (ErrP) is elicited when the subject's expectation does not match actual outcomes. An ErrP is generated in the anterior cingulate cortex (ACC) [4]. Its waveform's characteristics can be observed by averaging multiple trials, as shown in Fig. 1. There is a positive peak at around 200 ms after feedback presentation, followed by a large negative peak at about 250 ms, and a positive peak at around 320 ms [5].

H. Liu et al. (Eds.): ICIRA 2022, LNAI 13456, pp. 711–721, 2022.
https://doi.org/10.1007/978-3-031-13822-5_64

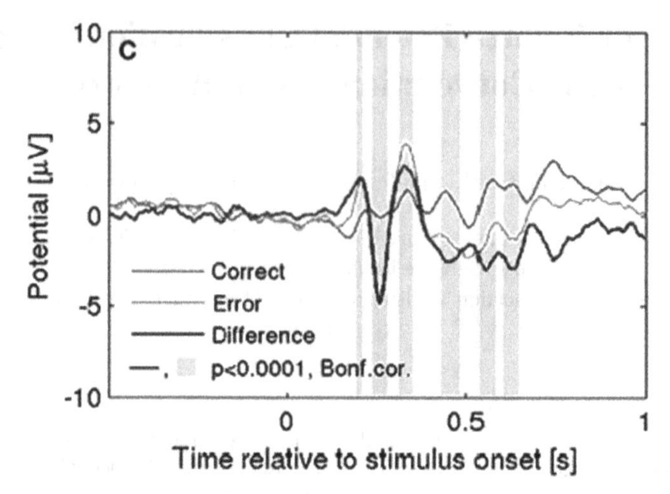

Fig. 1. Grand-average ERP at channel FCz for correct, error and difference (error minus correct) conditions. t = 0 corresponds to the stimulus presentation onset [5].

Studies have shown that ErrP is an inherent feedback mechanism of humans. An ErrP can combine with other BCIs to correct incorrect instructions of BCI and have the potential to improve the performance of BCI [6]. However, the existing ErrP decoding methods still have room for improvement. Recently, a great deal of researches has been done about the classification of ErrP signals.

Ricardo and Millan et al. [7] used the data of two channels (FCz and Cz) and input them into a Gaussian classifier for classification. Akshay Kumar et al. [8] applied a Linear Discriminant Analysis (LDA) classifier to classify correct and error trials through statistical mean features.

The above traditional machine learning methods have achieved certain achievements by manually selecting the time window, filter interval and channel of ErrP. However, due to the highly nonstationary nature of EEG, manual selection features may reduce the generalization ability of the algorithm. In recent years, more and more researchers have focused on deep learning methods that can automatically extract valid features.

Mayor Torres et al. [9] adopted GAN (generative adversarial network) for data augmentation to compensate for insufficient training data, and then trained DCNN (deep convolutional neural network) proposed by the authors. Swamy Bellary et al. [10] proposed two neural networks: ConvArch1 and ConvArch2. ConvArch2 was developed by adding batch normalization and drop out layers on the basis of ConvArch1, with shorter training time and better performance. Praveen K et al. [11] put forward a two-stage trained neural network classifier, which divides the neural network learning into two stages. The first stage aimed to capture the global features of ErrP, and the second stage finetune the model parameters to fit each subject.

The above methods are based on deep learning, which can automatically learn the internal characteristics of signals and achieve end-to-end classification. These approaches are generally superior to traditional machine learning methods and employ the great potential of deep learning in ErrP classification, but there is still room for improvement. At present, deep learning has entered the commercialization stage in image classification tasks. Some researchers converted 1 D EEG signals into 2-D images and then input them into 2-D CNN, which showed better performance than 1-D CNN [12, 13]. Similar to the electrocardiogram, ErrP can recognize the peak information on the feature period with eyes. If the 1D EEG signals are transformed into 2D EEG images and then classified by a specific convolutional neural network, it is expected to improve the ErrP classification accuracy.

In this paper, we propose an effective ErrP classification method using the proposed attention-based convolutional neural network (AT-CNN). The results of experiments demonstrate that our proposed method outperforms other state-of-the-art methods and achieve 83.42% accuracy, 66.17% sensitivity, and 88.48% specificity in the Monitoring Error-Related Potential dataset. The main contributions of this work can be outlined as follows:

1) To the best of our knowledge, this is the first attempt to apply 2D CNN to classify ErrP. The experimental results verify the effectiveness of our method.
2) We propose a deep neural network model named attention-based convolutional neural network (AT-CNN) to fully utilize the temporal and spatial characteristics of 2D EEG image.
3) In order to improve the generalization performance of the model, we use label smoothing instead of cross entropy as the loss function

The remainder of this paper is organized as follows: The proposed AT-CNN is introduced in Sect. 2. The dataset used in this paper is briefly described in Sect. 3. Section 4 contains experimental results and discussions. Finally, Sect. 5 presents the conclusions and future work.

2 The Proposed Method

2.1 Data Preprocessing

The raw EEG signal was spatially filtered using common average reference (CAR) and then filtered to [1–10] Hz using a third-order Butterworth band-pass filter. Because channel FCZ has the most significant ErrP feature, we chose only one channel ("FCZ") and crop the data of 0.2 s before and 0.8 s after stimulus presentation. After that, 1D EEG signal is transformed into a 2D EEG grayscale image with an image size of 224 × 224.

2.2 AT-CNN Architecture

The current ErrP classification method based on deep learning is to input 1D EEG signals into 1D-CNN for training. Because ErrP has large noise and less significant features under a single trail, the performance of 1D-CNN is not ideal. Since 2D convolution and pooling layers are more suitable for extracting temporal waveform changes and spatial locality of EEG images, we attempted to detect ErrP using 2D EEG images. The local small changes of EEG signals induced by ErrP can be extracted more effectively.

In order to fully extract the spatiotemporal features of the 2D EEG image, we propose the Attention-based Convolutional Neural Network (AT-CNN). A schematic diagram of the network structure is shown in Fig. 2. The first four convolutional modules of AT-CNN are used for extraction of feature in image, followed by a Convolutional Block Attention Module (CBAM) [14]. CBAM consists two independent sub-modules: channel attention module (CAM) and spatial attention module (SAM), which achieve feature fusion of channel dimension and spatial dimension respectively. The fully connected layers are followed by CBAM to obtain the final outcome. To reduce overfitting, Dropout is applied between fully connected layers. There are two benefits to adding CBAM. Firstly, CAM can help the model focus on the informative and important parts of 2D EEG image, which improve the detection performance. Second, SAM can help the model highlight the most informative regions in the 2D EEG image.

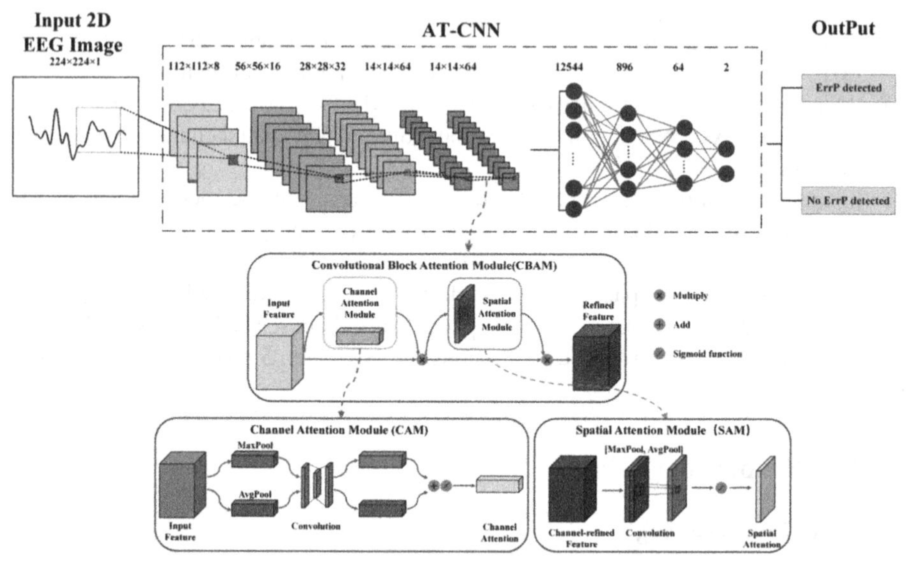

Fig. 2. The overview of proposed Attention-based Convolutional Neural Network architecture (AT-CNN). The Convolution Block Attention Module (CBAM) is added after four convolutional modules. The CABM has two sequential sub-modules: channel and spatial.

The architecture of AT-CNN is shown in Table 1. The convolution layer extracts feature on the input 2D-EEG-image. Different convolution kernels can capture different features. The operation of the convolutional layer is defined in Eq. (1):

$$x_j^l = f(b_j^l + \sum\nolimits_{i \in M_j} w_{ij}^l * x_i^{l-1}) \tag{1}$$

where x_j^l denotes the jth feature map of the lth layer, f represents the activation function. Usually ReLU is used, M_j is the set of input feature map, w_{ij}^l is the convolution kernel, and b_j^l is bias.

Bath normalization (BN) is used after each convolutional layer. BN normalizes the results of neuron output, which can effectively alleviate overfitting, and make the training process of the network more stable at the same time. BN is applied to each mini-batch, and the calculation formula for normalization by mini-batch is as follows

$$\hat{x}_i^{(k)} = \frac{x_i^{(k)} - \mu_B^{(k)}}{\sqrt{\acute{o}_B^{(k)^2} + \varepsilon}} * \gamma + \beta \tag{2}$$

where $x_i^{(k)}$ is the ith element of the k-th dimension, $\mu_B^{(k)}$ and $\acute{o}_B^{(k)^2}$ are the mean and variance of the input mini-batch respectively, γ and β are hyperparameters. The model learns to adjust the mean and bias of the normalized data.

Table 1. Architecture of AT-CNN.

Module/Features	Type	Input size	Filters	Kernel Size	Stride	Activation
M1	2D convolution	(224,224,1)	8	(3,3)	1	ReLu
	BN	(224,224,8)				
	Max polling	(224,224,8)		(2,2)	2	
M2	2D convolution	(112,112,8)	16	(3,3)	1	ReLu
	BN	(112,112,16)				
	Max polling	(112,112,16)		(2,2)	2	
M3	2D convolution	(56,56,16)	32	(3,3)	1	ReLu
	BN	(56,56,32)				
	Max polling	(56,56,32)		(2,2)	2	
M4	2D convolution	(28,28,32)	64	(3,3)	1	ReLu
	BN	(28,28,64)				
	Max polling	(14,14,64)		(2,2)	2	
M5	CBAM	(14,14,64)				
M5	Flatten	(1,12544)				
M6	Fully-connected	(1,12544)				ReLu
	Dropout	(1,12544)				
M7	Fully-connected	(1,896)				ReLu
	Dropout	(1,896)				
M8	Fully-connected	(1,64)				Softmax
	Dropout	(1,64)				

The pooling layer is usually followed by convolution operation. The main purpose is to retain the main features while reducing the parameters. In AT-CNN, the maximum pooling operation is used, and its definition is as follows:

$$x_j^l = \max\left(x_j^{l-1}\right), l \in P_j \tag{3}$$

where P_j is the set of input feature maps.

The fully connected layer is located after the convolutional layer and acts as a classifier in the entire convolutional neural network. Before inputting to the FC layer, the output features of the previous layer will be tiled into a one-dimensional vector form. After the fully connected layer, the prediction result of the output model is the prediction result of the problem. For the work in this paper, two nodes are output, representing the detection of ErrP from the EEG signal and the absence of detection of ErrP.

2.3 Loss Function

The loss function is used to measure the degree of inconsistency between the predicted value and the true value of the model. The smaller the loss function, the better the model fit. The optimizer is used to minimize the loss function, and the cross-entropy loss is commonly used in binary classification. In this paper we employ label smoothing [15]. As shown in Eq. (4), C is the cost function and the optimizer is used to minimize it, N is the number of all samples, p_c is the output value of the neural network (between 0 and 1) and y_c' is the target label. Here, instead of using the original target label y_c, we adjust the actual label y_c to y_c' and set ε to 0.1. (i.e. replace the exact classification target from 0 and 1 to ε and 1 − ε). The reason is that the data labels obtained may not be completely correct, and the wrong labels will adversely affect the model. Using label smoothing can prevent the model from pursuing exact probabilities without affecting the model learning to classify correctly. Thus, the prediction performance is improved and the overfitting of the model is reduced.

$$C = -\frac{1}{N}\sum\nolimits_{i=1}^{N}[y_c' Logp_c + \left(1 - y_c'\right)\log(1 - p_c)] \tag{4}$$

$$y_c' = (1 - \varepsilon) * y_c \tag{5}$$

2.4 Model Training

To meet the requirements of experimental setup, the obtained ErrP dataset is often unbalanced. To balance the dataset, we adopt random oversampling, the minority class is randomly copied the same as majority class.

While training AT-CNN, we use back-propagation to update the parameters of the model. During model training, pre-setting 100 epochs, the batch size is set to 32 and the initial learning rate is 0.001. We use Cosine learning rate decay to dynamically reduce the learning rate, and the optimizer uses Adam.We also employ Early stopping.Model training was terminated if no drop in validation loss was observed in 10 epochs to avoid overfitting. The model was implemented in Pytorch (Pytorch, 1.8.0), a freely available DL library from Facebook. For experiments, we use GeForce GTX 1080 GPU with 10 GB memory.

2.5 Performance Metrics

In our paper, we refer to the EEG signal with ErrP as a error case, and the EEG signal without ErrP as a correct case. We use accuracy, sensitivity, specificity to evaluate the method. These performance metrics are as follows:

$$Accuracy = \frac{TP + TN}{TP + FP + FN + TN} \tag{6}$$

$$Sepecifity = \frac{TN}{TN + FP} \tag{7}$$

$$Sensitivity = \frac{TP}{FN + TP} \tag{8}$$

where TP (true positives) are the number of error cases, which are predicted as error. FN (false negatives) are the number of correct cases, which are predicted as correct. TN (true negatives) are the number of correct cases that are predicted as correct and FP (false positives) are the number of error cases that are classified as error. Sensitivity refers to the proportion of correctly detected error from EEG signal. Specificity refers to the proportion of correctly identified correct from EEG signal.

3 Dataset

In this study, we use the BNCI2020 – Monitoring Error-Related Potential dataset [7] to evaluate our method. The experimental protocol is shown in Fig. 3. the subjects were asked to observe the movement of a cursor displayed on a monitor. In order to obtain the ErrP signal, the experiment controls the cursor with an error rate of 20%, that is, the cursor has a 20% probability of moving away from the target.

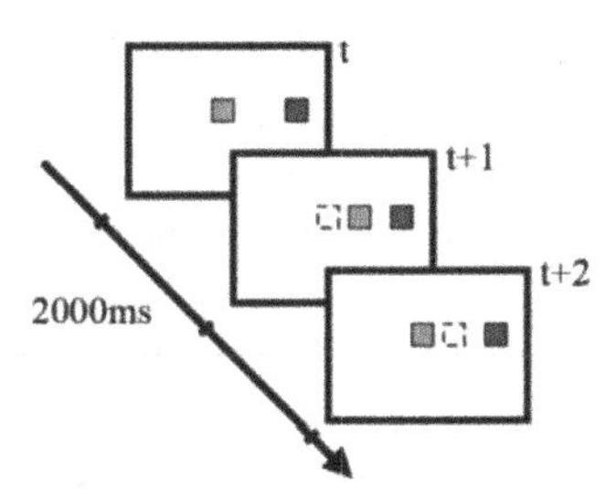

Fig. 3. Experimental protocol. Green cursor is moving to red square (target), dotted square is the cursor location at the previous time step [7].

The dataset contains 6 subjects and 2 sessions. Table 2 shows time difference (in days) between the two sessions for all subjects. In the experiment, the EEG signals of all subjects were recorded at a sampling rate of 512 Hz, and the acquisition process used 64 channels according to the 10–20 international standard.

As for the dataset segmentation of training and evaluation models, we simulated the practical application of BCI system. We use data from the first session for model training and validation, and data from the second session for testing.

Table 2. Time difference (In days) between the two sessions.

Subject	1	2	3	4	5	6
Days between two sessions	51	50	54	211	628	643

4 Experimental Results and Discussions

4.1 Comparison with Other State of Art Methods

To evaluate the proposed method, we compare it with other state-of-the-art methods, and all experiments are averaged using five-fold cross-validation. The reported results were calculated based on trial labels and presented as mean ± std deviation.

ConvNet [10], ConvArch [10] and ANN [11], these methods use the same dataset as this paper. But the data segmentation in the original paper is different from this paper. Therefore, we replicate their model and uniformly adopt the data segmentation method of this paper. The data segmentation method in this paper is the closest to the actual application scenario. Table 3 shows the experimental results and we obtained the following observations. First, the classification accuracy of deep learning-based methods, such as ConvNet (81.58%), ConvArch (79.61%), ANN (83.21%), and ours (83.42%)) outperforms traditional methods, such as Gaussian classifiers (70.41%) and LDA (78.91%), indicating that deep learning-based methods can learn from the data and learn more discriminative features. Second, we find that ANN outperforms ConvNet and ConvArch in accuracy, which suggests that learning global features using other topics and then fine-tuning model parameters to fit each topic can improve classification performance. However, it consumes a lot of time in practical applications. Third, the accuracy of our proposed method is 83.42%, the sensitivity is 63.17%, and the specificity is 88.48%, and our method achieves the best classification performance compared with other methods.

Table 3. Performance indices of the different methods using the same dataset.

Methods\ Metrics	Accuracy (%)	Sensitivity (%)	Specificity (%)
Gaussian classifier [7]	70.41 ± 7.40	63.21 ± 9.06	75.81 ± 6.84
LDA [8]	78.91 ± 5.43	**68.83 ± 7.46**	81.43 ± 5.95
ConvArch [10]	79.61 ± 1.78	60.84 ± 6.76	84.30 ± 4.22
ConvNet [9]	81.58 ± 2.81	65.10 ± 3.84	85.70 ± 3.64
ANN [11]	83.21 ± 3.47	64.31 ± 3.31	87.81 ± 2.34
Ours	**83.42 ± 2.64**	63.17 ± 4.25	**88.48 ± 1.82**

4.2 Contribution of Different Train Settings

To further verify the effectiveness of our proposed model, we evaluate the contribution of different train settings to AT-CNN. We tested it when some elements are omitted and employed the same hyperparameters (learning rate, batch size, etc.) for a fair comparison.

To evaluate the contribution of the label smoothing, we performed the training of the AT-CNN again while not using label smoothing instead of cross-entropy as loss function. Table 4 shows the results. We found that the accuracy of the whole model decreased from 83.42% to 82.71%. The sensitivity decreased greatly, from 66.17% to 58.45%. The results show that label smoothing also plays an important role in the learning of the model, for the acquired data and labels may and completely correspond. Using label smoothing can prevent overfitting data, thereby improving the generalization performance of the model.

To test the contribution of batch normalization (BN), we trained AT-CNN again without using it. Table 4 shows the results. We found that the accuracy, specificity and sensitivity all decreased, and the accuracy rate dropped from 83.42% to 81.88%. The result shows that the BN can effectively improve the performance of AT-CNN.

To show the importance of the convolutional block attention module (CBAM) of AT-CNN, we retrained the system without applying CBAM in AT-CNN. Table 4 shows the results. We obverse that the accuracy drops from 85.62% to 82.12%. We found that adding CBAM to the model can greatly improve the prediction performance of the model.

When we do not augment erroneous samples with over sampling, the accuracy of the AT-CNN decreased from 83.42% to 82.64%, and the specificity was slightly improved from 88.48% to 91.06%, shown in Table 4. However, the sensitivity decreased from 63.17% to 48.97%, showing the importance of over sampling. We speculate that the reason for the large drop in model sensitivity is that the samples in the original data are unbalanced and the samples of ErrP accounts for 20% of the total samples. AT-CNN will tend to predict samples as the majority class. After random oversampling, the model can improve the prediction performance.

Table 4. Performance of AT-CNN when some elements are omitted.

Metrics\technique	No label smoothing	No BN	No CBAM	No over sampling
Accuracy (%)	82.71	81.88	82.12	82.64
Specificity (%)	90.28	87.30	86.58	91.06
Sensitivity (%)	52.45	60.21	63.93	48.97

5 Conclusions and Future Work

In this paper, we proposed an effective ErrP classification method using 2D convolutional neural networks with EEG image as an input. The raw 1D EEG signals are converted into a 224 × 224 grayscale images after preprocessing. In order to effectively extract the spatiotemporal features of 2D-EEG images, a new deep learning model named AT-CNN was put forward. To the best of our knowledge, this is the first work that 2D-CNN is applied to classify ErrP. And the experimental results demonstrate that the proposed method has the best performance compared with the state-of-the-art methods. For the future work, we will consider utilizing signals from multiple channels in the anterior cingulate cortex (ACC) and integrating information from multiple EEG channels, which is expected to further improve the classification accuracy.

Acknowledgements. The work was supported by Key Research and Development Program of Shaanxi (Program No. 2020KWZ-003).

References.

1. Baniqued, P.D.E., Stanyer, E.C., Awais, M., Alazmani, A., Holt, R.J.: Brain-computer interface robotics for hand rehabilitation after stroke: a systematic review. J. NeuroEng. Rehabil. **18** (2021). https://doi.org/10.1101/2019.12.11.19014571
2. López-Larraz, E., et al.: Brain-machine interfaces for rehabilitation in stroke: a review. Neurorehabilitation **43**, 77–97 (2018). https://doi.org/10.3233/NRE-17239
3. Klonowski, W.: Everything you wanted to ask about EEG but were afraid to get the right answer. Nonlinear Biomed. Phys. **3**, 2 (2009). https://doi.org/10.1186/1753-4631-3-2
4. Pascual-Marqui, R.D.: Standardized low-resolution brain electromagnetic tomography (sLORETA): technical details. Methods Find. Exp. Clin. Pharmacol. **24**(Suppl. D), 5–12 (2002). https://doi.org/10.1002/med.10000
5. Chavarriaga, R., Sobolewski, A., Millán, J.d.R.: Errare machinale est: the use of error-related potentials in brain-machine interfaces. Front. Neurosci. **8**, 208 (2014). https://doi.org/10.3389/fnins.2014.00208
6. Parashiva, P.K., Vinod, A.P.: Improving direction decoding accuracy during online motor imagery based brain-computer interface using error-related potentials. Biomed. Signal Process. Control **74**, 103515 (2022). https://doi.org/10.1016/j.bspc.2022.103515
7. Chavarriaga, R., Millan, J.d.R.: Learning from EEG error-related potentials in noninvasive brain-computer interfaces. IEEE Trans. Neural Syst. Rehabil. Eng. **18**, 381–3888 (2010). https://doi.org/10.1109/TNSRE.2010.2053387
8. Kumar, A., Pirogova, E., Fang, J.Q.: Classification of error-related potentials using linear discriminant analysis. In: 2018 IEEE-EMBS Conference on Biomedical Engineering and Sciences, pp. 18–21 (2018). https://doi.org/10.1109/IECBES.2018.8626709
9. Torres, J.M.M., Clarkson, T., Stepanov, E.A., Luhmann, C.C., Lerner, M.D., Riccardi, G.: Enhanced error decoding from error-related potentials using convolutional neural networks. In: 2018 40th Annual International Conference of the IEEE Engineering in Medicine and Biology Society (EMBC), pp. 360–363 (2018). https://doi.org/10.1109/EMBC.2018.8512183
10. Swamy Bellary, S.A., Conrad, J.M.: Classification of error related potentials using convolutional neural networks. In: 2019 9th International Conference on Cloud Computing, Data Science & Engineering (Confluence), pp. 245–249 (2019). https://doi.org/10.1109/CONFLUENCE.2019.8776901
11. Parashiva, P.K., Vinod, A.P.: Improving classification accuracy of detecting error-related potentials using two-stage trained neural network classifier. In: 2020 11th International Conference on Awareness Science and Technology (iCAST), pp. 1–5 (2020). https://doi.org/10.1109/iCAST51195.2020.9319482

12. Ullah, A., et al.: A hybrid deep CNN model for abnormal arrhythmia detection based on cardiac ECG signal. Sensors **21** (2021). https://doi.org/10.3390/s21030951
13. Rohmantri, R., Surantha, N.: Arrhythmia classification using 2D convolutional neural network. Int. J. Adv. Comput. Sci. Appl. **11**, 201–208 (2020). https://doi.org/10.48550/arXiv.1804.06812
14. Woo, S., Park, J., Lee, J.-Y., Kweon, I.S.: CBAM: convolutional block attention module. In: Ferrari, V., Hebert, M., Sminchisescu, C., Weiss, Y. (eds.) ECCV 2018. LNCS, vol. 11211, pp. 3–19. Springer, Cham (2018). https://doi.org/10.1007/978-3-030-01234-2_1
15. Szegedy, C., et al.: Rethinking the inception architecture for computer vision. In: 2016 IEEE Conference on Computer Vision and Pattern Recognition (CVPR), pp. 2818–2826 (2016). https://doi.org/10.1109/CVPR.2016.308

A Multi-sensor Combined Tracking Method for Following Robots

Hao Liu, Gang Yu(✉), and Han Hu

Department of Mechanical Engineering and Automation, Harbin Institute of Technology, Shenzhen, Shenzhen 518055, China
gangyu@hit.edu.cn

Abstract. At present, the research on tracking methods is mainly based on visual tracking algorithm, which has reduced the accuracy at night or under the condition of insufficient light intensity. Therefore, this paper starts from the direction of multi-sensor combined tracking. Firstly, in order to verify the feasibility and performance of the multi-sensor combined tracking method proposed in this paper, a set of tracking robot system is designed. Secondly, aiming at the problem that the visual tracking method fails to track in scenes such as complete occlusion and insufficient illumination, the non-line-of-sight perception of the following target is realized based on the fusion of ultra-wide band (UWB) and inertial measurement unit (IMU) sensors. Besides, based on coordinate transformation and decision tree algorithm, this paper makes decisions on UWB and visual tracking targets to achieve combined tracking.

Keyword: Following robot · Target tracking · Multi-sensor combined tracking

1 Introduction

The Intelligent System Laboratory of the Central Research Institute of Toyota in Japan has developed a personal following robot to assist in handling and loading [1]. The following robot is mainly equipped with panoramic camera, Lidar and inertial measurement sensor for sensing, and the robot follows the target according to the robot's kinematic model. Among them, the panoramic camera is used for target tracking, Lidar is used for obstacle avoidance, and inertial sensors measure the acceleration and angular acceleration of the robot to control the robot to keep its balance.

The school of robotics engineering at Inha University in Korea has developed a following robot [2]. The following robot mainly integrates monocular camera and Lidar to track human targets. In the aspect of visual tracking, the particle filter method is used to track the morphological features extracted from the image. At the same time, the laser ranging sensor is used to measure the distance and angle of the target, and then the data of Lidar and visual tracking are fused to realize the reliable tracking of the target.

Han yang University in South Korea has developed a following robot for marathon athletes [3]. The robot mainly obtains the point cloud data of the surrounding environment through laser sensors. According to the support vector data description, the point

H. Liu et al. (Eds.): ICIRA 2022, LNAI 13456, pp. 722–734, 2022.
https://doi.org/10.1007/978-3-031-13822-5_65

cloud data is mapped to high-dimensional space for classification, and the target area is distinguished, the target position is tracked. At the same time, Kalman filter is used to estimate the state of the tracking human body and the optimal position of the tracking target, to realize the motion control of the robot.

The commercial version developed by Intel follows the Segway robot. An intelligent upgrade is made on the platform of the balance car, which senses the surrounding environment through the RGB-D camera, realizes gesture recognition, obstacle avoidance and following based on the visual algorithm, and also has the functions of speech recognition, mobile photography and home monitoring.

A humanoid robot developed by Shenyang Institute of automation, Chinese Academy of Sciences [4], which realizes target tracking based on three degrees of freedom redundant vision. It is equipped with a binocular camera composed of a laser based TOF camera and two CCD cameras. Its processing logic is very similar to human eyes, which is to find and track the target in a relatively large range. After finding the target, carefully observe the target and track it. Firstly, it looks for the target to be tracked through the time-of-flight camera, roughly locates the target, and then accurately locates it through the binocular camera. The time-of-flight camera does not have a high resolution, which can reduce the computation during the coarse localization phase. However, the binocular camera has a higher resolution, which allows precise measurement and localization of the target.

2 Following System Design

The method of human target tracking based on multi-sensor proposed in this paper is mainly used to solve the shortcomings of visual tracking. For example, in completely obscured and poorly illuminated scenes, the vision tracking algorithm is unable to re-identify and track the target when the human target leaves the camera's field of view. Therefore, this paper introduces Ultra Wide Band (UWB) and Inertial Measurement Unit (IMU) sensors, which mainly address the problem of how to provide reliable coordinate information for tracking targets in the presence of occlusion.

Based on the analysis of the above application scenarios and adopted technologies, the overall scheme design of the following robot in this paper is mainly divided into four parts, including core processing layer, hardware layer, control system layer and power module as shown in Fig. 1.

(1) The perception layer is the bottom part of the whole system, and it is also an important device for following robot to realize the perception of the surrounding environment. The 1080p camera sensor is installed on the 2-DOF camera PTZ (Pan/Tilt/Zoom) to provide the system with video stream in the following process. Two nine axis gyroscopes are respectively installed on the camera pan tilt and the chassis of the following robot to provide the system with the relative angle and pitch angle of the camera. The tags of IMU and UWB are fixed together with Bluetooth module to provide the acceleration information and angle information of human target for the system. The motor driver is responsible for controlling the motor following the robot motion.

(2) The core processing layer is built based on ROS system, which is mainly responsible for visual target detection and tracking and sensor combination tracking. In the initialization process, the target detection algorithm detects the human target closest to the robot in the picture as the target tracked in the subsequent process, and releases the detected target. The target tracking node subscribes to the target location data of the target detection node as the initialization target box of target tracking. The fusion node parses the data according to the communication protocol and obtains the UWB tracking data and the attitude data of the nine-axis gyroscope sensor. At the same time, it achieves the matching between the UWB tracking target and the visual tracking target according to the conversion relationship between the UWB coordinate system and the camera coordinate system, and achieves the effect of multi-sensor tracking fusion.
(3) The control system layer mainly collects the sensing data with the core processing layer through UART serial port and transmits it to the core processing layer for processing. There are two processing units in the control layer. The first processing unit STM32F103 is responsible for processing the measurement data of each UWB base station and label. Through the distance data between the three base stations and the label, the position information of the label relative to the robot can be obtained through the solution algorithm. Another processing unit STM32F407 is mainly responsible for the PTZ attitude control of the 2-DOF camera. It controls the motor speed of the robot chassis through the motor driver and receives the data from the core processing layer. At the same time, it receives the position information calculated from STM32F103 and sends the relevant data to the core processing unit through the serial port.

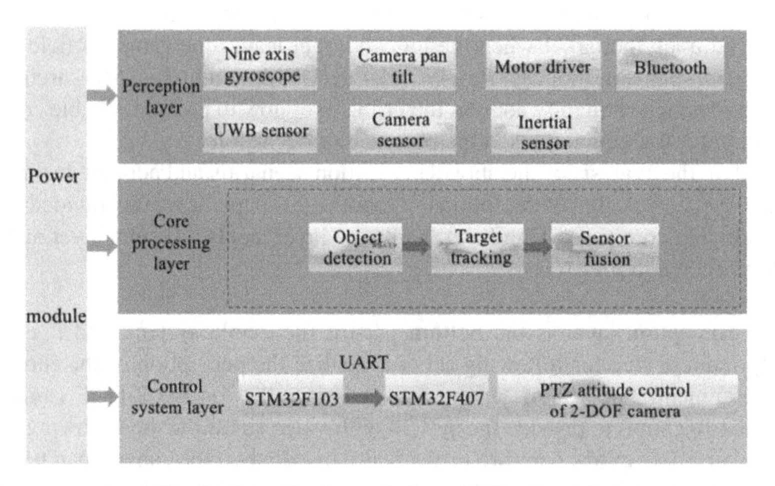

Fig. 1. Overall scheme design of following robot

The hardware layout of the following robot is shown in Fig. 2.

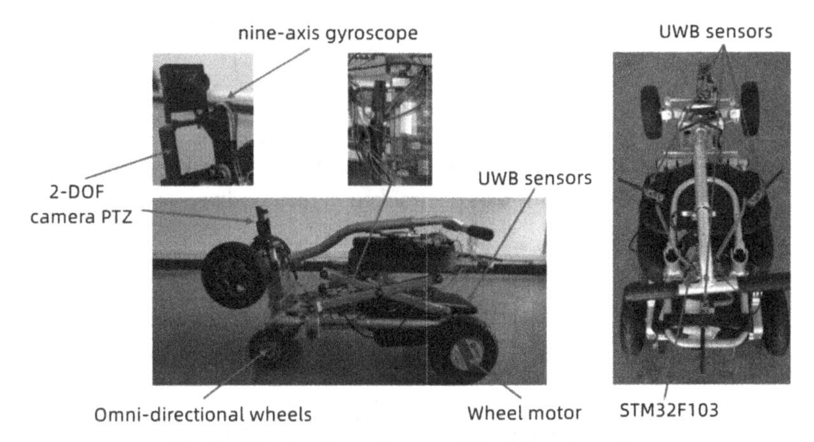

Fig. 2. The hardware layout of the following robot

3 Multi-sensor Combined Tracking

3.1 Fusion Model of UWB and IMU Based on Adaptive Kalman Filter

According to Kalman filter, for general linear system, the relationship between state equation and observation equation is as follows:

$$X_k = \Phi_{k,k-1}X_{k-1} + B_{k-1}U_{k-1} + \Gamma_{k-1}W_{k-1} \tag{1}$$

$$Z_k = H_kX_k + V_k \tag{2}$$

where X_k represents state vector at time k, B_{k-1} represents the influence of $k-1$ time input on the system, U_{k-1} represents $k-1$ time input, W_{k-1} represents dynamic noise of random system, H_k represents k-time measurement matrix, V_k represents measurement noise sequence at time k, Γ_{k-1} represents System noise matrix.

In this paper, the distance measured by three base stations $[d_1, d_2, d_3]^T$ is directly used as the observation variable, that is, the observation equation is nonlinear and needs to be processed by extended Kalman filter. According to the human target model in this paper, it can be obtained by Kalman filter. We can get the state vector describing human motion and the measurement of human target view obtained by using UWB solution model as follows:

$$\begin{pmatrix} x_k \\ y_k \\ \dot{x}_k \\ \dot{y}_k \end{pmatrix} = \begin{pmatrix} I_2 & \Delta TI_2 \\ 0 & I_2 \end{pmatrix} \begin{pmatrix} x_{k-1} \\ y_{k-1} \\ \dot{x}_{k-1} \\ \dot{y}_{k-1} \end{pmatrix} + \begin{pmatrix} \frac{1}{2}\Delta T^2 I_2 \\ \Delta TI_2 \end{pmatrix} \begin{pmatrix} \ddot{x}_{k-1} \\ \ddot{y}_{k-1} \end{pmatrix} + W_{k-1} \tag{3}$$

$$\begin{pmatrix} d_1 \\ d_2 \\ d_3 \end{pmatrix} = \begin{pmatrix} \sqrt{(x_k - x_1)^2 + (y_k - y_1)^2} \\ \sqrt{(x_k - x_2)^2 + (y_k - y_2)^2} \\ \sqrt{(x_k - x_3)^2 + (y_k - y_3)^2} \end{pmatrix} + V_k \tag{4}$$

where, ΔT represents the sampling time interval, I_2 represents the 2×2 unit matrix, the positions of the three UWB base stations are (x_1, y_1),(x_2, y_2) and (x_3, y_3), respectively, and (x_k, y_k) represents the position of the tracked human body.

$$h(X_k) = \begin{pmatrix} \sqrt{(x_k - x_1)^2 + (y_k - y_1)^2} \\ \sqrt{(x_k - x_2)^2 + (y_k - y_2)^2} \\ \sqrt{(x_k - x_3)^2 + (y_k - y_3)^2} \end{pmatrix} \tag{5}$$

Because $h(X_k)$ is a nonlinear function, there is no constant matrix H_k, which makes both sides of the equation hold. According to the extended Kalman filter, the nonlinear function is expanded by Taylor formula, and the approximate linearized equation is obtained by ignoring the higher-order terms of more than quadratic. Then you can get:

$$H_k = \frac{\partial h(X_k)}{\partial X_k} = \begin{pmatrix} \frac{\partial p_1}{\partial x_k} & \frac{\partial p_1}{\partial y_k} & \frac{\partial p_1}{\partial \dot{x}_k} & \frac{\partial p_1}{\partial \dot{y}_k} \\ \frac{\partial p_2}{\partial x_k} & \frac{\partial p_2}{\partial y_k} & \frac{\partial p_2}{\partial \dot{x}_k} & \frac{\partial p_2}{\partial \dot{y}_k} \\ \frac{\partial p_3}{\partial x_k} & \frac{\partial p_3}{\partial y_k} & \frac{\partial p_3}{\partial \dot{x}_k} & \frac{\partial p_3}{\partial \dot{y}_k} \end{pmatrix} \tag{6}$$

In the application of Kalman filter, it is necessary to ensure that the driving noise and measurement noise of the system must be white noise. In the process, the driving noise and measurement noise of the system are colored, and the change of the actual environment leads to the change of noise. Finally, the deviation between the estimated value and the real value becomes larger and larger. At this time, the filter can not make the optimal estimation of the state. Therefore, this paper uses adaptive weighted Kalman filter to estimate the parameters of the model.

Adaptive Kalman filter mainly introduces fading factor to modify the error covariance matrix online and update the noise matrix and state noise matrix to prevent the divergence of the filter and improve the robustness of the algorithm. The adaptive Kalman filter algorithm introduces the weighting coefficient to adjust the measurement noise and state noise. The weighting coefficient is calculated as:

$$d_k = \frac{1 - b}{1 - b^{k+1}} \tag{7}$$

where k represents the current k time, b is forgetting factor, value takes from 0 to 1. It can be seen that when k $\rightarrow \infty$, d_k tends to 1, that is, the Kalman filter algorithm of the same standard.

The weighting coefficient is calculated and the noise is dynamically adjusted by the residual at the last time as follow:

$$\begin{cases} r_k = (1 - d_k)r_{k-1} + d_k\left(Z_k - H_k\bar{X}_k^-\right) \\ R_k = (1 - d_k)R_{k-1} + d_k\left(\varepsilon_k\varepsilon_k^T - H_kP_{k-1}H_k^T\right) \\ q_k = (1 - d_k)Q_{k-1} + d_k\left(\bar{X}_k - \Phi_{k,k-1}\bar{X}_{k-1}\right) \\ Q_k = (1 - d_k)Q_{k-1} + d_k\left(K_k\varepsilon_k\varepsilon_k^TK_k^T + P_k - \Phi_{k,k-1}P_k^-\Phi_{k,k-1}^T\right) \end{cases} \tag{8}$$

where $\varepsilon(t)$ represents the residual as follow:

$$\varepsilon(t) = Y_k - H_k\overline{X}_k^- \tag{9}$$

where, r_k and q_k respectively represent the mean value of measurement noise and state noise, R_k and Q_k represent the variance of measurement noise and state noise.

3.2 Combined Tracking with Camera Sensor After UWB Fusion

Vision-based target tracking algorithms cannot be completely reliable in real scenarios, and there is a possibility of misidentification in some extreme conditions. However, although UWB sensors have large errors in the measured data due to multipath effects, the fusion of IMU sensors can limit the errors to a certain range. Therefore, when the difference between the visually tracked target and the UWB fused target is small, the position of the visually tracked target can be considered more accurate. And when the difference between the position of the tracked target after UWB fusion and the position of the visually tracked target is large, it may be due to the misidentification of the visual tracking, so the position of the tracked target after UWB fusion is more reliable at this time, so the position of the UWB tracked target is mainly used at this time.

In this paper, three characteristics are used to determine which primary tracking method is used to drive the robot to achieve following. These include whether there is a tracked target in the camera field of view and whether the position of the visually tracked target point and the position of the UWB tracking point are on the same side of the x-axis of the camera coordinate system. The fusion status can be obtained as shown in Table 1.

Table 1. Fusion status.

Features	Target in camera field of view A_1	On the same side of x-axis A_2	UWB coordinates are within the camera range A_3	Primary tracking methods
1	Yes	Yes	Yes	Visual tracking position
2	Yes	Yes	No	Visual tracking position

(continued)

Table 1. *(continued)*

Features	Target in camera field of view A_1	On the same side of x-axis A_2	UWB coordinates are within the camera range A_3	Primary tracking methods
3	Yes	No	Yes	Visual tracking position
4	Yes	No	No	UWB tracking position
5	No	No	Yes	UWB tracking position
6	No	No	No	UWB tracking position
7	No	Yes	Yes	UWB tracking position
8	No	Yes	No	UWB tracking position

The decision tree ID3 algorithm uses the information gain criterion to select features for classification on each node of the decision tree, calculates the information gain for all possible features, and selects the feature with the greatest information gain as the classification criterion.

As shown in the above table, A_1 indicates that there is a target in the camera field of view, A_2 indicates that it is on the same side of the x-axis, and A_3 indicates that the coordinates after UWB fusion are on the same side of the camera. Suppose there are k categories in data set D that can be classified, and these categories can be represented by $A_i(\mathrm{i} = 1,\ldots, \mathrm{k})$. The information gain of feature A to training data set D is expressed as $g(D \mid A)$ as shown below:

$$g(D \mid A) = H(D) - H(D \mid A) \tag{10}$$

It is defined as the difference between the empirical entropy $H(D)$ of set D and the empirical entropy of set D under the given characteristic A_k condition.

Empirical entropy $H(D)$ represents the uncertainty in set D, as shown below:

$$H(D) = -\sum_{k=1}^{K} \frac{|A_k|}{|D|} log\left(\frac{|A_k|}{|D|}\right) \tag{11}$$

The empirical entropy of set D under the given characteristic A_k condition as shown below:

$$H(D \mid A) = \sum_{k=1}^{n} \frac{|D_k|}{|D|} H(D_i) = -\sum_{i=1}^{n} \frac{|D_i|}{|D|} \sum_{k=1}^{K} \frac{|D_{ik}|}{|D_i|} log\left(\frac{|D_{ik}|}{|D_i|}\right) \tag{12}$$

The classification table of sensor fusion analyzed above in this paper is shown in Table 1. Then the fusion of UWB data and camera data is realized according to the ID3 algorithm of decision tree. According to Eq. (10), Eq. (11) and Eq. (12), the information gain based on different characteristics can be calculated, respectively, $g(D, A_1) = 0.548$, $g(D, A_2) = 0.0488$, $g(D, A_3) = 0.0488$, $g(D, A_1) > g(D, A_2) = g(D, A_3)$. Therefore, first selecting features for classification can reduce the uncertainty of class information more. According to $g(D, A_2) = g(D, A_3)$, it shows that selecting features A_2 and A_3 has the same effect in reducing the uncertainty of the set. Therefore, the decision tree model shown in Fig. 3 can be obtained. According to the decision tree model shown in Fig. 3, the fusion of UWB tracking data and camera tracking data after fusion with IMU can be realized.

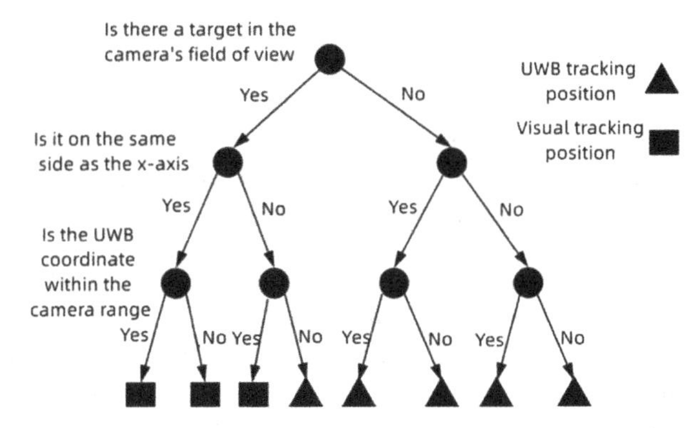

Fig. 3. Decision tree model of multi-sensor and camera fusion

4 Experiment Verification

This paper designs relevant verification experiments according to the design indexes of the following robot designed in this paper.

4.1 Following Distance Test

In order to verify the following distance, the experiment of following distance is carried out in this paper. The following scene is shown in Fig. 4, and the distance between the real-time following target and the following robot and the speed of the following car are recorded. As shown in Fig. 4 (a) and Fig. 4 (d), the robot can follow the target in a straight line or on a gentle slope; In Fig. 4 (b) and Fig. 4 (c), the robot follow the target to the right and the left respectively.

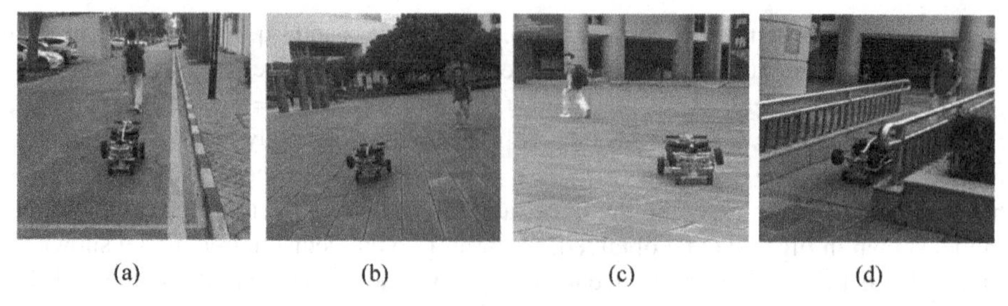

(a) (b) (c) (d)

Fig. 4. Follow distance experiment

4.2 Following Occlusion Experiment

In order to verify the design index of the extraction distance of the occluded target, the tracking occlusion experiment is designed in this paper. The experiment verifies that the tracking vehicle can sense the tracking target within a certain range, and achieves ultra-broadband hyper-visual distance perception to extract the target position under visual target occlusion.

As shown in Fig. 5 (a), when moving to the position shown in Fig. 5 (b), the tracker can perceive the position of the target. When moving from the position shown in Fig. 5 (c) to the position shown in Fig. 5 (d), the visual tracker has failed and cannot perceive the position of the tracked target. As shown in the visual index diagram in Fig. 6, since the target is no longer within the camera field of view, there is no pixel error and overlap rate with the calibration frame, so they are all 0. As can be seen from Fig. 7, although the tracking accuracy of UWB is not high, it can also sense the position of the target. Therefore, it can be concluded from this experiment that when the tracking target is completely blocked, the following robot can effectively perceive the tracking target position.

(a) (b) (c) (d)

Fig. 5. Human target tracking experiment

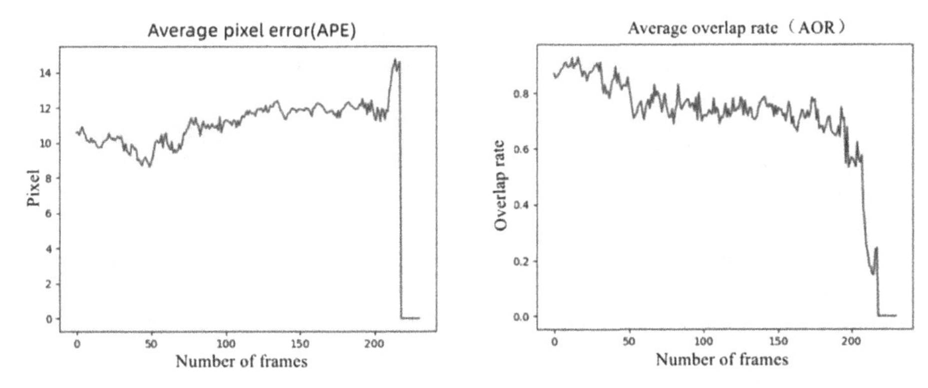

Fig. 6. Visual tracking index

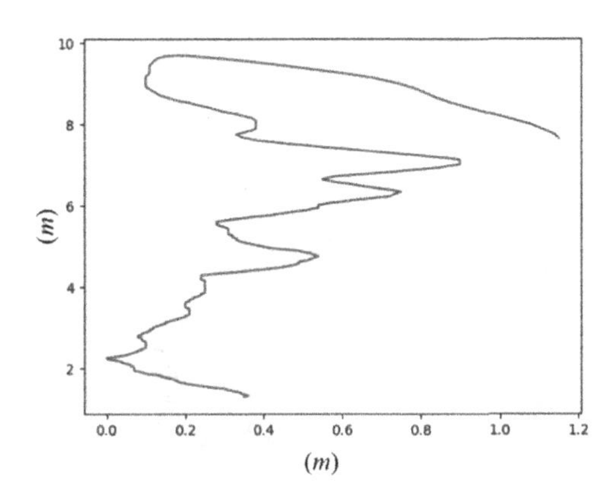

Fig. 7. Sage-huse adaptive Kalman filter fusion trajectory

4.3 PTZ Following Experiment

As shown in Fig. 8 (a) and (b), when the target leaves the camera field of view, the visual tracking algorithm fails and the target position cannot be perceived. As shown in Fig. 8 (c), when the target reappears within the camera field of view, the target tracking algorithm in this paper can continue tracking again. When moving from Fig. 8 (c) to Fig. 8 (d), the camera pan tilt can follow the target to ensure that the target is within the camera field of view. When moving the state of Fig. 8 (e), the target is partially obscured at this time, but the tracker can still track the target. After manually calibrating 1290 pictures, calculate the average pixel error (APE) and average overlap (AOR) with the tracking frame output by the tracking algorithm, as shown in Fig. 9.

APE is the error value based on the pixel distance between the predicted target center position and the real position, and the final result is averaged. AOR is the intersection ratio of the predicted area to the real area for each frame, and the final result is averaged.

As shown in Fig. 9 (a), there are four average errors of 0, which are in frames 85–101, 303–319, 837–900 and 1073–1092 respectively. This is because the target is completely obscured by obstacles in these frames. From the index of overlap rate, the overlap rate of tracking target frame and calibration frame can be maintained above 0.6, which shows that when the target is within the field of view, the tracking algorithm can track the target and has a certain accuracy.

In conclusion, this experiment can verify that the tracking effect of the lightweight tracking network in this paper can meet the tracking requirements, and the two degree of freedom camera platform can realize the continuous tracking of the target.

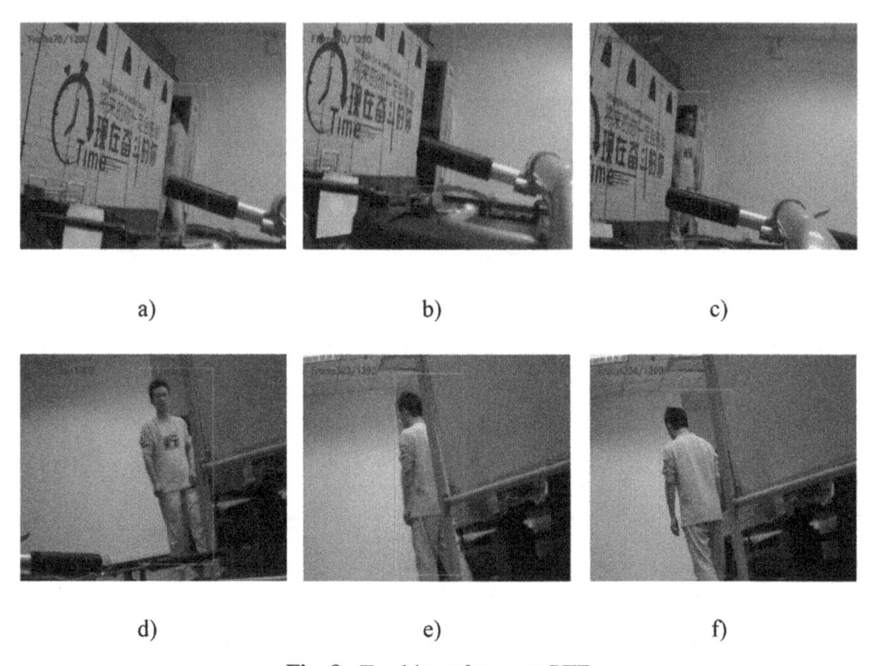

Fig. 8. Tracking of camera PTZ

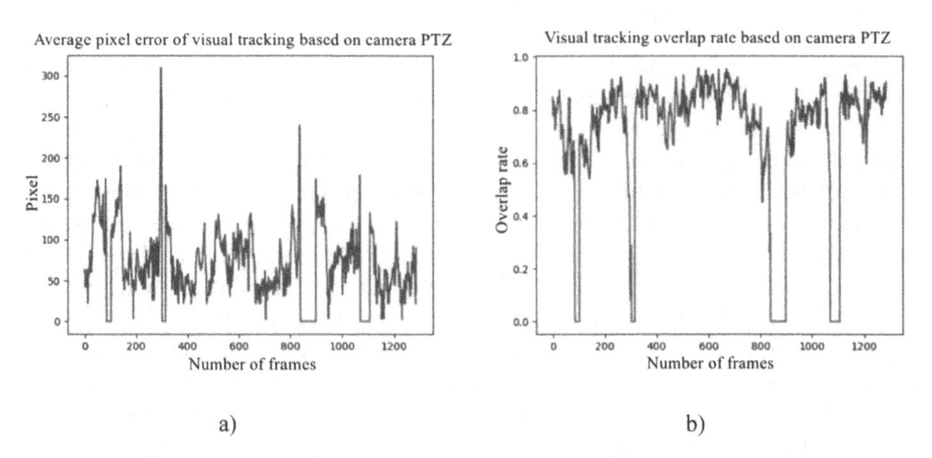

Fig. 9. APE and AOR indexes in camera PTZ following experiment

5 Conclusion

Taking the following robot as the application scenario, this paper designs the overall scheme of the following robot from the perspective of the reliability of the following robot, focuses on the research of human target tracking method based on multi-sensor, and the main results are summarized as follows:

(1) Aiming at the human target tracking method studied in this paper, a set of target following robot system is designed, including sensor type selection and so on. At the same time, the motion model of the robot is modeled to realize the motion control of the robot.
(2) Based on Kalman filter and decision tree, a tracking method combining UWB, IMU and monocular camera is proposed to realize that when the target is completely blocked by obstacles, the following robot can still perceive the problem of following the target so as to realize robust tracking.

References

1. Hirose, N., Tajima, R., Sukigara, K.: Personal robot assisting transportation to support active human life — human-following method based on model predictive control for adjacency without collision. In: 2015 IEEE International Conference on Mechatronics (ICM), pp. 76–81 (2015)
2. Kim, H., Lee, J., Lee, S., Cui, X., Kim, H.: Sensor fusion-based human tracking using particle filter and data mapping analysis in in/outdoor environment. In: 2013 10th International Conference on Ubiquitous Robots and Ambient Intelligence (URAI), pp. 741–744 (2013)
3. Jung, E., Yi, B.: Study on intelligent human tracking algorithms with application to omnidirectional service robots. In: 2013 10th International Conference on Ubiquitous Robots and Ambient Intelligence (URAI), pp. 80–81 (2013)
4. Wan, M., Zhang, H., Fu, M., Zhou, W.: Motion control strategy for redundant visual tracking mechanism of a humanoid robot. In: 2016 8th International Conference on Intelligent Human-Machine Systems and Cybernetics (IHMSC), pp. 156–159 (2016)
5. Jia, S., Zang, R., Li, X., Zhang, X., Li, M.: Monocular robot tracking scheme based on fully-convolutional siamese networks. In: 2018 Chinese Automation Congress (CAC), pp. 2616–2620 (2018)
6. Hare, S., Golodetz, S., Saffari, A., et al.: Struck: structured output tracking with kernels. IEEE Trans. Pattern Anal. Mach. Intell. **38** 2096–2109 (2015)
7. Kalal, Z., Mikolajczyk, K., Matas, J.: Tracking-learning-detection. IEEE Trans. Pattern Anal. Mach. Intell. **6**(1), 1409–1422 (2010)
8. Zhang, K., Zhang, L., Yang, M.: Fast compressive tracking. IEEE Trans. Pattern Anal. Mach. Intell. **36**(10), 2002–2015 (2014)
9. Zhong, W., Lu, H., Yang, M.H.: Robust object tracking via sparse collaborative appearance model. IEEE Trans. Image Process. **23**(5), 2356–2368 (2014)
10. Danelljan, M., Bhat, G., Khan, F.S., et al.: ATOM: tracking by overlap maximization. In: 2019 IEEE/CVF Conference on Computer Vision and Pattern Recognition (CVPR). IEEE (2020)

11. Ferrer, G., Zulueta, A.G., Cotarelo, F.H., et al.: Robot social-aware navigation framework to accompany people walking side-by-side. Auton. Robots **41**(4), 775–793 (2017)
12. Cifuentes, C.A., Frizera, A., Carelli, R., et al.: Human–robot interaction based on wearable IMU sensor and laser range finder. Robot. Auton. Syst. **62**(10), 1425–1439 (2014)
13. Jiang, B., Luo, R., Mao, J., et al.: Acquisition of localization confidence for accurate object detection (2018)

Extracting Stable Control Information from EMG Signals to Drive a Musculoskeletal Model - A Preliminary Study

Jiamin Zhao, Yang Yu, Xinjun Sheng(✉), and Xiangyang Zhu

State Key Laboratory of Mechanical System and Vibration, Shanghai Jiao Tong University, 800 Dongchuan Road, Shanghai, China
xjsheng@sjtu.edu.cn

Abstract. Musculoskeletal models (MMs) driven by electromyography (EMG) signals have been used to predict human movements. Muscle excitations of MMs are generally the amplitude of EMG, which shows large variability even when repeating the same task. The general structure of muscle synergies has been proved to be consistent across test sessions, providing a perspective for extracting stable control information for MMs. Although non-negative matrix factorization (NMF) is a common method for extracting synergies, the factorization result of NMF is not unique. In this study, we proposed an improved NMF algorithm for extracting stable control information of MMs to predict hand and wrist motions. Specifically, we supplemented the Hadamard product and $L2$-norm regularization term to the objective function of NMF. The proposed NMF was utilized to identify stable muscle synergies. Then, the time-varying profile of each synergy was fed into a subject-specific MM for estimating joint motions. The results demonstrated that the proposed scheme significantly outperformed a traditional MM and an MM combined with the classic NMF (NMF-MM), with averaged R and NRMSE equal to 0.89 ± 0.06 and 0.16 ± 0.04. Further, the similarity between muscle synergies extracted from different training data revealed the proposed method's effectiveness of identifying consistent control information for MMs. This study provides a novel model-based scheme for the estimation of continuous movements.

Keywords: Musculoskeletal model · Electromyography · Non-negative matrix factorization algorithm

1 Introduction

Electromyography (EMG) signals have been widely used to decode motion intentions in the control of prosthetics, exoskeletal systems, and robots [1–3]. To provide intuitive control of multiple degrees of freedom (DoFs), continuous and

H. Liu et al. (Eds.): ICIRA 2022, LNAI 13456, pp. 735–746, 2022.
https://doi.org/10.1007/978-3-031-13822-5_66

coordinated myoelectric control has attracted increasing attention. In the literature, there have been many attempts for achieving multi-DoF continuous and simultaneous control, such as linear regression [4], support vector regression [5], artificial neural network [6], and deep learning methods [7].

Recently, EMG-driven musculoskeletal models (MMs) have been proposed with the aim of better mimicking physiologic human movements by decoding explicit representations of anatomical structures of the neuromusculoskeletal systems [8–13]. EMG-based MM is a step towards understanding how neural commands are translated into mechanical outputs through modeling of muscle activation, kinematics, contraction dynamics, and joint mechanics [8,14]. Muscles in MMs are modeled as Hill-type actuators, in which subject-specific model parameters (e.g., maximum isometric force and optimal fiber length) are utilized to depict neuromuscular characteristics relating to muscle force production and can be estimated by the constrained global numerical optimization [10]. An important criterion for myoelectric control is whether the controller enabled consistent trial-to-trial performance under various movement conditions with the same customized model parameters. Crouch *et al.* [10,15] proposed a lumped-parameter MM to decode hand/wrist motions with high accuracy and their studies proved that the MM-based controller could enable relatively consistent control across two separate testing days. These studies revealed the stability of the subject-specific neuromuscular system. However, large variabilities were shown in the comparison of subject-specific model parameters of biceps and triceps brachii in repeatability experiments [16]. The variation highlighted the model's low repeatability and the strong non-stationarity of EMG signals even in the same task. The non-stationary EMG signals in across-trial usage will lead to the requirement for re-calibration of MM-based controllers. Therefore, to improve the robustness of MMs and reduce the training burden, it is important for the MM-based controller to extract stable and consistent control information from multi-channel EMG signals.

Recent evidence suggests that the performance of complex human movements such as walking and grasping may be accomplished using muscle synergies [17–19]. Kristiansen *et al.* [20] performed a cross-correlation analysis, in which the synergies extracted in different sessions were compared and the same general structure of these muscle synergies was present across sessions. Therefore, muscle synergy might provide a perspective for extracting consistent control information for MMs. Non-negative matrix factorization (NMF) [21] is often used to identify muscle synergies from EMG signals. However, it is reported that the extracted synergies based on the classic NMF are not unique. We suppose that, by supplementing Hadamard product and $L2$-norm regularization term to the objective function of NMF (NMF-HPWH-L2), muscle synergies would be unique and consistent across sessions and better reflect the neural coordinative structures. Thus, in this study, we aimed to investigate if the improved NMF could extract relatively consistent control information for MMs to improve the model performance. We proposed an MM combined with the improved NMF (NMF-HPWH-L2-MM) to predict hand and wrist flexion/extension joint angles.

The remainder of the paper is organized as follows: Sect. 2 describes the experimental setups, the details of NMF-HPWH-L2-MM, and the evaluation criteria. The estimation performance of different approaches is elaborated in Sect. 3. Section 4 discusses the effectiveness of the proposed method. Conclusions are drawn in Sect. 5.

2 Methodology

2.1 Subjects

Three able-bodied subjects (S1-S3, 2 females and 1 male, aged from 25 to 27 years old) participated in the experiment and received informed consent with respect to the experiment details. The experimental procedures were in accordance with the Declaration of Helsinki.

2.2 Experimental Protocol

Subjects were required to accomplish three tasks in a static upper limb posture with the hand and arm in neutral positions and the elbow flexed to 90°: (1) metacarpophalangeal (MCP) flexion/extension only, (2) wrist flexion/extension only, and (3) simultaneous MCP and wrist flexion/extension. Five trials were performed for each task. The duration of each trial was 20 s. And there was a 20-second break between trials to avoid fatigue. During the experiment, surface EMG signals and joint angles were recorded at the same time. Two angle sensors (Witmotion Ltd, China) were used to record MCP and wrist movements at 40 Hz. Then, joint angles were low-pass filtered (6 Hz). EMG signals were recorded by Delsys sensors (Delsys Trigno Wireless, Delsys Inc., USA) from six muscles (see Fig. 1): 1-flexor carpi radialis (FCR), 2-flexor digitorum superficialis (FDS), 3-flexor carpi ulnaris (FCU), 4-extensor carpi ulnaris (ECU), 5-extensor digitorum (ED), and 6-extensor carpi radialis longus (ECRL). EMG signals were sampled at 2000 Hz and bandpass filtered (29–450 Hz). A notching comb filter (50 Hz) was used to eliminate the power-line interference. The signals were then rectified and normalized by the maximum EMG values recorded during maximum voluntary contractions.

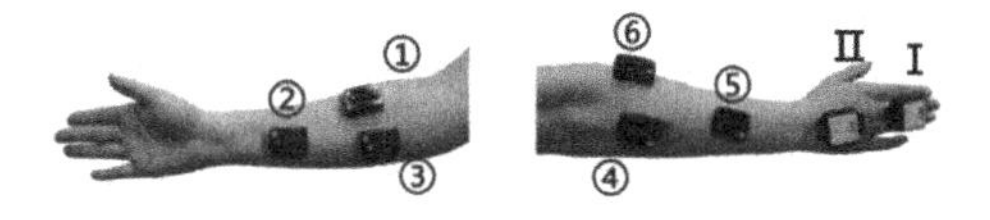

Fig. 1. Experimental setups. Two angle sensors and six surface EMG sensors: angle sensors I and II for MCP and wrist angle, and surface EMG sensors 1–6 for FCR, FDS, FCU, ECU, ED, and ECRL.

2.3 MM Combined with NMF-HPWH-L2 (NMF-HPWH-L2-MM)

Classic NMF. NMF is a factorization algorithm for multivariate analysis proposed by Lee and Seung [21]. Due to its non-negative elements, it is suitable to use NMF to extract muscle synergies from multi-channel EMG signals. NMF was applied to the time-domain features of EMG signals (Z) to identify muscle synergies with a set of muscle weightings (W) and the associated time-varying profiles (H) in Eq. 1. Z was a $6 \times N$ matrix indicating the root mean square (RMS) values of post-processed EMG signals; 6 and N were the numbers of muscle and trial frame, respectively. The dimensionality of the muscle weighting set (W) in this study was four with the Variation Accounted For (VAF) index exceeding 0.8 [22]. For each DoF (1 for MCP, 2 for wrist), there were two synergies, synergy "+" and synergy "–", representing synergies corresponding to joint flexion and extension.

$$Z = W \times H = \left[W_1^+, W_1^-, W_2^+, W_2^-\right] \times \begin{bmatrix} H_1^+(t) \\ H_1^-(t) \\ H_2^+(t) \\ H_2^-(t) \end{bmatrix} \tag{1}$$

NMF-HPWH-L2 For the improved NMF, we applied Hadamard product to the muscle weighting (W) and time-varying profile (H). The Hadamard product is defined as follows:

$$[X \circ Y]_{mn} = [X]_{mn} \times [Y]_{mn}, \tag{2}$$

where $[\cdot]_{mn}$ represented the mth row and nth column element of a matrix.

In NMF-HPWH-L2, there were two constraint matrices (S_w and S_h) applied to W and H, respectively. The matrix S_w was used to reduce the adverse effect of EMG crosstalk for the MM-based controller, which has been proved in our previous study [13]. The selection of S_w was detailed in [13] and briefly introduced as follows. In Eq. 1, $W_i^+ = \left[W_{1,i}^+, \ldots, W_{j,i}^+, \ldots, W_{6,i}^+\right]^{\mathrm{T}}$ and $W_i^- = \left[W_{1,i}^-, \ldots, W_{j,i}^-, \ldots, W_{6,i}^-\right]^{\mathrm{T}}$. They were both 6×1 matrices with $i = 1, 2$ and $j = 1 \sim 6$ representing the ith DoF and the jth muscle. Then, the weightings were normalized to the highest value in each column. The normalized weightings in Eq. 3, $\overline{W}_{j,i}^+$ and $\overline{W}_{j,i}^-$, determined the relative contribution of each muscle to the activation of synergy-related movement within each synergy.

$$\begin{aligned} \overline{W}_{j,i}^+ &= \frac{W_{j,i}^+}{max(W_{1,i}^+, \ldots, W_{6,i}^+)} \\ \overline{W}_{j,i}^- &= \frac{W_{j,i}^-}{max(W_{1,i}^-, \ldots, W_{6,i}^-)} \end{aligned} \tag{3}$$

Besides, we checked the difference of a muscle's contribution to four synergies by the percentages of weightings (P), which were calculated as follows: (1) each row of H was normalized to the maximum of each row, and the normalized profiles were $\overline{H}_i^+(t)$ and $\overline{H}_i^-(t)$, (2) W was converted to V (see Eq. 4) to ensure that $V \times \left[\overline{H}_i^+(t), \overline{H}_i^-(t)\right]^{\mathrm{T}}$ was equal to Z, and (3) P could be calculated by Eq. 5.

$$\begin{cases} V_{j,i}^+ = W_{j,i}^+ \times \max H_i^+(t) \\ V_{j,i}^- = W_{j,i}^- \times \max H_i^-(t) \end{cases} \tag{4}$$

$$\begin{cases} P_{j,i}^+ = \frac{V_{j,i}^+}{V_{j,1}^+ + V_{j,1}^- + V_{j,2}^+ + V_{j,2}^-} \\ P_{j,i}^- = \frac{V_{j,i}^-}{V_{j,1}^+ + V_{j,1}^- + V_{j,2}^+ + V_{j,2}^-} \end{cases} \tag{5}$$

In this study, only if both $\overline{W}_{j,i}^+$ and $P_{j,i}^+$ were greater than 0.2, the muscle was assumed to contribute to the excitation of joint flexion of DoF i and the corresponding element ($S_{w_{j,i}^+}$) in the matrix S_w was set to 1, otherwise $S_{w_{j,i}^+} = 0$. The same cut-off criterion was also applied to $\overline{W}_{j,i}^-$ and $P_{j,i}^-$ to determine $S_{w_{j,i}^-}$.

The matrix S_h was designed to represent the activation state of each muscle synergy. For example, the elements in S_h corresponding to MCP rotation were set to 1 during task (1) and other elements were 0. Equation 1 was converted to Eq. 6 by adding S_w and S_h.

$$\begin{aligned} Z &= (S_w \circ W) \times (S_h \circ H) \\ &= (S_w \circ W) \times \left(\begin{bmatrix} 1_{1\times\frac{N}{2}}, 0_{1\times\frac{N}{2}} \\ 1_{1\times\frac{N}{2}}, 0_{1\times\frac{N}{2}} \\ 0_{1\times\frac{N}{2}}, 1_{1\times\frac{N}{2}} \\ 0_{1\times\frac{N}{2}}, 1_{1\times\frac{N}{2}} \end{bmatrix} \circ \begin{bmatrix} H_{1,1}^+(t), H_{2,1}^+(t) \\ H_{1,1}^-(t), H_{2,1}^-(t) \\ H_{1,2}^+(t), H_{2,2}^+(t) \\ H_{1,2}^-(t), H_{2,2}^-(t) \end{bmatrix} \right) \end{aligned} \tag{6}$$

$H_{q,i}^+$ and $H_{q,i}^-$ represented the time-varying profiles corresponding to joint flexion and extension of DoF i during task q ($q = 1, 2$). Z was composed of two-trial data separately from task (1) and task (2). Thus, $N/2$ in matrix S_h was the number of trial frame in each task. $1_{1\times\frac{N}{2}}$ and $0_{1\times\frac{N}{2}}$ were $1 \times \frac{N}{2}$ matrices composed of 1 and 0, respectively.

The $L2$-norm regularization terms [23] of W and H were also complemented to the objective function of classic NMF in order to avoid the factorization relying too much on the training data. Thus, the objective function of NMF-HPWH-L2 was

$$\min_{W,H} \frac{1}{2} \|Z - (S_w \circ W) \times (S_h \circ H)\|_{Fro}^2 + \lambda \sum_{j=1}^{6} \|W(j,:)\|_2^2 + \beta \sum_{l=1}^{4} \|H(l,:)\|_2^2 \quad s.t. W, H \geq 0, \tag{7}$$

where the lower corner "Fro" was Frobenius norm, λ and β were regular terms. In this study, $\lambda = 5$ and $\beta = 1$ were empirical values, which enabled better performance of the proposed method. Then, Eq. 7 could be rewritten as:

$$\min_{W,H} \frac{1}{2} \left\| \begin{bmatrix} S_w \circ W \\ \sqrt{\lambda} I_{4\times4} \end{bmatrix} \times \left[S_h \circ H, \sqrt{\beta} I_{4\times4} \right] - \begin{bmatrix} Z, 0_{6\times4} \\ 0_{4\times N}, 0_{4\times4} \end{bmatrix} \right\|_{Fro}^2 \quad s.t. W, H \geq 0, \tag{8}$$

where I was a unit matrix. Equation 8 can be solved by alternating non-negative least-squares [24], which was detailed in Eqs. 9 and 10:

$$\begin{bmatrix} S_w \circ W \\ \sqrt{\lambda} I_{4\times4} \end{bmatrix}^{(k+1)} = arg \min_{W>0} \left\| \begin{bmatrix} S_w \circ W \\ \sqrt{\lambda} I_{4\times4} \end{bmatrix} \times \left[S_h \circ H, \sqrt{\beta} I_{4\times4} \right]^{(k)} - \begin{bmatrix} Z, 0_{6\times4} \\ 0_{4\times N}, 0_{4\times4} \end{bmatrix} \right\|_{Fro}^2 \tag{9}$$

$$\left[S_h \circ H, \sqrt{\beta} I_{4\times4} \right]^{(k+1)} = arg \min_{H>0} \left\| \begin{bmatrix} S_w \circ W \\ \sqrt{\lambda} I_{4\times4} \end{bmatrix}^{(k+1)} \times \left[S_h \circ H, \sqrt{\beta} I_{4\times4} \right] - \begin{bmatrix} Z, 0_{6\times4} \\ 0_{4\times N}, 0_{4\times4} \end{bmatrix} \right\|_{Fro}^2. \tag{10}$$

Finally, Eqs. 9 and 10 can be solved using multiplicative iteration algorithm [25].

NMF-HPWH-L2-MM. As described above, muscle weighting W was resolved with two-trial data using the proposed NMF-HPWH-L2 algorithm. Then, muscle excitation H_M for MMs was determined by minimizing $\|W \times H_M(t) - Z\|_2^2$ with the constraint $H_M(t) > 0$. Each row of H_M represented muscle excitations of the corresponding muscle synergy. In this study, each muscle synergy was modeled as a Hill-type muscle actuator. Muscle excitations were subsequently converted into muscle activations using the excitation-activation dynamics [10,13]. Then, muscle activations were fed into the subject-specific MM as inputs.

2.4 Other Algorithms for Comparison

To investigate the effectiveness of the proposed method, we made a comparison with a traditional MM [13] and an MM combined with the classic NMF algorithm (NMF-MM). For each method, the training data consisted of two-trial data separately from task (1) and task (2). The training data were the same as those of NMF-HPWH-L2-MM. The remaining data were used for model evaluation. Muscle excitations in the traditional MM were post-processed EMG signals (refer to Sect. 2.2).

2.5 Model Calibration and Testing

Customized model parameters were determined by the constrained global numerical optimization. Model parameters were optimized to minimize the mean square error between measured and predicted joint angles [10,13], including optimal fiber length, maximum isometric force, moment arms, passive elastic element stiffness, and fiber length at the neutral position.

2.6 Performance Evaluation and Statistical Analysis

The estimation performance of different methods was assessed by Pearson correlation coefficient (R) and normalized root mean square error (NRMSE) between measured and predicted joint angles. R and NRMSE during tasks (1) and (2) were the values at the excited DoF. Additionally, we implemented a one-way ANOVA to test if the proposed method had better performance than MM and NMF-MM. The independent variable was the algorithm (NMF-HPWH-L2-MM, MM, and NMF-MM). Dependent variables were R and NRMSE. Figure 2 demonstrated the procedures of comparing NMF-HPWH-L2-MM, MM, and NMF-MM. The significance level in the statistical analysis was set to 0.05.

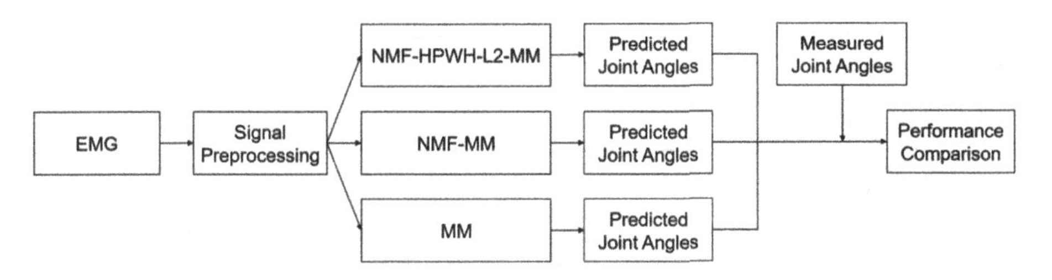

Fig. 2. Block diagram of the procedures of comparing the proposed NMF-HPWH-L2-MM, MM, and NMF-MM.

3 Results

3.1 Estimation Performance of NMF-HPWH-L2-MM

The estimation results of one subject (S2) are represented in Fig. 3, indicating the feasibility of the proposed model to predict hand and wrist movements during different tasks. The blue dashed lines were the measured joint angles, while the yellow ones represented the predicted angles. For the trials in Fig. 3, the R values in tasks (1), (2), and (3) were 0.96, 0.97, and 0.88, respectively. And the corresponding NRMSE values were 0.16, 0.08, and 0.16.

3.2 Comparison with MM and NMF-MM

Figure 4 illustrates the averaged R and NRMSE of three algorithms (NMF-HPWH-L2-MM, MM, and NMF-MM) across subjects, tasks, and DoFs. The results demonstrated that NMF-HPWH-L2-MM had the highest estimation accuracy, with averaged R values: NMF-HPWH-L2-MM: 0.89 ± 0.06, MM: 0.82 ± 0.09, and NMF-MM: 0.83 ± 0.09. The statistical analysis revealed that the R value of NMF-HPWH-L2-MM was significantly higher than those of MM and NMF-MM. Meanwhile, the NRMSE of NMF-HPWH-L2-MM, MM, and NMF-MM were 0.16 ± 0.04, 0.19 ± 0.05, and 0.18 ± 0.05, respectively. There was a significant difference between the NRMSE of NMF-HPWH-L2-MM and MM.

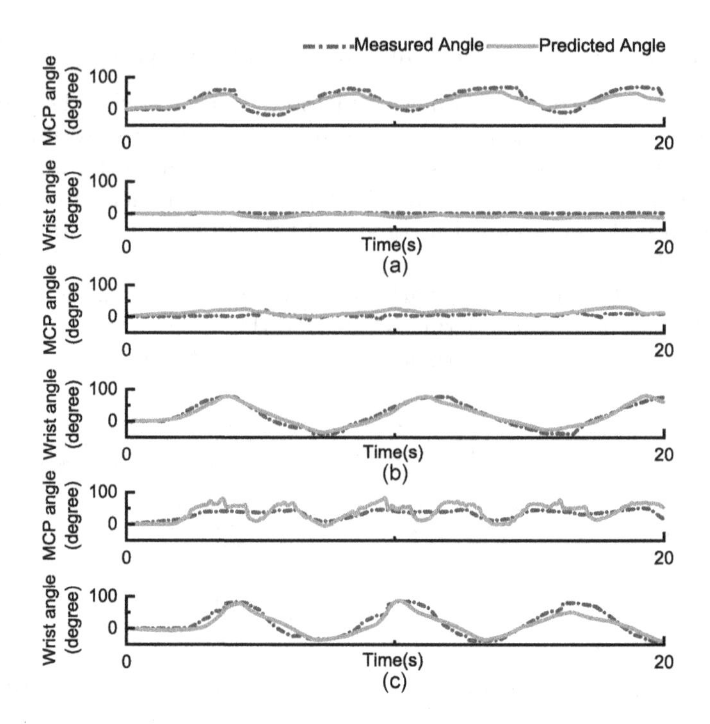

Fig. 3. Measured and predicted joint angles during (a) MCP flexion/extension only, (b) wrist flexion/extension only, and (3) simultaneous MCP and wrist rotation of S2. (Color figure online)

4 Discussion

4.1 Estimation Performance of NMF-HPWH-L2-MM

In this study, we proposed an MM combined with the improved NMF algorithm to improve the estimation performance of model-based approaches. Figure 3 showed the estimated and measured joint angles in different tasks. Obviously, the proposed scheme could achieve a promising estimation accuracy. To investigate the effectiveness of NMF-HPWH-L2-MM, we compared the estimation performance of the proposed scheme with a traditional MM and NMF-MM. In Fig. 4, NMF-HPWH-L2-MM had much higher estimation accuracy with averaged R and NRMSE: NMF-HPWH-L2-MM: 0.89 ± 0.06 and 0.16 ± 0.04, MM: 0.82 ± 0.09 and 0.19 ± 0.05, and NMF-MM: 0.83 ± 0.09 and 0.18 ± 0.05. Further, the proposed scheme significantly outperformed MM and NMF-MM ($p < 0.05$).

4.2 Analysis of the Effectiveness of NMF-HPWH-L2-MM

To improve the robustness of MMs, we proposed an improved NMF algorithm to extract stable control information for MMs. We supplemented Hadamard product to the muscle weighting (W) and time-varying profile (H). Besides,

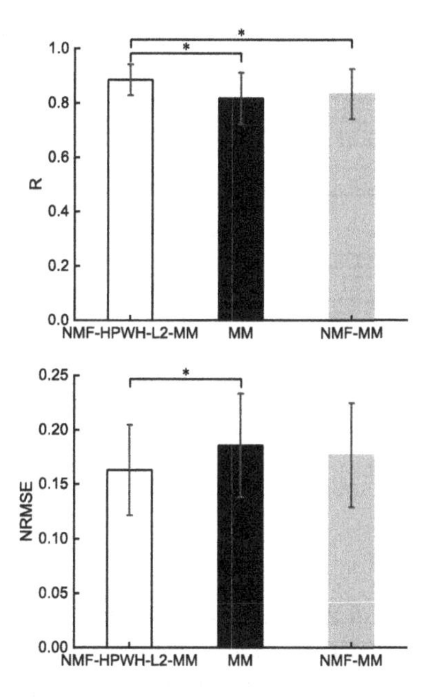

Fig. 4. Averaged R and NRMSE across subjects, tasks, and DoFs for NMF-HPWH-L2-MM, MM, and NMF-MM. The error bar means the standard deviation. Symbols "*" indicate significant differences in the statistical analysis ($p < 0.05$).

$L2$-norm regularization term was added to the objective function of the classic NMF. To better demonstrate the effectiveness of the proposed NMF-HPWH-L2 for extracting stable muscle synergies for model-based controllers, we compared the extracted muscle weighting matrices of subject S2 from two different training data (each training data consisted of two different trials separately from tasks (1) and (2)). In Fig. 5, the subfigures (a) and (b) depicted these two muscle weighting matrices, respectively. The histograms with different colors represented different muscle synergies, which were responsible for different movements at each DoF. In each histogram, the Y-axis represented various muscles (refer to Fig. 1), while the X-axis was the muscle weighting matrix. The comparison between subfigures (a) and (b) revealed that the general structure of muscle synergies across trials extracted by NMF-HPWH-L2 was highly similar. As such, the proposed NMF-HPWH-L2 algorithm enables the extraction of stable low-dimensional control information from multi-channel EMG signals for model-based controllers, which perhaps results in its excellent estimation performance.

Further, to evaluate the across-trial similarity of the extracted muscle synergies, cosine similarity was considered for the across-trial similarity measurement [26]. Cosine similarity is a measure between two vectors that measures the cosine angle between them and ranged from -1 and 1. -1 indicates the opposite and 1 indicates the identical exactly. For subject S2, we calculated the cosine similarities of muscle synergies, which were extracted from five different training data.

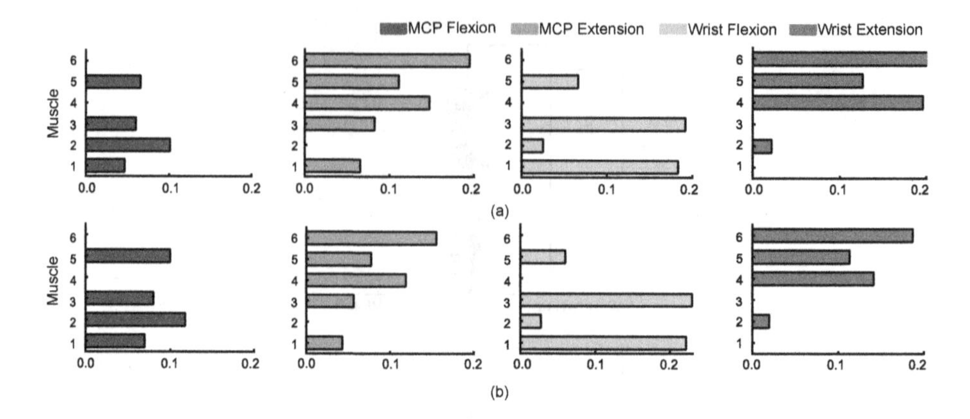

Fig. 5. The comparison of muscle weightings of two muscle synergies extracted from different training data.

As described in Sects. 2.3 and 2.4, each training data were composed of two-trial data from task (1) and task (2). Table 1 presented the cosine similarities between each two muscle synergies, which revealed the capacity of NMF-HPWH-L2 to extract stable muscle synergies and consistent control information for MMs. The cosine similarity between the synergies in Fig. 5 was 0.9811. Further, the extracted stable control information may provide a perspective for reducing the retraining burden of MMs in across-day usage, which will be investigated in our future work.

Table 1. Cosine similarities between extracted muscle synergies

Synergy number	1	2	3	4	5
1	1.0000	0.9978	0.9748	0.9811	0.9938
2	–	1.0000	0.9855	0.9860	0.9934
3	–	–	1.0000	0.9900	0.9832
4	–	–	–	1.0000	0.9925
5	–	–	–	–	1.0000

4.3 Limitations

However, there existed several limitations in the current study. The number of subjects was limited and the study was an offline analysis. Future work will demonstrate the feasibility of the proposed scheme with more subjects in online tests.

5 Conclusion

This study introduced an improved NMF algorithm for extracting stable control information of MMs to estimate hand and wrist motions, which could be potentially applied in neural control for prostheses. To investigate the effectiveness of the proposed scheme, we made a comparison with a traditional MM and NMF-MM. The outcomes of comparisons indicated that the proposed approach had significantly higher estimation accuracy than the traditional MM and NMF-MM. The proposed method paves the way for extracting stable muscle excitations and reducing the retraining burden for model-based controllers.

Acknowledgements. The authors would like to thank all the participants in the experiments. This work was supported by the National Natural Science Foundation of China under Grant 91948302 and Grant 51905339.

References

1. Piazza, C., Rossi, M., Catalano, M.G., Bicchi, A., Hargrove, L.J.: Evaluation of a simultaneous myoelectric control strategy for a multi-DoF transradial prosthesis. IEEE Trans. Neural Syst. Rehabil. Eng. **28**(10), 2286–2295 (2020)
2. Yao, S., Zhuang, Y., Li, Z., Song, R.: Adaptive admittance control for an ankle exoskeleton using an EMG-driven musculoskeletal model. Front. Neurorobot. **12**, 16 (2018)
3. Delpreto, J., Rus, D.: Sharing the load: Human-robot team lifting using muscle activity. In: 2019 International Conference on Robotics and Automation (ICRA), pp. 7906–7912 (2019)
4. Smith, L.H., Kuiken, T.A., Hargrove, L.J.: Evaluation of linear regression simultaneous myoelectric control using intramuscular EMG. IEEE Trans. Biomed. Eng. **63**(4), 737–746 (2016)
5. Ameri, A., Kamavuako, E.N., Scheme, E.J., Englehart, K.B., Parker, P.A.: Support vector regression for improved real-time, simultaneous myoelectric control. IEEE Trans. Neural Syst. Rehabil. Eng. **22**(6), 1198–1209 (2014)
6. Muceli, S., Farina, D.: Simultaneous and proportional estimation of hand kinematics from EMG during mirrored movements at multiple degrees-of-freedom. IEEE Trans. Neural Syst. Rehabil. Eng. **20**(3), 371–378 (2012)
7. Yu, Y., Chen, C., Zhao, J., Sheng, X., Zhu, X.: Surface electromyography image-driven torque estimation of multi-DoF wrist movements. IEEE Trans. Industr. Electron. **69**(1), 795–804 (2022)
8. Buchanan, T.S., Lloyd, D.G., Manal, K., Besier, T.F.: Neuromusculoskeletal modeling: estimation of muscle forces and joint moments and movements from measurements of neural command. J. Appl. Biomech. **20**(4), 367–395 (2004)
9. Lloyd, D.G., Besier, T.F.: An EMG-driven musculoskeletal model to estimate muscle forces and knee joint moments in vivo. J. Biomech. **36**(6), 765–776 (2003)
10. Crouch, D.L., Huang, H.: Lumped-parameter electromyogram-driven musculoskeletal hand model: A potential platform for real-time prosthesis control. J. Biomech. **49**(16), 3901–3907 (2016)
11. Sartori, M., Durandau, G., Dosen, S., Farina, D.: Robust simultaneous myoelectric control of multiple degrees of freedom in wrist-hand prostheses by real-time neuromusculoskeletal modeling. J. Neural Eng. **15**(6), 066,026.1-066,026.15 (2018)

12. Zhao, Y., Zhang, Z., Li, Z., Yang, Z., Xie, S.: An EMG-driven musculoskeletal model for estimating continuous wrist motion. IEEE Trans. Neural Syst. Rehabil. Eng. **28**(12), 3113–3120 (2020)
13. Zhao, J., Yu, Y., Wang, X., Ma, S., Sheng, X., Zhu, X.: A musculoskeletal model driven by muscle synergy-derived excitations for hand and wrist movements. J. Neural Eng. **19**(1), 016027 (2022)
14. Winters, J.M., Woo, S.LY.: Multiple Muscle Systems, pp. 69–93. Springer, New York (1990)
15. Crouch, D.L., Huang, H.: Musculoskeletal model-based control interface mimics physiologic hand dynamics during path tracing task. J. Neural Eng. **14**(3), 036008 (2017)
16. Jayaneththi, V.R., Viloria, J., Wiedemann, L.G., Jarrett, C., Mcdaid, A.J.: Robotic assessment of neuromuscular characteristics using musculoskeletal models: A pilot study. Comput. Biol. Med. **86**, 82–89 (2017)
17. Neptune, R.R., Clark, D.J., Kautz, S.A.: Modular control of human walking: a simulation study. J. Biomech. **42**(9), 1282–1287 (2009)
18. Sartori, M., Gizzi, L., Lloyd, D.G., Farina, D.: A musculoskeletal model of human locomotion driven by a low dimensional set of impulsive excitation primitives. Front. Comput. Neurosci. **7**, 79 (2013)
19. Pale, U., Atzori, M., Müller, H., Scano, A.: Variability of muscle synergies in hand grasps: Analysis of intra- and inter-session data. Sensors **20**(15), 4297 (2020)
20. Kristiansen, M., Samani, A., Madeleine, P., Hansen, E.A.: Muscle synergies during bench press are reliable across days. J. Electromyogr. Kinesiol. **30**, 81–88 (2016)
21. Lee, D.D., Seung, H.H.: Learning parts of objects by non-negative matrix factorization. Nature **401**, 788–791 (1999)
22. Gizzi, L., Nielsen, J.F., Felici, F., Ivanenko, Y.P., Farina, D.: Impulses of activation but not motor modules are preserved in the locomotion of subacute stroke patients. J. Neurophysiol. **106**(1), 202–210 (2011)
23. Wang, D., Liu, J.X., Gao, Y.L., Yu, J., Zheng, C.H., Xu, Y.: An NMF-L2,1-norm constraint method for characteristic gene selection. PLoS ONE **11**(7), e0158494 (2016)
24. Cichocki, A., Zdunek, R., Amari, S.: New algorithms for non-negative matrix factorization in applications to blind source separation. In: IEEE International Conference on Acoustics Speech and Signal Processing Proceedings (2006)
25. Cichocki, A., Amari, Si., Zdunek, R., Kompass, R., Hori, G., He, Z.: Extended SMART algorithms for non-negative matrix factorization. In: International Conference on Artificial Intelligence and Soft Computing (ICAISC), pp. 548–562 (2006)
26. Scano, A., Dardari, L., Molteni, F., Giberti, H., Tosatti, L.M., d'Avella, A.: A comprehensive spatial mapping of muscle synergies in highly variable upper-Limb movements of healthy subjects. Front. Physiol. **10**, 1231 (2019)

Construction of Complex Brain Network Based on EEG Signals and Evaluation of General Anesthesia Status

Zhiwen Xiao, Ziyan Xu, and Li Ma(✉)

School of Information Engineering, Wuhan University of Technology, Wuhan 430070, China
excellentmary@whut.edu.cn

Abstract. General anesthesia is now an important part of surgery, it can ensure that patients undergo surgery in a painless and unconscious state. The traditional anesthesia depth assessment mainly relies on the subjective judgment of the anesthesiologist, lacks a unified standard, and is prone to misjudgment. Since general anesthesia is essentially anesthesia of the central nervous system, the state of anesthesia can be monitored based on EEG analysis. Based on this, this paper proposes a method to reasonably construct a brain connection network system based on the characteristic parameters of EEG signals and combine machine learning to evaluate the state of anesthesia. This method extracts the EEG signals related to the depth of anesthesia, The knowledge of graph theory introduces the three functional indicators of Pearson correlation coefficient, phase-lock value and phase lag index to construct a complex brain network, and then perform feature selection based on the constructed brain network to generate a dataset, and use machine learning methods for classification. To evaluate the anesthesia state, the experimental results show that the accuracy of the method for evaluating the anesthesia state can reach 93.88%.

Keywords: EEG · Anesthesia assessment · Complex brain networks · Machine learning

1 Introduction

As a highly complex system, the traditional research methods cannot sufficiently reflect the internal operation principle of the brain [1]. It is necessary to find a suitable way to study it. The research based on graph theory shows that the structure and functional network of the human brain conforms to the complex network model. Using the complex network data analysis method to construct the corresponding brain network to analyze the brain helps understand the physiological mechanism of the brain, explore the relationship between different brain regions, and has significant advantages in the study of related physiological activities such as brain diseases.

Brain science research has become an important research direction today [2]. Countries worldwide have launched relevant plans to explore the mystery of the brain. Most studies on functional brain networks are based on fMRI, EEG, and MEG data. Among

H. Liu et al. (Eds.): ICIRA 2022, LNAI 13456, pp. 747–759, 2022.
https://doi.org/10.1007/978-3-031-13822-5_67

them, EEG has the advantage of high temporal resolution [3]. EEG signals to analyze the state of brain activity have become a common means to construct the EEG brain functional network. From the perspective of EEG, it can reveal the topological structure of the brain and find abnormal indicators of mental diseases, which provides a reference for the clinical diagnosis of brain diseases and the judgment of anesthesia depth.

Anesthesia is an essential part of the surgery, ensuring that patients receive surgery unconsciously. Since 1942, due to the clinical application of muscle relaxant drugs, the classical ether anesthesia depth staging is no longer applicable to the clinical judgment of anesthesia depth. Later, scholars such as Jcssop and Artusio conducted research on the mechanism of anesthesia and the relationship between anesthesia and consciousness [4]. However, the interaction between anesthesia drugs and the central nervous system is very complex, so there is no "gold standard" for the identification of different depths of anesthesia so far. The traditional evaluation of anesthesia depth mainly relies on the subjective judgment of anesthesiologists. Affected by the individual differences of patients and the psychological and physiological conditions of anesthesiologists, it is prone to misjudgment. To maintain a stable anesthesia level, an objective and reliable method is needed to evaluate the depth of anesthesia and assist anesthesiologists in better completing anesthesia. The primary role of anesthetics is the brain's central nervous system, and EEG has a solid sensitivity to anesthetics. Therefore, monitoring anesthesia depth based on EEG detection has a solid theoretical basis. Consequently, it is of great scientific value to reasonably construct the brain connection network system and conduct real-time monitoring to evaluate the anesthesia status.

This paper proposes a reasonable brain connection network system combined with machine learning to evaluate the anesthesia state. This method extracts the EEG time-frequency spectrum characteristics or parameters related to anesthesia depth. It makes statistical analysis and classification according to these characteristics or parameters to determine the final anesthesia depth index to achieve efficient and stable clinical anesthesia monitoring. This paper consists of four sections. Section 2 mainly introduces constructing a complex brain network based on the extracted EEG features. Section 3 presents the process and classification results using the machine learning method and records the experimental results in Sect. 4.

2 Construction of Complex Brain Network Based on EEG Signal

2.1 Acquisition of EEG Signals

EEG signals are the overall activity of brain nerve cells, including ion exchange, metabolism, and other comprehensive external manifestation. An in-depth study of the characteristics of EEG will promote the exploration and research process of people's brains and enhance their ability to diagnose diseases [5]. As an essential part of the brain, the cerebral cortex contains a variety of nerve cells. The electrical signals generated by various activities between nerve cells constantly adjust the potential on the scalp surface of the brain, and thus the EEG signals are generated. When the brain is in different states due to various factors such as sleep and anesthesia, the EEG signals generated in the brain are significantly different.

The EEG data used in this study came from the BCIcompetitionIV public dataset, which collected EEG signals from 14 patients undergoing general anesthesia surgery and 14 healthy controls in a resting state. All people did not suffer from any brain disease and were in deep anesthesia. The EEG signal acquisition method adopts the electrode placement method of the 10–20 system stipulated by the International Electroencephalogram Society. This method is simple in operation, efficient in the collection, and does not cause any damage to the collected person. Table 1 is the corresponding relationship table of the electrode system.

Table 1 .

Site	Electrode code
Front	Fp1, Fp2
Area	F3, F4, Fz
Central	C3, C4, Cz
Top area	P3, P4, Pz
Pillow area	O1, O2
Side	F7, F8
Temporal region	T3, T4
posterior temporal	T5, T6
Ears	A1, A2

2.2 EEG Signals Preprocessing Based on EEGLAB

Rejection Bad Channel. The collected EEG signals contain EOG signals and some larger artifacts, which can interfere with the experiment. These signals are bad leads. We use EEGLAB to remove bad leads from the original signal [6] (see Fig. 1(a)).

Re-reference. When dealing with continuous multi-channel signals such as EEG, often because of the interference between the signs, the voltage of the electrode with different distances will have other effects due to the other potential difference. Therefore, it is necessary to convert the position of the reference point, that is, to re-reference the signal to restore the accurate signal display of each channel of EEG. Some commonly used reference locations have a bilateral mastoid average reference, which refers to the average value of two mastoid data as the reference data, or the intermediate reference of the whole brain. It refers to using the average value of all data of the entire brain as the reference data. In this paper, the average data of the entire brain are selected for re-reference (see Fig. 1(b)).

Filtering. In the experimental study, we need to use the target frequency band we are interested in. At this time, filtering can help us filter the frequency band that we do not

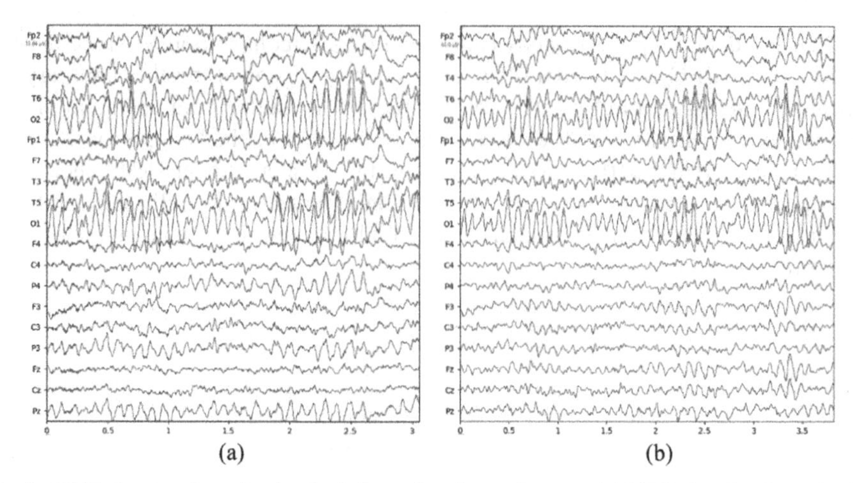

Fig. 1. EEG signals after rejecting bad channel and re-reference. (a) EEG signals after rejecting bad channel; (b) EEG signals after re-reference

care about, but also can filter some useless noise signals to make EEG signals smoother and improve the signal-to-noise ratio of the whole signal. Here we use a low-pass filter, and the cut-off frequency is 30 Hz (see Fig. 2 and Fig. 3).

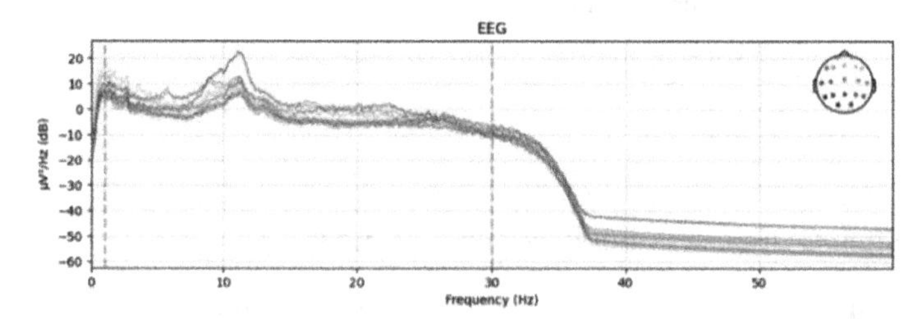

Fig. 2. Frequency domain filter graph of EEG signals

Independent Component Analysis. Independent component analysis is an effective method to solve the problem of blind source separation. The primary purpose of this step is to remove the artifacts such as eye electricity, ECG, and EMG and retain the accurate EEG signals, to obtain relatively clean data (see Fig. 4).

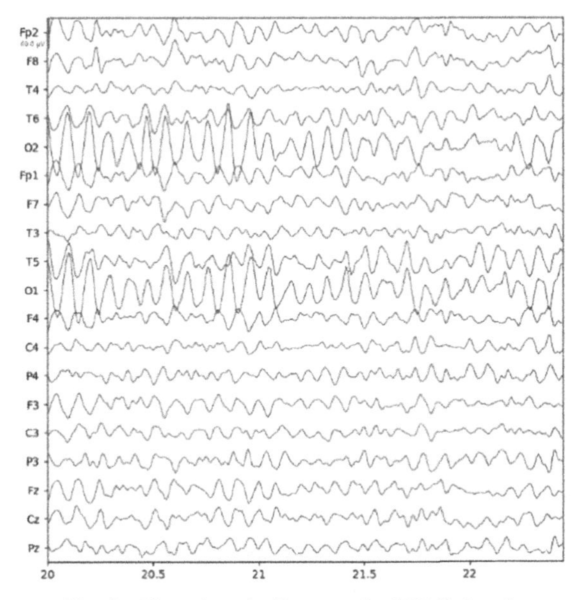

Fig. 3. Time domain filter graph of EEG signals

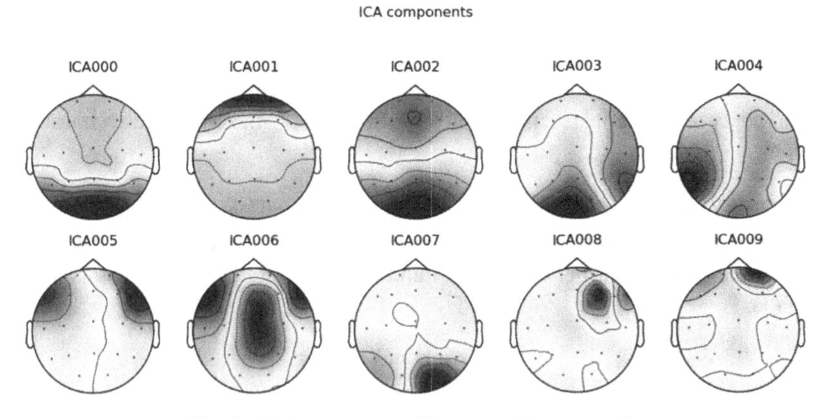

Fig. 4. ICA component 2D potential topography

After analysis, it is found that ICA01 and ICA08 contain artifacts and channel noise. Therefore, two ICA components are eliminated, and then the EEG signal sequence diagram after ICA processing is output (see Fig. 5).

2.3 Brain Network Construction

Brain Functional Network and Its Characteristic Parameters. As a method in EEG research, functional connectivity is essential in studying the correlation between different brain regions. The technique is to measure the dependence or correlation between the two signals to indicate the closeness of the connection between other brain regions.

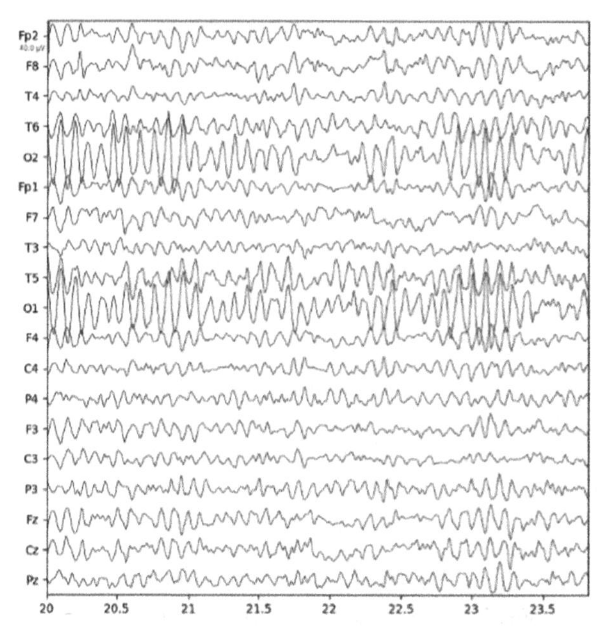

Fig. 5. EEG signals after ICA processing

Based on the graph theory analysis method, this paper introduces the Pearson correlation coefficient, phase locking value, and phase lag index functional connectivity index and analyzes and constructs complex brain networks based on these characteristic parameters.

Pearson Correlation Coefficient. Pearson correlation coefficient is defined as the quotient of covariance and standard deviation between two variables. This parameter is used to measure the correlation between two channels. The calculation process is shown in formula (1).

$$\rho_{X,Y} = \frac{cov(X, Y)}{\sigma_X \sigma_Y} \tag{1}$$

In which $cov(X, Y)$ represents the covariances of X and Y, σ_X and σ_Y represents the sample standard deviation of X and Y, respectively.

Formula (1) defines the overall correlation coefficient, which is usually represented by ρ. The Pearson correlation coefficient can be obtained by estimating the covariance and standard deviation of the sample, which is commonly represented by the English lowercase letter r, as shown in formula (2) and formula (3).

$$r = \frac{E((X - \mu_X)(Y - \mu_Y))}{\sigma_X \sigma_Y} \tag{2}$$

$$r = \frac{E(XY) - E(X)E(Y)}{\sqrt{E(X^2) - E^2(X)}\sqrt{E(Y^2) - E^2(Y)}} \tag{3}$$

The range of r is $[-1, 1]$, so the Pearson correlation coefficient can measure whether the two signals are positive or negative, and the greater the value, the stronger the correlation.

Phase Locking Value. Phase lock value is a phase-based functional connection method that measures the phase difference between two-channel signals. In brain network analysis, PLV is defined as amplitude $A(t)$ and instantaneous phase $\varphi(t)$ of$s(t)$, which are estimated by using Hilbert transform$H\{\}$, the following formula (4).

$$z(t) = s(t) + i \cdot H\{s(t)\} = A(t) \cdot e^{i \cdot \varphi(t)} \tag{4}$$

The analytic signal $z(t)$ can be regarded as the projection of one-dimensional time series in a two-dimensional complex plane.

Phase $\phi(t)$ is calculated by the formula (5).

$$\phi(t) = arctan\left(\frac{Im\{z(t)\}}{Re\{z(t)\}}\right) = arctan\left(\frac{H\{s(t)\}}{s(t)}\right) \phi \in [-\pi, \pi] \tag{5}$$

Phase synchronization is defined as the phase locking of two oscillators, as shown in formula (6).

$$\phi_{12}(t) = \phi_1(t) - \phi_2(t) = const \tag{6}$$

where $\phi_1(t)$ and $\phi_2(t)$ denote the phase of the oscillator, $\phi_{12}(t)$ is defined as the relative phase of the two signals.

The PLV is defined as follows, see formula (7) and formula (8).

$$PLV = \left| \frac{1}{N} \sum_{N-1}^{j=0} e^{i\phi_{12}(j\Delta t)} \right| \tag{7}$$

$$PLV = \left(\left[\frac{1}{N} \sum_{N-1}^{j=0} sin(\phi_{12}(j\Delta t)) \right]^2 + \left[\frac{1}{N} \sum_{N-1}^{j=0} cos(\phi_{12}(j\Delta t)) \right]^2 \right)^{1/2} \tag{8}$$

N is the total number of samples; Δt represents the time between consecutive sample j from 1 to $N - 1$.

The value range of PLV is [0, 1], 0 means the phase is not synchronized, and 1 means the phase is constant, that is, the signal synchronization is perfect. The greater the value of PLV, the stronger the phase synchronization between the two signals.

Phase Lag Index. The phase lag index is a phase-based functional connectivity method that can measure the synchronization of two-channel signals. The introduction of the phase lag index reduces the impact of familiar sources (such as volume conduction or active reference electrode in EEG) in phase synchronization estimation.

For the calculation of PLI, see formula (9).

$$PLI = |\langle sign(\Delta\phi_{rel}(t))\rangle| = \left|\frac{1}{N}\sum_{N}^{n=1} sign(\Delta\phi_{rel}(t_n))\right| \tag{9}$$

where N denotes the time point and ϕ_{rel} denotes the phase difference of the two-channel signals at time t_n, the sign is a symbolic function. When the independent variable is positive, the output is 1, and when the independent variable is negative, the result is – 1. For 0, the result is also 0. The function is shown in formula (10).

$$sign(x) = \begin{cases} 1, x > 0 \\ 0, x = 0 \\ -1, x < 0 \end{cases} \tag{10}$$

The range of PLI is [0, 1], and the larger the value, the stronger the phase synchronization between the two signals.

Threshold Calculation. Based on these functional connectivity indicators, a two-state brain network is constructed, and the selection of appropriate thresholds is essential for the successful construction of a complex brain network. If the selection threshold is too low, it may lead to many nodes that are judged to be connected are not relevant, resulting in complex network topology; if the selection threshold is too high, it may lead to the loss aof influential connections, resulting in more scattered points or sub-networks, and the network is no longer a whole. This paper uses the average threshold method to average the correlation coefficients between all different channels to obtain the appropriate threshold. See formula (11).

$$Threshold = \frac{\sum \varnothing_{x,y}}{M}(x \neq y) \tag{11}$$

Threshold represents the value of threshold, $M = C_N^2$ represents the total number of permutations and combinations between two different channels, N is the total number of EEG channels, $\varnothing_{x,y}$ represents the functional connectivity index parameters between two different channels.

Brain Network Construction Based on Characteristic Parameters. Visualization of thermal map based on characteristic parameters and brain network construction according to visualization and calculated threshold [7].

*Brai*n Network Construction Based on Pearso*n Correlation Coefficient.* According to formula (2), the PCC of 19 channels of EEG signals is calculated, and the PCC matrix of 19 × 19 is constructed. The thermal diagrams of the resting state and anesthesia state are drawn for visualization (see Fig. 6).

The transverse and longitudinal coordinates represent a channel of the EEG signal. The corresponding value of the transverse and longitudinal coordinates on the graph is the phase lag index between the two media. From the color state column of the thermal diagram, it can be found that the color is bluer, and the phase lag index between the two channels is more extensive. That is, the correlation between the two channels is stronger.

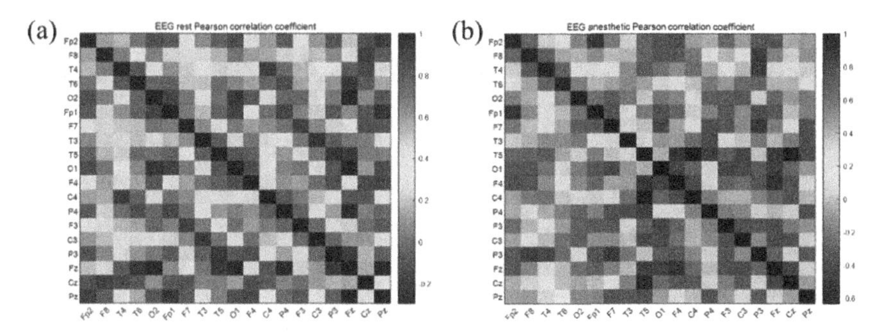

Fig. 6. The PCC correlation visualization. (a) resting state; (b) anesthesia state

On the contrary, the color is redder, and the phase lag index between the two channels is smaller. That is, the correlation between the two channels is weaker.

Based on the PCC matrix and the color depth of the thermal map, the brain network connection based on PCC can be constructed. According to formula (11), it can be obtained that the threshold value of the characteristic parameters of the constructed brain network is 0.7. The constructed brain network is shown in Fig. 7, which are the brain network connection diagrams of resting and anesthesia states, respectively.

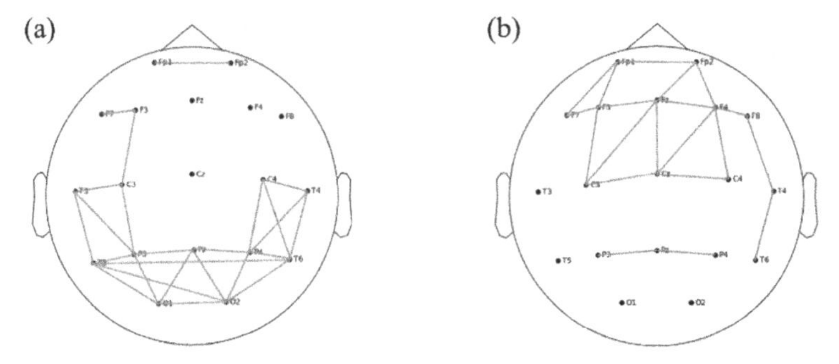

Fig. 7. Brain network connectivity map based on PCC. (a) resting state; (b) anesthesia state

Brain Network Construction Based on Phase Locking Value. According to the formula (7), the PLV of two channels of 19 channel EEG signals were calculated, and the PLV matrix of 19×19 was constructed. The thermal diagrams of resting state and anesthesia state were drawn for visualization. (see Fig. 8.)

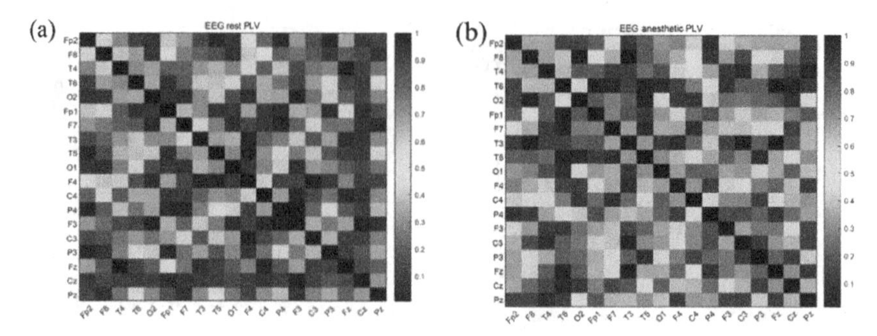

Fig. 8. The PLV correlation visualization. (a) resting state; (b) anesthesia state

According to formula (11), it can be obtained that the threshold value of the characteristic parameters of the constructed brain network is 0.6. The constructed brain network is shown in Fig. 9.

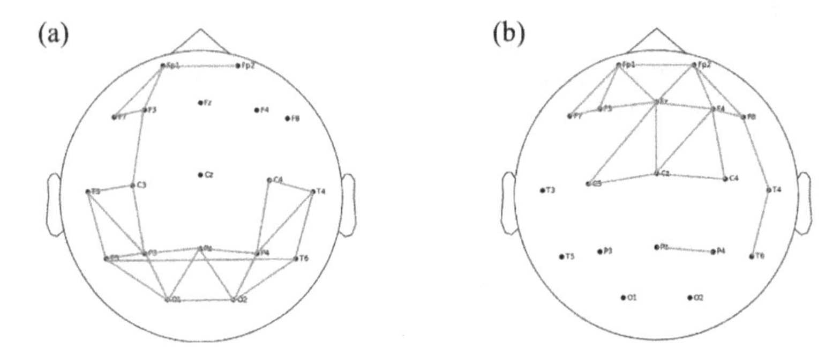

Fig. 9. Brain network connectivity map based on PLV. (a) resting state; (b) anesthesia state

Brain Network Construction Based on Phase Lag Index. According to the formula (9), the PLI of two channels of 19 channel EEG signals were calculated, and the PLI matrix of 19 × 19 was constructed. The thermal diagrams of resting state and anesthesia state were drawn for visualization. (see Fig. 10.)

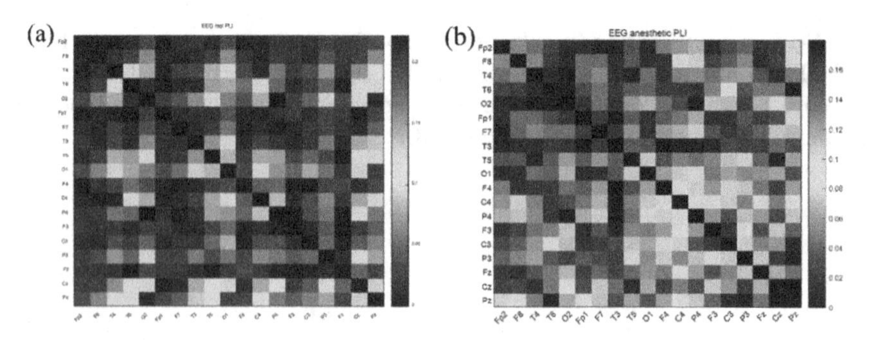

Fig. 10. The PLI correlation visualization. (a) resting state; (b) anesthesia state

According to formula (11), it can be obtained that the threshold value of the characteristic parameters of the con-structed brain network is 0.11. The constructed brain network is shown in Fig. 11.

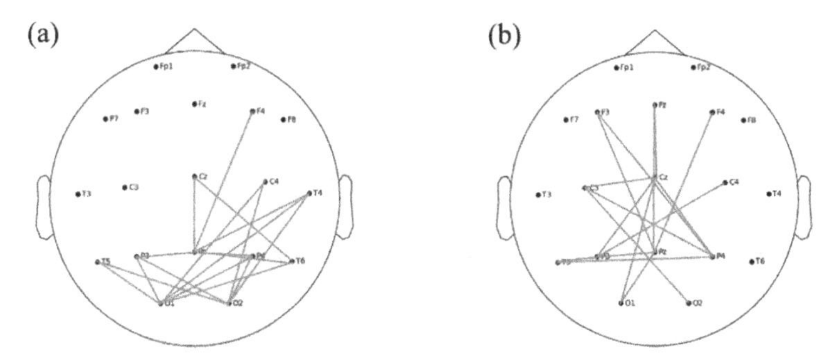

Fig. 11. Brain network connectivity map based on PLI. (a) resting state; (b) anesthesia state

3 Anesthesia State Evaluation Based on Machine Learning

Since the data after feature extraction of EEG signals have more dimensions and redundant features, it is necessary to select the feature from the original feature set and eliminate some irrelevant data sets, which not only reduces the running time of the classification algorithm and improves the classification accuracy of the classification model, but also reduces the complexity of the model and enhances the intelligibility of the model. After the successful training of the prediction model, combined with the concept of intelligent model, a new and efficient method for evaluating anesthesia status with various and uncertain standards is proposed. Feature selection is a complex brain network based on EEG feature extraction [8].

The research object of this paper is the EEG signals collected during clinical anesthesia. The collected data include resting state and anesthesia state. After feature extraction and feature selection, the dimension of the data set is finally constructed as 1124×247, that is, a total of 1124 groups of data, corresponding to 1124 state sets (only resting state and anesthesia state). Each state contains 247 attributes, namely 247 features.

Before classification, the data is labeled and divided into anesthesia state and awake state. To obtain higher classification accuracy, the classification model needs to be adjusted. This paper uses 10-fold cross-validation to improve the generalization ability of the model.

This paper uses five different classification algorithms [9], namely, linear discriminant analysis, support vector machine, decision tree, random forest, and k nearest neighbor algorithm. The flow chart of machine learning is shown in Fig. 12.

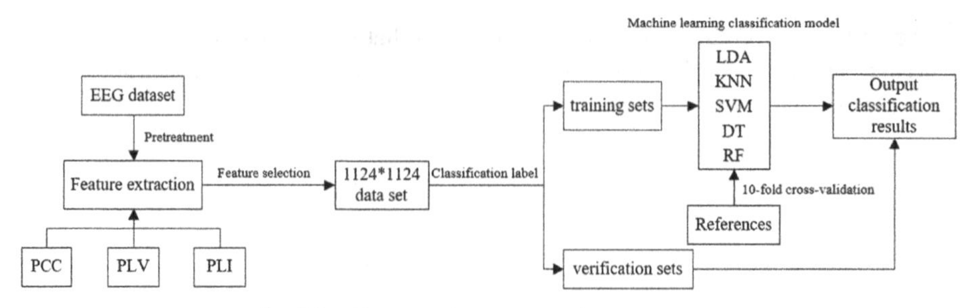

Fig. 12. Machine learning flow chart

We divided the states of the subjects into two categories, namely, "anesthesia state" and "awake state." According to the above, various machine learning methods classi-fied the collected EEG signals after processing (see Table 2).

Table 2. Machine learning classification results

		LDA	KNN	SVM		DT	RF
				linear	rbf		
Accuracy		0.875	0.510	0.910	0.837	0.869	0.939
Precision	Resting	0.85	0.46	0.87	0.82	0.86	0.97
	Anesthesia	0.9	0.54	0.95	0.85	0.87	0.91
Recall	Resting	0.89	0.4	0.95	0.83	0.85	0.89
	Anesthesia	0.86	0.61	0.88	0.84	0.89	0.98
F1 score	Resting	0.87	0.43	0.91	0.82	0.86	0.93
	Anesthesia	0.88	0.57	0.91	0.85	0.88	0.95

4 Conclusion

Anesthesia has almost become an indispensable part of the surgery. However, the evaluation of the anesthesia state is a complex process. In this paper, combined with the characteristics of significant changes in EEG during anesthesia, a method of anesthesia state evaluation is proposed to construct a complex brain network and combine it with machine learning classification evaluation. Train a fast and accurate evaluation model. Experiments show that this method has high accuracy in assessing anesthesia state, and further investigations are needed to verify the effectiveness for assessing clinical anesthesia state.

Acknowledgment. The author sincerely appreciates the support of Professor Ma Li and the School of Information Engineering of Wuhan University of Technology. National innovation and entrepreneurship training program for college students + S202110497213.

References

1. Brain's reward region helps to supply resilience in the face of stress. Nature **576**(7786) (2019)
2. Chen, M., Li, H., Wang, J., Dillman, J.R., Parikh, N.A., He, L.: A Multichannel deep neural network model analyzing multiscale functional brain connectome data for attention deficit hyperactivity disorder detection. Radiol. Artif. Intell. **2**(1) (2019)
3. Avcu, P., Sinha, S., Pang, K.C.H., Servatius, R.J.: Reduced avoidance coping in male, but not in female rats, after mild traumatic brain injury: implications for depression. Behav. Brain Res. **373**(C), 112064 (2019)
4. Duan, F., et al.: Topological network analysis of early Alzheimer's disease based on resting-state EEG. IEEE Trans. Neural Syst. Rehabil. Eng. **28**, 2164–2172 (2020). Publication of the IEEE Engineering in Medicine and Biology Society
5. Khosropanah, P., Ramli, A.R., Abbasi, M.R., Marhaban, M.H., Ahmedov, A.: A hybrid unsupervised approach toward EEG epileptic spikes detection. Neural Comput. Appl. **32**(7), 2521–2532 (2020)
6. Maren, S., Bauer Anna-Katharina, R., Stefan, D., Bleichner Martin, G.: Source-modeling auditory processes of EEG data using EEGLAB and brainstorm. Front. Neurosci. **12**, 309 (2018)
7. Zhang, M., Riecke, L., Fraga-González, G., Bonte, M.: Altered brain network topology during speech tracking in developmental dyslexia. NeuroImage **254**, 119142 (2022)
8. Bilucaglia, M., Duma, G.M., Mento, G., Semenzato, L., Tressoldi, P.: Applying machine learning EEG signal classification to emotion-related brain anticipatory activity. F1000Research **9**, 173 (2020)
9. Nathan, N.: Future tense: machine learning and the future of medicine. Anesth. Analg. **130**(5), 1114 (2020)

A Hybrid Asynchronous Brain-Computer Interface Combining SSVEP and EOG Signals for Rapid Target Recognition

Ximing Mai[1], Xinjun Sheng[1], Xiaokang Shu[1], Yidan Ding[2], Jianjun Meng[1(✉)], and Xiangyang Zhu[1(✉)]

[1] State Key Laboratory of Mechanical System and Vibration, Shanghai Jiao Tong University, 800 Dongchuan Road, Shanghai, China
{mengjianjunxs008,mexyzhu}@sjtu.edu.cn
[2] School of Mechanical Engineering, Shanghai Jiao Tong University, 800 Dongchuan Road, Shanghai, China

Abstract. Brain-Computer interfaces (BCIs) can help the disabled restore their ability of communicating and interacting with the environment. Asynchronous steady-steady visual evoked potential (SSVEP) allows the user to fully control the BCI system, but it also faces the challenge of the long erroneous state when the user switches his/her aimed target. To tackle this problem, a training-free SSVEP and EOG hybrid BCI system was proposed, where the spontaneous saccade eye movement occurred in SSVEP paradigm was hybridized into SSVEP detection by Bayesian approach to recognize the new target rapidly and accurately. The experiment showed that the proposed hybrid BCI had significantly higher asynchronous accuracy and short gaze shifting time compared to the conventional SSVEP-BCI. The results indicated the feasibility and efficacy of introducing the spontaneous eye movement information into SSVEP detection without increasing the user's task burden.

Keywords: Asynchronous BCI · Hybrid BCI · SSVEP · Saccade eye movement · Target recognition

1 Introduction

In the last few decades, brain-computer interfaces (BCIs) have gained tremendous attention and interest because they provide the disabled with an additional channel for communicating and interacting with the environment [1,2]. Among different types of BCI, the noninvasive BCI, especially electroencepalography (EEG)-based BCI, is favored by the researchers due to its advantages of portability, low cost and high temporal resolution [3–5]. Steady-state visual evoked

This work is supported in part by the China National Key R&D Program (Grant No. 2020YFC207800), the National Natural Science Foundation of China (Grant No. 52175023 & No. 91948302), Shanghai Pujiang Program (Grant No. 20PJ1408000).

H. Liu et al. (Eds.): ICIRA 2022, LNAI 13456, pp. 760–770, 2022.
https://doi.org/10.1007/978-3-031-13822-5_68

potential (SSVEP) [6–8] is one of the typically used modalities in EEG-based BCI, which has the metrics of high recognition accuracy, high information transfer rate (ITR), and multiple available commands. The SSVEP is elicited when the subject looks at a stimulus flickering at a frequency higher 6 Hz. The frequency response of SSVEP shows peaks at the flickering frequency and its harmonics.

The SSVEP-BCI can be divided into synchronous SSVEP-BCI and asynchronous SSVEP-BCI. In synchronous SSVEP-BCI, the stimuli in the subject's view perform flickering and non-flickering sequentially according to the fixed timeline set by the BCI system. Even though an extremely high ITR is achieved by synchronous SSVEP-BCI [9,10], the restriction of the fixed timeline to the subject's behavior limits its practical use. In asynchronous SSVEP-BCI [11–15], the BCI system continuously decodes the subject's intention, so the subject can output a command by gazing at the target stimulus or rest at any time he/she wants without any timeline.

One of the existing challenges of asynchronous SSVEP-BCI is the long erroneous state caused by target switching among the flickering stimuli. When the subject is delivering a new command, the new SSVEP induced by the new stimulus starts to append into the decoding epoch. Since the epoch from the latest moment to the previous amount of time is extracted to decode the subject's current intention, the epoch is combined with the mixture of SSVEP induced by the new stimulus and the SSVEP induced by the previous one. It leads to the recognition failure until the new SSVEP takes up a large portion of the epoch. During the erroneous state, the recognized results are inconsistent and incorrect. Previous studies utilize the dwell time strategy [11–15] to solve the problem, i.e., the recognized command is outputted only if it has been detected for a predefined number of consecutive times. While the dwell time strategy stabilize the system output, it further increases the period of the erroneous state.

A Hybrid BCI utilizing two or more physiological signals shows advantages over conventional single modality BCI in some applications [16,17]. In the existing hybrid BCI combining SSVEP and electrooculography (EOG), the subject performs the gazing task to evoke SSVEP and the eye movement task (blinks, winks) sequentially or simultaneously so that the recognition accuracy or the number of available commands is boosted [18–20] . However, the requirement of multiple tasks increases the subject's burden, leading to the physical and mental fatigue. It still remains an unsolved problem of how to integrate multiple physiological signals without increasing the subject's burden.

In this study, a training-free hybrid BCI combining SSVEP and EOG for asynchronous SSVEP-BCI is proposed. The subject gazes at the stimulus corresponding to the desired command as the conventional SSVEP paradigm does. However, when the subject shifts his/her gaze towards a new target, the spontaneous eye movement of saccade [21,22] is captured and hybridized into SSVEP decoding using Bayesian approach. By doing so, the problem of long erroneous state caused by target switching is diminished without introducing additional eye movement task. The experiment results showed that the proposed hybrid BCI significantly outperformed the conventional SSVEP-BCI in terms of accuracy and target switching time.

2 Methods

2.1 Subjects and Experimental Setup

Nine healthy subjects with no BCI experience of SSVEP or EOG task(2 females; average age 22.0 ± 1.58; range: 20–25) participated in the experiment. All procedures and protocols were approved by the Institutional Review Board of Shanghai Jiao Tong University, Protocol No. (IRB HRP AD03V02), date of approval (20210120). Informed consent was obtained from all subjects prior to their participation in the experiment. The subjects were sitting in a chair at 75 cm distance in front of the screen with a chin rest to fixate the head position. The stimuli used to evoke SSVEPs were presented on a 24.5-in. LCD monitor with a 1920×1080 pixels resolution and a refresh rate 240 Hz. A sampled sinusoidal stimulation method was used to present up, right, down, and left stimulus flickering at the frequency of 7.5, 9, 10.5, 12 Hz, respectively.

EEG and EOG signals were acquired at 1024 Hz using a 64 channel Biosemi Active Two system, a cap with active electrodes, and four additional EXG electrodes (Biosemi, Amsterdam, The Netherlands). Two electrodes near channel POz, named CMS and DRL, were used as reference and ground. The offsets of all electrodes were kept below ± 40 mV. The raw EEG and EOG signals were firstly filtered by 50 Hz notch filter. Nine channels in the parietal and occipital region, including Pz, POz, PO3, PO4, PO7, PO8, Oz, O1 and O2 were for the SSVEP decoding, as shown in Fig. 1(a). A bandpass filter encompassing 0.1 90 Hz was applied to EEG signals before any further processing. Four channels on the horizontal (A, B) and vertical (C, D) axis of eyes were used to record EOG data, as shown in Fig. 1(b), and were filtered 20 Hz low pass filter. A Tobii TX300 (Tobii, USA) eye tracker at a sampling rate 300 Hz was used to record eye movement during the experiments as the gold standard.

2.2 Experimental Design and Protocol

The experiment was designed for simulating the online asynchronous SSVEP-BCI. The experiment had 10 runs in total, with 12 trials for each run, as shown in Fig. 2. In each trial, the subject were instructed to gaze at the stimulus where the visual cue (the red square) was located. In the first 3 s of the trial, the visual cue was at the middle of the screen, and the subject needed to gaze at the blank. Next, the visual cue switched to one of the locations of the 4 stimuli, indicating the subject to gaze at the target, labeled as the 1st target. Then, the subject was guided by the visual cue to shift his/her gaze towards one of the 3 remaining stimuli, i.e., the 2nd target. Finally, the screen went blank for 2 s and the next trial began. The duration of cueing the subject to gaze at the 1st target and 2nd target was randomly choosen from 4, 5, or 6 s to prevent user's prediction

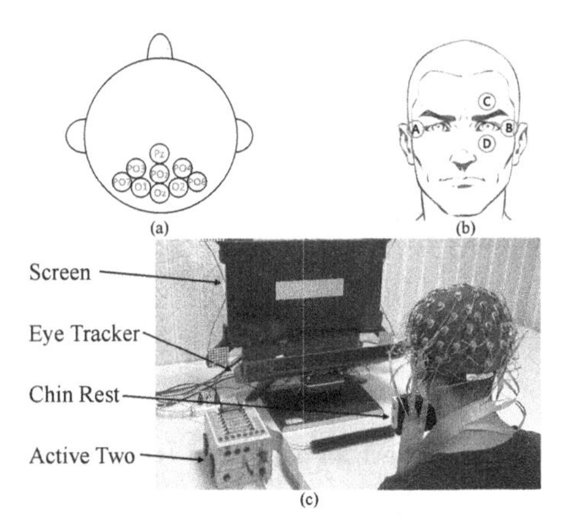

Fig. 1. Experimental setup. (a) Channel configuration for SSVEP (solid black circles). (b) Channel configuration for EOG (solid red circles). (c) Facilities used in this study. (Color figure online)

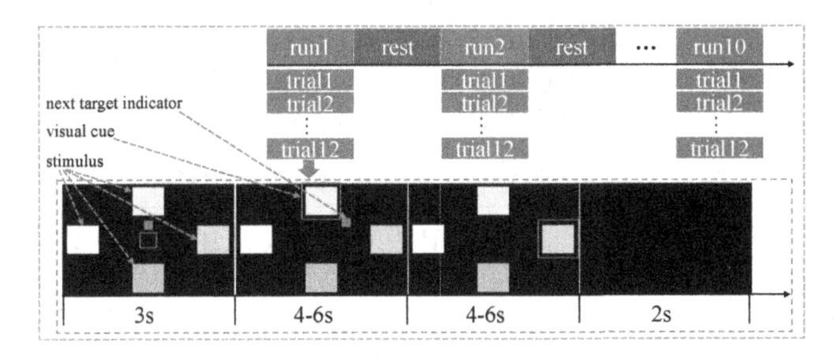

Fig. 2. Asynchronous experiment design. (Color figure online)

and adaptation. The number of all possible choice of 1st target and 2nd target was 12, corresponding to the 12 trials in each run. The sequence of the trials in each run was also randomly chosen, and a interval of several minutes between two consecutive blocks was provided with the subject to rest.

2.3 Signal Processing

The ground truth of the subject's gaze position was captured and decoded by the eye tracker signals. The horizontal gaze position of left side (0), center (0.5) and right side (1), and the vertical gaze position of upside (0), center (0.5) and downside (1) were obtained using a threshold to eye tracker signals by:

$$G_{adjust} = 0.5 \times round((G - 0.5) \times 3) + 0.5 \quad (1)$$

G was the raw value of gaze position, G_{adjust} was the projected value after setting the threshold. The target that the subject gazed at could be determined by the G_{adjust}. The valid gaze shifting time was obtained at the last rising or

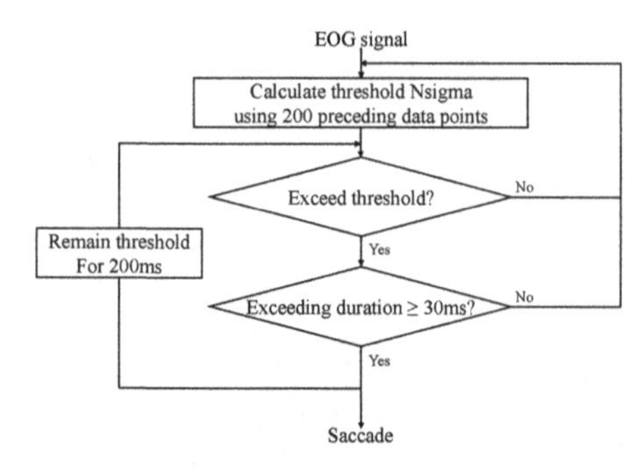

Fig. 3. Flow chart for saccade detection method.

falling edge of the adjusted curve if the gaze shift was correct both horizontally and vertically, or the gaze shifting time was invalid and the trial was discarded.

SSVEPs were analyzed using canonical correlation analysis (CCA) [23], a widely used method in detecting the frequency of SSVEPs. Given two multi-dimensional variables X and Y and their linear combinations $x = X^T W_X$ and $y = Y^T W_Y$, weight vectors W_X and W_Y were optimized to maximize the correlation between x and y by solving the following problem:

$$W_X, W_Y \max \rho(x, y) = \frac{E[W_X^T X Y^T W_Y]}{\sqrt{E[W_X^T X X^T W_X] E[W_Y^T Y Y^T W^Y]}} \tag{2}$$

For SSVEP detection, $X \in R^{N \times M}$ indicated the multi-channel EEG data epoch with the size of N channels and M samples, and $Y \in R^{Q \times M}$ was referred to reference signal:

$$Y = [sin(2\pi ft), cos(2\pi ft), ...sin(2\pi N_h ft), cos(2\pi N_h ft)]^T \tag{3}$$

where f was the stimulation frequency, N_h was the number of harmonics and Q was set as $2N_h$. CCA calculated the canonical correlation between multi-channel EEG data epoch and reference signals with respect to each stimulation frequency. The frequency with maximal correlation ρ was recognized as the target. SSVEPs were detected in different window lengths (from 0.5 s to 3.0 s in steps of 0.25 s) with a 31.25 ms' sliding window step in the experiment analysis.

Saccades were induced when the subject shifted his/her gaze to a new position, e.g., switching targets in SSVEP paradigm. The saccade detection algorithm was inspired by Behrens and colleagues' work [24]. Firstly, the 1st order derivative of EOG signal was calculated. Then, the adaptive threshold $NSigma$ was constantly calculated and updated by multiplying N by the standard deviation of the 200 preceding 1st derivative signal, N was set to 3 in this study to ensure the 1st order derivative signals were within the $[-NSigma, NSigma]$ when the subject was fixating on any target. When the subject switched towards a new target, the derivative signal would exceed the threshold due to the rotation of eyeball, i.e., the saccade. If the duration of exceeding threshold was longer

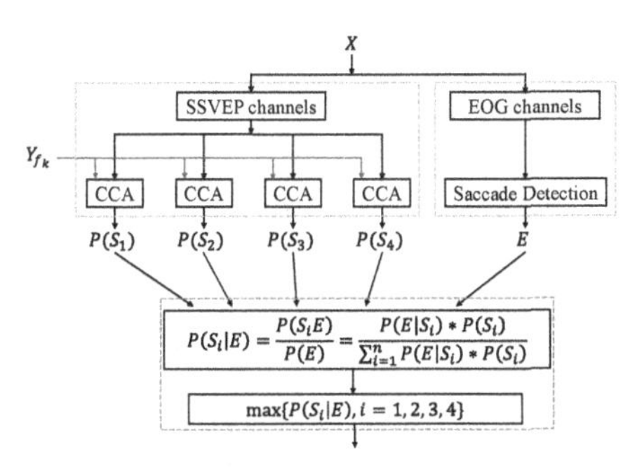

Fig. 4. The SSVEP-EOG-hybrid Bayesian approach by combining SSVEP decoding and saccade detection

than 30 ms, the event was labeled as a valid saccade, and the threshold was fixed for 200 ms for the possible incoming saccades, or it was labeled as a fluctuation and the threshold was updated. The detection procedure was shown in Fig. 3.

When the subject switched targets, the EEG signals were dynamically non-stationary, resulting in an erroneous state when using SSVEP detection methods alone. Therefore, the information of saccade events were introduced and the Bayesian approach was used to combine the SSVEP detection and saccade detection to diminish the erroneous state. Suppose the last target decoded by the hybrid BCI was S_{pre}. Firstly, the CCA was used to detect SSVEP and the canonical correlation ρ_i for each stimulus was interpreted as the probability $P(S_i)$ that the subject was fixating on this stimulus currently (Fig. 4).

$$P(S_i) = \frac{\rho_i}{\sum_{i=1}^{n} \rho_i} \tag{4}$$

where S_i was the event that the subject was currently gazing at the stimulus i ($i = 1, 2, 3, 4$ represented for stimulus up, right, down and left respectively), and n was 4. If no saccade was captured by EOG signal, the stimulus corresponding to the highest probability was the predicted target. If a valid saccade was detected, Then, the probability distribution of S_i was further inferred to posterior probabilities by the saccade event E:

$$P(S_i|E) = \frac{P(S_i E)}{P(E)} = \frac{P(E|S_i) \times P(S_i)}{\sum_{i=1}^{n} P(E|S_i) \times P(S_i)} \tag{5}$$

where $P(E|S_i)$ was the probability of the saccade event occurrence when the subject shifted his/her gaze from S_{pre} to S_i. The probability distribution of $P(E|S_i)$ was estimated empirically from statistics of saccade detection performance. In the condition of $S_i = S_{pre}$, which meant the subject didn't shift his/her gaze, detection of the saccade was a false positive event, and therefore the probability was equal to the false positive rate (FPR) of saccade detection. Suppose $S_i \neq S_{pre}$, which meant the subject shifted his/her gaze, the probability of detecting the correct saccade was equal to the accuracy of saccade detection,

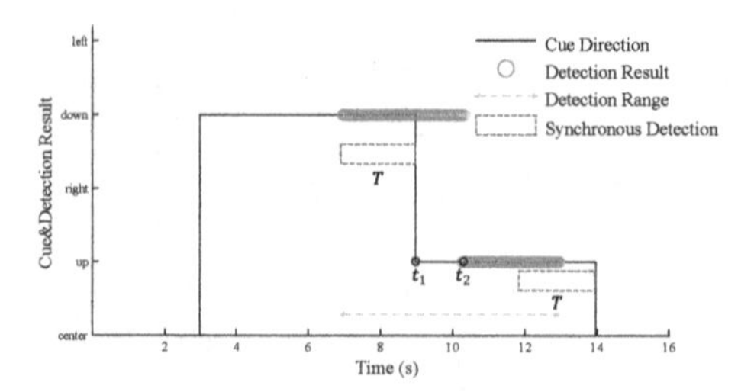

Fig. 5. Performance evaluation in a single trial. The x-axis represents the ongoing time, and the y-axis represents the targets. The black curve shows the visual cues, while the detection results at each step are shown by the red circles. (Color figure online)

while the probability of detecting a wrong saccade was equal to (1 - accuracy). Finally, is the stimulus k with the maximum probability $P(S_k|E)$:

$$k = \arg\max\{P(S_i|E), 1, 2, ..., n\} \tag{6}$$

was selected as the target.

2.4 Evaluation of Performance

The performance of the proposed hybrid BCI and conventional SSVEP-BCI were evaluated by the asynchronous accuracy and gaze shifting time (GST; unit: second) of the experiment. In the conventional SSVEP-BCI, CCA algorithm was used to detect the gazing target. Since the sliding window with the step of 31.25 ms was utilized into each trial, multiple detection results would be obtained from each trial. Therefore, the asynchronous accuracy was calculated as the number of correction detection points divided by the total number of detection points in each trial's detection period (shown in Fig. 5). The detection period of each trial was set from 2 s before the visual cue changed from 1st to 2nd target to 4 s after the change in order to focus on the evaluation of control state of asynchronous BCI. The GST was measured by the duration from the time of visual cue changing from 1st target to 2nd target, i.e., t1 as shown in Fig. 5, to the time of the first steady detection point occurrence, i.e., t2 as shown in Fig. 5. The steady detection point was defined as the current and the following N (20) results are all the same as the cueing target.

3 Results

3.1 Pseudo Synchronous SSVEP Performance

The pseudo synchronous SSVEP performance was first calculated to verify the CCA algorithm performance. Figure 6(a) showed the pseudo synchronous SSVEP accuracy in terms of different window length T (from 0.25 s to 3.0 s in steps of 0.25 s) and different number of harmonics. The accuracy was $>$ 90%

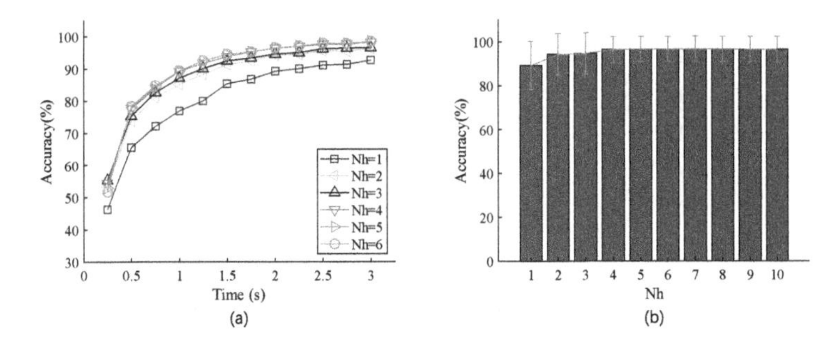

Fig. 6. Pseudo synchronous SSVEP-BCI performance. (a) Pseudo synchronous accuracy under different window lengths and different harmonics. (b) Pseudo synchronous accuracy under different harmonics when window length $T = 2\,\mathrm{s}$.

Actual Saccades \ Predicted Saccades	up	upRight	right	downRight	down	downLeft	left	upLeft	none
upLeft	1.724	0	0	0	1.149	0.5747	32.18	64.37	0
left	1.149	0	0	0	5.747	12.64	80.46	0	0
downLeft	0	0	0	0	0.5747	96.55	2.299	0	0.5747
down	2.299	0	0	0	96.55	0	0	0	1.149
downRight	2.299	1.149	18.39	77.01	0.5747	0	0	0	0.5747
right	0	0	100	0	0	0	0	0	0
upRight	0.5747	94.25	5.172	0	0	0	0	0	0
up	97.7	0	0	0	1.149	0	0	0	1.149

Fig. 7. Confusion matrix for saccade detection.

when the window length $T > 2\,\mathrm{s}$ and the harmonic $N_h > 2$. Figure 6(b) illustrated the accuracy in terms of different harmonics when $T = 2\,\mathrm{s}$. Repeated measures ANOVA indicated that there is a significant difference among accuracy of different harmonics $F(1.591, 72) = 7.598, p = 0.009$. Pairwise comparisons revealed that $N_h = 1$ had significantly lower accuracy than other N_h conditions. N_h was set to 5 empirically to achieve a high recognition accuracy and the relatively short calculation time.

3.2 Saccade Detection Performance

Figure 7 showed the saccade detection performance under different saccade directions. The accuracy and FPR of saccade detection were $87.04 \pm 2.15\%$ and $5.22 \pm 0.65\%$ respectively, indicating that the eye movement of saccade could be accurately detected by the proposed adaptive threshold algorithm.

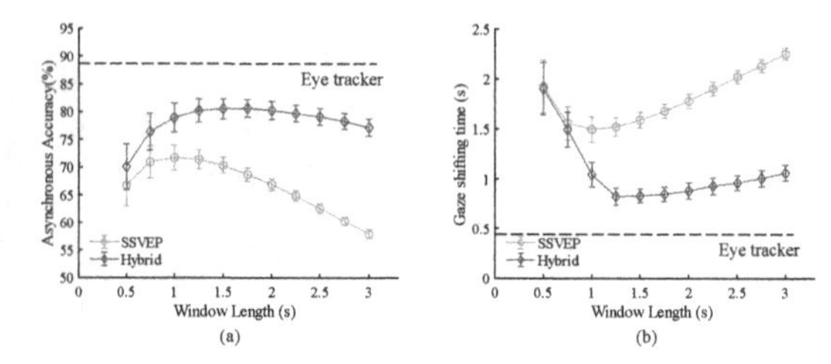

Fig. 8. Asynchronous performance under different window lengths and different types of BCI. (a) Asynchronous accuracy, (b) gaze shifting time (GST) of the proposed hybrid BCI and the conventional SSVEP-BCI using CCA detection algorithm.

3.3 BCI Performance

The performance of eye tracker, the proposed hybrid BCI and the conventional SSVEP-BCI was evaluated in terms of asynchronous accuracy and GST under different window lengths. As shown in Fig. 8, the eye tracker detection had the highest asynchronous accuracy ($87.57 \pm 0.67\%$) and the shortest GST (0.47 ± 0.01 s). The proposed hybrid BCI had the higher asynchronous accuracy and shorter GST than the conventional SSVEP-BCI in all window lengths. The highest asynchronous accuracy of the hybrid BCI was $80.6417 \pm 1.60\%$ when $T = 1.75$ s. Paired t-test showed a significant difference in asynchronous accuracy between the hybrid BCI and the SSVEP-BCI in all window length. The shortest GST of the hybrid BCI was 0.82 ± 0.08 s when $T = 1.25$ s. Paired t-test showed a significant difference in GST between the hybrid BCI and the SSVEP-BCI in all window length except for $T = 0.5$ s.

4 Discussion

In this study, a hybrid BCI combining SSVEP and EOG was proposed for enhancing the performance of asynchronous SSVEP control state detection. Switching target occurs frequently in asynchronous SSVEP-BCI control state, resulting in a long erroneous state when using SSVEP detection algorithm alone. The integration of detected saccade events into SSVEP detection helps diminish the erroneous state by optimizing the probability distribution of the potential target using saccade event as an additional event and Bayesian approach. The efficacy of the proposed hybrid BCI was verified by the significantly higher accuracy and shorter tgaze shifting time.

In the existing asynchronous SSVEP-BCI studies, the dwell time strategy [11–15] was introduced to solve the erroneous state. While the stability of the system output was improved it increases the delay period from the time when subject switched to a new target to the time when the system output the new target. For the proposed hybrid BCI, both the stability and the reaction speed of the system were improved.

While the eye movements could be accurately captured and recognized by EOG signals, it remains a challenge of obtaining the absolute gaze position using EOG signal alone. The proposed hybrid BCI introduced the saccade event which occurs spontaneously when the subject shifted his/her gaze towards new target, and the gaze position is predicted fast and accurately, while in the existing hybrid BCI the subject need to perform multiple tasks to evoke both SSVEP and eye movement, which is unnatural and increases the subject's workload.

5 Conclusion and Future Work

In this study, a novel training-free approach for asynchronous BCI was proposed by hybridizing SSVEP and EOG signal. The detected saccade event was integrated into the SSVEP detection to obtain a optimized posterior probability distribution of the possible target in SSVEP paradigm. The experiment showed significantly higher asynchronous accuracy and shorter gaze shifting time compared to the conventional asynchronous SSVEP-BCI.

In the future study, although the proposed SSVEP and EOG hybrid BCI is preliminarily testified in this work, the feasibility of the system for general population needs to be verified by recruiting more subjects into the experiment. Moreover, the online experiment will be conducted to further evaluate the efficacy of the proposed hybrid BCI.

References

1. He, B., Yuan, H., Meng, J., Gao, S.: Brain–computer interfaces. In: He, B. (ed.) Neural Engineering, pp. 131–183. Springer, Cham (2020). https://doi.org/10.1007/978-3-030-43395-6_4
2. Tonin, L., Millán, J.d.R.: Noninvasive brain–machine interfaces for robotic devices. Annu. Rev. Control Robot. Auton. Syst. **4**, 191–214 (2021)
3. Edelman, B.J., et al.: Noninvasive neuroimaging enhances continuous neural tracking for robotic device control. Sci. Robot. **4**(31), eaaw6844 (2019)
4. Schwarz, A., Höller, M.K., Pereira, J., Ofner, P., Müller-Putz, G.R.: Decoding hand movements from human EEG to control a robotic arm in a simulation environment. J. Neural Eng. **17**(3), 036010 (2020)
5. Xu, M., Han, J., Wang, Y., Jung, T.P., Ming, D.: Implementing over 100 command codes for a high-speed hybrid brain-computer interface using concurrent p300 and SSVEP features. IEEE Trans. Biomed. Eng. **67**(11), 3073–3082 (2020)
6. Wong, C.M., Wang, B., Wang, Z., Lao, K.F., Rosa, A., Wan, F.: Spatial filtering in SSVEP-based BCIS: unified framework and new improvements. IEEE Trans. Biomed. Eng. **67**(11), 3057–3072 (2020)
7. Chen, X., Huang, X., Wang, Y., Gao, X.: Combination of augmented reality based brain-computer interface and computer vision for high-level control of a robotic arm. IEEE Trans. Neural Syst. Rehabil. Eng. **28**(12), 3140–3147 (2020)
8. Cao, Z., Ding, W., Wang, Y.K., Hussain, F.K., Al-Jumaily, A., Lin, C.T.: Effects of repetitive SSVEPs on EEG complexity using multiscale inherent fuzzy entropy. Neurocomputing **389**, 198–206 (2020)

9. Chiang, K.J., Wei, C.S., Nakanishi, M., Jung, T.P.: Boosting template-based SSVEP decoding by cross-domain transfer learning. J. Neural Eng. **18**(1), 016002 (2021)
10. Nakanishi, M., Wang, Y., Chen, X., Wang, Y.T., Gao, X., Jung, T.P.: Enhancing detection of SSVEPs for a high-speed brain speller using task-related component analysis. IEEE Trans. Biomed. Eng. **65**(1), 104–112 (2017)
11. Yang, C., Yan, X., Wang, Y., Chen, Y., Zhang, H., Gao, X.: Spatio-temporal equalization multi-window algorithm for asynchronous SSVEP-based BCI. J. Neural Eng. **18**(4), 0460b7 (2021)
12. Wang, M., Li, R., Zhang, R., Li, G., Zhang, D.: A wearable SSVEP-based BCI system for quadcopter control using head-mounted device. IEEE Access **6**, 26789–26798 (2018)
13. Bi, L., Fan, X.A., Jie, K., Teng, T., Ding, H., Liu, Y.: Using a head-up display-based steady-state visually evoked potential brain-computer interface to control a simulated vehicle. IEEE Trans. Intell. Transp. System. **15**(3), 959–966 (2013)
14. Li, Y., Pan, J., Wang, F., Yu, Z.: A hybrid BCI system combining p300 and SSVEP and its application to wheelchair control. IEEE Trans. Biomed. Eng. **60**(11), 3156–3166 (2013)
15. Diez, P.F., Mut, V.A., Perona, E.M.A., Leber, E.L.: Asynchronous BCI control using high-frequency SSVEP. J. Neuroeng. Rehabil. **8**(1), 1–9 (2011)
16. Abiri, R., Borhani, S., Sellers, E.W., Jiang, Y., Zhao, X.: A comprehensive review of EEG-based brain-computer interface paradigms. J. Neural Eng. **16**(1), 011001 (2019)
17. Meng, J., Streitz, T., Gulachek, N., Suma, D., He, B.: Three-dimensional brain-computer interface control through simultaneous overt spatial attentional and motor imagery tasks. IEEE Trans. Biomed. Eng. **65**(11), 2417–2427 (2018)
18. Zhou, Y., He, S., Huang, Q., Li, Y.: A hybrid asynchronous brain-computer interface combining SSVEP and EOG signals. IEEE Trans. Biomed. Eng. **67**(10), 2881–2892 (2020)
19. Zhu, Y., Li, Y., Lu, J., Li, P.: A hybrid BCI based on SSVEP and EOG for robotic arm control. Front. Neurorobot. **14**, 95 (2020)
20. Saravanakumar, D., Reddy, R.: A brain computer interface based communication system using SSVEP and EOG. Procedia Comput. Sci. **167**, 2033–2042 (2020)
21. Constable, P.A., Bach, M., Frishman, L.J., Jeffrey, B.G., Robson, A.G.: ISCEV standard for clinical electro-oculography (2017 update). Doc. Ophthalmol. **134**(1), 1–9 (2017)
22. Larsson, L., Nyström, M., Stridh, M.: Detection of saccades and postsaccadic oscillations in the presence of smooth pursuit. IEEE Trans. Biomed. Eng. **60**(9), 2484–2493 (2013)
23. Lin, Z., Zhang, C., Wu, W., Gao, X.: Frequency recognition based on canonical correlation analysis for SSVEP-based BCIS. IEEE Trans. Biomed. Eng. **53**(12), 2610–2614 (2006)
24. Behrens, F., MacKeben, M., Schröder-Preikschat, W.: An improved algorithm for automatic detection of saccades in eye movement data and for calculating saccade parameters. Behav. Res. Methods **42**(3), 701–708 (2010)

Wearable Sensing and Robot Control

ZNN-Based High-Order Model-Free Adaptative Iterative Learning Control of Ankle Rehabilitation Robot Driven by Pneumatic Artificial Muscles

Xianliang Xie, Quan Liu, Wei Meng(✉), and Qingsong Ai

School of Information Engineering, Wuhan University of Technology, Wuhan 430070, China
{264254,quanliu,weimeng,qingsongai}@whut.edu.cn

Abstract. As a new lightweight actuator, pneumatic artificial muscles (PAMs) have been widely used in rehabilitation robots attributing to the compliant interaction and good safety features. However, it is challenging to precisely control such PAMs-driven devices due to the non-linear and time-varying nature. For complex controlled plants like PAMs, the model-based controllers are challenging to design and use for their complex structure, while the convergence of the mode-free method can be further enhanced. In this paper, a Zero Neural Network based high-order model-free adaptive iterative control (ZNN-HOMFAILC) method is proposed to realize high precise position control of the rehabilitation robot. Firstly, the model of PAMs is converted to an equivalent linearized data model along the iterative axis. Then, to achieve fast convergence speed, a high-order pseudo partial derivative (PPD) law is designed to improve the convergence performance under different initial PPDs. The control law based on ZNN is designed to improve the learning ability from errors during iterations. Finally, the control performance and convergence speed of the ZNN-HOMFAILC are validated by simulation and actual control experiments of the PAMs-driven ankle rehabilitation robot. Results show that ZNN-HOMFAILC can significantly improve the convergence speed by 54% compared with MFAILC and HO-MFAILC, and the average tracking error of PAM can be reduced to a small level (2% of the range) after 7 iterations.

Keywords: Iterative learning control · Zero Neural Network · Ankle rehabilitation robot

1 Introduction

With the improvement of living standards and the aging of society, the incidence of movement disorders caused by stroke and neurological injury also increases every year. According to the World Health Organization (WHO), stroke ranked second among the top 10 causes of death worldwide [1]. In traditional rehabilitation therapy, a physician provides one-on-one rehabilitation treatment to patients; due to time and resource constraints, manual physiotherapy methods cannot provide sufficient training frequency and intensity [2]. However, rehabilitation robots are not limited by these restrictions and can

H. Liu et al. (Eds.): ICIRA 2022, LNAI 13456, pp. 773–784, 2022.
https://doi.org/10.1007/978-3-031-13822-5_69

help patients recover quickly [3]. Unlike conventional rehabilitation robots that use the motor and hydraulic drives, PAMs are more compliant because of their lightweight and higher power-to-weight ratio [4]. So, rehabilitation robots driven by PAMs can safely assist patients in training.

Over the past few years, several advances have been made in model-based control methods for PAMs. For instance, Qian *et al.* [5] constructed the PAM model under a three-element and proposed an ILC control scheme incorporating the composite energy function. Zhang *et al.* [4] formulated a simplified three-element model for PAM under a specific range of pressure and developed a compensation control to reject modeling errors. Ai *et al.* [6] integrated the Hammerstein model and Modified Prandtl-Ishlinskii model to characterize the hysteretic of PAMs, and the validity of the model is verified by PID control. Khajehsaeid *et al.* [7] proposed a continuum mechanics-based model to calculate contraction force of PAMs and designed a fast adaptive sliding mode controller with back stepping. Although the above studies have developed PAM models, the models are only an approximation of the control process [8]. Compensators need to be designed for unmodeled features. Due to the complexity of the PAM model, the model-based controllers are difficult to design and analysis. Therefore, model-free control methods need to be introduced to improve the performance of PAM in practical applications [9].

The data-driven control method does not require the exact mathematical model of controlled plants and only uses output and input data to develop the controller [10]. Compared to model-based approaches, model-free controllers are versatile and straightforward, while modeling characteristics are included in input and output data. The robot repeatedly performs the same rehabilitation tasks in passive rehabilitation training [11]. Model-free iterative learning can be applied to the control of PAMs because of the repetitive nature of training. MFAILC has been applied in many fields. For instance, Ai *et al.* [12] introduced a high-order form into the PPD learning law of MFAILC to improve tuning the PPD and convergence speed. Yu *et al.* [13] combined an RBFNN into the PPD tunning law and achieved a faster convergence rate. Li *et al.* [14] introduced dynamic linearization methods to nonlinear plant and outer-loop nonlinear set-points, and higher-order indirect adaptive ILC is designed to improve control performance. Chi *et al.* [15] designed a high-order error form in the MFAILC control law to improve the convergence speed. Wu *et al.* [16] introduced the iterative feedback tuning to the optimization of MFAILC and applied the proposed method to the position control of a single PAM.

Considering the complexity of the PAM model and the inadequacy of model-free control methods in terms of control performance, the ZNN-HOMFAILC method with enhanced convergence is proposed to enhance the control accuracy of the ankle rehabilitation robot. The rest parts are organized as follows: Sect. 2 introduces the inverse kinematic of the ankle rehabilitation robot. Section 3 designs the ZNN-HOMFAILC controller. In Sect. 4, simulation and practical control experiments are performed, and the results are analyzed. The conclusion is in Sect. 5.

2 Inverse Kinematics of the Ankle Rehabilitation Robot

To realize the precise position control of ankle rehabilitation, the formula between the angle of rotatable support platform and displacement of PAMs should be derived. This paper's ankle rehabilitation robot as a control plant is redundantly driven by PAMs shown in Fig. 1(a). The structural and geometric model of the rehabilitation robot is shown in Fig. 1(b).

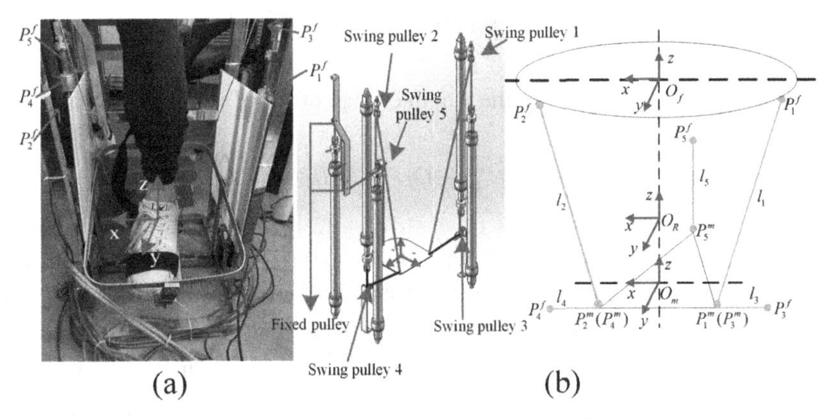

Fig. 1. (a) ankle rehabilitation robot, (b) structure and geometric model

$O_f - xyz$, $O_m - xyz$, $O_R - xyz$ are the fixed bracket platform, the rotatable support platform and the actual rotation coordinate system respectively. The connection points between the drive cable and fixed bracket platform are $P_i^f\ (i = 1, \cdots, 5)$, the connection points with the rotatable support platform are $P_i^m (i = 1, \cdots, 5)$. The distance between the center O_f of fixed bracket platform and the real center of rotation O_R of rotatable support platform is $H = 0.477\,\text{m}$, the distance between the center O_m of the rotatable support platform and O_R is $h = 0.106\,\text{m}$. The coordinates of P_i^f in $O_f - xyz$ and P_i^m in $O_R - xyz$ are shown as follows.

$$\begin{cases} P_1^m = P_3^m = \left[-0.06\ 0.05\ -0.106\right]^T \\ P_2^m = P_4^m = \left[0.06\ 0.05\ -0.106\right]^T \\ P_5^m = \left[0\ -0.074\ -0.106\right]^T \end{cases} \begin{cases} P_1^f = \left[-0.1875\ 0.089\ 0\right]^T \\ P_2^f = \left[0.1875\ 0.089\ 0\right]^T \\ P_3^f = \left[-0.1725\ 0.05\ -0.583\right]^T \\ P_4^f = \left[0.1725\ 0.05\ -0.583\right]^T \\ P_5^f = \left[0\ -0.074\ -0.345\right]^T \end{cases} \tag{1}$$

Let $\boldsymbol{q} = [\theta_x\ \theta_y\ \theta_z]^T$ be the rotation angle vector of the rotatable support platform, in which θ_x, θ_y and θ_z are rotation angles of the rotatable support platform around x, y, and z axes respectively. The vector of the real rotation center O_R of the rotatable support platform in the fixed bracket coordinate system is $\overrightarrow{O_f O_R} = [0\ 0\ H]^T$. The vector from P_i^f to P_i^m can be deduced from Fig. 1(b) as (2):

$$\boldsymbol{L}_i = \boldsymbol{R}\overrightarrow{O_R P_i^m} + \overrightarrow{O_f O_R} - \overrightarrow{O_f P_i^f}\,(\mathrm{i} = 1, 2, \ldots, 5) \tag{2}$$

$\overrightarrow{O_R P_i^m}$ is the position vector of P_i^m in O_R, $\overrightarrow{O_f P_i^f}$ is the position vector of P_i^f in O_f. $\boldsymbol{R}$ is the rotation matrix of the rotatable support platform, which is determined by $\boldsymbol{q}$, where a single-axis rotation matrix $\boldsymbol{R} = \boldsymbol{R_x R_y R_z}$ is expressed by (3).

$$\boldsymbol{R}_x = \begin{bmatrix} 1 & 0 & 0 \\ 0 & cos\theta_x & -sin\theta_x \\ 0 & sin\theta_x & cos\theta_x \end{bmatrix} \boldsymbol{R}_y = \begin{bmatrix} cos\theta_y & 0 & sin\theta_y \\ 0 & 1 & 0 \\ -sin\theta_y & 0 & cos\theta_y \end{bmatrix} \boldsymbol{R}_z = \begin{bmatrix} cos\theta_z & -sin\theta_z & 0 \\ sin\theta_z & cos\theta_z & 0 \\ 0 & 0 & 1 \end{bmatrix} \tag{3}$$

According to Eqs. (2) and (3), the length change of $\overrightarrow{P_i^f P_i^m}$ is:

$$\Delta l_i = \left\| \overrightarrow{O_f O_R} + \boldsymbol{R}\overrightarrow{O_R P_i^m} - \overrightarrow{O_f P_i^f} \right\| - l_{i0}, \; (i = 1, 2, \cdots, 5) \tag{4}$$

l_{i0} is the distance between P_i^f to P_i^m when $\boldsymbol{q} = [0\,0\,0]^T$. The displacement of each PAM actuator can be calculated according to any rotation angle of the rotatable support platform. By changing the displacement of P_5^f, P_1^f and P_2^f, the rotation around x-axis can be achieved; Similarly, P_1^f and P_2^f corresponds the rotation around y-axis, P_4^f and P_5^f corresponds the rotation around z-axis. It's noted that the forward kinematics model can be trained by neural networks used for active control which is not necessary for model-free control method.

3 Design of the ZNN-HOMFAILC Controller

3.1 MFAILC Algorithm

The nonlinear non-affine discrete SISO system can be expressed by (5):

$$y_{k+1}(t) = f(u_k(t), \cdots, u_k(t - o_u), y_k(t), \cdots, y_k(t - o_y)) \tag{5}$$

where $y_k(t)$ and $u_k(t)$ denote the control input and output at time t for kth iteration, where $t = 0 \cdots N - 1$, N is the desired output period of the system. o_u and o_y represent the system order respectively, $f(\cdot)$ is the uncertain nonlinear system function.

Assumption I: Each of components of $f(\cdot)$ has a continuous partial derivative for input $u_k(t)$.

Assumption II: System (5) is generalized Lipschitz, for all $t = 0 \cdots N - 1$ and iteration rounds k, if $|\Delta u_k(t)| \neq 0$, following condition is satisfied:

$$|\Delta y_k(t+1)| \leq b|\Delta u_k(t)| \tag{6}$$

where b is a positive number representing the uncertainty after linearization of the system. The above assumptions come from [17]. If (5) satisfies these assumptions, it can be transformed into a compact form dynamic linearization (CFDL) model:

$$\Delta y_k(t+1) = \phi_k(t)\Delta u_k(t) \tag{7}$$

where $\phi_k(t)$ is the PPD of system (5) meets $|\phi_k(t)| \leq b$, which has been demonstrated in detail in [14].

Consider the criterion function for control input:

$$J(u_k(t)) = \lambda|u_k(t) - u_{k-1}(t)|^2 + |y_d(t+1) - y_k(t+1)|^2 \tag{8}$$

in which λ is the weight parameter, y_d is the desired output. Let $\frac{\partial J(u_k)}{\partial u_k} = 0$, the MFAILC control method can be derived as:

$$u_k(t) = u_{k-1}(t) + \frac{\rho\phi_k(t)}{\lambda + |\phi_k(t)|^2} e_{k-1}(t+1) \tag{9}$$

Introducing step coefficient ρ into (9) to make the controller more universal, $e_k(t) = y_d(t) - y_k(t)$ is the control error. The accurate value of $\phi_k(t)$ cannot be calculated directly. Therefore, Eq. (10) is used to estimate the PPD.

$$J\left(\widehat{\phi}_k(t)\right) = \mu\left|\widehat{\phi}_k(t) - \widehat{\phi}_{k-1}(t)\right|^2 + \left|\Delta y_{k-1}(t+1) - \widehat{\phi}_k(t)\Delta u_k(t)\right|^2 \tag{10}$$

where $\mu > 0$ is weight factor, let $\frac{\partial J\left(\widehat{\phi}_k\right)}{\partial \widehat{\phi}_k} = 0$, we can get:

$$\widehat{\phi}_k(t) = \widehat{\phi}_{k-1}(t) + \frac{\eta\Delta u_k(t)}{\mu + |\Delta u_k(t)|^2}(\Delta y_{k-1}(t+1) - \widehat{\phi}_k(t)\Delta u_k(t)) \tag{11}$$

where η is step factor introduced in (11) same as ρ, and $\widehat{\phi}_k(t)$ is an estimated value of $\phi_k(t)$. To ensure the validity of CFDL model, the reset formula needs to be introduced:

$$\widehat{\phi}_k(t) = \widehat{\phi}_0(t),\ if\,\widehat{\phi}_k(t) \le \varepsilon\ \ or\ \ |\Delta u_k(t)| \le \varepsilon \tag{12}$$

where ε is a reset threshold and $\widehat{\phi}_0(t)$ is the initial value of PPD, which is determined empirically. It can be seen from (9), (11) and (12) that the MFAILC only utilizes the I/O data of the system, which is entirely model-free. However, initial value of PPD has a significant impact on the convergence speed of the controller, which is the critical issue to be addressed in the ZNN-HOMFAILC algorithm.

3.2 ZNN-HOMFAILC Algorithm

The error decay formula of linear time-varying equations by following the ZNN design methodology can be given as follow [18]:

$$\dot{e}(t) = -\gamma\mathcal{F}(e(t)) \tag{13}$$

where $\gamma > 0$ and $\mathcal{F}(\cdot)$ is an activation function composed of any monotonically increasing odd function, such as Linear function, Power-Sum function, and Hyperbolic-Sine function. The Linear function is used in this paper, that is $\dot{e}(t) = -\gamma e(t)$.

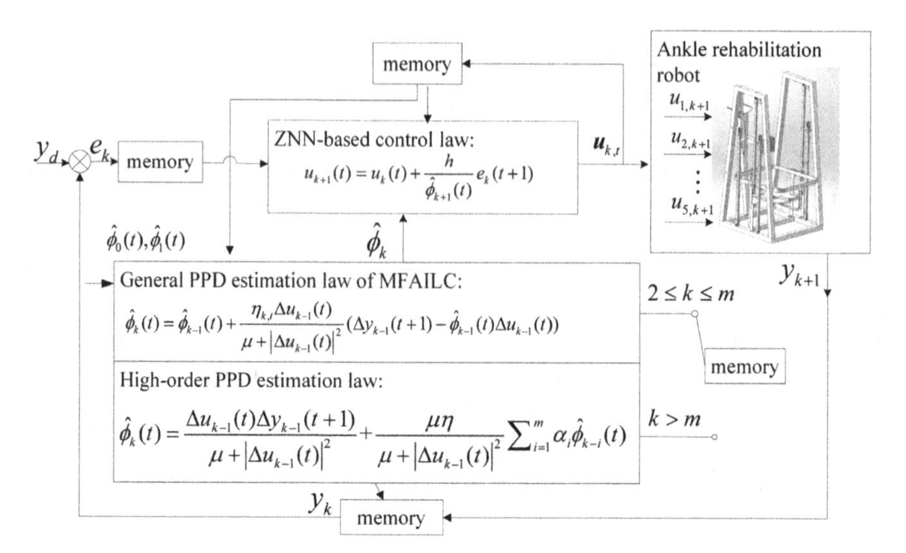

Fig. 2. The block diagram of ZNN- HOMFAILC

Tracking error is represented by the CFDL model (7) as:

$$e_k(t+1) = y_d(t+1) - y_k(t+1) - \widehat{\phi}_k(t)\Delta u_k(t) \tag{14}$$

Converting Eq. (13) to discrete form:

$$\frac{e_{k+1}(t) - e_k(t)}{T} = -\gamma e_k(t) \rightarrow e_{k+1}(t) = (1-h)e_k(t) \tag{15}$$

in which $h = \gamma T > 0$, is generalized parameter, substituting (14) into (15), we can derive the control law:

$$u_{k+1}(t) = u_k(t) + \frac{h}{\phi_{k+1}(t)} e_k(t+1) \tag{16}$$

The Eq. (16) is the ZNN-based controller law with an enhanced convergence speed. To address the impact of different initial values of PPD on the convergence speed, the high-order tunning law of PPD is introduced.

The high-order PPD tunning law is introduced in criterion function (10):

$$J\left(\widehat{\phi}_k(t)\right) = \left|\Delta y_{k-1}(t+1) - \widehat{\phi}_k(t)\Delta u_k(t)\right|^2 + \mu\left|\widehat{\phi}_k(t) - \sum\nolimits_{i=1}^{m} \alpha_i \widehat{\phi}_{k-i}(t)\right|^2 \tag{17}$$

where m is order of PPD and α_i is the forgetting factor with $\sum_{i=1}^{m} \alpha_i = 1$. Let $\frac{\partial J(u_k)}{\partial u_k} = 0$, the high-order tunning law of PPD can be derived as (18):

$$\begin{cases} \hat{\phi}_k(t) = \frac{\Delta u_{k-1}(t)\Delta y_{k-1}(t+1)}{\mu+|\Delta y_{k-1}(t+1)|^2} + \frac{\mu\eta}{\mu+|\Delta y_{k-1}(t+1)|^2} \sum_{i=1}^{m} \alpha_i \hat{\phi}_{k-i}(t) & 2 \le k < m \\ \hat{\phi}_k(t) = \hat{\phi}_{k-1}(t) + \frac{\eta \Delta u_k(t)}{\mu+|\Delta u_k(t)|^2}\left(\Delta y_{k-1}(t+1) - \hat{\phi}_k(t)\Delta u_k(t)\right) & k \ge m \end{cases} \tag{18}$$

The ZNN- HOMFAILC scheme of the system (5) consists of PPD reset Eq. (12), controller law (16), and PPD estimation law (18) shown in Fig. 2.

4 Experiment and Results Discussion

To verify the high convergence speed of ZNN-HOMFAILC, the simulation and actual ankle rehabilitation robot control of MFAILC, HOMFAILC [12] and ZNN-HOMFAILC are conducted.

4.1 Simulation Results

A nonlinear system (19) with time-varying parameters is designed for simulation, in which α(t) is a non-repetitive disturbance used to approximate the nonlinear characteristics of PAMs.

$$y(t) = \begin{cases} \frac{y(t-1)}{1+y^2(t-1)} + u^3(t-1) \; 1 \le t \le 50 \\ (y(t-1)y(t-2)y(t-3)u(t-2)\frac{y(t-3)-1+\alpha(t)u(t-1)}{1+y^2(t-2)+y^3(t-3)} \; 50 < t \le 100 \end{cases} \tag{19}$$

where $\alpha(t) = 0.1 * round(\frac{t}{50})$, the desired tracking trajectory is (20):

$$y_d(t+1) = \begin{cases} 0.5 * (-1)^{round(0.1t)} \; 0 \le t \le 50, \\ 0.5 * \sin(0.1\pi t) + 0.3 * \cos(0.1\pi t) \; 50 < t \le 100 \end{cases} \tag{20}$$

For three control method, the reset threshold $\varepsilon = 0.01$, the step factor of PPD law $\eta = 0.6$, weight coefficient $\mu = 2$. For ZNN-HOMFAILC the controller parameter $h = 8$, order $m = 3$ with $\alpha_1 = 0.6, \alpha_1 = 0.2, \alpha_1 = 0.2$. For MFAILC and HOMFAILC, the weight factor of controller law $\lambda = 1$, step factor $\rho = 1$. The trajectory tracking experiments are conducted separately under $\widehat{\phi}_0(t) = \{20, 30, 40\}$.

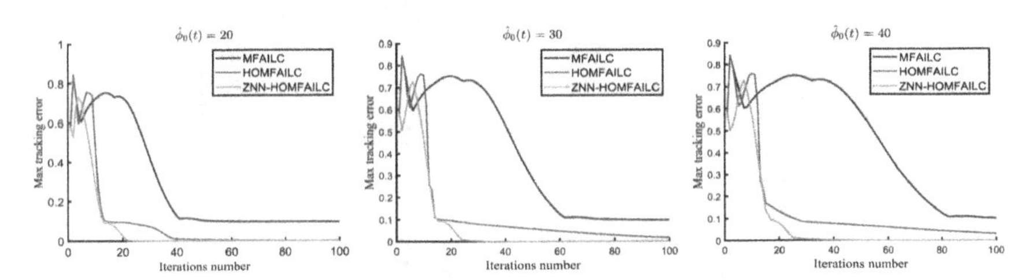

Fig. 3. Maximum tracking error of three methods

The maximum output error of three control methods is shown in Fig. 3. ZNN-HOMFAILC has the fastest convergence speed with different initial PPDs, and the control error is within 0.01. With the increase of the initial value of PPD, the convergence speed and error of MFAILC and HOMFAILC are deteriorating, while ZNN-HOMFAILC converges at 20 iterations with a maximum error around 0.01.

To compare the control performance of three control algorithms, the system output of three methods at the 10th, 20th, 30th and 40th iteration under $\widehat{\phi}_0(t) = 40$ is given in Fig. 4. For quantitatively analyzing the tracking performance of the three control algorithms, the RMSE at the 10th iteration was calculated. The RMSE of ZNN-HOMFAILC, HOMFAILC and MFAILC are 0.1456, 0.3396 and 0.4045. Therefore, ZNN-HOMFAILC can track desired trajectories with a minimum number of iterations. It's noted that the reason for the concentration of tracking bias in the first half is that the system function (19) in $0 \le t \le 50$ is more complex, which contains three terms of control input.

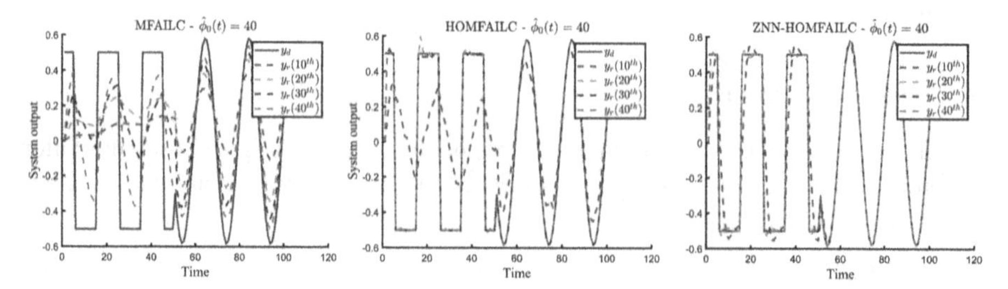

Fig. 4. Trajectory tracking of three control methods under $\widehat{\phi}_0(t) = 40$

The simulation results show that ZNN-HOMFAILC significantly improves the convergence speed and reduces the control error of MFAILC and HOMFAILC for complex control plants. Its control performance is not affected by the initial PPDs.

4.2 Experimental Results

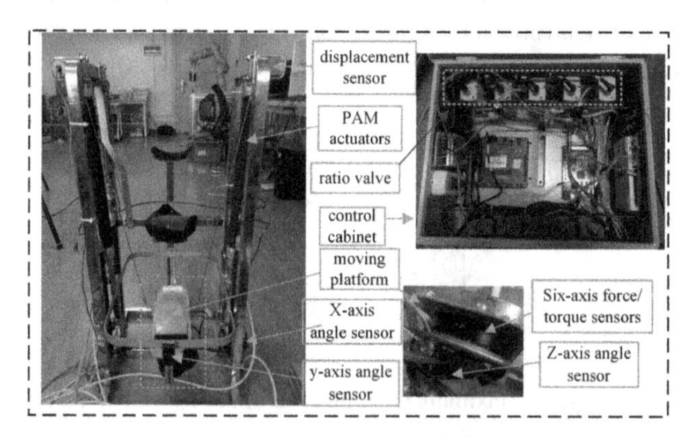

Fig. 5. Experiment platform of ankle rehabilitation robot

To prove that ZNN-HOMFAILC can improve the convergence performance of MFAILC and HOMFAILC in actual robot control, the control experiment is conducted on the ankle rehabilitation robot, shown in Fig. 5, driven by 5 PAM actuators (FESTO MAS-20-400N), and the control methods are running in LabVIEW. The data acquisition module (USB-7635) delivers the calculated control signal to the ratio valve (VPPM-6L-G18-0L6H), which controls the air pressure of PAMs. The displacement of PAMs is collected by displacement sensor (MLO POT-225-TLF), and the angle of the rotatable support platform is collected by angle sensor (EVB-MLX90316GO).

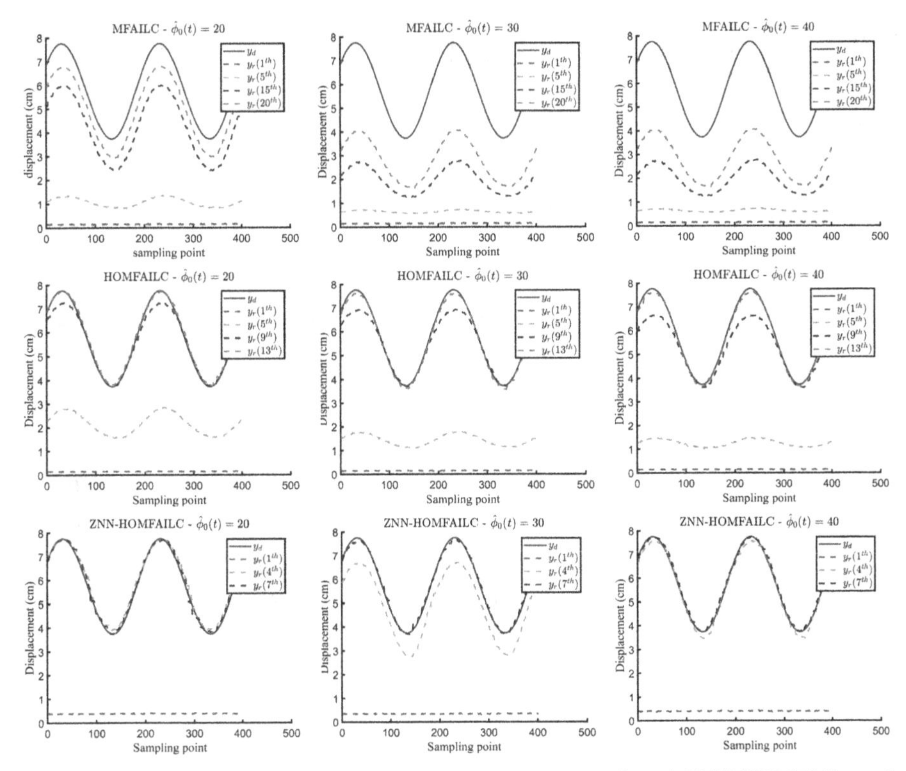

Fig. 6. PAMs displacement tracking of MFAILC, HOMFAILC and ZNN-HOMFAIL under $\widehat{\phi}_0(t) = \{20, 30, 40\}$

The desired rotation angle of the support platform is set $\theta = 0.2sin40\pi t$ in x, y axis and $\theta = 0.1sin40\pi t$ in z axis, the desired displacement of PAMs is calculated by the inverse kinematics model (4). The tracking of displacement of PAMs verifies control performance. The parameter of ZNN-HOMFAILC $h = 5$ and the order is 3 with $\alpha_1 = 0.7, \alpha_1 = 0.2, \alpha_1 = 0.1$ both for HOMFAILC and ZNN-HOMAILC. The other parameters are set to the same value, the step and weight factor of PPD law is set $\mu = 2, \eta = 0.8$, the weight and step factor of the control law is set $\lambda = 1, \rho = 1$. The experiments are conducted under $\widehat{\phi}_0(t) = \{20, 30, 40\}$.

Figure 6 shows the desired displacement and actual displacement of the PAM actuator for three control methods. It can be seen that ZNN-HOMFAILC achieves the tracking of the desired displacement with the least number of iterations under the same initial PPD. When the initial value of PPD changes, the performance of MFAILC will gradually decrease and cannot track the desired displacement. The AVME and RMSE are provided in Table 1.

Figure 7 shows the average error convergence curve of three control methods under the different initial values of PPD. Among the three control methods, ZNN-HOMFAILC has a faster convergence speed (5th iteration), and the displacement tracking bias is less than 0.1 cm. The practical robot control experiments show that ZNN-HOMFAILC can improve the convergence speed and control performance of MFAILC and HOMFAILC, which is applicable to PAM tracking control.

Table 1. The absolute value of maximum error and RMSE of three control methods

Iteration number		AVME (cm)			RMSE		
		3	5	7	3	5	7
$\widehat{\phi}_0(t) = 20$	MFAILC	4.8562	4.3659	3.7904	5.3210	4.8520	4.2656
	HOMFAILC	4.5456	3.3909	0.9950	5.0864	3.6930	1.2260
	Proposed	0.7129	0.2553	**0.2431**	0.3536	0.1417	**0.0988**
$\widehat{\phi}_0(t) = 30$	MFAILC	4.8665	4.6718	4.4184	5.4187	5.1953	4.8158
	HOMFAILC	4.7284	4.0388	1.9118	5.2748	4.4635	2.1957
	Proposed	2.1015	0.4281	**0.2030**	0.7928	0.1009	**0.0890**
$\widehat{\phi}_0(t) = 40$	MFAILC	4.8854	4.7556	4.5921	5.4335	5.2855	5.0891
	HOMFAILC	4.6431	4.1885	2.6236	5.1896	4.6750	2.8519
	Proposed	1.4305	0.1994	**0.1476**	0.8310	0.1089	**0.0975**

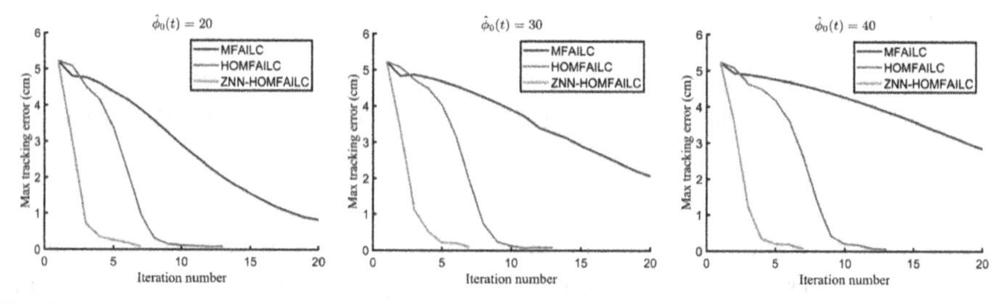

Fig. 7. Maximum error convergence trajectory profiles of the three algorithms with different initial PPDs

5 Conclusion

This paper proposes a ZNN-HOMFAILC to control an ankle rehabilitation robot powered by PAMs. First, inverse kinematics is derived. Then ZNN-based error decay formula and high-order PPD estimation law are introduced in the MFAILC scheme to enhance the control performance. Finally, the complex system function is used to verify that ZNN-HOMFAILC can effectively improve the convergence speed and reduce the control error for the controlled plant with time-varying parameters by simulation. In the practical robot control experiments, the convergence speed of ZNN-HOMFAILC improves by 54% compared to HOMFAILC. The maximum tracking error and RMSE decreased by 93% compared to MFAILC and 75% compared to HOMFAIL in the 7th iteration with $\widehat{\phi}_0(t) = 20$. The experimental results of the actual control show that ZNN-HOMFAILC can achieve the precise position control of PAMs. In addition to precise position control of PAMs, the ankle rehabilitation robot also needs to provide appropriate force assistance. Therefore, in the future, we intend to introduce interactive force information and position tracking errors into the ZNN-HOMFAILC controller for force and position control. Experiments on patients using the ankle rehabilitation platform are one of the future priorities.

Acknowledgments. This work is supported by the National Natural Science Foundation of China under Grant 52075398 and the Wuhan Application Frontier Project under Grant 2020020601012220.

References

1. World Health Organization (2018). https://www.who.int/news-room/fact-sheets/detail/the-top-10-causes-o
2. Liu, L.: Chinese stroke association guidelines for clinical management of cerebrovascular disorders: executive summary and 2019 update of clinical management of ischaemic cerebrovascular diseases. Stroke Vascul. Neurol. **5**(2), 10 (2020)
3. Khalid, Y.M.: A review on the mechanical design elements of ankle rehabilitation robot. Proc. Inst. Mech. Eng. **229**(6), 452–463 (2015)
4. Zhang, D.: Active model-based control for pneumatic artificial muscle. IEEE Trans. Industr. Electron. **64**(2), 1686–1695 (2016)
5. Qian, K.: Robusti iterative learning control for pneumatic muscle with uncertainties and state constraints. IEEE Trans. Industr. Electron. (2022)
6. Ai, Q., Peng, Y., Zuo, J., Meng, W., Liu, Q.: Hammerstein model for hysteresis characteristics of pneumatic muscle actuators. Int. J. Intell. Robot. Appl. **3**(1), 33–44 (2019). https://doi.org/10.1007/s41315-019-00084-5
7. Khajehsaeid, H., Esmaeili, B., Soleymani, R., Delkhosh, A.: Adaptive back stepping fast terminal sliding mode control of robot manipulators actuated by pneumatic artificial muscles: continuum modelling, dynamic formulation and controller design. Meccanica **54**(8), 1203–1217 (2019). https://doi.org/10.1007/s11012-019-01012-4
8. Hou, Z.: An overview of dynamic-linearization-based data-driven control and applications. IEEE Trans. Industr. Electron. **64**(5), 4076–4090 (2016)
9. Chakraborty, S.: Efficient data-driven reduced-order models for high-dimensional multiscale dynamical systems. Comput. Phys. Commun. **230**, 70–88 (2018)
10. Hou, Z.: Data-driven model-free adaptive control for a class of MIMO nonlinear discrete-time systems. IEEE Trans. Neural Netw. **22**(12), 2173–2188 (2011)
11. Dong, M.: State of the art in parallel ankle rehabilitation robot: a systematic review. J. Neuroeng. Rehabil. **18**(1), 1–15 (2021)
12. Ai, Q.: High-order model-free adaptive iterative learning control of pneumatic artificial muscle with enhanced convergence. IEEE Trans. Industr. Electron. **67**(11), 9548–9559 (2019)
13. Yu, Q.: RBFNN-based data-driven predictive iterative learning control for nonaffine nonlinear systems. IEEE Trans. Neural Netw. Learn. Syst. **31**(4), 1170–1182 (2019)
14. Li, H.: Double dynamic linearization-based higher order indirect adaptive iterative learning control. IEEE Trans. Cybern., 1–12 (2021)
15. Chi, R.: Computationally efficient data-driven higher order optimal iterative learning control. IEEE Trans. Neural Netw. Learn. Syst. **29**(12), 5971–5980 (2018)

16. Wu, W.: Iterative feedback tuning-based model-free adaptive iterative learning control of pneumatic artificial muscle. In: 2019 IEEE/ASME International Conference on Advanced Intelligent Mechatronics (AIM), Hong Kong, China, pp. 954–959. IEEE (2019)
17. Chi, R.: Adaptive iterative learning control for nonlinear discrete-time systems and its applications. Beijing Jiaotong University, Beijing (2006)
18. Xu, F.: Zeroing neural network for solving time-varying linear equation and inequality systems. IEEE Trans. Neural Netw. Learn. Syst. **30**(8), 2346–2357 (2018)

The Calibration of Pre-travel Error and Installation Eccentricity Error for On-Machine Probes

Jianyu Lin[1], Xu Zhang[1,2](✉), and Yijun Shen[3]

[1] School of Mechatronic Engineering and Automation, Shanghai University, Shanghai, China
[2] Wuxi Research Institute, Huazhong University of Science and Technology, Jiangsu, China
zhangxu@hust-wuxi.com
[3] School of Mechanical Engineering, Shanghai Jiao Tong University, Shanghai, China

Abstract. Touch-trigger probes are widely used in modern high-precision CNC machine tools as an on-machine inspection solution. The calibration of the probe pre-travel error is a significantly factor affecting the accuracy of the inspection system. However, the existing pre-travel calibration methods either have poor accuracy in practical application or require complicated extra equipments. In this paper, a simple but effective method for measuring the pre-travel error is proposed. With a high-precision standard ring gauge of known size, the center of the ring gauge is obtained by least square method fitting all the measured points. The eccentricity of the probe installation is separated by the reverse method. Then, the probe pre-travel table is established. In order to verify the accuracy of the pre-travel error table and the probe installation eccentricity, the spindle is rotated to touch the same point repeatedly on the high-precision plane. The experimental results show that the proposed novel calibration method can improve the accuracy considerably.

Keywords: On-machine inspection · Probe calibration · Pre-travel · Least square method

1 Introduction

In order to quickly obtain the feedback on the machining results of the machine tool in real time and realize high-speed and high-precision manufacturing, it is necessary to introduce on-machine inspection technology [1]. Probe works safely,accurately, and efficiently, which can help to improve production quality. Currently, a large number of Computerized Numerical Control (CNC) machine tools are equipped with probe [2]. The on-machine probe relies on the multi-axis coordinate system of the CNC machine tool body, which can align the workpiece positioning, check the processing quality of the parts without repeatedly clamping the workpiece, and correct the processing parameters by immediate deviation feedback [3, 4].

Compared with the Coordinate Measuring Machines (CMM), the on-machine inspection environment is complex. There are many factors affecting the measurement

H. Liu et al. (Eds.): ICIRA 2022, LNAI 13456, pp. 785–795, 2022.
https://doi.org/10.1007/978-3-031-13822-5_70

accuracy, making precision control more difficult. The error of the On-machine inspection mainly comes from the probe installation eccentricity error, the pre-travel error and the error of compensation method. Butler et al. proved that the pre-travel error accounts for nearly 60% of the measurement error [5]. The compensation of the pre-travel error of the probe is an important step to improve the accuracy of on-machine inspection. At present, there are two main methods of pre-travel error compensation, i.e., the theoretical analysis model and experimental calibration [6]. The theoretical analysis is based on the structure of the probe and the triggering process, considering the measurement direction, triggering speed, probe length and other factors. Li et al. proposed a dynamic model that treats the probe as a spring link unit, and calculated the pre-travel by the probe deformation during the measurement process [7]. Li et al. proposed a modeling and compensation method of probe pre-travel errors for a five-axis on-machine inspection system considering the influence of gravity [8]. However, there are many probe configurations, which makes it difficult for the theoretical analysis method to exhaustively enumerate all the factors that affect the pre-travel error, and the model accuracy is difficult to meet the requirements of practical engineering. The experimental calibration method is to detect the pre-travel error by means of a calibration device. Woźniak et al. designed a low trigger force and high resolution displacement sensor to measure static pre-travel [9, 10]. RenéMayer et al. used a high-magnification magnification camera to observe the contact position of the probe stylus tip and the workpiece, and used a micrometer to measure the pre-travel [11]. Although the measurement accuracy of these methods is not affected by the accuracy of the machine tool, the result is only the static pre-travel of the probe, and there is a certain deviation between the calibration process and the actual working conditions of the probe. Therefore, these methods are always used to evaluate the probe triggering performance, which is not suitable to measure the pre-travel in engineering because of the complicated equipments.

The existing pre-travel calibration methods either have poor accuracy in practical application or require complicated extra equipments. In this study, a simple but effective method is proposed to calibrate the probe pre-travel. First, a model is built to locate ring gauge center more accurately considering the anisotropy of probe. The probe installation eccentricity error is separated from pre-travel error to reduce deviation between the detected point and planed point. Second, based on the map of pre-travel, the number of detection points that can efficiently loacte the center is simulated and analyzed. Finally, the errors are verified by repeatedly measuring the same point by rotating the spindle. The experimental results show that the proposed novel calibration method can improve the accuracy considerably.

The rest of this paper is organized as follows. In Sect. 2, working principle of the probe is introduced. In Sect. 3, the probe calibration method is described. The pre-travel error and probe installation eccentricity error are verified by experiment in Sect. 4. The conclusions are drawn in Sect. 5.

2 Working Principle of Probe

The basic structure of the probe can be decomposed into the return spring, probe wing seat, ball contact pair, as shown in Fig. 1. The probe wing seat is connected with three evenly distributed positioning columns and the ball pair supports to form a pair of contact pairs. The normally closed circuit is disconnected when any contact pair is disconnected. The probe will send a trigger signal, which can be triggered in -Z direction compression and any direction in XY plane. The pre-travel of the probe is anisotropic, because the deflection of the probe is different in different directions. The probe is continuously bent, not only waiting for the trigger force to reach the threshold, but also waiting for the transmission of the measurement signal. There is an error between the saved coordinate positions and the actual center of the stylus tip. This error is called pre-travel error, which is relatively stable under the same conditions of use, so the probe needs to be calibrated before applying to the actual inspection.

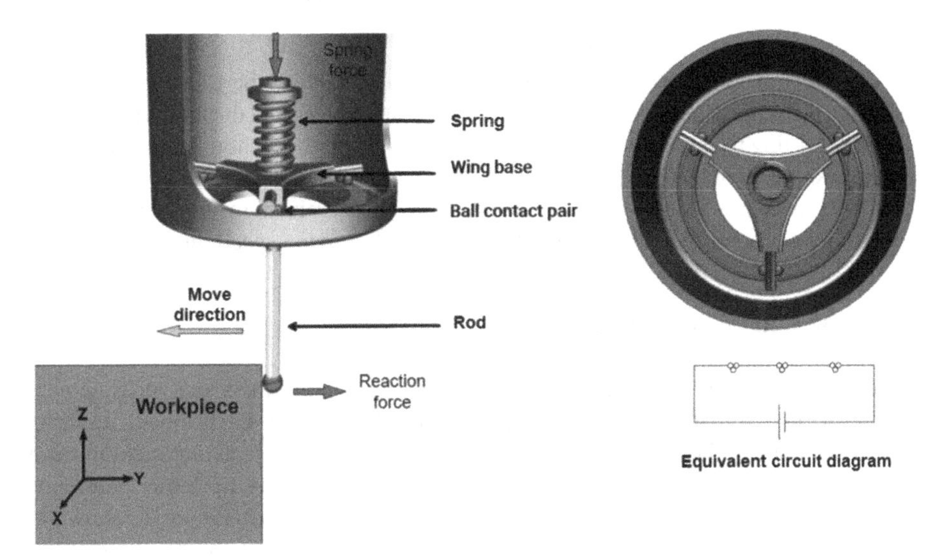

Fig. 1. Probe structure diagram.

The procedure of touch measurement is introduced as follows. At T0, the stylus tip just touches the object to be measured, the probe doesn't trigger. The probe continues to move in the set direction, and the probe rod is deflected by force. At T1, when the measurement force reaches a certain threshold, the positioning column leaves the ball support, the current circuit is disconnected, and the probe sends a trigger signal. At T2, the CNC system detects the probe trigger signal. At T3, the CNC system latches the machine tool coordinate position and sends out the servo motion stop signal. At T4, the machine tool decelerates and brakes to stop. The probe returns to the initial state under the action of the return spring and ready for the next measurement cycle. The measurement system timing diagram is shown in Fig. 2.

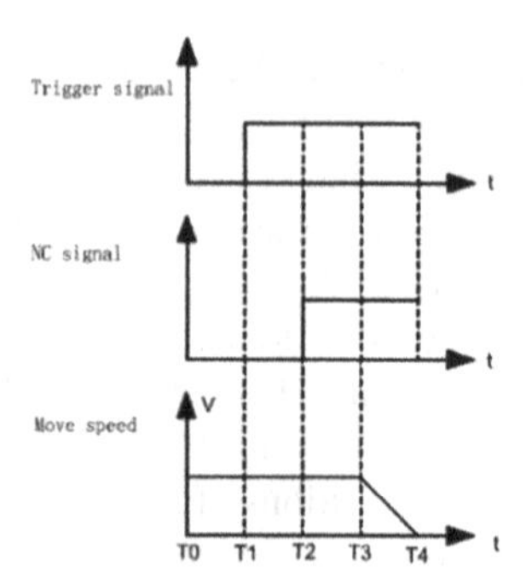

Fig. 2. Measurement system timing diagram.

The pre-travel error of the probe is mainly determined by the touch speed of the probe and the time difference between touching the workpiece and the numerical control system recording the coordinate position. It is quite beneficial to choose a smaller touch detection speed to reduce the measurement error. However, reducing the moving speed will correspondingly increase the uncertainty of the measurement results and reduce the measurement efficiency [12], so the appropriate touch moving speed must be selected.

3 Probe Calibration Method

Probe eccentricity error caused by probe installation will make the actual measurement position deviate from the measurement point planned based on the CAD model, resulting in misalignment of the pre-travel compensation direction and the actual measurement offset direction [13]. Therefore, it is necessary to adjust the stylus tip center before the measurement cycle, to reduce the eccentric error of the probe installation, as shown in Fig. 3. It is also difficult to adjust the probe stylus tip and the spindle rotation center to be concentric with a micrometer, so it is necessary to measure and calculate the eccentricity error of the probe. Eccentricity error compensation is performed before the measurement movement to reduce the offset of the actual measurement position.

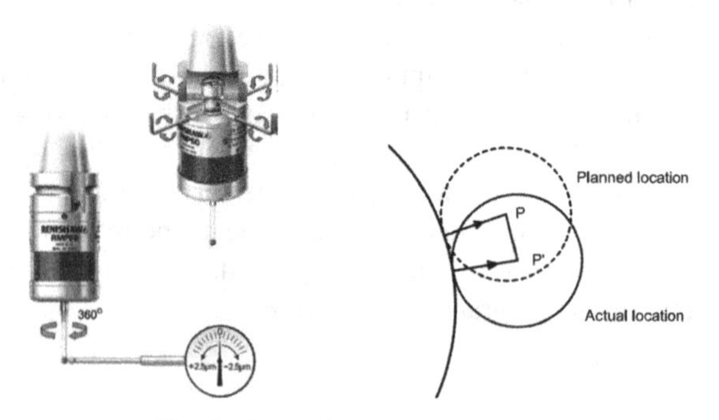

Fig. 3. Probe installation eccentric.

A high-precision standard ball or ring gauge of known diameter is commonly used for calibration, and its accuracy is much higher than the measurement accuracy requirement. Renishaw probes have a repeatability error of ± 1 μ for one-way triggering without changing the measurement environment. Taking a standard ring gauge as an example, points are uniformly collected on the inner diameter surface of the ring gauge in the XY plane, and least square method is used to calculate the distance from each measurement point to the optimal center of the circle [14], which is used as the reference point of the center of the ring gauge.

According to the measurement system timing diagram, the procedure for analyzing the sampling point of the probe at the inner diameter of the ring gauge is as follows Fig. 4. Taking the O' point as the assumed center of the ring, the probe touches the measurement points along the set directions. At this time, the stylus tip center is offset from the spindle center by $\Delta R_P(\theta)$, the tangent contact position of the stylus tip and the inner diameter ring gauge is $C(\theta)$, the position of the stylus tip center is $P(\theta)$, the machine continues to move in the set direction, and the probe rod is slightly deformed along the normal direction of the contact point. When the contact force reaches the threshold, the probe sends a trigger signal, the CNC latches the spindle center position $Q(\theta)'$, but the theoretical trigger position is $Q(\theta)$. $Q(\theta)'$ and $Q(\theta)$ can coincide only when the moving direction is consistent with the normal direction of the contact point. Since the stylus tip is tangent to the ring gauge, the normal of the contact point $C(\theta)$, points to the theoretical center O of the ring gauge, and $P(\theta)$ is on the $OC(\theta)$ line. The radius of the probe stylus tip is r, the radius of the ring gauge is R, and the position $P(\theta)$ of the probe stylus tip center is on the cylinder with the coaxial line with the ring gauge and the radius is $(R - r)$.

$\Delta R_O(\theta)$ is the trigger direction deviation caused by the error between O' and O. $r(\theta)$ is the stylus tip dynamic radius triggered at $Q(\theta)$. $r(\theta + \Delta\theta)$ is the stylus tip dynamic radius triggered at $Q(\theta)'$. The relationship is as Eq. (1). To reduce $\Delta R_O(\theta)$, it is equivalent to reduce $\Delta\theta$. We must first control the error between O' and O.

$$\Delta R_O(\theta) = r(\theta) - r(\theta + \Delta\theta) \tag{1}$$

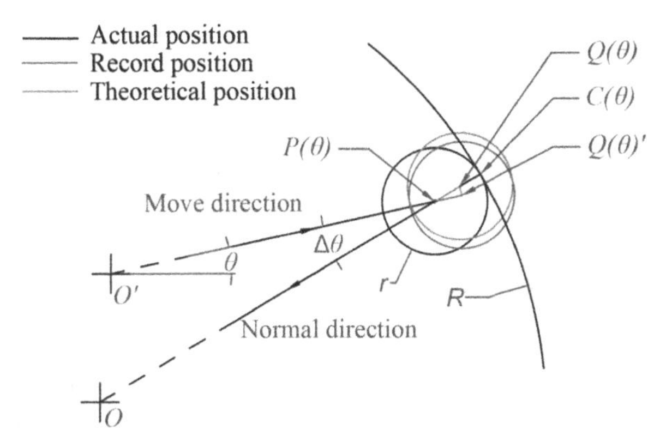

Fig. 4. Pre-travel calibration model.

The error magnitude of the probe anisotropy is larger than that of the radius size of the high-precision standard ring gauge. Based on the above assumption, the following iterative measurement and calibration method is proposed.

Step 1: The measurement program is executed with the spindle angle equal to 0° and 180° respectively. Setting O' (by visual inspection) as assumed center, probe moves in set directions from O' to target points, obtains a set of touch points coordinates $Q(\theta)'$.

Step 2: Using least square method to fit $Q(\theta)'$ and get the center O'', O' is replaced by O''. The fitting result can be used as a good initial value for the center O of the ring gauge, because the radius value of the ring gauge is much larger than the pre-travel.

Step 3: Taking the new O' as the center, probe moves in the set directions, obtains a new set of touch points coordinates $Q(\theta)'$.

Step 4: $\Delta R_P(\theta)$ is probe installation eccentricity. The pre-travel error $E(\theta)$ and the probe dynamic radius $r(\theta)$ can be calculated by following equations.

$$E(\theta) = Q(\theta)' - P(\theta) + \Delta R_P(\theta) + \Delta R_O(\theta) \tag{2}$$

$$r(\theta) = r - E(\theta) \tag{3}$$

Step 5: If the difference between the new fitted center O'' and O' is greater than the threshold of 1 μ, O'' takes place of O', then, step 3 and step 4 repeats. When the difference is smaller than the threshold, the influence of $\Delta R_O(\theta)$ is small enough to be ignored. The stylus tip center is close to the theoretical center of the ring gauge, the which is recorded as O''_0. Now, we can get the probe pre-travel $E(\theta)$ and dynamic action radius $r(\theta)$.

Step 6: The probe installation eccentricity error could be separated by reverse method [15]. The φ is the spindle rotation angle. O''_{180} can be obtained in the same way. Since the installation eccentricity error of the probe is centrally symmetric at 0° and 180°, the symmetrical center point is the probe installation eccentricity T. $(\Delta R_O(\theta + 180) - \Delta R_O(\theta))$ error is smaller than $(\Delta R_P(\theta + 180) - \Delta R_P(\theta))$ order of magnitude at this position, which can be ignored. The probe installation deviation $\Delta R_P(\theta)$ can be separated from $E(\theta)$. The installation eccentricity compensation should be performed for all subsequent measurement points.

$$E(\theta + \varphi) = Q(\theta + \varphi)' - P(\theta + \varphi) + \Delta R_P(\theta + \varphi) + \Delta R_O(\theta + \varphi) \tag{4}$$

$$T = \frac{O''_0 + O''_{180}}{2} = \frac{\Delta R_P(\theta) - \Delta R_P(\theta + 180)}{2} \tag{5}$$

4 Experimental Verification and Discussion

The experiment is conducted on SIEMENS CNC system test platform with LIECHTI five-axis machining tool and Renishaw OMP60 touch probe. After the probe is activated, the green signal light is on. The red signal light flashes and the programmable logic controller (PLC) signal DB10.DBX107.0 jumps when manually triggering the probe.

After the probe measurement cycle is activated, the original measurement movement stops when the probe is manually triggered. The stylus tip center coordinate is obtained through $AA_MM[AIXS], $AA_MW[AIXS].

During the experiment, it is necessary to keep the parameters consistent, the touch speed is 100 mm/min, the length of the measuring rod is 100 mm, and the standard ring gauge with a diameter of 50.001 mm is selected.

4.1 Probe Calibration by Proposed Method

In the experiment, the ring gauge is installed to keep the axis parallel to the spindle axis. $O_0^{''}$ and $O_{180}^{''}$ could be obtained by fitting 36 points evenly distributed across the ring gauge with the method described above, the center is shown in Fig. 5.

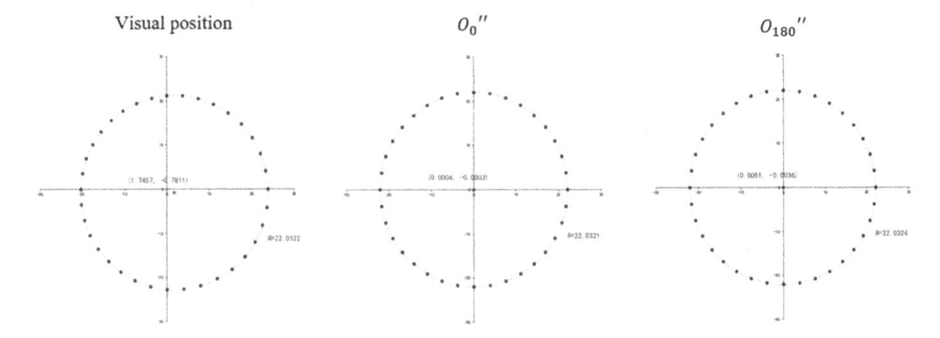

Fig. 5. The least square fitting measurement points.

The probe pre-travel error $E(\theta)$ in each measurement direction is calculated by the measurement points when the probe at $O_0^{''}$, as show in Fig. 6. The error at uncalibrated angle is predicted by the linear interpolation method. The map of probe dynamic radius $r(\theta)$ is shown in Fig. 7.

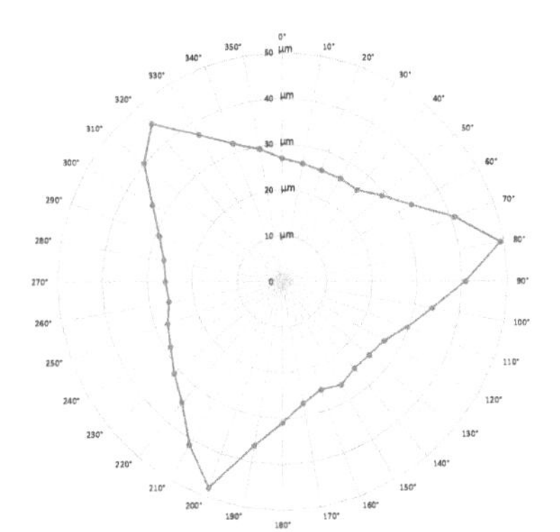

Fig. 6. The error map of pre-travel.

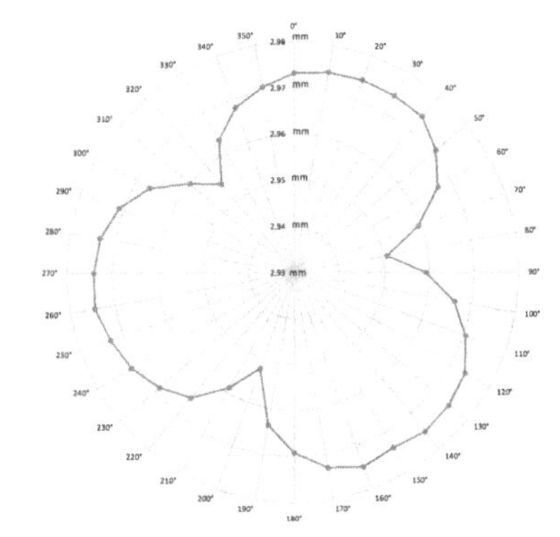

Fig. 7. The map of probe dynamic radius.

After compensating for probe installation eccentricity T, spindle center accuracy can be verified by measuring the ring gauge runout with a tool coupled micrometer. In this experiment, the runout measured by this method is about 2 μm, which is smaller than that 12 μm calibrated by Siemens CYCLE996.

4.2 Compensation Experiment

The probe dynamic radius compensation table and the probe installation eccentricity error results are verified by inspecting a high precision flat surface. Adjust the contact angle between the trigger stylus tip and the plane by rotating the spindle, probe touches the same target point along the -X direction in this experiment, sports sketch is shown in Fig. 8. The rotation angle of the spindle is β. The probe rotation error is $S(\beta)$, C is a constant, $L(\beta)$ is the X coordinate position of touch point located by PLC. $Error(\beta)$ is the measurement error after compensating the probe dynamic radius and probe rotation error. $Error(\beta)'$ is the compensation result without considering the probe installation eccentricity error.

$$S(\beta) = T + \Delta R_P(0) * cos(\beta + C) \tag{6}$$

$$Error(\beta) = L(\beta) - r(\beta) - S(\beta) \tag{7}$$

$$Error(\beta)' = L(\beta) - r(\beta) \tag{8}$$

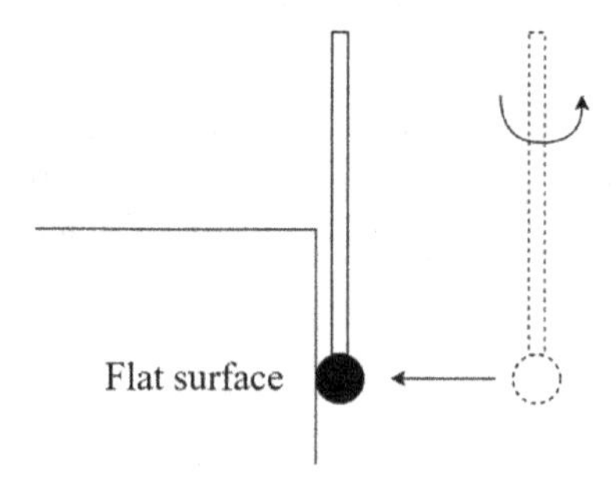

Fig. 8. Repeating measurement of specified points after rotating the spindle.

4.3 Experimental Results and Discussion

Because of the anisotropy of the probe, there is a certain variation in the center position when different numbers of fitting points are adopted. In order to improve the efficiency of iteratively fitting the circle center, the influence of the number of fitting points on the fitting accuracy of the circle center is simulated based on the obtained map of pre-travel error. The measurement points are evenly distributed, 12 sets of data are sampled from each group of measurement points, and the mean Euclidean distance of the fitted circle

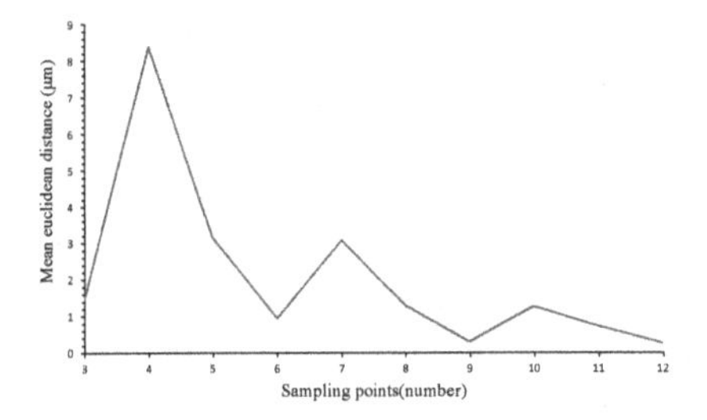

Fig. 9. Mean Euclidean distance of fitting circle centers with different sampling points number.

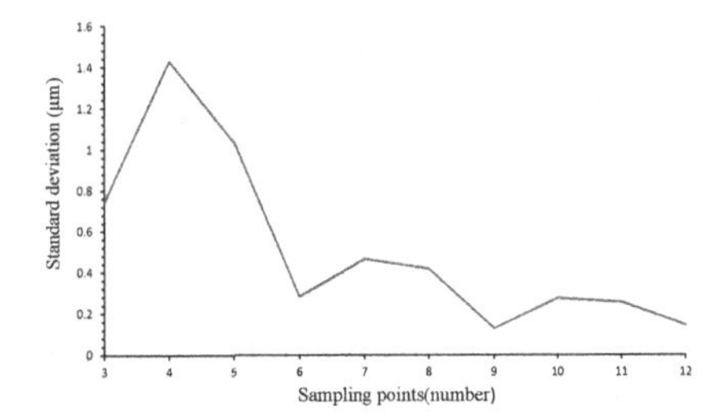

Fig. 10. Euclidean distance standard deviation of fitting circle center with different sampling points.

center with respect to the current center, as shown in Fig. 9. The Euclidean distance standard deviation of the fitted circle center relative to the current circle center, as shown in Fig. 10.

Analyzing the simulation results, it can be judged that the average Euclidean distance is relatively smaller when the number of sampling points is a multiple of 3. Generally, the accuracy increases with increase in the number of sampling points. When the number of sampling points is greater than 8, the degree of dispersion of the fitting center is relatively small. Considering the fitting accuracy and calibration efficiency comprehensively, it is most appropriate to select 9 touch detection points.

Due to the probe installation eccentricity error, there exists a deviation between the inspection point and planned point, which affects the accuracy of inspection result. The probe rotation error is caused by probe installation eccentricity error. The probe installation eccentricity error separated from pre-travel error can be verified by inspecting the

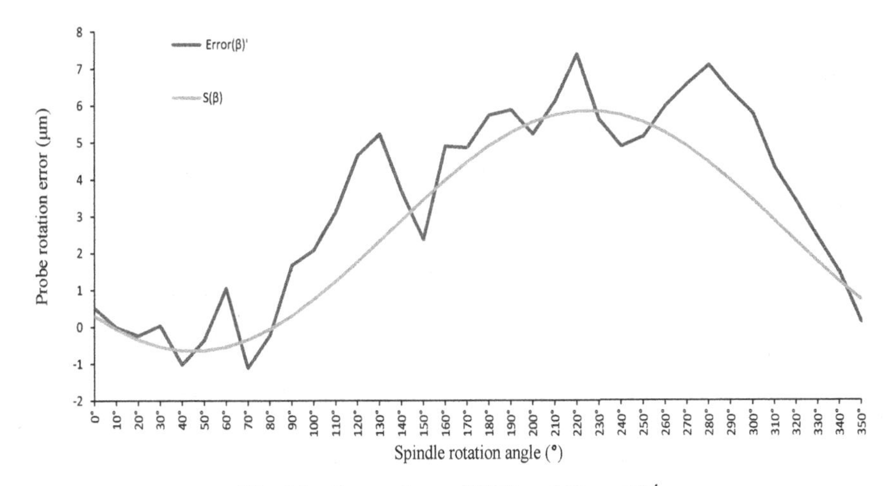

Fig. 11. Comparison of $S(\beta)$ and $Error(\beta)'$.

specified points repeatedly after rotating the spindle. If only the pre-travel is compensated, the inspection result will have a probe rotation error. The difference between $S(\beta)$ and $Error(\beta)'$ is shown in Fig. 11.

Error comparison is shown in Fig. 12. It is obvious that the probe pre-travel influences the accuracy of feature measurement seriously, so it is necessary to calibrate probe before applying to the actual inspection. The probe installation eccentricity error will not only offset the inspecting point, but also affect the result of pre-travel. Probe rotation error approximately 6 μ is contained in $Error(\beta)'$. $Error(\beta)$ fluctuates within 4 μ, which is much smaller than that before compensation. Therefore, this method is relatively accurate for the calibration of the pre-travel error and the probe installation eccentricity error.

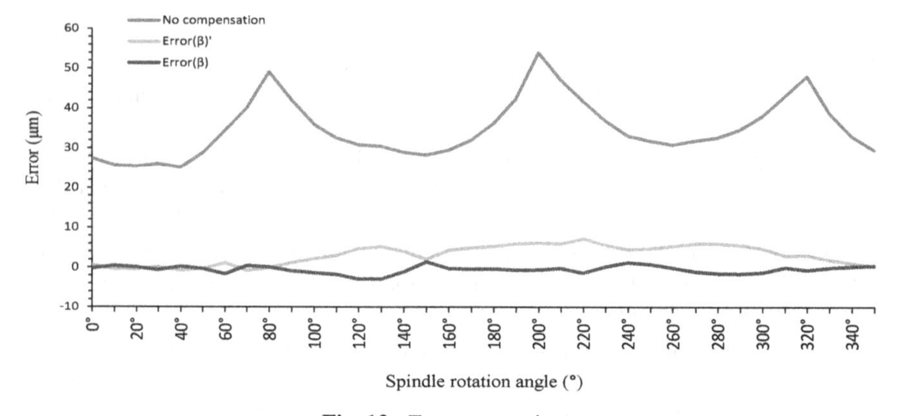

Fig. 12. Error comparison

5 Conclusion

Based on the probe's mechanical structure and working principle, this paper proposes a simple and accurate method to calibrate probe on-machine inspection. The proposed method improves the positioning accuracy of ring gauge center. The probe installation eccentricity error isolated from pre-travel error is calculated by the reverse method. The compensation of installation eccentricity error helps to reduce the deviation between inspection points and planning points. The positioning accuracy of ring gauge center increases with increase in the number of detecting points. In order to improve the efficiency of calibration, it is appropriate to evenly distribute 9 inspection points in the simulation. To validate the proposed method, compensation experiment is carried out. The compensation result indicates that the proposed method can significantly increase the inspection accuracy.

Acknowledgement. This research was supported by the National Natural Science Foundation of China (Grant No. 51975344).

References

1. Wang, P., Lei, Y., Zou, S., Cui, L.: Current situation and development trend of on-machine inspection technology. Combined Mach. Tools Autom. Process.Technol., 1–4 (2015)
2. Zhang, J., Luo, S., Pang, C., Du, H.: Research and development trend of on-machine measurement technology for modern machine tools. Aerospace Manufact. Technol. **09**, 43–49 (2016)
3. Guiassa, R., Mayer, J.R.R., St-Jacques, P., Engin, S.: Calibration of the cutting process and compensation of the compliance error by using on-machine probing. Int. J. Adv. Manuf. Technol. **78**(5–8), 1043–1051 (2014). https://doi.org/10.1007/s00170-014-6714-6
4. Wang, L., Huang, X., Ding, H.: Error analysis and compensation of trigger probe on-machine inspection System. China Mech. Eng. **23**(15), 1774–1778 (2012)
5. Butler, C.: An investigation into the performance of probes on coordinate measuring machines. Ind. Metrol. **2**(1), 59–70 (1991)
6. Gao, F., Zhao, B., Li, Y., Yang, X., Chen, C.: Novel pre-travel calibration method of touch trigger probe based on error separation. Chin. J. Sci. Instrum. **34**(7), 1581–1587 (2013)
7. Li, Y., Zeng, L., Tang, K., Li, S.: A dynamic pre-travel error prediction model for the kinematic touch trigger probe. Measurement **146**, 689–704 (2019)
8. Li, S., Zeng, L., Feng, P., Yu, D.: An accurate probe pre-travel error compensation model for five-axis on-machine inspection system. Precis. Eng. **62**, 256–264 (2020)
9. Woźniak, A., Dobosz, M.: Metrological feasibilities of CMM touch trigger probes part II: experimental verification of the 3D theoretical model of probe pretravel. Measurement **34**, 287–299 (2003)
10. Zhao, B., Gao, F., Li, Y.: Study on pre-travel behaviour of touch trigger probe under actual measuring conditions. In: Proceedings 13th CIRP Conference on Computer Aided Tolerancing, vol. 27 (2015)
11. RenéMayer, J.R., Ghazzar, A., Rossy, O.: 3D characterisation, modelling and compensation of the pre-travel of a kinematic touch trigger probe. Measurement **19**(2), 83–94 (1996)
12. Han, R., Zhang, J., Feng, P., Yu, D.: Analysis and modeling of radius error of touch-trigger probe on-machine inspection system. Combined Mach. Tools Autom. Process.Technol. **12**, 60–64 (2014)
13. Sepahi-Boroujeni, S., Mayer, J.R.R., Khameneifar, F.: Repeatability of on-machine probing by a five-axis machine tool. Int. J. Mach. Tools Manuf. **152**, 103544 (2020)
14. Wozniak, A., Jankowski, M.: Random and systematic errors share in total error of probes for CNC machine tools. J. Manuf. Mater. Process. **2**(1), 1–8 (2018)
15. Marsh, E.R., Arneson, D.A., Martin, D.L.: A comparison of reversal and multiprobe error separation. Precis. Eng. **34**(1), 85–91 (2010)

A Preliminary Tactile Conduction Model Based on Neural Electrical Properties Analysis

Xiqing Li and Kairu Li(✉)

Shenyang University of Technology, Liaoning 110879, China
Kairu.Li@sut.edu.cn

Abstract. The absence of tactile feedback leads to a high rejection rate from prostheses users and impedes the functional performance of dexterous hand prostheses. To effectively deliver tactile feedback, transcutaneous electrical nerve stimulation (TENS) has attracted extensive attention in the field of tactile sensation restoration, due to its advantages of non-invasive application and homology with neural signals. However, the modulation of electrotactile stimulation parameters still depends on operators' experience instead of a theoretical guidance. Thus, this paper establishes a preliminary tactile conduction model which is expected to provide a theoretical foundation for the adjustment of electrotactile stimulation parameters. Based on a review of studies about the electrical conduction properties of electrodes and upper-limb tissues which are related to tactile generation process, a tactile conduction model is established to describe the neural signal transduction path from electrodes to tactile nerve fibres and the influence of different stimulation parameters on subjects' sensation experience is briefly analysed.

Keywords: Tactile conduction model · Tactile feedback · Sensory nerve fibres · Transcutaneous electrical nerve stimulation (TENS)

1 Introduction

Limb loss which is mostly caused by accidents and trauma is projected to affect 3.6 million people in the USA by 2050, of which 35% of cases will affect the upper limb [1]. Prostheses may improve amputees' quality of life by restoring the function of their lost hands during daily life activities [2]. However, a highly rejection rate (30%–50%) of hand prostheses is found among amputees, partly due to the absence of tactile feedback [3–5]. Amputees can not be aware of tactile sensations, such as force and pressure, through their prosthetic fingers, which undermines their sense of perceptual embodiment [6]. The lack of tactile feedback also constrains the commercial promotion of hand prostheses. Therefore, it is important to restore tactile sensation feedback for hand prostheses.

As a promising tactile stimulation technique, transcutaneous electrical nerve stimulation (TENS) attracts extensive attention due to its advantages of non-invasive application and homology with neural signals. It stimulates cutaneous sensory afferent neurons with a local electric current [7, 8]. The adjustable parameters include stimulation frequency, amplitude, pulse width, electrodes distribution, etc. which have a significant impact on

H. Liu et al. (Eds.): ICIRA 2022, LNAI 13456, pp. 796–807, 2022.
https://doi.org/10.1007/978-3-031-13822-5_71

the stimulating effect [9]. However, a challenge entailed in TENS is that the electrical stimulation parameters are adjusted based on experience without a theoretical foundation [10], which results in poor stability and naturality [11, 12]. Improper parameter setting may even lead to unexpected feelings such as burning pain. Therefore, it is necessary to establish a tactile conduction model as a theoretical guidance for the modulation of electrotactile stimulation parameters.

Despite existing literature studying the electrical property of skin or tactile neurons separately, it is expected to have a comprehensive understanding of the whole electrical signal conduction process from electrodes, through skin, to tactile nerve fibres. This paper is organised as follows. Section 2 reviews existing studies about models or equivalent circuits of upper-limb tissues which are related to tactile generation process. Section 3 presents the establishment of the tactile conduction model and briefly analyses the influence of different stimulation parameters on subjects' sensation experience, then followed by a summary in Sect. 4.

2 Equivalent Models of Tissues Related to Sensory Generation

The impedance of the skin under the influence of electrical stimulation exhibits obvious dual resistance and capacitance features; hence, researchers also employ resistance and capacitance parallel networks to equate the skin impedance in model creation [13]. In 1989, Kaczmarek [14] proposed a dynamic impedance model of electrode skin that closely resembles the voltage-current properties of skin but is largely applicable to low frequency and low current situations. In 2015, Vargas Luna [15] investigated the ion diffusion effect of skin impedance and constructed a dynamic model in which the electrolyte takes a short time to fill the spaces between the cells of the stratum corneum, often within a few microseconds, due to the fact that the impedance and capacitive resistance generated by the ion flow of the cells of the stratum corneum vary with time, and the cells osmotically contract [16]. The model captures the dynamic properties of skin impedance.

Most studies on tactile nerve fibre models fail to describe details of ion channels, and models of tactile sensory afferent nerve fibres with ion channels are more consistent with the physiological properties of actual subcutaneous nerve fibres [17–20]. In 1907, Lapicque proposed the Integrate-and-Fire paradigm, the IF model [21]. It's a straightforward theoretical model with a formula such as Eq. (1):

$$C\frac{dV}{dt} = -\frac{V - V_{rest}}{\gamma} + I_{syn}(t) \tag{1}$$

where C is the membrane capacitance, V is the membrane potential, V_{rest} is the resting potential, γ is the membrane resistance, and $I_{syn}(t)$ is the input current.

The IF model is a simple one-dimensional neuronal model, a deep simplification of the biological neuron, but its variable is only the membrane potential V, which means that it is not possible to microscopically describe the relationship between parameters such as membrane potential and ion channel conductance, etc. The advent of the HH model has optimised this problem.

In 1952, Hodgkin and Huxley [22] jointly proposed a computational model to describe the changes in membrane potential on neurons as shown in Fig. 1.

In Fig. 1, I is the total current flowing through the cell membrane, which is composed of four branched currents together, namely the sodium channel current, the potassium channel current, the chloride ion-based leakage current and the current flowing through the membrane capacitance. The model successfully explains the principles of neuronal action potential generation. As a result of the Nobel Prize awarded in 1963, the study is considered one of the most effective models for describing neuronal electrophysiological activity [23, 24]. Later studies of transcutaneous electrical nerve stimulation were greatly influenced by the model.

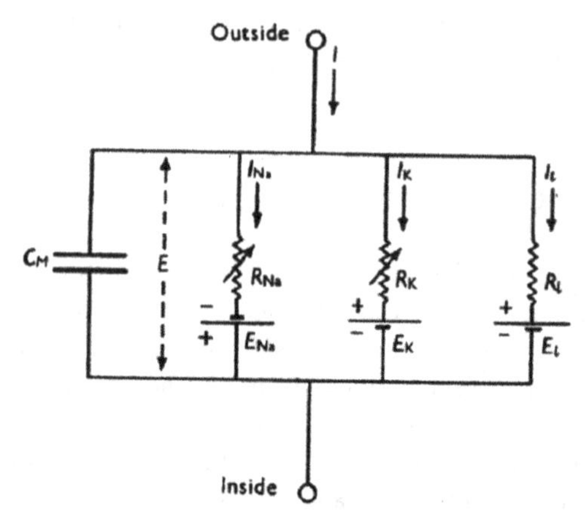

Fig. 1. Equivalent circuit diagram of cell membrane [21]

After the HH model was proposed, various neuronal models were proposed due to the computational complexity of the HH model, typically represented by the FHN model [25], the HR model [26] and the Morris-Lecar model [27].

The FHN model, a two-dimensional dynamical system, was proposed in 1961 by Fitz Hugh and Nagumo, who studied the link between the ion channel switching variables in the HH model to reduce it to two differential equations. The mathematical expression is given by Eq. (2):

$$\begin{cases} \dot{V} = f(V) - W + I + V_{xx} \\ \dot{W} = a(bV - cW) \end{cases} \tag{2}$$

where $f(V)$ is a cubic polynomial, V is the membrane potential, a, b, and c are constants, and W is the recovery variable. FHN model is obtained by downscaling and simplifying the HH model, which has fewer and more concise variables than the HH model, and it can also reflect the main characteristics of neuronal firing activity.

The Morris-Lecar model, a simplification of the HH neuron model, was proposed by Keyne et al. in 1973 to describe the relationship between muscle cell membrane voltage and the voltages associated with Ca^{2+} and K^+ ion channels, initially as a third-order nonlinear differential equation, and later as a second-order system of equations

(Morris-Lecar model), which is mathematically described by Eq. (3):

$$\begin{cases} C\frac{dV}{dt} = G_{Ca}M_{\infty}(V)(V_{Ca} - V) + G_k N(V_k - V) + G_L(V_L - V) + I \\ \frac{dN}{dt} = \lambda_N(V)(N_{\infty}(V) - N) \end{cases} \quad (3)$$

where N is the potassium channel opening probability, M is the calcium channel opening probability, I is the current, G_{ca} and G_k are the maximum conductance, respectively, and G_L is the leakage conductance. V_{ca}, V_k, and V_L are the reversal potentials, respectively. $M_{\infty}(V)$ and $N_{\infty}(V)$ are the values of M and N at steady state, respectively. $\lambda_N(V)$ is the potassium ion activation variable of the time constant.

The HR neuron model was first proposed in 1984 by J L Hindmash and R M Rose to represent the electrical activity of neurons. The HR model is straightforward in appearance, and one of its key aspects is its ability to represent the cluster firing pattern of neurons, which is commonly used to model repeating peaks and irregular behaviour in neurons and is mathematically described as Eq. (4):

$$\begin{cases} \frac{dx}{dt} = y - ax^3 + bx^2 - z + I \\ \frac{dy}{dt} = c - dx^2 - y \\ \frac{dz}{dt} = r\left(x - \frac{1}{4}(z - z_0)\right) \end{cases} \quad (4)$$

where the three variables x, y and z denote the membrane potential, fast recovery current, and slow adaptation current, respectively. I is the external stimulation current, a, b, c, d, and z_0 are constant values, z_0 is used to regulate the resting state, and r is a control parameter related to the calcium ion concentration and is used to control the slow adaptation current z.

The Morris-Lecar model contains only two variables, so it cannot provide a comprehensive description of neuronal electrical activity. However, it can describe the mechanism of action potential generation in some neurons, and it is easier to calculate than the HH model, so it is also widely used in neurology.

3 Establishment of a Preliminary Tactile Conduction Model

3.1 Physiological Basis of Tactile Feedback

As the leading edge of the human tactile system, the skin has the dual function of receiving and transmitting impulses. The features of the skin affect its ability to sense and transmit external stimuli, particularly electric current stimuli, and analysing the laws of the affected skin and constructing a model of electric conduction can be accomplished by analysing the structure of the skin.

The skin has a complex multi-layered structure consisting of three main layers: the epidermis, the dermis and the subcutaneous fat, plus skin derivatives such as sweat glands, hair, sebaceous glands and nails [28]. As the largest organ of the body, it is the part of the body that is in direct contact with the outside world.

There are many receptors with a density of sensory nerve endings close to 200/cm^2 under the skin [29]. Sensory nerve endings located under the epidermis, tactile vesicles in the dermis papillary layer, and deeper circumferential vesicles are capable of sensing touch, pain, temperature, and pressure. Ruffini vesicles, Meissner vesicles, Merkel discs, and Pacinian vesicles are the four mechanoreceptors in the skin that detect external touch [30]. The four touch receptors have diverse locations and responses to tactile stimuli. The design of the prosthesis focuses on the sensory function of the skin.

The number of tactile receptor vesicles under the skin, which correspond to vesicles scattered in the intermediate junction between the dermis and the epidermis, is mostly responsible for tactile generation. Meissner vesicles, which are found in the papillary layer of the dermis and can detect mechanical stimuli at frequencies of 20–50 Hz [31], underpin the smooth skin. Merkel's contact discs, which are found beneath smooth or hairy skin and are particularly sensitive to 5–15 Hz, detect tactile sensations caused by separation and touch between the body and objects. It is extremely sensitive to the material qualities and spatial attributes of the items it comes into contact with, such as the object's structural organization, pattern, and thickness. Figure 2 depicts the location and organization of different receptors. The electrode sheet is fitted to the forearm and delivers an electric current to the skin, which eventually stimulates the tactile nerve fibres (Fig. 3).

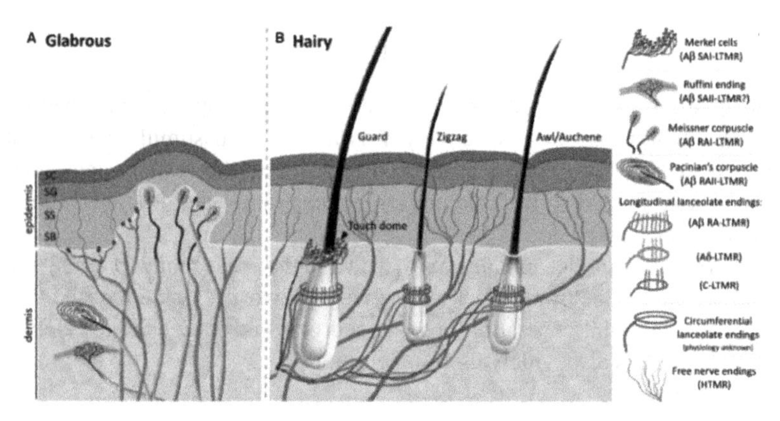

Fig. 2. Location and morphological distribution of subcutaneous mechanoreceptors [6]

Both hairy and hairless skin contain Pacinian vesicles and Ruffini terminals, which are sensitive to high frequency mechanical stimuli at 60–100 Hz. They are the most sensitive to mechanical stimulation of all the receptors, and can detect skin deformation at 10 nm.

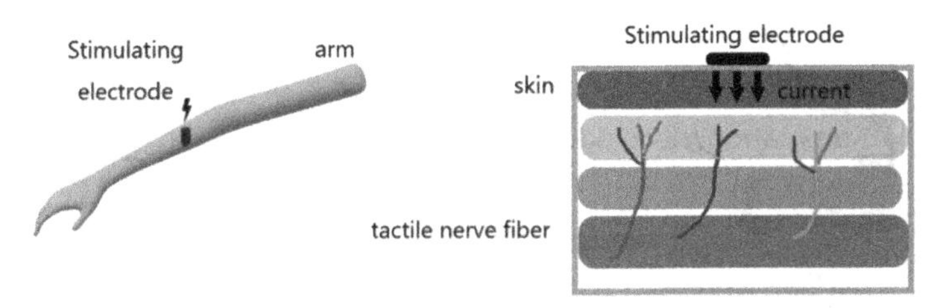

Fig. 3. A physiological demonstration of current transmission structure

When the human skin is electrically stimulated by delivering pulses directly to the skin with electrodes, the cell membranes of these mechanoreceptors are distorted, leading to the opening of certain ion channels via which ions (Na^+, K^+) depolarize the membrane [32, 33]. The action potential [34] is conveyed to primary sensory neurons via axons attached to the receptors [35] when these membrane potentials hit a specified threshold (depending on the biological features of the individual receptors). In addition, nerve impulses are transported by nerve fibres to the peripheral nerve-central nerve and, finally, to the associated functional regions of the brain in order to generate the corresponding perception.

3.2 A Preliminary Electrotactile Conduction Model

An impedance modelling approach was utilized to study how the sense of touch is generated during TENS, as well as the selection of electrical stimulation parameters and their effect on the excitation of subcutaneous nerve fibres. By constructing the proper impedance models, numerous aspects of the formation of electrical tactile sensations can be determined. The rationale behind the construction of the model is the focus of attention.

Among the factors that influence the electrical conduction process are electrode-skin impedance, skin impedance, and nerve fibres impedance; understanding these impedance models provides insight into behaviour when exposed to electrical stimulation. In the model review section, incomplete models of independent conduction are shown, the majority of them are only designed for a specific area of the skin. Figure 4 shows the equivalent circuit of the conduction passway of the electrotactile stimulation. Part A is the electrode-skin impedance under the action of electrical stimulation, part B is the skin impedance, and Part C shows the two-dimensional tactile neuron equivalent circuit.

C_e is the amount of charge between the electrode and skin, while R_L is the resistance created when a charge transition occurs between the skin and the electrode (charge migration resistance) [36, 37]. Both C_e and R_e rely on the current's density and frequency. C_s and R_{s1} are the analogous capacitance and resistance of the epidermis and sweat glands of the skin. R_{s2} is the equal dermal and subcutaneous tissue resistance to the front of the tactile neurons.

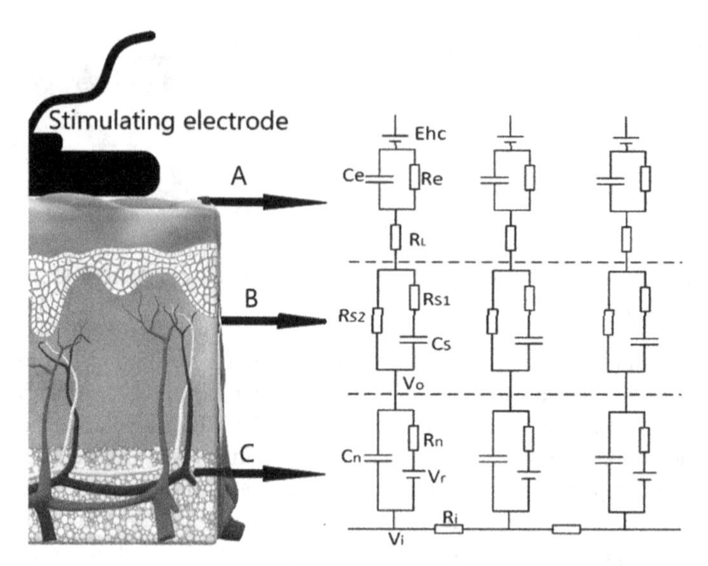

Fig. 4. Equivalent circuit of the electrotactile conduction process

The total impedance Z at the skin-electrode interface is shown as a function of frequency in Eq. 5:

$$Z = R_l + \frac{R_e}{1 + j\omega C_e} \tag{5}$$

where ω is the corner frequency, which can be expressed as $\omega = 2\pi f$ and f is the frequency. At high frequencies, the capacitor C_e conducts and the resistor R_e is short-circuited, and the impedance in the equivalent circuit is $Z = R_l$; at low frequencies, C_e is broken, no current passes, and the impedance of the equivalent circuit is $Z = R_l + R_e$; over the entire frequency range, the lower the frequency, the higher the impedance.

The current passes through the skin and eventually acts on the tactile nerve fibres. The nerve fibres are separated into myelinated and unmyelinated nerve fibres, with the myelinated nerve fibre being the primary one linked with the feeling of touch. An electric field is applied to myelinated nerve fibres, with the current penetrating from node. Figure 5 depicts the model of tactile nerve fibres impedance. A visual depiction of a tactile nerve fibre, comprised of node of Ranvier and myelin, is shown in the upper section of the figure. The equivalent circuit for the neuron is shown in the lower half. The current excites the subcutaneous tactile nerve fibres and creates the appropriate action potentials during electrical stimulation.

The cell membrane is denoted by n, Δx is the length of myelin, and in the myelinated nerve model, the length of Ranvier node is set to L. C_n is the value of the cell membrane's capacitance in the radial direction, R_n is the cell membrane's resistance in the radial direction, R_i is the cell's internal axial resistance, and D is the radius of the axon. The radial direction current is set to I_i, whereas the magnitude of the current going through R_n is $I_{\mathrm{i,\,n}}$. The external voltage of the n th membrane is $V_{o,n}$ and the internal voltage is $V_{i,n}$ and the internal static voltage is V_r, so by the nodal theorem we can obtain Eq. (6)

and Eq. (7):

$$C_n d\left(V_{i,n} - V_{o,n}\right)/dt + I_{i,n} + G\left(V_{i,n} - V_{i,n-1}\right) + G\left(V_{i,n} - V_{i,n+1}\right) = 0 \tag{6}$$

$$V_n = V_{i,n} - V_{o,n} + V_r \tag{7}$$

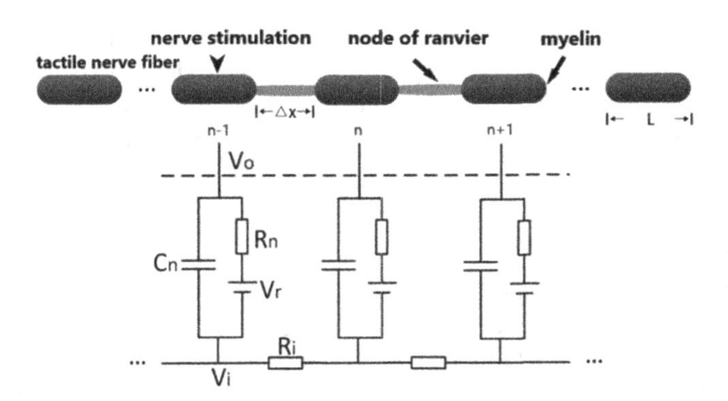

Fig. 5. Tactile nerve fibre impedance model

Combining Eq. (7) and Eq. (5), we can obtain Eq. (8):

$$dV_n/dt = \left(G\left(V_{n-1} - 2V_n + V_{n+1} + V_{o,n-1} - 2V_{o,n} + V_{o,n+1}\right) - I_{i,n}\right)/C_m \tag{8}$$

Equation (8) illustrates the effect of intra-membrane voltage on extra-membrane voltage, which is mostly created by the external electrode after passing through a distance of skin tissue, and is dependent on the position of the electrode and the conductivity of the skin.

Under the influence of the electrodes, the local surface skin generates an electric field resembling a circle. The produced electric field is supposed to have a hemispherical distribution within the skin, with the electrodes at the centre of the circle.

Assuming that the resistivity of the skin epidermis is $\rho(t)$ at moment t, the electric potential at radius r is $\psi(r, t)$, and the electric field is $E(r, t)$, the following relationship exists between the electric potential and the electric field Eq. (9):

$$E(r, t) = -\frac{\partial \psi(r, t)}{\partial r} \tag{9}$$

In addition, the relationship between the electric field and the current density $J(r, t)$ and resistivity $\rho(t)$ is as follows:

$$E(r, t) = \rho(t) J(r, t) \tag{10}$$

Transformation of Eq. (9) and Eq. (10) gives Eq. (10):

$$\psi(r, t) = -\int E(r, t) dr = -\int J(r, t) \rho(t) dr \tag{11}$$

According to $J(r,t) = \frac{i(t)}{S} = \frac{i(t)}{2\pi r^2}$, a three-dimensional model is introduced for the action of a single electrode at radius r, the resistivity in Eq. (12) is a time-dependent function.

$$\psi(r,t) = -\int \frac{i(t)}{2\pi r^2}\rho(t)dr = \frac{i(t)}{2\pi r}\rho(t) \quad (12)$$

The electrotactile conduction model develops based on previous models because it contains not only the skin impedance, but also the electrode-skin contact impedance. The skin surface is curved and difficult for the electrodes to fit completely, so the electrode-skin impedance due to the gaps of the non-complete-contact surfaces should be considered.

3.3 Influencing Factors of the Model

The model's impedance varies from 100 Ω to 1 MΩ depending on the frequency of the input current. The lower the frequency, the higher the impedance is [38]. The skin impedance varies from 10 KΩ to 1 MΩ at a frequency of 1 Hz, while the impedance is 100 Ω–220 Ω with a high frequency input of 100 Hz [16]. The stratum corneum is primarily responsible for the variation in capacitance of human skin, which ranges from 0.02 $\mu F/cm^2$ to 0.06 $\mu F/cm^2$. Sweat of fingers creates a space between the skin and the electrode, lowering the conductivity [39].

3.4 Description of Electrical Stimulation Parameters

At stimulation frequencies less than 60 Hz, people can clearly feel the vibration and a slight sense of pressure [40]. At pulse frequency range between 60 and 400 Hz, the subjective sensation is more natural and comfortable, and the needed stimulation pulse amplitude is minimal [41]. When the stimulation frequency is more than 400 Hz, people will have a strong tactile feeling; however, over a period of time, the tactile sensation will gradually fade, necessitating an increase in pulse amplitude [42]. For current pulses between 200 and 500 μs, people can feel a pleasant sense of touch.

4 Summary

The mechanism of tactile feedback induced by TENS (transcutaneous electrical nerve stimulation) is analysed with a particular focus on features of the skin and subcutaneous tactile receptors. Then, based on the literature review of existing electrophysiological models of electrode-skin and neurons, a preliminary electrotactile conduction model is developed to predict tactile receptors' reactions to external stimulations. It can also be viewed as a theoretical guidance for the parameter adjustment during the process of electrotactile stimulation.

Acknowledgments. This work is supported by the National Natural Science Foundation of China (Grant No. 62003222) and the Research Fund of Liaoning Provincial Department of Education (Grant No. LQGD2020018).

References

1. Ziegler-Graham, K., MacKenzie, E.J., Ephraim, P.L., Travison, T.G., Brookmeyer, R.: Estimating the prevalence of limb loss in the United States: 2005 to 2050. Arch. Phys. Med. Rehabil. **89**(3), 422–429 (2008)
2. Maimon-Mor, R.O., Makin, T.R.: Is an artificial limb embodied as a hand? Brain decoding in prosthetic limb users. PLoS Biol. **18**(6), 1–26 (2020)
3. Gholinezhad, S., Dosen, S., Jakob, D.: Electrotactile feedback outweighs natural feedback in sensory integration during control of grasp force. J. Neural Eng. **18**(5), 056024 (2021)
4. Johansson, R.S., Flanagan, J.R.: Coding and use of tactile signals from the fingertips in object manipulation tasks. Nat. Rev. Neurosci. **10**(5), 345–359 (2016)
5. Engeberg, E.D., Meek, S.G.: Adaptive sliding mode control for prosthetic hands to simultaneously prevent slip and minimize deformation of grasped objects. IEEE/ASME Trans. Mechatron. **18**(1), 376–385 (2013)
6. Abraira, V.E., Ginty, D.D.: The sensory neurons of touch. Neuron **79**(4), 618–639 (2013)
7. Cansever, L., et al.: The effect of transcutaneous electrical nerve stimulation on chronic postoperative pain and long-term quality of life. Turk. J. Thorac. Cardiovasc. Surg. **29**(4), 495–502 (2021)
8. Schmid, P., Bader, M., Maier, T.: Tactile information coding by electro-tactile feedback. In: Proceedings of 4th International Conference on Computer Computer-Human Interaction Research and Applications, CHIRA 2020, pp. 37–43 (2020)
9. Kajimoto, H., Kawakami, N., Maeda, T., Tachi, S.: Tactile feeling display using functional electrical stimulation. In: Proceedings of ICAT (1999)
10. Tarnaud, T., Joseph, W., Martens, L., Tanghe, E.: Dependence of excitability indices on membrane channel dynamics, myelin impedance, electrode location and stimulus waveforms in myelinated and unmyelinated fibre models. Med. Biol. Eng. Comput. **56**(9), 1595–1613 (2018). https://doi.org/10.1007/s11517-018-1799-y
11. Lee, J., Lee, H., Eizad, A., Yoon, J.: Effects of using TENS as electro-tactile feedback for postural balance under muscle fatigue condition. In: 2021 21st International Conference on Control, Automation and Systems (ICCAS), Piscataway, NJ, pp. 1410–1413. IEEE (2021)
12. Alonso, E., Giannetti, R., Rodríguez-Morcillo, C., Matanza, J., Muñoz-Frías, J.D.: A novel passive method for the assessment of skin-electrode contact impedance in intraoperative neurophysiological monitoring systems. Sci. Rep. **10**(1), 1–11 (2020)
13. Watkins, R.H., et al.: Optimal delineation of single c-tactile and c-nociceptive afferents in humans by latency slowing. J. Neurophysiol. **177**(4), 1608–1614 (2017)
14. van Boxtel, A.: Skin resistance during square-wave electrical pulses of 1 to 10 mA. Med. Biol. Eng. Comput. **15**(6), 679–687 (1977)
15. Hu, Y., Zhao, Z., Vimal, A., Hoffman, G.: Soft skin texture modulation for social robotics. In: 2018 IEEE International Conference on Soft Robotics, Piscataway, NJ, pp. 182–187. IEEE (2018)
16. Vargas Luna, J.L., Krenn, M., Cortés Ramírez, J.A., Mayr, W.: Dynamic impedance model of the skin-electrode interface for transcutaneous electrical stimulation. PLoS One **10**(5), 1–15 (2015)
17. Chizmadzhev, Y.A., Indenbom, A.V., Kuzmin, P.I., Galichenko, S.V., Weaver, J.C., Potts, R.O.: Electrical properties of skin at moderate voltages: contribution of appendageal macropores. Biophys. J. **74**, 843–856 (1998)
18. Yang, H., Meijer, H.G.E., Doll, R.J., Buitenweg, J.R., Van Gils, S.A.: Computational modeling of Adelta-fiber-mediated nociceptive detection of electrocutaneous stimulation. Biol. Cybern. **109**(4), 479–491 (2015)

19. Kuhn, A., Keller, T., Lawrence, M., Morari, M.: The influence of electrode size on selectivity and comfort in transcutaneous electrical stimulation of the forearm. IEEE Trans. Neural Syst. Rehabil. Eng. **18**(3), 255–262 (2010)
20. Dahl, C., Kristian, M., Andersen, O.K.: Estimating nerve excitation thresholds to cutaneous electrical stimulation by finite element modeling combined with a stochastic branching nerve fiber model. Med. Biol. Eng. Comput. **49**(4), 385–395 (2011)
21. FitzHugh, R.: Impulses and physiological states in theoretical models of nerve membrane. Biophys. J. **1**(6), 445–466 (1961)
22. Kajimoto, H.: Electro-tactile display with tactile primary color approach. In: International Conference on Intelligent Robots & Systems (2004)
23. Stefano, M., Cordella, F., Loppini, A., Filippi, S., Zollo, L.: A multiscale approach to axon and nerve stimulation modelling: a review. IEEE Trans. Neural Syst. Rehabil. Eng. **9**, 397–407 (2021)
24. Frankenhaeuser, B., Huxley, A.F.: The action potential in the myelinated nerve fibre of Xenopus laevis as computed on the basis of voltage clamp data. J. Physiol. **171**(2), 302–315 (1964)
25. Hindmarsh, J.L., Rose, R.M.: A model of neuronal bursting using three coupled first order differential equations. Proc. R. Soc. Lond. B. Biol. Sci. **221**(1222), 87–102 (1984)
26. Morris, C., Lecar, H.: Voltage oscillations in the barnacle giant muscle fiber. Biophys. J. **35**(1), 193–213 (1981)
27. Qinghui, Z., Bin, F.: Design of blind image perception system based on edge detection. In: Transducer Microsystem Technologies, pp. 2–4 (2019)
28. Goganau, I., Sandner, B., Weidner, N., Fouad, K., Blesch, A.: Depolarization and electrical stimulation enhance in vitro and in vivo sensory axon growth after spinal cord injury. Exp. Neurol **300**, 247–258 (2018)
29. Saal, H.P., Delhaye, B.P., Rayhaun, B.C., Bensmaia, S.J.: Simulating tactile signals from the whole hand with millisecond precision. Proc. Natl. Acad. Sci. U. S. A **114**(28), E5693–E5702 (2017)
30. Lee, W.W., et al.: A neuro-inspired artificial peripheral nervous system for scalable electronic skins. Sci. Robot **4**(32), eaax2198 (2019)
31. Raspopovic, S., et al.: Bioengineering: restoring natural sensory feedback in real-time bidirectional hand prostheses. Sci. Transl. Med. **6**(22), 222ra19 (2014)
32. Vallbo, A.B., Johansson, R.S.: Properties of cutaneous mechanoreceptors in the human hand related to touch sensation. Hum. Neurobiol. **3**(1), 3–14 (1984)
33. Greffrath, W., Schwarz, S.T., Büsselberg, D., Treede, R.D.: Heat-induced action potential discharges in nociceptive primary sensory neurons of rats. J. Neurophysiol. **102**(1), 424–436 (2009)
34. Morse, R.P., Allingham, D., Stocks, N.G.: Stimulus-dependent refractoriness in the Frankenhaeuser-Huxley model. J. Theor. Biol. **382**, 397–404 (2015)
35. McIntyre, C., Richardson, A.G., Grill, W.M.: Modeling the excitability of mammalian nerve fibers: influence of afterpotentials on the recovery cycle. J. Neurophysiol. **87**(2), 995–1006 (2002)
36. Zhao, L., Liu, Y., Ma, Z.: Research progress of tactile representation technology. Jisuanji Fuzhu Sheji Yu Tuxingxue Xuebao/J. Comput. Des. Comput. Graph **30**(11), 1979–2000 (2018)
37. Hodgkin, A.L., Huxley, A.F.: Action potentials recorded from inside a nerve fibre. Nature **6**, 710–711 (1939)
38. Chizmadzhev, Y.A., Indenbom, A.V., Kuzmin, P.I., Galichenko, S.V., Weaver, J.C., Potts, R.O.: Electrical properties of skin at moderate voltages: contribution of appendageal macropores. Biophys. J. **74**(2), 843–856 (1998)

39. Stephens-Fripp, B., Alici, G., Mutlu, R.: A review of non-invasive sensory feedback methods for transradial prosthetic hands. IEEE Access **6**, 6878–6899 (2018)
40. Tamè, L., Tucciarelli, R., Sadibolova, R., Sereno, M.I., Longo, M.R.: Reconstructing neural representations of tactile space. Neuroimage **229**, 117730 (2021)
41. Vance, C.G.T., Rakel, B.A., Dailey, D.L., Sluka, K.A.: Skin impedance is not a factor in transcutaneous electrical nerve stimulation effectiveness. J. Pain Res. **8**, 571–580 (2015)
42. Bütikofer, R., Lawrence, P.D.: Electrocutaneous nerve stimulation-II: stimulus waveform selection. IEEE Trans. Biomed. Eng. **26**(2), 69–75 (1979)

Author Index

Printed in the United States
by Baker & Taylor Publisher Services